TREATMENT WETLANDS

Robert H. Kadlec
The University of Michigan, Ann Arbor
and
Wetland Management Services
Chelsea, Michigan

Robert L. Knight
CH2M HILL
Gainesville, Florida

LEWIS PUBLISHERS
Boca Raton New York

Library of Congress Cataloging-in-Publication Data

Kadlec, Robert H.
Treatment wetlands / Robert H. Kadlec, Robert L. Knight.
p. cm.
Includes bibliographical references and index.
ISBN 0-87371-930-1
1. Sewage--Purification--Biological treatment. 2. Wetlands.
TD7556.K33 1995
628.3′5—dc 95-9492
CIP

International Standard Book Number 0-87371-930-1
Library of Congress Card Number 95-9492
Printed in the United States of America 2 3 4 5 6 7 8 9 0
Printed on acid-free paper

Frog pioneers

Larson's *Frog Pioneers* epitomize three important aspects of Treatment Wetlands: (1) the innovative (pioneering) aspect of the technology; (2) the fact that wetlands can and are being constructed in nearly any climate; and (3) the age-old truism that wetlands attract frogs and other wildlife. (Reprinted with permission, Universal Press Syndicate; drawing by Gary Larson.)

About the Authors

Robert H. Kadlec, holds BS, MS and PhD degrees in chemical engineering from the Universities of Wisconsin and Michigan, 1958–62. That era saw the culmination of the "unit operations" approach to chemical processing, and the transition to the use of principles of transport phenomena to describe transfer and reaction rates in a wide variety of chemical and biochemical processes. Those techniques and analytical tools are also the foundation of today's environmental engineering. Bob began applying engineering analysis to wetland processes in 1970, with the goal of managing wetlands for water quality improvement. The result was the Houghton Lake (Michigan) natural wetland treatment system, which is still operating successfully.

Research on that natural wetland, and on the ensuing twenty years of its operation for engineered treatment, formed the early framework for Dr. Kadlec's development of wetland process characterization. The technology has grown tremendously and so has Bob's involvement in treatment wetland projects. He has participated in over 100 projects, ranging from simple feasibility studies to comprehensive university research projects. His graduate and undergraduate students have contributed significantly to the engineering science of treatment wetlands. Those studies have focused primarily on wetland hydrology and water chemistry. In the course of those interdisciplinary studies, a good deal of knowledge of practical ecology was imparted by his colleagues.

Bob has worked on treatment wetlands in many states and several other countries. Major projects have included Houghton Lake, MI; Incline Village, NV; Des Plaines, IL; Columbia, MO; and the Everglades Stormwater Treatment Areas. He is currently chairman of the Macrophyte Specialist Working Group of the International Association on Water Quality. He has authored or co-authored over 60 publications on treatment wetlands, in addition to dozens of project reports. He was a proposer and developer of the USEPA North American Treatment Wetland Database.

Dr. Kadlec retired from his teaching duties in 1993 and is currently principal of Wetland Management Services, and provides specialty consulting services to a wide range of governmental and private organizations. His contributions to this book are an effort to consolidate over two decades of research and practical experience.

Robert L. Knight, PhD, began studying wetlands receiving wastewaters in 1970 when he worked under Elizabeth McMahan and H.T. Odum at the Institute for Marine Sciences in Morehead City, North Carolina. Bob's academic and field experience continued with research on *in situ* populations of lake algae as part of the algal assay test development at the University of North Carolina (UNC); masters research at UNC under Charles Weiss and Ed Kuenzler on effects of thermal pollution on phytoplankton; graduation with an MSPH at UNC in 1973; research on mesocosms receiving mercury and cadmium with Hank Kania and John Giesy at the Savannah River Ecology Laboratory (SREL) (1973–1977); and doctoral studies with H.T. Odum at the University of Florida, Gainesville, from 1977 through 1980. Dr. Knight's PhD research synthesized ecosystem control studies at Silver Springs, Florida and heavy metal studies at the SREL into a hypothesis of ecosystem control.

Since 1981, Dr. Knight has worked as an Environmental Scientist with CH2M HILL in Gainesville, Florida. Since his first month on the job, Bob has been involved with studying, planning, permitting, designing, and operating treatment wetlands for all types of wastewaters for clients across the U.S. and around the world. He has been actively involved in over 100 treatment wetland projects and currently is CH2M HILL's Wetland Technology and Services Leader. Dr. Knight served on the Water Pollution Control Federation's Natural Systems Technical Committee and was the primary author of *Chapter 9 — Wetlands Systems* in

Manual of Practice FD-16. He has written more than 70 technical reports concerning treatment wetlands and more than 25 published papers on the subject. He was co-principal investigator on the North American Wetland Treatment System Database project for the U.S. Environmental Protection Agency and primary author of Arizona's Guidance Manual for Constructed Wetlands. *Treatment Wetlands* represents the attainment of an important professional and personal goal for Dr. Knight.

Preface

Treatment wetlands comprise an amazing diversity of managed ecosystems and provide an impressive array of water quality improvement mechanisms. This technology is solidly wedged in the gap between ecological desires for wetland restoration and preservation and engineering concerns for reliable and economical methods for water quality improvement. We, as authors, have come from the extremes of "old-school" ecology and engineering, through a variety of wetland projects, to an understanding of the need for cross-education and communication between disciplines.

This book was undertaken with the intent of technology transfer and not as an attempt to define the last word in design procedures. Information is presently growing at an exponential rate. The number of newly issuing project reports, conference proceedings, books, and reviewed papers is so large that it is a full-time job to keep abreast of new developments. It is no longer feasible for an entry-level practitioner to find the scientific and operational background relevant to even a subset of the treatment wetlands technology.

In that environment, it is very tempting to seek out summaries and manuals that provide simplistic design guidelines. Several such manuals exist, but in our opinion these are often obsolete by the time they are published. Indeed, we know this to be true for this book as well. In the interest of brevity, previous design guidelines omit the ecological and engineering fundamentals and cannot include the necessary details for all potential areas of application. It is the data and the fundamentals that must survive to the next iteration, not the recipes. Methods of analysis, leading to design, are constrained by the data that support them. We fully anticipate improvements in methodology occurring at a rapid pace. The reader is encouraged to think critically about design bases and design calculations and to be certain that a new situation warrants the use of a particular method or estimate.

Wetlands are built for mitigation and are restored for purposes of improving and increasing habitat and species diversity. Some of the ideas in this book bear on those types of activities, but they are not the central theme. Many wetlands are built to improve the quality of acid mine drainage, but these too are outside the scope of this book. The principal focus is on transformation/removal of biodegradable materials, nutrients, trace metals, pathogens, and toxicants.

Of necessity, there is more emphasis on surface-flow (SF) wetlands than on subsurface-flow (SSF) systems. A disproportionate amount of research and monitoring effort in North America has gone into SF systems because of their cost effectiveness and ability to optimize multiple benefits. There is also more emphasis on domestic wastewater than some readers will like, but that too is due to the availability of information.

At several points in the text, values of rate constants, regression parameters, and other constants are given. It is to be emphasized that these represent current estimates of central tendencies over diverse populations of wetlands and are best estimates to be used only in the range of the data that generated them, and that they may undergo changes as additional data are analyzed. Adjustment of these parameters is often necessary for site-specific design projects and should be done on the basis of more current information, such as pilot or regional prototype projects. The practitioner is also well advised to study the available literature on specific issues in greater detail. To that end, a rather extensive list of references has been provided.

Wetlands contain assemblages of many plants and animals adapted to local climatic and geological conditions and influenced by initial conditions; therefore, wetlands exhibit the individuality characteristic of a living organism. The concept of precise replication is dubious even at a single site, and the designer is advised not to copy other systems without due attention to the primary variables that define the character and function of that ecosystem.

It is somewhat frustrating that only simple first-order and zero-order uptake and release models can be substantiated with data at this point in time. The level of detail is nevertheless fairly complicated, especially when describing the full sequential nitrogen conversion pathway. Detailed hydraulics also require nontrivial computations. The presumption is made that the reader is familiar with one or more of the many powerful spreadsheet numerical analysis programs available for personal computers. It is our view that this level of computational power is necessary for some treatment wetland design calculations and is useful in many other situations. In other words, final design will require something more than "back-of-the-envelope" calculations and rules of thumb.

Water mass balances form the basis for all reliable data analysis and design calculations. Data sets that do not include this vital information must be viewed with some suspicion because rainfall, evapotranspiration, and leakage can all have large effects on performance of treatment wetlands.

The assumption is made in this work that design will be performance based. This approach may appear exceptionally stringent when considering very small systems for domestic wastewater treatment, such as for individual residences. At this stage of the development of the technology, there is not a great deal of operating data on individual systems due to cost factors. It is therefore premature to offer guidelines that purport to codify the construction of a treatment wetland at any scale, but especially small ones, until significant regionalized databases are available.

In contrast, there is a very large amount of information for large wetlands, some of which have been assembled into computerized formats that facilitate analysis. Where possible, we have attempted to integrate the data and experiences from Scandinavia, the U.K., Europe, Australia, Asia, Africa, and New Zealand. In some instances, notably for SSF wetlands, these data outweigh those available in the U.S.

Even with these caveats and cautions, we believe that the careful reader will find that *Treatment Wetlands* can provide a firm foundation from which to embark on a new and exciting technological journey. This is an age when the environment is under increasing demands due to expanding populations and decreasing per capita energy supplies. Treatment wetlands can provide a double dose of environmental protection when appropriately designed to optimize assimilation, habitat, and public-use functions. We wish to welcome the growing number of individuals who are considering options to enhance the environment through ecological engineering.

Robert H. Kadlec
Robert L. Knight
March 1995

Acknowledgments

The authors wish to acknowledge the efforts of the hundreds of engineers and scientists who have had the courage to try something new, to solve endless challenges, and to ultimately move the treatment wetland technology to the forefront of providing a cost-effective and environmentally protective alternative to conventional wastewater treatment methods. These contributors include the pioneers in the field, the countless graduate students who have dedicated their academic lives and energies to experimental design and data collection, and the far-sighted administrators who gave their adventurous staff the responsibility to find the best solutions. Without the data assembled from these diverse resources, this book and the treatment wetland technology would be hollow.

In terms of their more direct impact on the completion of this book, the authors wish to thank the following individuals for review of portions of the draft manuscript: Jim Bays, Pat Cline, Michelle Girts, and Henry Sheldon. In the final determination, the authors must take responsibility for any errors or omissions in the text.

CH2M HILL provided a significant contribution to the completion of this book by providing editorial, word processing, and graphics assistance. Three CH2M HILL editors earned special recognition: Patti Garcia, Tara Boonstra, and Dianne Cothran.

CRC/Lewis project editor Gail Renard and other staff are gratefully acknowledged for their skill and professionalism in taking the manuscript to press. Skip DeWall, formerly with Lewis Publishers in Chelsea, Michigan, is sincerely thanked for his enthusiasm and support for the whole idea and beginning of this book.

Deepest appreciation is extended to our wives, Kelly Kadlec and Gail Knight, who put up resolutely with seemingly endless late hours and working weekends. Their contribution can best be measured by the observation that this book was finished and our marriages have been strengthened by the process.

Table of Contents

Chapter 10

Chapter 11

Chapter 12

Chapter 13

Chapter 14

Chapter 15

Chapter 16

Chapter 20

Chapter 21

Chapter 22

Chapter 23

Section 5 Wetland Treatment System Establishment, Operation, and Maintenance

Chapter 24

Chapter 25

Section 6 Wetland Data Case Histories

Chapter 26

Chapter 27

Section 1
Introduction and Scope

CHAPTER 1

Introduction To Wetlands For Treatment

WHY WETLANDS?

Wetlands are land areas that are wet during part or all of the year because of their location in the landscape. Historically, wetlands were called swamps, marshes, bogs, fens, or sloughs, depending on existing plant and water conditions and on geographic setting. Wetlands are frequently transitional between uplands (terrestrial systems) and continuously or deeply flooded (aquatic) systems. Wetlands are also found at topographic lows (depressions) or in areas with high slopes and low permeability soils (seepage slopes). In still other cases, wetlands may be found at topographic highs or between stream drainages when land is flat and poorly drained (blanket bogs and pocosins). In all cases, the unifying principle is that wetlands are wet long enough to alter soil properties because of the chemical, physical, and biological changes that occur during flooding, and to exclude plant species that cannot grow in wet soils.

Wetlands hold seemingly magical properties that make them unique among major ecosystem groups on the earth. Ample water is important for most forms of biological productivity, and wetland plants are adapted to take advantage of this abundant supply of water while overcoming the periodic shortage of other essential chemical elements such as oxygen. Because of this, wetlands are among the most biologically productive ecosystems on the earth (Figure 1-1). As such, they are frequently inhabited by jungle-like growths of plants and are home to a multitude of animals including mammals, birds, reptiles, amphibians, and fish that are uncommon in other ecosystems.

In addition, because wetlands have a higher rate of biological activity than most ecosystems, they can transform many of the common pollutants that occur in conventional wastewaters into harmless byproducts or essential nutrients that can be used for additional biological productivity. These transformations are accomplished by virtue of the wetland's land area, with its inherent natural environmental energies of sun, wind, soil, plants, and animals. These pollutant transformations can be obtained for the relatively low cost of earthwork, piping, pumping, and a few concrete structures. Wetlands are one of the least expensive treatment systems to operate and maintain. Because of the natural environmental energies at work in a wetland treatment system, minimal fossil-fuel energy and no chemicals are necessary.

Historically, wetlands have been the last areas of the landscape to be altered for traditional upland uses such as urban or industrial development, farming, and mining because they must be drained to become productive and because effective drainage and successful flood control may be rather expensive. With the development of a widespread environmental awareness in North America and worldwide in the last 20 years, the culturally and environmentally beneficial aspects of wetlands and other natural ecosystems have been discovered by the public. Because wetlands are less likely than other natural ecosystems to have been previously altered by humans, the remaining wetlands (only about 50 percent of the historical wetland area still exists in the continental U.S., exclusive of Alaska) have become a rallying point for restoration, conservation, and preservation.

Figure 1-1 Wetlands are renowned for their high plant and wildlife productivity.

A value-conscious society frequently asks, "What return can I realize on my privately owned wetlands that will compensate me to leave them in their natural state?" or "How can we find the money to pay for wetland restoration or construction of new wetlands in upland areas?". The environmentalists' appeals for wetland preservation on the basis of aesthetic and wildlife values are resulting in the protection of some significant wetland areas. In addition, the "no net loss of wetland functions" objective is being approached as society recognizes that compatible uses of wetlands also exist that augment their value and further warrant their protection and construction. Wetlands can be used as a low-cost, natural technology for water quality treatment by small towns and large cities and for passive nonpoint runoff management in agricultural areas. These advantages encourage the creation and restoration of wetlands for water treatment across the landscape and in a variety of climates and geographical settings. The net result may be an increasing wetland resource in some areas.

Wetland conservation as distinguished from preservation refers to a wise, sustainable use of one or more wetland functions which do not interfere or may actually enhance the other functional attributes of the wetland (such as biological productivity, groundwater recharge, or flood control). Wetlands used for water quality treatment, when practiced through a conservative and knowledgeable design, constitute this form of wetland conservation, while providing a measurable value to society to create or maintain wetlands in the landscape.

Agricultural, urban, and industrial development causes wastes to be concentrated, which can overwhelm traditional receiving waters used to transport and assimilate pollutants. To deal with this concentration of human activities and to protect and enhance receiving waters used for fish and wildlife propagation and for human recreation, society has developed a multitude of technologies to purify or treat waste streams before they are discharged. Many conventional wastewater treatment technologies require intensive inputs of concrete, steel, energy, and chemicals; however, a relatively newer group of technologies relies more on natural energies to accomplish this treatment. The use of natural or constructed wetlands to provide treatment is one of these land-intensive, natural technologies. Figure 1-2 shows how a wetland treatment system can serve as a natural or green-belt buffer between concentrated wastewaters and surface waters that must be kept clean for multiple uses.

When concentrated wastewaters are pretreated before they are discharged into a wetland, many of the residual pollutants remaining in the effluent become a resource for the wetland biota, which transforms these raw materials into biomass, new soil, or harmless atmospheric

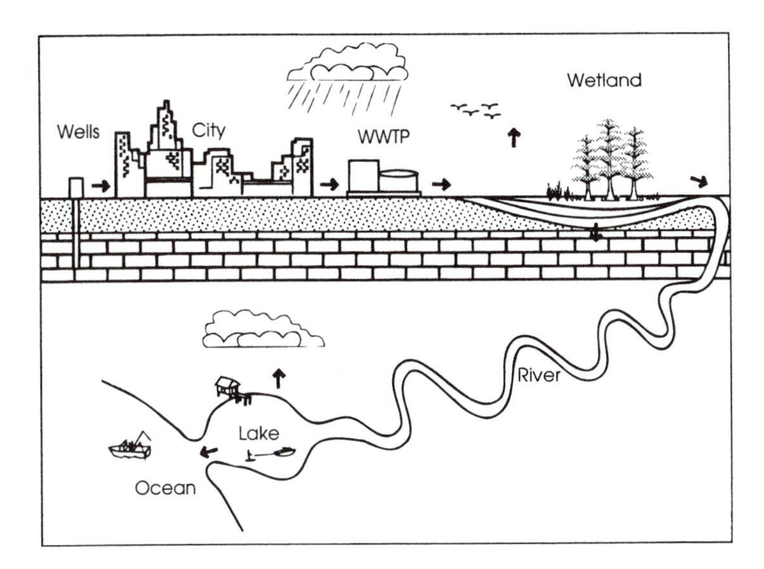

Figure 1-2 Wetland treatment systems can provide a cost-effective, seminatural buffer between urban areas and traditional recreational waters including lakes, rivers, estuaries, and oceans. Constructed and natural wetland treatment systems can augment recreational resources by incorporating nature centers, hiking trails, boardwalks, and other passive public use facilities.

gases. These transformations result in the final purification of the water and the beneficial recycling of the wastewater constituents into biological food chains, leading to support of natural wildlife populations or alternatively to harvestable biomass in aquaculture or forestry. For this reason, the use of wetlands for treatment should be viewed as a beneficial reuse of wastewaters, where this reuse benefits the environment as compared to reuse that more directly benefits society (such as landscape and crop irrigation or potable reuse). Wetland treatment systems, if designed and operated in a way that does not degrade natural wetland functions, result in an important environmental reuse of wastewaters as compared to discharges directly into streams, estuaries, oceans, or saline aquifers.

HISTORICAL PERSPECTIVE

Natural wetlands have been used as convenient wastewater discharge sites for as long as sewage has been collected (at least 100 years in some locations). Examples of old wetland sites include the Great Meadows natural wetland near the Concord River in Lexington, MA which began receiving wastewater in 1912; the Brillion Marsh in Wisconsin that has received municipal wastewater discharges since 1923; the Dundas sewage treatment plant which began discharging to the Cootes Paradise natural wetland near Hamilton, Ontario in 1919; and a discharge to a natural cypress swamp from the city of Waldo, FL since 1939.

More recently, the U.S. Environmental Protection Agency (EPA) found that there were about 324 "swamp discharges" in the 14 states in Regions IV and V in the mid-1980s (Table 1-1). None of these discharges was monitored for water quality or biological integrity within the receiving wetlands until the 1960s and 1970s. When monitoring was initiated at some of these existing discharges, an awareness of the water quality purification potential of wetlands began to emerge.

Table 1-2 presents an annotated chronology of some of the events leading to the acceptance of the use of natural and constructed wetlands for water quality management. The table lists selected research efforts, full-scale project initiation dates, and key technical conferences at

Table 1-1 Inventory of Wastewater to Wetland Discharges in the U.S. EPA Regions IV and V, 1981–1982

State	EPA Region	Wetland Discharges: Municipal	Industrial	Other or Unknown	Total
Alabama	IV	10	3	0	13
Florida	IV	16	12	30	58
Georgia	IV	8	2	0	10
Illinois	V	4	1	1	6
Indiana	V	4	2	3	9
Michigan	V	7	0	0	7
Minnesota	V	27	5	3	35
Mississippi	IV	10	2	28	40
North Carolina[a]	IV	20	1	40	61
Ohio	V	2	1	0	3
South Carolina	IV	17	1	16	34
Tennessee	IV	6	2	4	12
Wisconsin	V	28	7	1	36
Total		159	39	126	324

[a] North Carolina reported a total of 267 discharges to "swamp waters." Only 61 were greater than 376 m^3/d and discharged directly to natural wetlands.

Note: Data from Brennan and Garra, 1981; U.S. EPA, 1983c.

which the use of wetlands for water quality control was a featured topic. The worldwide spread of confidence in this technology originated from research conducted in Europe by Seidel and Kickuth at the Max Planck Institute in Plon, Germany starting in 1952 (Bastian and Hammer, 1993) and in the western hemisphere during the 1970s. Implementation of the technology has accelerated around the world since about 1985.

Between 1967 and 1972, Howard T. Odum and A.C. Chestnut of the University of North Carolina, Chapel Hill began a 5-year study using coastal lagoons (with marsh wetland littoral vegetation) for recycling and reuse of municipal wastewaters (Odum, 1985). Sixty-seven faculty and students examined a multitude of details related to augmenting these seminatural aquatic/wetland ecosystems with municipal effluent. The studies also included an examination of a natural *Spartina* salt marsh ecosystem that was receiving a discharge of secondarily treated wastewater (Marshall, 1970; McMahan et al., 1972; Camp et al., 1971; Stiven and Hunter, 1976).

In 1972, Odum, who had relocated to the University of Florida in Gainesville in 1970, and Katherine Ewel began work on another 5-year research effort directed at assessing the effectiveness of natural cypress wetlands for municipal wastewater recycling. The Florida Cypress Wetland Study was funded by the Rockefeller Foundation and the National Science Foundation's Division of Research Applied to National Needs (RANN). The study involved more than 50 graduate students and dozens of faculty members. In one portion of the study, from March 1974 until September 1977, secondarily treated municipal wastewater from a trailer park north of Gainesville was discharged to two isolated cypress wetlands (domes), while two control wetlands were monitored (Figure 1-3). The cypress domes were loaded with about 0.4 centimeters per day (cm/d) of effluent to assimilate all water through evapotranspiration and infiltration. Research studies measured nearly all aspects of the physical, chemical, and biological processes occurring in the wastewater and control cypress domes. This information, along with information from cypress wetlands throughout the southeastern U.S., was compared and summarized by Ewel and Odum (1984).

Concurrently with work at the University of Florida, Robert H. Kadlec and co-workers at the University of Michigan in Ann Arbor began the first in-depth study using engineered wetlands for wastewater treatment in a cold climate region. The Houghton Lake, MI project

Table 1-2 Wetland Treatment Technology Historical Timeline of Selected Events

Date	Location	Description
		Selected Early Research Efforts
1952–late 1970s	Plon, Germany	Removal of phenols and dairy wastewater treatment with bulrush plants by K. Seidel and R. Kickuth
1967–1972	Morehead City, NC	Constructed estuarine ponds and natural salt marsh studies of municipal effluent recycling by H.T. Odum and associates
1971–1975	Woods Hole, MA	Potential of natural salt marshes to remove nutrients, heavy metals, and organics was studied by I. Valiela, J.M. Teal and associates
1972–1977	Porter Ranch, MI	Natural wetland treatment of municipal wastewater by R.H. Kadlec and associates
1973–1974	Dulac, LA	Discharge of fish processing waste to a freshwater marsh was studied by J.W. Day and co-workers
1973–1975	Seymour, WI	Constructed marshes were planted with bulrush and pollutant removal was studied by Spangler and co-workers
1973–1976	Brookhaven, NY	Meadow/marsh/pond systems studied by M.M. Small and associates
1973–1977	Gainesville, FL	Studies of the use of cypress wetlands for recycling of municipal wastewaters by H.T. Odum, K. Ewel, and associates
1974–1975	Brillion, WI	Constructed and natural marsh wetlands were tested by F.L. Spangler and associates for phosphorus removal
1974–1988	NSTL Station, MS	Gravel-based, subsurface-flow wetlands tested for recycling municipal wastewaters and priority pollutants by B.C. Wolverton and co-workers
1975–1977	Trenton, NJ	Small enclosures in the Hamilton Marshes (freshwater tidal) were irrigated with treated sewage by Whigham and co-workers
1976–1979	Eagle Lake, IA	A natural marsh wetland was studied for assimilation of agricultural drainage and municipal wastewater nutrients by G.B. Davis, A.G. van der Valk, and co-workers
1976–1982	Southeast Florida	Natural marsh wetlands receiving agricultural drainage waters were studied for nutrient removal by F.E. Davis, A.C. Federico, A.L. Goldstein, S.M. Davis, and co-workers
1979–1982	Arcata, CA	Pilot wetland treatment system for municipal wastewater treatment by Gearheart and co-workers
1979–1982	Humboldt, SK	Batch treatment of raw municipal sewage in lagoons and wetland trenches by Lakshman and co-workers
1980–1984	Listowel, Ontario	Constructed marsh wetlands were tested for treatment of municipal wastewater under a variety of design and operating conditions by Herskowitz and associates
1981–1984	Santee, CA	Subsurface-flow wetlands were tested for treatment of municipal wastewaters by R.M. Gersberg and co-workers
		Selected Full-Scale Projects
1972	Bellaire, MI	Natural forested wetland receiving municipal wastewaters
1973	Mt. View, CA	Constructed wetlands for municipal wastewater treatment
1974	Othfresen, Germany	Full-scale root zone facility treating municipal wastewater based on the design method of Kickuth and co-workers
1975	Mandan, ND	Constructed ponds and marshes to treat runoff and pretreated process wastewater from an oil refinery
1976	Vermontville, MI	Volunteer wetlands created by flood irrigation of municipal wastewater in 1972
1977	Lake Buena Vista, FL	Natural forested wetland was used for year-round advanced treatment and disposal of up to 27,700 m^3/d of municipal wastewater

Table 1-2 (continued)

Date	Location	Description
	Selected Full-Scale Projects—continued	
1978	Houghton Lake, MI	Natural peatland receiving summer flows of municipal wastewater
1979	Drummond, WI	Sphagnum bog receiving summer flows from a facultative lagoon
1980	Show Low, AZ	Constructed wetland ponds for municipal wastewater treatment and wildlife enhancement
1984	Fremont, CA	Constructed marsh for urban stormwater treatment at Coyote Hill
1984	Incline Village, NV	Constructed wetlands for total assimilation (zero discharge) of municipal effluent
1986	Arcata, CA	Constructed marsh wetlands for municipal wastewater treatment
1987	Myrtle Beach, SC	Natural Carolina bay wetlands for municipal wastewater treatment
1991	Columbus, MS	First full-scale constructed wetland for advanced treatment of pulp and paper mill wastewater
1993	Everglades, FL	Treatment of phosphorus in agricultural runoff in a 1380-ha constructed filtering marsh
	Major Conferences	
May 1976	Ann Arbor, MI	Freshwater Wetland and Sewage Effluent Disposal (Tilton et al., 1976)
February 1978	Tallahassee, FL	Environmental Quality Through Wetlands Utilization (Drew, 1978)
November 1978	Lake Buena Vista, FL	Wetland Functions and Values (Greeson et al., 1978)
July 1979	Higgins Lake, MI	Freshwater Wetland and Sanitary Wastewater Disposal (Sutherland and Kadlec, 1979)
September 1979	Davis, CA	Aquaculture Systems for Wastewater Treatment (Bastian and Reed, 1979)
June 1981	St. Paul, MN	Wetland Values and Management (Richardson, 1981)
June 1982	Amherst, MA	Ecological Considerations in Wetlands Treatment of Municipal Wastewaters (Godfrey et al., 1985)
July 1986	Orlando, FL	Aquatic Plants for Water Treatment and Resource Recovery (Reddy and Smith, 1987)
June 1988	Chattanooga, TN	Constructed Wetlands for Wastewater Treatment (Hammer, 1989b)
September 1989	Tampa, FL	Wetlands: Concerns and Successes (Fisk, 1989)
September 1990	Cambridge, U.K.	Constructed Wetlands in Water Pollution Control (Cooper and Findlater, 1990)
October 1991	Pensacola, FL	Constructed Wetlands for Water Quality Improvement (Moshiri, 1993)
September 1992	Columbus, OH	INTECOL Wetlands Conference (Mitsch, Chairman)
December 1992	Sydney, Australia	Wetland Systems in Water Pollution Control (Pilgrim, Chairman)
November 1994	Guangzhou, China	4th International Conference on Wetland Systems for Water Pollution Control (Jiang and Kadlec, Co-Chairmen)

was an off-shoot of an interdisciplinary seminar generated in response to Earth Day activities. Work at the Porter Ranch peat-based wetland (peatland), located near the community of Houghton Lake, began in 1971 with the first introduction of water and nutrients to the peatland in 1972. After 2 years of discharges to mesocosm-size plots in the peatland, a 360-cubic meter per day (m^3/d) facility began operation and continued seasonal discharges for 3 years under the direction of university personnel. A full-scale system was initiated in 1978 with a design flow of 6410 m^3/d during the summer to the Porter Ranch peatland (Figure 1-4). This system continues to operate today, and information from the Houghton Lake

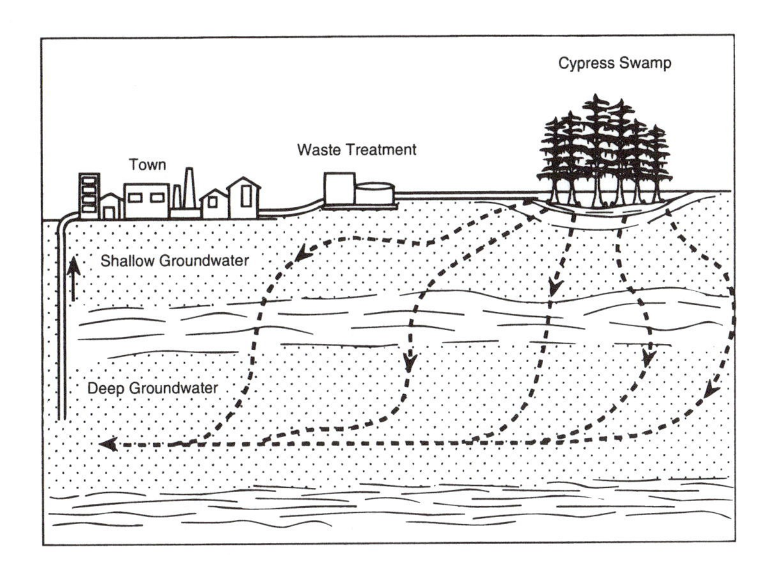

Figure 1-3 Conceptual diagram illustrating the use of natural cypress dome wetlands for wastewater recycling. The cypress dome treatment wetland concept was developed by Howard Odum and co-workers to provide cost-effective and environmentally protective effluent management for small communities throughout Florida.

Natural Peatland Treatment System represents one of the longest and most complete data sets currently available.

In 1973, the first intentionally engineered, constructed wetland treatment pilot systems in North America were constructed at Brookhaven National Laboratory near Brookhaven, NY. These pilot treatment systems combined a marsh wetland with a pond and a meadow in series and were designated as the meadow/marsh/pond (MMP) treatment system (Figure 1-5).

The public was not prepared to wait for results from these and other research efforts under way in the 1970s. In 1972, the city of Bellaire, MI began discharging stabilized municipal wastewater to a 16-hectare (ha) forested wetland. While research was conducted on this system, the wetland was the primary means of effluent disposal for the city. In 1973, the Mt. View Sanitary District in California constructed about 8.5 ha of wetland marshes for wildlife habitat and wastewater discharge.

Industrial stormwaters and process waters were also applied to constructed pond/wetland systems as early as 1975 at Amoco Oil Company's Mandan Refinery in North Dakota (Litchfield, 1989). In 1976, the communities of Pinetop and Lakeside and, in 1977, Show Low, AZ created a series of constructed lake/wetland areas for effluent evaporative disposal and wildlife production (Wilhelm et al., 1989). In 1978, the city of Houghton Lake, MI began full-scale discharge of treated municipal effluent to a natural peatland prior to eventual effluent flow to the nutrient-sensitive waters exiting Houghton Lake (Kadlec, 1993a).

The Reedy Creek Wetland Treatment System, constructed at Walt Disney World in 1977 (McKim, 1982; Knight et al., 1987), was an early natural wetland treatment system built in central Florida, somewhat on the model provided by the University of Florida cypress dome study.

The Reedy Creek system (Figure 1-6) used two wetlands, one with about 34 ha of natural mixed cypress/hardwood forested swamp and the other with about 0.2 ha of constructed marsh and 5.6 ha of natural swamp forest, to provide advanced wastewater treatment between 1977 and 1991 for monthly average flows as high as 22,700 m^3/d (7.2 cm/d). Flow to this

Figure 1-4 The Houghton Lake, MI wetland treatment system uses a 716-ha natural sedge/shrub peatland for seasonal discharges.

wetland was discontinued in 1991 when a zero surface discharge option was implemented through landscape irrigation and groundwater recharge.

Florida has had a number of other natural wetland wastewater discharges. Some of these discharges have been monitored, including Waldo (started 1931), Pottsburg Creek Swamp near Jacksonville (started 1967), Wildwood (started 1955), Gainesville (started 1930s), and Jasper (start date unknown). In addition, a number of other natural wetland treatment systems

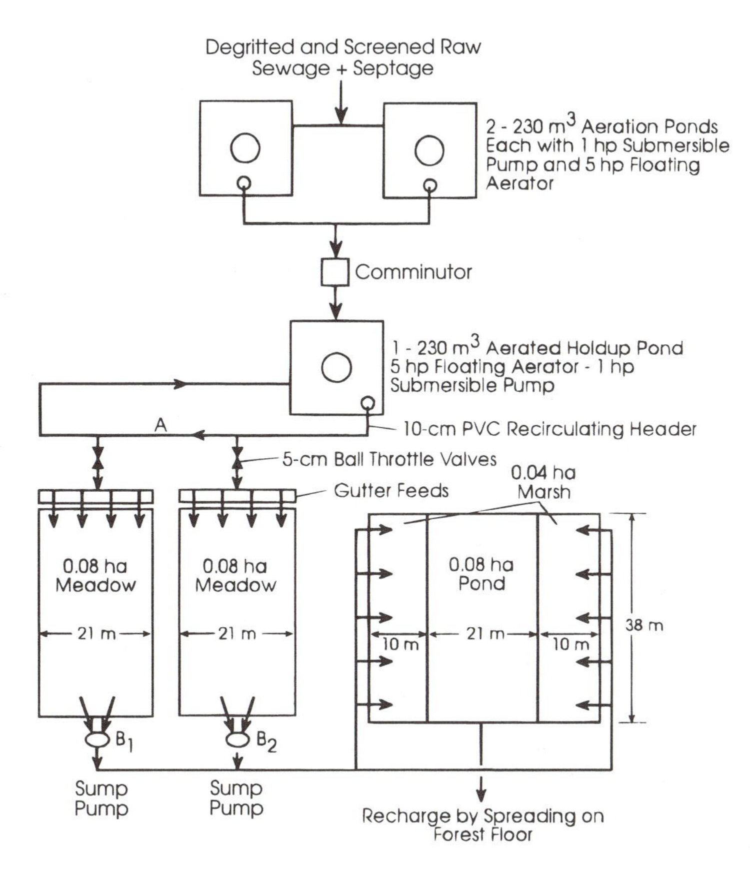

Figure 1-5 Diagram of the constructed meadow/marsh/pond pilot treatment system at Brookhaven, NY. (Redrawn from Small and Wurm, 1977.)

have been designed and permitted for treatment and disposal, including Pasco County's 50-ha Deer Park cypress wetlands (started 1988); Poinciana's 47-ha cypress dome wetland (started 1985), Appalachicola's titi swamp wetland (started 1985), and Buenaventura Lakes' cypress domes (started 1982).

Currently, Florida has several of the largest constructed wetland treatment areas in the world, including the Lakeland and Orlando constructed wetlands, both of which were started in 1987. Each wetland has about 500 ha for advanced treatment of municipal wastewater (Figure 1-7). Other large constructed wetlands in Florida are treating agricultural drainage waters at a number of sites in the Kissimmee River watershed (Boney Marsh, Armstrong Slough, Chandler Slough, and the three Water Conservation Areas), at Lake Apopka (100 ha) since 1990, and the Everglades Agricultural Area (1,490 ha) since 1994. An additional 16,000 ha of constructed wetlands are planned in the Everglades region for treatment of runoff from sugar cane operations.

The largest constructed treatment wetland is the 1800-ha Kis-Balaton project in Hungary, which has operated since 1985.

Florida, along with several other states, also has numerous smaller natural and constructed wetlands that treat urban stormwaters. The Lake Jackson system, a model constructed stormwater system in Tallahassee, has been operated and monitored by the Florida Department of Environmental Protection since 1984. A small urban stormwater system was built in Greenwood Park in Orlando in 1989 (Figure 1-8). About 50 natural wetlands, including

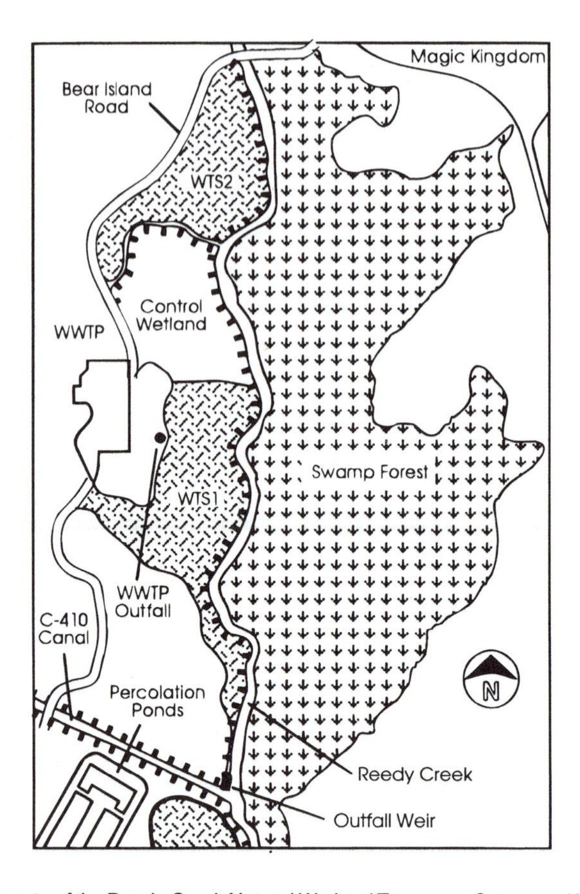

Figure 1-6 Components of the Reedy Creek Natural Wetland Treatment System at Walt Disney World, FL.

mostly cypress domes and a few wet prairie areas, receive urban runoff from residential and commercial developments in southwest Florida.

This historical perspective should help to illustrate the relatively recent development of wetland treatment technology and visualize the youth of even the oldest operating, full-scale engineered wetland treatment systems (about 20 years in 1995). This relatively short period of experience in the design and operation of wetland treatment systems is cause for reflection and understanding and is not unlike many of the other wastewater treatment technologies used today. For example, preliminary research on land treatment of municipal wastewaters was conducted during the 1960s, and full-scale implementation did not begin until the early 1970s. Biological nutrient removal techniques have been largely developed since 1970; rotating biological contactors and other attached growth treatment technologies were largely developed during the early 1970s; and the oxidation ditch concept was not in commercial use until the mid-1970s. In fact, the only "older" treatment technologies still in use are facultative and aerated lagoons, trickling filters, and activated sludge treatment systems. All of the modern technologies were developed in response to the Clean Water Act of 1972 and the rising awareness of the extent of water pollution occurring at that time.

It is not surprising that the value of wetlands for water quality improvement was "rediscovered" during the last 20 years of heightened environmental awareness. But it is important to remember how short our experience base is and to apply this knowledge by following conservative design approaches when permit limits are stringent and inflexible. At the same time, as with any young technology, many questions concerning optimizing wetlands for

(a)

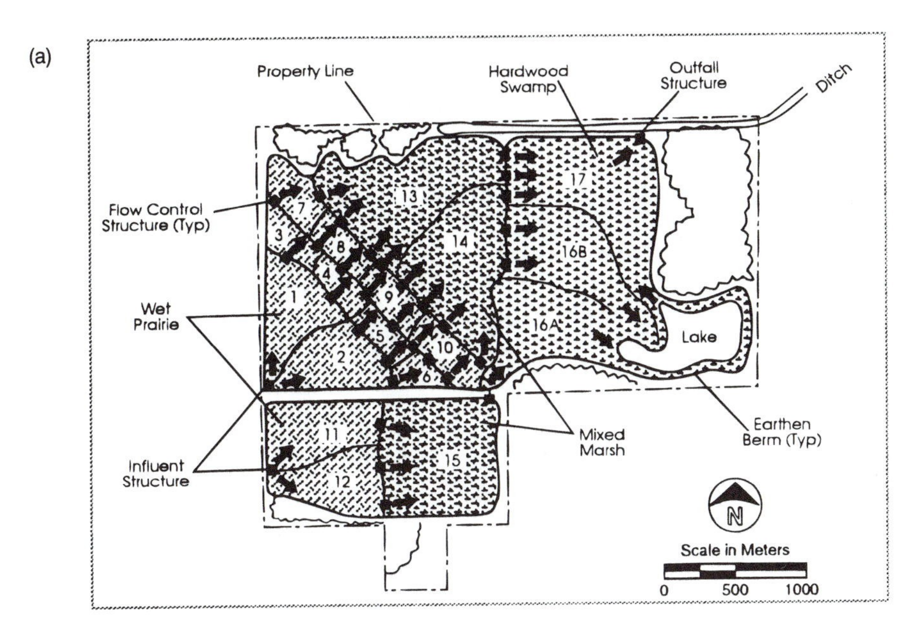

(b)

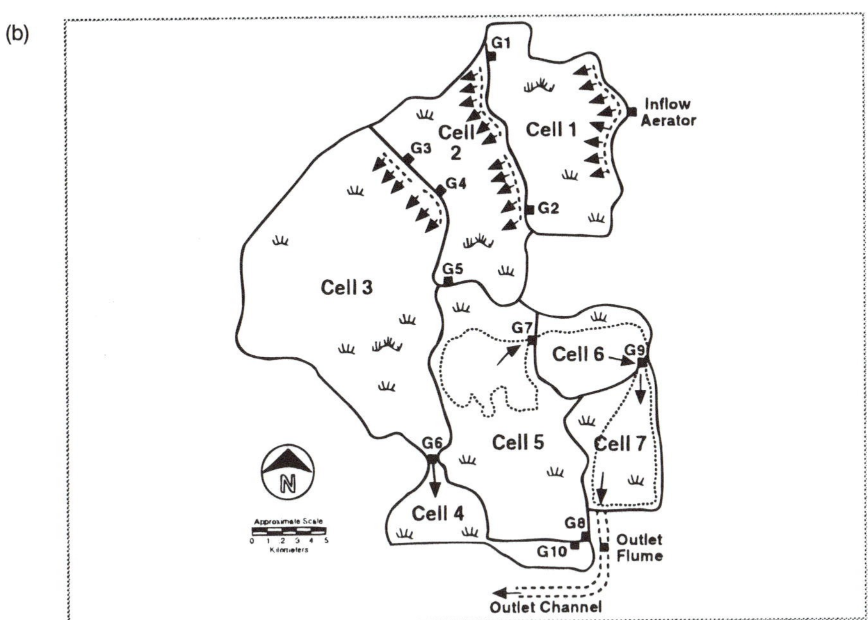

Figure 1-7 Two large constructed wetland treatment systems: (a) the Iron Bridge wetlands near Orlando, FL, and (b) the Lakeland Treatment Wetlands, Lakeland, FL.

treatment must be answered through research and analysis of the ever-increasing accumulation of wetland operational data. This book is intended to provide a solid basis for future projects that will address these important issues.

ORGANIZATION AND SCOPE OF THIS BOOK

Wetland treatment systems have taken their place among the proven technologies available for protecting surface water quality. Full-scale wetland treatment systems are being used

Figure 1-8 Orlando's Greenwood Park stormwater wetland treatment facility.

routinely to treat municipal, industrial, and agricultural wastewaters; agricultural and urban runoff; landfill leachate; and acid-mine drainage waters. However, the designs for these systems (more than 300 in North America by 1994 [excluding mine drainage] and over 1000 worldwide) are still highly individualistic and variable. There has been little consensus on approaches for successful wetland system operation and management.

Although information that standardizes wetland treatment system designs and summarizes successful techniques for project planning, construction, and operation exists, this information is scattered in many locations, making it difficult for engineers, scientists, and public officials to access it. This book summarizes and consolidates this information into one source that can be used by these diverse groups to implement wetland systems for water quality management.

This book provides background, design, construction, and operational guidance for wetlands that primarily treat nontoxic, organic, and inorganic wastewaters. It does not provide design guidelines for acid-mine drainage wetlands or for the loosely allied group of treatment technologies using floating aquatic plants. However, many of the principles and much of the information contained in this book will be helpful to designers and managers of those more specialized technologies. An overview of the contents of this book is provided below to help the reader locate information relevant to specific applications of wetlands technology.

SECTION 1

The three chapters in this section (Chapters 1 through 3) provide background information to help the reader understand the position of wetlands as a treatment technology. This background information includes an historical account of the development of the technology; a detailed discussion of the need for wastewater treatment; a description of wastewater characteristics and other, more conventional treatment practices; a discussion of how wetlands fit into the suite of natural treatment technologies that are available to the wastewater manager; and a discussion of how wetland systems can be used to provide cost-effective treatment and disposal.

SECTION 2

The five chapters in Section 2 (Chapters 4 through 8) summarize wetland geography, hydrology, and ecology issues that are relevant to wetland design and use. A general knowledge of these issues will enable flexibility during wetland planning, design, construction, and operation by describing the components of a wetland and how they normally interact. This section organizes information about wetland ecology into chapters describing wetland definitions, landform, and occurrence (Chapter 4); the properties and origin of wetland soils (Chapter 5); unifying principles of wetland hydrology and water quality interactions (Chapter 6); the diversity of wetland microbial and plant communities (Chapter 7); and the variety of wetland animal populations from invertebrates to mammals (Chapter 8). All of these chapters discuss their subject material on the basis of the individual species, groups, communities, or populations, as well as from the perspective of how they are an integral part of the overall wetland ecosystem.

SECTION 3

Section 3 (Chapters 9 through 17) examines existing data that show the effects of wetlands on water quality. This section describes the natural physical, chemical, and biological transformations and storages that occur in wetlands and alter the concentrations of water quality parameters of interest in conventional wastewaters. It also provides a combination of empirical and rational approaches that can be used to estimate the wetland area necessary to achieve a given quantitative water quality change. Section 3 is introduced by a detailed description of the hydraulic and chemical design tools that are used in treatment wetlands (Chapter 9). This chapter provides the reader with important background information needed for design of both surface- and subsurface-flow wetlands.

Chapter 10 describes wetland water temperature and the importance and prediction of dissolved oxygen concentrations and summarizes hydrogen ion dynamics (pH changes) in treatment wetlands. The physical and chemical processes attenuating concentrations of suspended solids in wetlands are described and quantified in Chapter 11. Chapter 12 describes the carbon cycle in treatment wetlands and ties this information into design methods for assimilation of biochemical oxygen demand. Chapter 13 summarizes the complex nitrogen transformations and storages that occur in wetlands and provides several alternate methods of sizing wetland treatment systems for nitrogen reduction. Chapter 14 describes the typical phosphorus cycle in wetlands and provides regression and mechanistic models for predicting phosphorus removal in wetland treatment systems. Chapter 15 summarizes information concerning the effects of wetlands on concentrations of other inorganic chemical constituents of wastewater including salts, micronutrients, and various metals. Chapter 16 provides a similar summary for the fate of some common trace organic wastewater constituents in treatment wetlands. Chapter 17 provides a review and summary of the fate of pathogens in wetlands, including the major bacterial indicators, viruses, and parasites.

SECTION 4

Successful project planning requires a careful analysis of the wetland treatment alternatives that are available and subsequent comparison of these wetland alternatives to other natural and conventional treatment technologies. Section 4 (Chapters 18 through 23) provides background for project planning and a summary of design methods for constructed and natural treatment wetlands. Preliminary and final design methods include guidance on site selection, system configuration, safety factors for sizing wetlands based on actual permit limitations, selection of appropriate wetland vegetation, system hydraulics, and water control/monitoring structures. Chapter 18 describes stormwater runoff and the typical range of pollutant concentrations found in municipal, industrial, and agricultural wastewaters. Chapter 19 provides an overview of the various wetland-based treatment technologies, including natural and constructed surface-flow wetlands (SF) and soil- or gravel-based subsurface-flow (SSF) wetlands. These technologies are compared to show their common attributes and the design features unique to each technology. Chapter 19 also provides suggestions concerning project planning and permitting, including determination of the technical, regulatory, and economic feasibility of a wetland treatment system project. Chapter 20 describes design of constructed SF systems, and Chapter 21 details design methods for constructed SSF systems. Natural SF system design is described in Chapter 22. Wetland design for ancillary environmental and human benefits is described in Chapter 23.

SECTION 5

Construction techniques to successfully implement constructed and natural treatment wetland projects are described in Chapter 24. Treatment wetland success depends on vegetation establishment and maintenance, as well as continuing operational control and maintenance. Chapter 24 also describes the art of successful wetland vegetation establishment and troubleshooting for the types of plant establishment problems that are sometimes encountered.

One of the newest aspects of wetland treatment technology is optimizing water quality treatment through operational controls. Chapter 25 provides guidance concerning hydraulic and constituent loading to the various wetland systems as well as information on vegetation maintenance, water level control, prevention of nuisance conditions, and general system maintenance.

SECTION 6

One of the best ways to learn is from experience. Although the user of this book may have limited personal experience concerning wetlands, Section 6 (Chapters 26 and 27) compensates for this by providing fairly detailed information concerning planning, permitting, design, construction, and operation of existing representative wetland treatment systems.

Chapter 26 summarizes the results of treatment wetland database efforts and provides information on the geographical occurrence of wetland treatment systems, their sizes, other design information, and their overall treatment performance. Chapter 27 provides case histories for 10 different full-scale wetland treatment systems and includes information concerning the successes and shortcomings of each system. Case histories are provided for five constructed SF systems which receive water from a variety of sources including municipal wastewater, industrial stormwater, and degraded river water. Subsurface-flow wetland case histories include two systems receiving municipal effluents, and the natural wetland case histories include three municipal wastewater systems from both northern and southern climatic areas.

SUMMARY

Treatment wetlands are providing a cost-effective, natural alternative for water quality management in a wide variety of climatic conditions worldwide. Although this technology is still somewhat innovative, long-term operational information now exists from many full-scale engineered wetlands. This information, when augmented by detailed process studies in a number of smaller pilot facilities and combined with operational data from dozens of younger full-scale systems, has led to the development of a significant understanding of the normal physical, chemical, and biological processes occurring in treatment wetlands and to the development of design and operational guidelines to harness this natural purification potential while maintaining or enhancing other beneficial wetland functions important to society.

The information summarized in this book represents the efforts of hundreds of scientists and engineers over the past 30 years. While a synthesis of this massive collection of information is necessary to carry the wetland treatment technology to a wider audience, it may still be beneficial for the reader to refer to the original sources and to examine the evolution of wetland engineering during this formative period. A list of the major publications dealing with the use of wetlands for wastewater management is provided at the end of this chapter. The reader is also encouraged to examine the detailed references cited throughout the text and provided at the end of this book.

KEY REFERENCES CONCERNING TREATMENT WETLANDS

Australian Water and Wastewater Association (AWWA) and International Association on Water Quality (IAWQ), *Wetland Systems in Water Pollution Control: Proceedings of an International Symposium,* Sydney, Australia, 1992.

Bastian, R. K. and S. C. Reed, Eds. *Proceedings of the Seminar on Aquaculture Systems for Wastewater Treatment,* U.S. EPA Publication No. MCD-67, 1979.

Behandlung von häuslichen Abwasser in Pflanzenbeeten, ATV, Gesellschaft zur Föderung der Abwassertechnik e.V., Postfach1160, Markt 71, D-5205 St. Augustin 1, Germany, 1989. English translation of this design guideline document available.

Chan, E., T. A. Bursztynsky, N. Hantshe, and Y. J. Litwin. *The Use of Wetlands for Water Pollution Control,* U.S. EPA600/2-82-036, 261 pp. 1982.

Cooper, P. F., Ed. *European Design and Operations Guidelines for Reed Bed Treatment Systems.* Prepared by EC/EWPCA Emergent Hydropohyte Treatment Systems Expert Contact Group. Swindon, U.K.: Water Research Centre, 1990.

Cooper, P. F. and B. C. Findlater, Eds. *Constructed Wetlands in Water Pollution Control.* Pergamon Press, London, 1990.

Davis, F. E., Ed. *Water: Laws and Management.* Bethesda, MD: American Water Resource Association, 1989.

Eisenberg, D. M. and J. R. Benneman. *An Overview of Municipal Wastewater Aquaculture,* U.S. EPA, Draft Final Report, Contract No. DM41USC252(C), 1982.

Etnier C. and B. Guterstam, Eds. *Ecological Engineering for Wastewater Treatment.* Gothenburg, Sweden: Bokskogen, 1991.

Fisk, D. W., Ed. *Wetlands: Concerns and Successes.* Bethesda, MD: American Water Resource Association, 1989.

Greeson, P. E., J. A. Clark, and J. E. Clark, Eds. *Wetland Functions and Values: The State of Our Understanding.* Minneapolis, MN: American Water Resource Association, 1979.

Hammer, D. A., Ed. *Constructed Wetlands for Wastewater Treatment.* Chelsea, MI: Lewis Publishers, 1989.

Hammer, D. E. and R. H. Kadlec, *Design Principles for Wetland Treatment Systems.* U.S. EPA600/2-83-026, 1983.

Hook, D. D., et al., Eds. *The Ecology and Management of Wetlands.* Portland, OR: Timber Press, 1988.

Hyde, H. C., R. S. Ross, and F. Demgen, Eds. *Technology Assessment of Wetlands for Municipal Wastewater Treatment,* U.S. EPA600/2-84-154, NTIS No. PB 85-106896, 1984.

Jiang, Ch., Ed. *Proceedings of the Fourth International Conference on Wetland Systems for Water Pollution Control.* Eds. Center for International Development and Research, South China Institute for Environmental Sciences, Guangzhou, China, 798 pp. 1994.

Kaynor, E. R., P. J. Godfrey, and J. Benforado, Eds. *Ecological Considerations in Wetland Treatment of Municipal Wastewaters.* New York: Van Nostrand Reinhold, 1989.

Knight, R. L., R. W. Ruble, R. H. Kadlec, and S. C. Reed. Database: North American Wetlands for Water Quality Treatment, Phase II Report. Prepared for U.S. EPA, September 1993, 824 pp. Diskette available from D.S. Brown, U.S. EPA, Risk Reduction Engineering Laboratory, Cincinnati, OH 45268, 1993.

Mitsch, W. J., Ed. *Global Wetlands: Old World and New.* Amsterdam, Elsevier. 1994.

Moerassen voor de Zuivering van Water (Wetlands for the Purification of Water). Post Academic Course, published as The Utrecht Plant Ecology News Report. University of Utrecht, Lange Nieuwstraat 106, 3512 PN Utrecht, The Netherlands, No. 11. October 1990 (about 50% in English),

Moshiri, G. A., Ed. *Constructed Wetlands for Water Quality Improvement. Proceedings of an International Symposium.* Chelsea, MI: Lewis Publishers, 1993.

Reddy, K. R. and W. H. Smith, Eds. *Aquatic Plants for Water Treatment and Resource Recovery.* Orlando, FL: Magnolia Publishing, 1987.

Reed, S. C., E. J. Middlebrooks, and R. W. Crites. *Natural Systems for Waste Management and Treatment.* New York: McGraw-Hill, 1988.

Reed, S. C. *An Inventory of Constructed Wetlands used for Wastewater Treatment in the United States.* Prepared for U.S. EPA, 1990.

Saurer, B., Ed. *Pflanzenkläranlagen: Abwasserreinigung mit Bepflantzen Bodenkörpern.* Wasserwirtschaft Land Steiermark, Graz, Austria, 172 pp. (in German), 1994.

Tennessee Valley Authority. General Design, Construction, and Operation Guidelines: Constructed Wetlands Wastewater Treatment System for Small Users Including Individual Residences. Technical Report TVA/MW—93/10. Chattanooga, TN, 1993.

Tilton, D. L., R. H. Kadlec, and C. J. Richardson, Eds. Freshwater Wetlands and Sewage Effluent Disposal. *Proceedings of NSF/RANN Conference.* Ann Arbor, MI: The University of Michigan. NTIS PB259305, 1976.

U.S. Environmental Protection Agency. *The Effects of Wastewater Treatment Facilities on Wetlands in the Midwest.* U.S. EPA 905/3-83-002, 1983.

U.S. Environmental Protection Agency. *The Ecological Impacts of Wastewater on Wetlands: An Annotated Bibliography.* U.S. EPA 905/3-84-002, 1984.

U.S. Environmental Protection Agency Office of Water. Subsurface Flow Constructed Wetlands for Wastewater Treatment: A Technology Assessment. EPA Report 832-R-93-001, 1993.

U.S. Environmental Protection Agency. *Freshwater Wetlands for Wastewater Management Handbook.* U.S. EPA 904/9-85-135, 1985.

U.S. Environmental Protection Agency. Report on the Use of Wetlands for Municipal Wastewater Treatment and Disposal. EPA Report 430/09-88-005, 1987.

U.S. Environmental Protection Agency. Phase I Report: Freshwater Wetlands for Wastewater Management. EPA Report 904/9-83-107, 1983.

U.S. Environmental Protection Agency. *Design Manual: Constructed Wetlands and Aquatic Plant Systems for Municipal Wastewater Treatment.* U.S. EPA 625/1-88/022, 1988.

Water Pollution Control Federation. *Manual of Practice: Natural Systems, for Wastewater Treatment.* Manual of Practice FD-16, Chapter 13: Wetland Systems—Alexandria, VA, 1990.

Wetland Systems in Water Pollution Control. *Water Sci. Technol.,* 29 (4):1–336. 1994.

Wittgren, H. B. and K. Hasselgren. Naturliga system för avloppsrening och resursutnyttjande i tempererat klimat. VA-VORSK, Stockholm, Sweden, 1993. 68 pp. (in Swedish).

CHAPTER 2

Review of Water Quality Treatment Needs

INTRODUCTION

Civilization has been highly successful at using the energy flows and natural resources of nature for its own nurture and growth. The primary and secondary productivity of the world has increasingly been directed toward feeding humans, primarily through cultural evolution, a process largely independent of genetic evolution in which environmental conditions are adapted through the use of tools, shelter, trade, extraction of natural resources, and ecosystem control. While cultural evolution has increased human populations and dominance, it has also led to the increase of concentrated solid and liquid wastes and has resulted in environmental pollution, which can be defined as the release of gaseous, liquid, and solid wastes that exceed the assimilative processes of the adjacent natural systems.

As human populations grow in size and density, the magnitude of environmental contamination and pollution problems grow even more rapidly (Figure 2-1). When human populations are relatively sparse and spread across the landscape, waste products are minor compared to the assimilative capacity of the surrounding natural environment. But as populations become more focused geographically and the borders between developed and natural systems are pushed back due to urban sprawl, the ratio of waste products to natural assimilative capacity (which is a function of natural landscape area) increases rapidly. The challenge of protecting the environmental health of human and natural ecosystems in a developed world has led to two major waste management processes: (1) sophisticated waste storage and treatment technologies and (2) increased pollution prevention through conservation, recycling, and byproducts use.

In many situations, pollution prevention is clearly the most cost-effective and environmentally sound method of dealing with environmental pressures exerted from overpopulation and urban sprawl. Pollution prevention includes the reduction of pollution by minimizing inputs to the production and manufacturing sectors by developing higher efficiencies, best management practices, and new processes. It also includes the conscious reduction of energy and material use (conservation). Recycling nutrients and solids from the waste stream or from manufacturing byproducts also reduces the waste stream volume and, in some cases, its concentration of pollutants. However, technological limits to pollution prevention and recycling, as well as economic forces, further influence decisions in manufacturing that are not based on environmental considerations.

Concentrated wastes that cannot be eliminated through pollution prevention must be treated to lower pollutant levels so that they are below threshold concentrations that will impair the receiving environment. Treatment levels must be continually increased as the available assimilative capacity of the remaining natural environment shrinks. Technological development must focus resources on continuously improving technologies for waste management. Also, as wastes become more concentrated and pollution more of a problem, existing waste sources that were not previously a problem will become a new focus for treatment.

(a)

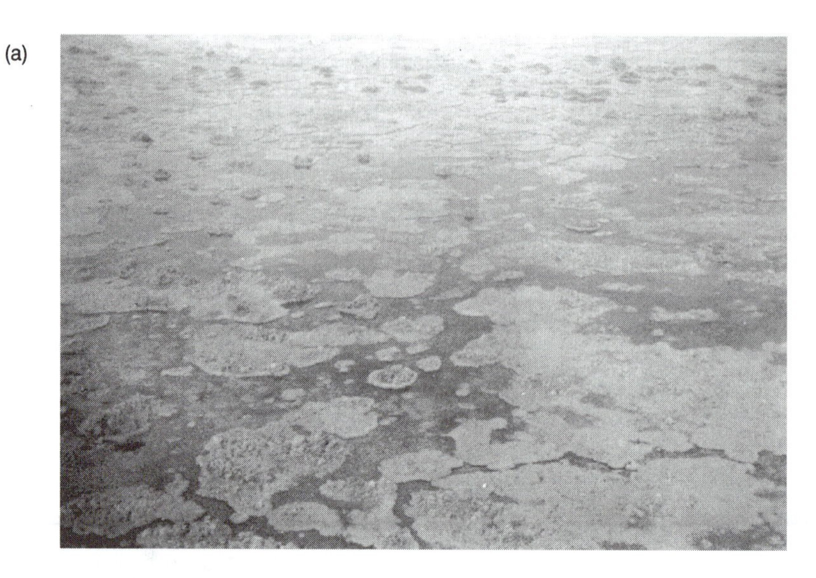

(b)

Figure 2-1 The ratio of concentrated urbanized development to natural landscapes is proportional to the cost per capita of sophisticated pollution treatment technologies. In the rural setting (a), there is little evidence of pollution in spite of a low cost per capita spent on wastewater control, while in the urban setting (b), the cost per capita spent on pollution control must increase to maintain a constant receiving water condition.

SOURCES OF POLLUTANTS

In this book, wastewater pollution sources are organized into the following categories: municipal wastewaters, industrial wastewaters, agricultural wastewaters, and stormwaters (Table 2-1). Although these sources nearly always overlap, this classification system will help to focus design on the typical flow regimes, constituent concentrations, and mass loadings from each source.

Municipal wastewater consists of a combination of domestic wastewaters, originating in households, offices, and public restrooms, and lesser contributions from many commercial and small industrial sources. Domestic wastewater solids include a mixture made of 33

Table 2-1 Categorization of Wastewater Sources Covered in This Book

Wastewater Category	Origin	Frequency	Description
Municipal wastewater	Primarily residential and commercial	Continuous flows	Dilute to concentrated mixture of urine, feces, paper, plastics, soaps, grease, and miscellaneous household and industrial chemicals
Industrial wastewater	Primarily industrial processes and leachates	Continuous to intermittent	Dilute to concentrated solution or slurry of a few to many biodegradable and nondegradable chemicals
Agricultural wastewater	Intensive agricultural practices such as milking or washing barns, feedlots, and slaughter houses	Continuous to intermittent	Dilute to concentrated mixture of biodegradable compounds
Stormwaters	Runoff from urban, suburban, and rural areas	Intermittent	Typically dilute mixtures of mineral and organic solids and dissolved salts, nutrients, and trace constituents

percent soaps and soil solids, 20 percent urine, 18 percent ground food wastes, 16 percent feces, and 7 percent paper (Metcalf and Eddy, 1991). The remaining 5 percent of the solids in domestic wastewater originate in the water supply itself. Commercial inputs to municipal wastewater sources include wastewater generated by car washes, restaurants, photofinishing shops, laundries, and bars. Industrial wastewaters, with or without pretreatment, frequently are a component of municipal wastewater flows.

Because of the residence time of the wastewaters in pipes and basins, municipal wastewater flow rates are relatively continuous, with hourly, daily, and seasonal fluctuations corresponding to the mix of residential and commercial sources. The typical ratio between the maximum month and the annual average flow rates is about 1.2 (Metcalf and Eddy, 1991). Ratios for the maximum week, day, and hour daily flow rates and the annual average flow are about 1.4, 1.8, and 2.7, respectively. The per capita flow rates used in the design of conventional wastewater treatment facilities are about 0.23 to 0.38 m^3/d.

Currently, there are more than 15,600 municipal wastewater treatment facilities in the U.S. treating a total wastewater flow of about 112,000,000 m^3/d (U.S. EPA, 1993c). Most of these systems are relatively small, with 80 percent treating less than 3786 m^3/d, and 96 percent treating less than 37,860 m^3/d. However, municipal treatment facilities greater than 37,860 m^3/d treat about 66 percent of the total flow. While wetland treatment systems can provide primary and secondary treatment of municipal wastewaters, they are more typically used for advanced treatment beyond the secondary level. Most of the wetland treatment systems in North America are being used to treat municipal wastewaters, generally at flows less than 3786 m^3/d (Knight et al., 1993a).

Industrial wastewaters originate from a wide variety of processes and facilities. Only a fraction of industrial facilities discharge wastewaters that might be amenable to use of wetland systems for treatment. The rest contain potentially toxic constituents at toxic concentrations that could be detrimental to wildlife attracted to natural treatment systems. The types of industrial wastewaters that are most amenable to treatment by wetlands are listed in Table 2-2 and include pulp and paper, petroleum refining, food processing, slaughtering and rendering, some landfill leachates, and some chemical and plastics manufacturing wastes. Nontoxic waste streams (including stormwaters) from industrial facilities also can be treated by wetlands. As with all wetland treatment options, pretreatment requirements for industrial wastes are determined from maximum sustainable wetland inflow concentrations and loads discussed in Chapters 18 and 19.

Table 2-2 Industrial Wastewater Sources Most Amenable to Treatment by Natural System Processes

Category	Typical Pollutants	Pretreatment Process
Pulp and paper	BOD, COD, TSS, TN, color	Primary sedimentation; aeration stabilization; activated sludge
Food processing	BOD, TSS, TN, salts	Flocculation; sedimentation; anaerobic digestion; activated sludge
Slaughtering and rendering	BOD, TSS, TN, grease, salts	Flocculation; sedimentation; anaerobic digestion, activated sludge
Chemical manufacturing	COD, TN, metals, salts, organics	Sedimentation chemical precipitation; aeration
Petroleum refining	Oil and grease, TSS, BOD_5, COD, salts, metals, organics	Oil/water separation; aeration stabilization; activated sludge
Landfill leachates	BOD_5, COD, TSS, TN, salts, metals, organics	Aeration stabilization; metals precipitation; activated sludge

Note: BOD, biochemical oxygen demand; COD, chemical oxygen demand; TSS, total suspended solids; TN, total nitrogen; and BOD_5, 5-day biochemical oxygen demand.

Agricultural wastewater sources include the wastewaters generated from intensive agricultural practices including dairies, feedlots, swine houses, and aquaculture facilities. Most of these flows currently receive minimal treatment, primarily through the use of facultative lagoons. Wetlands are becoming increasingly important for treatment of fairly concentrated or pretreated agricultural wastewaters.

The last wastewater source category identified in this book is stormwaters. Table 2-3 summarizes the general characteristics of stormwaters that may require water quality treatment. Runoff from storm events may be encountered as either point or nonpoint discharges. Stormwater runoff is generally intermittent with varying pollutant concentrations dependent on storm duration and frequency (Figure 2-2). Typically, urban and suburban area runoff is channelized to storm sewers where it enters receiving waters at a number of point discharges or is joined by municipal wastewaters and discharged as a combined sewer overflow (CSO).

Stormwater runoff from agricultural and natural lands is generally more diffuse, entering receiving waters as overland or sheet flow or at identifiable ditches and canals. The quality of rural stormwaters and required treatment needs may be very different from urban and suburban stormwaters. Included in this category are flows in canals, lakes, and rivers that are being diverted to wetlands for treatment.

Table 2-3 Stormwater Categories That May Require Water Quality Treatment

Category	Typical Pollutants	Frequency	Pretreatment Process
Agricultural runoff	Mineral solids, nutrients, pesticides	Intermittent	Typically none
Residential runoff	Mineral solids, organic matter, nutrients, pesticides, oil and grease	Intermittent	Vegetated swales, wet or dry detention ponds; oil and grease skimmers
Urban runoff	Mineral solids, garbage, nutrients, metals, oil and grease	Intermittent	Detention ponds; oil and grease skimmers
Land drainage	Mineral solids, organic matter, nutrients	Intermittent to continuous	Typically none
Industrial runoff	Mineral solids, priority pollutants	Intermittent	API oil/water separators, wet detention
Streams and drains	Mineral solids, nutrients, pesticides	Continuous to intermittent	Typically none

Note: API = American Petroleum Institute.

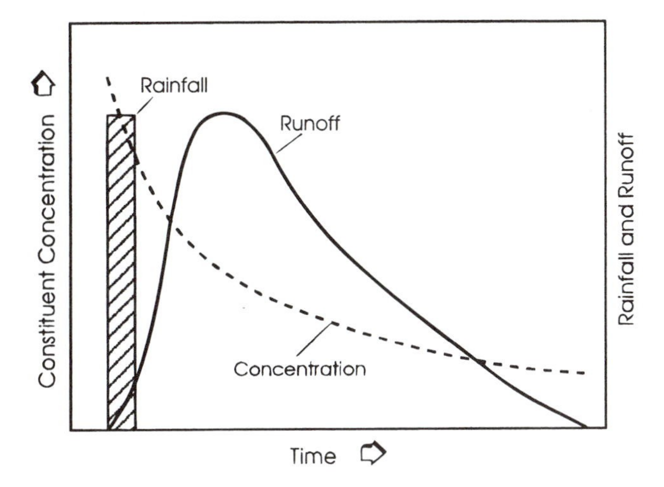

Figure 2-2 Generalized progression of runoff volume and pollutant concentrations during and following a rainfall event.

Until recently, most stormwater flows received no treatment. Now, stormwaters are being treated to further protect receiving waters that were previously affected by municipal wastewaters before the Clean Water Act of 1974 and its amendments were implemented. In many cases, the stormwater component is now the greatest contributor to water body impairment (U.S. EPA, 1989a). Therefore, wetland treatment systems are being increasingly used for new stormwater treatment systems (Strecker et al., 1990; Olson, 1992).

CONVENTIONAL TREATMENT TECHNOLOGIES

Conventional wastewater treatment is accomplished by physical, chemical, and biological processes. Many of these processes are general in nature and can function within a variety of treatment schemes. A review of these technologies is valuable for planning and designing a wetland treatment project for at least two reasons. First, pretreatment with conventional processes is usually advisable before discharge into a wetland because of the potential solids or oxygen demand overload that might create nuisance conditions within a wetland receiving raw or inadequately treated wastewaters. The wetland designer should be aware of the types of conventional processes that can be used to accomplish this pretreatment.

Second, a wetland treatment technology may not be the most cost-effective, environmentally sensitive, or technically reliable process for a given wastewater or project location. Conventional treatment technologies should be compared with wetlands and other land treatment technologies before final project planning and design is begun. A knowledge of conventional wastewater treatment approaches, as well as the other natural land technologies described in Chapter 3, is essential to make a sound evaluation of the most appropriate treatment technology or combination of technologies for a given application. For detailed references on conventional wastewater treatment technologies, see Metcalf and Eddy (1991) and the Water Environment Federation (WEF) (1992).

Table 2-4 summarizes the typical removal efficiencies obtained with the conventional treatment processes discussed later. These efficiencies are the estimated average unit efficiencies for typical municipal wastewaters and may be different for industrial wastewaters or for nontypical municipal wastewaters.

Table 2-4 Summary of Mass Removal Efficiencies (%) for Conventional, Energy Intensive, Wastewater Treatment Technologies

Treatment Process	Constituent						
	BOD_5	COD	TSS	TP	Org-N	NH_3-N	TN
Grit removal	0–5	0–5	0–10	<1	<1	<1	<1
Primary sedimentation (without coagulation)	30–40	30–40	50–65	10–20	10–20	0–20	10–20
Primary sedimentation (with coagulation)	40–70	30–60	60–90	70–90	—	—	—
Activated sludge	80–95	70–85	80–90	10–50	15–20	8–65	—
Trickling filters	65–80	55–80	60–85	8–12	15–50	8–15	—
Rotating biological contactors	80–85	80–85	80–85	10–25	15–50	8–15	—
Oxidation ditch	86–99	—	81–98	—	—	20–80[a]	—
Tertiary treatment[a]	80	80	80	60–95	0–20	80–95	80–85

[a] Estimated.
Note: Data obtained from Metcalf and Eddy, 1991; WEF, 1992; Williams, 1982.

PRIMARY TREATMENT

Primary treatment is considered "the first line of defense" in wastewater treatment (WEF, 1992) because it sets the stage for the majority of biological treatment technologies that follow. Primary treatment consists of screening, grit removal, and primary sedimentation (Figure 2-3). Screening and grit removal may be referred to as "preliminary treatment" because they remove larger solids from the wastewater and the heavier mineral solids that might otherwise erode mechanical equipment downstream in the treatment facility.

Screening of large solids removes gross pollutants that might otherwise interfere with other treatment processes or with mechanical equipment such as pumps, valves, and aerators. Coarse screens have openings of about 6 mm, while fine screens trap solids from about 1.5 to 6 mm. Coarse screens may be as simple as manually cleaned bar screens or trash racks or may consist of continuously or intermittently operated, mechanically cleaned, moving screens. A review of screening data from 29 wastewater treatment plants in the U.S. indicates that the average screenings recovered are about 10.9 cm^3/m^3 of effluent, with a range from 0.4 to 88 cm^3/m^3 (WEF, 1992).

An alternative to screening in preliminary treatment is the use of a comminutor or grinder to reduce the physical size of wastewater solids. Generally, these systems are not preferred because of the plastics and rag solids that remain in sludges in downstream processes.

Grit in raw wastewater primarily consists of inorganic and organic solids that enter the collection system and include materials such as sand, gravel, seeds, coffee grounds, and other minimally decomposable organic solids. Because grit is more settleable than more highly decomposable organic solids, it should be removed in the front end of the treatment plant to protect mechanical equipment from abrasion and prevent sedimentation in pipelines and basins. Grit removal processes include aerated grit chambers or other turbulent flow chamber designs. The grit is then concentrated and removed from the chamber with bucket

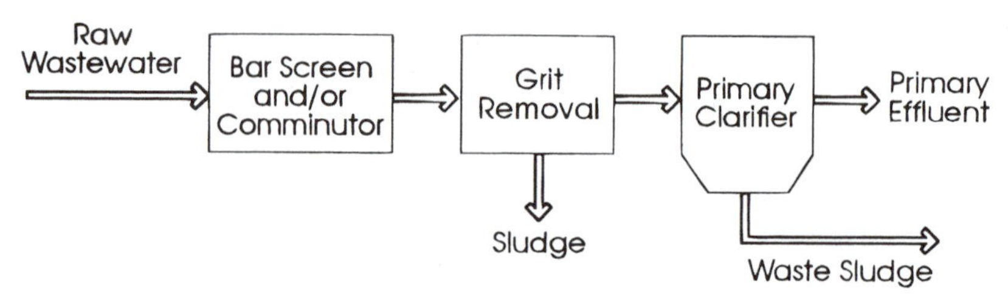

Figure 2-3 Diagram of typical primary treatment.

collectors, screw augers, or slurry pumps. After grit is removed, it is generally washed and landfilled. A review of data from 38 treatment plants indicates that typical grit production is about 21.7 cm^3/m^3 with a range from 2.9 to 78 cm^3/m^3 (WEF, 1992).

Primary sedimentation is used to initially reduce the high concentration of total suspended solids (TSS) present in raw wastewater. Sedimentation is accomplished by creating quiescent flow conditions within a fairly deep (typically 3 to 5 m) pond or concrete vessel known as a primary clarifier. Settled solids are removed as sludge for further treatment, dewatering, and disposal, while the supernatant is removed via weirs to undergo additional treatment or discharge. Primary sedimentation basins typically incorporate skimmers for removal of floatable materials called scum (grease, oils, waxes, hair, paper, and plastics). Scum is typically disposed of by landfilling, incinerating, or digesting with other sludges. Primary sedimentation may also be enhanced by preaeration to promote flocculation (aggregation of smaller particles into larger particles called flocs) or through chemical coagulation with iron salts, alum, or lime.

SECONDARY TREATMENT

Secondary treatment is the minimal level of municipal and industrial treatment that is required in the U.S. before discharge to most surface receiving waters. Secondary treatment requires a treatment level that will produce concentrations of 5-day biochemical oxygen demand (BOD_5) and TSS less than 30 mg/L and, in addition, a minimum percent reduction of 85 percent.

Secondary treatment generally consists of the removal of additional wastewater solids and dissolved organic matter through microbial uptake and growth. Thus, secondary treatment is essentially a biological process in which bacteria and fungi are encouraged to grow in lagoons, mixed tanks, and ponds or on fixed surfaces. The principal secondary treatment technologies are facultative ponds, aerated lagoons, aeration basins with solids recycling (activated sludge), trickling filters, and rotating biological contactors. The pond systems, generally classified as natural treatment systems, are described in Chapter 3. Brief descriptions of the other secondary treatment processes are provided below.

The basic process of activated sludge or suspended growth wastewater treatment consists of two basins: an aerated reactor basin with a hydraulic residence time between 0.5 and 24 h, followed by a settling basin or clarifier with a hydraulic residence time of about 3 to 7 h; and recycling and removal of sludge collected in the two basins (Figure 2-4). Suspended biological solids are grown heterotrophically in the natural wastewater "broth," thereby removing soluble organics and nutrients from the water column and allowing removal in a solid form as sludge. These microorganisms or microbes would be constantly washed out

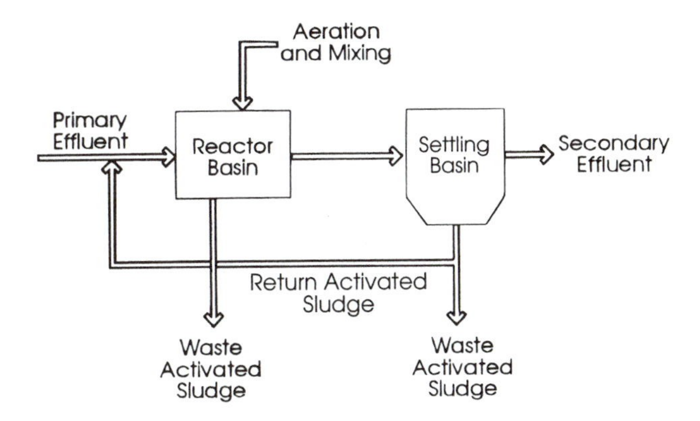

Figure 2-4 Diagram of activated sludge (suspended growth) treatment process.

of the reactor basin without a continuous recycle of additional populations concentrated in the settling process (return activated sludge). The biological solids in the reactor basin are kept suspended and aerobic by the addition of vigorous aeration.

The activated sludge process is highly efficient at removing residual biochemical oxygen demand and suspended solids remaining in primary wastewater and can be adapted to also reduce ammonia nitrogen, total nitrogen, and phosphorus. Some modifications of the conventional activated sludge treatment process include extended aeration in which hydraulic residence times are extended to about 16 to 24 h and the density of suspended microbial solids is increased through a higher recycle rate and lower wastage of sludge; contact stabilization which is a plug flow system with relatively short hydraulic residence time (0.5 to 3 h) and high inflow loading; and oxidation ditches which rely on an oval basin design with the extended aeration process and sometimes with internal clarification.

The second group of secondary treatment technologies relies on attached growth of microbial populations to extract soluble carbon and nutrients from primary effluents. These attached microbial populations are continually or intermittently exposed to the atmosphere to provide reaeration and to maintain aerobic conditions. Attached growth technologies can be very effective for removal of carbonaceous, oxygen-demanding constituents and can be easily adapted for simultaneous removal of nitrogenous oxygen demand (ammonia nitrogen). The trickling filter and rotating biological contactor technologies are two variations of this attached growth treatment process.

Trickling filters consist of a 1- to 8-m deep, typically round basin filled with a growth medium for microbial populations (Figure 2-5). The media most commonly used are rock (or slag) and plastics. Primary effluent is loaded over the surface of the trickling filter and allowed to move downward through the media by gravity, entraining air and nourishing the attached microbial populations with water, organic carbon, and nutrients. An underdrain system collects and transports the water and sloughed solids to a secondary clarifier. Part of

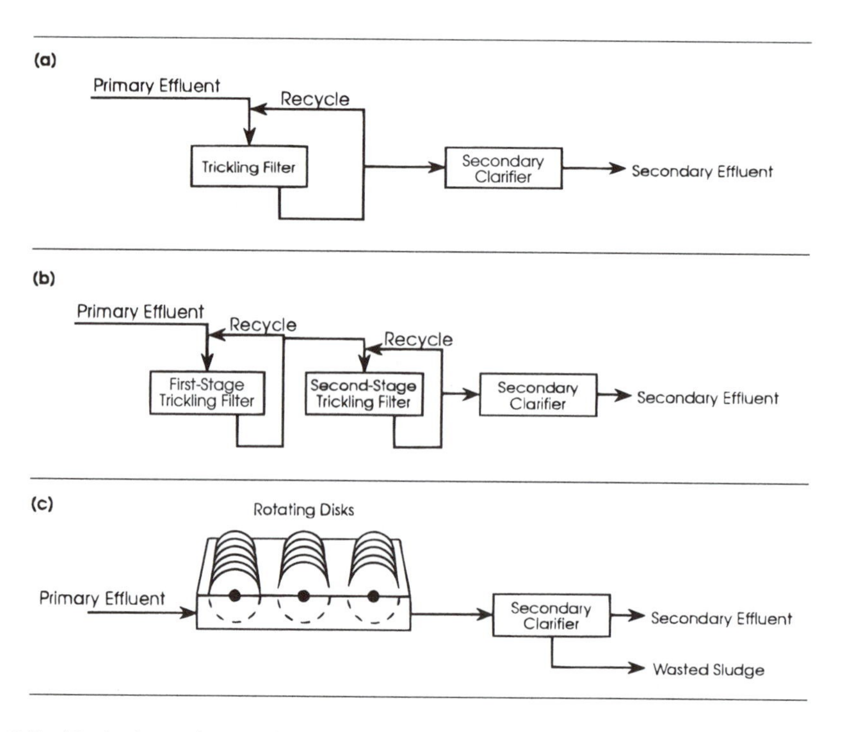

Figure 2-5 Attached growth secondary treatment system schematic diagrams: (a) single-stage trickling filter, (b) two-stage trickling filter, and (c) rotating biological contactor.

the water collected from the trickling filter is typically recirculated and mixed with the primary effluent to increase the hydraulic loading rate without increasing the constituent mass loading rates. Many trickling filter process configurations are used, and Figure 2-5 illustrates examples of typical single-stage and two-stage flow configurations.

Rotating biological contactors use circular plastic disks (media) mounted on a horizontal shaft and turning at about 1 to 2 rpm through a shallow wastewater-filled tank (Figure 2-5c). The lower 40 percent of the disks are suspended in the wastewater flow, while the upper portion of the disks are exposed to the air. The plastic disks are colonized by attached microbial growth which accomplishes secondary wastewater treatment. Each shaft containing disks is called a stage, while a unit may contain a number of stages. Multi-unit rotating biological contactors are frequently designed in series, parallel, or both. Covers are typically provided over rotating biological contactors to minimize variation in the physical environment of the treatment process.

ADVANCED WASTEWATER TREATMENT (AWT)

Secondary treatment technologies are evolving rapidly, primarily through the combination of existing process units into new combined systems. For example, aerobic and anaerobic stages are combined for biological nutrient removal, and dual biological treatment uses attached growth processes in combination with suspended growth processes. A number of these combined technologies can reduce concentrations of biochemical oxygen demand and suspended solids below the typical secondary treatment level and transform or remove nutrients (primarily nitrogen and phosphorus) from the wastewater stream. Reductions of biochemical oxygen demand, total suspended solids, nitrogen, and phosphorus beyond those typically accomplished by secondary treatment are called tertiary or advanced treatment.

Three forms of advanced wastewater treatment are briefly described in this section: nitrification, denitrification, and phosphorus removal. Refer to Metcalf and Eddy (1991) and WEF (1992) for detailed descriptions of conventional advanced wastewater treatment processes.

Nitrification can be accomplished in either suspended growth or in attached growth systems. Nitrification is an aerobic process (see Chapter 13 for a detailed description) in which bacteria oxidize ammonia to nitrate nitrogen. Standard activated sludge treatment can be modified to accomplish nitrification by increasing solids recycling (sludge age) and the overall hydraulic residence time of the treatment system. Extended aeration-activated sludge systems and oxidation ditch designs typically are capable of nitrification. Trickling filters and rotating biological contactor systems can also be designed for nitrification, especially through the use of multi-stage systems.

The total mass of nitrogen in the wastewater is not reduced by nitrification alone, but can be reduced by a second microbial transformation process called denitrification. In denitrification, nitrate nitrogen is microbially transformed into nitrogen gas which is lost to the atmosphere. This process is anoxic, lacking dissolved oxygen but containing other oxidized compounds such as nitrate nitrogen, and occurs to a limited extent in conventional aerated treatment processes such as activated sludge or trickling filter units. Wastewater treatment systems can be designed for denitrification by including an anaerobic process after effluent nitrification. This process has been added in separate units as well as within single-vessel units. Also, denitrifiers are dependent on a source of reduced organic carbon for heterotrophic growth, so methanol is frequently added to these treatment systems.

Normal microbial growth during secondary treatment results in a sludge with about 1.5 to 2 percent total phosphorus on a dry weight basis. Through sludge wasting, the total phosphorus content of the wastewater receiving secondary activated sludge treatment is reduced by about 10 to 25 percent. Biological phosphorus removal relies on a "luxury uptake" of phosphorus that occurs in microbial populations during growth in more vigorously aerated conditions. With higher uptake rates and increased sludge wastage, a higher percentage of the dissolved phosphorus can be removed from the wastewater. In addition, by sequencing through an anaerobic reactor before entry into the aerobic reactor, phosphorus content of the microbes is initially reduced, allowing a greater removal efficiency from solution in the second reactor.

Figure 2-6 illustrates three of the numerous proprietary biological nutrient removal processes that are accomplished through combinations of aerobic, anoxic, and anaerobic reactor vessels with recycles of water and solids. The A/O™ (licensed by I. Krüger AF, Denmark) process combines an anaerobic and an aerobic reactor in sequence to reduce phosphorus and carbon from secondary wastewaters. The A²/O™ (Kruger, Denmark) process incorporates an anoxic reactor in front of the aerobic vessel with a wastewater recycle to promote denitrification, resulting in a combined biological nitrogen and phosphorus removal system. The Virginia Initiative Process, or VIP (developed by Hampton Roads Sanitation District and CH2M Hill), is another combined process with additional wastewater recycles capable of accomplishing phosphorus removal and partial nitrogen removal. Typical biological phosphorus reduction treatment systems can reliably achieve total phosphorus concentrations lower than 1.0 mg/L.

Phosphorus removal from wastewaters is also frequently accomplished through several conventional chemical and physical processes. Chemical processes typically use aluminum (alum) or iron (ferric) salts to chemically precipitate dissolved phosphorus and remove it in

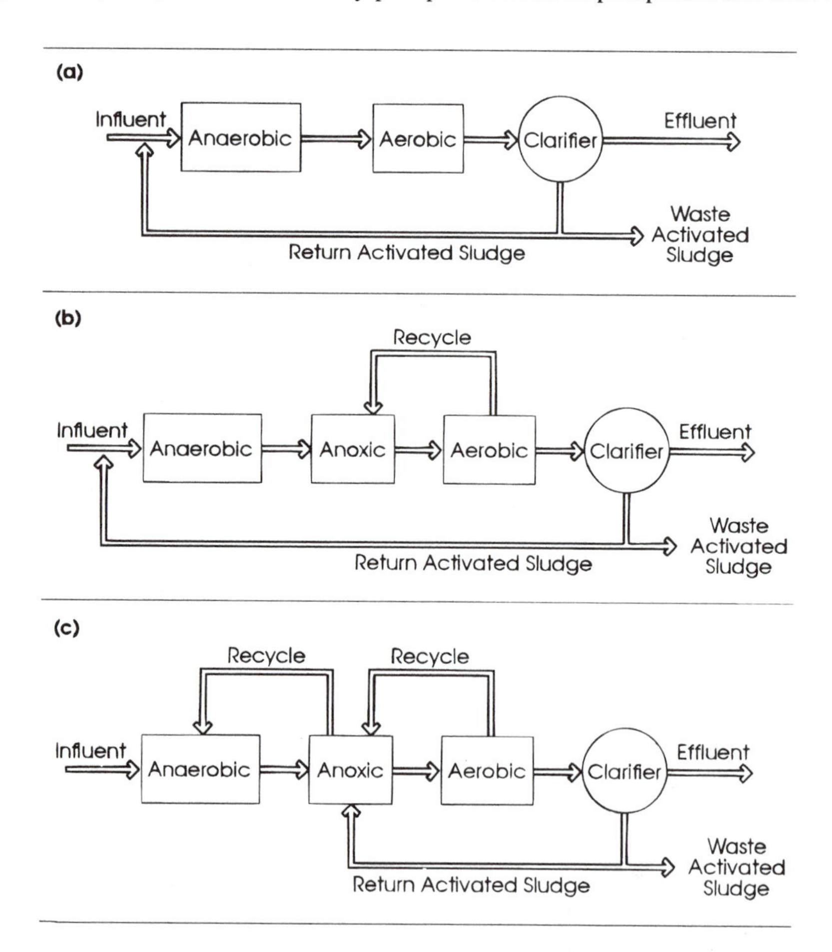

Figure 2-6 Diagrams of some advanced wastewater treatment processes for nitrogen and phosphorus removal: (a) A/O™ process, (b) A²/O™ process, and (c) VIP process.

a solid (sludge) form. Typical iron or aluminum concentrations for phosphorus removal are between 3 and 25 mg/L. This removal may be combined with the primary or secondary clarification process or included as an additional or tertiary process. Total phosphorus removal efficiency through chemical precipitation can exceed 90 percent in municipal effluents, resulting in final total phosphorus concentrations lower than 0.5 mg/L. Physical phosphorus removal processes include ultrafiltration, reverse osmosis, and ion exchange. The first two physical processes rely on filtration of colloidal and dissolved phosphorus with a membrane, while ion exchange relies on the electrical attraction between ionized forms of phosphorus and specific ion exchange resins. All three of these physical processes require extensive pretreatment for suspended solids reduction, and all generate a reject waste stream that may require additional chemical treatment for ultimate phosphorus removal.

SUMMARY

Environmental pollution is a consequence of the society we live in. Increasing population pressure combined with excessive consumption of stored fossil-fuel resources results in widespread exceedances of the natural assimilative capacity of remaining natural environments. As these beneficial processes are exceeded, human populations are limited from further growth and prosperity due to pollution-related loss of renewable natural resources and public health. Pollution control in response to excessive waste releases has developed into a social priority in many developed nations including the U.S.

Pollution control can be effective through prevention or treatment. Pollution prevention is preferable as a first line of defense, and practices such as recycling and energy conservation make good sense in times of short energy supplies and declining economic growth; but, at some point, pollution treatment is easier and more cost-effective in an expanding economy.

Although there are many pollutants commonly found in a variety of waste streams, different pollution sources have some important differences in chemical composition and concentrations and in flow periodicities. The major pollutant source categories discussed in this book include municipal, industrial, agricultural, and stormwater. Each of these general categories may also be conveniently divided into subcategories of more specific pollution sources.

Although this book focuses on the use of wetlands for treatment of wastewaters, a knowledge of conventional treatment methods is important in determining the most effective treatment technology and in design of multi-component treatment "trains" which include pretreatment, primary treatment, secondary treatment, and advanced treatment. In the U.S., wetland systems are most frequently used for advanced secondary or tertiary treatment, and conventional energy-intensive or natural treatment processes are most frequently used for secondary treatment.

Conventional treatment technologies are available and widely used for screening of coarse solids and grits; sedimentation of high suspended solids concentrations; and decomposition of particulate and soluble, oxygen-consuming organic matter. Advanced biological treatment processes are also available for varying levels of nitrogen transformation and phosphorus concentration reductions. Chemical methods may also be employed for phosphorus removal and for treatment of some industrial wastewaters containing excessive metal levels.

Detailed descriptions of wastewater categories and their typical pollutant composition are given in Chapter 18. Multiple conventional and wetland alternatives are available to treat these wastewater sources, and the rest of this book provides a rationale necessary to integrate these diverse technologies into successful, wastewater management systems.

CHAPTER 3

Natural Systems for Treatment

INTRODUCTION

Natural treatment systems for wastewater management are differentiated from conventional systems based on the source(s) of energy that predominates in the two treatment categories (Figure 3-1). In conventional wastewater treatment systems, nonrenewable, fossil-fuel energies predominate in the treatment process. While conventional treatment relies largely on naturally occurring, biological pollutant transformations, these processes are typically enclosed in concrete, plastic, or steel basins and are powered by the addition of forced aeration, mechanical mixing, and/or a variety of chemicals. Because of the power intensity in conventional treatment systems, the physical space required for the biological transformations is reduced considerably compared to the area required for the same processes in the natural environment.

Natural treatment systems require the same amount of energy input for every kilogram of pollutant that is degraded as conventional biological treatment systems; however, the source of this energy is different in natural systems. Natural treatment systems rely (to a greater or lesser extent) on renewable, naturally occurring energies, including solar radiation; the kinetic energy of wind; the chemical-free energy of rainwater, surface water, and groundwater; and storage of potential energy in biomass and soils. Natural treatment systems are land intensive, while conventional treatment systems are energy intensive.

Figure 3-2 summarizes and contrasts the estimated construction and operation and maintenance costs for a conventional activated sludge treatment system capable of achieving advanced secondary effluent quality and a natural treatment system incorporating a facultative lagoon and a constructed wetland, both with a treatment capacity of 3786 m^3/d and with final disinfection. This example does not include the raw wastewater collection and pumping system necessary to deliver wastewater to either of these two systems.

In this highly simplified analysis, the conventional system requires about 2 ha of land area, \$427/d of high-quality labor, energy, and chemical input, with a capital cost of about \$4,112,000. The natural treatment system requires about 36 ha of land, \$123/d of high-quality energies, and solar and wind energies that come with the land, with a capital cost of about \$3,664,000. A detailed comparison of these options would need to analyze the total energies focused into this treatment process, including energy losses occurring during fossil-fuel use (coal and oil) to produce electricity and chemicals. Generally, however, this example provides a good illustration of how conventional and natural treatment processes are different in their individual mixes of energy and land area uses.

Conventional technologies have been an attractive alternative for wastewater treatment in many locations because they provide a compact, controllable method of pollution abatement where large amounts of fossil-fuel energies can be focused to deal with increasing wastewater flows and mass loads. Conventional treatment systems will continue to be used to deal with pollution control in many highly urbanized areas; however, some negative aspects of these energy-intensive systems are increasingly evident.

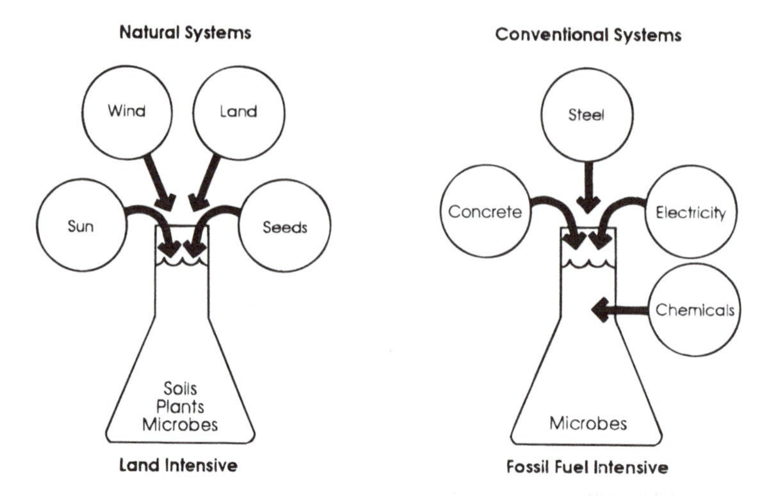

Figure 3-1 Comparison of the energy inputs to natural and conventional wastewater treatment technologies.

Three environmental consequences that are common to most conventional treatment systems include (1) depletion of nonrenewable resources, (2) ancillary environmental degradation associated with extraction and use of these nonrenewable resources, and (3) the fate of residual byproducts resulting from many conventional treatment technologies.

Fossil fuels are essentially nonrenewable resources and are being depleted over time. Any unessential use of fossil fuels will eventually eliminate their availability for more essential uses. For example, reaeration of wastewaters during secondary or advanced treatment can be accomplished by use of electricity to power mechanical aerators or alternatively by more land-intensive atmospheric diffusion. Use of fossil fuels (coal or oil) to generate electricity for aeration that could be provided naturally consumes a resource (electricity) that is irreplaceable for our electronic information society.

There is always an environmental effect associated with the extraction, refining, and transportation of fossil-fuel energies. Thus, use of electricity, plastics, concrete, and chemicals to reduce pollution at a conventional treatment facility results in some pollution elsewhere (Best, 1987). Many conventional treatment processes result in the formation of wastewater residuals or sludge, which in turn presents an environmental disposal problem. Thus, where natural treatment technologies are feasible, they offer the potential to reduce offsite and future environmental consequences associated with pollution control.

The goal of this chapter is to provide an overview of the natural treatment technologies that are currently in use. Treatment technologies included in the overall category of natural systems include onsite infiltration systems, slow-rate land application systems, rapid infiltration land treatment systems, overland flow treatment systems, wastewater stabilization pond systems, floating aquatic plant systems, and wetlands (Water Pollution Control Federation [WPCF], 1990b).

All of these natural treatment technologies are relatively land intensive; however, they have widely varying requirements for supplemental, fossil-fuel energy inputs; specific treatment capabilities; and different strengths and weaknesses for individual applications. Table 3-1 provides a comparison of design parameters and the cost of these natural wastewater treatment technologies.

This chapter contrasts wetland treatment techniques with those of the other land-intensive, natural treatment technologies to help the reader choose the most suitable alternative or group of alternatives for a given treatment need. Detailed information concerning the planning

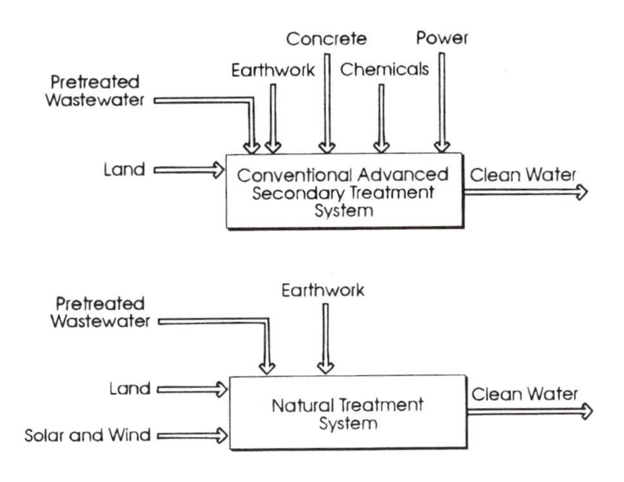

	Construction Costs ($)	
Cost Category	Conventional[a] WWTP	Natural[b] Treatment System
Mobilization & Administration	$ 95,000	$ 91,000
Earthwork (Cleaning, Grubbing, and Excavation)	381,000	1,336,000
Wetland Planting	0	309,000
Other Sitework (Electrical, Controls and Piping)	728,000	1,720,000
Conventional Primary	639,900	0
Conventional Activated Sludge	698,000	0
Sludge Handling	687,000	0
Biological Nitrification	476,000	0
Chlorination and Outfall	208,000	208,000
	$4,112,000	$3,664,000
Operation and Maintenance Costs ($/Year)		
Personnel	$ 63,000	$ 24,000
Utilities	23,000	5,000
Chemicals (including Disinfection)	23,000	11,000
Equipment/Supplies	47,000	5,000
	$156,000	$45,000

[a] Conventional activated sludge with nitrification and disinfection; costs from EPA (1978, 1983) adjusted to 1994.

[b] Faculative lagoon and constructed surface flow wetland with disinfection from EPA (1983) and West Jackson County, MS.

Figure 3-2 Generalized comparison of a conventional activated sludge nitrification advanced secondary treatment plant and a natural treatment system composed of a facultative lagoon and a constructed wetland, both treating 3786 m^3/d of secondary effluent to 10 mg/L BOD and TSS and 2 mg/L NH_4^+.

and design of these other natural treatment systems can be found in WPCF (1990b), Reed et al. (1988), Metcalf and Eddy (1991) Water Environment Federation (WEF) (1992), U.S. EPA (1981, 1984a), and others.

UPLAND NATURAL TREATMENT SYSTEMS

Onsite infiltration systems, slow- and high-rate land application systems, and overland flow systems all rely on the use of relatively well-drained upland areas for treatment (Figure

Table 3-1 Comparison of Natural Wastewater Treatment Technologies

			Design Parameters			Capital Costs					
Natural System Type	Pretreatment Requirements	Treatment Goals	Hydraulic Loading (cm/d)	Specific Treatment Area (ha/1000 m³/d)	Water Depth (m)	$1,000/ha	$/m³/d	O&M Costs $/m³	Disposal To	Advantages	Disadvantages
Onsite infiltration	Primary settling in septic or Imhoff tank	BOD_5 and TSS reduction (approximately secondary)	0.5–4.0	2.5–20	N.A.		1000–3000	0.01–0.10	Groundwater	Zero discharge; low energy use	Requires permeable, unsaturated soils; limited to small systems (<200 m³/d)
Slow-rate land application	Primary or secondary	BOD_5, TSS, and nutrient reductions	0.15–1.6	6–67	N.A.	60–150	800–2000	0.10–0.20	Groundwater	Zero discharge	Requires permeable, unsaturated soils; high energy cost
High-rate land application	Primary or secondary	BOD_5 and TSS reduction	1.6–25	0.4–6	<1	300–600	450–900	0.05–0.10	Groundwater	Zero discharge; low energy use	Requires highly permeable, unsaturated soils; potential nitrate contamination
Overland flow	Primary or secondary	BOD_5 and TSS reduction	1–10	1–10	<0.1	240–400	600–1000	0.08–0.15	Surface water	Aerobic treatment; moderate energy use	Crop maintenance; TSS breakthrough
Facultative ponds	Primary	BOD_5 and TSS reduction	0.7–3.4	3–14	1.2–2.5	80–160	500–1000	0.07–0.13	Surface water	Aerobic/anaerobic treatment; low energy use	High algal TSS in outflow; little operational control
Floating aquatic plant systems	Primary or secondary	BOD_5, TSS, and nutrient reduction	2–15	0.7–5	0.4–1.8	270	500–1000	0.12–0.14	Surface water	Phosphorus removal through harvesting	Anaerobic treatment; plant harvesting and disposal; pests
Wetlands	Primary, secondary, or advanced	BOD_5, TSS, and nutrient reduction	0.4–20	0.5–20	<0.6	25–250	500–1000	0.03–0.09	Surface water	Low energy; aerobic/anaerobic treatment; wildlife habitat	Maintenance of plant populations; hydraulics in subsurface-flow systems

Note: Data from Water Pollution Control Federation (1990b). N.A.—not available.

3-3). All of these technologies use an unsaturated soil layer to provide either direct filtration and assimilation of pollutants or a rooting medium for growth of upland plants which filter wastewater solids and absorb dissolved pollutants for eventual harvest and removal.

Onsite and land application systems provide wastewater treatment coupled with ultimate discharge to groundwater. These systems are called "zero discharge" systems because they typically do not discharge, or only seasonally discharge, to surface waters. Overland flow treatment uses lower permeability, upland soils planted with a grass cover crop. Only a small fraction of the wastewater infiltrates to the groundwater in overland flow treatment, so this technology normally includes a discharge to surface waters.

ONSITE INFILTRATION

Onsite infiltration systems are the most numerous wastewater treatment systems in the U.S. Onsite systems include residential septic tanks and their associated drain fields and

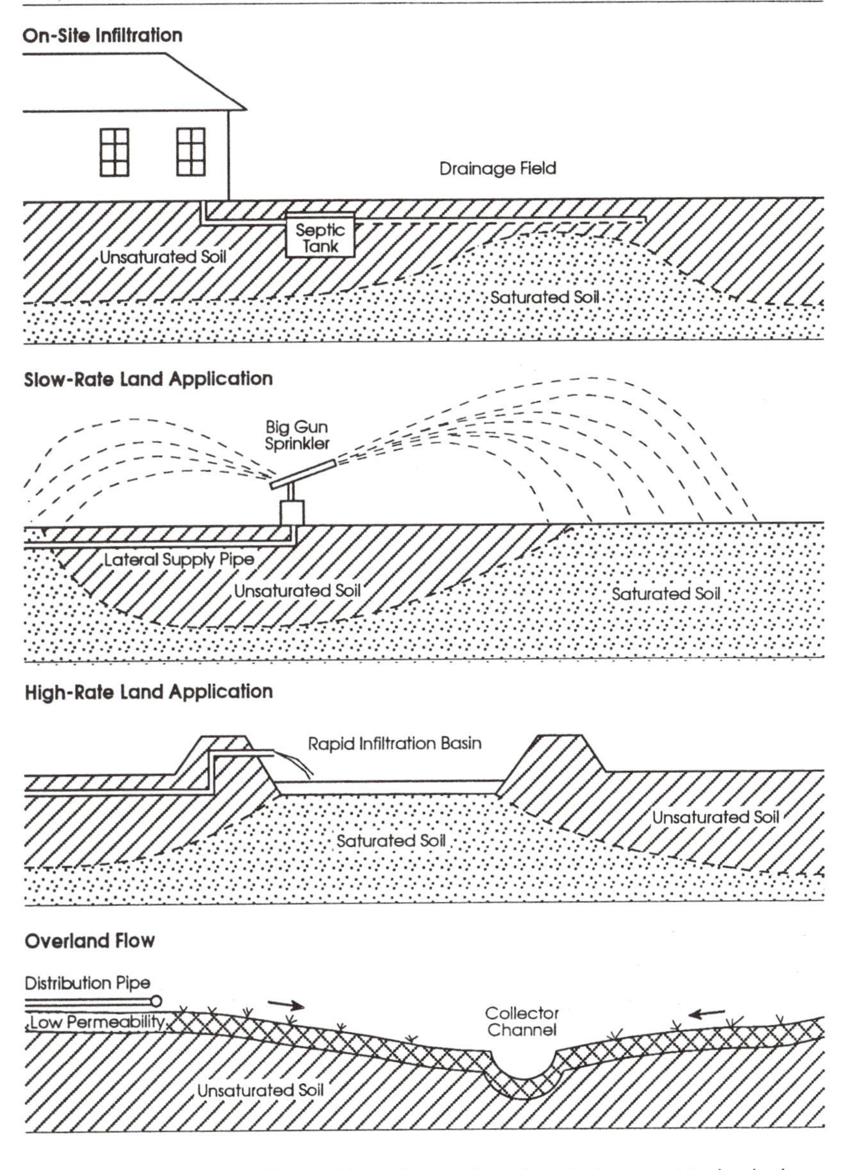

Figure 3-3 Diagrams of upland-based natural wastewater treatment technologies.

larger community systems consisting of a septic or Imhoff tank and a larger drainfield area. Typical flow rates to these systems are less than 200 m^3/d. Most single-family, onsite systems treat less than 1 m^3/d.

The septic tank provides a buried basin which is used for solids settling and anaerobic digestion of solids (Figure 3-4). Although only a small fraction of carbon and other wastewater constituents are removed by a septic tank, these constituents are partially transformed by anaerobic decomposition and converted to more stable particulate and dissolved forms before entering the leach lines.

The leach field consists of branched, perforated pipes surrounded by a highly porous media (typically coarse gravel) and buried in a permeable soil with a minimum of about 1.5 m of unsaturated zone above any existing shallow groundwater. The unsaturated zone can be constructed in areas with low permeability or high surficial groundwater by the use of a mound system using imported soil. The area necessary for a leach field is site specific and depends on existing soil and groundwater conditions. This area can be estimated by using Equation 3-1 from WPCF (1990b):

$$A = 1.5\ Q/k \qquad (3\text{-}1)$$

where A = leach field area, m^2
Q = average wastewater flow, m^3/d
k = soil permeability, $m^3/m^2/d$

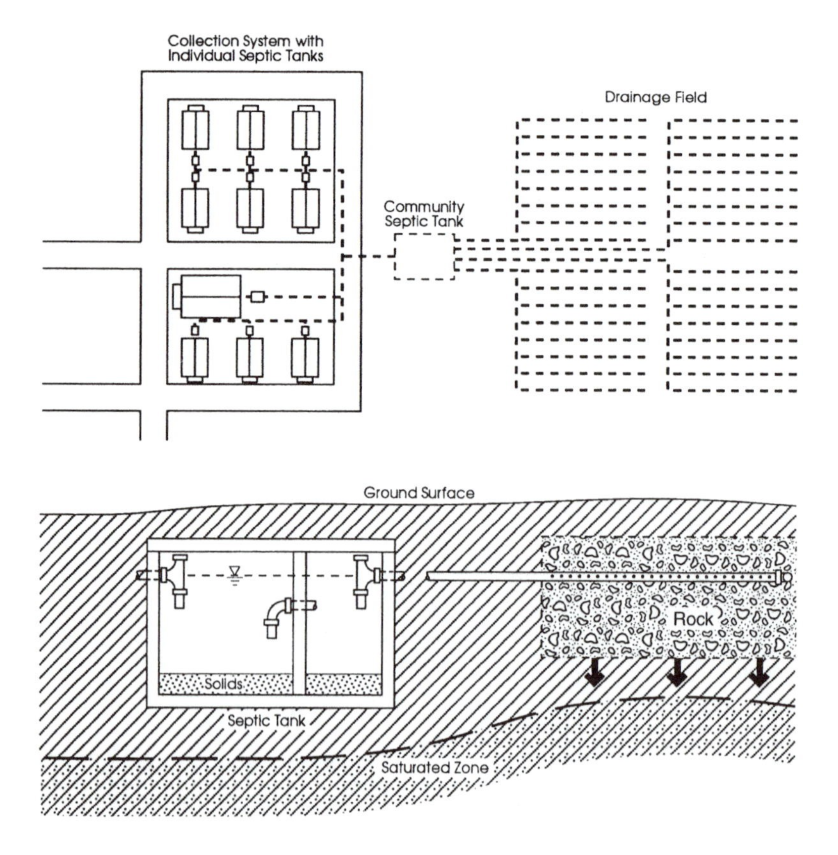

Figure 3-4 Schematic plan and section profiles of a small community onsite infiltration system.

Alternatively, WPCF (1990b) provides a range of hydraulic loading rates (cm/d) for onsite systems based on the texture of the upper 1 m of soil, ranging from 4 cm/d (2.5 ha/1000 m^3/d) for coarse to medium sand to 0.5 cm/d (20 ha/1000 m^3/d) for clays.

Hydraulic loading rate is directly related to the land area required for a given wastewater flow by the equation

$$A_s = 100/HLR \tag{3-2}$$

where A_s = specific treatment area for a given flow, $m^2/m^3/d$
HLR = hydraulic loading rate, cm/d

Onsite systems require relatively low capital investment and operational control. Typical capital cost is \$1000 to \$3000/m^3/d and operation cost is \$0.01 to \$0.1/m^3. However, onsite system design is more complicated and is subject to errors for larger systems because onsite systems typically operate continuously without resting and reestablishment of unsaturated soil conditions. Assumptions concerning the soil infiltrative capacity change radically when the application area is large compared to the wetted edge of the mound of applied wastewaters. Alternatively, slow- and high-rate land application system design accounts for this limitation by alternating application between different spray fields or infiltration basins.

SLOW-RATE LAND APPLICATION

Slow-rate land application of wastewaters uses irrigation of vegetated systems for wastewater polishing and ultimate disposal (Figure 3-5). Irrigation rates are generally low and intermittent, allowing reestablishment of aerobic soil conditions at regular intervals. These aerobic conditions are essential for growth of dry land vegetation which in turn is essential for nutrient removal, filtering of wastewater solids, and maintenance of permeable soil texture. Slow-rate systems are used to treat and dispose of both municipal and industrial wastewaters. More than 800 slow-rate land application systems currently exist in the U.S.

The slow-rate land application technology has a wide variety of process modifications and design criteria depending on project goals. In some cases, water disposal is the primary goal, and the maximum wastewater volume compatible with site characteristics and groundwater criteria is applied to a given land area. These systems frequently use cover crops for partial nutrient removal through harvesting and byproduct recovery. Commonly used cover crops include pasture grasses, corn, legumes, and pine trees. The hydraulic loading rate to this type of land application system is limited by either long-term sustainable soil permeability or by the concentration of the most limiting wastewater constituent at the point of compliance with groundwater standards. The design hydraulic loading rate can be increased by adding soil underdrains; however, underdrains significantly increase system cost and convert this zero discharge technology into an alternative with an intermittent or continuous surface discharge.

In other cases, slow-rate land application is used to irrigate golf courses and other human contact, landscaped areas following a high level of pretreatment. These systems use only enough water to satisfy the requirements of the cultivated plants and generally store or discharge excess wastewaters during periods of rainy weather. In areas with water shortages, treated wastewater becomes a valuable commodity to be conserved and is used sparingly for irrigation of crops or landscaped areas.

Slow-rate land application systems are typically designed with hydraulic loading rates between 0.15 and 1.6 cm/d (6 to 67 ha/1000 m^3/d). Detailed guidelines for calculating land areas for slow-rate land application systems are given by the U.S. EPA (1981), Reed et al. (1988), Metcalf and Eddy (1991), and WEF (1992). Wastewater is generally pumped to multiple irrigated areas and spread using sprinklers, center-pivot irrigators, or ridge and furrow irrigation techniques. Individual irrigation areas may receive water from less than

Figure 3-5 Two slow-rate land application systems.

once to three times per week. Irrigation is generally ceased if there is surface runoff observed from the application area.

The most common problems encountered with slow-rate land application systems are related to overestimation of the long-term soil infiltration capacity during periods of sustained irrigation. These problems are associated with the difficulty of establishing percolation rates during initial site investigations and with changes in soil structure that may occur during construction or operation of these systems. These types of problems are most commonly experienced in areas with clayey soils and in soils with higher seasonal groundwater levels that lack sufficient unsaturated capacity year round or develop low permeability strata through chemical and physical soil changes.

Because of the high land area requirements for slow-rate land application systems and because of the investment in piping and pumping necessary for wastewater distribution, these systems are generally the most costly of the natural system alternatives. Typical capital costs for system installation are about \$60,000 to \$150,000/ha (\$800 to \$2000/m^3/d), and operation and maintenance costs typically range from \$0.1 to \$0.2/m^3.

HIGH-RATE LAND APPLICATION (RAPID INFILTRATION)

High-rate land application systems use highly permeable soils for groundwater discharge (Figure 3-6). High-rate land application systems are generally designed as relatively small or narrow, shallow basins or ponds with berm heights less than 1.5 m. High-rate systems are typically loaded at hydraulic loading rates between 1.6 and 25 cm/d over the bottom area of the basins (0.4 to 6 ha/1000 m^3/d). Berm and buffer areas are additional.

Because of groundwater mounding that occurs beneath high-rate land application basins, a sustainable infiltration rate is a function of the ratio between the length of the basin edges and the bottom surface area. Smaller basin areas and higher length-to-width ratios increase this infiltration rate. Multiple basins are typically used to allow dry down and resting. A careful rotational schedule can eliminate problems occurring due to overlapping groundwater mounds beneath basins. During resting periods, basin permeability may be renovated by rototilling or harrowing. Alternatively, a water-tolerant ground cover crop can be planted in the basins to maintain soil texture and aeration.

At typical hydraulic loading rates, high-rate land application systems provide limited wastewater quality renovation. While a significant fraction of the particulate organic matter and nutrients present in the pretreated wastewater are removed, soluble fractions are generally not diminished. One of the potential problems that occurs with rapid infiltration systems is the oxidation of reduced nitrogen compounds in the aerobic soil zone with the potential for elevated nitrate nitrogen concentrations in receiving groundwaters. The other potential problem with high-rate land application is over optimism concerning long-term soil infiltration rates. A successful design requires careful measurement of infiltration capacity and conservative hydraulic loading rates.

Because of the potentially low land area requirements for high-rate land application systems and the relative ease of periodically applying wastewater to the basins, when technically and regulatorily feasible, this technology is less costly (on a flow basis) than slow-rate land application and most other natural treatment alternatives. Capital costs range from \$300,000 to \$600,000/ha or about \$450 to \$900/m^3/d. Operational and maintenance costs range from \$0.05 to \$0.10/m^3 (WPCF, 1990b).

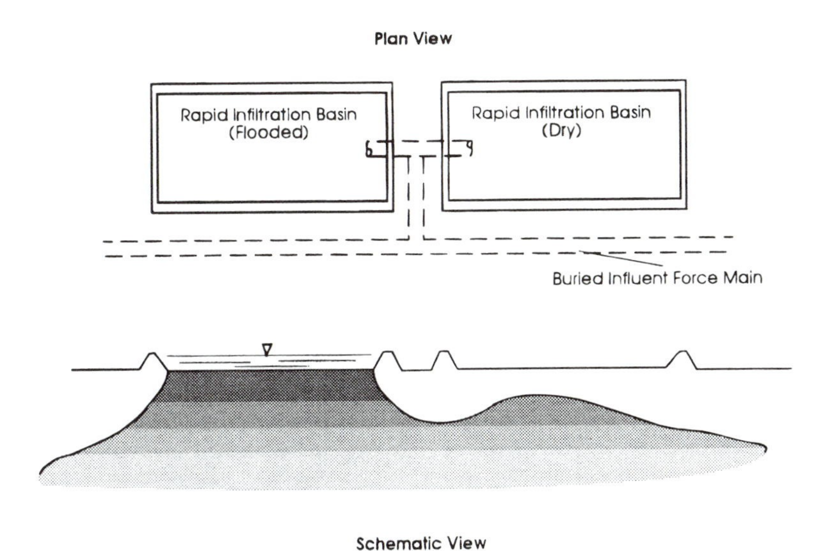

Figure 3-6 Diagram of a typical rapid infiltration system for municipal wastewater disposal.

OVERLAND FLOW SYSTEMS

Unlike other upland alternatives, overland flow treatment systems rely on low permeability soils to restrict infiltration and consequently have a surface discharge.

The conceptual basis of overland flow for treatment is illustrated in Figure 3-7. Pretreated (primary or secondary) wastewater is applied intermittently to the top of sloped, vegetated terraces by gated pipes or by spray nozzles and allowed to flow by gravity down the slopes to a series of collection channels. As water flows through the dense vegetation on the slope, particulate pollutants settle, and dissolved constituents are sorbed by plants and soils. Typically, wastewater application continues for 8 to 12 h out of every 24 h. During resting periods with no application, the organic fraction of the settled particulates is microbially oxidized, and sorbed nutrients are incorporated in biomass (primarily inorganic nitrogen and phosphorus), microbially transformed (nitrification of ammonia nitrogen to nitrate nitrogen), or bound in the soil layer.

Typically, overland flow slopes from 1 to 6 percent are graded by laser technology and are between 30 and 60 m in length. The width of slopes varies to provide the necessary wetted area to accomplish treatment goals. Typical average hydraulic application rates to overland flow systems range from 1 to 10 cm/d (1 to 10 ha/1000 m^3/d).

Overland flow systems are prone to operational problems in two areas: (1) maintenance of a viable cover crop and (2) violation of suspended solids criteria. Both of these problems can occur because of the difficulty of sustaining an even sheetflow on these slopes.

Ponding is likely to occur on overland flow terraces with low slopes, resulting in soil oxygen depletion and eventual death of desired cover crops. Alternatively, on higher slope terraces, erosion is likely to occur and result in high discharge concentrations of mineral sediments. A second factor that can contribute to suspended solids violations in overland flow systems is the relative inability of these systems to remove algal solids. When preceded by facultative or aerated lagoons with high algal production, overland flow systems have had difficulty consistently meeting total suspended solids limits.

Due to lower potential hydraulic loading rates and higher costs for plant maintenance and surface-discharge monitoring, overland flow systems are generally more expensive than high-rate land application systems. Typical costs for overland flow terrace construction are about \$240,000 to \$400,000/ha or \$600 to \$1000/m^3/d (WPCF, 1990b). Operation and maintenance costs range from \$0.08 to \$0.15/m^3.

AQUATIC AND WETLAND SYSTEMS

INTRODUCTION

Aquatic and wetland treatment systems are fundamentally different from upland systems because they are continuously flooded and typically develop an anaerobic sediment and soil

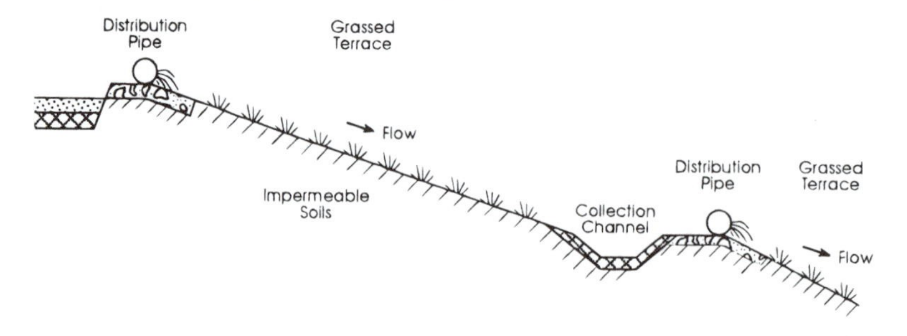

Figure 3-7 Diagram of a typical overland flow wastewater treatment system.

layer. This anaerobic condition excludes the growth of plant species that rely on abundant soil oxygen and results in the simultaneous occurrence of aerobic and anaerobic assimilation processes in a single, layered, natural treatment system. This section briefly describes and contrasts three types of natural, flooded, treatment systems: facultative ponds, floating aquatic plant-based systems, and wetland systems.

FACULTATIVE PONDS

Pond systems are one of the oldest and most widely used wastewater treatment technologies. Pond systems can be passive lagoons dominated by renewable energies from the sun, wind, and biota, or they can be highly sophisticated systems with liners and substantial forced aeration, in which case they are similar to conventional suspended growth treatment systems. This section only describes the lower energy, facultative (stabilization) pond approach to treatment (Figure 3-8).

Facultative ponds are designed to maintain a natural aerated surface layer over a deeper anaerobic layer. Natural aeration occurs because of the combined action of atmospheric oxygen diffusion and the release of oxygen during algal photosynthesis in the water column. Oxygen concentration may be highly variable over daily and seasonal periods within a facultative pond system. Excessive anaerobic conditions in a facultative pond are controlled by limiting the biochemical oxygen demand (BOD) loading rate. Typical design loading rates vary from about 14 to 50 kg BOD_5/ha/d with a detention time between 80 and 180 days (WEF, 1992).

Pond performance is typically a function of the effective hydraulic retention time, which in turn is related to flow dynamics and short circuiting. Multiple cell ponds typically are more effective, and flow curtains or cell configuration can be used to increase the ratio between actual and theoretical residence times. A typical depth for facultative ponds is about 1.2 to 2.5 m. Typical hydraulic loading rates range from about 0.7 to 3.4 cm/d (3 to 14 ha/m^3/d) (WEF, 1992).

Conservatively designed and carefully operated facultative ponds are effective at consistently achieving reductions of biochemical oxygen demand. However, because of their reliance on algal growth, ponds have a fundamental limitation on attaining low suspended solids

Figure 3-8 Photograph of a typical facultative pond wastewater treatment system.

outflow concentrations. These elevated levels of suspended solids (up to and exceeding 100 mg/L) contain a fraction of decomposable organics and nutrients, and, thus, facultative ponds do not produce tertiary quality water. Facultative ponds also have some potential for total nitrogen removal (Reed, 1985), but have little affect on total phosphorus concentrations.

Typical pond capital costs are about \$80,000 to \$160,000/ha, resulting in treatment costs of about \$500 to \$1000/m^3/d (WPCF, 1990b). Typical operation and maintenance costs range from \$0.07 to \$0.13/m^3.

FLOATING AQUATIC PLANT SYSTEMS

Pond systems can be purposely inoculated with floating aquatic plant species to provide wastewater treatment (Figure 3-9). Typical plant species that have been used in large-scale applications are water hyacinths (*Eicchornea crassipes*) and duckweed species (*Lemna, Spirodela,* and *Wolfiella*). Floating aquatic plant treatment systems are functionally different from facultative ponds because the photosynthetic component (floating aquatic plants as opposed to submerged planktonic algae) is releasing oxygen above the water surface, effectively reducing atmospheric oxygen diffusion. Consequently, floating aquatic plant systems are oxygen deficient, and aerobic processes are largely restricted to the plant root zone. The majority of the water column in floating aquatic plant systems is generally anaerobic, with the degree of oxygen depletion dependent on the organic loading rate.

Treatment occurs in floating aquatic plant systems through three primary mechanisms: (1) metabolism by a mixture of facultative microbes on the plant roots suspended in the water column and in the detritus at the pond bottom, (2) sedimentation of wastewater solids and of internally produced biomass (dead plants and microbes), and (3) incorporation of nutrients in living plants and subsequent harvest. Floating aquatic plant systems are typically effective at reducing concentrations of biochemical oxygen demand and total suspended

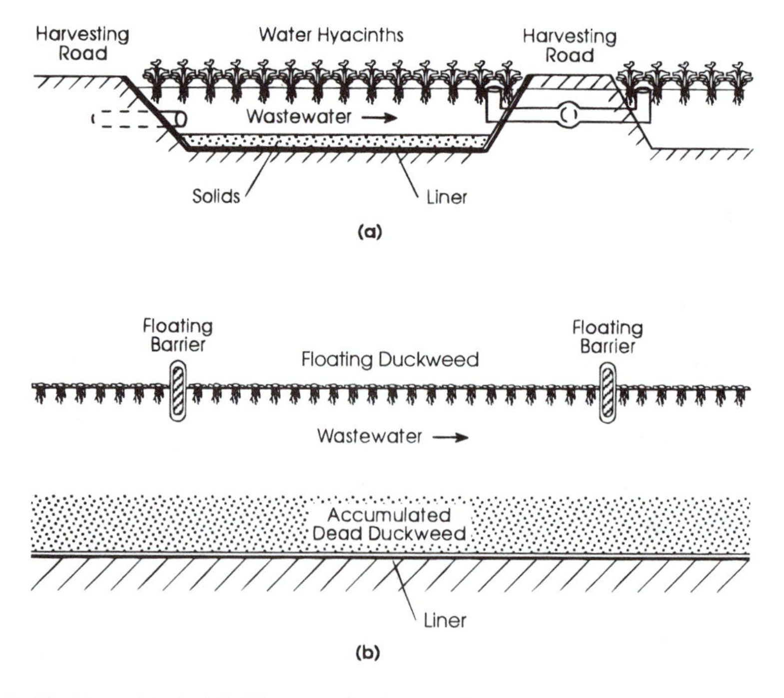

Figure 3-9 Diagram of typical floating aquatic plant treatment systems: (a) water hyacinth and (b) duckweed.

solids. Nitrate nitrogen may be effectively removed by denitrification. Total nitrogen and phosphorus removal can be consistently accomplished if the plants are harvested routinely.

Pond depth in floating aquatic plant systems is typically from 0.4 to 1.2 m for water hyacinth and 1.2 to 1.8 m for duckweed treatment systems. These systems can be used to provide secondary treatment, in which case biochemical oxygen demand mass loading should be limited to less than 100 kg/ha/d. When floating aquatic plant systems are used for advanced wastewater treatment and nutrient removal, organic loadings should be kept below 35 kg/ha/d. Typical hydraulic loading rates are in the range of 2 to 15 cm/d (0.7 to 5 ha/1000 m^3/d). Floating aquatic plant systems cost about $270,000/ha to build (capital costs are $500 to $1000/$m^3$/d), and operation and maintenance costs are about $0.12 to $0.14/$m^3$ (WPCF, 1990b).

Floating aquatic plant systems have some potential weaknesses that have limited their widespread use. Since these systems depend on one or just a few plant species for colonization of the pond surface, they are susceptible to catastrophic events which can kill part or all of these populations during a short time period. For example, water hyacinths are easily killed by cold weather and are attacked by numerous plant pest species. Duckweed is less sensitive to cold weather and pests, but it can also be killed by winter conditions. When plant cover is lost in a floating aquatic plant system, treatment effectiveness may be seriously impaired for a period of weeks or months as new plants are established.

A second potential problem with floating aquatic plant systems results from harvesting biomass for nutrient removal and for maintenance of plant growth at an optimum rate. These plants are more than 95 percent water when harvested so drying is required, and once dried there is typically a significant residual solids disposal problem.

WETLAND SYSTEMS

Wetland treatment systems use rooted, water-tolerant plant species and shallow, flooded, or saturated soil conditions to provide various types of wastewater treatment. The three basic types of wetland treatment systems include natural wetlands, constructed surface flow (SF) wetlands, and constructed subsurface-flow (SSF) wetlands (Figure 3-10).

While there are many types of naturally occurring wetlands, only those types with plant species that are adapted to continuous flooding are suitable for receiving continuous flows of wastewaters. Also, due to their protected regulatory status, discharges to natural wetlands must receive a high level of pretreatment (minimum of secondary). Constructed wetlands mimic the optimal treatment conditions found in natural wetlands, but provide the flexibility of being constructible at almost any location. They can be used for treatment of primary and secondary wastewaters as well as waters from a variety of other sources including stormwaters, landfill leachate, industrial and agricultural wastewaters, and acid-mine drainage.

Surface-flow wetlands (natural and constructed) are densely vegetated by a variety of plant species and typically have water depths less than 0.4 m. Open water areas may be incorporated into the design to provide for optimization of hydraulics and for wildlife habitat enhancement. According to the WPCF (1990b), typical hydraulic loading rates are between 0.4 to 4.0 cm/d (2.5 to 25 ha/1000 m^3/d) in natural wetlands and 0.7 to 5.0 cm/d (2 to 14 ha/1000 m^3/d) in constructed surface-flow wetlands.

Subsurface-flow wetlands use a bed of soil or gravel as a substrate for growth of rooted wetland plants. Pretreated wastewater flows by gravity, horizontally through the bed substrate where it contacts a mixture of facultative microbes living in association with the substrate and plant roots. Bed depth in SSF flow wetlands is typically less than 0.6 m, and the bottom of the bed is sloped to minimize that water flow overland.

Typical plant species used in SSF wetlands include common reed (*Phragmites australis*), cattail (*Typha* spp.), and bulrush (*Scirpus* spp.). Some oxygen enters the bed substrate by

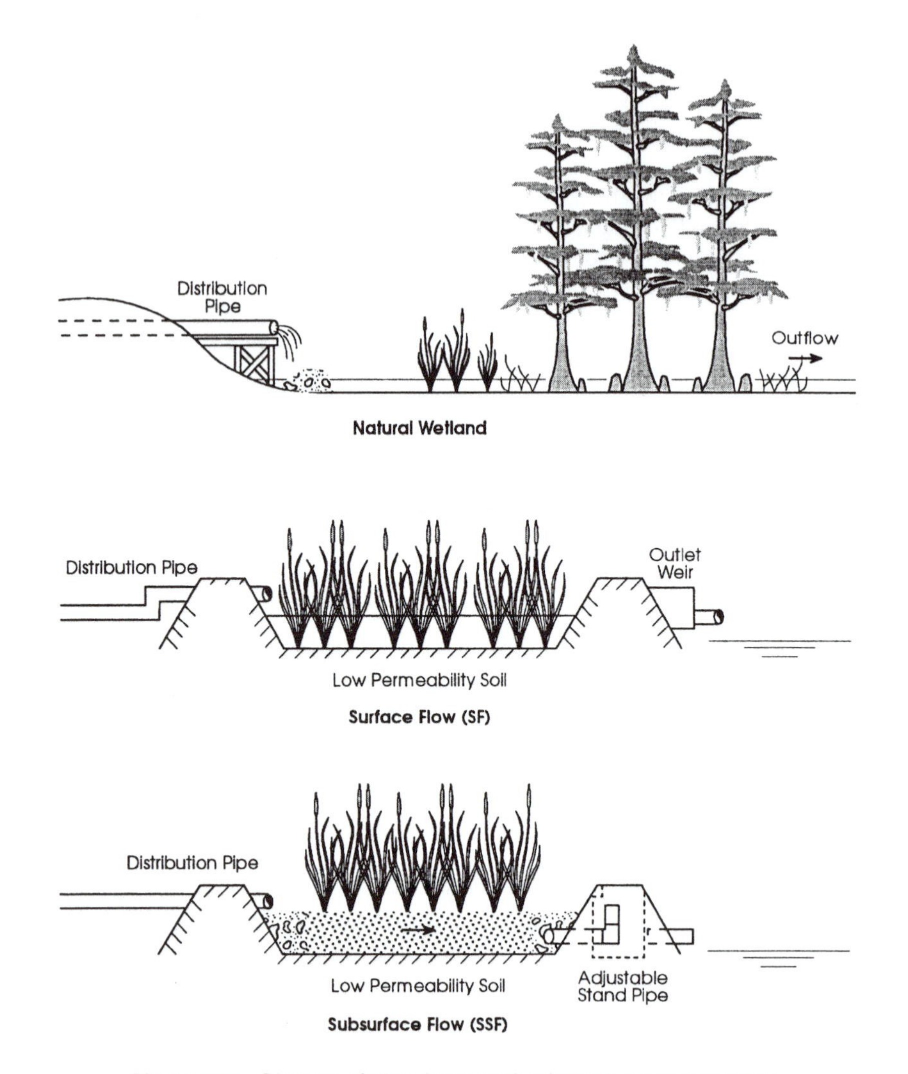

Figure 3-10 Diagram of three basic wetland treatment system types.

direct atmospheric diffusion and some through the plant leaves and root system, resulting in a mixture of aerobic and anaerobic zones. The majority of the saturated bed is anaerobic under most wastewater design loadings. According to the WPCF (1990b), typical hydraulic loading rates in SSF wetlands range from 2 to 20 cm/d (0.5 to 5 ha/1000 m^3/d).

Wetlands have been found to be effective in treating biochemical oxygen demand, suspended solids, nitrogen, and phosphorus, as well as for reducing metals, organics, and pathogens. Effective wetland performance depends on adequate pretreatment, conservative constituent and hydraulic loading rates, collection of monitoring information to assess system performance, and knowledge of successful operation strategies.

The most common difficulties experienced by wetland treatment systems have been related to maintaining partially aerated soil conditions. When wetland systems are overloaded by oxygen-demanding constituents or are operated with excessive water depth, highly reduced conditions occur in the sediments, resulting in plant stress and reduced removal efficiencies for biochemical oxygen demand and ammonia nitrogen. A common problem encountered in SSF constructed wetlands is inadequate hydraulic gradient and resulting surface flows.

Natural wetlands, when available, are typically the least expensive natural treatment alternative, requiring minimal capital expenditures for pumps, pipes, and water distribution structures in addition to the cost of the land itself. However, pretreatment and operational monitoring costs are typically higher for discharges to natural wetlands.

Constructed SF wetlands require a capital expenditure typically between $10,000 to $100,000/ha (20th and 80th percentile), primarily as a result of the earthwork costs. Subsurface flow wetlands are typically more expensive on a per area basis than SF systems, with capital costs from $100,000 to $200,000/ha (Knight et al., 1993a). Operation and maintenance costs for natural and constructed wetlands are primarily related to system monitoring and are generally low ($0.03 to $0.09/m^3) (WPCF, 1990b).

SUMMARY OF NATURAL TREATMENT TECHNOLOGIES

Table 3-1 summarizes and contrasts the principal features of natural wastewater treatment technologies. Each technology has strengths and weaknesses that must be considered during project planning and implementation. All of the natural treatment system technologies have the advantage of reducing the use of fossil fuels during construction and operation compared to conventional treatment systems. Where land is available, energy costs are expected to increase over time, and permit criteria do not preclude their use; natural treatment systems will often provide the most cost-effective and practicable alternatives.

Onsite infiltration systems have been the technology of choice for single households and small communities when soil percolation rates and groundwater levels are not limiting. These systems are relatively inexpensive, easy to install, and require little or no operation and infrequent maintenance. In some areas where groundwater levels are a constraint to percolation, the SSF wetland technology has been combined with septic tanks, resulting in an onsite system with periodic surface discharges. This alternative has been found to be preferable to mounded or failing drainfields where central sewage collection and treatment is not feasible.

Small- to medium-sized towns and cities have a number of natural treatment system options to consider. Where technically feasible and approved by regulating agencies, high-rate land application systems are generally the most cost-effective choice. They have moderate capital costs and low operation and maintenance costs. If suitable natural wetlands are available and approved, then they also represent a relatively low cost alternative for disposal, usually following a minimum of advanced secondary treatment. Natural wetland systems must be sized conservatively to minimize alterations of the existing biota (see Chapter 22 for a detailed approach to natural wetland treatment system design).

Facultative ponds, overland flow systems, and unharvested floating aquatic plant systems also offer a viable approach for small towns located adjacent to a receiving water with adequate assimilative capacity to accept secondarily treated wastewater. Where receiving waters do not have adequate capacity to directly discharge from a lagoon, overland flow system, or floating aquatic plant system, a constructed wetland can be added for advanced wastewater treatment. If surface discharge is not permittable and soils are only moderately permeable, a slow-rate land application system offers a final alternative for natural treatment at a reasonable cost.

Medium- to large-sized cities may believe that natural systems cannot be used for dealing effectively with their large wastewater flows. Medium- to large-sized cities, such as Arcata, CA; Orlando, FL; Lakeland, FL; and Columbia, MO have combined conventional technologies with natural systems to achieve very stringent discharge requirements in a cost-effective manner and also provide ancillary benefits to their citizens and surrounding environment by discharging to natural systems. When conventional technologies are used to provide a consistent, high-quality reclaimed water through tertiary treatment, this water can be used for beneficial reuse for humans (crop and landscape irrigation) and the environment (construction of habitat wetlands). When high levels of nutrient removal are required, harvested floating aquatic plant systems and constructed wetlands provide natural treatment technologies that do not create chemical sludges.

One last general point to make about natural treatment systems concerns both the designer and the regulator of these technologies. Land-intensive systems typically have longer hydraulic residence times (from about 3 to 200 days) than conventional systems (less than 1 to 2 days) and therefore are effective at modulation of erratic inflow volume and quality. However, because of their long hydraulic and solid residence times and because natural systems are typically outdoors and are spread over larger land areas that are susceptible to storms, wind, fires, insects, floods, and earthquakes, these natural systems are relatively slower to respond to operational changes and more apt to respond to natural events outside of the control of the system operator or owner. To achieve project success, both the engineer and the regulator must be aware of these differences between natural and conventional treatment systems.

The engineer who wishes to design natural systems should use conservative design criteria founded upon operational data from successful systems, and the regulator should provide realistic permit criteria that allow for normal daily, weekly, monthly, seasonal, or annual effluent quality variations typical of natural systems. The remainder of this book presents the information necessary to plan, design, and operate a successful wetland treatment system. The other texts referenced earlier provide information for the other natural treatment system technologies.

Section 2
Wetland Structure and Function

CHAPTER 4

Landform and Occurrence

INTRODUCTION

The term wetland describes a diverse spectrum of ecological systems. Scientific consensus of what constitutes a wetland has been subjectively influenced by definitions that attempt to encompass regulatory and environmental concerns. These concerns have been heightened by historic conversions of wetlands to dry lands and the resulting losses of a variety of natural functions originally provided by the former wetlands. Scientific definitions of wetland types have also been refined as the various structural and functional aspects of these ecosystems have been better described through accelerated research efforts.

A basic understanding of a wetland landform will increase an engineer's ability to successfully design natural or constructed wetlands as part of water pollution control systems. This chapter provides a general description of what natural wetlands are, where they occur, and how they can be constructed for water quality treatment.

WHAT IS A WETLAND?

WETLAND DEFINITION

The technical meaning of the term wetland includes a wide range of ecosystems from areas that are never flooded to areas that are deeply flooded all of the time. Areas that are not flooded may still be classified as wetlands because of saturated soil conditions where water is at or below the ground surface during part of a typical growing season. Wetland areas that are deeply flooded grade imperceptibly into aquatic ecosystems as water depth exceeds the growth limits of emergent or submergent vegetation. Figure 4-1 shows how wetlands lie on a continuum between dry lands (uplands) and deeply flooded lands (aquatic systems). Since this is a true continuum, with temporal and biological variability, there is no absolute hydrologic demarcation between these ecosystems, and all definitions are somewhat arbitrary.

The most consistent attribute of wetlands is the presence of water during some or all of an average annual period. Wetlands are areas where the soil is saturated with water or where shallow standing water results in the absence of plant species which depend on aerobic soil conditions. Wetlands are dominated by plant species that are adapted to growing in seasonally or continuously flooded soils with resulting anaerobic or low-oxygen conditions. At their upslope margin, wetlands can be distinguished from uplands by the latter's tendency to remain flooded or saturated for less than 7 to 30 days each year, a short enough period so that oxygen and other soil conditions do not limit plant growth. At their downgradient edge, wetlands grade into aquatic systems which are flooded to a depth or at a duration where

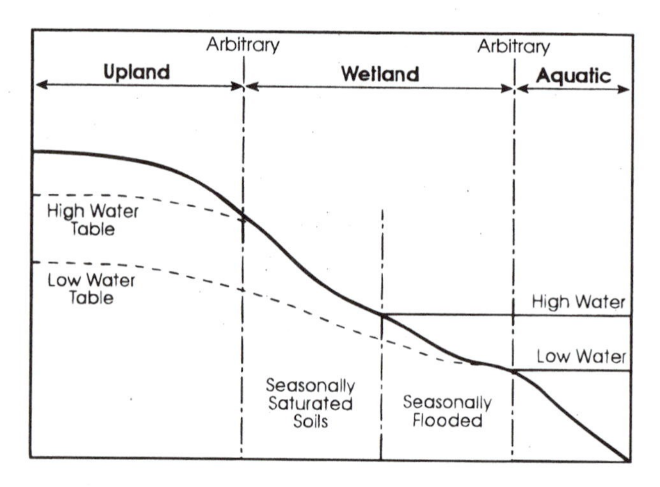

Figure 4-1 Wetlands are transitional areas between uplands, where excessive water is not a factor for plant growth, and aquatic ecosystems, where excessive flooding excludes rooted, emergent vegetation.

emergent, rooted plants cannot survive. The average water depth which typically separates wetlands from adjacent aquatic ecosystems is in the range of 1 to 2 m.

WETLAND CLASSIFICATION

Several terms are used most frequently to describe natural wetland systems. These terms are based primarily on the salinity of the water which influences the vegetation of the wetland and on the growth form of the resulting vegetation. These wetland classes are broken into two physical categories based on water salinity:

Freshwater wetlands—those that are inundated with fresh water (salinities less than about 1000 mg/L).

Saltwater wetlands—those that are inundated with brackish or saline waters (salinities greater than 1000 mg/L).

Natural saltwater wetlands have two major classes:

Salt marsh—saltwater wetlands dominated by emergent (rooted and extending above the level of the water), herbaceous (soft tissue, nonwoody) plant species.

Forested salt water or mangrove—saltwater wetlands dominated by woody plant species (typically mangrove trees).

Natural freshwater wetlands may also be conveniently separated into two general classes:

Freshwater marsh—freshwater wetlands dominated by emergent, herbaceous plant species adapted to intermittent to continuous flooding.

Freshwater swamp—freshwater wetlands dominated by tree species adapted to life in infrequent to prolonged flooded conditions.

Figure 4-2 provides examples of these four general wetland categories.

Because of the continuum of hydrologic conditions in wetlands, wetland plant species are found along gradients of increasing and decreasing dominance. Therefore, some wetlands fall between the commonly recognized marsh and swamp categories. Many commonly used colloquial terms describe local types of marshes and swamps. Some wetland terms in general usage are defined in greater detail in Table 4-1.

The U.S. Fish and Wildlife Service (USFWS) has developed the most widely used wetland and deepwater habitat classification scheme currently available (Cowardin et al., 1979). This technical classification system enables wetland habitat types to be mapped precisely. The overall structure of the USFWS wetland and deepwater habitat classification scheme is presented in Table 4-2.

The USFWS classification scheme provides hierarchical classes for wetlands with systems, subsystems, classes, and subclasses for finer definition of individual recognizable wetland habitats. Of particular relevance to this book are certain estuarine intertidal, lacustrine littoral (associated with lake margins), and palustrine classes. These wetland and deepwater habitat classes are listed in Table 4-3 along with their typical common names and vegetation types. Figure 4-2 illustrates the common wetland classes. Marshes and forested wetlands are the two types that are most commonly used for natural wetland treatment.

Freshwater swamp

Salt marsh

Mangrove wetland

Freshwater marsh

Figure 4-2 Typical natural wetland categories used for water quality management.

Table 4-1 Common Freshwater Wetland Terms and Descriptions

Wetland Term	Descriptions
Bog	An acidic wetland dominated by mosses which typically accumulates peat. Sometimes forested.
Bottomland	Floodplain wetlands typically dominated by wetland tree species.
Fen	A wetland occurring on low, poorly drained ground and dominated by herbaceous and shrubby vegetation. Soil is typically organic peat or marl.
Marsh	A wetland dominated by emergent soft-tissue macrophytes such as cattails, reeds, or bulrushes.
Pocosin	A freshwater wetland typically occurring on poorly drained hillsides and lowlands and dominated by ericaceous shrubs and trees that are adapted to periodic fires. Soil is typically organic peat.
Pothole	Shallow wetlands and ponds typically created by glacial action and vegetated with herbaceous and aquatic plant species.
Slough	A slow-moving creek or stream characterized by herbaceous and woody wetland vegetation.
Wet prairie	A shallow wetland dominated by sedge and grass species.
Reedswamp	A wetland dominated by common reed (*Phragmites australis*).

Table 4-2 USFWS Classification System for Wetlands and Deepwater Habitats

System	Subsystem	Class
Marine (open oceanfront)	Subtidal (continuously submerged)	Rock bottom; unconsolidated bottom; aquatic bed; reef
	Intertidal (exposed at low tide)	Aquatic bed; reef; rocky shore; unconsolidated shore
Estuarine (tidal embayments; variable salinity)	Subtidal (continuously submerged)	Rock bottom; unconsolidated bottom; aquatic bed; reef
	Intertidal (exposed at low tide)	Aquatic bed; reef; stream bed; rocky shore; unconsolidated shore; emergent wetland; scrub-shrub wetland; forested wetland
Riverine (associated with river channels)	Tidal (fluctuating flows)	Rock bottom; unconsolidated bottom; aquatic bed; rocky shore; unconsolidated shore; emergent wetland
	Perennial (continuously inundated)	Rock bottom; unconsolidated bottom; aquatic bed; rocky shore; unconsolidated shore; emergent wetland
	Intermittent (seasonally exposed)	Stream bed
Lacustrine (associated with lakes)	Limnetic (deep water)	Rock bottom; unconsolidated bottom; aquatic bed
	Littoral (shoreline, shallow water)	Rock bottom; unconsolidated bottom; aquatic bed; rocky shore; unconsolidated shore; emergent wetland
Palustrine (nontidal, emergent vegetation)	None	Rock bottom; unconsolidated bottom; aquatic bed; unconsolidated shore; moss-lichen wetland; emergent wetland; scrub-shrub wetland; forested wetland

Modified from Cowardin, L.M., et al., 1979.

Table 4-3 Comparison of Common Wetland Categories, USFWS Wetland Classes, and Typical Dominant Plant Species in the U.S.

Common Name	USFWS Classification	Typical Vegetation
Salt marsh	Estuarine intertidal emergent	*Distichlis spicata; Juncus roemerianus; Spartina alterniflora*
Mangrove wetland	Estuarine intertidal scrub-shrub or forested	*Avicennia germinans; Laguncularia racemosa; Rhizophora mangle*
Freshwater marsh	Riverine tidal emergent Riverine perennial emergent Lacustrine littoral emergent Palustrine emergent	*Carex* spp.; *Cladium jamaicense; Cyperus* spp.; *Eleocharis* spp.; *Juncus* spp.; *Panicum* spp.; *Phragmites communis; Polygonum* spp.; *Pontederia* spp.; *Sagittaria* spp.; *Scirpus* spp.; *Typha* spp.
Freshwater swamp	Palustrine scrub-shrub Palustrine forested	*Acer rubrum; Alnus* spp.; *Betula* spp.; *Cyrilla racemiflora; Decodon verticillatus, Fraxinus* spp.; *Ilex* spp.; *Larix larcina; Lyonia lucida; Nyssa* spp.; *Persea palustris; Picea* spp.; *Pinus* spp.; *Quercus* spp.; *Salix* spp.; *Taxodium* spp.; *Vaccinium* spp.

ANATOMY OF A WETLAND

Figure 4-3 shows structural components typical of wetland ecosystems. Starting with the unaltered sediments or bedrock below the wetlands, these typical components are

- Underlying strata—unaltered organic, mineral, or lithic strata which are typically saturated with or impervious to water and are below the active rooting zone of the wetland vegetation.
- Hydric soils—the mineral to organic soil layer of the wetland which is infrequently to continuously saturated with water and contains roots, rhizomes, tubers, tunnels, burrows, and other active connection to the surface environment.
- Detritus—the accumulation of live and dead organic material in a wetland which consists of dead emergent plant material, dead algae, living and dead animals (primarily invertebrates), and microbes (fungi and bacteria).
- Seasonally flooded zone—the portion of the wetland that is seasonally flooded by standing water and provides habitat for aquatic organisms including fish and other vertebrate animals, submerged and floating plant species that depend on water for buoyancy and support, living algae, and populations of microbes.
- Emergent vegetation—vascular, rooted plant species which contain structural components that emerge above the water surface, including both herbaceous and woody plant species.

Technically, wetlands exist where any one or more of these components occur. Natural wetlands usually have all of these attributes. Constructed wetlands may have less mature components, especially soil organic matter which forms over an extended period of time. The structural components of natural wetlands are highly variable and depend on hydrology, underlying sediment types, water quality, and climate. These components are described in more detail in Chapters 5 through 8.

DISTRIBUTION OF NATURAL WETLANDS

It is important to be familiar with the occurrence, distribution, and types of natural wetlands in an area when planning a wetland treatment project, because in some cases natural wetlands can be used to manage or treat wastewaters and stormwaters. An inventory of the natural wetland types available, the extent to which they can be loaded with additional water and nutrients without harm to their existing biology, and an understanding of the need for preservation or creation of threatened wetland habitat areas is important in project planning.

In some locations, it may be necessary to avoid natural wetland areas because of permitting constraints. Also, dominant plant types in local natural wetlands can provide the designer of constructed wetlands with important clues for selecting plant species. In some cases, natural wetlands may serve as donor sites for wetland plants or seed materials for constructed wetlands.

Natural wetlands do not occur with even frequency or area in all regions. The greatest concentrations of natural wetlands occur in the northeastern, midwestern, and Alaskan glaciated plains; the lower Mississippi alluvial plain; and the low coastal plains along both the Atlantic and Pacific Oceans, gulfs, and embayments. Roughly 42 million hectares of natural wetlands have been identified by the USFWS National Wetlands Inventory (NWI) in the lower 48 states (Dahl, 1990). Figure 4-4 shows the general occurrence of natural wetlands in the United States. An additional 69 million hectares of wetlands are estimated to exist in Alaska, resulting in a total estimated inventory of about 111 million hectares of natural wetlands in the U.S. An estimated area of 47 million hectares or 30 percent of the original wetland habitat within the U.S. has been converted to other uses since the late 18th century.

Table 4-4 summarizes the estimated wetland areas by states. Those with the highest percentage of their areas in natural wetlands are Alaska (45%), Florida (29%), Louisiana (28%), Maine (24%), and South Carolina (23%). The dominant wetland classes are freshwater swamp (61%) and freshwater marsh (34%) (Tiner, 1984; Dahl et al., 1991). Saltwater wetlands make up only 5 percent of the total natural wetland inventory.

These natural wetland classes include a range of hydrologies and therefore cannot be used to estimate the available area of natural wetlands that are compatible with water quality management. Furthermore, area is not a measure of wetland quality or importance. However, these areas and percentages generally reflect the percentage of the existing landscape that is not compatible with other, upland-dependent forms of water quality treatment. These data indicate where there may be a need to consider the interactions between wetlands and water

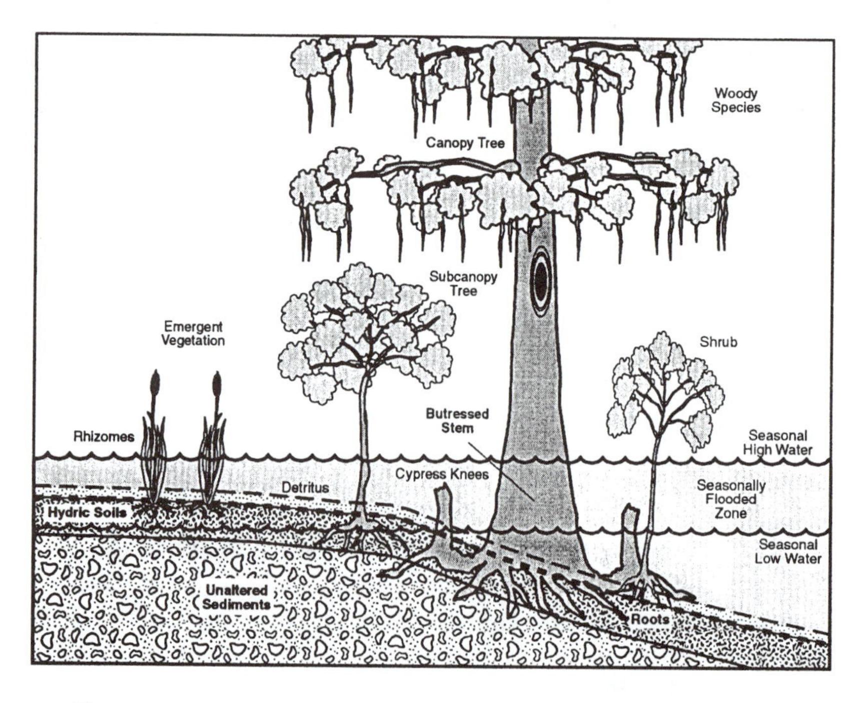

Figure 4-3 Typical structural components common to many wetland types.

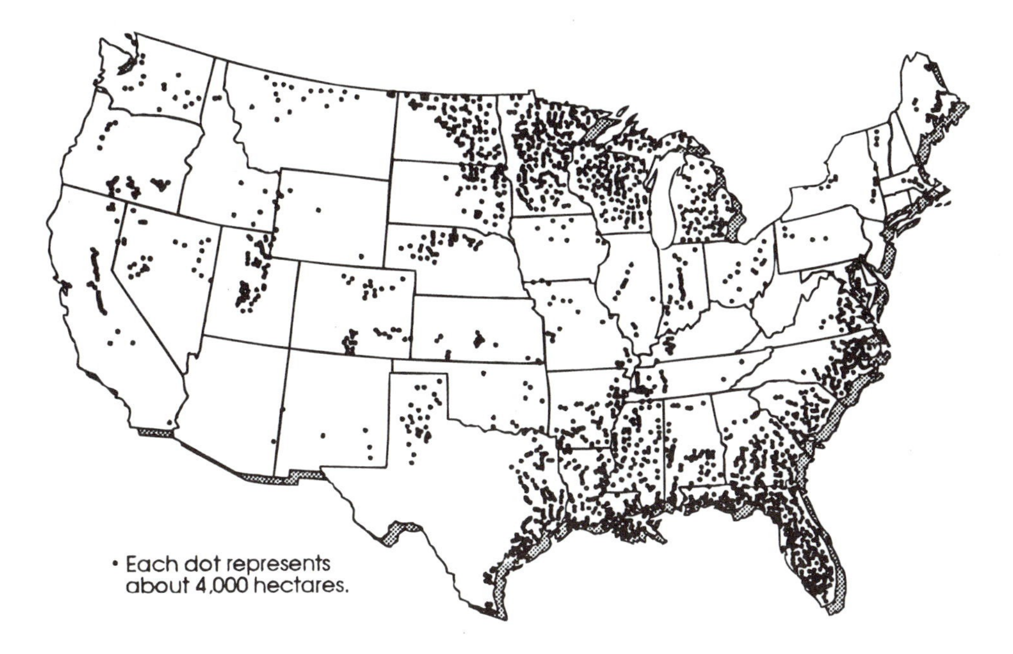

Figure 4-4 Distribution of natural wetlands in the coterminous U.S.

quality. Not surprisingly, the use of natural wetlands for wastewater management has been principally developed in several states with high natural wetland areas (Florida, Michigan, and South Carolina).

CONSTRUCTED WETLANDS

The purposeful construction of wetland ecosystems is a relatively new technology. Although wetlands have been created incidentally by human and animal engineering of ponds and lakes throughout history and prehistory (Figure 4-5), the intentional construction of wetlands to provide habitat and water quality functions began with the environmental movement in the 1970s.

Hammer (1992b) defines wetlands built expressly for habitat replacement and mitigation as "created" wetlands, and "constructed" wetlands as systems that are built expressly for water quality treatment. This terminology indicates the importance of distinguishing between "created" and "constructed" wetlands because of the regulatory limitations potentially placed on wetlands constructed primarily as wildlife habitat.

The proposed usage by Hammer is not followed in this book. The word "created" carries incorrect connotations when referring to wetlands that are engineered. However, it is important to distinguish between wetlands constructed for habitat that might be considered to be waters of the U.S. and wetlands constructed for treatment and thereby excluded from waters of the U.S. by the U.S. Army Corps of Engineers. This distinction is discussed in greater detail in Chapter 19.

Wetlands are engineered and constructed for four principal reasons as indicated by specific descriptive terminology: (1) to compensate for and help offset the rate of conversion of natural wetlands resulting from agriculture and urban development (constructed habitat wetlands); (2) to improve water quality (constructed treatment wetlands); (3) to provide flood control (constructed flood control wetlands); and (4) to be used for production of food and fiber (constructed aquaculture wetlands).

Table 4-4 Estimated Natural Wetland Areas in the U.S.

State	Natural Wetland Area (ac)	Percent of Total State Area
Alabama	3,783,800	11.5
Alaska	170,000,000	45.3
Arizona	600,000	0.8
Arkansas	2,763,600	8.1
California	454,000	0.4
Colorado	1,000,000	1.5
Connecticutt	172,500	5.4
Delaware	223,000	16.9
Florida	11,038,300	29.5
Georgia	5,298,200	14.1
Hawaii	51,800	1.3
Idaho	385,700	0.7
Illinois	1,254,500	3.5
Indiana	750,633	3.2
Iowa	421,900	1.2
Kansas	435,400	0.8
Kentucky	300,000	1.2
Louisiana	8,784,200	28.3
Maine	5,199,200	24.5
Maryland	440,000	6.5
Massachusetts	588,486	11.1
Michigan	5,583,400	15.0
Minnesota	8,700,000	16.2
Mississippi	4,067,000	13.3
Missouri	643,000	1.4
Montana	840,300	0.9
Nebraska	1,905,500	3.9
Nevada	236,350	0.3
New Hampshire	200,000	3.4
New Jersey	915,960	18.3
New Mexico	481,900	0.6
New York	1,025,000	3.2
North Carolina	5,689,500	16.9
North Dakota	2,490,000	5.5
Ohio	482,800	1.8
Oklahoma	949,700	2.1
Oregon	1,393,900	2.2
Pennsylvania	499,014	1.7
Rhode Island	65,154	8.4
South Carolina	4,659,000	23.4
South Dakota	1,780,000	3.6
Tennessee	787,000	2.9
Texas	7,612,412	4.4
Utah	558,000	1.0
Vermont	220,000	3.6
Virginia	1,074,613	4.1
Washington	938,000	2.1
West Virginia	102,000	0.7
Wisconsin	5,331,392	14.8
Wyoming	1,250,000	2.0

Note: Modified from Dahl et al., 1991.

HABITAT WETLANDS

Wetlands are being constructed at an accelerating rate to help offset the historical and continuing losses of natural wetland functions. In this book the term "constructed habitat wetlands" is used to indicate wetlands constructed primarily for creation or replacement of wildlife habitat. These wetlands may be built as part of overall environmental enhancement programs or as compensatory mitigation (replacement of wetland area and functions) for

"Sure, kid. You start by working for the ecosystem, but pretty soon you figure out how to get the ecosystem working for you!"

Figure 4-5 Beavers have one of the original design patents on constructed wetlands. *(Drawing by Ed Fisher; © 1991; The New Yorker Magazine, Inc.)*

specific conversions of natural wetlands. The federal Clean Water Act of 1974 and its amendments, as well as numerous state laws, require compensatory mitigation for many permitted wetland alterations. A growing number of habitat wetlands are also intended and designed for public use (bird watching, hiking, and other nature study). The technology of wetlands construction for habitat replacement is maturing rapidly through the political and environmental support for a national policy of "no net loss" voiced by former U.S. President George Bush.

Habitat wetlands are being constructed in all four of the major wetland categories: salt marsh, forested salt water, freshwater marsh, and freshwater swamp. The success of habitat wetlands and the development of a range of wetland functions has varied. Overall, the greatest success has been achieved in the establishment of salt marsh and freshwater marshes.

Salt marsh and forested saltwater wetlands constructed adjacent to estuarine waters provide water in the correct salinity range to encourage the established species. Upland areas adjacent to these tidal salt waters are excavated to about 0.2 to 1.0 m below the mean high-water line and are planted with salt marsh and mangrove plant species.

Freshwater habitat wetlands can be constructed in nearly any upland location or environment. The principal requirement is a predictable source of water for maintenance of the wetland plant communities that are established. Inadequate knowledge of the natural variation of water levels in habitat wetlands and biological naivete concerning plant growth requirements have resulted in a number of failures to establish desired plant communities in constructed freshwater wetlands. Water for freshwater habitat wetlands typically is provided by excavating to a depth at or below seasonal high groundwater levels or by constructing the wetlands adjacent to a lake or river with perennial water. There is growing consideration and utilization of treated wastewater as the water source. Freshwater wetlands frequently are constructed for recreational uses in addition to their use as wildlife habitat (Figure 4-6).

CONSTRUCTED TREATMENT WETLANDS

All constructed wetlands specifically engineered with water quality improvement as a primary purpose are called constructed treatment wetlands in this book. The waters to be treated are highly diverse and include municipal, industrial, and agricultural wastewaters; stormwaters (both urban and rural); and polluted surface waters in rivers and lakes.

Many details concerning the construction of wetlands for water quality treatment are provided elsewhere in this book; this chapter presents information concerning the general form of constructed treatment wetlands and their occurrence and importance in the existing landscape. Constructed treatment wetlands may contain some or all of the structural components shown in Figure 4-3. The main difference between natural and constructed treatment wetlands is the origin of their landform. The importance of landform to wetlands is immediately evident to the engineer attempting to design and build a wetland in an upland area and to the owner who has to pay the expense of this construction.

Constructed treatment wetlands may be built at, above, or below the existing land surface if an external source of water, such as wastewater, is to be added (Figure 4-7). When built at or above the ground surface, earthwork is primarily fill to create berms or levees. This fill material may be gathered onsite or may be hauled from an offsite location, depending on treatment wetland design. When treatment wetlands are built below the existing ground surface, earthwork construction may include some fill for berms and low areas, but it also may include varying degrees of excavation. A cost-effective design will attempt to balance cut and fill needs within a limited project area. For above- and below-ground situations, earthwork is not complete until the bottom shape of the treatment wetland has been graded to the appropriate elevation to allow controlled flooding. For constructed treatment wetlands, this final grading is extremely important to effectively distribute water in wetland treatment cells and to take advantage of the greatest portion of the constructed wetland area.

Liners or relatively impervious site soils are critical to the success of constructed treatment wetlands in areas where groundwater levels are typically below the ground surface. Most

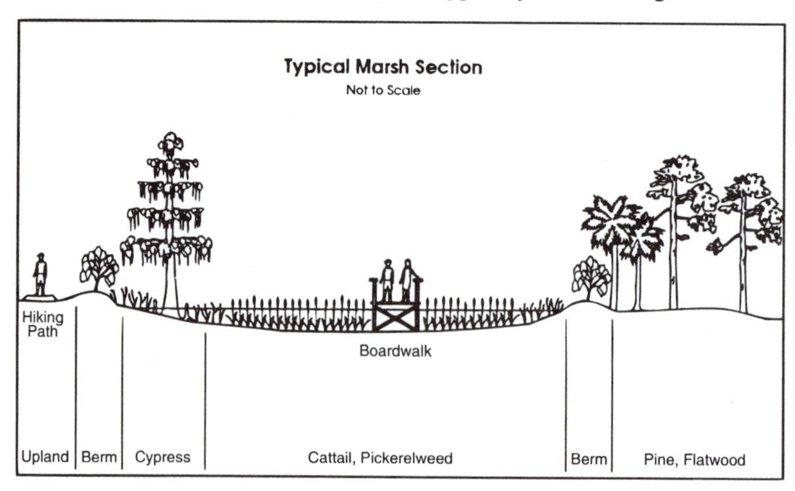

Figure 4-6 Diagram of a freshwater wetland design for inclusion in a park setting.

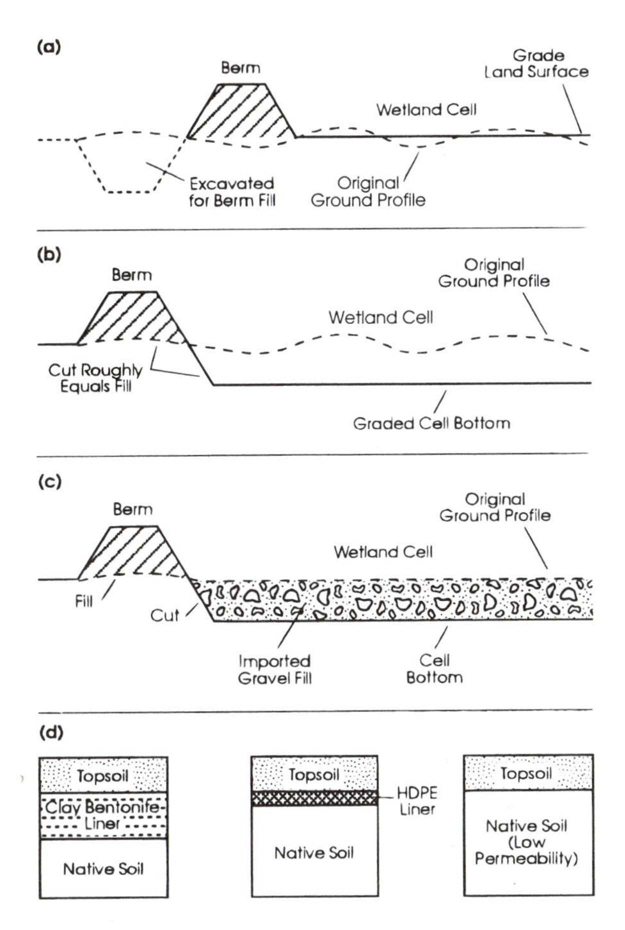

Figure 4-7 Some examples of constructed wetland landforms: (a) above grade, (b) below grade, (c) gravel filled, and (d) liner alternatives.

soils, except excessively drained sands and gravels, can be loaded with enough water to result in saturated conditions and establishment of wetland vegetation. However, if water inflow volume is limited or if the flow will be periodic, an unlined constructed treatment wetland can become desiccated and unable to support wetland vegetation. If site soils naturally have low permeability (clays, fine silts, and clayey loams) or if groundwater is naturally at or near the ground surface due to site topography, liners are not necessary for constructed treatment wetland success.

When they are necessary to hold water or to protect groundwater quality, the most frequently used liners for treatment wetlands are imported clay, clay-bentonite mixtures, polyvinylchloride (PVC), and high-density polyethylene (HDPE) synthetic materials. Synthetic liners must be well protected from construction and root development damage by overlying soils or gravels (typically 0.3 to 1.0 m). Synthetic liners are potentially expensive (approximately $32,000 to $100,000/ha) and are prone to more damage than clay or clay-bentonite liners.

One final ingredient in the landform of a constructed treatment wetland is an adequate rooting medium for the wetland plants. This rooting substrate must allow ample roots to develop for stability and nutrition of the mature plants. Most wetland plants will grow slowly or will die in dense clay soils or large, sharply angular rocks. A loamy or sandy topsoil layer (typically 0.2 to 0.3 m thick) is ideal for propagation of most wetland plant species. Topsoils, typically expensive to import, can usually be stockpiled during initial site grading and grubbing and redistributed over the wetland prior to final grading. Coarse sands and gravels often form the substrate (or media) for subsurface-flow wetlands.

The U.S. Army Corps of Engineers (USACE), which implements Section 404 of the Clean Water Act, has excluded treatment ponds and lagoons from its definition of regulated waters of the U.S. (40 CFR 123.12). This policy has generally been followed for wetlands as well; however, in cases where treatment ponds and constructed wetlands have been abandoned and have developed significant environmental and wetland functions, the USACE has ruled that these areas have become waters of the U.S. and are subject to the provisions of Section 404 of the Clean Water Act. Various issues related to the regulatory status of wetlands used for water quality treatment are presented in Chapter 19.

FLOOD CONTROL WETLANDS

Small to large impoundments are being used increasingly to offset losses of natural flood storage volume during urban and agricultural development. Many of these impoundments are designed as wet detention systems, and an increasing number of wet detention systems incorporate, purposely or fortuitously, areas colonized by emergent wetland plants. These wet detention systems are engineered to provide specific hydraulic functions for water storage and for bleed-down intervals following rainfall events. Flood control systems that include significant areas of wetlands vegetation are called constructed flood control wetlands.

Flood control wetlands are sited generally at low elevations in the landscape to allow gravity inflows from the adjacent upland system generating runoff. In some cases, these flood control wetlands must be located in former converted floodplain areas, resulting in restoration of lost wetland functions in addition to flood control. Flood control wetlands frequently must be designed to maximize storage volume, thereby necessitating incorporation of large and deep open water areas in addition to vegetated zones. Fluctuating water levels in constructed flood control wetlands may limit their ability to serve secondary functions such as providing habitat for plants and wildlife or for providing water quality treatment. All three of these wetland functions can be provided if the flood control wetland area is designed conservatively.

CONSTRUCTED AQUACULTURE WETLANDS

Aquaculture is the husbandry of food organisms in aquatic systems. Constructed and natural wetlands are being incorporated into aquaculture operations at an increasing rate. Shallow ponds used for rearing crayfish, shrimp, and commercial fish species have many characteristics in common with constructed wetlands. Shallow ponds and diked areas used for production of water-tolerant food crops such as rice, cranberries, and water chestnuts are technically wetlands in every respect, with the sole differentiator being the intensity of human management.

Significant potential exists for compatible aquaculture within wetlands constructed for water quality management. This potential is being exploited for rice culture in China and Brazil. Agricultural runoff treatment wetlands in Maine designed by the U.S. Department of Agriculture Natural Resources Conservation Service include bait fish as a possible aquaculture byproduct. Increased fish and wildlife productivity in wetlands receiving elevated nutrients in wastewater inflows can be harvested for economic purposes. Some other compatible activities currently occurring at wetland treatment systems in North America include waterfowl hunting, trapping of furbearing mammals, and timber production.

SUMMARY

The term wetland is a relatively recent invention used to describe ecosystems that are transitional between uplands and deeply flooded lands. Wetlands take on a variety of forms in response to their hydrology, soils, climate, and regional plant communities. Common wetland categories that incorporate most wetland treatment systems are salt marsh, mangrove wetland, freshwater marsh, and freshwater swamp. The distinguishing characteristics of these wetland categories are water salinity (salt water vs. fresh water) and plant form (herbaceous vs. woody).

Because of the seasonal or continuous presence of water in wetlands, they have come under increasing regulatory protection from alteration and impacts which might have offsite significance. In the U.S., the Clean Water Act and numerous other federal, state, and local ordinances protect the water quality and, to a growing extent, the biology of naturally occurring wetlands. These regulations greatly influence the use of natural wetlands for water quality treatment and, to a lesser but important degree, the construction of wetlands for pollution control.

Natural wetlands occur throughout the U.S. (and the world). Their frequency and distribution are dictated by regional differences in geology, landform, surface and groundwater hydrology, and net rainfall. There are currently about 111 million hectares of natural wetlands in the U.S.; however, about 47 million hectares or 30 percent of the original wetland area have been converted to nonwetlands. The percentage of wetland area in individual states ranges from a low of about 0.3 percent of the land area in Nevada to a high of about 45 percent in Alaska. A number of states in the southeastern and north central U.S. have significant percentages of their land areas occupied by natural wetlands, a factor which reduces the availability of well-drained upland areas suitable for other land-intensive forms of water quality treatment.

Wetlands can be engineered and constructed in nearly any location for a variety of purposes. Typically, wetlands are constructed for one or more of four primary purposes: creation of habitat, water quality management, flood control, and aquaculture. There is currently a poor distinction by many regulatory personnel between constructed wetlands that should be regulated as waters of the U.S. and wetlands that are part of privately or publicly owned treatment works and should not be burdened by excessive regulation.

Constructed wetlands incorporate many of the components typical of natural wetlands. Some of these components can be replicated relatively simply, but others may take a long time to develop. In all instances, the form and functions of natural wetlands are relatively expensive to duplicate through design and construction (typical costs for constructed marsh wetlands in the U.S. are between $10,000 and $100,000/ha). Also, valuable functions provided by the uplands present at a site prior to wetlands construction may be lost in the process. These factors should be given careful consideration when selecting, designing, and building constructed treatment wetlands.

CHAPTER 5

Wetland Soils

INTRODUCTION

Soils result from physical, chemical, and biological modifications of parent materials including bedrock, sedimentary deposits, and metamorphosed sediments and rocks. Wetland soils are typical of other soils, except for the influence of water which modifies soil characteristics by reducing exposure to the atmosphere with subsequent reduction of aerobic conditions. Flooding in wetlands also may concentrate or dilute the chemical constituents of the soils depending on the chemistry of the flood water, physical and chemical nature of the soils, and the surrounding environment. A basic knowledge of the physical and chemical properties of wetland soils is important in the planning, design, and operation of wetland treatment systems because these properties affect wetland plant growth and assimilation of some wastewater constituents.

This section provides an introduction to wetland soils and sediments, definitions of common terms, and a discussion of the importance of soils in the design and operation of wetland treatment systems. More detailed discussions of wetland soils can be found in Reddy and D'Angelo (1994), Faulkner and Richardson (1989), Gambrell and Patrick (1978), Good and Patrick (1987), Buol et al. (1980), and Mitsch and Gosselink (1993). Subsequent chapters discuss many of the other environmental variables that modify wetland soils and sediments and are, in turn, modified themselves.

WHAT ARE WETLAND SOILS?

Many wetland soils are characterized by a lack of oxygen induced by flooding. Oxygen diffusion in flooded soils is nearly 10,000 times slower than in aerobic soils (Armstrong, 1978). Well-aerated upland soils rapidly experience a decline in soil oxygen and redox potential when they are flooded. Continuous or seasonal inundation combined with the production of large amounts of dead organic matter (litterfall) results in nearly perpetual soil anaerobiosis in many wetlands. The resulting lower dissolved oxygen level results in the accumulation of organic matter in wetland soils because of a reduced level of microbial activity and organic decomposition. This condition leads to the formation of the primary taxonomic characteristics of wetland soils lumped under the term *hydric.*

Hydric soils are defined by the National Technical Committee for Hydric Soils (U.S. Soil Conservation Service, 1987) as

> soils that in their undrained condition are saturated, flooded, or ponded long enough during the growing season to develop anaerobic conditions that favor the growth and regeneration of hydrophytic vegetation

Hydric soils fall into two categories: (1) organic soils or histosols which typically contain at least 12 to 20 percent organic carbon (20 to 35 percent organic matter) on a dry-weight basis and (2) mineral soils including some entisols, ultisols, and inceptisols that typically have less than 12 to 20 percent organic carbon (20 to 35 percent organic matter).

The properties of organic and mineral wetland soils are compared and contrasted in Table 5-1. In fact, many natural and constructed wetlands have soil/sediment profiles that include organic and mineral horizons or layers, and also have spatial heterogeneity, so that some physical, chemical, and biological properties of each soil type may have an influence on a particular wetland's ecology.

Organic soils are defined by the U.S. Soil Conservation Service (1975) as (1) soils that are saturated with water and have greater than 18 percent organic carbon (if mineral fraction is greater than 60 percent clay), greater than 12 percent organic carbon (with no clay in the mineral fraction), or 12 to 18 percent organic carbon (0 to 60 percent clay in the mineral fraction); or (2) soils that are never saturated for more than a few days at a time and have more than 20 percent organic carbon (nonwetland soils that may form in tropical areas). All other soils are defined as mineral.

As indicated in Table 5-1, organic soils typically have lower bulk density (dry weight per volume less than 1 g/cm^3) and a higher porosity (volume of water held per volume of soil) and higher water holding capacity than mineral soils. Consequently, flooded organic soils can often hold more water than flooded mineral soils. In spite of higher porosity, organic soils may have low to high hydraulic conductivities, indicating that they may impede horizontal or vertical groundwater flows. Thus, organic soils can act as an aquitard to seal a wetland and increase hydroperiod. Organic soils typically have a lower stock of available plant nutrients than mineral soils and are less capable of storing some cations and anions in wastewater effluents.

A leaf or root mat may occur above the first soil horizon in many wetlands. This nonsoil layer consists of newly deposited or partially decomposed plant material (leaves, stems, and roots) and can be easily removed from the soil surface. This litter or detrital layer is not considered to be part of the wetland soils, but in many cases will be a precursor for future soils development in the wetland.

New wetland soils form in response to external and internal loadings of carbon and mineral sediments to the wetland. In closed systems without inputs and outputs, new sediment formation is dependent upon the net difference between carbon fixation by plants and the rate of carbon degradation. In wetland systems that are open to inflows of water (including wastewaters and stormwaters), sediments may accrete because of sedimentation of mineral solids and carbon accretion. In a flow-through wetland, influent sediments tend to accumulate near the front end of the system (Figure 5-1).

Figure 5-2 contrasts two stages in the idealized development of a constructed wetland treatment system on an upland mineral soil. The newly planted and flooded constructed wetland has no dead plant litter, and oxygen consumption in the sediments is limited to

Table 5-1 Comparison of Mineral and Organic Soils in Wetlands

	Mineral Soil	Organic Soil
Organic content, percent	Less than 20–35	Greater than 20–35
Organic carbon, percent	Less than 12–20	Greater than 12–20
pH	Usually circumneutral	Acid
Bulk density	High	Low
Porosity	Low (45–55%)	High (80%)
Hydraulic conductivity	High (except for clays)	Low to high
Water holding capacity	Low	High
Nutrient availability	Generally high	Often low
Cation exchange capacity	Low, dominated by major cations	High, dominated by hydrogen ion
Typical wetland	Riparian forest, some marshes	Northern peatland, southern swamps, and marshes

Modified from Mitsch and Gosselink, 1993.

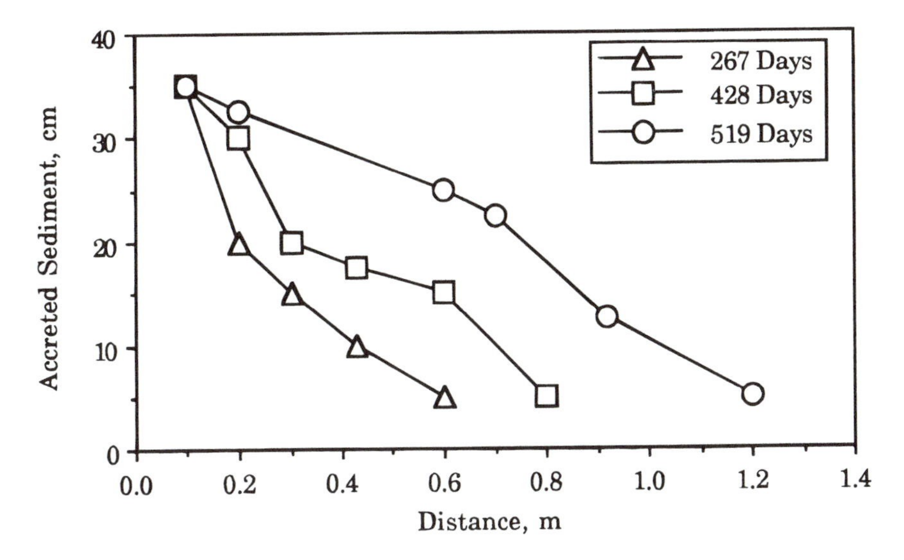

Figure 5-1 Accretion of new sediments in a treatment wetland. The incoming water had TSS = 321 mg/L, and 40% of this material remained in the bottom sediments after 610 days. (Adapted from van Oostrom, 1994).

existing low concentrations of soil organic matter. For this reason, the redox potential in the new wetland sediments (Figure 5-2a) is relatively high compared to the mature system shown in Figure 5-2b. Litter accumulation results in an additional diffusional barrier between atmospheric gases and the wetland sediments and exerts a very high, internally produced oxygen demand due to high rates of microbial respiration. Very low redox levels may be reached in mature wetland sediments, resulting in the formation of truly anaerobic conditions as chemically bound oxygen is utilized through microbial oxidation-reduction reactions.

ACCUMULATION RATES OF WETLAND SOILS

The buildup of mineral matter which settles from incoming stormwater, river water, or wastewater is often slow. At the low end of the spectrum are the clean wastewaters from advanced treatment plants, which may have less than 10 mg/L of suspended matter. If this material is all inorganic and undecomposible, it can accrete in the treatment wetland. Under most circumstances, this represents only a few millimeters per decade of solids buildup in the wetland. Typical loadings are less than 100 g/m^2/yr. At the high end of the spectrum, the turbid waters of a muddy river can contain several hundred mg/L of suspended matter, mostly mineral. This sediment load can cause a buildup of several cm/yr if it is concentrated in the inlet region of a treatment wetland. Indeed, it was this kind of sediment load that built the Mississippi delta wetlands of Louisiana and sustained and expanded it, despite soil subsidence, over the presettlement era. Channelization has diverted this sediment load and caused the conversion of large areas of wetland to open water (Patrick, 1994). In smaller watersheds, the accretion rates are lower; for instance, Kadlec and Robbins (1984) report a range of 0.05 to 0.9 cm/yr for sites in two Great Lakes coastal wetlands.

THE BUILDING OF PEATS AND MUCKS

Under oxygen-deficient conditions created by extended and deep innundation and a high consumption rate of available electron acceptors, there is a net accretion of organic matter

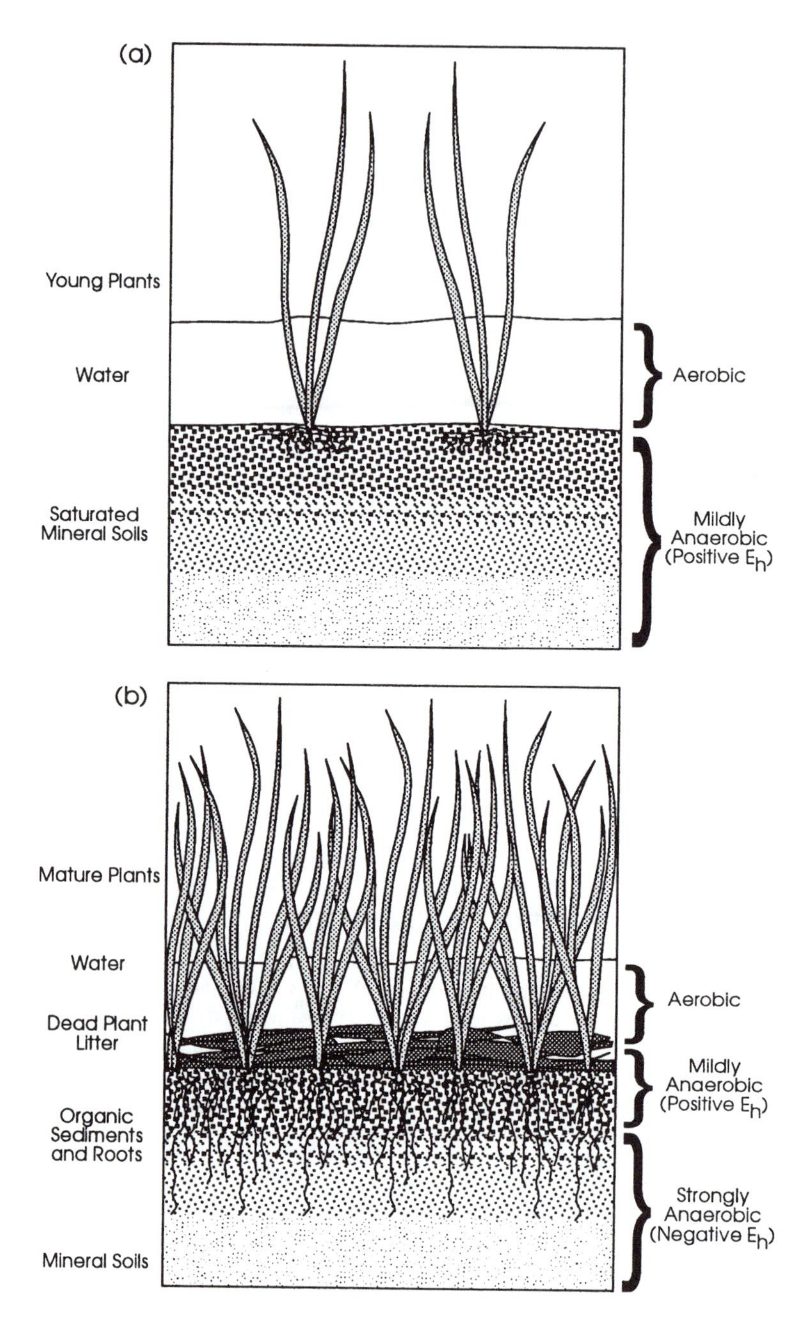

Figure 5-2 Two idealized stages in the maturation of constructed wetland soils: (a) newly planted and flooded wetland on mineral soils has low organic carbon accumulation and is mildly anaerobic due to microbial degradation of existing low organic carbon concentrations; (b) mature wetland with actively decomposing litter layer and surficial organic sediments where redox potential is lower due to higher rate of microbial respiration.

over and above any sedimentation of incoming suspended matter. This is the process of peat accumulation if most of the material originates from leaf and stem detritus of emergent macrophytes in marshes or from sphagnum mosses in bogs. This can range from 0.3 to 1.4 cm/yr in Louisiana freshwater marshes (DeLaune et al., 1978), from 0.3 to 1.1 cm/yr in the Everglades (Reddy et al., 1993), and from 0.1 to 1.1 mm/yr for northern bogs in the U.K. (Durno, 1961).

This accretion of organic matter generated within the wetland is stimulated by extra nutrients. Reddy et al. (1991a) show a decrease in accretion along the decreasing gradient of phosphorus in the Everglades.

CALCITIC MUDS AND MARL

Wetlands may develop with conditions of a short hydroperiod (the annual period of inundation). Dry periods subsequently promote the oxidation of organic detritus and, thus, preclude the formation of peat. Under flooded conditions, an open water system may develop dense algal populations that can quickly utilize nutrients. Given a supply of dissolved calcium, the algae can create calcium carbonate precipitates. In combination with silt, these solids become calcitic muds (Gleason and Stone, 1994).

If the calcium carbonate precipitates are mixed with sands and clay, the result is marl, a substrate in the alkaline fens of northern climates. These are fine-grained tan or whitish materials which may be hard or soft depending on exact composition and conditions of deposition. The hydraulic conductivity of soft marl is on the order of 0.02 m/d (Kadlec and Bevis, 1992), which places it in the category of poorly conductive soils.

PHYSICAL PROPERTIES OF HYDRIC SOILS

WETLAND MINERAL SOILS

The definition of hydric soils indicates that any upland soil utilized for construction of a wetland treatment system will become a hydric soil following a short to long period of flooding and continuous anaerobiosis. Since most wetlands are constructed in areas that were formerly uplands, most constructed wetlands are initially dominated by mineral soils. As constructed wetland treatment systems mature, the percent of organic matter in the soil generally increases, and in some systems, soils might eventually cross the arbitrary line between mineral and organic.

Mineral soils are classified by particle size distributions, color, depth, and a number of other factors. The three major mineral soil classes are clays, silts, and sands.

Clays are soils with very fine particles packed closely together. In some clay soils, these fine minerals become organized into crystalline lattices. Because of their very fine texture and low hydraulic conductivity, clays may function as aquitards. The existence of many natural wetlands depends on impermeable clay lenses in sedimentary or wind-blown (loess) deposits. Clays typically have the highest adsorption potential of any soils because of their high surface area to volume ratio resulting from their small particle size distribution. When water in a wetland is in contact with underlying clays or when water percolates through the bottom of a clay-lined wetland, the presence of the clays may greatly increase treatment potential for conservative ions such as phosphorus and metals.

Clays typically inhibit the flow of water and gases because of their dense structure. Although clays may have good nutrient availability, their dense structure can make root development difficult for many emergent wetland plants, resulting in poor or slow plant growth (Pierce, 1989).

Like clays, silts may also reduce the flow of water due to low hydraulic conductivities. However, silts have larger particle sizes than clays and do not typically exhibit the molecular (crystalline) properties of clays. Loamy soils consist of mixtures of clays, silts, and sands and typically have excellent plant growth characteristics because of adequate nutrient holding capacity and higher hydraulic conductivities and gaseous diffusion rates.

Sandy soils are made of larger mineral particles, typically with a preponderance of quartz crystals which are chemically inert compared to clays. Thus, sandy soils are less likely to be able to bind chemical nutrients important for plant nutrition and may require fertilization to insure healthy plant growth. The hydraulic conductivity of sandy soils may be quite high, depending on the percentage of silts and clays in the soil.

Wetland mineral soils are generally light to dark grey in color if they are continuously saturated or light tan to brown with darker mottles in intermittently flooded areas. The grey color typical of many saturated mineral soils is called *gleying* and results from the presence of reduced iron compounds in a clay matrix (Pierce, 1989).

WETLAND ORGANIC SOILS

Histosols occur in wetland environments where the rate of organic matter formation is greater than the rate of decomposition (Buol et al., 1980). This condition is most likely to occur in areas with relatively low dissolved oxygen, low-nutrient conditions, and high carbon fixation rates by woody or herbaceous emergent macrophytes, mosses, and/or algae.

Organic soils may be classified by their extent of decomposition. These organic soils may be called peat, muck, or mucky peat. Organic soils that have the least amount of decomposition (less than one third decomposed) are called fibrists (peat). Fibric peats have more than two thirds of the plant fibers still identifiable. Saprists or mucks have greater than two thirds of the original plant materials decomposed. Hemists (mucky peat or peaty mucks) are in between saprists and fibrists.

Bulk density of organic soils tends to increase with decomposition. Slightly decomposed organic soils (fibrists) have larger pore spaces and higher rates of saturated water movement compared to well-decomposed saprists which may have hydraulic conductivity rates lower than clay soils (Buol et al., 1980).

Due to their fibrous nature, histosols may shrink, oxidize, and subside when they are drained. Fire may also accelerate this oxidation process, and agricultural practices (drainage, cropping, harrowing, and burning) are known to result in soil subsidence in highly organic soils such as those in the Everglades Agricultural Area where subsidence rates have been estimated at about 3 cm/yr (Stephens, 1956).

Drying of organic soils promotes oxidation and gasifies carbon, but not the mineral nutrients associated with those soils. Although the available nutrient content of a peat or muck is often quite low, there are large amounts of nitrogen, phosphorus, sulfur, and other mineral constituents organically bound in unavailable forms. Oxidation destroys these recalcitrant organics and releases the associated substances. Upon reflooding, those substances can dissolve and provide relatively high concentrations of nutrients and other dissolved minerals.

Organic soils cannot easily be characterized by grain size because the necessary act of drying destroys the physical-chemical structure. The general range of hydraulic conductivity for soils found in sedge, reed, and alder wetlands is 0.1 to 10 m/d, placing these materials in the range of other mineral soils (Loxham and Burghardt, 1986). However, this is true only for fully saturated soils; even a slight degree of unsaturation lowers the hydraulic conductivity by two orders of magnitude. The reason is the extremely large capillary suction pressure created in the micropores. This means that organic soils and sediments are virtually undrainable; they retain a very high percentage of water.

Organic soils are typically dark in color, ranging from black mucks to brown peats.

HYDRIC SOIL CHEMICAL PROPERTIES

Soil chemical properties are primarily related to the chemical reactivity of soil particles and the surface area available for chemical reactions. Chemical reactivity is related to the surface electrical charge of the soil particles. Soil charge is typically highest in clays and organic soil particles.

CATION EXCHANGE CAPACITY

Wetland soils have a high trapping efficiency for a variety of chemical constituents; they are retained within the hydrated soil matrix by forces ranging from chemical bonding to physical dissolution within the water of hydration. The combined phenomena are referred to as *sorption.*

A significant portion of the chemical binding is cation exchange, which is the replacement of one positively charged ion, attached to the soil or sediment, with another positively charged ion. The humics substances found in wetlands contain large numbers of hydroxyl and carboxylic functional groups, which are hydrophilic and serve as cation binding sites. Other portions of these molecules are nonpolar and hydrophobic in character. The result is the formation of *micelles,* which are groups of humic molecules with their nonpolar sections combined in the center and their negatively charged polar portions exposed on the surface of the micelle (Wershaw et al., 1986). Protons or other positively charged ions may then associate with these negatively charged sites to create electrical neutrality.

Micelles are one form of *ligand* that can bind metal ions. The cumulative process of binding of a metal ion to a ligand (L) to form a complex may be described by a chemical equation; here, it will be illustrated for the binding of a divalent metal ion (M):

$$2HL + M^{2+} \Leftrightarrow ML_2 + 2H^+ \tag{5-1}$$

The number of ligands per gram of dry solid is determined from the number of metal ions that can be sorbed by a fully protonated sample (Peat Testing Manual, 1979). This is referred to as the cation exchange capacity (CEC) of the material, usually measured in milliequivalents per gram. Peats have CEC values of approximately 1.0 to 1.5 meq/g (Kadlec and Keoleian, 1986). For a heavy metal such as copper, this can translate to a large binding capacity, on the order of a few percent by weight (Kadlec and Rathbun, 1984).

Clearly, the pH of the soil or sediment has a large influence on the partitioning of a metal to the ligand because excess hydrogen ions drive Equation 5-1 toward the ionic form of the metal.

The complexation Equation 5-1 may operate in an equilibrium governed by the stability constant or selectivity coefficient (K):

$$K = \frac{C_{ML_2} C_{H^+}^2}{C_{HL}^2 C_{M^{2+}}} \tag{5-2}$$

The larger the value of K, the more stable the complex. Metal ligand formation constants less than 10^2 are considered weak, and formation constants greater than 10^8 are considered to be strong (Pankow, 1991). In some cases, metal ions have multiple protons available for complexation, and multiple ligands may be complexed by a single metal ion. This phenomenon leads to stepwise formation constants which generally decrease in strength with each additional ligand.

Determination of K from Equation 5-2 requires experimental measurement of the total number of ligands (the CEC), as well as the distribution of the metal between solution and solid as a function of pH (Kadlec and Keoleian, 1986). More often, the partitioning of a metal cation between solid and water is described by a partition coefficient or sorption isotherm graph, the latter sometimes expressed as a power law (Freundlich) equation:

$$C_{ML_2} = K_p(C_{M^{2+}})^n \tag{5-3}$$

An example of such an isotherm, for copper sorbing and exchanging to a sedge peat at pH 6.5, is (Kadlec and Rathbun, 1984):

$$C_s = 3.4(C_w)^{0.4} \tag{5-4}$$

where Cs = solid phase copper concentration, mg/g
Cw = water phase copper concentration, mg/l

Comparing Equations 5-2 and 5-3, the isotherm is expected to be strongly dependent on pH. There is 45-fold reduction in sorption for copper as pH is lowered from 6.5 to 1.2.

Anionic solutes such as phosphate, PO_4^{3-}, also sorb to wetland solids and soils. However, the ligand functional group will be positively charged. Iron and aluminum in the soil add to the numbers of such sorption sites (Reddy and D'Angelo, 1994). Phosphorus sorption isotherms have been represented by both power law (Freundlich) and saturating (Langmuir) isotherms (Richardson and Marshall, 1986; Hammer and Kadlec, 1980; Reddy and D'Angelo, 1994).

Drying of the organic material will destroy some of the character of the highly hydrated micellular chemical-physical structures, therefore destroying some of the sorption capacity of the material. The sorption capacity of dried, harvested peats are not as large as that of wet, living peats.

OXIDATION AND REDUCTION REACTIONS

Wetlands are ideal environments for chemical transformations because of the range of oxidation states that naturally occur in wetland soils. Free oxygen decreases rapidly with depth in most flooded soils because of the metabolism of microbes which consume organic matter in the soil and through chemical oxidation of reduced substances. This decline in free oxygen is measured as an increasingly negative electric potential between a standard platinum electrode and the concentration of oxygen in the soil. This measure of electric potential is called reduction-oxidation or redox potential (Eh) and provides an estimate of the soil's oxidation or reduction potential.

When Eh > 300 mV, conditions are termed aerobic because dissolved oxygen is available. When Eh < -100 mV, conditions are termed anaerobic because there is no dissolved oxygen. Some authors refer to intermediate conditions (near-zero DO) as anoxic.

Oxidation and reduction are chemical transformations involving the movement of protons and electrons between molecules. These transfers frequently result in striking differences in the chemical properties of the molecules being oxidized or reduced. A generalized oxidation-reduction reaction can be described as two half reactions where one half reaction involves the acceptance of electrons by an oxidized molecule to become a reduced molecule (*reduction*):

$$OX_1 + ne^- = RED_1 \tag{5-5}$$

and the second half reaction where a reduced molecule is *oxidized* by the loss of one or more electrons:

$$RED_2 = OX_2 + ne^- \tag{5-6}$$

Half reactions generally do not occur alone because free electrons are reactive in aqueous environments. When an oxidant and a reductant are combined in a saturated soil, the overall oxidation-reduction reaction is of the form:

$$OX_1 + RED_2 = RED_1 + OX_2 \tag{5-7}$$

where OX_1 is considered to be the oxidant and RED_2 is considered to be the reductant.

The reduction of ferric iron (Fe^{3+}) to ferrous iron (Fe^{2+}) is a redox half reaction typical of anaerobic sediments:

$$Fe^{3+} + e^- = Fe^{2+} \qquad K = \frac{Fe^{2+}}{(Fe^{3+})(e^-)} \tag{5-8}$$

where K is the stability or formation constant for this reaction using the same convention as used earlier for complexation reactions. The typical complementary half reaction for this reduction of ferric to ferrous iron is the oxidation of reduced sulfur (hydrogen sulfide, H_2S) to sulfate (SO_4^{2-}) by the half reaction

$$H_2S + H_2O = SO_4^{2-} + 10H^+ + 8e^- \tag{5-9}$$

for an overall oxidation-reduction reaction of

$$8Fe^{3+} + H_2S + H_2O = 8Fe^{2+} + SO_4^{2-} + 10H^+ \tag{5-10}$$

However, the real situation in treatment wetlands is far more complicated because there are more species present and more redox routes than those used here for illustration. For instance, Vile and Weider (1993) determined that iron reduction can be accompanied by sulfate reduction, not sulfate production.

As soils become increasingly reduced, chemicals other than oxygen provide electrons for further reduction. While aerobic soils generally have an Eh range from about 400 to 700 millivolts (mV) corrected to a pH of 7, wetland soils may have an Eh range from −300 mV or lower (strongly reduced) to about 700 mV (well oxidized) (Gambrell and Patrick, 1978).

Oxygen depletion in wetland soils is typically complete in the Eh range between +320 to +340 mV. Nitrate reduction (denitrification or ammonification—see Chapter 13) may begin before complete oxygen removal and is considered to be complete at about +220 mV. Manganic manganese (Mn^{4+}) is reduced to manganous manganese (Mn^{2+}) by +220 mV; ferric iron is reduced to the ferrous form by +120 mV; sulfate is reduced to sulfide at −150 mV; and carbon dioxide (CO_2) is reduced to methane (CH_4) from −250 to −300 mV (Gambrell and Patrick, 1978). Figure 5-3 illustrates the time sequence of oxidation-reduction reactions occurring in a newly flooded mineral or organic soil.

The redox potential of many wetland soils decreases with vertical depth into the sediments because the only source of free oxygen is from atmospheric diffusion at the top of the sediment layer. The typical oxygen gradient in wetland sediments includes a thin (less than a few centimeters) oxidized surface horizon at the sediment-water interface, underlain by increasingly reduced conditions with depth based on the amount of biological and chemical reducing activity in the sediments (Figure 5-4). Vertical redox gradients in treatment wetland soils will vary in response to distance from the point of wastewater loading (Figure 5-5).

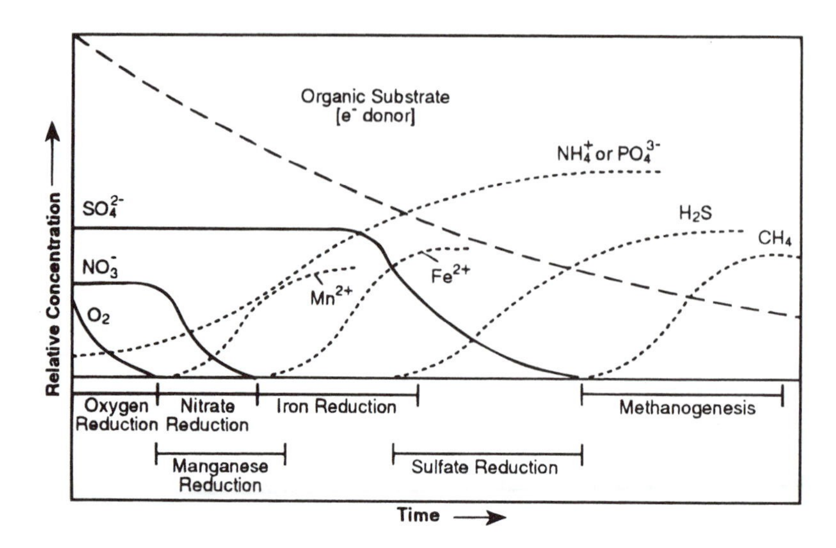

Figure 5-3 Time sequence for oxidation reduction in newly flooded wetlands soils. (Adapted from Reddy and D'Angelo, 1994, in *Global Wetlands: Old World and New*, W. J. Mitsch, Ed., Elsevier. With permission.)

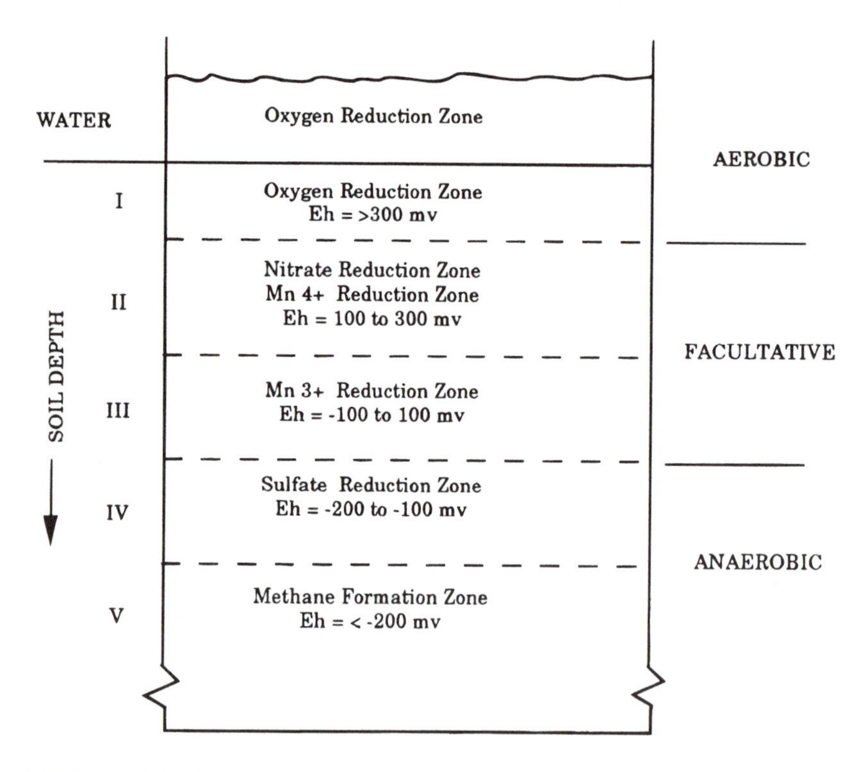

Figure 5-4 A typical depth profile for oxidation-reduction reactions and redox potential in a lightly loaded wetland. (Adapted from Reddy and D'Angelo, 1994, in *Global Wetlands: Old World and New*, W. J. Mitsch, Ed., Elsevier. With permission.)

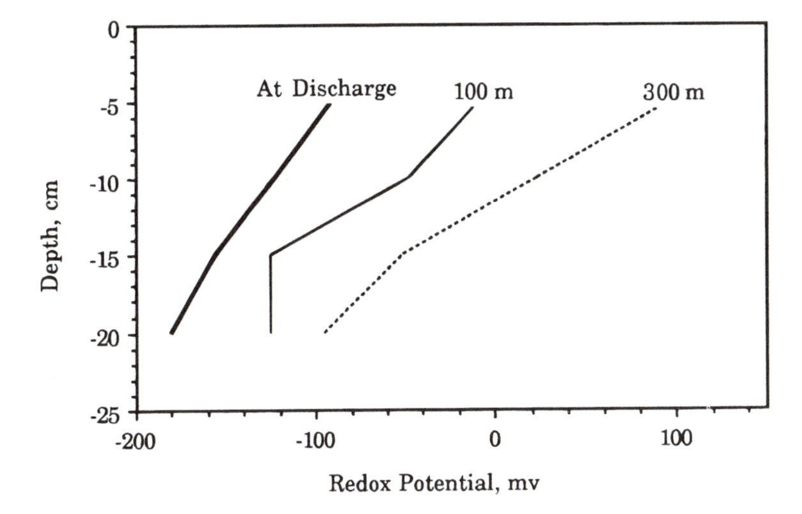

Figure 5-5 Vertical profiles in redox potential as a function of distance from discharge, and depth below water, in the peatland treatment system at Houghton Lake, MI. The soil surface is ill defined, but would be approximately 10 to 15 cm below the water surface. (Unpublished data from the Houghton Lake Treatment Wetland, 1993 field season.)

Northern wetlands are sometimes sealed by an ice cap in winter which prevents the supply of oxygen to the water and/or soils. This then shifts the redox profile to much lower Eh values, causing sulfate reduction and methanogenesis to dominate even the upper soil horizons. Gaseous sulfur compounds cannot escape and may reach lethal levels for wetland biota.

Treatment wetlands are often subjected to waters with higher oxygen demands exerted by both carbonaceous and nitrogenous compounds. This causes a greater depletion of electron acceptors such as oxygen, nitrate, sulfate, and iron in both the water column and the underlying soils. The redox potential in treatment wetlands is therefore typically lower than for natural wetlands, ranging from the nitrate reducing regime downward to the methanogenesis regime (Figure 5-6). Exceptions are treatment wetlands that receive very high-quality wastewaters.

Chemical conditions in the soil column are responsible for some of the variability in the overall performance of treatment wetlands. For instance, a sudden excursion of soil Eh to lower values may be accompanied by a release of coprecipitated phosphorus and a switch in the BOD reduction mechanisms (Herskowitz, 1986).

HYDROGEN ION (pH)

Prior to flooding, soils may have widely varying pH conditions ranging from about 3 to 10 units (Hammer, 1992). Following flooding, pH in wetland soils may initially decline due to aerobic decomposition liberating carbon dioxide into the interstitial water. However, this initial pH swing is generally transient and is followed by a typical trend in both acidic and alkaline soils toward pH neutrality (pH 6.7 to 7.2 units) over time (Figure 5-7). This typical trend is considered to be the result of ferric iron reduction under flooded soil conditions (Gambrell and Patrick, 1978). In some highly organic histosols, pH may remain very low, even following long periods of flooding. This result is likely due to the slow oxidation of organic sulfur compounds resulting in production of sulfuric acid and due to the presence of humic acids (Mitsch and Gosselink, 1993). Wetland treatment system hydrogen ion dynamics are described further in Chapter 10.

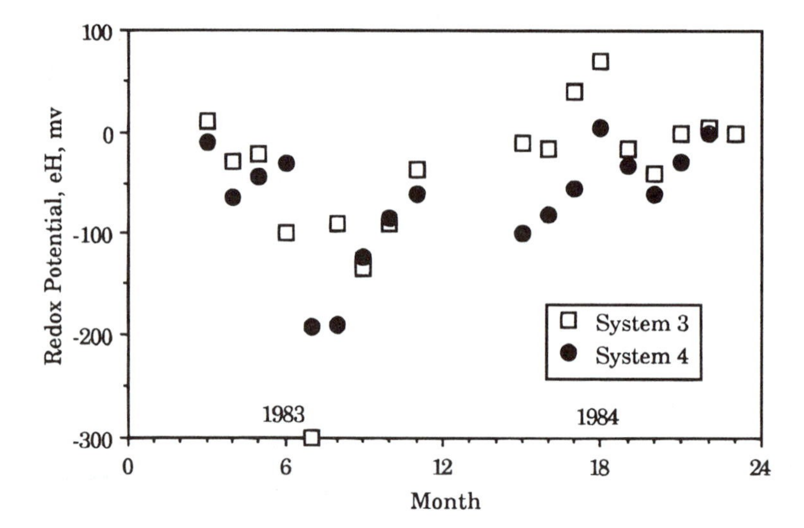

Figure 5-6 Listowel redox potential as a function of time. The measurement was taken at a depth of 1.5 cm. The points shown represent the average of three inlet and three outlet sampling points for each system. Measurements were not taken under the ice in winter. (Adapted from Herskowitz, 1986.)

PHOSPHORUS DYNAMICS

Phosphorus is frequently the most limiting macronutrient in upland and wetland ecosystems. Most municipal and agricultural wastewater discharges contain elevated phosphorus concentrations compared to levels that will stimulate excessive or nuisance algal growth in downstream receiving waters. Therefore, the phosphorus assimilation capacity of wetlands used for treatment is important. Chapter 14 provides a description of the typical phosphorus cycle in wetlands and how to estimate phosphorus removal potential of these systems.

Because of the biological importance of phosphorus, wetland ecosystems have developed mechanisms for trapping and recycling this element. Phosphorus entering a wetland ecosystem is readily interconverted from organic to inorganic forms and forms chemical complexes with organic and inorganic ligands which in turn may be adsorbed or precipitated within wetland soils. Phosphorus typically forms insoluble complexes with oxidized iron, calcium,

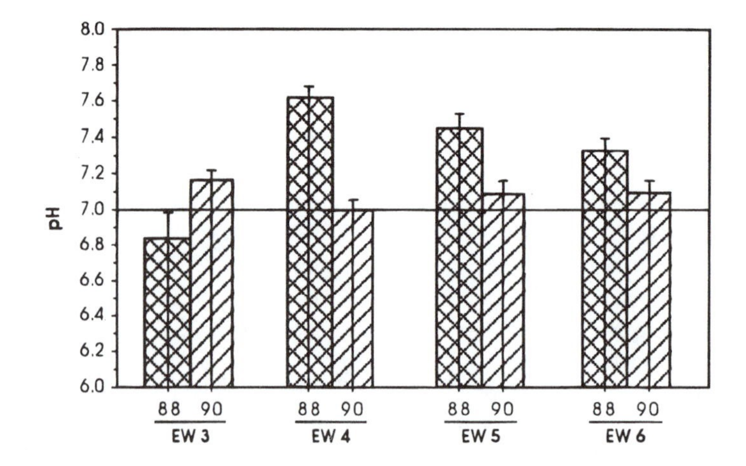

Figure 5-7 Changes in sediment pH following 2 years of operation of the Des Plaines River Constructed Wetland Treatment Systems near Wadsworth, IL. (Modified from Fennessy, 1992).

and aluminum in aerobic wetland soils. Under anaerobic conditions, much of this bound phosphorus becomes soluble and more available for plant uptake and for diffusion from the soil system. If an aerobic surface soil horizon occurs within the wetland, soluble phosphorus may be retained or trapped within the wetland sediments until it is biologically cycled (Gambrell and Patrick, 1978).

Phosphorus may be directly adsorbed onto clays located within and under wetlands. This adsorption is strongest under low pH conditions. In some instances, incoming P sorbed to particulates settles rapidly and is effectively buried as this form of the nutrient. However, exchangeable sorbed P is available for further cycling. Soluble reactive P often persists longer, but not in oligotrophic wetlands (Qualls et al., 1994). Some portion of the phosphorus bound in wetland sediments may eventually be recycled through plant uptake and either reburied as organic phosphorus or lost in the wetland outflow.

BIOLOGICAL INFLUENCES ON HYDRIC SOILS

Biological, chemical, and physical properties of wetland soils are interdependent. Microbial wetland fauna (see Chapter 8) make up a significant fraction of the organic carbon occurring in hydric soils. These tiny organisms are competing for sometimes limited and rapidly shifting supplies of energy-containing compounds and nutrients, and their growth and death have a very significant effect on the fate and transport of the majority of soil chemical constituents. In addition to the microbial populations, macrophytic plants diversify soil structure through the growth and death of roots and the creation of decaying plant litter, and wetland animals dig, burrow, scrape, and otherwise cause bioturbation of wetland sediments on an almost continuous basis. Some of the major interactions between wetlands biology and sediments are described below. Additional discussion of these processes can be found in other sections of this book.

MICROBIAL SOIL PROCESSES

Soil microbial populations have significant influence on the chemistry of most wetland soils. Important transformations of nitrogen, iron, sulfur, and carbon result from microbial processes. These microbial processes are typically affected by the concentrations of reactants as well as the redox potential and pH of the soil.

Nitrogen transformations in wetlands are described in detail in Chapter 13. Organic nitrogen is biologically transformed to ammonia nitrogen through the process of mineralization. Mineralization results as a consequence of organic matter decomposition resulting from the actions of both aerobic and anaerobic microbes. Ammonia is in turn converted to nitrite and nitrate nitrogen through an aerobic microbial process called nitrification. As indicated earlier in this chapter, nitrate nitrogen can be further transformed to nitrous oxide or nitrogen gas in anaerobic wetland soils by the action of another group of microbes (denitrifiers). Nitrogen gas can also be transformed to organic nitrogen by bacterial nitrogen fixation in some aerobic and some anaerobic wetland soils (nitrogen fixation).

Bacteria can transform reduced iron and possibly manganese to oxidized forms. These chemosynthetic processes utilize oxygen as an electron acceptor and typically are accelerated by acidic conditions typical of acidic coal mine drainage waters. The iron and manganese cycles typical of treatment wetlands are described in more detail in Chapter 15.

Sulfate can be reduced to sulfide by anaerobic bacteria in wetlands. The sulfate serves as an electron acceptor in the absence of free oxygen at low redox potentials. Sulfides can provide a source of energy for chemoautotrophic and photosynthetic bacteria in aerobic wetlands, resulting in the formation of elemental sulfur and sulfate. Chapter 15 provides a more detailed description of the typical wetland sulfur cycle.

Various aspects of the wetland carbon cycle are described in Chapters 7 and 12. Organic carbon is microbially degraded to carbon dioxide by aerobic respiration when oxygen is available and by fermentation under anaerobic conditions. Greater energy is released under aerobic respiration, resulting in more efficient assimilation of organic matter into microbial cellular material (Gambrell and Patrick, 1978). In fermentation, organic matter serves as the terminal electron acceptor, forming acids and alcohols. Methane can be formed in wetlands due to the action of bacteria using carbon dioxide as an electron acceptor at very low redox potentials.

WETLAND ALGAE AND MACROPHYTES

Chapter 7 provides a description of the ecology of algae and macrophytes in wetlands. These photosynthetic organisms may or may not be important in soil formation and chemistry in wetlands used for wastewater treatment. Organic matter accumulation in some wetlands is a direct or indirect result of the primary fixation of carbon from the atmosphere by these plants. Particulate macrophytic material and dead algal cells contribute organic carbon, nitrogen, and phosphorus to the wetland littersoil layer in the form of cellulose, hemicellulose, lignin, proteins, and phospholipids (Reddy and D'Angelo, 1994). In some low nutrient wetlands and in wetlands that are drained and exposed to the atmosphere, oxidation (aerobic respiration) may result in no net accumulation of organic matter.

Wetland macrophytes further modify soil texture, hydraulic conductivity, and chemistry by the growth of plant roots and rhizomes. These plant structures initially serve as pathways for increased gaseous diffusion into and out of the wetland sediments. Gas-filled aerenchyma in many wetland plants provide significantly less diffusional resistance, allowing some oxidation of soils in the immediate vicinity of the roots (rhizosphere) and diffusion of carbon dioxide, hydrogen sulfide, and even methane back to the atmosphere through the plants (Figure 5-8). Several of the important chemical transformations mentioned earlier occur upon or within the aerated rhizosphere and roots of wetland plants.

The top layer of soil contains the roots of the emergent macrophytes. These most often occupy the top 20 to 30 cm of soil. A dense macrophyte stand will have a large amount of below-ground biomass in the form of roots and rhizomes, often in the range 1000 to 2000 g/m^2. Although numerous, these make up only a tiny fraction, less than 1 percent, of the upper soil horizon. However, roots are extremely important to both the physical and chemical condition of this soil.

Roots propagate through the soil by pushing their way into pore space and physically moving the soil grains. This is possible in fine-grained media, either soils or fine gravels. It is not possible in coarse angular rocks, and those media inhibit root penetration (Brix, 1994c; Tanner and Sukias, 1994). Roots die, with a typical turnover time on the order of a few years. This leaves behind a mat of fibrous root litter or detritus. In some cases, dead rhizomes can offer new pore space for water and gas movement. The live and dead root mat provides stability and strength to low density organic soils such as peats; without roots there is no bearing strength in wetland histosols. Travel becomes difficult for humans and large mammals, and hence the origin of the term "mired down."

The chemical influences of the plant roots are also quite large. Very steep gradients in most chemical species occur in the root zone. These are attributable to the extraction of nutrients and other dissolved substances by the plants into their root system. This depletes the pore water and, in turn, allows release of sorbed materials which also can be ingested. The end result is a depletion of the available, extractable pool of biochemically ingestible substances. These materials are translocated to either above-ground or below-ground plant

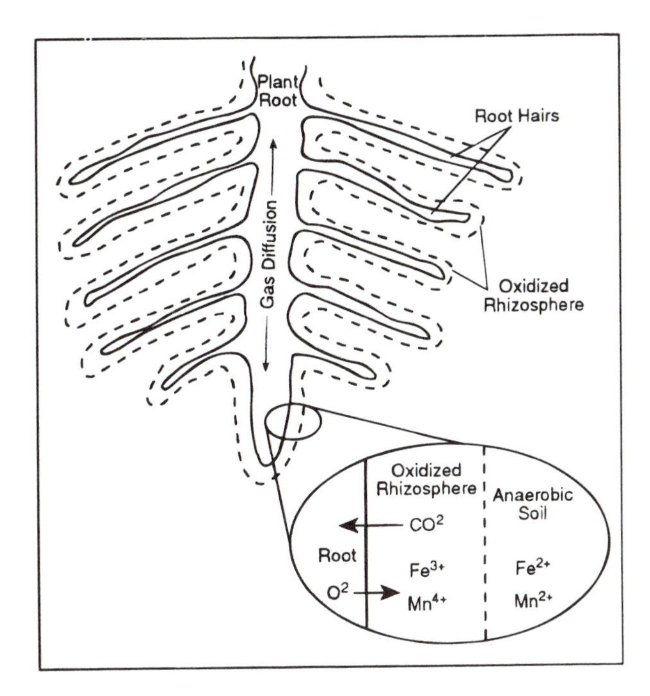

Figure 5-8 Idealized macrophyte root structure in a wetland soil illustrating development of an oxidized rhizosphere.

parts, where they are incorporated into tissues and reside for the lifetime of that plant part. When the biomass dies, these chemicals become subject to either release or burial with undecomposed detritus, forming new soil. Thus, it is seen that top soils are a vital and integral part of the biogeochemical cycle.

Plant roots are not the only living organisms in the rhizosphere. Seeds of the macrophytes lodge in this zone and make it the seed bank for possible regeneration of those species. Bacteria and fungi are numerous and perform a wide variety of chemical conversions. Many species of worms and larvae inhabit the top soil layer. The correct image is one of teeming life, much of which is not visible. This active surface layer of soil and its inhabitants are sometimes called the *acrotelm* in a peatland to distinguish it from the uninhabited, dead soil zones below, which are termed the *catotelm.*

EFFECTS OF ANIMALS ON WETLAND SOILS

Wetland fauna are important in the formation of soils and in the continual modification of soil properties. Much of the organic matter reaching the wetland litter compartment and ultimately contributing to soil organic matter has been processed by microscopic and macroscopic invertebrates living in the wetland water column, in the litter/detritus, and in the oxidized portion of the wetland sediments. Chapter 8 provides a description of the general classes of invertebrate organisms occurring in wetlands. Depending on size, location in the vertical wetland profile, and feeding strategy, wetland vertebrates and macroinvertebrates tear, shred, graze, and otherwise process large plant materials into smaller and smaller particles until they provide ample surface area for attachment of microscopic invertebrates and microbes.

In addition to the processing of plant materials into degraded organic matter, larger wetland animals may occasionally be important for bioturbation or the physical process of digging, stirring, and mixing sediments across vertical and horizontal profiles. Fish such as carp are known to stir lake and wetland sediments in their continual search for prey organisms. Wading birds also will feed in aerobic sediments on macroinvertebrates, and their resulting beak holes may number in the dozens per square meter in shallow wetland areas. Mammals may inhabit wetlands and either dig for crayfish, clams, or other sediment-colonizing food organisms or may build dens and burrows in or through the wetland sediments. When they occur, all of these faunal processes tend to increase the localized oxidation potential of wetland soils.

TREATMENT WETLAND SOILS

SURFACE-FLOW WETLANDS

The sediments that form in treatment wetlands are often different from those that form in natural wetlands, for a number of reasons. First, the enhanced activity of various microbes, fungi, algae, and soft-bodied invertebrates leads to a greater proportion of fine detritus compared to leaf, root, and stem fragments. There is a significant formation of low-density biosolids (sludge). Second, there may be a precipitation of metal hydroxides or sulfides, which add mineral flocs to the sediments. Finally, there is often a high ionic strength associated with effluents being treated, reflected in high dissolved salt content. The effect of high ionic strength is to alter the structure of the highly hydrated organic materials that comprise wetland sediments and soils.

The result of high nutrient waters on existing wetland sediment-soil profiles has often been observed to be a shift to low-density, mushy materials occupying the "water" column (Kadlec and Bevis, 1990; van Oostrom, 1994). This phenomena is illustrated in Figure 5-9 for the Houghton Lake peatland. Solids densities in the water column are on the order of 3% by weight in the wastewater discharge zone, which is the same as those observed in a constructed treatment wetland in New Zealand (van Oostrom, 1994). Sedge (*Cladium*) peats in the Everglades have solids densities of only 4 to 8% (Reddy et al., 1991a); those at Houghton Lake (*Carex*) reach 12 to 18%.

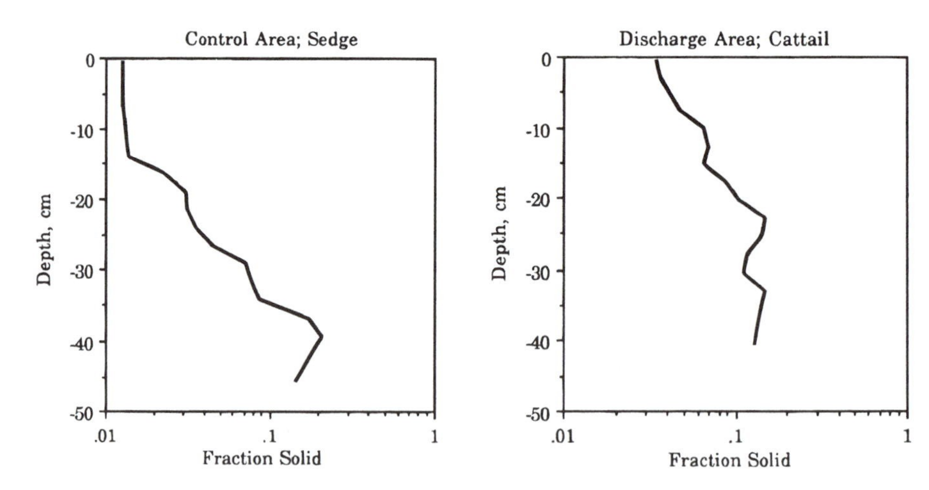

Figure 5-9 Vertical sediment-soil density profiles in a natural wetland (left) and in the wetland after wastewater discharge has converted it to cattail (right). On the left, density starts to increase at a depth of about 14 cm, which corresponds to the top of the litter layer. Solids then increase to the 15 to 20% range, which is the density of the peat substrate. On the right, even the top-most water has a few percent solids, and there is no definitive break in the profile to signal either litter or soil horizons.

This water-holding capacity of organic sedimentary materials and soils results from the hydration of the complex organic molecules that form these solids. Binding strengths range from simple pore-water retention in large pores to chemically bound water attached to hydration sites on the molecules. A single monolayer of water on humic or fulvic acid components of an organic soil requires a water content of about 5% by weight (Schnitzer, 1986). As a result, peat cannot be completely air dried because the more tightly bound water cannot evaporate. An air-dried peat still retains about 30% water by weight.

The considerations listed previously show that wetland soils are a vital and integral link in processes that govern water quality. It therefore seems reasonable to expect consideration of wetland soils to be an important part of treatment wetland design. However, that is not the case. Antecedent soils are altered and replaced by new organic soils.

The sorption capacity of the antecedent soils is reequilibrated with the new water quality of the incoming water, perhaps along a gradient from inlet to outlet. If there are leachable chemicals, they are depleted and exit the wetland.

Roots and rhizomes of new plantings (constructed wetlands) or replacement species (natural wetlands) repopulate the top 30 cm of the wetland and set up a new biogeochemical cycle.

The long hydroperiods of treatment wetlands are conducive to the buildup of organics: first litter and microdetritus, then the sediments formed from their decomposition, and finally the organic soils generated from those sediments and deposited mineral solids.

In short, the wetland rearranges itself to accommodate the environment created by the designer. The functioning of the wetland after such adaptation is no longer dependent upon the previous condition and type of soils, hydrology, and biota. It is totally dependent upon the new soils, hydrology, and biota. It is this new sustainable mode of wetland operation that is the target of most designs.

Available data indicate that the final state of a treatment wetland, and the accompanying suite of water quality functions, are largely independent of the initial condition of the real estate upon which it is built. During the interim period of adaptation, antecedent conditions are important because they dictate the short-term performance of the wetland. That period of adaptation appears to extend for up to 2 years for newly constructed wetlands and longer for alteration of natural wetlands to a treatment function.

SUBSURFACE-FLOW WETLANDS

Some measure of performance control can be exerted via the utilization of specially tailored bed media for subsurface-flow (SSF) wetlands. In some sense, these are the "soils" of this type of constructed wetland. If sands, soils, or gravels are borrowed from natural sources, there will be a period of adaptation as for surface-flow (SF) wetlands, and it appears to be of the same general duration as for SF wetlands. However, a bed material may be chosen that is manufactured to have a very large phosphorus sorption capacity, such as an expanded clay (Jenssen et al., 1994). The design philosophy is now quite different than for most existing treatment wetlands; the intent is to exhaust a short-term capacity, regenerate the wetland, and repeat the cycle. This may be a feasible strategy in some cases, provided that the expense of regeneration coupled with its frequency are within acceptable economic bounds.

SUMMARY OF THE IMPORTANCE OF WETLAND SOILS

Wetland soils are the foundation for and principal storage of all biotic and abiotic components that exist in wetlands. Wetland soils support plant and microbial growth that provide the substrates necessary for water quality enhancement in wetland treatment systems. If soils are unsatisfactory for plant or microbial growth, the wetland treatment system is liable to have inadequate plant cover. A knowledge of the physical and chemical composition of site soils is essential to accurately predict some internal chemical and biological processes in treatment wetlands. The rate of soil accretion in wetland treatment systems affects the potential removal of conservative elements such as phosphorus and heavy metals and also is an important consideration during design of berm height above the wetland substrate. Some sorbed materials can be released if exposed to lower concentrations in incoming water. Violent hydrologic events are capable of resuspending particulate deposits. It is therefore incumbent on the designer to minimize or prevent such occurrences.

The main environmental factor that influences the nature of wetland soils is dissolved oxygen concentration. Vertical oxygen gradients are typically established in wetland soils due to bacterial respiration and chemical oxidation demand and due to the greatly reduced rate of oxygen diffusion in saturated soils compared to unsaturated soils. These oxidation gradients result in a chain of oxidation-reduction reactions that provide many wetlands with their typical profile of declining redox potentials with depth. Redox, in turn, affects the microbial processes that are important in most aspects of wetland use for water quality improvement, including especially removal of organic carbon and nitrogen.

Wetland soils are as dynamic in character as all other aspects of the wetland ecosystem. Soils in constructed wetlands built on upland sites undergo gradual transformations, resulting in accumulations of organic carbon and reduced elements such as iron and sulfur that are typical of natural wetland soils. Many of the changes that occur during wetland development and succession are the result of biological factors such as growth of bacteria and fungi, algae and macrophytic plants, micro- and macroinvertebrates, and larger animals that occur in wetlands. While many of these natural biological processes are not within the control of the wetland treatment system designer and operator, their effects should be considered when trying to maximize chances for success with a wetland project.

CHAPTER 6

Wetland Hydrology and Water Quality

WETLAND HYDROLOGY

The water status of a wetland defines its extent and is a determinant of species composition in natural wetlands (Mitsch and Gosselink, 1993). Hydrologic conditions also influence the soils and nutrients, which in turn influence the character of the biota. The flows and storage volume determine the length of time that water spends in the wetland and, thus, the opportunity for interactions between waterborne substances and the wetland ecosystem.

Water enters natural wetlands via streamflow, runoff, groundwater discharge, and precipitation (Figure 6-1). These flows are extremely variable in most instances, and the variations are stochastic in character. Stormwater treatment wetlands generally possess this same suite of inflows. Treatment wetlands dealing with continuous sources of wastewater may have these same inputs, although streamflow and groundwater inputs are typically absent. The steady inflow associated with continuous source treatment wetlands represents an important distinguishing feature. A dominant steady inflow drives the ecosystem toward an ecological condition that is somewhat different from a stochastically driven system.

Wetlands lose water via streamflow, groundwater recharge, and evapotranspiration (Figure 6-1). Stormwater treatment wetlands also possess this suite of outflows. Continuous source treatment wetlands would normally be isolated from groundwater, and the majority of the water would leave via streamflow in most cases. Evapotranspiration (ET) occurs with strong diurnal and seasonal cycles because it is driven by solar radiation, which undergoes such cycles. Thus, ET can be an important water loss on a periodic basis.

Wetland water storage is determined by the inflows and outflows together with the characteristics of the wetland basin. Depth and storage in natural wetlands are likely to be modulated by landscape features such as the depth of an adjoining water body or the conveyance capacity of the outlet stream. Large variations in storage are therefore possible in response to the high variability in the inflows and outflows (Figure 6-2). Indeed, some natural wetlands are wet only a small fraction of the year, and others may be dry for interim periods of several years. Such periods of dryout have strong implications for the vegetative structure of the ecosystem. Constructed treatment wetlands, on the other hand, typically have some form of outlet water level control structure. Therefore, there is little or no variation in water level, except in stormwater treatment wetlands. Dryout does not normally occur, and only those plants that can withstand continuous flooding will survive.

Temporal changes in depth, combined with an uneven topography of the wetland bottom, lead to vegetaiive pattern effects in natural wetlands. Constructed treatment wetlands usually have nearly uniform bottoms. Combined with controlled, steady water levels, this means uniform hydrologic conditions and an absence of pattern effects. Pattern effects interact with water flows through the wetland, with preferential, sparsely vegetated channels carrying a disproportionately high fraction of the water. This in turn impairs the treatment potential because much of the wetland surface is not exposed to the water flow.

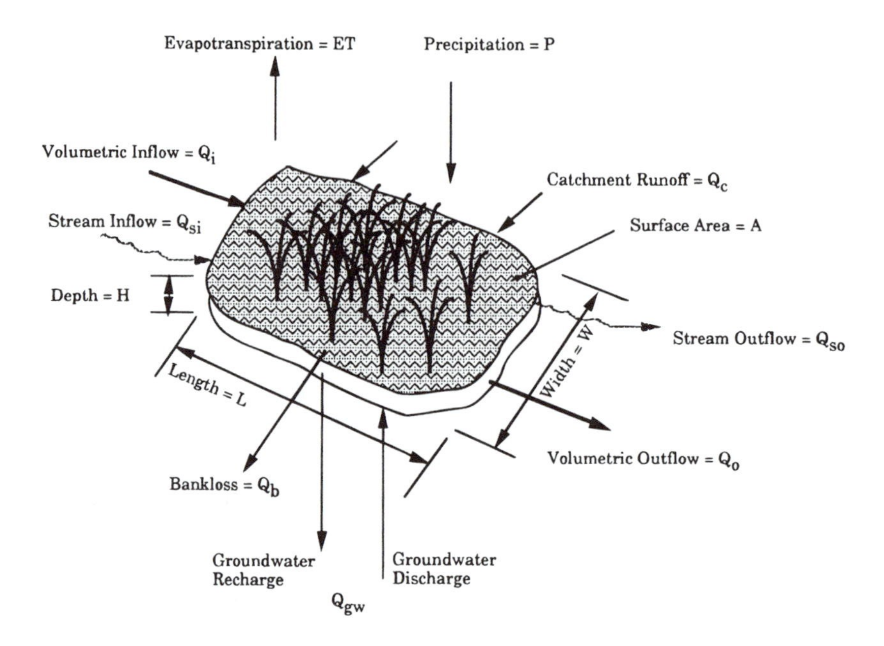

Figure 6-1 Components of the water budget and associated terminology.

The same processes are involved in the hydrology of both natural and constructed wetlands. However, different flows are dominant in the two situations, and the physical structure of the basin integrates these processes in a different manner. It is therefore possible to utilize the extensive body of knowledge on natural wetland hydrology, provided that care is taken to adapt it to the conditions of the treatment wetland.

The important features of wetland hydrology from the standpoint of treatment efficiency are those which determine the duration of water-biota interactions and the proximity of waterborne substances to the sites of biological and physical activity. There is a strong tendency in the wetland treatment literature to borrow the detention time concept from other aquatic systems such as "conventional" wastewater treatment processes. In purely aquatic environments, reactive organisms are distributed throughout the water, and there is often a clear understanding of the flow paths through the vessel or pond. However, wetland ecosystems are more complex and therefore require more descriptors.

DEFINITION OF HYDROLOGIC TERMS

Literature terminology is somewhat ambiguous concerning hydrologic variables. The definitions used in this book are specified in the pages that follow. The notation and parent variables are illustrated in Figure 6-1.

HYDRAULIC LOADING RATE

Symbol:	q
Alternative symbol:	HLR
Units:	cm/d;m/d;m/yr

(a)

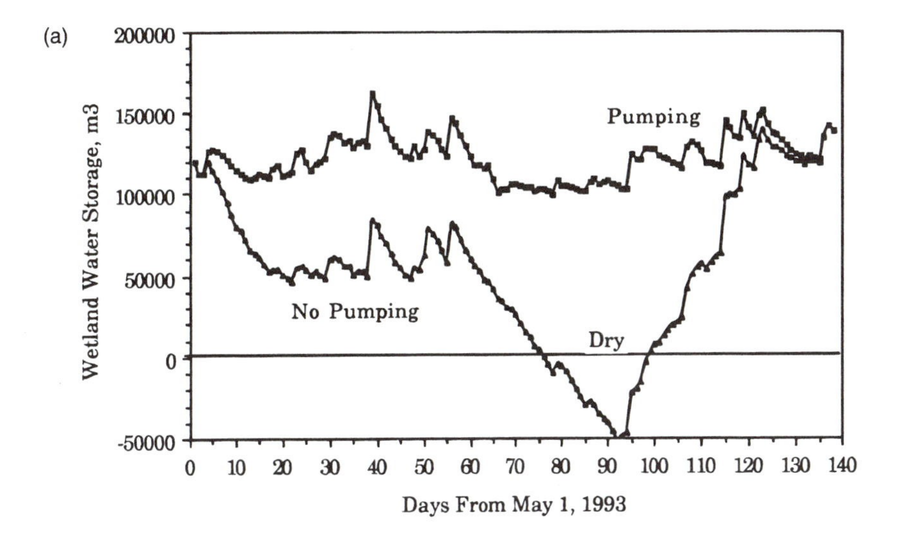

(b)

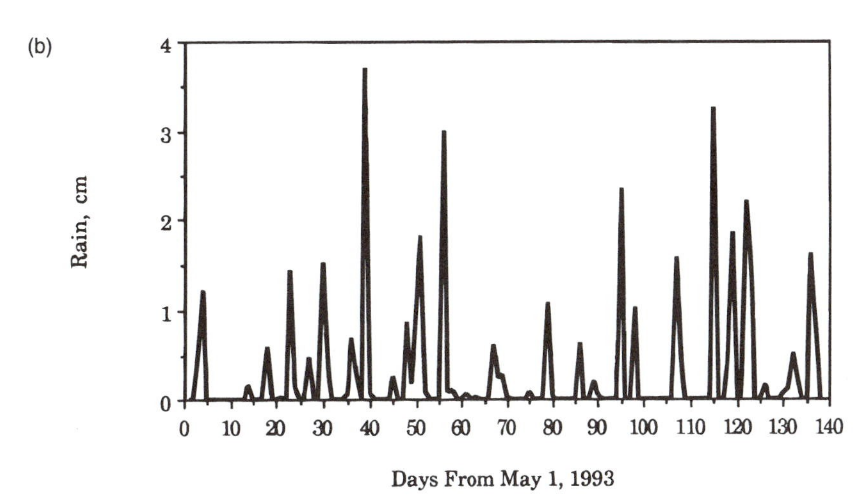

Figure 6-2 (a) The effect of added wastewater on water storage in the Houghton Lake wetland. Conditions of no pumping are calculated from a calibrated hydrology model and correspond well to preproject observations. ET would have dried out the wetland in mid-summer without pumped additions. (b) Rainfall at the Houghton Lake, MI wetland during the 1993 pumping season. Each rainstorm has a temporary effect on water storage.

Defining equation:

$$q = \frac{Q}{A} \tag{6-1}$$

where A = wetland area (wetted land area), m^2
Q = water flow rate, m^3/d

The hydraulic loading rate is the rainfall equivalent of whatever flow is under consideration. It does not imply the physical distribution of water uniformly over the wetland surface. The wetted area is usually known with good accuracy because of berms or other confining features. The definition is most often applied to the wastewater addition flow at the wetland inlet: $q_i = Q_i/A$. The subscript "i" is often omitted for simplicity. Care must be taken because the unsubscripted symbol q also may refer to the local volumetric flow rate at an interior location.

Some wetlands are operated with intermittent feed, notably, vertical flow wetlands. Under these circumstances, the term hydraulic loading rate refers to the time average flow rate. The loading rate during a feed portion of a cycle is the instantaneous hydraulic loading rate, also called the hydraulic application rate.

Some wetlands are operated seasonally, for instance, during warm weather conditions in northern climates. Although these are in some sense intermittently fed, common usage refers to the loading rate during operation and not to the average over the entire year.

MEAN WATER DEPTH

Symbol: $\tilde{h}$
Alternative symbol: d
Units: m,cm
Defining equation:

$$\tilde{h} = \frac{1}{A}\int_0^L \int_0^W h(x, y)dydx \qquad (6.2)$$

where x = longitudinal distance, m
y = transverse distance, m
h = water depth at coordinates (x,y), m

The mean depth calculation requires a detailed survey of the wetland bottom topography combined with a survey of the water surface elevation. The accuracy and precision must be better than normal because of the small depths usually found in treatment wetlands. The two surveys combine to give the local depth:

$$h = H - G \qquad (6\text{-}3)$$

where G = local ground elevation, m
H = local water elevation, m

As-built surveys under dry conditions may not suffice for determination of ground levels because of possible soil swelling and lift upon wetting. If the substrate is a peat or muck, there is not a well-defined soil-water interface. Common practice in that event is to place the surveyor's staff "firmly" into the diffuse interface. Water surface surveys are necessary in those situations where a head loss is incurred. That includes most subsurface-flow (SSF) wetlands and some larger, densely vegetated surface-flow (SF) wetlands. Local water depth is then determined as the difference between two field measurements and, hence, is subject to double inaccuracy.

The difficulties outlined previously have prevented accurate mean depth determinations in many treatment wetlands.

WETLAND WATER VOLUME

Symbol: V
Units: m^3

Defining equation:

$$V = \int_0^L \int_0^W \int_0^h e(x, y, z)dxdydx = \tilde{\varepsilon}V_T = \tilde{\varepsilon}A\tilde{h} \quad (6\text{-}4)$$

where ε = water volume fraction in the water column, m^3/m^3
V_T = total volume between water and ground surfaces, m^3
z = vertical distance coordinate, m

The void fraction ε, or porosity of the wetland, is also difficult to determine. In the case of an SF wetland, it may vary spatially in the x and y directions due to pattern effects. It also varies in the vertical direction, with lesser values near the bottom in the litter layer. However, the mean value is usually greater than 0.95, and many authors therefore consider $\varepsilon = 1.0$ as a good approximation. No one has ever directly measured the void fraction in an SF wetland.

The case of the SSF wetland presents more difficulties. The mean porosity of a clean sand or gravel media is apt to be in the range 0.30 to 0.45. But in an operational wetland, roots block some fraction of the pore space, as do accumulations of organic and mineral matter associated with treatment. Lateral pattern effects are typically minimal, but there may be gradients in both the vertical and flow directions. Roots block the upper horizons, and mineral matter preferentially settles to the bottom void spaces. Canister measurements of void fraction are not accurate because of vessel wall effects and compaction problems. Attempts to measure the water-filled void fraction by wetland draining have been thwarted by hold-up of residual water. Wetland filling is an unexplored option for porosity determination.

NOMINAL DETENTION TIME

Symbol: τ_n
Alternative symbol: t
Units: days
Defining equation:

$$\tau_n = \frac{V}{Q} = \frac{\tilde{\varepsilon}A\tilde{h}}{Q} \quad (6\text{-}5)$$

There is obviously a possible ambiguity that results from the choice of the flow rate that is used in this equation. The inlet flow rate is often used when there is no measurement or estimate of the outlet flow rate. Given the exit flow, some authors base the calculation on the average flow rate (inlet plus outlet divided by 2). When there are local variations in total flow and water volume, the correct calculation procedure must involve integration of transit times from inlet to outlet.

In any case, there is a good deal of inaccuracy in the calculation of nominal detention time because it combines the uncertainties in depth and porosity.

Nominal detention time is not necessarily indicative of the actual detention time because it is based on the presumption that the entire volume of water in the wetland is involved in the flow. This can be seriously in error, with the result that actual, measured detention times are much smaller than the nominal value. For example, the nominal detention time for the Boggy Gut treatment wetland was estimated to be 19 days; the measured value was 2 days (Knight, 1994). Careful consideration of the site characteristics showed that this effect is due to large zones of wetland that are not in the flow path.

The relation between nominal detention time and hydraulic loading rate is easily seen to be

$$q = \frac{\varepsilon \tilde{h}}{\tau_n} \tag{6-6}$$

Thus, it is seen that hydraulic loading rate is inversely proportional to nominal detention time for a given wetland depth. Therefore, hydraulic loading rate embodies the notion of contact duration, just as nominal detention time does.

ACTUAL VELOCITY

Symbol: v
Units: m/d
Defining equation:

$$v = \frac{Q}{(\varepsilon A)_c} \tag{6-7}$$

where $(\varepsilon A)_c$ = open area perpendicular to flow, m^2.

This is the velocity that would be measured by a probe or tracer.

SUPERFICIAL VELOCITY

Symbol: u
Units: m/d
Defining equation:

$$u = \frac{Q}{A_c} \tag{6-8}$$

where A_c = area perpendicular to flow, m^2

This quantity is also called the cross-sectional hydraulic loading rate. The cross-sectional area is just the width times the mean depth.

$$A_c = W\tilde{h}_c \tag{6-9}$$

where $\tilde{h}_c$ = mean depth perpendicular to flow, m.

The relation between superficial and actual velocities is

$$u = \varepsilon v \tag{6-10}$$

HYDRAULIC REGIME

Depths can vary, especially for natural and stormwater treatment wetlands. The term "water regime" is used as an umbrella name for a variety of representations of the depth and duration information. The parent information is the depth-time data, such as that shown in Figure 6-2 (storage is depth times area).

A derived representation is the stage-duration graph, or depth-duration graph, for the wetland. The time series of depths is converted to a distribution curve of depth vs. duration. This has the advantage of permitting several years' data to be rolled into one representation.

Figure 6-3 shows such a plot for Water Conservation Area 2A in the Florida Everglades (Burns and McDonnell, 1992). From such a plot, the time average depth at a given location may be determined; but that is not the mean depth for the entire wetland. When their use is required, time averages will be denoted by an overbar; thus, for gauge 2A-17 in Figure 6-3, $\bar{h} = 0.52$ m. Spatial averages will be denoted by an overtilde.

Constructed wetlands with outlet level control have very steep depth-duration curves.

HYDROPERIOD

This term is used in several ways in the wetlands literature. It is sometimes used to designate all attributes of the water regime. A second common usage is the number of days per year that there is surface water at a given wetland location; that is the definition adopted here. In Figure 6-3, gauge 2A-17 is inundated 93 percent of the time, and hence, the hydroperiod is 340 days.

Continuous source treatment wetlands usually have hydroperiods of 365 days. Stormwater wetlands may have lesser hydroperiods.

FLOODING FREQUENCY

Characterization of the wetland water time series is incomplete until the frequency with which standing water periods occur is specified. In general terms, there are significant ecological differences between systems with brief and frequent flooding vs. those with longer flood periods, even though the hydroperiods and depth-duration curves are identical.

Continuous source treatment wetlands usually are continuously flooded and, therefore, have a flooding frequency of 1.0. Stormwater wetlands may have greater flooding frequency.

GLOBAL WATER MASS BALANCES

GENERAL PRINCIPLES

Transfers of water to and from the wetland follow the same pattern for SF and SSF wetlands (Figure 6-1). In treatment wetlands, wastewater additions are normally the dominant

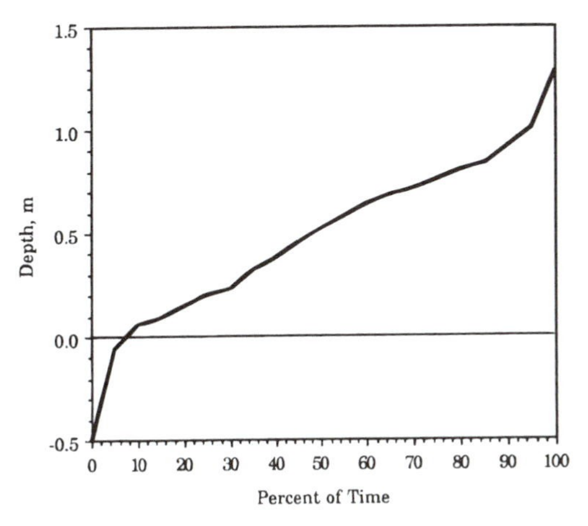

Figure 6-3 The depth-duration curve for gauge 2A-17 in WCA2A in the Everglades, compiled over a 10-year period of record. The wetland was dry about 7 percent of the time; thus, the hydroperiod was $0.93 \times 365 = 340$ days. Depth exceeded 1 m 6 percent of the time. The mean depth was 0.52 m (50th percentile). (Adapted from Burns and McDonnell, 1992.)

flow, but under some circumstances, other transfers of water are also important. The dynamic overall water budget for a wetland is

$$Q_i - Q_o + Q_c - Q_b - Q_{gw} + Q_{sm} + PA - ETA = \frac{dV}{dt} \qquad (6\text{-}11)$$

where
A = wetland top surface area, m^2
ET = evapotranspiration rate, m/d
P = precipitation rate, m/d
Q_b = bank loss rate, m^3/d
Q_c = catchment runoff rate, m^3/d
Q_{gw} = infiltration to groundwater, m^3/d
Q_i = input wastewater flow rate, m^3/d
Q_o = output wastewater flow rate, m^3/d
Q_{sm} = snowmelt rate, m^3/d
t = time, d
V = water storage in wetland, m^3

Over any budget period, each flow rate adds or removes a corresponding volume of water:

$$V_i - V_o + V_c - V_b - V_{gw} + V_{sm} + V_r - V_e = \Delta V_{stored} \qquad (6\text{-}12)$$

where
V_b = bank loss volume, m^3
V_c = catchment runoff volume, m^3
V_e = evaporated volume, m^3
V_{gw} = volume infiltration to groundwater, m^3
V_i = input wastewater volume, m^3
V_o = output wastewater volume, m^3
V_r = rain volume, m^3
V_{sm} = snowmelt volume, m^3
ΔV_{stored} = change in water storage in wetland, m^3

Each term in these water budgets may be important for a given treatment wetland, but rarely do all terms contribute significantly.

MASS BALANCE COMPONENTS

Inflows and Outflows

Most moderate- to large-scale facilities will have input flow measurement; but a smaller number of facilities will have the capability of independently measuring outflows as well as inflows. Therefore, the overall water budget (Equation 6-11 or 6-12) is often used to quantify the outflow. Usually, only rainfall is a significant addition and only ET is a significant subtraction to the inflow. This calculation is best done when there is no change in storage.

The change in storage (ΔV_{stored}) over a short averaging period T can be a significant quantity. For example, if the nominal detention time in the wetland is 10 days, then a 10 percent change in stored water represents 1 day's addition of wastewater. Because water depths are not large, changes of a few centimeters may be important over short averaging periods.

If there is significant infiltration, there are two unknown outflows, and the water budget alone is not sufficient to determine either outflow by difference.

Runoff

The total catchment area for a wetland is likely to be just the area enclosed by the containing berms and roads; that area is easily computed from site characteristics. Rainfall on the catchment area will in part reach the bed by overland flow, in an amount equal to the runoff factor times the rainfall amount and the catchment area (see Figure 6-4). A very short travel time results in this flow being additive to the rainfall:

$$Q_c = \psi \cdot P \cdot A_c \tag{6-13}$$

where the units of each term are cubic meters per day, m³/d;

A_c = catchment surface area, m²
ψ = catchment runoff coefficient,

For small- and medium-sized wetlands, the catchment area will typically be about 25 percent of the wetland area, as it is for the Benton, KY system, for example. About 20 percent of a site will be taken up by berms and access roads which may drain to the wetland. Runoff coefficients are high, since the berms are impermeable—a range of 0.8 to 1.0 might be typical. The combined result of impermeable berms, their necessary area, and the quick runoff is an addition, of about 20 to 25 percent to direct rainfall on the bed.

Bank Losses

Bank losses will be negligibly low if impermeable embankments have been used. However, there may arise situations where this is not the case. An empirical procedure may then

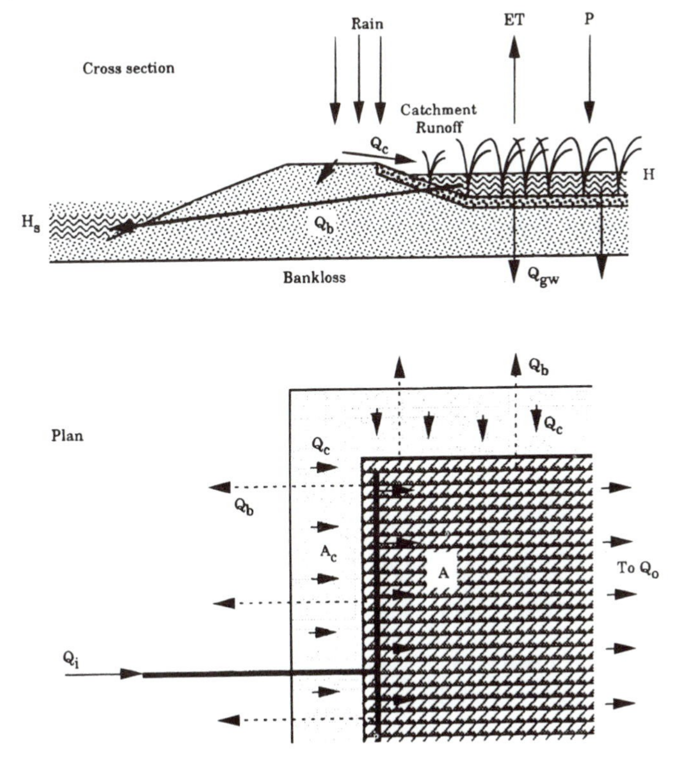

Figure 6-4 Water budget quantities.

be used in which the bank loss is calibrated to the head difference between the water inside and outside of the berm (Burns and McDonnell, 1992; Kadlec, J. A., 1986) (see Figure 6-4):

$$Q_b = \lambda L_b(H - H_s) \tag{6-14}$$

where H = wetland water elevation, m
H_s = external water elevation, m
L_b = length of the berm, m
λ = empirical coefficient, m/d

For instance, levees in the Everglades are typically built from the peat and limestone soils native to the area. Leakage is therefore significant and has been studied extensively in connection with many canal and storage projects. The value used is $\lambda = 15$ m/d (Burns and McDonnell, 1992), which represents a very leaky berm.

Rainfall

For conceptual design, some estimate of average annual precipitation is of use. For this purpose, Figure 6-5 may be used.

Rainfall amounts may be measured at or near the site for purposes of wetland design or monitoring. However, the gauging location must not be too far removed from the wetland because some rain events are extremely localized.

For most design purposes, historical monthly average precipitation amounts suffice. These may be obtained from archival sources such as *Climatological Data,* a monthly publication of the National Oceanic and Atmospheric Administration (NOAA), National Climatic Data Center, Asheville, NC.

For short-term simulation model calculations, rainfall amounts must be predicted from short-interval historical records. This entails three probabilistic data sets: the distribution of rain amounts, the distribution of storm durations, and the distribution of times between storms. Weather records are generally available for extended historical periods, but the three probability distributions must typically be constructed from the historical data. To provide some idea of the nature of these distributions, the 20-year period (1951–1970) for the Houghton Lake wetland treatment site is utilized here (Parker, 1974). Data from a weather station located about 7 km from the wetland were available (NOAA). A Weibull distribution was found to adequately represent both duration ($R^2 = 0.994$) and interstorm period ($R^2 = 0.997$) data:

$$F(\tau) = 1 - \exp\left[-\left(\frac{\tau - 1}{b}\right)^c\right] \tag{6-15}$$

where b = constant, days
c = constant,
F = cumulative probability that the duration is less than t
τ = time, days

The distributions were normalized so that 1 day was the minimum value of duration or interstorm period. The parameter values at Houghton Lake, MI were

Rain duration a = 0.262 b = 1.056
Interstorm duration a = 1.000 b = 0.979

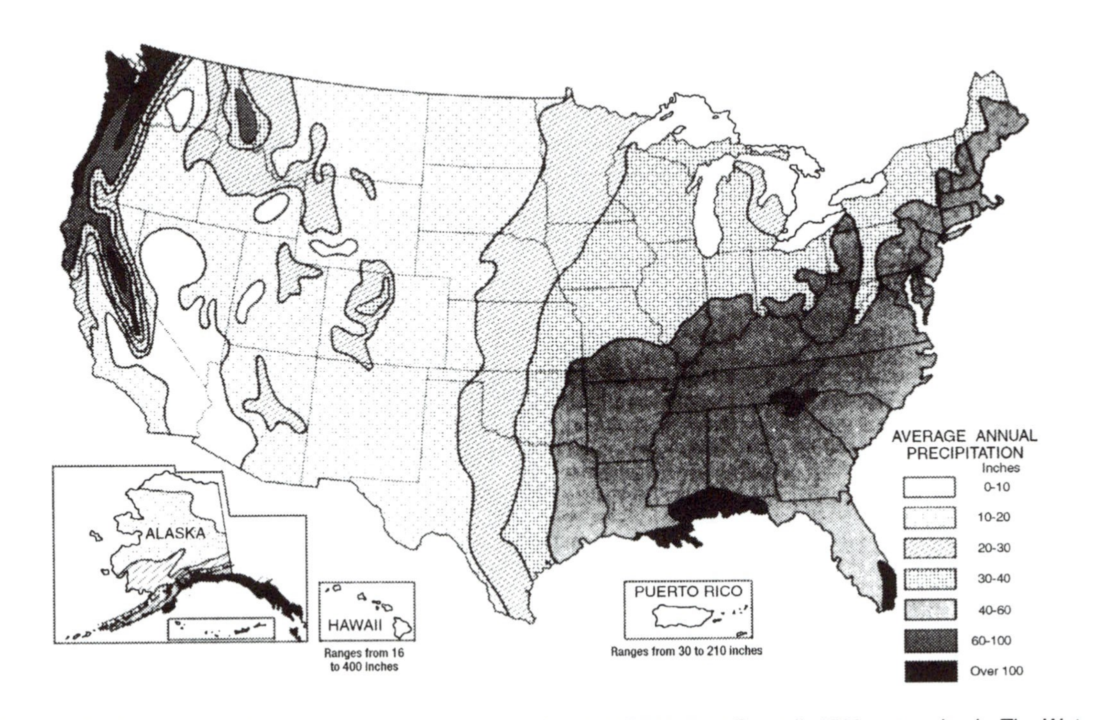

Figure 6-5 Annual precipitation in the U.S. (Data from U.S. Water Resources Council, 1968, appearing in *The Water Encyclopedia*, 2nd Ed., van der Leeden, F., Troise, F.L., and Todd, D.K., (Eds.), Lewis Publishers, Chelsea, MI, 1990. With permission.)

The distribution of rainfall rates for each month for this site was adequately fit (R^2 = 0.985) by

$$F(R) = 1 - \exp\left[-\left(\frac{P - B}{A}\right)\right] \tag{6-16}$$

where A = constant, m/d
B = constant, m/d
F = cumulative probability that the rain rate is less than R
P = rain rate, m/d

The parameters A and B varied from month to month; seasonal values in mm/day were

	A	B
Winter	2.939	3.269
Spring	4.444	4.189
Summer	6.655	7.422
Fall	5.064	4.645

Snowmelt

In northern climates, snowmelt is a possible component of the liquid water mass balance. In such a case, the end-of-season snow pack is melted over time in rough proportion to the temperature excess above freezing. The amount of the snowpack is documented in weather records, such as *Climatological Data* (NOAA). A procedure for determining snowmelt rate is given by Chow (1964):

$$Q_{sm} = (0.0254)[0.03(T_{avg} - 24) + 0.02(T_{max} - 27)] \cdot A \tag{6-17}$$

where T_{avg} = average daily temperature, °C
T_{max} = maximum daily temperature, °C

Melting proceeds at the calculated rate until the snowpack is depleted.

Example 6-1 Monthly Water Budget

The town of Ross River, Yukon Territories plans to discharge lagoon wastewater to natural wetlands rather than directly to the Pelly River. Those wetlands are situated on rather permeable soils interspersed with permafrost lenses. Therefore, some water infiltrates into the wetland soils, but the shallow receiving aquifer immediately discharges (vents) to the Pelly River. As part of the planning process, the water budget for the wetlands was projected for a summer discharge period of 105 days (June, July, August, and half of September). In the antecedent condition, no water leaves the wetlands via surface flow, and that condition was presumed to exist under discharge conditions as well. What monthly water levels and flows may be expected?

Solution

This site receives catchment runoff during the unfrozen months. During winter months, a snowpack builds, which then melts during the spring period. Exfiltration from the wetlands takes place during the warm months. Rain- and snowfall are available from historic records, and evapotranspiration was estimated. The imbalance between inputs and outputs results in a change in storage. The month-by-month analysis (Table 6-1) shows that water levels will not rise during the summer months because the combined evaporative and exfiltration losses are sufficient to compensate for the added water. The assumption of zero surface outflow is verified.

Infiltration

The soils under a treatment wetland may range in water condition from fully saturated, forming a water mound on the shallow regional aquifer, to unsaturated (Figure 6-6). If the wetland is lined with a relatively impervious clay layer, it is likely that the underlying strata will be partially dry with the regional shallow aquifer located some distance below (Figure 6-6). In this case, it is easy to estimate leakage from the wetland from

$$Q_{gw} = kA\left[\frac{H_w - H_{cb}}{H_{ct} - H_{cb}}\right] \qquad (6\text{-}18)$$

where
A = wetland area, m^2
H_{cb} = elevation of clay bottom, m
H_{ct} = elevation of clay top, m
H_w = wetland water surface elevation, m
k = hydraulic conductivity of the clay, m/d
Q_{gw} = infiltration rate, m^3/d

In practice, a leak test is often required to demonstrate that a clay liner in fact performs as designed. One such procedure is known as the Minnesota barrel test. The water loss from a bottomless barrel placed in the wetland is compared to the water loss from a barrel with a bottom. The barrels collect rain and evaporate water with equal efficiency, so any additional loss from the bottomless barrel must be due to vertical seepage.

Example 6-2 Wetland Leakage Calculation

The city of Columbia, MO plans to discharge secondary wastewater to 37 ha of constructed wetlands rather than directly to the Missouri River (Brunner and Kadlec, 1993). Those wetlands are sealed with 30 cm of clay, but are situated on rather permeable soils. The hydraulic conductivity of the clay sealant is 1×10^{-7} cm/s. Water is expected to be 30 cm deep, and there is 30 cm of topsoil above the clay as a rooting medium. What water leakage flow may be expected?

Solution

This amount of sealant should leave an unsaturated layer beneath the wetlands, as a worst case scenario. Therefore, Equation 6-18 may be used to estimate leakage:

$$Q_{gw} = \left[\frac{1 \cdot 10^{-7} \cdot 86{,}400}{100}\right](370{,}000)\left[\frac{90 - 0}{30 - 0}\right]$$

$$Q_{gw} = 96\ m^3/d = 25{,}330\ gal/day$$

$$Q_{gw}/A = 0.79\ cm/mo = 0.31\ inches\ per\ month$$

Because of the proximity of Columbia's drinking water supply wells, this leakage rate was experimentally confirmed prior to startup. Over a 27-day period, wetland unit one lost 0.21 cm more than the control (Figure 6-7), indicating a tighter seal than designed.

Table 6-1 Projected Monthly Water Budget Calculations for an Evaporative/Infiltrating Wetland Treatment System

	Jan	Feb	Mar	Apr	May	Jun	Jul	Aug	Sep	Oct	Nov	Dec	Total or Annual Average
Pump input, m^3	0	0	0	0	0	10,410	10,757	10,757	4,858	0	0	0	36,782
Total precip, cm	1.94	1.57	1.45	1.23	1.45	3.19	4.15	3.24	2.29	1.58	2.33	1.93	26.35
Rainfall, cm	0.00	0.00	0.00	0.11	1.45	3.19	4.15	3.24	2.29	0.53	0.00	0.00	14.96
Catchment runoff, m^3	0	0	0	323	2,452	3,286	1,909	1,490	1,053	242	0	0	10,755
ET, cm	0.00	0.00	0.00	4.54	12.90	16.83	15.67	11.02	3.97	0.00	0.00	0.00	64.93
ET output, m^3	0	0	0	4,994	14,190	18,513	17,237	12,122	4,367	0	0	0	5,952
Rain input, m^3	0	0	0	121	1,595	3,509	4,565	3,564	2,519	579	0	0	1371
Snowmelt, m^3	0.0	0	0	3,666	8,874	0	0	0	0	0	0	0	12,540
Exfiltration, m^3	0.0	0	0	1,100	1,100	1,100	1,100	1,100	1,100	1,100	0	0	7,700
Net input, m^3	0	0	0	−1,984	−2,369	−2,408	−1,106	2,589	2,963	−279	0	0	32,201
Output, m^3	0.00	0.00	0.00	0.00	0.00	0.00	0.00	0.00	0.00	0.00	0.00	0.00	0.00
Δ Storage, m^3	0	0	0	−1984	−2,369	−2,408	−1,106	2,589	2,963	−279	0	0	−2,594
Δ Level, cm	0.0	0.0	0.0	−1.8	−2.2	−2.2	−1.0	2.4	2.7	−0.3	0.0	0.0	−2.4
Water surface, cm	0.0	0.0	0.0	−1.8	−4.0	−6.1	−7.2	−4.8	−2.1	−2.4	−2.4	−2.4	−2.8

Note: Area, m^2 = 110,000; spring snowpack = 11.4 cm H_2O equivalent.

Adapted from the Ross River, Yukon system; David Nairne & Associates and NovaTec Consultants Inc., Engineers (1991).

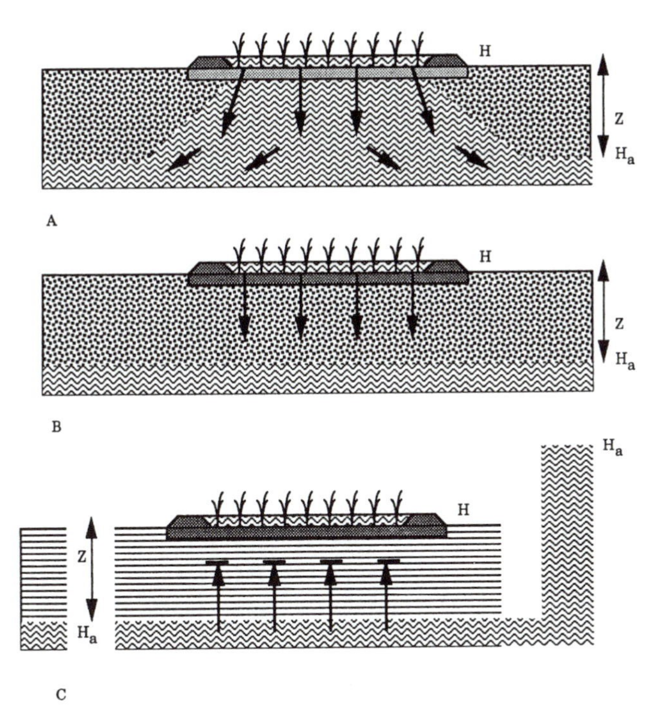

Figure 6-6 Three groundwater-wetland interactions. (A) large leakage, leading to groundwater mounding; (B) small leakage, unsaturated conditions beneath the clay seal layer; and (C) a wetland perched above an aquifer under positive pressure. H = head in wetland, H_a = head in aquifer, and Z = distance wetland to aquifer.

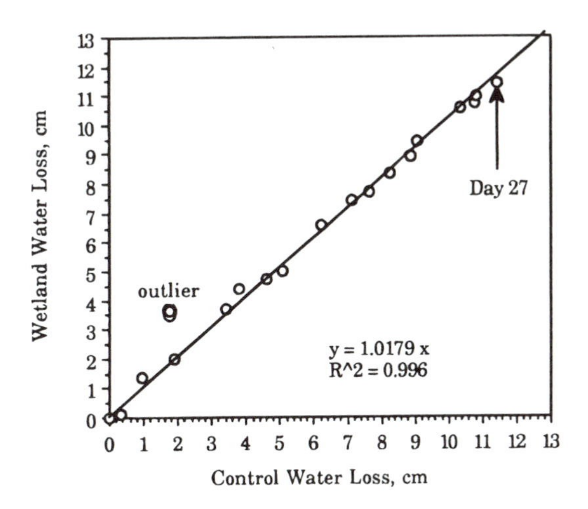

Figure 6-7 Seepage test results for wetland treatment unit one at Columbia, MO. In this Minnesota barrel test, water loss from a bottomless barrel placed in the wetland is compared to water loss from a control barrel with a bottom.

If there is enough leakage to create a saturated zone under the wetland (Figure 6-6A), then complex three-dimensional flow calculations must be made to ascertain the flow through the wetland bottom to groundwater. These require a substantial quantity of data on the regional water table, regional groundwater flows, and soil hydraulic conductivities by layer.

Such calculations are expensive and usually warranted only when the amount of seepage is vital to the design.

A third possibility is that the wetland is perched on top of, and is isolated from, the shallow regional aquifer. In some instances, such as the Houghton Lake site, the wetland may be located in a clay "dish" which forms an aquiclude for a regional shallow aquifer under pressure (Figure 6-6C). A well drilled through the wetland to the aquifer displays artesian character. The "in-leak" for this system is very small because the clay layer is many feet thick (Haag, 1979).

Water Storage

The computation of the volume of water stored in a wetland involves the stage-storage curve for the wetland. The derivative of this function is the water surface area.

$$A = \frac{dV}{dh} \quad (6\text{-}19)$$

where A = wetland area, m^2
h = wetland depth, m
V = wetland volume, m^3

In normal practice, no allowance is made for the volume occupied by vegetation because of the difficulty of measurement of the vegetation volume.

Some constructed wetlands have steeply pitched side slopes and may be regarded as constant area systems. This implies that the stage-storage curve is a straight line. For instance, Mierau and Trimble (1988) report a nearly linear stage-storage curve for a rectangular diked marsh treating river water (Figure 6-8). But some wetlands have more complicated topography, such as the treatment wetlands at Des Plaines (Figure 6-9).

This information permits computation of water elevation changes from a knowledge of changes in storage volume. Over any time period, the stage change (ΔH) is given by

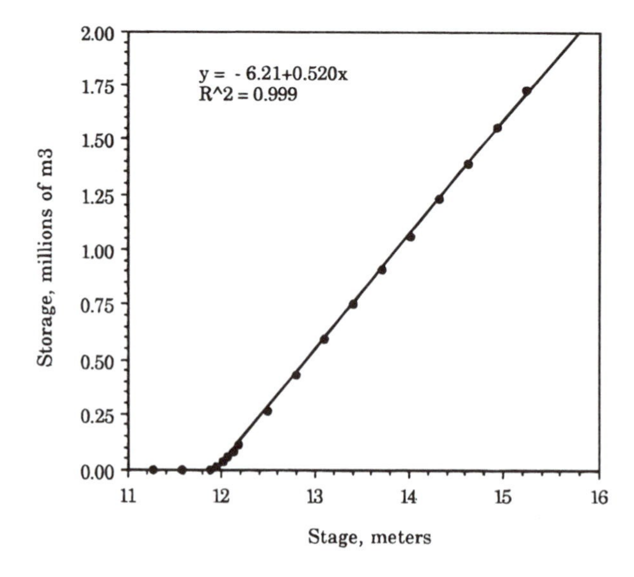

Figure 6-8 Stage-storage curve for Boney Marsh. (Data from Mierau and Trimble, 1988.)

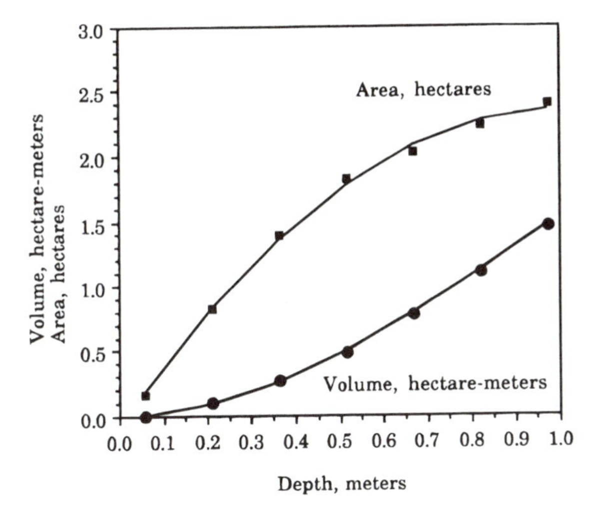

Figure 6-9 Stage-storage and stage-area curves for wetland EW3 at Des Plaines, IL. The curves are predicted by $V = 0.0022 - 0.104h + 2.44h^2 - 0.809h^3$

$$A = \frac{dv}{dh} = -0.104 + 4.88h - 2.43h^2$$

$$\Delta H = \int_{t1}^{t2} \frac{dV}{A} = \frac{\Delta V}{A_{avg}} \tag{6-20}$$

where

A_{avg} = mean water surface area over the period t_1 to t_2.

In the extreme, a wetland may evaporate much of the added water, such as at Incline Village, NV. The area of these wetlands responds by expanding and shrinking in response to added water and evapotranspiration (Figure 6-10).

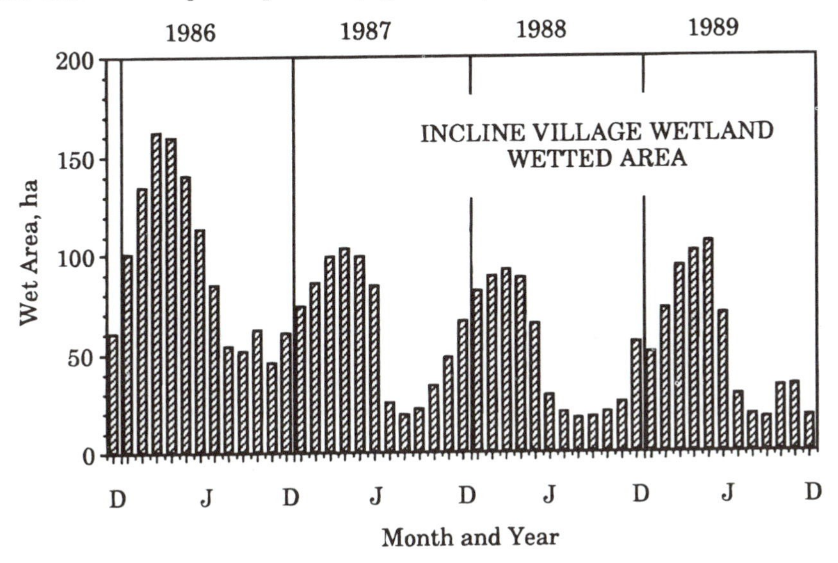

Figure 6-10 The expansion and shrinkage of the Incline Village, NV wastewater wetlands as a function of time. Summer water diversions to agricultural uses accelerate the dryout caused by arid conditions. (Data from Kadlec et al., 1990a.)

Evapotranspiration (ET)

Wetland treatment systems frequently operate with small hydraulic loading rates. For 100 SF wetlands in North America, 1.00 cm/d is the 40th percentile (Knight et al., 1993). ET losses approach a daily average of 0.50 cm/d in summer in the southern U.S.; consequently, more than half the daily added water may be lost to ET under those circumstances. But ET follows a diurnal cycle, with a maximum during early afternoon and a minimum in the late nighttime hours. Therefore, outflow can cease during the day for this extreme example.

The importance of evaporation and transpiration requires that methods be available for estimation. Several methods are explained in Chapter 9. The energy balance method will be used for determining the wetland temperatures in Chapter 10, as well as for estimating evaporative losses. This procedure is somewhat involved. At this juncture the two simplest estimators will be noted: wetland ET is roughly equal to lake evaporation, which in turn is roughly equal to 80 percent of pan evaporation. Figure 6-11 shows the distribution of annual ET across the U.S.

WATER MASS BALANCE IMPACTS ON POLLUTANT REDUCTIONS

For purposes of contaminant mass balancing, an overall water balance is required. The time period over which averaging is done will generally be dictated by the frequency of water quality sampling. For instance, weekly water quality results would normally be combined with weekly average flows to determine mass removal rates. In design, seasonally variable wastewater flows can combine with seasonally variable rain and ET to produce large differences in hydrologic functions.

Precipitation and ET are time variable transfers, but they combine over extended periods to either dilute or concentrate the wastewater during its passage through the wetlands. Periods of heavy rainfall can cause dilution, but may have an adverse effect on detention time and, hence, cause poor mass removals.

An illustration of seasonal water budget variability is given in Table 6-2. This illustration is based on actual data for rain, pan evaporation, and wastewater inflows for the treatment wetlands at Millersylvania State Park in Washington state. However, for simplicity, the SF wetland and the target contaminant are hypothetical; the park actually is considering a hybrid SSF-SF system. In this particular climatic zone, winter rains are heavy when park use is low, leading to a large amount of dilution. Conversely, rainfall is low in summer, ET is high, and park use is high, leading to high loadings and evaporative concentration. The combination produces a design "bottleneck" in midsummer. Annual averages do not give an adequate representation of the anticipated seasonal performance.

Example 6-3

Platzer and Netter (1992) report that the nominal detention time, based on inflow, for the SSF wetland at See, Germany is 20 days. There was a measured net loss of 70 percent of the water to ET in summer. What is the correct nominal detention time? Platzer and Netter report that the wetland removes 88 percent of the incoming nitrogen. What will be the exit concentration reduction?

Solution

The water slows as it travels through the constant cross-section system. The outflow rate will be only 30 percent of the inflow rate, and the average of inlet and outlet velocities is 65 percent of the inlet velocity. It is tempting to conclude that the time of travel is (1/0.65) equal to 1.54 times the inlet-based nominal detention time, but that is not quite correct.

The inlet nominal detention time is defined by

$$\tau_i = \frac{L}{v_i} \tag{6-21}$$

Recall the definition of inlet hydraulic loading:

$$q_i = \frac{Q_i}{A} \tag{6-22}$$

Define fractional atmospheric augmentation:

$$\alpha = \frac{[(1 + \psi r)P - ET]}{q_i} \tag{6-23}$$

The relation between travel time and travel distance is

$$\frac{dx}{dt} = v = \frac{Q}{Wh\varepsilon} = v_i\left[1 + \alpha\frac{x}{L}\right] \tag{6-24}$$

Integrating Equation 6-24 from x = 0 to x = L gives the time of transit, or actual residence time,

$$\tau = \int_0^\tau dt = \int_0^L \frac{dx}{v} = \tau_i\left[\frac{\ln(1 + \alpha)}{\alpha}\right] \tag{6-25}$$

For the parameters of the present example,

$$\frac{\tau}{\tau_i} = \left[\frac{\ln(1 + [-0.7])}{[-0.7]}\right] = 1.72 \tag{6-26}$$

In general, the actual contact time will be closely approximated by using an average of inlet and outlet flow rates. Here, the actual detention time would be 34.4 days; the use of an average flow rate gives 30.8 days.

A mass balance on nitrogen over the whole wetland requires that

$$Q_oC_o = 0.12 \cdot Q_iC_i \tag{6-27}$$

The water mass balance requires

$$Q_o = 0.30 \cdot Q_i \tag{6-28}$$

Combining these two produces the exit concentration ratio

$$C_o = 0.40 \cdot C_i \tag{6-29}$$

The fractional concentration remaining (40 percent) is much greater than the fractional mass remaining (12 percent).

WETLAND WATER QUALITY

Wetlands have been called the "kidneys" of the global water cycle because they often cause improvements in quality for flow-through waters. However, this idea should not be construed to mean that wetlands will produce perfectly pure water under any circumstances. It is useful to explore here the water quality to be expected in natural wetlands, since that forms a background for expectations from treatment wetlands.

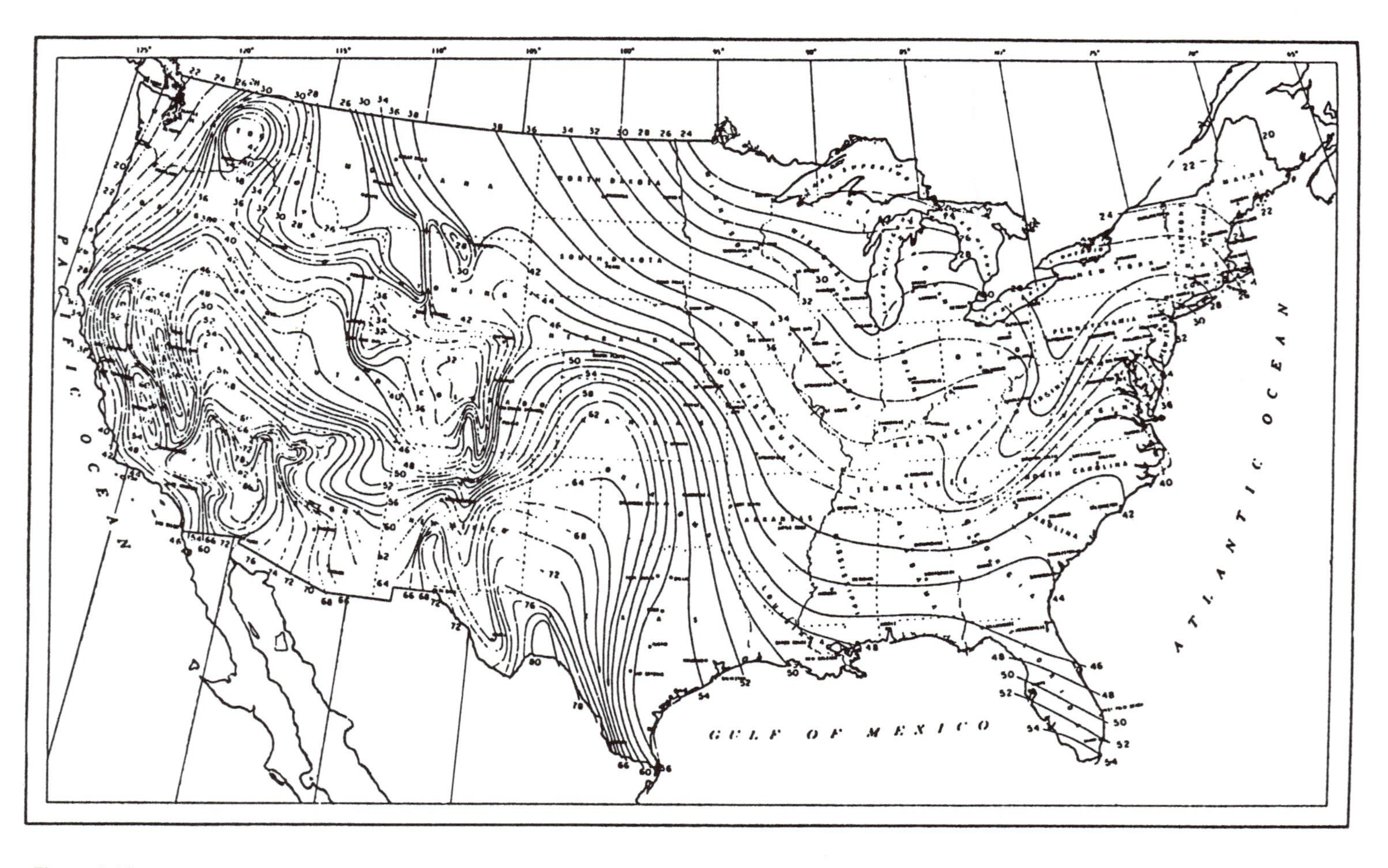

Figure 6-11 Annual lake evaporation in the U.S. (inches; 1946–1965). (Data from U.S. Water Resources Council, 1968, appearing in *The Water Encyclopedia*, 2nd Ed., van der Leeden, F., Troise, F.L., and Todd, D.K., (Eds.), Lewis Publishers, Chelsea, MI, 1990. With permission.)

Table 6-2 Projected Monthly Water Budget Calculations for a State Park Wetland Treatment System

	Jan	Feb	Mar	Apr	May	Jun	Jul	Aug	Sep	Oct	Nov	Dec	Annual Average
Pump input, gal/d	1,400	2,200	2,700	6,300	9,200	14,500	24,900	22,000	9,500	3,300	1,700	1,500	8,267
Pump input, m^3/d	5.3	8.3	10.2	23.9	34.8	54.9	94.3	83.3	36.0	12.5	6.4	5.7	31.3
Precip, in.	8.50	5.77	4.85	3.13	1.85	1.44	0.76	1.34	2.36	4.68	7.58	8.70	50.96
Catchment factor	1.20	1.20	1.20	1.20	1.20	1.20	1.20	1.20	1.20	1.20	1.20	1.20	1.20
EP, in.	0.39	0.59	1.20	2.14	3.35	4.26	5.02	4.40	3.13	1.83	0.82	0.52	27.65
ET factor	0.80	0.80	0.80	0.80	0.80	0.80	0.80	0.80	0.80	0.80	0.80	0.80	0.80
ET area factor	1.00	1.00	1.00	1.00	1.00	1.00	1.00	1.00	1.00	1.00	1.00	1.00	1.00
P input, m^3/d	33.4	25.1	19.1	12.7	7.3	5.9	3.0	5.3	9.6	18.4	30.8	34.2	17.1
ET output, m^3/d	1.0	1.7	3.1	5.8	8.8	11.5	13.2	11.5	8.5	4.8	2.2	1.4	6.1
Net input, m^3/d	37.7	31.7	26.2	30.8	33.3	49.2	84.1	77.0	37.1	26.1	35.0	38.5	42.2
Δ Storage, m^3	0	0	0	0	0	0	0	0	0	0	0	0	0
Output, m^3/d	37.7	31.7	26.2	30.8	33.3	49.2	84.1	77.0	37.1	26.1	35.0	38.5	42.2
Avg flow, m^3/d	21.5	20.0	18.2	27.3	34.1	52.1	89.2	80.2	36.5	19.3	20.7	22.1	36.8
Dilution factor	7.11	3.81	2.56	1.29	0.96	0.90	0.89	0.92	1.03	2.09	5.44	6.78	1.35
HRT, d =	75.3	80.9	89.1	59.3	47.5	31.1	18.2	20.2	44.4	83.9	78.2	73.3	44.1
Inlet HLR, cm/d =	0.13	0.21	0.26	0.60	0.87	1.37	2.36	2.08	0.90	0.31	0.16	0.14	0.78
Average HLR, cm/d =	0.54	0.50	0.45	0.68	0.85	1.30	2.23	2.00	0.91	0.48	0.52	0.55	0.92
Atmospheric augmentation, cm/d =	0.81	0.59	0.40	0.17	−0.04	−0.14	−0.25	−0.16	0.03	0.34	0.71	0.82	0.27
Concentration reduction =	1.00	1.00	1.00	0.99	0.96	0.86	0.67	0.72	0.95	1.00	1.00	1.00	0.96
Mass removal =	1.00	1.00	1.00	0.98	0.96	0.88	0.71	0.75	0.95	1.00	1.00	1.00	0.95

Note: The SF wetland is level controlled at all times. Area, m^2 = 4,000; depth, m = 0.45; porosity = 0.9; k-value, m/yr = 10.

AMBIENT CONCENTRATIONS IN NATURAL WETLANDS

Natural wetlands exist in a wide variety of types, as discussed in Chapter 4. In addition, each type may exist across a range of water chemistries, ranging from nutrient rich to nutrient poor. Some idea of the water chemistry to be expected in natural wetlands can be gained from the literature (Table 6-3). While there are clearly no absolutes, it is apparent that nutrients are typically present at low levels.

Phosphorus is often at levels well below regulatory ranges for most receiving waters. Most pristine natural wetlands have total phosphorus less than 0.1 mg/L and often below 0.05 mg/L. However, it must be recognized that many natural wetlands are not pristine because they are affected by point and nonpoint discharges of water from agriculture, industry, and municipalities. Therefore, the data from Theresa Marsh (Table 6-3) exhibit total phosphorus (TP) concentrations that are an order of magnitude higher, 0.1 to 0.7 mg/L. In this particular case, the source of the phosphorus was traceable to the incoming streamflows (Klopatek, 1978). At the opposite extreme, the Everglades ecosystem is adapted to TP of approximately 0.01 mg/L. Rainfall TP can range from 0.01 to 0.05 mg/L and depends on land use and industrial patterns in the region. Average values are more commonly 0.02 to 0.04 mg/L. Thus, when rainfall dominates the water budget, comparably low levels of total phosphorus can be expected in the wetland waters.

Organic nitrogen is formed in wetlands as a product of biomass decomposition. Proteins degrade to smaller organic species such as amines, which in turn degrade to ammonium nitrogen. In natural wetlands, the result is a pervasive low concentration of organic nitrogen, at concentrations of approximately 1 to 2 mg/L.

Ammonium nitrogen is the preferred form for plant growth. It is formed by decomposition of dead plant material, and in the absence of dissolved oxygen, it is not microbially converted to nitrate. During the growing season, ammonium is utilized in the formation of biomass. However, during the winter season in northern climates, plant growth is minimal, and an ice cap ensures anaerobic conditions in the soil-litter zone, as well as blocking transverse flow through this organic-rich zone. Decomposition processes are slower in winter, but the ammonium nitrogen formed has no outlet. As a consequence, ammonium nitrogen can build to 2 to 4 mg/L under the ice in northern wetlands in winter. This ammonium remains available to fuel the spring growth of the vegetation. Snowmelt usually runs off over the ice and does not flush this pool of nitrogen.

During the growing season, the vegetation will avail itself of stored nitrogen, as well as atmospheric inputs. But without streamflow inputs, the supply of nitrogen is too small to permit the vegetation to reach its full growth potential. All of the ammonium nitrogen released from decomposition is utilized, and the water concentrations drop to very low levels, approximately 0.05 to 0.10 mg/L. Plant growth increases as more ammonium is added, until full growth potential is realized.

Microbial processes also contribute to ammonium reduction under warm-weather conditions. The wetland usually contains a population of nitrifying bacteria which utilize dissolved oxygen to convert ammonium to nitrate.

If ammonium is supplied in excess of bacterial and plant growth requirements, the concentrations in wetland waters rise to reflect the subsidy. The same effect can be produced by excessive additions of organic nitrogen which can be mineralized to ammonium.

Nitrate and nitrite nitrogen are usually absent from natural wetland waters, in the sense that concentrations are frequently at or below the lower detection limit (about 0.1 mg/L for normal analytical procedures). Denitrification is efficient in the natural wetland environment because the necessary carbon source is present together with the anoxic conditions that favor the utilization of nitrate as an electron acceptor. Denitrification is a microbially mediated process and therefore slows at cold temperatures.

Table 6-3 Water Chemistry of Some Typical Natural Wetlands

Chemical Parameter		Theresa Marsh, WI (1)	Porter Ranch Peatland, MI (2)	Portage Fen, MI (3)	Seminole Ranch, FL (4)	Ombrotrophic Bog, Newfoundland (5)	Cypress Dome, FL (6)
Nitrate + nitrite	mg/L	0.1–1.7	0.04 ± 0.02	0.1	0.1		0.08
Ammonium	mg/L	0.1–1.6	0.73 ± 0.81	0.09	0.15		0.14
Organic N	mg/L	0.7–4.8		0.2	0.75		1.2
Total N	mg/L	1.5–6.8		0.4	1		1.6
Dissolved P	mg/L	0.1–0.5	0.02 ± 0.01		0.04		0.07
Total P	mg/L	0.1–0.7	0.04 ± 0.01	0.05	0.07		0.18
Na	mg/L	5–24		3.5		3.9	5
K	mg/L	0.9–9.1	0.7 ± 0.6			0.1	0.34
Ca	mg/L	56–168	19 ± 11	22		0.25	2.9
Mg	mg/L	23–73	4 ± 2	22		0.46	1.4
Fe	mg/L		0.5 ± 1.6				
Chloride	mg/L	17–54	28 ± 25	9			
Sulfate	mg/L	15–99					2.6
pH		7.4–8.2	6.0–7.5	7.7	6.9		4.5
Conductivity	μmho/cm	520–940	150–350	520	390		60
Alkalinity ($CaCO_3$)	mg/L	270–420		182			1.8

(1) Eutrophic, *Typha* + *Scirpus* — Klopatek, 1978
(2) Rich fen, *Carex* + *Salix* — Richardson et al., 1978
(3) Alkaline fen, *Carex* — Kadlec and Bevis, 1992
(4) *Spartina bakeri* — Post et al., 1992
(5) *Sphagnum* — Damman and French, 1987; as cited by Mitsch and Gosselink, 1993
(6) Cypress — Dierberg and Brezonik, 1984

Carbon compounds are prevalent in the wetland environment because of the large amounts of biomass that occupy the ecosystem. Atmospheric carbon fixation is the basis of plant growth. Humic substances are the products returned to the water after the growth-death-decomposition cycle. The amount of carbon in the water is often represented by the total organic carbon (TOC) content for natural wetlands, in contrast to the biological oxygen demand (BOD) and chemical oxygen demand (COD) that are used to characterize wastewaters. Because much of the humic material is not a suitable food source for bacteria, the COD of a natural wetland is higher than its BOD. The orders of magnitude for these parameters in a natural wetland are TOC $\sim$ 40 mg/L, COD $\sim$ 100 mg/L, and BOD $\sim$ 5 mg/L.

Total suspended solids (TSS) are not normally measured in natural wetland ecosystem studies. However, treatment studies sometimes extend to very high degrees of reduction of these parameters, to levels that may be regarded as characteristic of unimpacted wetlands. Background concentrations so deduced are in the range of 2 to 10 mg/L for densely vegetated marshes. However, zones of open water can foster large populations of planktonic algae, which contribute to TSS of up to 100 mg/L.

Measurement of TSS is a very difficult task in shallow water conditions. Wetlands contain easily suspended flocs and particulate detritus. The act of taking a grab sample can disturb these sediments, either by the feet of the sampler or by the currents which fill the sample bottle. Thus, a carefully taken surface sample may reflect 10 mg/L TSS, while a carelessly dipped sample at the same location may register 1000 mg/L.

Common metals often reflect the water source for natural wetlands. *Minerotrophic* wetlands are those which possess significant inputs of calcium, magnesium, sodium, and potassium from either streamflow or groundwater discharge. Concentrations of the monovalent cations may reach 10 to 20 mg/L, and those for divalent cations may reach 20 to 50 mg/L. *Ombrotrophic* wetlands are isolated from those water sources and receive their water exclusively from precipitation. As a result, common metals are much scarcer in the water and biomass. Concentrations of common metals may be in the range of 0 to 5 mg/L.

Heavy metals are not present in natural wetland waters in high concentrations. The organic sediments are good cation exchange sites for the divalent heavy metals (Ni, Cu, Pb, Cd), and hence, these are found bound to sediments in preference to the dissolved forms. Further, the anaerobic zones of the wetland soils can generate sulfide ions which precipitate the insoluble sulfides of these metals. Iron and manganese possess some of these chemical features as well and, in addition, possess a complex hydroxide chemistry that can lead to precipitation. As a consequence, the action zone for heavy metals is the soil-sediment compartment of the wetland.

pH reflects the hydrogen ion content of the wetland waters. In marshes and swamps, pH values are typically circumneutral (pH 6 to 8). However, in ombrotrophic bogs, the exchange of metal ions for hydrogen ions by plants, coupled with the decomposition of organic matter can lead to very low pH, typically 3.5 to 5.5. The pH of the water is strongly correlated with the calcium content; neutral wetland waters have Ca $>$ 20 mg/L (Glaser, 1987).

Algae can drive the pH of wetland waters to high values during periods of high productivity. Algal blooms can be associated with temporary periods of pH $\sim$ 8 to 9.

Dissolved oxygen (DO) and *redox potential* (Eh) are measures of the oxidation potential in the water or sediments. Dissolved oxygen can range from zero to more than twice the theoretical solubility in response to many ecosystem variables. Wetland surface waters typically have a vertical gradient in DO, with high DO at the air water interface and very low DO at the sediment-water interface. The water depth and vegetation density dictate whether this gradient is confined to a region near the sediment (high sediment oxygen demand) or whether the bulk of the surface water is oxygenated (low sediment oxygen demand). Mixing processes can erase this gradient, which is sustained by diffusion under quiescent conditions.

Algal processes can raise DO to very high levels during bloom conditions due to photosynthetic production. Because gases easily supersaturate with respect to their equilibrium solubility, this in-water production can cause significant supersaturation. Open water areas are required for this phenomenon; dense vegetation blocks the light necessary for the algae.

More often than not, natural wetlands are oxygen poor, and there are very low redox potentials in the sediments. The decomposition of decaying vegetation and microorganisms requires more oxygen than can be transported through the overlying water. Therefore, DO $\leq$ 2 mg/L is not an uncommon occurrence.

Sulfates are often reduced as a result of anaerobic wetland processes. The resultant sulfides are precipitated by divalent metals or released as hydrogen sulfide. Significant sulfur accumulations are not uncommon in freshwater wetland soils, and these can dominate the soil chemistry in marine wetland environments.

Some substances are nearly *conservative* in the wetland environment. Chloride ions are the best example, although total ion content, as measured by electrical conductivity, is also relatively unaffected by wetland processes. As a consequence, a chloride budget is often used as a check on the water budget for the wetland. A very important corollary of this conservatism is lack of treatment potential. Total dissolved solids (TDS) are not much affected by wetland processes and cannot be effectively reduced.

PARTICULATE AND DISSOLVED FORMS

Because wetlands are rich in plant detritus as well as living and dead microorganisms, wetland waters often transport and contain many chemicals associated with the resultant particulate matter. Many organic chemicals are hydrophobic and have an affinity for the organic sediments of the wetland. Wetland sediments and soils are good cation exchange media and, therefore, carry metals on the cation exchange sites.

Wetland sediments may be flocculent as a result of electrical forces controlled by the ionic strength of the water. Dead algae and bacteria form such flocs, as do the hydroxide precipitates of iron and aluminum compounds.

Inorganic ions form a large fraction of the dissolved substances in wetland waters. Electrolytes may total 100 to 500 mg/L in freshwater marshes, but only 30 to 100 mg/L in bogs (Shotyk, 1987). Inland wetlands in arid climates may display very high dissolved solids, measured in parts per thousand.

Dissolved organic compounds are characteristic of wetland waters. Humic substances, notably humic and fulvic acids, may be present. In northern bog waters, these substances comprise about one third of the total dissolved carbon, and their concentrations range up to 25 mg/L (Bourbonniere, 1987). These materials are excellent complexors for a variety of chemicals, including metals and small organic molecules. Therefore, the wetland water may contain organically bound soluble forms of pollutants.

It is important to note that most waterborne constituents are found both in and on suspended solids as well as in organic or inorganic dissolved forms. Wetland ecosystems often act to alter the proportions as well as the amounts of these various forms.

TEMPORAL PATTERNS

Diurnal variations in wetland water chemistry are confined to variables strongly influenced by the photosynthetic processes. For instance, DO can display large changes between the nighttime hours (low values) and the daytime hours (high values). For this reason, DO measurements must be placed in the context of the time of day that they are made.

Most wetland water chemistry variables are not strongly and directly linked to the solar cycle and, hence, do not display marked diurnal cycles (Kadlec, 1975). In any case, great

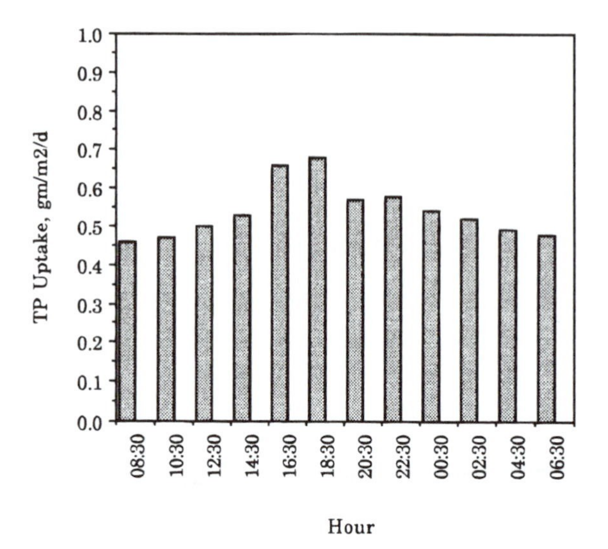

Figure 6-12 Diurnal fluctuations in phosphorus uptake in an experimental microcosm, *Phalaris arundinacea,* in sand. (Adapted from Adler, et al., 1994.)

care must be exercised to separate random sampling errors from the detection of the 24-hour cycle effect (Kadlec, 1988). This can be done under controlled laboratory conditions, and the cycle can be defined, as indicated in Figure 6-12 (Adler et al., 1994).

Seasonal changes in wetland water chemistry are more pronounced. Seasonal changes reflect temperature, photoperiod, hydroperiod, and growth status, which in total can exert a large effect on water chemistry. For instance, nitrogen chemistry can be significantly affected (Figure 6-13). In general, the growing season shows a tendency to deplete nutrients, while the winter season shows a slow anaerobic digestion of organic matter. Dry seasons can accentuate organic matter decomposition, while wet seasons reflect dilution. In general, the interactions among the driving forces is site specific and year specific.

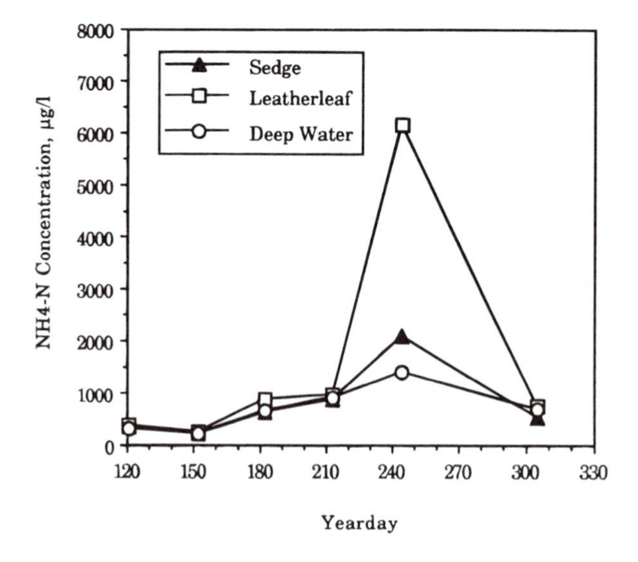

Figure 6-13 Seasonal variation in surface water ammonium in the natural Houghton Lake wetland in 1973. (Replotted from Kadlec, J. A., 1975).

SPATIAL PATTERNS

Horizontal

Natural wetlands display inhomogeneity across the width and length of the ecosystem. There are spatial patterns of water depth, which in turn relate to spatial patterns in vegetation type and density. Water quality may vary accordingly. The same is true of treatment wetlands, except that there is typically a gradient in water quality superimposed upon the spatial patterns of the wetland. Figure 6-14 illustrates the interior total phosphorus for a reconstructed wetland receiving pumped river water. This wetland contains large expanses of open water and submerged aquatics and is therefore subject to wind mixing. It is seen that concentrations vary from point to point and that cover type is not a sufficient predictor of the observations. There is no obvious regular progression from inlet to outlet.

However, if the wetland is spatially uniform in depth and vegetation type and density, as is the case for many constructed treatment wetlands, then transect measurements of water quality along the direction of flow will show monotonic gradient effects.

Vertical

Water quality also varies with depth in the water-soil column in wetlands. There may be measurable differences in water quality within the water column alone, but the differences become more apparent within the soil porewater (Table 6-4).

Chemistry in the vertical direction is mediated strongly by the effects of the roots of the macrophytes and by a strong vertical gradient in redox potential. However, it is difficult to sample steep vertical gradients in shallow water without resorting to special sampling equipment. Consequently, there is a tendency for results to be averaged over the free-standing water column. This does not mean that vertical gradients are absent, only that these are often ignored.

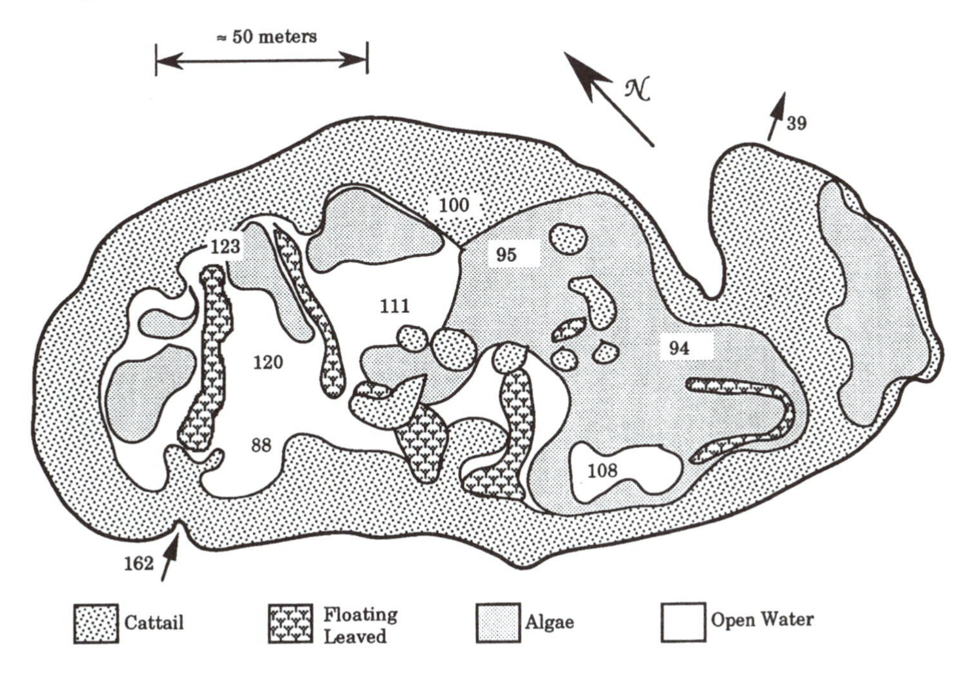

Figure 6-14 Principal vegetation types and distribution for Des Plaines, IL wetland EW3, 1990. Interior numbers represent TP (μg/L) values on three dates, time averaged. Exterior streamflow TP values represent averages over the same time span. (Data from WRI, 1992.)

Table 6-4 Variation of Water Quality with Depth in a Natural Peatland

Depth	Unit	Surface Water	15 (cm)	45 (cm)
Chloride	mg/L	27.9	28.5	23.3
NH_4-N	μg/L	728	2099	1889
TDP	μg/L	53	98	77
Iron	mg/L	0.50	1.80	1.80
Sodium	mg/L	6.70	5.50	5.10

Note: There is a significant difference for porewater, excepting chloride. N ≥ 127. (Data from Kadlec, J. A., 1975.)

VARIABILITY IN MEASUREMENTS

All sampling and measurement procedures have inherent variability. Sources include ecosystem variability, sampling errors, and analytical errors. Wetlands are certainly no better than other ecosystems with respect to inherent variability. Spatial and temporal inhomogeneity contribute to data scatter, even with perfect field and laboratory techniques. It is desirable to take an adequate number of replicates, thus ensuring accuracy of the mean, but that is often too expensive to be practical. Replicates are of necessity separated, either in distance or in time or both. Intuition would indicate that closer measurements in time or space would yield smaller variance, and that is the actual situation. Figures 6-15 and 6-16 show the magnitude of the sampling variability for a specific wetland (Kadlec, 1988a). These are not intended to be a reliable pair of reference graphs; rather, a few data sets have been selected from data for a natural (undisturbed) wetland to illustrate a trend. The chloride determination is quite accurate and precise in the laboratory, so the effects seen are nearly entirely due to the wetland. Two features are worthy of attention: not much is to be gained by sampling too frequently or too closely; and beyond a certain spacing or interval, single grab samples lose much of their meaning. For the wetland in question, monthly sampling and a 50-m grid seem to represent a reasonable compromise between effort and accuracy.

There are differences in coefficients of variation among the various water quality parameters, but none show small standard deviations.

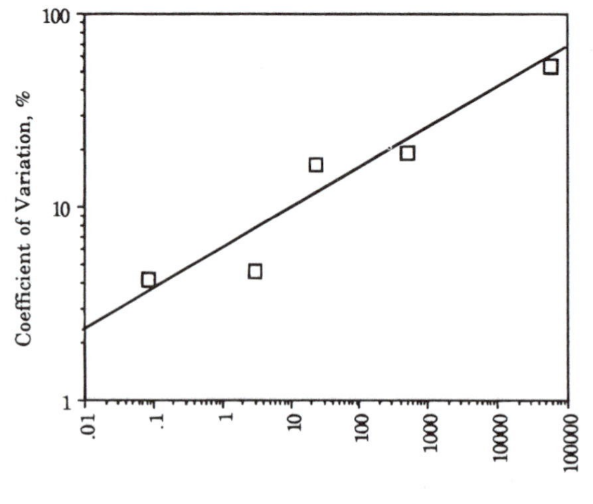

Figure 6-15 An example of the variability encountered as a function of sampling frequency. The parameter is chloride, and the data are from the Houghton Lake peatland. (Data from Kadlec, 1988a.)

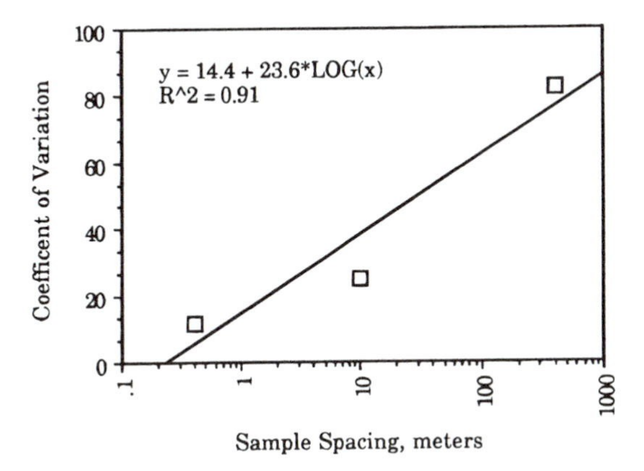

Figure 6-16 An example of the variability encountered due to spatial variability. The parameter is chloride, and the data are from the Houghton Lake peatland. (Data from Kadlec, 1988a.)

BIOGEOCHEMICAL CYCLING

The utilization of wetlands for water quality improvement has led to the involvement of several disciplines, each with its own perspective on the character of wetland chemical processing. The limnologist views the wetland as a shallow lake and therefore borrows concepts from the chemistry of the aquatic environment. The soil scientist views the wetland as waterlogged soil and applies ideas that heavily emphasize the wetland substrate. The engineer is used to microbial processes associated with both attached and suspended organisms; he visualizes the wetland as a reaction vessel, with the plants being likened to vessel internals for attached growth. The botanist focuses on the vegetation; to him, the wetland is a water garden, and plant processes occupy center stage. The hydrologist views the wetland as a vegetated channel. The ecologist is concerned with types and patterns of plant and animal species.

In point of fact, all points of view have merit, and there is great danger in a narrow view. All portions of the ecosystem play important roles in some phase of pollutant processing. Also, some concepts from related disciplines do not carry over to the wetland situation.

All the chemicals of interest in water treatment can pass into and through the micro- and macroflora within the wetland. As a plant grows, it incorporates nutrients and other chemicals into its tissues, both above and below ground. During senescence, some of these chemicals may be translocated from above-ground parts to below-ground roots and rhizomes. Upon death of a plant part, the chemicals become part of a pool associated with the standing dead and litter so formed. Decomposition of freshly dead material results in a partial return of the chemicals to the water, with the balance being retained within the accreting soils and sediments in the wetland (Figure 6-17).

Organisms of all sizes participate in such cycling. Microbes grow, die, and decompose on a much faster time scale than do soft-tissue macrophytes, but they can be just as important in pollutant immobilization. Woody plant parts may resist decomposition for periods of decades or centuries.

A few substances can be transformed into gaseous products by chemical or biochemical reactions and so be released from the cycle to the atmosphere. Under proper conditions, nitrate, ammonium, and sulfur may all be lost via this pathway.

Some materials may deposit within the wetland without passing through the cycle outlined previously. Incoming particulate matter may settle to the bottom and carry with it an associated suite of sorbed and structural chemicals. In addition, purely chemical reactions can form precipitates which also settle out, including calcium and iron minerals.

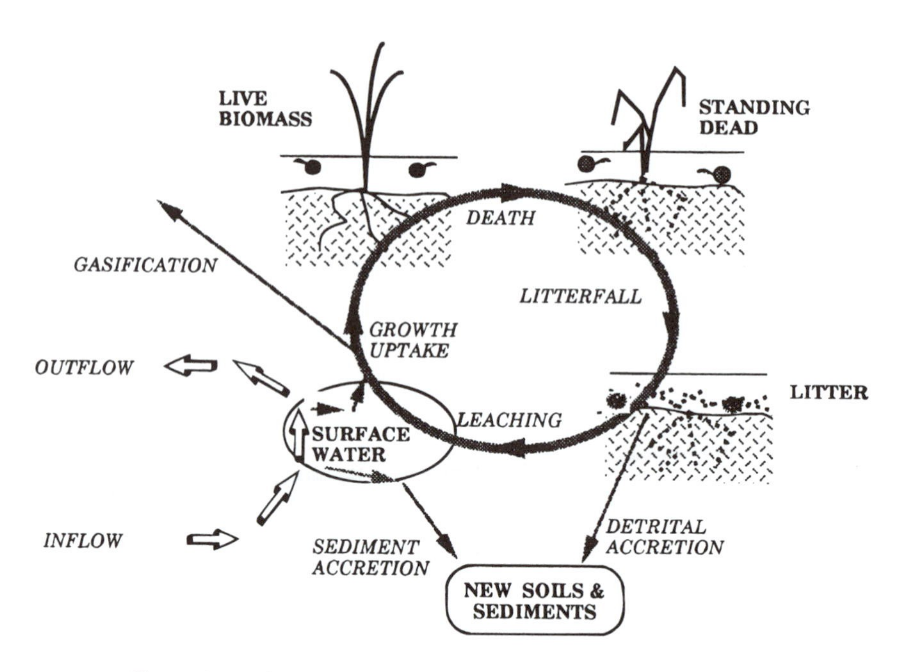

Figure 6-17 A simplified view of the wetland biogeochemical cycle.

It is important to recognize that plant uptake is not a measure of the net removal which is fostered by the vegetation and microbes. Initial uptake is counterbalanced by releases later in the cycle, with the effect that net uptake and burial as new sediments is only a fraction of the gross uptake. Johnston (1993) shows that only 26 to 55 percent of the annual nitrogen and phosphorus uptake is retained in above-ground tissues in herbaceous wetlands, the balance is lost to leaching and litterfall. But that is not the percentage of interest in long-term treatment design; rather, it is necessary to know the percentage of a chemical that will never return to the water.

As a hypothetical illustration, consider the nitrogen uptake into the above-ground tissues of an emergent macrophyte in a northern environment. The fate of a 1-year growth cohort will be considered. In a eutrophic marsh, annual growth may be 5000 g/m^2/year, which all dies back by the end of the fall season. The above-ground, end-of-season, standing live crop would be less than this figure, perhaps only 2500 g/m^2, because some leaf death occurs during the growing season and because of the contribution of roots. The nitrogen content of the live biomass is assumed to be 2.0 percent, which means a plant uptake for the year of 100 g/m^2. The standing dead plants begin to decompose, with a subsequent loss of both biomass and nitrogen. For simplicity, assume that decomposition consumes half the standing dead at constant nitrogen content. As a worst case, the nitrogen in the decomposed material (50 g/m^2) all winds up back in the surface water due to leaching by rain. The balance of 50 g/m^2 nitrogen is contained in the remaining dead biomass, which falls into the water as litter.

The soft-tissue portion of the litter decomposes according to a first-order mechanism, with a half life of about 1 year. The hard-tissue, lignin-rich portion does not decompose; that fraction is assumed to be 20 percent in this example. Therefore, 80 percent of the litter will decompose over a period of years, ultimately releasing its nitrogen, totaling 40 g/m^2. Twenty g/m^2 will be lost the first year, 10 the second; 5 the third, and so on until all soft tissues are gone.

Net burial of biomass is only 500 g/m^2/year, and for nitrogen it is only 10 g/m^2/year. That detritus is mixed with the material arriving with litter cohorts from succeeding years.

If the bulk density of the litter is 0.2 g/cm^3, then the detrital accretion rate may be calculated to be 2.5 mm/year. That is within the range measured in such wetlands.

This illustration shows that care must be taken to precisely define what is meant by "removal by plants." The content of the annual production of new biomass is a serious overestimate of the net removal that may eventually be attributed to that cohort.

In an actual wetland, this cycle is augmented by processes involving the roots, by cycling in the microphytes, and by microbial conversion and loss processes.

GLOBAL CHEMICAL BUDGETS

Measurements of chemical composition of wetland inflows and outflows are the most obvious method of characterizing water quality functions. However, such measurements by themselves can be very misleading. A much better characterization is achieved by computing the mass balance, or budget, for an individual chemical constituent.

A proper mass balance must satisfy several conditions:

1. The *system* for the mass balance must be defined carefully. A system in this context means a defined volume in space; this is often taken to be the surface water in the wetland in the case of a surface-flow wetland (SF) or the water in the media for a subsurface-flow wetland (SSF). A precise definition is needed to compute the change in storage. The mass balance is termed *global* when the entire wetland water body is chosen as the system. In later chapters, it will be useful to compute the *internal* mass balance, which is based on an internal element or subdivision of the water body.
2. The *time period* for totaling the inputs and outputs must be specified. It may be desirable to express inflows and outflows in terms of rates, but these must then be averaged over the time period chosen.
3. *All inputs and outputs* to the chosen system must be included. The concept of mass conservation may be invoked to calculate one or a group of material flows. A partial listing of some of the inflows and outflows does not constitute a mass balance.
4. Compounds undergo chemical reactions within a wetland ecosystem. Any *production or destruction reactions* that occur within the boundaries of the chosen system are to be included in the mass balance. Reactions outside the boundary are not counted because an outflow must occur to transport the chemical to the external reaction site and that is accounted as an outflow.
5. Waterborne chemical flows are determined by separate measurements of water flows and concentrations within those waters. Therefore, an accurate *water mass balance is a prerequisite* to an accurate chemical mass balance.
6. If at all possible, it is desirable to *demonstrate closure* of the mass balance. This is achieved by independently measuring every component of the mass balance. The degree of closure is often expressed as a percentage of total inflows. Unfortunately, closure has rarely been demonstrated for any chemical in any wetland.

It is difficult to establish detailed chemical mass balances over the wetland surface water because of the number and complexity of the possible transfers to and from the water and their nonsteady character (Figure 6-18). It is common practice to measure only the principal inflows and outflows, and to ascribe the difference to "removal," which may be positive or negative. This lumping of all transfers to and from the water body is often unavoidable due to economic constraints. It is possible to write a general mass balance equation for a generic chemical species:

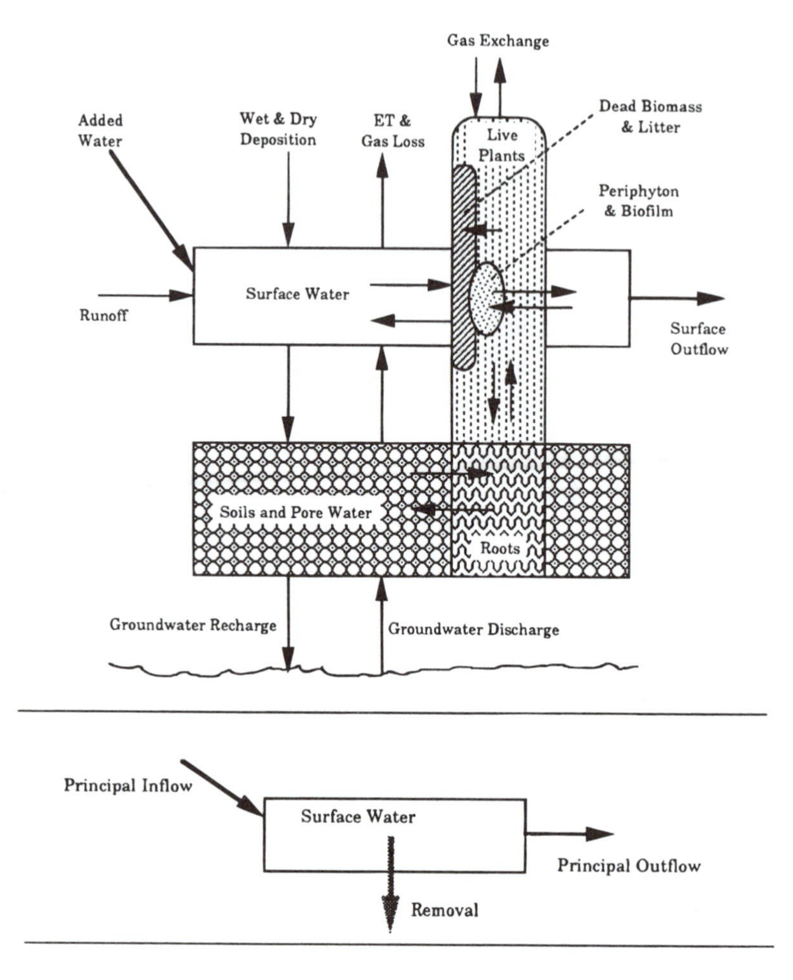

Figure 6-18 Major components of an SF wetland ecosystem and the principal routes of chemical transfer. Upper panel: actual transfers. Lower panel: common practice in the literature.

$$Q_iC_{k,i} - Q_oC_{k,o} + Q_cC_{k,c} \pm Q_{gw}C_{k,gw} + PAC_{k,p} - \tilde{J}_kA = \frac{d(VC_{k,s})}{dt} \tag{6-30}$$

where $C_{k,c}$ = concentration in catchment runoff, g/m^3
$C_{k,gw}$ = concentration in groundwater recharge or discharge, g/m^3
$C_{k,i}$ = concentration in inflow, g/m^3
$C_{k,o}$ = concentration in outflow, g/m^3
$C_{k,p}$ = concentration in precipitation, g/m^3
$C_{k,s}$ = concentration in wetland surface water, g/m^3
$\tilde{J}_k$ = spatially averaged removal rate, $g/m^2/d$

In Equation 6-30, bank losses and snowmelt have been omitted for the sake of simplicity. All of the transfers have been lumped into one removal rate. Flow rates are instantaneous. The removal rate is the average over the entire wetland area, and the system concentration is averaged over the entire water volume.

In most instances, this instantaneous mass balance will be averaged over a specified time period denoted by t_m. This is done by adding up all inflows and outflows and equating the difference to the change in inventory. The mathematical operation is the integration of Equation 6-30 over the time period.

$$\int_0^{t_m} \left[Q_i C_{k,i} - Q_o C_{k,o} + Q_c C_{k,c} \pm Q_{gw} C_{k,gw} + PAC_{k,p} - \tilde{J}_k A = \frac{d(VC_{k,s})}{dt} \right] dt \quad (6\text{-}31)$$

The result of this integration is

$$\overline{Q_i C_{k,i}} - \overline{Q_o C_{k,o}} + \overline{Q_c C_{k,c}} \pm \overline{Q_{gw} C_{k,gw}} + \overline{PAC_{k,p}} - \overline{\tilde{J}_k A} = [VC_{k,s}]_{t_m} - [VC_{k,s}]_0 \quad (6\text{-}32)$$

where the overbar indicates time averaging. An average mass flow, $\overline{QC}$, is the product of the average flow and the flow-weighted (or mass average) concentration:

Definition

$$\overline{C} = \frac{1}{t_m} \int_0^{t_m} C dt \quad (6\text{-}33)$$

Definition

$$\hat{C} = \frac{\frac{1}{t_m} \int_0^{t_m} QC dt}{\frac{1}{t_m} \int_0^{t_m} Q dt} = \frac{1}{\overline{Q} t_m} \int_0^{t_m} QC dt \quad (6\text{-}34)$$

Hence,

$$\overline{QC} = \overline{Q}\hat{C} \quad (6\text{-}35)$$

where the hat notation indicates a flow-weighted average.

Finally, these definitions lead to the mass balance over the time period t_m:

$$\overline{Q}_i \hat{C}_{k,i} - \overline{Q}_o \hat{C}_{k,o} + \overline{Q}_c \hat{C}_{k,c} \pm \overline{Q}_{gw} \hat{C}_{k,gw} + \overline{P} A \hat{C}_{k,p} - \overline{\tilde{J}}_k A = [VC_{k,s}]_{t_m} - [VC_{k,s}]_0 \quad (6\text{-}36)$$

This is a simple statement that inflows and outflows are balanced by removals and changes in storage. However, the rigorous mathematical treatment yields a very important result: it is necessary to use flow-weighted concentrations with time-average flow rates. One method of accomplishing this is the use of flow-proportional samplers to measure input and output concentrations. Of course, if either the flow rate or the concentration are constant, there is no difference between flow-weighted averages and time averages. But if there are significant event impacts on the wetland due to rain or pulsed inflow or runoff, then the difference is important.

Table 6-5 presents mass balances for water and phosphorus for a 1-year study of Wingra Marsh near Madison, WI (Perry et al., 1981). A somewhat different breakdown of the various streams and transfers was used by these authors in an attempt to identify and quantify some of the individual transfers shown in Figure 6-18. It may be seen that surface flows dominate this wetland because it is receiving stormwater flows.

The time period for the global mass balance is of critical importance because of the time scale of interior phenomena. Many wetlands, whether treatment or pristine natural, have long nominal detention times which usually reflect long actual detention times. A 2-week detention is not uncommon. If the wetland were in plug flow, an entering cohort of water would exit 2 weeks later. Clearly, same-day samples taken from inlet and outlet should not be used to compute "removals." In fact, wetland flow patterns are more complex than plug flow; the entering cohort breaks up, and pieces depart at various times after entry, some earlier and some later than the implied 2-week detention.

This difficulty of synchronous sampling may be alleviated in the mass balancing process by selecting a mass balance period that spans several detention times.

The change in storage is important for short-term mass balances and for periods of significant water level change. For example, if the detention time is 2 weeks, the wetland can hold 14 day's inflow. In reverse, that means that a change in water level of 1/14 = 7 percent is equivalent to 1 day's inflow. Simple estimating procedures can thus be used to ascertain the need for tracking chemical inventory in the wetland water body.

The removal term is the result of transfers to and from the soil and biomass compartments in the wetland, as well as of transfers to and from the atmosphere. Those biomass and soil

Table 6-5 Phosphorus Mass Balances for Wingra Marsh, WI, 1975–1976

Season				Autumn	Winter	Spring	Summer	Annual
Water budgets (cubic meters)								
	Inputs	Precipitation		20,344	1,648	27,126	7,963	57,081
		Stormwater		176,993	15,656	509,959	54,415	757,023
		Groundwater		6,103	3,296	5,425	3,982	18,806
		Total		203,440	20,600	542,510	66,360	832,910
	Outputs	ET		16,421	0	16,300	26,390	59,111
		Groundwater		30,789	8,976	21,734	9,802	71,301
		Surface outflow		158,050	9,724	505,306	39,208	712,288
		Total		205,260	18,700	543,340	75,400	842,700
	Δ Storage			−1,820	1,900	−830	−9,040	−9,790
Phosphorus budgets (kilograms)								
	Inputs	Stormwater	Dissolved	38.3	7.8	147.5	12.5	206.1
		Stormwater	Particulate	65.2	1.6	166.4	47.8	281.0
		Groundwater		0.1	0.1	0.1	0.1	0.4
		Wet fall		0.3	0.1	0.8	0.1	1.3
		Dry fall		0.3	0.1	0.3	0.3	1.0
		Total		104.2	9.7	315.1	60.8	489.8
	Outputs	Infiltration		5.7	3.8	5.4	5.5	20.4
		Exchange		1.8	0.2	7.6	0.5	10.1
		Sedimentation		48.8	0.8	126.8	33.2	209.6
		Surface outflow		48.4	4.6	175.3	22.0	249.9
		Total		104.7	9.4	315.1	61.2	490.0
	Δ Storage			−0.5	0.3	0.0	−0.4	−0.6
	%Removal			54	53	44	64	49

Adapted from Perry, et al. 1981.

compartments dominate the overall wetland storage and transformations for most chemicals. Therefore, the water body mass balance is very sensitive to small changes in transfers, reactions, and storages in biomass and soils. The removal rate depends very strongly on events in these solids compartments and, hence, is determined in major part by the changing ecological state of the wetland. Because wetland biological processes are more or less repetitive on an annual cycle, the long-term performance of the wetland is best characterized by global mass balances which span an integer number of years. Seasonal effects require a time period of 3 months, which is usually long enough to avoid storage errors and detention time offset.

The removal is an areal average. However, in most flow-through wetlands, there is a strong gradient in the unaveraged removal in the direction of flow. Water Conservation Area 2A (WCA2A) provides an example of this nonuniformity (Figure 6-19). WCA2A is a very large wetland for which phosphorus removal has been very carefully measured. Mass balance closure has been demonstrated (Walker, 1995). As the wetland system "boundary" is moved successively further from the inlet, the areal average removal rate decreases. The average removal rate depends on the size of the portion of the overall wetland that is chosen for the global mass balance.

This weakness of the global mass balance can be corrected by using the internal mass balance which reflects distance effects. That procedure will be explained in later chapters.

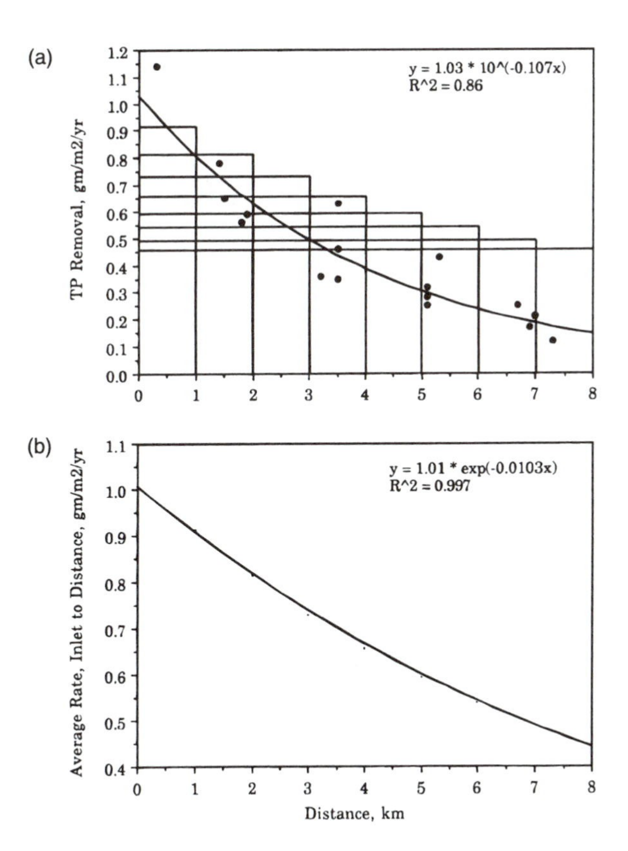

Figure 6-19 (a) Local phosphorus uptake rate, j_k, as a function of distance from the water inlet, WCA2A. (Adapted from Walker, 1995.) Horizontal bars indicate the average value from inlet to distance. (b) Phosphorus uptake rate averaged over the wetland from inlet to distance, J_k, as a function of distance from the water inlet, WCA2A.

CHEMICAL TERMINOLOGY

It is important to distinguish among the various measures of global wetland chemical removal. The definitions used in this book are specified in the following sections. The parent variables are identified in Figure 6-20.

Mass Input Rate, Mass Output Rate

Symbol:	M_i
Alternative symbol:	
Units:	g/d; kg/d; kg/year
Defining equation:	

$$M_i = Q_i C_i \tag{6-37}$$

The mass output rate is defined in the analogous manner, with the subscript "o" reflecting the outlet measurement point.

Input Mass Loading Rate, Output Mass Loading Rate

Symbol:	m_i
Alternative symbol:	□ LRI, □ LI
Units:	$g/m^2/d$; kg/ha/d; kg/ha/yr
Defining equation:	

$$m_i = \frac{M_i}{A} = \frac{Q_i C_i}{A} = q_i C_i \tag{6-38}$$

The box in the alternate symbol is a letter designating the chemical; PLI would denote *P*hosphorus *L*oading Rate *I*n, for instance. A chemical loading rate is a measure of the distributed, "rainfall" equivalent of a chemical mass flow. It does not imply the physical distribution of water uniformly over the wetland. The subscript "o" reflects the outlet measurement point.

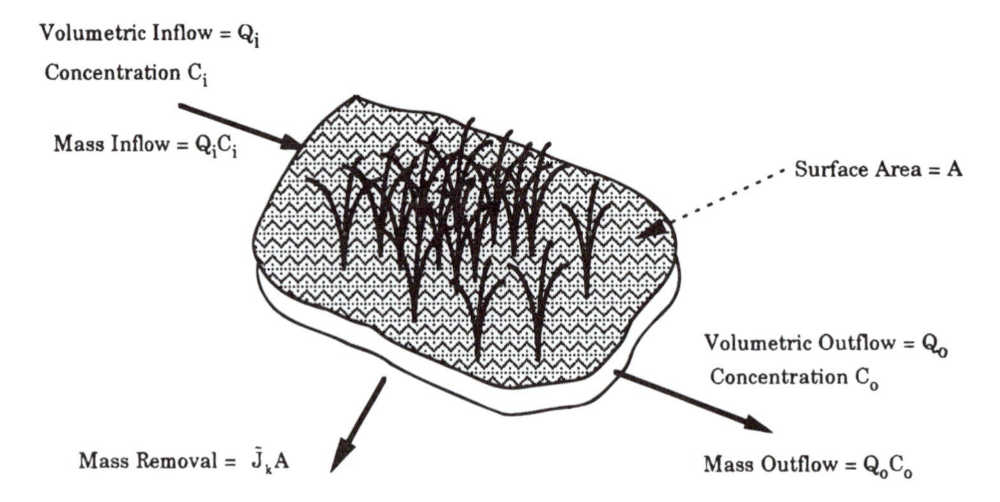

Figure 6-20 Components of a chemical budget and associated terminology.

Mass Removal Rate

Symbol: $\tilde{J}A$
Units: g/d; kg/d; kg/yr
Defining equation:

$$\tilde{J}A = M_i - M_o \tag{6-39}$$

Specific Mass Removal Rate

Symbol: $\tilde{J}$
Units: $g/m^2/d$; kg/ha/d; kg/ha/yr
Defining equation:

$$\tilde{J} = m_i - m_o \tag{6-40}$$

This single-number measure of wetland performance can be misleading in the common event of strong concentration gradients and removal gradients. See the discussion in the preceding section.

Percent Concentration Reduction

Defining equation:

$$\%\ \text{Concentration Reduction} = 100\,\frac{C_i - C_o}{C_i} \tag{6-41}$$

This term is quite ambiguous because it usually refers to the average of one or more synchronous samples for selected streamflows. As noted earlier, such contemporaneous measures do not reflect the internal dynamics of the wetland. Further, dilution or concentration due to rain, ET, or other unaccounted flows renders this an imperfect measure of true removal. Nevertheless, this terminology is frequently used in the literature.

Percent Mass Removal

Defining equation:

$$\%\ \text{Mass Removal} = 100\,\frac{M_i - M_o}{M_i} = 100\,\frac{m_i - m_o}{m_i} \tag{6-42}$$

This term is less ambiguous than concentration reduction because it traces only the chemical of interest and displays no effect of the quantity of water in which that chemical is located. However, the difficulties of contemporaneous measurement remain.

The Trap of Small Reductions

In some instances, the incoming concentration of a particular chemical may be small for some period of time. Then, due to measurement errors or small transfers from wetland storages and productions, it may give outflow concentrations that are greater than the incoming values. A one-time calculation of a "reduction efficiency" will properly reflect that condition

Table 6-6 Nitrate Reduction in Listowel System 5

	Inflow Concentration (mg/L)	Outflow Concentration (mg/L)	Percent Concentration Reduction	Average Percent Reduction
FA 80	0.34	0.367	−8	
WI 80/81	0.01	0.297	−2867	
SP 81	0.01	0.020	−100	
SU 81	0.16	0.003	98	−719
Annual 80/81	0.13	0.17	−33	

as a (large) negative percent reduction. At other times, a larger inflow concentration may be reduced by the wetland, leading to a positive percentage removal. If the removal percentages are then averaged, the large negative value improperly dominates the calculation. Data from Listowel Ontario clearly demonstrate this effect (Table 6-6). The averaged percent reductions yield −719 percent; the averaged concentrations yield −33 percent.

SUMMARY

This chapter has provided an overview of wetland hydrology and water quality oriented toward the treatment wetland concept. The importance of the water budget in understanding wetland characteristics cannot be understated. Hydrologic information forms the basis for mass balances for chemical constituents. Hydraulic loading rate and nominal detention time are the one-number parameters often used to categorize the wetland. Key features of the water budget have been introduced.

Wetland water quality is variable with respect to time, space, and wetland type. Therefore, great care must be exercised in interpretation of water quality data. Averaging periods should account for water and chemical dynamics and storages in water and biomass. Global mass balances are the first step in understanding the water quality function of wetlands, but are generally not sufficient to form a design basis for wetland construction or utilization.

The interactions of wetland waters with biota and soils are very strong. Many chemical processes are dominated by the vegetation and sediments and soils. Therefore, the wetland must not be viewed as an aquatic system carrying out only microbial and algal reactions.

CHAPTER 7

Wetland Microbial and Plant Communities

INTRODUCTION

Because of the presence of ample water, wetlands are typically home to a variety of microbial and plant species. The diversity of physical and chemical niches present in wetlands results in a continuum of life forms from the smallest viruses to the largest trees. This biological diversity creates interspecific interactions, resulting in greater diversity, more complete utilization of energy inflows, and ultimately to the emergent properties of the wetland ecosystem.

The study of organisms and their populations is a convenient way to catalog these life forms into groups with general similarities. However, the genetic and functional responses of wetland organisms are essentially limitless and result in the ability of natural systems to adapt to changing environmental conditions such as the addition of wastewaters. Genetic diversity and functional adaptation allow living organisms to use the constituents in wastewaters for their growth and reproduction. In using these constituents, wetland organisms mediate physical, chemical, and biological transformations of pollutants and modify water quality. In wetlands engineered for water treatment, design is based on the sustainable functions of organisms that provide the desired transformations.

The wetland treatment system designer should not expect to maintain a system with just a few known species. Such attempts frequently fail because of the natural diversity of competitive species and the resulting high management cost associated with eliminating competition or because of imprecise knowledge of all the physical and chemical requirements of even a few species. Rather, the successful wetland designer creates the gross environmental conditions suitable for groups or guilds of species; seeds the wetland with diversity by planting multiple species, using soil seed banks, and inoculating from other similar wetlands; and then uses a minimum of external control to guide wetland development. This form of ecological engineering results in lower initial cost, lower operation and maintenance costs, and the most consistent system performance.

This chapter presents an overview of the floristic diversity that naturally develops in treatment wetlands, as well as some details of the growth requirements of key microbial and plant species that generally occur in wetland treatment systems. These microbial and plant species are typically the dominant structural and functional components in treatment wetlands. An understanding of their basic ecology will provide the wetland designer or operator with insight into the mechanics of their "green" wastewater treatment unit. For more detailed information on aquatic and wetland microbial communities, the reader is referred to Portier and Palmer (1989), Pennak (1978), and WPCF (1990a). For more detailed information on the ecology of the vascular plant species found in wetlands, the reader is referred to Hutchinson (1975), Mitsch and Gosselink (1993), and Guntenspergen et al. (1989).

WETLAND BACTERIA AND FUNGI

Wetland and aquatic habitats provide suitable environmental conditions for the growth and reproduction of microscopic organisms. Two important groups of these microbial organisms are bacteria and fungi. These organisms are important in wetland treatment systems primarily because of their role in the assimilation, transformation, and recycling of chemical constituents present in various wastewaters.

Bacteria and fungi are typically the first organisms to colonize and begin the sequential decomposition of solids in wastewaters (Gaur et al., 1992). Also, microbes typically have first access to dissolved constituents in wastewater and either accomplish sorption and transformation of these constituents directly or live symbiotically with other plants and animals by capturing dissolved elements and making them accessible to their symbionts or hosts.

Taxonomy of the microbes is complex and frequently revised, but the general groups of bacteria and fungi are commonly recognized. Bacteria are classified in the Procaryotae (Buchanan and Gibbons, 1974). Procaryotes are distinguished by their lack of a defined nucleus with nucleic material present in the cytoplasm in a nuclear region. Cyanobacteria or blue-green algae are also classified as procaryotes, but they are discussed with algae later. Fungi are classified as eucaryotes because they have a nucleus separated from the cytoplasm by a nuclear membrane.

BACTERIA

Bacteria are unicellular, procaryotic organisms classified by their morphology, chemical staining characteristics, nutrition, and metabolism. *Bergey's Manual* (Buchanan and Gibbons, 1974) places bacteria into 19 associated groups with unclear evolutionary relationships. The groups that are important to water quality treatment and wetlands are listed and briefly described in Table 7-1.

Most bacteria can be classified into four morphological shapes: coccoid or spherical, bacillus or rod like, spirillum or spiral, and filamentous (Figure 7-1). These organisms may grow singly or in associated groups of cells including pairs, chains, and colonies. Bacteria typically reproduce by binary fission, in which cells divide into two equal daughter cells. Most bacteria are heterotrophic, which means they obtain their nutrition and energy requirements for growth from organic compounds. In addition, some autotrophic bacteria synthesize organic molecules from inorganic carbon (carbon dioxide [CO_2]). Some bacteria are sessile, while others are motile by use of flagella. In wetlands, most bacteria are associated with solid surfaces of plants, decaying organic matter, and soils.

FUNGI

Fungi represent a separate kingdom of eucaryotic organisms and include yeasts, molds, and fleshy fungi. All fungi are heterotrophic and obtain their energy and carbon requirements from organic matter. Most fungal nutrition is saprophytic, which means it is based on the degradation of dead organic matter. Fungi are abundant in wetland environments and play an important role in water quality treatment. For general information about fungi, refer to Ainsworth et al. (1973).

Yeasts are unicellular fungi that grow on fruits and in aquatic habitats. Yeasts reproduce by budding, which differs from bacterial fusion because budding can result in numerous daughter cells being formed from a single parent cell. Yeasts can degrade organic matter to carbon dioxide and water through aerobic respiration or can live as facultative anaerobes by using organic compounds as terminal electron acceptors. Byproducts of anaerobic respiration

Table 7-1 Classification of Bacteria Important in Wetland Treatment Systems

Group	Representative Genera	Comments
Phototrophic bacteria	*Rhodospirillum, Chlorobium*	Members of these genera are nonsymbiotic N fixers
Gliding bacteria	*Beggiatoa, Flexibacter, Thiothrix*	Filamentous bacteria found in activated sludge; *Beggiatoa* oxidizes hydrogen sulfide
Sheathed bacteria	*Sphaerotilus*	Filamentous bacteria implicated in reduced sludge settling rates in sewage treatment plants and common in polluted waters
Budding and/or appendaged bacteria	*Caulobacter, Hyphomicrobium*	Aquatic bacteria growing attached to surfaces with a hold fast
Gram-negative aerobic rods and cocci	*Pseudomonas, Zooglea, Azotobacter, Rhizobium*	*Pseudomonas* spp. denitrifies NO_2^- to N_2 under anaerobic conditions and can oxidize hydrogen gas; *P. aeruginosa* causes a variety of bacterial infections in humans; *Azotobacter* spp. is a nonsymbiotic N fixer; *Rhizobium* is a symbiotic N fixer
Gram-negative facultatively anaerobic rods	*Escherichia, Salmonella, Shigella, Klebsiella, Enterobacter, Aeromonas*	*E. coli* is the predomiant coliform in feces; *Salmonella* spp. cause food poisoning and typhoid fever; *Shigella* spp. causes bacillary dysentery; species in the genera *Klebsiella* and *Enterobacter* are nonsymbiotic N fixers and are in the total coliform group; *K. pneumoniae* is important in human and industrial wastes and can cause bacterial infections in humans
Gram-negative anaerobic bacteria	*Desulfovibrio*	Reduces sulfate to hydrogen sulfide
Gram-negative, chemolithotrophic bacteria	*Nitrosomonas, Nitrobacter, Thiobacillus*	*Nitrosomonas* catalyze the conversion of NH_4^+ to NO_2^-; *Nitrobacter* oxidize NO_2^- to NO_3^-; *T. ferrooxidans* oxidize ironsulfides producing Fe^{+3} and SO_4^{-2}
Methane-producing bacteria	*Methanobacterium*	Anaerobic bacteria of wetland sediments that convert carbonate to methane
Gram-positive cocci	*Streptococcus*	Fecal streptococci include human species (*S. faecalis* and *S. faecium*) and animal species (*S. bovis, S. equinus, S. avium*)
Endospore-forming rods and cocci	*Clostridium, Bacillus*	*C. botulinum* survives in soils and bottom sediments of wetlands and causes avian botulism; some *Clostridium* spp. are nonsymbiotic N fixers; *B. thuringiensis* is an insect pathogen; *B. licheniformis* denitrifies NO_2^- to N_2O
Actinomycetes and related organisms	*Nocardia, Frankia, Streptomyces*	Filamentous bacteria occurring aquatically and in soils; *Nocardia* is implicated in sludge bulking in sewage treatment; *Frankia* is a symbiotic N fixer with alder trees

include alcohols. Examples of yeasts found in aquatic ecosystems include *Cryptococcus* and *Dactyella,* which prey on aquatic nematodes.

Molds and fleshy fungi are filamentous and produce long chains of cells called mycelia. Reproduction may be vegetative through spore production or sexual between adjacent mating strains. Most aquatic fungi are molds and conduct their life cycle in association within the

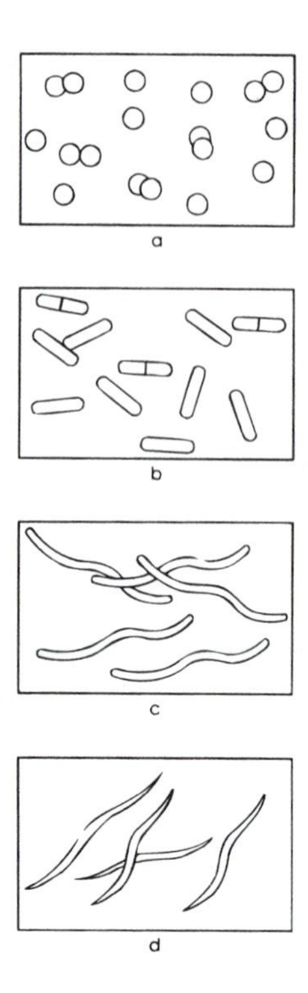

Figure 7-1 Typical bacterial growth forms: (a) coccoid, (b) bacillus, (c) spirillum, and (d) filamentous.

detritus and sediment layers of the water body or wetland. In freshwater streams colonized by riparian vegetation, aquatic hyphomycetes are frequently the dominant fungi that degrade leaf litter (Suberkropp et al., 1988). Water molds (Phycomycetes) include parasitic genera such as *Rhizophydium* and *Saprolegnia* and filamentous, saprophytic forms such as *Allomyces.* Fleshy fungi or Basidiomycetes are uncommon in wetlands and aquatic habitats except in association with wet and decaying wood, such as *Aureobasidium* in streams.

Fungi are ecologically important in wetlands because they mediate a significant proportion of the recycling of carbon and other nutrients in wetland and aquatic environments. Aquatic fungi typically colonize niches on decaying vegetation made available following completion of bacterial use. Saprophytic fungal growth conditions dead organic matter for ingestion and further degradation by larger consumers. Fungi live symbiotically with species of algae (lichens) and higher plants (mycorhizzae), increasing their host's efficiency for sorption of nutrients from air, water, and soil. If fungi are inhibited through the action of toxic metals and other chemicals in the wetland environment, nutrient cycling of scarce nutrients may be reduced, greatly limiting primary productivity of algae and higher plants. In wetlands, fungi are typically found growing in association with dead and decaying plant litter.

MICROBIAL METABOLISM

Most of the important chemical transformations conducted by microbes are controlled by enzymes, genetically specific proteins that catalyze chemical reactions. To a varying extent, bacteria and fungi are classified by their ability to catalyze certain reactions. Microbial metabolism includes the use of enzymes to break complex organic compounds into simpler compounds with the release of energy (catabolism) or the synthesis of organic compounds (anabolism) by the use of chemically stored energy. Microbial metabolism not only depends on the presence of appropriate enzymes, but also on environmental conditions such as temperature, dissolved oxygen (DO), and hydrogen ion concentration (pH). Also, the concentration of the chemical substrate undergoing the transformation is of primary importance in determining reaction rate.

Microbes can be classified by their metabolic requirements. Photoautotrophic bacteria such as the green and purple sulfur bacteria use light as an energy source to synthesize organic compounds from CO_2. Reduced sulfur compounds such as hydrogen sulfide or elemental sulfur serve as electron acceptors in oxidation-reduction reactions. Photoheterotrophs use light as an energy source and organic carbon as a carbon source for cell synthesis. The organic carbon sources most typically used by photoheterotrophs are alcohols, fatty acids, other organic acids, and carbohydrates. Because photosynthetic bacteria do not use water to reduce CO_2 they do not produce O_2 as a byproduct of metabolism, as do the algae and higher plants.

Chemoautotrophic bacteria derive their energy from the oxidation of reduced inorganic chemicals and use CO_2 as a source of carbon for cell synthesis. A number of the bacteria which are important in wetland treatment of wastewater are chemoautotrophs. Bacteria in the genus *Nitrosomonas* oxidize ammonia nitrogen to nitrite, and *Nitrobacter* oxidize nitrite to nitrate, deriving energy which is used in cell metabolism. The genus *Beggiatoa* derives energy from the oxidation of H_2S, *Thiobacillus* oxidizes elemental sulfur and ferrous iron, and *Pseudomonas* oxidizes hydrogen gas.

Chemoheterotrophs derive energy from organic compounds and also use the same or other organic compounds for cell synthesis. Most bacteria and all fungi, protozoans, and higher animals are chemoheterotrophs.

During microbial metabolism, carbohydrates are broken into pyruvic acid with the net production of two pyruvic acid molecules and two adenosine triphosphate (ATP) molecules for each molecule of glucose and the subsequent decomposition of pyruvic acid through fermentation or respiration. Fermentation by substrate-level phosphorylation does not require oxygen and results in the formation of a variety of organic end products such as lactic acid, ethanol, and other organic acids.

Aerobic respiration is the process of biochemical reactions by which carbohydrates are decomposed to CO_2, water, and energy (38 ATP molecules for each glucose molecule fully oxidized). The Krebs Cycle results in the loss of carbon dioxide (decarboxylation) and energy storage (two molecules of ATP per molecule of glucose). For complete oxidation to occur, oxygen and hydrogen ions must be available as the final electron acceptor in a chain of reactions called the electron transport chain. The overall reaction for aerobic respiration can be summarized as follows:

$$C_6H_{12}O_6 + 6H_2O + 6O_2 + 38\ ADP + 38\ P = 6CO_2 + 12H_2O + 38\ ATP \quad (7\text{-}1)$$

Also, approximately 60 percent of the energy of the original glucose molecule is lost as heat during the complete aerobic respiration process.

Anaerobic respiration is an alternative catabolic process that occurs in the absence of free oxygen gas. In anaerobic respiration, some other inorganic compound is used as the final electron acceptor. A variable and lower amount of energy is derived during the process of anaerobic respiration. This form of respiration is important to several groups of bacteria which occur in wetlands and aquatic habitats. Bacteria in the genera *Pseudomonas* and *Bacillus* use nitrate nitrogen as the final electron acceptor, producing nitrite, nitrous oxide (N_2O), or nitrogen gas (N_2) by the process termed denitrification. *Desulfovibrio* bacteria use sulfate (SO_4^{-2}) as the final electron acceptor, resulting in the formation of H_2S. *Methanobacterium* uses carbonate (CO_3^{-2}), forming methane gas (CH_4).

MICROBIAL GROWTH

Anabolism is the enzymatically controlled biosynthesis of complex organic compounds from less complex organic and inorganic molecules. Some microbial anabolic processes are simply the reverse of the catabolic, respiratory processes described previously. For example, ATP generated from aerobic respiration is used with intermediates of the Krebs Cycle to synthesize organic molecules (carbohydrates), which are the building blocks of the specific microbial organisms. The nutrients essential for microbial growth (and production of protoplasm in general) are summarized in Table 7-2. The molar ratios of major elements in protoplasm are 106C:180H:460:16N:1P + 25 percent trace minerals by weight.

Because microbial cells do not change size significantly during growth, the growth of microbial populations is generally described in terms of cell numbers. The specific growth rate, u_s of a microbial population is a function of the concentrations of the growth substrates as well as the physical environment of the microbe. Where all growth factors are present in excess of the required levels except for one which limits growth, S, a useful formulation to describe microbial growth is the Monod equation:

$$u_s = (u_{max}S)/(K + S) \qquad (7\text{-}2)$$

where u_s = the specific growth rate, cells/h
u_{max} = the maximum specific growth rate, cells/h
S = the substrate concentration, mg/L
K = the half saturation constant, mg/L

If more than one growth requirement is limiting, then the Monod equation can be expanded to show multiple limitations:

$$u_s = (u_{max})(S_1/K_1 + S_1)(S_2/K_2 + S_2) \qquad (7\text{-}3)$$

where S_2 = the concentration of the second limiting substrate, mg/L.

If u_s = constant, microbial growth can generally be described as an exponential increase in cell numbers (and in dry weight). This growth form is approximated by the following equation:

$$dN/dt = u_s N \qquad (7\text{-}4)$$

Table 7-2 Mineral Composition (% of Dry Weight) of *Escherichia coli* and Average Protoplasm

Element	*E. coli*	Protoplasm
Carbon	50	39.0
Oxygen	20	22.6
Nitrogen	14	6.9
Hydrogen	8	5.5
Phosphorus	3	1.0
Sulfur	1.0	
Potassium	1.0	
Magnesium	0.5	
Calcium	0.5	
Iron	0.2	

Note: Data from VanDemark and Batzing, 1987.

where N = the number of cells
t = the time since growth started
u_s = the specific growth rate, h^{-1}

Equation 7-4 describes the exponential phase of microbial growth, and the population at a given time can be estimated from the equation

$$N_t = N_o e^{u_s t} \quad (7\text{-}5)$$

where e = the base for the natural logarithm (2.71828)
N_t = the population size at time t
N_o = the starting population size

When microbial growth is studied in the laboratory in a closed container, the typical growth curve is more complicated than the simple logarithmic increase (Figure 7-2a). There are three components of microbial growth. The first is a lag phase, during which enzymes are synthesized, cells sorb luxury quantities of growth nutrients, and individual cell sizes increase in preparation for cell division. The second growth phase is the logarithmic phase, during which exponential growth occurs in response to the asexual division of microbial cells because of relatively unlimited supplies of raw materials necessary for growth. The third phase in the typical microbe growth cycle is the stationary phase where nutrient limitation and the accumulation of waste byproducts eventually results in a zero net increase in cell numbers. The final phase observed in a closed container is the death or exponential decay phase in which the size of the population declines at an exponential rate.

Microbial growth in nature may differ from what is observed in the laboratory (Figure 7-2b) because in flowing water systems, nutrients and growth requirements are replenished

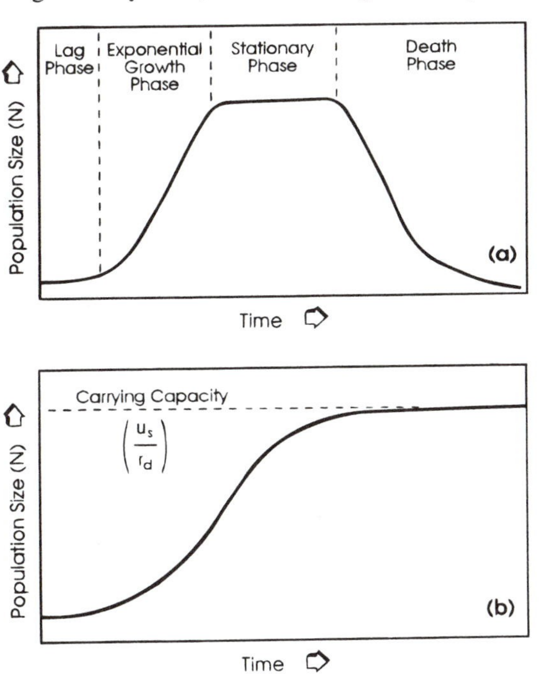

Figure 7-2 Typical growth pattern for microbial populations with (a) not renewed and (b) renewed growth substrate.

and cell waste products are removed. When growth requirements are supplied in a steady manner, steady state populations of adapted microbial species will predominate. New organism growth is generally exponential until population size reaches a growth plateau where the population size levels off. The height of this plateau or the steady state population is called the carrying capacity of that organism. The carrying capacity is proportional to the chemical and physical factors in the wetland that regulate growth of that species.

One model of this logistic or sigmoidal growth is given by the equation

$$dN/dt = u_s N - r_d N^2 \quad (7\text{-}6)$$

where r_d is the death rate (d^{-1}).

This population levels off at a steady state carrying capacity equal to the following:

$$N_s = u_s/r_d \quad (7\text{-}7)$$

Counting the cells in microbial populations can be difficult and time consuming. To determine the growth of specific indicator species such as *Escheriella coli,* cell numbers are monitored through plate count techniques or microbial examination. A useful way to estimate population size for mixed microbial populations is through dry weight and volume determinations. Turbidity and volatile suspended solids measurements can be used to monitor population sizes in aquatic microbial populations.

In a wetland treatment system, microbial activity is frequently more important than population size. Functional parameters used to assess microbial activity include the rates of appearance or loss for chemical constituents of concern, such as DO, NH_4^+, NO_3^-, and organic carbon. One of the most common measurements of microbial activity is the 5-day biochemical oxygen demand (BOD_5) test (American Public Health Association, 1992), which measures the amount of oxygen lost in a water sample under controlled temperature conditions (see Chapter 12 for a more complete discussion of BOD). *In situ* measurement of microbial activity together with other autotrophic and heterotrophic organisms is accomplished by changes in ambient concentrations of DO. Community metabolism is measured as the oxygen change, corrected for diffusion, occurring in the dark.

The decline of pathogenic microbial organisms is of interest in some wetland treatment systems receiving domestic and municipal wastewaters. This decline is generally exponential because of unsuitable conditions for growth of human pathogens in a wetland environment.

$$dN/dt = -r_d N \quad (7\text{-}8)$$

and

$$N_t = N_o e^{-r_d t} \quad (7\text{-}9)$$

The half life of the microbial population (the time when one half of the initial population is remaining during a period of exponential death) can be calculated from the relationship

$$t_{1/2} = 0.693/r_d \quad (7\text{-}10)$$

WETLAND ALGAE

INTRODUCTION

Algae are unicellular or multicellular, photosynthetic bacteria and plants that do not have the variety of tissues and organs of higher plants. Algae are a highly diverse assemblage of species that can live in a wide range of aquatic and wetland habitats. Many species of algae are microscopic and are only discernable as the green or brown color or "slime" occurring on submerged substrates or in the water column of lakes and ponds. Other algal species develop long, intertwined filaments of microscopic cells that look like mats of hair-like "sea weed" submerged or floating in ponds and shallow water environments. Figure 7-3 illustrates some of the major algal life forms typical in wetland environments.

For the most part, algae depend on light for their metabolism and growth and serve as the basis for an autocthonous food chain in aquatic and wetland habitats (Figure 7-4). Organic compounds created by algal photosynthesis contain stored energy which is used for respiration or which enters the aquatic food chain and provides food to a variety of microbes and other heterotrophs. Alternatively, this reduced carbon may be directly deposited as detritus to form organic peat sediments in wetlands and lakes.

Algae also depend on an ample supply of the building blocks of growth, including carbon, typically extracted from dissolved carbon dioxide in the water column, and on macro- and micronutrients essential to all plant life. When light and nutrients are plentiful, algae can create massive populations and contribute significantly to the overall food web and nutrient cycling of an aquatic or wetland ecosystem. When shaded by the growth of macrophytes, algae frequently play a less important role in wetland energy flows.

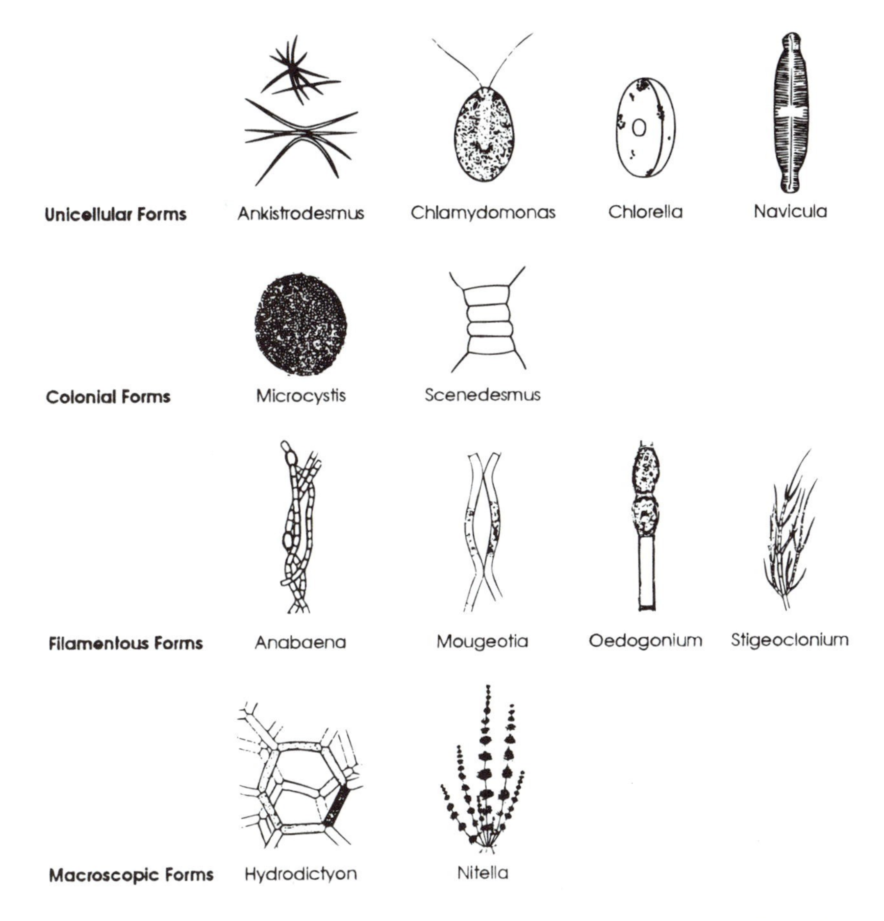

Figure 7-3 Algal life forms typically found in wetland and aquatic environments (From *Algae of the Western Great Lakes Area*, G.W. Prescott. Copyright © W.C. Brown Communications, Inc., 1962. Reprinted by permission of W.C. Brown Communications, Inc.).

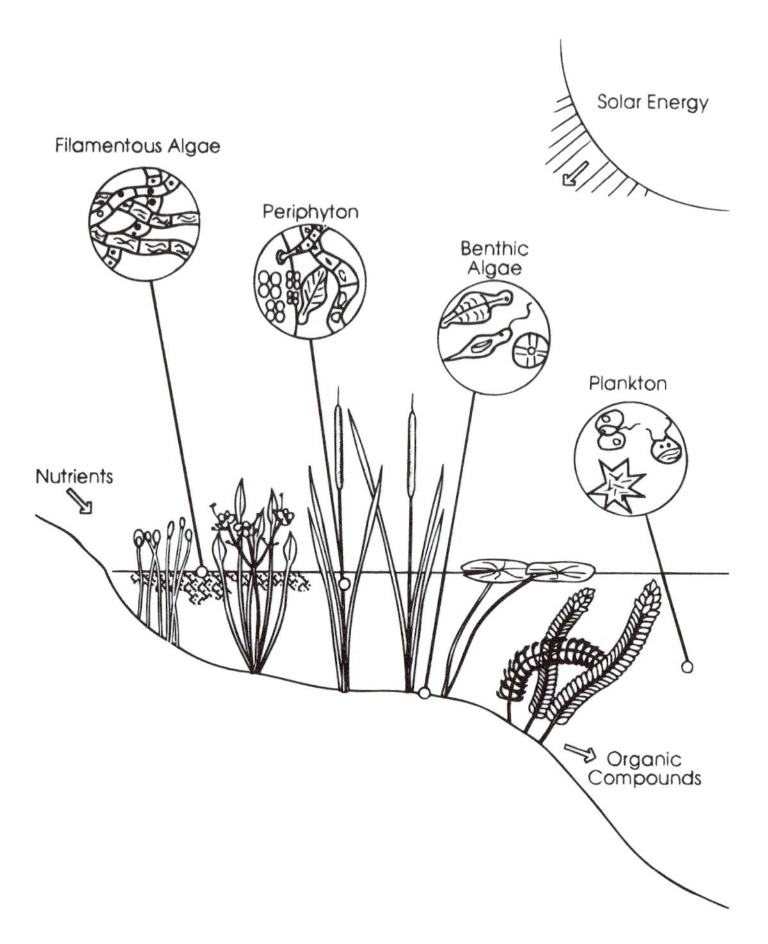

Figure 7-4 Major energy sources and ecological niches affecting the occurrence of algae in wetlands.

Most species of algae need ample water during some or all of their life cycles. Because water quality and climatic variables such as air and water temperature and light intensity are the principal determinants of algal species distribution, the algal flora of wetlands is generally similar to the regional algal flora living in ponds, lakes, springs, streams, rivers, and similar aquatic environments. The algal flora of wetlands differs from the flora of more aquatic environments, primarily in response to varying water chemistry, water depth, light inhibition by emergent macrophytes, and seasonal desiccation which is more likely in shallow water environments.

CLASSIFICATION

Algae may be classified by evolutionary or genetic relationships, morphological adaptations, or ecological functions. Taxonomic identification of algae in wetlands is rarely required to design or operate wetland treatment systems. For detailed taxonomy of this phylum, the reader is referred to Lee (1980) and South and Whittick (1987).

Table 7-3 lists the major algal classes that occur in wetland and aquatic habitats. Of principal occurrence in terms of volume and importance in natural wetlands are the diatoms (Chrysophyta), green algae (Chlorophyta), and blue-green algae (Cyanophyta). The desmids, a subgroup of the green algae, are typically a major part of the flora of acidic, freshwater

Table 7-3 Classification of Algae Found in Wetland and Aquatic Habitats

Group	Growth Form	Typical Growth Habitats	Typical Genera
Cyanophyta (blue-green algae)	Unicellular, filamentous, colonial, sheathed, or unsheathed	All habitats	*Anabeana, Phormidium, Lyngbya, Microcystis, Oscillatoria*
Chlorophyta (green algae)	Unicellular, filamentous, colonial, motile, and attached	All habitats	*Chlamydomonas, Chlorella, Stigeoclonium, Oedogonium, Spirogyra, Hydrodictyon*
Euglenophyta (euglenoids)	Unicellular, motile, facultatively heterotrophic	Primarily benthic, occassional in periphyton and plankton	*Euglena, Trichomonas, Phacus, Strombomonas*
Pyrrophyta (dinoflagellates)	Mostly unicellular and motile	Primarily planktonic in lakes and open water	*Gymnodinium, Peridinium*
Chrysophyta (yellow-green algae)	Unicellular and colonial motile	Occurs in all habitats infrequently	*Chromulina, Ochromonas, Dinobryon, Mallomonas, Synura*
Bacillariophyta (diatoms)	Unicellular and filamentous, motile, and sessile	Occurs in all habitats, most common in benthic algae and periphyton	*Navicula, Fragillaria, Pinnularia, Melosira, Cyclotella*
Cryptophyta (crytomonads)	Unicellular and motile	Primarily planktonic, some in periphyton	*Cryptomonas, Rhodomonas, Chilomonas*
Rhodophyta (red algae)	Mostly filamentous, some unicellular	Primarily benthic or periphytic, mostly marine and estuarine	*Batrachiospermum, Bangia*
Phaeophyta (brown algae)	Macroalgae (seaweeds)	Primarily marine and estuarine	*Gracilaria, Fucus, Laminaria, Sargassum*

wetlands. Motile algae in the Euglenophyta are typically more important in wetlands and other aquatic environments with high concentrations of dissolved organic carbon. Each of these algal classes has different photosynthetic pigment ratios and different requirements for nutrients and light. These differences result in a wide range of adaptability to microhabitats within wetlands and aquatic environments, both spatially in three dimensions as well as temporally through the seasons.

Because the functional role of algae is most important in understanding wetlands, this section focuses on the functional algal groups that are found in wetlands. Three overlapping functional terms used to describe algae important in wetland ecosystems are periphyton, aufwuchs, or benthic algae. The term "periphyton" describes the community of organisms that grows attached to emergent and submerged plants in littoral and other marsh wetlands. Although periphyton usually begin colonization of new plant surfaces by attached algal growth of filamentous and unicellular species, this functional component also includes a variety of free-living algae (not attached to the surface), fungi, bacteria, and protozoans following a period of maturation. Aufwuchs is a more general term than periphyton and includes all algae and associated microscopic life attached to all surfaces in an aquatic or wetland system. These surfaces frequently include living vascular plants as well as dead plants, leaves, branches, trunks, stones, and exposed substrates. Benthic or attached algae are more specific terms which refer only to the algal component of the periphyton or aufwuchs.

Planktonic or free-floating algae are generally not important in wetland ecosystems unless open or deep water areas are present. Plankton spend most of their life cycle suspended in the water column and are the most important algal component in lakes and some ponds. Tychoplankton are algae that initially grow as attached species and which subsequently break free from their substrate and live planktonically for part of their life cycle. Tychoplanktonic algal species are most common in streams and in littoral wetlands. The chlorophyll content of the water is commonly used as a measure of algal density. The degradation product, pheophytin, is a measure of algal remains.

ECOLOGY

Filamentous algal mats are sometimes a dominant component of the plant biomass in wetland systems. The mats are made of a few dominant species of green or blue-green filamentous algae in which individual filaments may include thousands of cells. These microscopic or barely visible filaments occur in such numbers as to produce extensive, macroscopic mats. Although filamentous algal mats are generally dominated by one or a few species of algae, they also often contain from a few to dozens of other filamentous, colonial, and unicellular algal species and a diversity of bacteria and fungal species.

Filamentous algal mats first develop below the water surface on the substrate of the wetland in areas with little emergent vegetation. During the day, entrained gas bubbles (primarily pure oxygen resulting from photosynthesis) may cause the mats to move up through the water column and float at the surface. During the night, the mats sink again to the wetland substrate.

Filamentous algae that occur in wetlands as periphyton or mats may dominate the overall primary productivity of the wetland, controlling DO and CO_2 concentrations within the wetland water column. Diurnal DO profiles in wetlands and other aquatic environments with substantial populations of submerged plants undergo major changes in relation to the daily gross and net productivity. Figure 7-5 illustrates these changes in a partially vegetated, shallow, freshwater marsh. Wetland water column DO can fluctuate from near zero during the early morning following a night of high respiration to well over saturation (>20 mg/L) in high algal growth areas during a sunny day. Dissolved carbon dioxide and consequently the pH of the water varies proportionally to DO because of the corresponding use of CO_2 by plants during photosynthesis and release at night during respiration. As CO_2 is stripped from the water column by algae during the day, pH may rise by 2 to 3 pH units (a 100- to 1000-fold decrease in H^+ concentration). These daytime pH changes are reversible, and the production of CO_2 at night by algal respiration frequently returns the pH to the previous day's value by early morning.

Algae also are important for storing and transforming essential growth nutrients in wetlands and aquatic habitats. Table 7-4 summarizes the typical elemental composition of algae with carbon (35 to 50 percent), nitrogen (2 to 9 percent), and phosphorus (0.2 to 0.8

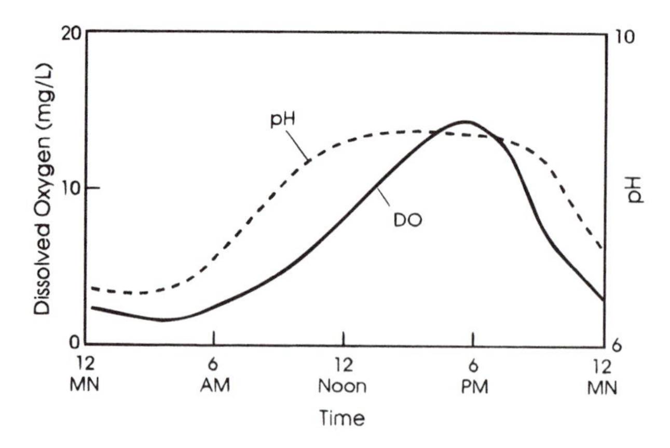

Figure 7-5 Typical diurnal plots of DO concentration and pH in a wetland dominated by filamentous algae.

Table 7-4 Mineral Composition of Some Wetland Algal Species as Percent of Dry Weight

Constituent	Average (Range) for 5 to 9 Species
Ash	22.1 (11.7–41.2)
Calcium	5.23 (0.27–24.64)
Carbon	41.9 (35.4–49.7)
Copper	0.0122 (0.0019–0.065)
Crude protein	18.3 (4.5–31.3)
Fiber	20.8 (9.3–40.9)
Iron	0.325 (0.104–0.940)
Magnesium	0.40 (0.01–0.92)
Manganese	0.345 (0.093–1.64)
Nitrogen	5.1 (2.57–9.43)
Percent solids	5.6 (0.4–14.9)
Phosphorus	0.23 (0.05–0.56)
Potassium	2.46 (0.21–6.09)
Sodium	0.26 (0.04–1.42)
Sulfur	0.69 (0.15–1.58)
Zinc	0.0128 (0.0010–0.0395)

percent) as some of the important elements cycling through this wetland trophic level. Because of their relatively low contribution to the overall fixed carbon in wetlands, algae do not constitute a major storage reservoir for these elements in wetlands. However, because of their high turnover rates in some aquatic habitats, algae may be important for short-term nutrient fixation and immobilization with subsequent gradual release and recycling. The functional result of this nutrient cycling is that intermittent high inflow concentrations of pollutants used by algae for growth may be immobilized and transformed more effectively than would be possible without these components, thereby reducing the amplitude of wetland constituent outflow concentrations.

Algae also may be important in the long-term fixation of some pollutants into sediments through sorption, settling, and burial as organic peat. This phenomenon has been documented primarily in low-nutrient wetland systems such as the Everglades marshes in Florida, the Okefenokee Swamp in Georgia, and in northern oligotrophic bogs and marshes. In higher-nutrient systems, microbial decomposition of dead algae is more rapid and may exceed the rate at which these solids accumulate in the sediments, while in low-nutrient wetlands the microbial decomposition rate may not be able to keep pace with the production of algal carbon and associated minerals.

For a detailed description of the importance of algae in wetlands, see Vymazal (1995).

WETLAND MACROPHYTES

INTRODUCTION

The term macrophyte includes vascular plants that have tissues that are easily visible. Macroalgae were discussed earlier. Vascular plants differ from algae through their internal organization into tissues resulting from specialized cells. A wide variety of macrophytic plants occur naturally in wetland environments. The USFWS has more than 6700 plant species on their list of obligate and facultative wetland plant species in the U.S. Godfrey and Wooten (1979, 1981) list more than 1900 species (739 monocots and 1162 dicots) of wetland macrophytes in their taxonomy of the southeastern U.S. Obligate wetland plant species are defined as those which are found exclusively in wetland habitats, while facultative species are those that may be found in upland or in wetland areas. There are many guidebooks that illustrate wetland plants (for example, Hotchkiss, 1972; Niering, 1985). Lists of plant species that occur in wetlands are available (for example, RMG, 1992).

Wetland macrophytes are the dominant structural component of most wetland treatment systems. A basic understanding of the growth requirements and characteristics of these wetland plants is essential for successful treatment wetland design and operation.

CLASSIFICATION

The plant kingdom is divided taxonomically into phyla, classes, and families, with certain families either better represented or occurring only in wetland habitats. The major plant phyla are the mosses and clubmosses (Bryophyta) and the vascular plants (Tracheophyta). In the vascular plant phylum, there are three important classes of plants: ferns (Filicinae), conifers (Gymnospermae), and flowering plants (Angiospermae). The flowering plants are further divided into the monocots (Monocotyledonae) and dicots (Dicotlyedonae). This section describes the major groups of plants that occur in wetlands used for water quality treatment.

Since plant taxonomic families were developed to provide insight into the evolutionary affinity of plant species, it is not surprising that some families are well represented by multiple obligate wetland species. Table 7-5 lists the plant groups and families that are principally represented by wetland species or that have a few highly important wetland plant species. The interested reader may wish to take advantage of the instructional program for aquatic plant identification developed by the University of Florida in Gainesville. This consists of seven videotapes covering floating, submersed, and emersed plants, as well as grasses, sedges, and rushes (IFAS, 1991).

Vascular plants including wetland plants may also be categorized morphologically by descriptors such as woody, herbaceous, annual, perennial, emergent, floating, and submerged. Woody species have stems or branches that do not contain chlorophyll. Since these tissues are adapted to survive for more than 1 year, they are typically more durable or woody in texture. Herbaceous species have above-ground tissues that are leafy and filled with chlorophyll-bearing cells that typically survive for only one growing season. Woody species include shrubs that attain heights up to 2 or 3 m and trees that generally are more than 3 m in height when mature.

Annual plant species survive for only one growing season and must be reestablished annually from seed. Perennial plant species live for more than 1 year and typically propagate each year from perennial root systems or from perennial above-ground stems and branches. Nearly all woody plant species are perennial, but herbaceous species may be annual or perennial.

The terms emergent, floating, and submerged refer to the predominant growth form of a plant species (Figure 7-6). Some species have one or more of these growth forms; however, there is usually a dominant form that enables the plant species to be classified. In emergent plant species, most of the above-ground part of the plant emerges above the water line and into the air. These emergent structures may be self-supporting or may be supported by other emergent species or physical structures such as in the case of vines. Emergent wetland plant species are the primary concern of this section because they provide surface area for microbial growth which is important in many of the chemical assimilation processes in wetland treatment systems.

Both floating and submerged vascular plant species may occur in wetland treatment systems. Floating species have leaves and stems buoyant enough to float on the water surface. Submerged species have buoyant stems and leaves that fill the niche between the sediment surface and the top of the water column. Floating and submerged species are more typical of deeper, aquatic habitats than of wetlands, but they may occur in wetlands when water depth exceeds the tolerance range for rooted, emergent species.

Table 7-5 Classification of Major Macrophytic Plant Groups Occurring in Wetlands Used for Water Quality Treatment

Group	Family	Typical Genera	Typical Growth Habit
Mosses	Fontinalaceae	*Fontinalis*	Attached, floating, or submersed
	Sphagnaceae	*Sphagnum*	Rooted emergent or floating
Ferns	Blechnaceae	*Blechnum Woodwardia*	Rooted emergent
	Osmundaceae	*Osmunda*	Rooted emergent
	Salviniaceae	*Salvinia, Azolla*	Floating
Conifers	Cupressaceae	*Chamaecyparis*	Tree
	Pinaeae	*Pinus*	Tree
	Taxodiaceae	*Taxodium*	Tree
Monocots	Alismataceae	*Sagittaria*	Emergent herbs
	Araceae	*Colocasia, Peltandra, Pistia*	Emergent or floating herbs
	Cannaceae	*Canna*	Emergent herbs
	Cyperaceae	*Cyperus, Carex, Cladium, Eleocharis, Rhynchospora, Scirpus*	Emergent herbs
	Graminae	*Panicum, Paspalum, Phragmites, Spartina, Zizaniopsis*	Emergent herbs (occasionally rooted with floating leaves)
	Hydrocharitaceae	*Hydrilla, Egeria, Elodea, Limnobium*	Rooted submergents to floating
	Juncaceae	*Juncus*	Emergent herb
	Lemnaceae	*Lemna, Spirodela, Wolffia, Wolffiella*	Small, floating herbs
	Pontederiaceae	*Eichhornia, Pontederia*	Floating and emergent herbs
	Typhaceae	*Typha*	Emergent herb
Dicots	Aceraceae	*Acer*	Tree
	Amaranthaceae	*Alternanthera*	Emergent to floating herb
	Aquifoliaceae	*Ilex*	Tree or woody shrub
	Betulaceae	*Betula, Carpinus*	Trees
	Compositae	*Bidens, Mikania*	Emergent herbs
	Cyrillaceae	*Cliftonia, Cyrilla*	Woody shrubs to trees
	Ericaceae	*Lyonia*	Woody shrub
	Fagaceae	*Quercus*	Trees
	Guttiferae	*Hypericum*	Woody shrubs
	Lauraceae	*Persea*	Tree
	Lythraceae	*Decodon*	Herbaceous/woody shrub
	Magnoliaceae	*Liriodendron, Magnolia*	Trees
	Nymphaceae	*Nuphar, Nymphaea*	Floating, rooted herbs
	Nyssaceae	*Nyssa*	Tree
	Oleaceae	*Fraxinus*	Tree
	Onagraceae	*Ludwigia*	Herbaceous or woody shrub
	Polygonaceae	*Polygonum*	Emergent herb
	Rubiaceae	*Cephalanthus*	Shrub or small tree
	Salicaceae	*Populus, Salix*	Trees
	Saururaceae	*Saururus*	Emergent herb
	Umbelliferae	*Hydrocotyle, Lilaeopsis*	Emergent to floating herbs

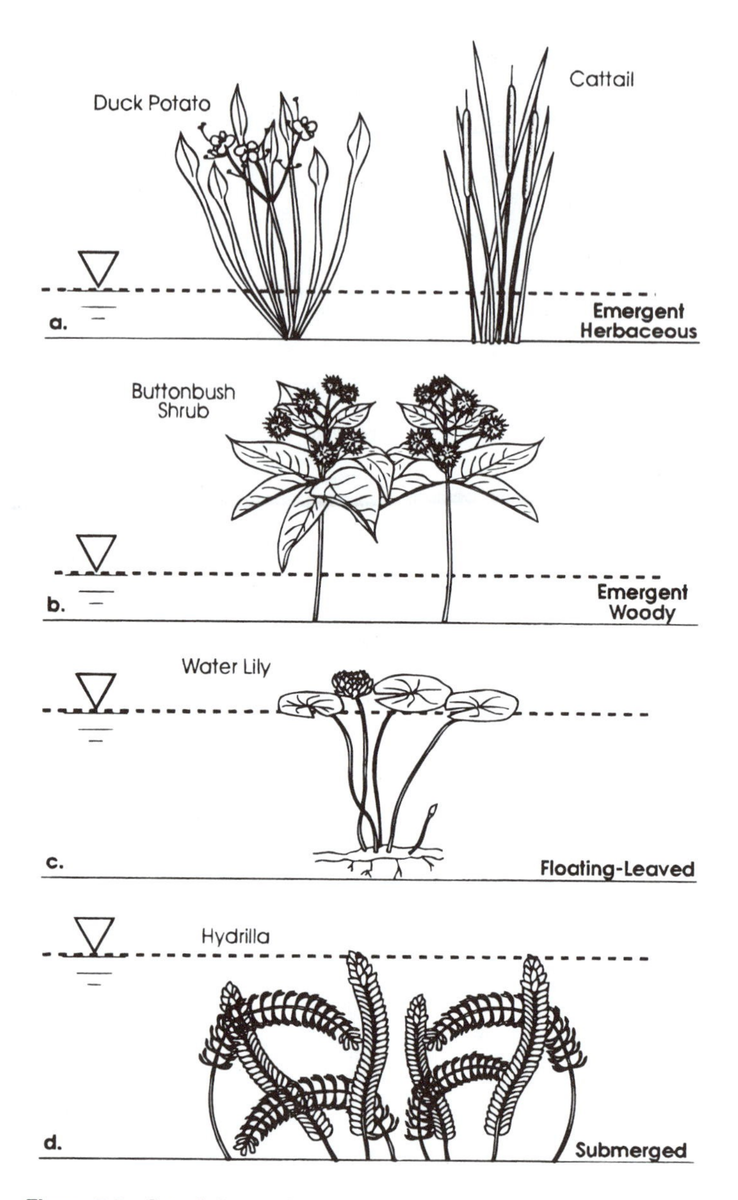

Figure 7-6 Growth forms of rooted wetland and aquatic vascular plants.

ADAPTATIONS TO LIFE IN FLOODED SOILS

All vascular plant roots require gaseous exchange to supply oxygen for cell respiration and to exhaust gases such as CO_2 that might accumulate during metabolic processes (Vartapetian, 1978). All plants also require water for numerous biochemical processes including photosynthesis and transpiration, which assist with intercellular transport of nutrients and metabolites. However, excess water presents a stress because it inhibits diffusion of gases to and from plant roots (oxygen diffusion is about 10,000 times faster in air than in water) and because of the presence of oxygen-demanding constituents dissolved in the water which tend to lower the DO available to supply root metabolism (Whitlow and Harris, 1979). Flooding and oxygen limitation also quickly change soil properties, allowing accumulation of potentially toxic gases such as H_2S and ions such as iron and manganese (Gambrell and Patrick, 1978).

One adaptation to flooding is the development of aerenchymous plant tissues (Figure 7-7a) that transport gases to and from the roots through the vascular tissues of the plant above water and in contact with the atmosphere, providing an aerated root zone and thus lowering the plant's reliance on external oxygen diffusion through water and soil (Armstrong, 1978; Jackson and Drew, 1984; Zimmerman, 1988). Lenticels or small openings on the above-water portions of these plants provide an entry point for atmospheric oxygen into this aeranchymous tissue network. Lenticel surface area may be increased through plant growth, height increases, or the formation of swollen buttresses in trees and woody herbs and in cypress knees (Figures 7-7b and c).

Oxygen transport into the root zone through lenticels and aerenchymous tissue has been measured between 2.08 g $O_2/m^2/d$ (Brix and Schierup, 1990; Figure 7-8) and 5 to 12 g $O_2/m^2/d$ (Armstrong et al., 1990) in *Phragmites australis* grown in gravel beds. In some cases, this oxygen transport appears to be sufficient only to offset root metabolism and not to result in additional aeration of the surrounding sediments (Brix, 1990), while in other cases there appears to be some "wastage" of O_2 with release to the surrounding microbial populations.

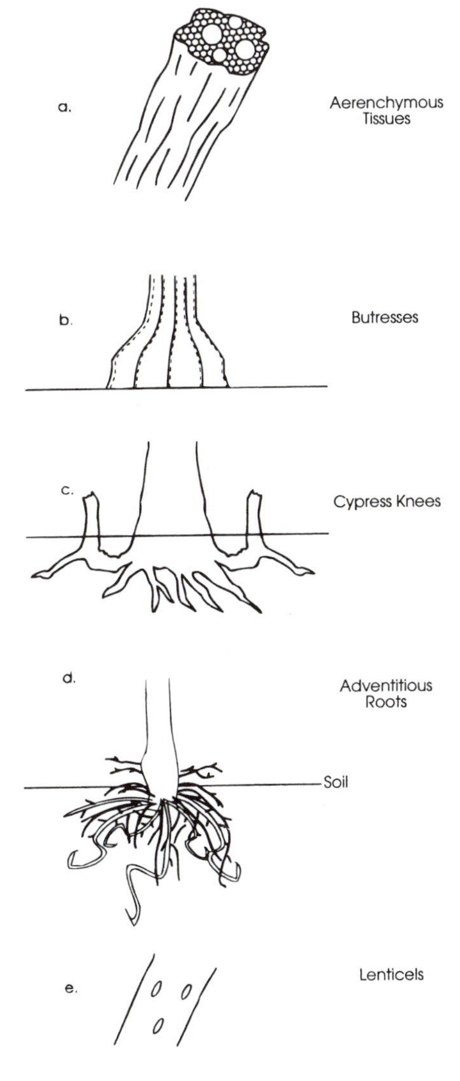

Figure 7-7 Vascular plant morphological adaptations to life in wetlands.

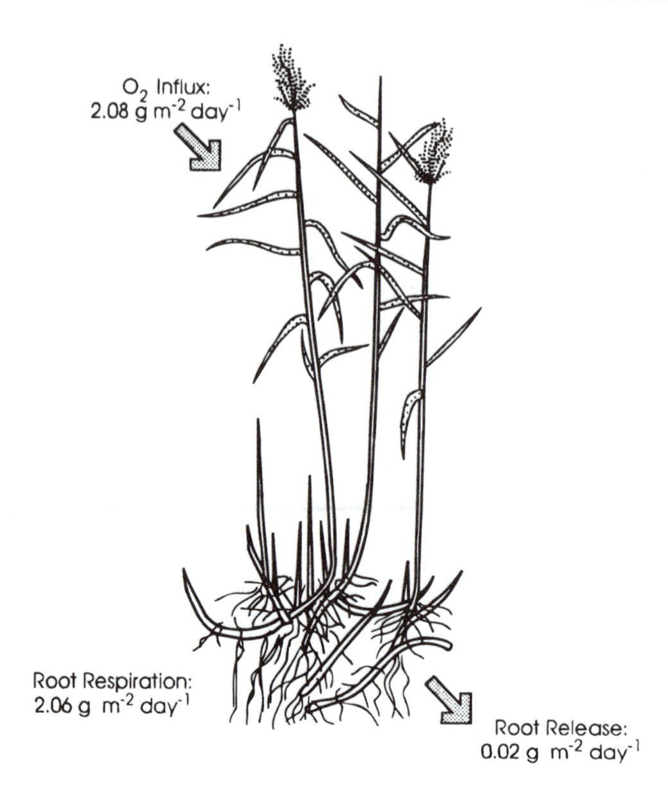

Figure 7-8 Oxygen flux through *Phragmites australis* to the root zone (Reprinted from "Proceedings of the International Conference on the Use of Constructed Wetlands in Water Pollution Control," Brix and Schierup, 1990, with kind permission from Elsevier Science Ltd, The Boulevard, Langford Lane, Kidlington OX5 1GB, UK).

The flow of O_2 through this "snorkel" system (Zimmerman, 1988) is greater than can be explained by gaseous diffusion alone and appears to be enhanced through convective flow driven by temperature and humidity gradients between the above-water and root portions of the plants (Brix and Schierup, 1990).

A second adaptation to flooding shared by some hydrophytes is the generation of adventitious roots (Figure 7-7d) from flooded stem tissue (Kozlowski, 1984). These roots have the potential to extract DO and plant nutrients from water where gases and nutrients may be more available than in the anoxic soil zone.

It has been observed that some hydrophytes can withstand much lower levels of anoxia in the root zone than nonhydrophytic species, allowing an increased survival rate under extreme conditions such as peak flood levels, rainy periods, or dormant periods. Metabolic explanations of this adaptability are currently inconclusive (Jackson and Drew, 1984), and actual mechanisms imparting all aspects of flood tolerance remain unexplained.

Plant survival in flooded environments is a balance between the severity of oxygen limitation and the adaptations available to overcome this oxygen shortage (Figure 7-9). Thus, hydrophytic plants may be adapted to survive and even grow in specific flooded conditions, such as 3 months each year, or in "clean" or flowing water which might have higher *in situ* DO concentrations (Gosselink and Turner, 1978). However, these same plants may not be able to grow or survive during 5 months of flooding or in stagnant or "dirty" water conditions. Likewise, plants may have adaptations that allow prolonged survival in 1 ft of water, but not at 2 ft. It may be hypothesized that this balance is tilted unfavorably at higher water levels because of reduced aerial plant stem surface area to provide oxygen to the roots through the lenticels and aerenchymous tissues. This proposed explanation is supported by

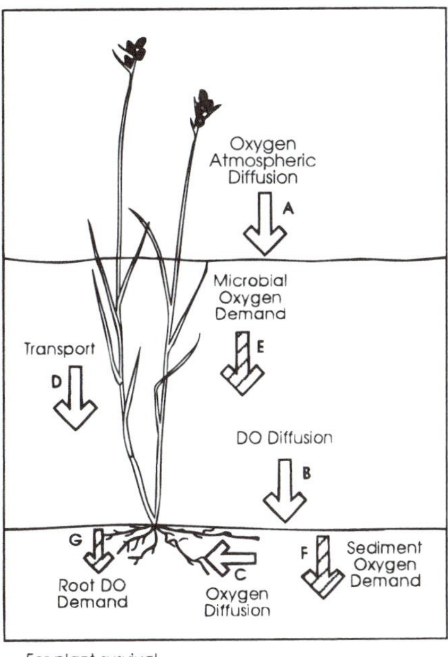

For plant survival
$D + C > G$

If plant is not adapted to flooding, then for survival
$C > G$, $B > C + F$, and $A > B + E$

Figure 7-9 Summary of dissolved oxygen balance in wetland plant growth. Although wetland plants can survive for a time with no significant aerobic respiration in the root zone, this time may be short and is limited by existing storages of sugars and other organic compounds in the energy-bearing, above-ground stems and leaves. For plant survival, there must be a net oxygen surplus to the roots allowing respiration. Thus, oxygen transport to the roots via internal transport mechanisms through aerenchymous tissues (D) plus diffusion through the soils (C) must exceed the oxygen required for root metabolism (G) or the plant will die prematurely. If a plant is not adapted to flooding (inadequate internal gas transport, $D \approx 0$), then root metabolism depends upon adequate oxygen diffusion from the atmosphere (A), through the water column (B), and through the sediments (C) to overcome microbial sediment oxygen demand (F) and still provide enough oxygen for root metabolism (G). These conditions are rare in continuously flooded wetland environments, and only occur when dissolved oxygen-demanding substances (E) and sediment organic matter are low.

the finding that hydrophytes generally respond to flooding by growing taller, a growth response that allows a more favorable balance between emergent and submerged plant organs (Grace, 1989).

HYDROPERIOD AND WATER REGIME

While the presence of water separates uplands from wetlands and aquatic ecosystems, the hydroperiod of the water is the most important contributor to wetland type or class (Gosselink and Turner, 1978; Gunderson, 1989). The importance of this factor in wetland treatment system design and operation cannot be overstated, since incorrect understanding of the hydroperiod and water regime limitations of wetland plant species is the most frequent cause of vegetation problems in natural and constructed wetlands. Measuring the hydroperiod is relatively easy. However, selecting the optimal hydroperiod for wetland treatment design and performance is complex. For that reason, this section describes the dynamics of the hydroperiod and the implications for treatment systems.

The concepts of hydroperiod and water regime include two interdependent components: (1) the duration of flooded or saturated soil conditions (the hydroperiod as a percentage of

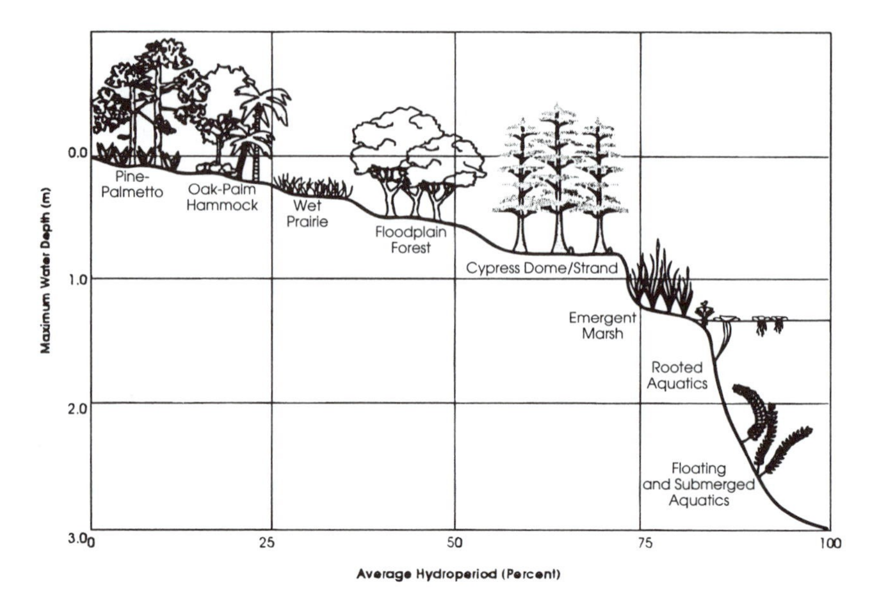

Figure 7-10 A generalized diagram of natural Florida wetland ecosystems along a hydroperiod gradient.

time with flooding) and (2) the depth of flooding (Gunderson, 1989). While hydroperiod refers to the duration of flooding, the term water regime refers to hydroperiod as well as to the combination of water depth and flooding duration (depth-duration curve).

The duration and depth of flooding affect plant physiology because of soil oxygen concentration, soil pH, dissolved and chelated macro- and micronutrients, and toxic chemical concentrations. The importance of the water regime in regulating the occurrence of natural plant communities is illustrated in Figure 7-10 with an example from Florida. Plant types classified as wetlands by the USFWS have as few as 7 days of flooding or soil saturation each year (hydroperiod = 2 percent duration). In Florida, the change in the hydroperiod results in the disappearance of the palmetto (*Sabal minor*) in wet pine flatwood forests and gives rise to pine-cypress forests, wet prairies, and at longer durations and greater water depth, cypress-gum forests, mixed emergent marshes, and, ultimately, aquatic habitats.

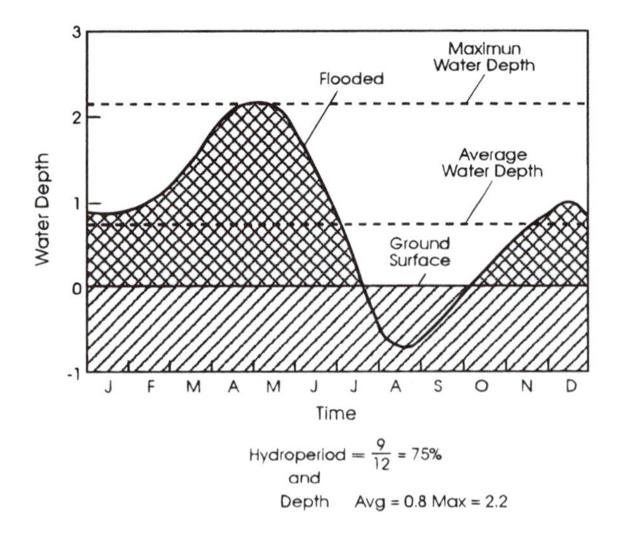

Figure 7-11 Components of wetland hydroperiod and water regime.

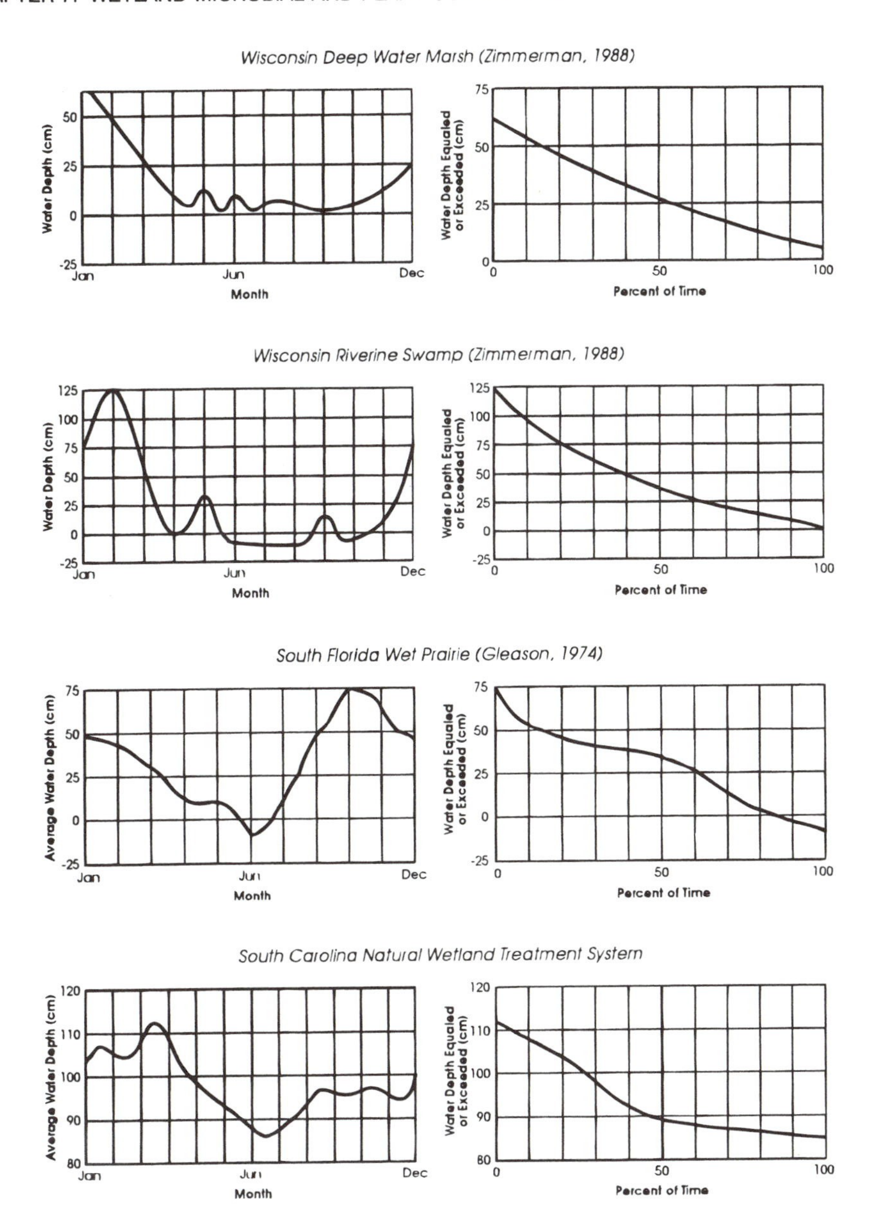

Figure 7-12 Water level charts and depth-duration curves for several wetland ecosystems.

Figure 7-11 uses a graph of water level within a wetland over an annual period to illustrate these two aspects of hydroperiod and water regime. Duration of flooding refers to the percentage of time that a wetland site is flooded or saturated, and depth of flooding refers to the minimum, average, and maximum depths of water at a given or typical spot within a wetland.

For any specific location within a wetland, a depth-duration curve can be prepared to summarize the water regime and hydroperiod. Figure 7-12 illustrates water level and depth-duration curves for a variety of wetland ecosystems including a natural wetland treatment system. A water stage-duration curve also can be used to summarize the effects of hydroperiod on vegetation within a larger wetland ecosystem (Figure 7-13). Hydroperiod curves provide a convenient method for estimating the percentage of time that a wetland is flooded at any water depth and can summarize water level data over any period of record. Note that water

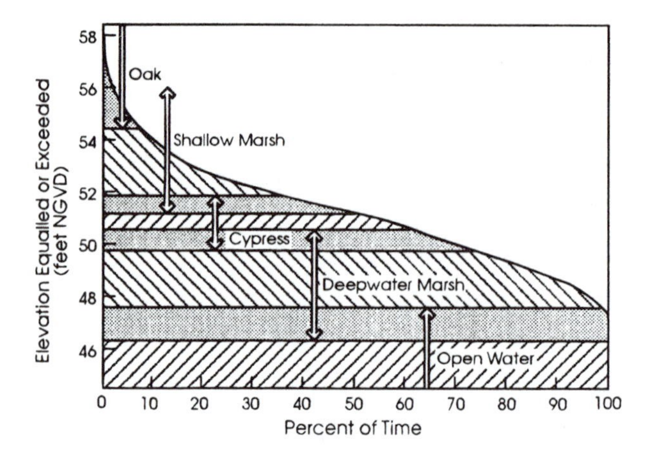

Figure 7-13 Stage-duration curve for wetland zones fringing Lake Hatchineha, FL, based on daily water level measurements from 1942 to 1964. (Adapted from Duever et al., 1988).

level charts and depth-duration curves also can summarize the time and depth that water is located below the ground.

An appropriate water regime and hydroperiod is particularly important in the design of wetland treatment systems because of its effect on soil oxygen levels and subsequently on plant community structure and dominance. Figure 7-14 illustrates idealized water level charts for several wetland classes. Many natural wetlands will undergo significant biological changes if their natural hydroperiod tolerance limits are exceeded through the addition of intermittent or continuous wastewater flows. The effect of excessive hydroperiods on natural wetland treatment systems is discussed in more detail in Chapter 22.

Typical hydroperiod tolerance ranges for wetland plant species important in wastewater treatment are presented in Table 7-6. Few quantitative studies are available on the effect of water quality and water depth on the growth of wetland plants, so Table 7-6 has been developed from the authors' experience with wetland plant occurrence and from available published information. Hydroperiod ranges are given when necessary to reflect either uncertainty or variation in plant survival based on other factors such as genetic variability or water quality.

The emergent wetland plant species listed in Table 7-6 are organized into four tolerance groups: (1) herbaceous species that can withstand continuous inundation, (2) woody species that can withstand continuous inundation, (3) herbaceous species that cannot tolerate continuous inundation; and (4) woody species that cannot tolerate continuous inundation. Clearly, only species from the first two groups can be used successfully in treatment wetlands that are continuously inundated by wastewater. Species from the third and fourth groups may be compatible with treatment of intermittent stormwater or in wetland treatment systems where adequate area is available to seasonally rest portions of the system from operation. Plant species in the third and fourth groups may be used in wetland treatment systems if they are isolated from flooding by living along the wetland's margins or are situated on higher ground or on logs and stumps within the wetland.

PLANT REPRODUCTION

As with all plant species, wetland plants increase their numbers and density through asexual and sexual reproduction. Asexual reproduction refers to an increase in the number of individuals of a plant species through vegetative growth and typically occurs through growth of the roots or rhizomes with subsequent emergence of new above-ground stems and

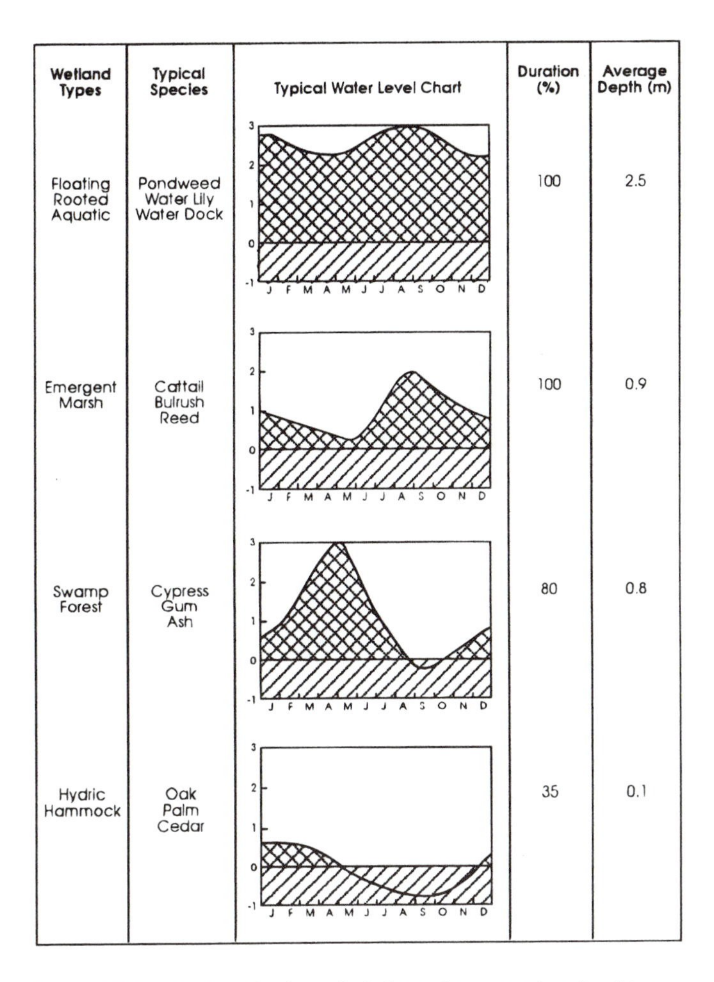

Figure 7-14 Idealized hydroperiod charts for general wetland types.

leaves. Technically, a cattail bed that developed vegetatively from a single-parent plant is a single plant. However, when these rhizomes are cut or decay, the individual daughter plants may remain viable and continue to spread vegetatively. A number of woody wetland plant species can spread vegetatively through coppice growth from viable root systems. This phenomenon is frequently observed following logging activities when new sprouts emerge from the roots or from the stump of the parent tree.

Most wetland plant species also may increase their numbers through sexual reproduction. In sexual reproduction, two individual plants, or male and female flowers from a single plant, contribute gametes to form seeds with new combinations of genetic material. Sexual reproduction is important in providing alternative strategies for plants to survive from year to year through seasonal extremes, to propagate the species over large distances, to rapidly colonize new habitats, and to provide genetic variants that can adapt to changing environmental and competitive conditions. The great diversity of plant adaptations for sexual reproduction is described in detail by Willson (1983) and Doust and Doust (1988).

Because of the potential for year-to-year hydrological variations in natural herbaceous wetlands with large numbers of annual plant species, many species produce seeds which remain viable for years. These seeds accumulate during productive years and remain dormant until conditions are favorable for germination, frequently after a period of desiccation and rewetting. This storage of viable seeds is known as a seed bank and has been studied in a

Table 7-6 Approximate Hydroperiod Tolerance Ranges for Typical Wetland Plant Species Found in Wetland Treatment Systems

Scientific Name	Common Name	Maximum Water Depth (m)	Flooding Duration (%)
	Submerged Aquatic Species		
Fontinalis spp.	Water mosses	0.1–1.5	80–100
Egeria densa	Brazillian waterweed	0.1–3	90–100
Elodea spp.	Elodea	0.1–3	90–100
Hydrilla verticillata	Hydrilla	0.1–3	90–100
Vallisneria spp.	Eel grasses	0.25–3	90–100
Najas spp.	Naiads	0.25–2	90–100
Myriophyllum spp.	Water milfols	0.25–3	90–100
Utricularia spp.	Bladderworts	0.1–1	70–100
Cabomba caroliniana	Fanwort	0.25–3	90–100
	Floating Aquatic Species		
Azolla caroliniana	Mosquito fern	None	70–100
Salvinia rotundifolia	Water fern	None	70–100
Pistia stratiotes	Water lettuce	None	70–100
Lemna spp.	Duckweeds	None	90–100
Spirodela spp.	Giant duckweeds	None	90–100
Wolffia spp.	Water meals	None	90–100
Wolffiella spp.	Bog mats	None	90–100
Eichhornia crassipes	Water hyacinth	None	80–100
	Floating Rooted Aquatic		
Limnobium spongia	Frog's bit	0.2–2	70–100
Potamogeton spp.	Pondweeds	0.5–3	90–100
Brasenia schreberi	Water shield	0.25–2	80–100
Nymphoides spp.	Floating hearts	0.1–3	80–100
Nelumbo spp.	Lotuses	0.25–1.5	70–100
Nuphar spp.	Spatterdocks	0.5–3	90–100
Nymphaea spp.	Water lilies	0.5–3	90–100
Hydrocotyle spp.	Pennyworts	<0.05–1.0	25–100
	Emergent, Woody Shrubs, and Trees (Continuous Inundation)		
Taxodium spp.	Cypresses	<0.05–1.0	30–100
Decodon verticillatus	Water-willow	0.1–0.5	70–100
Nyssa spp.	Tupelo gums	0.1–1.0	50–100
Cephalantus occidentalis	Buttonbush	0.1–0.5	50–100
Salix spp.	Willows	0.1–0.5	50–100
Rhizophora mangle	Red mangrove	<0.05–0.5	50–100
Gordonia lasianthus	Loblolly bay	<0.05–0.25	50–100
	Emergent, Woody Shrubs, and Trees (Less Than Continuous)		
Pinus serotina	Pond pine	<0.05–0.1	<10–50
Sabal palmetto	Cabbage palm	<0.05–0.1	<10–50
Acer spp.	Maples	<0.05–0.2	0–70
Ilex spp.	Hollys	<0.05–0.1	<10–30
Avicennia germinans	Black mangrove	0.1–0.3	20–70
Alnus serrulata	Hazel alder	<0.05–0.1	<10–50
Betula nigra	River birch	<0.05–0.1	<10–50
Cornus foemina	Swamp dogwood	<0.05–0.1	<10–50
Cyrilla racemiflora	Titi	<0.05–0.1	25–50
Lyonia lucida	Fetterbush	<0.05–0.1	<10–50
Quercus spp.	Oaks	<0.05–0.1	0–30
Hypericum spp.	St. John's wort	<0.05–0.1	<10–50

Table 7-6 Continued

Scientific Name	Common Name	Maximum Water Depth (m)	Flooding Duration (%)
	Emergent, Herbaceous (Continuous Inundation)		
Sphagnum spp.	Sphagnum mosses	<0.05–0.1	75–100
Sagittaria spp.	Arrowheads	0.2–0.5	50–100
Colocasia esculenta	Wild taro	0.1–0.5	25–100
Peltandra spp.	Spoon flowers	<0.05–0.25	50–100
Canna spp.	Canna lilies	<0.05–0.25	50–100
Carex spp.	Sedges	<0.05–0.25	50–100
Cladium jamaicense	Sawgrass	0.1–0.25	50–100
Cyperus spp.	Sedges	<0.05–0.50	50–100
Eleocharis spp.	Spikerushes	<0.05–0.5	50–100
Rhynchospora spp.	Beak rush	<0.05–0.5	50–100
Scirpus spp.	Bulrush	0.1–1.5	75–100
Hydrocloa	Watergrass	<0.05–1.0	75–100
Panicum hemitomon	Maidencane	0.1–0.3	50–100
Panicum repens	Torpedo grass	<0.05–0.5	50–100
Phragmites spp.	Common reed	<0.05–0.5	70–100
Zizania aquatica	Wild rice	0.1–1.0	70–100
Zizaniopsis milacea	Southern wild rice	0.1–1.0	70–100
Iris spp.	Blue flag iris	<0.05–0.2	50–100
Juncus spp.	Rushes	<0.05–0.25	50–100
Thalia geniculata	Arrowroot	0.1–0.75	70–100
Pontederia spp.	Pickerelweeds	0.1–0.25	70–100
Sparganium americanum	Bur reed	0.1–0.5	70–100
Typha spp.	Cattails	0.1–0.75	70–100
Alternanthera philoxeroides	Alligator weed	0.1–1.0	70–100
Ludwigia spp.	Water primroses	0.1–0.5	70–100
Polygonum spp.	Smartweeds	<0.05–0.25	50–100
Saururus cernuus	Lizard's tail	<0.05–0.2	50–100
Phalaris arundinacea	Reed canary grass	<0.05–0.30	13–100
Glyceria spp.	Mannagrass	<0.05–0.30	0–100
	Emergent, Herbaceous (Less Than Continuous Inundation)		
Blechnum serrulatum	Swamp fern	<0.05–0.1	30–70
Woodwardia spp.	Chain ferns	<0.05–0.1	30–50
Osmunda spp.	Royal ferns	<0.05–0.1	30–50
Arundinaria gigantea	Cane	<0.05–0.1	<10–20
Paspalum distichum	Knot grass	<0.05–0.1	50–70
Spartina spp.	Cordgrasses	<0.05–0.25	20–70

number of marsh ecosystems (Leck et al., 1989; Pederson, 1981). In some cases, soil from a natural wetland with a seed bank can be used to establish a new constructed wetland. Most seeds in seed banks are annuals, but in some marshes up to 50 percent may be perennials. Numbers of seeds range from <100 to >375,000/m^2 (Leck et al., 1989). A typical freshwater marsh in Manitoba had 4582 seeds/m^2 and 34 species in emergent areas and only 93 seeds/m^2 in open water areas (Pederson, 1981).

MINERAL COMPOSITION OF PLANTS

The mineral composition of plants is important in treatment wetlands for two reasons: (1) uptake and release of chemical compounds by plants changes water quality and (2) because plants are an integral part of wetland treatment systems, their mineral nutrition must be adequate for survival and growth.

Because of their diverse genetic and phenotypic manifestations, plants may have a wide range of average elemental compositions (Boyd, 1978). Also, a single species' mineral composition may vary greatly by nutritional history, age, and physical and chemical environment. A detailed knowledge of plant mineral composition is generally important only in a few wetland systems, where plant harvest is a principal method of nutrient removal and in detailed mass balance research studies which are intended to determine internal component wetland treatment processes.

Table 7-7 summarizes the average mineral composition and percent ash for several plants used in wetlands treatment and provides typical ranges. The nutritional elements that comprise the dry weight of herbaceous, wetland plants are carbon (41 percent), potassium (2.6 percent), nitrogen (2.3 percent), calcium (1.34 percent), sodium (0.51 percent), sulfur (0.41 percent), and phosphorus (0.25 percent). The remaining dry weight comprises oxygen, hydrogen, and trace elements. Ash weight, the amount of mineral weight that remains after the dried plant sample is ignited at about 550°C, generally consists of noncarbonaceous plant compounds. In addition to interspecies variability, there are large differences (factor of 4) in chemical composition of different plant parts, and these may vary greatly with season (factor of 4) (Kühl and Kohl, 1993).

Because of their potentially high biomass, wetland tree species may contain a relatively high percentage of the nutrients stored in forested wetland ecosystems. As can be seen in Table 7-7, the nutrient content of woody tissues in trees is very much less than in leafy tissues and herbaceous plant species. However, since woody tissues make up a higher proportion of the total plant biomass (about 87 percent of total dry weight for bald cypress and 84 percent for black gum [Phillips et al., 1989]), they also contain the greatest mass of nutrients (about 47 to 55 percent of the nitrogen and 58 to 61 percent of the phosphorus) and represent the largest nutrient storage compartment in some natural wetlands.

Reddy and DeBusk (1987) have estimated the maximum nitrogen and phosphorus removals that can be obtained through plant growth, harvest, and removal from wetland treatment systems (Table 7-8). The range of nitrogen standing stock in wetland plants varies from 4 to 50 kg/ha in small, floating plant species (*Lemna*) to more than 1200 kg/ha for herbaceous and woody emergents. For plant species that can be maintained in an exponential growth stage (especially floating aquatics such as water hyacinth and duckweed), higher nitrogen removal rates can be achieved. For plant species that require significant structure for growth, the estimated annual uptake rates in Table 7-8 could not be sustained with harvesting.

Because of its lower concentration in plant tissues, the phosphorus standing stock in wetland plant communities is generally about one tenth that of nitrogen. Phosphorus standing stock ranges from 1 to 16 kg/ha in *Lemna* to 375 kg/ha in a *Typha* marsh.

In spite of the high standing stock of nutrients in woody species in forested wetlands, mass balance studies have demonstrated that this storage accounts for a relatively small portion of the nutrient flux created by municipal wastewater discharge (Dierberg and Brezonik, 1984; Mitsch et al., 1979; Reddy and DeBusk 1987). In the cypress dome study reported by Dierberg and Brezonik (1984), only about 20 percent of the total nitrogen and 1 percent of the total phosphorus contained in the pretreated sewage was actually incorporated into long-term woody storage compartments.

The relatively high nutrient uptake rates in wetland plants and the potential accumulation of these nutrients have caused wetland treatment enthusiasts to suggest that routine harvesting of plant material might optimize nutrient removal potential (Breen, 1990). Nutrient storage in wetland plants is frequently significant, but the regular harvest of wetland plants from treatment systems has not been successful in full-scale applications because of cost and

Table 7-7 Summary of the Average and Range (in parentheses) of Mineral Composition (percent dry weight) of Plants Typically Used in Wetland Treatment Systems

Category	C	K	N	Ca	Na	S	P	Ash
All wetland plants[a]	41.0 (29–50)	2.61 (0.42–4.56)	2.26 (1.46–3.95)	1.34 (0.20–8.03)	0.51 (0.07–1.52)	0.41 (0.11–1.58)	0.25 (0.08–0.63)	14.0 (6.1–40.6)
Typha latifolia	45.9 (43.3–47.2)	2.38 (0.91–4.39)	1.37 (0.86–2.12)	0.89 (0.35–1.62)	0.38 (0.05–1.09)	0.13 (0.05–0.53)	0.21 (0.08–0.41)	6.75 (3.96–10.2)
Hydrocotyle umbellata	—	1.73	2.56	1.85	0.98	0.16	0.18	—
Scirpus americanus	—	2.83	1.22	0.50	0.09	0.59	0.18	—
Juncus effusus	—	0.89	1.24	0.38	0.40	0.26	0.27	—
Pontederia cordata	—	2.58	1.40	0.96	0.83	0.22	0.24	—
Phragmites communis	—	1.47 (1.0–1.7)	2.57 (1.6–4.2)	0.28 (0.21–0.36)	0.17 (0.08–0.26)	—	0.18 (0.16–0.20)	—
Taxodium distichum								
Whole tree	—	0.06	0.14	0.22	—	—	0.01	—
Foliage	—	0.65	1.37	0.78	—	—	0.13	—
Branches	—	0.14	0.34	0.69	—	—	0.04	—
Stem bark	—	0.13	0.46	1.42	—	—	0.03	—
Stem wood	—	0.04	0.09	0.10	—	—	0.01	—
Nyssa biflora								
Whole tree	—	0.10	0.13	0.21	—	—	0.01	—
Foliage	—	0.39	1.44	0.64	—	—	0.12	—
Branches	—	0.15	0.28	0.41	—	—	0.03	—
Stem bark	—	0.16	0.34	0.99	—	—	0.03	—
Stem wood	—	0.08	0.07	0.11	—	—	0.01	—

[a] Remaining major elements include oxygen (44%), hydrogen (6%), and magnesium (0.1–0.8%).

Note: Data from Boyd, 1978; Prentki et al., 1978; Reddy and DeBusk, 1987; Raven et al., 1981; Phillips et al., 1989.

Table 7-8 Nutrient Removal Potential of Wetland Plant Species Used in Wetland Treatment Systems

Species	Biomass		Nitrogen		Phosphorus	
	Stock (kg/ha)	Growth (kg/ha/year)	Stock (kg/ha)	Uptake (kg/ha/year)	Stock (kg/ha)	Uptake (kg/ha/year)
Typha	4,300–22,500	8,000–61,000	250–1,560	600–2,630	45–375	75–403
Juncus	22,000	53,300	200–300	800	40	110
Scirpus	6,000	7,130	175–530	125	40–110	18
Phragmites	6,000–35,000	10,000–60,000	140–430	225	14–53	35
Eichhornia crassipes	20,000–24,000	60,000–110,000	300–900	1,950–5,850	60–180	350–1,125
Pistia stratiotes	6,000–10,500	50,000–80,000	90–250	1,350–5,110	20–57	300–1,100
Hydrocotlye	7,000–11,000	30,000–60,000	90–300	540–3,200	23–75	130–770
Alternanthera philoxeroides	18,000	78,000	240–425	1,400–4,500	30–53	175–570
Lemna minor	1,300	6,000–26,000	4–50	350–1,200	1–16	116–400
Salvinia	2,400–3,200	9,000–45,000	15–90	350–1,700	4–24	92–450
Cypress swamp			996–1,219	10.4–213	25–86	3.3–23
Hardwood swamp			815		86	0.7–8.7

Note: Data from Reddy and DeBusk, 1987.

sustainability. For phosphorus removal, floating aquatic plant systems have been designed to harvest vegetation without disrupting plant growth and water treatment (USEPA, 1988).

PRIMARY PRODUCTIVITY

Primary productivity is the fixation of solar energy by plants in the form of organic chemical bonds. This energy fixation provides an energy source for metabolism and growth of the plants and subsequently provides an energy source for heterotrophs which consume these plant tissues. Photosynthesis is the biological process that captures solar energy in chemical bonds. The general stoichometric equation for photosynthesis is

$$6CO_2 + 12H_2O + \text{sunlight} \xrightarrow[\text{enzymes}]{\text{chlorophyll}} C_6H_{12}O_6 + 6H_2O \tag{7-11}$$

Photosynthesis requires solar energy and the chemical building blocks provided by CO_2 and H_2O. It also requires plant growth nutrients and micronutrients, plant pigments including chlorophyll, metabolic enzymes, and plant structures to bring together these essential ingredients. The plant kingdom shows the multitude of adaptations that nature has made to accomplish this fairly complex process.

Several photosynthetic processes have evolved to provide competitive advantages to various plant groups. The two most common photosynthetic processes in higher plants are the C_3 and C_4 pathways, which refer to the number of carbon atoms in the first step in the carbon fixation process. In C_3 plants, the first stable organic compound resulting from CO_2 fixation is a three-carbon compound, phosphoglycerate. In C_4 plants, CO_2 is fixed to form oxaloacetate, a four-carbon organic acid. The more highly evolved C_4 process gives plants an advantage because they are not limited by typical ambient atmospheric concentrations of O_2 and CO_2 dioxide and are not as susceptible to photosynthetic saturation at high light intensities. Many wetland plants, including cattails, bulrushes, duckweed, water hyacinth, and soft rush, are C_3 plants. Examples of wetland C_4 plants include pigweed (*Amaranthus* spp.) and torpedograss (*Panicum repens*). Some floating leaved and submerged plant species are known to be able to increase photosynthesis by using bicarbonate ion in the C_4 pathway (Madsen and Sand-Jensen, 1991).

In spite of their use of the less efficient carbon fixation method, several aquatic C_3 plants attain high rates of photosynthesis. This high productivity appears to be related to the almost unlimited supply of water and growth nutrients available to plants growing directly in water and to morphological adaptations such as leaf area and structural position (Bowes and Beer, 1987).

Ecosystem primary productivity is a complicated process of photosynthesis by one or more plant species. Most wetland ecosystems contain numerous species of photosynthetic plants including algae and vascular plants. Thus, ecosystem primary productivity is the collective photosynthesis of multiple species competing for a finite resource base available in the wetland. The total carbon fixed in this photosynthetic process is called the gross primary productivity. However, some of this fixed energy is expended as respiration for cell maintenance and growth during the process of photosynthesis in the daytime and nighttime. Net primary productivity refers to the difference between gross primary productivity and this metabolic respiration rate. Definitions of gross and net primary productivity vary, and methodologies for their estimation may actually measure different processes. Figure 7-15 illustrates some details of these metabolic processes and how they are commonly defined in wetland ecosystems.

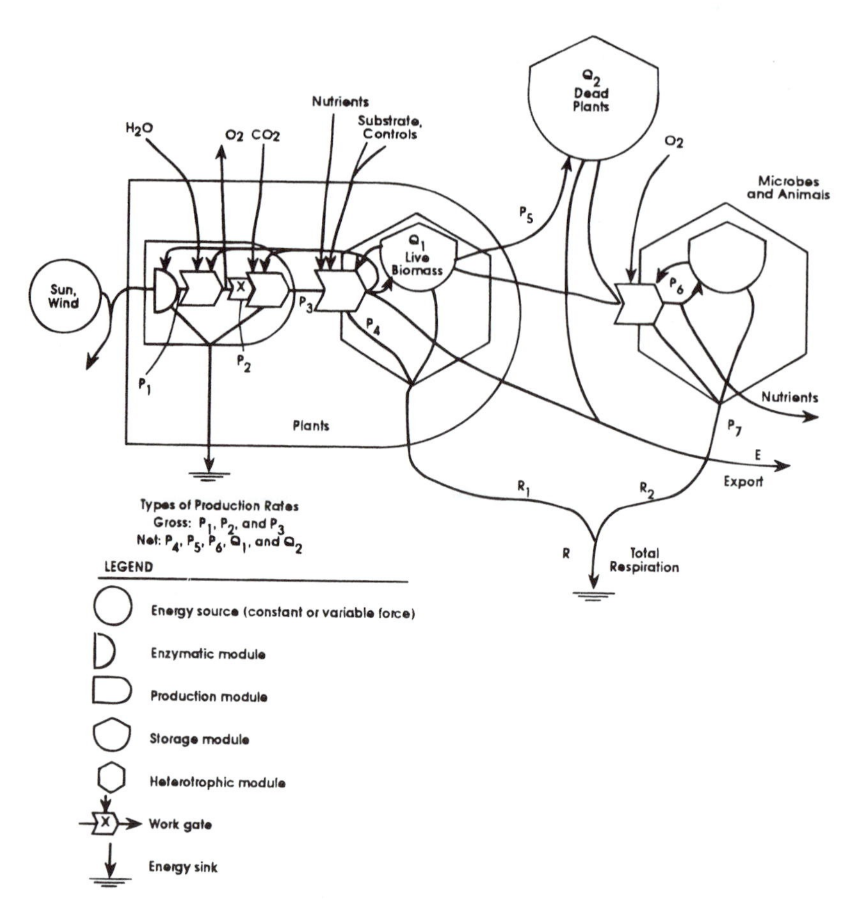

Figure 7-15 Diagram of generalized wetland photosynthesis and respiratory pathways using Odum's energy circuit symbols. (Adapted from Odum, 1978).

Net primary productivity (NPP) can be estimated instantaneously, varying from positive during daylight to negative at night, or more frequently, it is estimated on a daily, seasonal, or annual basis. During any period when metabolism outpaces photosynthesis and NPP is negative, cell and plant resources are declining. During an extended period of negative NPP, individual plant populations and plant communities are declining and eventually will die. In a stationary ecosystem that does not accumulate additional plant material from one period to another, the gross primary productivity is equal to respiration and the NPP is equal to zero. If organic matter is accumulating as is frequently the case in wetlands, then the NPP is clearly positive.

Wetland ecosystems are often sites of long-term positive NPP and develop in accumulations of buried organic matter in the form of peat and eventually coal. This net accumulation of organic matter is primarily because of the reduced metabolic rate of microbes in flooded wetland sediments compared to metabolic rates in well-aerated, upland soils. When living and dead plant material sinks to the level of anaerobic sediments, it is protected from abundant free oxygen and from the higher rates of degradation typical of an oxygenated system.

Reduced metabolism of dead organic matter in wetlands may also result from the relatively low nutrient content of water in many natural wetlands that depend on rainfall for water. Under low-nutrient conditions, microbial metabolism may be reduced because of a lack of essential growth nutrients rather than because of a shortage of available oxygen. Under high-nutrient conditions such as in wetland treatment systems, increased primary productivity may not result in increased net production of peat or increased export downstream because of the increased metabolism of the wetland microbes.

Net primary productivity of freshwater marshes is estimated most frequently through the harvest of annual peak standing stocks of live and dead plant biomass. When root biomass is measured, it is usually an important part of net annual plant production. Some researchers consider NPP estimates by peak standing stock to be underestimates because they do not account for biomass turnover during the growing season (Pickett et al., 1989). Table 7-9 summarizes some typical estimated net production data from wetland ecosystems. The range of net production rates in natural wetlands which are not subject to obvious anthropogenic nutrient enrichments vary from about 50 g/m^2/year in arctic tundra to 3500 g/m^2/year, in southeastern marshes. Most temperate freshwater marshes have net primary production rates of 600 to 3000 g/m^2/year.

Figure 7-16 illustrates the typical pattern of living and standing dead biomass accumulation in a freshwater marsh dominated by *Zizaniopsis miliacea, Pontederia cordata,* and *Alternanthera philoxeroides.* Total live biomass peaks during the late growing season and is followed by a fall peak in standing dead biomass and a winter peak in fallen plant litter in the marsh. There is essentially no change in the live biomass component annually, and thus, the annual NPP is much lower than the seasonal maximum productivities estimated in Table 7-9. Long-term net productivity in wetlands can be estimated by measuring peat accretion (including net increases in root biomass), the increase in woody biomass, and system export. In northern regions, the seasonal growth period is obviously much shorter than that in Figure 7-16, as is the decline of live biomass in fall. Standing dead biomass increases quickly during northern fall senescence and is typically collapsed into the water by the weight of snow.

Primary productivity of wetland plants is increased by the availability of water, light, and nutrients. Adding wastewater to wetlands generally increases the availability of water and nutrients and consequently results in the stimulation of gross and net primary productivity

Table 7-9 Net Primary Production Estimates From A Variety of Wetland Ecosystems

Description	Aboveground Annual NPP (g dry wt/m^2/year)	Ref.
Typha spp. (freshwater tidal)	956–1868	Whigham et al., 1978
Typha spp. (brackish tidal)	626–1668	Whigham et al., 1978
Typha spp. (managed stands)	1697–6941	Pratt and Andrews, 1981
Typha angustifolia, TX	2560–2895	Hill, 1987
Typha domingensis, FL	1077–3035	Davis, 1989
Typha spp. (temperate wetlands)	754–2760	Davis, 1989
Phragmites australis (freshwater tidal)	1367–2066	Whigham et al., 1978
Phragmites australis, DE	3550	Kibby, 1979
Phragmites australis, MS	2330	Kibby, 1979
Pontederia/Peltandra (freshwater tidal)	650–1126	Whigham et al., 1978
Zizania aquatica (freshwater tidal)	560–1589	Whigham et al., 1978
Acorus calamus (freshwater tidal)	605–1071	Whigham et al., 1978
Sagittaria latifolia (freshwater tidal)	214–1071	Whigham et al., 1978
Northern bog marshes	101–1539	Reader, 1978
Cladium jamaicense, FL	802–2080	Davis, 1989
Freshwater tidal marsh, SC	2562	Pickett et al., 1989
Freshwater tidal marsh, NJ	2346	Whigham et al., 1978
Freshwater marsh receiving municipal effluent, FL	1460	Dolan et al., 1978
Swamp forest, VA	1227	Fowler and Hershner, 1989
Swamp forest	764–1074	Brown et al., 1979
Cypress/tupelo swamp forest, IL	1963	Mitsch, 1979
Cypress/tupelo swamp forest, LA	1140–1516	Conner and Day, 1976
Swamp forest, GA	692	Schlesinger, 1978
Cypress dome receiving municipal effluent, FL	1530	Odum et al., 1977

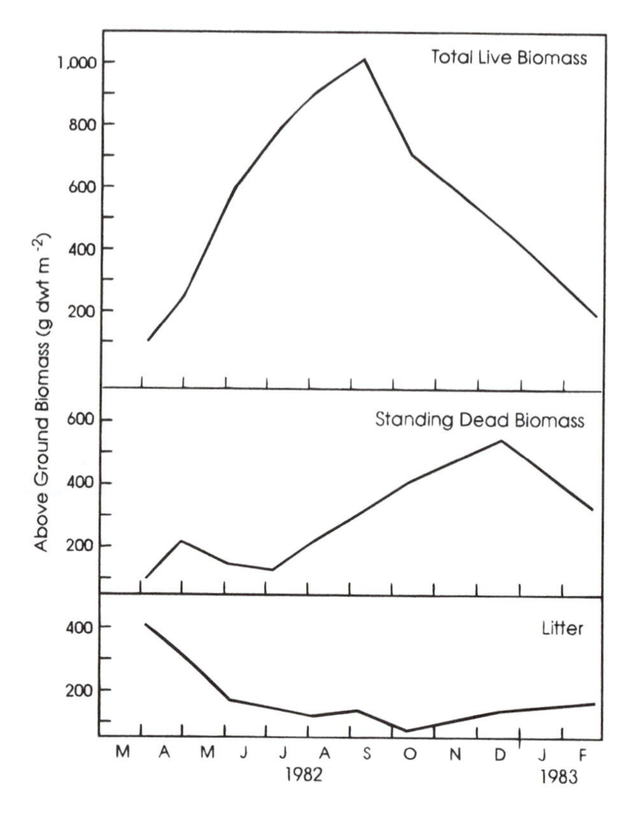

Figure 7-16 Typical seasonal pattern of biomass and litter in a mixed herbaceous tidal marsh in South Carolina. (Modified from Pickett, et al., 1989).

of these ecosystems (Guntenspergen and Stearns, 1981; Reader, 1978; Nixon and Lee, 1986; Knight, 1992a).

Water is a subsidy to primary productivity as long as it does not limit the availability of other essential ingredients in the photosynthetic process. As described earlier, all wetland plants have an optimal range of hydroperiods and water depths. To maximize productivity, wastewater additions must be controlled within this optimum range.

Nutrients also affect wetland plant growth. Figure 7-17 illustrates the typical plant growth response curve to increased concentrations of nitrogen and phosphorus. The maximum rate

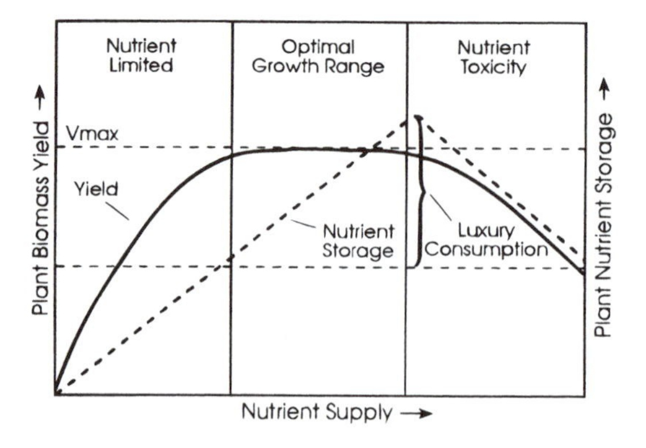

Figure 7-17 General relationship between plant biomass yield (solid line) and nutrient accumulation (dashed line) as a function of nutrient concentration in the water column. (Adapted from Reddy and DeBusk, 1987).

of plant growth is attained as nutrient levels are initially increased. However, at higher nutrient levels, plant growth levels off, while luxury nutrient uptake continues, and at higher nutrient concentrations, phytotoxic responses are observed. Figure 7-18 illustrates the effect of nitrate N on the growth of cattails; similar results occur for ammonium nitrogen and phosphorus.

Different wetland plant species have varying adaptations to optimize their use of sunlight for primary production. For instance, the response of light-adapted species to increasing light energy is to increase gross productivity, while shade-adapted species have higher productivity at lower light intensities. Data from two types of saltmarsh wetlands in Florida illustrate that these species have slightly different efficiencies when using light and that the ecosystem adapts to seasonally higher and lower light intensities by shifting its photosynthetic response (Figure 7-19).

LITTERFALL AND DECOMPOSITION

Over the life cycle of a vascular plant, all plant tissues are either consumed, exported, or eventually recycled back to the ground as plant litter. Litterfall and the resulting decomposition of organic plant material is an ecologically important function in wetlands.

Wetland plant tissues fall at variable rates depending on the survival strategy of the individual plant species. Herbaceous plant species typically recycle the entire above-ground portion of the plant annually in temperate environments (Figure 7-16). The growth season may vary from 10 or more months in subtropical regions to less than 3 months in colder climates. Also, most herbaceous species lose a fraction of living leaf and stem material as litter throughout the growing season, so there is a continuous rain of dead plant tissues throughout the year with seasonal highs and lows of litterfall.

Woody plant species also participate in this production of plant litter through a natural pruning of small branches throughout the annual period. In the northern hemisphere, large amounts of flowers are shed during the spring, and leaves and fruiting bodies are lost during

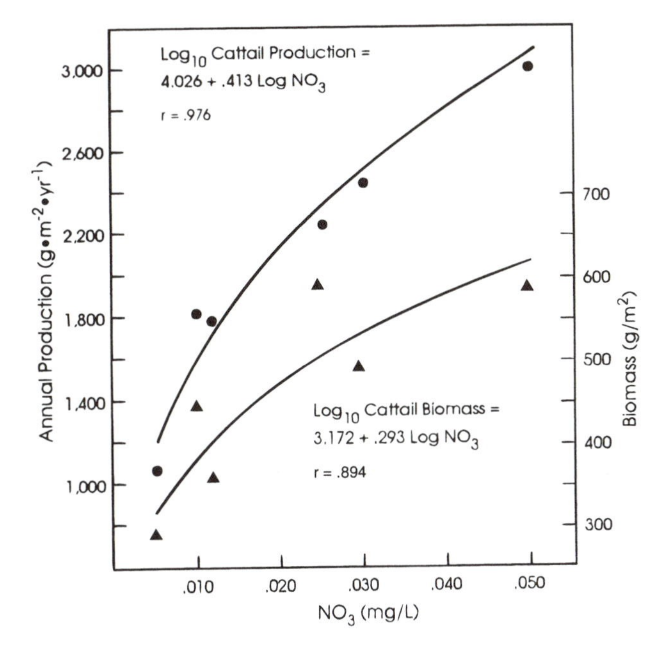

Figure 7-18 Example of the effect of a plant growth nutrient (nitrate N) on growth of a wetland emergent, *Typha domingensis* (Adapted from Davis, 1989).

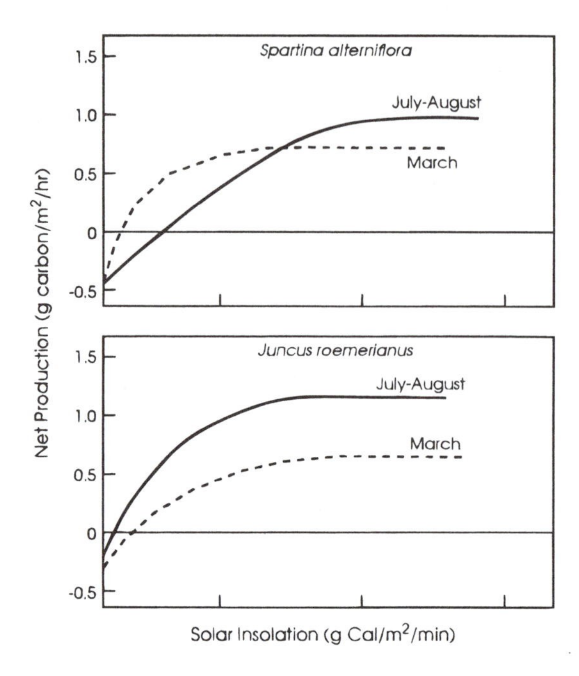

Figure 7-19 Relationship between solar insolation and net production of *Spartina* and *Juncus* salt marshes. (Adapted from Odum, 1978).

the fall. Figure 7-20 illustrates the overall litter production rates in a forested wetland in South Carolina receiving municipal effluent.

Plant litter production is a direct measure of the macrophyte component of NPP. Just as NPP varies widely among different wetland ecosystem types, so does plant litter production vary among ecosystem types. Table 7-10 summarizes litterfall rates for a number of typical wooded wetland ecosystems on an annual basis.

Plant litter contributes directly to the amount of organic matter a wetland treatment system must process and the resulting oxygen consumption and ambient dissolved oxygen levels in the wetland. The process of litter decomposition is sequential, depending on the starting point of the litter (woody trunk, woody branch, twigs, herbaceous stems, herbaceous leaves, coriaceous leaves, flowers, or fruits). Litter decomposition is largely mediated by vertebrates, invertebrates, and microbes living in wetlands. New litter is typically conditioned by fungi and bacteria for several weeks before it is shred into smaller particles by aquatic macroinvertebrates (Merritt and Lawson, 1979). The organisms important in litter decomposition in wetlands are discussed elsewhere in Chapters 7 and 8.

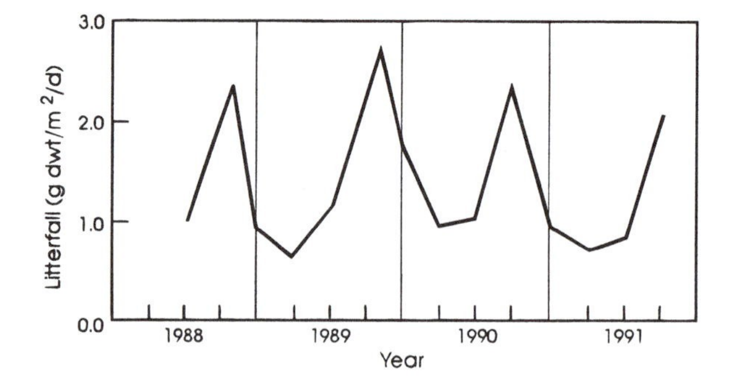

Figure 7-20 Seasonal pattern of litterfall in a mixed hardwood swamp forest receiving municipal wastewater in South Carolina.

Table 7-10 Litterfall Rates in a Variety of Forested Wetland Ecosystems

Description	Litterfall (g dry wt/m²/yr)	Ref.
Cypress/tupelo swamp forest, IL	348	Mitsch, 1979
Cypress/tupelo swamp forest, LA	620	Conner and Day, 1976
Cypress swamp forest, GA	310	Schlesinger, 1978
Mixed hardwood swamp forest receiving municipal effluent, SC	508	CH2M HILL, 1992
Mixed hardwood swamp forest, VA	252	Fowler and Hershner, 1989
Shrub birch-willow peatland, MI	90	Chamie, 1976
Black mangrove swamp forest, FL	485	Lugo and Snedaker, 1974

Litter decomposition can be measured with litter decomposition bags. Nylon mesh bags are filled with plant material and suspended in the wetland at various sites. Litter bags are collected after varying time intervals, oven or air dried, and reweighed to measure the weight loss per unit time. Detailed litter decomposition study methods used in wetlands research are summarized by Chamie and Richardson (1978), Davis and van der Valk (1978), and Odum and Heywood (1978).

Litter decomposition can be monitored in a wetland treatment system using a standard plant material. This method provides a normalized rate of litter decomposition during differing seasonal periods or between years. Figure 7-21 summarizes different decomposition rates for the leaves of an evergreen shrub, fetterbush (*Lyonia lucida*), in a natural wetland used for wastewater management in the coastal plain of South Carolina. Research has concluded that litter decomposition is accelerated by aerated standing water and varies seasonally in response to air temperature (Godshalk and Wetzel, 1978). Flooding in wetlands has been found to increase the litter decomposition rate through physical leaching of inorganic and organic compounds from the plant tissues (Day, 1989; Whigham et al., 1989) and by providing habitat for aquatic microbes and invertebrates which are important mediators in this process. However, if flood waters are anaerobic, biological activity is greatly reduced (Tupacz and Day, 1990) and only the leaching mechanisms and anaerobic respiration will occur. The differential mobility of mineral elements during the leaching process is typically K>Mg>P>N>Ca (Chamie and Richardson, 1978).

Litter decomposition rates for plant species and tissues from the same species vary widely because of different structure and chemical composition. Table 7-11 summarizes some of the reported first-order decay rates estimated for wetland plant litter and provides the calculated halflife for these litter types. Typical decay rates for the leaves of wetland vegetation

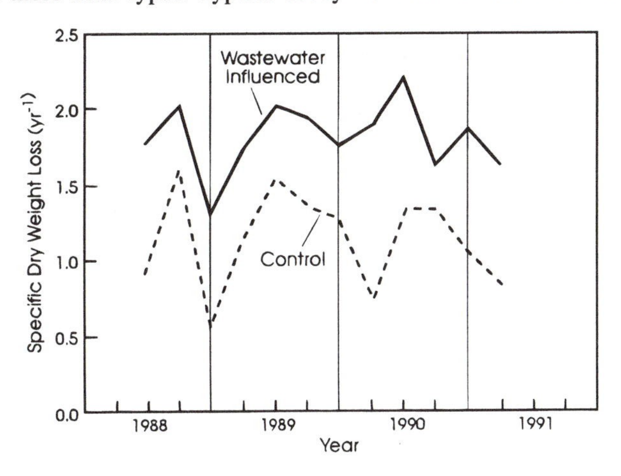

Figure 7-21 Seasonal variation in decomposition rate of *L. lucida* leaves in a South Carolina wetland receiving municipal effluent.

vary based on nutrient recycling strategy and have been measured from 0.24 to 25 $year^{-1}$ with halflives from 10 to 1054 days. Twigs, branches, and roots of woody species have lower decay rates (0.18 to 0.30 $year^{-1}$) and halflives from 2.3 to 3.8 years.

SUMMARY

Structurally and functionally, most wetlands are dominated by their populations of microbes and plant life. A general understanding of the nature of these components is important in the design and operation of wetlands used for water quality treatment. A detailed knowledge about the ecology of these organisms is essential for refining this new technology and for solving existing problems.

Microbial populations in wetlands include the diverse flora of bacteria, fungi, and algae that are important for nutrient cycling and pollutant transformations. The variety of microbial species in wetlands function in a wide range of physical and chemical conditions. Because of this variety of species and the niches they occupy, wetland ecosystems can operate consistently to treat water. Since many of these organisms are the same as those important in conventional treatment systems, their growth requirements and characteristics are known to many engineers. Overall ecosystem parameters such as DO, water temperature, and influent constituent concentrations must be controlled through design and system operational control to keep this microbial symphony in tune.

Macrophytic plants provide much of the visible structure of wetland treatment systems. There is no doubt that they are essential for the high quality water treatment performance of most wetland treatment systems. The numerous studies measuring treatment with and without plants have concluded almost invariably that performance is higher when plants are present. This finding led some researchers to conclude that wetland plants were the dominant source of treatment because of their direct uptake and sequestering of pollutants. It is now known that plant uptake is the principal removal mechanism only for some pollutants and only in lightly loaded systems. During an initial successional period of rapid plant growth, direct pollutant immobilization in wetland plants may be important. For many other pollutants, plant uptake is generally of minor importance compared to microbial and physical transformations that occur within most wetlands. Macrophytic plants are essential in wetland treatment

Table 7-11 Estimated Plant Litter Decomposition Rates in Wetland Ecosystems

Description	Decay Rate ($year^{-1}$)	Half Life (days)	Ref.
Cypress needles	0.28–0.59	429–903	Day, 1989
Maple/gum leaves	0.29–0.67	378–872	Day, 1989
Mixed hardwood leaves	0.24–0.42	602–1054	Day, 1989
Maple branches	0.18–0.30	843–1405	Day, 1989
Herbaceous marsh species	1.86–19.5	13–136	Whigham et al., 1989
Herbaceous marsh species	25.2	10	Odum and Heywood, 1978
Cattail and bulrush	0.48–1.12	225–525	Davis and van der Valk. 1978
River ash leaves, MI	1.05–1.41	180–240	Merritt and Lawson, 1979
Fetterbush leaves, SC	1.50	169	CH2M HILL, 1992
Wetland tree roots, VA	0.175–0.365	693–1444	Tupacz and Day, 1990
Black mangrove leaves, FL	0.52	486	Lugo and Snedaker, 1974
Willow leaves, MI	0.459	551	Chamie and Richardson, 1978
Willow stems, MI	0.182	1391	Chamie and Richardson, 1978
Sedge, MI	0.448	566	Chamie and Richardson, 1978
Water hyacinth leaves, India	7.66	33	Gaur et al., 1992
Common reed leaves, Austria	0.359–1.125	225–705	Hietz, 1992
Common reed stalks, Austria	0.135–0.171	1479–1874	Hietz, 1992
Tidal marsh species, MA	0.99–1.40	180–255	Morris and Lajtha, 1986

systems because they provide structure for the microbes that mediate most of the pollutant transformations that occur in wetlands.

The diversity of wetland plant adaptations provides the wetland treatment system designer with numerous options and potential problems. Some plant species produce large amounts of carbon that are able to support heterotrophic microbes important in nutrient transformations. Other plant species provide shading of the water surface, in turn controlling algal growth and suspended solids levels in the discharge from the wetland treatment system. Many wetland plant species cannot withstand continuous inundation, preventing the use of certain natural wetlands for water quality treatment. An understanding of the ecological properties of these wetland plant species is essential for successful wetland treatment system design, construction, and operation.

CHAPTER 8

Wetland Wildlife

INTRODUCTION

Animals frequently focus and control energy flows in wetland ecosystems (Figure 8-1). Without heterotrophic animals and microbes (consumers), plant populations would achieve a high biomass and productivity based on available nutrients and then stop growing and die, similar to the microbial growth curve illustrated in chapter 7, Figure 7-2. Fortunately, consumers recycle the limiting elements necessary to sustain a continuous, high level of primary productivity.

In natural wetlands, as in most naturally occurring ecosystems, the relative populations of producers and consumers remain balanced through highly complex biological and environmental control mechanisms (Odum, 1978). Through ecosystem-level adaptations, consumers have evolved feedback control behaviors to regulate their food or prey populations at optimal levels through each link in the food chain. The populations of top consumers (generally large, carnivorous species) are in turn controlled by environmental factors such as temperature, storms, floods, fires, and other "catastrophes" (Figure 8-2).

Because of the energy-focusing aspects of the food chain in wetlands and other ecosystems, wildlife populations are biological "integrators" of ecosystem structure and function. These wildlife species frequently provide a convenient focus for human attention on the health and value of wetland ecosystems. Larger wildlife species such as fish, birds, herptiles (reptiles and amphibians), and mammals have a variety of commercial and aesthetic uses in society. Wetlands provide habitat for a vast array of animal species, some of which are rare or do not occur in other habitats. For example, it has been estimated that about 900 species of wildlife in the U.S. require wetlands as habitat for a significant portion of their life cycle (Feierabend, 1989).

As discussed in Chapter 7, because of ample water, wetlands are among the most productive ecosystems on the earth. Much of this primary productivity is used in the secondary production of animal populations. Using wetlands to treat wastewaters may stimulate wildlife production and typically results in both a change in species diversity and in numbers of organisms. In the long term, diversity in heavily loaded treatment wetlands is likely to be lower than for comparable natural wetlands. In a growing number of applications of the wetland treatment technology, wildlife enhancement is an important secondary goal. In other systems designed to treat potentially toxic industrial wastewaters, exclusion of certain wildlife species may be an important consideration during planning and design. Wildlife species such as destructive insects, burrowing mammals (nutria and muskrats), bottom-feeding fish, and high bird populations (with associated nutrient enrichment) have been found to be detrimental to the operation of wetland treatment systems. A basic understanding of the types of wildlife that occur in treatment wetlands and the conditions necessary to either favor or discourage their occurrence is important for project success.

This chapter provides an introduction to the diverse wildlife populations that occur in wetlands, including invertebrates, fish, herptiles, birds, and mammals. The range of diversity

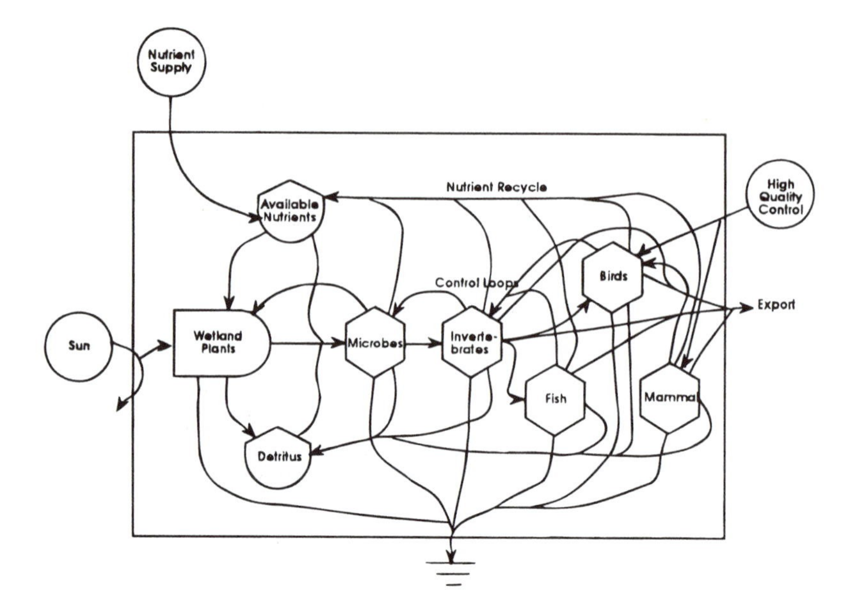

Figure 8-1 Generalized model of a wetland food chain illustrating consumer control through nutrient recycle, feed forward control, and feedback control. (See Figure 7-15 for an explanation of symbols.)

for each of these groups, their typical population densities, and the environmental factors that influence these populations are briefly summarized. References are provided for more detailed background information on each group.

INVERTEBRATES

Most aquatic invertebrate groups are represented in wetland ecosystems. The classification of these invertebrate groups and their general characteristics are summarized in Table 8-1. The invertebrate types that occur most frequently in wetlands include protozoans, aschelminthes, and aquatic arthropods. The ecology of each of these groups is briefly described.

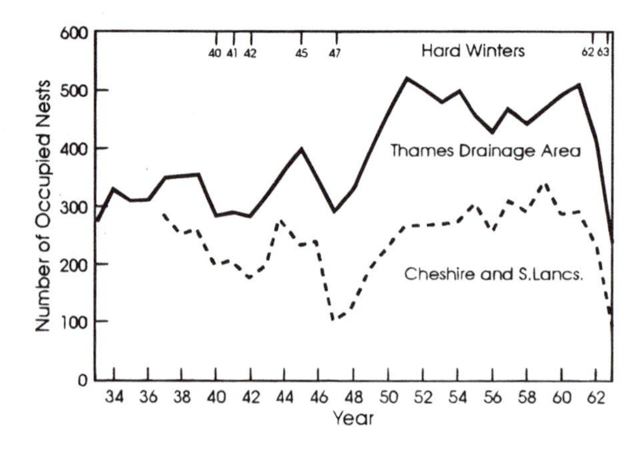

Figure 8-2 Example of apparent regulation of top consumer populations by abiotic control (global weather system). Changes in heron (*Ardea cinerea*) populations in two areas of Great Britain (Adapted from Lack, 1966).

Table 8-1 Phylogenetic Classification of Invertebrate Animals Characteristic of Freshwater Wetlands

Phylum	Class
Protozoa (unicellular)	Mastigophora (flagellates)
	Sarcodina (amoeboid forms)
	Ciliata (ciliates)
Porifera (sponges)	Demospongea (bath sponges)
Coelenterata (hydroids)	Hydrozoa (hydra)
Platyhelminthes (flatworms)	Turbellaria (planaria)
Aschelminthes (cavity worms)	Gastrotricha (gastrotrichs)
	Rotifera (rotifers)
	Nematoda (nematodes)
Tardigrada (water bears)	Heterotardigrada
	Eutardigrada
Bryozoa (moss animalcules)	Gymnolaemata
	Phylactolaemata
Annelida (segmented worms)	Oligochaeta (aquatic worms)
	Hirudinea (leeches)
Arthropoda (arthropods)	Arachnida (spiders, water mites)
	Crustacea (crayfish, water fleas, ostracods, copepods, isopods)
	Insecta (aquatic insects)
Mollusca (molluscs)	Gastropoda (snails)
	Pelecypoda (clams)

PROTOZOANS

Protozoans are unicellular animals that represent a multitude of forms and species in aquatic and wetland environments (Figure 8-3). Protozoans can live in colonies, but they also

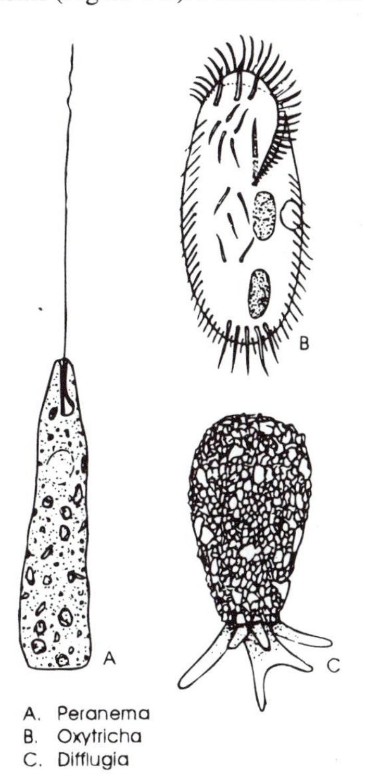

Figure 8-3 Representative protozoan groups that occur in wetlands (From *Fresh-Water Invertebrates of the United States,* R.W. Pennak. Copyright © 1978 John Wiley & Sons, Inc. Reprinted by permission of John Wiley & Sons, Inc.)

function as individual organisms and commonly reproduce by binary fission. Functionally, protozoans play an important, though microscopic, role in nearly every microhabitat within wetlands. Individual species are photoautotrophs, discussed in Chapter 7 (*Euglena, Chlamydomonas, Volvox*), and heterotrophs. Heterotrophic species include holozoic species that ingest a variety of particulate foods (including other protozoans) and saprozoic species that absorb dissolved nutrients and salts. Protozoans can be predaceous or parasitic on other animal species. Because of their variety of functional adaptations, protozoans are important in many chemical and particulate transformations that occur in wetland treatment systems.

Protozoan species are typically microscopic (30 to 300 μm), but may occur in macroscopic "blooms" under specific environmental conditions in wetlands. After a dry wetland is flooded, resulting in increased decomposition of dead plant material, various protozoan forms may be observed along a temporal gradient, starting with small flagellates, paramecia, and amoeba. Although most protozoan species favor aerobic conditions, as many as 90 percent of the freshwater species can survive at dissolved oxygen concentrations as low as 10 percent of saturation (Pennak, 1978). Anaerobic and sewage environments also typically have an assemblage of protozoan species including genera such as *Euglena, Peranema, Paramecium, Amoeba, Vorticella,* and *Bodo.*

In contrast, disease-causing protozoans such as *Giardia* spp., *Cryptosporidium* spp., and *Entamoebae* spp. may enter the wetland with wastewater. Not only the organisms, but their cysts are of concern, especially in warm climates. Experience has shown that their numbers are reduced upon contact with the wetland environment (Rivera et al., 1994).

Ecologically, protozoa function as primary producers, herbivores, carnivores, and omnivores and facilitate nutrient cycling in wetlands. They are an important link in the food chain, leading to the small crustaceans and insects in wetlands and aquatic habitats. Protozoan populations are significantly reduced during periods of low temperature, but species diversity does not appear to be sensitive to winter temperatures (Pennak, 1978). Optimum temperatures for protozoan growth are between 10 and 20°C. Most protozoan species are adapted to an optimum pH range between 6.5 and 8.0 (Pennak, 1978). Protozoan ecology is described in greater detail in Pennak (1978) and Noland and Gojdics (1967).

Typically, protozoans are monitored by suspending polyurethane foam substrates in aquatic habitats and allowing natural colonization to occur (Cairns and Yongue, 1974). Following incubation, the foam cubes are removed from the water and the entrapped water and invertebrates are squeezed into a jar for concentration, identification, and enumeration. Also, the researcher may wish to examine natural substrates, including plant surfaces, detritus, and sediments, to observe the natural diversity and abundance of protozoan species.

ASCHELMINTHES

Aschelminthes (cavity worms) include important groups of wetland invertebrates such as gastrotrichs, rotifers, and nematodes (Figure 8-4).

Gastrotrichs, which range in size from 100 to 300 μm, are found in puddles, marshes, wet bogs, and the littoral zone of ponds. Frequently, they are found in environments with abundant detritus and decaying organic matter and where the dissolved oxygen content is less than 1 mg/L. Gastrotrichs are browsers that feed on bacteria, algae, detritus, and small protozoans. In turn, gastrotrichs are prey for amoeboid protozoans, hydras, oligochetes, and nematodes.

Rotifers occur in a variety of wetland and aquatic environments. They are microscopic organisms, typically 100 to 500 μm, that are sessile on substrates or mobile through the water column. Rotifers feed on detritus, algae, and living invertebrates. As with all invertebrates, rotifers are an important part of the heterotrophic food chain in aquatic and wetland ecosystems. They function as controllers of algal and protozoan populations and as nutrient

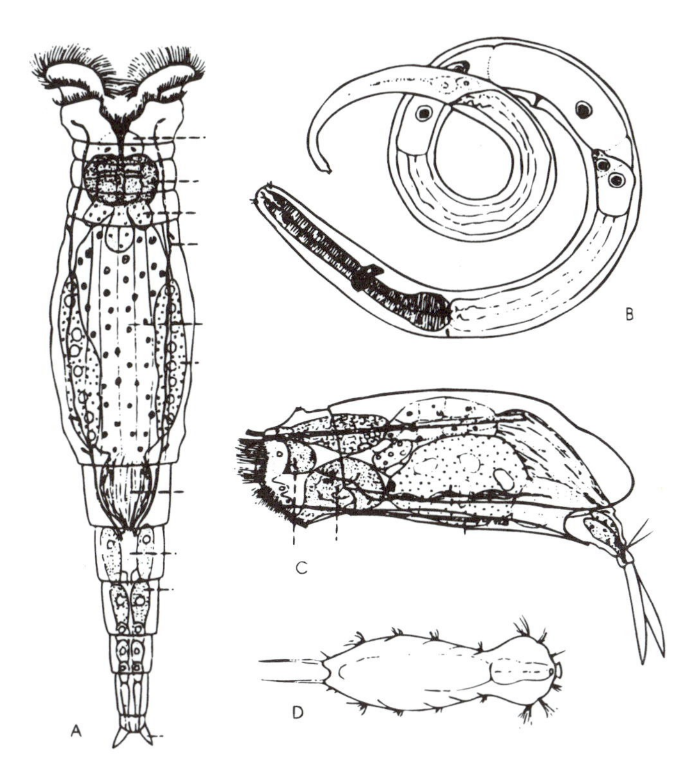

Figure 8-4 Representative aschelminthes that occur in wetlands: (A) bdelloid rotifer, *Philodina;* (B) nematode, *Achromadora;* (C) loricate rotifer, *Euchlanis;* and (D) gastrotrich, *Dasydytes* (From *Fresh-Water Invertebrates of the United States*, R.W. Pennak. Copyright © 1978 John Wiley & Sons, Inc. Reprinted by permission of John Wiley & Sons, Inc.)

recyclers, and by cropping the outer layer of attached microbial growth, they help stimulate oxygen penetration in fixed-film wastewater treatment systems (WPCF, 1990a).

Nematodes are narrow, worm-like organisms that are generally less than 1 cm in length. Nematodes comprise more than 4000 freshwater species. They have a variety of food preferences and function as saprovores, herbivores, and carnivores. Also, many species are parasitic on plants and animals. Nematodes may be found in large numbers in wetland soils, and dense populations of genera including *Mononchus, Ironus,* and *Tripyla* occur in sewage treatment facilities. Although adult nematodes can only withstand anaerobic conditions for a few weeks at a time, their eggs are highly resistant to anerobiosis, desiccation, and freezing.

Some nematodes, for example, *Ascaris* spp., present human health problems, particularly in warm climates. Their eggs may enter a treatment wetland. Experience has shown wetlands reduce the number of eggs (Rivera et al., 1994).

AQUATIC ARTHROPODA

Arthropoda include most wetland invertebrate species including the Class Crustacea (cladocerans, ostracods, copepods, isopods, and decopods) and the Class Insecta which includes 11 aquatic orders. Arthropods are "macroinvertebrates" because they can be observed by the unassisted eye. They are used as an indicator of the physical and chemical environment of aquatic and wetland environments. As numerous studies have shown, these macroinvertebrates are sensitive to a variety of natural and polluted water conditions. For that reason, the "macroinvertebrate species diversity index" is the most widely used biological criterion for pollution control.

The Shannon index of general diversity is defined as (Odum, 1971)

$$\overline{H} = -\sum \left(\frac{n_i}{N}\right)\log\left(\frac{n_i}{N}\right) \tag{8-1}$$

where n_i = number of individuals or biomass of species i
N = total number of individuals or total biomass for all species

Shannon diversity is commonly reported as $\log_{10}$, $\log_e$, or $\log_2$ (bits per individual).

Class Crustacea includes microscopic to macroscopic invertebrates that are found in aquatic environments, breath through gills or the body surface, have two pairs of antennae, and have jointed appendages on most body segments. Representative crustacea that occur in wetland environments are shown in Figure 8-5 and include the cladocerans (water fleas), ostracods, copepods, isopods, and decopods (crayfish).

Cladocerans typically range in size from 0.2 to 3 mm and are enclosed in a folded shell which is open ventrally. Cladocerans feed on detritus including bacterial and algal components

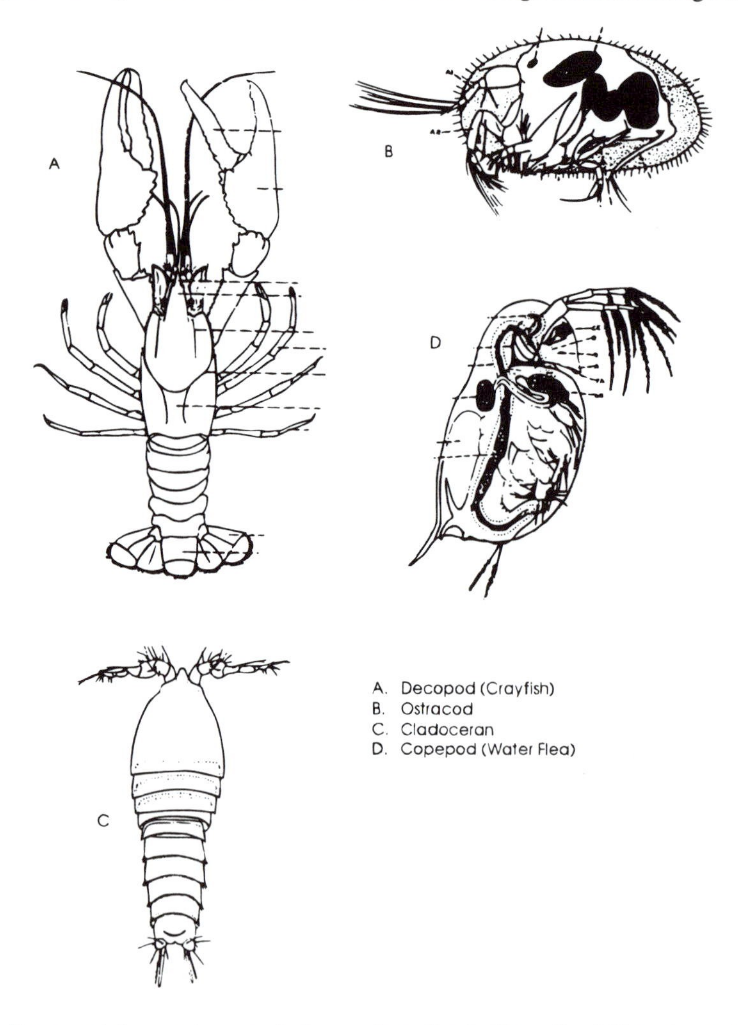

Figure 8-5 Representative crustaceans that occur in wetlands (From *Fresh-Water Invertebrates of the United States*, R.W. Pennak. Copyright © 1978 John Wiley & Sons, Inc. Reprinted by permission of John Wiley & Sons, Inc.)

as well as on smaller invertebrates such as protozoa. Many species of cladocerans can live in low-oxygen conditions typical of wetlands (less than 1 mg/L) and are tolerant of pHs from 6.5 to 8.5. A few species of cladocerans are of great importance to wastewater treatment technology because they are used to measure whole water and compound-specific toxicity. *Ceriodaphnia dubia* is the most frequently used cladoceran species for toxicity testing. It is used for short-term (24 to 96 h), acute toxicity tests and to measure chronic toxicity during a 7-day testing period (US EPA, 1989e).

Ostracods, copepods, amphipods, and isopods are small crustaceans that are generally less important in freshwater wetland environments because of their low diversity and numbers. Amphipods and isopods typically are not observed in freshwater wetlands. Ostracods are generally less than 1 mm long and are enclosed in an opaque carapace or shell. Ostracods feed mostly on bacteria, fungi, algae, and small detrital particles and are characterized as omnivorous scavengers. Ostracods are tolerant of a wide range of environmental conditions including low dissolved oxygen levels. Copepods, typically less than 2 mm in length, occur in a variety of benthic and littoral aquatic environments. Copepods are an important food item for larger macroinvertebrates and fish. Members of the suborder Harpacticoida are most typical of wetland environments and feed by raking, seizing, and scraping bacteria, fungi, and algae from submerged substrates.

The crustacean order Decopoda includes the freshwater crayfish. Crayfish frequently inhabit freshwater wetlands. They are important consumers because they graze on plants and continually process detritus to consume bacteria, algae, and dead plant material. Crayfish are obligate aerobes and normally extract dissolved oxygen by use of gills. Many crayfish species can live outside of the water column and inhabit anaerobic areas by intermittently moving into more highly oxygenated water or by constructing burrows or "chimneys" which extend from below the water table up to the atmosphere. Typical crayfish densities in lakes are less than 100 kg/ha; however, when cultured as an economic crop, densities can range from 500 to 1500 kg/ha. Crayfish are prey for a number of carnivorous fish and for many herptiles, birds, and mammals.

Aquatic insects comprise a diverse assemblage of 11 orders in freshwater systems and fill a wide spectrum of ecological niches in wetlands. Representatives of the most commonly found insect groups in wetlands are illustrated in Figure 8-6 and include the nymphal and/or adult stages of the Odonata (dragonflies and damselflies), Hemiptera (true bugs such as back swimmers, creeping water bugs, and water scorpions), Coleoptera (aquatic beetles), and Diptera (mosquitoes, horseflies, deer flies, and midges).

Aquatic insects have a variety of food preferences and environmental requirements in wetland ecosystems. Odonate nymphs prey on other macroinvertebrates including chironomids (midges) and larval fish. Most Hemiptera are carnivorous, and some species such as the creeping water bug (*Pelocoris femoratus*) are capable of painful bites to humans wading in densely vegetated wetlands. Dragonflies and damselflies typically have grasping and chewing mouth parts, and true bugs typically have piercing and sucking mouth parts. Aquatic beetles are found in wetlands in both larval and adult stages and feed on a variety of foods including plant material, detritus, and other invertebrates.

Aquatic diptera are the most important macroinvertebrates in wetlands in terms of number of species and population size. Mosquitoes (Family Culicidae) and midges (Families Chironomidae and Heleidae) are common wetland inhabitants. Only the larvae and pupae of these groups actually occur in the wetland water column because the adults emerge from their pupal skins and lead a terrestrial life feeding on nectar or blood. Mosquitoes may attain very high populations in polluted wetlands (over 2400 larvae and pupae in a single 350 mL dipper) and are tolerant of low dissolved oxygen conditions because they can breath atmospheric oxygen at the water surface. The nuisance potential of mosquitoes in wetlands and normal control methods are discussed in Chapter 25.

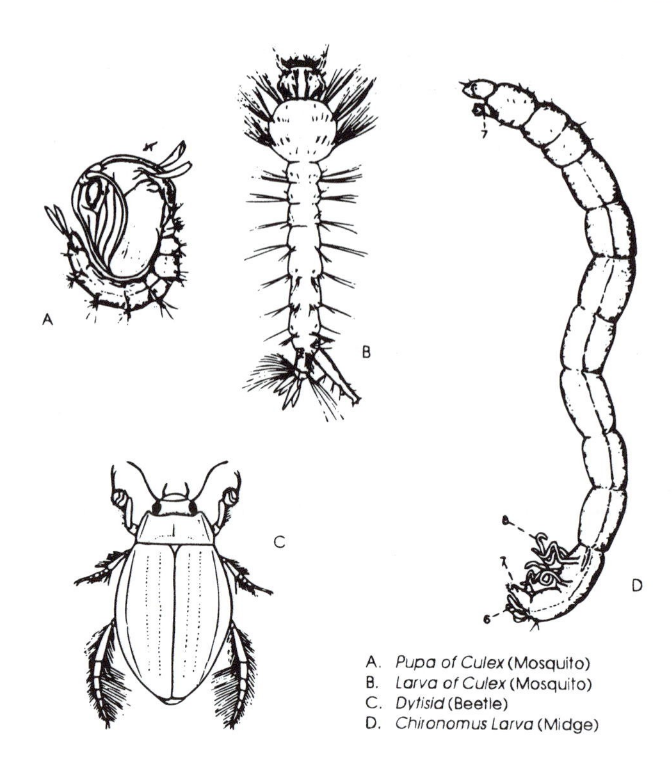

Figure 8-6 Representative aquatic insects that occur in wetlands (From *Fresh-Water Invertebrates of the United States*, R.W. Pennak. Copyright © 1978 John Wiley and Sons, Inc. Reprinted by permission of John Wiley & Sons, Inc.)

Information concerning wetland macroinvertebrate populations is infrequent in the published literature. A number of specific findings from wetland treatment systems are described later to provide an indication of typical macroinvertebrate population densities and diversities in those systems.

Benthic macroinvertebrate populations have been studied in a variety of wetland ecosystems receiving wastewater. In the Florida cypress dome study, Brightman (1984) found fewer individuals, less biomass, and lower species diversity of macroinvertebrates in domes receiving treated wastewater (Table 8-2). These reductions were attributed to low dissolved oxygen concentrations in the wastewater-treated cypress domes and to reduced atmospheric oxygen diffusion and increased organic loading due to development of a duckweed cover on the treated domes. Microarthropod studies at the same site (McMahan and Davis, 1984) found that adult organisms collected from the vegetation in the domes were not negatively impacted compared to a control dome and that species diversity and biomass were higher in the sewage domes compared to the groundwater-augmented dome. Mosquito population studies at this site (Davis, 1984) indicated that wastewater treatments in the cypress domes caused no immediate or unusual increase in mosquito abundance and that populations of floodwater mosquito species in the genera *Aedes* and *Psorophora,* which are known human pests, declined in the treatment domes compared to the control because of stabilized water levels.

Central Slough is a natural, forested, floodplain wetland near Conway, SC that began receiving secondarily treated municipal wastewater in 1986. No benthic macroinvertebrates were present in this swamp forest in the year before discharge because of low water levels in the adjacent Waccamaw River. After effluent inflows were initiated, this system became continuously inundated, and macroinvertebrate populations were established. Beginning in 1988, these macroinvertebrate populations were monitored annually during May by using

Table 8-2 Summary of Macroinvertebrate Data from Cypress Domes

	Average Sewage Domes		Groundwater Domes		Control Dome	
Taxonomic Group	No. of Species	No. of Individuals	No. of Species	No. of Individuals	No. of Species	No. of Individuals
Annelida	1	60	1	83	1	37
Arthropoda						
Arachnoida	0	0	0	0	1	17
Crustacea						
Amphipoda	1	1	0	0	1	2
Decapoda	1	1	1	8	1	3
Isopoda	1	2	0	0	2	13
Insecta						
Coleoptera	5	4	0	0	3	26
Diptera	12	296	11	786	12	947
Hemiptera	2	1	1	6	1	14
Odonata	4	9	7	64	8	43
Mollusca	0	0	1	46	0	0
Total	27	374	22	993	30	1102
Average biomass (kg/ha)[a]		2.9		13.9		6.7
Shannon-Weaver diversity (bits/indiv)[b]		0.3		1.3		0.6

[a] kg/ha: Kilograms dry weight/hectare.
[b] bits/indiv: Bits per individual.

Note: Data from Brightman, 1984.

soil cores. A similar control area known as South Slough was also monitored. Table 8-3 summarizes these data for a 4-year period. Macroinvertebrate population density increased in Central Slough during this period from about 420 to 7144 organisms per square meter, while the control wetland population averaged about 980 organisms per square meter. The number of species of macroinvertebrates (from 3 to 22 species) was relatively low in both systems, but generally was lower in the wetland receiving effluent (average 9 species in Central Slough compared to 13 in the control wetland). Average dissolved oxygen levels were typically less than 1.5 mg/L in Central Slough during the May sampling events and were 1 to 2 mg/L lower in Central Slough than in the control area.

FISH

All freshwater fish are included in the Phylum Cordata, Subphylum Vertebrata, Class Osteichthyes, or "bony fish." Since fish are largely restricted to aerobic waters, many species

Table 8-3 Summary of Benthic Macroinvertebate Collections from Central Slough and a Control Slough, Conway, SC

	1988		1989		1990		1991	
Density (No./m^2)[a]	Central	Control	Central	Control	Central	Control	Central	Control
Annelida	0	12	0	0	0	0	0	0
Oligochaeta	6	0	0	0	112	212	25	447
Hirudinea	0	0	0	0	0	6	0	13
Crustacea	0	314	57	399	56	312	0	170
Mollusca	0	100	0	0	0	0	0	0
Insecta	414	1105	456	57	1543	400	7119	372
Total	420	1531	513	456	1712	931	7144	1002
Number of Species	9	21	7	3	5	7	14	22

[a] No./m^2: Number per square meter.

do not commonly occur in wetland ecosystems. As shown in Table 8-4, the list of fish families that typically occur in wetland environments in North America is rather brief. All of the fish species typical of wetlands are either adapted to obtaining oxygen in low oxygen waters or only visit wetlands on a seasonal or shorter basis.

Fish fill a variety of ecological niches in wetlands depending on the individual species adaptations as well as the geographical location and life stage. As they grow, most fish species shift food preferences from small macroinvertebrates to larger macroinvertebrates, small fish, and small herptiles, birds, or mammals. Most fish are observed to be "conscious consumers," selecting prey in a nonrandom pattern and rejecting unsuitable prey organisms. In this manner, fish are important controllers of energy flows in wetlands and aquatic ecosystems. The factors affecting food selection in most fish species are poorly understood, but are likely based on competitive selection of stable and productive food chains (Knight, 1980).

Boltz and Stauffer (1989) divide wetland fish in Pennsylvania into three groups based on habitat preferences: (1) those occupying swamps, marshes, and bogs; (2) those inhabiting littoral areas of lakes and streams; and (3) those visiting floodplain areas of low-gradient streams. These three habitat types share some environmental traits and common fish species. Although floodplains, swamps, and marshes frequently have seasonally low dissolved oxygen conditions, they also tend to have high primary productivity that can translate into abundant food supplies for consumers such as fish.

Isolated wetlands such as swamps and marshes typically have lower dissolved oxygen concentrations (to less than 1 mg/L) on a seasonal basis than floodplain areas. Therefore, fish species that regularly inhabit these areas are typically adapted to life in a low dissolved oxygen environment. Mud minnows (Umbridae), bowfin (Amiidae), catfish (Ictaluridae), killifish (Cyprinodontidae), and top minnows (Poeciliidae) can all "gulp" atmospheric oxygen

Table 8-4 Representative Wetland Fish Families and Species in North America

Family/Species	Habitat
Amidae	Swamps, slow streams, lake littoral zones
Amia calva (bowfin)	
Umbridae	Low-gradient streams and swamps
Umbra pygmaea (eastern mud minnow)	
U. limi (central mud minnow)	
Esocidae	Weedy lakes, streams, and marshes
Esox americanus (redfin pickerel)	
E. niger (chain pickerel)	
Cyprinidae	Small streams and backwater areas
Notropis spp. (shiners)	
Phoxinus eos (redbelly dace)	
Cyprinus carpio (carp)	
Ictaluridae	Littoral areas of lakes and streams, swamps, and marshes
Ictalurus spp. (bullheads)	
Noturus gyrinus (tadpole madtom)	
Aphredoderidae	Vegetated streams and swamps
Aphredoderus sayanus (pirate perch)	
Cyprinodontidae	Lake and stream littoral areas
Fundulus spp. (killifish)	
Poeciliidae	Shallow areas of marshes, lakes, and streams
Gambusia affinis (mosquitofish)	
Heterandria formosa (least killifish)	
Centrarchidae	Low-gradient streams and backwater areas
Acantharcus pomotis (mud sunfish)	
Enneacanthus spp. (sunfish)	
Lepomis gulosus (warmouth)	
Percidae	Low-gradient streams, swamps, and littoral areas
Etheostoma spp. (darters)	
Lepisosteidae	Low-gradient streams, swamps, and marshes
Lepisosteus spp. (gar)	

at the water surface (Figure 8-7) and also are known to preferentially inhabit wetland areas with slightly higher dissolved oxygen conditions, such as near a water inflow or in open water. Magnuson et al. (1989) observed that wetland fish species including mud minnows have multiple adaptations to low-oxygen conditions, including small body size (low overall oxygen consumption), low oxygen respiratory requirements, avoidance behaviors, air breathing, reduced activity levels, and the ability to recolonize rapidly after local extinction.

A second group of fish do not have specific adaptations to breathe atmospheric oxygen, but they have physiological tolerance to lower-oxygen conditions. These fish include some species of pickerel (Esocidae), shiners (Cyprinidae), chubsuckers (Castostomidae), sunfish (Centrarchidae), and darters (Percidae). These species can be killed by sudden oxygen depletion in wetlands caused by pollution or natural conditions such as rapid temperature changes and algal die offs.

The most widely distributed fish species in warmer temperate, subtropical, and tropical wetland treatment systems is the mosquitofish (*Gambusia affinis*) in the top minnow family (Figure 8-8). This species has been widely disseminated in North America and abroad in shallow wetlands and ponds because of its noted ability to consume the larvae and pupae of mosquitoes. In more northern climates, the black striped top minnow (*Notrophus fundulus*) is a mosquito predator. The adult mosquitofish is about 12 to 15 mm in length and bears live young. Estimated densities of mosquitofish in wetland treatment systems range from 5 to 50 individuals per square meter (5 to 50 kg wet weight/ha). These densities can be achieved from low initial inoculations of adult fish in as few as 3 to 6 months. Mosquitofish are the prey of many larger consumers in wetlands, including other fish, snakes, and birds.

In northern temperate areas, the central mud minnow (*Umbra limi*), shown in Figure 8-7, is an important fish species in treatment wetlands. Typically 51 to 102 mm in length, the central mud minnow can live in waters with summer temperatures up to 29°C and with dissolved oxygen concentrations as low as 0.25 mg/L. However, mud minnows can be killed by low-oxygen levels during the winter (Scott and Crossman, 1973).

Figure 8-7 Mud minnow gulping air at the water surface under simulated anoxic conditions. (From Magnuson, J.J., et al. 1989.)

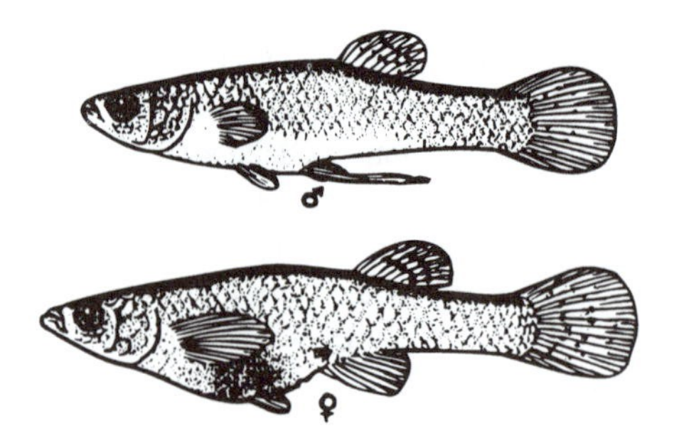

Figure 8-8 Male and female adult mosquitofish (*Gambusia affinis*) (From Samuel Eddy and James C. Underhill, *How to Know the Freshwater Fishes,* 3d ed. Pictured Key Nature Series. Copyright © 1978 Wm. C. Brown Communications, Inc., Dubuque, Iowa. All rights reserved. Reprinted by permission.)

Larger fish species not specifically adapted to life in low dissolved oxygen conditions also have been found in wetland treatment systems in a number of cases. These species, which include genera such as *Lepomis, Essox,* and *Micropterus,* are important in the control of mosquitofish populations and provide food for larger wetland consumers such as snakes, large wading birds, and mammals such as otters.

Fish population studies have been conducted at only a few natural and constructed wetlands receiving wastewater. Three studies are summarized to give the reader a general idea of fish density and diversity in treatment wetlands. At Central Slough, SC, the average fish density over a 5-year period was about 20 organisms per square meter with an average of two species present, while in two adjacent control areas the fish density averaged 13 organisms per square meter with 2.2 species per sampling event. The most common fish species observed at this site was the mosquitofish, but other fish collected were the banded sunfish (*Enneacanthus obesus*), spotted sunfish (*Lepomis punctatus*), flier (*Centrarchus macropterus*), dusky shiner (*Notropis cummingsae*), eastern mud minnow, and redfin pickerel (*Esox americanus*).

The Champion pilot constructed wetlands near Pensacola, FL were designed to test the effectiveness of wetlands for final polishing of pulp mill effluent (Knight et al., 1994). Mosquitofish were introduced shortly after system startup in May 1991. After 6 months of colonization, the mosquitofish population in these wetlands had increased to an average density of about 7.3 organisms per square meter. In addition to mosquitofish, channel catfish and sunfish were introduced into these constructed wetlands and were found to survive despite low dissolved oxygen conditions (typically less than 2 mg/L) and high seasonal water temperatures (daily values above 35°C).

At the Reedy Creek natural wetland treatment system in central Florida, fish monitoring discovered five species including mosquitofish, least killifish (*Heterandria formosa*), Everglades pygmy sunfish (*Elassoma evergladei*), sailfin molly (*Poecilla latipinna*), and spotted sunfish. Total fish densities ranged from 6 to 48 fish per square meter, and biomass values were between 1.8 and 9.0 g/m^2. These figures represent stable fish populations after more than 10 years of wastewater discharge to this natural forested wetland.

Fish populations can influence treatment. For instance, carp stir bottom sediments and impair suspended solids removal.

AMPHIBIANS AND REPTILES

Much of the wetland mystique comes from the variety of creeping or crawling critters that typically abound in these ecosystems. Amphibians and reptiles frequently provide much of the sound and part of the sights that greet wetland visitors. Frogs are observed or heard splashing and calling throughout the night and during the morning and evening hours. Turtles may be abundant in aerobic wetlands and associated treatment lagoons. Snakes are generally rare, but always are looked for because of their potential danger and because of a general human fascination with them. In subtropical and tropical wetlands, alligators or other larger reptiles may occupy a top carnivore role. All of these amphibians and reptiles (known jointly as herptiles) may play important roles as consumers and recyclers in wetlands and as regulators of lower level consumer populations. Because of the higher primary and secondary productivity in treatment wetlands compared to natural wetlands, herptile populations and all other higher consumer groups are frequently abundant.

Table 8-5 lists some of the principal groups of herptiles that are typically found in treatment wetlands. The four most important groups (frogs, turtles, snakes, and alligators)

Table 8-5 Representative Amphibian and Reptile Species Encountered in Wetlands Used For Water Quality Treatment

Class	Order	Family	Species
Class Amphibia			
	Order Caudata (salamanders)		
		Family Amphiumidae (amphiumas)	
			Amphiuma means (two-toed amphiuma)
	Order Anura (toads and frogs)		
		Family Hylidae (treefrogs)	
			Acris gryllus (cricket frog)
			Hyla cinerea (green treefrog)
			Pseudacris triseriata (chorus frog)
		Family Ranidae (true frogs)	
			Rana catesbeiana (bullfrog)
			R. pipiens (leopard frog)
Class Reptilia			
	Order Crocodylia (crocodilians)		
		Family Alligatoridae (alligators and caiman)	
			Alligator mississippiensis (American alligator)
	Order Testudines (turtles)		
		Family Chelydridae (snapping turtles)	
			Chelydra serpentina (snapping turtle)
		Family Kinosternidae (mud and musk turtles)	
			Stenotherus odoratus (stinkpot)
			Kinosternum subrubrum (mud turtle)
		Family Emydidae (box and water turtles)	
			Clemmys guttata (spotted turtle)
			Chrysemys floridana (cooter)
			C. scripta (pond slider)
			C. rubriventris (red-bellied turtle)
			C. picta (painted turtle)
		Family Trionychidae (soft-shell turtles)	
			Trionyx spiniferus (spiny softshell)
	Order Squamata Suborder Lacertilia (lizards)		
		Family Iguanidae (iguanids)	
			Anolis carolinensis (green anole)
	Order Squamata Suborder Serpentes (snakes)		
		Family Colubridae (colubrids)	
			Natrix spp. (water snakes)
			Liodytes alleni (striped swamp snake)
			Seminatrix pygaea (black swamp snake)
			Farancia abacura (mud snake)
			Elaphe spp. (rat snakes)
			Lampropeltis spp. (king snakes)
		Family Viperidae (pit vipers)	
			Agkistrodon contortrix (copperhead)
			A. piscivorus (cottonmouth)
			Sistrurus catenatus (massasauga)
			Crotalus spp. (rattlesnakes)

are described here. Stebbins (1966) contains more detailed information on identification and life history of these organisms (see also Duellman and Trueb, 1986 and Siegel et al., 1987).

AMPHIBIANS

Amphibians (Class Amphibia) are cold-blooded vertebrates with moist, glandular skin. Typically, amphibians are aquatic early in life and then metamorphose into terrestrial or aquatic adults. They do not have clawed toes. Amphibians include salamanders, frogs, and toads, but only frogs are well represented in wetlands used for wastewater treatment.

Frogs and toads are part of the Order Anura and have about 2700 species worldwide. Frogs are distinguished by smooth skin and long legs adapted for leaping. Treefrogs (Family Hylidae) are well represented in many wetland systems and reach relatively high densities in treatment wetlands because of the abundance of food (primarily flying insects). Members of this family can live on the ground, on floating and emergent vegetation, or in trees. True frogs (Family Ranidae) live on the ground adjacent to wetlands and spend much of their time in the shallow littoral zone in the water or on floating plants. These frogs typically feed on flying insects, but can also feed on aquatic invertebrates and even small fish.

Little is known about the energy flow through the trophic level occupied by wetland frogs or of their importance in regulating insect populations. However, their occasionally large populations, observed abilities to devour a wide range of insects with few other natural enemies, and importance as prey for other consumers such as birds and fish indicate that frogs have a necessary and important role in the function of wetland ecosystems.

REPTILES

The Class Reptilia includes cold-blooded vertebrates characterized by skin covered with scales, shields, or plates and by toes that bear claws. In general, reptiles lay eggs with leathery skins. The young resemble the adults in general form with no metamorphose process. Reptiles that most frequently inhabit wetlands in North America associated with water quality treatment include the American alligator (Order Crocodylia), turtles (Order Testudines), and snakes (Order Squamata, Suborder Serpentes). Each of these groups is briefly described.

The alligator (*Alligator mississipiensis*) occurs across the southern coastal plain of North America from coastal North Carolina to southern Texas. Alligators have been observed at relatively high densities in wetland treatment systems with open water areas and high fish populations. For example, more than 25 adult alligators resided in a 20-ha constructed wetland/pond receiving secondarily treated municipal wastewater near Ocala, FL. Although individual alligators are sometimes observed in densely vegetated wetlands, they are typically transient unless deeper, open water habitat areas are provided.

Turtles are found primarily in wetland treatment systems that have areas of deeper, open water and especially in pretreatment lagoons. Species of turtles frequently found in these habitats include snapping turtles, mud turtles, cooters or sliders, and soft-shelled turtles. Most of these species feed on a variety of plant materials, but some also eat meat, fish, and invertebrates. Soft-shell and snapping turtles are largely carnivorous and feed on invertebrates, fish, and small birds.

A wide variety of snakes are associated with wetlands and wetland treatment systems and play an important ecological role by feeding on fish, invertebrates, birds, and small mammals. In North American treatment wetlands, the snakes that receive the most interest are the venomous species including the cottonmouth or water moccasin (*Agkistrodon piscivorus*), the copperhead (*A. contortrix*), and a few rattlesnakes (*Sistrurus catenatus* and *Crotalus* spp.) that occasionally inhabit wetlands. In fact, most snakes in wetlands are nonpoisonous species of water snakes (Genus *Natrix*), swamp snakes (*Liodytes* and *Semina-*

Table 8-6 Amphibian Structure and Function in Cypress Domes Receiving Secondarily Treated Wastewater and in a Control Dome Receiving Groundwater

	Average Annual Biomass (kg/ha)		Net Migration(+)/Immigration (−) (g/m of edge)	
Genus	**Sewage Dome**	**Control Dome**	**Sewage Dome**	**Control Dome**
Rana	17.2	13.8	−352	+137
Acris	4.2	4.0		
Hyla	0.4	0.5	−148	+71
Bufo	2.5	2.7		
Total	24.3	21.0	−499	+208

Note: Data from Harris and Vickers, 1984.

trix), rat snakes (*Elaphe*), and king snakes (*Lampropeltis*). The water and swamp snakes feed voraciously on wetland fishes, and rat snakes feed primarily on small mammals and on birds and their eggs. King snakes have the interesting adaptation of feeding on other snakes and are immune to the venom of our native poisonous snake species.

Few quantitative herptile studies have been conducted on wetlands used for wastewater treatment. Harris and Vickers (1984) studied the structure and function of amphibian populations in Florida cypress domes receiving secondarily treated municipal wastewater. Four genera (*Rana, Acris, Bufo,* and *Hyla*) were dominant in these systems on a biomass basis (Table 8-6). Although annual average amphibian biomass was slightly higher in the cypress domes receiving sewage, the difference was not statistically significant. Amphibian biomass from a cypress dome that was not receiving any water augmentation (and thus would normally dry out during parts of most years) was not reported for comparison. However, amphibian productivity in the sewage domes was found to be essentially zero, possibly because of mortality of larval stages in the mostly anaerobic water (average dissolved oxygen was 0.17 to 0.29 mg/L). Table 8-6 also indicates that there was a net immigration of amphibians into the treatment domes, while the control dome had a net migration out.

BIRDS

To humans, birds are typically the most important visual feature of wetlands. This importance is not surprising considering that about 600 different bird species (one third of the total resident bird species) are either partially or wholly dependent on wetlands for some part of their life history in North America (Kroodsma, 1978). These bird species represent many shapes and colors, a variety of habitat preferences, and large populations. The diversity and abundance of birds in and around wetlands attracts many bird watchers who repeatedly have observed that their species lists are longer and their counts higher when they include wetlands in their counting areas. Also, where these water bodies are enriched by nutrients and organic matter from wastewater and stormwater discharges, bird watchers usually find even better success in their sport.

In addition, a number of wetland bird species are game species and attract hunters. Because treatment wetlands are augmented by greater water inflows than adjacent natural wetlands, they frequently provide waterfowl habitat that is otherwise in short supply. In a few cases, these treatment wetlands have been designed with duck hunting as a recognized secondary benefit. In other cases, wastewater is used to create waterfowl breeding and feeding habitat as the primary purpose with water quality treatment as a secondary or nonessential side effect.

Table 8-7 lists the major groups of birds associated with treatment wetlands in North America. Twenty-one families of birds are most frequently encountered in these systems.

Table 8-7 Representative Birds Important in North American Wetlands Used For Water Quality Treatment

Family Podicipedidae (grebes)
- *Podiceps nigricollis* (eared grebe)
- *Podilymbus podiceps* (pied-billed grebe)

Family Pelecanidae (pelicans)
- *Pelecanus erythrorhynchos* (white pelican)

Family Anhingidae (anhingas)
- *Anhinga anhinga* (anhinga)

Family Phalacrocoracidae (cormorants)
- *Phalacrocorax auritus* (double-crested cormorant)

Family Ardeidae (herons and egrets)
- *Ixobrychus exilis* (least bittern)
- *Botaurus lentiginosus* (American bittern)
- *Nycticorax* spp. (night herons)
- *Butorides striatus* (green-backed heron)
- *Egretta* spp. (herons and egrets)
- *Bubulcus ibis* (cattle egret)
- *Ardea herodias* (great blue heron)

Family Ciconiidae (storks)
- *Mycteria americana* (wood stork)

Family Gruidae (cranes)
- *Grus americana* (whooping crane)
- *G. canadensis* (sandhill crane)

Family Threskiornithidae (ibises)
- *Plegadis* spp. (ibises)
- *Eudocimus albus* (white ibis)

Family Anatidae (swans, geese, ducks)
- *Cygnus* spp. (swans)
- *Branta canadensis* (Canada goose)
- *Anas* spp. (dabbling ducks)
- *Aix sponsa* (wood duck)
- *Aythya* spp. (diving ducks)
- *Mergus* spp. (mergansers)

Family Aramidae (limpkin)
- *Aramus guarauna* (limpkin)

Family Rallidae (rails, gallinules, coots)
- *Rallus* spp. (rails)
- *Porzana carolina* (sora)
- *Gallinula chloropus* (common moorhen)
- *Fulica americana* (American coot)

Family Recurvirostridae (stilts and avocets)
- *Himantopus mexicanus* (black-necked stilt)

Family Scolopacidae (sandpipers)
- *Tringa* spp. (yellowlegs)
- *Actitis macularia* (spotted sandpiper)
- *Gallinago gallinago* (common snipe)
- *Scolopax minor* (American woodcock)

Family Accipitridae (kites, hawks, eagles)
- *Haliaeetus leucocephalus* (bald eagle)
- *Rostrhamus sociabilis* (snail kite)
- *Circus cyaneus* (northern harrier)
- *Buteo lineatus* (red-shouldered hawk)
- *Pandion haliatus* (osprey)

Family Strigidae (owls)
- *Strix varia* (barred owl)
- *Otus asio* (eastern sreech owl)

Family Alcedinidae (kingfishers)
- *Ceryle alcyon* (belted kingfisher)

Family Hirundinidae (swallows)
- *Tachycineta bicolor* (tree swallow)

Family Troglodytidae (wrens)
- *Cistothorus palustris* (marsh wren)

Table 8-7 Continued

Family Vireonidae (vireos)
Vireo griseus (white-eyed vireo)
Family Emberizidae (warblers, sparrows)
Protontaria citrea (prothonotary warbler)
Wilsonia citrina (hooded warbler)
Geothlypis trichas (common yellowthroat)
Melospiza georgiana (swamp sparrow)
Family Icteridae (blackbirds)
Agelaius phoeniceus (red-winged blackbird)
Xanthocephalus xanthocephalus (yellow-headed blackbird)
Quiscalus spp. (grackles)

Most of these groups have specialized environmental requirements that allow them to occupy specific ecological niches in the wetlands. Families that feed on invertebrates, fish, and amphibians along the wetland edge include the herons and egrets (Ardeidae), woodstork (Ciconiidae), ibises (Threskiornithidae), limpkin (Aramidae), stilts (Recurvirostridae), and sandpipers (Scolopacidaea). Bird families that perch on the water surface and feed primarily on aquatic organisms include pelicans (Pelecanidae); cormorants (Phalacrocoracidae); swans, geese, and ducks (Anatidaea); and rails, gallinules, and coots (Rallidae). Birds that perch in trees near wetlands or hunt in flight include anhingas (Anhingidae); kites, hawks, and eagles (Accipitridae); owls (Strigidae); and kingfishers (Alcedinidae). Bird families that feed primarily on insects over wetlands include swallows (Hirundinidae) and vireos (Vireonidae). Warblers and sparrows (Emberizidae) and blackbirds (Icteridae) typically feed on invertebrates or seeds within the emergent plant zone near the water surface. Figure 8-9 shows silhouettes of representative species in some of the most typical wetland bird families.

Each bird species in these families has a different feeding strategy in terms of water depth, substrate type, vegetative cover, or other environmental variables. For example, Baskett (1988) studied preferred feeding water depths at the Grand Pass Wildlife Area in Missouri and found that the spotted sandpiper preferred moist, unflooded sites; the solitary sandpiper preferred 0 to 5 cm of water; and the dowitcher preferred 2 to 10 cm of water. Dabbling ducks at this same site (pintail, blue-winged teal, and mallard) preferred water depths of about 10 to 20 cm for feeding. Kushlan (1989) found that invertebrate populations used by feeding birds were stimulated in a marsh after drawdown and resubmergence. Thus, wetland treatment systems may be more attractive to birds after their initial flooding than in a continuously flooded state. Management for wildlife enhancement may involve periodic rotation of flows with drawdowns in marsh water levels. Chapter 23 contains more information on enhancing the wildlife benefits of treatment wetlands.

A few quantitative studies of bird populations in wetlands treating wastewaters are available. A brief summary of these data is provided to assist the reader with estimating typical bird usage during project planning. Harris and Vickers (1984) studied bird populations in Florida cypress domes receiving secondarily treated sewage, in a cypress dome augmented by groundwater, and in an unaugmented cypress dome used as a control. They found that bird populations and diversity were increased slightly in the sewage domes compared to the controls, but that densities were not significantly different.

Shifts in bird populations, both species and numbers, have been documented at the Houghton Lake site over a 23-year period. Fluctuations have been noted, partly due to changing habitat caused by wastewater. Through the course of the project, species numbers at control sites (unimpacted by wastewater) have remained stable. In the discharge area, numbers of species increased, then decreased (Table 8-8). A similar course of events has occurred at the Biwabik, MN treatment wetlands (Hanowski and Niemi, 1993).

If the wetland is constructed on previous upland, increases in bird species and abundance have been observed, compared to the previous upland habitat (Hickman, 1994).

Figure 8-9 Representative wetland bird family silhouettes (Illustrations by Arthur Singer from *Birds of North America.* Copyright © 1966 Western Publishing Company, Inc. Used by permission.)

Table 8-8 Numbers of Bird Species in the Houghton Lake Wetland at Treatment and Control Sites over the Course of the Project

Year	Control Area	Discharge Area	Ref.
1973	20	20	Bergland, 1974
1978	19	27	Rosman, 1978
1979	21	33	Rabe, 1979
1983	17	34	Rabe, 1984
1987	21	30	Rabe, 1988
1989	19	25	Rabe, 1990
1991	18	20	Ludwig, 1991
1993	—	20	Dorset et al., 1994

Note: Discharges began in 1978.

During the summer of 1991, the U.S. EPA conducted a study of the environmental condition of six constructed wetland treatment systems in the U.S. (McAllister, 1992, 1993a, 1993b). This inventory included two systems in the arid west (Incline Village, NV and Show Low, AZ), two systems along the coastal plain of Mississippi (Ocean Springs and Collins), and two systems in peninsular Florida (Orlando and Lakeland). Bird surveys were conducted by local ornithologists using slightly different methods. Table 8-9 summarizes the major findings of this research.

The U.S. EPA bird surveys at constructed wetland treatment systems confirmed that these systems have high species richness and population densities compared to control wetlands. The total number of bird species observed at each site during the 1991 study ranged from 33 to 63, with average daily population densities of all wetland-dependent species between about 7 and 19 birds/ha. Densities of wading birds at the two central Florida constructed wetland treatment systems averaged 0.29 birds/ha at Orlando and 0.38 birds/ha at Lakeland. These densities were as high as or higher than comparison data from marshes along the St. Johns River, the central Everglades, and Central America. Highest total bird densities were noted at the arid region sites where there are few natural wetlands available to compete as alternative habitat. The U.S. EPA studies also concluded that the constructed treatment wetlands are important habitat for a number of endangered or threatened wetland-dependent bird species.

Bird usage of the 16-ha DUST (demonstration urban stormwater treatment) marsh in Coyote Hills Regional Park near Freemont, CA was studied by Duffield (1986). Weekly or biweekly bird censuses were conducted in the DUST marsh and a nearby 18.3-ha control from mid-January 1984 through mid-June 1985. The DUST marsh consists of three cells that vary from primarily open water to mud flats to a vegetated channel colonized by cattails (*Typha* spp.), alkali bulrush (*Scirpus robustus*), and pickleweed (*Salicornia virginica*). The control marsh is about 41 percent open water and 59 percent emergent vegetation dominated by cattails, bulrush, and pickleweed.

Mean abundance of wetland birds in the marsh areas ranged from 100 to 300 (6.25 to 18.75 birds/ha) in the DUST marsh and 90 to 420 (4.91 to 23.0 birds/ha) in the control marsh. Mean number of species per census ranged from 14 to 23 in the DUST marsh and 10 to 18 in the control area. Highest mean species counts were observed in both marshes during the winter and spring seasons. During all seasons, dabbling and diving ducks preferred the control marsh, and shorebirds, gulls, and terns preferred the DUST marsh. Mallards (*Anax platyrhychos*) and American coots (*Fulica americana*) were the most abundant dabbling species at the DUST marsh. Individual fish-eating wading birds were generally present in greater numbers in the DUST marsh than in the control area. These populations were enhanced by the presence of roosting black-crowned night herons (*Nycticorax nycticorax*). Shorebirds

Table 8-9 Results from Constructed Wetland Treatment System Bird Surveys by the U.S. EPA During 1991

Site	Constructed Wetland Area	Total Species	Density #/ha average (range)
Incline Village, NV	198	47	19.1 (0.8–42.2)
Show Low, AZ	284	42	13.8 (7.8–21.7)
Collins, MS	4.5	35	7.2 (5.9–8.5)
Ocean Springs, MS	22	35	10.4 (6.4–14.5)
Orlando, FL	494	141[a]	0.29 (0.11–0.54)[b]
Lakeland, FL	498	63	0.38 (0.04–0.72)[b] (7.7–13.5)[c]

[a] U.S. EPA observed 33 species during 1991, and the remainder of the species had been observed by site operators since 1987.
[b] Wading birds only.
[c] All species.

preferred the mud flat areas of the overland flow cell in the DUST marsh. The most common shorebirds were American avocets (*Recurvirostra americana*), black-necked stilts (*Himantopus mexicanus*), and marbled godwits (*Limosa fedoa*). Other shorebirds that regularly used the mud flat area include long-billed dowithchers (*Limnodromus scolopaceus*), least sandpipers (*Calidris minutilla*), and killdeer (*Charadrius vociferus*).

MAMMALS

Although the number of mammal species that inhabit treatment wetlands is typically quite low, a few important species are usually associated with both natural and constructed wetland systems. Table 8-10 lists the mammals typically found in North American wetlands. All of these species are found periodically in treatment wetlands.

The opossum (*Didelphis marsupialis*), an omnivorous member of the Class Marsupialia (marsupials), is sometimes found in North American wetlands. Shrews (Class Insectivora, Family Soricidae) are found along the edges of wetlands and moist fields and feed on insects. By far, the largest and most important group of mammals associated with wetlands are rodents. The Class Rodentia includes mice, voles, and rats (all Family Cricetidae), most of which are herbivorous species that graze on plants and seeds and are prey to fish, wading birds, and raptors. Marsh and swamp rabbits (Family Lepoidae) are also frequently encountered in wetlands in the southern U.S. Three other species of rodents that play an important role in wetlands are the muskrat, beaver, and nutria.

Muskrat (*Onadatra zibethica*) cut large numbers of emergent herbaceous plants, primarily cattails (Lacki et al., 1990), and build feeding platforms and nests (mounds). This grazing

Table 8-10 North American Mammal Species Sometimes Associated With Treatment Wetlands

- Order Marsupialia
  - Family Didelphiidae
    - *Didedelphis marsupialis* (opossum)
- Order Insectivora
  - Family Soricidae
    - *Sorex* spp. (shrews)
- Order Carnivora
  - Family Ursidae
    - *Ursus americanus* (black bear)
  - Family Procyonidae
    - *Procyon lotor* (raccoon)
  - Family Mustelidae
    - *Mustela* spp. (weasels and minks)
    - *Lutra canadensis* (otter)
  - Family Felidae
    - *Lynx rufus* (bobcat)
- Order Rodentia
  - Family Castoridae
    - *Castor canadensis* (beaver)
  - Family Cricetidae
    - *Peromyscus* spp. (mice)
    - *Synaptomys* spp. (bog lemmings)
    - *Microtus* spp. (voles)
    - *Neofiber alleni* (water rat)
    - *Ondatra zibethica* (muskrat)
    - *Myocastor coypus* (nutria)
- Family Leporidae
  - *Sylvilagus* spp. (swamp and marsh rabbits)
- Order Artiodactyla
  - Family Cervidae
    - *Odocoileus virginianus* (white-tailed deer)
    - *Alces alces* (moose)

can change treatment wetland areas from densely vegetated to a patchwork of open and emergent areas. Berg and Kangas (1989) quantified this effect in Michigan marshes and found that although muskrats only consumed about 2 percent of the annual net primary productivity, their mounds represented about 20 percent of this production. Placing this cut biomass in mounds was found to accelerate litter decomposition rates in these marshes compared to litter which falls into the water column. Muskrats are also problematic in constructed wetlands because they burrow into dikes, creating operational headaches and potential for system failure. Muskrat densities reported in Iowa's Wall Lake ranged from almost zero during some years to a high population of 3.6/ha, with an average density of about 1.2/ha (Errington, 1963).

Nutria (*Myocastor coypus*) are an introduced species from South America. Muskrats are prey for a number of raptors and mammalian predators (foxes, coyotes, and bobcat), but mature nutria are larger (Figure 8-10) and have fewer natural predators (typically only the alligator). Nutria are strictly herbivorous and feed on a broad range of plants in treatment wetlands including cattails, alligator weed, grasses, pickerelweed, water hyacinth, duckweed, and young tree seedlings. Nutria cut vegetation in a manner similar to muskrats and build feeding platforms and nest mounds. They are prolific and commonly achieve damaging population densities unless they are controlled by trapping or shooting.

Beavers (*Castor canadensis*) typically are found in natural wetland treatment systems, but they also appear in constructed wetlands with trees in/or nearby the wetland. Beavers feed on bark, twigs, and leaves from a variety of trees and also use sticks, branches, and whole trees to build lodges and dams. Consequently, beavers may cause undesirable damage in wetland treatment systems when they kill trees. Also, when beavers find a reliable source

Figure 8-10 Comparison of muskrat, nutria and beaver, three wetland grazers (Excerpt from *A Field Guide to the Mammals* by Burt and Grossenheider. Copyright © 1952, 1964, 1976, by William Henry Burt and Richard Philip Grossenheider. Reprinted by permission of Houghton Mifflin Company. All rights reserved.)

of flowing water such as in a natural wetland receiving a wastewater discharge, they construct a dam to create a deep water habitat that will protect their nesting lodge from predators. Increased water levels may augment treatment through increased residence time in the wetland, but this potential advantage may be offset by the low-oxygen conditions which result when waters rich in organic matter and ammonia nitrogen are stagnated. Low-oxygen conditions also may cause tree mortality in the deep water areas and therefore reduce treatment effectiveness, especially for ammonia nitrogen.

Other mammalian species characteristic of North American wetlands include the following: the river otter (*Lutra canadensis*), which prefers deeper, open water areas and feeds on crustacea, mollusks, and fish; the raccoon (*Procyon lotor*), which primarily visits wetlands in search of food and less for cover; the black bear (*Ursus americanus*), an omnivorous denizen of shrub-scrub and forested wetlands in local areas of North America (Hellgren and Vaughan, 1989); the bobcat (*Felis rufus*), which feeds primarily on small birds in wetlands; and the white-tailed deer (*Oedocoileus virginianus*), elk (*Cervus canadensis*), and moose (*Alces alces*), which are herbivorous and feed seasonally along the edges of natural wetlands.

SUMMARY

Animals play a sometimes subtle, but important role in wetlands used for water quality enhancement. From the tiniest microscopic protozoans to the largest mammalian representatives such as otters or bears, animals consume energy-yielding biomass, convert part of this energy into new biomass, and recycle unused organic matter and nutrients. Nutrients spiral their way up the food chain and are continuously used and transformed so they may be used again. Consumers keep nutrients in circulation and regulate the populations of lower trophic levels in a manner that maximizes system function (Odum, 1983). Wetland ecosystems exposed to toxins or other factors that eliminate consumer populations have reduced nutrient cycling functions, which may in turn affect the performance of biological water quality treatment.

In most cases, the wetland designer does not need to be concerned with the nutritional and habitat requirements of the animal populations present in a wetland treatment system. The diversity of adjacent wetlands and aquatic systems is frequently adequate to provide faunal colonizers for constructed wetland treatment systems. When these natural colonizers are present, a diverse assemblage of organisms will establish in a new constructed wetland in a few years or less and will create a balanced wetland ecosystem which has essential self-regulating functions. However, if a wetland treatment system is to be constructed where adjacent sources of adapted species are not present, the designer may need to promote colonization artificially through importation of water, sediments, and plants containing microscopic and minute wetland animals and microbes from more distant sources.

In cases where the wetland designer wants to achieve significant wildlife benefits in addition to water quality treatment benefits, greater consideration must be given to wildlife populations during design, construction, and operation of wetland treatment systems. Chapter 23 summarizes the current state of the knowledge on enhancing wildlife use of constructed and natural wetlands. The ancillary benefits potentially achieved when treatment wetlands are built to attract wildlife may be an added value with relatively little capital and operating cost.

Section 3
The Effects of Wetlands on Water Quality

CHAPTER 9

Hydraulic and Chemical Design Tools

INTRODUCTION

The preceding chapters have shown that wetlands are complex, dynamic ecosystems. Many different types of wetlands exist, with different soils, vegetation, fauna, and hydrology. When such ecosystems are considered for treatment purposes, it becomes desirable to predict the degree of treatment to be expected. That information is used to project the treatment potential of natural wetlands and for the design of constructed treatment wetlands.

It has been supposed by some that wetlands are easily conceived and built and are "low-tech" devices; therefore, they should be easily described in terms of simple equations. In fact, they are exceedingly complex ecoreactors that require complex descriptions if they are to meet expectations. The reader is warned that the successful use of equations in this book, or any other on the subject, requires understanding of their origins and conditions for applicability. This chapter is intended to provide a foundation for the quantitative aspects of treatment wetland design.

A design model must first do an adequate job of predicting wetland hydraulics. The basic tool is the interior water budget. Stream inflows and outflows are typically measurable and controllable in design. The wetland often will not communicate with groundwater due to existing or constructed clay seals or liners. The uncontrollable elements of the water budget are precipitation and evapotranspiration. Weather records may be used as the basis for design data for precipitation to use in water mass balances. However, evapotranspiration is not so easily estimated and hence will be discussed in greater detail.

The fundamental hydraulics question for treatment wetlands is conveyance capacity and the related issues of depths and slopes. These issues are answered by the interior water budget, together with data on the flow resistance of the wetland. Surface-flow and subsurface-flow wetlands are a bit different with respect to frictional characteristics and will be considered separately. The hydraulic sizing of the wetland is then set so that operation will occur within the desired parameters of depth and flow.

Other tools are necessary to determine the size of the wetland to achieve the desired chemical reductions. These are a description of internal flow paths, the interior chemical mass balance, and the reaction rate expression. Chapter 6 has set forth some of the complexity of the transformations and transfers that occur in wetlands. A conceptually ideal model would incorporate all such transfers, but at great computational expense. Further, the existing information is insufficient to calibrate a highly detailed model. This chapter and others in this section will set forth models at the greatest degree of complexity that can be supported by the available data. That data will be from wetland systems; extrapolations from companion technologies are used for confirmation when available.

WETLAND EVAPOTRANSPIRATION

Water losses to the atmosphere from a wetland occur from the water and soil (evaporation) and from the emergent portions of the plants (transpiration). The combination of the two processes is termed evapotranspiration (ET). There are many reports of studies of these wetland water losses, including the reviews of Linacre (1976) and Ingram (1983). All the studies have attempted to correlate data with ecosystem and meteorological variables and, in most cases, have compared data to open water evaporation. The results form a large and confusing literature, but a coherent set of principles may be drawn from these efforts. Attention is focused on marsh wetlands: those which contain nonwoody emergent macrophytes such as cattails, bulrush, and reeds.

It is worth noting that ET is the air conditioning system for the treatment wetland. Without the attendant loss of the latent heat of vaporization of water, the summer wetland temperature would increase to a desert-condition temperature and bake the vegetation. Some wetland plants, notably cattails, can in fact withstand desiccated conditions for many months.

SURFACE-FLOW WETLANDS

The presence of vegetation retards evaporation. This is logically to be expected for a number of reasons including shading of the water surface, increased humidity near the surface, and reduction of the wind at the surface. The presence of a litter layer can create a mulching effect. The reported magnitude of this reduction is on the order of 50 percent. A sampling of percentages of open water evaporation is Bernatowicz et al., 1976: 47 percent; Koerselman and Beltman, 1988: 41 to 48 percent; and Kadlec et al., 1987: 30 to 86 percent. However, these data should not be interpreted as meaning that the wetland conserves water, because transpiration can make up the difference or more.

Wetland ET and lake evaporation are roughly equal. Roulet and Woo (1986) report this equality for a low arctic site, and Linacre's (1976) review concludes: "In short, rough equality with lakes is probably the most reasonable inference for bog evaporation." Eisenlohr (1966) found that vegetated potholes lost water 12 percent faster than open water potholes, but Virta (1966) (as cited by Koerselman and Beltman, 1988) found 13 percent less water loss in peatlands. There is a seasonal effect which can invalidate this general observation in the short term.

The seasonal variation in ET shows the effects of both radiation patterns and vegetation patterns. The seasonal pattern of ET resembles the seasonal pattern of incoming radiation. During the course of the year, the wetland reflectance changes, the ability to transpire is gained and lost, and a litter layer fluctuates in a mulching function. Agricultural water loss calculations include a crop coefficient to account for the vegetative effect. This is in addition to effects due to radiation, wind, relative humidity, cloud cover, and temperature and may be viewed as the ratio of wetland ET to lake evaporation. The result is a growing-season enhancement, followed by winter reductions.

Very small wetlands will display strong reactions to the surrounding microclimate. Linacre (1976) refers to this as the "clothesline effect" and cites several studies which show enhanced ET for what amount to potted plants. Since constructed water treatment wetlands tend to be small, it is reasonable to inquire at what size this effect becomes important. There is very little information available on the size effect. The Koerselman and Beltman (1988) study was on a wetland of "less than one hectare" and displayed no large differences from similar studies on larger wetlands. Studies at Listowel, Ontario (Herskowitz et al., 1987) indicated that lake evaporation was a reasonable estimator of wastewater wetland ET for 0.1- and 0.4-ha wetlands. However, as size is decreased, the advective terms in the energy balance become important at some point, and conventional methods are no longer adequate. Ratios to pan and lake evaporation, and to radiation, would not be expected to hold. For instance, Bavor et al. (1988) found ET enhanced by a factor of two over pan evaporation in an open water, unvegetated wetland 4 m wide by 100 m long.

The type of vegetation is not a strong factor in determination of water losses. Bernatowicz et al. (1976) found relatively small differences among several reed species, including *Typha.* Koerselman and Beltman (1988) similarly found little difference among two *Carex* species and *Typha.* Linacre (1976) concludes: "... it appears that differences between plant types are relatively unimportant ...". Lafleur (1990) recommended using the energy balance ET estimate as the independent variable in linear regression for specific vegetation types. In agriculture, this approach leads to crop coefficients which influence ET at a specific site. That approach has the advantage of retaining the energy balance used in other ecosystems, but modifying it slightly for site specific circumstances. To date, such site regressions do not exist for treatment wetlands.

Because ET is driven by solar radiation, there is a strong diurnal cycle in water loss. A water level recorder placed in a wetland will show a drop in water level during the daytime hours, superimposed on longer-term drainage or filling trends. If the hydraulic loading from other sources is low enough, this drop may be used to infer wetland ET (Figure 9-1). At

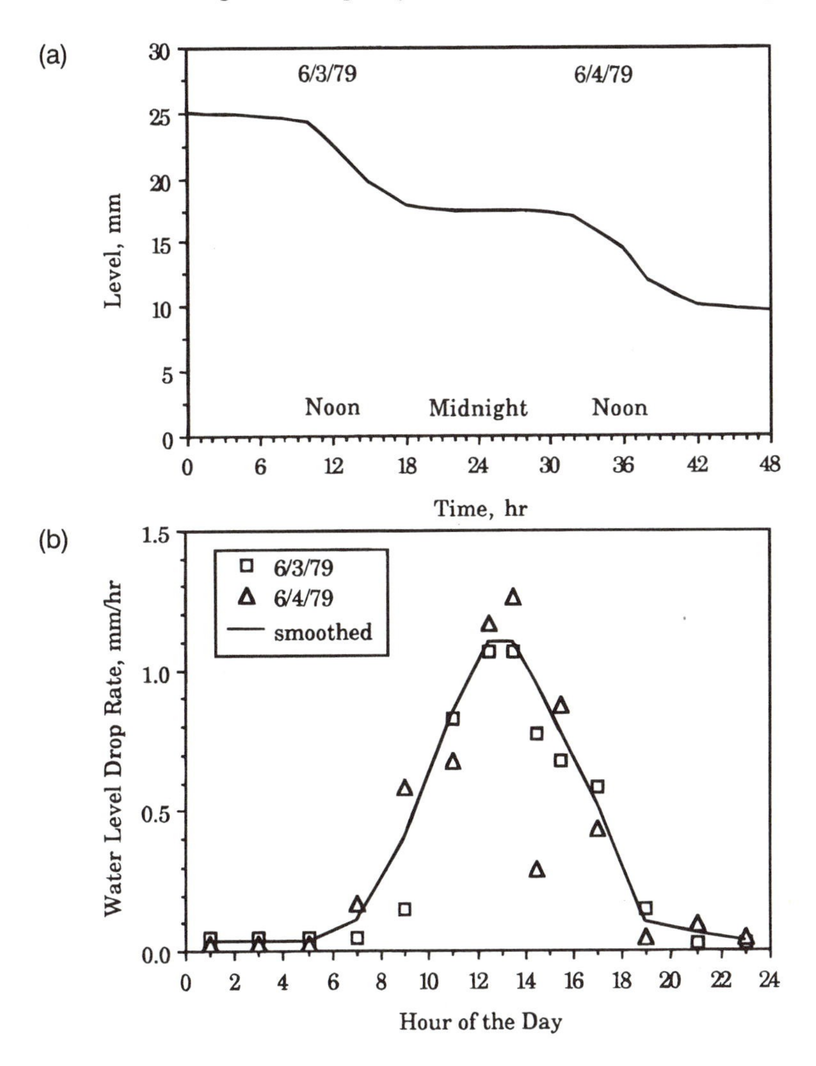

Figure 9-1 (a) Continuous recorder measurement of water level drops in the Houghton Lake, MI wetland over the period 6/3/79 through 6/6/79. Daytime drops are due to evapotranspiration. (b) Water loss rates from the Houghton Lake wetland. Peak loss occurs in conjunction with the peak daily radiation, at 1200 to 1400 hours.

higher loadings or in the event of strong outflow structure controls, the technique may not be accurate.

Predictive Tools

The available techniques for estimating ET from the wetland water are generally restricted to the unfrozen season. For systems operating at air temperatures below freezing, there will be an insulating layer above the wetland. This may be either an ice and snow blanket over the flowing water (Herskowitz, 1986) or an artificial layer such as straw covered by snow (Jenssen et al., 1992). It is likely that very little water evaporates under these conditions. Solar radiation would not penetrate to the water, and the air layer would likely be saturated. The vapor pressure of water at temperatures near zero Celsius is low, but not zero; so the potential for vapor transport is much reduced. The wetland can lose water from the snow blanket, however. Sublimation and melting can create over-ice flows and some evaporative losses. These effects do not influence the under-ice water body.

There are two basic procedures for estimating ET: relating ET to pan evaporation (EP) and relating ET to meteorological conditions, with numerous modifications of each. Complete descriptions and evaluations of these procedures are given in ASCE Manual of Practice No. 70 (ASCE, 1990). Both estimating procedures use regressed equations fit to data for a fully wet condition and therefore predict potential evapotranspiration (ET_0). Because treatment wetlands are typically wet 100 percent of the time, this potential is realized, and $ET = ET_0$.

Pan Factor Methods

The pan evaporation method uses the data from a Class A evaporation pan as the reference. Each state operates pans at few stations, and data are reported in *Climatological Data,* a publication of the National Oceanic and Atmospheric Administration (NOAA), National Climatic Data Center, Asheville, NC. The pan is placed on a platform above the ground, and therefore, it evaporates more water than a lake or wetland. Wetland ET, over at least the growing season, is well represented by about 0.70 to 0.80 times Class A pan evaporation from an adjacent open site. The Class A pan integrates the effects of many of the meteorological variables, with the notable exception of advective effects. This result has been reported in several studies, including northern Utah (Christiansen and Low, 1970), western Nevada (Kadlec et al., 1987), and southern Manitoba (Kadlec, J. A., 1986). The stipulation of a time period in excess of the growing season is important because the short-term effects of the vegetation can invalidate this simple rule of thumb. The effect of climate is apparently small, since Florida data of Zoltek et al. (1979) for a wastewater treatment wetland at Clermont are represented by 0.78 times the Class A pan data from the nearby station at Lisbon on an annual basis. This multiplier is the same as that recommended by Penman (1963) for the potential ET from terrestrial systems.

Refinements to the 70 to 80 percent rule of thumb have been proposed by Christiansen (1968) in the form of multipliers:

$$ET_o = 0.755 \cdot EP \cdot C_T \cdot C_W \cdot C_H \cdot C_S \tag{9-1}$$

where

$$C_T = 0.862 + 0.179\left[\frac{T}{20}\right] - 0.041\left[\frac{T}{20}\right]^2 \tag{9-2}$$

$$C_W = 1.189 - 0.24\left[\frac{W}{1.86}\right] + 0.051\left[\frac{W}{1.86}\right]^2 \tag{9-3}$$

$$C_H = 0.499 + 0.62\left[\frac{H}{60}\right] - 0.119\left[\frac{H}{60}\right]^2 \quad (9\text{-}4)$$

$$C_S = 0.904 + 0.008\left[\frac{S}{80}\right] + 0.088\left[\frac{S}{80}\right]^2 \quad (9\text{-}5)$$

where C_H = humidity coefficient,
C_S = sunshine coefficient,
C_T = temperature coefficient,
C_W = wind coefficient,
EP = pan evaporation, mm/d
ET = evapotranspiration, mm/d
H = relative humidity, percent
S = percentage of possible sunshine, percent
T = temperature, °C
W = wind speed, m/s (1.86 m/s = 100 miles/day)

The regression coefficients were not developed for wetlands, but for well-watered grass surfaces. However, the Christiansen approach has been found to adequately represent data from wetlands in Nevada and Michigan (Kadlec et al., 1987).

Energy Balance Methods

The principal driving force for ET is solar radiation. A good share of that radiation is converted to the latent heat of vaporization. Therefore, the energy balance is the proper framework to interpret and predict not only evaporative processes, but also wetland water temperatures. The energy balance equations are more complex than the regression equations and rules of thumb in the preceding section and are not necessarily more accurate. However, the concepts will form the basis for estimating wetland water temperatures in Chapter 10.

Energy balance methods have been used to predict marsh ET in many studies. Water balances generally close adequately under this procedure, as for example in the Des Plaines River Demonstration wetlands (Hey et al., 1992). Further, direct calibrations and checks have been conducted in wetland environments (see, for example, Scheffe, 1978 or Lafleur, 1990).

About half the net incoming solar radiation is converted to water loss on an annual basis. Reported values include 0.49 (Bray, 1962), 0.47 (Christiansen and Low, 1970), 0.51 (Kadlec et al., 1987), and 0.64 (Roulet and Woo, 1986). If the radiation data from the central Florida area is used to test the concept for the Clermont wetland (Zoltek et al., 1979), the value is 0.49. However, much more refined energy balances may be utilized to estimate wetland ET. A modification of the Penman (1948) approach is described here (Wright, 1982 and 1987).

The energy balance for an interior element of wetland is depicted in Figure 9-2. Over some balance period, the net incoming radiation is dissipated to heat and evaporated water:

$$R_N = \rho\lambda_m \cdot ET + H_a + G + (U_o - U_i) + \Delta S \quad (9\text{-}6)$$

where G = conductive transfer to ground, $MJ/m^2/d$
H_a = convective transfer to air, $MJ/m^2/d$
ET = water lost to ET, m/d
R_N = net radiation reaching the ground, $MJ/m^2/d$
ΔS = energy storage change in the wetland, $MJ/m^2/d$
U_i = energy entering with water, $MJ/m^2/d$

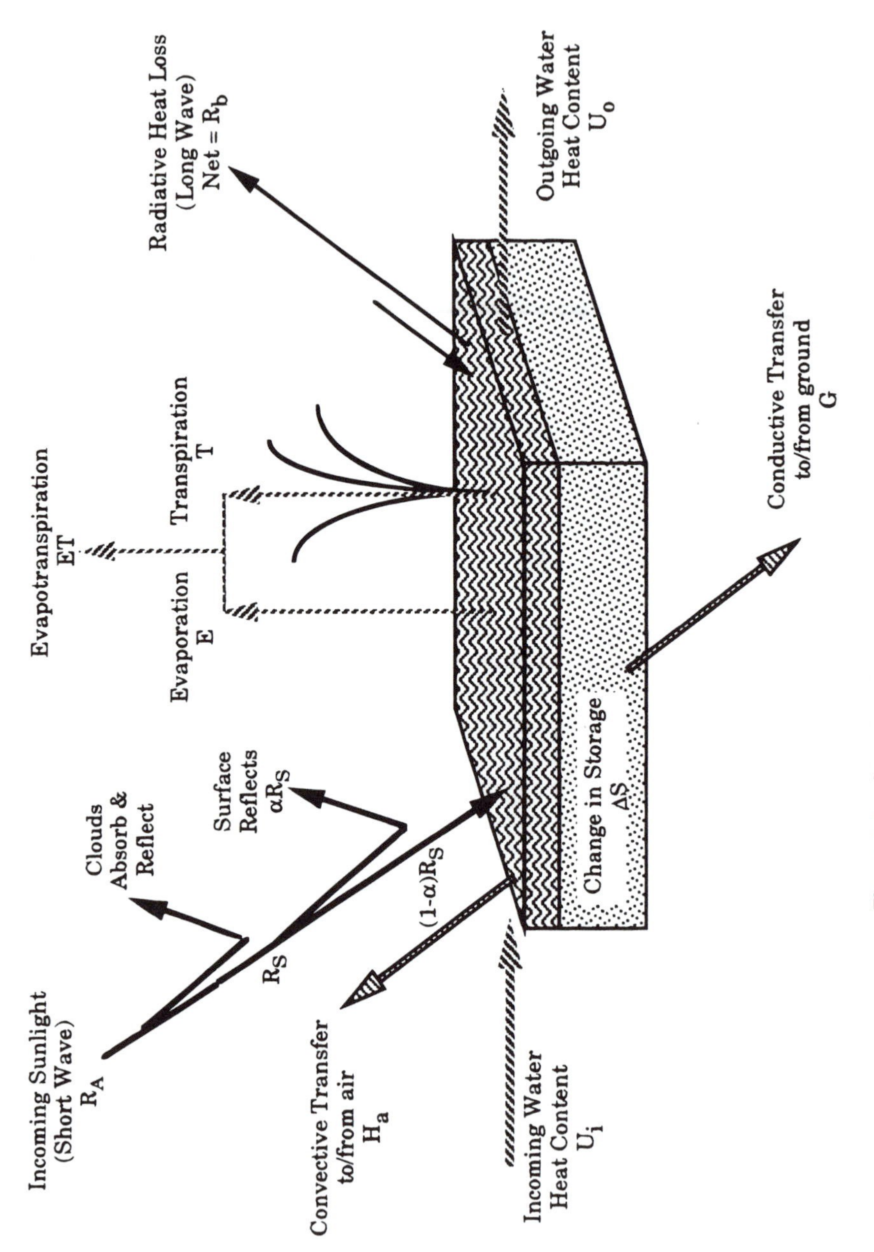

Figure 9-2 Components of the wetland energy balance.

U_0 = energy leaving with water, $MJ/m^2/d$
λ_m = latent heat of vaporization of water, MJ/kg (2.453 MJ/kg at 20°C)
ρ = density of water, kg/m^3

The value of R_N at the height of summer is on the order of 10 to 20 $MJ/m^2/d$. The maximum temperature gradients in the wetland soil column are on the order of 0.1°C/cm (see Chapter 10), which leads to an estimate of G = 0.5 $MJ/m^2/d$. Peak storage changes occur in spring (positive) and fall (negative); these are on the order of 0.5 $MJ/m^2/d$. At a hydraulic loading rate of 2.5 cm/d and a temperature change of 5°C, $(U_0 - U_I)$ = 0.5 $MJ/m^2/d$. Therefore, in the wetland environment, these terms are often negligible compared to R_N, and Equation 9-6 reduces to

$$R_N = \lambda \cdot ET + H_a \tag{9-7}$$

where $\lambda = \rho\lambda_m$, volumetric latent heat of water, = 2.453 $MJ/m^2/mm$ at 20°C.

This equation and its variants are widely used in the literature to predict ET. Its use is dependent on equations relating the quantities in Equation 9-7 to meteorological and environmental variables which are addressed here.

Net Radiation

The net radiation reaching the surface of the wetland is calculated through a series of steps which estimate the absorptive and reflective losses from incoming extraterrestrial radiation, R_A, given in Figure 9-3. The amount of radiation which makes it through the outer atmosphere is solar radiation:

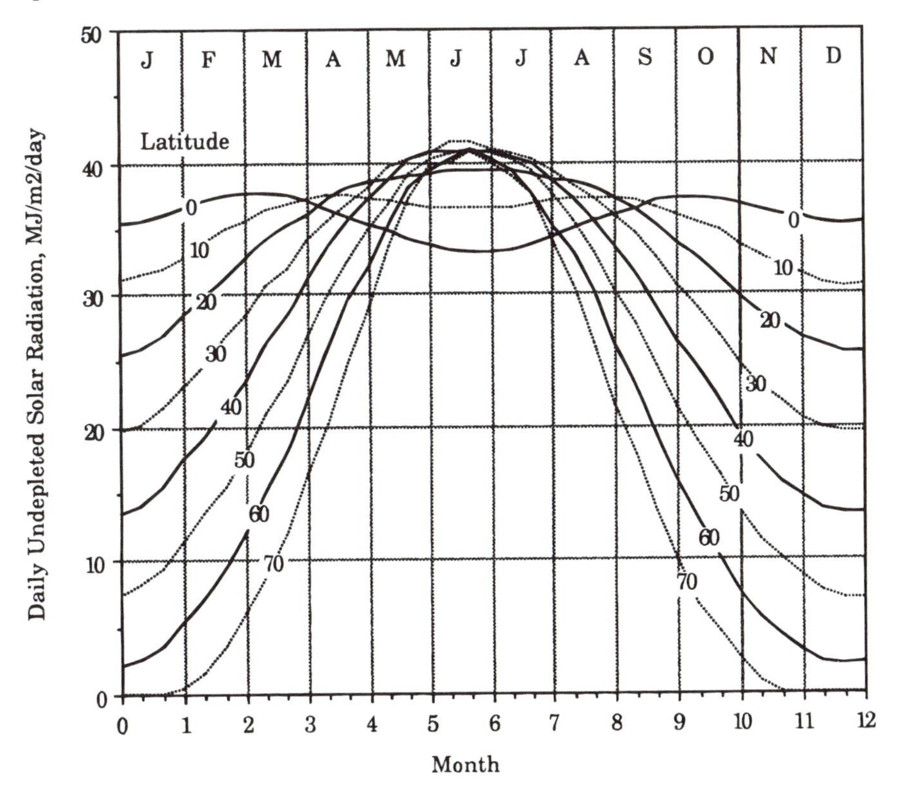

Figure 9-3 Undepleted solar radiation as a function of latitude and season. (Reprinted from ASCE, Manuals and Reports on Engineering Practice, No. 70, *Evapotranspiration and Irrigation Requirements*, Jensen, M.E., Burman, R.D., and Allen, R.G. (Eds.), 1990. With permission.)

$$R_S = \left(0.25 + 0.5\frac{S}{100}\right) \cdot R_A \tag{9-8}$$

where R_A = extraterrestrial radiation, $MJ/m^2/d$
R_S = solar radiation, $MJ/m^2/d$
S = percent daily sunshine

A fraction α, the wetland albedo, of this radiation is reflected by the wetland. A value of $\alpha = 0.23$ is commonly used for green crops (ASCE, 1990).

Net outgoing long wave (heat) radiation is computed based on atmospheric characteristics of cloud cover, absolute temperature, and moisture content:

$$R_b = \left[0.1 + 0.9\left(\frac{S}{100}\right)\right](0.34 - 0.139\sqrt{P_w^{sat}(T_{dp})})\sigma(T + 273)^4 \tag{9-9}$$

where R_b = net outgoing long wave radiation, $MJ/m^2/d$
$P_w^{sat}(T_{d_p})$ = water vapor pressure at the dew point, kPa*
T = air temperature, °C
σ = Boltzman's constant, $= 4.903 \times 10^{-9}$ $MJ/m^2/d$

In combination, the net incoming radiation is

$$R_N = 0.77R_S - R_b \tag{9-10}$$

Heat Transfer and Water Convective Mass Transfer

The saturation pressure of water (vapor pressure) is a function of temperature. It is well represented ($R^2 = 0.99999$) by

$$\ln[P_w^{sat}(T)] = 19.0971 - \frac{5{,}349.93}{T + 273.16} \tag{9-11}$$

This vapor pressure is exerted at the air-water interface.

The vapor flow is calculated as a mass transfer coefficient times the water vapor pressure difference between the water surface and the ambient air above the wetland (Figure 9-4):

$$ET = K_e[P_w^{sat}(T_w) - P_{wa}] = K_e \Delta P_w \tag{9-12}$$

where K_e = water vapor mass transfer coefficient, m/d/kPa
P_{wa} = ambient water vapor pressure, kPa
$P_w^{sat}(T_w)$ = saturation water vapor pressure at T_w, kPa
T_w = water temperature, °C

Typically, the amount of water in the ambient air is a known quantity; calculated as the relative humidity times the saturation pressure of water at the ambient air temperature:

* The metric unit of pressure is the kiloPascal (kPa). 1 Atmosphere = 101.3 kPa = 760 mmHg.

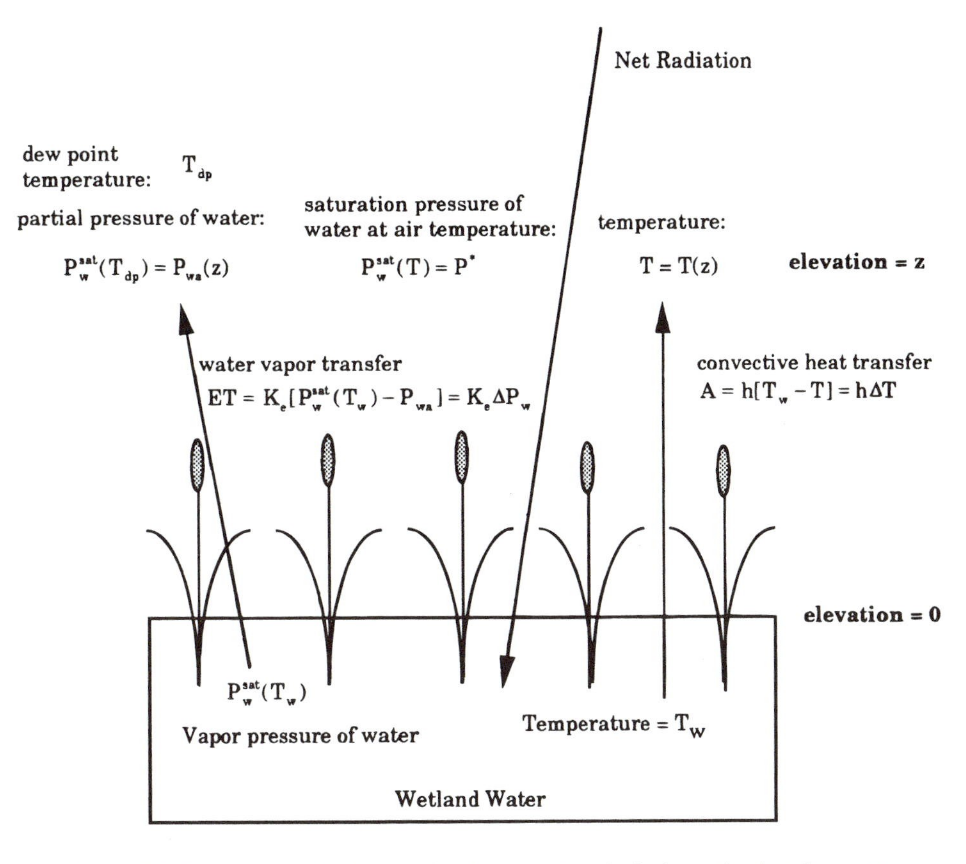

Figure 9-4 Driving forces for heat and water vapor transfer in the wetland environment.

$$P_{wa} = RH \cdot P_w^{sat}(T) \tag{9-13}$$

where RH = relative humidity fraction
T = air temperature, °C

The water transport coefficient has been found to be a linear function of the wind velocity, with the following correlation being one of several in common use (ASCE, 1990):

$$K_e = \frac{(4.82 + 6.38u)}{\lambda} = (10^{-3})(1.965 + 2.60u) \tag{9-14}$$

where u = wind speed at 2 meters elevation, m/s
$\lambda = \rho\lambda_m$ = volumetric latent heat of vaporization of water, MJ/m³ (2453 MJ/m³)

The convective heat transfer from the water to the air is likewise represented as a heat transfer coefficient times the temperature difference:

$$H_a = h[T_w - T] = h\Delta T \tag{9-15}$$

where h = heat transfer coefficient, MJ/m²/d/°C.

The relation between heat and mass transfer in the air-water system has resulted in an accurate, calibrated relation between the heat and mass transfer coefficients (ASCE, 1990):

$$h = \gamma\lambda K_e = (0.0666)(2453)K_e = 163.3K_e \tag{9-16}$$

where $\gamma = \dfrac{c_pP}{[0.622]\lambda}$ = the psychometric constant, kPa/°C
$\gamma = 0.0666$ at 20°C and 1 kPa
(0.622 = 18/29 = molecular weight ratio, water/air)
c_p = heat capacity of air, MJ/kg°C
P = ambient air pressure, kPa

thus,

$$h = (0.0666)(4.82 + 6.38u) = 0.321 + 0.425u \tag{9-17}$$

These equations may be used to estimate evaporative loss. The procedure is to find the water temperature and corresponding vapor pressure that satisfy the energy conservation Equation 9-6. Storage change and conduction losses are taken to be negligible (ASCE, 1990). The required information includes the air temperature humidity, and wind speed, together with the net incoming radiation. The full equation set is Equation 9-7 plus Equations 9-11 through 9-17.

These calculations may be easily done via an iterative spreadsheet, which searches for the water temperature (and corresponding vapor pressure, via Equation 9-13) that balances the energy flows. In the past, such spreadsheets were unavailable, and ad hoc methods were developed by Penman (1956) and others to achieve the same energy balance computations. A brief description of that method follows.

Penman Estimator

The slope of the vapor pressure curve at any temperature is Δ, and this permits a connection between the water T and dew point T and the corresponding water saturation pressures:

$$\Delta = \frac{dP_w^{sat}}{dT} \cong \frac{P_w^{sat}(T_w) - P_w^{sat}(T_{dp})}{T_w - T_{dp}} \tag{9-18}$$

After considerable manipulation (see ASCE, 1990), the Penman equation results:

$$\lambda \cdot ET = \phi(R_N - G) + (1 - \phi)\lambda K_e[P_w^{sat}(T) - P_w^{sat}(T_{dp})] \tag{9-19}$$

where

$$\phi = \left[\frac{\Delta}{\Delta + \gamma}\right] \tag{9-20}$$

Equation 9-19 requires the use of an average value of the saturation pressure of water:

Table 9-1 Variation of $\Delta/(\Delta + \gamma)$ with Elevation and Temperature[a]

Elevation (m)=		−0	500	1000	1500	2000	2500	3000
Elevation (ft) =		−0	1600	3281	4921	6562	8202	9843
°C	°F							
0.0	32	0.403	0.418	0.433	0.449	0.465	0.482	0.499
5.0	41	0.479	0.495	0.511	0.527	0.543	0.559	0.576
10.0	50	0.553	0.568	0.584	0.599	0.615	0.630	0.646
15.0	59	0.621	0.636	0.651	0.665	0.680	0.694	0.708
20.0	68	0.683	0.696	0.710	0.723	0.736	0.748	0.761
25.0	77	0.736	0.748	0.760	0.772	0.783	0.794	0.805
30.0	86	0.782	0.792	0.803	0.813	0.822	0.832	0.841
35.0	95	0.820	0.829	0.838	0.846	0.854	0.862	0.870
40.0	104	0.852	0.859	0.866	0.873	0.880	0.887	0.893
45.0	113	0.877	0.884	0.890	0.896	0.901	0.907	0.912
50.0	122	0.898	0.904	0.909	0.914	0.918	0.923	0.927

[a] $\gamma/(\Delta + \gamma) = 1 - [\Delta/(\Delta + \gamma)]$. Assumption $c_p = 1.013$ kJ kg^{-1} °C^{1} (0.242 cal g^{1} °C^{1}) at 0°C^{1} and at sea level.

$$P_w^{sat}(T) = \frac{1}{2}[P_w^{sat}(T_{max}) + P_w^{sat}(T_{min})] \tag{9-21}$$

Values of the parameter ϕ are given in Table 9-1. Vapor pressures of water may be calculated via Equation 9-13.

Example 9-1

A peatland at latitude 45°N and elevation 500 m (Houghton Lake, MI) is exposed to the following August meteorological conditions:

Maximum air temperature	23.3°C
Minimum air temperature	8.9°C
Mean air temperature	16.1°C = 289 K
Relative humidity	40 percent
Percent sunshine	60 percent
Wind velocity	0.43 m/s

From this information, the balance of the meteorological data can be calculated:

Water pressure if saturated at 16.1°C	1.826 kPa	Equation 9-11
Actual water pressure	0.730 kPa	0.4 = RH
Dew point temperature	2.4°C	Equation 9-11
Saturation water pressure at T_{max}	2.861 kPa	Equation 9-11
Saturation water pressure at T_{min}	1.139 kPa	Equation 9-11
Average saturation water pressure	2.000 kPa	Equation 9-21
Extraterrestrial radiation	32.21 MJ/m²/d	Figure 9-2
Solar radiation	17.72 MJ/m²/d	Equation 9-8
Net long wave back radiation	4.86 MJ/m²/d	Equation 9-9
Net incoming radiation	8.78 MJ/m²/d	Equation 9-10
Penman factor, ϕ	0.636	Table 9-1

Conduction losses to the ground are assumed negligible for reasons given earlier. The Penman Equation 9-19 may now be exercised to obtain the estimated ET:

$$\lambda ET = 0.636(8.78 - 0) + (0.364)[4.82 + 6.38(0.43)][2.000 - 0.730]$$

$$\lambda\ ET = 9.08\ MJ/m^2/d$$

$$ET = 9.08/2.453 = 3.70\ mm/d$$

The Penman Estimate

Soil temperature profiles were measured on this date and may be used to estimate the effect of neglecting this energy transfer. The conduction heat equation is

$$G = k\frac{dT}{dz} = 0.0518\frac{(-5)}{0.6} = -0.43\frac{MJ}{m^2d} \tag{9-22}$$

where k = soil thermal conductivity, $MJ/d/m^2/(°C/m)$ (see Table 10-1)
z = depth into soil, m

The new Penman estimate of ET = 3.59 mm/d or 3 percent lower than the previous estimate.

The nearest Class A evaporation pan is at the Lake City, MI weather station. Pan evaporation at the end of August has averaged 4.83 ± 0.25 mm/d. From this reference, the Christiansen method (Equation 9-1) may be applied. The coefficients are

$$C_T = 0.862 + 0.179\left[\frac{16.1}{20}\right] - 0.041\left[\frac{16.1}{20}\right]^2 = 0.980 \tag{9-2}$$

$$C_W = 1.189 - 0.24\left[\frac{0.43}{1.86}\right] + 0.051\left[\frac{0.43}{1.86}\right]^2 = 1.136 \tag{9-3}$$

$$C_H = 0.499 + 0.62\left[\frac{40}{60}\right] - 0.119\left[\frac{40}{60}\right]^2 = 0.859 \tag{9-4}$$

$$C_S = 0.904 + 0.008\left[\frac{60}{80}\right] + 0.088\left[\frac{60}{80}\right]^2 = 0.957 \tag{9-5}$$

Thus, the estimated ET is

$$ET = 0.755 \cdot 4.83 \cdot 0.980 \cdot 1.136 \cdot 0.859 \cdot 0.957 \tag{9-1}$$

$$ET = 0.691 \cdot 4.83 = 3.34 \text{ mm/d}$$

The Christiansen Estimate

The quickest estimate of 70 to 80 percent of pan evaporation leads to an estimate of 3.48 to 3.86 mm/d.

FORESTED WETLANDS

Agricultural forest canopies of sufficient areal extent do not display radically different crop coefficients than those for nonwoody crops (ASCE, 1990). It is therefore anticipated that the predictive approaches for marshes will also hold approximately for forested wetlands, but calibrations and comparisons for forested wetlands are not available.

SUBSURFACE-FLOW WETLANDS

When the water surface is below the ground, a key assumption in the energy balance approach is no longer valid; the transfers of water vapor and sensible heat are no longer similar. Water vapor must first diffuse through the dry layer of gravel and then be transferred by swirls and eddies up through the vegetation to the air above the ecosystem. Heat transfer to the water must now pass through a porous media in addition to the eddy transport in the

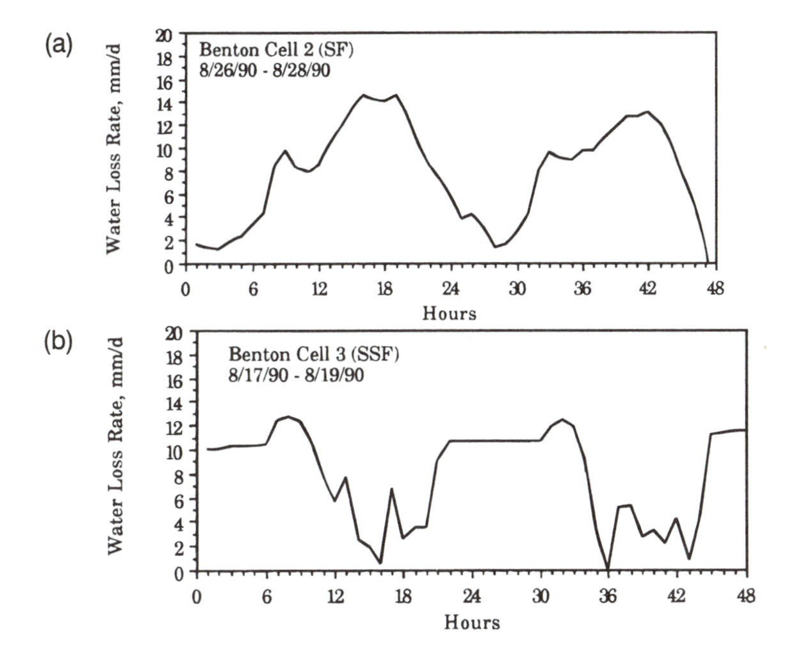

Figure 9-5 (a) Diurnal flow variation over a 2-day period for an SF wetland at Benton, KY. (b) Diurnal flow variations over a 2-day period for an SSF wetland at Benton, KY. The mean loss rate was about 7.5 mm/d in both cases.

air for convective transport or in addition to radiative transport to the gravel surface. The heat storage capacity of the media is also directly involved, since it is in the water.

The energy balance approach is still valid, but there are no estimates of the transport coefficients within the porous media. It is therefore necessary to rely on wetland-specific information. Treatment wetland data collected at Benton, KY provide some insights concerning ET in SSF wetlands (Figure 9-5). Both the SF and SSF wetlands show a strong diurnal swing in outflow, presumably in response to the diurnal radiation pattern, and the resultant pattern in ET. The SF wetland shows a maximum loss rate at about 1800 hours, which is in agreement with results from other sites (Kadlec et al., 1987). The SSF wetland shows a maximum in loss rate during the middle of the night. The amplitude of the cycle is larger for the SF wetland. Both of these observations are consistent with the larger heat storage capacity of the water plus gravel system.

Water budgets were used by Bavor et al. (1988) to estimate gravel bed wetland ET. The correlations were

Cattail/gravel

$$\begin{gathered} ET = 1.128\ EP + 0.072\ mm/d \\ R^2 = 0.72 \\ 12 < T_{air} < 25°C \end{gathered} \tag{9-23}$$

Bulrush/gravel

$$\begin{gathered} ET = 0.948\ EP - 0.0027\ mm/d \\ R^2 = 0.93 \\ 12 < T_{air} < 25°C \end{gathered} \tag{9-24}$$

Gravel(no plants)

$$\begin{gathered} ET = 0.0757\ EP - 0.028\ mm/d \\ R^2 = 0.15 \\ 12 < T_{air} < 25°C \end{gathered} \tag{9-25}$$

The transpiration component of ET is clearly important in SSF systems.

OVERLAND FLOW IN WETLANDS

There is not a long history of research and development related to overland flow in wetlands. Mathematical descriptions are often adaptations of turbulent, open channel flow formulae. These are discussed in detail in a number of texts, for example, the work of French (1985). The general approach is utilization of mass, energy, and momentum conservation equations, coupled with an equation for frictional resistance. Perhaps the most common friction equation is Manning's equation, which will be discussed more fully later in this section. There is a fundamental problem with the utilization of Manning's equation to wetland surface water flows: Manning's equation is a correlation for turbulent flows, while SF wetlands are nearly always in a laminar or transitional flow regime based on open channel flow criteria. Under these conditions, Manning's n is strongly velocity dependent (Hosokawa and Horie, 1992). There is also a difficulty with the extension of open channel flow concepts to densely vegetated channels. The frictional effects that retard flow in open channels are associated primarily with drag exerted by the channel bottom and sides. Wetland friction in dense macrophyte stands is dominated by drag exerted by the stems and litter, with bottom drag playing a very minor role.

As a consequence, overland flow parameters determined from open channel theory are not applicable to wetlands. In particular, Manning's coefficient is no longer a constant; it depends upon velocity and depth as well as stem density. Predictions from previous information on nonwetland vegetated channels are seriously in error (Hall and Freeman, 1994). Unfortunately, virtually all existing information on wetland surface flow has been interpreted and reported via Manning's equation, and so it cannot be avoided.

This section provides methods to predict depths and velocities in surface-flow wetlands.

THE CONSERVATION LAWS

Models of channel water flow can be quite complex when all possibilities of bed slope, water flow rate, and depth are considered. The situation is even more complex if two lateral directions and transient behavior are considered. Fortunately, wetland surface flows form a somewhat simpler subset: they are gradually varied flows on very mild slopes. Kinetic energy changes are also usually negligible compared to potential energy changes. As a result, the energy and momentum balances do not contribute to calculations; these simply state that potential energy is converted solely to frictional work and that flow proceeds with gravitational forces balanced by frictional forces.

The mass balance for water flow in two dimensions is

$$\frac{\partial(h\varepsilon)}{\partial t} = -\frac{\partial(hu_x)}{\partial x} - \frac{\partial(hu_y)}{\partial y} + P - ET \tag{9-26}$$

where ET = evapotranspiration rate, m/d
h = water depth, m
P = precipitation rate, m/d
t = time, d
u_x = superficial water velocity, in x-direction, m/d
x = distance from inlet, m
y = transverse distance, m
ε = volumetric fraction water in the wetland, m^3/m^3
u_y = superficial water velocity, in y-direction, m/d

A one-dimensional flow description suffices for many applications, and that will be further developed here. A schematic representation of the notation is shown in Figure 9-6. The plants and litter in the wetland typically occupy some small fraction of the total volume, on the order of $\varepsilon = 2$ to 10 percent. Wetlands are rarely in true steady flow, but if a long averaging period is considered, changes in storage become less important, and periodic ET and stochastic rain events may be replaced by their time averages. The high-frequency events are not described, but time average depths and flows are properly modeled. The time-averaging, overbar symbol will not be used in this section, but it is implicit in every time varying quantity.

These conditions of steady flow in one dimension lead to substantial simplification:

$$\frac{d(hu)}{dx} = \frac{d(Q/W)}{dx} = (P - ET) \tag{9-27}$$

where Q = volumetric flow rate, m^3/d
W = wetland width, m

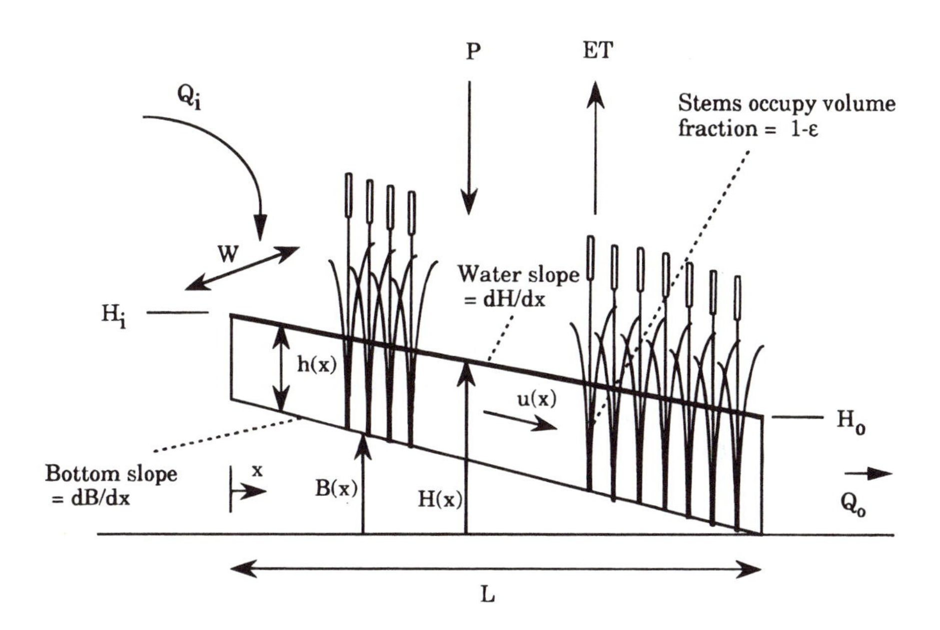

Figure 9-6 Notation used to describe overland flow in SF wetlands. Inlet is denoted by subscript "i"; outlet is denoted by subscript "o."

B(x) = ground elevation above datum, m
h(x) = water depth, m
H(x) = water surface elevation above datum, m
L = wetland length, m
Q = volumetric flow rate, m^3/d
u(x) = superficial water velocity, m/d
W = wetland width, m
x = distance from inlet, m
ε = volume fraction water, m^3/m^3

or, the equivalent:

$$\frac{d\Lambda}{dx} = (P - ET) \tag{9-28}$$

where Λ = Q/W, volumetric flow per unit width, $m^3/m/d$.

The inlet boundary condition that normally accompanies Equation 9-27 for treatment wetlands is a specification of the inlet flow rate:

$$Q(x = 0) = Q_i \tag{9-29}$$

One integration of Equation 9-27 is easy because P-ET is not a function of position:

$$Q = Q_i + (P - ET)Wx \tag{9-30}$$

Equation 9-30 represents the simple fact that rain increases the volumetric flow, while ET decreases it, as the water progresses through the wetland.

However, it is necessary to do more because, as yet, there is no connection with depth. The volumetric flow rate is related to linear velocity via geometry:

$$Q = uWh = v\varepsilon Wh \tag{9-31}$$

where v = actual water velocity, m/d.

Water elevation is water depth plus bed bottom elevation profile (Figure 9-6):

$$H = B + h \tag{9-32}$$

where B = elevation of the bed bottom above datum, m.

One more piece of information is required, that being the relation between velocity and water slope, which is the friction equation for the wetland.

FRICTION EQUATIONS

Water flow through the wetland is associated with a local frictional head loss:

$$u = ah^{b-1}S^c \tag{9-33}$$

where a,b,c = constants
S = −dH/dx = negative of the water surface slope, m/m

When a = 1/n, b = 5/3, and c = 1/2, Equation 9-33 becomes Manning's equation (see French, 1985, for example):

$$u = \frac{1}{n} h^{2/3}S^{1/2} \tag{9-34}$$

where n = Manning's coefficient, $s/m^{1/3}$.

Note that a unit conversion is necessary to convert to the mass balance unit of days.

Equations 9-30, 9-31, and 9-33 combine to provide an equation for the depth of flow as a function of position:

$$aWh^b\left(-\frac{d(h+B)}{dx}\right)^c = Q_i + (P - ET)Wx \tag{9-35}$$

The boundary condition necessary to solve Equation 9-35 is typically a specification of the exit water level, as determined by a weir or receiving pool:

$$H(x = L) = H_o \tag{9-36}$$

Equations 9-35 and 9-36 cannot be solved analytically to a closed form answer, but numerical solution is quite easy via any one of a number of methods.

A great deal of important information can be obtained from a special case, the case of negligible (P-ET), for which

$$aWh^b\left(-\frac{d(h+B)}{dx}\right)^c = Q_i \tag{9-37}$$

The numerical solution of Equation 9-37 is still necessary for wetlands with any sort of nonlevel bottom slope or bottom elevation profile. The use of average depths and slopes can lead to significant errors because of the extremely nonlinear character of Equation 9-37. The solution is easily done by spreadsheet.

There is a special depth which forms a dividing line between distance-thinning and distance-thickening flows. If the depth in Equation 9-37 is constant, it is termed the *normal depth,* and then flow is parallel to the bottom. This occurs for the case of a flat, inclined bottom.

$$h_n^b = \frac{\Lambda}{a\left(-\frac{dB}{dx}\right)^c} \tag{9-38}$$

Example 9-2

A 25 × 100 m wetland is to be operated with the exit water level set at 20 cm. The bottom is inclined at a slope of 0.01 percent. The wetland is densely vegetated, and the Manning's coefficient is estimated to be constant at $n = 1.0\ m^{-1/3}\ s$. What water surface profile may be expected at flow rates of approximately 200 m^3/d?

Solution

The possible contribution of rain and ET will not be considered in this example. First, calculate the normal depth for the bottom slope in question. The required values of the constants for Equation 9-38 are

$$a = 1/n = 1.0\ m^{1/3}/s \cdot 86{,}400\ s/d = 86{,}400\ m^{1/3}/d$$

$$b = 5/3$$

$$c = 1/2$$

Then,

$$h_n^b = \frac{200/25}{86{,}400(0.0001)^{1/2}} = 0.00926$$

$$h_n = 0.060\ m$$

Thus, this wetland can carry the design flow at a constant depth of 6 cm. By setting the exit weir above this level, the wetland will operate in the mode of distance-thickening flow.

Next, calculate the operational surface elevation profile from Equation 9-37. This calculation requires a numerical integration, which is readily accomplished on a spreadsheet. The result is a nearly horizontal water surface (gradient = 0.0000020). This wetland can comfortably carry the design flow without any measurable friction head loss.

Example 9-3

The Houghton Lake, MI treatment wetland is approximately 1000 m wide and extends to an impoundment located approximately 3 km from the discharge line. The slope of the ground is 0.00014. In August 1979, depths were observed to be about 15 cm in the treatment zone, which extended only a fraction of the distance to the outlet pool. Under these conditions, what influence does the level of the outlet pool have on the depths in the treatment zone? The average flow was 4000 m^3/d. The values of the power law friction coefficients have been calibrated to be

$$a = 1.0E + 7\ m^{-1}d^{-1}$$

$$b = 3$$

$$c = 1$$

Solution

The possible contribution of rain and ET need not be considered in this example. In fact, for the data period of August 1979, ET was measured to be 8.1 cm and precipitation totaled 9.0 cm. The difference is about 3 percent of the applied water. First, calculate the normal depth for the bottom slope in question. From Equation 9-38,

$$h_n^3 = \frac{4000/1000}{1.0E7 \cdot (0.00014)^1} = 0.00286$$

$$h_n = 0.142\ m$$

Thus, this wetland can carry the design flow at a constant depth of 14.2 cm.

Beavers, man, and P-ET control the level in the receiving impoundment. To answer the question posed, the depth at the entrance to the impoundment is set to values below and above the normal depth: 10 and 30 cm, respectively. Next, calculate the operational surface elevation profile from Equation 9-37. This calculation requires a numerical integration, which is readily accomplished on a spreadsheet. The result in the treatment zone is a water surface parallel to the ground slope at the normal depth for both pool elevations (Figure 9-7). The effect of pool elevation is not appreciable for upstream zones with ground elevation higher than the receiving pool elevation. It is of interest to calculate the Manning's coefficient for this same system:

$$\left[\frac{1}{n} h^{5/3}\left(-\frac{dB}{dx}\right)^{1/2}\right] = \frac{Q}{W} = \Lambda \tag{9-37}$$

$$\left[\frac{1}{n} 0.142^{5/3}(0.00014)^{1/2}\right] = \frac{4000}{1000}$$

$$n = 0.0001143\ d/m^{1/3} = 9.8\ s/m^{1/3}$$

This high value of n pertains to a shallow depth of flow, which is heavily influenced by litter and microtopography.

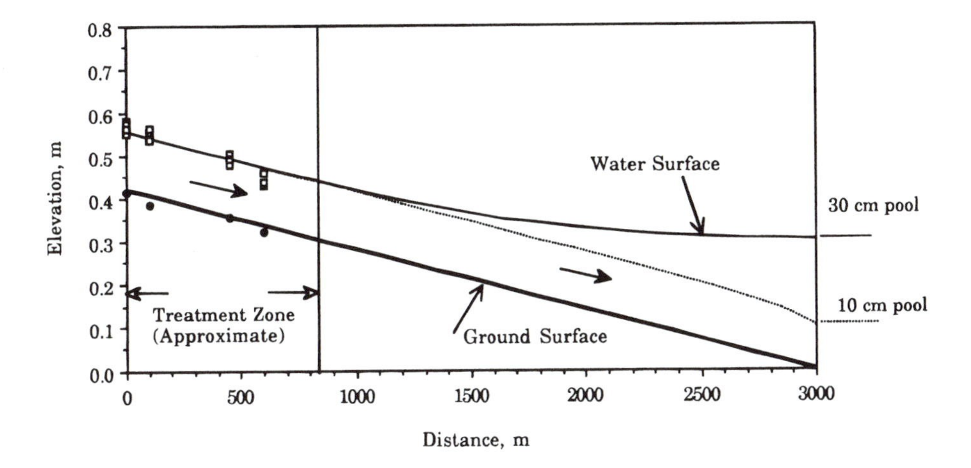

Figure 9-7 The typical water surface profile for the Houghton Lake wetland treatment system. The downstream pool control on water elevation is far removed from the treatment zone; flow in that zone is at the normal depth. Weekly data points are shown for August 1979.

In the preceding two examples, the wetlands were operated at the two limits of the water surface profile: a level pool in the wetland and a water surface slope equal to the ground slope. Intermediate cases obviously can exist, as illustrated in Figure 9-8.

FRICTION EQUATION COEFFICIENTS

It would be desirable to have predictive methods for the parameters a, b, and c in friction equations. At the present time, data exist for only a few types of wetlands. Site-specific factors are known to be very important, and it is not possible to extrapolate from nonwetland information.

Manning's Coefficients

Suppose that open channel information were to be used to estimate Manning's n for a wetland. Guidance may be found in estimation procedures in the hydraulics literature, for instance, French (1985). The value of n may be estimated from information on the channel

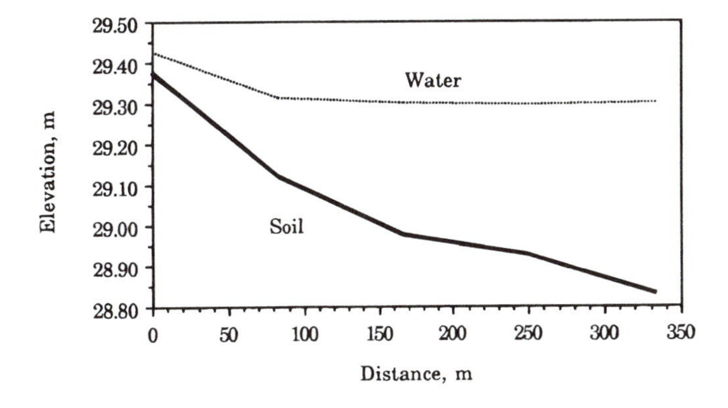

Figure 9-8 Profiles of soil and water elevations for Benton, KY cell 2. The Manning's coefficient for this profile is 3.2 $s/m^{1/3}$. Note that there are influences of both the outlet weir elevation and the bed slope.

character, type of vegetation, changes in cross-section, surface irregularity, obstructions, and channel alignment. Using the highest value of every contributing factor, the maximum open channel n value is 0.29 $s/m^{1/3}$ (French, 1985). This is approximately one order of magnitude less than values determined from actual wetland data. Clearly, nonwetland information is inadequate.

Florida emergent marsh studies comprise a large fraction of the available wetland friction information. These serve to provide general guidelines for site-specific factors. Generally, Manning's n is strongly depth dependent, decreasing as depth increases. The nature of this dependence is illustrated in Figure 9-9 for three Florida marsh studies. A general equation for this dependence is

$$\frac{n}{n_l} = \left(\frac{h_l}{h}\right)^m \tag{9-39}$$

Use of Equation 9-39 requires a single value of n at a known depth. For sparse vegetation and depths greater than 20 cm, the value of m ranges from 0.33 (MacVicar, 1985) to 0.5 (Shih and Rahi, 1982). At smaller depths, the value of m is greater, in the range of 1.0 to 2.0.

Likewise, n values are dependent on vegetation density because stems and litter provide the dominant drag surfaces. The relationship is linear for *Scirpus validus* (Figure 9-10). Therefore, it is not surprising to find a strong seasonal dependence (Figure 9-11). Since both litter and live stems are involved, the relation is not easily predictable; it depends on litterfall events.

The progress of a constructed system from an initial sparse vegetation to a more densely vegetated condition is accompanied by increases in the friction coefficient. Boney Marsh, FL received pumped river water over several years beginning in 1976. Hydrologic studies produced weekly values of Manning's n (Mierau and Trimble, 1988). The biological dynamics of the Boney Marsh operation produced considerable scatter in these n values, but the year-to-year trend line was upward from 0.6 to 2.7 $s/m^{1/3}$ (Figure 9-12).

As a result of the Florida work and work at the Houghton Lake, MI site (Kadlec et al., 1981; Hammer and Kadlec, 1986; Kadlec, 1990a), it is possible to place a general lower bound on Manning's coefficients for flow-through emergent marshes with relatively low vegetation densities. Figure 9-13 indicates this lower bound as a boomerang-shaped area. This estimate would apply to sedge meadows. More heavily fertilized systems, such as those at Benton, KY, display larger values of Manning's n. For a treatment marsh receiving nutrient-

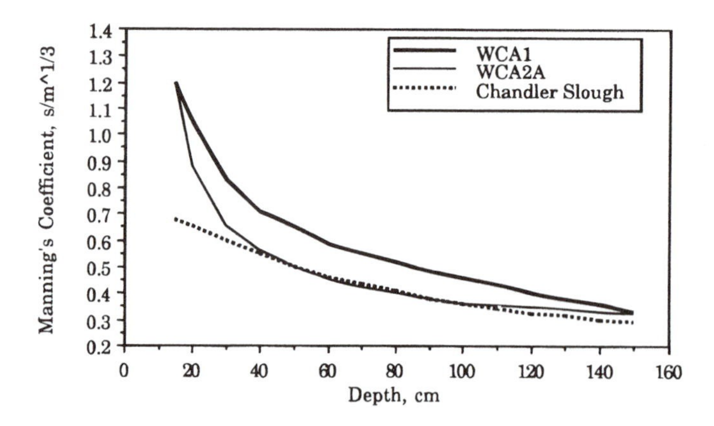

Figure 9-9 Manning's coefficient vs. depth for three Florida marshes. These wetlands store and convey runoff water. (Replotted from Shih, et al., 1979).

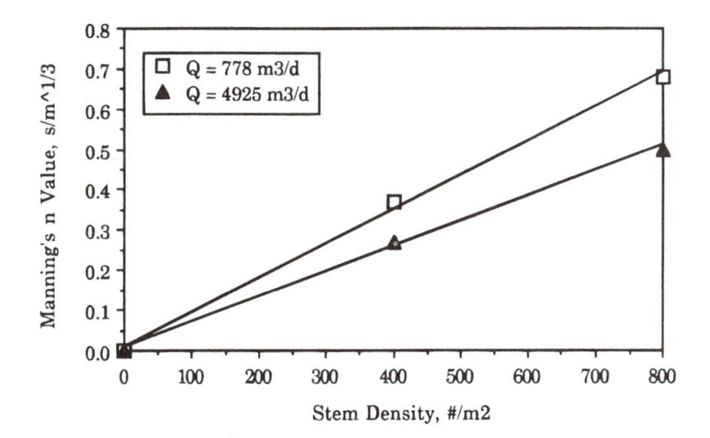

Figure 9-10 The dependence of Manning's n value on bulrush stem density in experimental channels at high velocity. The depths were on the order of 30 cm. (Replotted from Hall and Freeman, 1994).

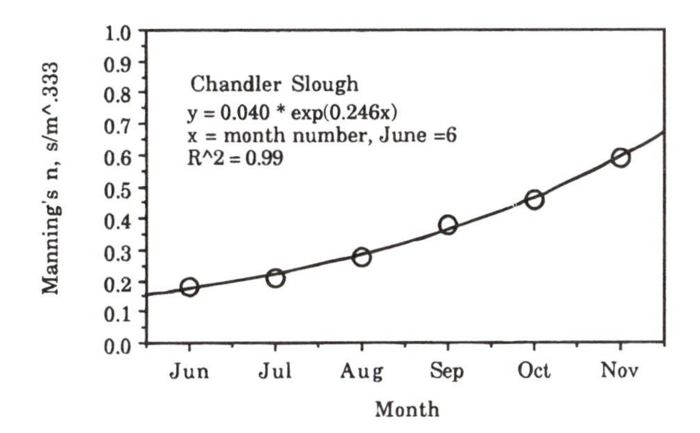

Figure 9-11 The variation of Manning's coefficient at a depth of 50 cm in Chandler Slough. (Redrawn from Shih and Rahi, 1982).

rich water, the vegetation is dense. A preliminary suggestion may be made for estimating n, as shown by the "D" line on Figure 9-13:

$$n \approx 1.0\ h^{-1.7} \tag{9-40}$$

$$0.1 \leq h \leq 1.0\ \text{m}$$

If vegetation is sparse, with no litter, then the "S" line is a better estimator:

$$n \approx 0.2\ h^{-1.7} \tag{9-41}$$

$$0.05 \leq h \leq 1.0\ \text{m}$$

The Power Law Model

Manning's "constant" is clearly not constant for the wetland environment, and it would be preferable to utilize a model which describes the depth variability. There are sound reasons

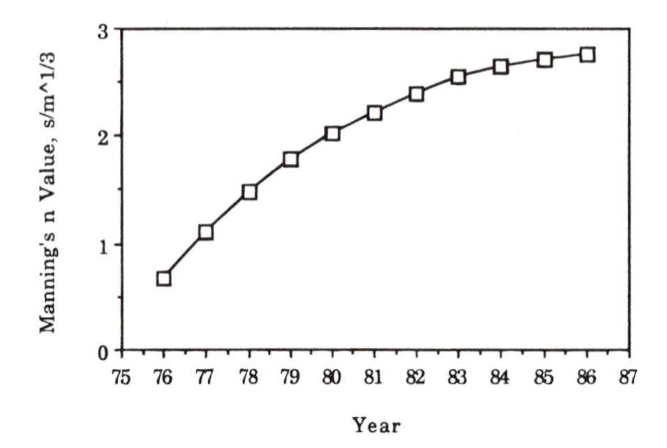

Figure 9-12 The increase of friction in a developing wetland as shown by the central tendency of each year's data. Vegetation densities in Boney Marsh may have been only part of the reason for increases; increases in litter and sediment accumulation could have altered these measurements. (Data from Mierau and Trimble, 1988).

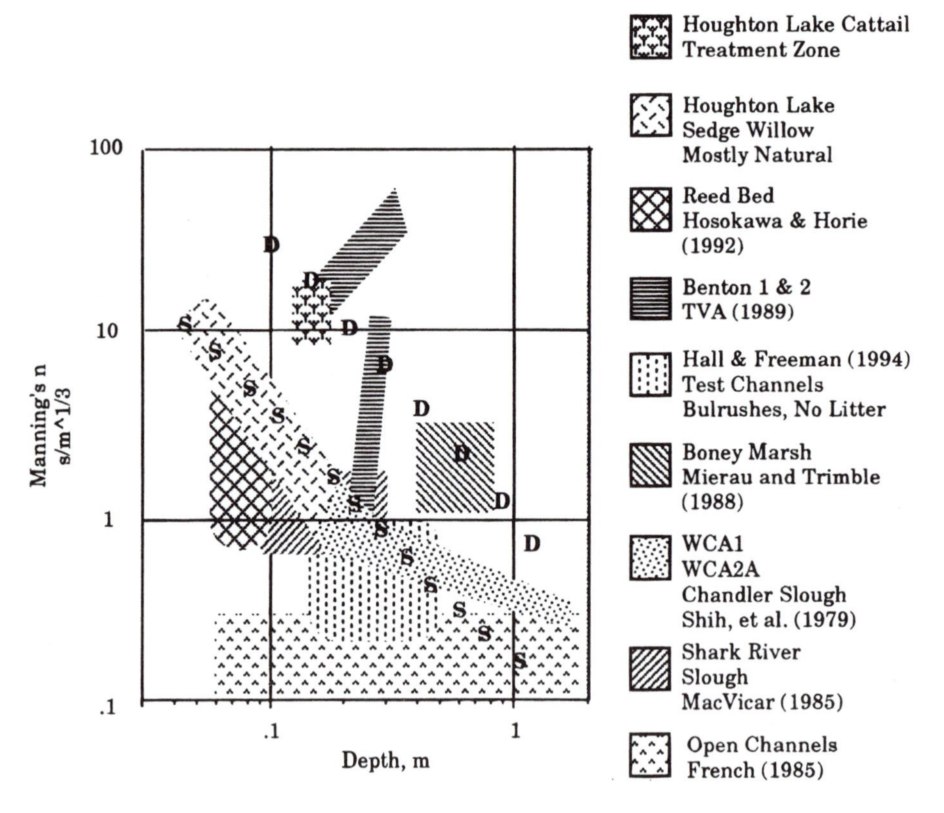

Figure 9-13 Ranges of Manning's coefficient. Values for open channels are too low for emergent marshes even after adjustment for vegetation and roughness. The clusters representing Houghton Lake, Shark River Slough, WCA1, WCA2A, Chandler Slough, and test channels are for a low-nutrient condition with relatively sparse vegetation. Boney Marsh and Benton 1 & 2 are more densely vegetated. The line of letter "D" is a speculative line for dense emergent vegetation; letter "S" is for sparse stands of plants.

for adopting a different form of the friction equation (Kadlec, 1990a). First, Manning's equation applies to turbulent flows controlled by bottom friction, whereas wetland flows are transition flows controlled by vegetation resistance. Second, the depth of flow is typically so shallow that the topography of the wetland bottom is of importance. At very shallow depths, portions of the bottom are not submerged, and at typical operating conditions, the distribution of depths is fairly wide. Third, there is a vertical distribution of the density of submerged vegetation and litter. Thus, the bottom layer at any depth is apt to have more resistance than upper layers. These factors are easily and completely described by the simple expedient of relaxing the exponents in the Manning's formulation to yield Equation 9-33.

$$u = ah^{b-1}S^{c} \tag{9-33}$$

or

$$Q = aWh^{b}S^{c} \tag{9-42}$$

The exponent "c" is 0.5 in the turbulent open channel formulation. However, wetland investigations indicate a higher value. Kadlec et al. (1981) and Hammer and Kadlec (1986) found reasonable data fits for $c = 1.0$ for sedge wetlands. Roig and King (1993) presented tidal marsh model flume data, for which $c = 0.80$. Kadlec (1990a) showed that a typical bottom topography coupled with a measured vertical stem density and a conventional correlation for stem drag produced an exponent of 0.71. As a limiting value, laminar flow around a uniform array of submerged objects over a flat bottom is theoretically described by $c = 1.0$. Until more data becomes available, a value of $c = 1.0$ is recommended.

The exponent "b" is 1.67 in the turbulent open channel formulation. But the depth variability measured for wetlands increases this value by an amount ranging from 0.33 (MacVicar, 1985) to the value 1.7 in Equation 9-40, although the latter is for use with $c = 0.5$. The end result is a range of $2.0 \leq b \leq 3.4$. Until more data becomes available, a value of $b = 3.0$ is recommended. As noted by Kadlec et al. (1981) and Hammer and Kadlec (1986), water surface profiles are not sensitive to these parameters over usual operating ranges.

The coefficient "a" remains a function of vegetation and litter density. Calibrations of Equation 9-42 have been made for treatment wetlands in the startup mode (Hammer and Kadlec, 1983). But because of the prevalence of Manning's equation as an analytical tool, the comparison of available data and Equation 9-42 is best accomplished via Figure 9-13. Because the depth dependence of the flow rate is embedded in Equation 9-42, a single value of "a" describes the trends in the n values. Until more data become available, a value of $a = 1.0 \times 10^{7}\ m^{-1}d^{-1}$ is recommended for densely vegetated wetlands.

In summary, for densely vegetated marshes,

$$Q = (1 \times 10^{7})Wh^{3}S \tag{9-43}$$

And for sparsely vegetated marshes,

$$Q = (5 \times 10^{7})Wh^{3}S \tag{9-44}$$

where the units are meters and days.

HEAD LOSS CALCULATIONS

The implementation of Equations 9-35 or 9-37 requires numerical integration, which is inconvenient in conceptual design calculations. Because of the extreme nonlinearity of the equations, it is very inaccurate to use average values. Accordingly, it is better to use precalculated values of the head loss for the intended design conditions. To accomplish this, the case of a rectangular constructed wetland is considered, with a negligible loss or gain of water due to P and ET. Equation 9-37 is dedimensionalized using the wetland length and the outlet water depth:

$$y = \frac{h}{h_o} \qquad y_B = \frac{B}{h_o} \qquad z = \frac{x}{L} \qquad S_1 = \frac{dy_B}{dz} \tag{9-45}$$

$$y^3\left(-\frac{d(y + y_B)}{dz}\right) = y^3\left(-\frac{dy}{dz} - S_1\right) = \frac{qL^2}{ah_o^4} = M_1 \tag{9-46}$$

where h_o = water depth at outlet, m
L = wetland length, m
q = hydraulic loading rate, m/d

and the rest of the new variables are defined in Equations 9-45 and 9-46. It is presumed that the outlet water depth is fixed, since a weir is frequently used at outlets. Integration of Equation 9-46 yields the inlet water depth and hence the head loss for a given wetland. Solutions to Equation 9-46 depend on two parameters: S_1 which represents the bed slope and M_1 which contains the friction coefficient, the hydraulic loading rate, outlet depth, and wetland length.

Figure 9-14 presents the solution of Equation 9-46 for different parameter values. It may be used to estimate head losses in SF wetlands.

Example 9-4

An SF wetland is to be built to treat 200 m^3/d of secondary municipal wastewater. The appropriate hydraulic loading rate has been determined to be 2 cm/d. Site considerations indicate that a length of 400 m is desirable. A bed slope of 20 cm over the 400 m length is to be used to provide drainage. The outlet weir is to be set at 20 cm depth. What is the estimated head loss?

Solution

The possible contribution of rain and ET are not be considered in this example. The constants needed to use Figure 9-14 are

$$a = 1 \times 10^7\ d^{-1}m^{-1} \qquad h_o = 0.20\ m \qquad L = 200\ m \qquad q = 0.02\ m/d$$

$$dB/dx = 0.20/400 = 0.0005 \qquad S_1 = 1.0 \qquad M_1 = \frac{(0.02)(400)^2}{(0.2)^4(1 \times 10^7)} = 0.2$$

Referring to Figure 9-14, the ratio of inlet depth to outlet depth is 0.6. Therefore,

$$h_i = (0.6)(0.20) = 0.12\ m = 12\ cm \qquad H_i = B_i + h_i = 0.20 + 0.12 = 0.32\ m$$

$$\Delta H = 0.32 - 0.20 = 0.12\ m = 12\ cm$$

If the complete profile is desired, integration of Equation 9-37 is required. This profile is shown in Figure 9-15A. Note that the normal depth is given by Equation 9-38:

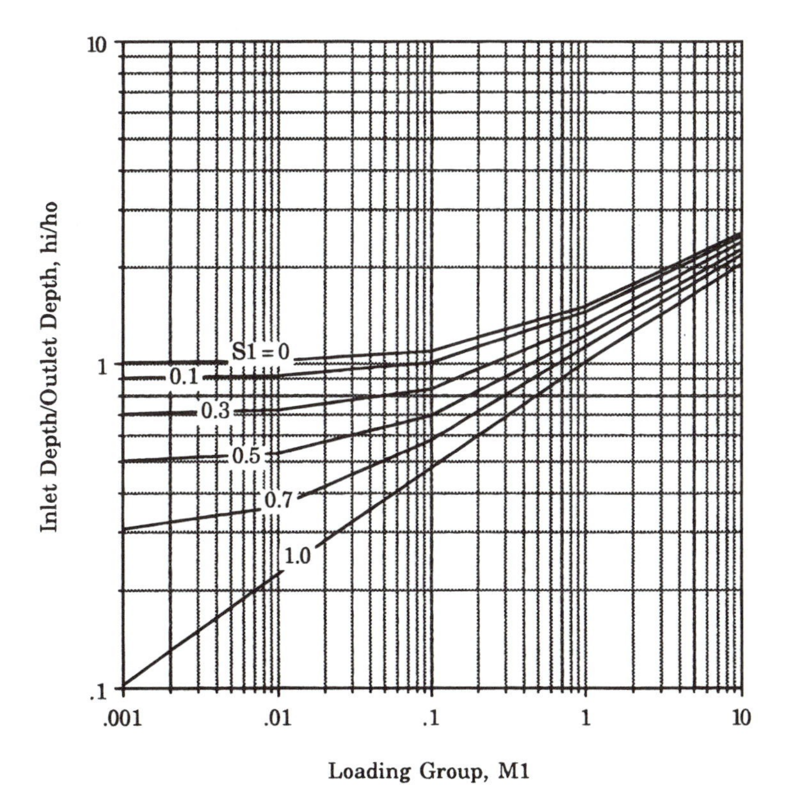

Figure 9-14 Inlet to outlet depth ratio for SF wetlands of different slops and different loading rates. The friction power law is used, with b = 3 and c = 1.
Note:

$$S_1 = \frac{L}{h_o}\left(-\frac{dB}{dx}\right) \qquad M_1 = \frac{qL^2}{h_o^4 a}$$

$$h_n^b = \frac{200/25}{(1 \times 10^7)(0.0005)}$$

$$h_n = 0.117 \text{ m}$$

This turns out to be the inlet depth in this example.

Example 9-5

Repeat Example 3 for a flat bottom.

Solution

The constants needed to use Figure 9-14 are the same, except $S_1 = 0$. Referring to Figure 9-14, the ratio of inlet depth to outlet depth is 1.16. Therefore,

$$H_i = h_i = (1.16)(0.20) = 0.232 \text{ m} = 23.2 \text{ cm}$$

$$\Delta H = 0.232 - 0.200 = 0.200 \text{ m} = 0.032 \text{ m} = 3.2 \text{ cm}$$

If the complete profile is desired, integration of Equation 9-37 is required. This profile is shown in Figure 9-15b. Note that there is no normal depth for the case of a flat bottom.

THE EFFECTS OF RAIN AND ET

Evaporation and rain do not have a large effect on time-averaged water depths for typical conditions in SF wetlands. This is true whether the wetland is close to the weir-controlled operating condition or the slope-controlled operating condition. For instance, if the ET rate in Example 9-4 is taken to be 0.5 cm/d, 25 percent of the water is lost to ET. However, the average depth of flow drops only 0.3 cm, from 14.6 cm to 14.3 cm. There are exceptions, such as the Incline Village, NV wetland treatment system, which evaporates nearly all the applied water (Kadlec et al., 1990a).

However, evaporation slows the water, and this can cause significant increases in detention time or corresponding decreases in actual hydraulic loading rate. In Example 9-4 with 0.5 cm/d ET, the detention time increases from 7.3 to 8.3 days. In addition, evaporative water losses cause concentration increases, thus partly counterbalancing the extra detention time. In Example 9-4 with 0.5 cm/d ET, an inert material would increase in concentration by 33.3 percent. So it is necessary to consider rain and ET in pollutant mass balances, even though there is a negligible effect on depth.

Sudden rain events can increase the depth of flow by an amount limited by the total amount of the rain event, but the actual increase will be less, depending upon the duration

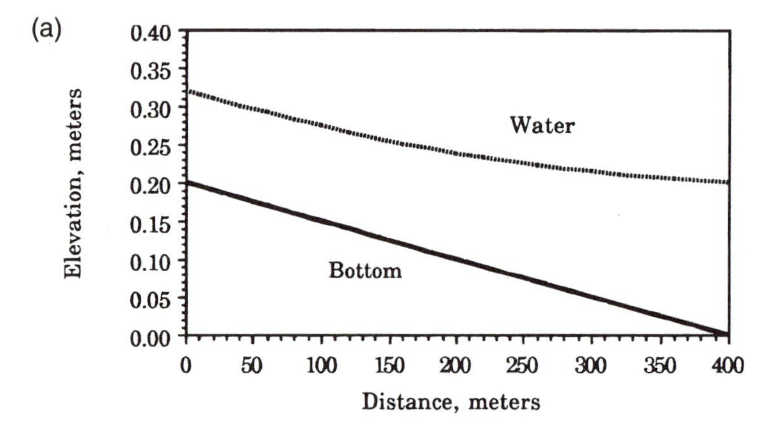

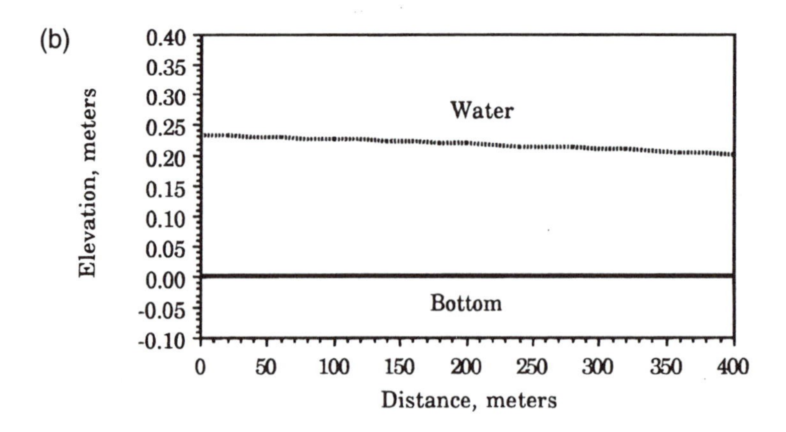

Figure 9-15 Water surface profiles for Examples 3 and 4: (a) a sloped bottom; (b) a flat bottom.

of the rain event. As an example, consider the wetland in Example 9-4. A 10-cm rain is imposed over a 4.8-h (0.2 days) period. The simulated results in Figure 9-16 indicate that there is a steady increase in water level to a maximum of 8.5 cm average depth increase at the end of the rain. About 15 percent of the added rain is lost during the ramp-up due to increases in outflow during the rain period. Interestingly, the added depth is slow to dissipate; there is still an extra 1 cm of water 2 days after the rain stops. The preceding are simulation results, but these effects have been confirmed in the field (Hammer and Kadlec, 1986).

SUBSURFACE WETLAND HYDRAULICS

INTRODUCTION

The idea of flowing water through a planted bed of porous media seems simple enough; yet numerous difficulties have arisen in practice. Sometimes these problems have been traced to incorrect design calculations; at other times problems have resulted from changes in the conditions in the bed. A great deal of confusion has been evidenced regarding the movement of water through SSF wetlands. Rules of thumb abound in the literature, many of which do not acknowledge the simple physics of water movement. The literature is replete with misapplications of the fundamental relations between head loss and flow rate. In this section, relevant calculations are examined and bounds are placed on the variables governing the ability to operate in subsurface flow with rooted macrophytes.

Gravel bed SSF wetlands in the U.S. are frequently observed to be overflooded. The two probable causes are clogging of the media with particulates and improper hydraulic design. The same appears to be true for other countries as well (Brix, 1994b), especially the SSF wetlands that use soil for the medium. The underlying cause of such hydraulic failure is the ad hoc procedure of designing to guessed values of hydraulic parameters.

The SSF technology has been rescued by the fact that the hydraulically failed mode of flooded operation is the SF wetland, which is nearly as efficient as the SSF wetland.

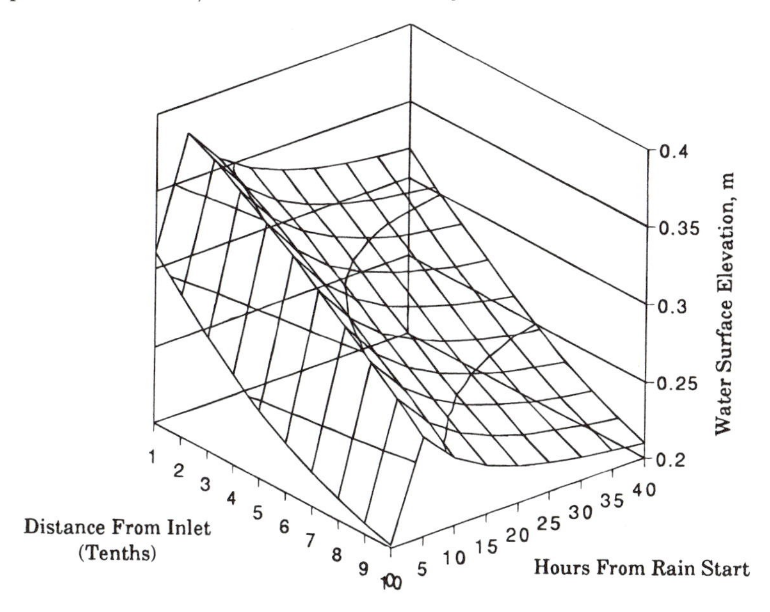

Figure 9-16 The response of an SF wetland to a rain event for the wetland in Example 9-4. The 10-cm rain occurs over 4.8 h. The ground slopes 20 cm from inlet to outlet and is at 0.0 m elevation at the outlet.

The flow direction is most often horizontal, although there are some downflow and some upflow systems. The same fundamentals apply to all orientations, but final equations differ somewhat. Here, the focus is on horizontal flow only.

THE GEOHYDROLOGICAL BACKGROUND

There is a very long history of research and development related to flow in porous media. Descriptions of flow phenomena started with the propositions of Darcy in 1856 and have grown to include several texts on the subject. These fall in two major categories: groundwater hydrology, as typified by the text of Freeze and Cherry (1979), and the movement of hydrocarbon fluids in petroleum reservoirs, as typified by the text of Greenkorn (1983). Several types of flow can occur in general; here, the concern is solely for the case of a fully saturated flow with an unconfined top interface with air, either in or above the bed. Full saturation refers to the absence of a capillary fringe, in which both air and water occupy the voids between particles. As noted earlier, capillary phenomena can be quite significant in the soils of free-water surface wetlands. Such two-phase behavior generally requires very fine pore size, which implies a very low hydraulic conductivity. Consequently, the feasibility of subsurface flow at realistic rates generally precludes capillary fringe effects.

The most complete model of water flow in porous media accounts for the possibility of transient behavior and flow in all three directions. Enormous effort has been devoted to this subject over a long history, motivated by the desire to model groundwater movement. The basic equations and assumptions used to simplify them are named in honor of early researchers.

The mass balance for water flow in three dimensions is

$$\frac{\partial s}{\partial t} = -\vec{\nabla} \cdot \vec{u} \tag{9-47}$$

where s = volumetric fraction water in the porous media, m^3w/m^3
t = time, d
$\vec{u}$ = superficial water velocity, m/d = $m^3/d/m^2$

Water flow through the media is frequently associated with the local pressure head according to Darcy's (1856) equation:

$$\vec{u} = -k\vec{\nabla}h \tag{9-48}$$

where h = local pressure head, m
k = hydraulic conductivity, m/d

These combine by substitution to yield the Richards equation:

$$\frac{\partial s}{\partial t} = \vec{\nabla} \cdot (k\vec{\nabla}h) \tag{9-49}$$

SSF wetlands operate in thin sheetflow, with a free (but underground) upper surface. Equation 9-49 may be averaged over the vertical (thin) dimension, for the case of the upper surface exposed to the atmosphere, to yield the Dupuit-Forcheimer equation:

$$\frac{\partial(\varepsilon H)}{\partial t} = \frac{\partial}{\partial x}\left(kH\frac{\partial H}{\partial x}\right) + \frac{\partial}{\partial y}\left(kH\frac{\partial H}{\partial y}\right) + P - ET \tag{9-50}$$

where ET = evapotranspiration loss, m/d
P = rain, m/d
H = elevation of the free-water surface, m
x = longitudinal distance, m
y = transverse distance, m
ε = water-filled porosity

It is important to note that this equation embodies the assumption that the driving force for flow is a tilt to the water surface.

Much simpler versions of this theory will suffice for wetland design purposes. However, the full theory is of considerable importance in understanding operating variability.

Adaptations for SSF Wetlands

In the following, it is presumed that the porous medium is isotropic. This is probably not true, due to the presence of plant roots and other introduced particulates. The variability in the vertical and transverse directions is accounted for by averaging. Longitudinal variations in hydraulic conductivity are also present after the wetland has been in operation for a time.

The following developments presume that the wetland is in a steady state condition, but later it will be shown that this is rarely the case. The representation will therefore be for long-term average performance. With that understanding, the overbar, used in earlier sections to indicate time averaging, will be omitted.

Most SSF wetlands are rectangular, and so that feature is added to the list of restrictions. Notation is outlined on Figure 9-17.

The mass balances and geometrical definitions have been presented in Equations 9-27 through 9-32, which also hold for SSF wetlands. The porosity is lower, usually in the range 0.35 to 0.45 m^3/m^3 for sands and gravels, and there is the added geometry of a bed surface to consider. The elevation of the top surface of the media is

$$G = B + \delta \tag{9-51}$$

where G = elevation of the bed top above datum, m
δ = thickness of the bed media, m

The headspace is defined to be the distance from the top surface of the media down to water:

$$f = \delta - h \tag{9-52}$$

where f = headspace, m.

In general, the variables h, H, G, δ, f, and B are all dependent on distance from the bed inlet.

BED FRICTION AND HYDRAULIC CONDUCTIVITY

The simplest friction relationship states that superficial velocity is proportional to the slope of the water surface:

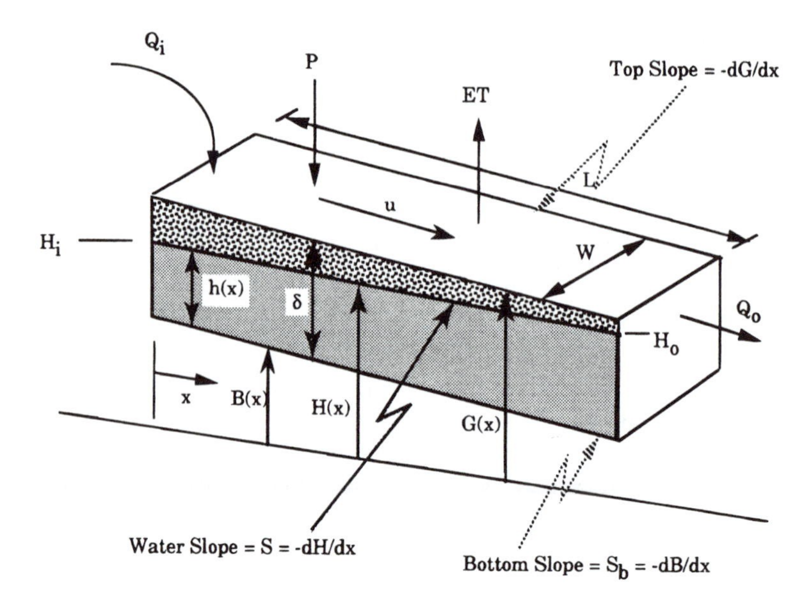

Figure 9-17 Notation for SSF bed hydraulic calculation for the simplest case. The actual velocity of water is $v = u/\varepsilon$. The subscripts "i" and "o" stand for inlet and outlet, respectively.

Note:
B(x) = elevation of bed bottom, m
ET = evapotranspiration, m/d
G(x) = elevation of bed surface, m
h(x) = water depth, m
H(x) = elevation of water surface, m
L = bed length, m
P = precipitation, m/d
Q = volumetric flow rate, m^3/d
x = distance from inlet, m
δ = water depth, m

$$u = -k \frac{dH}{dx} \tag{9-53}$$

where H = elevation of the water surface, m
k = hydraulic conductivity, m/d

This is the one-dimensional version of Darcy's law (Equation 9-48). It is restricted to the laminar flow regime.

A more general correlation spans both laminar and turbulent flow. The laminar term in Equation 9-53 is preserved, and a turbulent term is added:

$$-\frac{dH}{dx} = \frac{1}{k} u + \omega u^2 \tag{9-54}$$

where ω = turbulence factor, d^2/m^2.

The turbulent contribution ωu^2 is negligible when the particle Reynolds number is less than 1.0, and it may be ignored with small error at Reynolds numbers up to 10. The Reynolds number is defined as

$$\text{Re} = \frac{D\rho u}{(1 - \varepsilon)\mu} \tag{9-55}$$

where D = particle diameter, m
ρ = density of water, kg/m^3
μ = viscosity of water, kg/m/d

Sand media will typically be in the laminar range, but rock media will often be in the transition region between laminar and turbulent, with significant contributions from the turbulent term.

Simple rearrangement of Equation 9-54 gives

$$u = -k_e \frac{dH}{dx} \tag{9-56}$$

where k_e = effective hydraulic conductivity, m/d.

Comparison of Equations 9-54 and 9-56 indicates that

$$\frac{1}{k_e} = \frac{1}{k} + \omega u \tag{9-57}$$

When velocity is beyond the laminar range, the effective hydraulic conductivity will depend on velocity.

Correlations for Hydraulic Conductivity

The hydraulic conductivity and turbulence factor for a particulate media depend on the characteristics of the media:

1. Mean particle diameter
2. Variance of the particle size distribution
3. Particle shape
4. Porosity of the bed
5. Arrangement of the particles

Of these, the effects of particle size and porosity have been quantified in the form of equations. For instance, the Ergun equation (Ergun, 1952) is widely accepted for random packing of uniform spheres:

$$-\frac{dH}{dx} = \frac{150(1 - \varepsilon)^2\mu}{\rho g \varepsilon^3 D^2} u + \frac{1.75(1 - \varepsilon)}{g \varepsilon^3 D} u^2 \tag{9-58}$$

where g = acceleration of gravity, m/d^2.

Comparison with Equation 9-54 indicates that

$$\frac{1}{k} = \frac{150(1 - \varepsilon)^2\mu}{\rho g \varepsilon^3 D^2} \tag{9-59}$$

$$\omega = \frac{1.75(1 - \varepsilon)}{g \varepsilon^3 D} \tag{9-60}$$

Equation 9-58 works for spheres of a single size, but gravel bed wetlands do not utilize such media. Hu (1992) applied Equation 9-58 to an SSF system at Bainikeng, China and found that Ergun-predicted depths were about 10 cm too large.

The effects of a nonspherical shape are indeed significant (Brown and Associates, 1956). Idelchik (1986) gives a correlation for crushed, angular materials which predicts conductivities about three times lower than those for spheres of the same size:

$$k_l = \frac{\rho g \varepsilon^{3.7} D^2}{127.5(1 - \varepsilon)\mu} \tag{9-61}$$

where k_l = hydraulic conductivity for uniformly sized particles, m/d.

Note that Equation 9-61 is quite similar to Equation 9-59, except for the porosity effect and the size of the numerical coefficient. The overall porosity dependence is approximately the fifth power, which indicates the extreme importance of this variable. Unfortunately, there is not a simple technique to measure *in situ* porosity, and measurements on extracted samples are prone to significant errors. For instance, if a 25-cm diameter bucket is used to measure the porosity of 2.5 cm media, a +7 percent error will result from the edge effect of the bucket wall (Leva, 1947). That error propagates through Equation 9-59 to predict a hydraulic conductivity that is 35 percent too high.

Most media possess a distribution of sizes. The effect of a size distribution is to lower the hydraulic conductivity. This occurs because small particles have a disproportionately large amount of surface area which causes drag on the water and because the small particles can fit in the spaces between the larger particles. For instance, Freeze and Cherry (1979) present a technique based on the work of Masch and Denny (1966), which utilizes the variance of the particle size distribution to estimate a correction factor for the hydraulic conductivity of large sand particles. For a variance of 50 percent of the mean particle size, the reduction is a factor of two.

Most wetland basins will be filled by processes that lead to a random arrangement of particles because it is not practical or necessary to use any technique other than dumping and leveling. However, hydraulic conductivity is very sensitive to the arrangement of the particles within the bed. Martin (1948) studied the head loss through different arrangements and orientations of spheres in packed beds. Spheres may be packed in cubical, orthorhombic, tetragonal, or rhombohedral arrays, or they may be arranged randomly. The resultant porosities vary from 0.2595 to 0.476. The resulting spread in hydraulic conductivity is about one order of magnitude.

Given all the previous uncertainties, each of which can greatly influence the hydraulic conductivity of the clean media, it is prudent to measure the conductivity of the candidate media for a proposed project. Correlations may be used to guide the initial selection, but should not be trusted for final design purposes. For the laminar contribution, the following correlation is recommended:

$$k = \phi \frac{\rho g \varepsilon^{3.7} D^2}{127.5(1 - \varepsilon)\mu} \tag{9-62}$$

where ϕ = laminar size distribution factor, dimensionless
≈ 0.5

Hence,

$$k = \frac{\rho g \varepsilon^{3.7} D^2}{255(1 - \varepsilon)\mu} \tag{9.63}$$

The turbulence size distribution factor is closer to unity, and a value of 0.88 is chosen based on data fitting of a series of flume experiments carried out on 1.0- to 4.5-cm media (Hansen, 1993):

$$\omega = \frac{1.75(1 - \varepsilon)}{0.88g\varepsilon^3 D} = \frac{2(1 - \varepsilon)}{g\varepsilon^3 D} \tag{9.64}$$

Combining these two terms gives the guidance formula for clean media:

$$\left[\frac{255(1 - \varepsilon)\mu}{\rho g\varepsilon^{3.7}D^2} + \frac{2(1 - \varepsilon)u}{g\varepsilon^3 D}\right] = \frac{1}{k_e} \tag{9-65}$$

Validation of Equation 9-65 can be accomplished *a posteriori* for very few gravel bed wetlands since the gradient, porosity, and velocities have seldom been reported. Data for media from four treatment wetland sites with porosities in the range 0.35 to 0.38 are displayed along with Equation 9-65 on Figure 9-18. Data for clean sands and soils is lacking in the SSF literature.

It is very important to recognize that Figure 9-18 is valid only for porosities near 0.35 and size variances near 50 percent.

Measurement of Hydraulic Conductivity

Either trenches or pipes filled with media may be used to measure the conductivity of a candidate media; the trench technique will be explained here. The low resistance to flow produces very small head losses through small test systems, and specialized experimental procedures and data analysis are required. These procedures are difficult for media greater than 3 cm. The media is placed in a sealed horizontal trench with a cross-section large enough to contain more than ten particles across the smallest dimension. This is necessary to avoid edge effect errors. The length should be at least ten times the largest transverse dimension to avoid entrance and exit effects. Provisions for flow measurement are necessary, such as the bucket-and-stopwatch procedure. Head loss is measured across the test length by a sensitive procedure such as differential manometry.

Head losses in the laminar region will typically be immeasurably small. The trench is therefore operated at a sequence of flow rates in the turbulent region for which head loss is accurately measurable. A rearrangement of Equation 9-54 is then used to determine the values of k and ω:

$$\frac{1}{u}\left(-\frac{\Delta H}{\Delta x}\right) = \frac{1}{k} + \omega u \tag{9-66}$$

The left-hand side term is determined experimentally and regressed against the superficial velocity, u. Equation 9-66 shows that a straight line is expected, with slope ω and intercept l/k.

Data from a laboratory channel are shown in Figure 9-19 for spheres of diameter 1.415 cm in a nearly regular, orthorhombic array. Head losses were measured using a differential manometer, accurate at greater than 0.15 cm of water. Over a trough length of 137 cm, this permitted measurements of gradients greater than 0.001 m/m. The intercept was at $1/k = 6.7 \times 10^{-6}$ or k = 149,000 m/d. The porosity was lab measured at $\varepsilon = 0.344$, for which the Ergun equation predicts k = 107,000 m/d. The difference may be due to the regular,

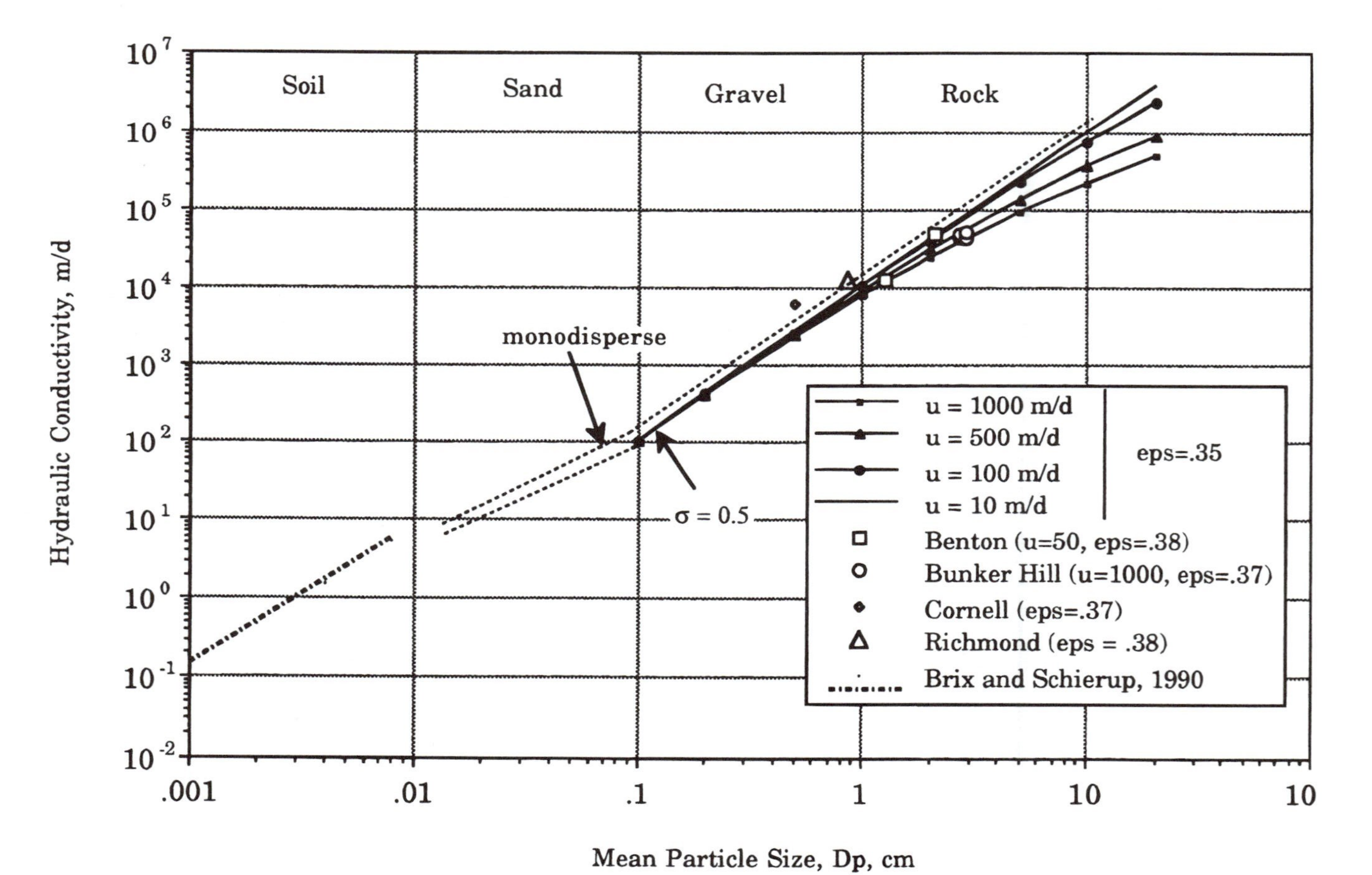

Figure 9-18 The dependence of hydraulic conductivity on media grain size. This plot is approximate; it is based on a porosity of 35% and a 50% variance in the particle size distribution. Other size distributions and deviation in particle shape and packing will influence values for specific media.

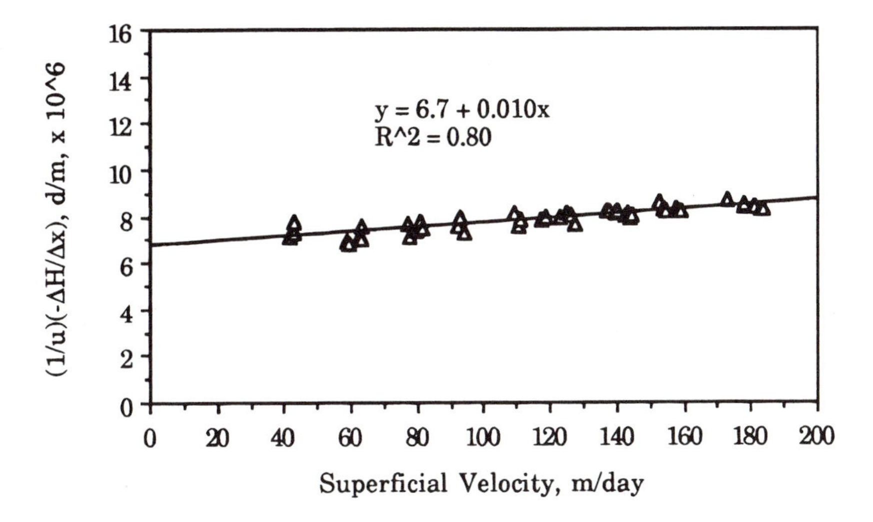

Figure 9-19 Determination of hydraulic conductivity parameters from trough test data.

rather than random, packing in the trough. A change in porosity to $\varepsilon = 0.376$ gives an Ergun $k = 149{,}000$ m/d.

The turbulence factor determined from the slope of the plot is 0.010 d^2/m^2. The Ergun prediction at $\varepsilon = 0.344$ is 0.02 d^2/m^2, which drops to 0.015 at $\varepsilon = 0.376$.

The extreme sensitivity of either a correlation or an experiment to the porosity of the media has strong implications for design. It is apparent that the hydraulic conductivity of the as-placed media cannot be measured or predicted accurately in advance of construction. Therefore, designs must be robust enough to withstand this uncertainty.

Soil-Based Systems

As particle sizes become smaller, it becomes more and more difficult to characterize the media on the basis of size. Soil classifications are based on the amounts of sand, silt, and clay (Figure 9-20). Soils used in most SSF systems in Denmark have more than two thirds sand. Sand is graded by a rough scale from very fine to coarse. The U.S. EPA (1980) has provided a means of estimating hydraulic conductivity based on soil type (Table 9-2), which in turn may be keyed to a rough estimate of size through the USDA classes of soil separates (Table 9-3).

Clogging

The bed will not maintain the clean media conductivity because of the deposition of solids and the blockage of pore space by plant roots. If one third of the pore space is blocked, the hydraulic conductivity will decrease by one order of magnitude according to Equation 9-61. This possibility must be acknowledged in design if the potential for flooding is to be minimized.

Solids deposition can occur for a variety of reasons, beginning with the placement of the media. Unwashed media will carry a load of fine dust or soil. Mud on the wheels of vehicles can add to the dirt supply during placement. Also, those beds which are constructed with a layer of fine media on top of coarse media can be subject to the penetration of the lower layer by that finer material. Planting activities can introduce soils associated with the roots of the plants. All of these factors can combine to cause problems right from startup.

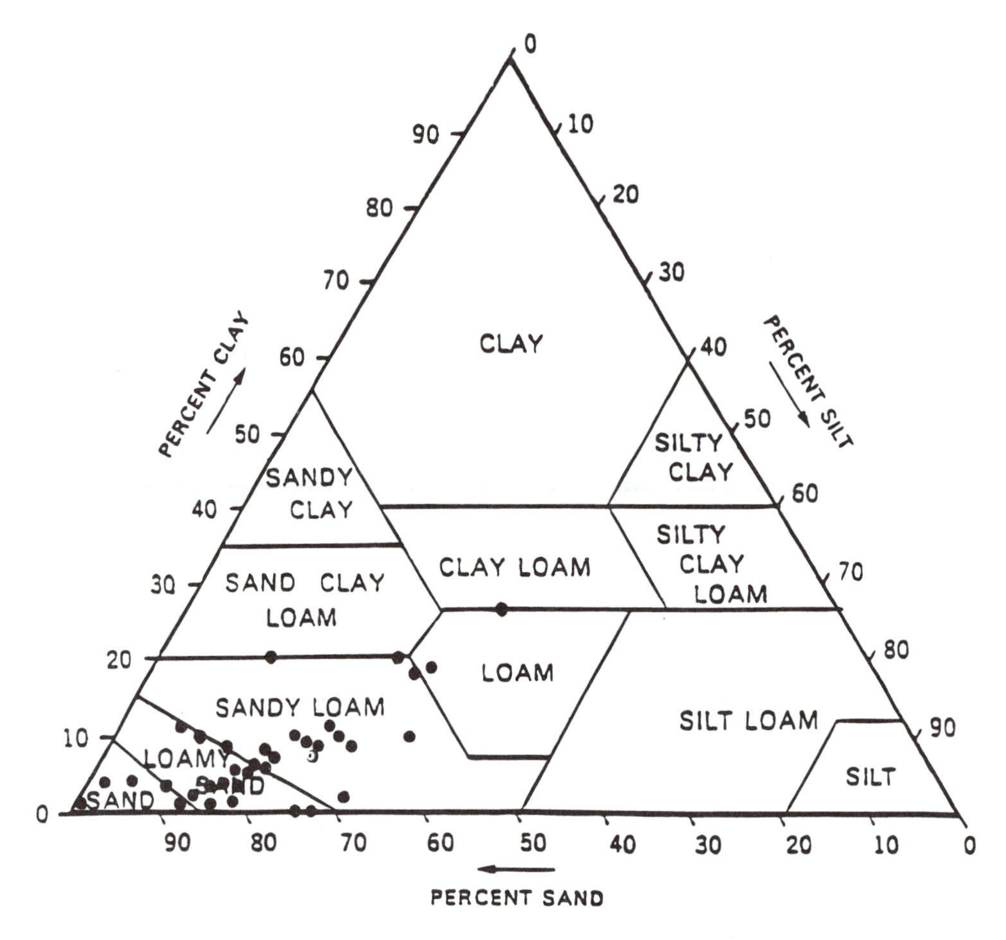

Figure 9-20 Soil classification according to sand, silt, and clay content. The soils of 42 Danish SSF wetlands are shown. (From Schierup, et al., 1990a.)

Table 9-2 Typical Values of Hydraulic Conductivity for Soil Materials

Soil Texture	Hydaulic Conductivity (m/d)
Gravel, coarse sand	36–
Coarse, medium sand	6–36
Fine sand, loamy sand	2.4–6
Sandy loam, loam	1.2–2.4
Loam, porous silt loam	0.6–1.2
Silty clay loam, clay loam	0.3–0.6

Note: Data from U.S. EPA, 1980.

Table 9-3 Size Groups of Soil Separates for Fine Earth

Soil Component	Size Range (mm)
Very coarse sand	2.0–1.0
Coarse sand	1.0–0.5
Medium sand	0.5–0.25
Fine sand	0.25–0.10
Very fine sand	0.10–0.05
Silt	0.05–0.002
Clay	<0.002

Note: Data from SCS, 1981.

For instance, Zachritz and Fuller (1993) note that "Clogging . . . has been an operational problem since plant startup" at the Carville, LA facility.

Irrespective of construction problems, a gravel bed will become clogged with organic sediments and detritus and possibly with roots. The biofilms that achieve treatment occupy some pore space. The result is a severe decline in bed hydraulic conductivity, especially in the front end of the bed. The magnitude of such decline may be as much as a factor of ten (Figure 9-21). Most of the decline is apparently associated with sediments and detritus since unplanted gravel beds show declines similar to those found in planted systems (Fisher, 1990; Sanford et al., 1995). Fine sand systems also display much lower conductivities than predicted (Netter and Bischofsberger, 1990).

The microbial populations associated with nutrient cycling and BOD reduction are highest in the inlet section of the bed in response to the elevated contaminant concentrations in that region. These organisms, together with their detritus, reduce the pore volume in the entrance region of the bed to a greater extent than the downstream sections. In turn, this implies a greater reduction of hydraulic conductivity in the inlet region. This effect has been measured by several investigators (Fisher, 1990; Kadlec and Watson, 1993). Establishment of the low inlet conductivity occurs over the first months of operation and stabilizes thereafter (Figure 9-21).

Soil-based systems may not display clogging simply because their as-placed conductivities are already very low and may match those of the deposited solids. Haberl and Perfler (1990) found no evidence of reduced conductivities over 5 years for any of three SSF soil-based wetlands at Mannersdorf, Germany. They found erratic changes in k, with minimum and maximum values of 0.37 ± 0.34 and 3.8 ± 2.2 m/d, respectively. Coombes (1990) reports conductivities for several U.K. reed beds in the range of 0.2 to 9.8 m/d, which is the same as the range shown in Table 9-2. The system at Acle, U.K., had essentially the same conductivity at year four as at startup. However, because of low conductivity, soil-based systems are extremely prone to flooding.

The implications of these phenomena in design are very important. The SSF wetland must be designed to operate properly and to establish and sustain plant growth, in the face of large changes in hydraulic characteristics.

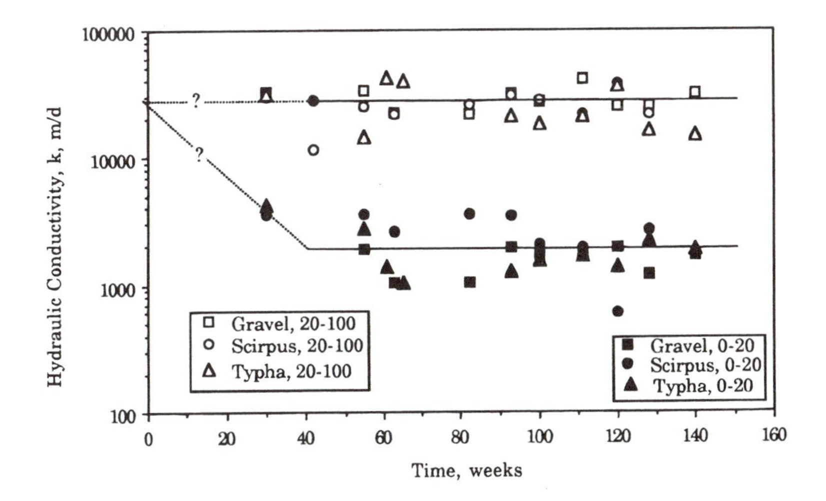

Figure 9-21 The dependence of hydraulic conductivity on time. Three beds are represented: an unplanted gravel bed, gravel planted with *Scirpus,* and gravel planted with *Typha.* Data fall into two groups: the inlet sections, 0 to 20 m from the inlet, and the rest of the bed, 20 to 100 m from the inlet. (Adapted from Fisher, 1990.)

SURFACE WATER ELEVATION PROFILES

More than just Darcy's law is required to calculate flow rates and depths in SSF wetland. Equations 9-27, 9-31, 9-32, and 9-54 provide the ability to calculate h, H, u, and Q as functions of distance down the bed. Since two of these equations are differential equations, an integration procedure must be implemented, and boundary conditions must be specified. Combining Equations 9-27, 9-31, and 9-32 yields one equation in elevation and velocity; Equation 9-54 provides another:

$$\frac{d[u(H - B)]}{dx} = (P - ET) \tag{9-67}$$

$$-\frac{dH}{dx} = \frac{1}{k}u + \omega u^2 \tag{9-54}$$

From a knowledge of u and H profiles, it is then easy to compute the volume flow and depth profiles from

$$Q = uWh \tag{9-31}$$

and

$$h = H - B \tag{9-32}$$

The boundary conditions would most often be a specification of the exit water elevation (set by a structure) and a specification of the inlet flow rate (set by the delivery system):

$$H(x = L) = H_o \tag{9-68}$$

$$Q(x = 0) = Q_i \tag{9-69}$$

The required input information must also include bed width W, the bottom elevation profile B(x), the hydraulic conductivity profile k(x), and the turbulence factor profile $\omega(x)$.

There is a very important constraint to be met in the course of solution of these model equations: flow must be underneath the media surface, or the hydraulic conductivity Equation 9-54 does not apply. Mathematically, this means satisfying the inequality

$$0 < h < \delta \tag{9-70}$$

These model equations may be solved on a desktop computer without great difficulty.

Some sense of the validity and sensitivity of the model can be gained from its calibration to field data at Benton, KY. The inlet zone of this crushed limestone SSF cell was flooded at the time of the study, which was about 3 years after startup. Measurements included detailed surveys of the water surface elevation and of flow rates. The media were tested in the field to determine the *in situ* conductivity (TVA, 1989). The washed media was also tested in the laboratory (Kadlec and Watson, 1993). Although the lab conductivity was roughly comparable to the field values, the lab value was clearly on the high side (67 percent above field mean). That kind of difference can easily arise from differences in void fraction resulting from packing factors. Further, the media in the bed contained a considerable amount of dirt from work vehicles. A 12 percent void fraction difference would account for the difference in k's. Nevertheless, the model fits the water surface profile correctly in the SSF zone (Figure 9-22).

At the scale of Figure 9-22, there does not appear to be any large problem, because the mean depth in the wetland is only about 15 percent greater than predicted. However, the effects on vegetation were huge: the inlet section (20 percent) was an SF *Scirpus* wetland, and the remaining 80 percent of the bed contained only sparse terrestrial vegetation. The plants respond to headspace (δ), which is very sensitive to model parameters.

This level of computational effort is probably not warranted in design. Some further simplifying assumptions will offer aid in that task. A very much simplified hydraulic model appears in the literature, but it does not serve a very useful purpose in design. An alternate design procedure is subsequently presented.

THE SIMPLEST CASE

Most existing SSF beds have flat, but inclined bottoms. Many operate in the laminar region and do not experience large atmospheric gains or losses. Most are intended to operate at constant water depth. In their initial startup condition, the hydraulic conductivity will not be a function of distance from the inlet. Under these ideal conditions, the model reduces to

$$\frac{H_i - H_o}{L} = \frac{[Q/W\tilde{h}]}{\tilde{k}} = \frac{\tilde{u}}{\tilde{\tilde{k}}} \tag{9-71}$$

where the overtilde indicates spatial averaging. It is important to note that the gradient on the left side of this equation is the slope of the water surface, not the slope of the bottom of the bed. As is the case for lakes, if there is not a slope to the water surface, there is not a flow of water. Details of notation for this case are shown in Figure 9-17.

The use of this equation *by itself* for design can lead to serious errors (and has done so). The average water depth $\tilde{h}$ is not adequate for design; the longitudinal depth *profile* is required. The problems are related to the large changes in k, to the violation of the constraint (Equation 9-70) (flooding), and to the sensitivity of vegetation to the headspace (f = δ − h). Basically, the design goal is to keep the water below ground, but high enough for plant roots to reach it.

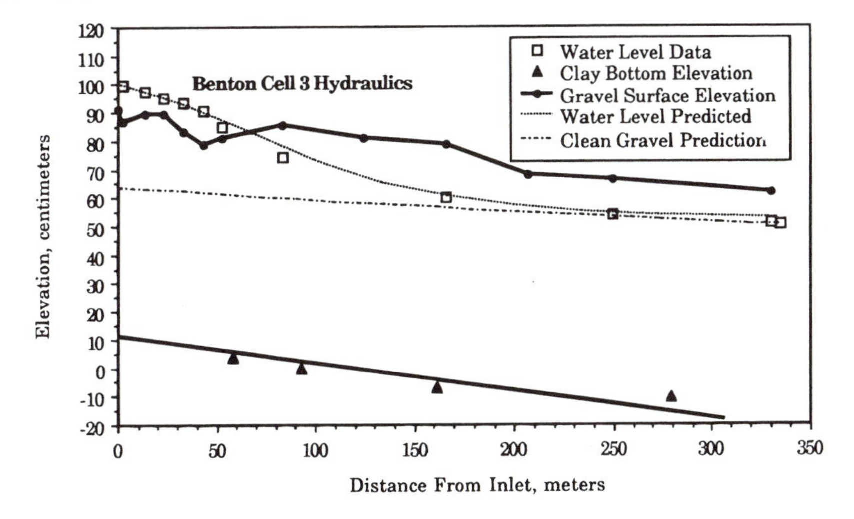

Figure 9-22 The dependence of hydraulic conductivity on distance for an SSF wetland. At the time of the study, the inlet section of the bed was clogged significantly. The "prediction" is a calibration of Equations 9-54 and 9-67. (Adapted from Kadlec and Watson, 1993).

RAPID ASSESSMENT OF PROFILES

As a benchmark case, attention will be restricted to rectangular SSF systems with uniform media depth. For simplicity, attention will be focused upon negligible atmospheric additions via rain and negligible losses via ET. The hydraulic conductivity will be presumed to be constant, with no turbulent contribution, along with width, bed depth and slope, and flow rate. However, the equations previously presented may be used to examine other cases as well.

For these conditions, Equations 9-54 and 9-31 reduce to

$$-kWh\frac{dH}{dx} = Q \tag{9-72}$$

with the boundary condition

$$h(x = L) = h_0 \tag{9-73}$$

Dedimensionalization of Equation 9-72 results in

$$y\frac{dy}{dz} - G_1\, y = -G_3 \tag{9-74}$$

with the boundary condition

$$y(z = 1) = 1 \tag{9-75}$$

where the definitions of Equation 9-45 have been used, and where the geometry number G_1 is given by

$$G_1 = \frac{S_b}{(h_0/L)} \tag{9-76}$$

and the loading number G_3 is given by

$$G_3 = \frac{(q/k)}{(h_0/L)^2} \tag{9-77}$$

Based on solution of the model, it may be shown that the condition $G_3 = G_1$ corresponds to the situation of the water surface paralleling the bed bottom. For $G_3 < G_1$, and the water depth increases with distance from the inlet; for $G_3 > G_1$, the water depth decreases with distance from the inlet. Thus, the parameters G_3 and G_1 are useful to determine if a particular flow is distance thinning or distance thickening. They will also be useful in setting design constraints.

FLOODED OPERATION

A combination of clogging and inappropriate design has produced overland flow in what may be the majority of existing SSF systems.

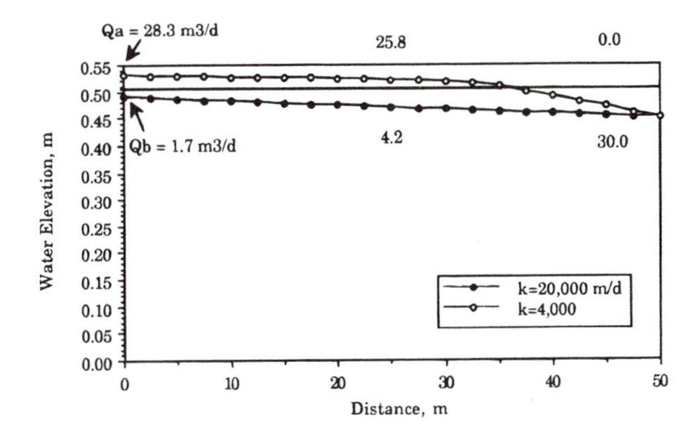

Figure 9-23 Magnitudes of below-ground and above-ground flows for a hypothetical SSF bed. A fivefold reduction in conductivity causes flooding. Where surface flow exists, it comprises the majority of the flow.

Flooding is usually confined to the inlet region of the bed. Overland flow carries the excess water until the gradient over the remaining travel distance is sufficient to permit the flow to be carried below ground (Figure 9-23).

The amount of water carried in overland flow may be estimated from equations describing flow in vegetated channels, such as Manning's equation. However, the presence of the permeable bed below the channel affects the value of the frictional coefficient (Manning's n, for instance). Das Gupta and Paudyal (1985) compared overland flow over permeable and impermeable media of the same size and found that shallow flows were subject to more resistance (Figure 9-24). As the flow depth increased, the magnitude of the effect diminished, but it was significant for the flooding depths encountered in SSF wetland beds, *circa* 2 to 5 cm.

If the bed is densely vegetated, friction is due principally to stems and litter, in which case the magnitude of this correction is negligible.

If overflooding depths are as much as a few centimeters, most of the water will be carried by the overland flow. As an illustration, suppose that a level bed has been designed using the conductivity of the clean media and later clogs to a lower value, as discussed in the following example.

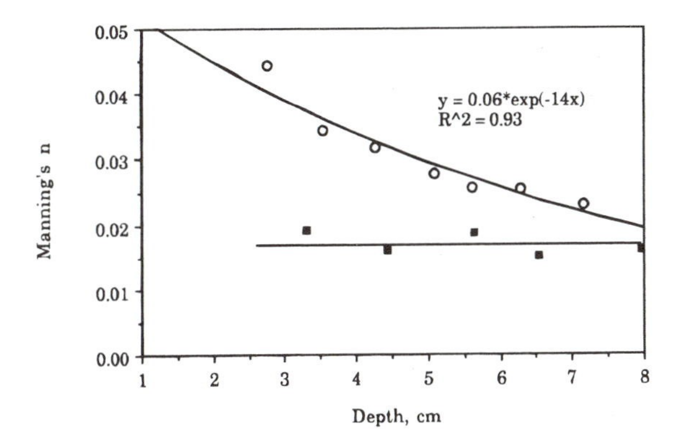

Figure 9-24 The effect of a permeable boundary on overland flow frictional resistance. (From Das Gupta and Paudyal, 1985.)

Example 9-6

A hypothetical SSF wetland is designed for clean media as follows:

$S_b = 0$	$L = 50$ m
$Q = 30$ m^3/d	$\delta = 0.50$ m
$W = 4$ m	$h_0 = 0.45$ m
$q = 15$ cm/d	$k = 20{,}000$ m/d
$G_1 = 0.075$	

As a result of operation, the conductivity drops to 4000 m/d. What water surface profile is expected, and how much water is carried above ground?

Solution

The water surface profile, when entirely below ground is calculated from model Equations 9-54 and 9-67. These are integrated backwards from the bed outlet to the point where the surface flow ends. The combined SF and SSF region requires calculations of the SF component from considerations in the previous SF section and the subsurface component from equations presented in this SSF section. The water surface elevation is the bed surface elevation plus the free-water depth, and the same gradient drives both flows.

Figure 9-23 shows the results of such calculations. Two thirds of the bed length is flooded, and over the first 50 percent of the bed, surface flow comprises more than 90 percent of the flow. The depth of surface water does not exceed 3 cm.

Several features of flooded performance are quite different from totally subsurface flow. In the flooded zone, the majority of the water flows over the surface of the bed. In the Example 9-6, more than 90 percent of the flow was overland, despite the shallow depth of flow. The volume of water in the wetland was increased compared to the design condition, and consequently, the average detention time is greater than that for the design condition. However, the surface water moves at high velocity compared to subsurface water and thus shifts the distribution of detention times to a shape containing an early peak.

All features described here for the hypothetical example were in fact reported by Spangler et al. (1976): "The main flow of water, however, was horizontally across the surface to the area immediately above the outlet, and then vertically down to the outlet. . . ."

Example 9-6 illustrates the need for careful hydraulic design if an SF wetland is to live up to its name. The design must be robust enough to withstand the reduction of conductivity as the system matures.

DYNAMIC RESPONSES: RAIN AND ET

Most gravel bed wetland systems are fed a constant flow of wastewater. There is therefore a strong tendency to visualize a relatively constant set of system operating parameters — depths and outflows in particular. This is not necessarily true in practice. Effects of rain and evaporative losses are magnified by the porosity of the bed to yield larger depth changes, and flows are very sensitive to depth and gradients. Thus, there may be significant outflow variability due to precipitation and ET.

As a case in point, Cell #3 at Benton, KY was operated in September 1990 at an HLR of 1.7 cm/d, corresponding to a nominal detention time (HRT) of approximately 13 days. Evapotranspiration at this location and at this time of year was estimated to be about 0.5 cm/d. Consequently, ET forms a significant fraction of the hydraulic loading. Because ET is driven by solar radiation, it occurs on a diurnal cycle. The anticipated effect is a diurnal variation in the outflow from the bed, with amplitude mimicking the amplitude of the combined (feed plus ET) loading cycle. This was the observation at Benton, KY (Figure 9-25).

In such an instance, because the night outflow peak is nearly double the daytime minimum outflow, it is important to use diurnal timed samples of the outflow and to appropriately flow weight them for determination of water quality.

It is perhaps easier to visualize the effect of rain. A sudden rain event, such as a summer thunderstorm, will raise water levels in the bed. The amount of the level change is magnified by porosity and catchment effects: there will be a threefold magnification due to a bed porosity of 33 percent, and some further increase due to bank runoff. Thus, a 3-cm rain can raise bed water levels by more than 10 cm — if there is that much gravel headspace. Overflooding of the bed may occur if there is insufficient headspace. In any case, outflows from the system increase greatly as the rainwater flushes from the system.

As an illustration, again consider Cell #3 at Benton, KY in September 1990. Figure 9-26 shows a rain event of about 2 cm occurring at noon on 9/10/90 for antecedent conditions given previously. The bed was subjected to a surplus loading of over 100 percent of the daily feed in a brief time period. The result was a sudden increase in outflow of about 300 percent, which subsequently tapered off to the original flow condition.

The implications for water quality are not inconsequential. In this example, samples taken during the ensuing day represent flows much greater than average. Water has been pushed through the bed, exits on the order of 1 day early, and has been somewhat diluted. Velocity increases are great enough to move particulates that would otherwise remain anchored. Internal mixing patterns will blur the effects of the rain on water quality.

Sampling intervals are not normally small enough to define these rapid fluctuations. For instance, weekly sampling of Benton Cell #3 would have missed all of the details of the rain and ET effects in the previous illustrations. It is therefore important to realize that compliance samples may give the appearance of having been drawn from a population of large variance, despite the fact that the variability is in large part due to deterministic responses to atmospheric phenomena.

SHADOW ZONES

Vertical Shadow Zones

There has been some speculation in the literature that there are potentially large geometrical "shadows" in the longitudinal cross-section in which there is zero flow. Hu (1992)

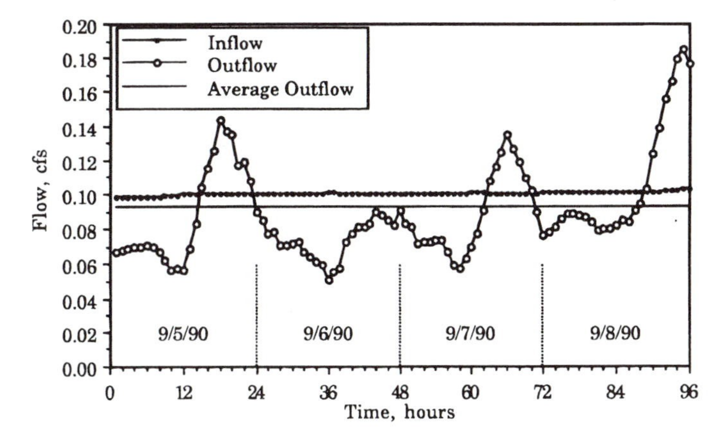

Figure 9-25 Flows into and out of Benton Cell #3 vs. time. Flows were measured automatically via data loggers; the values were stored as hourly averages. The data points on this graph are 6-h running averages, which smooth out short-term "noise" and emphasize the diel trends. (Data from TVA, unpublished data.)

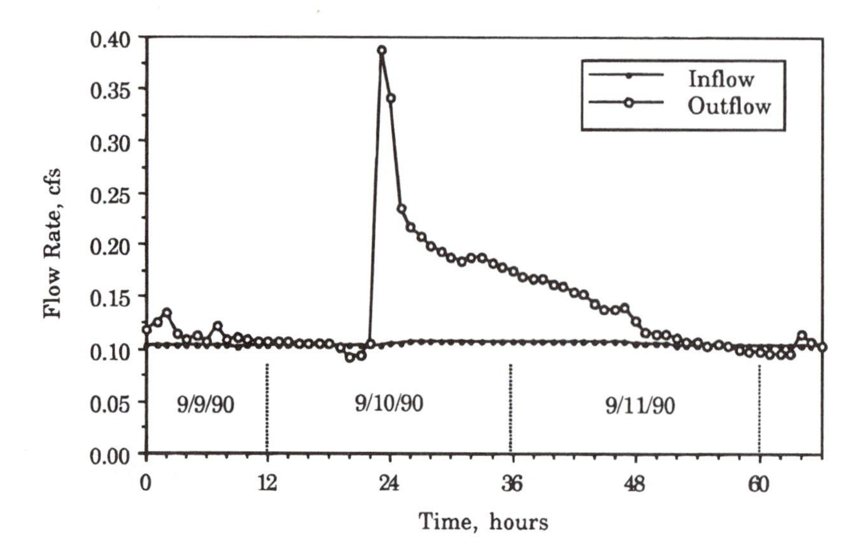

Figure 9-26 Flows into and out of Benton Cell #3 vs. time during a rain event period. Flows were measured automatically via data loggers; the values were stored as hourly averages. The rain event totaled approximately 0.75 in. (Data from TVA, unpublished data.)

postulated that only parallel flow lines could exist in a parallelogram configuration; therefore, triangular dead zones would occupy any extra space. Pennison (1993b) qualitatively estimated flow lines, allowing for curved flow lines which expand after the inlet and contract to the outlet. There is no need to guess at these flow lines, since interior velocities may be calculated, based on Richards' Equation 9-49.

Solution of Richards' equation shows that such stagnant shadows are much more limited than the Hu and Pennison ideas. The assumptions in this equation are mass conservation and Darcian flow. In a sample calculation for the expansion of a top feed into a clean, bare bed, the velocity profile is essentially uniform after a horizontal distance equal to the bed depth (Figure 9-27). A similarly scaled contraction is predicted at the downstream end of the bed if there is a top take-off.

Data from the Spangler et al. (1976) study confirm that top-fed tracer reached the bottom of the bed (maximum detected concentration) at the first sampling point, which was about 3 m from the entrance of the 0.75-m deep bed.

The presence of vegetative blockage is a much more important factor: the root zone impedes flow more than the bare media below it. Fisher's (1990) studies document this phenomenon. And there is some suggestion that accumulated solids selectively occupy a bottom layer in the media (Sanford et al., 1995).

Corner Zones

The Spangler et al. (1976) tracer data demonstrate that single-point feed, at the center of the inlet width, can lead to sizable dead zones in the inlet corners for SSF gravel beds. This is correctly predicted by Richards' equation. The result is similar to that shown in Figure 9-27, but with the bed width W replacing the bed depth H. The effect is thus scaled differently, and the edge velocity is only 50 percent of the average after a travel distance of one half the bed width. If this criterion is used to define dead space, there is a loss of area $= 0.5W^2$. For small L to W ratios (R_z), this can be a significant fraction of the bed.

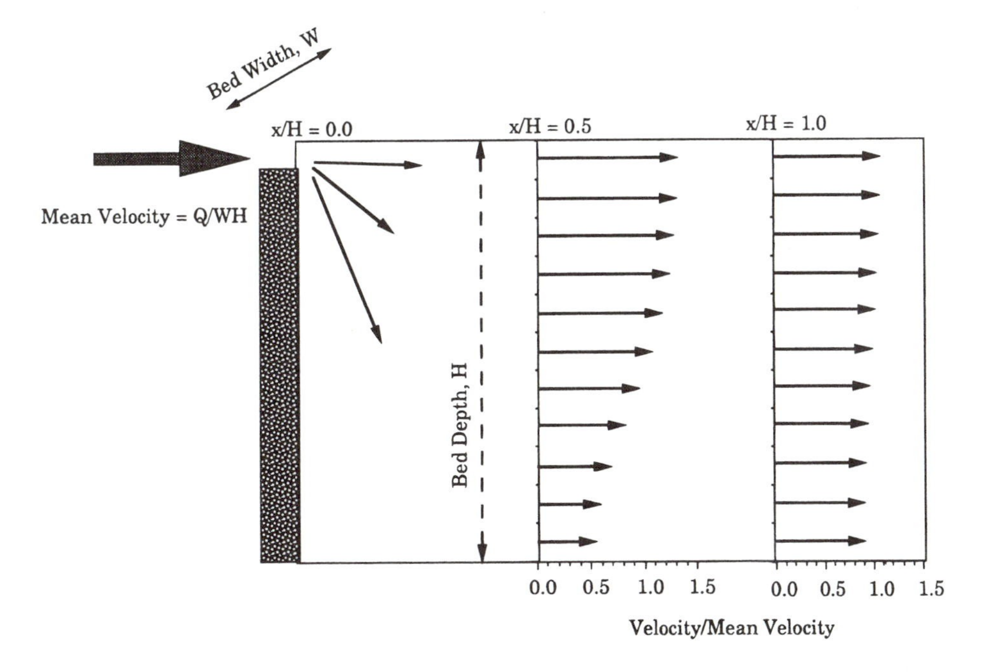

Figure 9-27 Flows rearrange quickly in the inlet region of the bed. This sketch shows the velocities computed for the inlet region of a bed fed at the top inlet edge. After one bed depth of travel, velocities have evened out to within 8% of the mean.

HYDRAULIC DESIGN OF SSF WETLANDS

The purpose here is to explore the effects of design parameters on the water surface and its relation to the media surface. The goals of hydraulic design for SSF wetlands are to convey the design flow below ground while providing an acceptable environment for wetland plants. Two philosophies may be used to achieve these goals: a management-intensive scheme, in which operators compensate for hydraulic upsets, and a design-intensive scheme, in which the design is made robust enough to eliminate most operational adjustments. In either case, the surface area ($L \cdot W$) is presumed to be set from the treatment requirements.

LARGE BOTTOM SLOPES, OUTLET LEVEL ADJUSTMENT

Management-intensive schemes may be set in a number of different ways. A Danish alternative for soil-based phragmites beds is briefly discussed here as an example of the use of the describing equations set forth earlier (Johansen, 1994). This and similar schemes do add to operation costs.

1. The depth of the bed is taken to be 0.6 m and is uniform. This is determined by the observation that phragmites can root to the full 60 cm bed depth.

$$\delta = 0.6\text{m} \tag{9-78}$$

2. The drop in the bed surface must not exceed 0.3 m. This ensures that inlet zone roots will likely touch the water even at no flow.

$$\Delta H = (H_i - H_o) \leq 0.3\text{m} \tag{9-79}$$

3. The bed cross-sectional area is set to convey the maximum instantaneous daily flow at the design bed slope.

$$W\delta = \frac{Q_{daily\ max}}{k\left(\frac{\Delta H}{L}\right)} \tag{9-80}$$

4. The soil media is determined by availability, but should contain sufficient sand to provide a "reasonable" conductivity, e.g., $k \geq 250$ m/d.
5. The width of any one cell should not exceed about 30 m. This ensures good lateral flow distribution at the cell inlet.

Example 9-7 The Danish Criteria

A small SSF wetland has been set at 1534 m^2 to treat an average of 80 m^3/d of effluent. The maximum daily flow is 145 m^3/d. What should be the dimensions of the bed?

Solution

The media is assumed to have a conductivity of 250 m/d. The bed slope is set to the maximum so that $\Delta H = 0.3$ m. The bed is set to a depth $\delta = 0.6$ m.

Then

$$LW = 1534$$

and

$$W(0.6) = \frac{(145)}{250\left(\frac{0.3}{L}\right)}$$

or

$$W = 3.222\ L$$

Solving

$$L = 21.8 \text{ m and } W = 70.3 \text{ m}$$

Two cells in parallel should be used, 35 m wide by 22 m long. This also allows a degree of flexibility in operation.

The expected water surface profiles, with no exit level adjustment, are shown in Figure 9-28.

The operator must use some judgment in setting the exit bed level in such a design. Under conditions of low seasonal flow due to source variations or high ET, the operator raises the exit water level to flood the exit end of the bed and to wet the entire root column in the inlet end. During higher source flows or in protracted periods of high rainfall, the exit level is lowered in an attempt to minimize inlet zone flooding. In the Danish setting, bed flooding is not perceived to be a serious problem of human or wildlife contact. There is also a limitation on media availability, leading to a strong preference for cheap, locally available soils.

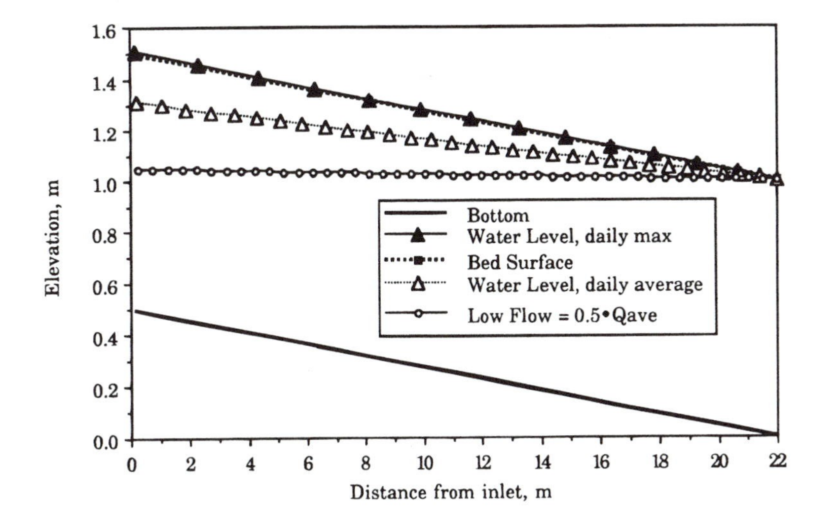

Figure 9-28 Water levels for clean media for the Danish criteria. The conditions for daily maximum and daily average flows and for one half the daily average flow are shown.

ROBUST DESIGN PROCEDURE

The specialized Danish procedure works well for its intended purposes. However, the plants of choice are not always *Phragmites*, and other species do not root to as great a depth. In other settings, flooding should be avoided if at all possible, and operators may not always react to changing source flows or changing environmental conditions. In general, media choice is more open than in the preceding example. The bed design can be set to operate within design constraints without operator attention.

The bed depth is usually selected to be in the range of 30 to 60 cm, based upon assumptions on plant rooting depth and its effect on treatment potential. At the time of this writing, such depth "criteria" remain speculative. However, most candidate plants do root to at least 30 cm, and there needs to be some room for sediment accretion in the bottom of the bed. Therefore, the upper half of the range, 45 to 60 cm, seems to be the best choice.

Normally, the requirement for concentration reduction will provide a specification of either hydraulic loading rate or detention time. In turn, these combine with the required volumetric flow rate to determine either the volume of water in the bed or the area of the bed. The requirements for stable and controllable water flow and for proper vegetation conditions serve to further restrict the geometry of the bed and the size of the media.

These requirements are

1. Anticipated flows must pass through the bed without overland flow or flooding.
2. Anticipated flows must pass through the bed without stranding the plants above water; i.e., there must not be protracted, excessive headspace.
3. Operation should remain acceptable in the likely event of changing hydraulic conductivity. As the bed clogs with roots and sediments, it should not flood.
4. The bed should be drainable.
5. The bed should be floodable.
6. Water levels within the system should be fully controllable through the use of inlet and outlet structures.
7. The configuration must fit the site in terms of project boundaries and in terms of hydraulic profiles.

There are often series and parallel arrangements of individual wetland cells within a system. In the present discussion, attention is restricted to individual cells and to simple rectangular geometries. The variables that may be chosen to satisfy the previous considerations are length (L), width (W), bottom slope (S_b), and the media (D or k). Bed depth (δ) is usually in the narrow range specified previously and is therefore set by conditions other than hydraulics.

There is no theoretical need for a slope of the top of the bed. If water level control is to include the ability to totally inundate the bed, for vegetation management, then a top slope is detrimental.

Because the media is likely to undergo significant changes in hydraulic conductivity, its frictional resistance cannot be relied upon for control of the water depth. This implies that the conductivity of the media must be high enough, or the water travel distance short enough, to sustain a nearly level pool condition.

The width of the system is scaled by the need to carry the full design flow. Thus, the planar aspect ratio (L:W) is not a controlling hydrologic parameter.

The bottom slope should be set to provide for complete bed drainage. Normally, a few centimeters of elevation differential allows for this requirement. *Bottom slope should not be considered as the design driving force for water movement.* The reason is that designs based on bed slope are excessively sensitive to changing conditions of flow and hydraulic conductivity; dryout or flooding are virtually certain to occur with such designs. An example is provided to illustrate this point.

Example 9-8

A wetland is to be designed to carry 300 m^3/d of water through a bed 0.6 m deep. The aspect ratio has been chosen according to a rule of thumb to be L:W = 10. The nominal detention time in the media is to be 4 days, and the void space is thought to be about 40 percent. A cheap gravel is available with a nominal size of 0.5 cm.

Solution

The overall size of the wetland can be computed from the detention time requirement:

$$\tau = \frac{LW\delta\varepsilon}{Q} = \frac{(10W)W\delta\varepsilon}{Q}$$

$$4 = \frac{(10)W^2(0.6)(0.4)}{300}$$

$$W = 22.4 \text{ m}, \qquad L = 224 \text{ m}$$

The superficial velocity through the wetland is

$$u = \frac{L\varepsilon}{\tau} = \frac{224(.4)}{4} = 22.4 \text{ m/d}$$

Equation 9-65 gives a hydraulic conductivity estimate of k_e = 10,500 m/d. The simplest model of flow and head loss, Equation 9-56, predicts the necessary water surface gradient;

$$\frac{dH}{dx} = -\frac{u}{k_e} = -\frac{22.4}{10{,}500} = -0.00213$$

The head loss over 224 m will therefore be

$$\Delta H = 0.00213(224) = 47.8 \text{ cm}$$

In an effort to keep the water surface parallel to the bed bottom, the bottom slope would be set at 0.213 percent. A 6-cm headspace will be controlled at the exit end to ensure subsurface flow.

This design leads to trouble! Suppose the available water flow at startup is only 150 m^3/d. Equations 9-54 and 9-67 may be solved for the new operating condition, with the result shown in Figure 9-29a. There is then a headspace of 22 cm at the inlet end, which endangers wetland vegetation establishment.

If later in the project life the full design flow is realized and the conductivity drops to one fourth the bare media value, flooding will occur over 90 percent of the bed surface.

If the design is for this reduced conductivity, the bed must be tilted four times as much. There will be a much greater inlet headspace at full conductivity and reduced flow; the inlet headspace would be greater than 50 cm. Wetland plants would not grow. The water profile is shown in Figure 9-29b.

By using bed tilt to control flow, control of headspace is lost.

DESIGN CRITERIA

Bed Slope

Excessive headspace is prevented by limiting the bed slope. For instance, a value of 10 percent of the bed depth might be chosen, based on the desire to utilize most of the bed depth for treatment and to encourage proper rooting. The worst-case water surface is the level pool created by low flows. This criterion is purely geometrical:

$$\Delta B = B_i - B_o \tag{9-81}$$

$$\Delta B < 0.1 \cdot \delta \tag{9-82}$$

This criterion translates easily to dimensionless variables by dividing by bed length:

$$S_b = \frac{\Delta B}{L} < 0.1 \cdot \frac{\delta}{L} \tag{9-83}$$

To a first approximation, $\delta \approx h_o$, thus,

$$G_1 = \frac{S_b}{(h_o/L)} < 0.1 \tag{9-84}$$

The value of G_1 must be less than about one tenth to ensure that the inlet end of the bed does not have too much headspace, leading to dryout conditions that strand plant roots above water (Figure 9-30). The variables in Figure 9-30a are fractional distance, x/L, and fractional elevations, H/h_o, B/h_o, and G/h_o.

Loading

Excessive loading or low hydraulic conductivity lead to excessive gradients of the water surface and flooding. A limit should be placed on the head loss through the bed; a value of 10 percent of the bed depth for instance.

This is also a geometrical constraint:

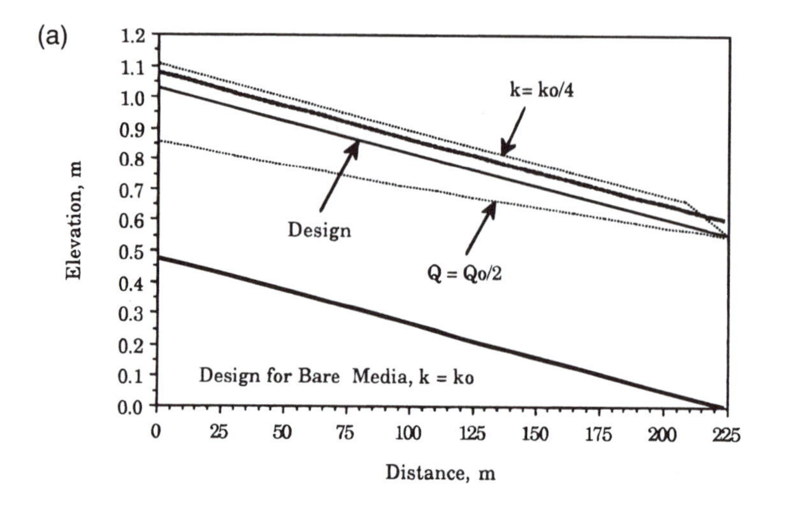

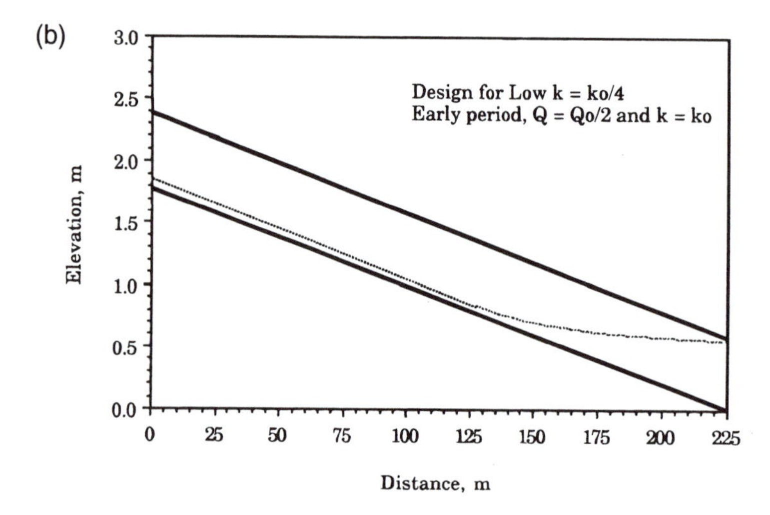

Figure 9-29 The change in water surface profiles resulting from departures from basis of design: (a) design for bare media; (b) design for clogged bed.

$$\Delta H = H_i - H_o \tag{9-85}$$

$$\Delta H < 0.1 \cdot \delta \tag{9-86}$$

Design should not be in the regime of severe distance thickening or thinning flows. Therefore, to a first approximation, Equation 9-71 applies, and

$$\Delta H = \frac{QL}{\tilde{k}\tilde{h}W} < 0.1\delta \tag{9-87}$$

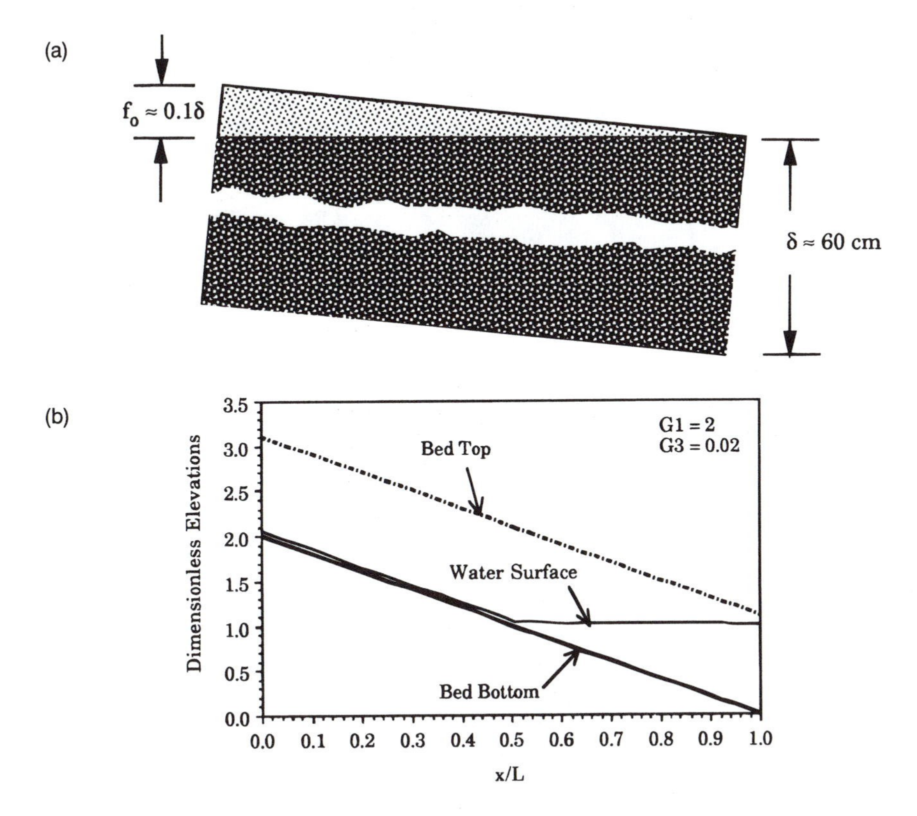

Figure 9-30 (a) Schematic of headspace criterion. At very low flow rates, the level pool should not leave plant roots far above water. (b) Calculated water surface for excessive bed slope, with G_1 at 20 times the recommended criterion of $G_1 \leq 0.1$. Dryout occurs over the majority of the bed in a low-flow situation. Raising the outlet level to prevent dryout would lead to flooding of the outlet end.

This criterion also has a counterpart in dimensionless variables. If $h_o \approx \delta \approx \tilde{h}$, then

$$G_3 = \frac{(q/k)}{(h_o/L)^2} < 0.1 \tag{9-88}$$

For a level bed with outlet depth set at 90 percent of bed depth, this criteria ensures that the inlet end will not flood (Figure 9-31).

Other criteria may be implemented according to the rooting requirements of plants or flooding allowances. The means for interpreting the design to anticipated water surfaces is Equation 9-74, or its equivalent derived from Equations 9-54 and 9-67, for conditions of high P or ET. Likewise, beds with distance-variable media depth will require calculations using Equations 9-54 and 9-67.

Such constraints must be met for all anticipated operating conditions, including initial and clogged conductivity, and the range of expected operating flows.

The Denham Springs, LA, SSF rock-reed filter will serve to illustrate the application of these constraints.

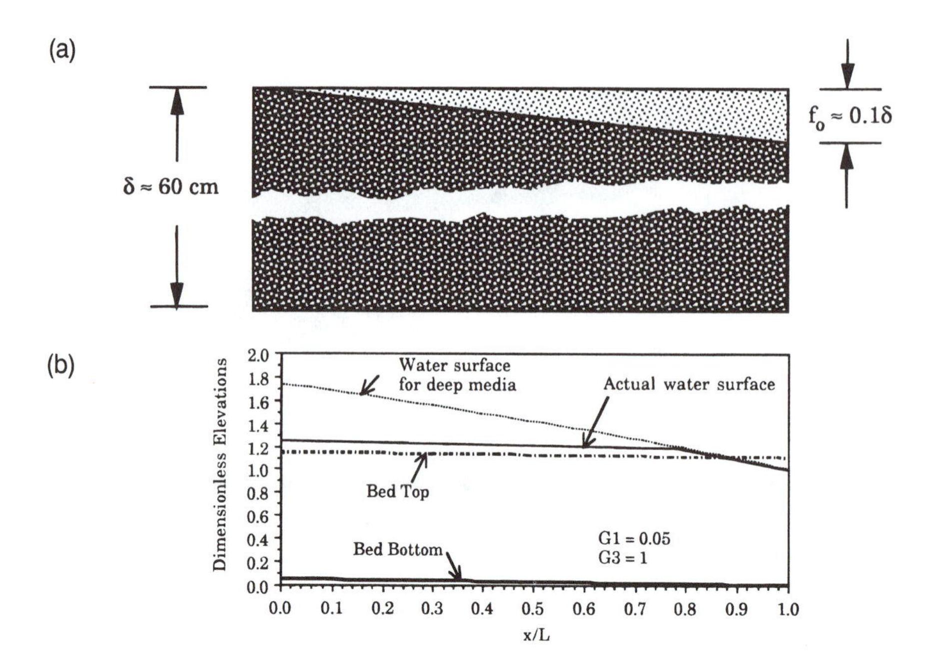

Figure 9-31 (a) Schematic of loading criterion. At high-flow rates, the high gradient should not cause flooding of the inlet end. This schematic is for a level bed. If the wetland is designed with a bed slope, the outlet headspace may be reduced accordingly. (b) Calculated water surface for excessive loading, with G_3 at ten times the recommended criterion of $G_3 \leq 0.1$. Flooding occurs over the majority of the bed. Lowering the exit level would lead to dryout at the exit end.

Example 9-9

Cell #1 at Denham Springs is reported (U.S. EPA North American Database [NADB], 1993) to operate approximately as follows:

$q = 0.10$ m/d $\quad S_b = 0$
$L = 320$ m $\quad D = 6.35$ cm
$h_o = 0.6$ m

Is this SSF wetland within the recommended design constraints?

Solution

First, there is no bottom slope, so $G_1 = 0 < 0.1$, and the constraint in Equation 9-83 is satisfied. The bed is expected to be hard to drain, however.

The exit drains from the system were set at the bed surface, thus, any gradient of the water surface causes flooding (and has done so). However, even if the exit level were dropped to 90 percent of the bed depth, this wetland would be on the brink of flooding. This may be demonstrated from the second constraint in Equation 9-88 for the estimated bare media case at startup. The superficial linear velocity is $u = Lq/h = 320 \times 0.10/0.6 = 53$ m/d. From Figure 9-18, at $D = 6.35$ cm, the estimated $k = 400{,}000$ m/d. Therefore,

$$G_3 = \frac{(q/k)}{(h_o/L)^2} = \frac{(0.1/400{,}000)}{(0.6/320)^2} = 0.071 < 0.1$$

If the hydraulic conductivity drops as plant roots grow and solids deposit, it would be expected to flood.

Example 9-10

The gravel-only trench at Richmond, New South Wales, Australia was operated at hydraulic loading rate $q = 0.027$ m/d. Other operating parameters were

$S_b = 0$	$L = 100$ m
$D = 0.75$ cm	$h_o = 0.24$ m
$\varepsilon = 0.45$	$\delta = 0.35$ m

Was this SSF wetland within the recommended design constraints in Equations 9-83 and 9-88?

Solution

There is no bottom slope, so $G_1 = 0 < 0.1$, and the constraint in Equation 9-83 is satisfied. The bed is expected to be hard to drain.

Check the second constraint in Equation 9-88. The superficial linear velocity is $u = Lq/h_b = 100 \times 0.027/0.24 = 11.3$ m/d. Figure 9-18 does not apply, since the bed porosity is considerably higher than 0.35. From Equation 9-65,

$$\left[\frac{255(1-\varepsilon)\mu}{\rho g \varepsilon^{3.7} D^2} + \frac{2(1-\varepsilon)u}{g\varepsilon^3 D}\right] = \frac{1}{k_e}$$

The necessary property numbers are

$g = 980$ cm/s^2	$\mu = 0.01$ g/cm/s
$\rho = 1.00$ g/cm^3	$u = 11.3$ m/d $= 0.013$ cm/s
$D = 0.75$ cm	$\varepsilon = 0.45$

$$\left[\frac{255(1-0.45)(0.01)}{1.0(980)(0.45)^{3.7}(0.75)^2} + \frac{2(1-0.45)(0.013)}{980(0.45)^3(0.75)}\right] = \frac{1}{k_e}$$

$$[0.0488 + 0.0002] = \frac{1}{k_e} = 0.0490$$

$$k_e = 20.4 \text{ cm/s} = 17{,}620 \text{ m/d}$$

The operating conditions allow for a head loss of (0.35 − 0.24 = 0.11 m) before flooding of the inlet end. Thus, the loading criterion becomes

$$\Delta H = \frac{L^2 q}{k\bar{h}} < 0.11\text{m}$$

Checking:

$$\Delta H = \frac{100^2(0.027)}{(17{,}620)(0.24)} = 0.064 \text{ m} < 0.11 \text{ m}$$

This clean-media, unplanted bed should not display front end flooding.

After wastewater application, the conductivity dropped to 2000 m/d (Figure 9-21). This condition predicts flooding:

$$\Delta H = \frac{100^2(0.027)}{(2000)(0.24)} = 0.56 \text{ m} > 0.11 \text{ m}$$

Actual operation of the Richmond gravel bed displayed flooding in the inlet (clogged) region, but subsurface flow in the downstream (clean) region (Figure 9-32).

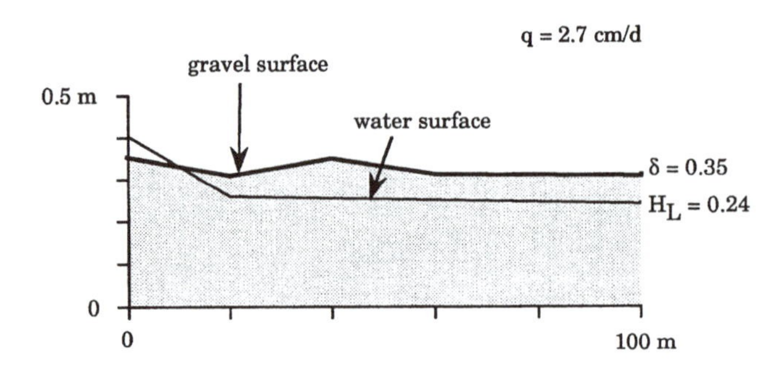

Figure 9-32 The water surface profiles for a gravel control trench at Richmond, NSW, Australia. (Adapted from Bavor, H.J., et al., 1988).

A HYDRAULIC DESIGN EXAMPLE

This hypothetical example is used to illustrate the complete hydraulic design procedure. The basis is the same as for Example 9-8.

Example 9-11

A wetland is to be designed to carry 300 m^3/d of water at full capacity, but only half that amount will be available at startup. The nominal detention time in the media is to be 4 days. A cheap gravel is available with a nominal size of 0.5 cm and a porosity of 40 percent; other sources of media are not as cheap.

Solution

The upper limit on loading sets the wetland volume required, since the largest flow must be treated. The depth of media is selected as 0.6 m to allow for a bottom sediment buildup and to match the rooting requirement of the vegetation. The porosity of the gravel is presumed to be 40 percent.

The wetland area will be selected to meet the treatment requirement:

$$\tau = \frac{LW\delta\varepsilon}{Q}$$

$$4 = \frac{LW(0.6)(0.4)}{300}$$

$$A = LW = 5000 \text{ m}^2$$

The hydraulic loading is therefore

$$q = Q/LW = 300/5000 = 0.06 \text{ m/d} = 6 \text{ cm/d}$$

The length to width ratio and media conductivity must meet the hydraulic constraint. A ten-fold reduction in the conductivity will be presumed to occur due to clogging; k = 10,500/10 = 1050 m/d. The loading constraint is

$$G_3 = \frac{(q/k)}{(\delta/L)^2} < 0.1$$

$$G_3 = \frac{(0.06/k)}{(0.6/L)^2} < 0.1$$

$$\frac{L^2}{k} < 0.6 \qquad \text{units: meters, days}$$

Equation 9-63 gives a hydraulic conductivity estimate of k = 1050 m/d for the cheap media for laminar flow. Therefore,

$$L^2 < 630$$

$$L < 25 \text{ m}$$

Selecting L = 25 m, the width is then W = 5000/25 = 200 m.

The velocity is u = Q/Wh. The depth will be approximately 90 percent of the media depth or 55 cm. Hence u = 300/(200)(0.55) = 2.73 m/d, which is well within the laminar range. This is established by checking the Reynolds number:

$$\text{Re} = \frac{D\rho u}{(1 - \varepsilon)\mu} = \frac{(0.5)(1.0)[(2.73)(100/6{,}400)]}{(1 - 0.4)(0.01)} = 0.26 < 10$$

The drainability condition sets the bottom slope to avoid dryout at reduced loadings:

$$G_l = \frac{S_b}{(\delta/L)} < 0.1$$

$$S_b < 0.1(\delta/L) = 0.1(0.6/25) = 0.0024$$

For the 25-m length, this means a drop in bottom elevation of no more than 6 cm. Select a drop of 5 cm (2 in.) to keep the American surveyors happy.

The aspect ration of 200/25 = 0.125 is conducive to short circuiting and should be corrected by using several cells in parallel. This is consistent with a design strategy which allows for the possibility of cell shutdown. An aspect ration of 1.0 or greater is reasonable. This leads to eight parallel cells in the current design, each 25 × 25 m.

A single bed can also be designed if a different media is obtained. Suppose the aspect ratio is to be L:W = 2.0. Then L = 100 m and W = 50 m. The slope of the bottom can be no higher than 0.0006. The flooding constraint requires

$$k > L^2/(0.6) = (100)^2/(0.6) = 16{,}700 \text{ m/d}$$

Allowing for a tenfold reduction due to clogging, the bare media should have k = 167,000 m/d. In the laminar range, conductivity is proportional to the square of the particle size; hence, we need a size of

$$k = 167{,}000 = \left(\frac{D}{0.5}\right)^2 (10{,}500)$$

$$D = 2.0 \text{ cm}$$

A decision must now be made on the media to be used. The extra cost of the larger media must be compared to the extra cost of more separation dikes and more inlet distribution piping. Inlet distribution must be along the entire width of the cell to avoid corner dead zones. The six-cell system requires 625 m of diking; the one-cell system requires only 300 m. The six-cell system requires 200 m of inlet distribution; the one-cell system requires 50 m. The cost of the media is likely to be more than half the cost of the system, so this decision is economically important. Conductivity measurements on the candidate media are therefore warranted.

Next examine the response of this system to the anticipated variations in operating conditions. Figure 9-33 shows that water level control is virtually completely due to the positioning of the exit water level control point.

Further, the water surface is stable with respect to evaporative losses, as long as the water fed exceeds ET. Protection against rainfall events of all magnitudes is not feasible. The system as designed has approximately 5 cm of headspace at 40 percent porosity, which can contain a 2-cm rain event without surface flow.

NONIDEAL FLOW PATTERNS

Three types of hydraulic inefficiencies may occur in treatment wetlands: one caused by internal islands and other topographical features, a second due to preferential flow channels at a large-distance scale, and a third caused by mixing effects, such as water delays in litter layers and transverse mixing. The first is characterized by a gross areal efficiency; the second and third are characterized by a dispersion number or an equivalent set of well-mixed units in series. All three influence the ability of the wetland to improve water quality.

Nonideal flow patterns can have very large effects upon the removal of pollutants in wetland treatment systems (Kadlec et al., 1993; Chen and Wang, 1994). It is therefore necessary to consider flow pattern effects and the related mixing in the design of wetland treatment systems.

GROSS AREAL EFFICIENCY

In some instances, there are zones within a wetland that do not interact with water entering at a particular influent location. These zones can occur simply due to the presence of islands that have no standing water. Large, measurable, permanent islands would typically be excluded from the measure of total wetland areal extent. But relatively small islands are difficult to quantify, and some are out of the water for only a portion of the time. Internal channels may intercept added water flows and prevent them from reaching some portions of the wetland.

In other circumstances, wetland margins may be wetted by marginal inflows and not by the particular discharge under consideration. This feature would normally not be associated with constructed wetlands. Natural wetlands, and those constructed with contoured margins, can be susceptible to areal segregation from the introduced water flow. Recalling that the

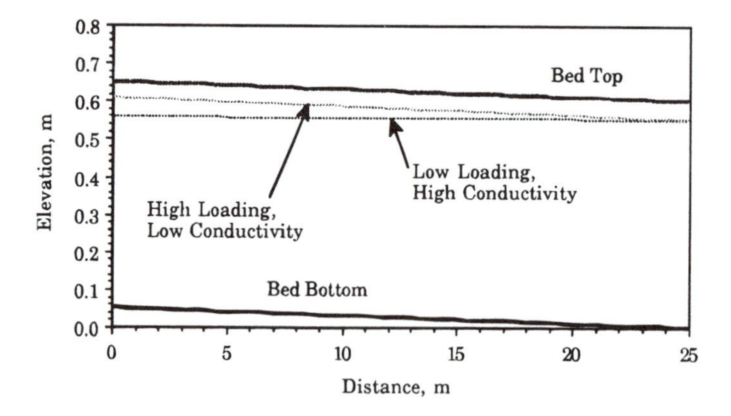

Figure 9-33 Water surface profiles under robust design conditions for Example 9-11. Extremes of operation cause neither flooding nor dryout. Exit level control is at 55 cm.

jurisdictional definition of wetland margins can involve fairly brief (2 weeks) periods of inundation, it is especially inaccurate to use jurisdictional delineation to define wetland treatment area for natural wetland treatment systems. The area within a wetland that has the potential to treat a particular inflow might be determined from areal tracer studies, but this has been done in very few wetlands. In concept, an inert tracer is added to the incoming water, and the wetland is surveyed to find all the areas to which the tracer travels.

In practice, it is not generally feasible to conduct preproject tracer studies, even for existing natural wetlands. In such cases, it is necessary to interpret site topography and the presence of existing channels to form estimates of the potential contact area for a particular inflow. As an example, consider the Wingra marsh study site (Perry et al., 1981) shown in Figure 9-34. This study focused on the efficiency of the marsh in phosphorus removal. The dominant water inflow (90 percent) was urban stormwater. This water flowed through numerous channels inside the wetland and exited through six ditches to two receiving creeks. Flows were measured during one year and were used to estimate the water budget in the succeeding year during which water chemistry was determined. Water and phosphorus budgets were compiled. An area of 9.8 ha was assigned to the wetland. Two flow bands leading downhill to the outflow locations total considerably less area.

Based on the data in the Perry et al. (1981) study, it is possible to estimate an areal uptake coefficient for phosphorus for the year of the study (1975–1976). The methods involved will be described in detail in Chapter 14. The first-order areal rate constant so calculated is 4.5 m/yr. In Chapter 14, it will be shown that a large number of wetlands display rate constants that are considerably larger, averaging 12.1 ± 6.1 m/yr (N = 82). The disparity in the case of Wingra marsh may well be explained on the basis of gross areal

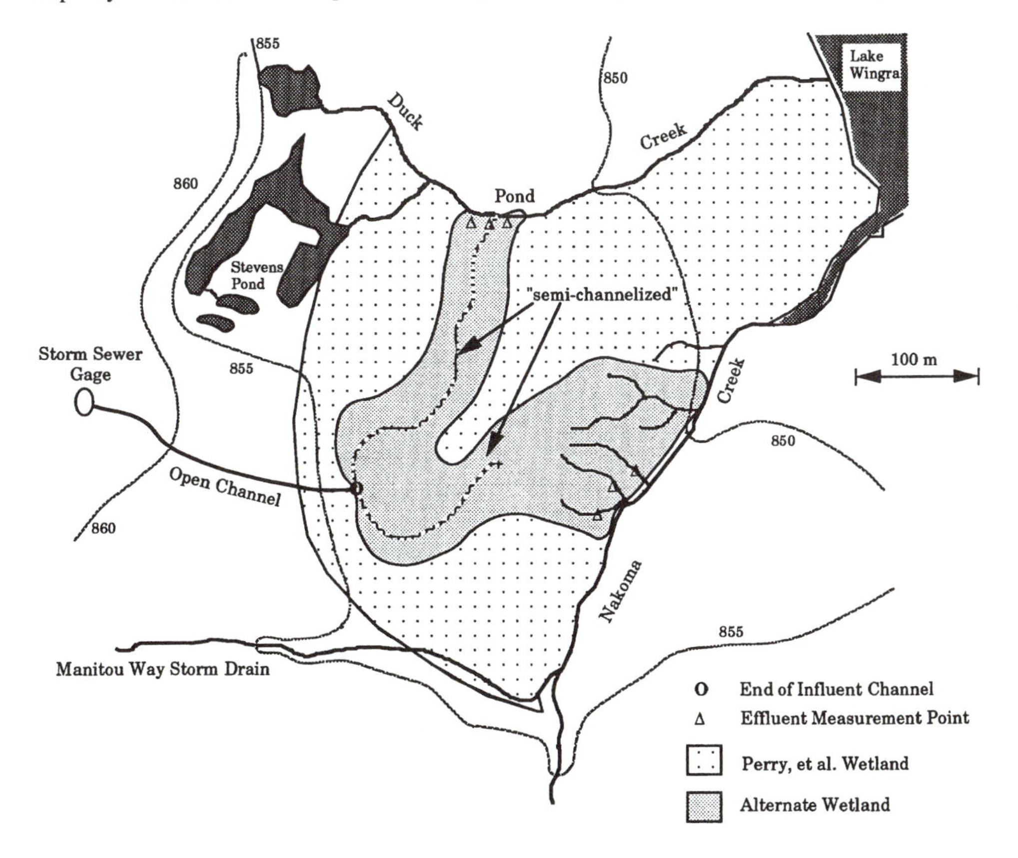

Figure 9-34 Wingra marsh area studied by Perry et al., 1981.

efficiency. Some of the area counted by Perry et al. is separated from the stormwater flow by major streams; other portions are uphill from the discharge point (Figure 9-34). In addition, internal streams convey water in a direct manner to the bounding receiving streams. A considerable portion of the marsh, as defined by Perry et al., is further from the discharge than all the observed and monitored effluent points. Considering the internal channels and the locations of inflows and outflows, it seems plausible that a smaller area is involved in the treatment of the stormwater. An alternative area is estimated to be 3.6 ha (Figure 9-34). When phosphorus removal is assigned to this smaller area, the uptake rate constant is calculated to be 12.2 m/yr, which is central to the distribution of rate constants from other marsh studies.

A secondary method for gaining partial information on gross areal efficiency is the tracer determination of the volumetric efficiency of the wetland. Volumetric efficiency is defined to be the fraction of the wetland water volume that is traversed by an inert tracer added to the incoming water flow of interest. Without detailed internal tracer information, it is not possible to know what surface area is associated with the volume traversed by the tracer.

In the absence of information on gross areal efficiency at data-producing sites, only qualitative assessments may be made. Constructed wetlands, with designer-controlled topography, should not be designed to the pessimistically low standards of some inefficient natural wetlands.

VERTICAL AND TRANSVERSE MIXING

Mixing due to swirls, eddies, stagnant zones, and segregated velocity profiles has been extensively studied due to necessity for reactor design for the chemical, petrochemical, and biochemical process industries. The discipline of chemical reaction engineering has matured to the point that this information is standard in many textbooks which address several levels of audiences (see Levenspiel, 1972, or Fogler, 1992, for instance). Standard procedures are available for the blending of hydrodynamics and intrinsic chemical or biochemical process rates. The principal ingredients of these procedures are the development of models for flow from tracer tests and mass balances and the development of localized models of chemical interactions. In concrete-and-steel process reactors, flow models may often be inferred from past research and the geometry of the system and from operating conditions such as flow rate and reactor volume.

In the SF wetland environment, there are mixing processes on a number of different distance scales. Expanses of open water permit surface, wind-driven currents to develop, which are matched by return flows in lower water layers. Deeper parallel zones in the SF wetland carry more flow because of the depth effect on hydraulic resistance. These preferential channels may also be due to a lower vegetation density along some flow paths. A tracer impulse added to the incoming water provides a way to find such preferential paths, since the tracer will later be found preferentially in those wetland zones. Both natural and constructed SF wetlands display such flow variability (Figure 9-35). In particular, the results for constructed wetlands indicate that it is not possible to avoid such flow irregularities even with extreme care in construction.

There are also mixing effects in the vertical direction in SF wetlands. Water may be moving more slowly near the bottom because of the increased drag of the dense litter layer. Those slow moving zones exchange chemical constituents with adjacent faster-moving layers and thus create vertical mixing. Dense plant clumps can effectively block flow even though these are of very high void fraction. Water in these clumps can exchange constituents with the adjacent microchannels by diffusive processes. All of these effects combine to form a complicated overall mixing pattern. The result of such mixing is evidenced in the blurring of a tracer impulse added to the incoming water. If water moved through the wetland in

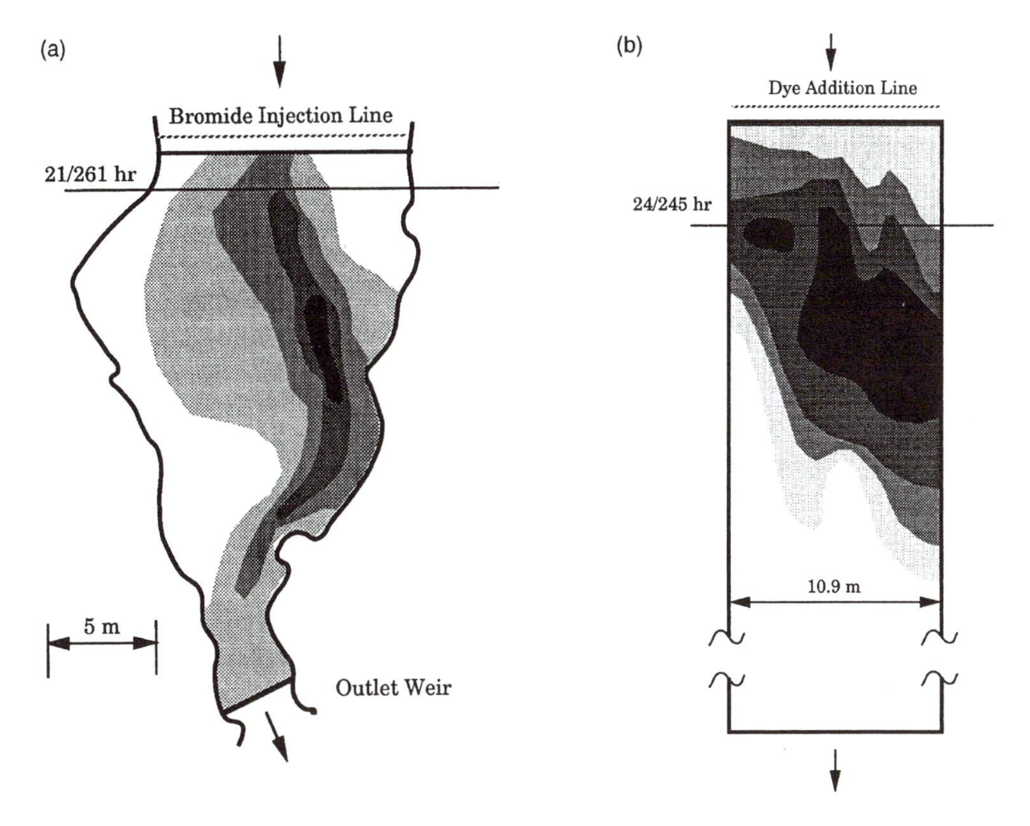

Figure 9-35 Tracer isopleths in natural and constructed wetlands. In both cases, tracer was added uniformly across the inlet width. The theoretical location of the pulse centroid is shown by the horizontal line labeled with elapsed time/theoretical detention time. (a) *Typha orientalis* natural wetland in New Zealand (data from Cooper, 1902); (b) *Typha latifolia* constructed wetland in Ontario (data from Herskowitz, 1986).

lock-step, then such a tracer pulse would exit as an impulse (a sharp spike of concentration). This result has never been observed in a wetland tracer test; the exit tracer is always a blurred, skewed bell-shaped curve (Figure 9-36). Mixing is seldom if ever observed to be complete within the SF wetland. This extreme would lead to an exponentially decaying tracer response to the inlet impulse.

In an SSF wetland, large-scale eddies and wind mixing are absent. However, preferential flow channels can occur on a large scale. Evidence of this was found for the SSF wetland at Benton, KY by internal sampling of tracer tests (Figure 9-37). An impulse of tracer (Rhodamine WT) was added to the inlet flow to this SSF wetland. Water was distributed across the entire width of the rectangular wetland. If a uniform front were to move down the length of the gravel bed, approximately equal tracer responses would be anticipated at sampling points located at equal distances from the inlet. The observed responses were considerably different at equidistant sampling points, indicating subsurface preferential paths. It is further noted that the tracer impulse is blurred to a skewed bell-shaped curve just as in the case of the SF wetlands. Previous research on dispersion in packed beds has provided the explanation: at the scale of an individual particle of the media, there exist microchannels and dead water microzones. A dissolved constituent can diffuse into a backwater microzone, remain for a duration, and then diffuse back into the main stream of the adjacent microchannel. In the SSF bed, an extremely large number of such subsystems exist in series to form the passages through the wetland bed. Therefore, the amount of mixing per unit length in an

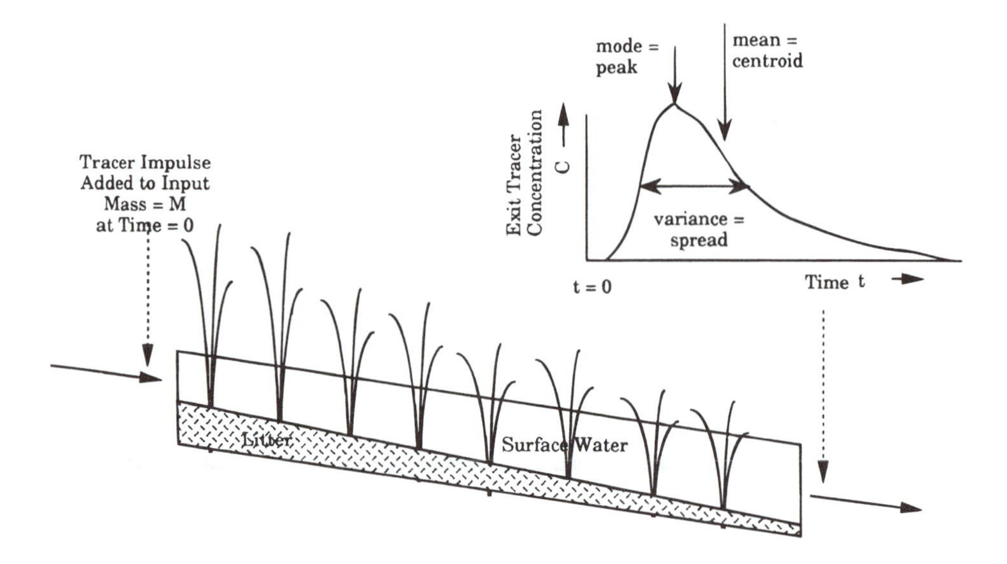

Figure 9-36 The impulse tracer addition experiment and the tracer response curve.

SSF wetland can be, and often is, the same as for an SF wetland with a few large side pools interacting with one main flow channel.

For both types of wetland, the intensity of mixing increases as the water velocity increases and often in nearly direct proportion. The transit time through the wetland is inversely proportional to the water velocity. The combination of these effects means that an element of water undergoes about the same amount of mixing per unit length during its travel through the wetland. This provides the means for correlating mixing in a large number of wetland geometries, as will be demonstrated later in this section. It is first necessary to establish the definitions which form the framework of interpretation for quantitative studies of wetland mixing.

THE THEORETICAL BACKGROUND

A very large body of knowledge about mixing in reactors has been available for several decades. The necessary elements of that knowledge are reproduced here to permit utilization in treatment wetland design.

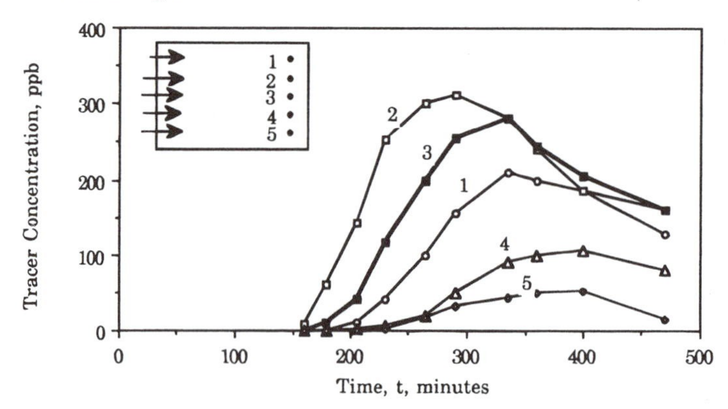

Figure 9-37 Tracer concentrations at five stations normal to the flow direction in gravel bed wetland 3 at Benton, KY. Although these traces are not complete, it is clear the more tracer arrives sooner at station 2 than at some stations. (Data from TVA, unpublished data.)

Most of the literature on wetland hydrology defines a nominal residence time (also called the detention time) to be the ratio of the volume of water in the wetland divided by the volumetric flow rate of water through the wetland:

$$\tau_n = V/Q \tag{9-89}$$

where V = volume of water in the wetland, m^3
τ_n = nominal residence time, d

The value of τ_n is often not known with a high degree of accuracy. Normally, there is a good deal of uncertainty about the depth of water in the wetland. All wetlands possess a depth distribution, due to unavoidable irregularities in the as-built condition or due to the microtopography of a natural wetland. It is rarely possible to perform a quantitative fill or drain experiment to quantify the water volume. A well-performed tracer study can determine the actual mean detention time for the wetland.

The ideas used here were first proposed by Danckwerts (1953) who used the residence time distribution (RTD) to characterize chemical reactors. More recently, the theory of RTDs is elucidated in the textbooks of Levenspiel (1972) and Fogler (1992). These basic chemical reaction engineering principles can be applied to wetlands since these are in fact chemical reactors. However, these principles can often be circumvented in "conventional" concrete and steel devices: the designer can create a good approximation to the desired flow pattern by using vessel internals and judicious choices of geometry. Many of those options are not available to a wetland designer, such as the addition of stirrers.

The RTD represents the time various fractions of fluid (water in the case of the wetland) spend in the reactor; hence, it is the contact time distribution for the system. In a broader context, the RTD is the probability density function for residence times in the wetland. This time function is defined by

$$f(t)\Delta t = \text{fraction of the incoming water which stays in the wetland for a length of time between t and t} + \Delta t \tag{9-90}$$

where f = RTD function, 1/d
t = time, d

The RTD function may be measured by injecting an impulse of dissolved inert tracer material into the wetland inlet and then measuring the tracer concentration as a function of time at the wetland outlet. All of the tracer material enters at time zero. Some elements of this material follow high-speed routes through the wetland and arrive well ahead of the actual mean detention time. Other elements are delayed in slow-moving regions and exit the wetland much later than the mean detention time. Typically, the last water leaves after about three mean detention times (Figure 9-38).

For an impulse input of tracer into a steadily flowing system, the function [f(t)] is

$$f(t) = \frac{QC(t)}{\int_0^\infty QC(t)dt} = \frac{C(t)}{\int_0^\infty C(t)dt} \tag{9-91}$$

where C(t) = exit tracer concentration, g/m^3
Q = water flow rate, m^3/d

The first numerator is the mass flow of tracer in the wetland effluent at any time t after

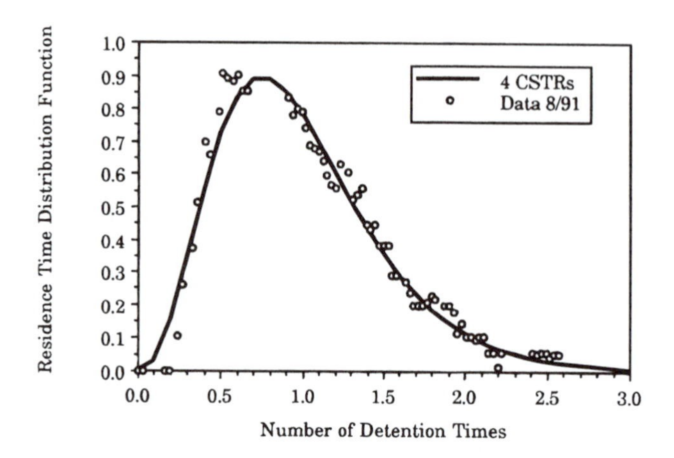

Figure 9-38 The RTD for SF wetland EW3 at Des Plaines, IL in August 1991. Four well-mixed units model the response.

the time of the impulse addition. The first denominator is the sum of all the tracer collected and thus should equal the total mass of tracer injected. The tracer concentration can be measured at interior wetland points as well as at the outlet. Equation 9-91 may then be used to determine the distribution of transit times to that internal point. In this broader sense, the RTD becomes a function of internal position, f(x,t).

The *moments* of the RTD define the key parameters which characterize the wetland; the two most important being the actual detention time and the spreading of a concentration pulse due to mixing (variance of the pulse). The *nth* moment is defined by

$$M_n = \int_0^{\infty} t^n f(t)dt \tag{9-92}$$

The *zeroth moment* represents the definition of the fractional character of the RTD function. Since the term $f(t)\Delta t$ represents the fraction of tracer that spends between time t and $t + \Delta t$ in the system, the sum of these fractions is unity:

$$\int_0^{\infty} f(t)dt = 1 \tag{9-93}$$

However, if the parent concentration distribution is used, the zeroth moment becomes a mass balance check on recovery of tracer:

$$\text{Mass Added} = \int_0^{\infty} QC(t)dt \tag{9-94}$$

Recovery of tracer is an extremely important quality check for the measured RTD. Failure to recover 100 percent of the tracer may mean that the tracer adsorbed or was degraded during passage through the wetland. Organic compounds, including the fluorimetric dyes often used in water tracing in mineral systems, are notorious in this respect; they disappear in wetlands. Noninteractive, inorganic substances are to be preferred; lithium ions and bromide ions being two of the more popular. Batch microcosm tests can help to establish the degree of interaction between a specific tracer and the wetland in question. Interactive, disappearing tracers produce serious errors in the inferred mixing parameters.

Often, the "tail" of the measured exit concentration distribution is poorly defined. Sometimes the sampling is terminated too soon; sometimes the final baseline does not return to the starting zero. Under these circumstances, the tail may be determined as an exponentially decreasing function, extrapolated from the data points past the second inflection of the response.

The *first absolute moment* is the tracer detention time (τ), which is the average time that a tracer particle spends in the wetland. This value defines the centroid of the exit tracer concentration distribution. This average time is also called actual residence time:

$$\int_0^\infty tf(t)dt = \tau \tag{9-95}$$

A wetland may have internal excluded zones which do not interact with flow, such as the volume occupied by plant materials. In a steady state system without excluded zones, the tracer detention time (τ) equals the nominal residence time (τ_n). This is true whether the flow patterns are ideal (plug flow or well mixed) or nonideal (intermediate degree of mixing). An adsorbing tracer will produce an artificially short detention time, which may then be erroneously presumed to result from a large excluded zone. An incorrect topography may be due to either positive or negative differences between τ and τ_n.

A second parameter which can be determined directly from the residence time distribution is the variance (σ^2), which characterizes the spread of the tracer response curve about the mean of the distribution, which is τ. This is the *second central moment*:

$$\int_0^\infty (t - \tau)^2 f(t)dt = \sigma^2 \tag{9-96}$$

where σ^2 = RTD variance, d^2.

The variance of the RTD is created by mixing of water during passage or equivalently by a distribution of velocities of passage. This can be lateral, longitudinal, or vertical mixing. An adsorbing tracer will lead to a narrowed response pulse and, hence, to an erroneously low degree of mixing. This measure of dispersive processes may be rendered unitless by dividing by the square of the tracer detention time:

$$\sigma_\theta^2 = \frac{\sigma^2}{\tau^2} \tag{9-97}$$

The new parameter is σ_θ^2, the *dimensionless variance* of the tracer pulse.

CHEMICAL REACTION CALCULATIONS

The RTD is useable to determine the extent of pollutant reduction in the wetland, provided there is a known reaction rate equation. In chemical technology, a first-order volumetric rate equation finds frequent application, and this expression has also been used for wetland applications. The remaining concentration for a first-order, homogeneous batch reaction is given by

$$\frac{C}{C_i} = \exp(-k_v t) \tag{9-98}$$

where C_i = pollutant concentration at start, g/m^3

C = pollutant concentration at time t, g/m^3
k_v = reaction rate constant, 1/d
t = reaction time, d

For a flow-through wetland reactor, entering elements of water are presumed to move through the system as individual, noninteracting "packets." The length of time any packet spends in the wetland before departing is specified by the RTD. The average remaining concentration is determined by averaging over the distribution of residence times within the wetland to yield

$$\frac{C_e}{C_i} = \int_0^\infty f(t)\exp(-k_V t)dt \tag{9-99}$$

where C_i = pollutant concentration at inlet, g/m^3
C_e = pollutant concentration at outlet, g/m^3

A similar procedure may be used to compute concentrations for a first-order areal rate equation. It is sufficient to use a generalized version of Equation 9-99 involving the Damköhler number:

$$\frac{C}{C_i} = \exp(-Da) \tag{9-100}$$

where $Da = k_v t$ for a first-order volumetric reaction
$Da = k/q$ for a first-order areal reaction

Some care must be used in applying Equation 9-100 to an areal reaction because the hydraulic loading rate, and hence the Damköhler number, may depend on detention time in a nonlinear way.

It will be shown later in this chapter that wetland pollutant reduction may often be modeled with first-order, area-based equations of the form

$$J = k(C - C^*) \tag{9-101}$$

where C = pollutant concentration, g/m^3
C^* = background pollutant concentration, g/m^3
J = reduction rate, $g/m^2/yr$
k = rate constant, m/yr

For illustration, assume that atmospheric processes are in balance and that the wetland is in plug flow, leading to the pollutant mass balance equation

$$q\frac{dC}{dy} = k(C - C^*) \tag{9-102}$$

where q = hydraulic loading rate, m/yr
y = fraction of distance from inlet to outlet

The fraction remaining F of the total possible change in pollutant is termed the *concentration approach.* From inlet to outlet ($y = 1.0$) for plug flow, it is

$$F = \frac{(C_o - C^*)}{(C_i - C^*)} = \exp\left(-\frac{k}{q}\right) = \exp\left(-\frac{k\tau}{\varepsilon h}\right) \qquad (9\text{-}103)$$

where C_o = outlet concentration, g/m^3
C_i = inlet concentration, g/m^3
h = wetland depth, m
ϵ = wetland porosity, m^3/m^3

The RTD averaging technique may also be applied to the concentration approach:

$$F = \int_0^{\infty} f(t)\exp\left(-\frac{kt}{\varepsilon h}\right)dt \qquad (9\text{-}104)$$

The knowledge of f(t) permits the adaptation of simple kinetic expressions to the nonideal flow patterns in a wetland for the prediction of average behavior in the steady flow system. Transient flow patterns, with the accompanying transient storages of water and pollutants, require transient RTD functions, which may be quite complex, as shown by Kadlec et al. (1993). It is therefore convenient to construct flow models which embody the RTD and the dynamic hydrological mass balance for water. There are dozens of published methods for constructing such flow models (see the review of Call, 1989, for example). Many of these consist of series and parallel combinations of the two ideal extremes of mixing: totally mixed zones (corresponding to a continuous stirred tank reactor or CSTR in the reaction literature) and totally unmixed zones (corresponding to a plug flow reactor or PFR in the reaction literature). The PFR may in turn be represented by an infinite series of CSTRs or approximated by a smaller number, usually less than ten (Levenspiel, 1972).

MODELS FOR WETLAND MIXING WITH REACTION

The need to be concerned about flow profiles will be established first by considering the situation in which different elements of incoming water take different lengths of time to pass through the wetland. This has the effect of blurring a tracer impulse and is one type of non-ideal flow pattern in wetlands. This implies the existence of noninteracting parallel paths through the wetland. Other types of mixing are presumed to be absent, which means that each parallel channel is in plug flow. This condition will serve to illustrate the potential for effects of the commonly observed RTD.

Suppose that a wetland has multiple "parallel" noninteracting paths for water movement from inlet to outlet, as indicated in Figure 9-39. For illustration, assume that atmospheric processes are in balance. Each path may be regarded as a plug flow path, but with different loading rates and residence times. The pollutant reduction along each path is obtained from Equation 9-103. For instance, along channel 1, the fraction remaining F_1 of the total possible change in pollutant is

$$F_1 = \frac{(C_o - C^*)_1}{(C_i - C^*)} = \exp\left(-\frac{k\tau_1}{\varepsilon h}\right) \qquad (9\text{-}105)$$

where C_{o1} = outlet concentration from channel 1, g/m^3
C_i = inlet concentration, g/m^3
τ_1 = detention time down channel 1, d
= V_1/Q_1

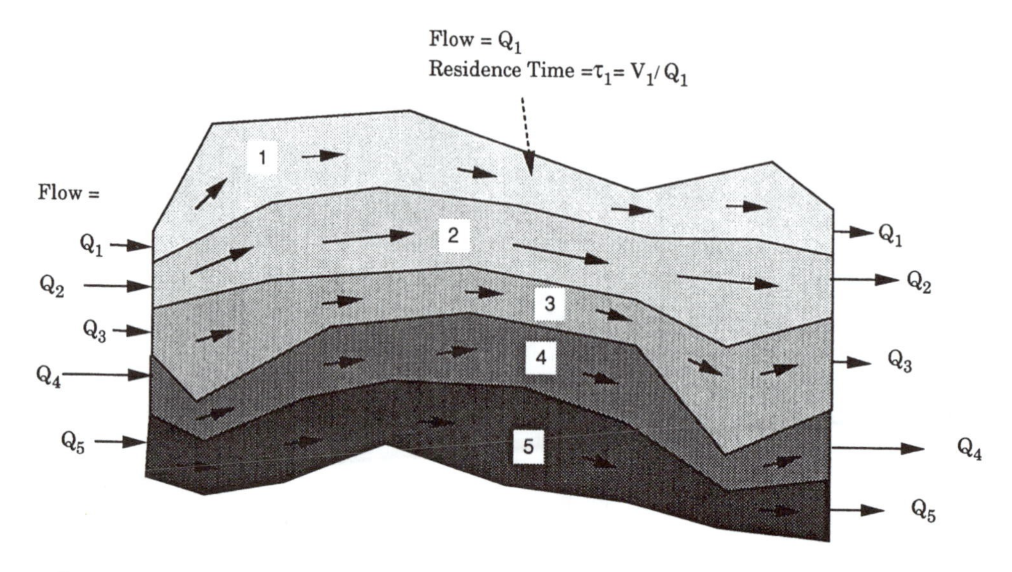

Figure 9-39 Hypothetical wetland with five flow paths of different flow rates and detention times.

Let $X_1 = Q_1/Q_T$ be the fraction of water passing through section 1, where Q_T is the total flow rate through all paths. Then the overall exit concentration approach is then the flow-weighted average:

$$F = \sum_i X_i \exp\left(-\frac{k\tau_i}{\varepsilon h}\right) \tag{9-106}$$

This entire wetland possesses a nominal detention time, calculated from Equation 9-89:

$$\tau_n = \frac{\sum_1^5 V_k}{\sum_1^5 Q_k} \tag{9-107}$$

This detention time is often used to compute the plug flow exit concentration according to Equation 9-103, which is different from that computed using Equation 9-106.

Example 9-12 illustrates the difference in design results for a typical case. The effect of the RTD on the design area is quite large.

Example 9-12

A wetland is to be designed to carry 139 m^3/d of water and to lower the concentration of a pollutant from 100 to 7 mg/L. The kinetic constants are known to be k = 35 m/yr and C* = 5 mg/L. Site constraints limit the length to 140 m. The wetland will behave as five parallel channels (Figure 9-39), with the RTD shown in Figure 9-40. Size the wetland assuming plug flow, and compare to the size required for the actual RTD.

Solution

Each channel is calculated separately, and the results are flow weighted to obtain the average outlet concentration. The area is adjusted until the desired target of 7 mg/L is achieved, the result being 9520

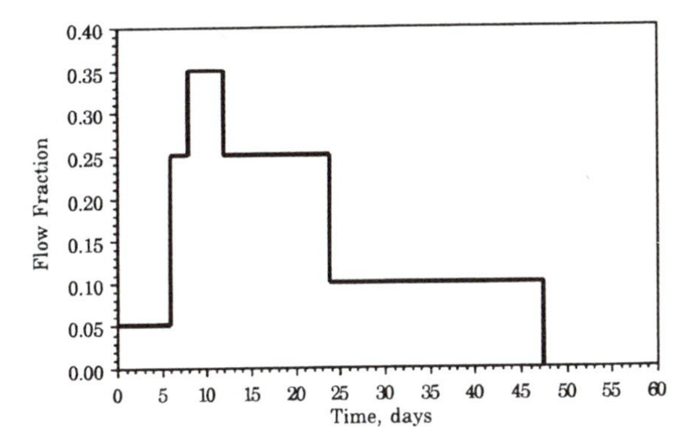

Figure 9-40 RTD for the wetland in Example 9-12. This is a stepwise approximation to the actual RTDs observed, such as Figure 9-38.

m^2. Computations are shown in Table 9-4. The plug flow calculation, using the nominal detention time, yields an area of 5579 m^2, which is 70 percent smaller than that required.

Examination of the computation reveals that the difference is due to the rapid passage of some pollutant to the exit without sufficient time for reduction. When mixed with the slower-moving material, with a greater reduction, the average reduction is more difficult to achieve.

DATA ANALYSIS, THE RTD, DESIGN, AND THIS BOOK—IMPORTANT!

Intrinsic rate constants are to be used in RTD-based conversion equations such as Equation 9-105. These are rate constants determined from data for which the degree of flow nonideality is known. This means that the rate constant must have been derived from a tracer-tested wetland, with a known RTD. At this point in the development of wetland technology, a large number of wetlands have produced data on pollutant reductions, but a very small proportion have been tracer tested.

It is not reasonable to abandon large amounts of useful information from operating systems based on the lack of knowledge of the RTDs. Accordingly, an assumption must be

Table 9-4 Five Channel Flow Example

Width	= 68 m	PFR: $\dfrac{C_o - C^*}{C_i - C^*} = \exp\left(-\dfrac{k\tau}{H}\right)$
Length	= 140	
Depth	= 0.25 m	
C^*	= 5 mg/L	
C_i	= 100 mg/L	$\dfrac{7-5}{100-5} = \exp\left(-\dfrac{0.96\tau}{.25}\right)$
C_o	= 7 mg/L	
k	= 35 m/yr	
	= 0.096 m/d	
Q	= 138.73 m^3/d	
actual τ	= 17.16 days	nominal τ = 10.05 days
Area	= 9520 m^2	Area = 5579m^2

Flow Fraction	Velocity m/d	Residence Time d	Weighted τ d	Outlet Conc. mg/L	Weighted Outlet Conc. mg/L
0.05	23.53	5.95	0.298	14.7	0.735
0.25	17.65	7.93	1.983	9.5	2.383
0.35	11.76	11.90	4.165	6.0	2.096
0.25	5.88	23.80	5.950	5.0	1.253
0.10	2.94	47.60	4.760	5.0	0.500
			17.16		7.0

made about the degree of mixing in the data-producing wetlands. For first-order processes, the smallest value of the rate constant corresponds to the least degree of mixing—the PFR assumption. Rate constants are determined with this assumption in this book, based on the authors' wish to provide a conservative interpretation of the available information from operating wetland systems. The use of the rate constants presented in this book, therefore, must recognize this fact.

It is recommended that the user adopt one of two philosophies related to wetland design when using the PFR rate constants given in this book. The easier of the two is to presume that the wetland to be built or used has less than or equal mixing compared to the wetlands from which the rate constants were derived. If the new wetland is more hydraulically efficient, this procedure will result in overdesign; if less efficient, the result will be underdesign.

The second procedure is to adopt a presumed degree of mixing for both the data-generating wetlands and for the new wetland. This is desirable in those cases where other site and design considerations indicate the new wetland will be of very low hydraulic efficiency. The first step is to correct the PFR rate constant to the presumed degree of mixing in the database wetlands. For instance, if the three-tank assumption is made for data reduction, then the PFR rate constant must be modified by the ratio of the three-tank Damköhler number to the PFR Damköhler number to yield a higher intrinsic rate constant.

The second step is to determine the required Damköhler number (Da = k/q) for the wetland under design, from the figures and equations in the following chapters, for the required pollutant reduction. The design hydraulic loading or detention time is then determined from the rate constant and the design Damköhler number.

Two procedures are commonly used to describe partially mixed flows and associated reactions: the tanks-in-series model and the dispersion model.

THE TANKS-IN-SERIES MODEL (TIS)

This model has been advocated by the WPCF (1990b) for wastewater lagoons and is also a popular model for many other applications. The wetland may be conceptually partitioned into a number of equal sized pieces (N), each of which is presumed to be completely mixed (Figure 9-41). Therefore, the concentration departing each is equal to the uniform, internal concentration.

The number of tanks which characterizes a particular wetland may be determined in a number of ways, all of which derive from the solution of the transient tracer mass balance (Levenspiel, 1972). If a tracer impulse is added to the inlet of a wetland, the form of the outlet pulse for the TIS model is given by

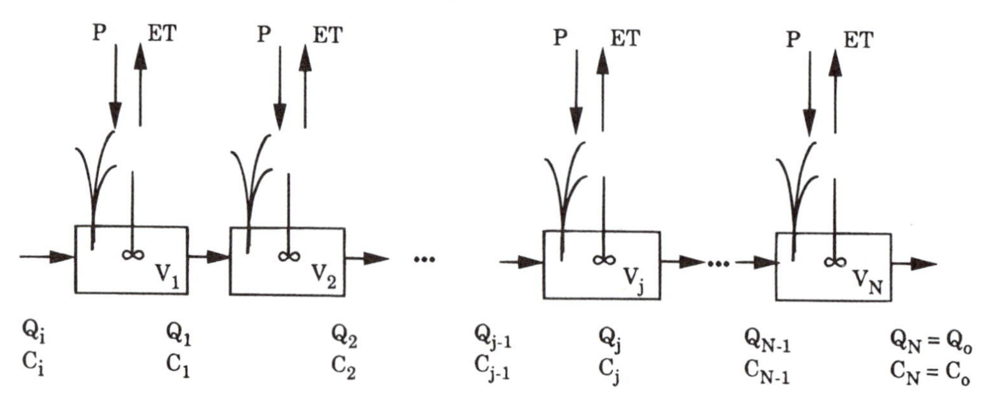

Figure 9-41 The tanks-in-series model for mixing in a treatment wetland. The bottom area of each unit is A_j, and the depth is h_j.

$$f(t) = \frac{N}{(N-1)!}\left(N\frac{t}{\tau}\right)^{N-1}\exp\left(-N\frac{t}{\tau}\right) \tag{9-108}$$

The mean of this distribution is the mean residence time, τ.

Special cases of the TIS are the single continuous stirred tank reactor (CSTR) ($N = 1$) and the plug flow reactor (PFR) ($N = \infty$). These have the following RTDs:

CSTR:

$$f(t) = \exp\left(-\frac{t}{\tau}\right) \tag{9-109}$$

PFR:

$$f(t) = \delta(t - \tau) \tag{9-110}$$

where $\delta\ (t-\tau)$ is the Dirac delta function, which is a "spike" of area = 1.0 located at $t = \tau$. The *mode* of this RTD is the time of the peak of the distribution. It is given by

$$\frac{\tau - \tau_{peak}}{\tau} = \frac{1}{N} \tag{9-111}$$

This immediately presents an interesting result. The peak of the tracer distribution does not occur at the mean residence time, it precedes it; for a small value of N, the lead time is a significant fraction of the mean detention time.

If the exit distribution has a broad crest, Equation 9-111 is not an accurate way to determine N. It is better determined from the dimensionless variance (Levenspiel, 1972)

$$\sigma_\theta^2 = \frac{1}{N} \tag{9-112}$$

Tracer test data from operating SF treatment wetlands indicate that the value of N ranges from 2 to 5. A value of $N = 1.0$ corresponds to a totally mixed wetland; a value of $N = \infty$ is the plug flow extreme. The corresponding range for σ_θ^2 is from 1.0 to 0.0. It is not necessary that N be an integer; fractional values are allowable.

Figure 9-42 indicates that many SF wetland geometries yield similar tracer responses and that $N = 3$ is a fair approximation to the tracer data. Several investigators have reported small numbers of tanks, $2 < N < 8$, fitting tracer response data for SF wetlands (see, for instance, Stairs, 1993; Kadlec, 1994). However, open water systems are subject to wind-driven mixing and therefore may be quite well mixed even if the channel is very long and slender (Bavor et al., 1988). Figure 9-43 demonstrates that even an aspect ratio of 25:1 may be representable by one or two well-mixed units.

Chemical Reactions in Tanks in Series (TIS)

For a CSTR, the mass balance equation for the *jth* tank is

$$Q_{j-1}C_{j-1} - Q_jC_j = R_j = kA_j(C_j - C^*) \tag{9-113}$$

For simplicity, suppose that rain and ET are in balance, so that all flows are equal, $Q_j = Q$. Then C^* may be subtracted and added on the left side:

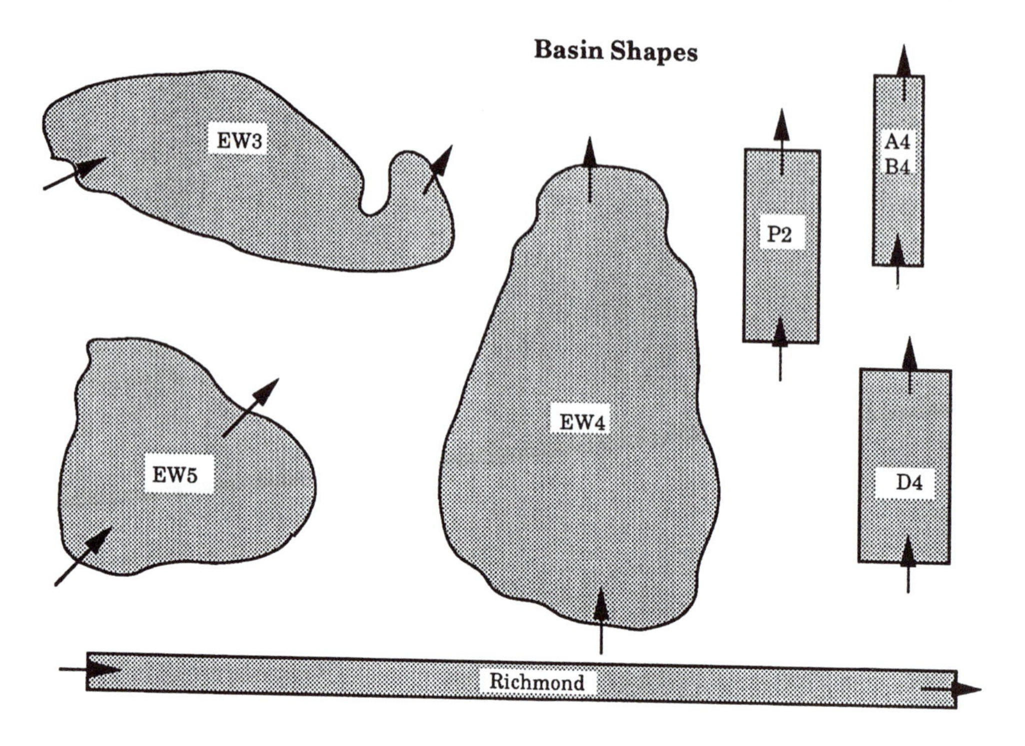

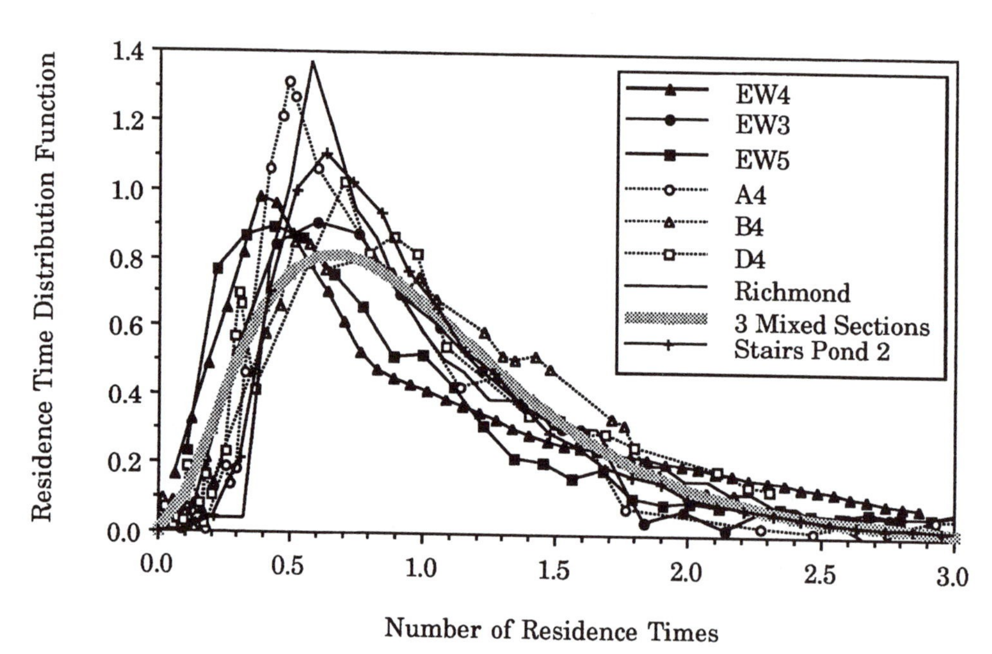

Figure 9-42 Tracer response curves from several SF wetlands. The response of three well-mixed units is shown for comparison. EW3, 4, and 5 are at Des Plaines; A4, B4, and D4 are at Champion. The Richmond wetland was a *Myriophyllum* bed. Stairs pond 2 was at Pope and Talbot.

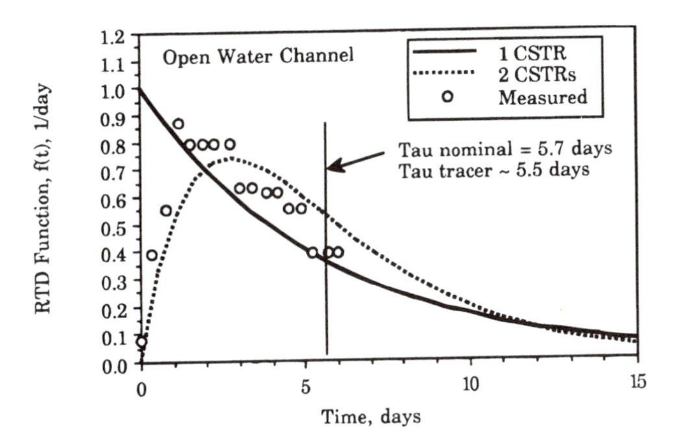

Figure 9-43 Tracer response for an open water, unvegetated SF wetland at Richmond, NSW (Data from Bavor, et al. 1988). The aspect ratio was L to W = 25:1. The channel mixing is the equivalent of between one and two CSTRs.

$$C_j - C^* = \frac{C_{j-1} - C^*}{1 + kA_j/Q} = \frac{C_{j-1} - C^*}{1 + k/q_j} \qquad (9\text{-}114)$$

or

$$F_j = \frac{C_j - C^*}{C_{j-1} - C^*} = \frac{1}{1 + k/q_j} \qquad (9\text{-}115)$$

However, $A_j = A_T/N$, so

$$F_j = \frac{F_{j-1}}{\left[1 + \frac{kA_T}{NQ}\right]} = \frac{F_{j-1}}{\left[1 + \frac{(k/q)}{N}\right]} = \frac{F_{j-1}}{\left[1 + \frac{Da}{N}\right]} \qquad (9\text{-}116)$$

Combining the mass balances for all N tanks produces the relation between inlet and outlet concentrations:

$$\frac{(C_o - C^*)}{(C_i - C^*)} = \left[1 + \frac{k}{Nq}\right]^{-N} \qquad (9\text{-}117)$$

If in addition $C^* = 0$, the relation between inlet and outlet concentrations is

$$C_o = \frac{C_j}{\left[1 + \frac{Da}{N}\right]^N} \qquad (9\text{-}118)$$

Although the tanks-in-series model may reproduce the correct overall effect for a given wetland, the measured internal distance profiles will not necessarily be a stepwise sequence of flat concentration profiles. Rather, this model represents continuous profiles in the direction of flow as such a sequence.

Example 9-12

A wetland is to reduce a pollutant by 90 percent. The reduction follows a first-order areal model, with an intrinsic rate constant of 20 m/yr. The wetland is expected to behave as three CSTRs. What is the required (maximum) hydraulic loading rate?

Solution

Equation 9-117 gives the required Damköhler number:

$$0.1 = \frac{1}{\left[1 + \frac{Da}{3}\right]^3} \tag{9-118}$$

$$Da = 3.46$$

For the areal model, Da = k/q; hence,

$$q = k/Da = 20/3.46 = 5.77 \text{ m/yr} = 1.58 \text{ cm/d}$$

TANKS-IN-SERIES MODEL WITH A DELAY

Subsurface-flow wetlands are not as well represented by a series of CSTRs. Most SSF tracer responses reflect a delay during which no tracer reaches the outlet of the bed (see Figure 9-44). Such a delay is describable by the addition of a PFR element of a flow network. Suppose the delay time is t_d. Then the RTD is given in two pieces:

$$f(t) = 0 \qquad t < t_d \tag{9-118}$$

$$f(t) = \frac{N}{(N-1)!}\left(N\frac{(t-t_d)}{\tau}\right)^{N-1} \exp\left(-N\frac{(t-t_d)}{\tau}\right) \qquad t > t_d$$

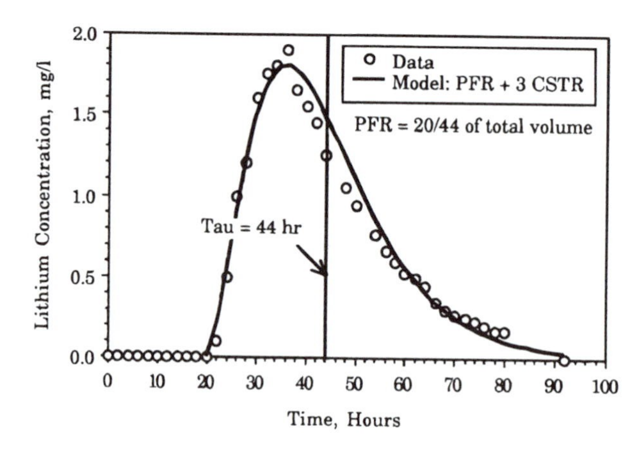

Figure 9-44 Tracer response for the Carville, LA SSF wetland. A network model fits the data, with a PFR of 45 percent and 3 CSTRs of 18 percent of the volume each. (Data are from the U.S. EPA 1993b.)

The conversion in the plug flow component of this nonideal configuration is described by an exponential relation:

$$\frac{C_{o,PFR}}{C_{i,PFR}} = \exp(-Da_{PFR}) \tag{9-119}$$

and the concentration relation for the CSTRs is given by Equation 9-118 written for the remaining portion of the bed:

$$C_{o,CSTRs} = \frac{C_{i,CSTRs}}{\left[1 + \frac{Da_{CSTRs}}{N}\right]^N} \tag{9-120}$$

The Damköhler numbers are written for the corresponding portion of the total detention time:

$$Da_{PFR} = Da \cdot \frac{t_{delay}}{\tau} \tag{9-121}$$

$$Da_{CSTRs} = Da \cdot \frac{1 - t_{delay}}{\tau} \tag{9-122}$$

For the particular case of a first-order reaction model, it does not matter whether the PFR precedes or follows the set of CSTRs.

Example 9-13

An SSF wetland is to reduce a pollutant by 90 percent. The reduction follows a first-order areal model, with an intrinsic rate constant of 20 m/yr. The wetland is expected to behave as 50 percent in PFR, 50 percent in three CSTRs. What is the required (maximum) hydraulic loading rate ($C^* = 0$)?

Solution

A combination of Equations 9-119 and 9-120 gives

$$\frac{C_o}{C_i} = \frac{\exp(-Da_{PFR})}{\left[1 + \frac{Da_{CSTRs}}{3}\right]^3}$$

The required Damköhler numbers are related to the system Da via Equations 9-121 and 9-122:

$$Da_{PFR} = Da_{CSTRs} = (0.5) \cdot Da$$

Then,

$$\frac{C_o}{C_i} = 0.10 = \frac{\exp(-Da/2)}{\left[1 + \frac{Da/2}{3}\right]^3}$$

$$Da = 2.03$$

For the areal model, Da = k/q; hence,

$$q = k/Da = 20/2.03 = 9.85 \text{ m/yr} = 2.70 \text{ cm/d}$$

PLUG FLOW MODIFIED BY DISPERSION

A second technique is the use of a dispersion process superimposed on a plug flow model (PFD). Mixing is presumed to follow a diffusion equation. A one-dimensional spatial model is chosen because analytical expressions are available for computation of pollutant removal for the one-dimensional case (Fogler, 1992). A two-dimensional version requires the two-dimensional velocity field, which has yet to be determined for any operating treatment wetland. The tracer mass balance equation includes both spatial and temporal variability:

$$D \frac{\partial^2 C}{\partial x^2} - \frac{\partial (uC)}{\partial x} = \frac{\partial C}{\partial t} \tag{9-123}$$

where u = velocity, m/d
D = dispersion constant, mm^2/d
x = distance from inlet toward outlet, m

The appropriate wetland boundary conditions for this mass balance are known as the *closed-closed* boundary conditions (Fogler, 1992). These imply that no tracer can diffuse back from the wetland into the inlet pipe, nor back up the exit structure at the wetland outlet. These are different from the *open-open* boundary conditions that are appropriate for river studies. There are analytical, closed-form solutions to the later case, which has led to their repeated misapplication to wetlands (Bavor et al., 1988; Stairs, 1993). There are no closed-form solutions to the former case, but numerical solutions to the closed-closed tracer mass balance have been available for more than 20 years (Levenspiel, 1972). Fortunately, there are correlations for the moments of the resulting RTD, and hence, it is possible to easily calculate the dispersion constant that fits a particular data set.

The dimensionless parameter which characterizes Equation 9-123 is the Peclet number or its inverse, the wetland dispersion number:

$$\mathscr{D} = \frac{D}{uL} = \frac{1}{Pe} \tag{9-124}$$

where Pe = Peclet number, dimensionless
$\mathscr{D}$ = wetland dispersion number, dimensionless
L = distance from inlet to outlet, m

The two results of interest from modeling of the pulse test are the tracer detention time and the dimensionless variance:

$$\tau = L/u \tag{9-125}$$

$$\sigma_\theta^2 = 2\mathscr{D} - 2\mathscr{D}^2(1 - e^{-1/\mathscr{D}}) \tag{9-126}$$

REGRESSION OF NUMERICAL MODEL

The wetland dispersion number may therefore be calculated from the moments of the tracer response curve (Equations 9-93, 9-94, and 9-95). All the caveats discussed in the fitting of the TIS model also apply here, such as tracer mass conservation, sorption, and treatment of "tails."

Values of D typically range from 0.07 to 0.33 (Table 9-5).

Chemical Reactions in PFD

The mass balance for a first-order reactive substance in steady flow through a PFD wetland is a second-order ordinary differential equation first solved by Danckwerts in 1953 (Fogler, 1992):

$$\frac{C_o}{C_i} = \frac{4b \exp\left(\frac{Pe}{2}\right)}{(1 + b)^2 \exp\left(\frac{bPe}{2}\right) - (1 - b)^2 \exp\left(\frac{-bPe}{2}\right)} \tag{9-127}$$

where

$$b = \sqrt{1 + 4\frac{Da}{Pe}} \tag{9-128}$$

This equation has been advocated for wastewater stabilization ponds (WPCF, 1990b). Figure 9-45 gives a graphical representation of Equations 9-127 and 9-128.

Example 9-14

Performance of an existing wetland is described by a first-order areal rate constant of 10 m/yr for conversions from 75 to 90 percent under the plug flow approximation. It is suspected that the wetland really obeys a PFD model, with D = 0.15. What is the value of the intrinsic rate constant?

Table 9-5 Mixing Parameter Values for SF Wetland and Aquatic Systems

System	Source	Normalized Variance σ_θ^2	System Dispersion Number $\mathscr{D} = \frac{D}{uL} = \frac{1}{Pe}$	Depth Dispersion Number $\frac{D}{uh}$
SF Wetlands				
Cattail/open water	Kadlec (1994)	0.40 ± 0.10	0.26 ± 0.10	89 ± 34
Cattail	Stairs (1993)	0.45 ± 0.32	0.250	34.2
Cattail	TVA (1990a)	0.22	0.12	136
Cattail	Herskowitz (1986)	0.13	0.07	
Open water	Bavor et al. (1988)	0.40	0.27	63
Myriophyllum	Fisher (1990)	0.26	0.15	30
Sawgrass	Rosendahl (1981)	0.14	0.07	50
SSF Wetlands				
Schoenoplectus	Bavor et al. (1988)	0.19	0.11	
Soil	Schierup et al. (1990a)	0.46 ± 0.03	0.35 ± 0.03	
Ponds				
Nssuka, Nigeria	Agunwamba et al. (1992)	0.265 ± 0.035	0.157 ± 0.025	65 ± 10
Dredge ponds	Thackson et al. (1987)	0.40 ± 0.28	0.48 ± 0.64	77 ± 38
Rivers				
22 Rivers and channels	Fischer et al. (1979)			60 ± 41
5 Rivers	Day (1975)	0.019–0.076	0.009–0.038	58 ± 30
Experimental channel	Seo (1990)	0.31 ± 0.05	0.19 ± 0.04	

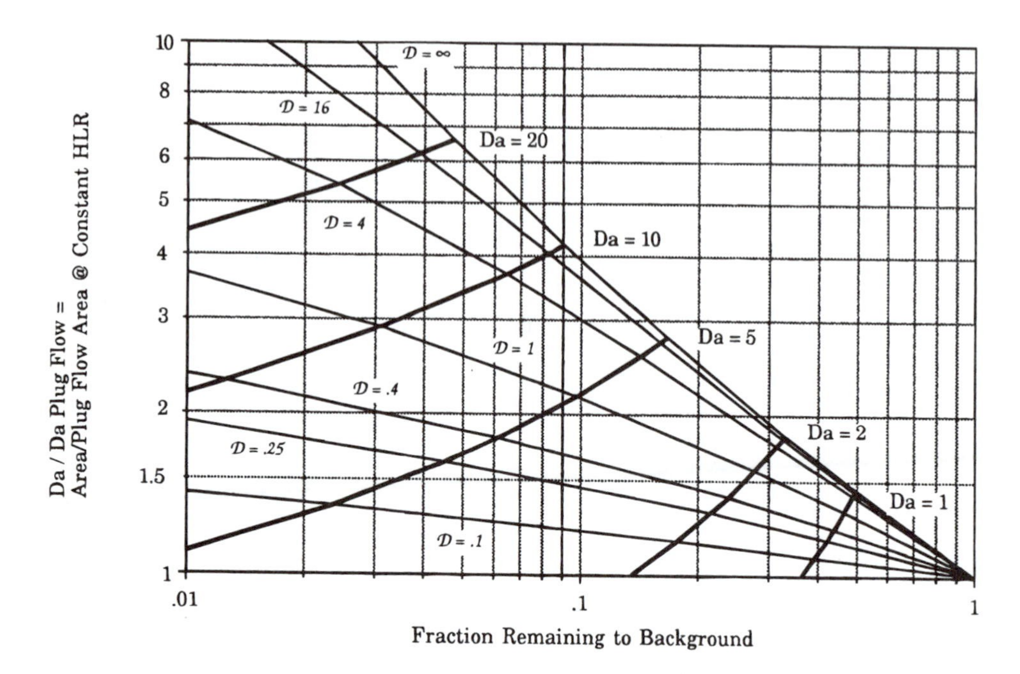

Figure 9-45 Comparison of plug flow and partially mixed wetlands for first-order reactions. The model is plug flow with dispersion. The vertical axis is the ratio of the Damköhler number for a PFD to the Damköhler number for a PFR for a first-order k-C* model. At any given hydraulic loading rate, the vertical axis is the ratio of area to the corresponding plug flow area.

Note: Fraction Remaining to Background $= \dfrac{C_o - C^*}{C_i - C^*}$

Solution

From Figure 9-45, it is seen that the ratio of actual to plug flow Damköhler numbers ranges from 1.20 to 1.30 over the specified conversion range. According to the definition of Da = k/q, and the fact that q is the same for both models, the ratio of k values is also 1.20 to 1.30. Hence, the intrinsic rate constant is in the range of 12 to 13 m/yr.

CHEMICAL REACTIONS IN PARTIALLY MIXED WETLANDS: AN APPROXIMATION

Equations 9-127 and 9-128 are cumbersome and may be approximated by a simple procedure. The same approximation may be used for the TIS model, as discussed here.

The extremes of mixing are represented by the PFR and CSTR models, for which wetland exit concentrations are easily calculated from the first-order equations:

$$\frac{C_e}{C_i} = \exp(-\mathrm{Da}) \tag{9-129}$$

$$\frac{C_e}{C_i} = \frac{1}{1 + \mathrm{Da}} \tag{9-130}$$

Actual wetlands will produce exit concentrations intermediate to these two extremes, as predicted by Equations 9-127 and 9-128. The dimensionless variance σ_θ^2 also finds its extremes

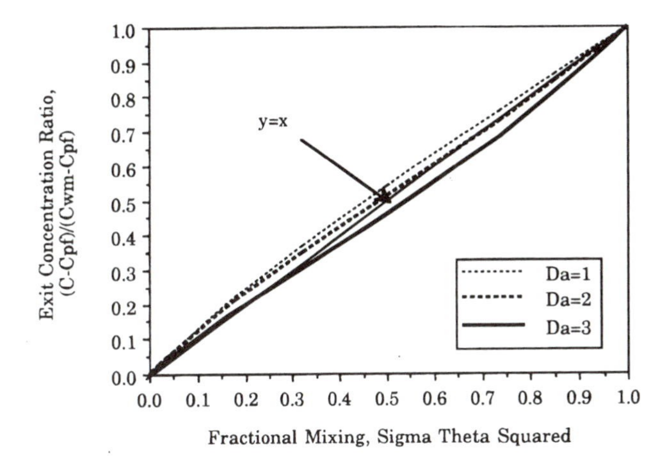

Figure 9-46 The predicted effect of mixing on a concentration remaining for first-order reactions. (Data from Danckwerts, 1953.) Exit concentrations are for partially mixed reactor, C; for the well-mixed reactor, C_{wm}; and for the plug flow reactor, C_{pf}. Mixing is quantified by the dimensionless variance, σ_θ^2. The speed of the reaction is characterized by the Damköhler number, Da = k/q. The range shown here, Da < 3, corresponds to plug flow concentration fractions remaining from 0.05 to 1.00 or percent reductions from 0 to 95 percent. Values of Da < 1 are also represented by the line for Da = 1.

for the PFR and the CSTR, the values zero and unity, respectively. To a good approximation, σ_θ^2 may be used as a linear interpolator between PFR and CSTR exit concentrations. Thus, both the TIS and PFD models may be implemented in a simple way. The parameters from these are simply related to σ_θ^2:

$$\sigma_\theta^2 = 2\mathcal{D} - 2\mathcal{D}^2(1 - e^{-1/\mathcal{D}}) = \frac{1}{N} \tag{9-131}$$

The interpolation is

$$\frac{C_{ACTUAL}/C_i - C_{PFR}/C_i}{C_{CSTR}/C_i - C_{PFR}/C_i} \cong \sigma_\theta^2 \tag{9-132}$$

Equations 9-127 and 9-128 are presented graphically in Figure 9-46, illustrating the degree of fit of the approximation in Equation 9-132.

Example 9-15

A wetland receives a hydraulic loading of 1.58 cm/d. The reduction of a pollutant follows a first-order areal model, with an intrinsic rate constant of 20 m/yr. The wetland is expected to behave as three CSTRs. What is the fraction pollutant remaining ($C^* = 0$)?

Solution 1

Use Equation 9-118 to determine the answer.

$$\frac{C_o}{C_i} = \frac{1}{\left[1 + \frac{(20/(1.58)(3.65)}{3}\right]^3} = 0.0998 \tag{9-118}$$

Solution 2

If the wetland were a PFR,

$$\frac{C_o}{C_i} = \exp(-Da) = \exp(-k/q) = \exp(-20/(3.65 \cdot 1.58) = 0.0312$$

If the wetland were a CSTR,

$$\frac{C_o}{C_i} = \frac{1}{[1 + Da]} = \frac{1}{\left[1 + \frac{20}{3.65 \cdot 1.58}\right]} = 0.224$$

The interpolator is

$$\sigma_\theta^2 = \frac{1}{N} = 0.333$$

The actual exit concentration is

$$\frac{C_{ACTUAL}/C_i - 0.0312}{0.224 - 0.0312} = 0.333$$

$$C_{ACTUAL}/C = 0.0954$$

Note that this is just Example 9-13 in reverse, for which the fraction remaining was 0.100. The approximation is within 5 percent.

The tools presented here permit corrections for nonideal flow patterns. Implementation requires more data, namely, tracer tests on prototype wetlands. Until such data become more common, the designer may well opt for a conservative approach, such as the plug flow data interpretation followed by an estimate of the degree of mixing in the design wetland. In any case, it is clear that high degrees of pollutant reduction make it necessary to understand the degree of mixing.

A MASS BALANCE DESIGN MODEL

RATIONALE

The design of treatment wetlands centers on the question of size. How big does the wetland need to be to accomplish a specified pollutant reduction? To answer this, the wetland is viewed as a chemical reactor. The principles of chemistry and chemical reactor engineering may be applied to these complex "ecoreactors," but care must be taken not to overstep the existing data bounds. The procedure in this section will start from the complex characterization of the well-known suite of wetland processes and simplify it to a point which matches current intersystem data availability. The result will be a model for the calculation of internal and outlet concentrations, given the wetland size, water flow rates, and other driving forces on the ecosystem.

The framework will be the internal mass balances for water and the pollutant of interest, written for the water within the wetland. This approach acknowledges mass conservation for water and the pollutant. The pollutant interacts with the soils and biota of the ecosystem via transfers to and from those solid compartments. The result is an altered exit concentration, which is to be predicted by the design model.

Regression equations are an alternate way to describe relations between inlet and outlet concentrations and some measure of wetland size and water flow. That approach cannot account for effects such as dilution by rain or concentration by ET, nor can it provide information on the internal spatial distributions of pollutants. Utilization of mass balances does not remove the requirement to fit constants to a model; rather, it shifts that requirement to parameters in a local uptake equation.

Pollutant concentration profiles in treatment wetlands are typically decreasing with distance and often display exponentially decreasing character. That feature is characteristic of systems described by first-order removal rate equations coupled with plug flow through the system. This may be partially due to the fact that the active ecosystem arranges itself along the gradient from inlet to outlet in treatment wetlands in response to the gradients in waterborne substances, especially nutrients.

This model will lump together all the transfers of a substance within the active ecosystem. Modeling begins with a consideration of those transfers.

COMPARTMENTAL ANALYSIS

The wetland ecosystem at any location may be partitioned into pieces, referred to as *compartments*. Above-ground plant parts, roots, litter, biofilms, soils, and water are the major compartments (Figure 9-47). The rates of movement of a chemical between these compartments, together with rates of accumulation of that chemical within them, form the basis for rational design equations. If all rates are known, it is possible to calculate the interaction between the wetland ecosystem and the chemicals and water passing through it.

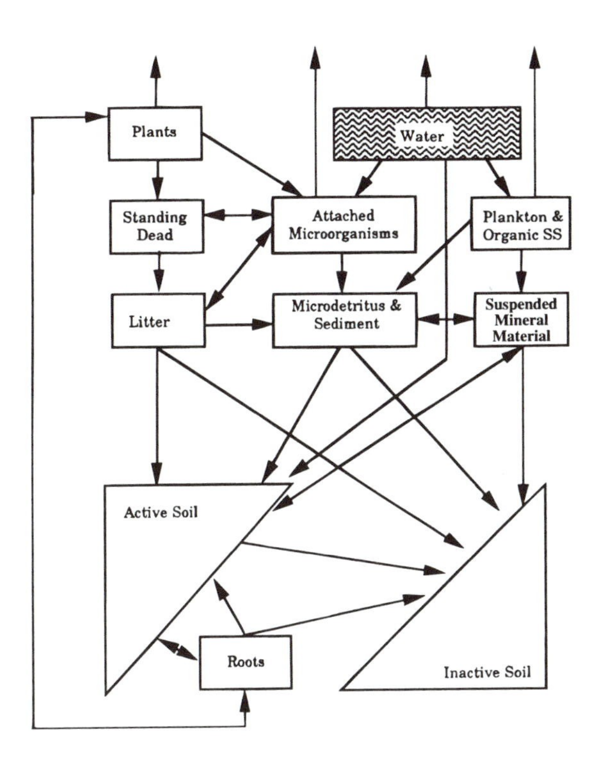

Figure 9-47 Transfer routes for a waterborne chemical in a wetland.

All compartments are static (nonmoving) with the exception of the water compartment. The water compartment is designated as number 1; the static compartments are numbered from 2 to as many such compartments as are desired.

The Water Compartment

Wetland depths are small in relation to their width and length. It is therefore reasonable to average all variables, such as concentrations, velocity, temperature, etc., in the vertical direction in the water body. We shall presume that such averaging has been done, and represent the result with the parent symbol, such as "C," "u," "T," etc., without overmarks or subscripts. However, many wetlands are of irregular shape, and there is often variability in the two planar dimensions, which will be denoted by "x" parallel to flow and "z" in the cross-flow direction.

Consider a small zone of the wetland water located at an interior point (Figure 9-48). The water mass balance is

$$\frac{\partial(\varepsilon h)}{\partial t} = -\frac{\partial(u_x h)}{\partial x} - \frac{\partial(u_z h)}{\partial z} + P - I - ET \tag{9-133}$$

The mass balance for a chemical in the water is

$$\frac{\partial(\varepsilon h C_1)}{\partial t} = \sum_{j=1}^{N} r_{j1} - \sum_{j=1}^{N} r_{1j} - \frac{\partial(u_x h C_1)}{\partial x} + D_x \frac{\partial}{\partial x}\left(h \frac{\partial C_1}{\partial x}\right) - \frac{\partial(u_z h C_1)}{\partial z} + D_z \frac{\partial}{\partial z}\left(h \frac{\partial C_1}{\partial z}\right) + PC_P - IC_1 - r_g - r_c \tag{9-134}$$

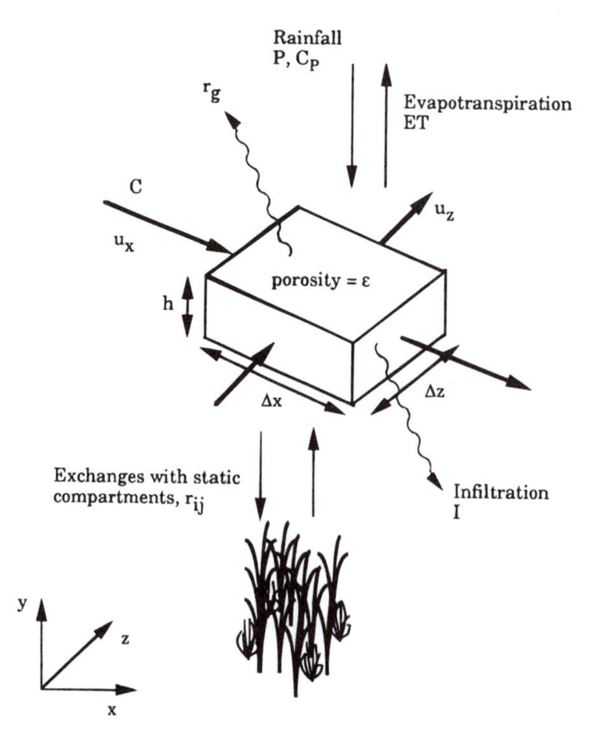

Figure 9-48 An element of water interior to the wetland.

where C_1 = concentration in surface water, g/m^3
C_p = concentration in rain, g/m^3
D_x = dispersion coefficient parallel to flow, m^2/d
D_z = dispersion coefficient normal to flow, m^2/d
ET = evapotranspiration, m/d
h = water depth, m
I = infiltration rate, m/d
P = rain rate, m/d
r_{ij} = rate of chemical transfer from compartment i to j, g/m^2/d
r_g = rate of chemical transfer from water to atmosphere, g/m^2/d
r_c = rate of chemical conversion in water, g/m^2/d
t = time, d
u_x = velocity in the principal flow direction, m/d
u_z = velocity in the cross-flow direction, m/d
x = distance parallel to flow, m
z = distance perpendicular to flow, m
ε = wetland water filled void fraction, m^3/m^3

Dispersion terms are next deleted because of the plug flow presumption. Methods for dealing with mixing were discussed in the previous section. As a result,

$$\frac{\partial(\varepsilon h C_1)}{\partial t} = \sum_{j=1}^{N} r_{j1} - \sum_{j=1}^{N} r_{1j} - \frac{\partial(u_x h C_1)}{\partial x} - \frac{\partial(u_z h C_1)}{\partial z} + PC_P - IC_1 - r_g - r_c \tag{9-135}$$

Often, there is no chemical reaction in the water itself. Inorganic precipitation reactions are an exception. There is rarely gas release of the chemical directly from water, although ammonia release is a potential exception. Therefore, water-phase reactions and atmospheric release are omitted from further consideration in this development.

It is also very difficult to calibrate models which include the cross-flows, since that calibration requires internal velocity measurements. No wetland project has yet accomplished that task. One alternative is to average this mass balance across the wetland width, which produces the following result:

$$\frac{\partial(\varepsilon h C_1)}{\partial t} = \sum_{j=1}^{N} r_{j1} - \sum_{j=1}^{N} r_{1j} - \frac{\partial(QC_1)}{\partial A} + PC_P - IC_1 \tag{9-136}$$

where A = wetland surface area upstream of the location, m^2 ($A = Wx$)
Q = volumetric flow rate of water, m^3/d ($Q = Whu_x$)
W = wetland width, m (width may vary with x, i.e., $W = W(x)$)

All of the variables in this equation have now been averaged across the wetland perpendicular to flow. This mass balance now accounts for dynamic variability and one-dimensional spatial variability. It includes transfers from any specified number of static compartments and allows for an input of rainfall chemicals and an output of chemical to infiltration.

It is convenient to combine all possible transfers to and from the water into one instantaneous uptake rate defined by

$$J_U = -\left\{\sum_{j=1}^{N} r_{j1} - \sum_{j=1}^{N} r_{1j}\right\} \tag{9-137}$$

where J_U = instantaneous uptake rate, g/m^2/d.

The Static Compartments

It is not difficult to isolate a number of nonmoving biomass and soil compartments. Figure 9-47 shows one such set of compartments. This particular configuration is predicated in major part by sampling and analytical methods. Live plants (green) and standing dead (brown) are above water and can be clipped, separated, dried, and weighed. Suspended matter is determined by filtration, and the mineral component is determined from ash content, with the balance being organic. Litter can be raked from the bottom. Microdetritus and sediments are the suspendable material remaining after clipping and raking. A soil core can be rinsed of all mineral and organic matter, leaving the live roots. The remaining soil material may be fractionated according to the extractant used to measure the chemical and regrouped into active and inactive portions. For instance, for phosphorus, the inactive portion might be considered to be that which cannot be extracted with water, KCl, NaOH, or HCl.

The mass balances for all static compartments are both spatially and temporally variable; for active biomass and soil, these are

$$\frac{\partial(m_i X_i)}{\partial t} = \sum_{j=1}^{N} r_{ji} - \sum_{j=1}^{N} r_{ij} + r_{pi} - r_{di} \qquad i = 2, N - 1 \tag{9-138}$$

where i = compartment number (i=1 designates water; i=N designates inactive soil)
m_i = total mass in compartment i, g/m^2
N = number of compartments, including water
r_{di} = rate of chemical destruction in compartment "i," $g/m^2/d$
r_{pi} = rate of chemical production in compartment "i," $g/m^2/d$
X_i = mass fraction of chemical in compartment i, dimensionless

and for the inactive soil compartment,

$$\frac{\partial(m_N X_N)}{\partial t} = \sum_{j=1}^{N} r_{jN} - \sum_{j=1}^{N} r_{Nj} = \sum_{j=1}^{N} r_{jN} \qquad i = N \tag{9-139}$$

The r_{Nj} are all zero because there are no return fluxes to any active compartment from the inactive soil, by definition. Further, there is no transfer from a compartment to or from itself, so $r_{ii} = 0$ as well.

In a truly dynamic model calculation, it is necessary to compute the amounts of chemical in each compartment, which means finding equations to calculate the m_i and h (total biomass and water mass balances) to go with calibrated values for the concentration in that biomass. It also means finding calibrated rate equations for each of the r_{ij} and r_{pi} and r_{di}. This is indeed a daunting task if one considers even a small number of biomass compartments. For instance, consider plankton, suspended matter, attached microorganisms, microdetritus, above-ground macrophytes, standing dead macrophytes, roots, litter, and active soil as the nine biomass compartments that interact with water and the inactive soil, then it is required to track total biomass via mass conservation for each, requiring nine biomass balance equations. In addition, it requires determination of about 30 of the 100 possible rate equations for the non-zero r_{ij} in Equations 9-137 through 9-139, since only about one third of the 100 are major, feasible pathways. In most instances, wetland science is not advanced enough to provide the necessary answers, although a few reasonably successful attempts have been made (Dixon, 1974; Hammer, 1984). Data requirements for calibration and validation are extreme, since there are in excess of 30 parameters to be determined.

LUMPING*

Some simplification clearly needs to be made in order to proceed with mass balance modeling. When adding all the chemical mass balances for active wetland compartments to obtain the chemical mass balance over the entire active wetland ecosystem at a specific location, most of the intercompartment transfers (r_{ij}) cancel, since they appear in two mass balances with opposite signs. The exceptions are transfers to the inactive soil compartment. A much simpler, overall active wetland chemical mass balance is

$$\sum_{i=2}^{N-1} \frac{\partial(m_i X_i)}{\partial t} + \frac{\partial(\varepsilon h C_1)}{\partial t} = \left\{ \sum_{j=1}^{N} (r_P - r_d)_j - \sum_{j=1}^{N} r_{jN} \right\} - \frac{\partial(QC_1)}{\partial A} + PC_P - IC_1 \tag{9-140}$$

or

$$\sum_{i=2}^{N-1} \frac{\partial(m_i X_i)}{\partial t} + \frac{\partial(\varepsilon h C_1)}{\partial t} = -J - \frac{\partial(QC_1)}{\partial A} + PC_P - IC_1 \tag{9-141}$$

where J = net chemical reduction rate, $g/m^2/d$.

The net chemical reduction rate is seen to be the result of transfers to inactive soils plus the net destruction rate of that chemical by reaction in all compartments:

$$J = \left\{ \sum_{j=1}^{N} (r_d - r_p)_j + \sum_{j=1}^{N} r_{jN} \right\} \tag{9-142}$$

Equations 9-139 and 9-141 represent a two-compartment model of the entire wetland, as depicted in Figure 9-49. These equations simply state that a chemical stripped from the water must appear as storage in either the active or inactive wetland biomass compartments or be chemically converted.

However, care must be taken to distinguish between the uptake rate and the net reduction rate. Uptake includes storage in active wetland components. Comparison of Equations 9-136, 9-137, and 9-141 shows that

$$J_U = J + \sum_{i=2}^{N-1} \frac{\partial(m_i X_i)}{\partial t} \tag{9-143}$$

The distinction between these two rates is quite important during a developmental phase for treatment wetlands. As vegetation density increases, so does the storage of tissue chemicals, thus causing a high uptake rate. However, the net reduction rate, which is the sum of permanent burial and chemical destruction, may be significantly lower. Conversely, if antecedent soils contain excesses of the chemical, the developmental phase may show a negative uptake.

It is now necessary to determine and calibrate an equation for the "lumped" rate "J" for overall chemical reduction and "J_U" for uptake. Some general statements may be made concerning the individual transfer rates, based on the conceptual model of Figure 9-47.

1. The first step in transfer of a dissolved chemical is the convection and diffusion of that chemical from the bulk of the water to solid surfaces. Such solid surfaces contain biofilms and sorption sites.

* The term "lumping" is in common use for the description of a set of multiple chemical reactions. Here, we adopt it for naming the combination of multiple transfers of a waterborne pollutant, phosphorus for example.

2. There are no direct transfers from the above-ground plants and standing dead, plankton, or periphyton to the inactive soil. These compartments transfer to the active soil, sediments, suspended matter, roots, litter, and possibly water.
3. Transfers of stable chemical precipitates are directly to the inactive soil.
4. Suspended solids transfer by sedimentation to the soil surface. The stable mineral and organic fraction of this settled material enter the inactive soil compartment.
5. Dead roots are considered part of the soil matrix. Root structural chemicals are added to the inactive soil due to dead root decomposition.
6. Structural chemicals, as a component of inactive soil, are formed as one product of the decomposition of litter, microdetritus, dead roots, and active soil.
7. Increased nutrients in the surface water will lead to increased biological activity, including increased plant growth. An end result is an increase in the amount of structural chemicals added to the inactive soil compartment.

The combined rates J and J_U are presumed to follow the law of mass action, which in this context means that more chemical in the active ecosystem results in a larger rate of transfer to the inactive part, higher chemical conversion rates, and a larger rate of net

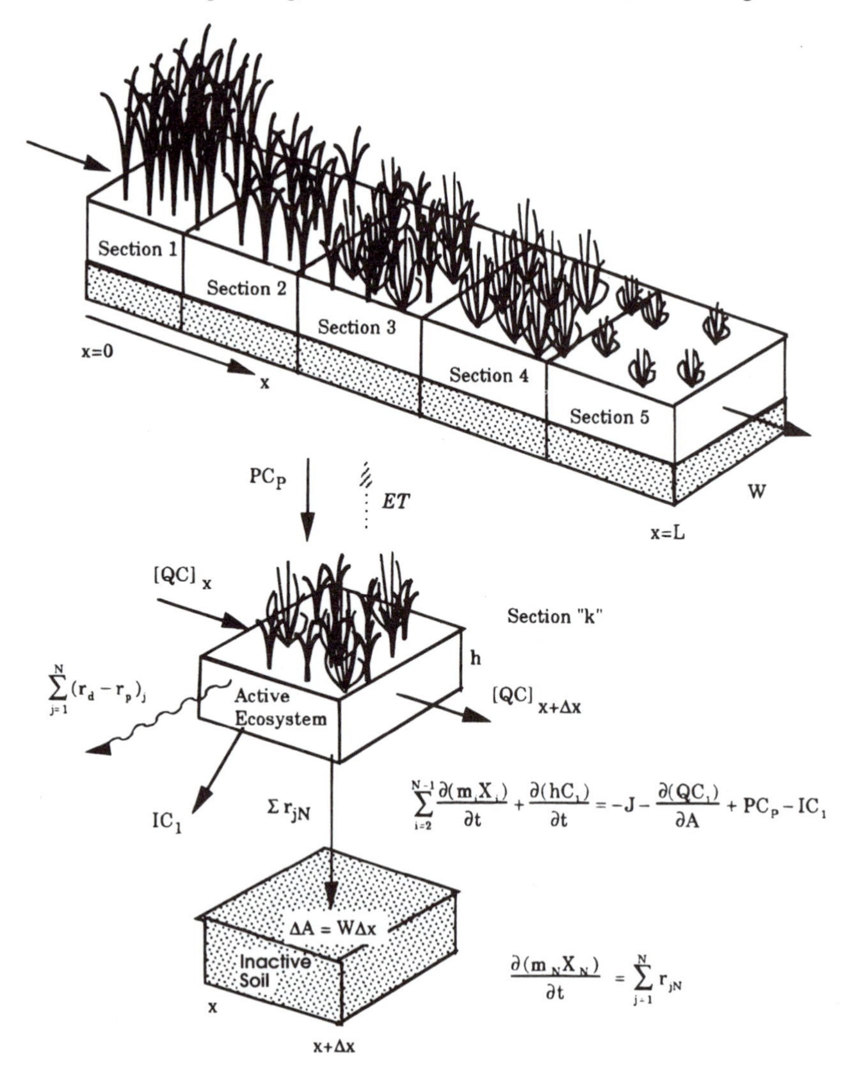

Figure 9-49 The two-compartment model of chemical transfers in a wetland ecosystem.

removal. Further, data from several wetland treatment systems shows an apparent first-order dependence on water column chemical concentration at high concentrations.

For the case of chemicals that undergo formation as well as destruction via chemical reactions, a multi-step mechanism is required. For instance, this is true for ammonium nitrogen, which may be formed from the mineralization of organic nitrogen as well as destroyed by processes such as nitrification. In this important case, all the terms in Equation 9-142 are required: there is chemical production, chemical destruction, and intercompartment transfers.

For many chemicals, the return rate to the water from the static compartments of the ecosystem—the soils and biomass—can be a significant (negative) contribution to the net rate. There is presently no scientific study to provide guidance on modeling this transfer. Therefore, the simplest option is used here: a constant (zero-order) return rate.

The lumped rate equation for the net reduction of a chemical with no precursors is therefore written as

$$J = kC - r^* = k(C - C^*) \tag{9-144}$$

where kC^* = chemical return rate from static compartments, $g/m^2/d$
k = removal rate constant, m/d
(or with unit conversion, m/yr)
r^* = return rate of chemical, $g/m^2/d$

In the terminology of reaction engineering, the model is first order in the forward direction and zero order in the reverse direction. It will be applicable to BOD, TSS, phosphorus, metals, and individual chemicals that are not themselves decomposition products of other chemicals. It will not always be applicable to nitrogen compounds, which can be chemically formed as well as destroyed.

For consistency, the uptake rate is defined in a similar way:

$$J_U = k_U(C - C^*) \tag{9-145}$$

This k-C* model will be used with the interior mass balances as the design model for pollutant removal in treatment wetlands in this book. Its validation and calibration will be presented in subsequent chapters.

SIMPLIFYING SITUATIONS

Equation 9-141 relates temporal changes in storage, spatial changes in advective chemical flow, and transfer. Many of the storages in the wetland can and do fluctuate over time. Two distinct types of time variation exist: probabilistic variation and adaptation trends.

Probabilistic variations occur due to the vagaries of climate and weather and the influences of animal populations. Periods of warm weather can speed decomposition and other microbial processes. Excessive rainfall, or conversely, periods of drought, can alter wetland function and in turn affect the biomass and P content of ecosystem compartments. As a means of defining probabilistic behavior, consider a time period over which there are no climatic trends or changes in human use patterns. Further suppose that wetland species have sorted themselves out into relatively stable patterns, as typified by relatively invariant vegetation patterns. Then, over a sufficiently long period of time, this mature wetland ecosystem will change in many ways, but will eventually return to a state closely approximating the starting condition at the beginning of the time period.* Ecosystem state variables, such as macrophyte biomass or tissue concentration of an element, will vary from time to time in response to environmental variables such as temperature and solar energy. Such variations will center on a mean with no long-term time trend. Figure 9-50 illustrates this type of variability for the phosphorus entering and leaving Boney Marsh, a constructed SF wetland system that consistently removed phosphorus from a portion of the Kissimmee River in south central Florida.

Adaptation trends occur in response to a sustained change in the set of ecosystem driving forces or in response to the construction of a wetland ecosystem that is not in a mature, stable state. The latter is often called *initial condition forcing.* A wetland constructed on a previously upland site will undergo change as the ecosystem adapts to the water and nutrient inputs. Vegetation, which is almost certain to be incomplete in areal coverage, will fill in and subsequently undergo changes in species composition. If a new input of water and nutrients is begun and sustained for an existing wetland, there may be an alteration of existing species composition. In either case, wetland soils will change to new chemical storages, consistent with the surface water concentrations and the new suite of chemical and microbial reactions.

The term *stationary state* is used here to describe a wetland which is not undergoing adaptation trends. Such a wetland displays only probabilistic variability.

SIMPLIFICATION FOR THE STATIONARY STATE

Wetland design will most often be for conditions after the startup period or, in other words, after adaptation trends have ceased. This includes storages in water and all other active wetland compartments.

Time averaging of the performance of the wetland provides a means of avoiding the task of describing the details of the short-term fluctuations. This is accomplished by applying the definition of time averaging to the water and chemical mass balances. This process generates the flow-weighted average concentration, which is essentially equal to the mass average concentration, since the density of water is nearly constant for wetland operating temperatures. The definition of time average concentration is

$$\overline{C} = \frac{1}{t_m}\int_0^{t_m} C\,dt \tag{6-33}$$

The definition of flow-weighted (mass average) concentration is

$$\hat{C} = \frac{\frac{1}{t_m}\int_0^{t_m} QC\,dt}{\frac{1}{t_m}\int_0^{t_m} Q\,dt} = \frac{\int_0^{t_m} QC\,dt}{\overline{Q}t_m} \tag{6-34}$$

where C = concentration, g/m^3
Q = volumetric flow rate, m^3/d
t = time, d
t_m = time period for averaging, d
$^-$ = indicates time average value
$\hat{}$ = indicates flow weighted average value

* In other branches of science, such as statistical thermodynamics, this hypothesis has proven very useful. There it is called the *ergodic hypothesis.*

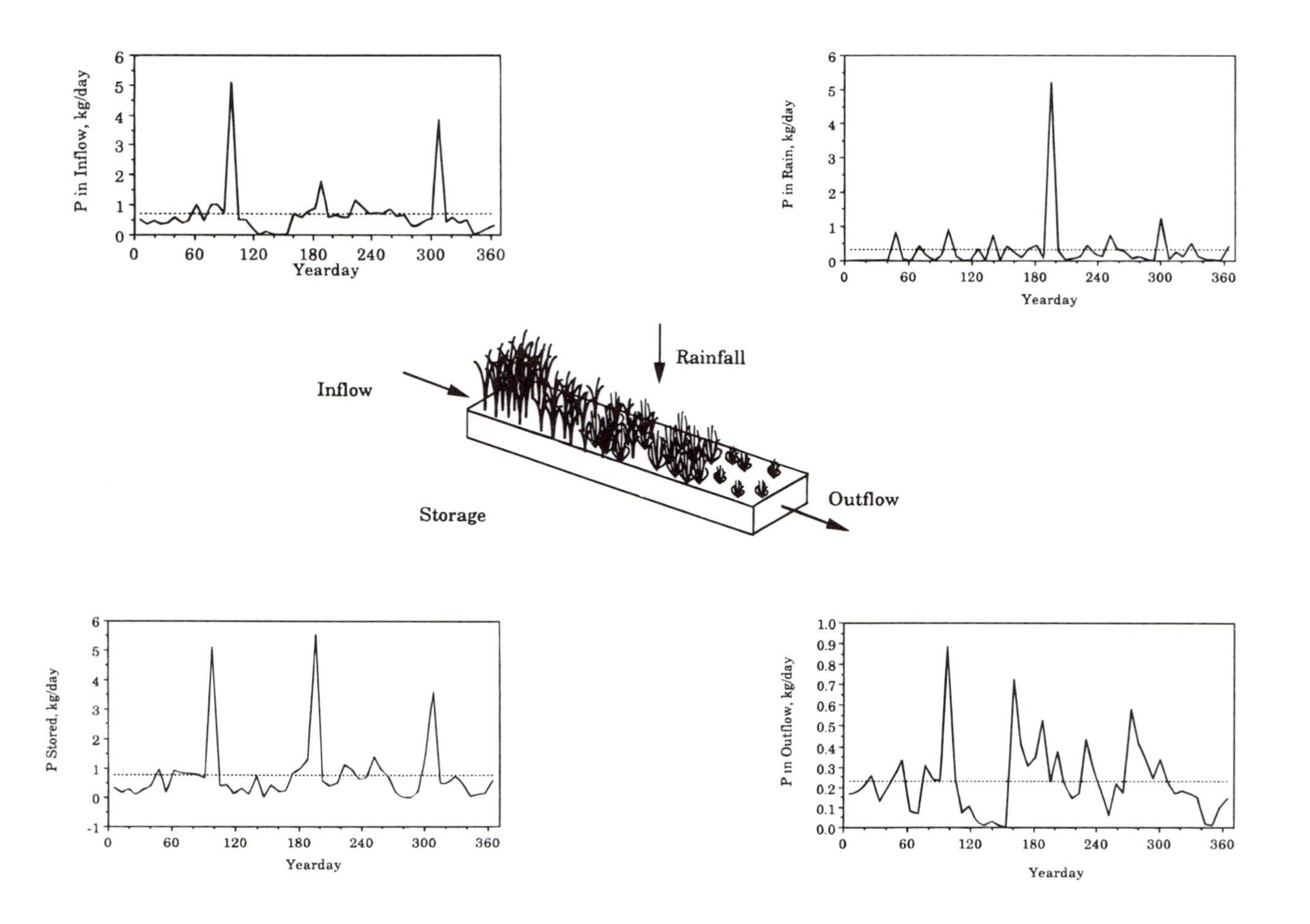

Figure 9-50 The variability of phosphorus flows and storages in Boney Marsh, 3/85–2/86.

The mass balance (Equation 9-141) is integrated over the averaging period and divided by the length of the averaging period to yield

$$\left\{\frac{1}{t_m}\sum_{i=2}^{N-1}\Delta(m_i X_i) + \frac{\Delta(HC_1)}{t_m}\right\} + \frac{d(\overline{Q}\hat{C}_1)}{dA} = -\bar{J} + \overline{P}\hat{C}_P \qquad (9\text{-}146)$$

where Δ is the value at end of the period minus value at beginning of the period.

If the averaging period is long enough or if it begins and ends with the same wetland state, the terms in braces in Equation 9-146 will become negligible. The averaging period t_m is in the denominator and grows large; the changes in active storage fluctuate, but do not have a growth trend. If operation is in the long-term stationary state, the changes in active storage will remain within the probabilistic band. In contrast, the change in storage within the inactive soil compartment may continue to grow, perhaps with fluctuations.

If rainfall chemicals are negligible compared to other inputs, then the last term in Equation 9-146 may be omitted.

The averaging period must also be long enough to account for the time of passage of water through the ecosystem. On average, water and chemicals entering the wetland will exit one actual residence time later. Earlier in this chapter, it was shown that it takes about three nominal detention times to completely flush an element of entering water. The averaging period must therefore be long enough to make transit time delays negligible, especially if there are significant temporal changes in the flow or concentration entering the wetland. Since nominal detention times are often 1 week or longer, averaging periods should not be shorter than monthly or quarterly for time varying inputs. If inflows and inlet concentrations are constant, shorter averaging periods may be used.

The resulting simplified active ecosystem mass balance is

$$\frac{d(\overline{Q}\hat{C})}{dA} = -\bar{J} = -k(\overline{C} - C^*) \qquad (9\text{-}147)$$

The subscript "1," referring to compartment 1, water, has been dropped since there are no other concentrations in the model.

If the wetland operates under relatively steady flow conditions, there is negligible difference between the flow-weighted average $\hat{C}$ and the time average $\bar{C}$. For instance, the values for the inflow for Listowel System 4 over a 4-year period were $\bar{C}$ = 3.165 mg/L and $\hat{C}$ = 3.146 mg/L. For this common case, normal practice is to drop the averaging designation:

$$\frac{d(QC)}{dA} = -J = -k(C - C^*) \qquad (9\text{-}148)$$

If the volumetric flow does not vary through the wetland, which occurs when precipitation and ET are in balance over the averaging period, then Q is constant and Equation 9-148 may be written as

$$Q\frac{dC}{dA} = -J = -k(C - C^*) \qquad (9\text{-}149)$$

Many wetlands are rectangular. The area upstream of a given point in the wetland, A, is then simply the constant width times the distance from the inlet end:

$$A = Wx \tag{9-150}$$

where W = wetland width, m
x = distance from inlet end, m

The introduction of Equation 9-150 into Equation 9-149 gives

$$Q\frac{dC}{dA} = \Lambda\frac{dC}{dx} = -k(C - C^*) \tag{9-151}$$

where Λ = flow rate per unit width, = Q/W, $m^3/d/m = m^2/d$.

Next, introduce the fractional distance from inlet to outlet, y = x/L:

$$\frac{Q}{W}\frac{dC}{d(yL)} = \frac{Q}{A_T}\frac{dC}{dy} = -k(C - C^*)$$

$$q\frac{dC}{dy} = -k(C - C^*) \tag{9-152}$$

where A_T = total wetland area, = LW, m^2
q = hydraulic loading rate, m/d
(or with unit conversion, m/yr)

Application of Equation 9-152 requires integration from the wetland inlet, where the concentration is C_i, to an intermediate distance y, where the concentration is C. The result is the *concentration profile* through the wetland:

$$\ln\left(\frac{C - C^*}{C_i - C^*}\right) = -\frac{k}{q}y \tag{9-153}$$

At the outlet, where the concentration is C_o, the *input-output relation* is

$$\ln\left(\frac{C_n - C^*}{C_i - C^*}\right) = -\frac{k}{q} = -Da \tag{9-154}$$

where Da = Damköhler number, k/q, dimensionless.

We have now simplified the model to the fullest extent possible and therefore designate Equations 9-153 and 9-154 as the simple *first-order areal model.*

A recapitulation of the assumptions and presumptions behind Equation 9-153 is in order. A distinction is made on the basis of whether a check is possible; an assumption may be checked by means independent of the model, whereas a presumption may only be checked by exercising the model.

Assumptions

A. No adaptation trends, as implied by a stationary state for all active wetland storages
B. Long-term time average performance
C. Spatially invariant time-averaged flow, as implied by time average rainfall equaling time average ET
D. No infiltration

E. Time average concentrations equal flow-weighted concentrations
F. No chemical arrives via atmospheric deposition
G. Rectangular wetland
H. No backmixing or bypassing; i.e., plug flow
I. No variation in the cross-flow direction

Presumptions

J. Chemical conversion plus removal to the inactive wetland compartment follows a first-order areal rate Equation 9-144.

Equation 9-154 may be applied to any set of input and output concentrations and the associated hydraulic loading rate. For each set of such data, a value of "k" may be calculated. Such calculations give no information on the appropriateness of the model. The values of "k" so determined are simply a rearrangement of operating data, unless the assumptions are checked, and the presumptions tested. In other words, it is necessary to verify that exponentially decreasing concentration profiles (Equation 9-153) do in fact exist in wetland treatment systems. Such tests will be applied in following chapters.

AVERAGING PULSE DRIVEN SYSTEMS

In a wetland subject to episodic inflows, $\hat{C}$ and $\overline{C}$ may differ significantly, especially if the concentration pulse is out of phase with the flow pulse. As an illustration, the inflow average concentrations corresponding to Figure 9-50 are $\overline{C} = 68.9$ ppb and $\hat{C} = 71.6$ ppb.

As a second illustration, consider the flow entering Water Conservation Area 2A in south Florida through structure S10C:

Flow (arith. mean ± SD)	1,041,000 ± 1,678,000	m^3/d
P Concentration (arith. mean ± SD)	94 ± 63	ppb
P Concentration (flow wtd. mean ± SD)	110 ± 237	ppb

In this pulse-driven wetland, there is about a 15 percent difference between time average and flow-weighted average P concentrations. The difference is greater at some interior locations in this wetland (Walker, 1992). A large variance in flow or concentration is a signal of possible significance for this effect.

Calculations may take this difference into account via the introduction of a conversion factor between time average and flow-weighted average concentrations:

$$\overline{C} = \beta\hat{C} \tag{9-155}$$

$$k_e = \beta k \tag{9-156}$$

$$\frac{d(\overline{Q}\hat{C})}{dA} = -k\overline{C} = -\beta k\hat{C} = -k_e\hat{C} \tag{9-157}$$

where k_e = effective permanent P storage rate constant, m/yr
β = averaging factor

There is no *a priori* method available to estimate β. It may be calculated from complete, transect water and phosphorus mass balances on an existing wetland. In design, presumptions must be made about the nature of the flow and concentration sequences which reach the wetland.

MASS TRANSFER WITH REACTION: BIOFILM PROCESSES

Transfer of a chemical from the water to immersed solid surfaces is the first step in the overall removal mechanism. Those surfaces contain the biofilms responsible for microbial processing, as well the binding sites for sorption processes. Roots are the locus for nutrient and chemical uptake by the macrophytes. Mass transfer takes place both within the biofilm and in the bulk water phase. The following sections present an analysis of this potentially rate-limiting process.

THICK BIOFILMS

The following discussion analyzes the transport of dissolved constituents to reaction sites located within the biofilms that coat all wetland surfaces. The sediment-water interface is but one such active surface; the litter and stems within the water column comprise the dominant wetted area in SF wetlands, and the media surface is the dominant area in SSF wetlands.

In any case, dissolved materials must move from the bulk of the water to the vicinity of the solid surface, then diffuse through a stagnant water layer to the surface, and then penetrate the biofilm while undergoing chemical transformation (Figure 9-51). This sequence of events has been described and modeled in the text of Bailey and Ollis (1986) and is outlined here. The case of zero wetland background concentration will be described here, but extension to the case of non-zero background is possible.

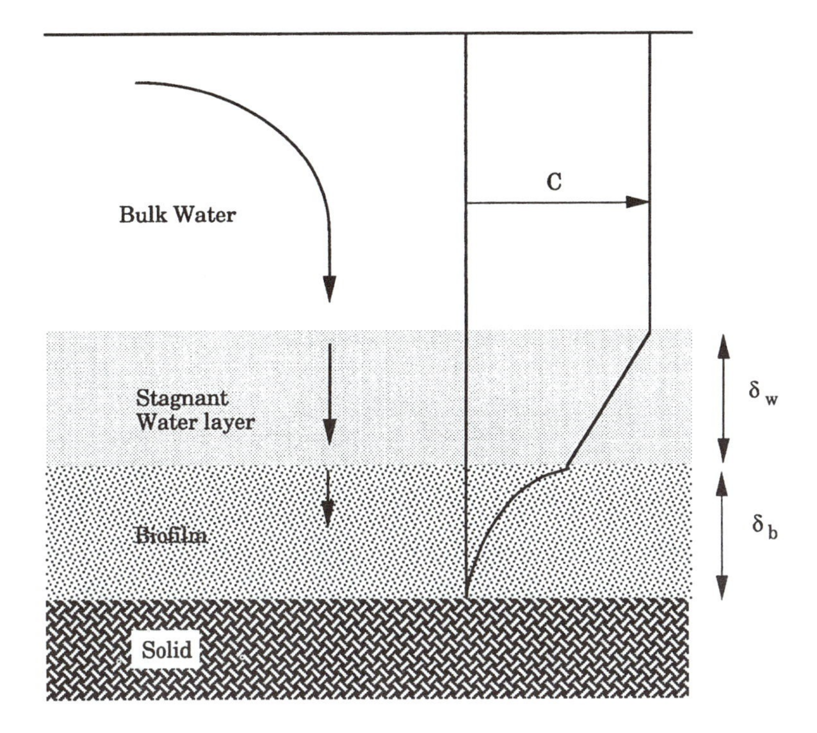

Figure 9-51 Pathway for movement of a pollutant from the water across a diffusion layer and into a reactive biofilm. The solid may be a sediment, a litter fragment, or a submerged portion of a live plant.

The rates of transfer across the two films are

$$J_{mt} = \frac{D_w}{\delta_w}(C - C_i) \quad (9\text{-}158)$$

$$J_{mt} = \left[\frac{\tanh(\phi)}{\phi}\right] k_b \delta_b C_i = E k_b \delta_b C_i \quad (9\text{-}159)$$

where C = concentration in the bulk water, mg/L = g/m^3
C_i = concentration at the biofilm surface, mg/L = g/m^3
D_w = diffusion coefficient in water, m^2/d
D_b = diffusion coefficient in biofilm, m^2/d
δ_b = thickness of the biofilm, m
δ_w = thickness of the stagnant boundary layer, m
E = $\tanh(\phi)/\phi$, biofilm effectiveness factor,—
J_{mt} = mass transfer rate, g/m^2/d
k_b = reaction rate constant inside biofilm, d^{-1}

and where

$$\phi = \delta_b \sqrt{\frac{k_b}{D_b}}$$

Eliminating C_i from Equations 9-158 and 9-159 gives

$$C_i = \frac{C}{1 + \frac{E k_b \delta_b \delta_w}{D_w}} = \frac{C}{1 + M} \quad (9\text{-}160)$$

where

$$M = \frac{E k_b \delta_b \delta_w}{D_w}$$

dimensionless.

Finally, the rate of transport of the pollutant from the bulk water to the biofilm is then

$$J_{mt} = \left[\frac{E k_b \delta_b}{1 + M}\right] C = k_i C \quad (9\text{-}161)$$

where k_i = intrinsic first-order areal reaction rate constant, m/d.

VERY FAST REACTION IN A VERY THIN FILM

If the chemical is consumed or converted immediately upon reaching the solid surface, the concentration at the surface is zero, and only Equation 9-158 is required:

$$J_{mt} = \frac{D_w}{\delta_w} C = k_i C \quad (9\text{-}162)$$

In this limiting case, mass transfer to the surface is the controlling mechanism.

SPECIFIC SURFACE AREA

In a field situation, it is also necessary to know the area of biofilms that occupy a given area of wetland (Figure 9-52). The overall removal rate from a wetland area A_w occurs from a biofilm area of A_b, and hence, the rate of removal is

$$JA_w = k_i A_b C$$

$$J = k_i \frac{A_b}{A_w} C = k_i a_s C = kC \tag{9-163}$$

where a_s = biofilm area per unit wetland area, m^2/m^2
A_w = wetland area, m^2
A_b = biofilm area, m^2
k = first-order areal reaction rate constant, m/d

Data for a_s has not been obtained for any SF wetland system, thus only a rough idea of the magnitude may be estimated. If there is no vegetation, and only the wetland bottom serves as the potential location of biofilms, the value of $a_s \leq 1.00$. If the emergent vegetation is considered as additional biofilm area, a dense stand of plants can yield $a_s \approx 5$. Inclusion of the litter can further increase the value to $a_s \approx 10$.

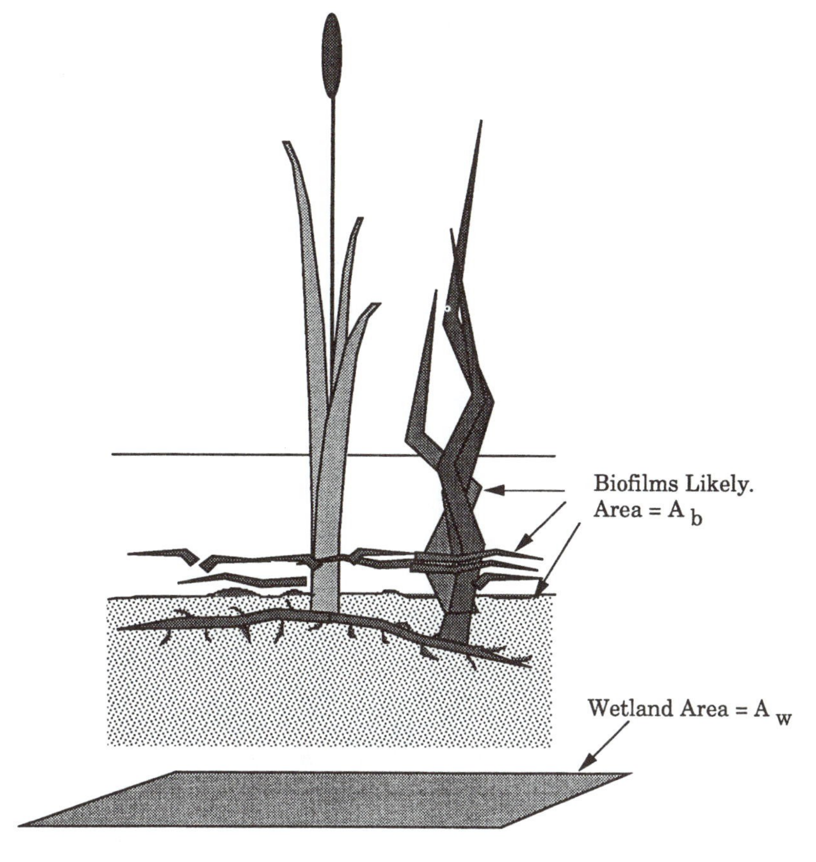

Figure 9-52 Biofilms dominate the sediment-water interface and the surfaces of litter and standing dead material.

The specific surface area of an SSF bed can be estimated from the grain size. For example, if the particles are roughly spherical, then the value of a_s is given by

$$a_s = \frac{6(1 - \varepsilon)h}{D_p} \tag{9-164}$$

where ε = void fraction, m^3/m^3
D_p = particle diameter, m
h = bed depth, m

For $d = 0.01$ m, $\varepsilon = 0.4$ and $H = 0.45$ m, $a_s = 162\ m^2/m^2$. It is this area availability that creates the potential for greater biological activity in the SSF wetland.

It is also possible to define the area per unit volume:

$$a_v = \frac{6(1 - \varepsilon)}{D_P} = \frac{a_s}{h} \tag{9-165}$$

For the data above, the value of $a_v = 360\ m^2/m^3$.

Example 9-16

Polprasert* and Agarwalla (1994) have measured or estimated all the component model parameters for BOD in a facultative lagoon in Bangkok at 20°C:

$D_w = 52.6 \times 10^{-6}\ m^2/d$ $D_b = 23.4 \times 10^{-6}\ m^2/d$
$\delta_w = 200 \times 10^{-6}$ m $\delta_b = 1538 \times 10^{-6}$ m
$k_b = 151.2\ d^{-1}$ $\phi = 3.91$

The efficiency of a biofilm under these conditions is

$E = 0.256$ $M = 0.226$

Therefore, the estimate for k_i is

$$k_i = \frac{Ek_b\delta_b}{1 + M}$$

$$k_i = \frac{(0.256)(151.2)(1538 \times 10^{-6})}{1 + 0.226} = 0.0486\ m/d$$

$$k_i = 17.7\ m/yr \tag{9-161}$$

Multiplication by a specific area factor of 2.0 m^2/m^2 would then yield a first-order areal rate constant of 35.4 m/yr.

This mechanistic approach is intuitively satisfying, but there are too many unknown and uncalibrated parameters for it to be useful at the time of this writing. However, some interesting features emerge with respect to the anticipated variability of the overall rate constant (k).

* This reference details the application of this model to wastewater stabilization ponds. The reader is cautioned that the original version contains an error in the partially mixed reactor model.

Temperature Dependence

Many of the variables that go into the mechanistic model are temperature dependent, such as diffusion coefficients and the biofilm rate constant (k_b). However, the apparent rate constant (k) is a combination of those parent variables and therefore exhibits a different temperature dependence. The θ model is used here to explore the consequences. For illustration, assume

1. The rate constant $k_b = k_{b20}\,(\theta)^{(T-20)}$. The value of $\theta = 1.05$, after Polprasert and Agarwalla (1994).
2. The diffusion coefficients have $\theta_w = 1.025$ and $\theta_b = 1.00$, after Polprasert and Agarwalla (1994).
3. The relative amounts of biofilm surfaces are greater in the winter after litterfall has occurred. Assume there is 25 percent more litter during the winter than during the spring and 25 percent less litter during the summer.
4. Climatic conditions give a winter temperature of 5°C and a summer temperature of 25°C.
5. The values of the 20°C parameters are those determined by Polprasert and Agarwalla (1994).

The resulting θ-value for the overall first-order areal uptake coefficient (k) is then 0.999. The temperature dependencies cancel each other.

Velocity Dependence

The reaction and diffusion within the biofilm are not expected to depend on the velocity in the bulk of the wetland water because the film is not subject to flushing. However, the transfer of a constituent from the bulk water to the surface of the biofilm is expected to be influenced by velocity, with higher velocities giving more turbulence and better transfer. In the limit of very rapid transfer from water to biofilm surface, the overall rate is dependent solely on processes within the biofilm. Wetlands will not typically operate at such high velocities, and hence, the water speed is a variable of interest.

For the case of the zero water film resistance, $M = 0$ and $C_i = C$.

The overall areal rate constant (k) is predicted to vary weakly with water velocity, as required by the mass transfer to the biofilm.

THE MASS TRANSFER STEP

Surface-Flow Wetlands

The value of the mass transfer coefficient $k_w = D_w/\delta_w$ can be estimated from data acquired in nonwetland contexts. At one limit, the scale of the convection currents that move dissolved constituents can be set at the scale of the wetland water depth. This corresponds to the situation of shallow, slow-moving streams. The O'Connor and Dobbins (1958) approach gives

$$k_w = \left[\frac{D_w u}{\pi h}\right]^{1/2} \tag{9-166}$$

where π = 3.14
h = wetland water depth, m
u = superficial water velocity, m/d

At $h = 0.3$, $D_w = 86$ e-6 and $u = 20$, this gives $k_w = 0.043$ m/d = 15.6 m/yr. The value is proportional to the square root of the velocity.

The scale of the convection currents that move dissolved constituents can be set at the scale of plant stems. This is analogous to the situation of mass transfer to groups of cylinders (Perry and Green, 1984). The estimating equation is

$$k_w = 0.9\left(\frac{D_w}{d}\right)\left[\frac{d\rho u}{\mu}\right]^{0.41} \tag{9-167}$$

where d = stem diameter, m
μ = water viscosity, kg/m/d
ρ = water density, kg/m^3

At d = 0.01, D_w = 86 e-6 and u = 20, this gives k_w = 0.011 m/d = 4.0 m/yr. The value is proportional to velocity to the 0.41 power.

Laboratory reactors, with controlled flow past biofilms, yield values somewhat higher (Rittman and McCarty, 1980), on the order of 0.3 m/d = 100 m/yr.

These mass transfer coefficients must be multiplied by the wetland area ratio, a_s, to obtain the limit of the overall areal rate constant for an infinitely fast biofilm reaction. In other words, if the biofilm produces a zero concentration of the pollutant at its surface, the only rate step is the transfer of the substance from the bulk water to the surface of the biofilm. The rate of disappearance is then limited only by how fast the pollutant arrives at the surface.

Subsurface-Flow Wetlands

The transport considerations discussed previously apply equally well to SSF wetlands, with the differences being the nature and amount of the solid substrate available for biofilms and the effectiveness of mass transfer in the packed bed. Generally speaking, the bed will have surface area determined by the media and the roots. Therefore, the surface area of the clean media will form a lower limit on the surface available for biofilms. However, that surface may not all be active, especially in air space above the water surface, but below the bed surface.

The mass transfer to particle surfaces inside the bed is predictable from the very large literature that has developed for describing packed bed chemical reactors of other sorts. A broad range correlation is (Fogler, 1992)

$$\frac{k_w d\varepsilon}{D_w} = 7.65\ Re^{0.18} + 3.65\ Re^{0.614} \tag{9-168}$$

where Re = Reynolds number, = $d\rho u/\mu$.

This equation is for water (Schmidt number = $\mu/\rho D_w$ = 1000) and is limited to Re > 0.01. At d = 0.01, D_w = 86 e-6, e = 0.4, and u = 20, this gives k_w = 0.322 m/d = 118 m/yr.

When multiplied by the large area factor associated with the media, the limiting value of the rate constant for mass transfer control is very, very large (118 × 162 m/yr). It is easy to see that external mass transfer cannot be the controlling factor in the removal of a pollutant in this SSF wetland.

In fact, not all the internal surface area can be available and active, or the observed field efficiencies of the SSF systems would be much greater. Availability of carbon and oxygen can and do become limiting. If these materials must be transported from the top surface of the water in the bed to the particle surface, then the travel distance for the limiting reactant is much longer, and a good share of the potential advantage of the internal surface area is lost.

No information is presently available on the rate of transport of materials from one side of a packed bed to its interior, as for oxygen supply in the SSF wetland.

Closure

The previous discussion has elucidated some simple and mechanistically appealing concepts that form a framework for interpreting the performance of wetlands in pollutant removal via the biofilms in the system. No one step is complicated, either conceptually or mathematically. However, the entire model construct contains many unknown parameters, and there are currently no means of accurately predicting several of them.

But several important items emerge from the exercise. The concept of a mass transfer controlled limit to wetland performance suggests that overall removal constants cannot exceed a value of a few hundred meters per year. There is reason to expect that temperature effects will be partially compensated by specific area effects. Any potential mass transfer limitation can be reduced by increasing the linear velocity of the water through the wetland. Thick biofilms are of lesser efficiency than thinner ones because of the importance of the diffusional resistance within the film.

The notion of a multiple-step process is strongly suggestive of the first-order areal model for pollutant removal. The value of this mechanistic interpretation is the ability to gain some small insights into the factors affecting the global removal constants.

WETLANDS WITH WATER LOSSES OR GAINS

Some treatment wetlands will gain significant water via precipitation or will lose significant water via ET or infiltration. In those cases, flow is not constant, and the water balance is more complicated. Here, it is assumed that averaging has been done over a time period for which there is a negligible change in inventories within the wetland, and the rest of the assumptions from the simplest case are adopted as well. Infiltration is assumed to be uniform and at the locally prevailing concentration.

THE PLUG FLOW CASE

The water mass balance is

$$\frac{dQ}{Wdx} = \frac{dQ}{dA} = P - I - ET \qquad (9\text{-}169)$$

where
- I = infiltration rate, m/yr
- A = surface area, m^2
- ET = evapotranspiration, m/yr
- Q = water flow rate, m^3/yr
- P = rainfall rate, m/yr
- W = wetland width normal to flow path, m
- x = distance along flow path, m

The inlet flow is specified to be $Q = Q_i$ at $x = 0$. Equation 9-169 may be integrated to give

$$Q = Q_i + (P - I - ET)Wx \qquad (9\text{-}170)$$

The mass balance on a chemical in the water is

$$\frac{d(QC)}{Wdx} = PC_P - IC - k(C - C^*) \tag{9-171}$$

where C = concentration in water, g/m^3
C_p = concentration in rain, g/m^3

The inlet concentration is specified to be $C = C_i$ at $x = 0$. Equation 9-170 may be substituted into Equation 9-171 to give

$$\frac{d([Q_o + (P - I - E)Wx]C)}{Wdx} = [PC_P + kC^*] - IC - kC \tag{9-172}$$

Equation 9-172 may be separated and integrated; a few intermediate steps are shown here.

Product rule for differentiation on left side:

$$[Q_i + (P - I - Et)Wx]\frac{dC}{Wdx} + (P - I - ET)C = [PC_P + kC^*] - IC - kC \tag{9-173}$$

Collect terms:

$$\frac{Wdx}{[Q_i + (P - I - ET)Wx]} = \frac{dC}{[(P - ET)C - [PC_P + kC^*] + kC]} \tag{9-174}$$

Integrate from $x = 0$ to x; corresponding to $C = C_i$ to C:

$$\frac{[(P - ET)C - [PC_P + kC^*] + k\dot{C}]}{[(P - ET)C_i - [PC_P + kC^*] + kC_i]} = \left(\frac{[Q_i + (P - I - ET)Wx]}{Q_i}\right)^{-\left(1+\frac{k+I}{P-I-ET}\right)} \tag{9-175}$$

or, with parameter grouping:

$$C = C_a + (C_i - C_a)\left(\frac{q_i + ay}{q_i}\right)^{-(1+(k+I)/a)} \tag{9-176}$$

where

$$a = P - ET - I \tag{9-177}$$

and

$$C_a = \frac{[PC_P + kC^*]}{(P - ET + k)} \tag{9-178}$$

and

$$q_i = \frac{Q_i}{A_T} \tag{9-179}$$

and

$$y = \frac{x}{L} \tag{9-180}$$

When rain equals ET plus infiltration, Equation 9-176 becomes indeterminate. The flow rate Q becomes constant, and integration gives the following:

$$C = C_b + (C_i - C_b)\exp\left(-\frac{(k + I)y}{q_i}\right) \tag{9-181}$$

and

$$C_b = \frac{[PC_P + kC^*]}{k + I} \tag{9-182}$$

A few concentration profiles are shown in Figure 9-53 for different combinations of ET and infiltration. An exponential fit is still reasonably good in many cases, but the apparent rate constant will vary in response to changes in the water balance for these nonconstant flows.

SUMMARY

This chapter has presented a synthesis of tools necessary to predict hydraulics and chemical transfers and reaction rates in treatment wetlands. Hydraulic and chemical processes in SF and SSF wetlands are similar in many respects, but there are significant differences. Adequate prediction methods are critical for treatment wetland design and successful operation.

Wetland water budgets are dominated by surface inflows and outflows, ET, and precipitation. Groundwater interactions are normally slight. Surface flows are generally measurable with sufficient precision. Precipitation may be projected from historical weather data, with the possibility of some error due to changing climatic conditions. Stochastic variability is

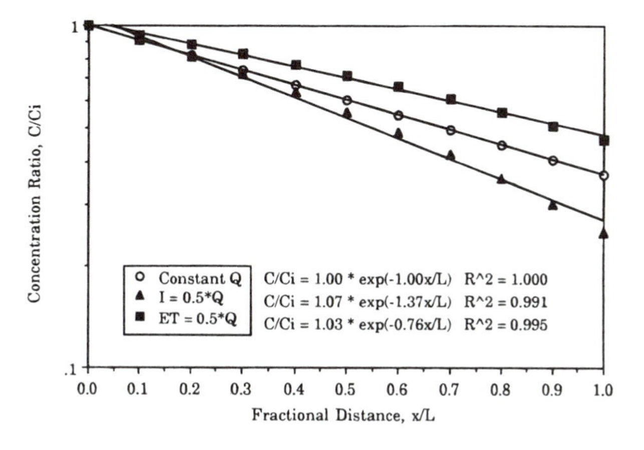

Figure 9-53 The effects of infiltration and ET on concentration profiles. Rainfall has been fixed at 10% of the inlet hydraulic loading rate. If half the added water infiltrates, the apparent rate constant rises by 37%. If half the water evaporates, the apparent rate constant falls by 24%. The exponential correlation remains excellent.

large, however, on several times scales of interest. Evapotranspiration is predictable by several techniques described in this chapter. In lightly loaded wetlands in warm seasons, this contribution may be very important in design calculations, so methods are presented for modifying pollutant reduction computations.

The internal water budget, or mass balance, for a treatment wetland is required for both conveyance calculations and pollutant reduction models. These equations, which have been detailed for SF and SSF wetlands, allow calculations of water depths and elevations and flow rates at interior points in the treatment wetland. Head losses in SF wetlands have sometimes caused operational problems and have often caused such difficulties in SSF systems. Procedures for estimating frictional effects in both types of wetlands have been presented, along with shortcut methods for estimating the necessary design parameters to ensure adequate conveyance.

The internal water mass balance also provides the framework for design models based on pollutant mass balances. The constraint of mass conservation is built into the design equations proposed for treatment wetlands. Recognition of the internal patterns of flow provides the ability to compute and calibrate to internal observations. Part of the description accounts for losses and gains as water passes through the system; a second part accounts for nonideal flow patterns. Transit time distributions and mixing processes are quantified by the RTD model. Tracer studies are used to establish parameters for this part of the description, which may have a large effect on wetland performance.

The transfers and reactions of chemicals involve many identifiable compartments in the wetland, such as biomass, microbes, and soils. The availability of operating data dictates a simplification in the modeling of the transfer and conversion network to the point where the model may be adequately calibrated to existing data. That condensation has been detailed and results in the k-C* model, which acknowledges a return flow to the water body as well as a first-order removal from the water. The principal remaining variables consist of concentrations and flows; the parameters are rate constants and background concentrations. The parameters may be deduced from or tested against both internal (transect) or terminal (input-output) information. A good share of the site-specific variability is accounted for by the reliance on mass balances because these reflect the vagaries of the external influences of meteorology.

These tools will be calibrated and used in later chapters to describe the calculations of reductions of the most common pollutants.

CHAPTER 10

Temperature, Oxygen, and pH

INTRODUCTION

The physical and chemical environment of a wetland affects all biological processes. In turn, many wetland biological processes modify this physical/chemical environment. Three of the most widely fluctuating and important abiotic factors are temperature, dissolved oxygen (DO), and hydrogen ion concentration ($pH = -\log_{10}C_{H+}$). Temperature is highly variable over daily, seasonal, and latitudinal gradients; however, it is affected very little by biology. Because temperature exerts a strong influence on some chemical and biological processes, it is important to wetland design. Oxygen, although abundant in the atmosphere, has a limited solubility in water. It is frequently a limiting factor for the growth of plants and animals in wetlands. Wetland plants have physiological adaptations that allow growth in low-oxygen soils. Hydrogen ion concentration, measured as pH, influences many biochemical transformations. It influences the partitioning of ionized and un-ionized forms of acids and bases and controls the solubility of many gases and solids. Hydrogen ions form part of the total cation content of wetland waters and are active in cation exchange processes with wetland sediments and soils.

These variables may be understood by examining the normal ranges of variation in natural and treatment wetlands. Successful design also requires that accurate predictions be made for intended operating conditions, which in turn implies prediction rules and equations.

WETLAND WATER TEMPERATURES

GENERAL

Temperatures are of interest for determination of the potential thermal condition of water leaving the treatment wetland. Receiving waters may be temperature sensitive, as in the case of a stream supporting a cold-water fishery. Some biochemical processes, notably the microbially mediated nitrogen processes, are temperature sensitive. Fog formation can accompany the discharge of warm waters into the wetland during cold periods.

The temporal pattern of wetland water temperatures is that of cycles within cycles: diurnal cycles reflecting the influence of solar radiation and annual cycles reflecting the seasonal changes in insolation. The amplitude of the daily water temperature swing depends on the type of wetland in question (Figure 10-1). A shallow, surface flow (SF) wetland strongly mimics the air temperature swing, with perhaps a slight delay in timing due to the thermal inertia of the surface soils and vegetation. A rooted aquatic system displays a moderately dampened cycle, and a subsurface-flow (SSF) wetland displays a strongly dampened cycle.

Differences in annual temperature cycles present a different story. The side-by-side comparison of the wetlands at Benton, KY shows very little difference in the monthly mean

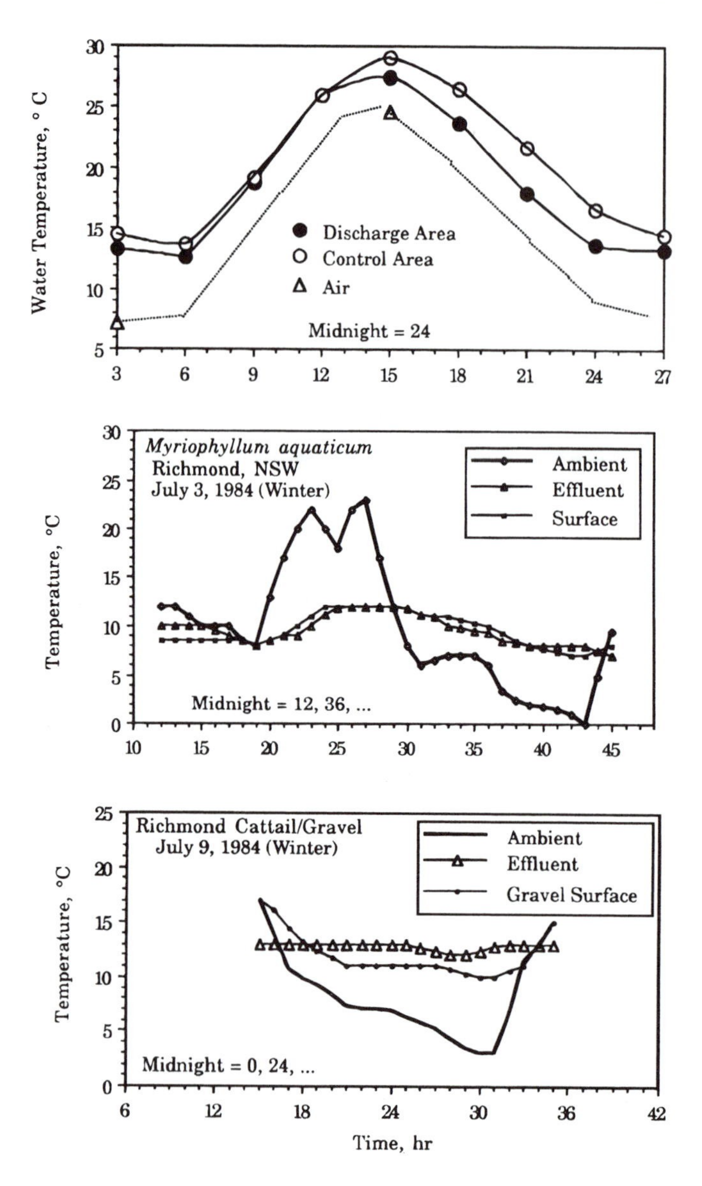

Figure 10-1 Temperature vs. hour for the surface water at the Houghton Lake, MI natural treatment wetland (top). Data are mean values spanning June, July, and August 1984. Temperatures for a floating aquatic bed at Richmond, NSW (middle) are dampened somewhat; those for a SSF wetland are dampened to a nearly constant value (bottom). (Data from Bavor et al., 1988.)

effluent water temperatures for SF and SSF wetlands (Figure 10-2). On this time scale, the thermal inertia of the wetland plays an insignificant role in the energy balance.

The transfers of energy to and from a wetland were illustrated in Chapter 9, Figure 9-2. The energy balance, Chapter 9, Equation 9-6, an expression of the principle of energy conservation, may be used to estimate wetland water temperatures as well as evapotranspiration losses. During spring, summer, and fall in northern climates and year-round in southern climates, the dominant processes are solar radiation and convective water and energy loss to the air.

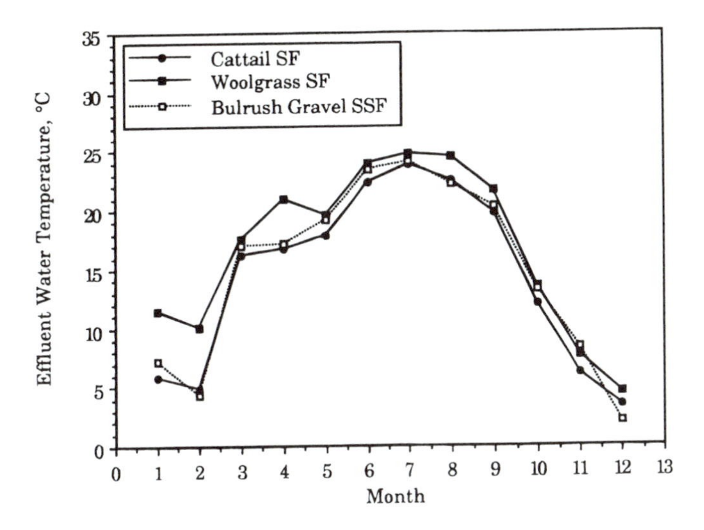

Figure 10-2 Effluent water temperatures for SSF and SF wetlands at Benton, KY. Note the absence of major differences due to wetland type.

WETLAND TEMPERATURES IN SPRING, SUMMER, AND FALL

SURFACE-FLOW WETLANDS

The Balance Point Temperature

Various energy transfer processes drive the wetland water toward an "equilibrium" temperature. This balance point temperature occurs when radiation gain is balanced by sensible heat loss and evaporative cooling. This is the temperature that will be established after a sufficiently long distance of travel in the wetland. In many cases, data show that this equilibrium temperature is reached in a relatively short distance. It is also the temperature reached in a wetland with no overland flow. For the case of a spatially uniform wetland with no change in energy storage, net incoming radiation is convected back to the air, conducted to the underlying soils, and used to evaporate and transpire water. Under these conditions, Chapter 9, Equation 9-6 reduces to

$$R_N = \rho\lambda_m \cdot ET + H_a + G \tag{10-1}$$

where G = conductive transfer to ground, MJ/m²/d
H_a = convective transfer to air, MJ/m²/d
ET = water lost to evapotranspiration, m/d
R_N = net radiation reaching the ground, MJ/m²/d
λ_m = latent heat of vaporization of water, MJ/kg (2.453 MJ/kg at 20°C)
ρ = density of water, kg/m³

Losses to the ground are typically quite small compared to atmospheric transfers. Methods for calculating R_N were presented in Chapter 9, the ET section.

The Penman calculation of ET disguises the calculation of the balance point temperature for the sake of a direct computation of ET. However, this temperature is easily retrieved from any estimate of ET. The ET loss depends on the difference in water partial pressures between the water surface and the ambient air above (see Chapter 9, Figure 9-4):

$$ET = K_e[P_w^{sat}(T_w) - P_{wa}] = K_e \Delta P_w \quad (10\text{-}2)$$

where K_e = water vapor mass transfer coefficient, m/d/kPa
P_{wa} = ambient water vapor pressure, kPa
$P_w^{sat}(T_w)$ = saturation water vapor pressure at T_w, kPa
T_w = water temperature, °C

The air transport coefficient depends on wind speed and has been represented as a linear function of the wind velocity (ASCE, 1990):

$$K_e = \frac{(4.82 + 6.38u)}{\lambda} = (10^{-3})(1.965 + 2.60u) \quad (10\text{-}3)$$

where u = wind speed at 2 meters elevation, m/s
$\lambda = \rho\lambda_m$ = volumetric latent heat of vaporization of water, MJ/m^3

Equations 10-2 and 10-3 combine to give

$$\lambda \cdot ET = (4.82 + 6.38u)[P_w^{sat}(T_w) - P_w] \quad (10\text{-}4)$$

Given the ET loss from the wetland and the water content of the air, Equation 10-4 is solved for the vapor pressure of water at the water temperature:

$$P_w^{sat}(T_w) = P_w + \frac{\lambda \cdot ET}{(4.82 + 6.38u)} \quad (10\text{-}5)$$

The saturation temperature corresponding to the vapor pressure may then be determined from Chapter 9, Equation 9-11.

Example 10-1

Find the balance point water temperature corresponding to Chapter 9, Example 9-1. In that example, ET = 3.70 mm/d, corresponding to $\lambda ET = 9.08\ MJ/m^2/d$:

$$P_w^{sat}(T_w) = 0.730 + \frac{9.08}{(4.82 + 6.38(0.43))} = 1.93\ \text{kPa} \quad (10\text{-}5)$$

From Chapter 9, Equation 9-11, $T_w = 17.0°C$.

This represents the mean daily equilibrium water temperature needed to evaporate the predicted amount of water.

Approach to the Balance Point Temperature

The energy balance may be extended to include the change in storage of energy within a parcel of flowing water. The transfers of energy illustrated in Chapter 9, Figure 9-2 produce a change in the energy content of the flowing water. The steady state flow energy balance is

$$\rho cuh \frac{dT}{dx} = R_N - G - H_a - \lambda ET \tag{10-6}$$

where c = heat capacity of water, MJ/kg/°C (1.003×10^{-3})
G = conductive transfer to ground, $MJ/m^2/d$
ET = water lost to ET, m/d
h = depth of water, m
H_a = convective heat transfer to air, $MJ/m^2/d$
R_N = net radiation reaching the ground, $MJ/m^2/d$
T = water temperature, °C
u = superficial velocity of water, m/d
λ_m = latent heat of vaporization of water, MJ/kg; (= λ/ρ)
λ = latent heat of vaporization of water, MJ/m^3 (2.453 MJ/kg at 20°C)
ρ = density of water, kg/m^3; (=1000) (ρc = 4.186 $MJ/m^3/°C$)

Note that

$$uh = \frac{Q}{W} = \Lambda \quad \text{and} \quad uh = \frac{Q}{LW} \cdot L = qL$$

where L = wetland length, m
q = hydraulic loading rate, m/d
W = wetland width, m
Λ = flow per unit width, $m^3/d/m$

Calculations with Equation 10-6 require the inlet water temperature and are very sensitive to the energy transfer terms, because the change in energy content of the water (the left-hand side) is equal to the difference between large numbers (the right-hand side). This difficulty is overcome by referencing the energy transfers to the balance point condition. Energy conservation for the balance condition is written as

$$0 = (R_N - G) - H_a^* - \lambda ET^* \tag{10-7}$$

Subtracting Equation 10-7 from Equation 10-6 gives

$$\rho cuh \frac{dT}{dx} = H_a^* - H_a + \lambda ET^* - \lambda ET \tag{10-8}$$

where the starred quantities refer to the balance point condition. Heat losses to the earth are presumed to be the same for the inlet zone and for the outlet, equilibrated zone. This is the presumption that the heat loss to the ground is nearly the same in the inlet region as in the outlet region. The starred quantities are available from the Penman ET calculation procedure (Chapter 9). The ratios of terms are given by the ratios of the driving forces:

$$\frac{H_a}{H_a^*} = \frac{h_c(T - T_a)}{h_c(T^* - T_a)} = \frac{(T - T_a)}{(T^* - T_a)} \tag{10-9}$$

$$\frac{\lambda ET}{\lambda ET^*} = \frac{\lambda_m K_e[P_w^{sat}(T) - P_{wa}]}{\lambda_m K_e[P_w^{sat}(T^*) - P_{wa}]} = \frac{[P_w^{sat}(T) - P_{wa}]}{[P_w^{sat}(T^*) - P_{wa}]} \tag{10-10}$$

where h_c = heat transfer coefficient, MJ/m²/°C/d
P_{wa} = water partial pressure in air, kPa
P_w^{sat} = vapor pressure of water, kPa
T_a = air temperature, °C
T = water temperature, °C
T^* = balance point water temperature, °C

The only remaining unknown in Equations 10-9 and 10-10 is the water temperature, T. It is now possible to solve Equations 10-8, 10-9, and 10-10 for the wetland water temperature as a function of distance. However, it is perhaps a good enough approximation to calculate the initial rate of change of temperature, since the drop to the balance point is rapid and occurs in a short distance.

$$\rho cuh \left.\frac{dT}{dx}\right]_{x=0} = H_a^*\left(1 - \frac{H_a(T_i)}{H_a^*}\right) + \lambda ET^*\left(1 - \frac{\lambda ET(T_i)}{\lambda ET^*}\right) \qquad (10\text{-}11)$$

This initial slope, combined with the balance point condition, gives a good approximation to the temperature profile.

Example 10-2

Example 10-1 is continued. The Houghton Lake wetland flowed at a rate of L = 10 m³/d/m on August 30, 1978. Pumped water entered at 20.7°C. The remaining conditions have been detailed in Examples 9-1 and 10-1.

In particular, the evaporative energy loss was

$$\lambda ET^* = 9.08 \text{ MJ/m}^2\text{/d}$$

The incoming solar radiation was

$$R_N = 8.78 \text{ MJ/m}^2\text{/d}$$

Consequently, the transfer of heat from the air to the water was

$$H_a^* = R_N - G - \lambda E^* = 9.08 - 8.78 = 0.30 \text{ MJ/m}^2\text{/d}$$

The balance point temperature, at which radiation equals convection and evaporative losses, was

$$T^* = 17.0 \text{ °C}$$

Now, it is possible to compute the inlet temperature change rate:

$$\rho c = 4.186 \text{ MJ/m}^3\text{/°C}$$

$$\rho c\Lambda = 4.186 \cdot 10 = 41.86 \text{ (MJ/m}^2\text{/d)/(°C/m)}$$

$$\frac{H_a}{H_a^*} = \frac{(T_i - T_a)}{(T^* - T_a)} = \frac{(20.7 - 16.1)}{(17.0 - 16.1)} = 5.11 \qquad (10\text{-}9)$$

$$\frac{\lambda E}{\lambda E^*} = \frac{[P_w^{sat}(T_i) - P_w(T_a)]}{[P_w^{sat}(T^*) - P_w(T_a)]} = \frac{[2.442 - 0.730]}{[1.930 - 0.730]} = 1.427 \tag{10-10}$$

$$41.86 \left.\frac{dT}{dx}\right]_{x=0} = 0.3(1 - 5.11) + 9.08(1 - 1.427) \tag{10-11}$$

$$\left.\frac{dT}{dx}\right]_{x=0} = -0.122 \frac{°C}{m}$$

The results of the complete integration of Equations 10-8, 10-9, and 10-10, and of the initial slope approximation, are shown on Figure 10-3.

SUBSURFACE-FLOW WETLANDS

SSF wetlands behave in a similar fashion: the incoming exotherm is rapidly dissipated by ET cooling. In this type of wetland, the transpiration component may dominate the transfer of water vapor, but the same energy balance must prevail. Data from a cattail-gravel wetland illustrates the rapid decline to a balance point temperature (Figure 10-4).

EMPIRICAL OBSERVATIONS

The net effect of the energy transfer processes produces the intuitively obvious result: more sunshine causes air, water, and vegetation to heat up and increases ET losses. Therefore, although the meteorological variables range through annual cycles and experience diurnal and daily variability, a strong correlation between air and water temperatures is to be expected. Monte Carlo calculations may be made, in which the variables in the Penman energy balance determination are allowed to take on a distribution of values. Input distributions of wind speed, relative humidity, maximum and minimum air temperatures, and incoming radiation are used to determine the input values. Large numbers of such calculations produce the

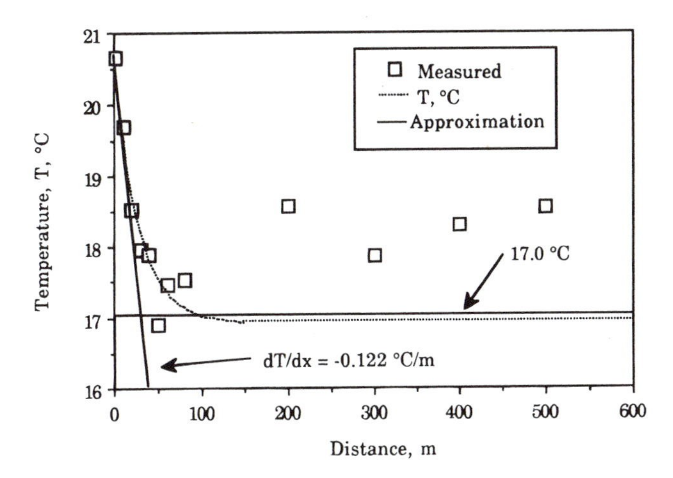

Figure 10-3 Energy balance and measured temperatures along a transect in the Houghton Lake wetland treatment system. The two-piece straight line approximation is a reasonable estimate of the more detailed energy balance integration. The discrepancy in the balance point temperature is probably due to the fact that data were taken in the photoperiod, whereas the balance point temperature is a daily mean.

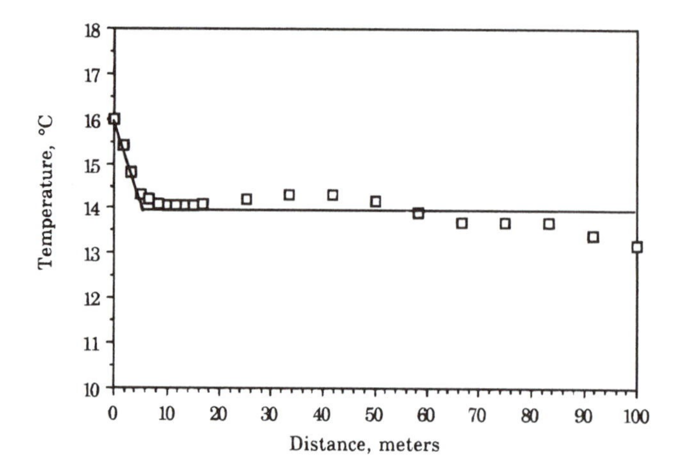

Figure 10-4 Longitudinal temperature profile in a cattail SSF wetland at Richmond, NSW. (Data from Bavor et al., 1988.)

distribution of expected ET and water temperatures which satisfy the energy balance equation. The resulting water temperatures display variability which centers on the mean daily air temperature for the temperature latitudes.

It is therefore not surprising that wetland data show precisely this same result: mean daily water temperatures are very close to mean daily air temperatures. Information from 15 treatment wetlands produces the following correlation:

$$T_w = (0.99 \pm 0.08) \cdot \hat{T}_a \tag{10-12}$$

where $R^2 = 0.87$ $N = 15$

Standard Error in $T_w = 2.1$ °C

$O < T_w < 27$ °C

$O < T_a < 27$ °C

This regression equation applies equally to SF and SSF data. The same energy balance applies to both types of wetland. Further, both systems are similarly vegetated. The presence of a dry gravel top layer in the SSF wetland would provide an additional barrier to both heat and water vapor transfers, but that extra resistance to transport is characterized by diffusive processes through air in the pore spaces. Above the surface of the gravel, transport is also characterized by diffusive processes through air.

Figure 10-5 illustrates the similarity between wetland types and the close approximation of balance point water temperatures to mean daily air temperatures.

WETLANDS IN WINTER

SURFACE-FLOW WETLANDS

General

Several SF wetland systems have been operated in the winter in northern climates. These include constructed SF wetlands at Cobalt, Ontario (Miller, 1989); Listowel, Ontario (Herskowitz, 1986); and Carson City, NV (Kadlec et al., 1990a) and natural SF wetlands at

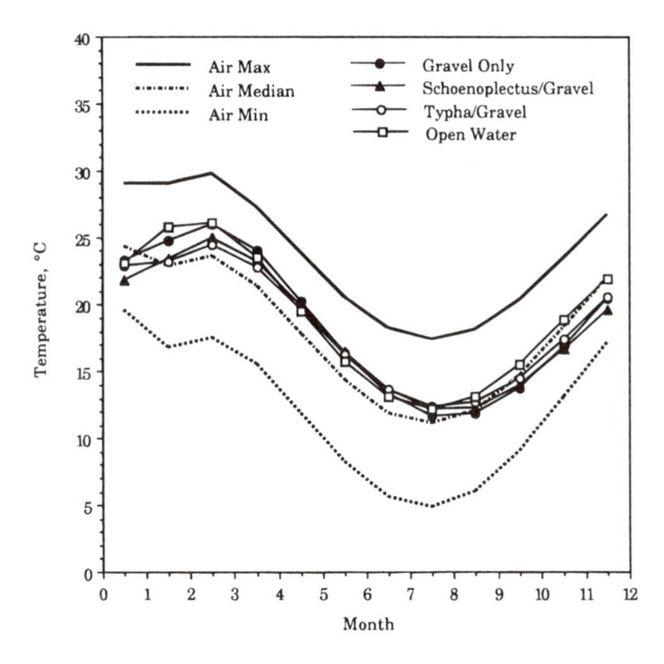

Figure 10-5 The annual variation in effluent water temperatures for various treatment wetlands at Richmond, NSW. Vegetated and unvegetated gravel beds do not differ from an open water channel. (Adapted from Bavor et al., 1988.)

Kinross, MI (Kadlec and Bevis, 1990); Concord, MA (Yonika et al., 1979); and Brillion WI (Spangler et al., 1976). SF wetlands have operated successfully under 2 m of snow in the Alps (Navarra, 1992).

The water in natural northern swamps and marshes often does not freeze in winter, due to the presence of an insulating layer of snow. In marsh environments, the standing dead vegetation is an effective snow trap which collects drifting and falling snow. Thus, the snow depth is frequently greater than the accumulated snowfall (Figure 10-6). If snow accumulates before a significant ice layer forms, subsequent freezing is strongly inhibited. If very cold temperatures precede snow accumulation, ice may form. However, ice thicknesses in vegetated

Figure 10-6 Emergent marshes trap snow in excess of that found on fields and on frozen lakes.

wetlands are very much less than lake ice thicknesses because of the insulating value of captured snow.

In densely vegetated wetlands, the plant stems hold the ice layer in place. If water levels subsequently drop, an air gap can form under the ice (Figure 10-7). This procedure was intentionally utilized at Listowel, Ontario (Herskowitz, 1986) and northern China (Yin and Shen, 1994) to allow winter operation. The insulation of the air gap, snow, and ice was sufficient to allow winter wetland water flow. However, this anchoring of the ice layer can also lead to over-ice flow if water levels are raised or the under-ice water pressure is increased (Kadlec, 1987). Water flows upward through holes or cracks, emerges, and flows downgradient over the ice and under the snow.

The point discharge introduction of warm water into either a constructed or natural treatment wetland causes an unfrozen, un-snow-covered inlet area to persist even in the event of extremely cold air temperatures. As the water moves out into the wetland, the incoming exotherm is dissipated, and a snow and ice cover becomes possible. This cover may consist of snow, ice, or a combination, depending on the vegetation density. If the discharge is into an unvegetated inlet zone, snow trapping is not possible, and ice covers the inlet pond in areas away from the discharge point (Figure 10-8). Flow then proceeds away under the ice. If the inlet zone is densely vegetated, snow may be held up above the water by standing dead vegetation or by a floating litter layer. In that event, flow from the unfrozen, uncovered inlet area proceeds away under a snow blanket.

Energy Balance Calculations

The winter energy balance may be broken into two pieces: a balance on the top layer of the snow and a balance on the water beneath the ice. The first energy balance determines the snow surface temperature, and the second determines the rate of freezing of the underlying water (Figure 10-7).

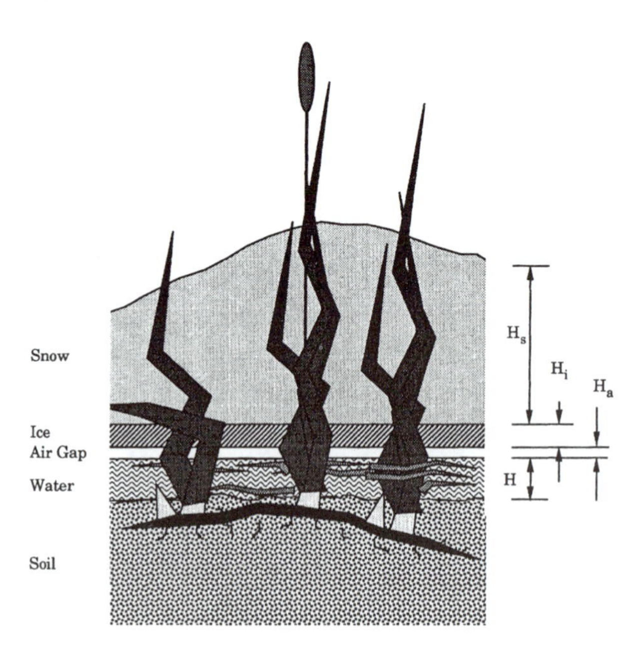

Figure 10-7 Cross-section of a marsh in winter. The layers shown may not all exist at all times in all places in all years. Cattail rhizomes will have set shoot buds for emergence in the following spring.

Figure 10-8 Inlet structure in winter: Incline Village, NV wetland treatment system.

Radiation, sublimation of snow, and convective heat transfer to the air dominate the snow surface energy balance. The insulating effect of the snow and ice layers prevents significant heat flow from below the snow surface; the heat arriving from below is one order of magnitude less than the surface fluxes mentioned above. The Penman calculation is modified to include a higher reflectance of the snow, and the heat of vaporization of liquid water is replaced by the latent heat of sublimation of the snow (2.834 MJ/kg). The net effect of these processes is a snow surface temperature that is not far different from the air temperature.

The second energy balance states that the net heat loss from the water must be balanced by temperature decline or by ice formation if the water is already at the freezing point. There are winter heat gains from deep soils, and losses by conduction through the snow, ice, and air layers that may exist above the water. Two calculations are of interest: the rate of freezing of stagnant water below the snow and the rate of temperature decline of water flowing beneath the ice. Both require a calculation of the heat loss through the air/snow/ice "sandwich."

$$E_{loss} = \frac{T_w - T_s}{R} \tag{10-13}$$

where E_{loss} = heat loss from water to air, MJ/m²/d
R = heat resistance of layers, $[\text{MJ/m}^2\text{/d/°C}]^{-1}$
T_s = snow surface temperature, °C
T_w = water temperature, °C

The resistance of each layer is its thickness divided by its thermal conductivity, and the resistances are additive:

$$R = \sum R_i = \frac{H_s}{k_s} + \frac{H_i}{k_i} + \frac{H_a}{k_a} \tag{10-14}$$

where H_a = air gap thickness, m
H_i = ice thickness, m
H_s = snow thickness, m
k_a = thermal conductivity of air, MJ/m/d/°C

k_i = thermal conductivity of ice, MJ/m/d/°C
k_s = thermal conductivity of snow, MJ/m/d/°C

There is energy gain from warm soils below the water. Its calculation is based on the vertical temperature gradient below ground:

$$E_{gain} = k_g\left(-\frac{dT}{dz}\right) \quad (10\text{-}15)$$

where E_{gain} = energy gain rate, MJ/m²/d
k_g = thermal conductivity of ground, MJ/m/d/°C
z = vertical distance upward, m

The thermal conductivities of the various layers are listed in Table 10-1.

Some winter referents make it easier to understand the orders of magnitude involved in winter energy balances. The latent heat of fusion of water is 0.334 MJ/kg, and the freezing of 1.0 mm of water therefore requires removal of 0.334 MJ. The volumetric heat capacity of water is 4.186 MJ/m³/°C. In a 30-cm deep SF wetland, 1.26 MJ must be removed to lower the water temperature 1.0°C.

These values may be compared to the heat extracted from deep soils during winter months. Data on soil temperature gradients are available for the Houghton Lake treatment wetland (Figure 10-9). The maximum winter gradient is about −6°C/m. The soils are saturated peats, with a thermal conductivity of about 0.6 W/m/°C. Therefore, the maximum heat extracted from deep soil is

$$E_{gain} = (0.6)(0.0864)(6) = 0.31 \text{ MJ/m}^2\text{/d} \quad (10\text{-}15)$$

Consequently, heat transfer from below counteracts heat losses that would otherwise produce freezing at the rate of about 1.0 mm/d.

A typical mid-winter mean daily temperature at 45° north latitude is −15°C (5°F), which would approximate the snow top temperature. A typical winter profile is 30 cm of snow on top of 5.0 cm of ice on top of water (Figure 10-7). The water temperature would be close to the freezing point. Under these conditions,

Table 10-1 Thermal Conductivities of Wetland Solids

Material	Thermal Conductivity W/m/°C
Air	0.024
New snow	0.080
Old snow	0.25
Dry litter	0.10
Dry gravel	0.30
Dry sand	0.35
Soil	0.52
Water	0.59
Saturated peat	0.60
Clay	1.3
Ice	2.2

Note: These are generic materials with considerable variability in property values, and the numbers are therefore approximate. To obtain values in MJ/m/d/°C, multiply by 0.0864.

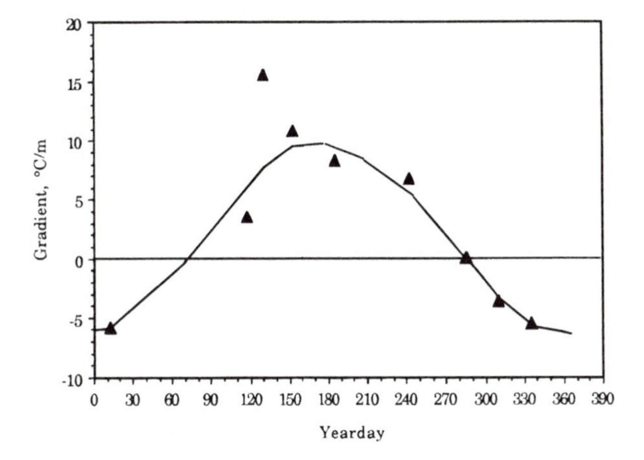

Figure 10-9 Vertical temperature gradients in the soils under the Houghton Lake treatment wetland through the course of the year. Negative values indicate temperatures increasing downward.

$$R = \frac{0.3}{(0.0864)(0.15)} + \frac{0.05}{(0.0864)(2.21)} + \frac{0.0}{k_a} \tag{10-14}$$

$$R = 23.42\ [\mathrm{MJ/m^2/d/°C}]^{-1}$$

$$E_{loss} = (0 - (-15))/23.42 = 0.64\ \mathrm{MJ/m^2/d}$$

There would therefore be a net loss of $0.64 - 0.31 = 0.33$ MJ/m²/d, which could freeze 1.0 mm/d of nonflowing water.

If water flows beneath the ice and snow without freezing, a slight modification of Equation 10-6 may be used to calculate the water temperature profile in the direction of flow under quasi-steady conditions:

$$\rho cuh \frac{dT_w}{dx} = E_{gain} - E_{loss} \tag{10-16}$$

This energy balance requires an inlet temperature to calculate by integration down the wetland. If there is a negligible heat gain from deep soils, an expression for the temperature profile may be developed by integration of

$$\rho cuh \frac{dT_w}{dx} = -E_{loss} = -\frac{T_w - T_s}{R} \tag{10-17}$$

which integrates to

$$\ln\left[\frac{T_w - T_s}{T_{wi} - T_s}\right] = -\frac{x}{R\rho cuh} = -\frac{x}{L_T} \tag{10-18}$$

$$T_w \geq 0$$

where T_{wi} = inlet water temperature, °C
L_T = characteristic thermal accommodation length, m
x = distance from inlet, m

Equation 10-18 predicts the wetland effluent water temperature as a function of the wetland length.

Example 10-3

Predict the wetland effluent water temperature for the following operational conditions:

Wetland length = 334 m	Snow depth = 0.15 m
Wetland width = 4 m	Ice thickness = 0.05 m
Water depth = 0.21 m	Air gap = 0.05 m
Flow rate = 27.5 m^3/d	Air temperature = −9°C
Inlet water T = 4.3°C	

Assume no heat gain from deep soil.

Solution

The thermal properties of the three layers are

Snow thermal conductivity = 0.15 W/m/°C = 0.01296 MJ/m/d/°C
Ice thermal conductivity = 2.21 W/m/°C = 0.191 MJ/m/d/°C
Air thermal conductivity = 0.024 W/m/°C = 0.00207 MJ/m/d/°C
Snow thermal resistance = 0.15/0.01296 = 11.57 $[MJ/m^2/d/°C]^{-1}$
Ice thermal resistance = 0.05/0.191 = 0.26 $[MJ/m^2/d/°C]^{-1}$
Air thermal resistance = 0.05/0.00207 = 24.15 $[MJ/m^2/d/°C]^{-1}$
Total thermal resistance = 36.0 $[MJ/m^2/d/°C]^{-1}$
Water velocity = 27.5/(4 × 0.21) = 32.7 m/d
ρcuh = (4.186)(32.7)(0.21) = 28.78 MJ/m/d/°C
Thermal accommodation length = (28.78)(36) = 1036 m

$$\ln\left[\frac{T_{wo} - T_s}{T_{wi} - T_s}\right] = -\frac{L}{L_T} \tag{10-18}$$

$$\ln\left[\frac{T_{wo} - (-9)}{4.3 - (-9)}\right] = -\frac{334}{1,036}$$

$$T_{wo} = 0.63\ °C$$

The conditions in this example are the 4-year January averages for the operation of Listowel System 4. The measured 4-year January average outlet water temperature was 0.55°C.

SUBSURFACE-FLOW WETLANDS

There is only a limited amount of information presently available concerning SSF performance in winter. A few SSF systems have been successfully operated through severe winter cold (Jenssen et al., 1992, 1994; Lemon and Smith, 1993). The same energy balance principles apply, but the over-water layers are different (Figure 10-10).

The low thermal conductivity of an under-ice air layer is not available in the SSF wetland, but the combined litter and dry gravel layers compensate. Further, it is possible to augment the litter layer with straw or other mulch materials.

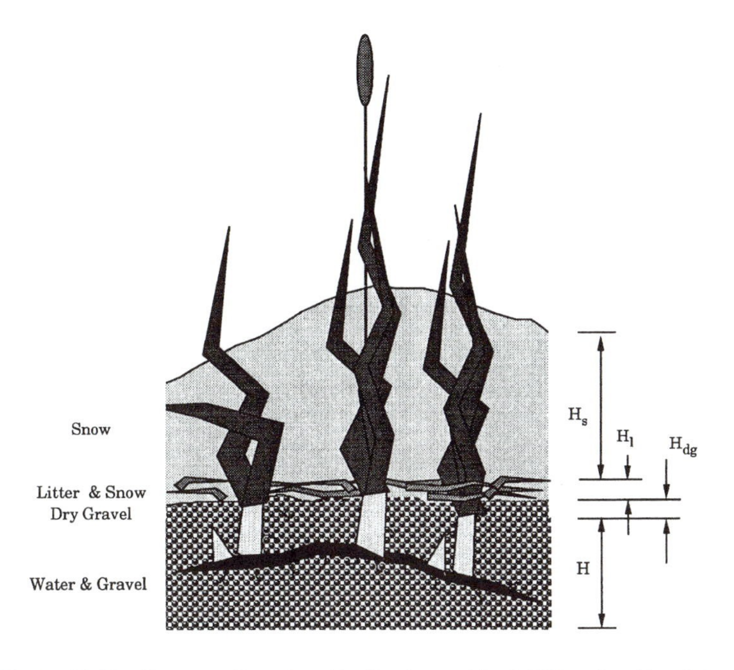

Figure 10-10 Cross-section of an idealized gravel bed SSF wetland in winter.

Example 10-4

Predict the wetland effluent water temperature for the following operational conditions:

Wetland length = 334 m	Snow depth = 0.15 m
Wetland width = 4 m	Dry gravel thickness = 0.05 m
Water depth = 0.5 m	Litter thickness = 0.10 m
Porosity = 0.42	Air temperature = −9°C
Flow rate = 27.5 m^3/d	
Inlet water T = 4.3°C	

Assume no heat gain from deep soil.

Solution

The thermal properties of the three layers are

Snow thermal conductivity = 0.15 W/m/°C = 0.01296 MJ/m/d/°C
Dry gravel thermal conductivity = 0.30 W/m/°C = 0.0259 MJ/m/d/°C
Litter thermal conductivity = 0.10 W/m/°C = 0.00864 MJ/m/d/°C
Snow thermal resistance = 0.15/0.01296 = 11.57$[MJ/m^2/d/°C]^{-1}$
Dry gravel thermal resistance = 0.05/0.0259 = 1.93 $[MJ/m^2/d/°C]^{-1}$
Litter thermal resistance = 0.10/0.00864 = 11.57 $[MJ/m^2/d/°C]^{-1}$
Total thermal resistance = 25.1 $[MJ/m^2/d/°C]^{-1}$
Water velocity = 27.5/(4 × 0.42 × 0.50) = 32.7 m/d
ρcuh = (4.186)(32.7)(0.21) = 28.78 MJ/m/d/°C
Thermal accommodation length = (28.78)(25.1) = 722 m

$$\ln\left[\frac{T_{wo} - T_s}{T_{wi} - T_s}\right] = -\frac{L}{L_T} \quad (10\text{-}18)$$

$$\ln\left[\frac{T_{wo} - (-9)}{4.3 - (-9)}\right] = -\frac{334}{722}$$

$$T_{wo} = -0.63\ °C$$

Because the predicted outlet temperature is less than the freezing point, ice formation would occur near the outlet end of the wetland. This example is the SSF analog of Listowel System 4, with the same size, water volume, and flow rate. Examination of the differences reveals that the thermal resistance of gravel and litter are somewhat lower than the thermal resistance of an air gap, leading to greater heat loss.

OXYGEN TRANSFER TO WETLAND WATERS

INTRODUCTION

Oxygen (O_2) makes up approximately 21% of atmospheric gases by volume. The concentration of oxygen now in the atmosphere is thought to have originated from biological processes that occurred over the past billion years. Oxygen is produced by photosynthesis and consumed by respiration. The apparent steady state concentration of oxygen that currently exists is probably the result of a balance between these two opposing processes. Although carbon dioxide concentrations have risen over the past century, no similar change in atmospheric oxygen has been observed.

The concentration of dissolved oxygen (DO) in water varies with temperature, dissolved salts, and biological activity. The effect of temperature on the equilibrium solubility of oxygen in pure water exposed to air has been widely studied and can be calculated from the following regression equation (Elmore and Hayes, 1960):

$$C_{DO}^{sat} = 14.652 - 0.41022T + 0.007991T^2 - 0.00007777T^3 \qquad (10\text{-}19)$$

where C_{DO}^{sat} = equilibrium DO concentration at 1.0 atmosphere, mg/L
T = water temperature, °C

DO solubility also varies with dissolved solids content. Figure 10-11 shows the effect of salinity and temperature on the equilibrium solubility of oxygen in water.

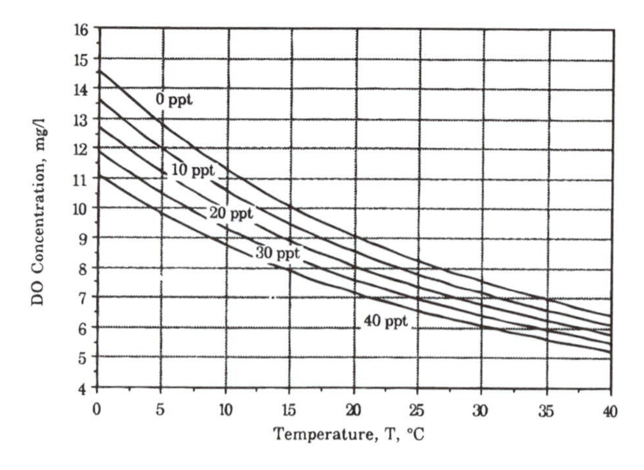

Figure 10-11 Effect of water temperature on DO saturation concentration at salinities between 0 and 40 ppt. (Plotted from data in Metcalf & Eddy, 1991.)

Water entering the treatment wetland has carbonaceous and nitrogenous oxygen demand (CBOD and NOD). After entering the wetland, several competing processes affect the concentrations of oxygen, biochemical oxygen demand (BOD), and nitrogen species (Figure 10-12). DO is depleted to meet wetland oxygen requirements in four major categories: sediment-litter oxygen demand, respiration requirements, dissolved carbonaceous BOD, and dissolved NOD. The sediment oxygen demand is the result of decomposing detritus generated by carbon fixation in the wetland, as well as decomposition of precipitated organic solids which entered with the water. The NOD is exerted primarily by ammonium nitrogen, but ammonium may be supplemented by the mineralization of dissolved organic nitrogen. Decomposition processes in the wetland also contribute to NOD and BOD. Microorganisms, primarily attached to solid, emersed surfaces, mediate the reactions between DO and the oxygen-consuming chemicals. Plants and animals within the wetland require oxygen for respiration. In the aquatic environment, this effect is seen as the nighttime disappearance of DO. Oxygen transfers from air and generation within the wetland supplement any residual DO that may have been present in the incoming water. Three routes have been documented for transfer from air: direct mass transfer to the water surface, convective transport down dead stems and leaves, and convective transport down live stems and leaves. The latter two combine to form the plant aeration flux (PAF). These transfers are largely balanced by root respiration, but may contribute to other oxidative processes in the root zone.

BIOCHEMICAL PRODUCTION OF OXYGEN

Oxygen is the byproduct of photosynthesis. When photosynthesis takes place below the water surface, as in the case of periphyton and plankton, oxygen is added to the water internally. A large algal bloom can raise oxygen levels to 15 to 20 mg/L, more than double the saturation solubility, as a result of wastewater addition (Schwegler, 1978). This process requires sunlight, and algal photosynthesis is suppressed in wetlands with dense covers of emergent macrophytes.

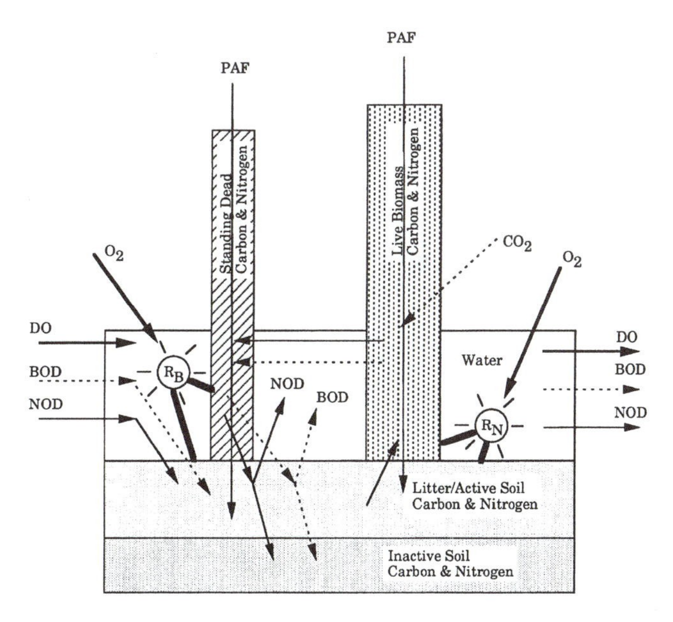

Figure 10-12 Oxygen interactions in the wetland environment. Note that both BOD and NOD are subject to consumption and generation within the litter and active soil and that atmospheric carbon eventually contributes to a BOD return to the water via decomposition.

Nonshaded, aquatic microenvironments within the wetland therefore display a large diurnal swing in DO due to the photosynthesis-respiration cycle. Nutrients stimulate the algal community and increase the DO mean and amplitude. When large amounts of nutrients are added to the wetland and water depths are shallow enough for emergent rooted plants, other components of the carbon cycle are increased, such as photosynthesis by macrophytes. It is then possible for other wetland processes to become dominant in the control of DO. The effect is typically a depression of average DO and a decrease in the amplitude of the diurnal cycle (Figure 10-13).

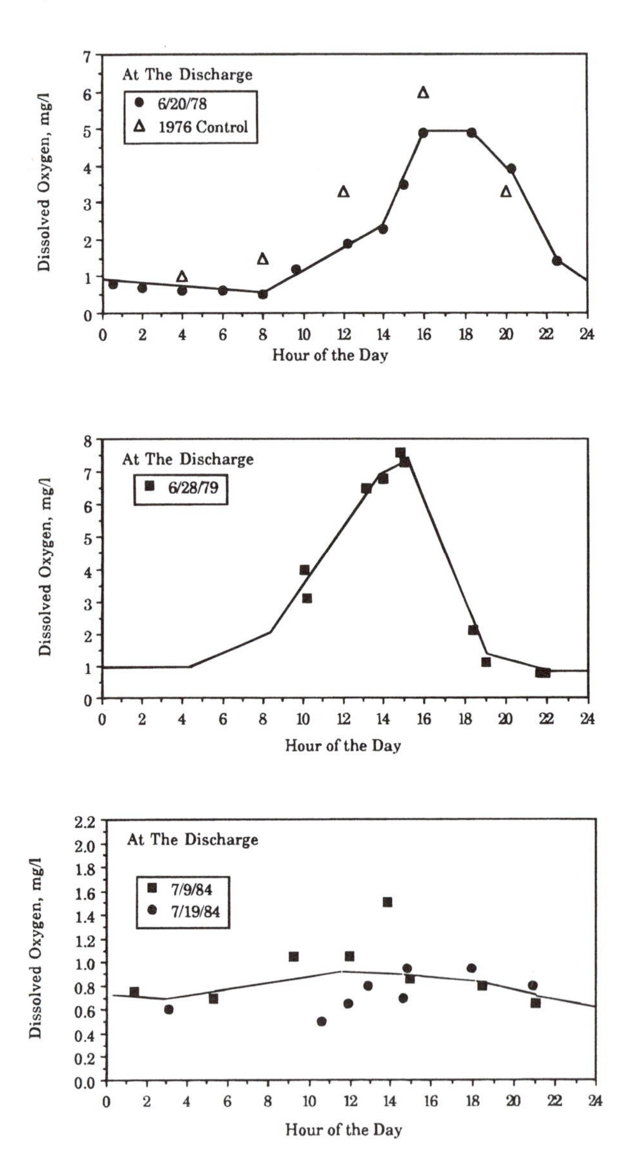

Figure 10-13 The progression of diurnal oxygen cycles at the Houghton Lake wetland treatment system. The full-scale project commenced on June 18, 1978, at which time the DO cycle still strongly resembled the 1976 control location (top). After 1 year, in late June 1979, the DO cycle at the discharge was enhanced (middle). After 5 years of operation, macrophyte communities had changed to cattail, and a large litter layer had developed. The DO cycle was lowered and damped (lower).

In wetlands dominated by macrophytes, oxygen processing is more complicated. Macrophytes and periphyton contribute to respiration and photosynthesis. The decomposition of litter and microdetritus returns ammonium nitrogen and BOD to the water and to the root zone. Oxygen transfer to the root zone occurs through plants as well as from mass transfer. BOD can degrade via anaerobic processes in the wetland litter and soil horizons.

OXYGEN CONSUMPTION IN THE WATER

The oxidative reduction of BOD may be crudely characterized by the reaction

$$BOD + O_2 \xrightarrow{\text{bacteria}} CO_2 + H_2O \tag{10-20}$$

This is a generic version of the aerobic respiration reaction. In the oxygenated environment of the BOD test, this is the dominant reaction. However, carbon compounds may be converted via a number of other pathways in the wetland environment. Nitrate and sulfate reduction occur in anoxic and mildly anaerobic zones, and methanogenesis occurs in strongly anaerobic zones. In the wetland environment, it is too simplistic to think the sole BOD reaction is that of DO with carbon compounds.

When a water is tested for BOD, there is preferential oxidation of BOD, followed by the disappearance of NOD (Metcalf and Eddy, 1991). The reason cited is the need to develop a community of nitrifying bacteria in the sample bottle, which takes several days. Therefore, BOD_5 is hopefully a measure of BOD with very little interference from NOD. If nitrification is inhibited, the test reflects carbonaceous compounds and is termed $CBOD_5$.

Five days does not complete the reduction of BOD. If the BOD test were allowed to continue, the oxidation process would go to completion. For wastewaters, the ultimate carbonaceous oxygen demand (UCBOD) is approximately 1.5 times the $CBOD_5$ value (U.S. EPA, 1985a). In a wetland, UCBOD represents an upper limit to the required oxygen transfer because of the existence of other carbon conversion pathways. At a minimum, the oxygen supplied by nitrate and sulfate should also be considered.

Oxygen associated with organic compounds (carbohydrates) is generally not available for the redox reactions which affect BOD and oxygen.

PHYSICAL OXYGEN TRANSFER TO WATER

Oxygen transfer from the atmosphere occurs because of DO levels that are below saturation in the water. Oxygen must be transported to the air-water interface and then into the bulk of the wetland water column. Resistance to mass transfer in the air is negligible, and the concentration adjacent to the interface is approximately 21% oxygen. This results in a saturation DO concentration in an extremely thin layer of water at the surface. That concentration is determined by temperature for pure water and further by the amount of dissolved solids for natural waters (Figure 10-11).

Transfer into the water below the surface is by a combination of molecular diffusion and bulk mixing. Diffusion is the dominant mechanism in totally stagnant water. However, wind and water currents create vertical mixing, even at the very low water velocities which occur in wetlands. Rain carries DO and promotes mixing.

The molecular diffusion process represents a lower limit to the transfer rate of oxygen into water. However, this process is accelerated by the chemical consumption of oxygen in the water column. The mass balance equation for vertical diffusion accompanied by a first-order volumetric consumption reaction is (Danckwerts, 1970)

$$D \frac{\partial^2 C}{\partial z^2} = \frac{\partial C}{\partial t} + kC \qquad (10\text{-}21)$$

where C = DO concentration, mg/L
D = diffusion coefficient, m^2/d
k = reaction rate constant, L/d
t = time, d
z = vertical distance, m

This equation may be solved to obtain the oxygen flux into stagnant water initially at zero DO; the closed form result is given by Danckwerts (1970). Transfer rates under these circumstances are quite low and would not provide for significant reductions in oxidizable pollutants.

In the absence of studies of vertical mixing or reaeration in wetlands, the rate of bulk mixing can only be estimated. Longitudinal mixing in wetland waters has been shown to be comparable to longitudinal mixing in lakes and rivers (Kadlec, 1994). Therefore, it seems reasonable to use information on lake and stream reaeration to gain some estimation of bulk mass transfer of oxygen into wetland waters. The mass transfer equation is

$$N_{O_2} = K(C_{DO}^{sat} - C_{DO}) \qquad (10\text{-}22)$$

where C_{DO}^{sat} = saturation DO concentration at water surface, mg/L = g/m^3
C_{DO} = DO concentration in the bulk of the water, mg/L = g/m^3
K = mass transfer coefficient, m/d
N_{O_2} = oxygen flux from air to water, $g/m^2/d$

The parameter K has been the subject of dozens of research studies in lakes and streams and in shallow laboratory flume studies (U.S. EPA, 1985). Lake mixing is driven primarily by wind shear. The value of K ranges from 0.05 to 5.0 m/d as the wind speed ranges from 1 to 10 m/s; a median value is K = 0.5 m/d. Wind shear can be a factor in wetlands that possess significant reaches of open water.

Mixing in rivers and streams is driven by water current speed with influences of wind shear. Parameter estimation for correlations for K do not extend to the low velocities found in treatment wetlands, so an extrapolation is involved, which is always hazardous. However, the dependence of K on velocity, diffusion coefficient, and depth in a variety of situations validates the form of the correlation over a wide enough range (Cussler, 1984). The O'Connor and Dobbins (1958) correlation produces mid-range estimates:

$$K = \sqrt{\frac{DU}{h}} \qquad (10\text{-}23)$$

where D = molecular diffusivity of oxygen in water, m^2/d ($D = 1.76 \times 10^{-4}$ m^2/d @ 20°C)
h = water depth, m
U = water speed, m/d

For a depth of 0.3 m and a water velocity of 30 m/d, the predicted value of K = 0.13 m/d. The rate of oxygen transfer to a bulk water at zero DO at 20°C is then estimated to be

$$N_{O_2} = K(C_{DO}^{sat} - C_{DO}) = 0.13(9.08 - 0.00) = 1.2 \text{ g/m}^2\text{/d}$$

This is one order of magnitude larger than predicted from molecular diffusion alone.

A temperature correction factor (*circa* $\theta = 1.020$) is sometimes used to adjust the value of K, but the uncertainty in extrapolation far exceeds the minor effects of such a correction for estimating wetland reaeration. Of more importance is the effect of wind, which is important in lakes and has been shown to be important in augmenting the mass transfer coefficient in streams (Eloubaidy and Plate, 1972; Mattingly, 1977). The Mattingly correction is given by

$$\frac{\Delta K}{K_o} = 0.24V_w^{1.64} \tag{10.24}$$

where K_o = mass transfer coefficient at zero wind velocity, m/d
V_w = near-surface, above-boundary layer, wind velocity, m/s

Thus, a 3 m/s (6.7 mi/h) wind increases K by a factor of 2.5 to K = 0.32 m/d; and oxygen transfer is increased accordingly, from 1.2 to 3.0 $g/m^2/d$.

It is to be reemphasized that this discussion of physical oxygen transfer is speculative because it is not based upon data collected in wetlands.

GRAVEL BED MASS TRANSFER

It may seem that the placement of the air-water interface below ground would form an additional impediment to oxygen transfer. However, rates of oxygen transfer to the water are extremely low and do not deplete the air space of oxygen. Differently stated, diffusion and mass transfer in the air space are many orders of magnitude faster than in water. Thus, the underground interface is exposed to air at essentially 21% oxygen.

The mechanisms of mixing within the water in the SSF media are different, in that the scale of turbulence is on the order of the size of the media. Therefore, different correlations are expected to apply. There is no specific study of oxygen transport from air to water moving in a gravel bed. It is possible to gain some understanding of the magnitude from information on internal mass transport (see Chapter 9).

PLANT OXYGEN TRANSFER

Wetland plants are rooted in soils and sediments that are frequently rendered anoxic by the overlying floodwater. In common with other organisms, plants require oxygen in the root zone for respiration. Consequently, many wetland plants have equipped themselves with airways from their above-water parts to their roots. These air ducts are called aerenchyma, and they transport gases to and from the root zone of the plant. Dead and broken shoots and stubble also form air pipes to the root zone. Of interest here is the fact that significant quantities of oxygen pass down through the airways to the roots (Brix, 1993b; Brix and Schierup, 1990) and that significant quantities of other gasses, such as carbon dioxide and methane, pass upward from the root zone.

Some—perhaps most—of the oxygen passing down the plant into the root zone is used in plant respiration (Brix, 1990). However, there is a great deal of chemical action in the microzones near the roots of wetland plants. Figure 10-14 shows that the oxygenated microzone around a rootlet can conduct oxidation reactions, while reduction reactions can occur only microns away in the anaerobic bulk soil. Diffusion easily connects these zones because of the close proximity. The excess supply of oxygen over that required for plant respiration, or plant aeration flux (PAF), has been the subject of several research endeavors (Brix, 1990; Gries et al., 1990; Armstrong et al., 1990). The difficulty of measuring processes and concentrations in the root microzones has been a major factor in the widely disparate estimates of PAF (Table 10-2).

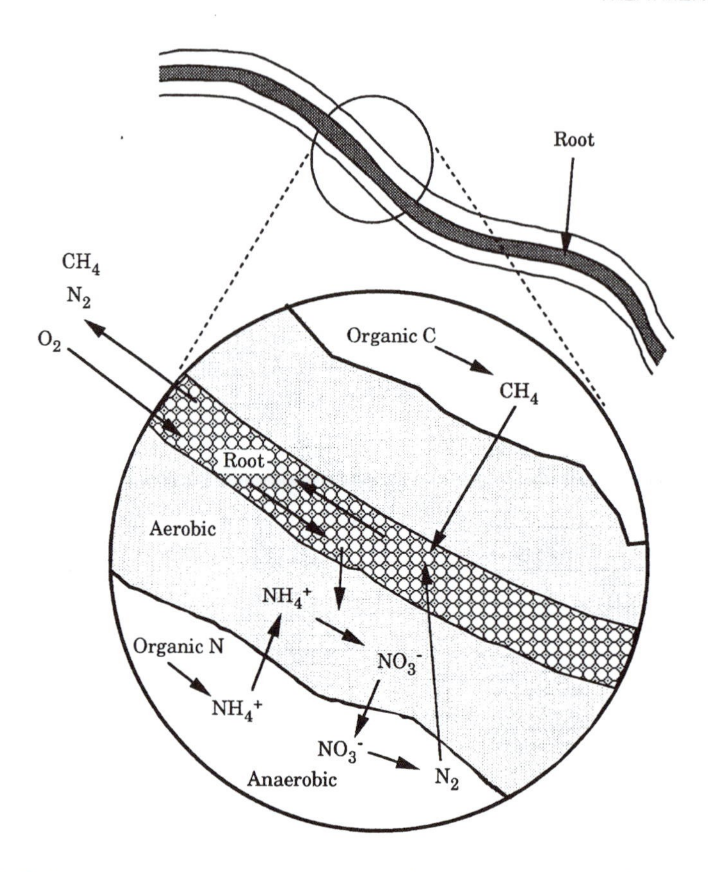

Figure 10-14 Pathways of nitrogen transformations in the immediate vicinity of a wetland plant root. (Adapted from Reddy and Graetz, 1988.)

The only direct field study on a 4-year-old operating treatment wetland is Brix and Schierup (1990) (Table 10-2). Both SF and SSF conditions were studied for *Phragmites* growing in soil at Kalø, Denmark. Portions of the wetland were enclosed in clear plastic chambers, and gas flows were measured, including oxygen inflows and carbon dioxide outflows. The latter permitted calculation of the respiration requirements of the plants. Several significant findings emerged from this study:

1. There was not a large difference between SF and SSF oxygen uptake (4.4 ± 3.0 vs. 3.3 ± 0.6 g $0_2/m^2/d$, respectively).
2. Oxygen transport through the plants (2.08 g $O_2/m^2/d$) was almost exactly balanced by respiration (2.06 g $O_2/m^2/d$). By difference, root release was estimated as 0.02 g $O_2/m^2/d$.
3. Chamber measurements of oxygen uptake in vegetated and unvegetated areas also showed plant uptake (4.5 ± 1.1 vs. 3.1 ± 1.9 g $O_2/m^2/d$, respectively).
4. The oxygen transfer calculated from system data for BOD and NH_4-N reduction, with stoichiometric coefficients of 1.5 and 4.5, respectively, was 5.2 g $O_2/m^2/d$.
5. Methane was emitted and at a greater rate in SF operation.
6. Only 23% of the lost ammonium exited the wetland as nitrate.

These direct measurements strongly suggest that it is not accurate to infer oxygen transfer from BOD and ammonium losses. Further, the results suggest that oxygen transfer to the water is dominated by air-water interfacial transfer, with the plants ingesting oxygen only to support their respiration.

Table 10-2 Reported Values of Plant Aeration Fluxes

Study	Plant and Substrate	Plant Aeration Flux g $O_2/m^2/d$	Inferred from	Comment
Armstrong et al., 1990	*Phragmites;* hydroponic	5–12	Water reoxygenation	Laboratory study on individual roots, extrapolated based on assumed root size and density distributions
Brix and Schierup, 1990	*Phragmites;* soil	2.08–2.06 = 0.02	O_2 gaseous uptake	Field study which measured uptake and respiration
Gersberg et al., 1989b	Cattails; gravel	2.1	NH_4-N disappearance	Field study
(as quoted by U.S. EPA, 1993b)	Bulrushes; gravel	5.7	NH_4-N disappearance	Assumes 4.5 mg O_2 per mg N, but there was no nitrate in effluent
	Phragmites; gravel	4.8	NH_4-N disappearance	
(as quoted by Burgoon, 1993)	Bulrushes; gravel	6.4	BOD + NH_4-N disappearance	Assumes 1.5 mg O_2 per mg BOD, 4.3 mg O_2 per mg NH_4-N
	Phragmites; gravel	4.4	BOD + NH_4-N disappearance	Assumes no other reduction mechanism or electron acceptors
Bavor et al., 1988	Cattails; gravel	0.8	NH_4-N disappearance	Field study
	Bulrushes; gravel	0.8	NH_4-N disappearance	Assumes 4.5 mg O_2 per mg N, but there was no nitrate in effluent
	Phragmites; gravel	0.8	BOD disappearance	Assumes no other reduction mechanism or electron acceptors
	Bulrushes; gravel	0.5	BOD disappearance	Assumes 1.5 mg O_2 per mg BOD
Burgoon, 1993				
Nitrate-reducing conditions	Bulrushes; plastic media	0.0	Carbon balance	Lab study; assumes 1.5 mg O_2 per mg BOD
Sulfate-reducing conditions	Bulrushes; plastic media	9.5–10.3	Carbon balance	Includes other reduction mechanism and electron acceptors
Nitrate-reducing conditions	Bulrushes; plastic media	0.8	$^{15}NH_4$-N balance	Lab study
Sulfate-reducing conditions	Bulrushes; plastic media	0.0	$^{15}NH_4$-N balance	Assumes 4.5 mg O_2 per mg N, but there was no nitrate in effluent
$t < 24$ h	Bulrushes; gravel	28.6	BOD disappearance	Mesocosms; assumes 1.5 mg O_2 per mg BOD; zero after 12 h.
$t > 24$ h	Bulrushes; gravel	0.0	BOD disappearance	
$t < 24$ h	Bulrushes; gravel	2.4	NH_4-N disappearance	Assumes 4.5 mg O_2 per mg N, but there was no nitrate in effluent
$t > 24$ h	Bulrushes; gravel	0.0	NH_4-N disappearance	
As referenced by Brix, 1993				
Lawson, 1985	*Phragmites*	4.3		
Greis et al., 1990	*Phragmites*	1–2		
Moorhead and Reddy, 1988	Floating-leaved plants	2.4–9.6		
Kemp and Murray, 1986; Sand-Jensen et al., 1982; Caffrey and Kemp, 1991	Submerged aquatic plants	0.5–5.2		

Chemical conditions in the root zone are important determinants of the potential for significant PAF. Hydroponic studies create root environments that do not include a significant sediment oxygen demand. Roots are numerous under such conditions and exchange oxygen along much of their length (Armstrong et al., 1990). The morphology and physiology of roots is very different in the anaerobic environment often associated with treatment wetland soils. Under treatment conditions, the number of roots is significantly less than in clean soil or hydroponic conditions. Roots become armored along much of their length, and oxygen losses to the soil and water occur only in a small apical region (Brix, 1994c).

Some studies infer the value of PAF from side-by-side tests for nitrogen loss (Gersberg et al., 1989b). These inferences are based on three presumptions: (1) that an unvegetated bed conducts precisely the same reactions that occur in an adjacent vegetated bed receiving the same water, (2) that plant uptake plays a minor role in nitrogen disappearance, and (3) that the oxygen stoichiometry of carbon disappearance and ammonium disappearance are the same in a wetland as they are for wastewater treatment plants. None are particularly good assumptions.

SF wetlands replicate well for BOD and total suspended solids (TSS) (Kuehn and Moore, 1994), but there has been no study to demonstrate the degree to which physically similar wetlands actually replicate performance for oxygen and nitrogen. However, it is clear that there are wide swings in all performance attributes of wetland treatment systems. When the difference between input and output for the gravel-only bed is subtracted from that difference for the adjacent vegetated gravel bed, the stochastic variability propagates through the calculation. For data from Richmond, NSW (Bavor et al., 1988), the standard deviations in quarterly averaged BOD and hydraulic loading rate (HLR) are about 40%. The resultant standard deviation in inferred oxygen uptake is about 100%.

If the nitrogen reaction sequence does not stop at nitrate, but goes on to include denitrification, then some of the oxygen consumed in nitrification may be recycled. This can reduce the oxygen consumption for nitrogen processing from 4.5 to 1.7 mg O_2/mg N_2, at the expense of the consumption of a carbon source to fuel the denitrification (see Chapter 13). The stoichiometric coefficient of 1.5 mg O_2/mg BOD presumes that BOD disappears solely due to reaction with DO, but there are many other demonstrated BOD reduction pathways in the wetland environment, such as methanogenesis, nitrate reduction, and sulfate reduction (Burgoon, 1993).

OXYGEN SAG CURVES IN SF WETLANDS

As water travels through the wetland, oxygen is consumed by carbonaceous and nitrogenous constituents of the water, soils, and detritus. After a sufficient travel distance, the oxygen demand exerted within the water may become depleted to the point where reaeration from external sources can begin to restore the DO to background wetland levels. This depletion and recovery phenomenon is characteristic of point source pollution of streams and is also observed in treatment wetlands. The oxygen sag curve for the SF treatment wetland at West Jackson County, MS is shown in Figure 10-15.

An overall oxygen sag analysis may be applied to wastewater discharges to rivers (Metcalf and Eddy, 1991). This analysis is predicated on the assumption that oxygen is increased in the flow direction by mass transfer from the air above and by photosynthesis occurring within the water column and is decreased by consumption of BOD and ammonium nitrogen oxidation and by consumption of sediment oxygen demand (SOD) and respiration. In the wetland environment, both sediments and litter consume oxygen during decomposition. Decomposition processes also release carbon and nitrogen compounds to the overlying water, which can exert an oxygen demand. It is therefore appropos to designate the sum as decomposition oxygen demand (DOD). Plants transfer oxygen to their root zone to satisfy

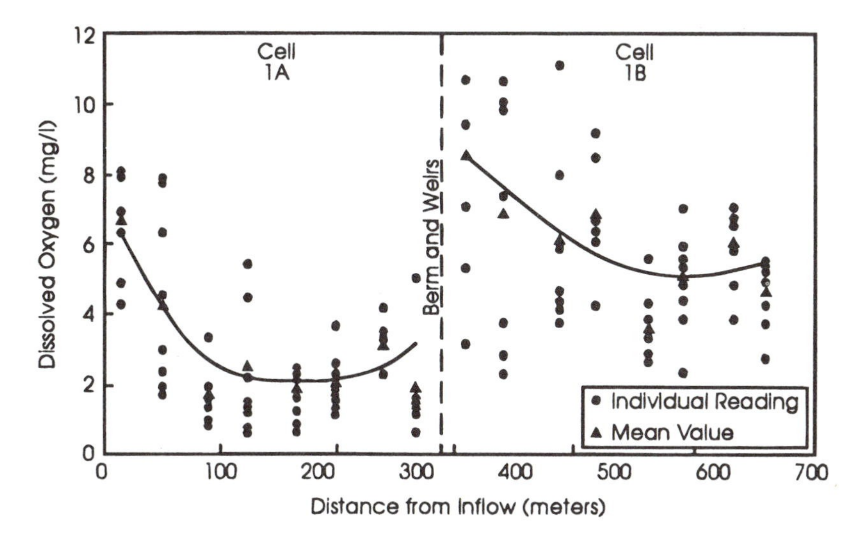

Figure 10-15 Observed oxygen profile in a constructed wetland treatment system in West Jackson County, MS.

respiratory requirements and may, in some instances, transfer a surplus to control the oxygen environment around the roots. The balance on DO in the wetland over a length L is written as

$$Q[C_{DO}(L) - C_{DO}(0)] = (LW)\left\{\begin{array}{l} K(C_{DO}^{sat} - \hat{C}_{DO}) + [r_{O,photo} + r_{PAF} - r_{O,res} - r_{O,DOD}] \\ + a_N Q[C_N(L) - C_N(0)] + a_B Q[C_{BOD}(L) - C_{BOD}(0)] \end{array}\right\} \tag{10-25}$$

where
a_N = stoichiometric coefficient for NH_4-N oxygen demand
a_B = stoichiometric coefficient for BOD
C_{DO} = DO concentration, mg/L
C_{BOD} = BOD concentration, mg/L
C_N = NH_4-N concentration, mg/L
$\hat{C}_{DO}$ = average DO concentration average over length L, g/m^3 = mg/L
LW = surface area, m^2
Q = flow rate, m^3/d
$r_{O,photo}$ = rate of DO generation by photosynthesis, $g/m^2/d$
$r_{O,res}$ = rate of DO consumption by respiration, $g/m^2/d$
$r_{O,DOD}$ = rate of DO consumption by decomposition, $g/m^2/d$
r_{PAF} = rate of DO addition by plant aeration flux, $g/m^2/d$

There is no treatment wetland data with which to separately evaluate photosynthesis, respiration, PAF, and DOD. It is necessary to lump these into wetland oxygen demand (WOD):

$$r_{O,WOD} = r_{O,DOD} + r_{O,res} - r_{PAF} - r_{O,photo} \tag{10-26}$$

where $r_{O,WOD}$ = net wetland oxygen consumption rate, $g/m^2/d$.

Further, there is often no data from which to estimate the reaeration coefficient K. Therefore, all transfer rates to and from the atmosphere and to and from the biomass in the wetland are lumped into a single term, the wetland net oxygen supply rate.

$$r_{NOSR} = K(C_{DO}^{sat} - \hat{C}_{DO}) - r_{O,WOD} \quad (10.27)$$

where r_{NOSR} = net oxygen supply rate, g/m^2/d.

It is then possible to simplify Equation 10-25:

$$q[\Delta C_{DO}] = -a_N q \Delta C_N - a_B q \Delta C_{BOD} + r_{NOSR} \quad (10\text{-}28)$$

where q = hydraulic loading rate, m/d.

The net oxygen supply rate can be positive (supply), negative (consumption), or zero. The data of Stengel et al. (1987) provide values of net oxygen consumption rates for *Phragmites* gravel bed wetlands. Fully oxygenated tap water with zero BOD and zero total kjeldahl nitrogen (TKN) was fed to the wetland, and the DO was found to decrease with distance in the inlet region. The SSF wetland was consuming oxygen in the absence of incoming BOD or NOD, with strong seasonal variations (Figure 10-16). The interpretation is simply that WOD exceeded the transfer of oxygen from air, and DO was depleted. Photosynthetic production of oxygen was likely zero in the gravel bed, and no mass transfer would be expected at the inlet because the water was saturated. Consequently, the rates shown in Figure 10-16 are $r_{O,WOD}$.

Stengel (1993) also found that after the initial drop in DO, reaeration did not occur; rather, DO reached a stable (constant) value with increasing distance along the bed. Cattails provided a stable root zone DO of about 1 to 2 mg/L in summer, whereas *Phragmites* stabilized at essentially zero DO. The implication is that in the downstream portions of the wetland, all oxygen uptake was consumed by respiration and SOD. If the wetland oxygen demand is longitudinally uniform,

$$K(C_{DO}^{sat} - 0) = r_{O,WOD} \quad (10\text{-}29)$$

In the summer, C_{DO}^{sat} = 10 mg/L, and the value of K = 0.16 m/d (Figure 10-16, June). This is in general agreement with the values of the reaeration coefficient predicted from lake and stream information.

It is important to note that this zero-loaded wetland was not able to sustain a high oxygen concentration in the water; the wetland processes consumed all transferred oxygen.

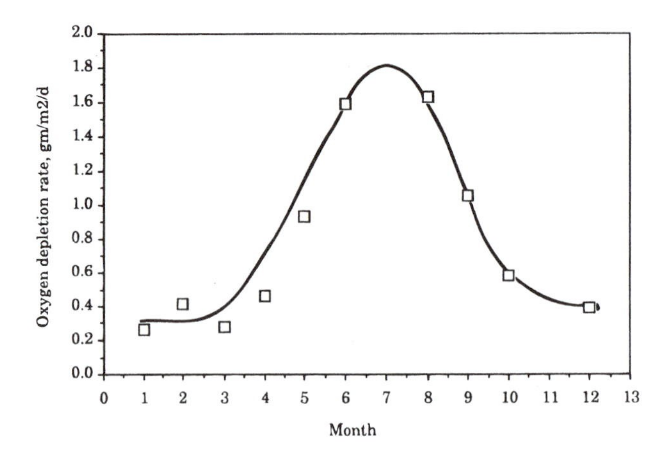

Figure 10-16 Oxygen-depletion rate in the inlet zone of a *Phragmites* gravel bed wetland receiving oxygenated tap water with nitrate at 30 ± 2 mg/L. (From data in Stengel et al., 1987.)

The stoichiometric coefficients in Equation 10-28 are often taken to be $a_B = 1.5$ and $a_N = 4.5$, as explained previously. However, there exist several wetland data sets that contain inlet and outlet concentrations of DO, NH_4-N, and BOD. It is then possible to let regression determine the suitable values for the net oxygen supply rate and the stoichiometric coefficients.

When Equation 10-28 is regressed for a wetland with BOD and NH_4-N at typical levels, the stoichiometric coefficients are very much smaller than $a_B = 1.5$ and $a_N = 4.5$. For example, for 3 years of I/O data for Listowel cattail SF wetland #4 (Herskowitz, 1986),

$$q\Delta C_{DO} = -0.11q\Delta C_N - 0.003q\Delta C_{BOD} - 0.027 \tag{10-30}$$

$$R^2 = 0.11$$

Almost none of the variability in DO levels in this wetland were attributable to changes in the waterborne BOD and NOD. The biomass compartments were dictating the oxygen level.

For 2 years of I/O data for the Richmond cattail gravel bed wetland (Bavor et al., 1988),

$$q\Delta C_{DO} = 0.080q\Delta C_N + 0.008q\Delta C_{BOD} + 0.028 \tag{10-31}$$

$$R^2 = 0.80$$

This wetland was a net supplier of oxygen, with $r_{NOSR} = 0.028$ g/m^2/d, which value represents the difference between supplies and WOD. Again, the stoichiometric coefficients are small.

There are a few wetland data sets that follow the progression of the key factors in oxygen consumption on transects through the wetland. The data presented by Butler et al. (1993) fit extremely well (Figure 10-17):

$$q\Delta C_{DO} = +0.199q\Delta C_{NH4-N} + 0.674q\Delta C_{SBOD} + 1.276 \tag{10-32}$$

$$R^2 = 0.99$$

The wetland in question was a gravel bed planted with *Phragmites* and subjected to periodic flow pulses with intermediate time for draining. Thus, the wetland was alternately exposed to air in the root zone and flooded conditions. The value of $r_{NOSR} = 1.276$ g/m^2/d presumably reflects the increased opportunity for mass transfer. Again, the stoichiometric coefficients regressed to wetland data are not 1.5 for BOD and 4.5 for NH_4-N.

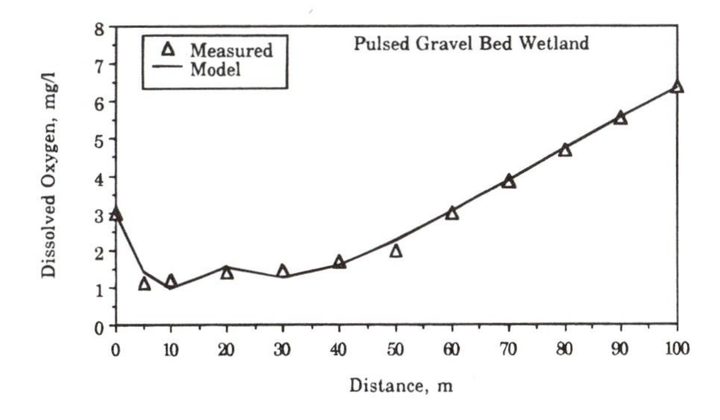

Figure 10-17 Fit of an oxygen sag model to averaged bed profiles for *Phragmites* in gravel. (Data from Butler et al., 1993.)

PREDICTION OF WETLAND EFFLUENT DISSOLVED OXYGEN

Oxygen sag regressions do not improve if an interwetland data set is used. Information from 14 SF wetlands produces the following:

$$q\Delta C_{DO} = 0.047q\Delta C_{TKN} - 0.91q\Delta C_{BOD} - 0.14 \quad (10.33)$$

$$R^2 = 0.16$$

Other independent variables may be tried in such a regression, such as TKN. Also, nitrate production may be added to the increase of oxygen in the water, but to no avail. The fundamental problem with this approach is that the component pieces of the net wetland oxygen supply are strongly regulated by the wetland biogeochemical processes, most importantly the chemical processes involving other electron acceptors.

The amount of oxygen in the incoming water, together with the loadings of ammonium and BOD in that water, may be used to determine rough estimates of the DO in the wetland effluent water. For average annual performance for 13 SF wetlands,

$$C_{DO,OUT} = 2.52 + 0.126C_{DO,IN} - 2.41qC_{NH_4-N,IN} \quad (10\text{-}34)$$

$R^2 = 0.62$
Standard error in $C_{DO,OUT} = 0.77$ mg/L
$0.9 < C_{DO,IN} < 9.9$ mg/L
$0.02 < qC_{NH4\text{-}N,IN} < 0.83$ g/m²/d

It is to be stressed that this regression Equation 10-34 does not apply to the instantaneous behavior in a wetland. For instance, it predicts the average behavior of Listowel wetland 4 as DO = 2.4 mg/L, compared to a 3-year average measured value of 1.7 mg/L. But it does not predict the measured time sequence on a monthly frequency. There are strong seasonal variations in DO in this wetland (Figure 10-18); the regression equation predicts a nearly constant outlet DO level.

Equation 10-34 was tested for three forested wetlands and predicted outlet DO concentrations with an average absolute error of 0.98 mg/L.

SSF wetlands in continuous flow discharge water with very low DO, often less than 1.0 mg/L (Table 10-3). Intermittent flow provides for direct air oxidation of solid materials, including adsorbed substances such as ammonium nitrogen. Such wetlands are analogs of overland flow and rapid infiltration systems, which operate more efficiently in the intermittent mode. During the air-filled period, sorbed ammonium and organic compounds can oxidize to nitrate and other products, and respiratory requirements may be met. The following water-filled period rinses out nitrate and other products of oxidation and cleans the surfaces for the next sorption cycle. Under these operating conditions, water reaeration can occur (Figure 10-17).

WETLAND HYDROGEN ION CONCENTRATIONS

GENERAL

Wetland water chemistry and biology are affected by pH. Many treatment bacteria are not able to exist outside the range $4.0 < pH < 9.5$ (Metcalf and Eddy, 1991). Denitrifiers operate best in the range $6.5 < pH < 7.5$, and nitrifiers prefer pH = 7.2 and higher. The same principles apply to other wetland biota; the acid-bog vegetation is adapted to low pH and differs greatly from the vegetation of an alkaline fen.

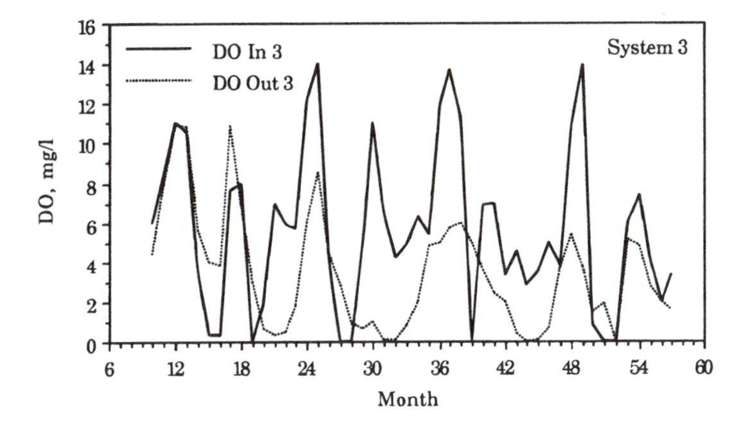

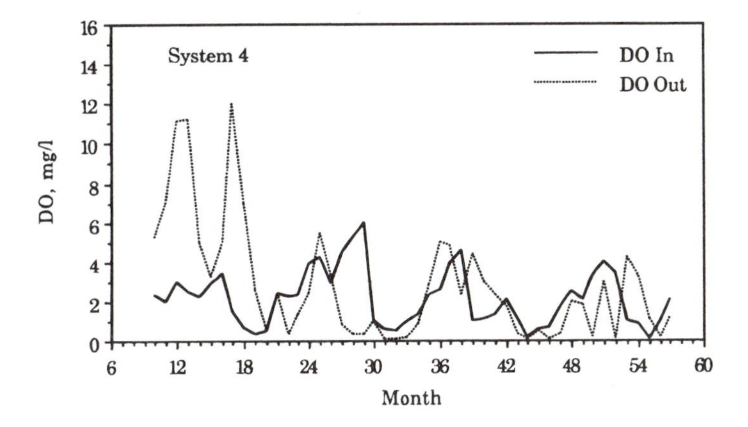

Figure 10-18 Four-year patterns of inlet and outlet DO for two wetlands at Listowel. System 3 was lightly loaded, receiving stabilization lagoon effluent. System 4 was more heavily loaded, receiving aeration cell effluent.

In addition to controlling various biological processes, pH is also a determinant of several important chemical reactions. Aluminum phosphate precipitates best at a theoretical pH of 6.3; iron phosphate precipitates best at a theoretical pH of 5.3. Ammonium changes to free ammonia at pH above neutral and at higher temperatures (see Chapter 13). The protonation of phosphorus changes with pH (see Chapter 14), and the hydroxide and oxyhydroxide precipitates of iron and aluminum are pH sensitive.

Table 10-3 DO Entering and Leaving Several SSF Wetlands

Wetland	Inlet DO	Outlet DO
Steady flow		
Benton, KY #3	8.20	1.00
Hardin, KY #1	5.20	1.20
Hardin, KY #2	5.20	0.70
Rector, AR	7.23	0.97
Waldo, AR	8.87	0.10
Richmond *Typha*	1.01	0.04
Richmond Cattail	1.01	0.00
Richmond Gravel #1	1.01	0.25
Richmond Gravel #2	1.20	0.13
Intermittent flow		
Portsmouth, U.K.	3.00	6.40
Phillips High School, AL	6.10	5.37

Because of its important influences on other chemistry, pH is an important regulatory parameter. Many states require effluents to meet a circumneutral pH range.

Natural wetlands exhibit pH values ranging from slightly basic in alkaline fens (pH = 7 to 8) to quite acidic in sphagnum bogs (pH = 3 to 4) (Mitsch and Gosselink, 1993). Natural freshwater marsh pH values are generally slightly acidic (pH = 6 to 7). Open water zones within wetlands can develop high levels of algal activity, which in turn create a high pH environment. Indeed, Bavor et al.'s (1988) data on an open water, unvegetated treatment "wetland" displayed high pH during some summer periods (pH $>$ 9) with circumneutral influent (7.0 $<$ pH $<$ 7.4).

In aquatic systems, algal photosynthetic processes peak during the daytime hours, creating a diurnal cycle in pH (Figure 10-19). Photosynthesis utilizes carbon dioxide and produces oxygen, thereby shifting the carbonate-bicarbonate-carbon dioxide equilibria to a higher pH. During nighttime hours, photosynthesis is absent, and algal respiration dominates, producing carbon dioxide and using oxygen. Open water areas in wetlands can exhibit these phenomena (Figure 10-19). Diffusive processes and calcium chemistry act as modifiers of this daily cycle.

Diurnal pH fluctuations are not evident in areas with dense emergent vegetation.

The organic substances generated within a wetland via growth, death, and decomposition cycles are the source of natural acidity. The resulting humic substances are large, complex molecules with multiple carboxylate and phenolate groups. The protonated forms have a tendency to be less soluble in water, and precipitate under acidic conditions. As a consequence, wetland soil-water systems are buffered against incoming basic substances. They are less well buffered against incoming acidic substances, since the water column contains a limited amount of soluble humics.

TREATMENT WETLANDS

Treatment wetland effluent hydrogen ion concentrations are typically circumneutral to slightly acidic. The notable exceptions are those wetlands receiving acid-mine drainage, which reflect the low pH of the incoming waters.

Continued application of circumneutral wastewater to a naturally acidic wetland can eventually alter the pH of the surface waters in the wetland. This was the case for an acid sphagnum-black spruce bog which received circumneutral wastewater for approximately 25

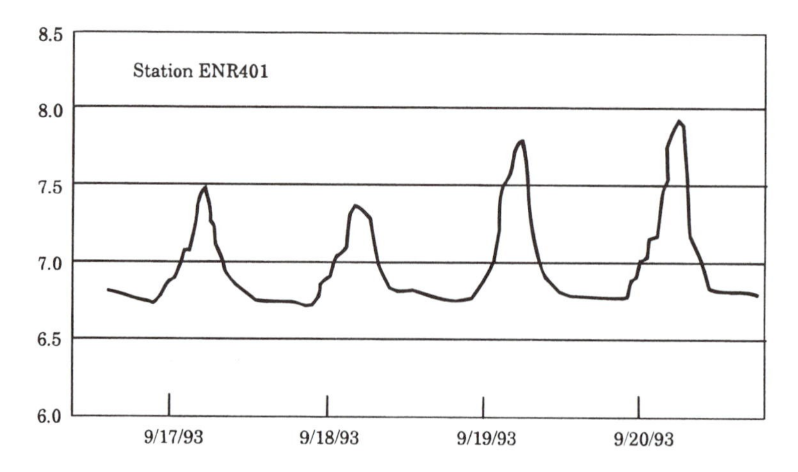

Figure 10-19 The diurnal fluctuations in wetland pH in an open water wetland environment. Algal photosynthetic activity drives large increases during the photoperiod. Data are from the Everglades Nutrient Removal wetlands. (Data from South Florida Water Management District, unpublished data.)

years prior to the measurements shown in Figure 10-20, as well as for a slightly acid peatland at Houghton Lake receiving slightly basic lagoon water. The effect on the peatland is the partial solubilization of the solid humic substances that existed under more acidic conditions. In addition to the chemical effect of humic solubilization, those decomposition processes that were acid inhibited can resume under the less acidic conditions.

The performance of the Byron Bay, NSW constructed SF wetlands illustrates the fact that long-term trends in acidic feed water pH are mimicked by similar long-term trends in wetland effluent pH (Figure 10-21). The amount of buffering is not great, averaging about one half a pH unit. In contrast, a Listowel constructed treatment wetland (System 3) received lagoon water, which periodically exhibited high pH due to algal activity in the lagoon (Figure 10-22). In the initial phases of the development of this wetland, during the first year of operation, little or no buffer capacity was evident. This was evidently due to the startup conditions, during which the vegetation spread to cover the wetland and litter formation and decomposition became operative. In later years, high incoming pH values were effectively damped out by the wetland.

Subsurface-flow wetlands also moderate and buffer the pH variations and levels of incoming basic waters (Figure 10-23). This is most likely due to the interactions between the substrate and its biofilms, rather than to the macrophytes. Bavor et al.'s (1988) data

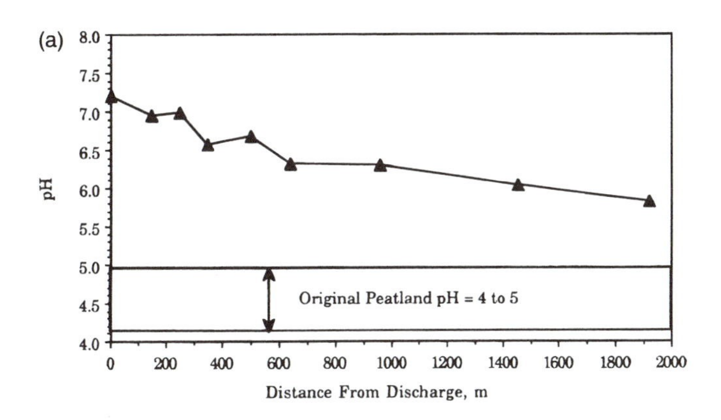

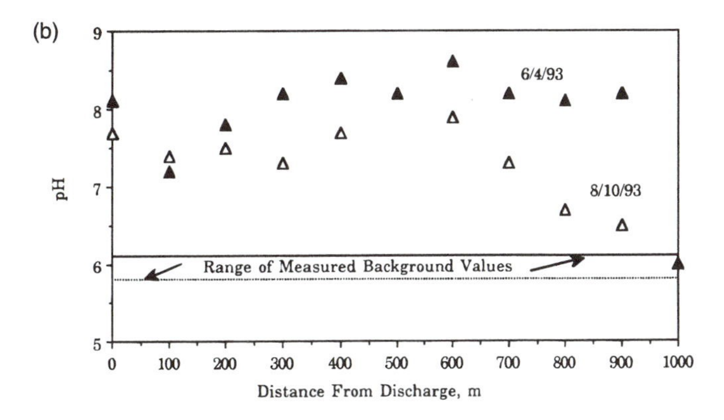

Figure 10-20 (a) Decline of pH along a transect in the direction of water flow at the Kinross, MI wetland. The original peatland pH range is inferred from contemporary data from close-by (500 m) remnant pristine peatlands. (b) pH along transects in the direction of water flow at the Houghton Lake, MI wetland in 1993, the 16th year of wastewater addition. The original peatland pH range was determined in predischarge investigations.

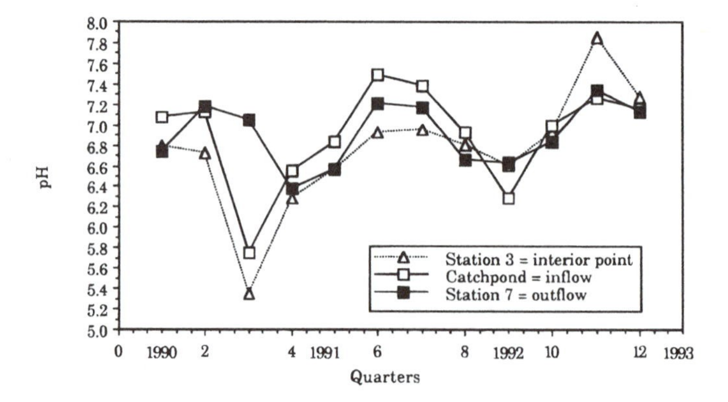

Figure 10-21 Long-term trends in the pH of waters in the Bryon Bay, NSW SF treatment wetlands.

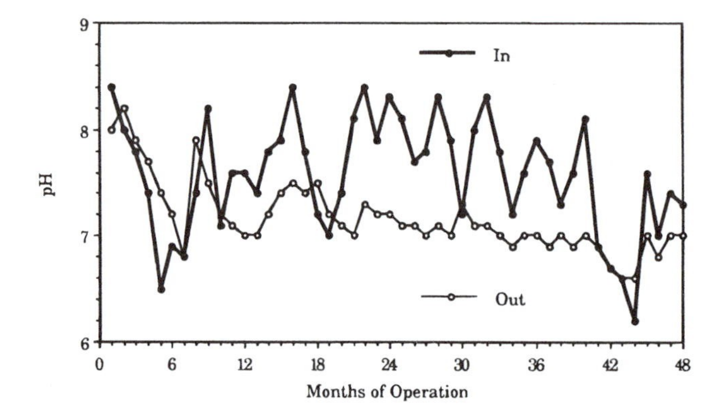

Figure 10-22 The pH of incoming and effluent waters from wetland 3 at Listowel. (Data from Herskowitz, 1986.) The inlet pulses are due to algal activity in the lagoon which feeds the wetland.

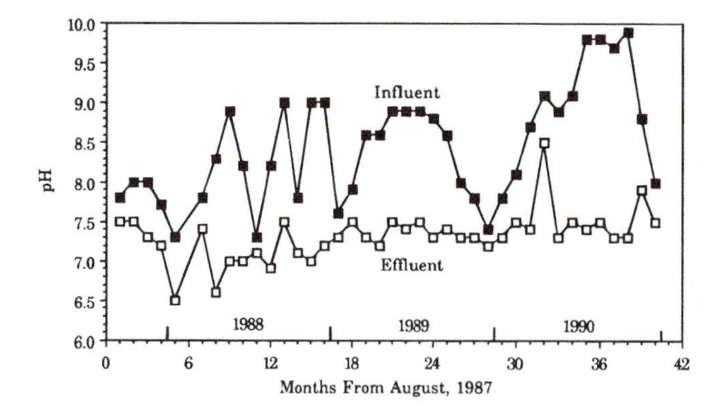

Figure 10-23 The variation of pH in the Benton, LA rock reed filter input and output. The feed is from a lagoon, which displays large excursions to basic pH.

support this idea, since their unplanted gravel bed produced the same pH as the planted SSF gravel systems.

Root zone (RZ) systems are built on sandy soil substrates and exhibit periodic overland flow. The pH phenomena in root zone treatment wetlands is the same as for SF and SSF wetlands. Incoming pH ranged from 6.9 to 8.1, and outgoing levels were in the range 6.6 to 8.1, for 62 root zone wetlands in Denmark (Schierup et al., 1990b).

SUMMARY

Many treatment wetlands exhibit a strong "buffer" capacity with respect to temperature, DO, and pH.

Wetland exit water temperatures are approximately equal to the mean daily air temperature during unfrozen seasons. This represents a balance between the dominant transfers: incoming solar energy gains and evaporative energy losses. The adjustment of the incoming water temperature to this balance is rapid because of the high energy content of evaporated water relative to typical water flow rates. In the winter, the insulation provided by snow, ice, mulch, and air gaps is enough to prevent water from freezing under north temperate conditions. The water temperature is then determined by losses upward through the insulating layers as well as vertically downward into the earth by conduction. In either case, energy balance equations permit calculation of wetland water temperatures.

Some natural wetlands operate with high levels of DO and others at low levels. Most treatment wetlands receive enough BOD and NOD loading to drive the DO level down to about 1 to 2 mg/L. Exceptions are oversized wetlands receiving very clean effluents. The soils, sediments, and biota in the wetland exert a strong influence on the DO concentrations in the water. Therefore, it is not accurate to assume that BOD and NOD disappearance is a measure of oxygen transfer. Most oxygen transfer is probably due to interfacial aeration, because there is no conclusive evidence that plants provide a significant aeration flux to the water or soil in excess of their respiratory demands.

Treatment wetlands operate at circumneutral pH for influents that are not strong acids or bases. This is true for constructed wetlands and for natural wetlands with acid antecedent conditions (peatlands).

CHAPTER 11

Suspended Solids

INTRODUCTION

A major function performed by wetland ecosystems is the removal of suspended sediments from water moving through the wetland. These removals are the end result of a complicated set of internal processes, including the production of transportable solids by the wetland biota.

Low water velocities, coupled with the presence of vegetation or a gravel substrate, promote fallout and filtration of solid materials. This transfer of suspended solids from the water to the wetland sediment bed has important consequences both for the quality of the water and the properties and function of the wetland ecosystem. Many pollutants are associated with the incoming suspended matter, such as metals and organic chemicals, which partition strongly to suspended matter. The accretion of solids contributes to pore space blockage in subsurface-flow (SSF) wetlands and to a gradual bottom elevation increase in surface-flow (SF) wetlands.

SOLIDS MEASUREMENT

Total suspended solids (TSS) are measured gravimetrically after filtration and drying (see APHA, 1992). The organic content is characterized as volatile suspended solids (VSS), determined from the weight loss on ignition at 550°C.

Turbidity in water is caused primarily by suspended matter, although soluble, colored organic compounds can contribute. Therefore, turbidity is sometimes used as a surrogate for gravimetric measurement of suspended matter. The measurement technique involves light scattering. The instrument is the turbidimeter, consisting of a nephelometer, light source, and photodetector. The standard unit is the nephelometric turbidity unit (NTU). The correlation is often good for a specific wetland system, but care must be taken in the extrapolation from one site to another. The rough correlation for an activated sludge process is $NTU \approx 0.42 \cdot TSS$ (Metcalf and Eddy, 1991). However, for river-borne and wetland sediments in northeastern Illinois, $NTU = 0.83 \cdot TSS$, $R^2 = 0.77$ (Figure 11-1).

It is virtually impossible to sample interior wetland waters for TSS because of the disturbance of sediments caused by sampling. Errors of one to two orders of magnitude can easily occur. This is the case in shallow zones of vegetated SF wetlands and in the pore space of SSF wetlands. If the water is deeper than about 20 cm, accurate sampling is possible, but not easy. Immersion of a sampler may cause disturbance of bed sediments, or the currents caused by water rushing into a sample bottle may disturb them. Ideally, the sample should flow into the sample bottle at the local velocity of the water in the wetland. This is termed isokinetic sampling and is necessary to prevent extraneous resuspension. It is often not possible to achieve undisturbed sampling for TSS, and therefore, it is difficult to obtain proper flow-weighted or volume-weighted values of TSS at interior points. For this reason, nearly all available TSS data from wetland treatment systems consists of input and output measurements in pipes and at structures.

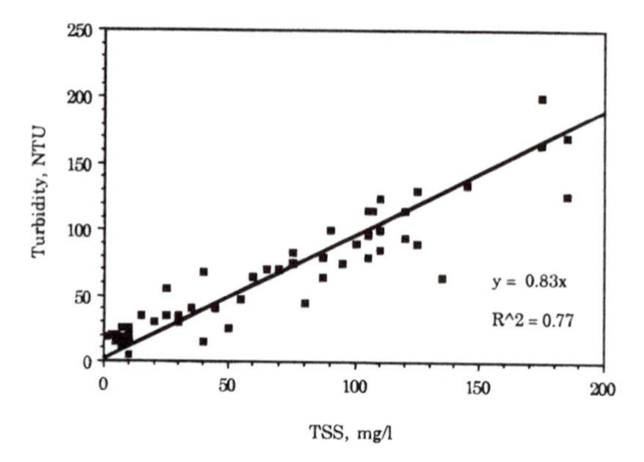

Figure 11-1 The correlation between suspended material and turbidity for wetland EW3 at Des Plaines. Data are from 1990–1991. (Data from WRI, 1992.)

This difficulty carries over to those chemical constituents which partition strongly to the solids. Any interior water sample will likely contain an unrepresentative proportion of the locally agitatible, or transportable, sediments and particulates. Subsequent analysis for the total amount of a partitioned substance will yield an inaccurately high value.

PARTICULATE PROCESSES IN SURFACE-FLOW WETLANDS

COMPONENT PROCESSES

The SF wetland processes sediments and TSS in a number of ways (Figure 11-2).

After the suspended material reaches the wetland, it joins large amounts of internally generated suspendable materials, and both are transported across the wetland. Sedimentation and trapping and resuspension, occur en route, as does "generation" of suspended material by activities both above and below the water surface. For example, algal debris may form at one location and deposit downgradient in the wetland.

Settling of Particulates

The slow moving waters in the SF wetland environment often permit time for physical settling of TSS. The settling velocity of the incoming particulates, combined with the depth of the wetland, gives an estimate of the time and travel distance for those solids.

Solids sink in water due to the density difference between the particle and water. For single, isolated spherical particles, the terminal velocity is reached quickly:

$$w^2 = \frac{4}{3}\frac{gd}{C_D}\left(\frac{\rho_s - \rho}{\rho}\right) \tag{11-1}$$

where d = particle diameter, m
C_D = drag coefficient, –
g = acceleration of gravity, m/s^2
w = terminal velocity, m/s
ρ = density of water, kg/m^3
ρ_s = density of solids, kg/m^3

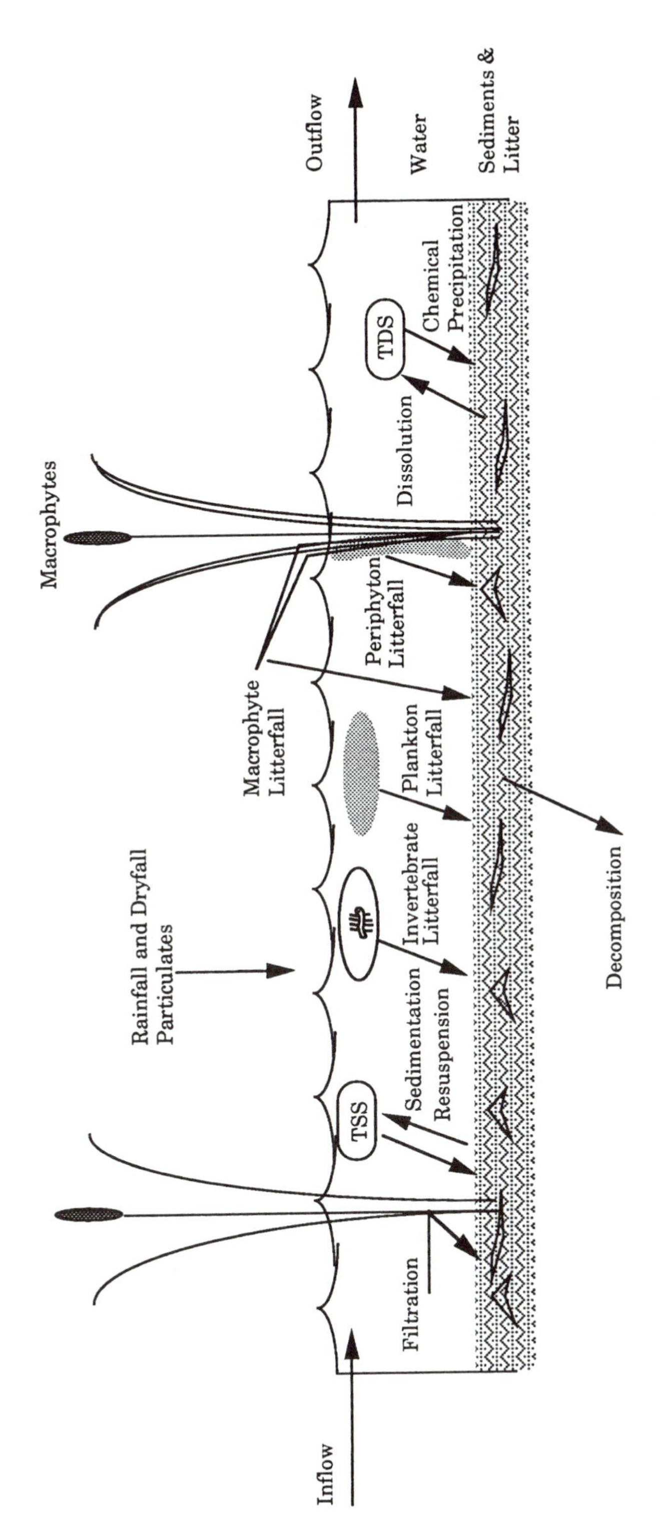

Figure 11-2 Suspended solids storages and transfers in the wetland environment.

In turn, the drag coefficient is a function of the particle Reynolds number (Figure 11-3), where the Reynolds number is

$$Re_p = \frac{d\rho w}{\mu} \tag{11-2}$$

where Re_p = particle Reynolds number, –
μ = viscosity of water, kg/m s (=0.001 × μ in centipoise)

If all physical properties are known, Equations 11-1 and 11-2 combine with Figure 11-3 to determine the settling velocity. This is an iterative calculation, in which there is a guess for w, calculation of Re_p from Equation 11-2, lookup of C_D from Figure 11-3, then calculation of w from Equation 11-1. The process is repeated until the value of w satisfies all conditions.

In the laminar flow region, $Re_p < 1.0$, the drag coefficient is inversely proportional to the particle Reynolds number (Figure 11-3), and the settling velocity of the particle is then calculable from Stokes law:

$$w = \frac{gd^2}{18\mu}(\rho_s - \rho) \tag{11-3}$$

A convenient, but very approximate, representation of the nature of the sizes and types of particles and their settling velocities is given in Figure 11-4.

In the wetland environment, neither the density nor the particle diameter are known, and the particles are not spheres or disks (Figure 11-5). There is also a distribution of particle sizes. While it is possible to correct for nonspherical shapes (Dietrich, 1982), there is not a convenient method for determination of the particle density.

Further, particles may agglomerate to larger size or be subject to interference from neighboring particles. Consequently, settling rates must be determined experimentally. Typi-

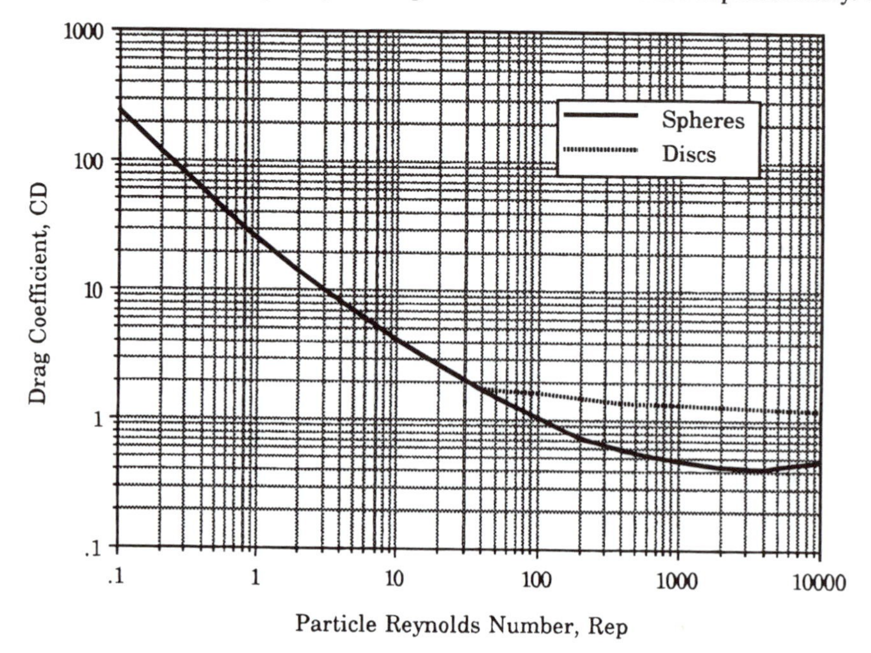

Figure 11-3 Drag coefficients for discs and spheres as a function of Reynolds number (Data from ASCE, 1975).

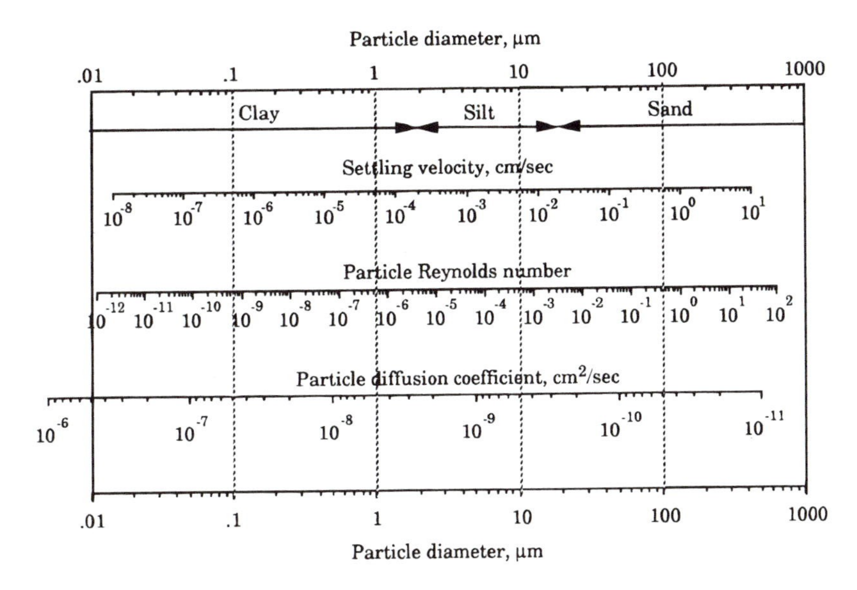

Figure 11-4 Characteristics of particles in 25°C water, (density = 2.0 gm/cc). (Adapted from Perry and Green, 1984.)

cally, a large diameter column of water is charged with a well-stirred suspension of particles, and the concentration is measured at a sequence of times at a series of depths below the water surface. Vertical profiles of TSS exist in differing shapes, depending on flocculation and interference. A number of analytical techniques may be applied to such data (Font, 1991). Only the mean water column concentration of TSS will be considered here. That concentration decreases as time progresses. Settling column data, for example, wetland waters

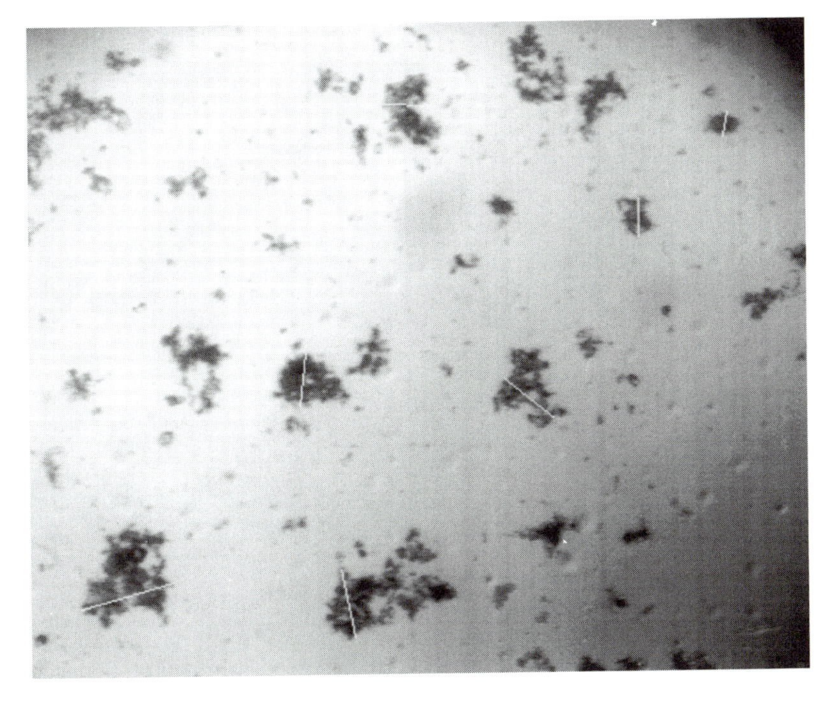

Figure 11-5 Photomicrograph of particulate suspended matter in the effluent from wetland EW4 at Des Plaines.

and other sources, indicate an exponential decrease in concentration with time, and a time scale of a few hours for the majority of settling to occur (Figure 11-6). The settling velocities shown on Figure 11-6 range from w = 2.75 to 26.3 m/d.

Caution must be used in those applications where colloidal materials may be present in the inflow because these materials are stable or very slow to settle. Very fine clay suspensions and some food processing wastewaters fall in this category. The settling velocity for planktonic solids was found to be on the order of w = 0.1 m/d for Wind Lake, WI (R.A. Smith and Associates, 1994).

Column settling data provide estimates of the removal time for TSS in the absence of dense vegetation. For instance, the water velocity in wetland EW3 at Des Plaines in 1991 was on the order of 30 m/d. The inlet zone was essentially unvegetated (Figure 11-7). Settling column data (Figure 11-6) suggest that solids should be essentially gone in 8 h or after a travel distance of about 10 m. Transect information confirms this estimate (Figure 11-8).

Resuspension

Much is known about the resuspension of particulates from flat surfaces (ASCE, 1975). Most interpretations are made in terms of the force per unit area (shear stress) required to tear a particle loose from the sediment surface. The concepts involve purely physical forces and apply most readily to mineral substrates and aquatic systems. Most results are for planar sediment bed bottoms and no extraneous objects. Vegetated wetland bottoms do not fit these conditions.

In the treatment wetland environment, physical resuspension is not a dominant process. Water velocities are usually too low to dislodge a settled particle from either the bottom or a position on submerged vegetation. However, in design, it is necessary to avoid wetland aspect ratios that produce high linear velocities. The potential for erosive velocities exists for highly loaded wetlands with high length to width ratios. Estimation of the velocity required to foster resuspension may be based on the settling characteristics of the solids and

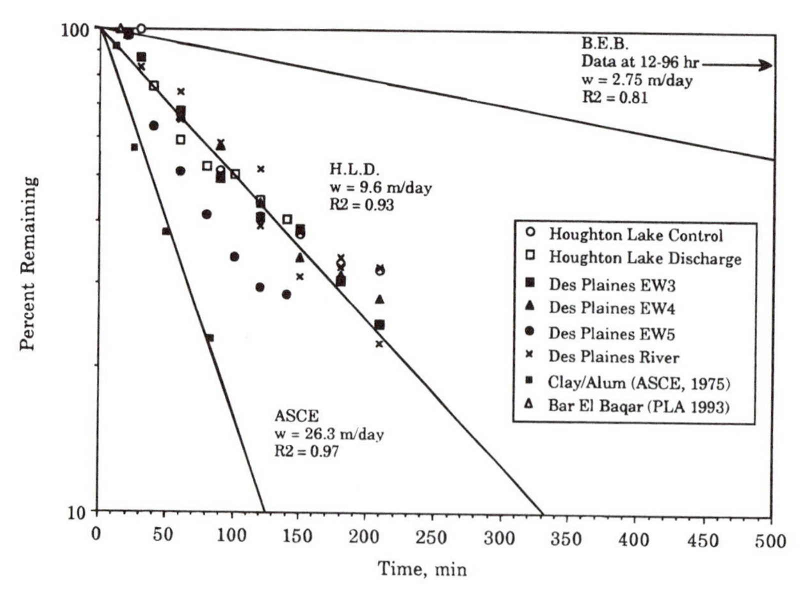

Figure 11-6 Examples of settling characteristics of TSS derived from wetlands and other natural contributing sources. (Data from sources are indicated, plus author's unpublished data.) The mean settling velocities range from 2.75 m/d for the Bar El Baqar (a branch of the Nile) TSS to 26.3 m/d for the clay alum mix.

Figure 11-7 Inlet zone of wetland EW3 at Des Plaines. Particulate settling was mostly complete at the second circle of stakes.

the frictional characteristics of the wetland combined with known correlations of the critical shear stress for particle dislodgment (ASCE, 1975).

The wetland environment provides an opportunity for three other mechanisms of resuspension: wind-driven turbulence, bioturbation, and gas lift. In open water areas, wind-driven currents cause surface flow in the wind direction and return flows along the bottom in the opposite direction. These velocities can exceed the net velocity from inlet to outlet.

Animals of all types and sizes can cause resuspension to occur. Feeding carp (Kadlec and Hey, 1994) and nesting shad (APAI, 1994) have been observed to cause problems. The carp rooted in the sediments for food and thus resuspended large amounts of sediments. Control was by drawdown and freezing. The shad fanned nests on the wetland bottom and thus resuspended sediments. Control was by drawdown and avian predation. Beaver activity can cause stirring, often at the outlet of the wetland, in conjunction with attempts to dam the outlet. Human sampling activities in the interior of treatment wetlands may also result in locally elevated concentrations of suspended solids.

Gas lift occurs when bubbles of gas become trapped in or attached to particulate matter. Wetland sediments are often of near-neutral buoyancy, so a small amount of trapped gas can cause "sinkers" to become "floaters". There are several gas-generating reactions in a wetland

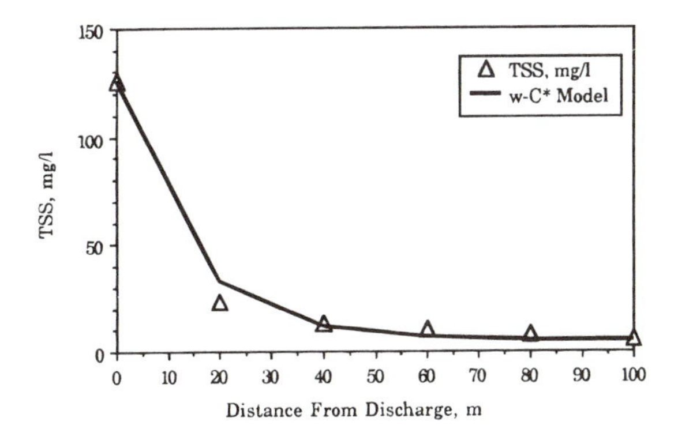

Figure 11-8 Decrease in TSS along a transect parallel to flow at Des Plaines EW3.

environment. Most important are photosynthetic production of oxygen by algae and production of methane in anaerobic zones.

In fully vegetated wetlands, the litter and root mats provide excellent stabilization of the wetland soils and sediments. This limits, but does not eliminate, resuspension.

Resuspension: The Extreme Case of Planar Sediment Beds

The theory of plain sedimentation predicts that as water velocity increases, there comes a point at which the shear stresses tear loose particles. Those shear stresses, in open channel flow, are reflected by a value for Manning's coefficient. The force necessary to dislodge the particle is related to the force that holds it down, which is in turn reflected in the settling velocity of the particle. These relations are explained in detail in ASCE (1975).

Wetland flows offer some simplification, since they are typically slow enough to be in the laminar regime with respect to particle settling and with respect to the water layers next to the top of the sediment. In actuality, wetland sediment resuspension is inhibited by the presence of obstacles such as the litter mat. However, some idea of the potential for resuspension may be gained by calculating the average water velocity that would cause incipient resuspension of an unobstructed bed of particles. This velocity places a lower limit on the resuspension velocity in the SF wetland; it will take more shear to dislodge particles from within local traps.

To prevent resuspension in this low-velocity region of the theory, it is required that

$$u \leq 7.2\left(\frac{w^{1/3}H^{1/6}}{nd^{2/3}}\right) \tag{11-4}$$

where d = particle diameter, m
H = water depth, m
n = Manning's coefficient for open channel, $s/m^{1/3}$
u = water velocity, m/d (=u m/s × 86,400 s/d)
w = particle settling velocity, m/s

Because this result is based on laminar theory, it is necessary that two Reynold's number criteria be met:

$$Re_1 = \frac{d\rho u}{\mu} \leq \left(\frac{2H^{1/6}}{n\sqrt{g}}\right) \tag{11-5}$$

$$Re_2 = \frac{d\rho w}{\mu} \leq 1 \tag{11-6}$$

where μ = water viscosity, kg/m·s (about 0.001 at 20°C)
ρ = water density, kg/m^3 (about 1000 at 20°C)

Example 11-1

Particles of 100 μm are observed to settle at a rate of 2.75 m/d. What is an estimate for the water velocity that would resuspend these particles from a flat, unvegetated bottom in a flow 30 cm deep?

Solution

First check the criterion in Equation 11-6):

$$Re_2 = \frac{(0.0001)(1000)(2.75/86{,}400)}{(0.001)} = 0.0032 \leq 1 \qquad (11\text{-}6)$$

OK, within laminar range for settling

Compute velocity for incipient resuspension. A value of the Manning's coefficient for the unobstructed flow over the sediment bed is needed. This is considerably lower than for the vegetated case. Following procedures outlined by French (1985), a value of 0.05 s/m$^{1/3}$ is assumed.

$$u \leq 7.2\left(\frac{(2.75/86{,}400)^{1/3}(0.3)^{1/6}}{(0.05)(0.0001)^{2/3}}\right) = 1{,}733 \text{ m/d} \qquad (11\text{-}4)$$

Next, check the criterion in Equation 11-5:

$$Re_1 = \frac{d\rho u}{\mu} \leq \left(\frac{2H^{1/6}}{n\sqrt{g}}\right) = \left(\frac{2(0.3)^{1/6}}{0.05\sqrt{9.8}}\right) = 10.45 \qquad (11\text{-}5)$$

$$\frac{(0.0001)(1000)(1{,}733/86{,}400)}{(0.001)} = 2.01 \leq 10.45$$

OK, within laminar range for resuspension theory

This analysis suggests that resuspension will not occur for these particles until water velocities exceed 1733 m/d.

"Filtration"

Conventional wisdom has it that the presence of dense wetland vegetation causes settling to be augmented by filtration. This is not true in the usual sense of the term filtration. It is the trapping of sediments in the litter layer that prevents resuspension and thus enhances the net apparent suspended sediment removal. The macrophytes and their litter form a nonhomogeneous "fiber bed" in the wetland context. The void fraction in the stems and litter is quite high, and straining and sieving are thus not dominant mechanisms. The principal mechanisms of fiber bed filtration are well known and documented in handbooks (see, for instance, Perry and Green, 1984, or Metcalf and Eddy, 1991). These include

1. Inertial deposition or impaction—particles moving fast enough that they crash head-on into plant stems rather than being swept around by the water currents.
2. Flow line interception—particles moving with the water and avoiding head-on collisions, but passing close enough to graze the stem and its biofilm and sticking.
3. Diffusional deposition—random processes at either microscale (Brownian motion) or macroscale (bioturbation) which move a particle to an immersed surface.

The efficiencies of collection for these mechanisms depend on the water velocity, particle properties, and water properties. A typical wetland "fiber" is a bulrush stem of about 1 cm diameter. A typical particle might be on the order of 1 to 100 μm. A typical water velocity is on the order of 10 to 100 m/d. Under these conditions, the collection efficiencies of all three mechanisms are predicted to be vanishingly small.

The measurements of Hosokawa and Horie (1992) on reed-field wetlands lend support to the previous concepts: they found no difference between the TSS removal efficiency of vegetated and unvegetated cells. They further confirmed the lack of filtration mechanisms in laboratory flumes with and without simulated vegetation.

The submerged vegetation does, however, trap sediment in sheltered microzones. The potential for resuspension is thereby lessened.

Biological Sediment Generation

Wetlands produce sediments via processes of death, litter fall, and litter attrition. This occurs for biota at a number of different size scales, ranging from macrophytes on down to bacteria. Algal productivity can be a major generator of suspended solids. A second set of processes adds pollen and seeds to the water. The TSS produced is organic in character, resulting in a high carbon content and a high proportion of VSS. The chlorophyll and pheophytin (dead chlorophyll) content is high if the algal pathway is dominant.

Some TSS originates from leaf and stem litter. For instance, annual leaf litterfall in a natural sedge-shrub peatland was found to be 60 to 70 g/m^2 (Chamie, 1976). Some part of this material contributes to TSS, either via direct attrition or via microbial decomposition.

The generation of sedimentary material is a very important internal process in nutrient-rich treatment wetlands. The generous supply of nutrients assures a large production of a wide variety of transportable organisms and associated dead organic material. Such wetlands are characterized by high water chlorophyll content and high sediment accumulation. Bacterial and algal growth is promoted, and decomposition products form a new pool of suspendable material. A host of wetland invertebrates, such as *Daphnia* and waterboatman (*Corixidae*), also die and contribute to the sediments, and they may be present in pumped lagoon water.

These processes are virtually impossible to predict and quantify. But, it is important to recognize that they exist because they contribute to an irreducible background level of TSS in a wetland.

Chemical Precipitation and Dissolution Reactions

Several chemical reactions can produce wetland sediments under the proper circumstances. Some of the more important are the oxyhydroxides of iron, calcium carbonate, and divalent metal sulfides. The hydroxides are typically flocs with the possibility of coprecipitates.

As conditions of chemical composition, pH and redox change in the wetland. These and other compounds may undergo dissolution and be removed from the sediment bed.

SOME INTERNAL DETAILS

The processes above combine to determine the amount of sediment at various locations within the wetland as a function of time and the TSS concentration in the wetland effluent. Cup collectors may be placed on the wetland bottom (Jordan and Valiela, 1983; Fennessy et al., 1992); these normally intercept the downward vertical flux of sediment, but prevent shear-induced resuspension. Horizon markers may be used to measure the accretion of sediment at a specific location. One technique involves the elevation of a blunt-footed rod which is lowered to the sediment surface. A reference rod, driven deep into stable soils, provides the local datum (Reeder, 1992). A second procedure involves placement of a flat feldspar plate on the wetland bottom, followed by sediment harvest above that horizon at a later time. Sediment accretion may also be determined on a longer time scale via conventional surveying and atmospheric radioisotope deposition (Kadlec and Robbins, 1984).

An area of wetland may be isolated and stirred to resuspend the sediment, yielding a measure of the potentially transportable solids. TSS may be measured at inlet, outlet, and interior wetland locations. Each procedure provides only a portion of the information needed to close the sediment budget for the wetland. The application of these and other field measurement techniques has produced estimates of the internal cycling of wetland sediments as well as their lateral transport.

Models of sediment transport have been developed and verified for estuaries (Nakata, 1989, for example). These are two- and three-dimensional models that allow for dispersion, and generation is not usually an important term. These models may be adapted to the wetland situation. The vertically averaged mass balance for TSS in a linear plug flow wetland is (see Figure 11-9)

$$\frac{\partial(hC)}{\partial t} + \frac{\partial(uhC)}{\partial x} = G + R - S \tag{11-7}$$

$$\frac{\partial B}{\partial t} = S - R - A - D \tag{11-8}$$

where A = consolidation rate of transportable solids, $g/m^2/d$
B = transportable solids bed, g/m^2
C = concentration, g/m^3 = mg/L
D = decomposition rate of transportable solids, $g/m^2/d$
G = generation rate, $g/m^2/d$
h = water depth, m
R = resuspension rate, $g/m^2/d$
S = settling rate, $g/m^2/d$
t = time, d
u = superficial water velocity, m/d
x = distance, m

Rate Equations

The settling rate (S) describes gravitational deposition; it is a lumped removal term. Gravitational deposition often follows a first-order areal removal rule:

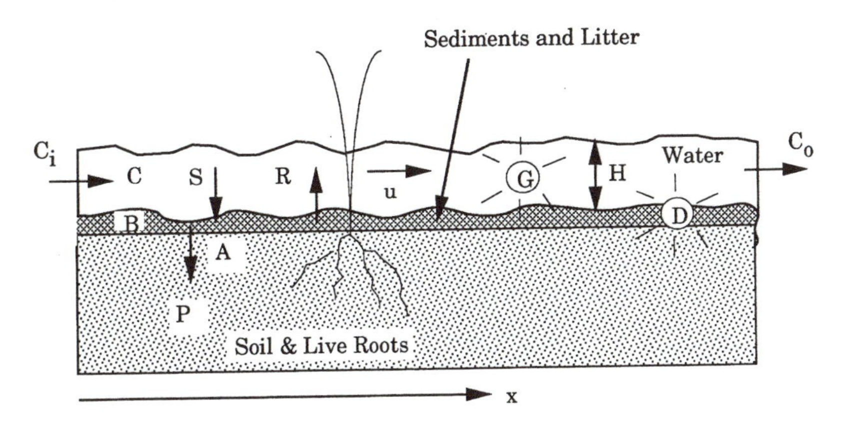

Figure 11-9 Framework for mass balances on suspended materials in the wetland environment.

$$S = wC \tag{11-9}$$

where w = solids settling velocity, m/d.

Column tests on a number of different suspended materials are shown in Figure 11-6. Equations 11-7 and 11-9 predict an exponential decrease in average solids concentration in the batch column test, which fits the example wetland data sets reasonably well. Settling velocities for those materials are on the order of 10 m/d, indicating the potential for high removal rates.

The detachment and resuspension of solids is a complicated process. Particles of the sediment bed can move in the direction of the water flow by a hopping mechanism driven by the shear stress on the top layer of particles. The vertical displacement is typically no more than a few particle diameters; thus, most lateral transport takes place in a thin layer next to the sediment bed surface. In a planar, unvegetated system, there exist methods for predicting the vertically averaged mean TSS concentration in this layer (Wiberg and Smith, 1989). There are no such methods for the wetland situation; therefore, there is no *a priori* method for computing (R). It seems probable that detachment would not be affected by the TSS concentration in the overlying water, but would depend on the amount of suspendable material in the sediment bed.

The generation rate (G) has likewise yet to be quantified in terms of wetland state variables such as nutrient status, vegetation type, and density.

There is also not a predictive equation for the consolidation rate (A) or the decomposition rate (D).

Averaging

In the short term, there are significant fluctuations in TSS storage within the water column in response to the variations in settling, resuspension, and generation. Childers and Day (1990) state: "Our results affirm the variability of short-term sediment transport and depositional processes. . . ." Over a long period, however, changes in water column storage are negligible compared to other inputs and outputs. The water column TSS mass balance then assumes the character of a steady state model. There is an accompanying sediment bed balance, in which the change in storage is the dominant feature. The long-term, time average profiles are calculated from:

$$uh\frac{dC}{dx} = G + R - S \tag{11-10}$$

$$\frac{\partial(B + P)}{\partial t} = S - R - D \tag{11-11}$$

where P = new immobile soil, g/m^2.

In a spatially uniform wetland, $dC/dx = 0$, and $S = G + R$. If there is no growth of the suspendable sediment pool (B), then $\partial B/\partial t = 0$, and $A = G - D$.

Using Equation 11-9 for (S) yields

$$C^* = \frac{G + R}{w} \tag{11-12}$$

where the asterisk refers to the spatial uniformity. This is the background TSS concentration that would prevail in the uniform wetland functioning in a stationary or steady state mode.

If it is assumed that generation and resuspension are constant over the entire wetland, Equation 11-10 may then be written as

$$uh\frac{dC}{dx} = w(C^* - C) \tag{11-13}$$

Integration of Equation 11 gives

$$\ln\left[\frac{C - C^*}{C_i - C^*}\right] = -\frac{wx}{uh} = -\frac{wx}{qL} = -\frac{w}{q}y = -\frac{w\tau}{h}y \tag{11-14}$$

where C_i = inlet TSS concentration, g/m^3
L = length of wetland, m
q = hydraulic loading rate, m/d
y = fractional distance through wetland

Equations 11-13 and 11-14 are restricted to long-term average performance. A short-term stochastic variability is superimposed upon this mean performance.

Equation 11-14 contains a subtle message that bears on the removal of nearly all pollutants in wetlands, not just TSS. The right-hand numerator contains the settling velocity times the wetland length. An increase in either will cause a faster approach to C*. The denominator contains the water velocity times the depth. An increase in either of those will cause a slower approach to C*. The detention time does not appear directly in this simplified mechanistic model, and the reason is easy to understand. If the water depth is doubled, for the same incoming volumetric flow rate and wetland area, the detention time will be doubled. But the particles do not fall any faster and now have twice as far to travel to the bottom. The extra detention time is used up by a greater travel time. On the other hand, doubling the area of the wetland, all else being equal, will also double the detention time. The vertical settling distance is not increased, and the extra time causes greater removal.

The model feature that creates this situation is the area-specific nature of the removal rate Equation 11-9: the amount of mass settling is proportional to the cross-sectional area of a settling column or to the surface area of a wetland at a given location. Other pollutant removals also obey such area-specific rate equations.

At the wetland outlet, y = 1, and

$$\ln\left[\frac{C_o - C^*}{C_i - C^*}\right] = -\frac{w}{q} = -Da_{SS} \tag{11-15}$$

where C_o = outlet TSS concentration, g/m^3
Da_{ss} = Damköhler number for TSS, = w/q, dimensionless

Equation 11-15 predicts long-term mean performance; stochastic processes will affect short-term results.

Model Verification

Data from the Des Plaines site support this model (Figure 11-8). Wetland EW3 was heavily loaded when the pump was operating and contained relatively sparse emergent vegetation. Independent measurements were made in settling columns, yielding w = 9.7 m/d. At the time of data acquisition, q = 1.3 m/d, from which the predicted value of Da_{ss} =

7.5. Independent measurements of R were also made utilizing sediment cups and I/O data, which gave R = 46.0 g/m²/d. Estimates of G = 1.6 g/m²/d (WRI, 1992). Accordingly, from Equation 11-12, the expected value of C* = 4.9 g/m³. Thus, both C* and Da_{ss} were estimated independently from the transect data for TSS. The predicted drop in TSS agrees quite well with the measurements.

The TSS data in Figure 11-8 may be regressed to produce values of $Da_{ss} \approx 9.1$ and C* ≈ 7 mg/L.

Information from a few other SF wetlands indicates similar TSS profiles. Listowel studies (Herskowitz, 1986) on densely vegetated channels at low loading rates show a fast decline to a spatially invariant background TSS of about 5 to 10 g/m³ (Figure 11-10). No settling velocity data are available for the Listowel wastewater.

Most available SF wetland TSS data consist of input-output information. Because of the rapid removal that typifies these systems, the output is reflective of the background concentration C* and not of the settling process that occurs in the inlet section. The 4-year study at Listowel pointed out this fact nicely (Herskowitz, 1986). Systems 3 and 4 were identically configured and hydraulically loaded wetlands. System 3 reduced TSS from 22.8 ± 3.1 to 9.2 ± 4.6 mg/L; System 4 reduced TSS from 111.1 ± 5.8 mg to 8.0 ± 2.4 mg/L. In fact, the TSS leaving the five Listowel wetlands over the 4-year test period were not significantly different ($\alpha = 0.05$), regardless of inlet TSS, wetland configuration, or operating variables.

Model Parameters

A single value of w cannot be recommended for design. The reader is referred to Figure 11-6, which shows a range of w from 3 to 30 m/d. For a new project, settling tests should be performed. The background C* will be presented in the Data Regression section.

Sediment Buildup

Trapped TSS, plus material generated within the wetland, will accrete as either movable sediment or the consolidated immovable new soil produced from the sediments. If Equations 11-10 and 11-11 are added and integrated from wetland inlet to outlet, it is seen that accretion is due to removal plus generation less decomposition:

$$\frac{\partial(\tilde{B} - \tilde{P})}{\partial t} = qC_o - qC_i + \tilde{G} - \tilde{D} \qquad (11\text{-}16)$$

It is an easy calculation to allocate the removed TSS to the buildup of new solids in the SF

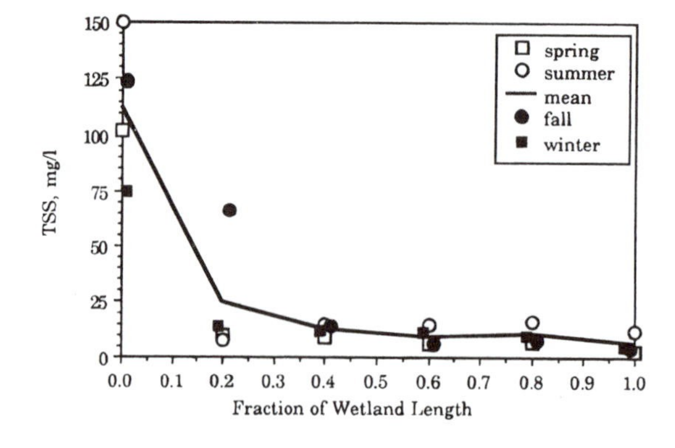

Figure 11-10 The TSS in the incoming waters at Listowel were effectively reduced to background levels in the first portion of the wetland. (Data from Herskowitz, 1986.)

wetland. For municipal wastewater polishing, typical operations lead to an accumulation of 1 to 2 mm/yr of new solids (50 mg/L removed at q = 5.5 cm/d at a bulk density of 0.5 g/cm^3 yields 2.0 mm/yr). But, as Equation 11-16 indicates, that material is augmented by internally generated solids and decreased by decomposition of the organic portion of sediments and soils. The net increase due to G − D may total up to 10 mm/yr in a highly eutrophic marsh (Reddy et al., 1991a).

It should be noted that buildup occurs preferentially in the inlet section of the wetland. Therefore, the deposition will exceed the average in that region (see Chapter 5, Figure 5-1).

No municipal wastewater polishing marsh has yet had to be serviced for solids removal. Two of the oldest, Vermontville (20 years, constructed) and Houghton Lake (17 years, natural), have experienced accretions in the range given earlier, but this has not jeopardized containment or operability. The accretions are slow enough to permit root zone adaptation.

Food processing wastewaters can contain very high TSS concentrations, which in turn can fill a treatment wetland with solids. Van Oostrom (1994) reported that one third of the volume of a floating *Glyceria* mat wetland was filled after 20 months of operation. The wastewater was a nitrified meat processing effluent with incoming TSS of 269 mg/L, and the removal rate was 5300 g/m^2/yr. Accreted sediments totaled 40 percent of the removed solids, 2100 g/m^2/yr, and these were concentrated near the inlet end of the wetland. The density of the solids was very low, *circa* 0.03 g/cm^3.

Application of the mass balance calculations to information from cup collectors and input-output information allows estimates of the various fluxes of TSS within a wetland. An example is given in Figure 11-11. The wetland is an effective sediment trap, but large amounts of sediments cycle within the wetland.

DATA REGRESSION

In the design of a wetland for treatment, it is not possible to accurately estimate G and R; thus, it is necessary to make a prediction of the expected background, C*. Information from the North American Database (NADB) shows few differences between forested and marsh wetlands for TSS reduction. The median reductions for all data are 67 (N = 164

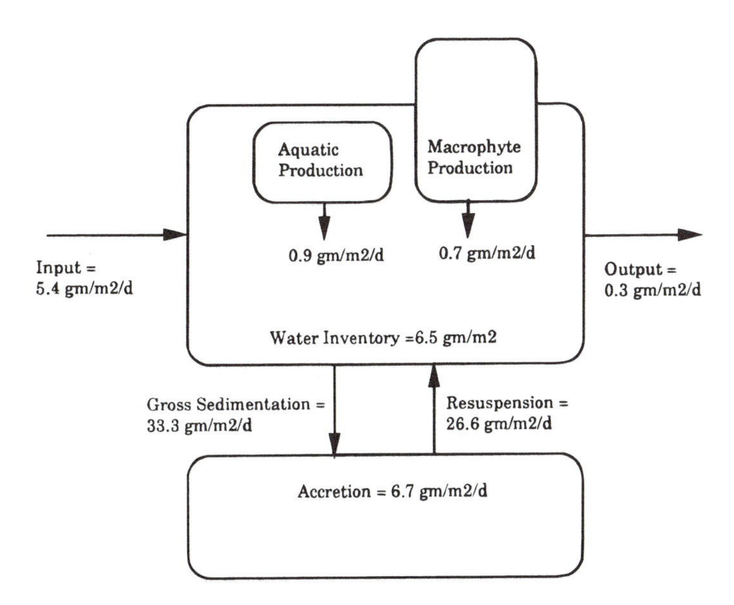

Figure 11-11 Components of the sediment mass balance for wetland EW3 at Des Plaines in 1991. The balance period is the 23-week pumping season. (Data from WRI, 1992.)

quarterly data points) and 66 percent (N = 371), respectively. Accordingly, these two types of SF wetlands are considered jointly for data analysis. For reasons stated earlier, the effluent concentrations are representative of C*. Regression of NADB information produces the following correlation:

$$C^* = C_o = 5.1 + 0.16\ C_i \tag{11-17}$$

$R^2 = 0.23$, N = 1582
Standard error in $C_o = 15$
$0.1 < C_i < 807$ mg/L
$0.0 < C_o < 290$ mg/L

The inclusion of C_i in the regression indirectly accounts for the enhanced TSS production of SF wetlands that receive a stronger influent. Mechanistically, it is the accompanying nutrients that cause an elevated internal production. For instance, the Gustine, CA bulrush wetlands had a median influent of 88 mg/L and a median effluent of 28 mg/L (LWAI, 1990). The average total Kjeldahl nitrogen (TKN) level in the incoming water was approximately 35 mg/L, which is the likely cause of the relatively high effluent TSS.

Percent Removal

A popular method of TSS data representation is the quotation of percentage removal or removal efficiency. Equation 11-17 may be used to calculate a percent removal. However, the presence of a background TSS level constrains removal efficiency to be below a level dictated by the inlet and background concentrations. This effect is illustrated with data from four systems in Figure 11-12.

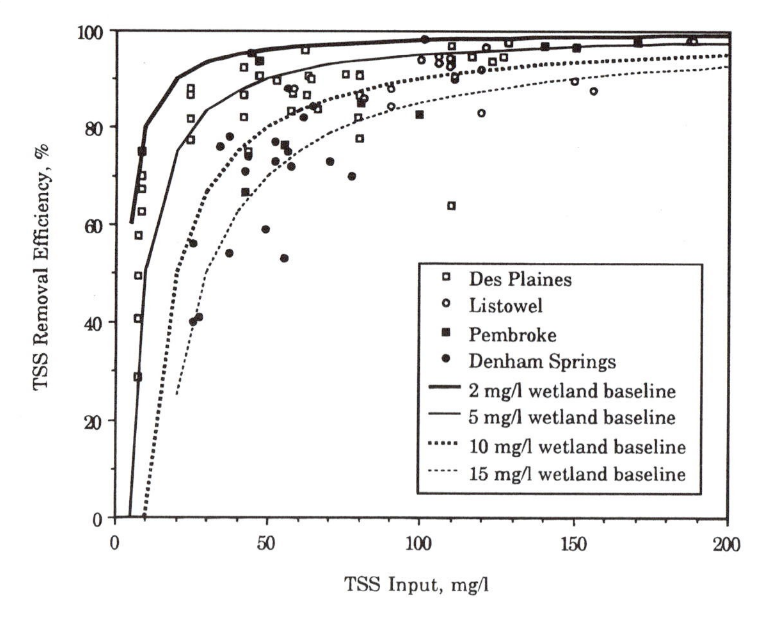

Figure 11-12 Removal efficiency for TSS as a function of inlet TSS concentration. The lines indicate the maximum removal efficiency which could be achieved for a wetland operating at the specified background level of TSS. Four wetland systems are selected for illustration of data.

Table 11-1 Overall Performance Efficiency of Wetlands for Reduction of Suspended Solids

Water Source	Type	Number of Wetlands	Median Percent Reduction	Maximum Percent Reduction	Data Source
Wastewater	SF marsh	39	66	93	NADB, 1993
Wastewater	SF forested	6	72	85	NADB, 1993
Wastewater	Duckweed	5	83	92	WEF, 1992
Wastewater	Hyacinths	4	82	92	WEF, 1992
Wastewater	SSF	22	75	96	NADB, 1993
Wastewater	SSF	3	92	96	Bavor et al., 1988
Wastewater	SSF	5	83	88	Green and Upton, 1992
Wastewater	SSF	77	86	97	Brix, 1994
Wastewater	SSF	7	86	94	Miscellaneous, NADB, 1993
Stormwater	SF natural	11	76	95	Strecker et al., 1992
Stormwater	SF constructed	14	81	98	Strecker et al., 1992
		193 Total	79 Weighted Median		

Table 11-1 presents the results from a large number of studies weighted at one point per wetland. For the types of solids normally encountered in polishing wetlands, the results may be summarized by a rough rule of thumb: a treatment wetland removes about three quarters of the incoming TSS, provided incoming TSS > 20 mg/L.

EFFECTS OF TEMPERATURE AND SEASON

Temperature appears to be a factor in the reduction of TSS. Figure 11-13 indicates the increase in the effluent TSS from SF marshes during the summer months, compared to the winter effluent concentrations. This is to be expected for the generation processes under consideration. Conversely, the effect of temperature on the properties of water is not sufficient to significantly influence the settling velocity of the particulate matter. Because some reduction of the generation processes would be expected at cold temperatures, the value of C* is expected to be less in cold seasons. The other contributing processes—settling and resuspension—remain operative and relatively unaffected by cold. A temperature coefficient for C* may be estimated from Listowel data:

$$C^* = C^*_{20}\theta^{(T-20)} \tag{11-18}$$

$$\theta = 1.065$$

Since the mean temperature for the NADB is close to 20°C, C^*_{20} may be taken from Equation 11-17. Agreement is within 10 percent for Listowel Systems 4 and 5.

Equation 11-18 represents mean behavior of the wetland and would be expected to account for temperature effects when applied to the average performance by season over a period of several years. Within any given year, this effect can be masked by stochastic variation. For instance, the temperature of wetland EW3 at Des Plaines changes from about 5°C in April to over 25°C in June. The effect of bioturbation by beaver in EW3 in 1992 was quite large at times, leading to very high turbidity for short periods of time (Figure 11-14). The summer increase in C* is difficult to see unless this "noisy" data is subjected to a filter, consisting of a 10-day rolling average.

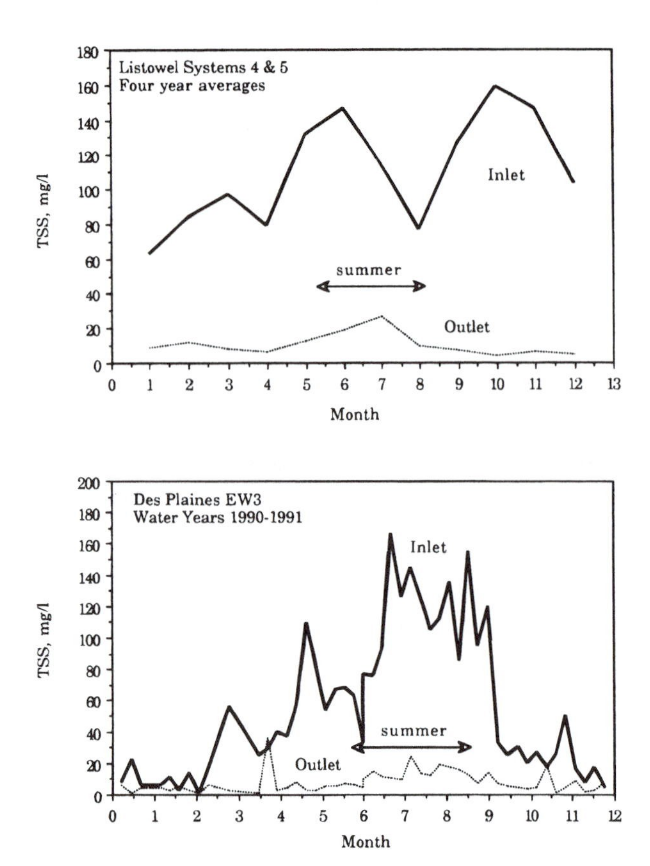

Figure 11-13 Response of SF wetlands to seasonal variables. Water temperature varied from 0 to 27°C at Des Plaines and from 0 to 19°C at Listowel.

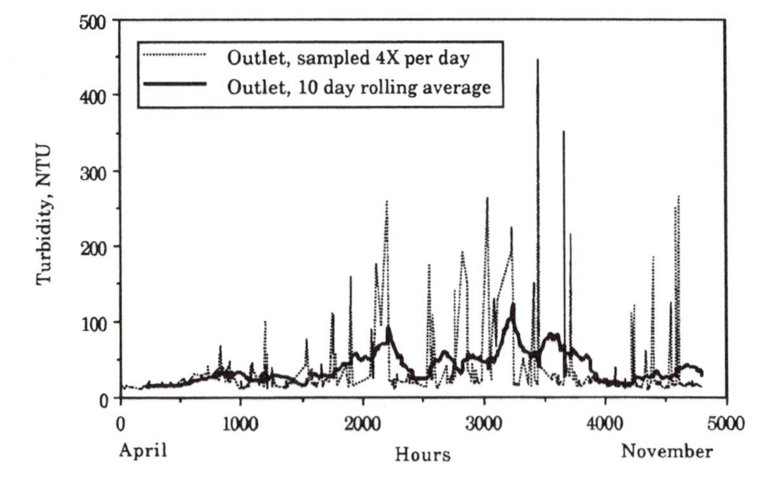

Figure 11-14 Variability in turbidity for wetland EW3 at Des Plaines. For this system, TSS ≈ NTU. Rolling averaging smooths out the sharp spikes, which were partly the result of beaver activity in the wetland.

PARTICULATE PROCESSES IN SUBSURFACE-FLOW WETLANDS

The fact of a subsurface air-water interface causes sediment processing in the SSF wetland to differ considerably from that in SF wetlands. Macrophyte leaf and seed litter is mostly contained on the surface of the bed and does not interact with the water flowing in the interstices below. Most vertebrates and invertebrates do not interact with the water. Resuspension is not caused by wind or vertebrate activities.

However, many particulate processes do operate in the water-filled voids. Particles settle into stagnant micropockets or are strained by flow constrictions (Figure 11-15). They may also impinge upon substrate granules and stick as a result of several possible interparticle adhesion forces. These physical processes are termed *granular medium filtration* in the literature (Metcalf and Eddy, 1991). Higher velocities can dislodge adhering or deposited material, which forms the basis for the backwashing method of filter regeneration.

Generation of particulate material can occur via all the mechanisms shown for SF wetlands. Below-ground macrophyte parts—roots and rhizomes—die, decay, and produce fine detrital fragments. Many other organisms are present in the bed that can contribute to TSS via the

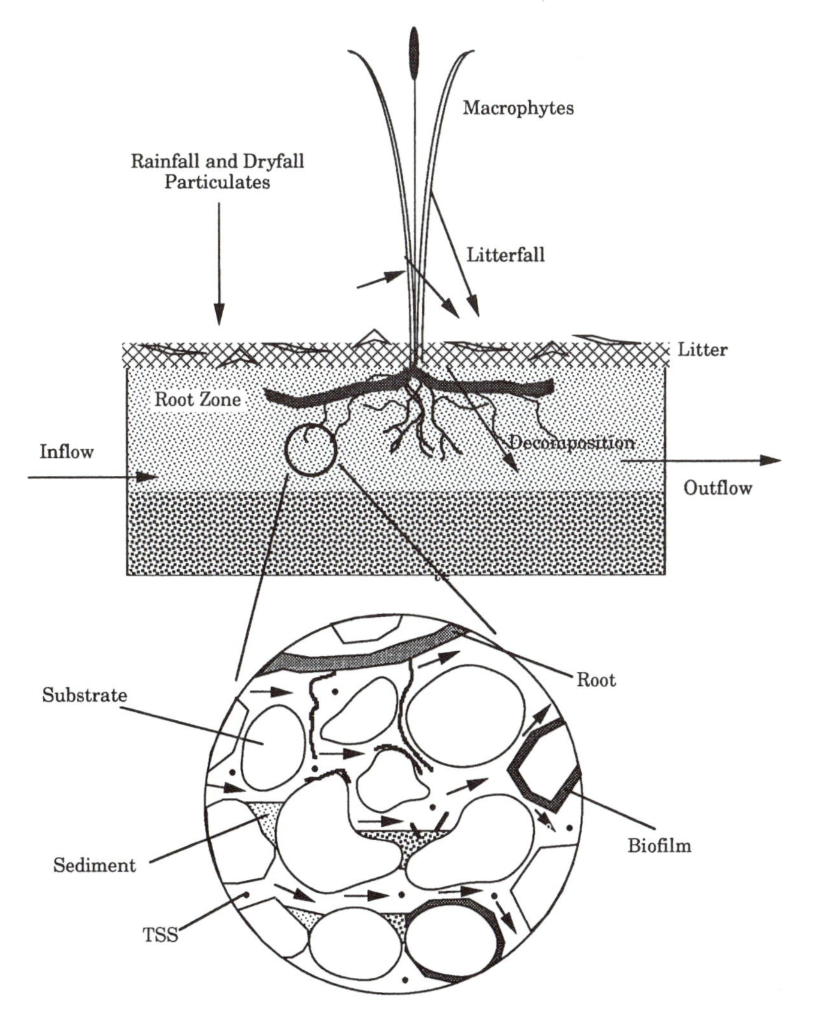

Figure 11-15 Particulates in the SSF environment. Generation is via chemical precipitation, biofilm decomposition, and root litterfall. Settling is in crevices; resuspension can be from pockets or biofilm sloughing.

Table 11-2 Biological Assay of Bed Solids and Litter, SSF Cell 3, Benton, KY

	Distance, m =	Inlet 2	58	93	156	279	Outlet 330
Bacteria millions of cfu/ml	Litter layer	4.40	41.00				
	Small rock, top		1.10	0.47	0.260	0.140	0.39
	Middle depth		0.60				
	Large rock, bottom		4.30	0.85	0.250	0.180	
Algae millions/ml	Litter layer	5.20	16.00				
	Small rock, top		0.46	0.094	0.130	0.076	0.032
	Middle depth		0.57				
	Large rock, bottom		5.40	0.042	0.076	0.015	
Fungi thousands of cfu/ml	Litter layer	0.90	100.00				
	Small rock, top		2.00	3.20	1.00	0.80	10.00
	Middle depth		5.00				
	Large rock, bottom		80.00	7.00	0.80	4.00	

Note: Data from Kadlec and Watson, 1993.

same route: algae, fungi, and bacteria all die and contribute particulate matter to the water flowing in the pore space. These microorganisms are unevenly spatially distributed within the gravel bed with more organisms located near the inlet and near the bottom (Bavor et al., 1988; see also Table 11-2).

There has been no study of internal solids transfer and generation processes in SSF wetlands, due to the near-impossibility of measurements below the gravel surface. However, transect measurements of TSS in SSF systems display features much like those in SF wetlands: a sharp decline, followed by little further change (Figure 11-16).

It is postulated that removal mechanisms are more numerous (settling plus straining) and more effective (shorter settling distances) in the gravel media.

REGRESSION EQUATIONS

Presently, insufficient data exist to support a regression equation for the initial removal of TSS in the bed inlet zone, nor are there mechanistic estimates such as those based on particle settling in SF systems. Accordingly, only input-output regressions can be offered at this point in time. Outputs probably reflect a balance between removal, resuspension, and

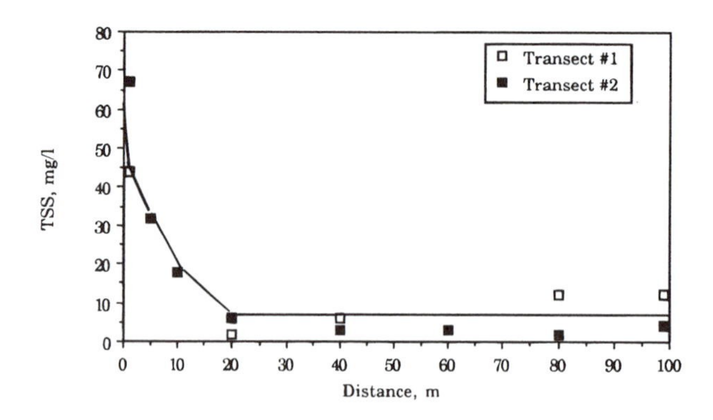

Figure 11-16 Profile of TSS through a gravel bed wetland vegetated with cattail. (Adapted from Bavor et al., 1988.)

generation, as in the SF wetlands. Regression of NADB information from 22 SSF wetlands produces the following correlation:

$$C_o = 7.8 + 0.063\ C_i \qquad (11\text{-}18)$$

$R^2 = 0.09$, $N = 107$
Standard error in $C_o = 10$
$0.1 < C_i < 253$ mg/L
$0.0 < C_o < 52$ mg/L

The mean hydraulic loading rate for this data was 12 cm/d.

Similar conclusions have been reached for soil-based SSF wetlands (Brix, 1994). Regression of data from 77 wetlands in the U.K. and Denmark produced

$$C_o = 4.7 + 0.09\ C_i \qquad (11\text{-}19)$$

$R^2 = 0.67$, $N = 77$
$0 < C_i < 330$ mg/L
$0 < C_o < 60$ mg/L

The soils in this group are mostly sands, with a finer grain size than the gravel beds in the NADB. The mean hydraulic loading rate for those data was 6 cm/d.

The corresponding removal efficiencies are given in Table 11-1.

Temperature and seasonal effects appear to be nil for SSF wetlands, although the database is not large. No effect can be found in the data from Richmond, NSW (Bavor et al., 1988) or in the data from Kalø, Denmark (Schierup et al., 1990b).

CLOGGING

The deposition of TSS within the SSF wetland produces the potential for bed clogging, especially near the inlet end of the system. The volatile portion of the TSS may partially decompose to gas and soluble matter, but the mineral portion of the deposit can accumulate. This can lead to pore blockage and a reduction in the hydraulic conductivity of the bed.

Root growth reduces the available pore space in SSF wetlands. Studies on beds with bulrushes have shown that roots and rhizomes are typically located in the upper 30 cm of the bed (U.S. EPA, 1993b). *Phragmites* roots and rhizomes have been reported to penetrate further in some instances (Gersberg et al., 1986), but other investigations show only 20 to 40 cm penetration (Schierup et al., 1990b; Saurer, 1994). The below-ground biomass of *Phragmites* is on the order of 2000 g/m^2, which approximates a quarter of the void volume in a 30-cm root zone.

The end result of subsurface biological and vegetative activity is the buildup of solids within the pore spaces of the media. That buildup is larger near the inlet and near the top of the bed (Tanner and Sukias, 1994; Kadlec and Watson, 1993). A significant portion of the pore volume can be blocked by accumulated organic matter, leading to increased hydraulic gradients and decreased retention times (Tanner and Sukias, 1994). The deposits consist of low-density biosolids together with fine mineral particulates, which can have a very low bulk density.

Tanner and Sukias (1994) measured organic depositions in the inlets of *Schoenoplectus* gravel beds on the order of 5 kg/m^2 over a 2-year period, exclusive of live and dead roots and rhizomes. The tracer detention time at the end of the period was about half the nominal detention time, suggesting that about 50 percent of the voids were blocked. The bulk density of the organic materials was measured to be approximately 0.2 g/cm^3. These organic deposits would block about 18 percent of the voids; the balance of the blockage was attributable to roots, rhizomes, and inorganic materials. Other field data indicate total void blockage accrual up to 10 percent per year, including root growth (Thut, 1989).

Estimates may be made of the potential accumulation rate of solids in SSF systems. Conley et al. (1991) estimate a service life of 100 years. However, these calculations were based on a solids density of 2.65 g/cm^3 (Tanner and Sukias, 1994). If live roots and rhizomes block a third of the pores, and the density is 0.2 g/cm^3, this estimate drops to less than 5 years.

Accretion of organic solids is only weakly correlated with organic loading to the wetland because of the strong contribution of plants to the total organic loading (Figure 11-17). However, there are strong distance profiles in vegetated gravel beds, with higher accretions occurring near the inlet and near the top of the bed. This implies that front end flooding is a possible result of solids accretion.

The decline of hydraulic conductivity associated with solids accumulation is magnified by the extreme sensitivity of hydraulic conductivity to void fraction, which is about a fifth-power dependence (see Chapter 9). For instance, a 2 percent void blockage can result in a 10 percent conductivity drop.

POND-WETLAND COMBINATIONS

Because incoming TSS is rapidly settled and filtered in the wetland environment, it is possible and desirable to provide a first element of the treatment wetland complex that traps the fastest settling fraction of the suspended material. A pond provides for that presettling and is more easily cleaned than an emergent macrophyte bed. It is further desirable to collect solids and their partitioned metals and chemicals in a location that is not foraged by sediment-feeding vertebrates. This presettling pond may require infrequent dredging to remove the accumulated deposits.

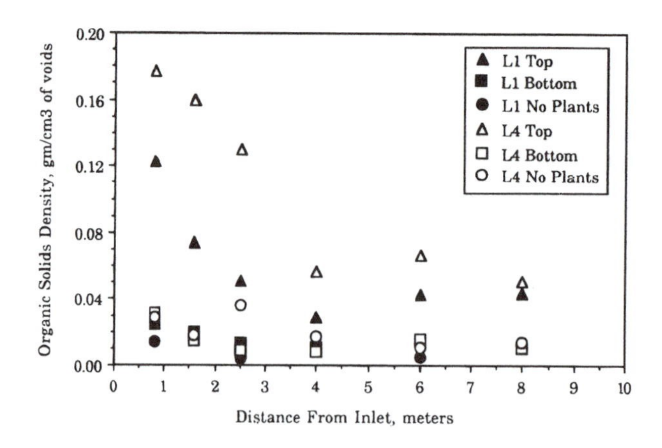

Figure 11-17 Solids concentration in the interstices of SSF wetlands. Solids do not include live or dead roots or rhizomes. Loading L1 was approximately 20 kg/ha/d (0.032 kg/m^2/d on the flow cross-section). Loading L4 was approximately 65 kg/ha/d (0.032 kg/m^2/d on the flow cross-section). Top refers to the top 10 cm; bottom refers to the bottom 30 cm. (Replotted from Tanner and Sukias, 1994.)

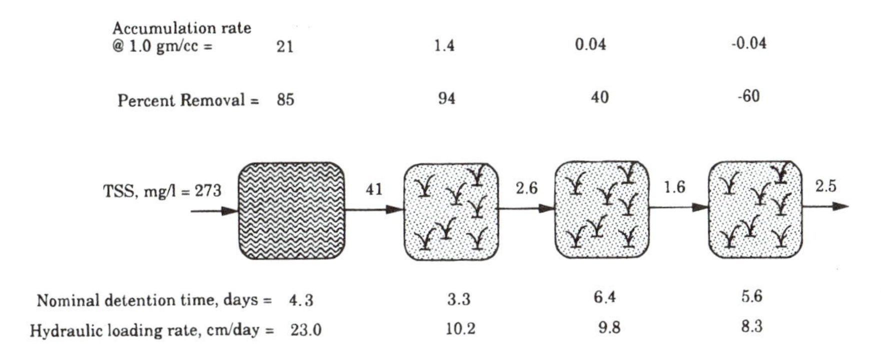

Figure 11-18 Performance of settling ponds preceding macrophyte cells in the Tarrant County, TX system. Values shown are averages for the system over an 8-week winter period. (Data from APAI, 1994.)

The performance of such pond-wetland complexes has been studied in the Tarrant County, TX system (APAI, 1994). Figure 11-18 illustrates the mean performance of three parallel marsh wetland cell trains of three cells each, following two parallel unvegetated settling ponds. The settling ponds occupy 15 percent of the area, but accounted for 86 percent of the solids removal. The first wetland cell completes the solids removal; the remaining two cells do not reduce TSS any further. The last wetland cells are, however, needed for phosphorus removal. The buildup of sediments in the ponds is relatively rapid (21 cm/yr if bulk density = 1.0 g/cm^3), and this implies that the settling ponds will require cleaning every few years. This is a desirable trade-off against dredging the first wetland cell, which would entail a period of re-establishment of vegetation.

A settling pond will not achieve the efficiency of a laboratory settling column because of flow nonidealities. Efficiencies are likely to be in the range of 60 to 80 percent (Perry and Green, 1984).

The placement of a pond as the final element in a wetland treatment system is generally not desirable from the standpoint of TSS reduction. The planktonic production in such a pond is typically quite high, leading to the reintroduction of high-chlorophyll microdetritus, much of which remains in suspension. An example of this phenomena is the Lakeland, FL system (Figure 11-19). Entering TSS is reduced in the first marsh cells, but is regenerated in later cells, because of planktonic activity (Bays, 1994).

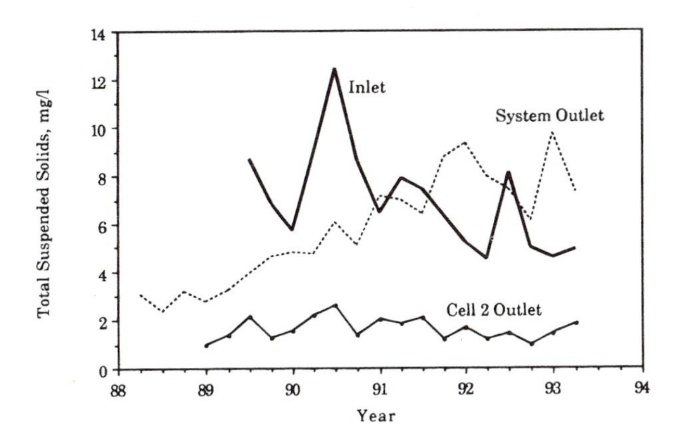

Figure 11-19 Suspended solids entering and leaving the Lakeland, FL treatment wetland. Cells 1 and 2 are marsh ecosystems; the final cells contain large proportions of open water. (Data from NADB, Knight et al., 1993b.)

DESIGN EXAMPLES

The settling characteristics of the solids in the incoming water are often unknown. In that event, it is necessary to extrapolate from known performance characteristics and apply a safety factor. This will usually not be unduly punitive, since TSS removal is often not the controlling factor in wetland design.

Example 11-2

An SSF wetland has been designed to operate at a hydraulic loading rate of 10 cm/d in order to reduce BOD to an acceptable level. The incoming TSS is 150 mg/L. What effluent TSS can be expected?

Solution

This loading rate is slightly lower than the mean for the data which support Equation 11-18. The inlet TSS is also within range for this regression. Therefore,

$$C_o = 7.8 + 0.063 \cdot 150 = 17.25 \qquad (11\text{-}18)$$

The estimated effluent TSS is 17 mg/L. This represents a removal efficiency of

$$\%\ \text{removal} = 100 \cdot \frac{150 - 17}{150} = 89\%$$

Since this is close to the average median removal, 84 percent, in Table 11-1, the estimate is confirmed.

The settling velocity may be measured for the particular water to be treated. If this information is available, the size of the wetland for TSS removal can be optimized.

Example 11-3

An SF wetland is to be designed to reduce incoming TSS from 100 mg/L to no more than 30 mg/L. The solids have been measured to settle at 10 m/d. What hydraulic loading rate is acceptable?

Solution

The design specification does not include temperature information, but summer temperatures can easily reach 25°C even in a temperate climate. The predicted value of C* at 20°C is

$$C^*_{20} = 5.1 + 0.16 \cdot 100 = 21.1\ \text{mg/L} \qquad (11\text{-}17)$$

The temperature factor for the assumed 25°C is

$$C^* = 21.1 \cdot 1.065^{(25-20)} = 21.1 \cdot 1.37 = 28.9\ \text{mg/L} \qquad (11\text{-}18)$$

The wetland may easily be sized to achieve the *long-term average* concentration of 30 mg/L:

$$\ln\left[\frac{30 - 28.9}{100 - 28.9}\right] = -\frac{10}{q} \qquad (11\text{-}15)$$

$$q = 2.4\ \text{m/d}$$

Even with an overall safety factor, the wetland can be loaded up to 1 m/d and achieve the removal goal in the long-term average sense. The exit concentration will be close to C^*. However, stochastic variability will add a probabilistic component to the long-term average, which may be on the order of 50 percent of the mean. Frequent summer violations of a 30-mg/L cap on performance are liable to occur no matter how big the wetland. The influent would need to be nutrient poor to have any chance of meeting the goal.

In fact, this example corresponds closely to the situation at Gustine, CA, which displayed a long-term average effluent TSS from test cells of 29.7 mg/L (Walker and Walker, 1990).

SUMMARY

Treatment wetlands are consistently effective at reducing elevated concentrations of suspended solids. Since most treatment wetlands are overdesigned in terms of TSS reduction, existing data are primarily useful for estimating background TSS outflow concentration variability and are not helpful in estimating an area-based reaction rate constant for this parameter. Rather, design for TSS removal should rely on measured settling rates for specific wastewaters, while taking due caution to recognize the inevitable internal wetland processes that will result in irreducible background TSS concentrations and stochastic variability in response to factors outside the treatment wetland operator's control.

CHAPTER 12

Biochemical Oxygen Demand

INTRODUCTION

Carbon compounds interact strongly with wetland ecosystems. The carbon cycle in wetlands is vigorous and typically provides carbon exports from the wetland to receiving ecosystems. Many internal wetland processes are fueled by carbon imports and by the carbon formed from decomposition processes.

Treatment wetlands frequently receive large external supplies of carbon in the added wastewater. Any of several measures of carbon content may be made, with biochemical oxygen demand (BOD) being the most frequent in the treatment of municipal wastewater. Degradable carbon compounds are rapidly utilized in wetland carbon processes. At the same time, a variety of wetland decomposition processes produce available carbon. The balance between uptake and production provides the carbon exports. In general, the amounts of carbon cycled in the wetland far exceed the quantities added in wastewater.

The growth of wetland plants requires carbon dioxide (CO_2) for photosynthesis. A variety of organisms release CO_2 as a product of respiration. Many pathways lead to the microbial production of CO_2, as well as methane (CH_4). Both gases dissolve in water to a limited extent, so there are active transfers of carbon to and from the atmosphere.

In terms of treatment, it is therefore not surprising to find good carbon reductions for the water added accompanied by non-zero background levels of various carbon compounds and the related BOD. For purposes of wetland design for BOD removal, the problem is to find relatively simple models for an enormously complex set of wetland functions.

GENERAL CONCEPTS

MEASURES OF CARBON CONTENT

In water, a wide spectrum of carbon compounds exists in either dissolved or particulate forms. The usual dividing line is a 0.45-μm filter. The following distinctions are made as a result of analytical methods:

- TC = total carbon, including all dissolved and suspended forms
- PC = particulate carbon, including organic and inorganic forms
- DC = dissolved carbon, including organic and inorganic forms
- IC = inorganic carbon, including all dissolved and suspended forms
    - DIC = dissolved inorganic carbon, usually comprised of carbon dioxide, carbonate, and bicarbonate
- TOC = total organic carbon, including all dissolved and suspended forms
    - DOC = dissolved organic carbon
    - NDOC = nondissolved organic carbon
- VOC = volatile organic carbon

In soils or biomass, samples are subjected to combustion and dissolution, followed by analysis for total carbon.

BOD, COD, AND TOC

Different analytical techniques are used to measure the amount of organic material in the wastewater. Biochemical Oxygen Demand (BOD) is a measure of the oxygen consumption of microorganisms in the oxidation of organic matter. It is measured as the oxygen consumption in an air-tight incubation of the sample. This test normally runs for 5 days, and the result is then more properly designated as BOD_5. Some oxygen may be used in nitrification if the necessary organisms are present in the sample. If this potential nitrogenous oxygen demand is inhibited chemically during the test, the result is carbonaceous biochemical oxygen demand, $CBOD_5$.

Chemical Oxygen Demand (COD) is the amount of a chemical oxidant, usually potassium dichromate, required to oxidize the organic matter. This measure is larger than BOD because the strong oxidant attacks a larger group of compounds. Either may be measured before or after filtration, leading to measures of total and soluble BOD and COD. In the wetland environment, the presence of humic materials leads to COD values that are much larger than BOD values. In northern peatlands, the ratio is approximately BOD_5 = 5 mg/L to COD = 100 mg/L (unpublished data from the Houghton Lake peatland). In municipal wastewaters, the ratio is typically 0.4–0.8 (Metcalf and Eddy, 1991). Industrial wastewaters may have lower ratios.

Total Organic Carbon (TOC) is measured by chemical oxidation followed by analysis for CO_2. In northern peatlands, the ratio BOD_5 to TOC is approximately BOD_5 = 5 mg/L to TOC = 25 mg/L (unpublished data from the Houghton Lake peatland). In municipal wastewaters, the ratio is 1.0–1.6 (Metcalf and Eddy, 1991). The interpretation of these ratios is that natural wetlands cycle at low levels of biologically usable carbon compounds, whereas municipal wastewaters are rich in usable carbon compounds.

Wetlands are efficient users of external carbon sources, manifested by excellent reductions in BOD_5 and COD. However, wetlands possess non-zero background levels of both BOD and COD, which depend on the type and status of the wetland. Typical ranges for background concentrations are 1 to 6 mg/L for BOD_5 and 30 to 100 mg/L for COD.

WETLAND WATER CHEMISTRY OF CARBON

Inorganic Carbon

Of the hundreds of carbon compounds which may occur in the wetland environment, relatively few are inorganic. Dissolved inorganic carbon consists primarily of carbon dioxide, carbonate, and bicarbonate.

In pure water solution, the principal carbonate species are related to atmospheric CO_2 by the temperature- and pH-dependent dissolution and dissociation series:

Henry's Law:

$$H_2CO_3^* \Leftrightarrow H_2O + CO_2\,(gas) \qquad K_H = \frac{[H_2CO_3^*]}{P_{CO_2}} \tag{12-1}$$

where

$$[H_2CO_3^*] = [H_2CO_3] + [CO_2] \tag{12-2}$$

Hydration

$$H_2CO_3 \Leftrightarrow H_2O + CO_2 \qquad K = \frac{[CO_2]}{[H_2CO_3]} \tag{12-3}$$

First dissociation

$$H_2CO_3 \Leftrightarrow HCO_3^- + H^+ \qquad K_{H_2CO_3} = \frac{[HCO_3^-][H^+]}{[H_2CO_3]} \tag{12-4}$$

Second dissociation

$$HCO_3^- \Leftrightarrow CO_3^= + H^+ \quad K_2 = \frac{[CO_3^=][H^+]}{[HCO_3^-]} \tag{12-5}$$

and where, as a result of Equation 12-2,

$$K_1 = \frac{K_{H_2CO_3}}{K + 1} \tag{12-6}$$

The notation of Pankow (1991) has been adopted. Brackets indicate the concentration of the chemical species, in molarity, and all are in water except for atmospheric carbon dioxide. The value of the equilibrium constant $K \approx 650$, and hence, most of the dissolved carbon dioxide is present as CO_2. Equations 12-1 through 12-6 may be solved for concentrations, given the partial pressure of CO_2 and the various equilibrium constants.

$$[H_2CO_3^*] = K_H P_{CO_2} \tag{12-7}$$

$$[HCO_3^-] = \frac{K_1}{[H^+]} K_H P_{CO_2} \tag{12-8}$$

$$[CO_3^=] = \frac{K_1 K_2}{[H^+]^2} K_H P_{CO_2} \tag{12-9}$$

The equilibrium constants, and hence the various concentrations, are all pH and temperature dependent (Table 12-1). These forms are distributed in water at 25°C as shown in Figure

Table 12-1 Temperature Dependence of the Carbonate System Equilibrium Constants

Temp. (°C)	log K_1	log K_2	log K_H (M/atm)
0	−6.58	−10.63	−1.11
5	−6.53	−10.56	−1.19
10	−6.46	−10.49	−1.27
15	−6.42	−10.43	−1.32
20	−6.38	−10.38	−1.41
25	−6.35	−10.33	−1.47
30	−6.33	−10.29	−1.53
35	−6.31	−10.25	—
40	−6.30	−10.22	−1.64
50	−6.29	−10.17	−1.72

From Pankow, J. F. 1991. *Aquatic Chemistry Concepts*. Chelsea, MI, Lewis Publishers. With permission.

12-1 (Pankow, 1991). However, it must be noted that wetland waters are more complex than the pure water system and therefore will not follow such idealized chemistry precisely. Modifications of the calculation (APHA, 1992) deal with expected deviations due to dissolved solids, but not the full suite of biological variations. Production and consumption of carbon dioxide in the wetland may significantly alter the chemical balance in the water.

A variety of cations can precipitate carbonates under certain conditions. The most important is calcium carbonate, $CaCO_3$. A major process in periphyton-dominated wetlands is chemical precipitation of $CaCO_3$ under conditions of high pH created by the algae (Gleason, 1972).

Analytical methods determine total alkalinity and phenophthalein alkalinity, from which carbonate, bicarbonate, and hydroxyl alkalinity are calculated.

A variety of cations can precipitate carbonate under certain conditions. Some important mineral precipitates in the wetland environment are

Calcite	$CaCO_3$
Aragonite	$CaCO_3$
Magnesite	$MgCO_3$
Dolomite	$CaMg(CO_3)_2$

The overall C-mineral chemistry is very complex; consequently, accurate calculations of solubilities are generally not possible. Calcium carbonate saturation indices may be calculated in a number of ways (APHA, 1992).

Organic Carbon

Biomass: Growth, Death, Decomposition. The general concepts of plant growth have been set forth in Chapter 7. In terms of carbon cycling within the wetland, there is a strong correlation between biomass and carbon content. The carbon content of wetland macrophytes is typically about 41 percent on a dry weight (dw) basis.

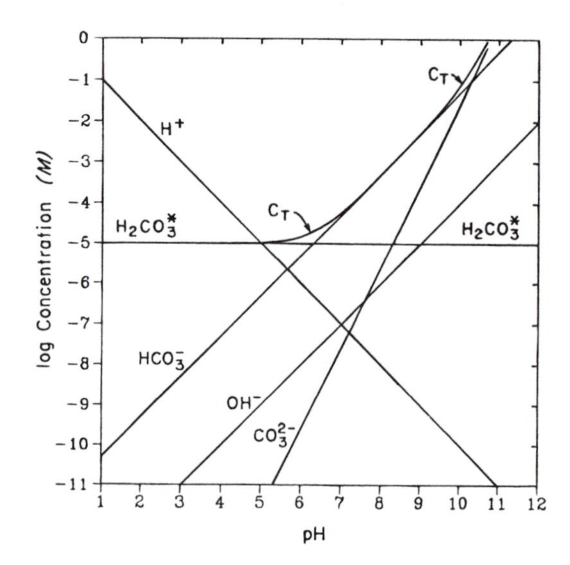

Figure 12-1 Distribution of carbonate species in water at 25°C. The partial pressure of CO_2 in the air is taken as 3.16×10^{-4} atm. (From Pankow, J.F. 1991. *Aquatic Chemistry Concepts.* Chelsea, MI: Lewis Publishers. With permission.)

The wetland cycle of growth, death, and partial decomposition uses atmospheric carbon and produces gases, dissolved organics, and solids (Figure 12-2). Decomposition involves the sugars, starches, and low molecular weight celluloses in the dead plant material. Gaseous products include methane and regenerated carbon dioxide. A spectrum of soluble, large, organic molecules, collectively termed humic substances, are released into the water. The solid residual of plant decomposition is peat or organic sediment, which originated as celluloses and lignins in the plants. These wetland soil organics are broadly classified as fulvic material, humic material, and humin, based upon whether they are acid soluble, base soluble, or insoluble (Peat Testing Manual, 1985).

The internal wetland carbon cycle is large. A general idea of the magnitudes of the various carbon transfers in a northern treatment marsh may be gained from considering the annual growth and decomposition patterns. A eutrophic treatment marsh grows about 3000 g/m^2 of above-ground dw biomass each year with a carbon content of about 41 percent. This translates to a requirement for 1230 $g/m^2/yr$ (34 kg/ha/d) of carbon. Decomposition of the resultant litter returns a significant portion of that carbon to the wetland ecosystem.

CARBON PROCESSING IN WETLAND SOILS

A rough representation of the decomposition "reactions" may be set down (Mitsch and Gosselink, 1993; Burgoon, 1993). These occur in different horizons in the wetland, as indicated in Figure 12-3.

Respiration occurs in aerobic zones:

$$\underset{\text{carbohydrates}}{C_6H_{12}O_6} + 6\,O_2 = 6\,CO_2 + 6\,H_2O \qquad (12\text{-}10)$$

Fermentation occurs in anaerobic zones:

$$\underset{\text{carbohydrates}}{C_6H_{12}O_6} = 2\,\underset{\text{lactic acid}}{CH_3CHOHCOOH} \qquad (12\text{-}11)$$

$$\underset{\text{carbohydrates}}{C_6H_{12}O_6} = 2\,\underset{\text{ethanol}}{CH_3CH_2OH} + 2\,CO_2 \qquad (12\text{-}12)$$

Methanogenesis occurs in anaerobic zones:

$$4H_2 + CO_2 = CH_4 + 2\,H_2O \qquad (12\text{-}13)$$

$$\underset{\text{acetate}}{CH_3COO^-} + 4\,H_2 = 2\,CH_4 + H_2O + OH^- \qquad (12\text{-}14)$$

Sulfate reduction occurs in anaerobic zones:

$$2\,\underset{\text{lactate}}{CH_3CHOHCOO^-} + SO_4^= + H^+ = 2\,\underset{\text{acetate}}{CH_3COO^-} + 2\,CO_2 + 2\,H_2O + HS^- \qquad (12\text{-}15)$$

$$\underset{\text{acetate}}{CH_3COO^-} + SO_4^= + 2\,H^+ = 2\,CO_2 + 2\,H_2O + HS^- \qquad (12\text{-}16)$$

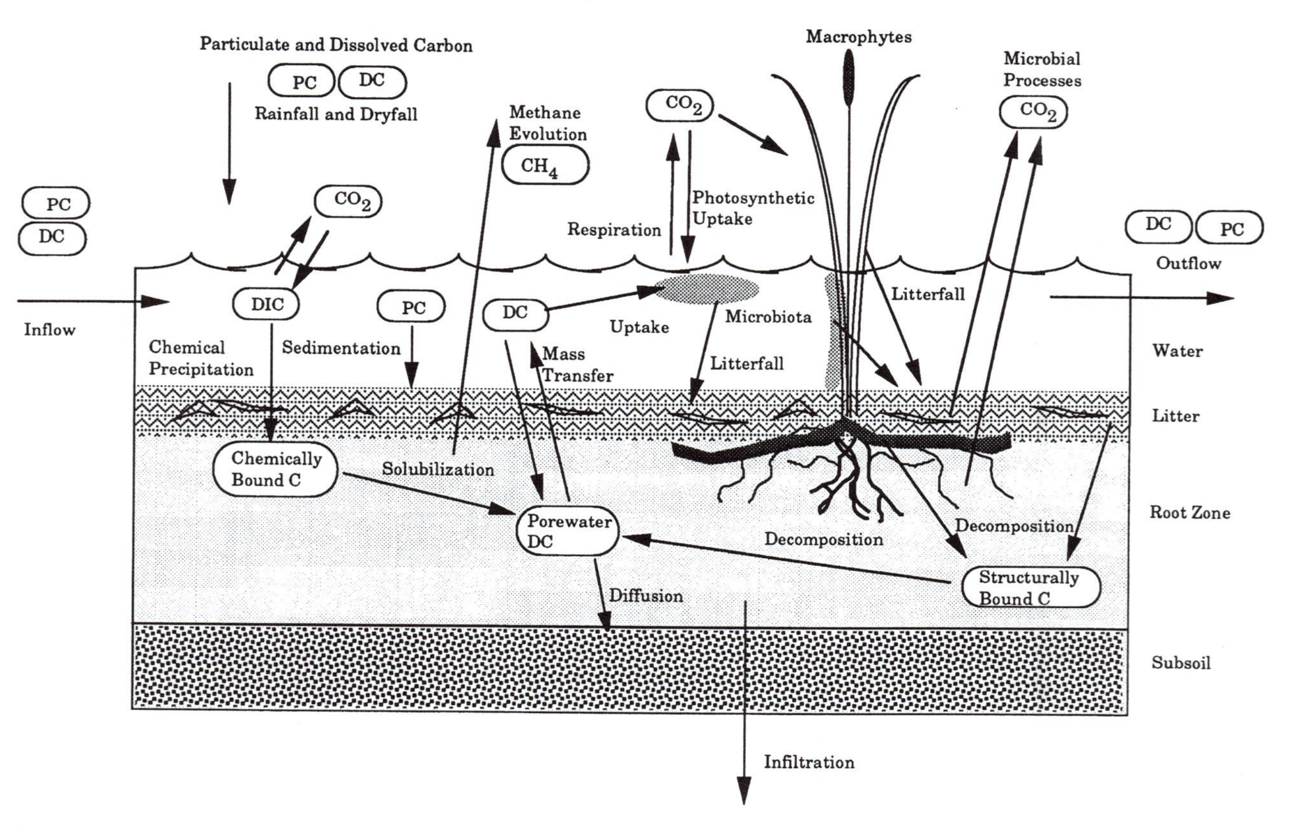

Figure 12-2 Carbon storages and transfers in the wetland environment. DC = dissolved carbon, PC = particulate carbon, DIC = dissolved inorganic carbon, DOC = dissolved organic carbon, CH_4 = methane, and CO_2 = carbon dioxide. Biomass carbon consists of living and dead biomass and organic decomposition products.

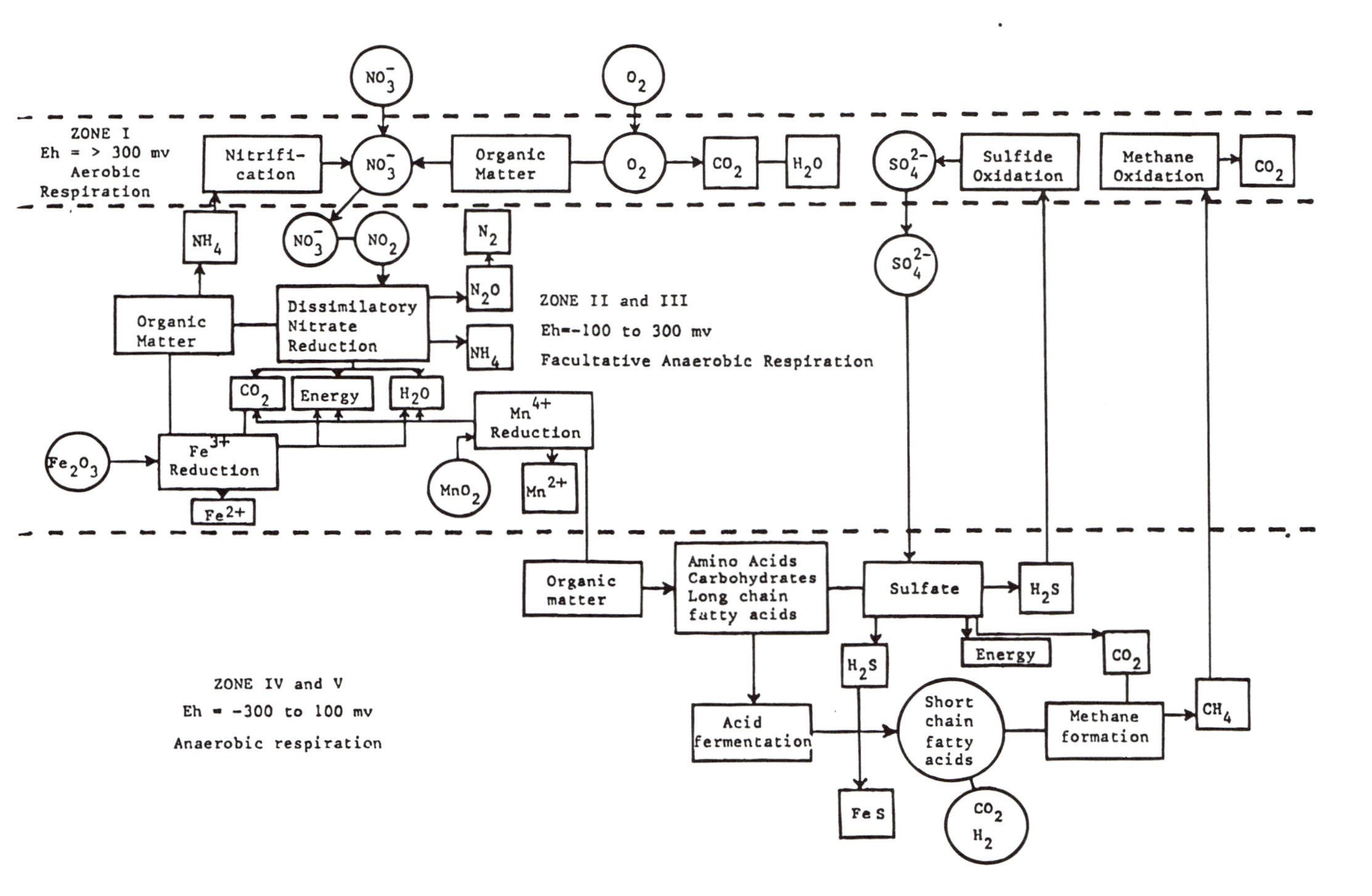

Figure 12-3 Pathways of organic carbon decomposition in wetland soils. Aerobic, facultative anaerobic, and obligate anaerobic processes are all typically present at different depths in the soil. (From Reddy, K.R. and D. A. Graetz. 1988. *The Ecology and Management of Wetlands.* London: Croom Helm. With permission.)

Nitrate reduction (denitrification) occurs in anaerobic zones:

$$\underset{\text{carbohydrate}}{C_6H_{12}O_6} + 4\ NO_3^- = 6\ CO_2 + 6\ H_2O + 2\ N_2 + 4\ e^- \qquad (12\text{-}17)$$

Iron reduction occurs in anaerobic zones:

$$\underset{\text{acetate}}{CH_3COO^-} + 8\ Fe^{+++} + 3\ H_2O = 8\ Fe^{++} + CO_2 + HCO_3^- + 2\ H_2O + 8\ H^+ \qquad (12\text{-}18)$$

The relative percentages of these reactions were investigated in controlled subsurface-flow (SSF) wetland microcosms by Burgoon (1993) using acetate as the carbon source. His results (Table 12-2) demonstrate that all routes can be important, depending upon physical and chemical conditions.

It is apparent that the wetland provides a spectrum of potential pathways for the utilization of organic carbon compounds. Sufficient information is not available to quantify both the complex chemistry and the spatial distribution of chemical compounds. Therefore, the interactions must be described via correlations and rate equations which are supportable by wetland performance data.

OVERALL INPUT/OUTPUT CORRELATIONS

The most commonly controlled parameter in wastewater is total BOD_5, and this will be described here. However, the same principles apply to COD and soluble BOD_5.

At the black-box level of data interpretation, correlations may be made between overall system operating variables. The most important are concentrations in the entering and departing flows (C_i and C_o), the flow rates in and out (Q_i and Q_o), the size of the wetland (A_T), and the hydraulic loading rate (q = HLR). Detention time is an alternative to hydraulic loading rate ($\tau = q/\epsilon h$).

Table 12-2 Percent Acetate Oxidized Via Various Pathways

	High Carbon Loading		Low Carbon Loading	
	Nitrate-Rich Environment			
Reaction	Plants	No Plants	Plants	No Plants
Oxidation	23.2	25.6	36.1	32.8
Nitrate reduction	70.6	69.3	51.7	56
Sulfate reduction	3.0	3.1	2.3	2.3
Ferric iron reduction	0.1	0.0	0.1	0.1
Methane formation	0.0	0.0	0	0
Bacterial biomass formation	3.1	2.0	9.8	8.8
Total	100.0	100.0	100.0	100.0
	Sulfate-Rich Environment			
Reaction	Plants	No Plants	Plants	No Plants
Oxidation	40.7	31.7	44.5	13.5
Nitrate reduction	0	0	0	0
Sulfate reduction	37.8	34.1	50.6	82.7
Ferric iron reduction	0.1	0.1	0.2	0.2
Methane formation	19.6	32.1	0	0
Bacterial biomass formation	1.8	2	4.7	3.6
Total	100.0	100.0	100.0	100.0

Note: Scirpus validus was planted in plastic media. Data from Burgoon, 1993.

Combinations of the principal variables may also be made and used in correlations. The most frequently used of these are hydraulic loading rate (q_i or q_o) and the loading rate ($LI = q_iC_i$ or $LO = q_oC_o$). Also in use is the removal rate, $RR = q_iC_i - q_oC_o$. Percentage reductions may be based on concentration units, $(C_i - C_o)/C_i$; or upon mass units, $(q_iC_i - q_oC_o)/q_iC_i$.

Two factors determine the utility of data correlations based on black-box information: how much of the variablility in the data is represented (eg., R^2 value) and the range of that data. Correlations may not be safely used outside of the range of the data that produced them.

There are not large differences in degree of fit among correlations that use different variable combinations from the previous list. There are differences based on wetland type.

REGRESSION EQUATIONS

Surface-Flow Wetlands

Data from 45 SF wetlands are presented in Figure 12-4. Most data were averaged seasonally, over 3-month periods, or quarterly. A total of 440 data points were included, most quarterly averages, with some monthly averages in cases where quarterly averaging was inappropriate. For inlet $BOD_5 \geq 10$ mg/L, the linear correlation of inlet and outlet concentrations is

SF BOD_5 correlation $$C_o = 0.173\ C_i + 4.70 \tag{12-19}$$

$R^2 = 0.62$, $N = 440$
Standard error in $C_o = 13.6$
$0.27 < HLR < 25.4$ cm/d
$10 < C_i < 680$ mg/L
$0.5 < C_o < 227$ mg/L

Adding the hydraulic loading rate to the linear correlation does not improve it, nor is a logarithmic correlation better.

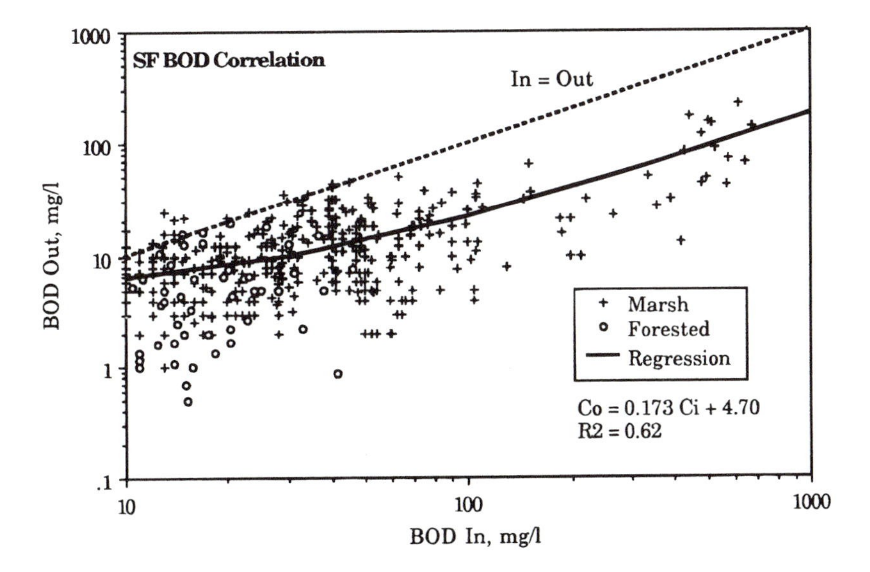

Figure 12-4 Input-output BOD_5 relation for SF wetlands.

When the inlet BOD_5 drops below 10 mg/L, there is no correlation of outlet BOD_5 with either inlet BOD_5 or hydraulic loading rate (Figure 12-5).

Many wetland variables are implicit in Equation 12-19: vegetation type, water depth, climate, wetland size and shape, and flow rates all differ from wetland to wetland. Emergent marshes and forested wetlands are included; SSF wetlands are excluded.

Subsurface-Flow Wetlands

Data from 73 SSF soil-based wetlands, one point each representing average performance, was correlated by Brix (1994b) in the same manner as for SF wetlands. These systems were reed beds (*Phragmites* spp.) in Denmark and the U.K. The average hydraulic loading rate was approximately 5 cm/d. The correlation of inlet and outlet concentrations was (Figure 12-6)

Soil-based SSF BOD_5 correlation $\quad C_o = 0.11\ C_i + 1.87 \quad$ (12-20)

$$R = 0.74,\ N = 73$$
$$1 < C_i < 330 \text{ mg/L}$$
$$1 < C_o < 50 \text{ mg/L}$$
$$0.8 < q < 22 \text{ cm/d}$$

The North American Database contains information on 24 SSF wetlands at 19 sites. Data were averaged quarterly for most wetlands; some were reported at other frequencies. Regression of this information gives

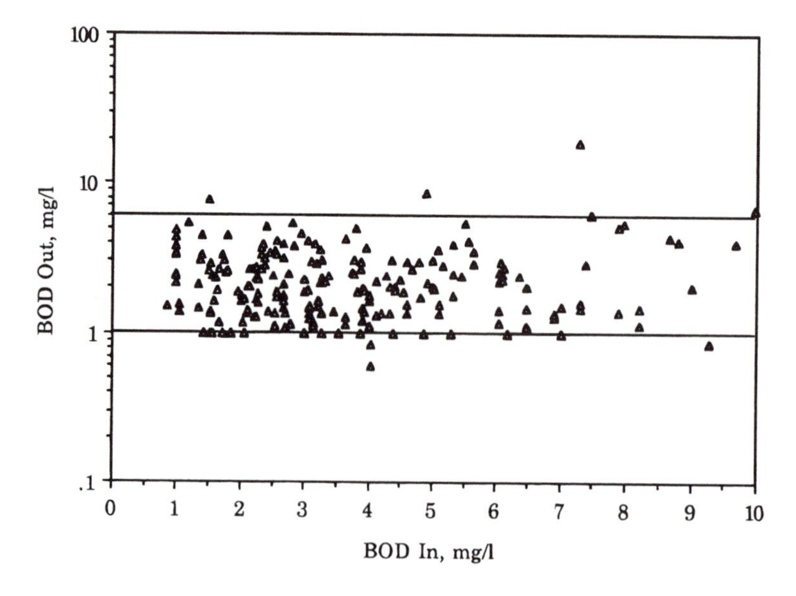

Figure 12-5 Outlet BOD_5 for U.S. SF wetlands receiving less than 10 mg/L of input BOD_5. There is no trend with input BOD_5 and most of the data lies between 1 and 6 mg/L; 1 mg/L is close to the minimum detection limit for the analytical procedure.

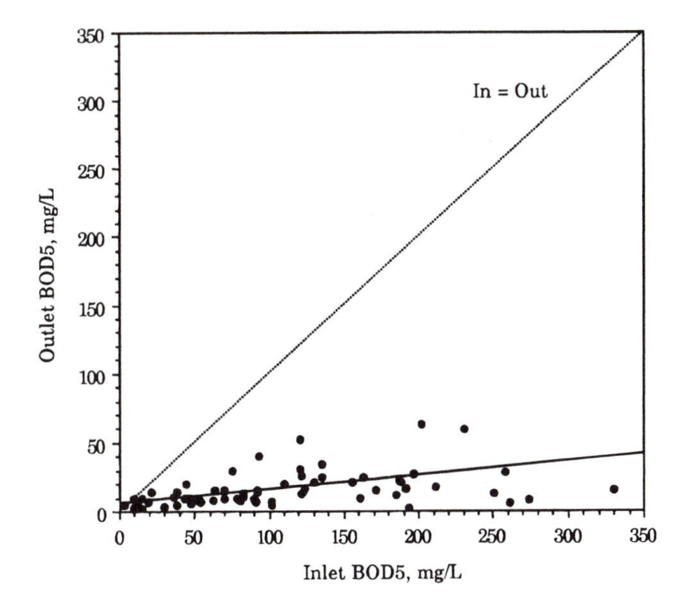

Figure 12-6 Input-output BOD_5 relation for soil-based SSF wetlands. Data shown are for 69 Danish systems. (Data updated from Brix, 1994b; additional data courtesy of H. Brix.)

U.S. gravel bed SSF BOD_5 correlation $\quad C_o = 0.33\ C_i + 1.4 \quad$ (12-21)

$R^2 = 0.48$, $N = 100$
Standard error in $C_o = 5.0$
$1 < C_i < 57$ mg/L
$1 < C_o < 36$ mg/L
$1.9 < q_{avg} < 11.4$ cm/d

Adding the hydraulic loading to the correlating variables produces no improvement, nor does restricting the inlet BOD_5 to more than ten.

INTERNALIZATION

Internal behavior of BOD_5 concentrations in a wetland may not be projected from correlations such as Equations 12-19 through 12-21 because the hydraulic loading rate is not involved. It is informative to examine internal profiles to verify a mass balance model that includes rate equations.

MASS BALANCE DESIGN MODEL

RATIONALE

The framework of the water mass balance provides the information necessary to calculate the interactions of BOD_5 with the wetland as the water passes through the ecosystem. Removal mechanisms operate via a number of different paths, and the wetland sediments and litter decompose to form BOD_5 (see Chapter 9). Mass accounting for water and BOD_5 requires a

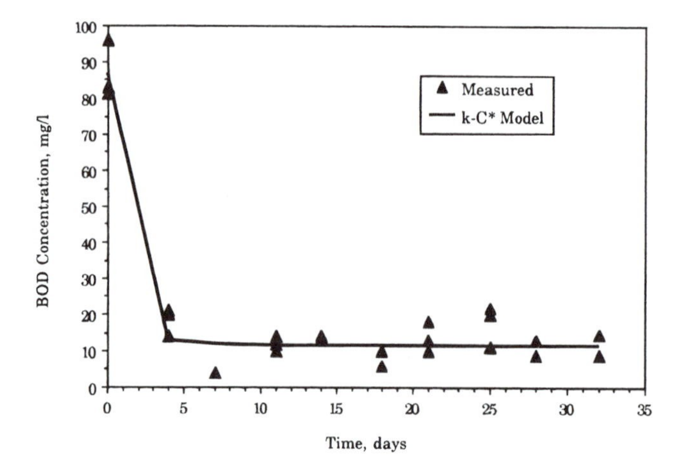

Figure 12-7 The progression of BOD concentrations in three wetlands operated in the batch mode. (Data from Lakshman, 1981.) The parameter values are k = 51 m/yr, C* = 11.3 mg/L; the fit is $R^2 = 0.956$.

mass balance equation for each. The BOD_5 mass balance contains the uptake and release rates, which in turn require parameters or rate constants.

A design model must describe the features of BOD_5 processes that have been measured in wetlands. Some idea of wetland response to BOD_5 additions may be gained by examining the results of Lakshman (1981) for batch wetland treatment of lagoon effluents. A set of wetlands were charged with wastewater and then closed in with no water additions or withdrawals. Typical response data showed a sharp decrease in BOD_5 to a non-zero, fluctuating background (Figure 12-7). There is not an exponential decrease to a zero BOD_5 with time of exposure to the wetland. The decrease is steep—perhaps exponential—but to a non-zero background BOD_5. Subsurface flow wetlands manifest the same characteristics (Figure 12-8). The same phenomena are also present on a distance scale for flow-through wetlands at Listowel Ontario (Figure 12-9). The rate equations selected for the process description should reflect this behavior.

It will be shown that the k-C* modification of a first-order uptake model can account for the observations; a BOD_5 generation rate must be added to the model.

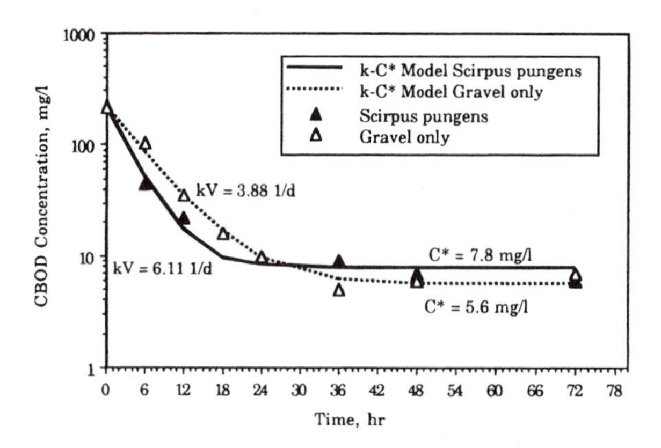

Figure 12-8 The progression of CBOD concentrations in two gravel bed wetlands operated in the batch mode. (Data from Burgoon, 1993.) The fits are $R^2 = 0.976$ for *Scirpus* and $R^2 = 0.969$ for gravel only.

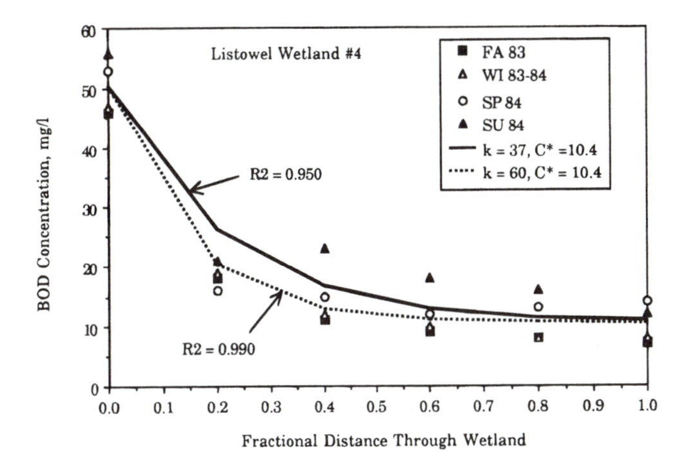

Figure 12-9 The progression of BOD_5 concentrations with distance through a wetland operated in the continuous flow mode. (Data from Herskowitz, 1986.) The model lines were independently determined from input-output data. The difference between the dashed and solid model lines corresponds to a 62 percent increase in k, but only a 1 percent decrease in the parameter search objective function.

MODEL EQUATIONS

Surface-Flow Wetlands

The compartmental view of the wetland is adopted, in which compartment 1 is the flowing wetland water body; all other compartments are static, including soils, sediments, litter, and living and dead biomass (see Chapter 9).

The water mass balance on the water compartment for time-averaged conditions in a non-infiltrating, one-dimensional flow is

$$\frac{d(hu)}{dx} = \frac{dQ}{Wdx} = (P - ET) \qquad (9\text{-}27)$$

The BOD_5 balance on the water compartment (without backmixing) is

$$\frac{d(QC)}{Wdx} = -J = -k(C - C^*) \qquad (9\text{-}148)$$

where C = BOD_5 concentration in surface water, g/m^3
C* = background BOD_5 concentration in surface water, g/m^3
ET = evapotranspiration, m/d
h = water depth, m
J = first-order areal removal rate, $g/m^2/d$
k = first-order areal rate constant for BOD_5, m/d (or m/yr)
P = rain rate, m/d
Q = water flow, m^3/d
u = water superficial velocity, m/d
W = wetland width, m
x = distance parallel to flow, m

The net return of BOD_5 from the wetland solids to the water is the result of many complex processes involving soils and litter. The generation of soluble and particulate carbon compounds which comprise BOD_5 is presumed to result from solid decomposition, and therefore not to involve the BOD_5 in the surface water. The lumped form of the net return is presumed to be a constant (zero order):

$$r_r = kC^* \tag{12-22}$$

where r_r = net production rate for BOD_5, $g/m^2/yr$.

The degree of biological activity in the wetland ecosystem may well influence r_r, with more active ecosystems returning a larger amount of BOD_5 by decomposition.

If there is insignificant atmospheric augmentation, the flow will be constant, and

BOD_5 Areal k-C* Model

$$q\frac{dC}{dy} = -k(C - C^*) \tag{12-23}$$

Application of Equation 12-23 requires integration from the wetland inlet, where the BOD_5 concentration is C_i, to an intermediate distance y, where the BOD_5 concentration is C:

$$\ln\left(\frac{C - C^*}{C_i - C^*}\right) = -\frac{k}{q}y \tag{12-24}$$

At the outlet, where the BOD_5 concentration is C_o,

$$\ln\left(\frac{C_o - C^*}{C_i - C^*}\right) = -\frac{k}{q} = -Da \tag{12-25}$$

where Da = Damköhler number for BOD_5, k/q, dimensionless.

The non-zero background BOD_5 has necessitated the use of a second parameter, C*, in addition to a first-order rate constant, k, to describe the field observations. Background levels are important in design if reductions are contemplated to reach such low levels.

If there is significant atmospheric augmentation, the more complicated Chapter 9, Equations 9-169 through 9-182 must be used.

Subsurface-Flow Wetlands

For BOD_5 in SSF wetlands, the volumetric rate constant may also be used. Profiles are then described by modified versions of Chapter 9, Equations 9-153 and 9-154:

$$\ln\left(\frac{C - C^*}{C_i - C^*}\right) = -\frac{k_v \varepsilon h}{q}y = -k_v \tau\, y \tag{9-153a}$$

and the input-output relation is

$$\ln\left(\frac{C_o - C^*}{C_i - C^*}\right) = -k_v \tau \tag{9-153b}$$

where τ = nominal detention time, d.

As a reminder of the transition from the areal to the volumetric versions of the model, substitute the definition of nominal detention time:

$$\ln\left(\frac{C_o - C^*}{C_i - C^*}\right) = -\frac{k_v h \varepsilon A}{Q} = -\frac{kA}{Q} \tag{12-26}$$

Volumetric vs. Areal Models

Equation 12-26 forms the basis for comparing the general utility of areal rate constants vs. volumetric rate constants. The calculation of concentrations (or areas) requires a value of $k = k_v h\epsilon$. In data analysis, calibration may be for k or k_v. If k_v is chosen, the values of water depth h and void fraction ϵ must be known. As discussed in Chapter 9, these are not accurately measurable, nor are they spatially uniform or constant in time. Therefore, it is reasonable to lump $k_v h\epsilon = k$ for purposes of model calibration.

PARAMETER VALUES

BOD_5 Reduction Parameters for SF Wetlands

Table 12-3 lists values of the rate constant and background for a number of SF wetlands. The average k value is 34 m/yr, and the average background is 6.2 mg/L for marshes and

Table 12-3 Rate Constants for BOD_5 Reduction for Some SF Wetland Systems

Site		k Value m/yr	Background C* mg/L	BOD_5 Return Rate kg/ha/d
Listowel, Ontario (Herskowitz, 1986)	System 1	13.8	4.3	1.63
	System 2	6.5	3.3	0.59
	System 3	12.4	4.6	1.56
	System 4	36.9	10.4	10.51
	System 5	42.8	13.9	16.30
Gustine, CA (Walker and Walker, 1990)	Marsh 1A	18.1	11.6	5.75
	Marsh 1B	13.7	6.4	2.40
	Marsh 1C	9.4	13.0	3.35
	Marsh 1D	28.7	5.9	4.64
	Marsh 2A	22	7.8	4.70
	Marsh 2B	41.6	5.5	6.27
	Marsh 6A	33.3	3.5	3.19
	Pilot marsh	21.6	4.7	2.78
Cobalt, Ontario (Miller, 1989)	Marsh	54.2	4.7	6.98
Iron Bridge, FL (NADB, 1994)	Marsh	22.5	2.1	1.29
Benton, KY (TVA, 1990a)	Marsh 1	93.7	5.4	13.86
	Marsh 2	59.6	7.9	12.90
Pembroke, KY (TVA, 1990a)	Marsh	51.4	3.3	4.65
West Jackson County, MS (Knight, 1994b)	Marsh	54	4.7	6.95
Lakeland, FL (NADB, 1994)	Marsh 1	47.9	1.1	1.44
Average	**Marsh**	**34.0**	**6.2**	**5.6**
Standard Deviation	**Marsh**	**22.0**	**3.5**	**4.5**
Cannon Beach, OR (NADB, 1994)	Forested	17.7	3.8	1.84
Bear Bay, SC (NADB, 1994)	Forested	6.8	1.9	0.36
Reedy Creek, FL (NADB, 1994)	Forested	34.2	1.7	1.56

1.9 mg/L for forested wetlands. The range of hydraulic loadings is approximately 0.9 to 13.5 cm/d, and the range of inlet BOD is approximately 4 to 600 mg/L.

The k-C* model describes the time series of batch wetland phenomena quite well. Figures 12-7 and 12-8 represent calibrations of the model to batch wetland data for SF and SSF systems, respectively. The calibration of the Humboldt SF data (Figure 12-7) produces $R^2 = 0.956$; calibration of the Burgoon SSF data (Figure 12-8) produces $R^2 = 0.976$ and 0.979.

There is a more stringent test that may be applied. For wetlands that have produced both input-output (I/O) data and transect data, one set may be used for *calibration* and the other used for *validation.* In general, the calibration should be from I/O data, since that represents the majority of available wetland information. However, there is considerable uncertainty in the value of k determined from I/O data, because most wetlands are operated to produce effluents close to C*. Searches for k values that minimize error may be insensitive to changes in k and may fail for some data sets.

Data from Listowel, Ontario illustrate this difficulty (Figure 12-9). Monthly I/O data from System 4 produced best values k = 37 m/yr and C* = 10.4 mg/L. Data acquisition at Listowel included weekly sampling along transects from inlet to outlet. Those weekly values were averaged to provide the quarterly transect data in Figure 12-9. The transect data are reasonably well represented by these parameter values ($R^2 = 0.95$), but a better fit of the transect data is obtained for k = 60 m/yr and C* = 10.4 mg/L ($R^2 = 0.99$). Close examination of the parameter estimation procedure for fitting the I/O data reveals that there is little difference in the I/O fit for k = 60 m/yr; the mean square error increases by only 1 percent.

It is important to recognize that the k-C* model, as developed for the flow-dominated hydrology (Equations 12-24 and 12-25), is a representation of average performance. On a short time scale, random events involving meteorology and wetland biota can affect the outlet concentration. Thus, the average performance of Listowel System 4, as characterized by effluent BOD_5 concentrations, is well calibrated by the k-C* model (Figure 12-10). However, short interval sampling produces values that *scatter around mean performance.* This scatter is important for regulatory purposes and will be discussed further in subsequent sections of this chapter.

The values of k and C* are expected to vary from one wetland to another, depending on site-specific factors such as vegetation type and density and hydraulic variables such as mixing intensity. Data from a second treatment wetland site located in Gustine, CA provides

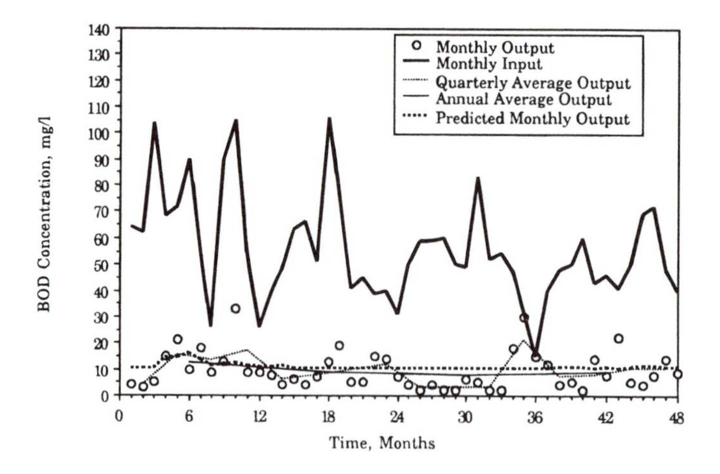

Figure 12-10 Time series of BOD_5 concentrations for Listowel System 4. (Data from Herskowitz, 1986.) The parameters k = 37 m/yr and C* = 10.4 mg/L yield the minimum squared error and represent long-term average behavior quite well. Short interval sampling shows scatter about this mean behavior.

verification of the k-C* model (data from Walker and Walker, 1990). Analysis of their I/O data from seven test wetlands produced k = 27 m/yr and C* = 16.4 mg/L. Independent transect data are well represented by these parameter values (Figure 12-11). Analysis of I/O data from a 1983 pilot study at Gustine produced k = 21 m/yr and C* = 4.7 mg/L, which is an imprecise prediction of the full-scale performance of 1989–1990. This effect is at least partly due to *wetland-to-wetland variability.*

When each of the eight Gustine wetlands data sets are analyzed individually, a range of k and C* values results, as is the case for the five Listowel wetlands. Table 12-3 lists the individual parameter values and includes information from 20 wetlands at 8 sites. The data from these sites was averaged over monthly periods before parameter estimation, except for the very large wetlands at Iron Bridge and Lakeland, for which quarterly averages were employed. In general, averaging should cover several detention times in order to avoid transport delay errors. The range of hydraulic loadings is approximately 0.1 to 10 cm/d, and the range of inlet BOD_5 is approximately 1 to 680 mg/L. The mean values are k = 34 m/yr and C* = 6 mg/L. The coefficients of variation (C.V.) are 65 and 56 percent for k and C*, respectively. This interwetland variability is important for design purposes and will be discussed further in subsequent sections of this chapter.

The implied BOD_5 return from the ecosystem to the water in an SF wetland, r = kC*, is also shown in Table 12-3. The mean value of 5.6 kg/ha/d is approximately the 50th percentile of the BOD_5 input loading rates to SF wetlands in the NADB. Consequently, the return BOD_5 is of approximately equal importance to the wastewater input for those wetlands.

Forested wetlands show similar values of k and C*. The database is not large because relatively few forested wetlands have been analyzed for BOD_5 removal. Table 12-3 lists results regressed from quarterly performance data from Bear Bay, SC; Reedy Creek, FL; and Cannon Beach, OR.

BOD_5 Reduction in SSF Wetlands

Several studies on horizontal SSF wetlands have produced I/O data from which k_V and C* values may be regressed, and transect data are available from a few. The k_V-C* model describes internal and sequential phenomena quite well, as evidenced by Figure 12-12. As for SF wetlands, calibration of the model is to Bavor et al.'s (1988) I/O data, and validation is the excellent fit to transect data.

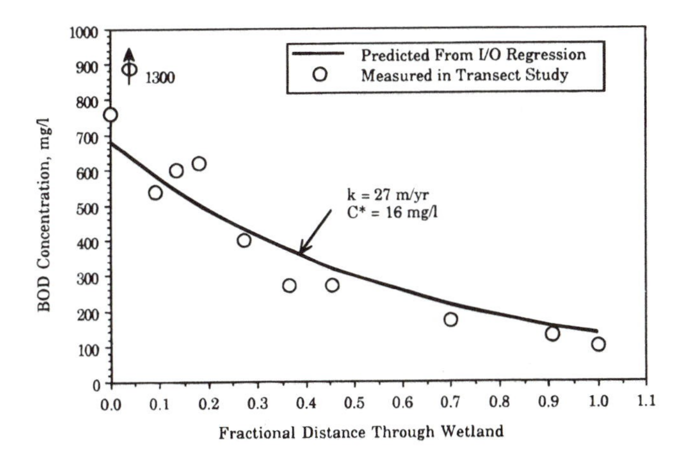

Figure 12-11 The progression of BOD concentrations through an SF wetland at Gustine, CA. (Data from Walker and Walker, 1990.) The line was fit to I/O data for seven separate wetlands cells at the site, not to the data points shown.

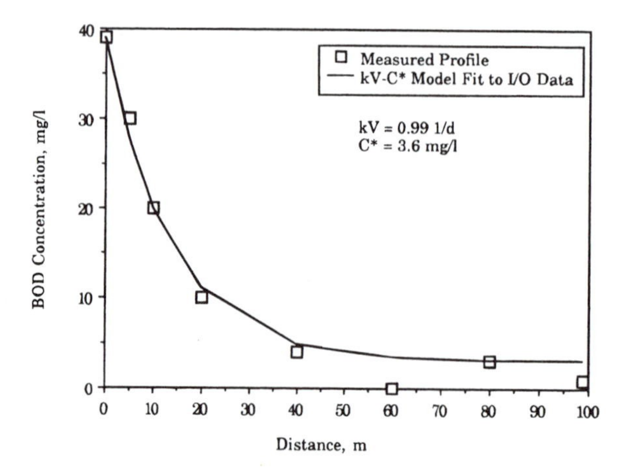

Figure 12-12 Transect values of BOD_5 in a cattail in gravel SSF wetland at Richmond, NSW. (Data from Bavor et al., 1988.) The model was calibrated to the I/O data for this wetland, not to the data points shown.

Data was analyzed by minimizing the mean square error for the two parameter models. Table 12-4 indicates ranges centered on about k = 180 m/yr (k_V = 1.96 d^{-1}) and C* = 9.8 mg/L, but there is a good deal of intersystem variability. The range of hydraulic loadings is approximately 3 to 40 cm/d, and the range of inlet BOD_5 is approximately 14 to 200 mg/L.

Values of the areal k are about five times higher than those for SF wetlands.

The value of C* depends weakly on the inlet BOD_5 concentration (Figure 12-13) for both SF and SSF systems. The regression equation is

$$C^* = 3.5 + 0.053\ C_i$$

$$R^2 = 0.67 \tag{12-27}$$

$$0 \leq C_i < 200 \text{ mg/L}$$

A possible explanation lies in the fact that a stronger wastewater also typically contains greater amounts of nutrients, which stimulate a larger biochemical cycle in the wetland. Waters in the more highly fertilized wetland are therefore likely to be more eutrophic, which in turn implies a higher BOD_5.

The implied BOD_5 return from the ecosystem to the water in SSF wetlands shows a mean value of 48 kg/ha/d (Table 12-4), which is considerably higher than the value of 5.6 kg/ha/d for SF wetlands. This number also depends very strongly on the strength of the incoming wastewater, because C* does; but, in addition, k is higher for the SSF wetland.

Limited information in the literature indicates that vertical flow SSF systems perform about the same as horizontal SSF wetlands for BOD_5 reduction (Table 12-5). Typically, incoming BOD_5 is reduced from over 100 mg/L to less than 10 mg/L at hydraulic loading rates comparable to those for horizontal SSF wetlands. These wetlands display enhanced nitrogen reduction compared to horizontal wetlands, so their use is anticipated to increase in the future.

BOD_5 Reduction in Soil-Based SSF Wetlands

The soil-based wetlands used in Denmark and other European countries are intermediate to SF and SSF wetlands. The hydraulic conductivity of many soils is insufficient to pass the required hydraulic loading, and some significant fraction of the water passes in overland flow. It is therefore not surprising to find that the performance of these soils in BOD_5 removal

Table 12-4 Rate Constants for BOD_5 Reduction for Some Horizontal SSF Systems

Site		k_V Value 1/d	Background C* mg/L	BOD_5 Return Rate g/m^3/d	k Value m/year	Background C* mg/L	BOD_5 Return Rate kg/ha/d
Santee, CA (Gersberg et al., 1986)	Gravel	0.30	16.1	4.8	31	14.5	12.3
	Bulrush	3.54	2.8	9.9	365	3.0	30.0
	Cattail	2.39	18.2	43.5	246	18.2	122.7
	Phragmites	2.92	14.8	43.2	300	14.8	121.6
Richmond, NSW (Bavor et al., 1988)	Gravel	1.27	3.4	4.3	96	4.2	11.0
	Bulrush	3.59	5.4	19.4	318	5.8	50.5
	Cattail	0.99	3.6	3.6	88	4.1	9.9
Florida (Burgoon, 1993)	Gravel	3.88	5.6	21.7	158	5.6	24.3
	Bulrush	6.11	7.8	47.7	250	7.8	53.4
Benton, KY3 (TVA, 1990a)	Bulrush	0.54	4.7	2.5	122	6.1	20.4
Carville, LA (Zachritz and Fuller, 1993)	*Sagittaria*	0.70	1.7	1.2	115	1.7	5.4
Fourteen Wetlands (U.S. EPA, 1993b)	Various	2.17	7.0	15.2	217	7.0	41.6
Five Wetlands[a] (Green and Upton, 1992)	*Phragmites*	1.22	0.8	1.0	97	1.8	4.8
Eighteen Wetlands[b] (Saurer, 1994)	*Phragmites*	1.75	15.5	27.1	168	15.5	71.3
Weighted Mean		**1.96**	**9.8**	**19**	**180**	**9.9**	**48**
Standard Deviation		**0.90**	**5.6**	**12**	**61**	**5.4**	**29**

[a] Depth assumed = 0.60 m; porosity assumed = 0.40.
[b] Depth assumed = 0.75 m; porosity assumed = 0.35.

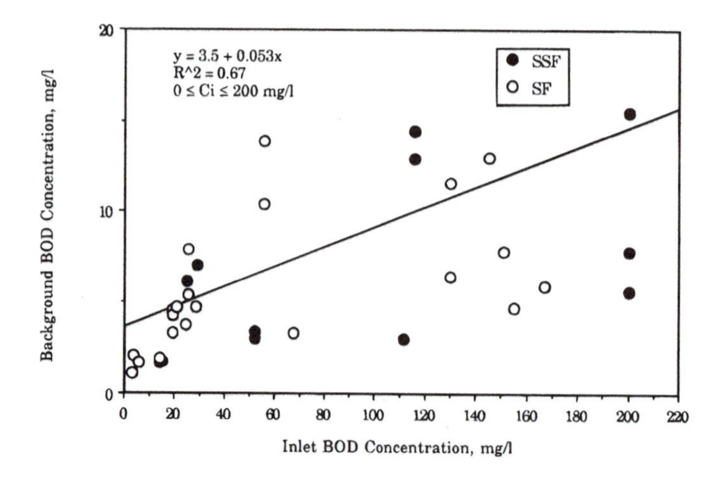

Figure 12-13 There is a slight increase in the apparent BOD background depending upon the inlet BOD concentration.

is intermediate between SF and SSF. Brix (1994c) has correlated the performance of over 70 soil-based SF wetlands to the k-C* model, resulting in k = 47.5 m/yr and C* = 3.0 mg/L.

Findlater et al. (1990) analyzed data from 14 soil-based wetlands assuming C* = 0; the irreversible first-order areal rate constant was found to be k_1 = 23.2 ± 1.9 m/yr. Schierup et al. (1990b) analyzed data from 39 soil-based wetlands assuming C* = 0; the irreversible first-order areal rate constant was found to be k_1 = 30.3 ± 3.2 m/yr.

FIRST-ORDER MODELS WITHOUT C*: IRREVERSIBLE MODELS

Equations 12-24, 12-25, and 12-26 become first-order irreversible models when C* = 0, because there is a zero return of BOD_5 from the ecosystem to the water. Under these circumstances, the values of the rate constant are designated by k_1 and k_{V1}, referring to the fact that these are one-parameter models.

Table 12-5 BOD_5 Performance of Vertical Flow Wetlands

Ref.	System	Period	HLR cm/d	BOD_5 In mg/L	BOD_5 Out mg/L
Haberl (1994);	System A	11/91–12/92	2.85	121	9
Haberl et al. (1994)		4/93–6/94	2.85	117	5
Haberl (1994)	System B	11/91–12/92	2.96	78	4
Haberl et al. (1994)		4/93–6/94	2.96	46	4
Pietsch (1992), as reported by Brix (1994b)	Phytofilt		3	200	<10
Haberl (1994)	Haider	1.5 years	10		
Burka and Lawrence (1990)	Oaklands Park System 1	6/88–1/90	16	108	1.6
Burka and Lawrence (1990)	Oaklands Park System 2	8/89–1/90	15	246	13.8
Bahlo and Wach (1990)	Plant I	6/88–12/90	4.8	147	<4
Bahlo and Wach (1990)	Plant II	8/88–1/90	2.7	289	<8
Rijs and Veenstra (1990)	Lauwersoog	1976–1988	1.4	222	5.2
Butler et al. (1993)	Egypt		14.4		
Butler et al. (1993)	England		14.4	22.5	2
Mean			**7.2**	**145**	**6.1**
Standard Deviation			**5.8**	**85**	**3.7**

$$\ln\left(\frac{C_o}{C_i}\right) = -k_{V1}\tau = -\frac{k_{V1}Ah\varepsilon}{Q} \qquad (12\text{-}28)$$

This model cannot be recommended because the rate constant is not constant, but a strong function of the detention time or hydraulic loading rate, as reflected by the distance progression through the wetland. This is illustrated for one of the Listowel, Ontario wetlands in Figure 12-14. The rate constant is determined from the slope of the concentration vs. distance graph, with a logarithmic scale for concentration, according to Equation 12-28. If the "outlet" is moved sequentially further from the inlet, the slope decreases markedly. The corresponding k_1 drops from 39.3 to 12.5 m/yr.

A comparison of the k_V-C^* model and the U.S. EPA (1988) model was made for the Denham Springs, LA SSF wetland by Skipper and Tittlebaum (1991). The mean square error for the former was one half the mean square error for the U.S. EPA (1988) model.

Design guidelines are prevalent in the literature which list values of k_{V1} to be used in design. For instance, the U.S. EPA (1993b) proposes $k_{V1} = 1.104\ d^{-1}$ at 20°C, with the caveat that this is intended for use in designs at an inlet BOD_5 loading ($qC_i = QC_i/A$) of 110 kg/ha/d. The area of the wetland is then to be calculated from Equation 12-32, solved for area

$$A = \frac{Q \ln\left(\frac{C_i}{C_o}\right)}{k_{V1}h\varepsilon} \qquad (12\text{-}29)$$

For design calculations, wetland area scaling according to Equation 12-29 is incorrect in many circumstances. The use of the irreversible first-order equation for design, with a constant k_{V1}, can lead to significant errors.

To illustrate, consider the SSF wetlands operated at Baxter, TN (George et al., 1994). Cell F of that study was operated at conditions nearly identical to the stated design point in the U.S. EPA (1993b) recommendation; i.e., an inlet BOD_5 loading of 112 ± 13 kg/ha/d compared to the recommendation of 110 kg/ha/d. Therefore, it is not surprising to find a good verification of the value of k_{V1}. Wetland F was operated at 0.3 days nominal detention

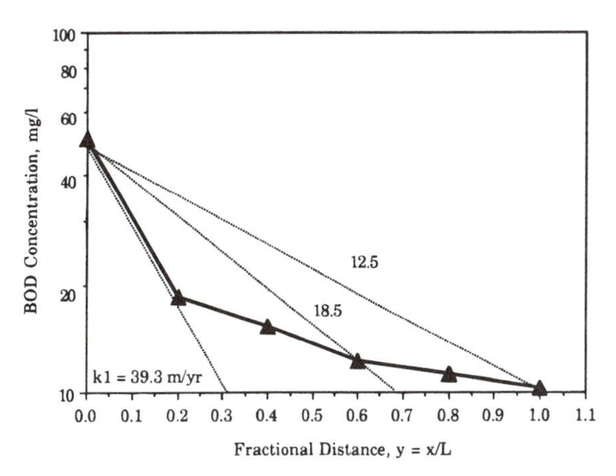

Figure 12-14 The progression of BOD_5 concentrations along Listowel Wetland 4. Data points are the averages of weekly samples during September 1983 through August 1984. The irreversible rate constant k_1 is determined from the inlet and outlet concentrations and hydraulic loading rate. The outlet is successively considered to be at fractional distances of 0.2, 0.6, and 1.0.

time and input $C_i = 44 \pm 5$ mg/L. Equation 12-28 predicts a long-term average outlet concentration of $C_o = 31 \pm 4$ mg/L; the observed average annual outlet concentration was $C_o = 32 \pm 9$ mg/L. This wetland achieved a 27 percent concentration reduction.

Now, suppose that a design calls for a long-term average effluent BOD_5 of 16 mg/L instead of 32 mg/L. Equation 12-30 is used to calculate the required new detention time, which is

$$\tau = \frac{\ln\left(\frac{C_i}{C_o}\right)}{k_{v1}} \tag{12-30}$$

$$\tau = \frac{\ln\left(\frac{44}{16}\right)}{1.104} = 0.92 \text{ days}$$

Data are available at the Baxter, TN site for detention times in the range 0.3 to 4.9 days (Figure 12-15). The data from eight side-by-side wetlands is well described by the k_V-C* model, with $k_V = 2.08$ and $C^* = 15.4$ mg/L. In contrast, the U.S. EPA (1993b) model underpredicts the observed concentrations over most of the range in nominal detention times. In practice, the design value of 16 mg/L is not reached until about 2 to 3 days' detention.

The literature contains a considerable number of references which set forth and reiterate the first-order, irreversible volumetric model for BOD_5 reduction in wetlands, both SF and SSF (Reed et al., 1988; U.S. EPA, 1988, 1993; WPCF, 1990b). The model was adopted for wetlands on the basis that "conditions for biological treatment are similar to those for the overland flow concept and to trickling filters and similar attached growth systems" (Reed et al., 1988). As data became available from many detailed and careful wetland studies, it became apparent to researchers that the model was not correct for wetlands. This recognition has come from those studying performance of SF systems: "The Gustine data are poorly correlated with the first order removal model" (Walker and Walker, 1990). The same evaluation has been expressed by those studying SSF wetlands: "This implies that after 18 hours the assumption of first-order kinetics is not valid, and a different kinetic model is required" (Burgoon, 1993).

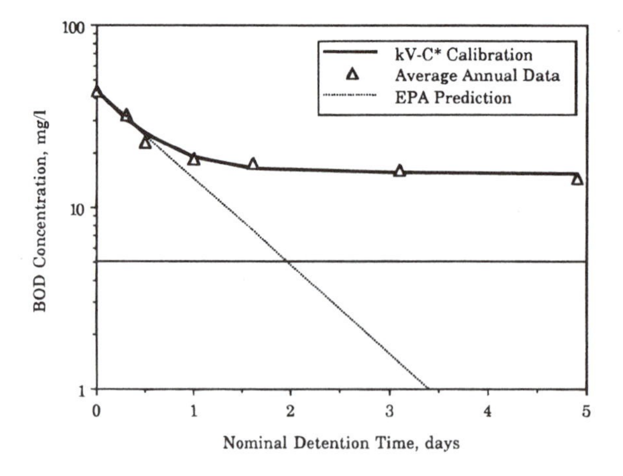

Figure 12-15 Annual average performance for BOD reduction for the Baxter, TN SSF wetlands. Original data were from George et al. (1994). The k_V-C* model fits the data better than the U.S. EPA (1993b) model.

Ponds

In the literature, pond reduction of BOD_5 has been presumed to be first order and to follow Equation 12-30 for plug flow (WPCF, 1990b).

The rate constant may be estimated in a two-step procedure: determination of the k value at 20°C from a correlation with BOD_5 loading, followed by a temperature correction.

$$k_{v20} = 0.027 + 0.00086 \cdot BLI \tag{12-31}$$

$$22 < BLI < 112$$

$$R^2 = 0.97$$

$$k_{VT} = k_{V20} \cdot (1.09)^{T-20} \tag{12-32}$$

where BLI = inlet BOD_5 loading, kg/ha/d
k_V = BOD_5 rate constant, 1/d
T = temperature, °C

The rate constants for ponds are significantly lower than those for wetlands. It is likely that the extra solid surfaces in the wetland contain active biofilms that enhance performance. The addition of such surfaces to ponds has been advocated (Polprasert and Agarwalla, 1994).

The correlations of Equations 12-31 and 12-32 cannot be used for high loadings. High-strength potato water has been studied in nonaerated lagoons (U.S. EPA, 1969). The values of the k-C* rate constants for those potato ponds were

BOD_5: C_i = 1600 mg/L k_{V20} = 0.025 d^{-1} C^* = 100 mg/L
COD: C_i = 2850 mg/L k_{V20} = 0.034 d^{-1} C^* = 180 mg/L

Floating Aquatic Beds

This type of "wetland" is a pond covered with one of several floating leaved plants, such as duckweed (*Lemna* spp.), pennywort (*Hydrocotyle* spp.), or water hyacinths (*Eichhornia crassipes*). A k_V-C* model appears to be well suited to these aquatic systems. Data from 15 "floating leaved wetlands" is represented by $k_V = 0.55\ d^{-1}$ and $C^* = 10.8$ mg/L, if the systems are considered to be plug flow (WPCF, 1990b; Hayes et al., 1987; Poole and Ngo, 1992; Hancock and Buddhavarapu, 1993). If the ponds are modeled as two continuous stirred tank reactors (CSTRs), $k_V = 0.98\ d^{-1}$. The Disney World facility treated high-strength wastewater (BOD_5 = 230 mg/L) with hyacinths and varied the detention time (Hayes et al., 1987). Those data are well represented by a two-CSTR model with $k_V = 1.50\ d^{-1}$ and $C^* = 26.6$ mg/L.

It appears that the degree of mixing in these deep water systems is quite important in the calibration of models. In addition, it is evident that a BOD_5 return must also be included in modeling floating leaved wetlands.

TEMPERATURE EFFECTS ON BOD_5 REDUCTION

BOD_5 reduction has been found to be temperature dependent in other water treatment processes, notably in attached growth microbial systems and in pond systems. These previous experiences led early workers to presume a similar temperature dependence in wetlands (Reed et al., 1988). Subsequent publications presented the Reed et al. (1988) presumption as fact (U.S. EPA, 1988, 1993b; WPCF, 1990b; Conley et al., 1991; Reed and Brown, 1992; Crites, 1994). Temperature dependence was expressed as

$$k_{V1} = k_{V1,20}(\theta)^{(T-20)} \tag{12.33}$$

where: $k_{V1,20}$ = BOD_5 rate constant at 20°C, 1/d
$k_{V1,T}$ = BOD_5 rate constant at T°C, 1/d
T = temperature, °C
θ = temperature factor, –

A value of $\theta = 1.1$ was initially used (Reed et al., 1988; U.S. EPA, 1988). Subsequently, the value of $\theta = 1.06$ was advanced (WPCF, 1990b; Conley et al., 1991; Reed and Brown, 1992; U.S. EPA, 1993b; Crites, 1994). At the time of this writing, no data analysis has been reported to support these numerical values of θ for treatment wetlands.

Surface-Flow Wetlands

The stochastic nature of wetland behavior tends to mask potential temperature trends, and it is necessary to examine several years of performance records to find minor trends. For those few sites where such data exist, temperature trends in C* are confused by the possibility of covariant seasonal trends. Growth patterns of the macrophytes, as well as seasonal patterns in litterfall and litter decomposition, are added features of a wetland treatment system compared to attached growth microbial systems.

Figure 12-16 shows that effluent BOD_5 peaks at low temperature at the Listowel, Ontario site, but also peaks at high temperature. Outlet concentrations at Listowel are dominated by the C* "tail" of the k-C* model for all seasons. Other SF wetlands show similar behavior. For instance, lightly loaded wetlands in Halsey, OR performed least well in the summer (Kuehn and Moore, 1994).

The temperature dependence of k may be estimated from transect studies. The initial decline in BOD_5 in the front end of the wetland is strongly dependent on k rather than C*.

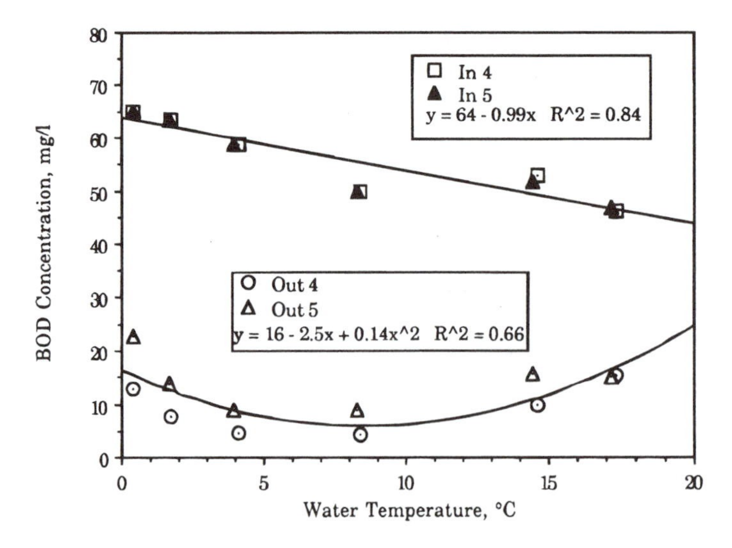

Figure 12-16 BOD reduction for Listowel Systems 4 and 5. Original data were from Herskowitz (1986). The aeration cell effluent that formed the wetland inflow shows a greater BOD at cold temperatures and a consistent decrease with increasing temperature. There is not a similar monotonic trend for the wetlands.

Channel studies at Listowel were conducted weekly for over 1 year, and water temperatures were measured. From those data, the values were for System 3, $\theta = 0.964 \pm 0.005$, and for System 4, $\theta = 1.010 \pm 0.020$, based on analysis of data from Herskowitz (1986).

The temperature dependence of k_1 was estimated for all five of the Listowel wetlands. This may be done for individual monthly data points because of the ease of parameter estimation for the one-parameter model. The result was $\theta = 0.976 \pm 0.031$ (n = 225); the water temperature range was $0.00 \leq T \leq 18.7$°C. Data analysis for the Iron Bridge; FL treatment wetlands (PBSJ, 1993) yielded $\theta = 0.969 \pm 0.025$ (n = 55); the water temperature range was $12.3 \leq T \leq 29.0$°C. These values should be considered more indicative of the temperature dependence of C* than of k, because these wetlands operated with outlet concentrations near C*.

In light of the data presently available, it appears that temperature effects for BOD_5 reduction are negligible for SF wetlands.

Soil-Based Wetlands

Data do not indicate a lessening of reduction efficiency for these systems at lower temperatures. The Findlater et al. (1990) study of 14 soil-based wetlands found no effect of temperature on BOD_5 reduction in the U.K. Schierup et al. (1990b) found no difference between winter and summer operation of 39 soil-based wetlands in Denmark for BOD_5.

Subsurface-Flow Wetlands

SSF wetlands display the same lack of temperature sensitivity. Bavor et al. (1988) attempted to find temperature dependence for the first-order volumetric rate constant in the data from five SSF wetlands operated on secondary effluent for 2 years, but could not find any such dependence. This led Bavor to conclude ". . . a temperature variation factor was found to be negligible, and could be omitted from the 1st order rate equation for prediction of removal of BOD_5 . . . ."

Gumbricht (1992) studied an SSF wetland at Snogeröd, Sweden for 3 years. The temperature coefficient inferred from those data was $\theta = 1.003$, which covers a temperature range from 2 to 21°C.

The SSF system at Hatzendorf, Austria showed an 88.7 ± 4.2 percent reduction in BOD_5 in the summer and 85.7 ± 3.2 percent in the winter (Saurer, 1994).

The Baxter, TN site (George et al., 1994) showed a 58.9 ± 0.8 percent reduction in BOD_5 for eight SSF wetlands, averaged across the wetlands: summer 58.1 percent, fall 59.0 percent, winter 60.0 percent, and spring 58.4 percent.

Bahlo and Wach (1990) found no temperature effect on BOD_5 removal in the vertical flow wetland at Springe-Eldagsen, Germany over a 2-year study period. The same result was reported by Lemon and Smith (1993) for wetlands located at Niagara-On-The-Lake, Ontario.

The conclusion from the available case study data is a negligible temperature effect on BOD_5 reduction in SSF wetlands.

Speculation on Contributing Factors

There has been no research attempt to elucidate the mechanistic reasons for the noneffect of temperature, but it is not difficult to rationalize them. BOD_5 is a lumped chemical category consisting of a multitude of carbon compounds. The wetland processes that consume BOD_5 involve several different vertical zones, in which a large number of chemical and biochemical processes take place. Some of those processes generate BOD_5 by decomposition, others consume BOD_5. As seen previously, wetlands frequently are of sufficient size to reach the background BOD_5 level created by the balance between generation and consumption of BOD_5. The net consumption is therefore the difference between two competing process categories. It is known that decomposition slows to a marked degree at cold temperatures. It is also intuitively plausible that BOD_5 destruction processes are also slow, since these are a combination of aerobic and anaerobic microbial reactions. The difference between slow destruction and slow generation, as manifested by net removal, need not differ significantly from that difference under warmer, more rapid, process conditions.

Not all BOD_5 removal processes slow with decreasing temperature. A significant fraction of wetland BOD_5 may be particulate and therefore susceptible to removal by particulate settling. This physical process is very weakly temperature dependent. Another important factor is the aerobic destruction of carbon compounds, which is dependent on the amount of oxygen available for the oxidation reactions. The required oxygen comes, in major part, from the dissolution of atmospheric oxygen. The solubility of gaseous oxygen in water is higher at lower temperatures, with a twofold increase as the temperature drops from 30°C to 0°C.

It would be hazardous to extrapolate this behavior to specific carbon compounds, especially those which are not normally found in wetlands. A compound that is not generated in the wetland will be subject to depletion alone and may also require a specific microorganism. Under these conditions, a temperature dependence is a distinct possibility.

SHORT-TERM VARIATIONS IN BOD_5 CONCENTRATIONS

The reductions calculated from either regression equations or from k-C* models pertain to conditions averaged over relatively long periods of time—several months or more. Over a shorter term, the many contributing wetland processes are subject to excursions in response to a wide variety of wetland variables. Consequently, neither of these models should be used to predict short-term dynamic behavior of the treatment wetland.

As an example of the short-term variance, consider the performance of the SSF wetland at Hansen's Disease Center in Carville, LA. Zachritz and Fuller (1993) present data on flow and BOD_5 concentrations over a 100-day study period in 1988. The regression model (Equation 12-21) provides an estimate of the average BOD_5 leaving the wetland (6.1 mg/L) in response to the average BOD_5 entering the wetland (14.1 mg/L) over this period. The measured average effluent BOD_5 was 5.9 mg/L. The prediction is well within the standard error of the regression equation. The k-C* model (Equation 12-24) predicts a wetland outlet concentration of 5.2 mg/L, also a reasonably close estimate. There is no way to model the daily behavior of this system using the regression equation, since the regression does not include any operating variables, nor can the inclusion of such variables be statistically justified from the available data.

The k-C* model appears to be usable on a short time step, since it may formally be used in conjunction with the daily water mass balance for the wetland. This is not a simple task since the travel time of water through this wetland is on the order of 2 days. It becomes necessary to model processes with sufficient temporal and spatial resolution to describe the movement and modification of concentration and flow changes ocurring at the wetland inlet. This is accomplished by using the one-dimensional version of the water and BOD_5 mass balance equations (Chapter 9, Equations 9-27 and 9-148):

$$\frac{\partial h}{\partial t} = -\frac{\partial (v_x h)}{\partial x} + P - ET = -\frac{\partial \Lambda}{\partial x} + P - ET \qquad (12\text{-}34)$$

$$\frac{\partial (hC)}{\partial t} = -k(C - C^*) - \frac{\partial (v_x hC)}{\partial x} = -k(C - C^*) - \frac{\partial (\Lambda C)}{\partial x} \qquad (12\text{-}35)$$

Solution of these equations may be accomplished by a numerical procedure and produces the effluent BOD_5 concentrations on a daily basis in response to daily inlet concentrations and flows, as well as daily precipitation and evapotranspiration estimates. The best-fit model parameters for the period in question are k = 115 m/yr and C* = 1.7 mg/L. Results of the mass balance calculations and the data are not in good agreement (Figure 12-17). The two problems are bad predictions of peak times and a low prediction of event amplitudes. On an average basis, the results are not bad. It is noteworthy that the first-order, irreversible volumetric model does an even worse job of predicting short-term behavior: upward and downward trends are not correctly reproduced.

A mechanistic explanation can be offered, but not substantiated, because there are insufficient data to do so. The k-C* model lumps together many processes, some of which are reduction processes, others being BOD_5 generation and return processes. The return of BOD_5 from the wetland biomass compartments is determined by conditions in the media, sediments, and litter, which are driven by factors other than water flow. Short-term storage and release processes are not included in the model. Further, large rainfall events can produce overland flow, which causes poor treatment for short periods.

These considerations mean that long-term models should not be used over time periods shorter than some multiple of the characteristic process times in the wetland. One such characteristic time is the nominal detention time of water within the wetland; another is a period representative of the stochastic chatter generated in the biomass compartments. As a general rule, the shortest model time step should be three to five times the wetland detention time. This confines the potential error caused by storage variation inside the wetland water body and minimizes the error associated with the time delay between input and output. The longer the averaging period, the less important are changes in storage.

BOD_5 concentrations in wetland waters are temporally variable even if there are no large changes in inlet conditions. This is evidenced in data from the Lakeland, FL SF treatment wetland (Figure 12-18). The frequency of effluent "chatter" has not been the subject of research, but the variance is reduced to the 20 percent level by using an averaging period of more than 1 month for Lakeland. Seasonal trends account for most of the remaining variance.

It thus appears that monthly averaging is the minimum to be used for detention times of 1 week or less and that longer averaging periods may be required for longer detention

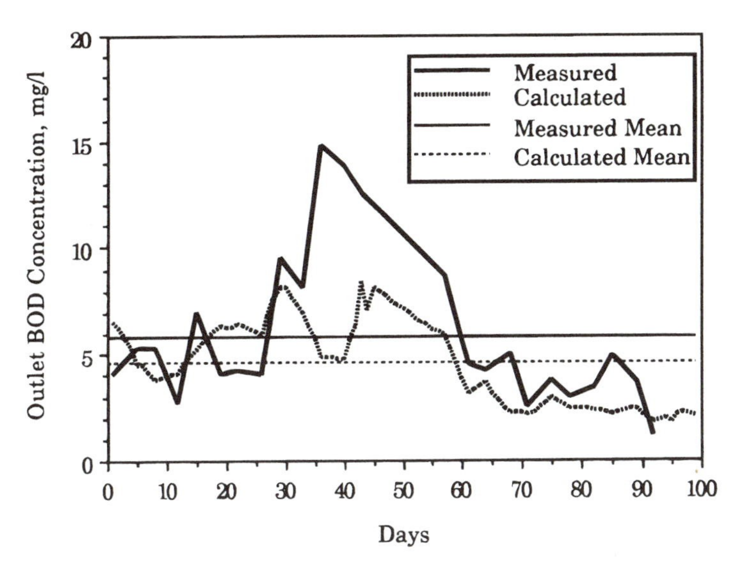

Figure 12-17 Measured and calculated daily BOD_5 performance of the Carville, LA SSF treatment wetland. (Data from Zachritz and Fuller, 1992.)

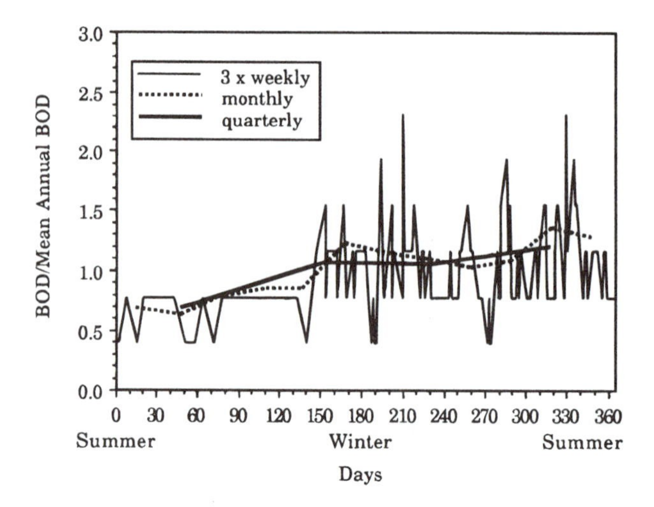

Figure 12-18 Averaging of BOD_5 concentrations in the effluent from the Lakeland treatment wetlands. Note that the high frequency "chatter" in concentrations is "data filtered" by averaging. Either monthly or quarterly averaging preserves the seasonal variations, but eliminates the high-frequency events. (Data from Hill, 1990.)

wetlands. Therefore, the mass balance equations, with the k-C* or irreversible removal models, should not be used for periods shorter than monthly or quarterly.

ADAPTATION TRENDS

Wetlands require a period of adaptation to reach a stationary state in which monotonic time trends are absent. This period includes vegetative areal fill-in, root and rhizome development, litter development, and microbial community establishment. The presence of full-sized mature plants is a necessary, but not sufficient condition for the realization of the stationary state for BOD_5 reduction. Operating data are the best indicator of the presence or absence of adaptation trends.

The concept of a wetland as a "microbial filter" creates the impression that establishment of the microbial population is the sole determinant of adaptation. Microbial populations are known to adapt rather quickly to their environment, and hence, a short adaptation period would be expected. In contrast, the litter decomposition that contributes to the return flux of BOD_5 may require 1 or 2 years to stabilize. Therefore, a newly constructed treatment wetland would be expected to require many months, including at least one full set of seasons, to stabilize.

Data from several locations indicate that this is the case in practice. Data from SF wetlands at Listowel, Ontario indicated weakly decreasing performance over a period of about 1 year, with more effect in the wetlands receiving the stronger effluent (Herskowitz, 1986). Other SF wetlands have displayed no adaptation trends over the first few years, such as Iron Bridge, FL and West Jackson County, MS (NADB, 1993).

Reed beds in coarse media (n = 16) and in soils (n = 14) were shown to be experiencing adaptation trends into their third year of operation, with performance improving (Findlater et al., 1990). Ten Danish soil-based wetlands displayed a monotonic decrease in performance over 3 years (Schierup et al., 1990b). The first 3 months of operation of the Baxter, TN SSF wetlands was more efficient than the ensuing operations (George et al., 1994). Other SSF wetlands appear to have stabilized more quickly. The Richmond, NSW gravel beds experienced little change in performance after decreasing for about 6 months operation (Bavor et al., 1988).

These observations indicate that some weak adaptation effects may be expected for a period of about 1 to 2 years and that performance may decrease during that period. The effect is presumably due to the development of the return flux associated with biomass decomposition.

STOCHASTIC VARIABILITY

The present lack of detailed internal data from which to estimate rapid transfer effects on rate constants and return rates results in the need to absorb these into the overall variability of BOD_5 performance. Mean values estimated from regression or the k-C* model are central tendencies; average values over short time periods are subject to variation from the long-term mean. The longer the averaging period, the closer the short-term mean value is to the long-term mean value.

For both SF and SSF wetlands, average effluent BOD_5 concentrations are distributed approximately according to the log normal distribution. Examples of these distributions are given in Figure 12-19 for SF wetlands and in Figure 12-20 for SSF wetlands. It is clear that the averaging period has a very strong influence on the higher percentiles. This means that regulatory specifications should acknowledge the high ratio of weekly to monthly values for low exceedance frequency. The usual "maximum weekly" and "maximum monthly" limits will be quite different. If the 98th percentile (probability variable = 2) is taken as the maximum value to be expected in an SF wetland, the anticipated maximum weekly value is more than twice the anticipated maximum monthly value (Figure 12-19).

Impact of Variability on Design

The values of k and C* vary from time to time for any specific wetland, and the average values over long periods will be different for different wetlands. The variables that enter the (simplest) mass balance design equation are hydraulic loading rate (q) and inlet concentration (C_i) and k and C*. Each is subject to variability. An understanding of the possibilities for the exit concentrations may be gained by exploring the design Equation 12-25 over the possible combinations of values for these four variables.

Example 12-1

As an illustration, consider an SSF wetland subjected to the following design spectra:

Hydraulic loading	Log normal	Mean = 25 cm/d	SD = 2 cm/d
Inlet concentration	Log normal	Mean = 100 mg/L	SD = 20 mg/L
Background C*	Normal	Mean = 8.8 mg/L	SD = 3.2 mg/L
Rate constant k	Log normal	Mean = 180 m/yr	SD = 60 m/yr

The design calculation for the mean conditions yields an outlet concentration of 21.5 mg/L. Exploration of the response space for the probability distributions above gives

Outlet concentration	≈Log normal	Mean = 24.0 mg/L	SD = 9.4 mg/L

The outlet distribution of concentrations (Figure 12-21) is close to log normal. For this illustration, the design area would yield a long-term average BOD_5 of 24 mg/L, but about 20 percent of the samples would exceed 31 mg/L and 10 percent would exceed 36 mg/L.

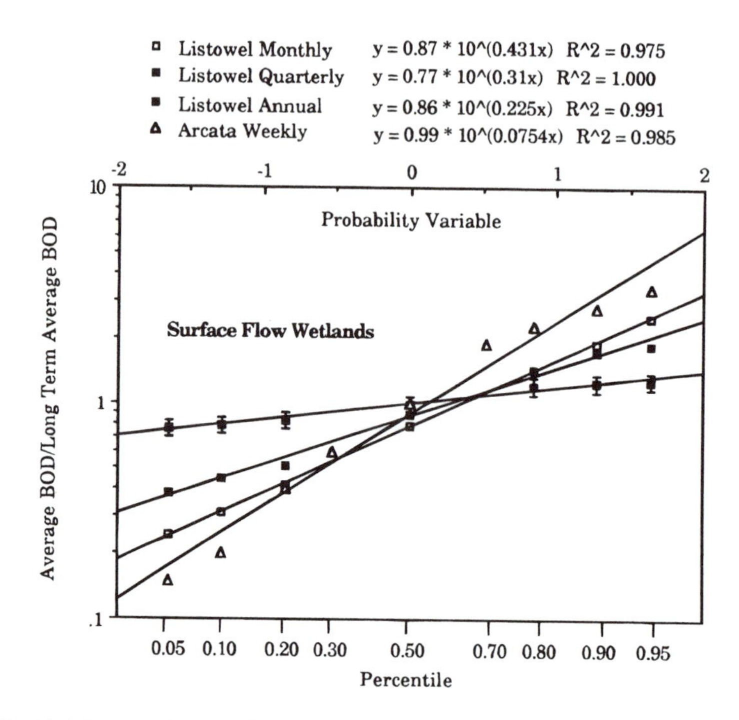

Figure 12-19 Variability in effluent BOD_5 percentiles as a function of the averaging period at Listowel and Arcata SF wetlands. Each Listowel point is the mean for five wetlands over 4 years, with a mean effluent BOD_5 = 10.1 mg/L. The Arcata points represent weekly values from one wetland over 4 years, with a mean effluent BOD_5 = 10 mg/L. Error bars are shown only for the monthly distribution; these represent the standard deviation in the percentile data. Read the graph as follows: at the 95th percentile, for quarterly averages, read 1.99 as the ratio of quarterly averages to the long-term (4-year) average. Interpretation: 95 percent of the quarterly averages will be less than 1.9 times the long-term average. (Data from Herskowitz, 1986, and Gearheart, 1992.)

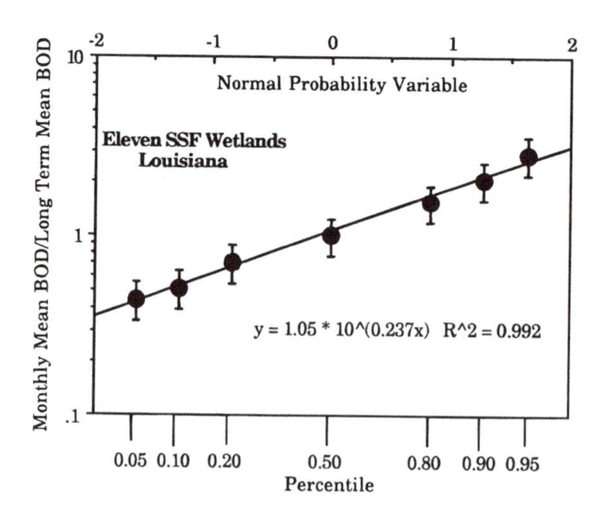

Figure 12-20 The variability of monthly averages of effluent BOD_5 for 11 SSF wetlands in Louisiana. The error bars shown represent the standard deviation in the percentile data. The long-term averages span an average of 3.1 years, and the average long-term BOD_5 was 7.1 mg/L. (Data from Pennison, 1993a.)

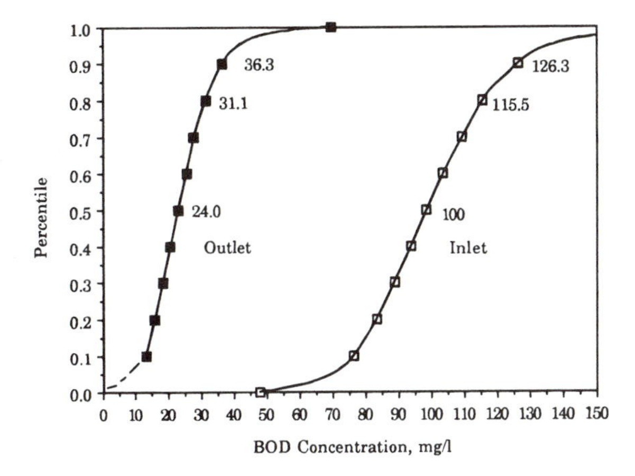

Figure 12-21 Hypothetical probability distributions showing the propagation of uncertainty in parameter values and operating conditions through the k-C* model for BOD_5 reduction. Both distributions are close to log normal; the lines appear as S-curves on these linear coordinates. Points on the outlet curve are not computed from the corresponding inlet point; distributions in four variables are involved.

The data from actual operating SSF systems in Louisiana displays this same character (Figure 12-20). However, inlet concentrations are lower than the 100 mg/L of the preceding illustration, and the variability is somewhat higher. The 80th percentile (one sample out of five) is 54 percent higher than the long-term mean for these wetlands.

Design calculations will normally have to allow for regulatory requirements that impose percentile restrictions on performance. Increased areas are the inevitable result.

SUMMARY

Wetlands are effective in the reduction of BOD_5, as long as incoming BOD_5 exceeds the natural level at which the wetland operates. A wealth of carbon conversion processes operate in wetlands, some which consume BOD_5 and others which produce it. Both anaerobic and aerobic processes have been measured to consume carbon compounds in the wetland environment. Litter and sediment decomposition produce soluble carbon compounds. As a consequence, the simplest mass balance model must include both consumption and generation of these substances. An irreversible first-order model does not fit wetland data. The two parameters of the recommended model are an areal uptake rate constant (k) and a background BOD_5 concentration (C*) for the wetland. A volumetric uptake constant is also appropriate for SSF wetlands, but its use is confounded by lack of good information on depth and porosity.

The background BOD_5 depends somewhat on season of the year. Neither the rate constant nor the background BOD_5 depends strongly on temperature. Variations in these parameters with temperature cannot be quantified adequately from the existing database, and consequently, all variability must be absorbed into the performance spectrum. The model parameters allow projection of long-term average behavior; the distribution of short-term averages displays widening variability as the averaging period is shortened.

Subsurface-flow wetlands have faster uptake than SF wetlands, by a factor of about five. Both types display a background BOD_5 of about 6 mg/L at an inlet BOD_5 of 50 mg/L as an intersystem average. Most operating wetlands are overdesigned for BOD_5 removal, and hence, effluent concentrations are at or near background levels.

CHAPTER 13

Nitrogen

INTRODUCTION

Nitrogen (N) compounds are among the principal constituents of concern in wastewater because of their role in eutrophication, their effect on the oxygen content of receiving waters, and their toxicity to aquatic invertebrate and vertebrate species. These compounds are also of interest because of the beneficial role that they can play in augmenting plant growth which in turn stimulates the production of wildlife.

The nitrogen cycle is very complex, and control of even the most basic chemical transformations of this element is a challenge in ecological engineering. This chapter describes the wetland nitrogen cycle, summarizes current knowledge about environmental factors that control nitrogen transformations, and provides alternative approaches that can be used to design wetland treatment systems to treat nitrogen.

NITROGEN FORMS AND STORAGES IN WETLANDS

Elemental nitrogen has an atomic weight of 14.01 g/mol with five electrons in the outer electron shell of its atomic structure. Because three electron positions are available in its outer shell, nitrogen can form compounds with varying stability that have oxidation states ranging from +5 to −3. These compounds include a variety of inorganic and organic nitrogen forms that are essential for all biological life.

The most important inorganic forms of nitrogen in wetlands are ammonia (NH_4^+), nitrite (NO_2^-), nitrate (NO_3^-), nitrous oxide (N_2O), and dissolved elemental nitrogen or dinitrogen gas (N_2). Nitrogen may also be present in wetlands in many organic forms including urea, amino acids, amines, purines, and pyrimidines. A basic understanding of the chemistry of these forms of nitrogen is critical for the design and operation of wetland treatment systems.

INORGANIC NITROGEN COMPOUNDS

Ammonia

Ammonia nitrogen is made up of a single, chemically reduced nitrogen atom (oxidation state of −3) that has either three or four hydrogen atoms, depending on water temperature and pH:

$$NH_3 + H_2O = NH_4^+ + OH^- \qquad (13\text{-}1)$$

where NH_3 is un-ionized ammonia and NH_4^+ is ionized ammonia (ammonium ion). Total ammonia is equal to the sum of the un-ionized and the ionized ammonia. The ionized form of ammonia is predominant in most wetland systems and is designated as ammonium nitrogen in this book. Figure 13-1 illustrates the percent of un-ionized ammonia for a range of temperature and pH conditions typical of wetland systems.

For an "average" environmental condition of 25°C and a pH of 7, un-ionized ammonia is only 0.6 percent of the total ammonia present. At a pH of 9.5 and a temperature of 30°C, the percentage of total ammonia present in the un-ionized form increases to 72 percent. At lower pH and temperature values, this percentage decreases significantly. The volatility of un-ionized ammonia results in ammonia losses from lagoons (Reed 1985) and from wetlands under high pH and temperature conditions.

Ammonia nitrogen is important in wetlands and other surface waters for three reasons: (1) ammonia is the preferred nutrient form of nitrogen for most wetland plant species and for autotrophic bacteria species; (2) ammonia is chemically reduced and therefore can be readily oxidized in natural waters, resulting in significant oxygen consumption (about 4.3 g of oxygen per gram of ammonia nitrogen oxidized); and (3) un-ionized ammonia is toxic to many forms of aquatic life at low concentrations (typically at concentrations greater than 0.2 mg/L). Because ammonia is one of the principal forms of nitrogen found in many wastewaters, and because of its potential role in degrading the environmental condition of wetlands and other receiving waters, reducing ammonia concentration drives the design process for many wetland treatment systems.

Nitrite

Nitrite (NO_2^-) is an intermediate oxidation state of nitrogen (oxidation state of +3) between ammonia (−3) and nitrate (+5). Because of this intermediate energetic condition, nitrite is not chemically stable in most wetlands and is generally found at very low concentrations. Detectable levels of nitrite in wetlands frequently indicate incomplete nitrogen assimilation and the presence of an anthropogenic nitrogen source.

Nitrate

Nitrate (NO_3^-) is the most highly oxidized form of nitrogen (oxidation state of +5) found in wetlands. Because of this oxidation state, nitrate is chemically stable and would persist unchanged if not for several energy-consuming biological nitrogen transformation processes that occur.

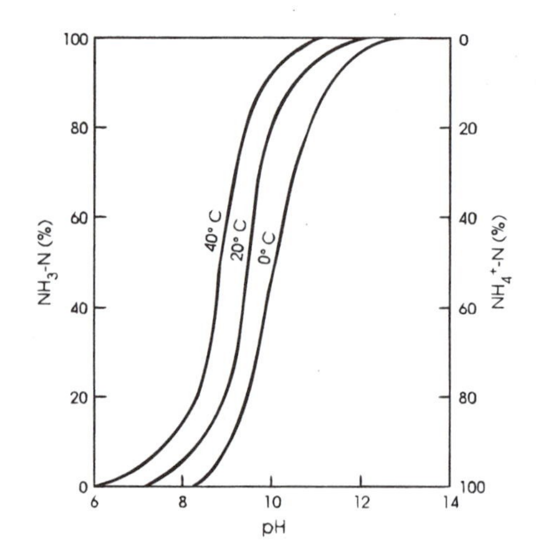

Figure 13-1 Effect of pH and water temperature on the fraction of total ammonia in the un-ionized (NH_3-N) and ionized (NH_4^+-N) forms.

Nitrate can also serve as an essential nutrient for plant growth, and in excess, nitrate leads to eutrophication of surface water. Nitrate and nitrite are also important in water quality control because they are toxic to infants (they result in a potentially fatal condition known as methylglobanemia) when present in drinking waters derived from polluted surface or groundwater supplies. The current regulatory criteria for nitrate in groundwater and drinking water supplies in the U.S. is 10 mg/L.

Gaseous and Atmospheric Forms of Nitrogen

Gaseous nitrogen may exist as dinitrogen (N_2), nitrous oxide (N_2O), nitric oxide (NO_2 and N_2O_4), and ammonia (NH_3). Under normal environmental conditions, dinitrogen is the only significant gaseous component in the atmosphere and is about 78 percent of air by volume. Nitrous oxide is an intermediate product of microbial denitrification and can be used in quantifying that process in wetlands via the acetylene blockage technique (Brodrick et al., 1988). Reddy and Patrick (1984) reported that typical concentration ratios for dinitrogen and nitrous oxide formed in laboratory test columns were between 100:1 to less than 10:1.

Dinitrogen gas and nitrous oxide are nearly inert for all purposes concerning eutrophication (except for nitrogen fixation discussed later) and the health of wetland organisms.

Atmospheric nitrogen also includes dissolved forms associated with cloud droplets and, hence, rainfall. Nitric oxide and ammonia are water soluble and react with water to form the soluble, ionized forms (nitrate, NO_3^-, and ammonium, NH_4^+). Ammonia volatilization and fossil-fuel combustion send the gaseous forms into the atmosphere; rainfall returns them to the earth's surface.

Like oxygen, dinitrogen gas and nitrous oxide occur in dissolved forms in surface waters in proportion to their partial pressure in the atmosphere. However, unlike dissolved oxygen, dissolved nitrogen gas has a very low biological activity compared to its abundance, and concentrations typically vary only in response to temperature changes. Dinitrogen gas saturation concentrations range from about 19 ml/L at 0°C to 12 ml/L at 25°C (Hutchinson, 1975).

ORGANIC NITROGEN COMPOUNDS

Organic nitrogen is made up of a variety of compounds including amino acids, urea and uric acid, and purines and pyrimidines. The amount of organic nitrogen in a water sample is functionally estimated by subtracting the NH_4-N concentration from the total Kjeldahl nitrogen (TKN), which is a measure of organic and ammonia nitrogen.

Amino Acids

Amino acids are the main components of proteins, which are a group of complex organic compounds essential to all forms of life. Amino acids consist of an amine group ($-NH_2$) and an acid group ($-COOH$) attached to the terminal carbon atom of a variety of straight carbon chain and aromatic organic compounds. The amine group is crucial in forming peptide chains that make up proteins. Only 20 different amino acids are combined to make the thousands of proteins found in living organisms. Nitrogen, primarily as amino acids, typically makes up from 1 to 7 percent of the dry weight of plants and animals, depending on protein content.

Urea and Uric Acid

Urea (CNH_4O) and uric acid ($C_4N_4H_4O_3$) are among the simplest forms of organic nitrogen in aquatic systems. Urea is formed by mammals as a physiological mechanism to dispose of ammonia that results when amino acids are used for energy production. Because ammonia is toxic, it must be converted to a less toxic form, urea, by the addition of carbon dioxide. Uric acid is produced by insects and birds for the same purpose. These organic forms of nitrogen are important in wetland treatment because they are readily hydrolyzed, chemically or microbially, resulting in the release of ammonia.

Pyrimidines and Purines

Pyrimidines and purines are heterocyclic organic compounds in which nitrogen replaces two or more of the carbon atoms in the aromatic ring. Pyrimidines consist of a single heterocyclic ring, and purines contain two interconnected rings. These compounds are synthesized from amino acids to become the main building blocks of the nucleotides that make up DNA in living organisms.

TOTAL NITROGEN

The mass of the various nitrogen forms can be added to estimate the total mass of nitrogen present in a wetland. In the water column, total nitrogen (TN) is calculated by adding the TKN value (organic and ammonia nitrogen) and the concentrations of nitrate and nitrite nitrogen (NO_x-N). In detritus, soils, and biological tissue, nitrogen is predominantly present as soluble and insoluble organic nitrogen. Total nitrogen in these wetland storages is approximately equal to TKN.

WETLAND NITROGEN STORAGES

Organic nitrogen compounds are a significant fraction of the dry weight of wetland plants, detritus, microbes, wildlife, and soils. In addition, nitrogen forms are a significant fraction of the total dissolved and suspended solids in many wetlands. The mass of these nitrogen storages varies in different wetland types, as does the accumulated nitrogen mass. A general idea of the sizes of these different storage compartments is necessary to understand the nitrogen fluxes discussed later (Figure 13-2).

Peaty sediments typically have nitrogen contents between 1 and 3 percent of dry weight, resulting in a nitrogen storage of about 1000 to 30,000 kg TN/ha in the upper 20 to 50 cm of wetland sediments (Table 13-1). The TN content of living biomass in marsh wetlands varies considerably among species, among plant parts, and among wetland sites. There is little variation from location to location within a homogeneous stand (Boyd, 1978). Example ranges of dry weight nitrogen percentages in natural wetlands are 0.93 to 2.56 percent for emergent plants; 1.86 to 3.79 percent for floating leaved plants; and 2.35 to 2.86 percent for submersed plants (Boyd, 1978).

Treatment wetlands are often nutrient enriched and display higher values of tissue nutrient concentrations than natural wetlands. For instance, cattail leaves in the discharge area of the Houghton Lake, MI wetland averaged 2.0 percent N; those in nutrient-poor control areas averaged 1.6 percent N.

Plant parts often show changes in nitrogen content during their various stages of existence. For instance, cattail standing dead leaves averaged 1.1 percent N and litter 2.2 percent N in the discharge area at Houghton Lake.

Different plant parts may show large differences in nitrogen content, and the seasonal variability may be very large. The extent of this variability is shown in Figure 13-3 for *Phragmites australis* for a reed stand in the margin of Templiner See, a heavily loaded eutrophic shallow lake in Germany (Kühl and Kohl, 1993). Biomass collected at the end of the growing season displays much lower nitrogen content than in the spring. Klopatek (1978)

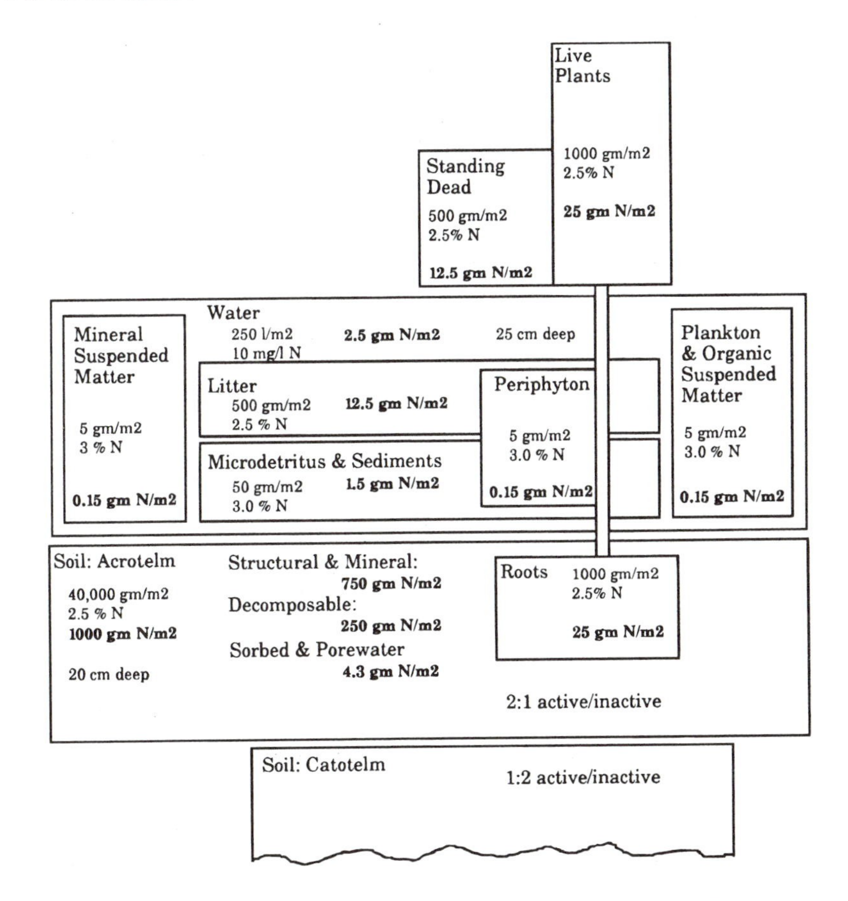

Figure 13-2 Illustration of nitrogen storages in a peat-based wetland.

has shown trends of the same magnitude for cattail roots and shoots. It is apparent that the timing and location of vegetation samples can greatly affect subsequent calculations of nitrogen storage in biomass.

The range of nitrogen content of above- and below-ground tissues is normally between 100 and 500 kg/ha in freshwater marshes and up to about 3000 kg/ha in wooded swamps. The nitrogen standing stock in the litter (detritus) and standing dead compartments in wetlands are typically between 20 and 200 kg/ha each in marshes and up to about 500 kg/ha in forested wetlands. For comparison, the nitrogen stored in the water column of a 0.25-m-deep wetland with a total nitrogen concentration of 10 mg/L is only 25 kg/ha.

These storages are seasonal, reflecting the growth cycle of the plant in question. The processes of growth, death, litterfall, and decomposition operate year-round and with different speed and seasonality depending on climatic conditions and genotypical habit. Even in northern climates, the total annual growth is slightly larger than the end-of-season standing crop, by about 20 percent (Whigham et al., 1978). In southern climates, measurements show 3.5 to 10 turnovers of the live above-ground standing crop in the course of a year (Davis, 1994). Decay processes release most of the nitrogen uptake, with the residual accreting as new sediments and soils.

NITROGEN TRANSFORMATIONS IN WETLANDS

Figure 13-4 shows the principal components of the nitrogen cycle in wetlands. The various forms of nitrogen are continually involved in chemical transformations from inorganic

Table 13-1 Typical Nitrogen Values for Wetland Substrates

Location	Total Nitrogen (mg/kg)	Exchangeable N (mg/kg)	Description	Ref.
Horry Co., SC	1,611	—	Forested Carolina bay receiving municipal wastewater	CH2M HILL, 1991
Waldo, FL	—	214	Forested wetland receiving partially treated municipal wastewater for 50 years	DeBusk and Reddy, 1987
Everglades Water Conservation Areas, Florida: Cattail *Cladium*	24,200–33,200 21,300–41,400	42–80, 30–106	Impounded cattail and sawgrass marsh receiving agricultural drain water	Reddy et al., 1991a
Holeyland Wildlife Management Area, Florida	13,100–28,100	25–149	Former Everglades sawgrass marsh, hydrologically altered, histosols	Reddy et al., 1991b
Southeastern Cypress Wetlands	3,100–23,200	—	Histosols	Coultas and Duever, 1984
Southeastern Cypress Wetlands	200–5,000	—	Utisols, Entisols, albisols, inceptisols	Coultas and Duever, 1984
Pasco Co., FL	8,300–26,200	—	Cypress domes, histosols	CH2M HILL, 1985
Wisconsin	13,600–19,400	—	Riverine marsh (*Scirpus*)	Klopatek, 1978
Scandinavia, Europe, Canada, Worldwide (N = 12)	12,800 ± 3,600	—	Rich/extremely rich fens	Waughman and Bellamy, 1980
Scandinavia, Europe, Canada, Worldwide (N = 16)	15,200 ± 6,950	—	Poor/intermediate fens	Waughman and Bellamy; 1980
Scandinavia, Europe, Canada, Worldwide (N = 17)	12,300 ± 3,700	—	Bogs	Waughman and Bellamy, 1980
Top 5 cm Houghton Lake treatment zone (rich fen, eutrophying)	22,500 ± 500	22 ± 7	Cattail/sedge over peat	Houghton Lake, unpublished data

to organic compounds and back from organic to inorganic. Some of these processes require energy (typically derived from an organic carbon source) to proceed, and others release energy, which is used by organisms for growth and survival. All of these transformations are necessary for wetland ecosystems to function successfully, and most chemical changes are controlled through the production of enzymes and catalysts by the living organisms they benefit.

The several nitrogenous chemical species are interrelated by a reaction sequence. Nitrogen is speciated in several forms in wetlands and partitioned into water, sorbed, and biomass phases. The surface-flow (SF) wetland is also stratified vertically into zones which promote different nitrogen reactions. And, as a further complicating factor, microenvironments around

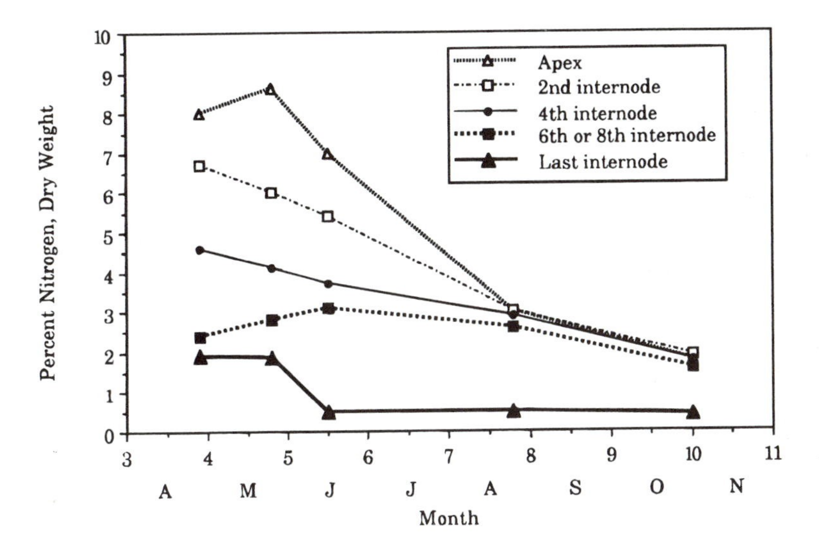

Figure 13-3 Nitrogen content in *Phragmites australis* as a function of season and position above ground. (Redrawn from the data of Kühl and Kohl, 1993.) The site was a highly productive reed stand which generated 1500 g/m^2 of biomass over the June to August period.

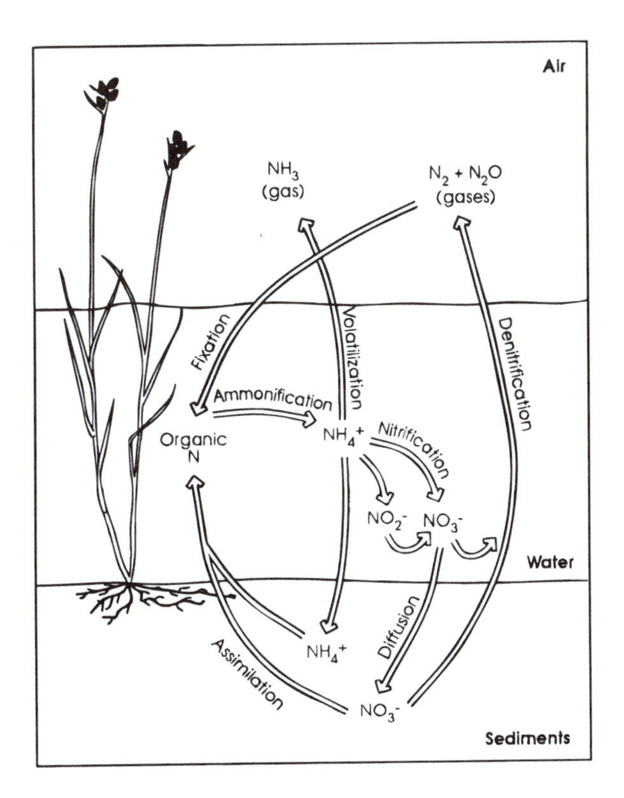

Figure 13-4 Simplified wetland nitrogen cycle.

individual plant roots may differ from the bulk surroundings (Reddy and D'Angelo, 1994). Although the detailed processes are well known, they have not been adequately quantified in many cases.

A number of processes transport or translocate nitrogen compounds from one point to another in wetlands without resulting in a molecular transformation. These physical transfer processes include the following: (1) particulate settling and resuspension, (2) diffusion of dissolved forms, (3) plant uptake and translocation, (4) litterfall, (5) ammonia volatilization, (6) sorption of soluble nitrogen on substrates, (7) seed release, and (8) organism migrations.

In addition to the physical translocation of nitrogen compounds in wetlands, five principal processes transform nitrogen from one form to another: (1) ammonification (mineralization), (2) nitrification, (3) denitrification, (4) nitrogen fixation, and (5) nitrogen assimilation.

A detailed understanding of these nitrogen translocation and transformation processes is important when designing wetland treatment systems. The following sections describe these processes and the environmental factors that regulate the transformations. Later in this chapter, empirical and theoretical design methods are presented for predicting the treatment wetland area necessary to accomplish the given nitrogen transformations.

AMMONIFICATION (MINERALIZATION)

Ammonification is the biological transformation of organic nitrogen to ammonia and is the first step in mineralization of organic nitrogen (Reddy and Patrick, 1984). This process occurs through microbial breakdown of organic tissues containing amino acids, through excretion of ammonia directly by plants and animals, and by hydrolysis of urea and uric acid.

Ammonia is converted from organic forms through a complex, energy-releasing, multistep, biochemical process. In some cases, this energy is used by microbes for growth, and ammonia is directly incorporated into microbial biomass. A large fraction (up to 100 percent) of the organic nitrogen in many wastewaters is readily converted to ammonia, requiring the designer to include enough treatment capacity to oxidize all of this reduced nitrogen. Kinetically, ammonification proceeds more rapidly than nitrification, thus creating the potential for increasing ammonia concentrations along the flow path of a wetland and requiring design for nitrogen removal to be based on the slower nitrification process.

Ammonifying bacteria were found mainly associated with roots in gravel bed systems *circa* 10^7 gram for roots vs 10^4 per gram for gravel (May et al., 1990). The activity for ammonification, measured in water and for biofilms on rhizomes and gravel, yields the same conclusion (Williams et al., 1994). There were not large distance effects found for a 100m long gravel bed in the United Kingdom (Williams et al., 1994).

Ammonification proceeds more slowly in anaerobic than in aerobic conditions because of the reduced efficiency of heterotrophic decomposition in anaerobic environments. Nevertheless, ammonia nitrogen is more likely to accumulate in anaerobic systems because of decreased nitrification rates, and ammonia nitrogen may be greater in low-oxygen wetlands. In aerobic wetland environments, the ammonia resulting from ammonification of organic nitrogen is more likely to undergo nitrification, and TN typically will be reduced through denitrification. Under nonflooded conditions, ammonification is substantially reduced (Reddy and Patrick, 1984).

The rate of ammonification in flooded soils also depends on temperature and pH (Reddy and Patrick, 1984). The ammonification rate increases with a doubling of the rate constant for a temperature increase of 10°C ($\theta = 1.07$). Optimum temperatures for ammonification range from 40 to 60°C, which typically are not encountered in wetland treatment systems. The optimum pH range for ammonification is between 6.5 and 8.5 (Reddy and Patrick, 1984).

Measurements of ammonification rates in natural wetlands range from 3 to 35 mg N/m^2/d (annual average 1.5 g/m^2/yr) in a swamp forest in central Minnesota (Zak and Grigal,

1991) and 4.3 to 5.9 $g/m^2/yr$ in a Minnesota bog (Urban and Eisenreich, 1988). Higher nitrogen mineralization rates reported in organic soils in Florida by Reddy (1982) were 41 to 125 and 22 $g/m^2/yr$ by Messer and Brezonik (1977).

In conventional biological treatment system design, ammonification of soluble organic nitrogen is typically treated as a first-order process of the form

$$R_{AMMON} = k_{VAMMON} C_{SON} \tag{13-2}$$

where R_{AMMON} = ammonification rate, mg/L soluble organic nitrogen per day
k_{VAMMON} = volume-based ammonification rate constant, d^{-1}
C_{SON} = concentration of soluble organic N, mg/L

A commonly used value for k_{VAMMON} in conventional biological treatment systems is 0.1 d^{-1}.

R_{AMMON} can be converted to an area-based organic nitrogen removal rate (J_{ON}) with units of $g/m^2/yr$ by multiplying by water depth, h in m:

$$J_{ON} = 365\, R_{AMMON} h \tag{13-3}$$

so,

$$J_{ON} = 365\, k_{VAMMON} h C_{SON} \tag{13-4}$$

Treatment wetlands data display decreases in organic nitrogen with contact time, which are consistent with first-order reduction kinetics, but show a non-zero background concentration. For long detention times, corresponding to large distances from the inlet, small concentrations of organic nitrogen persist.

For wetland treatment systems, the volume-based rate constant, k_{VAMMON}, can be replaced with an area-based rate constant, k_{ON} with units or m/yr, to give Equation 13-5:

$$J_{ON} = k_{ON}(C_{ON} - C^*_{ON}) \tag{13-5}$$

where C^*_{ON} = background wetland organic nitrogen concentration, mg/L.

Assuming an average wetland depth of 30 cm and using the value of k_{VAMMON} from biological treatment systems of 0.1 d^{-1}, the area-based ammonification rate constant would be 10.9 m/yr.

Actual first-order, area-based ammonification rate constants can be calibrated to operational wetland treatment system data and are listed in Table 13-2. The overall average value for k_{ON} for SF wetland treatment systems is 10.4 m/yr; for subsurface-flow (SSF) wetlands it is 31.3 m/yr. The conservative data reduction premise of a plug flow hydraulic pattern has been made in developing Table 13-2. Therefore, the rate constants shown are the minimum values that could be extracted from the data. For instance, the open water trench at Richmond, NSW was tracer tested, and the test indicated a hydraulic pattern intermediate between one and two well-mixed units in series (tank-in-series [TIS] model, see Chapter 9). Parameter estimation shows k_{ON} = 4.3 m/yr under the plug flow presumption; 5.2 m/yr for two TIS; and 6.4 for one well-mixed unit.

The average value of C^*_{ON} is approximately 1 to 2 mg/L for SF wetlands and less for SSF wetlands.

Ammonification rates depend on temperature, among many other influences in treatment wetlands. For example, Figure 13-5 illustrates the average k_{ON} values for the Listowel,

Table 13-2 Example First-Order, Area-Based Rate Constants for Ammonification for Treatment Wetlands

Location	Wetland	Type	Data Quarters	HLR (cm/d)	ORG N In (mg/L)	ORG N Out (mg/L)	k (m/yr)	C* (mg/L)	T °C	Data Ref.
Surface-flow marshes										
Listowel, Ontario	1	SF con marsh	16	2.67	5.10	2.89	7.3	1.5	8.0[a]	Herskowitz, 1986
	2	SF con marsh	16	2.84	5.10	3.23	4.3	0.4	8.0[a]	Herskowitz, 1986
	3	SF con marsh	16	1.92	5.10	2.54	7.9	1.5	7.84	Herskowitz, 1986
	4	SF con marsh	16	1.95	9.97	2.79	18.0	2.1	8.02	Herskowitz, 1986
	5	SF con marsh	16	2.60	9.97	3.49	12.5	1.4	8.0[a]	Herskowitz, 1986
Cobalt, Ontario	1	SF con marsh	5	5.31	3.04	1.15	21.7	0.2	0.4[a]	Miller, 1989
Benton, KY	1	SF con marsh	4	1.71	8.71	2.61	11.7	1.7	16.14	Choate et al., 1990a
	2	SF con marsh	4	1.71	8.71	3.08	8.9	1.5	17.75	Choate et al., 1990a
Gustine, CA	1A	SF con marsh	4	4.19	15.50	9.52	7.5	[b]	17[a]	Walker and Walker, 1990
	1B	SF con marsh	4	2.09	16.22	6.08	7.5	[b]	17[a]	Walker and Walker, 1990
	1C	SF con marsh	4	1.05	14.39	8.08	2.2	[b]	17[a]	Walker and Walker, 1990
	1D	SF con marsh	4	4.19	18.05	8.09	12.3	[b]	17[a]	Walker and Walker, 1990
	2A	SF con marsh	4	4.46	26.85	8.89	18.0	[b]	17[a]	Walker and Walker, 1990
West Jackson Co., MS	T1-7	SF con marsh	11	2.42	7.77	2.22	[b]	2.6	18[a]	NADB, 1993
Richmond, NSW	1	SF open water	8	6.41	8.29	6.90	4.3	0.0	19.70	Bavor et al., 1988
Forested surface flow							10.75			
Reedy Creek, FL	1	SF forested	42	3.51	1.86	0.80	34.3	0.7	22.30	NADB, 1993
Vereen, SC	1	SF forested	17	0.36	2.46	2.24	[b]	2.6	17.40	NADB, 1993
Floating/submergent aquatics										
New Zealand	2 Wetlands	SF con FAP	7	5.52	16.00	2.70	50.2	1.5	15	van Oostrom, 1994
Richmond, NSW	4	SF con SAB	8	7.34	8.29	1.59	45.5	0.0	17.16	Bavor et al., 1988
Soil-based reed beds										
Denmark	44 Wetlands	Soil-based Phrag	8 Typical	4.53	13.58	3.88	20.8	2.3	6.5[a]	Schierup et al., 1990b
Subsurface-flow reed beds										
United Kingdom	11 Wetlands	SSF Phrag	8 Typical	18.30	3.20	1.70	56.8	1.3	9.5[a]	Green and Upton, 1993
Gravel beds										
Benton, KY	3	SSF bulrush	4	5.22	9.42	1.53	28.3	0.4	16.74	Choate et al., 1990a
Hardin, KY	1	SSF cattail	4	8.81	11.67	1.83	29.0	0.0	16.15	Choate et al., 1990a
	2	SSF bulrush	4	7.34	11.96	2.09	30.8	0.0	13.36	Choate et al., 1990a
North America	9 Wetlands	SSF	Various	4.66	9.83	4.62	40.5	3.1		NADB, 1993
Richmond, NSW	2	Gravel only	8	3.83	8.29	0.68	24.9	0.0	18.57	Bavor et al., 1988
	3	SSF bulrush	8	4.61	8.29	1.56	34.8	0.0	18.20	Bavor et al., 1988
	5	SSF cattail	8	5.08	8.29	0.01	30.8	0.0	18.37	Bavor et al., 1988

[a] Estimated as mean annual, mean daily air temperature

[b] Data range precludes accurate parameter estimate.

Note: HLR—define, ORG N-organic nitrogen.

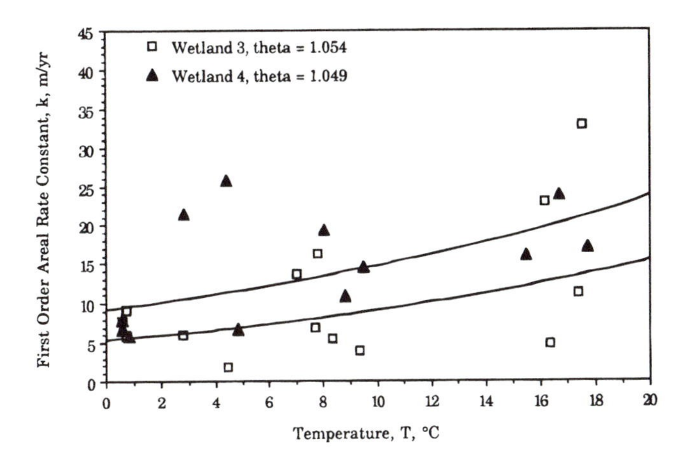

Figure 13-5 The temperature dependence of the ammonification rate constant for two Listowel wetlands. The value of C* = 1.6 and 1.7 mg/L, respectively, have been assumed based on regression over 4 years of data. (Data from Herskowitz, 1986.)

Ontario SF constructed wetlands (Systems 3 and 4, for which temperature was measured) as a function of temperature. The expected pattern of increasing ammonification rate constants is seen, and the temperature coefficients $\theta = 1.049$ and $\theta = 1.054$ are close to those measured in laboratory columns (Reddy and Patrick, 1984).

In view of the information from treatment wetlands and related technologies, an estimate of the temperature factor for organic reduction is $\theta = 1.05$. When the rate constants in Table 13-2 are adjusted to 20°C and averaged, the results are

k_{ON20} = 17 m/yr SF (includes soil-based reed beds and aquatics)
k_{ON20} = 35 m/yr SSF (includes U.K. reed beds and U.S. gravel beds)
k_{VON20} = 0.40 d^{-1} SSF (60 cm at 40 percent porosity)
C^* = 1.5 mg/L

These values are to be regarded as central tendencies for the systems listed. The intersystem variability is large in response to site factors not included in the simplified design model. For instance, septage-fed SSF wetlands may have higher concentrations and lower k-values.

Wetland data are not presently adequate to determine seasonality or temperature dependence of C* for ammonification, but the central tendency is $C^*_{ON} = 1.5$ mg/L.

NITRIFICATION

Nitrification Reaction Chemistry

Nitrification is the principal transformation mechanism that reduces the concentration of ammonia nitrogen in many wetland treatment systems by converting ammonia nitrogen to nitrates. Nitrification is a two-step, microbially mediated process in wetlands shown in Figure 13-6 and summarized by Equations 13-6 and 13-7 (Reddy and Patrick, 1984):

$$NH_4^+ + 1.5\ O_2 = 2\ H^+ + H_2O + NO_2^- \tag{13-6}$$

$$NO_2^- + 0.5\ O_2 = NO_3^- \tag{13-7}$$

The first step summarized in Equation 13-6 is mediated primarily by bacteria in the genus *Nitrosomonas* and the second step by bacteria in the genus *Nitrobacter.* Both steps can

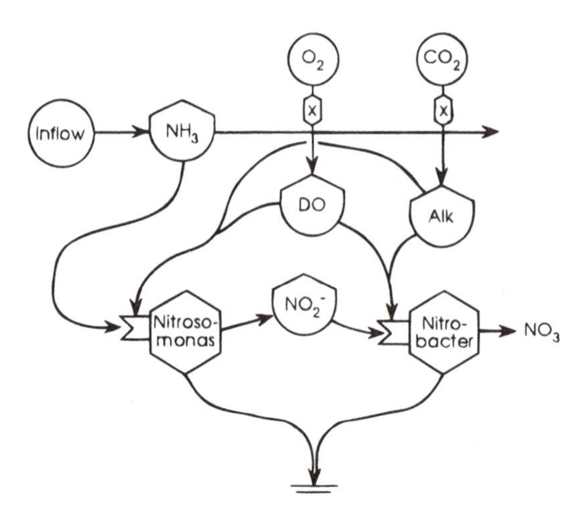

Figure 13-6 Simplified diagram of the nitrification process in wetlands. (See Figure 7-15 for an explanation of symbols.)

proceed only if oxygen is present; however, nitrification will still occur down to about 0.3 mg/L of dissolved oxygen (Reddy and Patrick, 1984). The actual nitrification rate may be controlled by the flux of dissolved oxygen into the system, which typically is mass transfer from atmospheric sources, through water, in wetland treatment systems. Since mass transfer is a first-order process, the nitrification rate in wetlands also might be considered first order.

The overall process of nitrification can be summarized by a single expression:

$$NH_4^+ + 2.0\ O_2 = NO_3^- + 2\ H^+ + H_2O \quad (13\text{-}8)$$

Based on this stoichiometric relationship, the theoretical oxygen consumption during the nitrification reaction is about 4.6 g of O_2 per gram of NH_3-N oxidized.

The oxidation reactions summarized as Equations 13-6 and 13-7 both release energy used by *Nitrosomonas* and *Nitrobacter* for cell synthesis. The combined processes of cell ($C_5H_7NO_2$) synthesis and oxidation reduction occurring during nitrification are summarized by Equations 13-9 and 13-10 from the U.S. EPA (1993e):

$$55\ NH_4^+ + 76\ O_2 + 109\ HCO_3^- = C_5H_7NO_2 + 54\ NO_2^- + 57\ H_2O + 104\ H_2CO_3 \quad (13\text{-}9)$$

for synthesis of Nitrosomonas cells and the following for Nitrobacter cell synthesis:

$$400\ NO_2^- + NH_4^+ + 4\ H_2CO_3 + HCO_3^- + 195\ O_2 = C_5H_7NO_2 + 3\ H_2O + 400\ NO_3^- \quad (13\text{-}10)$$

Equations 13-9 and 13-10 can be combined to determine the overall oxidation of ammonia and the synthesis of cell biomass during nitrification:

$$NH_4^+ + 1.83\ O_2 + 1.98\ HCO_3^- = 0.021\ C_5H_7NO_2 + 1.04\ H_2O + 0.98\ NO_3^- + 1.88\ H_2CO_3 \quad (13\text{-}11)$$

Note that 1 mol of ammonia nitrogen only produces about 0.021 mol of microbial biomass (0.17 g of dry weight biomass per gram of ammonia nitrogen consumed). This conversion is only a small fraction of heterotrophic biomass production because of the smaller amount of energy that can be derived from these oxidation and reduction reactions.

Equation 13-11 can be used to examine the major environmental requirements and consequences of the nitrification process. Nitrification of ammonia to nitrate consumes oxygen and bicarbonate ion and releases water and carbonic acid in addition to biomass and nitrate nitrogen. This overall reaction predicts that about 3.22 g of O_2 are consumed per gram of NH_4^+-N oxidized and 1.11 g of O_2 are consumed per gram of NO_2^- oxidized for a total use of about 4.3 g of O_2 for each gram of ammonia nitrogen nitrified. Thus, the actual oxygen consumption rate during nitrification is less than the 4.6 value predicted in Equation 13-8 because of the contribution of some oxygen from carbonate consumed during cell synthesis.

Approximately 7.14 mg/L (as $CaCO_3$) of alkalinity are consumed for each nitrified mg/L of ammonia nitrogen, and 1.98 mol of H^+ are released for each mole of ammonia nitrogen consumed. A high rate of nitrification lowers the alkalinity and pH of the water body.

Results From Other Technologies: Monod Kinetics and Environmental Factors

Nitrification has been extensively studied in connection with other wastewater treatment technologies, such as suspended growth and attached growth units. Treatment wetlands imbed the same microbiology in a complex ecological setting. The factors that are known to affect the nitrification in concrete and steel devices are also operative in wetlands, but they are combined with, and may be overwhelmed by, other competing wetland processes. It is useful to review the results from these other technologies because studies thus far have not fully delineated the effects of environmental variables in wetlands.

Because nitrification is mediated by microbes, the rate of nitrification is directly proportional to the growth of nitrifier bacteria. The Monod expression (introduced in Chapter 7) provides a convenient relationship for predicting bacterial growth rate (u) based on the concentration of a limiting substrate (S), the maximum potential growth rate when that substrate is not limiting (u_{max}), and the half saturation constant K_s for that substrate:

$$u = (u_{max}S)/(K_s + S) \qquad (13\text{-}12)$$

Besides oxygen and micronutrients, the limiting substrate for nitrifier growth is ammonia nitrogen for *Nitrosomonas* and nitrite nitrogen for *Nitrobacter.* Nitrite generally converts to nitrate faster than ammonia nitrogen converts to nitrite, so the sequential oxidation of ammonia nitrogen to nitrate is controlled kinetically by the slower first step, the *Nitrosomonas* bacterial growth rate. Based on this observation, the nitrifier growth rate (u_{NITR}) in yr^{-1}, can be described by the expression:

$$u_{NITR} = (u_{NITRmax}C_{AN})/(K_{NITR} + C_{AN}) \qquad (13\text{-}13)$$

where $u_{NITRmax}$ = maximum nitrifier growth rate, yr^{-1}
C_{AN} = the ammonium nitrogen concentration, mg/L
K_{NITR} = the nitrification half saturation constant, mg/L of NH_4-N

Reddy and Patrick (1984) summarize values of K_{NITR} applicable to flooded soils in the range of 1 to 10 mg/L of NH_4-N at temperatures between 20 and 30°C. Weber and Tchobanoglous (1986) estimated K_{NITR} = 0.8 mg/L for a pilot scale hyacinth wetland. K_{NITR} values as low as these would translate to a near-zero-order dependence of nitrification on ammonium concentration for many treatment wetlands, namely those with inlet ammonium greater than about 10 to 30 mg/L. In that case, over most of the range of concentrations during the reduction of ammonium, $C_{AN}/(K_{NITR} + C_{AN}) = 1$, and the rate would be zero order. There is confirmation of this effect for the low-concentration data of Weber and Tchobanoglous (1986), but not for high concentrations, because both Monod and first-order fits are nearly indistinguishable at elevated ammonium concentrations. Wetland microcosm data (Gale et al., 1993; Zhu and Sikora, 1994) do not support the Monod model, because ammonium disappearance is apparently first order rather than zero order at higher concentrations.

The nitrifier growth rate is directly proportional to the nitrification rate, U_{NITR}, based on a nitrifier yield coefficient Y_{NITR}:

$$U_{NITR} = u_{NITR}/Y_{NITR} \qquad (13\text{-}14)$$

where U_{NITR} = nitrification rate in gm of NH_4-N removed/gm nitrifiers per year
Y_{NITR} = nitrifier yield coefficient, gm nitrifiers grown/gm NH_4-N removed

A typical value of Y_{NITR} based on pure cultures of nitrifiers is 0.15 g/g (WPCF, 1983; Brown and Caldwell, 1975). This yield coefficient was developed for the design of conventional biological nitrification processes.

From an empirically derived effective nitrifier mass per wetland surface area V_{NITR} (g/m^2), this expression for nitrification can be presented as the following:

$$J_{NITR} = U_{NITR}V_{NITR} = (u_{NITR}V_{NITR})/Y_{NITR} \qquad (13\text{-}15)$$

where J_{NITR} = the nitrification rate in gm NH_4-N/m^2/yr.

Nitrifying bacteria were enumerated at the Listowel SF wetlands (Herskowitz, 1986), in the top 2 cm of sediment near the inlet and near the outlet, for three of the five wetlands. Numbers of both were typically in the general range of 10^4 to 10^5 per gram of sediment. There were spring and fall increases, attributed to increases in sediment redox potential. Numbers were typically lower in winter.

Nitrifiers were found at lower levels in a U.K. gravel bed, approximately 10^4 per gram for ammonium oxidizers and 10^3 per gram for nitrite oxidizers; most were associated with roots rather than the gravel (May et al., 1990).

Although nitrifying bacteria populations have been enumerated in wetlands, their biomass has not been reported. However, operational data from some wetland treatment systems can help estimate empirical values for V_{NITR}. Data from three typical, southeastern wetland treatment systems resulted in an average estimate of V_{NITR} of 0.05 g nitrifiers/m^2. The value of V_{NITR} can be used with estimates of u_{NITR} and Y_{NITR} to estimate the nitrification rate in wetland treatment systems. Calibration of u_{NITR} to treatment wetland data has not yet been attempted.

Environmental Factors Affecting Nitrification in Suspended Growth

Considerable effort has been expended to research the factors that influence nitrification in conventional wastewater treatment systems. Most work has examined nitrification in suspended growth, activated sludge systems. However, nitrification kinetics also have been documented in conventional attached growth systems such as trickling filters and rotating biological contactors (RBCs). Less experimental work has been published concerning nitrification kinetics in wetlands. This book reviews information from various nonwetland biological treatment systems to investigate potential environmental effects.

Temperature has a significant effect on both the maximum nitrifier growth rate ($u_{NITRmax}$) and on the nitrification half saturation constant (K_{NITR}) in suspended growth treatment systems based on the following empirically derived expressions from U.S. EPA (1993e) and converted to a yr^{-1} basis:

$$u_{NITRmax} = 172\ e^{0.098(T-15)} \tag{13-16}$$

where T = water temperature, °C, and

$$K_{NITR} \cong 1.0\ mg/L \tag{13-17}$$

These expressions are valid for water temperatures between about 5 and 30°C. Examination of Equation 13-16 indicates that $u_{NITRmax}$ increases significantly at higher water temperatures and decreases at lower water temperatures. The equivalent temperature factor for nitrification in Equation 13-16 is $\theta = 1.10$. The optimum temperature range for nitrification in pure bacterial cultures is from 25 to 35°C. Temperatures lower than 15°C are reported to lower nitrifier growth rates drastically (Reddy and Patrick, 1984). At water temperatures above 30°C, the nitrification rate begins to decline sharply. It is not currently known whether these same relationships adequately describe temperature effects on nitrification in treatment wetlands.

Dissolved oxygen is an important environmental factor that can limit nitrification rates in wetlands and conventional treatment systems. Equation 13-11 predicts a stoichiometric dissolved oxygen requirement of 4.3 g of O_2 per gram of NH_4-N that is fully nitrified to NO_3-N. Thus, oxidation of as little as 2 mg/L of ammonia nitrogen can consume more dissolved oxygen (8.6 mg/L) than is typically present in saturated surface waters. This dissolved oxygen must be constantly renewed by mass transfer from the atmosphere in wetland treatment systems, perhaps supplemented by a small amount of radial oxygen loss from roots.

The effect of dissolved oxygen concentration on the rate of nitrification in suspended growth treatment systems is frequently modeled by a second Monod limiting factor expression:

$$u_{NITR} = (u_{NITRmax} C_{DO})/(K_{DO} + C_{DO}) \tag{13-18}$$

where C_{DO} = the dissolved oxygen concentration, mg/L
K_{DO} = the dissolved oxygen half saturation constant, mg/L

Reported values of K_{DO} range from as low as 0.15 to 2.0 mg/L (U.S. EPA, 1993e). The U.S. EPA (1993e) suggests a design value of 1.0 mg/L for extended aeration systems, but lower values may be appropriate for wetland systems where dissolved oxygen concentrations are typically less than 2 mg/L. This half saturation constant needs to be the subject of additional research in wetland treatment systems. Observations at a number of SF and SSF wetland treatment systems indicate that rate of ammonia loss is very low when there is less than about 0.5 mg/L of dissolved oxygen present.

Alkalinity is consumed during the process of nitrification. Measurements of alkalinity reduction in nonwetland treatment systems are between 6.3 and 7.4 mg/L of alkalinity as $CaCO_3$ per mg/L of ammonia nitrogen oxidized for attached growth treatment systems (U.S. EPA, 1993e). These estimates compare favorably with the stoichiometric estimate of this ratio as 7.14. Alkalinity of surface waters is not immediately replenished from internal or external sources.

Adequate alkalinity in the untreated water must exist for a given level of ammonia nitrogen oxidation to occur. For example, if an ammonia nitrogen concentration of 20 mg/L is to be reduced to less than 2 mg/L, the total alkalinity of the water must exceed 128 mg/L. Alkalinity over this amount (at least 50 mg/L as $CaCO_3$) helps to buffer the pH changes that result from adding H^+ ions to the water column during nitrification. Some of the alkalinity lost during nitrification may be returned to the water column as nitrate nitrogen is denitrified. Assurance of adequate alkalinity is a priority in design of wetlands for nitrification.

Nitrification tends to lower pH in the water column by forming carbonic acid; however, this pH reduction can be buffered by the loss of carbon dioxide from the water column to the atmosphere. In quiescent wetlands or where diffusion through the water surface is impeded by a floating aquatic plant cover, pH may be lowered enough to limit the nitrification rate. The optimal pH range observed for nitrification in suspended growth treatment systems is between about 7.2 and 9.0 (Metcalf and Eddy, 1991). Other studies in nonwetland treatment systems have indicated that lower pH values (as low as 5.5 to 6.0) are not toxic to nitrifiers, but only inhibitory, and that nitrifier populations can adapt and achieve near optimum nitrification rates at a pH as low as 6.6. The WPCF (1983) nutrient control manual recommends that pH be maintained above 7.2 for nitrification stability, and the U.S. EPA (1993e) provides a formula for estimating the nitrification rate at pH values between 6.0 and 7.2:

$$u_{NITR} = u_{NITRmax}(1 - 0.833(7.2 - pH)) \qquad (13\text{-}19)$$

Treatment wetlands almost always operate at circumneutral pH (see Chapter 10); consequently, this factor should be a minor influence on nitrification in those systems.

Monod-type limiting equations can be multiplied to determine the reaction rate based on multiple variables (see Chapter 7). Combining Equations 13-13 and 13-16 through 13-19 provides an expression for nitrification rate in suspended growth treatment systems based on ammonia nitrogen concentration, dissolved oxygen concentration, water temperature, and pH:

$$u_{NITR} = 172\ e^{0.098(T-15)}[1 - 0.833(7.2 - pH)]\left(\frac{C_{AN}}{C_{AN} + 1}\right)\left(\frac{C_{DO}}{C_{DO} + 1.3}\right) \qquad (13\text{-}20)$$

This value for u_{NITR} in units of yr^{-1} can be substituted into Equation 13-15 to estimate the nitrification rate in a suspended growth system for a given combination of environmental conditions.

Attached Growth Treatment Systems

Wetlands possess some of the same characteristics as conventional attached growth treatment systems, such as trickling filters and rotating biological contactors. There is a higher availability of oxygen in the conventional processes, as well as a higher surface area for attachment, compared to wetlands. There is competition for attachment sites between nitrifying bacteria and the heterotrophic bacteria responsible for carbon oxidation. As a consequence, nitrification is impeded until biochemical oxygen demand (BOD) levels are reduced to approximately the same level as ammonium nitrogen. Design of these systems is based upon loading criteria (Metcalf and Eddy, 1991).

The temperature coefficients for attached growth processes are on the order of $\theta = 1.07$, based on limited information on trickling filters (U.S. EPA, 1993e).

Relation to Treatment Wetlands

The application of Equations 13-15 and 13-20, or the attached growth loading criteria, to treatment wetlands is impractical for several important reasons. First, the prediction of the specific yield per unit area of wetland (V_{NITR}/Y_{NITR}) is not presently possible. Second, wetlands promote attached growth processes to the near exclusion of suspended growth processes. Finally, treatment wetlands house other processes by which ammonium losses and gains may occur: nitrification in oxygenated microzones in the rhizosphere, plant uptake and biomass decay, and the above-water decomposition of plant material. Mineralization of organic nitrogen produces ammonium nitrogen. Thus, even if wetland biofilm processes are describable as an attached growth nitrification system, other processes can prevent accurate predictions of ammonium loss.

The rates of ammonium reduction in treatment wetlands are presently predictable only from wetland data.

Ponds

A second source of information on ammonia loss rates may be found in the literature on facultative ponds. Pano and Middlebrooks (1982) analyzed data from several ponds. They calibrated a first-order areal, well-mixed model to some of the data and verified the model with the remainder of the data. The precise mechanism for ammonium reduction was not studied, but the model was chosen based on previous modeling of ammonia stripping ponds. Two temperature regions were identified, in which

$$1 \leq T \leq 20: k_{av} = (.0038 + .000134T_{[°C]})\exp[(1.041 + .044T_{[°C]})*(pH - 6.6)] \quad (13\text{-}21)$$

$$20 \leq T \leq 25: k_{av} = (.0005035)\exp[1.540*(pH - 6.6)] \quad (13\text{-}22)$$

Data ranges:

$6.4 \leq pH \leq 9.6$
$0.7 \leq T \leq 28.1$ °C
$4.1 \leq C_{in} \leq 68$ mg/L
$2.9 \leq C_{out} \leq 83$ mg/L

The values of rate constants for this model are shown in Table 13-3. Predictions are for k_{av} in the range of 1 to 10 m/yr and for pH in the range 6.5 to 8.0, which is the range in treatment wetlands. The equivalent temperature coefficients are in the range $\theta = 1.045$ to 1.069 and for pH in the range 7.0 to 7.5.

Table 13-3 Rate Constants for Ammonia Loss from Ponds (m/yr)

T (°C)	pH = 6.5	pH = 7.0	pH = 7.5	pH = 8.0
0	1.25	2.10	3.54	5.96
5	1.44	2.70	5.08	9.53
10	1.62	3.39	7.11	14.92
15	1.79	4.19	9.80	22.95
20	1.95	5.10	13.33	34.82
25	1.58	3.40	7.35	15.87

Note: Units are in meters per year. Based upon data of Pano and Middlebrooks, 1982.

Area-Based, First-Order Ammonia Disappearance Model

An alternate expression that can be used to model the disappearance of ammonia nitrogen (nitrification and other processes combined) in wetland treatment systems is the area-based, first-order model, analogous to that presented above for organic nitrogen removal (Equation 13-5). There appears to be a very small background ammonia nitrogen outlet concentration for SF wetlands: in the range of 0.05 to 0.10 mg/L. For this reason, C* is assumed to equal zero, and the overall process of ammonia loss can be expressed as the following:

$$J_{AN} = k_{AN}C_{AN} \tag{13-23}$$

where k_{AN} = area-based, first-order ammonia disappearance rate constant, m/yr.

The integrated form of the ammonia nitrogen mass balance for this uptake rate and plug flow conditions is

$$\ln\left[\frac{C_{AN,o}}{C_{AN,i}}\right] = -\frac{k_{AN}}{q}y = -k_{VAN}\tau y \tag{13-24}$$

where k_{VAN} = first-order volumetric $NH_4 - N$ reduction rate constant, 1/d
q = hydraulic loading rate, m/yr
τ = detention time, days
y = fractional distance through the wetland

Transect water chemistry has been acquired at the Houghton Lake treatment wetland over a 16-year period, including measurements of ammonium nitrogen. These data show an exponential decrease in NH_4-N concentrations with distance from the discharge, thus supporting a first-order model (Figure 13-7). Summer season operation shows very low background concentrations (*circa* 0.05 mg/L), both for the wetland prior to wastewater discharges and for the unaffected zones of the present-day wetland. Wastewater concentrations are much higher (*circa* 10 mg/L), therefore, it is accurate to represent the data with a first-order model with a zero background.

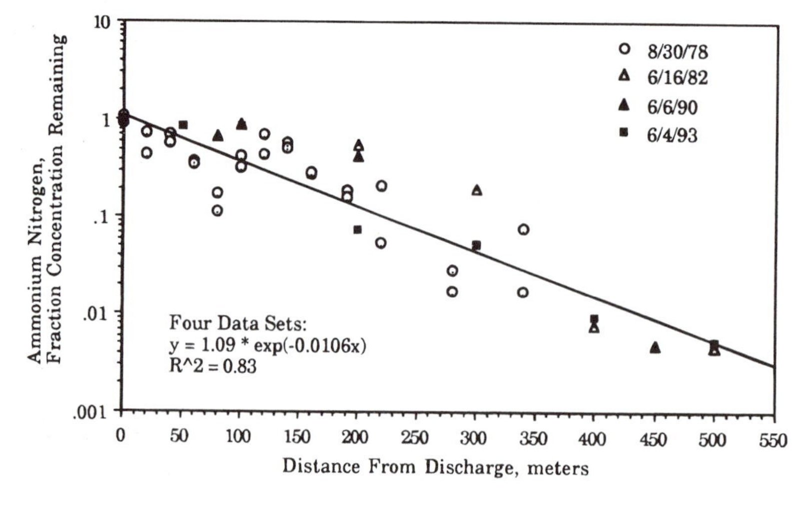

Figure 13-7 Transect data for ammonium nitrogen decline in the Houghton Lake treatment wetland. Data span a 16-year period over which there is not a noticeable change in the profiles.

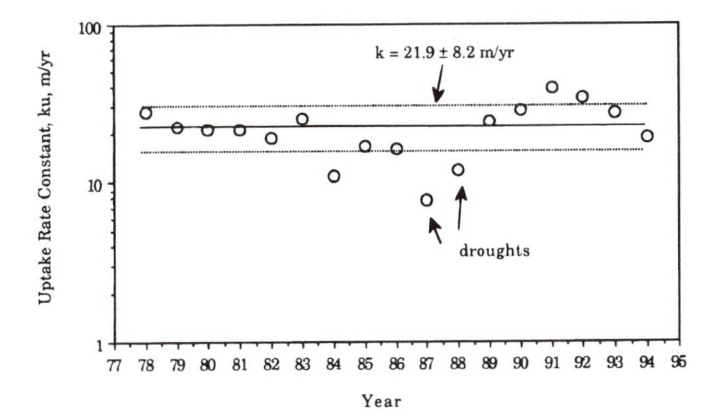

Figure 13-8 The first-order areal rate constant for ammonium reduction for the Houghton Lake wetland treatment system. Values are obtained from multiple transect data for each year. The effects of rain and evapotranspiration are included via a variable flow and depth hydrology model (see Chapter 9).

The hydrology of this wetland is not dominated by the discharge, thus making it necessary to use a dynamic water budget that includes the effects of rain, evapotranspiration, and variable water depth (see Chapter 9). Transect data were analyzed by that method to determine the areal rate constant for ammonium reduction. The rate constant varies from season to season, but there is not a time trend (Figure 13-8).

Growth of vegetation is an important factor in the removal of nitrogen in the Houghton Lake wetland. The wetland vegetation adapted to the new nutrient load over about a 5-year period, 1978–1982. The increase in biomass during this period provided some nitrogen storage. The wetland then became relatively stable in terms of biomass expansion.

The average total inorganic nitrogen loading (TIN) over the stable five-year period 1989–1993 was 0.113 ± 0.004 g/m^2/d during the summer pumping season which averaged 133 ± 7 days. The total application was therefore 15.0 ± 1.3 gN/m^2 during each season. Over the 5-year period, there was a net growth of aboveground live biomass of 1050 ± 150 g/m^2 during each summer season of about 150 days. This biomass contained approximately 1.9 ± 0.3% nitrogen. The nitrogen requirement for this growth was therefore about 20 ± 6 gN/m^2. On an instantaneous basis, the applied nitrogen was not quite sufficient for the new above-ground tissues formed during the growing season.

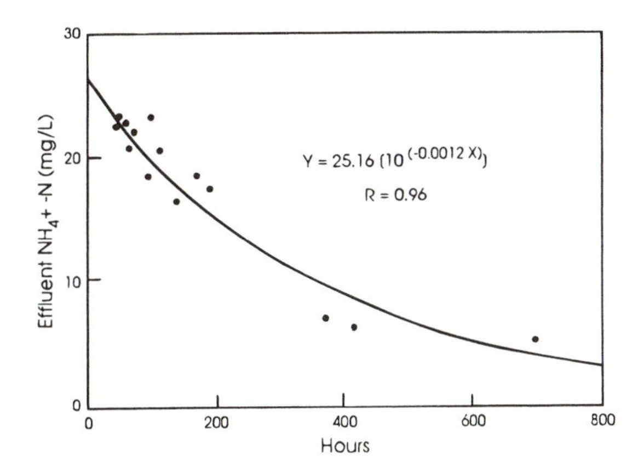

Figure 13-9 The effect of retention time on the removal of ammonia nitrogen at the Arcata pilot wetlands. (Data from Gearheart, 1990.) .

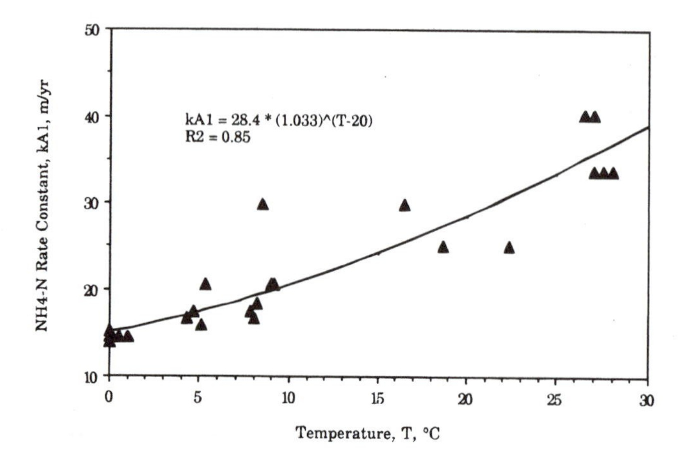

Figure 13-10 Temperature dependence of the rate constant for reduction of total nitrogen at Grumeth, Austria. (Data from Saurer, 1994.)

The remainder of the nitrogen requirement for roots and microbiota was provided by the return flux from decomposing litter and sediments, possibly augmented by atmospheric nitrogen fixation.

The net accumulation of nitrogen is in the new soils and sediments, which are the undecomposable residual from litter and microdetritus. Those new residuals total about 10 mm/yr at the point of discharge and about 2 mm/yr at remote locations. The mean accretion of 6 mm/yr contains approximately 2 percent nitrogen at a bulk density of 0.12 g/cc. This produces a net deposition of nitrogen of 14.4 $gN/m^2/yr$, which is nearly equal to the application rate of 15 $gN/m^2/yr$.

Similar verifications of the exponential reduction have been reported by Bavor et al. (1989) for SSF wetlands and by Gearheart (1990) for SF wetlands (see Figure 13-9).

If Equation 13-24 is calibrated directly from wetland operating data, low or negative values of k_{AN} sometimes result because of the production of ammonium nitrogen from ammonification of organic nitrogen. Rate constants so determined are termed *apparent rate constants.* Table 13-4 lists values of the apparent ammonium disappearance rate constant determined from operating data.

Temperature has an effect on apparent rate constants for ammonium reduction. Figure 13-10 shows this for a sand/*Phragmites* SSF wetland in Grumeth, Austria. The temperature factor for this system was approximately $\theta = 1.033$. Bavor et al. (1989) reported temperature coefficients averaging $\theta = 1.035$.

When a significant concentration of organic nitrogen is initially present, apparent rate constants are no longer adequate. The reaction calculations must be corrected for the proportion of organic nitrogen that mineralizes to ammonia in the wetland. It is important to note that Equation 13-24 may predict unrealistic internal ammonium nitrogen concentrations and inaccurate outlet concentrations in wetlands highly loaded with organic nitrogen. The kinetics of the sequential nitrogen conversion are next addressed.

Sequential Ammonium Kinetics

The reaction network shown in Figure 13-11 is assumed to consist of interconversions of nitrogen in the water accompanied by exchanges with the sediments and biomass and the atmosphere. This simplified nitrogen reaction network permits the exchange of ammonium nitrogen (AN) between the water and the live and dead biomass (B). Uptake may occur into living plants and sorption onto sediments or media. These are accounted by an uptake flux

Table 13-4 Example First-Order, Area-Based Rate Constants for Ammonium Reduction for Treatment Wetlands

Location	Wetland	Type	Data Quarters	HLR (cm/d)	NH_4-N In (mg/L)	NH_4-N Out (mg/L)	ORG N In (mg/L)	ORG N Out (mg/L)	Sequential k (m/yr)	Apparent k (m/yr)	T (°C)	Data Ref.
Surface-flow marshes												
Listowel, Ontario	1	SF con marsh	16	2.67	7.40	5.30	5.10	2.89	14.1	3.3	8.0[a]	Herskowitz, 1986
	2	SF con marsh	16	2.84	7.40	5.66	5.10	3.23	6.8	2.8	8.0[a]	Herskowitz, 1986
	3	SF con marsh	16	1.92	7.40	4.37	5.10	2.54	7.9	3.7	7.84	Herskowitz, 1986
	4	SF con marsh	16	1.95	8.58	6.43	9.97	2.79	6.0	2.1	8.02	Herskowitz, 1986
	5	SF con marsh	16	2.60	8.58	8.45	9.97	3.49	5.0	0.1	8.0[a]	Herskowitz, 1986
Houghton Lake, MI		SF nat marsh	16							22.1	16[a]	Kadlec, 1979–1994
Benton, KY	1	SF con marsh	4	1.71	5.04	7.89	8.71	2.61	1.7	−2.8	16.14	Choate et al., 1990a
	2	SF con marsh	4	1.71	5.04	6.43	8.71	3.08	2.6	−1.5	17.75	Choate et al., 1990a
Gustine, CA	5 Wetlands	SF con marsh	4	3.20	17.65	19.97	18.20	8.13	1.9	−1.4	17[a]	Walker and Walker, 1990
Richmond, NSW	1	SF open water	8	6.41	35.20	17.50	8.29	6.90	16.9	15.4	19.70	Bavor et al., 1988
Floating/submergent aquatics												
New Zealand	2 Wetlands	SF con FAP	7	5.52	60.00	37.00	16.00	2.70	15.3	9.7	15	van Oostrom, 1994
Richmond, NSW	4	SF con SAB	8	7.34	35.20	24.00	8.29	1.59	14.7	10.3	17.16	Bavor et al., 1988
Hyacinth Pilot	1	SF con FAP	High Cl	40.80	27.42	8.25			37.3	178.9	ca. 25	Weber and Tchobanoglous, 1986
Hyacinth Pilot	1	SF con FAP	Low Cl	44.00	6.45	1.65			ca. 300	218.9	ca. 25	Weber and Tchobanoglous, 1986

Table 13-4 Continued

Location	Wetland	Type	Data Quarters	HLR (cm/d)	NH_4-N In (mg/L)	NH_4-N Out (mg/L)	ORG N In (mg/L)	ORG N Out (mg/L)	Sequential k (m/yr)	Apparent k (m/yr)	T (°C)	Data Ref.
Soil-based reed beds												
Denmark	44 Wetlands	Soil-based *Phragmites*	8 Typical	4.53	21.00	14.10	13.58	3.88	14.9	6.6	6.5[a]	Schierup et al., 1990b
Subsurface-Flow Reed Beds												
U.K.	11 Wetlands	SSF *Phragmites*	8 Typical	18.30	4.60	2.80	3.20	1.70	27.3	33.2	9.5[a]	Green and Upton, 1993
Gravel beds												
Benton, KY	3	SSF bulrush	4	5.22	4.84	8.62	9.42	1.53	11.2	−11.0	16.74	Choate et al., 1990a
Hardin, KY	1	SSF cattail	4	8.81	7.95	9.32	11.67	1.83	8.6	−5.1	16.15	Choate et al., 1990a
	2	SSF bulrush	4	7.34	6.86	5.92	11.96	2.09	32.5	3.9	13.36	Choate et al., 1990a
North America	8 Wetlands	SSF	Various	4.66	5.70	3.20	9.83	4.62	25.8	9.8		NADB, 1993
Richmond, NSW	2	Gravel only	8	3.83	35.20	19.20	8.29	0.68	12.9	8.5	18.57	Bavor et al., 1988
	3	SSF bulrush	8	5.08	35.20	19.40	8.29	1.56	16.0	11.0	18.20	Bavor et al., 1988
	5	SSF cattail	8	4.61	35.20	18.80	8.29	0.01	22.7	10.6	18.37	Bavor et al., 1988

[a] Estimated as mean daily air temperature.

[b] Data range precludes accurate parameter estimate.

Note: The value of C* is taken as 0.0.

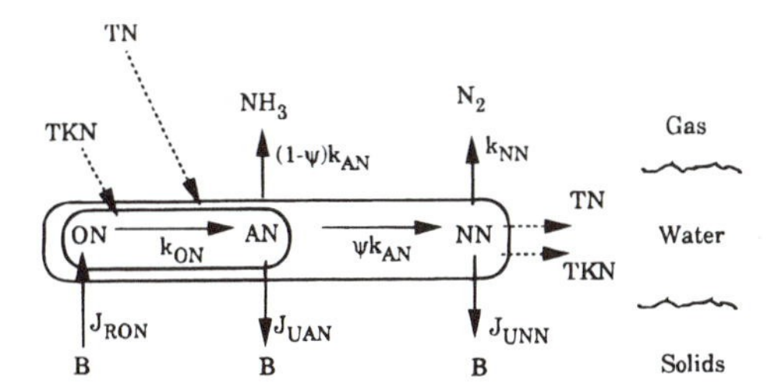

Figure 13-11 A simplified reaction sequence and transfer network for nitrogen in the wetland environment. Organic nitrogen (ON) may be ammonified to ammonium nitrogen (AN). The wetland contributes organic nitrogen from decomposition of biomass (B). Ammonium may be lost via volatilization of ammonia, nitrification, sorption, and plant uptake. Nitrate (NN) is formed by nitrification and lost by denitrification and uptake. Transfers to and from the wetland sediments and biomass are denoted by the fluxes (J). The first-order areal rate constants are denoted by (k).

J_{UAN}. Uptake of nitrate nitrogen (NN) by living plants is thought to be less important than uptake of ammonium nitrogen, but is still an allowed transfer route (J_{UNN}). Biomass decomposition can also release organic nitrogen (ON) into the water via decomposition (J_{RON}). There is no available mechanistic model for releases from decomposition or uptake by plants and sorption or releases of organic N; therefore, these are retained as zero-order rates.

Ammonium may be converted to ammonia gas (NH_3) under suitable conditions (see a later section of this chapter), and a first-order rate is ascribed to that process. It is irreversible because gaseous ammonia diffuses or is convected from the vicinity of the air-water interface. The fraction of the ammonium nitrogen that is nitrified is y, and the remaining fraction is gasified $(1 - \psi)$. Nitrate may be converted to nitrogen gas (N_2) via microbial denitrification. That process has been found to be first order (Phipps and Crumpton, 1994). In some natural wetlands, atmospheric nitrogen may be fixed by wetland plants and algae, but that process is usually assumed to be negligible in treatment wetlands.

This reaction network will be presumed to consist of zero- and first-order reactions which represent the simplest possible rate equations that may be chosen. The resulting model for the progress of concentrations as water moves through the wetland is tractable, but necessarily more complicated than for a single species such as BOD_5. For a steady state, plug flow wetland, the mass balance equations for the three dissolved species are

$$q\frac{dC_{ON}}{dy} = -k_{ON}C_{ON} + J_{RON} \tag{13-25}$$

$$q\frac{dC_{AN}}{dy} = +k_{O}C_{ON} - k_{A}C_{AN} - J_{UAN} \tag{13-26}$$

$$q\frac{dC_{NN}}{dy} = +\psi k_{AN}C_{AN} - k_{NN}C_{NN} - J_{UNN} \tag{13-27}$$

where q = hydraulic loading rate, m/yr
y = fractional distance through wetland,
ψ = fraction ammonium that is nitrified,—
and the rate notation is shown in Figure 13-7.

In batch tests, these equations still apply, with $q/y = \varepsilon h/t$, where εh = depth of free water and t = time.

Only the first two mass balances are required to describe ammonium concentrations, because dissimilatory nitrate reduction has been presumed negligible (see a later section of this chapter).

Solution for the Plug Flow Case

Equations 13-25 and 13-26 may be solved sequentially to provide the concentration profiles for the three single species. The solution of Equation 13-25 gives the k-C* model for organic nitrogen:

$$C_{ON} = C^*_{ON} + (C_{ONi} - C^*_{ON})e^{-(k_{ON}/q)y} \tag{13-28}$$

where C_{ONi} = inlet concentration of organic nitrogen, mg/L.

Next, use the result for C_{ON} to solve for the ammonium concentration profile:

$$C_{AN} = C^*_{AN} + (C_{ANi} - C^*_{AN})e^{-(k_{AN}/q)y} + \left(\frac{k_{ON}}{k_{AN} - k_{ON}}\right)(C_{ONi} - C^*_{ON})(e^{-(k_{ON}/q)y} - e^{-(k_{AN}/q)y}) \tag{13-29}$$

where C_{ANi} = inlet concentration of ammonium nitrogen, mg/L.

Data Fits

Equation 13-29 predicts internal concentration profiles as well as the outlet concentration of ammonium nitrogen (at y = 1.0). If there is only background organic nitrogen present in the feed water, the isolated k-C* model for ammonium reduction applies (Equation 13-24). This situation is rare in operating treatment wetlands, but has been tested in laboratory microcosms (Gale et al., 1993). Example depletion profiles in the field are exponential in shape (Figure 13-12), thus confirming first-order ammonium disappearance.

Very few wetland studies have gathered sufficient data to calibrate this model from internal transects. It is very desirable to have transect information because the interactions among nitrogen species cause local concentration maxima within the wetland. The project

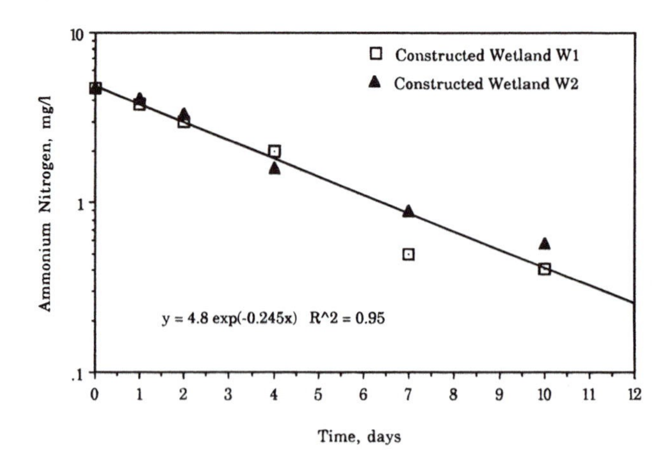

Figure 13-12 Ammonium nitrogen disappearance from water-soil microcosms from the Orange County Eastern Service Area treatment wetland. Conditions were well-mixed, aerobic (DO = 4.4 to 5.2 mg/L) and neutral pH (7.3 to 7.6). (Data replotted from Gale et al., 1993.)

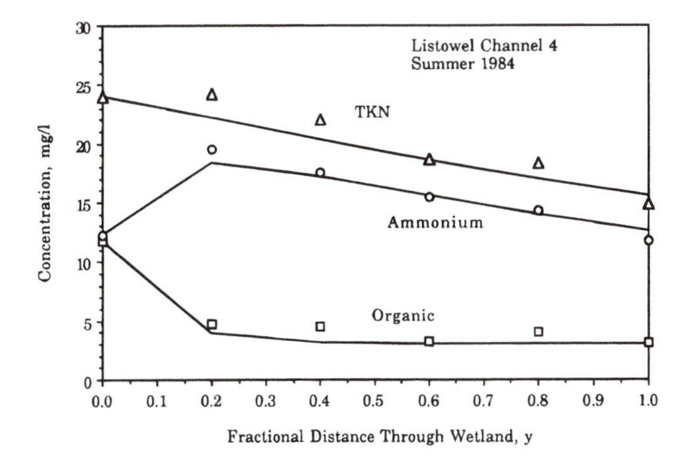

Figure 13-13 Profiles of major dissolved nitrogen species in Listowel Wetland 4 in summer 1984. Lines are model calibration, and symbols denote data points which are averages of biweekly data over 3 months.

at Listowel, Ontario obtained the requisite data by sampling along the length of the wetlands every 2 weeks in 1983–1984.

The data fit with Equations 13-28 and 13-29 is good, which is anticipated for a model with adjustable constants (Figures 13-13 and 13-14). In both summer and winter, organic nitrogen is quickly mineralized to ammonium, which subsequently decreases through most of the wetland travel distance. However, the background levels of organic nitrogen require a significant release of this constituent from the biomass and sediment, as well as a compensatory uptake of ammonium. This may be determined from the overall mass balance information produced by the integration of the model (Figure 13-15).

In the summer, a significant portion of that uptake is utilized in the formation of macrophyte tissues; the balance is attributable to the microflora such as algae. Thus, the incoming nitrogen species are processed in a large internal cycle of nitrogen: ammonium uptake by plants, organic N release from decay, and mineralization of organic N to ammonium. The summer background concentration of ammonium is calibrated to be zero, which is consistent with observations at other unfertilized wetlands in the vicinity.

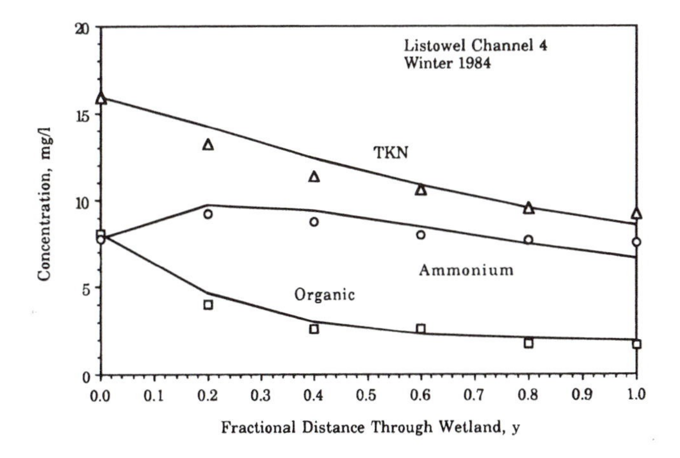

Figure 13-14 Profiles of major dissolved nitrogen species in Listowel Wetland 4 in winter 1984. Lines are model calibration, and symbols denote data points which are averages of biweekly data over 3 months.

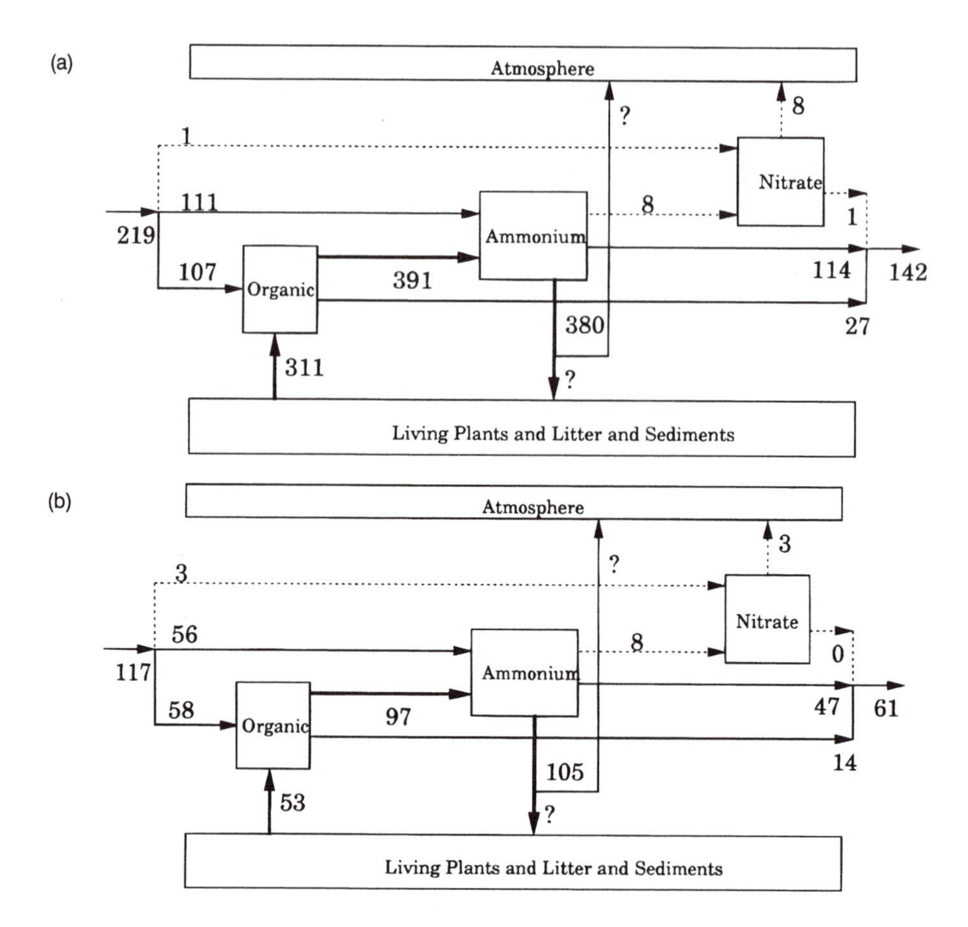

Figure 13-15 (a) Processing of nitrogen in Listowel Wetland 4 in summer 1984. Flows are in grams per square meter per year (g/m²/year). (b) Processing of nitrogen in Listowel Wetland 4 in winter 1984. Flows are in grams per square meter per year.

In the winter, there is release of a lesser amount of organic nitrogen from the wetland solids (Figure 13-15), and consumption of ammonium nitrogen is smaller, since there is no macrophyte growth operative during this period of the year in Ontario. There is also a much reduced rate of mineralization of organic nitrogen to ammonium. The result is a lesser overall efficiency of nitrogen removal by the system. The winter background concentration of ammonium is calibrated to be high (4 mg/L), but this is consistent with observations in unfertilized northern wetlands in winter.

Similar profiles of the nitrogen species have been observed for SSF flow wetlands, such as the Little Stretton, U.K. wetland in Figure 13-16 (Green and Upton, 1993).

At the present time, there are not enough wetland transect data to provide a firm, multi-system calibration of the nitrogen reaction sequence. Input-output data must be used to calibrate the values of k_{AN}. The outflow value of ammonium nitrogen is not sensitive to whether or not there is a maximum in the ammonium profile internal to the wetland, and hence, these data are not as satisfactory for parameter estimation as transect data. Estimates of these *sequential rate constants* for the multi-step model are given in Table 13-4.

There are also various possibilities for calibration (parameter estimation): k_{AN} may be determined from quarterly data subsets and then averaged or the whole data set may be averaged and then k_{AN} values calibrated to the average data set. The latter technique smooths the data over longer time intervals, thus evening out effects such as rain and evapotranspiration, seasonality, and vagaries of feed flows and concentrations.

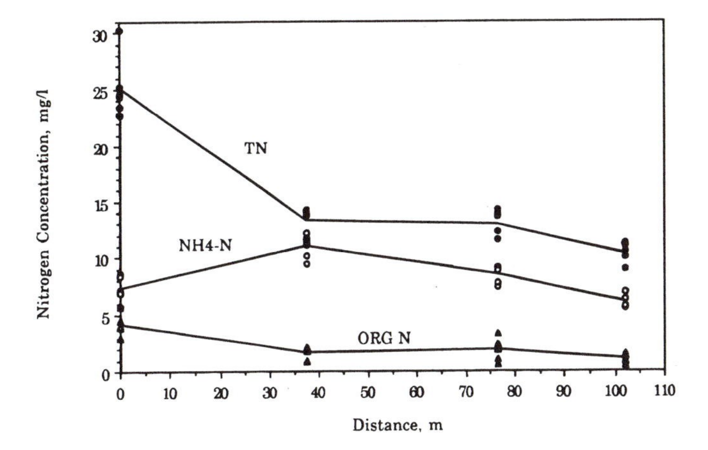

Figure 13-16 Progression of nitrogen species through the Little Stretton SSF wetland. Organic and oxidized nitrogen disappear quickly, with the former causing an increase in ammonium. These are followed by a slower decline in ammonium nitrogen. (Data from Green and Upton, 1993.)

Environmental Factors in Operating Wetlands

Some of the effects of environmental factors are evidenced by the data in Table 13-4 in light of other operating variables. The Gustine wetlands display low ammonium rate constants. They had very high BOD loads, ranging up to 600 mg/L in the feed water. It is possible that nitrifiers were present at low densities in the competitive environment dominated by heterotrophic carbon oxidation.

The open water trench at Richmond, NSW was well mixed, and possessed near-saturation dissolved oxygen. In general, cold-climate rates are lower than warm-climate rates. Both temperature and dissolved oxygen are relatively high in the Houghton Lake wetland.

Figure 13-17 illustrates the effects of temperature, dissolved oxygen, and season on the sequential area-based, first-order ammonia disappearance rate constant at the Listowel, Ontario pilot wetland marshes treating municipal wastewaters. The ammonia rate constant is very low during the coldest months, but is not zero even at freezing water temperatures. It rises during the spring and reaches a maximum during the warmest months. This maximum of k_{AN} during the summer may be lessened due to low dissolved oxygen during the summer

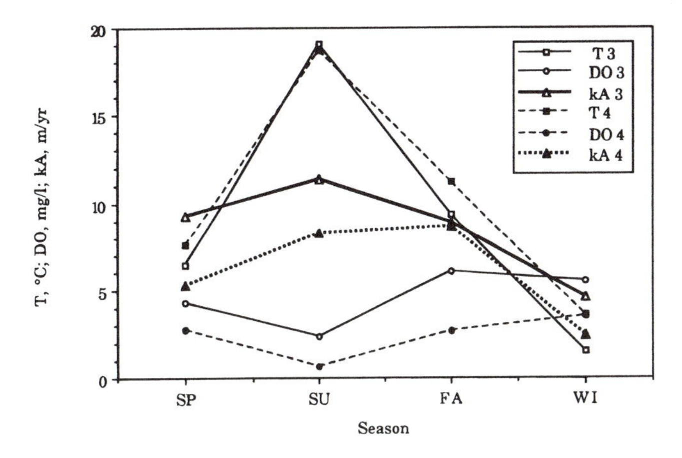

Figure 13-17 Temperature, dissolved oxygen, and sequential ammonium rate constants for Listowel Wetlands 3 and 4. Data are averages of 3 months for 4 years. (Data from Herskowitz, 1986.)

months. The dissolved oxygen (DO) correction (Equation 13-18) for K_{DO} = 1.3 mg/L ranges from 0.33 to 0.83 for these wetlands. The DO-unrestrained summer rate constant can be estimated at about 25 m/yr at Listowel at a temperature of 20°C. The estimated θ factors are 1.06 to 1.13.

In their review of nitrogen transformations in flooded soils, Reddy and Patrick (1984) summarized estimated first-order nitrification measurements from soil studies. The first-order, volume-based nitrification rates they summarized ranged from 0.003 to 3.1 d^{-1}, with a mean of 0.29 d^{-1}. Assuming an effective nitrification depth of 30 cm in a typical wetland treatment system, these values are equivalent to area-based, first-order nitrification rate constants between 0.33 to 339 m/yr with an average of 32 m/yr. It is likely that these values represent typical nitrification rate constants in systems that are not limited by a shortage of oxygen.

In view of the information from treatment wetlands and related technologies, an estimate of the temperature factor for ammonium oxidation is θ = 1.04. When the rate constants in Table 13-4 are adjusted to 20°C and averaged, the results are

k_{AN20} = 18 m/yr SF (includes soil-based reed beds and aquatics)
k_{AN20} = 34 m/yr SSF (includes U.K. reed beds and U.S. gravel beds)
k_{VAN20} = 0.39 d^{-1} SSF (60 cm at 40 percent porosity)

These values are to be regarded as central tendencies for the systems listed. The intersystem variability is large in response to site factors not included in the simplified design model. For instance, septage-fed SSF wetlands may have higher concentrations and lower k-values.

DENITRIFICATION

The biogeochemical processes that cause nitrogen fixation would ultimately deplete the atmosphere of nitrogen gas if the nitrogen cycle were not closed by the microbial liberation of oxidized nitrogen via the denitrification process. Denitrification is an energy-requiring reduction process where electrons are added to nitrate or nitrite nitrogen, resulting in the production of nitrogen gas, nitrous oxide (N_2O), or nitric oxide (NO).

Denitrification is an essential and complementary process that accompanies heterotrophic metabolism in aquatic and soil environments when dissolved or free oxygen is absent (anoxic). As described in Chapter 7, aerobic heterotrophic metabolism uses oxygen as the electron acceptor in the electron transport chain after the Krebs cycle. In denitrification, the enzyme nitrate reductase allows certain genera of bacteria to use the more tightly bound oxygen atoms in nitrate and nitrite molecules as this final electron acceptor. The most common facultative bacterial groups that accomplish denitrification include *Bacillus, Enterobacter, Micrococcus, Pseudomonas,* and *Spirillum.* These genera can switch easily from anoxic to aerobic metabolism because of the biochemical similarities of the two processes. However, because the use of free oxygen as a final electron acceptor yields more energy (about 686 kcal/mol of glucose) than the use of nitrate (about 570 kcal/mol of glucose), these organisms will typically not denitrify nitrate in the presence of free oxygen.

Nitrate and nitrite nitrogen ($NO_3 + NO_2 - N$) are summed together because they have not been considered separately in permits and because nitrite is generally a transient form of nitrogen in most wetlands. The combination is commonly called oxidized nitrogen, abbreviated as NO_X-N or total oxidized nitrogen (TON). Because nitrite is normally present in very low concentrations, the combination is also referred to as "nitrate nitrogen."

Chemical Reactions

The overall stoichiometric nitrate dissimilation reaction based on methanol (CH_3OH) as a carbon source is summarized by the following (U.S. EPA, 1993e)

$$NO_3^- + 0.833CH_3OH = 0.5N_2 + 0.833CO_2 + 1.167H_2O + OH^- \quad (13\text{-}30)$$

At low soluble ammonia concentrations, some nitrate is also used by these bacteria for cell synthesis. Denitrification, cell synthesis ($C_5H_7NO_2$), and the effect of these processes on total alkalinity are summarized by the following:

$$NO_3^- + 1.08CH_3OH + 0.24H_2CO_3 = 0.056C_5H_7NO_2 + 0.47N_2 + 1.68H_2O + HCO_3^- \quad (13\text{-}31)$$

From the stoichiometry of Equation 13-30, 2.47 g of methanol, or another equivalent carbon source, are required to support the denitrification of 1 g of nitrate nitrogen. In the absence of this carbon source, denitrification is inhibited.

As indicated by Equation 13-31, denitrification produces alkalinity. The observed rate of bicarbonate production by this process is about 3.0 g as $CaCO_3$ per gram of NO_3-N reduced. This increase in alkalinity is accompanied by an increase in the pH of the wetland surface water.

Theoretically, denitrification does not occur in the presence of dissolved oxygen. However, denitrification has been observed in suspended and attached growth treatment systems that have relatively low measured DO concentrations. This observation is explained by the presence of microscopic, anoxic zones that are likely to occur in bacterial films.

Denitrification has been observed in numerous wetland treatment systems which have measurable dissolved oxygen in their surface waters (Phipps and Crumpton, 1994; van Oostrom, 1994). Oxygen gradients occur between surface waters and bottom sediments in wetlands, allowing both aerobic and anoxic reactions to proceed. Also, nitrate formed by nitrification in wetlands tends to diffuse into anaerobic soil layers where it is effectively denitrified (Reddy and Patrick, 1984).

Denitrifying bacteria are more abundant than the nitrifiers, in both SF and SSF treatment wetlands. Listowel results show higher populations in the sediments in spring and summer, *circa* 10^6 per gram, vs 10^5 per gram in fall and winter (Herskowitz, 1986). Denitrifiers were found at higher levels in a U.K. gravel bed, approximately 10^7 to 10^8 per gram, and most were associated with roots rather than the gravel (May et al., 1990).

Denitrifying bacteria are more abundant in treatment wetlands than in natural wetlands. At the Houghton Lake wetland (unpublished data), there was a measured increase of about a factor of 100 in the number of denitrifiers in the wastewater discharge area after the discharge commenced (10^4 increased to 10^6 per gram for litter; 10^4 increased to 10^7 per gram for soil). At Clermont, FL, there was a similar increase, from 10^5 to 10^7 per gram of soil (Zoltek et al., 1979).

The adaptation of the sediment denitrification rate constant to higher values at higher nitrate loadings has been documented for sewage-impacted freshwater estuaries (King and Nedwell, 1987). Their study of the River Colne estuary, Essex, U.K., showed k_{V1} decreased by a factor of about three as a function of distance from the wastewater treatment plant.

Nitrate loss in treatment wetlands is often assigned to denitrification in the absence of proof that this mechanism is indeed the operative one. Other known and studied candidate mechanisms in wetlands include assimilation by plants, assimilation by microbiota, and dissimilatory reduction to ammonium nitrogen.

These alternative reduction routes have been documented to comprise from 1 to 34 percent of the total nitrate loss (Bartlett et al., 1979; Stengel et al., 1987; Cooke, 1994; van Oostrom, 1994). Bartlett et al., (1979) measured production of ammonium, dinitrogen, and nitrous oxide for microcosms with soils from a treatment wetland, but with no plants. From 1 to 6 percent of the product was ammonium nitrogen, the balance was measured as dinitrogen, with only trace amounts of nitrous oxide. Cooke (1994) measured ^{15}N-labeled nitrate, ammonium, and organic nitrogen in unvegetated microcosms in a treatment wetland. He found 34, 6, and 60 percent of $K^{15}NO_3$ converted by dissimilatory processes, microbial assimilatory processes, and denitrification, respectively, at one site and 25, 5, and 70 percent at a second site. Stengel et al. (1987) used the acetylene blockage technique to establish that 75 to 90 percent of the nitrate loss in a flow-through, *Phragmites*/gravel SSF unit was due to denitrification. van Oostrom (1994) measured 16 percent dissimilatory nitrate reduction in microcosms containing *Glyceria maxima* mats.

Nitrate uptake by wetland plants is presumed to be less favored than ammonium uptake. But in nitrate-rich waters, nitrate may become a more important source of nutrient nitrogen. Aquatic macrophytes utilize enzymes (nitrate reductase and nitrite reductase) to convert oxidized nitrogen to useable forms. The production of these enzymes decreases when ammonium nitrogen is present (Melzer and Exler, 1982). In the Santee, CA study of an SF *Scirpus*/gravel wetland (Gersberg et al., 1984), the entire nitrate loss was ascribed to plant uptake in the absence of an exogenous carbon source and with essentially no ammonium in the nitrified influent.

Results From Other Technologies: Monod Kinetics and Environmental Factors

Denitrification has been extensively studied in both attached and suspended growth treatment systems (U.S. EPA, 1993e). The results are reviewed here to provide guidance and a context within which to examine treatment wetland data.

Monod Kinetics

Monod factors for nitrate and the carbon source are recommended for suspended growth systems:

$$u_{DENITR} = u_{DENITRmax}[(C_{NN}/(K_{DENITR} + C_{NN})) \times (C_{ORGC}/(K_{ORGC} + C_{ORGC}))] \tag{13-32}$$

where
u_{DENITR} = actual denitrifier growth rate, d^{-1}
$u_{DENITRmax}$ = maximum denitrifier growth, d^{-1}
C_{NN} = nitrate nitrogen concentration, mg/L
$K_{DENITRN}$ = denitrification half-saturation constant, mg/L
C_{ORGC} = organic carbon concentration, mg/L
K_{ORGC} = organic carbon half-saturation constant, mg/L

The actual rate of denitrification, U_{DENITR} (kg NO_3-N/g biomass/d), can be determined by dividing the denitrifier growth rate, u_{DENITR}, by the denitrifier yield coefficient, Y_{DENITR} (kg biomass/kg NO_3^--N removed):

$$U_{DENITR} = u_{DENITR}/Y_{DENITR} \tag{13-33}$$

Experimental studies indicate that the nitrate half-saturation constant K_{DENITR} is in the range of 0.1 to 0.2 mg/L (U.S. EPA, 1993e). This very low half-saturation constant results in zero-order reaction kinetics (no effect of nitrate concentration) for denitrification at nitrate concentrations above about 1 to 2 mg/L. At lower concentrations, nitrate removal approaches first order. However, treatment wetlands are most often best described by first-order denitrification kinetics (with respect to nitrate), even at very high nitrate concentrations (Crumpton et al., 1993; Gale et al., 1993).

The empirical value for K_{ORGC} is also quite low (about 0.1 mg/L for methanol), resulting in low dependence of the denitrification process on available organic carbon when concentrations are above 1 mg/L. The measured range of values for Y_{DENITR} in conventional treatment systems is 0.6 to 1.2 g biomass per gram of nitrate nitrogen. The maximum denitrification rate, $U_{DENITRmax20}$, was found to be 0.38 and 0.45 g NO_3-N/gm biomass/d in suspended and attached growth systems, respectively (U.S. EPA, 1993e).

Temperature can significantly affect the rate of denitrification in attached and suspended growth treatment systems. A modified Arrhenius relationship describes the effect of temperature on denitrification:

$$K_{TDENITR} = K_{20DENITR}\ \theta^{T-20} \tag{13-34}$$

where $K_{TDENITR}$ is the denitrification reaction rate coefficient at temperature T, $K_{20DENITR}$ is the reaction rate coefficient at 20°C, and θ is the denitrification temperature coefficient (about 1.1 to 1.2 in attached growth systems and 1.08 to 1.10 in attached growth systems [Metcalf and Eddy, 1991; U.S. EPA, 1993e]). Brodrick et al. (1988) found that significant denitrification occurred at temperatures as low as 5°C in laboratory incubations.

The optimum pH for denitrification measured in conventional wastewater treatment systems ranges from 7 to 8 (Figure 13-18). Experimental studies in conventional systems indicate that denitrification is considerably reduced at a pH less than 6 and greater than 8. Treatment wetlands at circumneutral pH provide ideal environments in terms of hydrogen ion concentrations.

As with the nitrification process described earlier, prediction of the denitrification rate in treatment wetlands from these companion technologies is not recommended.

First-Order Models

Nitrate-nitrite disappearance has been modeled effectively in wetland treatment systems as a first-order reaction based on nitrate + nitrite nitrogen concentration (Phipps and Crumpton, 1994). The volume-based, first-order equation describing the loss of nitrate and nitrite nitrogen in wetland treatment systems follows:

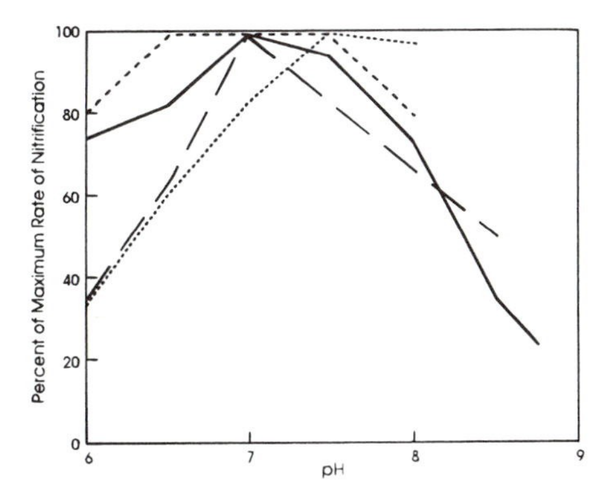

Figure 13-18 The effect of pH on denitrification rate in suspended growth wastewater treatment systems. (Modified from U.S. EPA, 1975.)

$$R_{NN} = k_{VNN}C_{NN} \tag{13-35}$$

where R_{NN} = denitrification rate, mg/L per day
k_{VNN} = volume-based denitrification rate constant, day^{-1}
C_{NN} = concentration of nitrate + nitrite nitrogen, mg/L

Half lives for nitrate and nitrite nitrogen in wetland microcosms not controlled by mass transfer or by carbon reported by Kadlec (1988c) were between 0.3 and 1.5 days, which are equivalent to k_{VNN} values between 0.46 to 2.31 d^{-1}. With increasing depth of the water column (10 to 30 cm) and increasing resistance to diffusion, the expected half life increased to 14 to 21 days (k_{VNN} = 0.03 to 0.05 d^{-1}).

Reddy and Patrick (1984) reported first-order denitrification rate constants for wetland soils and sediments between 0.004 and 2.57 d^{-1}, with an average value of 0.53 d^{-1}. Graetz (1980) measured denitrification rates in 14 flooded soils in Florida. Measured first-order rate constants varied from 0.04 to 0.192 d^{-1}, with an average value of 0.088 d^{-1}. Graetz confirmed that the denitrification rate constant in wetland soils is a function of organic carbon and pH (the rate declines below a pH of 6.5 units).

An area-based, first-order model for nitrate-nitrite loss in wetlands can be written as

$$J_{NN} = k_{NN}(C_{NN} - C^*_{NN}) \tag{13-36}$$

where J_{NN} = the area-based nitrate + nitrite loss rate, $g/m^2/yr$
k_{NN} = area-based, first-order nitrate-nitrite rate constant, m/yr
C^*_{NN} = irreducible background concentration of nitrate + nitrite N, mg/L

The value for C^*_{NN} is approximately zero because no investigation has shown a lower limit to the reduction of nitrate.

The integrated form of the NN mass balance for this uptake rate and plug flow conditions is

$$\ln\left[\frac{C_{NN,o}}{C_{NN,i}}\right] = -\frac{k_{NN}}{q}y = -k_{VNN}\tau y \tag{13-37}$$

where k_{VNN} = first-order volumetric TN reduction rate constant, 1/d
q = hydraulic loading rate, m/yr
τ = detention time, days
y = fractional distance through the wetland

Long-term system average values of k_{NN} for treatment wetlands are summarized in Table 13-5. This table contains apparent denitrification rate constants for wetlands receiving influents dominated by nitrate nitrogen. SF wetlands average k = 15 m/yr; SSF average k = 29 m/yr. The plug flow assumption has been used in the calculation of these apparent rate constants. The magnitude of the difference may be assessed at the Des Plaines, IL site. These wetlands have been measured to behave like three stirred tanks (Kadlec, 1994). If data are analyzed with that mixing pattern, the k values are approximately 15 percent higher.

Temperature Coefficients from Wetland Data

Crumpton and Phipps (1992) determined that denitrification in wetlands is controlled primarily by the rate of diffusion through a very thin layer of water near the sediments and

Table 13-5 Example Apparent, First-Order, Area-Based Rate Constants for Nitrate Reduction for Treatment Wetlands

Location	Wetland	Type	Data Quarters	HLR (cm/d)	NO_x-N In (mg/L)	NO_x-N Out (mg/L)	Apparent k (m/yr)	T (°C)	Data Ref.
Surface-flow marshes									
Des Plaines, IL	EW3	SF con marsh	6	9.66	1.66	0.75	28.3	20.2	Hey et al., 1994
	EW4	SF con marsh	6	1.40	1.65	0.21	10.5	19.8	Hey et al., 1994
	EW5	SF con marsh	6	6.02	1.66	0.73	18.1	21.1	Hey et al., 1994
Lakeland, FL	1	SF con marsh	21	3.71	6.62	1.14	23.8	22.6[a]	NADB, 1993
Sea Pines, SC	1	SF nat marsh	15	2.73	6.92	0.52	25.8	18.7[a]	NADB, 1993
Ames, IA	40 Mesocosms	SF con marsh		Batch	3 to 20	0.00	63.0	20[a]	Crumpton et al., 1993
Forested									
Reedy Creek, FL	WTS1	SF nat swamp	34	3.57	3.66	0.42	28.2	22.3[a]	NADB, 1993
Reedy Creek, FL	OFWTS	SF nat swamp	28	5.63	3.82	0.25	56.0	22.3[a]	NADB, 1993
Floating/submergent aquatics									
New Zealand	2 Wetlands	SF con FAP	7	5.52	121.00	64.00	12.8	15	van Oostrom and Russell, 1994
New Zealand	Pilot	SF con FAP	8	7.34	35.20	24.00	10.3	15	van Oostrom and Cooper, 1990
Soil-based reed beds									
Denmark	44 Wetlands	Soil-based Phrag	8 Typical	4.53	4.10	2.00	11.9	6.5[a]	Schierup et al., 1990b
Subsurface-flow reed beds									
U.K.	11 Wetlands	SSF Phrag	8 Typical	18.30	18.95	13.15	31.3	9.5[a]	Green and Upton, 1993
Gravel beds									
Santee, CA									
F-W, 1980–81	5 Wetlands	SSF vegetated	2	16.80	17.30	14.48	10.9	18[a]	Gersberg et al., 1983
Sp-Su 1981	5 Wetlands	SSF vegetated	2	16.80	17.30	10.76	29.1	18[a]	Gersberg et al., 1983
F-W, 1980–81	2 Wetlands	SSF unvegetated	2	16.80	17.30	14.11	12.5	18[a]	Gersberg et al., 1983
Sp-Su 1981	2 Wetlands	SSF unvegetated	2	16.80	17.30	11.92	22.8	18[a]	Gersberg et al., 1983
1980–81	2 Wetlands	SSF vegetated + methanol	4	16.80	17.30	0.52	214.9	18[a]	Gersberg et al., 1983
Sp-Su 1981	1 Wetland	SSF vegetated + mulch	2	8.40	17.30	1.56	73.8	18[a]	Gersberg et al., 1983
F-W, 1981–82	1 Wetland	SSF bulrush	2	18.50	18.30	16.65	6.4	18[a]	Gersberg et al., 1984
Sp-Su 1982	1 Wetland	SSF bulrush	2	18.50	18.30	14.82	14.2	18[a]	Gersberg et al., 1984
F-W, 1981–82	1 Wetland	SSF bulrush + mulch	2	10.50	18.30	5.49	46.1	18[a]	Gersberg et al., 1984
Sp-Su 1982	1 Wetland	SSF bulrush + mulch	2	10.50	18.30	2.01	84.7	18[a]	Gersberg et al., 1984

[a] Estimated as mean daily air temperature.

Note: The value of C* is taken as 0.0.

into the anaerobic wetland sediments. In turn, diffusion rate is controlled by temperature. In sediment core microcosms, they found that an area-based, first-order expression similar to Equation 13-28 could be corrected by introducing a temperature correction factor resulting in this model:

$$J_{NN} = k_{NN}C_{NN}\theta^{(T-20)} \tag{13-38}$$

where θ = denitrification temperature coefficient
T = water temperature, °C

Crumpton and Phipps (1992) estimated values for k_{NN} and θ of 15.44 m/yr and 1.069, respectively. Using data presented by Crumpton et al. (1993) for 48 tank mesocosms (9 m^2) operated in a batch mode, a value of k_{VNN} of about 0.57 d^{-1} (half life of about 1.2 days) is determined (Figure 13-19). This is equivalent to a k_{NN} of 63 m/yr for the 30-cm water column depth.

Studies in New Zealand have focused upon nitrate reduction at high strength (van Oostrom and Russell, 1994; van Oostrom, 1994). In one study, a single floating-mat wetland of *Glyceria maxima* was built and fed a nitrified high-strength effluent. Most of the incoming nitrogen (67 percent) was nitrate, but the ammonium level was still moderately high, *circa* 35 mg/L. The mean nitrate removal rate constant was 7.8 m/yr. This value was interpreted to be carbon-limited, based on estimates of the carbon production rate of the plants. The temperature coefficient, established from these data, is $\theta = 1.147 \pm 0.045$.

In view of the information from treatment wetlands and related technologies, an estimate of the temperature factor for nitrate reduction is $\theta = 1.09$. When the rate constants in Table 13-5 are adjusted to 20°C and averaged, the results are

k_{NN20} = 35 m/yr SF (includes soil-based reed beds and aquatics)
k_{NN20} = 50 m/yr SSF (includes U.K. reed beds and U.S. gravel beds)
k_{VNN20} = 0.57 d^{-1} SSF (60 cm at 40 percent porosity)

These values are to be regarded as central tendencies for the systems listed. The intersystem variability is large in response to site factors not included in the simplified design model.

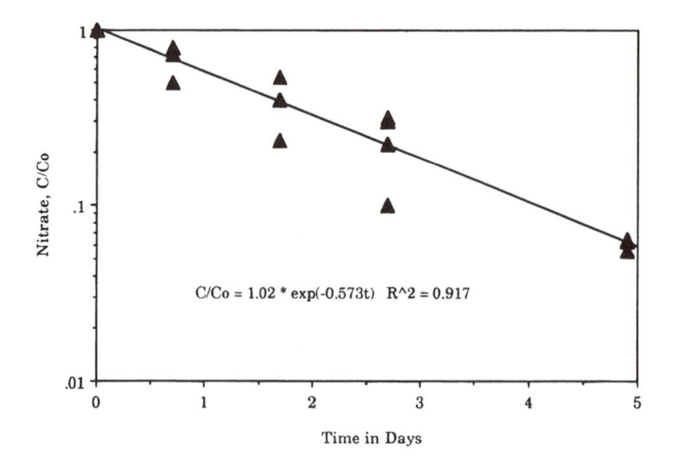

Figure 13-19 Batch nitrate reduction in SF cattail mesocosms. Each point represents the mean performance of ten mesocosms; each group of ten received different starting concentrations, from 3 to 221 mg/L nitrate nitrogen. (Data from Crumpton et al., 1993.) The slope corresponds to a rate constant of about 60 m/yr.

Accounting for Nitrate Production

The sequential reaction sequence may be extended to include nitrate. For the plug flow case, the profile for C_A is the basis for calculation of nitrification, and a first-order denitrification follows. Integration yields the nitrate concentration profile:

$$C_{NN} = (C_{NNi})e^{-(k_{NN}/q)y} + \psi\left\{\begin{aligned}&\left(\frac{k_{AN}}{k_{NN}-k_{AN}}\right)(c_{ANi})\left(e^{-(k_{AN}/q)y} - e^{-(k_{NN}/q)y}\right)\\&+\left(\frac{k_{ON}}{k_{AN}-k_{ON}}\right)\left(\frac{k_{AN}}{k_{NN}-k_{ON}}\right)(c_{ONi} - C^*_{ON})\left(e^{-(k_{ON}/q)y} - e^{-(k_{NN}/q)y}\right)\\&-\left(\frac{k_{ON}}{k_{AN}-k_{ON}}\right)\left(\frac{k_{AN}}{k_{NN}-k_{AN}}\right)(c_{ONi} - C^*_{ON})\left(e^{-(k_{AN}/q)y} - e^{-(k_{NN}/q)y}\right)\end{aligned}\right\} \quad (13\text{-}39)$$

where ψ = fraction of ammonium nitrified
= 1 − fraction volatilized

Equations 13-28, 13-29, and 13-39 represent the k-C* model for the three interconverting nitrogen species.

This model requires more information for nitrate calibration than is usually available from data reports. Further, it involves differences in k values, and differences in exponentials of k values, which magnify uncertainties. Significant inlet and outlet concentrations of organic, ammonium, and nitrate nitrogen are required for reasonable model calibration. In other words, it is not possible to estimate a denitrification rate constant when there is no nitrate in either the influent or the effluent. Consequently, there are presently few opportunites for calibrating the internal nitrate mass balance. One such case is the treatment wetlands of van Oostrom (1994) which were fed a partially nitrified, high-strength wastewater. Calibration of the full sequential model, presuming $\psi = 1.0$, yields

	Inlet C (mg/L)	Outlet C (mg/L)	C* (mg/L)	k(m/yr)
Organic	16	2.7	1.5	50
Ammonium	60	36	0.0	15
Nitrate	121	63	0.0	21

All three of these rate constants are within the range of values estimated from other treatment wetlands.

Calculations of nitrate reduction using apparent rate constants in Table 13-5 together with inlet nitrate concentrations will yield a lower bound to the effluent nitrate concentration. Calculations using apparent rate constants in Tables 13-2, 13-4, and 13-5 and the sequential model will yield an upper bound.

Nitrogen Fixation

Biological nitrogen fixation is the process by which nitrogen gas in the atmosphere diffuses into solution and is reduced to ammonia nitrogen by autotrophic and heterotrophic bacteria, blue-green algae, and higher plants. All photosynthetic bacteria are apparently capable of nitrogen fixation, as are some aerobic heterotrophs such as *Azotobacter,* some anaerobic bacteria such as *Clostridium,* and many facultative bacteria under anoxic conditions. Algal nitrogen fixers include filamentous, blue-green species in the genera *Anabaena, Gloeotrichia,* and *Nostoc.* Also, the aquatic fern, *Azolla,* and a few transitional, wetland vascular plant species in the genera *Alnus* and *Myrica* have been observed to fix atmospheric nitrogen (Waughman and Bellamy, 1980). Nitrogen also can be fixed from the atmosphere during lightning strikes. This electrical fixation process has been adapted for use in industrial production of nitrogen fertilizers.

Biological nitrogen fixation is an adaptive process that provides nitrogen for organisms to grow in conditions that are otherwise depleted of available nitrogen. Although nitrogen fixation apparently is not inhibited by high concentrations of available nitrogen (Brezonik, 1972), fixation generally is not observed in nitrogen-rich ecosystems. Because nitrogen fixation uses stored energy from either autotrophic or heterotrophic sources, it is not an adaptive process when nitrogen is otherwise available for growth. The presence of ammonium nitrogen is reported to inhibit nitrogen fixation (Postgate, 1978; as referenced by van Oostrom, 1994).

Nitrogen fixation rates have been measured in a number of lakes (Brezonik, 1972) with values that range from 0 up to 2.45 mg NH_4-N/m^3/hr, with most values less than 0.14 mg NH_4-N/m^3/hr. Fixation rates in wetlands receiving wastewater high in nitrogen are probably much lower or essentially negligible compared to other nitrogen transformation rates. Estimates of nitrogen fixation in a cypress dome receiving municipal wastewater ranged from 0.012 to 0.19 g/m^2/yr (Dierberg and Brezonik, 1984) and were concluded to be an insignificant component of the total nitrogen loading to this treatment wetland.

Under anaerobic conditions, microbial assemblages in the root zone of *Typha* spp. and *Glyceria borealis* were shown to fix considerable quantities of atmospheric nitrogen (Bristow, 1974). The majority of the activity was shown to be associated with the plants rather than the soils. Fixation rates at 20°C were determined to be 33.6 and 353 mg/kg roots per day for *Typha* and *Glyceria,* respectively. The measured rates of nitrogen fixation were estimated to be able to supply 10 to 20 percent of the growth requirement for *Typha* and 100 percent for *Glyceria.* Under aerobic conditions, fixation dropped by one order of magnitude. The effect of temperature was very strong, corresponding to $\theta = 1.16$.

The nitrogen fixation potential for the soil-microbe assemblage was studied for 45 sites in 17 peatlands in 8 countries by Waughman and Bellamy (1980). A total of 1297 assays were made. The appropriate subset in the context of treatment wetlands was the rich or extremely rich fen category, with $6.5 \leq pH \leq 7.6$, for which $N = 12$ sites. These showed fixation potentials averaging 0.622 mg/L/day of soil. A 30-cm root zone would then fix 187 mg/m^2/d. An indirect comparison with Bristow (1974) may be made for the assumption of 2000 g/m^2 of *Typha* roots, which gives 67 mg/m^2/d (245 kg/ha/yr) for Bristow's data.

These results do not permit quantification of the fixation ocurring in treatment wetlands, but do indicate the ability of wetland plants and soils to fix nitrogen. One of the factors contributing to variation in nitrogen mass balances for ammonium-poor treatment wetlands may well be the fixation of atmospheric nitrogen.

Nitrogen Assimilation

Nitrogen assimilation refers to a variety of biological processes that convert inorganic nitrogen forms into organic compounds that serve as building blocks for cells and tissues. The two forms of nitrogen generally used for assimilation are ammonia and nitrate nitrogen. Because ammonia nitrogen is more reduced energetically than nitrate, it is a preferable source of nitrogen for assimilation.

Although ammonia is the preferred nitrogen source, nitrate also can be used by some plant species. Nitrate uptake by wetland plants for plant growth is less favored than ammonium uptake under most conditions. In nitrate-rich waters, nitrate may become a more important source of nutrient nitrogen. Aquatic macrophytes utilize enzymes (nitrate reductase and nitrite reductase) to convert oxidized nitrogen to useable forms. The production of these enzymes decreases when ammonium nitrogen is present (Melzer and Exler, 1982). This process may be important in some treatment wetlands. For instance, a ^{15}N study of several SSF gravel wetland mesocosms (Zhu and Sikora, 1994) showed 70 to 85 percent of the entire nitrate loss was plant uptake—in the absence of an exogenous carbon source and with essentially no ammonium in the nitrified influent. Different species responded differently: 70 percent of the nitrate was taken up by *Phragmites australis,* 75 percent by *Typha latifolia,* and 85 percent by *Scirpus atrovirens georgianus.* Those rates correponded to uptake of 3 to 3.5 g N/m^2/d. The addition of acetate as a carbon source reduced plant uptake to about half those values. The balance of the nitrogen loss was determined to be about 25 percent immobilization and 75 percent denitrification.

Biota utilize nitrate and ammonium, and decomposition processes release organic nitrogen and ammonium back to the water. As noted in Chapter 7, typical biological material contains from 1.5 to 7 percent nitrogen on a dry-weight basis, with an average value of about 2.3 percent. One annual turnover of 2500 g/m^2 of biomass (typical standing crop) at 2 percent nitrogen represents 50 g/m^2 of nitrogen transfer (0.14 g/m^2/d annual average). This value is significant in comparison with most wastewater nitrogen loadings, which typically range from 0.01 to 1.0 g/m^2/d. There are typically one to five turnovers per year, depending on the climatic region: just over one in northern climates, five in southern climates. For five turnovers per year, uptake or release is therefore about $5 \cdot 50/365 = 0.68$ g/m^2/d. The uptake of 0.14 g/m^2/d is the 68th percentile of the distribution of TN inputs in the NADB; 0.68 g/m^2/d is the 98th percentile.

In temperate climates, macrophyte uptake is a spring-summer phenomenon. Both above- and below-ground plant parts grow during this period, but death phenomena are different. Plants such as *Typha* spp. and *Phragmites* in northern climates have an obvious annual cycle of above-ground live biomass: new shoots start from zero biomass in early spring and grow at a maximum rate in spring and early summer. Late summer is a period of reduced growth, and complete shoot die back occurs in the fall. In central Wisconsin, the process takes 6 months from start to finish for *Typha latifolia* shoots (Prentki et al., 1978); in Texas, it takes 9 months for *Typha angustifolia* shoots (Hill, 1987).

Roots comprise only a minor pool of biomass and nitrogen compared to rhizomes in plants such as *Typha* spp. and *Phragmites.* However, rhizome biomass, and hence nitrogen storage, is comparable to above-ground biomass (Whigham et al., 1978; Adcock et al., 1994). For *Typha* spp., the root/rhizome to shoot ratio is considerably smaller in treatment wetlands than in natural wetlands (Kadlec and Alvord, 1989). This may be due to the larger availability of nutrients which could influence the size of the nutrient gathering and storage organs. Rhizome turnover times (life expectancies) have been reported to be 1.5 to 2 years for *Typha* spp. and more than 3 years for *Phragmites* (Prentki et al., 1978). Cycling through the below-ground biomass is therefore slower than through the above-ground biomass.

The relative importance of plant uptake depends upon the nitrogen loading to the treatment wetland. For low nitrogen loadings, the plants cycle a significant fraction of the applied nitrogen. Nitrogen storage in the roots and rhizomes in the inlet zone of a SF *Phragmites*/*Typha* treatment wetland in Byron Bay, Australia was 35 g/m^2 and in the leaves and stems it was 92 g/m^2 (Adcock et al., 1994). Approximately 65% of the nitrogen added to this treatment wetland was found in the macrophyte biomass, due to a low nitrogen loading (*circa* 25 to 40 g/m^2/yr).

Similar results have been reported by Rogers et al. (1991) for gravel-based *Schoenoplectus validus* microcosms. In short-term (10 week) experiments, over 80 percent of the applied nitrogen was found in the plants at loading rates of 0.7 to 2.7 $gN/m^2/d$. Busnardo et al. (1992) found similar results for for gravel-based *Scirpus californicus* mesocosms. That study showed 50 percent of the added nitrogen, applied at 1.41 $gN/m^2/d$, was found to be in the plants after more than 1 year of operation. The corresponding uptake values (0.6 to 2.0 $gN/m^2/d$) are probably overestimates, due to factors such as container edge effects.

Macrophyte growth is not the only potential biological assimilation process; microorganisms and algae also utilize this nutrient. Ammonia is readily incorporated into amino acids by many autotrophs and microbial heterotrophs. The Krebs Cycle provides organic acids (such as pyruvic acid) for amination to produce amino acids. In turn, these amino acids are transformed into other amino acids or proteins, purines, and pyrimidines. The magnitude of this uptake process has not been quantified for treatment wetlands.

Permanent burial is only a small fraction of the nitrogen cycled by the plants and microbes. The magnitude of this accretion is discussed in the following section.

Most of the biomass, and its contained nitrogen, decomposes to release carbon and nitrogen. The locus of this release is a critical variable in the wetland nitrogen cycle: some portion of the nitrogen is released back to the wetland waters, some fraction is subjected to aerobic processes in above-water standing dead plant material and litter, and some may be translocated to rhizomes. The above-water decomposition favors oxygenative processes such as nitrification. The below-ground storage results in added burial of the residual from rhizome death and decomposition. Thus, the movement of nitrogen through the vegetation results in the enhancement of processes other than those in the water column and the associated biofilms.

OTHER NITROGEN FLUXES

The wetland nitrogen cycle includes a number of other pathways that do not result in a molecular transformation of the affected nitrogen compound. Some of the processes that may be important when designing wetland treatment systems include (1) atmospheric nitrogen inputs through rainfall and dryfall, (2) ammonia volatilization, (3) nitrogen release from biomass decomposition, (4) burial of organic nitrogen, and (5) ammonia adsorption. Each of these processes is briefly described.

Atmospheric Nitrogen Inputs

Atmospheric deposition of nitrogen contributes measurable quantities of nitrogen to receiving land areas. All forms are involved: particulate and dissolved, inorganic and organic. Wetfall contributes more than dryfall, and rain contributes more than snow (Tables 13-6 and 13-7). Atmospheric sources are almost always a negligible contribution to the wetland nitrogen budget for all but ombrotrophic systems.

Table 13-6 Atmospheric Deposition of Nitrogen on Boney Marsh in South Central Florida

Year March–February	P (mm/yr)	NO_x-N + NH_4-N (mg/L)	Dissolved Organic N (mg/L)	Particulate N (mg/L)	Total N (mg/L)	N Load (kg/ha/yr)
1979	846	0.40	0.94	0.87	2.20	18.6
1980	756	0.86	1.71	0.52	3.10	23.4
1981	788	1.49	2.07	2.64	6.20	48.8
1982	1,679	0.63	0.77	0.75	2.15	36.0
1983	1,002	0.72	1.20	0.72	2.64	26.5
1984	1,113	0.53	0.51	0.39	1.43	15.9
1985	1,021	0.61	0.68	0.68	1.96	20.0

Note: N = 278. Data from the unpublished data of South Florida Water Management District.

Table 13-7 Atmospheric Deposition of Nitrogen in North America

Site	Estimated Precipitation (mm)	Total Inorganic Nitrogen NO_x-N+NH_4-N (mg/L)	Total Nitrogen (mg/L)	TIN Load (kg/ha/yr)	TN Load (kg/ha/yr)
Ottawa, Ontario					
Rain	724	2.15		15.6	
Snow	147	0.85		1.3	
Hamilton, Ontario					
Wetfall	818		0.49		4.0
Dryfall					2.5
Geneva, NY					
Total	993	1.10		10.9	
Coshocton, OH					
Total	939	0.80	1.17	7.5	11.0
Cincinnati, OH					
Total	1020	0.69	1.27	7.0	12.9
Seattle, WA					
Dryfall					0.7
Boney Marsh, FL					
Dissolved	818	0.75	1.88	6.1	15.4
Particulate	818		0.94		7.7

Note: Data from Nitrogen Control and Phosphorus Removal in Sewage Treatment, 1978; Unpublished Data, South Florida Water Management District.

The nitrogen concentration of rainfall is highly variable depending on atmospheric conditions, air pollution, and geographical location. A typical range of total nitrogen concentrations associated with rainfall is 0.5 to 2.0 mg/L, with about half of this present as ammonia and nitrate nitrogen. These concentrations can be used with local rainfall amounts to estimate rainfall inputs in nitrogen mass balances (see Tables 13-6 and 13-7). Some dryfall of nitrogen also occurs through adsorption of atmospheric (volatilized) ammonia and through deposition of organic dust containing organic nitrogen. Typical dryfall nitrogen inputs that have been measured in a variety of ecosystems are between 0.07 and 0.25 g total nitrogen m^2/yr. These estimates can be used to examine the importance of dryfall to the nitrogen balance in a specific wetland.

Ammonia Volatilization

Un-ionized ammonia is relatively volatile and can be removed from solution to the atmosphere through diffusion through water upward to the surface and through mass transfer from the water surface to the atmosphere. Un-ionized ammonia generally is a small fraction of total ammonia in wetland waters, comprising less than 1 percent for circumneutral pH and $0 < T < 25°C$ (see Figure 13-1).

Mass transfer in the water may be limiting in SSF wetlands, but would not be in SF wetlands with patches of open water. In the latter case, measured dispersion coefficients are of sufficient magnitude to mix the upper water layers during the normally slow transit from inlet to outlet (Kadlec, 1994).

Measurements of ammonia gasification have been made for ammonium-fertilized wetlands vegetated in rice (Freney et al., 1985). Data correlated well with a first-order areal model based on mass transfer control in the atmosphere above the wetland, and calibration at one site predicted the second site ($R^2 = 0.96$). The model consisted of

$$J_v = K_{av}u_z(C_{GA,0} - C_{GA,z}) \tag{13-40}$$

$$\log_{10}\left[\frac{C_{GA,0}}{C_{LA,0}}\right] = 1.6937 - 1477.7/T \tag{13-41}$$

$$\log_{10}\left[\frac{C_{LA,0}}{C_{LA,0} + C_{LIA,0}}\right] = 0.09018 + 2729.92/T - pH \tag{13-42}$$

where C = concentration, g/m^3
J_v = ammonia emission rate, g N/m^2/d
K_{av} = ammonia volatilization rate constant, dimensionless
T = temperature, °K
u = wind speed, m/d

subscripts:

A = ammonia (un-ionized)
AI = ammonium (ionized)
G = gas phase
L = liquid phase
0 = at the water surface
z = at height z (meters) above the water

Equation 13-40 is the gas-phase mass transfer relation. Equation 13-41 represents the dissociation equilibrium for ammonia in solution, which is graphically represented in Figure 13-1. Equation 13-42 represents the Henry's Law equilbrium for gaseous ammonia over solution. Because temperature varies considerably on the diurnal cycle, ammonia losses are expected to reflect that same diurnal cycle, and did in the Freney et al. (1985) study (Figure 13-20). The loss rates in Figure 13-20 are comparable to plant uptake for dense stands of macrophytes.

Values of K_{av} were reported to be in the range 0.75×10^{-3} to 2.95×10^{-3}, with $K_{av} = 0.885 \times 10^{-3}$ providing the best fit for the calibration wetland.

For purposes of comparison, the model equations may be combined to calculate an overall areal rate constant based upon the liquid-phase ammonium concentration:

$$J_v = k_{av}C_{LA,0} \tag{13-43}$$

where k_{av} = apparent ammonia volatilization rate constant, m/d.

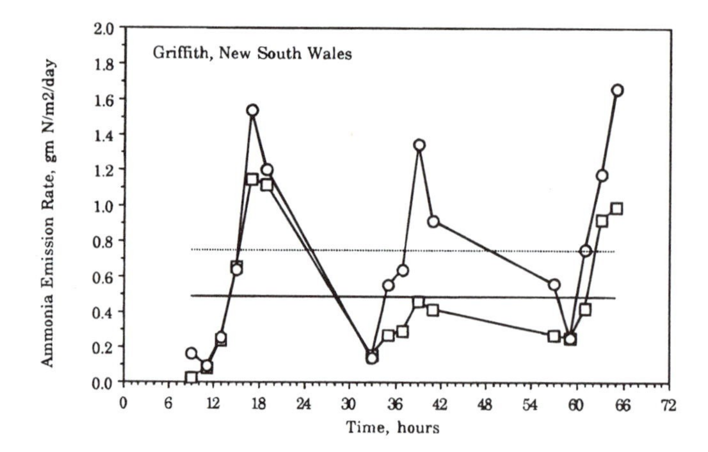

Figure 13-20 Ammonia emission rates from flooded rice fields at two sites. Soil fertility and crop density were different for the two 0.2-ha sites; water depth was 0.11 m. (Data from Freney et al., 1985.) Horizontal lines are the 3-day means.

Estimates of k_{av} may be made from the Freney model, for assumed wind speeds and atmospheric ammonia concentrations set equal to zero. The second assumption provides an upper bound to the potential ammonia volatilization. Table 13-8 shows ammonia loss rate constants ranging from about 0.1 to 10 m/yr, which are one order of magnitude lower than the ammonium reduction rate constants inferred from input-output data for treatment wetlands. The Freney data were acquired in warm temperatures and at pH > 7; therefore, Table 13-8 shows no values for low pH and low temperature.

It is noteworthy that these losses occur for extremely low concentrations of un-ionized ammonia, typically less than 1 percent of the total ammonium plus ammonia in solution. Nevertheless, the predicted emission rate constants are on the order of 5 to 10 percent of the overall loss rate constants for treatment wetlands.

Measurements of ammonia volatilization have been made for a wastewater wetland (Billore et al., 1994). At 35°C and an unspecified pH, the loss rate constants were 2.6 ± 1.9 m/yr for cattail, 6.5 ± 5.6 for duckweed, and 4.2 ± 1.8 for open water. These correspond to values in the circumneutral pH range in Table 13-8, which is expected to prevail in a wastewater wetland (see Chapter 10).

It is also noteworthy that the gas-phase mass transfer coefficients in the air above the wetland, inferred from the Freney model, are in reasonable agreement with the mass transfer coefficients which accurately predict water loss by evaporation under the same conditions (Freney et al., 1985). Discrepancies were attributed to small, water-side, mass transfer resistances.

The applicability of purely physical-chemical equilibrium in the wetland environment was challenged by Hemond (1983), who measured ammonia concentrations in the atmosphere above an acid (pH = 3.8) bog in Massachusetts. The concentrations immediately above the bog surface were three orders of magnitude greater than those predicted by Equations 13-40 and 13-41. Nevertheless, estimated vertical fluxes of ammonia were not a significant component of the nitrogen budget of this natural bog.

Therefore, volatilization typically has limited importance, except in specific cases where ammonia is present at concentrations greater than 20mg/L. Also, ammonia volatilization is accelerated by high water flow rates or vigorous mixing, neither of which are typical of most quiescent wetland ecosystems.

Biomass Decomposition

The nitrogen assimilated by macrophytes and microflora and microfauna is in part released during decomposition. Many wetland studies have addressed the question of the rates of decrease of biomass and the incorporated nitrogen. Turnover times for leaf litter vary from several months in warm climates to 2 years in cold climates, but rates of decomposition during warm months do not vary much with geographical region. The decomposition process is typified by a rapid initial weight loss, followed by an exponential loss of the remaining weight to an irreducible residual that contributes to sediment and soil building. The initial

Table 13-8 Rate Constants for Ammonia Volatilization from Flooded Rice

T, (°C)	pH = 7.0	pH = 7.5	pH = 8.0
15	0.14	0.44	1.36
20	0.25	0.78	2.39
25	0.43	1.34	4.10
30	0.73	2.28	6.85
35	1.23	3.79	11.16

Note: Wind speed = 5 m/s = 11.2 mi/h. Units are in meters per year. (Data from Freney et al., 1985.)

weight loss is about 10 percent for *Typha* spp. in natural wetlands and about 20 percent in wastewater wetlands (Kadlec, 1989b). Initial weight loss is greater (*circa* 50 percent) for "soft-tissue" plants such as *Sagittaria* spp. After initial weight loss, half lives of typical marsh plant leaf litter range from 0.4 to 1.2 years (Kadlec, 1989b).

Woody stems decay much more slowly. Wood litter from various wetland shrubs showed half lives from 5 to 13 years (Chamie and Richardson, 1978). Decomposition is typically faster in air, for the standing dead plant material, than it is for emersed litter. Acceleration of the decomposition process has been noted in nutrient-enriched water (Davis, 1984).

Decomposition may not be accompanied by a proportionate change in the nutrient content of the biomass. In the case of above-water decomposition, the fractional nitrogen loss typically exceeds the fractional biomass loss. The difference is attributable to leaching (Davis and van der Valk, 1978). The oxidation state of the nitrogen in the leachate is not known; it may be organic, ammonium, or nitrate nitrogen. However, the aerial origin of the leachate argues for nitrate as the form of nitrogen.

Underwater litter decay initially involves a transfer of nitrogen in plant tissues to nitrogen incorporated in decomposer microbial biomass. Consequently, the pool of particulate nitrogen is not reduced in the early stages of decomposition, which may last many months. For example, particulate nitrogen remained constant over 180 days for decomposing *Scirpus* spp. litter (Godschalk and Wetzel, 1978). S. M. Davis (1984) found constant nitrogen content over 2 years of decomposition of cattail (*Typha orientalis*) leaves in Florida, during which time biomass went down by 50 to 65 percent and nitrogen percentage more than doubled.

The importance of these findings is both the amount of nitrogen returned from the litter and the time delay in its release. Each year's cohort of litter may contain 15 to 25 g/m^2 of nitrogen each in below- and above-ground detritus. Allowing for accretion of half of this nitrogen (see below, Burial of Organic Nitrogen), the return rate is on the order of 15 to 25 $g/m^2/yr$. However, the return of a cohort of nitrogen apparently does not commence for a period of about 1 year and is then spread over the ensuing few years. Therefore, this component of the wetland nitrogen cycle may not stabilize for 2 or more years.

The form of nitrogen released upon decomposition of the underwater litter is also an unknown, but presumably it is comprised of organic and ammonium nitrogen, rather than nitrate.

Burial of Organic Nitrogen

Some fraction of the organic nitrogen incorporated in detritus in a wetland may eventually become unavailable for additional nutrient cycling through the process of peat formation and burial (see Chapters 11 and 12 for a discussion of the wetland carbon cycle). In natural, rich peatlands, the nitrogen content of these accumulated carbon reserves is typically about 2.5 to 3.0 percent on a dry-weight basis (Richardson et al., 1978; Reddy et al., 1991a). Under the more eutrophic conditions of many treatment wetlands, the nitrogen content exceeds the top of this range. Typical peat formation rate in natural peatlands and bogs ranges from 10 to 180 g dry matter m^2/yr yr (Richardson, 1989). However, under eutrophic treatment wetland conditions, accretion is much larger. For example, a lightly fertilized zone of Water Conservation Area 2A in Florida (inlet total phosphorus *circa* 0.2 mg/L) reported values range from 460 $g/m^2/yr$ (Richardson and Craft, 1990) to 1130 $g/m^2/yr$ (Reddy et al., 1991a). At 3 percent nitrogen, this accretion buries 14 to 34 g $N/m^2/yr$ (0.38 to 0.93 kg N/ha/d). By comparison, the nitrogen loading to a treatment wetland receiving 5 cm/d of 20 mg/L of total nitrogen is 365 g $N/m^2/yr$. Thus, nitrogen accretion and burial may be important for conditions of light nitrogen loading, but becomes insignificant for high nitrogen loadings.

Ammonia Adsorption

Ionized ammonia may be removed from solution through a cation exchange adsorption reaction with detritus and inorganic sediments in SF wetlands or the media in SSF wetlands.

The adsorbed ammonia is bound loosely to the substrate and can be released easily when water chemistry conditions change. At a given ammonia concentration in the water column, a fixed amount of ammonia is adsorbed to and saturates the available attachment sites. When the ammonia concentration in the water column is reduced, some ammonia will be desorbed to regain an equilibrium with the new concentration. If the ammonia concentration in the water column is increased, the adsorbed ammonia also will increase. Figure 13-21 illustrates this sorption process for the Houghton Lake, MI natural wetland treatment system; Water Conservation Area 2A in the Everglades; and for the gravel bed wetlands at Muscle Shoals, AL.

It is important to note that a large mass of ammonia nitrogen will not be adsorbed to detritus and sediment in a wetland and that this ammonia is very labile. The top 20 cm of the wetland substrate may contain up to 20 g N/m^2 in exchangeable form for a peat exposed to 10 mg/L ammonium nitrogen. This pool of nitrogen is quickly established at moderate nitrogen loadings (see Chapter 14 for an analogous discussion of sorption saturation times for phosphorus). At light nitrogen loadings, a short startup period may be influenced by this storage.

If the wetland substrate is exposed to oxygen, perhaps by periodic draining, sorbed ammonium may be oxidized to nitrate. This concept forms one basis for intermittently fed and drained, vertical flow treatment wetlands.

TOTAL NITROGEN REMOVAL RATES IN WETLANDS

BACKGROUND

Ponds and natural wetlands provide some insights about total nitrogen removal in aquatic and lightly loaded wetland ecosystems.

Reed (1984) analyzed treatment pond data and considered total nitrogen. The database was essentially that analyzed by Pano and Middlebrooks (1982) for ammonium nitrogen. A

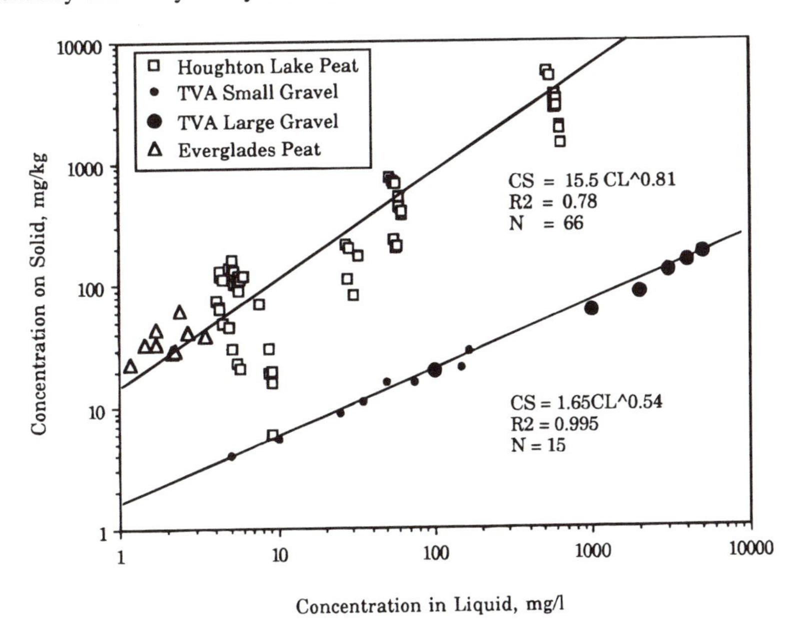

Figure 13-21 Ammonium adsorption on SF and SSF wetland substrates. The TVA data is from Sikora et al., 1994; Everglades data from Reddy et al., 1991; and Houghton Lake data from unpublished results.

pseudo-first-order, volumetric, plug flow model was fit and verified by Reed (1984). These pond data reflect ammonia volatilization, bacterial immobilization, and nitrification/denitrification. Bacteria are mostly associated with the pond sediments and are proportional to pond area. Volatilization is an areal loss from the upper water surface. Nitrification requires oxygen transfer, another surficial process. Therefore, areal models, such as that of Pano and Middlebrooks (1982), are suitable for ponds. Therefore, the k_{av} values corresponding to the Reed (1984) analysis were computed and are shown in Table 13-9.

Nixon and Lee (1986) reviewed the fluxes of total nitrogen in a variety of North American wetlands that had a complete spectrum of physical, chemical, and biological conditions. Figure 13-22 shows the overall estimated total nitrogen mass removal rates for several wetlands in this geographic area and demonstrates that nitrogen removal rates vary greatly in natural wetlands. Factors that affect the total nitrogen removal rate of these natural wetlands include (1) total nitrogen loading rate, (2) climate, (3) plant community composition, and (4) soil characteristics.

The individual process considerations discussed previously may be combined to form the integrated concept of nitrogen fluxes in treatment wetlands. An illustration of the relative magnitudes of those fluxes is shown in Figure 13-23 for a moderate nitrogen loading. A few important points emerge from this integrated view of nitrogen processing. First, the magnitude of the vegetative nitrogen cycle is by no means trivial; uptake represents 25 percent of the net removal in this example. However, net burial is only a fraction of plant uptake. Second, the influence of the biomass decay causes the true amount of ammonification to significantly exceed the apparent rate based only on water analyses: 220 vs. 160 $g/m^2/yr$. The k-C* model incorporates the appropriate correction as a return flux, kC*. Third, the true amount of nitrification greatly exceeds the amount based only on input-output (I/O) water analyses: 200 vs. 80 $g/m^2/yr$. The sequential nitrogen kinetic model corrects for the production of ammonium from organic nitrogen and calibrates to have higher rate constants accordingly. Finally, the rate of denitrification far exceeds the rate based only on input-output (I/O) water analyses: 280 vs. 80 $g/m^2/yr$. The contribution of nitrification means that apparent denitrification is much smaller than the true value.

It should also be noted that the cycle time for the biomass cycle is long: the returning organic nitrogen is from storages dating back an average of many months to over 1 year.

With the understanding that total nitrogen processing reflects this combination of transfers, the following analysis examines treatment wetland data.

NORTH AMERICAN TREATMENT WETLANDS

A database of design and operational data from North American treatment wetlands includes nitrogen data from 17 constructed SF wetlands, 26 natural SF wetlands and 9 constructed SSF wetlands (Knight et al., 1992, 1993b). Table 13-10 summarizes the average concentrations and estimated mass removals for total Kjeldahl nitrogen, nitrate, ammonia, and total nitrogen from the three types of systems. The number of systems and the number

Table 13-9 Rate Constants for Total Nitrogen Loss from Ponds

T, (°C)	pH = 6.5	pH = 7.0	pH = 7.5	pH = 8.0
0	1.33	2.14	2.94	3.75
5	1.61	2.59	3.57	4.54
10	1.95	3.13	4.32	5.50
15	2.36	3.79	5.23	6.66
20	2.85	4.59	6.33	8.07
25	3.45	5.56	7.66	9.77

Note: Units are in meters per year.
Based on Reed, 1984.

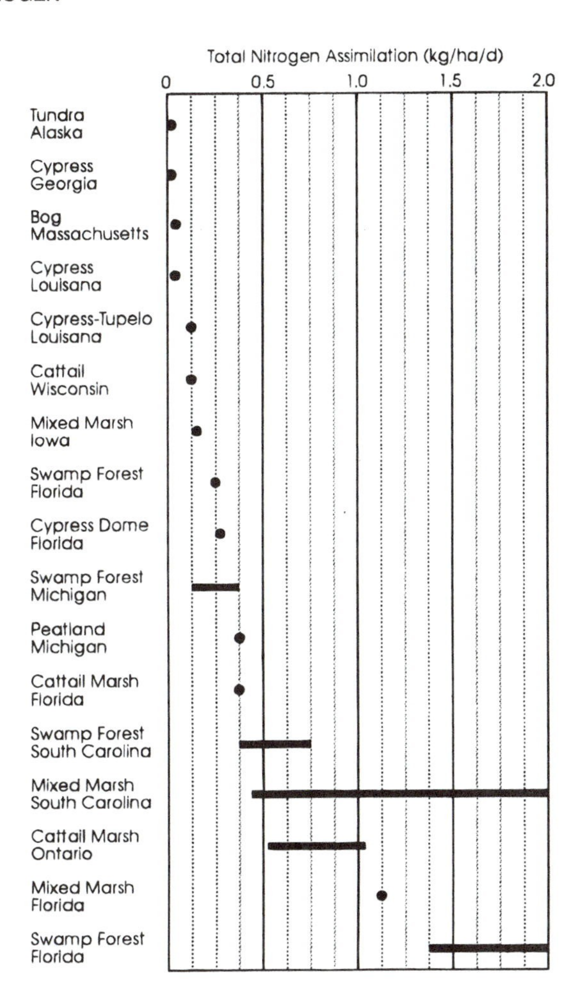

Figure 13-22 Range of total nitrogen assimilation in natural wetlands (Adapted from Nixon and Lee. 1986).

of cells at a given system represented by each average are variable, so these numbers must be used cautiously to develop generalizations.

As previously reported (Knight et al., 1985a), total nitrogen is reduced in most wetland treatment systems, with average mass reductions in Table 13-10 between 46 and 72 percent. Average total nitrogen mass removal rates are reported in Table 13-10 as 1.89 kg/ha/d for natural SF wetlands, 3.46 kg/ha/d for constructed SF wetlands, and 15.63 kg/ha/d for SSF wetlands. The total nitrogen mass removal rates in Table 13-10 are typically much higher than the rates summarized in Figure 13-22 for natural wetlands not receiving wastewaters.

DESIGN FACTORS AFFECTING TOTAL NITROGEN REMOVAL

The total nitrogen mass removal rate in natural wetlands is highly variable and depends on the form of nitrogen in the inflow, water depth, dissolved oxygen, and the total nitrogen mass loading rate (Figure 13-24). The absolute total nitrogen mass removal rate has been found to correlate consistently with total nitrogen mass loading rate up to loading rates of 30 kg/ha/d. However, at higher mass loading rates, the mass removal efficiency for total nitrogen removal generally declines (Knight et al., 1985a). Figure 13-24 illustrates that the

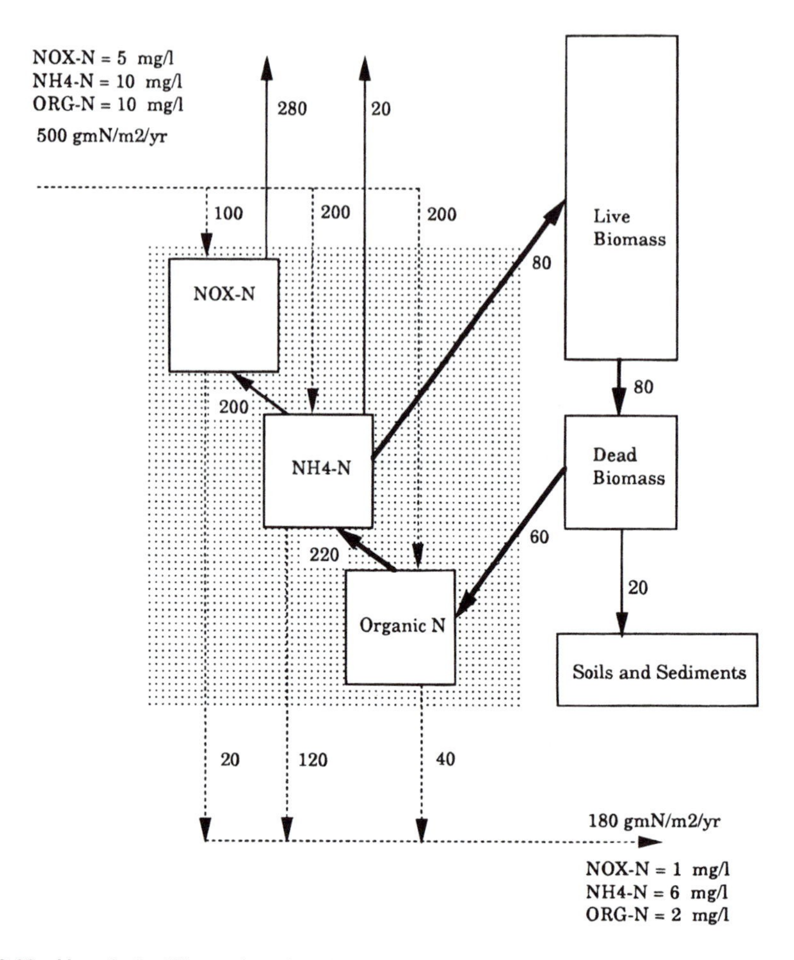

Figure 13-23 Hypothetical illustration of stationary nitrogen fluxes in a treatment wetland. The hydraulic loading is set at 20 m/yr = 5.5 cm/d. Apparent rate constants are 32 m/yr for nitrate, 18.3 m/yr for ammonium, and 49 m/yr for organic N ($C^* = 1.2$ mg/L). The return rate of organic nitrogen is also equal to $k_{ON} \cdot C_{ON}^* - 49 \cdot 1.2 = 60$ g/m^2/yr.

higher average total nitrogen mass removal rate reported in Table 13-10 for SSF systems is partly due to the typically higher total nitrogen mass loadings to these systems and not solely because of an inherently higher total nitrogen assimilation rate in SSF systems (Reed, 1992).

The efficiency of total nitrogen removal also is reduced by low total nitrogen inflow concentration because of internal nitrogen processes. At low inflow concentrations, the internal production and release of total nitrogen is greater than assimilation, resulting in negative calculated total nitrogen removal efficiencies. The lowest observed total nitrogen outflow concentrations from constructed wetlands used for treatment are typically between 1 and 2 mg/L. Outflow total nitrogen concentrations under these conditions are inversely correlated with the hydraulic loading rate (HLR).

The hydraulic loading rate affects total nitrogen removal efficiency, with reduced efficiency at higher loadings. Figure 13-25 illustrates the effect of the hydraulic loading rate on total nitrogen removal efficiency for a specific wetland. Scatter on this graph is caused by other factors: for example, total nitrogen inflow concentration and vegetative uptake and release.

Nitrogen reduction in constructed wetlands can be increased by optimizing flow distribution and residence time by incorporating deeper water zones arranged perpendicular to the

Table 13-10 Average Nitrogen Concentrations and Mass Balances for North American Treatment Wetlands

	Total Kjeldahl N					Nitrite + Nitrate N				
Wetland Type	In (mg/L)	Out (mg/L)	LR (kg/ha/d)	RR (kg/ha/d)	Eff (%)	In (mg/L)	Out (mg/L)	LR (kg/ha/d)	RR (kg/ha/d)	Eff (%)
Constructed SF	17.5	12.3	8.23	4.11	49.9	1.90	1.21	0.81	0.36	44.4
Natural SF	6.34	1.77	4.08	2.32	56.9	4.30	0.29	1.42	1.10	77.5
Constructed SSF	22.0	12.9	24.9	8.11	32.5	109	94.5	158	15.0	9.45
	Ammonia N					**Total N**				
	In (mg/L)	Out (mg/L)	LR (kg/ha/d)	RR (kg/ha/d)	Eff (%)	In (mg/L)	Out (mg/L)	LR (kg/ha/d)	RR (kg/ha/d)	Eff (%)
Constructed SF	8.84	8.19	4.93	1.67	33.9	8.08	4.58	7.59	3.46	45.6
Natural SF	4.09	0.69	2.21	1.58	71.5	10.2	2.30	2.63	1.89	71.9
Constructed SSF	7.89	3.58	10.2	2.21	21.7	41.4	12.1	29.0	15.6	53.8

Based on analysis of the North American Treatment System Database (Knight et al., 1993b).

Note: LR = Mass loading rate
RR = Removal rate
Eff = Effluent
SF = Surface-flow wetland
SSF = Subsurface-flow wetland

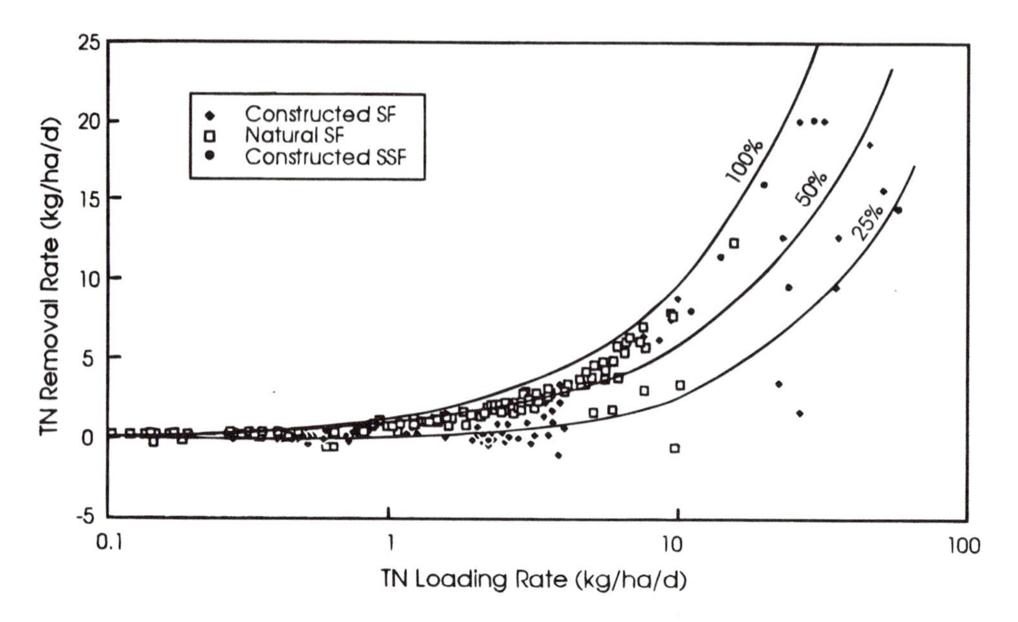

Figure 13-24 North American wetland treatment system total nitrogen mass removal rate as a function of total nitrogen mass loading rate. (Data from Knight et al., 1992.)

flow path (Knight and Iverson, 1990). These deep zones remain free of rooted emergent vegetation, but frequently are colonized by floating aquatics such as duckweed. Comparisons of test results from SF constructed wetlands with and without deep zones have shown enhanced reductions in total nitrogen, NH_3, total Kjeldahl nitrogen, and other wastewater pollutants (Knight et al., 1994; Hammer and Knight, 1994).

AREA-BASED, FIRST-ORDER k-C* MODEL FOR TOTAL NITROGEN

Because total nitrogen removal in wetland treatment systems is an integrative measure of individual nitrogen transformations which can be approximated by area-based, first-order rate expressions, the loss of total nitrogen between the inflow and outflow of a wetland treatment system can also be predicted with a simple model of the form

$$J_{TN} = k_{TN}[C_{TN} - C^*_{TN}] \tag{13-44}$$

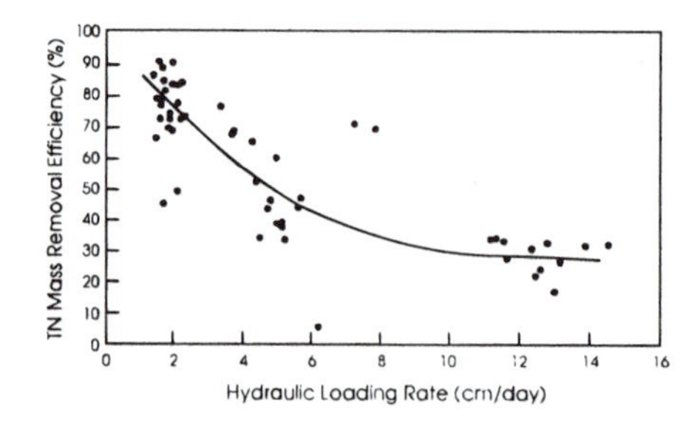

Figure 13-25 Effect of hydraulic loading rate on monthly average total nitrogen mass removal efficiency in a pilot wetland receiving pulp mill effluent.

where J_{TN} = the area-based, first-order total nitrogen reduction rate, $g/m^2/yr$
k_{TN} = area-based, first-order total nitrogen rate constant, m/yr
C^*_{TN} = irreducible background wetland TN concentration, mg/L

The integrated form of the total nitrogen mass balance for this uptake rate and plug flow conditions is

$$\ln\left[\frac{C_{TN,o} - C^*_{TN}}{C_{TN,i} - C^*_{TN}}\right] = -\frac{k_{TN}}{q} y = -k_{VTN}\tau y \quad (13\text{-}45)$$

where k_{VTN} = first-order volumetric total nitrogen reduction rate constant, 1/d
q = hydraulic loading rate, m/yr
τ = detention time, days
y = fractional distance through the wetland

The value of C* for total nitrogen in this model derives from background values for the components of total nitrogen, which are zero for ammonium and nitrate, and about 1.5 mg/L for organic nitrogen. Consequently, the value of C* for total nitrogen is approximately 1.5 mg/L. Whenever total nitrogen remains well above this level—above 20 mg/L, for instance—C* has little effect on calculations.

Treatment wetland data verify the decreasing trend of total nitrogen concentrations contained in this model. Bavor et al. (1989) calibrated this model to internal transect data for seven flow-through SF, SSF, and hybrid wetlands. Lakshman (1981) obtained data on batch operation of five SF wetlands, which may be fit by the model ($y = 1$ and τ is batch run time, see Figure 13-26). Tanner et al., (1995) obtained total nitrogen data from eight SSF wetlands run side-by-side at different hydraulic loading rates (Figure 13-27). That data is well described by the k-C* model.

The reader is again reminded that the plug flow model has been used for data reduction in this and other chapters on the premise that rate constants so estimated will be conservatively low. In the case of Tanner et al. (1995), tracer studies demonstrated nonplug flow conditions (Tanner and Sukias, 1994). A precise flow model was not developed, but the extreme of total mixing is shown for comparison on Figure 13-27. The rate constant for the plug flow model is lower by nearly a factor of two.

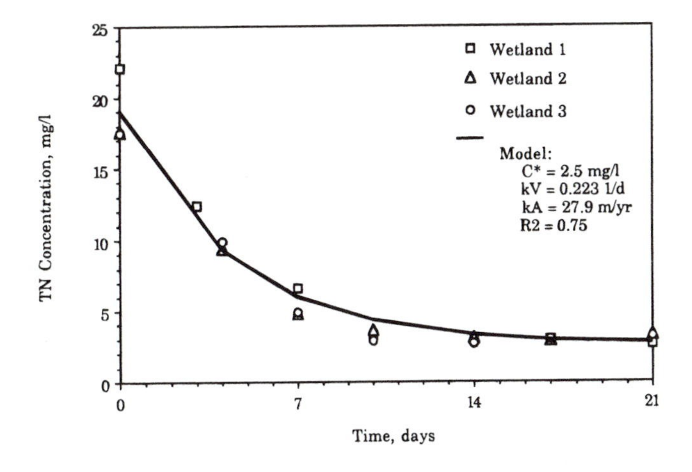

Figure 13-26 Reduction of total nitrogen in three SF wetlands operated in the batch mode. These data were taken in July in Humboldt, Saskatchewan (Lakshman, 1981).

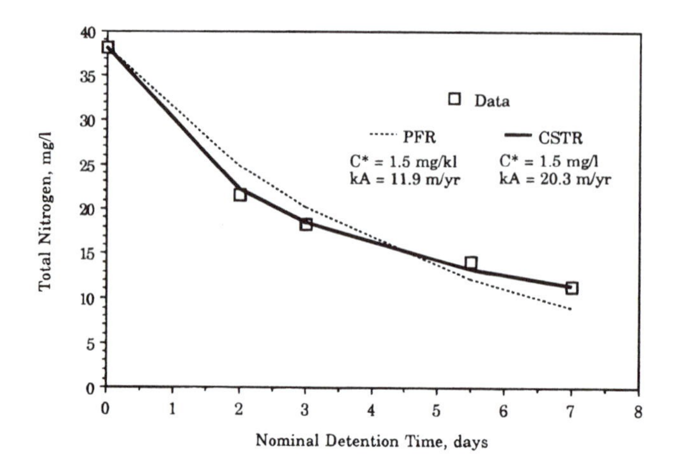

Figure 13-27 The reduction of total nitrogen in side-by-side SSF treatment wetlands in New Zealand. Note that a large degree of mixing provides a better data fit than no mixing. Also note that the parameter estimate for the rate constant is sensitive to the degree of mixing. Data are annual averages of fortnightly samples from four *Scirpus validus* wetlands. (Data from Tanner et al., 1995.)

The total nitrogen removal rate depends on total nitrogen concentration and approaches zero when total nitrogen inflow concentration is less than about 2 mg/L. The intersystem average of values for k_{TN} from Table 13-11, which represents systems receiving inflow total nitrogen concentrations above 2 mg/L, is 14.3 m/yr (N = 116). The average rate constant for SF marsh systems is 15.3 m/yr (N = 30), for SF forested systems is 16.0 m/yr (N = 6), and for SSF is 15.1 m/yr. An average value of C^*_{TN} = 1.5 mg/L has been used in Table 13-11 whenever the data do not support a lower value.

Effects of Season and Temperature

The component processes of ammonification, nitrification, and denitrification have all been shown to be temperature dependent in treatment wetlands; therefore, rates of total nitrogen reduction will also be temperature dependent. Wetland studies confirm this expectation and provide temperature factors.

Four wetlands in Sweden were used to model total nitrogen disappearance (Arheimer and Wittgren, 1994). Nitrate comprised about half of the total, which was in the range $5 \leq TN \leq 20$ mg/L. The calibration spanned two calendar years. The model was a temperature-sensitive, first-order areal model with a zero background concentration:

$$J_{TN} = [rT_{10}]C_{TN} \qquad (13\text{-}46)$$

where C_{TN} = total nitrogen concentration, mg/L
J_{TN} = total nitrogen removal flux, g/m²/d
r = retention calibration factor, m/°C/d
T_{10} = mean temperature for the last 10 days, °C

The product $[rT_{10}]$ is equal to the first-order irreversible rate constant [k] for total nitrogen reduction. The value of r = 0.0023 m/°C/d calibrated data from the four wetlands over 2 years with R^2 = 0.92. Over the range $5 \leq T \leq 25$°C, the equivalent k and θ values are

$$k_{20} = 16.2 \text{ m/yr}$$

$$\theta = 1.081$$

Table 13-11 First-Order, Area-Based Rate Constants for Total Nitrogen Reduction

Site	Wetland	Type	Data Quarters	HLR (cm/d)	TN In (mg/L)	TN Out (mg/L)	C* (mg/L)	kTN (m/yr)	Ref.
Surface-flow marshes									
Lakeland, FL	1	SF con marsh	21	3.71	11.01	3.72	1.5	19.70	NADB, 1993
	2	SF con marsh	16	3.37	3.83	2.54	1.3	8.77	NADB, 1993
	3	SF con marsh	15	1.44	2.54	1.6	1.3	7.46	NADB, 1993
Orange Co., FL	1	SF con marsh	11	1.38	2.3	1.39	0.8	4.70	NADB, 1993
Iron Bridge, FL	1–12	SF con marsh	16	2.68	3.41	1.09	0.7	18.15	NADB, 1993
Fort Deposit, AL	1–2	SF con marsh	1	0.82	10.88	3.6	1.5	4.48	NADB, 1993
West Jackson Co., MS	1–7	SF con marsh	2	2.41	8.26	3.9	1.5	9.11	NADB, 1993
Leaf River, MS	1	SF con marsh	2	9.72	11.71	8.3	1.5	14.42	NADB, 1993
	2	SF con marsh	2	12.65	14.58	9.26	1.5	24.11	NADB, 1993
	3	SF con marsh	1	13.32	14.58	7.34	1.5	39.20	NADB, 1993
Santa Rosa, CA	1	SF con marsh	1	16.83	16.17	9.57	1.5	36.71	NADB, 1993
	2	SF con marsh	1	16.83	9.57	4.24	1.5	66.36	NADB, 1993
	3	SF con marsh	1	16.83	4.24	2.59	1.5	56.62	NADB, 1993
	4	SF con marsh	1	16.83	2.59	2.51	1.5	4.68	NADB, 1993
Des Plaines, IL	EW3	SF con marsh	5	8.14	2.72	1.68	0.7	21.49	NADB, 1993
	EW4	SF con marsh	5	1.44	2.71	1.21	0.7	7.21	NADB, 1993
	EW5	SF con marsh	5	6.32	2.71	1.53	0.7	20.40	NADB, 1993
Cobalt, Ontario	1	SF con marsh	4	7.5	7.33	2.84	1.5	40.25	Miller, 1989
Benton, KY	1	SF con marsh	7	6.23	13.3	9.81	1.5	7.97	Choate et al., 1990a
	2	SF con marsh	7	6.23	14.05	9.44	1.5	10.41	Choate et al., 1990a
Listowel, Ontario	1	SF con marsh	16	2.67	12.77	8.70	1.5	4.36	Herskowitz, 1986
	2	SF con marsh	16	2.83	12.77	9.18	0.4	3.55	Herskowitz, 1986
	3	SF con marsh	16	2.01	12.77	7.14	1.5	5.09	Herskowitz, 1986
	4	SF con marsh	16	1.95	18.92	9.49	1.5	5.56	Herskowitz, 1986
	5	SF con marsh	16	2.60	18.92	12.15	1.4	4.63	Herskowitz, 1986
Gustine, CA	1A	SF con marsh	4	4.19	31.36	25.28	1.5	3.48	Walker and Walker, 1990
	1B	SF con marsh	4	2.09	31.36	26.26	1.5	1.43	Walker and Walker, 1990
	1C	SF con marsh	4	1.05	35.61	30.94	1.5	0.56	Walker and Walker, 1990
	1D	SF con marsh	4	4.19	40.52	33.77	1.5	2.90	Walker and Walker, 1990
	2A	SF con marsh	4	4.46	47.67	34.95	1.5	5.25	Walker and Walker, 1990

Table 13-11 Continued

Site	Wetland	Type	Data Quarters	HLR (cm/d)	TN In (mg/L)	TN Out (mg/L)	C* (mg/L)	kTN (m/yr)	Ref.
Forested surface flow									
Reedy Creek, FL	WTS1	SF forested	37	3.57	7.88	1.88	0.6	22.65	NADB, 1993
	OFWTS	SF forested	28	5.63	8.24	1.24	0.6	50.10	NADB, 1993
Central Slough, SC	1	SF forested	17	0.56	16.12	4.16	1.5	3.48	NADB, 1993
Poinciana, FL	1	SF forested	9	0.20	6.24	2.80	1.0	0.78	NADB, 1993
Vereen, SC	1	SF forested	12	0.37	16.69	2.81	1.0	2.92	NADB, 1993
Drummond, WI	1	SF forested	6	2.30	9.06	2.13	0.9	15.91	NADB, 1993
Floating/submergent aquatics									
New Zealand	2 Wetlands	SF con FAP	7	5.52	197.00	102.00	1.5	13.41	van Oostrom, 1994
Richmond, NSW	4	SF con SAB	8	7.34	44.07	26.04	1.5	14.76	Bavor et al., 1988
Soil-based reed beds									
Denmark	44 Wetlands	Soil-based Phrag	8 Typical	4.53	36.70	21.04	1.5	9.73	Schierup et al., 1990b
Subsurface-flow reed beds									
U.K.	11 Wetlands	SSF Phrag	8 Typical	18.30	26.73	17.66	1.5	29.76	Green and Upton, 1993
Subsurface flow									
Benton, KY	3	SSF bulrush	4	5.22	14.55	10.50	0.4	6.42	Choate et al., 1990a
Hardin, KY	1	SSF cattail	4	9.70	20.23	13.23	0.0	15.04	Choate et al., 1990a
	2	SSF bulrush	4	8.70	19.30	8.44	0.0	26.27	Choate et al., 1990a
Phillips High School, AL	1	SSF cattail	2	3.72	43.00	4.5	1.5	35.67	NADB, 1993
Mayo Pennisula, MD	1	SSF bulrush	3	10.63	31.36	13.07	1.5	36.79	NADB, 1993
Utica, MS	4 Wetlands	SSF bulrush	3	3.15	14.15	7.88	1.5	7.87	NADB, 1993
Richmond, NSW	2	SSF gravel only	8	3.83	44.07	21.59	0.0	9.98	Bavor et al., 1988
	3	SSF bulrush	8	4.61	44.07	20.12	0.0	13.19	Bavor et al., 1988
	5	SSF cattail	8	5.08	44.07	21.69	0.0	13.14	Bavor et al., 1988
Baxter, TN	A	SSF bulrush	5	8.3	25.6	22.8	1.5	3.74	George et al., 1994
	C	SSF bulrush	5	2.7	25.6	18.0	1.5	3.73	George et al., 1994
	F	SSF bulrush	5	25.5	25.6	20.1	1.5	24.11	George et al., 1994
	G	SSF bulrush	5	8.3	25.6	18.2	1.5	11.11	George et al., 1994
	H	SSF bulrush	5	8.3	25.6	16.7	1.5	13.96	George et al., 1994
	J	SSF bulrush	5	2.7	25.6	19.8	1.5	2.71	George et al., 1994
	M	SSF bulrush	5	25.5	25.6	18.2	1.5	34.14	George et al., 1994
	N	SSF bulrush	5	8.3	25.6	20.1	1.5	7.85	George et al., 1994
Hamilton, New Zealand	4 Wetlands	SSF gravel only	4	3.74	38.2	21.4	1.5	8.36	Tanner et al., 1995
	4 Wetlands	SSF bulrush	4	3.74	38.2	16.4	1.5	12.34	Tanner et al., 1995

Note: The value C* = 1.5 is used, unless the range of outlet concentrations shows less.

Glyceria maxima wetlands were studied by van Oostrom (1994). The incoming water was very high nitrogen wastewater, TN ~ 200 mg/L, of which 62 percent was nitrate nitrogen and 30 percent was ammonium nitrogen. Two wetlands were operated once-through; one was operated with 50 percent recycle. Only the once-through wetlands are discussed here. The hydraulic loading rate averaged 5.5 cm/d over a 610-day study period.

Fortnightly data showed a seasonal variation in the rate constant for total nitrogen reduction, which may be ascribed to a temperature effect (Figure 13-28). Analysis of the data yields the following approximate values:

$$k_{20} = 18.4 \text{ m/yr}$$

$$\theta = 1.084$$

Bavor et al. (1989) examined the effect of SSF gravel-based wetlands on nitrogen removal from primary and secondary quality, domestic wastewater at Richmond, NSW. These researchers found that the removal of total Kjeldahl nitrogen and ammonia nitrogen could be described adequately by first-order decay models based on hydraulic retention time, influent concentration, and corrected water temperature. Richmond data cover 2 years of complete speciation of nitrogen and include transect data. The incoming water contained little nitrate, and virtually no nitrate was present in the wetland effluents; consequently, total nitrogen and total Kjeldahl nitrogen were nearly identical (±5 to 10 percent). The influent total Kjeldahl nitrogen was about 80% ammonium and 20% organic nitrogen. From the transects, it was learned that total Kjeldahl nitrogen (total nitrogen) disappears monotonically, and the rate constant is temperature sensitive (Bavor et al., 1989). Rate constants and temperature coefficients derived from the transect data were

	k_{A20}, m/yr	k_{V20}, d^{-1}	θ
Gravel only	11.0	0.182	1.023
Bulrush	10.2	0.117	1.033
Cattail	10.5	0.121	1.024

It is clear that temperature is one factor leading to the large spread in TN k values in Table 13-11.

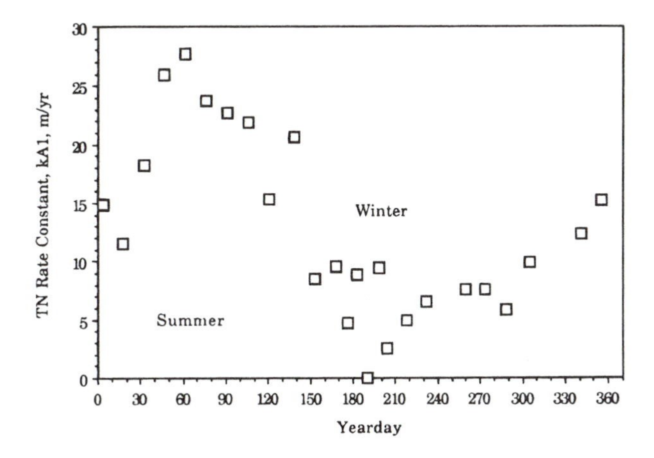

Figure 13-28 Seasonal variation in the areal rate constant for total nitrogen reduction in a *Glyceria maxima* wetland. (Data from van Oostrom, 1994.)

In view of the information from treatment wetlands and related technologies, an estimate of the temperature factor for total nitrogen reduction is $\theta = 1.05$. When the rate constants in Table 13-11 are adjusted to 20°C and averaged, the results are

k_{TN20} = 22 m/yr SF (includes soil-based reed beds and aquatics)
k_{TN20} = 27 m/yr SSF (includes U.K. reed beds and U.S. gravel beds)
k_{VTN20} = 0.31 d^{-1} SSF (60 cm at 40 percent porosity)

These values are to be regarded as central tendencies for the systems listed. The intersystem variability is large, in response to site factors not included in the simplified design model. For instance, septage-fed SSF wetlands may have higher concentrations and lower k-values.

OTHER FACTORS THAT AFFECT TOTAL NITROGEN REMOVAL

The distribution of the chemical forms of nitrogen in wastewater is an important factor in design for total nitrogen concentration and mass reduction. To be permanently removed from solution, organic nitrogen must be converted to ammonia, then to NO_X-N, and then to gas. An inflow of nitrogen in the nitrate form must only undergo denitrification to be removed from solution. Clearly, longer reaction times and multiple conversions must occur for a treatment wetland to assimilate an organic nitrogen inflow, compared to an inflow of nitrate.

Most wetland treatment systems are effective at total nitrogen removal because they provide diverse physical, chemical, and biological environments and have relatively long detention times. Total nitrogen removal efficiency may be reduced if limiting conditions exist. Those limiting conditions include the possibility of

- Short detention times = high hydraulic loadings
- Low temperatures
- pH too low or too high (inhibits both nitrification and denitrification)
- Significant contributions of organic nitrogen from decaying biomass
- Insufficient oxygen transfer to support nitrification
- Oxygen depletion due to preferential carbon oxidation
- Insufficient alkalinity to support nitrification
- Insufficient carbon source to support denitrification

SF treatment wetlands typically have aerated zones, especially near the water surface because of atmospheric diffusion, and anoxic and anaerobic zones in and near the sediments. In heavily loaded SF wetlands, the anoxic zone can move quite close to the water surface. Biomass decay provides a carbon source for denitrification, but that same decay competes with nitrification for oxygen supply. Low winter temperatures enhance oxygen solubility, but slow microbial activity.

The mechanisms for some of these processes are partially understood. For example, the observed first-order reduction in nitrate is likely due in part to the first-order diffusive transport of nitrate from the surface water to anoxic sediments. In winter, surface dissolved oxygen is higher, and mass transfer is lessened because of reduced diffusion coefficients and lowered convective dispersion. Thus, nitrate outflow concentrations may not be reduced as much in SF wetlands during winter months.

The transfer of dissolved oxygen from the atmosphere to SF wetlands was analyzed in Chapter 10, and estimated rates were found to be in the range of 1.2 to 3.0 $g/m^2/d$. That transfer is controlled by dispersive processes in the water. This estimate agrees with field measurements of oxygen uptake in SF treatment wetlands (Brix and Schierup, 1990).

Subsurface flow wetlands possess the same limitations for total nitrogen reduction. If the top of the bed is not capped with a layer of sediment and detritus, the resistance to oxygen transport through the air spaces in the gravel does not impede oxygen transport. Transfer of oxygen from the water surface into the bulk of the interstitial flowing water is then controlled by dispersive mixing within the water in the gravel bed. Literature correlations of dispersion in packed beds and open channels provide comparable estimates of dispersion in the two environments. This concept is reinforced by the tracer measurements of comparable dispersion in SF and SSF wetlands (see Chapter 9).

Oxygen transport by plants through their aerenchyma and roots may be insufficient to satisfy heterotrophic oxygen demands, resulting in anaerobic conditions and severe limitations on nitrification of NH_3. Oxygen transport to the root zone does not appear to exceed the respiratory requirements of the plant by any significant amount (see Chapter 10). Thus, SSF wetlands are capable of approximately the same oxygen transfer as SF wetlands. Brix and Schierup (1990) measured approximately 3 $g/m^2/d$ in excess of respiratory needs. This in turn agreed with the uptake necessary to produce the observed concentration reductions in the test wetland.

This rate limits nitrification to a maximum rate of about $3/4.3 \approx 0.8$ g $N/m^2/d$, which is in the range actually observed in European SSF systems (WPCF, 1990b).

A carbon source is necessary to provide the energy necessary for denitrification. Decay of litter and microdetritus can provide an internal source of that carbon. Approximately 2.5 g of carbohydrate are required to denitrify 1 g of nitrate nitrogen (see earlier discussion). Biomass release is limited to the decomposition of the annual gross production of the wetland, which may be as high as 5000 $g/m^2/yr$ in a SF wetland. The resulting denitrification rate, supportable from internal carbon sources, is then 5.5 $g/m^2/d$ (2000 $g/m^2/yr$). This limit corresponds to a 40 mg/L reduction at a hydraulic loading rate of 20 m/yr (5.5 cm/d).

Subsurface flow wetlands are at a disadvantage, with respect to providing an internal carbon source, because litter is deposited above the water surface. Carbon availability to the water is then limited by leaching of the litter.

DESIGN APPROACHES TO REMOVE NITROGEN

Because nitrogen removal is a goal common to many treatment wetlands, designers must consider nitrogen transformation processes carefully. This section reviews current knowledge on nitrogen removal rates and variability and introduces three design approaches: quick estimates, input/output, and mass balances. Examples are provided to illustrate the performance of these design methods at removing nitrogen from wastewater.

REGULATORY LIMITS AND STOCHASTIC VARIABILITY

In many treatment systems, nitrogen must be reduced to regulatory limits. For example, according to information in the North American Wetland Treatment Systems Database (Knight et al., 1993b), 50 percent of the permits for these systems include nitrogen, in some form, as a permit criterion. Of these permits, 90 percent have permit limits for ammonia nitrogen typically in the range of 1 to 10 mg/L, and only 8 percent have limits for total nitrogen in the range of 2 to 7.5 mg/L on an annual average basis. Currently, there are not many design examples from full-scale wetland treatment systems for nitrogen removal.

Regardless of the design method for nitrogen removal, the method should ensure regulatory compliance. For example, peak flows should be used to determine hydraulic loading rate, and minimum temperature or seasonal rate constants should be used to incorporate temperature dependence. The models should incorporate the variability of effluent nitrogen

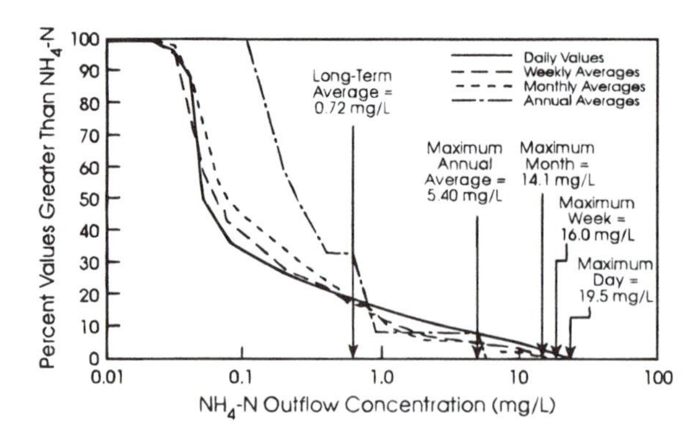

Figure 13-29 Cumulative frequency of outflow total ammonia nitrogen concentration at the Reedy Creek, FL natural wetland treatment system.

concentrations. As an example, Figure 13-29 illustrates the variability of ammonia nitrogen outflow concentrations from a wetland treatment system that was monitored 5 days each week for more than 10 years. In this case, maximum effluent concentrations over short time periods exceeded longer-term average values. For example, for the data in Figure 13-29, the long-term average (greater than 10 years) outflow ammonia nitrogen concentration was 0.72 mg/L, and the maximum annual average value over this period was 5.40 mg/L for a ratio (maximum annual to long-term average) of 7.5. In this example, the maximum monthly outflow was 16.67 mg/L, and the maximum daily outflow was 18.90 mg/L.

A second way of examining the effects of variability is to compare the probability distributions of the influent and effluent concentrations. For the Arcata full-scale SF marsh system, the percentile curves for total inorganic nitrogen (ammonium plus nitrate) are shown in Figure 13-30. A specified exceedence frequency defines a design target concentration.

Most wetland treatment systems are designed to meet either a maximum month or annual average limit for effluent nitrogen. Using nitrogen data from the North American Wetland Treatment System Database (Knight et al., 1993b), typical ratios of maximum month to annual average outflow concentrations were developed for each major nitrogen component important in wetland treatment systems. The maximum month to annual average ratios for the major water quality constituents are summarized in Chapter 19. Average ratios for nitrogen forms are 1.5 for total Kjeldahl nitrogen, 2.5 for total ammonia nitrogen, 2.5 for nitrite n+ itrate nitrogen, 1.8 for organic nitrogen, and 1.6 for total nitrogen. However, it must be

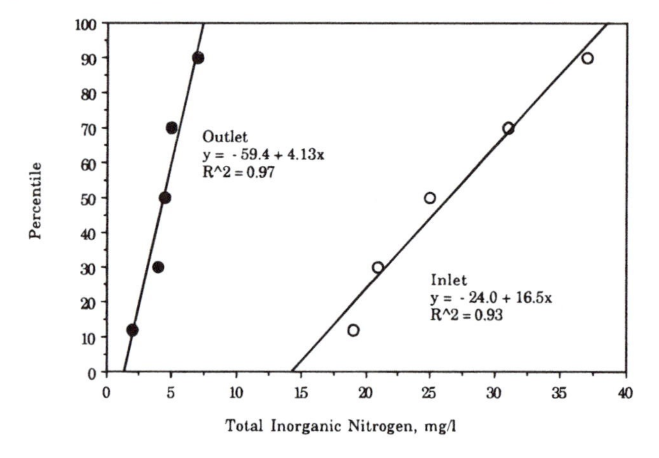

Figure 13-30 Percentile points of the distribution of total inorganic nitrogen entering and leaving the full-scale Arcata Marsh. (Replotted from Gearheart, 1992.)

noted that these ratios are not the 100th percentile for compliance, because of the finite number of months involved in the selection.

In fact, wetland data often contain sufficient intersystem variability to place the 100th percentile above the median inlet concentration! This is illustrated for the Danish soil-based reed bed's in Figure 13-31. The median inlet total nitrogen concentration for 66 wetlands was 30.0 mg/L; the median outlet was 17.1 mg/L. On average, a design for this reduction will be correct for annual averages, which formed the basis for the statistics. The (extrapolated) 100th percentile of the outlet concentrations is 52.3 mg/L, which is 75% higher than the design inlet concentration.

To incorporate this variability into design, the wetland designer must oversize the wetland. Suppose, for example, it is necessary to meet a monthly limit of 3.0 mg/L of total nitrogen. The designer must target an annual average outlet concentration 1.6 times smaller than the max monthly limit, 3.0/1.6 = 1.9 mg/L. The k-C* parameter estimates in this book are typically for data averaged over many quarters and should be taken to represent long-term average performance.

PRELIMINARY ESTIMATES OF NITROGEN REMOVAL

Detailed design calculations for nitrogen reduction are complex by necessity. Shortcut estimates are often useful for preliminary screening of concepts. Table 13-12 summarizes several bounding estimates for nitrogen removal in wetlands. The estimates are convenient to quickly compare alternatives, but they are based on different data ranges and system variables in the databases and do not account for a multitude of potential site-specific parameters that might affect nitrogen removal rates. For detailed conceptual or final design for nitrogen removal in wetlands, other empirical and preferably rational design approaches should be used.

When land area is not a major concern, total nitrogen loading rates and hydraulic loading rates can be conservative, and total nitrogen reduction is essentially certain to occur. These conservative loadings would include total nitrogen mass loading rates less than 3 to 5 kg/ha/d and hydraulic loading rates less than 1 to 3 cm/d. Based on the North American Wetland Treatment System Database, these conservative values have resulted in average total nitrogen outflow concentrations of less than 5 mg/L in at least 95 percent of the systems examined, including natural wetlands that usually do not benefit from even flow distribution or flexible water level control. Further examination of this data set indicates that outflow total nitrogen

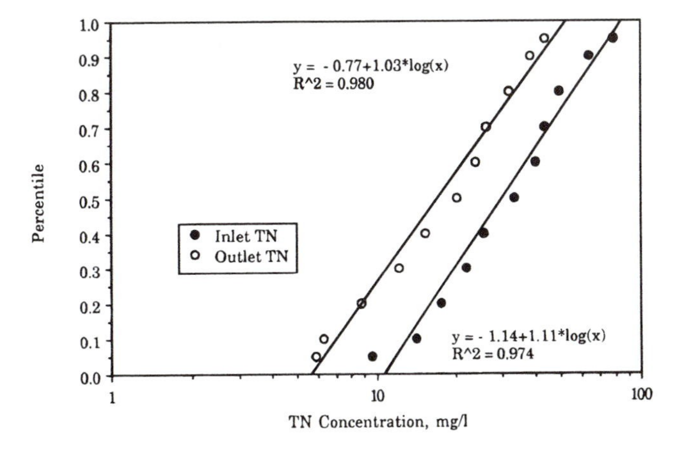

Figure 13-31 The distributions of inlet and outlet total nitrogen for 66 Danish soil-based wetlands. (Data from Schierup et al., 1990.)

Table 13-12 Limits and Reductions for Nitrogen Species in Treatment Wetlands

		Organic N	Ammonium N	Nitrate N	Total Nitrogen
Surface Flow	Reductions				
30 cm deep	Rate constant = k_{20} =	17 m/yr	18 m/yr	35 m/yr	22 m/yr
	Theta (θ) =	1.05	1.04	1.09	1.05
N loading @	Background = C* =	1.5 mg/L	0 mg/L	0 mg/L	1.5 mg/L
q = 5.0 cm/d	90% to C* in τ =	15 d	14 d	7 d	11 d
& 10 mg/L = 0.50 g N/m^2/d	90% to C* at q =	2.0 cm/d	2.1 cm/d	4.2 cm/d	2.6 cm/d
	Limits				
N loading @ τ = 7 days	Carbon supply limit =			5 g N/m^3/d	
& 10 mg/L = 0.43 g N/m^2/d	Oxygen supply limit =		0.8 g N/m^2/d		
	Burial and volatilization =				0.1 g N/m^2/d
	Temporary plant uptake =		0.3 g N/m^2/d		0.3 g N/m^2/d
Subsurface flow	Reductions				
60 cm deep	Rate constant = k_{20} =	35 m/yr	34 m/yr	50 m/yr	27 m/yr
0.40 porosity	Theta (θ) =	1.05	1.04	1.09	1.05
N loading @	Background =	1.5 mg/L	0 mg/L	0 mg/L	1.5 mg/L
q = 5.0 cm/d	90% reduction in τ =	6 d	6 d	4 d	7 d
& 10 mg/L = 0.50 g N/m^2/d	90% reduction at q =	4.2 cm/d	4.0 cm/d	5.9 cm/d	3.2 cm/d
	Limits				
N loading @ τ = 7 days	Carbon supply limit =			3 g N/m^2/d	
& 10 mg/L = 0.34 g N/m^2/d	Oxygen supply limit =		0.8 g N/m^2/d		
	Burial and volatilization =				0.1 g N/m^2/d
	Temporary plant uptake =		0.3 g N/m^2/d		0.3 g N/m^2/d

Note: The numbers in this table are for preliminary estimates, for T = 20°C. Reductions are for pure feeds, not mixtures of species.

concentrations less than 3 mg/L typically are achieved at total nitrogen inflow concentrations less than 10 mg/L.

A few of the early wetland treatment systems receiving nitrogen effluent limits were designed based on rule-of-thumb methods. Two of these systems are the Lakeland and Orlando, FL constructed wetland treatment systems. Both of these systems have consistently met their total nitrogen limit goals. The Lakeland constructed SF wetland treatment system was originally designed for nitrogen permit limits of 4 mg/L total nitrogen and 1.0 mg/L NH_4-N at a flow of 52,700 m^3/d (Jackson, 1989). Shortly after startup, the total nitrogen permit limit was reduced to 3.0 mg/L because of a regional water quality initiative (the Grizzle-Figg Bill). The basis for design of the Lakeland system was a literature review of the then-existing wetland data.

Table 13-13 summarizes some of the key nitrogen data from the Lakeland system. At the current hydraulic loading rate of about 0.63 cm/d, this wetland is providing excellent assimilation of total nitrogen and consistently achieves low outflow concentrations of NH_4-N. The mass loading data for Lakeland indicate that this system has a very low (conservative) nitrogen loading rate compared to many other treatment systems. Also, the dominant source of influent nitrogen at Lakeland is NO_3-N, which is typically removed in the first few cells of the wetland. To date, these conservative design conditions have resulted in consistent total nitrogen outflow concentrations less than 3.0 mg/L and NH_4-N less than 0.5 mg/L.

The Iron Bridge, FL constructed wetland also has a low outflow total nitrogen limit (2.31 mg/L on a monthly average basis). This system also was designed prior to the development of mass balance or empirical expressions for nitrogen removal. The design was based on existing data, primarily from natural wetlands, that indicated a probable nitrogen removal efficiency of 80 percent at a hydraulic loading rate less than 1 cm/d (100 ac/million gallons per day). Figure 13-32 illustrates the Iron Bridge total nitrogen performance through August 1991. Note that the highest monthly average total nitrogen loading rate to the Iron Bridge constructed wetland is 0.9 kg/ha/d. Total nitrogen concentration is lowered to an average of 3.89 mg/L prior to discharge into the wetland system. This total nitrogen is approximately 20 percent organic, 50 percent ammonia, and 30 percent nitrate + nitrite. Total nitrogen removal efficiency in the wetland averaged about 72 percent resulting in an average outflow total nitrogen of 0.89 mg/L and a monthly maximum of 1.74 mg/L.

Both of the previous examples are extremely large treatment wetlands, operated at very low loading rates, and correspondingly large detention times. In most instances, it is necessary to tighten the design to the smallest wetland area consistent with the desired degree of performance and risk. The tools for this are performance correlations and mass balance design models.

CORRELATIVE INPUT/OUTPUT DESIGN APPROACHES

The first quantitative design methods used for providing closer estimates of the actual treatment wetland area necessary to achieve nitrogen limits were based on statistical correlations derived from empirical data from operational wetland treatment systems. These correlative input-output design methods evolved and became more accurate as new data accumulated. This section describes two historical correlative nitrogen design models and presents new correlations based on the North American Wetland Treatment System Database (NADB, 1993).

Some Historical Perspective

The Fort Deposit SF constructed wetland design was sized based on nitrogen as the area-limiting constituent (Knight and Iverson, 1990). An empirical approach was used to determine

Table 13-13 Summary of Annual Nitrogen Concentrations and Loadings for the Inflow and Final Outflow at the Lakeland, Florida, Constructed Wetland Treatment System

			Total N					TKN					
Year	Flow (m^3/d)	HLR (cm/d)	In (mg/L)	Out (mg/L)	LR (kg/ha/d)	RR (kg/ha/d)	Eff (%)	In (mg/L)	Out (mg/L)	LR (kg/ha/d)	RR (kg/ha/d)	Eff (%)	NH_3 Out (mg/L)
1988	30,242	0.61	8.0	1.56	0.49	0.40	82	1.91	1.17	0.12	0.05	43	0.37
1989	24,905	0.50	10.0	2.10	0.50	0.40	80	3.14	1.69	0.16	0.07	47	0.35
1990	26,100	0.53	10.6	2.14	0.56	0.50	89	3.23	1.51	0.17	0.13	74	0.20
1991	31,718	0.64	12.2	1.76	0.78	0.68	88.1	2.97	1.29	0.19	0.12	64.2	0.15
1992	30,507	0.63	9.84	1.48	0.599	0.542	90.6	2.50	1.43	0.158	0.104	65.5	0.193
1993	33,136	0.68	9.62	1.22	0.612	0.531	86.8	3.37	1.09	0.224	0.151	67.4	0.194

Note: HLR = Hydraulic loading rate
LR = Mass loading rate
RR = Mass removal rate
Eff = Mass removal efficiency

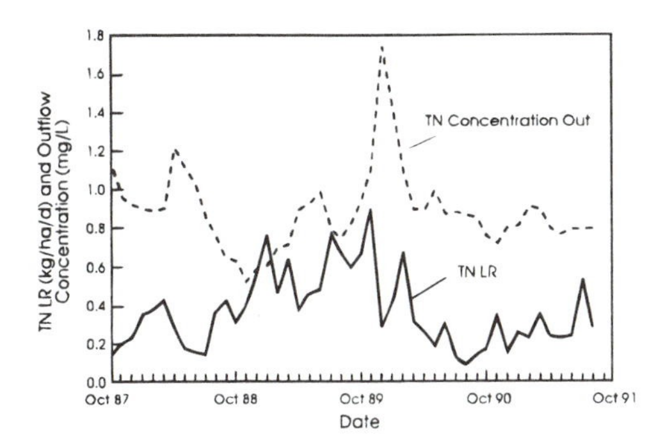

Figure 13-32 Monthly average total nitrogen loading and outflow concentration at the Iron Bridge Constructed Wetland Treatment System near Orlando, FL.

the wetland area necessary to satisfy the state criterion (a monthly average of 2 mg/L for NH_3-N) on a year-round basis. The minimum monthly water temperature was assumed to be 12°C, and the design flow was 908 m^3/d. A limiting factor (Monod)-type curve was fitted to long-term (annual average) total nitrogen removal data from a variety of wetland systems summarized by Knight et al. (1985a) to develop the design equation

$$A = [QC_oM_w]/[(QC_oN_{max}) - (1000M_wK_n)] \quad (13\text{-}47)$$

where
A = area, ha
Q = flow rate, m^3/d
C_o = total nitrogen inflow, mg/L
M_w = total nitrogen mass removal, kg/d
N_{max} = maximum specific total nitrogen removal rate, kg/ha/d
K_n = total nitrogen half saturation constant, kg/ha/d

The Fort Deposit design was based on the assumption that NH_4-N would be less than 2 mg/L when total nitrogen was less than 4 mg/L. Based on a design inflow total nitrogen concentration of 20 mg/L, an N_{max} of 16.8 kg/ha/d, and a K_n of 17.9 kg/ha/d, the calculated treatment area was found to be 6.0 ha. The system was enlarged to 6.1 ha during final design.

The total nitrogen design loading rate at Fort Deposit was 2.98 kg/ha/d, and Figure 13-33 illustrates the response of outflow NH_4-N concentration to variable total Kjeldahl nitrogen loadings. Outflow NH_4-N concentration did exceed the permit limit during 1 month in 1991, possibly in response to a total Kjeldahl nitrogen inflow spike about 3 months earlier. Dissolved oxygen concentrations in the wetlands were found to be extremely low at the time of the spike, and water levels were subsequently dropped to enhance atmospheric reaeration. Outflow ammonia concentrations have remained well below the permit limits since the water level was lowered.

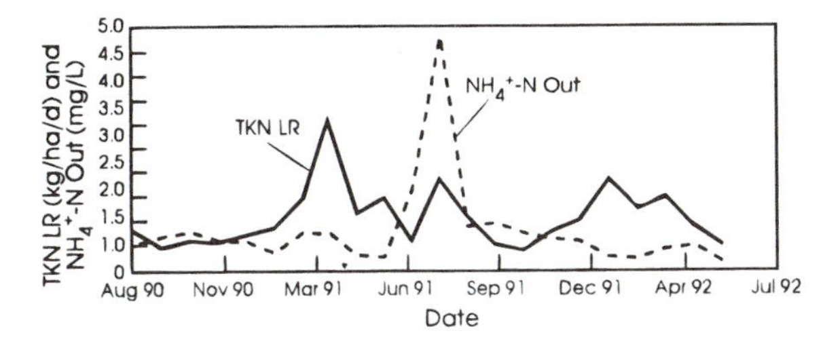

Figure 13-33 Total Kjeldahl nitrogen loading rate and outflow NH_4^+ concentration in the Fort Deposit, AL constructed wetlands.

The West Jackson County constructed wetlands near Ocean Springs, MS were designed using the same empirical design basis as the Fort Deposit system. However, the design loading was checked by the empirical design equation for ammonia presented in WPCF (1990b):

$$A = (0.01Q)/\exp[(1.527\ln C_e - 1.050\ln C_o + 1.69)] \qquad (13\text{-}48)$$

where C_e = outflow concentration, mg/L.

For the design conditions of 3785 m^3/d, an inflow total nitrogen of 20 mg/L, and an outflow limit of 2 mg/L ammonia (assumed outflow total nitrogen of 4 mg/L), Equation 13-47 predicts an area of 24.4 ha, while Equation 13-48 predicts 27.2 ha based on an assumed ammonia nitrogen inflow concentration of 10 mg/L.

WPCF (1990b) also provides an empirical equation for predicting the SF wetland area necessary to reduce a given total nitrogen inflow concentration to a desired outflow concentration:

$$A = (0.01Q)/(0.645C_e - 0.125C_o + 1.129) \qquad (13\text{-}49)$$

Using total nitrogen values of 20 and 4 mg/L for the West Jackson County example, Equation 13-49 predicts a wetland area requirement of 53 ha, considerably larger than the area predicted by Equations 13-43 and 13-44.

Ultimately the system was sized at 22 ha for a variable flow rate of 3785 to 8327 m^3/d depending on inflow total Kjeldahl nitrogen concentration to maintain a total Kjeldahl nitrogen loading rate of 3.4 kg/ha/d. Average inflow and outflow total nitrogen concentrations at West Jackson County between August 1990 and December 1992 were 11 and 4.4 mg/L at an average flow rate of 5800 m^3/d. Episodic violations of the design goal of 2.0 mg/L maximum monthly ammonium nitrogen have occurred, but the annual average performance is approximately as predicted.

These examples underscore the need to consider the seasonal and stochastic variability which is superimposed on the mean performance predicted by correlations.

Regressions for Wetland Nitrogen Removal

The North American Wetland Treatment System Database (Knight et al., 1993a; Knight, 1994c) provides a basis for empirical input-output regressions that describe SF wetland water quality treatment performance. However, the NADB is not adequate for regressions of SSF wetland performance because of the shortage of SSF data. The SSF database has been extended to include more data from both North American wetlands and those from other regions of the world.

Table 13-14 provides a summary of these regressions and appropriate concentration and hydraulic loading rate boundaries from which they were derived. Typically, about 500 data points are included for about 30 SF wetlands and about 100 points for about 30 SSF wetlands. The variables considered were inlet and outlet concentrations and hydraulic loading rate. Log-transformed data has also been considered, and the regression with the highest correlation coefficient is reported in Table 13-14. The standard errors in the regressed outlet concentrations are high, and the R^2 values are low, probably reflecting the fact that many site-specific factors are not included in the regressions. As has been shown in previous sections, there is a strong sequential interrelation among the nitrogen species, which implies the need to include precursor species as potential influences on outlet concentration of a given nitrogen form.

These SF nitrogen regressions may be extended to include forested wetlands, with almost no change in the parameters.

Table 13-14 Summary of Treatment Wetland Regression Equations for Nitrogen Species

		R^2	N	Standard Error C_2	q cm/d	Data Range (Median) C_1 mg/L	C_2 mg/L
Surface-flow marshes							
Parameter							
Organic N	$C_2 = 1.00\ C_1^{0.476}$	0.52	243	1.8	0.02–27.4 (2.9)	0.09–19.9 (2.8)	0.16–15.5 (1.4)
Ammonium N	$C_2 = 0.336\ C_1^{0.728} q^{0.456}$	0.44	542	4.4	0.1–33.3 (2.9)	0.04–58.5 (2.2)	0.01–58.4 (0.6)
Nitrate N	$C_2 = 0.093\ C_1^{0.474}\ q^{0.745}$	0.35	553	4.9	0.02–27.4 (2.7)	0.01–24.5 (1.7)	0.01–21.7 (0.2)
TKN	$C_2 = 0.569\ C_1^{0.840} q^{0.282}$	0.74	419	1.9	0.1–24.3 (2.9)	0.2–97.0 (8.7)	0.15–48.0 (3.0)
TN	$C_2 = 0.409 C_1 + 0.122 q$	0.48	408	3.5	0.2–28.6 (2.5)	2.0–39.9 (9.1)	0.4–29.1 (2.2)
Subsurface-flow wetlands							
Parameter							
Organic N	$C_2 = 0.1 C_1 + 1.0$	0.07	89	1.9	0.7–48.5 (6.2)	0.6–21.8 (6.9)	0.1–11.1 (1.1)
Ammonium N	$C_2 = 0.46 C_1 + 3.3$	0.63	92	4.4	0.7–48.5 (5.5)	0.1–43.8 (6.7)	0.1–26.6 (6.1)
Nitrate N	$C_2 = 0.62 C_1$	0.8	95	2.4	0.7–48.5 (5.5)	0.01–27.0 (0.3)	0.01–21.0 (0.4)
TKN	$C_2 = 0.752\ C_1^{0.821} q^{0.076}$	0.74	92	1.7	0.7–48.5 (5.5)	0.7–58.2 (15.2)	0.6–36.1 (8.2)
TN	$C_2 = 0.46 C_1 + 0.124 q + 2.6$	0.45	135	6.1	0.7–48.5 (7.1)	5.1–58.6 (21.0)	2.3–37.5 (13.6)

MASS BALANCE DESIGN FOR NITROGEN REMOVAL

Rational design for total nitrogen removal in wetlands must account, on a mass balance basis, for all of the nitrogen entering and leaving a wetland system. Formulas to quantify nitrogen fluxes and transformations must account for varying environmental conditions, or these environmental conditions must be shown to have minor impact on design. This section uses the typical wetland nitrogen cycle as a framework for integrating these various processes and estimating their impact on final wetland outflow nitrogen concentration. The complexity of calculations increases as the complexity of the nitrogen speciation increases.

TOTAL NITROGEN

Total nitrogen in the wetland water is a combination of all species. The ecosystem utilizes and transforms these, but also adds to them from desorption and decay processes. The plug flow mass balance Equation 13-35 may be solved for the wetland area that will yield the necessary outlet concentration:

$$A = \frac{Q\ln\left[\frac{C_{TN,i} - C^*_{TN}}{C_{TN,o} - C^*_{TN}}\right]}{k_{TN}} \tag{13-50}$$

where A = wetland area, m^2
Q = water flow rate, m^3/yr
k_{TN} = total nitrogen rate constant, m/yr

Values of k_{TN} are listed in Table 13-11, which are long-term mean values for the various wetland types. Intersystem averages and temperature correction factors are also given. The recommended value of $C^* = 1.5$ mg/L.

The assumptions underlying Equation 13-50 include a water balance that is dominated by the wastewater flow. In the event of significant evapotranspiration or rainfall augmentation, appropriate modifications are necessary (see Chapter 9). Another assumption is that the degree of flow non-ideality is the same for the system under design as it is for the database systems. In the event of anticipated better or worse mixing conditions, corrections must be made (see Chapter 9). And, as mentioned earlier, the nature of the regulatory requirement may indicate the need for a modified design target; i.e., a lower annual average total nitrogen for the outlet to meet a maximum monthly limit.

This mass balance approach is calibrated to the long-term performance of a wetland treatment system as opposed to short-term performance during an initial period. Depending on loading rate and environmental factors such as temperature and sunlight, the necessary startup period of successional development of plant and litter biomass, sediment nutrient concentrations, relatively constant outflow water quality characteristics may require from 1 to 2 years. Using steady state conditions for design is often overly conservative at first when storages are building within the treatment wetland. However, the opposite may be true if antecedent soils are overloaded with mobile nitrogen species.

The reader is further reminded that Equation 13-50 is valid only over time intervals sufficiently long to average out the asynchronousness of the input and output and the temporary storages in the wetland water body and biota.

The mass balance method of design is comparable to, but not identical to, the regression equation approach. The data fits of the two are compared in Figure 13-34. The NADB data set in this figure is the basis for the regression, but not for the values of k and C*, which were derived from the separate, intersystem basis of Table 13-11.

Example 13-1

Total nitrogen is to be reduced from a long-term average of 20 to 5 mg/L in a flow of 3786 m^3/d. What area of SF wetland is necessary?

Solution

The values of k_{TN} = 15 m/yr and C* = 1.5 mg/L will be used, since there is no site information on temperature given. Equation 13-50 yields

$$A = \frac{(3786 \cdot 365)\ln\left[\frac{20 - 1.5}{5 - 1.5}\right]}{15} = 153{,}400 \text{ m}^2 \qquad (13\text{-}50)$$

An area of 15.34 ha (38 ac) is predicted. Note that the effect of ignoring the background (return flow) would give 12.8 ha (31 ac).

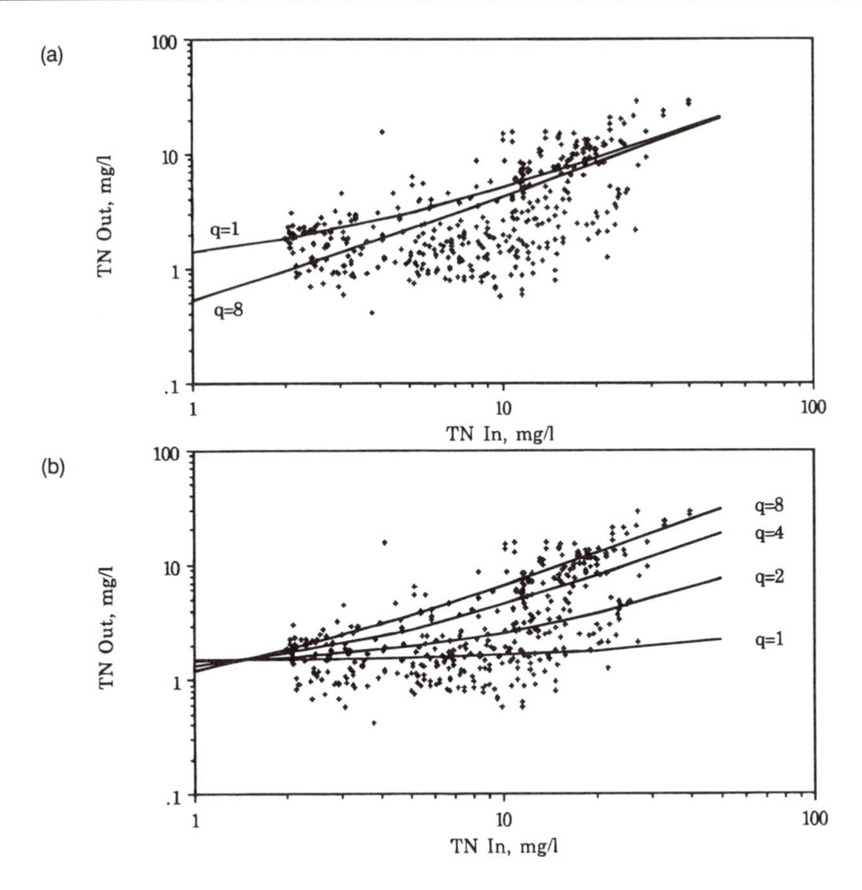

Figure 13-34 (a) Regression equation coverage of the NADB marsh data for TN > 2 in the inlet. The lines represent different values of q, the hydraulic loading rate (cm/d). (b) Coverage of the NADB marsh TN > 2 data by the separately calibrated k-C* model. The lines represent different values of q, the hydraulic loading rate (cm/d).

Nitrogen Component Transformations—Organic Nitrogen

Individual design equations for each of the principal nitrogen components were developed earlier. Only organic nitrogen is independent of the remaining species and may be treated as non-interactive in design. It is therefore possible to follow the simple approach shown above for total nitrogen.

$$A = \frac{Q\ln\left[\dfrac{C_{ON,i} - C^*_{ON}}{C_{ON,o} - C^*_{ON}}\right]}{k_{ON}} \tag{13-51}$$

where A = wetland area, m^2
Q = water flow rate, m^3/yr
k_{ON} = TN rate constant, m/yr

The caveats listed above for total nitrogen also apply to the reduction of organic nitrogen.

Example 13-2

Organic nitrogen is to be reduced from a long-term average of 30 to 5 mg/L in a flow of 3786 m^3/d. What area of SF wetland is necessary to accomplish this under winter conditions with a mean temperature of 5°C?

Solution

The value of k_{ON20} = 17 m/yr and C^* = 1.5 mg/L will be used. Correction to the winter temperature gives

$$k_{ON5} = 17(1.05)^{(5-20)} = 8.2 \text{ m/yr}$$

Equation 13-51 yields

$$A = \frac{(3786 \cdot 365)\ln\left[\dfrac{30 - 1.5}{5 - 1.5}\right]}{8.2} = 353{,}400 \text{ m}^2 \tag{13-51}$$

An area of 35.34 ha (87 ac) is predicted.

Nitrogen Component Transformations—Ammonium Nitrogen

Ammonium nitrogen is both produced and consumed in wetlands. If the incoming water contains large proportions of organic nitrogen, then ammonification will predominate, and effluent ammonium may be higher than influent. On the other hand, if ammonium dominates the feed, then a monotonic disappearance may result. If ammonium is regarded as an independent species, subject to a first-order reduction model, then both positive and negative rate constants may be representative of field data. These apparent rate constants are listed in Table 13-4. These are unusable in design because they represent an unacceptable oversimplification of the interaction between organic and ammonium nitrogen.

In some circumstances, there may be little or no organic nitrogen in the feed water. In that case, the first-order areal model may be applied as in the preceeding section for organic nitrogen, but with C* = 0 and the appropriate sequential rate constant.

The combined production-destruction (ammonification-nitrification) model cannot be circumvented by *ad hoc* methods; design must be based on the two-step reaction model given in Equation 13-29, written at $\Psi = 1.0$, the wetland outflow:

$$C_{AN} = (C_{ANi})e^{-k_{AN}/(Q/A)} + \left(\frac{k_{ON}}{k_{AN} - k_{ON}}\right)(C_{ONi} - C^*_{ON})(e^{-k_{ON}/(Q/A)} - e^{-k_{AN}/(Q/A)}) \quad (13\text{-}52)$$

where C_{ANi} = inlet concentration of ammonium nitrogen, mg/L
C_{ONi} = inlet concentration of organic nitrogen, mg/L

Unfortunately, this design equation is implicit in the hydraulic loading rate (Q/A). Therefore, design calculations require an iterative solution for the area necessary to achieve a given effluent ammonium concentration. Spreadsheet techniques are available for this purpose, such as the solver routine on Excel™ (Microsoft Corporation).

Example 13-3

A water with 10 mg/L ammonium and 20 mg/L organic nitrogen is to be reduced to 5 mg/L of ammonium nitrogen in a flow of 3786 m^3/d. What area of SSF wetland is necessary to accomplish this? T = 20°C.

Solution

From the foregoing sections, the value of k_{ON} = 35 m/yr and C^*_{ON} = 1.5 mg/L will be used; k_{AN} = 34 m/yr.

Solution of Equation 13-52 yields

$$A = 96{,}450 \text{ m}^2$$

An area of 9.65 ha (24 ac) is predicted. Ammonium peaks at 12 mg/L at 20% of the travel distance and does not go below 10 mg/L for the first 50% of the travel distance (Figure 13-35). Roughly speaking, the first half of the wetland is devoted to ammonification and the second half to nitrification.

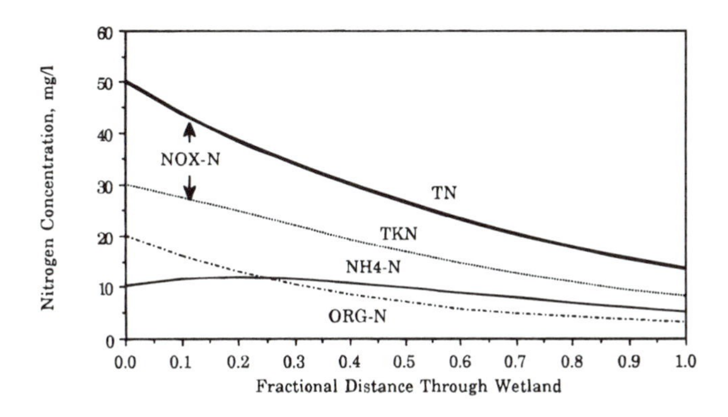

Figure 13-35 Concentration profiles for Examples 13-3 and 13-4.

Nitrogen Component Transformations—Nitrate Nitrogen

Oxidized nitrogen presents the same difficulty as ammonium: it is produced (nitrification) as well as consumed (nitrate reduction). It is also possible that this constituent is utilized in plant growth in the absence of significant ammonium nitrogen.

In the event that nitrate dominates the feed—as it would for a highly nitrified wastewater—it is possible to use the first-order areal model in isolation. The appropriate sequential k value is to be used.

In the event of significant ammonium and/or organic nitrogen in the feed, the sequential model should be used to estimate the combined effects of all processes on nitrate concentrations. The three-step plug flow kinetic model (Equation 13-39) is quite complicated and requires spreadsheet computations.

$$C_{NN} = (C_{NNi})e^{-(k_{NN}/q)y} + \psi \left\{ \begin{aligned} &\left(\frac{k_{AN}}{k_{NN} - k_{AN}}\right)(c_{ANi})\left(e^{-(k_{AN}/q)y} - e^{-(k_{NN}/q)y}\right) \\ &+ \left(\frac{k_{ON}}{k_{AN} - k_{ON}}\right)\left(\frac{k_{AN}}{k_{NN} - k_{ON}}\right)(C_{ONi} - C^{*}_{ON})\left(e^{-(k_{ON}/q)y} - e^{-(k_{NN}/q)y}\right) \\ &- \left(\frac{k_{ON}}{k_{AN} - k_{ON}}\right)\left(\frac{k_{AN}}{k_{NN} - k_{AN}}\right)(C_{ONi} - C^{*}_{ON})\left(e^{-(k_{AN}/q)y} - e^{-(k_{NN}/q)y}\right) \end{aligned} \right\} \quad (13.39)$$

where C_{Ni} = inlet concentration of nitrate nitrogen, mg/L.

Example 13-4

Suppose that the feed water in Example 13-3 also contained 20 mg/L of nitrate nitrogen. What nitrate would be expected in the effluent?

Solution

From the foregoing sections, the value of k_{NN} = 50 m/yr and C^{*}_{ON} = 0.0 mg/L will be used, with the other parameters used in Example 13-3.

Solution of Equation 13-39 yields the concentrations of nitrate. At the wetland outflow, the nitrate concentration is calculated to be 5.35 mg/L. The complete concentration profile is shown in Figure 13-31. It is interesting to note that this effluent concentration reflects nothing of the incoming nitrate, which is quickly denitrified (95% would be gone in the first half of the wetland). If there were zero nitrate in the feed, the predicted outlet nitrate would be 4.76 mg/L.

It is also interesting to note that the predicted total nitrogen profile is logarithmic (R^2 = 1.000). This shape is consistent with the first-order areal model for total nitrogen. The rate constant regressed from the calculation is 18.6 m/yr, compared to the mean value of 27 m/yr for SSF wetlands in Table 13-11.

SUMMARY OF NITROGEN REMOVAL DESIGN CONSIDERATIONS

The wetland designer chooses between two methods to achieve low nitrogen outflow: (1) extensive pretreatment (solids removal, nitrification, and denitrification) or (2) fairly large wetland treatment areas (typically greater than 5 ha/1000 m^3/d). Because the major nitrogen transformation mechanisms vary seasonally, conservative design must be based on specific permit limits, with different assumptions used for design to meet annual, monthly, or daily limits. Data from North American wetland treatment systems indicates that if design is based on methods derived from annual averages, target outflow concentrations should be divided by a factor from 1.6 to 2.5, as compensation for the usual maximum monthly limit vs. annual averages, to estimate the required wetland area necessary for satisfactory treatment.

Wetland design for nitrogen removal can be based on rule-of-thumb, regression equations or mass balance equations with first-order transformations. Limiting conditions provide for quick, but overly conservative estimates of size requirements.

Regression equations derived from input-output data from treatment wetlands provide a better basis for determination of wetland sizes. However, the residual variability is large, in part due to the limitations on the available variables. These regressions typically do not correctly predict the internal profiles of the various nitrogen species, such as the potential maximum in the ammonium profile. They cannot be adjusted for environmental variables such as temperature, evapotranspiration, or precipitation. Importantly, regressions are constrained to the limits of the independent variables in the generating data sets. Most regressions cannot be extrapolated beyond those limits without engendering serious errors. The advantage of these equations lies in their simplicity.

First-order, area-based nitrogen loss models provide a suitable method for design of wetland treatment systems in most circumstances. These have the advantage of correctly describing internal phenomena in flow-through wetlands, as well as describing batch wetland operation. Studies on side-by-side wetlands confirm the effects of the principal variables of inlet concentrations and hydraulic loading rates (or the equivalent detention times). The parent mass balance equation for water movement may be adjusted to fit extreme environmental conditions of precipitation or evapotranspiration. The rate equations account for return fluxes from the wetland biomass and thus can fit the entire range of hydraulic loadings.

In parameter estimation, the sequential nature of the nitrogen transformations cannot be ignored. Apparent rate constants, which do ignore species generation, can be seriously misleading, even to the point of having the wrong sign. Their use in design is usually inappropriate, unless applied to exactly the same feed speciation from which they were derived.

The quality of the first-order rate constants for wetland treatment systems will improve with time as additional operation data are collected and analyzed. These rate constants incorporate effects of pH, dissolved oxygen, and other physical, chemical, and biological processes that affect rates of ammonification, nitrification, and denitrification in wetland treatment systems. Estimates of temperature coefficients are available for the individual transformations. More complex Monod limiting-factor models appear to be applicable to wetland treatment system design, but wetland-derived rate and half saturation constants currently are lacking. The disadvantage of the sequential mass balance approach lies in the (necessary) complexity of description for ammonium and nitrate.

Several features of nitrogen processing are known with some degree of certainty. These are

1. The profiles of total Kjeldahl nitrogen and total nitrogen are typically monotonic decreasing when the input to the wetland is above background values. These groupings may therefore be described by a simple mass balance and first-order kinetics.
2. The background level of organic nitrogen is approximately 1.5 mg/L in SF wetlands and perhaps somewhat lower for SSF wetlands.
3. Organic nitrogen is the first species in sequence and therefore may be described by a simple mass balance and first-order kinetics.
4. Ammonium calculations must acknowledge the ammonification process and, hence, may require more complex equations. The background level of ammonium nitrogen is approximately zero in summer, but is non-zero in winter.

5. Nitrate calculations are the most complex, because these compounds are last in the reaction sequence and thus depend strongly on precursor reactions. The background level of nitrate nitrogen is approximately 0.0 mg/L. This implies that there is no return of nitrate nitrogen from the sediments to the overlying water.

In summary, wetland treatment systems consistently reduce total nitrogen concentrations in many wastewaters. The magnitude of these reductions depends on many factors including inflow concentrations, chemical form of the nitrogen, water temperature, pH, alkalinity, organic carbon, dissolved oxygen, water depth, and biota. Although these factors can be incorporated with some success into design of wetland treatment systems, precise nitrogen reaction rates and performance under different environmental variables are not known. These observations will lead the engineer to design conservatively.

CHAPTER 14

Phosphorus

INTRODUCTION

Treatment wetlands are capable of phosphorus (P) removal from wastewaters on both short-term and long-term bases. This chapter defines those relations which permit design and optimization of that capability.

Phosphorus is a nutrient required for plant growth and is frequently a limiting factor for vegetative productivity. Thus, the introduction of trace amounts of this element into receiving waters can have profound effects on the structure of the aquatic ecosystem. A measure of relative ecosystem requirements is the proportion among the nutrient elements in the biomass, which is often represented as a molar proportion of C to N to P = 106:16:1 or 41 to 7 to 1 on a mass basis (the Redfield ratio). Wastewaters do not have this ratio except by rare chance, and therefore, the addition of wastewater places a nutrient imbalance stress on a receiving ecosystem. Most often, there is excess phosphorus in the wastewater.

Wetland ecosystems may also be altered by the addition of phosphorus-rich waters. Phosphorus is utilized in the wetland in a complicated biogeochemical cycle which involves many pathways and diverse temporary and permanent sinks and sources. Natural wetlands—those with no added wastewater—function with a spatial and time varying balance between imports, exports, and storages. The wetland literature is replete with examples of natural wetland phosphorus functioning: net removal and net release and seasonal combinations of both. Further, a particular natural wetland adapts to the timing and amounts of natural phosphorus available to it, as well as other site factors.

When a wetland, either natural or constructed, is given a new supply of water and phosphorus, it responds by readjusting storages, pathways, and structure. If those new supplies are variable and within the stochastic band of historic inputs, a mature ecosystem will not change in character or function. Treatment of water for phosphorus removal implies that additions will significantly exceed the historic stochastic band of the natural wetland. A newly constructed wetland will require a successional period to adapt to the intended inputs. In both cases, a period of adaptation and change is expected and, in fact, occurs. Thereafter, the wetland functions in a long-term sustained mode, which is tuned to the inputs, but displays probabilistic variation as well.

Phosphorus removal is a difficult task in any water treatment technology, and wetland technology is no exception. On a per unit area basis, wetlands are not efficient in phosphorus reduction. Treatment wetlands in general are area intensive compared to "conventional" technologies, and the land requirement for wetland phosphorus reduction is typically the largest of all wetland requirements.

GENERAL CONCEPTS

PROCESS CHARACTERIZATION

For the purpose of understanding phosphorus cycling, wetlands may be visualized as consisting of several compartments: water, plants, microbiota, litter, and soil (Figure 14-1).

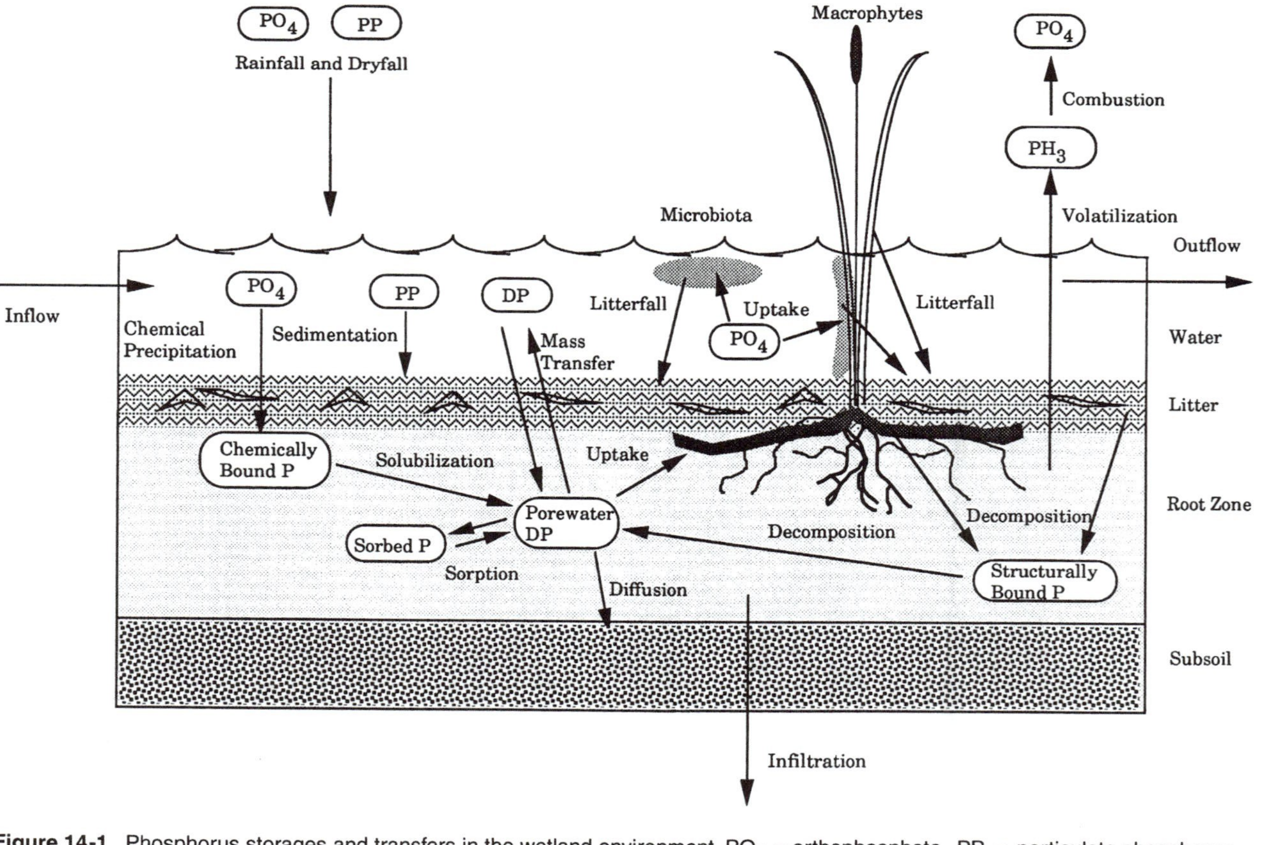

Figure 14-1 Phosphorus storages and transfers in the wetland environment. PO_4 = orthophosphate, PP = particulate phosphorus, DP = dissolved phosphorus, and PH_3 = phosphine. PP may consist of all the forms shown in the root zone.

Naturally occuring inputs of phosphorus are from surface inflows and atmospheric deposition, which consists of both wet- and dryfall. Outputs may be in surface outflows and in infiltration to groundwater. Inputs from groundwater and gaseous release to the atmosphere are less common or probable. Animal migration, ranging from insect movement to fish and bird travel, has been identified as a potential contribution to the phosphorus budget, but to date no quantification of this process exists.

A large number of transfer and alteration processes occur, as indicated in Figure 14-1, but only soil building provides a net long-term storage. Sediment and soil accretion provides phosphorus storage that can alternate between deposition and erosion on a short-term basis. The wetland environment provides appropriate conditions for net long-term buildups because inundation slows oxidative processes. Historical accumulations in natural wetlands are the genesis of peatlands. Natural accretions are on the order of a few millimeters per year (Mitsch and Gosselink, 1993).

Phosphorus removal by harvesting biomass has not thus far proven feasible. It is difficult to harvest rooted emergent macrophytes in wetlands, and when successful, relatively tiny amounts of phosphorus have been reclaimed in the harvested biomass. Herskowitz (1986) reported an average of 2.5 percent of the total phosphorus removal in surface-flow (SF) wetlands was achieved by harvest. Floating aquatic plants are somewhat easier to harvest; over 20 percent of the total phosphorus removal was achieved by water hyacinths (Fisher and Reddy, 1987). Harvesting is labor intensive and costly, which is antithetical to the passive character of wetlands technology. The problem of biomass utilization exacerbates the difficulties.

WETLAND WATER CHEMISTRY OF PHOSPHORUS

Wetland science has evolved to focus on categories of phosphorus compounds that are defined by methods of analysis (Table 14-1). In every case, the analytical procedure is reported as the elemental phosphorus content of the category.

Wetlands provide an environment for the interconversion of all these forms of phosphorus. Soluble reactive phosphorus is taken up by plants and converted to tissue phosphorus or may become sorbed to wetland soils and sediments. Organic structural phosphorus may be released as soluble phosphorus if the organic matrix is oxidized. Insoluble precipitates form under some circumstances, but may redissolve under altered conditions.

The principal phosphorus compounds in the wetland environment are dissolved phosphorus, solid mineral phosphorus, and solid organic phosphorus.

In solution, the principal inorganic species are related by the pH-dependent dissociation series:

$$H_3PO_4 \Leftrightarrow H_2PO_4^- + H^+ \tag{14-1}$$

$$H_2PO_4^- \Leftrightarrow HPO_4^= + H^+ \tag{14-2}$$

$$HPO_4^= \Leftrightarrow PO_4^{\equiv} + H^+ \tag{14-3}$$

These forms are distributed in water at 25°C as shown in Figure 14-2 (Freeze and Cherry, 1979).

A variety of cations can precipitate phosphate under certain conditions. Some important mineral precipitates in the wetland environment are (Reddy and D'Angelo, 1994)

Table 14-1 Forms of Phosphorus in the Wetland Environment

Dissolved in water (aqueous phase)
Procedures are performed on filtered (0.45 μm) samples:
- Orthophosphate (PO_4-P). This is the common ionic form of phosphorus.
- Condensed phosphates. These consist primarily of pyro-phosphate, meta-phosphate and poly-phosphates.
- Soluble reactive phosphorus (SRP). Primarily, PO_4-P, together with some condensed phosphates that are hydrolyzed in the analytical method.
- Dissolved acid hyrolyzable phosphorus (DAHP). Orthophosphate plus condensed phosphates. Determined after acid hydrolysis at boiling water temperature.
- Total dissolved phosphorus (TDP). Phosphorus that is convertable to PO_4-P upon oxidative digestion.
- Dissolved organic phosphorus (DOP). Phosphorus, in forms other than DAHP, that is convertable to PO_4-P upon oxidative digestion (= TDP-DAHP).

Dissolved or suspended in water (water plus associated suspended solids)
The procedures above may be performed on unfiltered samples to yield, by analogy,
- Total reactive phosphorus (TRP)
- Total acid hyrolyzable phosphorus (TAHP)
- Total phosphorus (TP)
- Total organic phosphorus (TOP) (= TP-TAHP)

Sorbed to the surface of soil particles
- Sorbed phosphorus is removed using gentle extracts of wet soil samples. Typical extractants include water or solutions of KCl or bicarbonate. Analysis may be for any of the dissolved forms.

Contained in the structure of biomass
- Total phosphorus may be found by analyzing for PO_4-P in digests of biomass samples.
- Digestion may involve dry or wet ashing, followed by redissolution.

Contained in the structure of soil particles:
- Structural, internal forms of phosphorus in the solid are removed (solubilized) using harsh extracts of wet soil samples. Typical extractants include:
  - Sodium hydroxide (0.1 *M*). The SRP in the extract is representative of iron- and aluminum-bound phosphorus. The balance of the total phosphorus in the extract (TP-SRP) is representative of organic phosphorus associated with humic and fulvic acids.
  - Hydrochloric acid (0.5 *M*). The SRP in the extract is representative of calcium-bound phosphorus.
- Total phosphorus may be found by analyzing for PO_4-P in digests of soil samples.
- Digestion may involve dry or wet ashing, followed by redissolution.

Apatite	$Ca_5(Cl,F)(PO_4)_3$
Hydroxylapatite	$Ca_5(OH)(PO_4)_3$
Variscite	$Al(PO_4)\cdot 2H_2O$
Strengite	$Fe(PO_4)\cdot 2H_2O$
Vivianite	$Fe_3(PO_4)_2\cdot 8H_2O$
Wavellite	$Al_3(OH)_3(PO_4)_2\cdot 5H_2O$

In addition to direct chemical reaction, phosphorus can coprecipitate with other minerals such as ferric oxyhydroxide and the carbonate minerals such as calcite (calcium carbonate), $CaCO_3$ (Reddy and D'Angelo, 1994). The overall phosphorus mineral chemistry is very

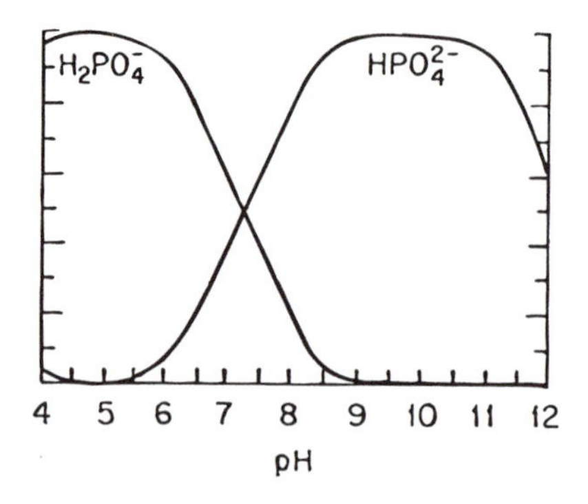

Figure 14-2 Distribution of major species of inorganic phosphorus in water at 25°C. (From Freeze, R.A. and J.A. Cherry, *Groundwater*, Englewood Cliffs, NJ, Prentice Hall, 1979. With permission.)

complex; consequently, quantitative calculations of solubilities are generally not possible. Trends for wetland soils are (Reddy and D'Angelo, 1994)

1. In acid soils, phosphorus may be fixed by aluminum and iron, if available.
2. In alkaline soils, phosphorus may be fixed by calcium and magnesium, if available.
3. Reducing conditions lead to solubilization of iron minerals and release of phosphorus coprecipitates. If free sulfide is present due to sulfate-reducing conditions, iron sulfide can form and preclude iron mineralization of phosphorus.

A gaseous form of phosphorus, phosphine (PH_3) has been identified as a potential compound of significance in wetland environments (Gassman and Glindemann, 1993). Phosphine is soluble in water, but has a high vapor pressure. It may be emitted from regions of extremely low redox potential, together with methane. Devai et al. (1988) measured PH_3 emissions from a constructed wetland (1.0 ha, *Phragmites* and bulrushes) in Hungary and estimated that 1.7 g/m^2/yr of phosphorus was being lost by this chemical route. To date, all North American wetland phosphorus mass balance studies have ignored this possibility.

PLANT CHEMISTRY OF PHOSPHORUS

Phosphorus is also incorporated in the tissues of all living organisms and is therefore present in all wetland biota and the corresponding detritus. The range of concentrations of phosphorus in the live leaves of various plant species is from approximately 0.1 to 0.4 percent on a dry-weight (dw) basis (Table 14-2). For 35 wetland species, under natural conditions, the range of phosphorus concentrations was 0.08 to 0.63 percent (mean $\pm$ SE = 0.25 $\pm$ 0.02; CV = 58%) (Boyd, 1978). The range in phosphorus content for a given species from site to site is not large, as might be expected. For instance, the phosphorus concentrations in *Typha latifolia* at three northern wetland treatment sites were as follows:

Listowel, Ontario	0.24 $\pm$ 0.06 percent
Kinross, MI	0.38 $\pm$ 0.14 percent
Houghton Lake, MI	0.33 $\pm$ 0.06 percent

Table 14-2 Examples of Phosphorus Concentrations in Wetland Plant Tissues (in percent dry weight)

Plant	Common Name	Wetland Status	Live	Dead	Litter	Source
Carex spp.	Sedge	Oligotrophic	0.08		0.07	Kadlec, 1988b
Carex spp.	Sedge	Eutrophic	0.30		0.22	Kadlec, 1988b
Cladium jamaicense	Sawgrass	Oligotrophic	0.04	0.02	0.02	Davis, 1990
C. jamaicense	Sawgrass	Eutrophic	0.08	0.04	0.12	Davis, 1990
Typha domingensis	Cattail	Oligotrophic	0.14	0.05	0.02	Toth, 1990; Davis, 1990
T. domingensis	Cattail	Eutrophic	0.20	0.07	0.16	Toth, 1990; Davis, 1990
T. latifolia	Cattail	Oligotrophic	0.09	0.04	0.07	Kadlec, 1988b
T. latifolia	Cattail	Eutrophic	0.28	0.29	0.24	Kadlec, 1988b
Scirpus californicus	Bulrush		0.13			Anonymous, 1992
Eleocharis sp.	Spikerush	Oligotrophic	0.18	0.08		Walker et al., 1988
Eleocharis sp.	Spikerush	Eutrophic	0.26	0.14		Walker et al., 1988
Panicum spp.	Maidencane	Oligotrophic	0.13	0.07		Walker et al., 1988
Panicum spp.	Maidencane	Eutrophic	0.16			Walker et al., 1988
Sagittaria sp.	Arrowhead	Oligotrophic	0.40	0.10		Walker et al., 1988
Sagittaria sp.	Arrowhead	Eutrophic	0.41	0.20		Walker et al., 1988
Utricularia spp.	Bladderwort	Oligotrophic	0.11			Walker et al., 1988
Utricularia spp.	Bladderwort	Eutrophic	0.16			Walker et al., 1988
Salix spp.	Willow leaves	Oligotrophic	0.12		0.10	Chamie, 1976
Salix spp.	Willow leaves	Eutrophic	0.31			Kadlec, 1988b
Betula pumila	Bog birch leaves	Oligotrophic	0.12		0.08	Chamie, 1976
B. pumila	Bog birch leaves	Eutrophic	0.33			Kadlec, 1988b

If the nutrient status of a wetland is increased from low ("oligotrophic") to high ("eutrophic"), there is a pronounced increase in tissue phosphorus concentration. The sample data in Table 14-2 shows an average increase of a factor of two for the ten species listed. The standing dead leaves have lesser phosphorus concentrations than their live counterparts. Litter may have slightly greater or slightly lesser concentrations.

Data are scant, but it is likely that the phosphorus content of periphyton and plankton is higher than for the macrophytes in phosphorus-rich waters. Schwegler (1978) reported values ranging from 0.06 to 0.48 percent dw for a treatment wetland, with a decreasing gradient with distance from the discharge.

The phosphorus content of the microdetritus formed by the death of microorganisms is also high compared to soils and vegetation; for example, it was found to be approximately 0.57 ± 0.39 percent dw in a sedge-cattail peatland (Kadlec, 1988b).

BIOMASS: GROWTH, DEATH, DECOMPOSITION

Due to the general scarcity of phosphorus in most natural environments and the absence of a significant atmospheric source, natural ecosystems, including wetlands, have numerous adaptations to scavenge and sequester this element. Consequently, phosphorus cycling is efficient and extensive in wetlands.

The amount of phosphorus sustainably removed by a wetland is generally much less than the phosphorus taken up by plants during a single growing season. All wetland biota undergo a cycle of growth, death, and partial decomposition (Figure 14-3). Total plant biomass—live, standing dead, litter—was found to be relatively constant across the seasons for a northern treatment wetland (Kadlec and Hammer, 1985). For cattails in the north, the live plant material in summer is replaced by dead material and extra rhizome biomass in winter.

At any given moment, live above-ground plant material is produced at a rate designated as the gross primary production rate (g/m^2/yr). At the same point in time, some of the live above-ground plant material is dying. The net primary production rate is the gross rate less

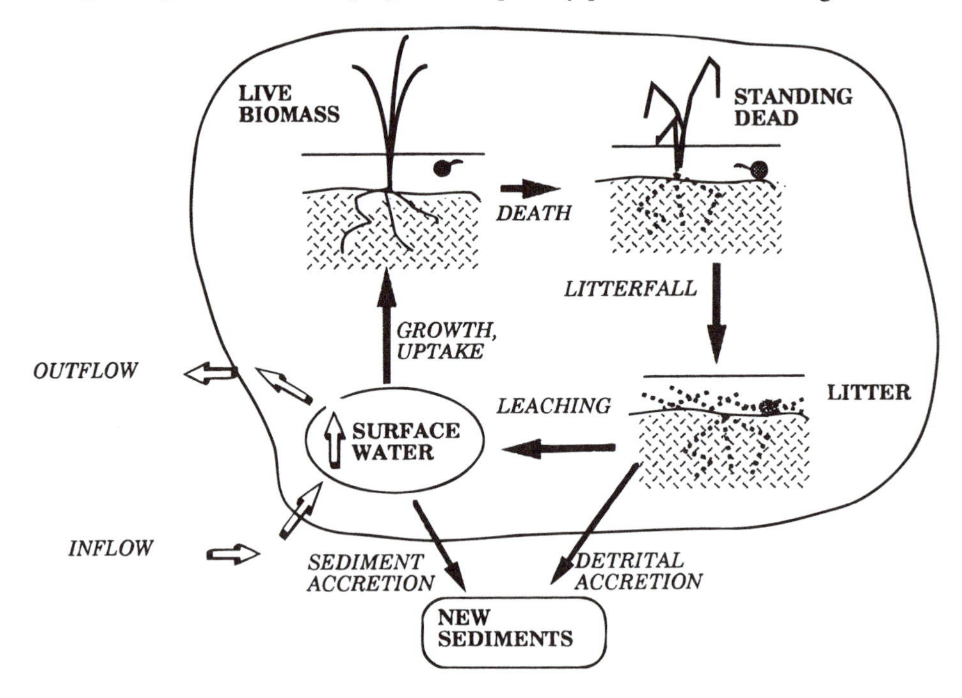

Figure 14-3 Phosphorus cycling in a wetland.

the death rate. Standing crop is the total amount of live or dead plant material to be found at a given moment. In northern wetlands, nonwoody growth occurs from a zero starting point in early spring. Most above-ground leaves and stems persist through the growing season and are measureable as the end-of-season standing crop. Under these circumstances, the turnover of plant material is the ratio of the end-of-season standing crop to the gross primary production rate. Macrophyte turnover in the northern environment is usually in the range of 1.0 to 2.0 reciprocal years, which means the live above-ground biomass is replaced between one and two times per year (Mitsch and Gosselink, 1993). However, growth occupies only about one third of the year.

Turnover is much higher in a southern climate because growth continues over a longer growing season. In the subtropical environment of the Florida Everglades, a eutrophic marsh may display turnovers of 3 to 6 reciprocal years (Figure 14-4). Thus, the speed of the cycle depicted in Figure 14-3 is roughly the same during the growing season, regardless of geographic location. But, the cycle is inoperative in winter in northern latitudes.

Leaf litter decomposes to a stable residual in a time span of 12 to 24 warm weather months; litter from plankton and periphyton decomposes much faster. The residual is a small fraction of the parent biomass, so there is no huge buildup of those residual solids. However, *the accretion of biomass residuals and minerals is the only sustainable storage mechanism for phosphorus removal.*

The biomass of either live or dead microbiota is virtually impossible to measure, but it is small compared to the biomass of live or dead macrophytes or of macrodetritus (litter). If these micro-biomass compartments are characterized by the amount of suspendable material exclusive of macrophyte litter, the standing "crop" has been measured to be in the range of 20 to 80 g/m^2 for a northern peatland receiving treated wastewater (Kadlec, 1989a).

Figure 14-5 provides a frame of reference for quantities of biomass and phosphorus that are present in a treatment wetland. This figure does not represent any specific wetland and is intended only to convey an idea of a general size of the various storages in a marsh ecosystem. These storages can, and do, grow and shrink in response to environmental variables and anthropogenic inputs.

UPTAKE AND STORAGE BY BIOTA

Organisms within the wetland require phosphorus for growth and incorporate it in their tissues. The most rapid uptake is by microbiota (bacteria, fungi, algae, microinvertebrates, etc.) because these organisms grow and multiply at high rates. Radio phosphorus (P-32) studies on wetland microcosms (Richardson and Marshall, 1986) indicated that phosphorus

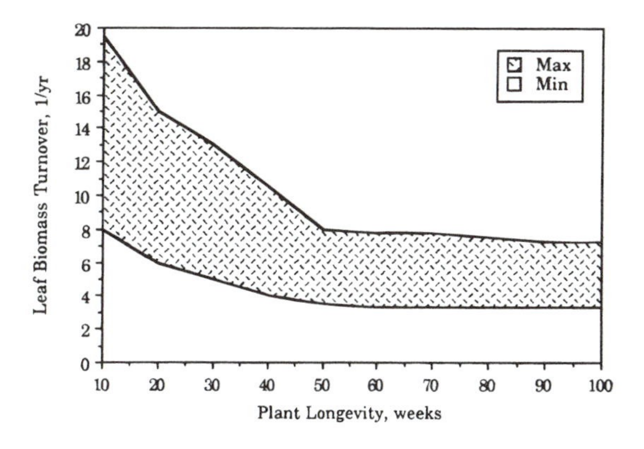

Figure 14-4 Leaf turnover rate vs. plant longevity for cattail (*Typha domingensis*) in the Florida Everglades. Data were obtained by tagging leaves. (Replotted from Davis, 1994.)

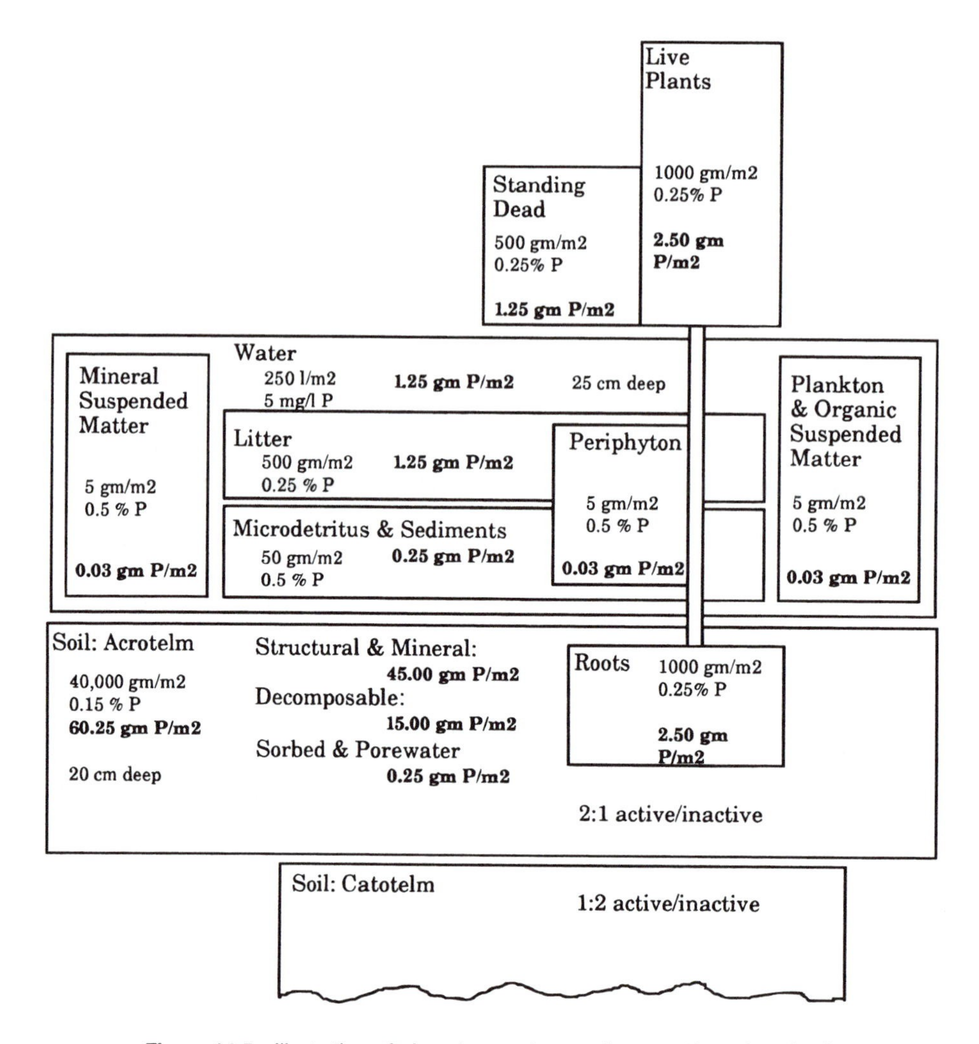

Figure 14-5 Illustration of phosphorus storages in a peat-based wetland.

uptake by microbiota occurs on a time scale of less than 1 h. However, more than 90 percent was released within the next 6 h.

At the other end of the size spectrum, the macrophytes obtain and use phosphorus much more slowly. Some is obtained by adventitious roots, which are located in or near the surface water. More is obtained from roots located below the soil surface, thus requiring that phosphorus move downward into the soil. Uptake response times for P-32 are the order of weeks (Richardson and Marshall, 1986).

Because phosphorus is a nutrient, the addition of this element to the wetland stimulates growth and causes increases in the amount of biomass. Phosphorus is used not only to make the plant, it is used to make the plant bigger. This in turn generates more litter. The increase in the biomass cycle is even slower than the intial uptake: one must wait for a cohort to undergo an entire decomposition cycle, which takes many months. The increase in the pool of biomass phosphorus is a short-term process and is presumably reversible. *Biomass increases should not be counted as part of the long-term sustainable phosphorus removal capacity of wetlands.*

Plant roots are an important part of the biomass and comprise a significant fraction of the active phosphorus storage. They are situated in the upper soil layer and extract phosphorus from it as it becomes available. Phosphorus may become available via desorption, reversal of chemical binding, or diffusion through porewater. This upper soil layer is sometimes called the *acrotelm*. In contrast, soils below the root zone are relatively inactive and are sometimes called the *catotelm*. However, both zones contain available and unavailable forms of phosphorus.

PHYSICAL PROCESSES

The two important physical processes for phosphorus removal in wetlands are sedimentation of particulate phosphorus and sorption of soluble phosphorus.

Incoming particles may contain phosphorus in available and unavailable forms. If the particulate matter is planktonic, then it may subsequently decompose to release soluble phosphorus. The particles may also contain weakly sorbed phosphorus, which may subsequently desorb. But if the particles contain phosphorus as insoluble minerals or refractory organophosphorus complexes, it may be permanently removed by the process of sedimentation.

All wetland soils have a capacity to sorb phosphorus, but that capacity is quite variable. This storage may be quickly exhausted in many SF treatment wetlands. In contrast, the particulate media in SSF wetlands may be designed to possess a large phosphorus storage via sorption. Iron- and aluminum-rich materials, limestone media, and specially prepared clays have all been employed to enhance this removal mechanism.

SUMMARY

Phosphorus cycling and storage involves a complex set of processes and several forms of phosphorus. Plant uptake is not a suitable measure of the net removal rate in a wetland because most of the stored phosphorus is returned to the water by decomposition processes. Long-term storage results from the undecomposed fraction of the litter produced by the various elements of the biogeochemical cycle, as well as the deposition of refractory phosphorus-containing particulates.

SOIL WATER PHOSPHORUS PROCESSES

STORAGE POTENTIAL: SOILS

There is a misconception that wetlands provide phosphorus removal only through sorption processes on existing soils. It is true that most soils do have sorptive capacity for phosphorus, but this storage is soon saturated under any increase in phosphorus loading. In this section, the capacity of this storage compartment, and the speed with which it fills are analyzed.

Figure 14-6 illustrates a section of soil in the wetland, consisting of soil particles and water-filled voids between them. The phosphorus storage is in porewater, in the solid as part of its chemical structure and on the surface of the solid as sorbed phosphorus compounds.

Porewater phosphorus may be partitioned into the various fractions described earlier: SRP, DOP, PO4-P, and TDP. By definition, there can be no particulate phosphorus associated with porewater; all particles are considered part of the soil.

Wetland soils often display steep vertical gradients in total phosphorus content, with a large reduction occurring through the first 30 cm. This corresponds roughly to the root zone for emergent macrophytes such as cattail (*Typha* spp.). Figure 14-7 shows examples of such gradients. The Jackson Bottoms system was newly built on phosphorus-rich mineral soils at the time the samples were taken.

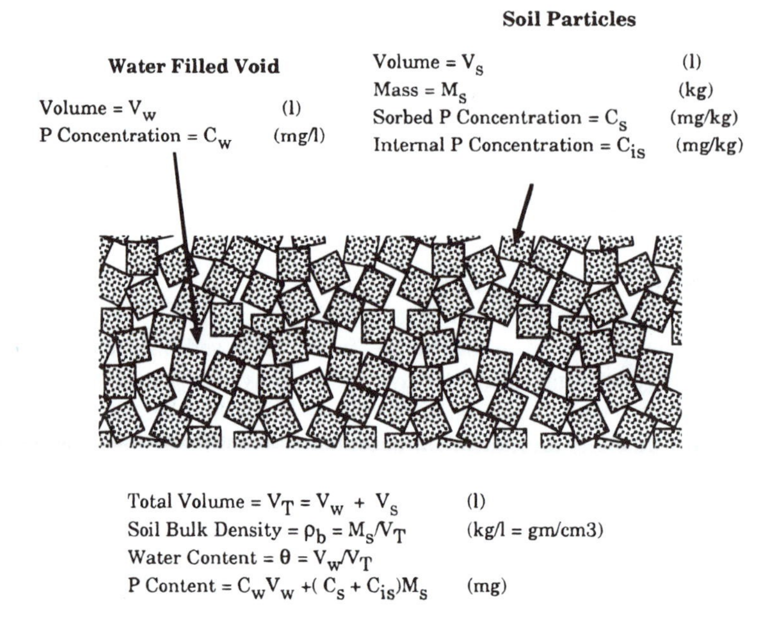

Figure 14-6 Microscopic section of soil showing terminology for phosphorus storage.

Most of the phosphorus in the soil column is structural phosphorus, both organic and inorganic. Very small fractions are found in porewater or as sorbed phosphorus. Table 14-3 illustrates the relative amounts for an Everglades peat. A peat has been selected for illustration, since most wetland treatment systems will build an organic sediment and soil layer with time. The first column in Table 14-3, for 0 to 10 cm, represents deposits which formed over

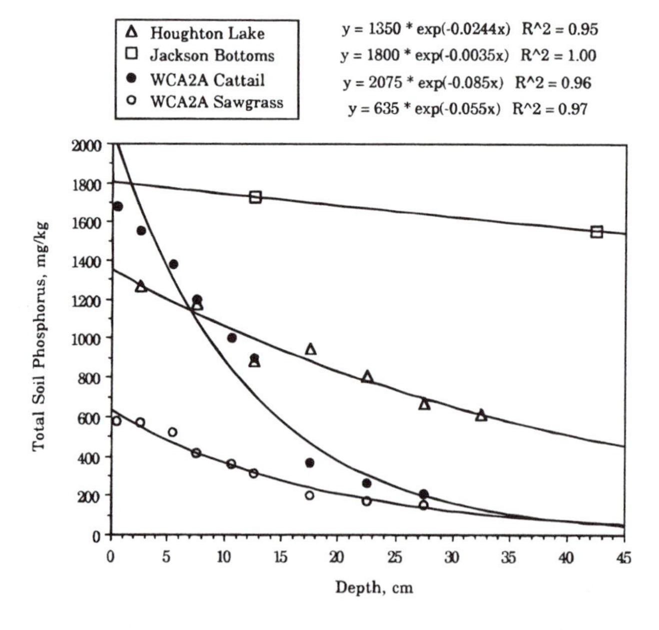

Figure 14-7 Vertical variation in total soil phosphorus expressed on a dry-weight basis. The Houghton Lake soil is a Houghton muck (Kadlec and Hammer, 1981); WCA 2A cattail is a recent cattail sediment/peat, WCA 2A sawgrass is a Loxahatchee peat (Reddy, et al., 1991a); and Jackson Bottoms is a mixture of Wapato and Cove soil types (CES, 1991). Curve fits are empirical exponentials.

Table 14-3 Example Vertical Profiles of Phosphorus Species in Everglades Peat (units: mg/L wet soil)

		Depth, cm			
Compartment	**Type**	**0–10**	**10–20**	**20–30**	**30–36.5**
Porewater	SRP	0.62	0.24	0.07	0.00
Soil surface	Sorbed Inorganic	0.82	0.29	0.20	0.01
Soil surface	Sorbed Organic	0.63	0.13	0.08	0.10
Soil surface	Sorbed Total	1.45	0.42	0.28	0.12
Soil structural	Fe + Al Bound	9.97	1.87	1.23	0.36
Soil structural	Ca Bound	15.02	2.36	0.63	0.18
Soil structural	Inorganic Total	26.37	4.39	1.96	0.61
Soil structural	Humic/fulvic	16.96	5.63	3.87	1.64
Soil structural	Organic Residual	30.79	11.26	6.66	3.81
Soil structural	Organic Total	47.75	16.89	10.54	5.45
Soil structural	Total	74.12	21.28	12.50	6.06
Total		76.19	21.94	12.84	6.19

the past 25 years, as determined by radio-cesium dating (Reddy et al., 1991a). An important point of Table 14-3 is that only a few percent of the phosphorus bound in the recently formed soils are typically available to the wetland phosphorus cycle.

SORPTION ISOTHERMS

There is a direct correlation between porewater and sorbed phosphorus for the data in Table 14-3. A linear relation explains inorganic phosphorus for that soil profile (a peat soil):

$$C_s = k_p C_w = 19.1 C_w \tag{14-4}$$

$R^2 = 0.82$
$0.009 < C_w < 0.689$ mg/L
$0.03 < C_s < 12.4$ mg/kg

where C_s = sorbed phosphorus concentration, mg/kg
C_w = porewater phosphorus concentration, mg/L
k_p = phosphorus capacity factor, l/kg

Equation 14-4 is a linear relation, the simplest of the phosphorus sorption isotherms which may be used to describe the relation between porewater and surface concentrations. A better fit of data is frequently found via a power-law fit, called the Freundlich isotherm. For SRP on Jackson Bottoms Wapato soils (mineral), this relation is

$$C_s = a_p C_w^b = 58\ C_w^{0.833} \tag{14-5}$$

$R^2 = 0.85$
$0.08 < C_w < 10.33$ mg/L
$0 < C_s < 274$ mg/kg

For SRP on Houghton Lake wetland soil (muck), this relation is

$$C_s = a_p C_w^b = 340\ C_w^{0.48} \tag{14-6}$$

$R^2 = 0.99$
$0.04 < C_w < 450$ mg/L
$73 < C_s < 5800$ mg/kg

where a_p = Freundlich phosphorus capacity factor, [mg P/kg]/[(mg P/l)^b]
b = Freundlich exponent, dimensionless

At a water concentration of 1.0 mg/L, these soils would hold 19.1, 58, and 340 mg/kg of phosphorus respectively.

The storage of phosphorus in porewater and on associated particle surfaces may be related to the porewater concentration via such isotherms. In a unit volume of wet soil, the total phosphorus storage may be represented by

$$V_T C_T = V_w C_w + M_s(a_p C_w^b) \tag{14-7}$$

$$C_T = \theta C_w + \rho_b(a_p C_w^b) = [\theta + \rho_b(a_p C_w^{b-1})]C_w = fC_w \tag{14-8}$$

or

$$f = [\theta + \rho_b(a_p C_w^{b-1})] \tag{14-9}$$

and

$$S = fC_w\delta \tag{14-10}$$

where f = phosphorus soil storage factor, dimensionless
S = phosphorus soil storage capacity, g/m^2
δ = soil depth, m

and where the rest of the terminology is defined on Figure 14-6.

Values of θ for wetland soils are typically in the range of 0.3 to 0.9. Bulk densities range from low values, circa 0.1 g/cm^3, for peats, to higher values, circa 1.5 g/cm^3, for mineral soils. The factor f is a multiplier on the volumetric storage capacity of water by itself and represents the added capacity of the sorption sites on the soil surface. Typical values of f range from 5 to 50.

The relevant interpretation of this storage is in the context of phosphorus added to the wetland. It is possible to gain some understanding of the relative importance of storage on existing soils by comparing the amount of phosphorus added to a wetland to the amount stored. For purposes of this discussion, complicating factors that are associated with storage in vegetation and that are critically important to storage in the ecosystem as a whole are ignored. As the basis for an order-of-magnitude calculation, consider a layer of soil 10 cm in thickness. The time to saturate a storage capacity of S g/m^2 at a phosphorus removal rate of J g/m^2/yr is equal to S/J. Sample calculations follow.

For the WCA 2A soil,

$f \approx 5$
$S \approx 0.2$ g/m^2 $\quad$ $J \approx 1.6$ g/m^2/yr $\quad$ $t \approx 1.5$ months

For the Houghton Lake wetland soil,

$f \approx 30$
$S \approx 7.5\ g/m^2$ $\qquad J \approx 20\ g/m^2/yr$ $\qquad t \approx 4.4$ months

For the Jackson Bottoms soil,

$f \approx 50$
$S \approx 6\ g/m^2$ $\qquad J \approx 50\ g/m^2/yr$ $\qquad t \approx 1.4$ months

Phosphorus sorption capacities for existing wetland soils are short-lived in these cases. Economic utilization of this storage in a wetland would require greatly enhanced sorption capacity or impossibly large wetland areas. Such enhancement is possible in SSF wetlands, if the media is tailored for phosphorus sorption capacity.

RATES OF MOVEMENT OF PHOSPHORUS IN SOILS WITHOUT VEGETATION

Capacity is not the only factor which needs to be considered in determining the sorptive lifetime of existing wetland soils. The phosphorus must move through the soil to reach unsaturated sorption sites at depth. Again, there are complicating factors associated with biological activity which are important to the rates of vertical phosphorus movement in the wetland. This section considers order-of-magnitude calculations based solely on a soil-water system.

Frontal Movement

The movement of solutes through sorbing media has been the subject of a considerable amount of research, primarily because of the importance of chromatography. Specialized applications to phosphorus movement in soils have been published (see, for example, Gerritse, 1993). If there is water movement through the soil, then there will be a movement of a phosphorus front in the same direction over a period of time. That front will move more slowly than the water due to the storage capacity of the solid surfaces. In the case of Freundlich isotherms with exponents less than 1.0, that front will be relatively abrupt (Ruthven, 1988).

The velocity of the phosphorus front may be estimated from

$$v = \frac{u}{f} \tag{14-11}$$

where u = water velocity, m/d
v = phosphorus front velocity, m/d

For the example systems, the factor f ranges from about 5 to 50. If a wetland is prevented from leaking, the upper soil horizon is still subject to a water flow to the root zone to satisfy the transpiration demand. A lower limit to vertical downward water movement for sealed wetlands is equal to the transpiration rate. That rate is in turn bounded by the evapotranspiration rate, which is on the order of 1 to 2 m/yr for the north temperate climate zone. The vertical downward velocity of a phosphorus front would be on the order of a few centimeters to as much as 1 m/yr.

Downward Diffusion

Solutes can move through sorbing media in the absence of water flow in that direction due to the presence of a vertical concentration gradient. Table 14-3 indicates that such gradients do exist. Diffusion is governed by Fick's equation:

$$N = D\left[-\frac{dC}{dx}\right] \tag{14-12}$$

where D = diffusion coefficient, m^2/yr
N = downward phosphorus diffusion flux, $g/m^2/yr$
x = distance downward from soil surface, m

Order-of-magnitude estimates of vertical downward diffusion fluxes may be made from measures of the vertical concentration profiles in porewater phosphorus. The diffusion coefficient may be estimated to be on the order of 0.5×10^{-5} cm^2/s = 0.016 m^2/yr (about half the free-water diffusion coefficient). The data in Table 14-3 yield the following estimate:

$$N \approx (.016)\left[\frac{0.62 - 0.24}{0.10}\right] = 0.06\ g/m^2/yr \tag{14-13}$$

Phosphorus does not penetrate to lower soil depths at a fast rate.

Fick's law may be combined with a mass balance to determine transient responses of the vertical phosphorus gradient. The result is

$$D\frac{\partial^2 C}{\partial x^2} = f\frac{\partial C}{\partial t} \tag{14-14}$$

If the starting condition is known, and the new phosphorus concentration at the surface is specified, this equation may be solved to ascertain the length of time required to establish a new vertical concentration gradient. Such calculations for the WCA 2A data given by Reddy et al. (1991a) produce times on the order of 1 year for gradient establishment. This transient result reinforces the idea that abiotic processes cannot move phosphorus at high vertical rates.

PHOSPHORUS RELEASES

In some circumstances, a wetland may be constructed or reestablished on a site that has a large, mobile, phosphorus storage already in place. Prime examples are previous peatlands that have undergone drying and peat oxidation. Under such circumstances, the available phosphorus may exceed the storage potential under the new water phosphorus concentrations. The result is the discharge of phosphorus from the antecedent soils into the new overlying water.

A specific case is the reflooding of drained forested peatlands in southern Florida at Walt Disney World. The site had been isolated from natural surface inflows and outflows for a period of several years. Many tens of centimeters of peat were lost to oxidation during this period, leaving a considerable residual of nonvolatile nutrients, including phosphorus. Wastewater discharges were commenced in 1988. The response of the wetland was to release stored phosphorus, creating a higher concentration of total phosphorus in the outflow than the inflow (Figure 14-8). Over the course of many weeks, new conditions were established which displayed phosphorus removal.

SUMMARY

Phosphorus in the wetland soil column is almost all in the form of tightly bound phosphorus, both inorganic and organic. This phosphorus is unavailable for use in normal biotic cycles. Order-of-magnitude calculations can be made for soil column response based on data available from wetland soils research. Wetland soils can hold only a few months of a new phosphorus loading to a wetland. Rate estimates show that the corresponding new soil storage would be achieved in a period of approximately 1 year for the upper soil horizon. Sorption on preexisting wetland soils does not typically provide a long-term or lasting phosphorus storage of consequence in treatment wetlands.

OVERALL INPUT-OUTPUT CORRELATIONS

INTRODUCTION

At the black-box level of data interpretation, correlations may be made between overall system operating variables via regression equations. The most important are concentrations in the entering and departing flows (C_i and C_o), the flow rates in and out (Q_i and Q_o), and the size of the wetland (A_T).

An alternate to the hydraulic loading rate is the detention time, and this variable has great intuitive appeal. The detention time has been used extensively to characterize reacting systems in which processes occur throughout the water body. Typical applications include biological wastewater treatment vessels, which conduct microbial reactions throughout the water; and treatment ponds, which rely upon suspended organisms, such as bacteria and plankton, for their effectiveness. The situation is different for SF wetlands, because most of the organisms responsible for pollutant reduction are associated with macrophytes and their litter. These in turn are proportioned to the surface area of the wetland and not to the volume of water within it. In addition, the depth and volume of water in the wetland are not accurately

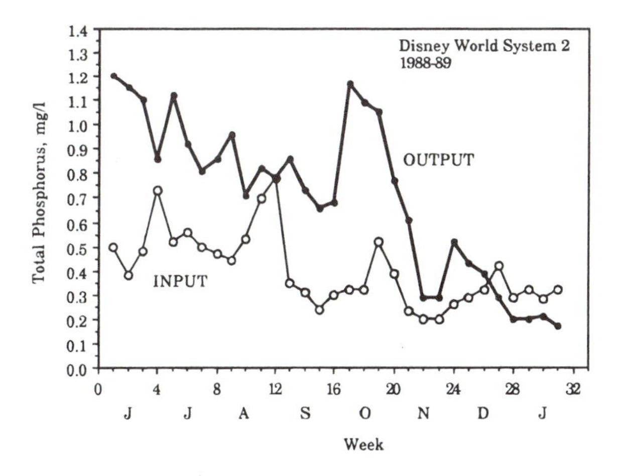

Figure 14-8 The progression of total phosphorus concentrations through the startup period of the natural forested wetland treatment system. The wetland had been damaged by drying and had stored the nutrients from the oxidation of half a meter of peat. (Replotted from DeBusk and Merrick, 1989.)

known in many cases; consequently, there is often large uncertainty in the nominal (calculated) detention time. For these reasons, detention time is not a satisfactory alternative.

Combinations of the principal variables may also be made and used in correlations. The most frequently used of these are hydraulic loading rate (q_i or q_o) and the phosphorus loading rate (PLI = q_iC_i or PLO = q_oC_o). Also in use is the phosphorus removal rate, PRR = $q_iC_i - q_oC_o$. percentage reductions may be based on concentration units, $(C_i - C_o)/C_i$; or upon mass units, $(q_iC_i - q_oC_o)/q_iC_i$.

Two factors determine the utility of data correlations based on black-box information: how much of the variablility in the data is represented (eg., R^2 value) and the range of those data. Correlations may not be safely used outside of the range of the data that produced them. For phosphorus, there are not large differences in degree of fit among correlations that use different variable combinations from the previous list. There are differences based on wetland type.

EMERGENT MARSHES

Data from a number of wetlands are presented in Figure 14-9. Information covers 49 wetland cells. Data were averaged seasonally, over 3-month periods, or quarters. A total of 373 quarterly data points were included. The power-law (logarithmic) correlation of inlet and outlet concentrations was

$$C_o = 0.34\ C_i^{0.96} \tag{14-15}$$

$R^2 = 0.73$, N = 373
Standard Error in $\ln C_o = 1.09$
$0.02 < C_i < 20$ mg/L
$0.009 < C_o < 20$ mg/L

A large number of characteristics of the wetland are included in such a correlation: vegetation type, water depth, climate, wetland size and shape, and flow rates, to name a few.

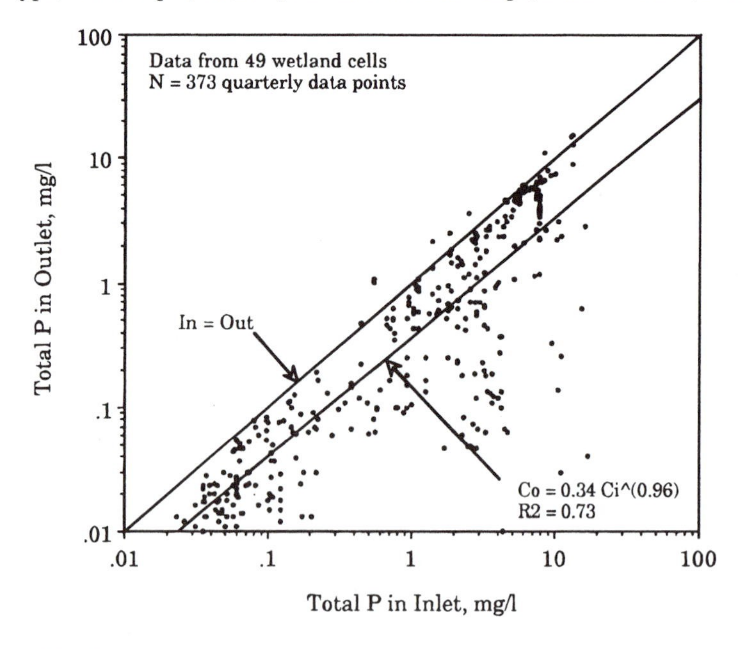

Figure 14-9 Black-box correlation of phosphorous removal in emergent marsh wetlands.

These range from reed canary grass in Oregon to cattails in Florida to sphagnum/labrador tea in Quebec. Equation 14-15 excludes forested wetlands and SSF wetlands.

The hydraulic loading may be added to the correlating variables, with the result

$$C_o = 0.195\, q_{avg}^{0.53} C_i^{0.91} \tag{14-16}$$

$R^2 = 0.77$, $N = 373$
Standard Error in $\ln C_o = 1.00$
$0.02 < C_i < 20$ mg/L
$0.009 < C_o < 20$ mg/L
$0.1 < q_{avg} < 33$ cm/d

Not much improvement has been gained by the addition of the hydraulic loading rate.

FORESTED WETLANDS

Data from nine natural wetlands in the U.S., including 166 quarters of data, was correlated in the same manner as for emergent marshes. The power-law (logarithmic) correlation of inlet and outlet concentrations is (Figure 14-10)

$$C_o = 0.48\, C_i^{0.46} \tag{14-17}$$

$R^2 = 0.10$, $N = 166$
Standard Error in $\ln C_o = 1.29$
$0.05 < C_i < 20$ mg/L
$0.02 < C_o < 9$ mg/L

A large number of characteristics of the wetland are included in such a correlation: vegetation type, water depth, climate, wetland size and shape, and flow rates, to name a few. These range from cypress strands in Florida to black spruce bogs in Minnesota. In Equation 14-17, marshes and SSF wetlands have been excluded.

The hydraulic loading may again be added to the correlating variables, with the result

$$C_o = 0.37\, q_{avg}^{0.53} C_i^{0.70} \tag{14-18}$$

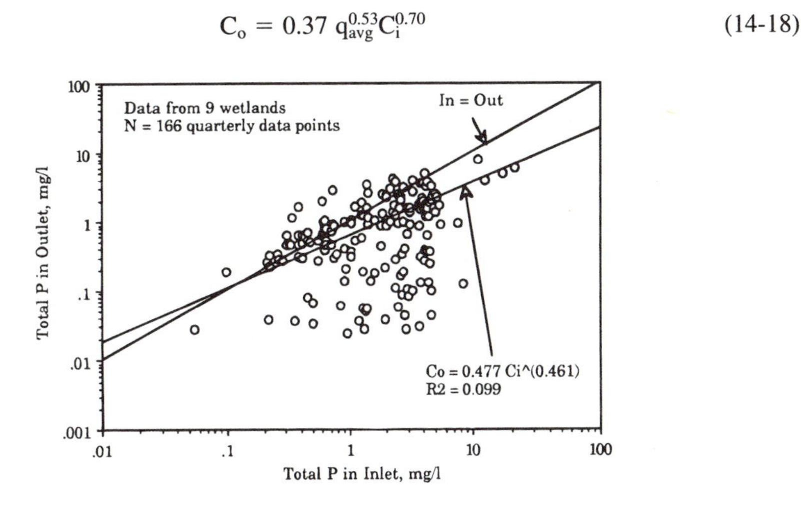

Figure 14-10 Black-box correlation of phosphorus removal in forested wetlands.

$R^2 = 0.33$, $N = 166$
Standard Error in $\ln C_o = 1.13$
$0.05 < C_i < 20$ mg/L
$0.02 < C_o < 9$ mg/L
$0.03 < q_{avg} < 13$ cm/d

Some improvement has been gained by the addition of the hydraulic loading rate.

SUBSURFACE-FLOW WETLANDS

Data from ninety wetlands located in the U.S., Australia, the U.K., and Denmark, each represented by one point representing average performance, were correlated in the same manner as for emergent marshes. The power-law (logarithmic) correlation of inlet and outlet concentrations is (Figure 14-11)

$$C_o = 0.51\, C_i^{1.10} \tag{14-19}$$

$R^2 = 0.64$, $N = 90$
Standard Error in $\ln C_o = 0.59$
$0.5 < C_i < 20$ mg/L
$0.1 < C_o < 15$ mg/L

A large number of characteristics of the wetland are included in such a correlation: vegetation type, water depth, climate, wetland size and shape, media type, and flow rates, to name a few. Most of the systems are reed beds (*Phragmites* spp.) in Denmark and the U.K.; the U.S. systems are more often planted with bulrushes (*Scirpus* spp.) In Equation 14-19, marshes and forested wetlands have been excluded.

The hydraulic loading may again be added to the correlating variables. However, this was done only for quarterly information from the eight U.S. wetlands for which the hydraulic loading rate was known, so no direct comparison should be made between Equations 14-19 and 14-20.

$$C_o = 0.23\, q_{avg}^{0.60} C_i^{0.76} \tag{14-20}$$

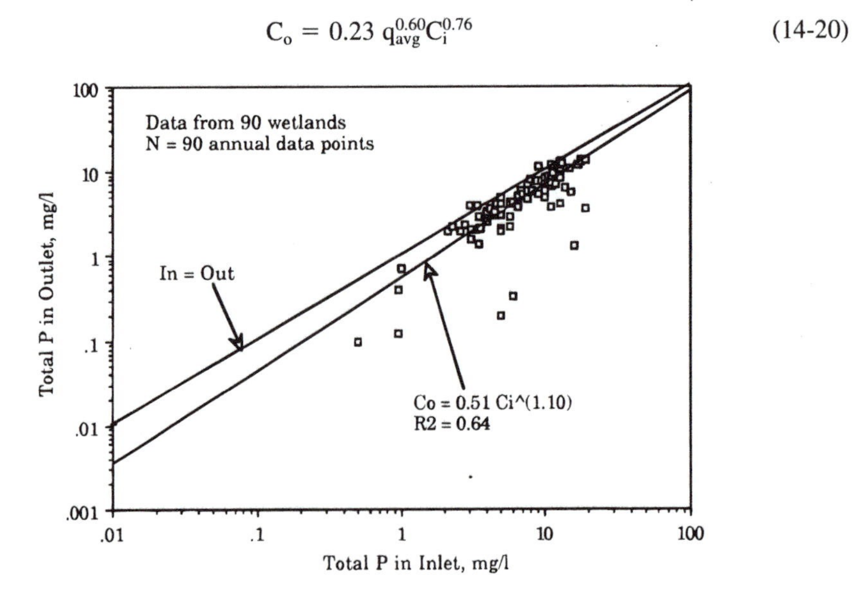

Figure 14-11 Black-box correlation of phosphorus removal in SSF wetlands.

$R^2 = 0.60$, $N = 23$
Standard Error in $\ln C_o = 1.01$
$2.3 < C_i < 7.3$ mg/L
$0.1 < C_o < 6$ mg/L
$2.2 < q_{avg} < 44$ cm/d

SUBMERGED AQUATIC BEDS

There are scant data for such wetlands. Gumbricht (1993) studied the performance of *Elodea canadensis* in 60-cm deep wetlands over 33 months of continuous operation. The simplest representation of the data for phosphorus was

$$PRR = 0.59\ PLI \qquad (14\text{-}21)$$

$R^2 = 0.78$, $N = 136$
$0.01 < C_o < 0.32$ mg/L

INTERNALIZATION

Internal behavior of phosphorus concentrations in a wetland may be projected from correlations. The hydraulic loading rate to any point in a wetland is the inlet hydraulic loading rate divided by the fractional area from the inlet, $q = q_i/y$. Equation 14-16, for instance, then becomes

$$C_o = 0.195\left(\frac{q_{avg}}{y}\right)^{0.53} C_i^{0.91} \qquad (14\text{-}22)$$

The profiles so calculated correspond nicely to data from some individual wetland systems, but are not at all accurate for others (Figure 14-12). Some wetlands happen to be close to the median behavior represented by the overall correlations; others are not. The scatter in the data is the result of site factors as well as the probabilistic nature of the controlling processes. In the case of the Lakeland system (Figure 14-12), there are several possible reasons for deviation from the mean. This wetland was constructed in an abandoned phosphate mine and has a configuration conducive to bypassing of large areas.

It is noteworthy that internal profiles in these systems are well represented by an exponential decrease through the wetland.

GENERAL CONSIDERATIONS

The previous power-law correlations (Equations 14-15 through 14-22) present the overall expectations of phosphorus removal, but do not permit acknowledgement of regional and site-specific factors. Nor do these correlations include temporal factors, such as the age of the wetland or the season of the year. Therefore, they have a large standard error in the prediction of effluent phosphorus concentration. Narrower (more specific) correlations would require a set of wetlands in the same geographical region, with the same density and types of plants, the same depth and duration of flooding, and the same age.

There is a disadvantage to attempting to narrow such correlations: they contain two or three *parameters* consisting of one multiplier and one or two exponents. It would be necessary to determine these from the narrower data set. The operating data presently available do not allow multiple parameter evaluation on a narrow basis, except at a few locations.

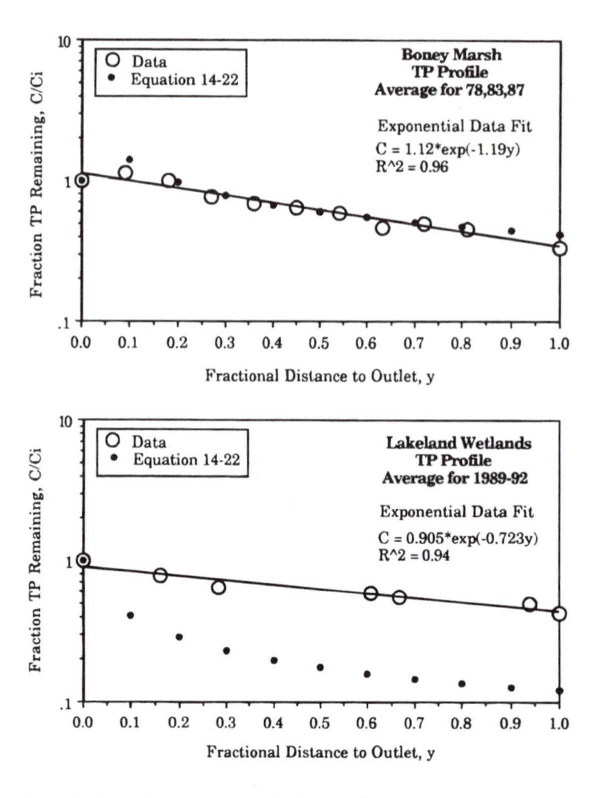

Figure 14-12 Internal predictions from the correlation of phosphorus removal in marsh wetlands (Equation 14-16).

TEMPERATURE AND SEASONAL DEPENDENCE

Temperature and solar radiation are primary determinants of the activity of the wetland photosynthetic processes. It is therefore intuitively plausible that these variables would be of importance in the removal of phosphorus. In this instance, intuition fails, for there are not large differences in phosphorus removal associated with temperature or season when viewed from the overall correlative perspective. To illustrate, consider the seasonal regression parameters for the general form of Equation 14-23:

$$C_o = aq_{avg}^{n}C_i^{m} \tag{14-23}$$

Partitioning the marsh data by season gives the parameters (±SE) listed in Table 14-4. The variation results in predicted seasonal swings in percent phosphorus reduction that are generally less than 10 percent.

Table 14-4 Seasonal Regression Parameters for Phosphorus Reduction in Marshes

Season	a	n	m	R^2
Spring	0.17 ± 0.03	0.58 ± 0.14	0.90 ± 0.06	0.80
Summer	0.17 ± 0.03	0.55 ± 0.12	0.90 ± 0.07	0.67
Fall	0.24 ± 0.04	0.41 ± 0.12	0.97 ± 0.05	0.84
Winter	0.22 ± 0.04	0.53 ± 0.16	0.90 ± 0.06	0.82
Whole year	0.20 ± 0.02	0.53 ± 0.06	0.91 ± 0.03	0.77

SUMMARY

This section contains general regression equations for phosphorus removal in various types of wetlands. These equations represent data from multiple operating systems with many site-specific differences, and that data scatters greatly. Therefore, the regression equations have large standard errors.

The simplest correlations relate inlet and outlet concentrations (Equations 14-15, 14-17, 14-19, 14-21) and predict phosphorus concentration reductions in the range of 40 to 70 percent.

If the hydraulic loading rate is included, the correlations become slightly better (Equations 14-16, 14-18, 14-20).

It is recommended that these equations be used only for preliminary design calculations and that site factors be considered in final design calculations. To do so, a rational design approach is required, which relies more strongly on mass balances for water and phosphorus and acknowledges site-specific factors.

THE MASS BALANCE MODEL WITH FIRST-ORDER AREAL UPTAKE

INTRODUCTION

The general mass balance and lumping concepts of Chapter 9 apply particularly well for phosphorus removal in treatment wetlands. Phosphorus moves from compartment to compartment and changes form readily (Figure 14-1), with a net accretion resulting in the inactive sediments and soils of the wetland.

Wetlands are capable of survival at very low phosphorus concentrations, compared to typical wastewater concentrations. Therefore, the background concentration in the k-C* model is close to zero. There has been no definitive study of natural phosphorus levels by wetland type, but oligotrophic marshes, such as the Florida Everglades or northern *Sphagnum* bogs, are conditioned to less than 20 μg/L. Cattail marshes often operate at less than 50 μg/L in northern climates. For purposes of treatment wetland design, these phosphorus background levels are usually negligible, and the assumption of $C^* = 0$ will be used here.

Chapter 9, Equations 9-152 through 9-154 may be written for this case:

$$q \frac{dC}{dy} = -k(C - C^*) = -kC \qquad (9\text{-}152)$$

Integration from the wetland inlet, where the concentration is C_i, to an intermediate distance y gives the *internal profile prediction:*

$$\ln\left(\frac{C}{C_i}\right) = -\frac{k}{q} y \qquad (9\text{-}153)$$

or alternatively:

$$C = C_i \exp\left(-\frac{k}{q} y\right) \qquad (14\text{-}24)$$

At the outlet, where the concentration is C_o, the *input-output relation* is

$$\ln\left(\frac{C_o}{C_i}\right) = -\frac{k}{q} = -Da \quad (9\text{-}154)$$

or alternatively:

$$C_o = C_i \exp\left(-\frac{k}{q}\right) \quad (14\text{-}25)$$

These "simplest case" phosphorus removal equations apply to average performance over a long term, after adaptation trends are over. Refer to Chapter 9 for the further details of assumptions and limitations. This mathematical description is useful in specific contexts, but the lumping process virtually guarantees that the rate constant "k" will vary with other ecosystem variables.

VERIFICATION OF THE SIMPLEST MODEL: SOME EXAMPLES

Figure 14-13 demonstrates that the logarithmic profiles required by the simplest model do in fact exist for wetlands at Lakeland, FL; Listowel, Ontario; and an 115-km^2 zone in the north end of Water Conservation Area 2A (WCA 2A), located south of Lake Okeechobee in Florida. The example wetlands were chosen to span a fair range of concentrations, wetland

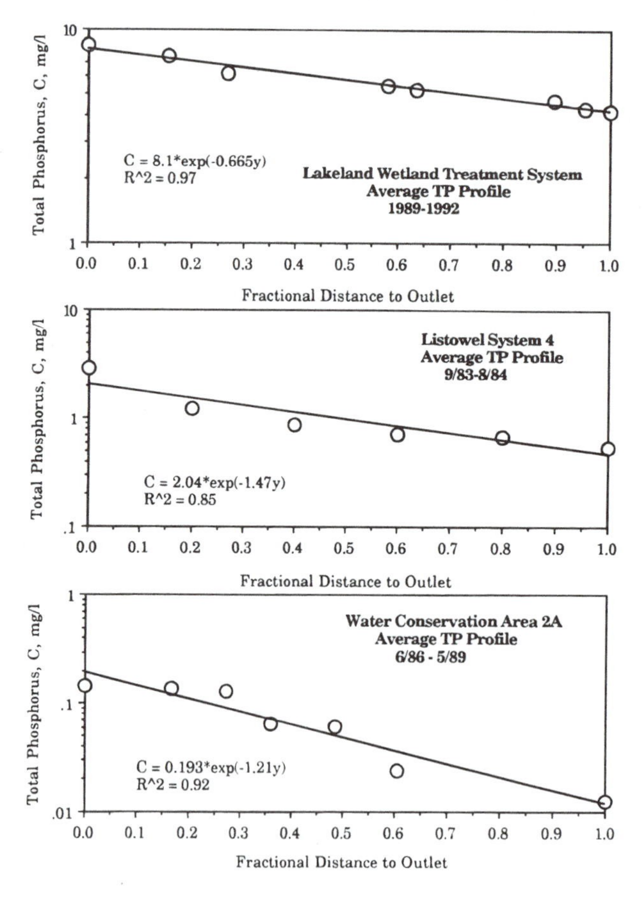

Figure 14-13 Examples of exponential decreases in phosphorus concentrations for varying levels of total phosphorus. Averaging is over several years and spans integer numbers of years.

types, and climate, while meeting the assumptions to differing degrees. Data for these systems comes from Hill (1992), Herskowitz (1986), and Walker (1992a, 1992b, 1992c), respectively.

The assumptions implicit in the simplest model were listed in Chapter 9 and are briefly discussed here in connection with Figure 14-13. The three ecosystems had undergone differing periods of adaptation: 1.5 years for Lakeland and 2.75 years for Listowel, both constructed; and in excess of 20 years for WCA 2A, a natural peatland. No trends remained in the data at the start of the averaging period. The conditions of assumption A are met for all three.

The averaging period was 4 years for Lakeland, 1 year for Listowel, and 3 years for WCA 2A. The number of times during the averaging period that phosphorus measurements were taken along transects were *circa* 200, 43, and 50, respectively. The averaging periods were also chosen in multiples of 12 months to span any seasonal effects exactly. The conditions of assumption B are met for all three.

Water budgets were compiled for these example systems. For Lakeland, a net (ET-R) of 4 percent of the incoming water was lost to the atmosphere and 23 percent was lost to seepage. Listowel System 4 was sealed against seepage and gained 13 percent. WCA 2A lost 11 percent to the atmosphere and gained 8 percent from seepage and cross-flow. Thus, the conditions of assumptions C and D are imperfectly met for all three; however, all are dominated by added water.

Lakeland and Listowel received a relatively steady flow of water at relatively uniform concentration and meet condition E, which requires that flow-weighted average concentrations coincide with time average concentrations. WCA 2A was event driven, and the two averages differed by as much as 40 percent in some cases.

Rainfall phosphorus was negligible for Lakeland and Listowel because these two systems received high phosphorus in pumped waters. For WCA 2A, 5.5 percent of the total phosphorus loading came from rain. Condition F was therefore met.

Listowel System 4 was rectangular, with an aspect ratio (L:W) of 84:1. The Lakeland system consists of seven cells in series, with extremely irregular shape. The length of the presumed flow path is about ten times the average width. This degree of compartmentalization into cells would lead to the expectation of plug flow approximation for a first-order reaction (Levenspiel, 1972). The inlet zone of WCA 2A was taken to be a rectangular parallelopiped with an aspect ratio of about 1.0. Confirmation was made by examination of soil phosphorus levels (Reddy, 1991). Condition G was thus imperfectly met for Lakeland and WCA 2A.

Mixing patterns in these wetlands were not measured, and therefore, assumption H was not tested. For Listowel, tracer tests were run, but only the actual detention times were reported (these matched calculated detention times, with an average difference of only 6 percent).

Assumption I was examined for WCA 2A by acquiring data on parallel transects across the width of the wetland (Reddy, 1991). Soil porewater phosphorus showed great similarity along three lines all parallel to flow. This assumption was untested for the other two wetlands.

Figure 14-13 shows good correlations for the exponential decrease predicted by the simplest model. These data provide strong support for the presumption (J) that a first-order areal model is reasonable for the stated assumptions.

CALIBRATION OF THE FIRST-ORDER MODEL: INTERSYSTEM VARIABILITY

So far it has been shown that the first-order areal phosphorus model describes some features of internal wetland phosphorus phenomena. The next step is the determination of model parameters, of which there is only one: the long-term average first-order areal phosphorus removal rate constant, "k." This parameter may be determined from either transect data or from input-output data. Most wetland treatment systems have measurements of inlet flows and inlet and outlet concentrations, but not transect concentration measurements. In those

Table 14-5 First-Order Phosphorus Rate Constant for Emergent Marshes

Site	No. of Wetlands	Years of Operation	Data Years	HLR cm/d	TP In mg/L	TP Out mg/L	k Value m/yr
Des Plaines, IL	4	6	6	4.77	0.10	0.02	23.7
Jackson Bottoms, OR	17	3	2	6.34	7.51	4.14	14.2
Lakeland, FL	6	7	7	7.43	6.54	5.69	3.4
Pembroke, KY	2	6	2	0.77	3.01	0.11	9.3
Great Meadows, MA	1	ca. 70	1	0.95	2.00	0.51	5.7
Fontanges, Quebec	1	2	2	5.60	4.15	2.40	11.2
Houghton Lake, MI	1	16	16	0.44	2.98	0.10	11.0
Cobalt, Ontario	1	2	2	7.71	1.68	0.77	20.9
Brookhaven, NY	1	3	3	1.50	11.08	2.33	8.9
Leaf River, MS	3	5	5	11.68	5.17	3.96	11.2
Clermont, FL	1	3	3	1.37	9.14	0.15	23.4
Sea Pines, SC	1	9	8	20.20	3.94	3.36	11.7
Benton, KY	2	6	2	4.72	4.54	4.10	2.4
Listowel, Ontario	5	4	4	2.41	1.91	0.72	8.2
Humboldt, SAS	5	3	3	3.04	10.16	3.24	12.8
Tarrant County, TX	9	2	1	9.44	0.29	0.16	20.1
Iron Bridge, FL	16	7	7	2.69	0.43	0.10	13.5
Boney Marsh, FL	1	11	11	2.21	0.05	0.02	14.2
WCA 2A	1	30	14	0.93	0.12	0.02	10.2
OCESA	4	6	6	0.83	0.27	0.16	6.4
	82	131	105			Average	12.1
						SD	6.1

cases, Equation 14-25 is used to calculate k. If infiltration or rainfall or evapotranspiration are important, appropriate modifications must be made (see Chapter 9). Rate constants are next determined from the data sets used to calibrate regression equations.

The data will be partitioned into the same subsets as for the regression approach: marshes, forested SF and SSF wetlands. There are several possible methods of weighting the data; in the following, each wetland is represented by one time average value of the rate constant.

EMERGENT MARSHES

The values of the uptake rate constant (Table 14-5) fall within a band; k = 12.1 ± 6.1 m/yr (mean ± standard deviation) when averaged at one point per site. The distribution of values is k = 13.1 ± 8.5 m/yr, when averaged at one point per wetland (Figure 14-14). Given the large number of significant differences in site-specific factors, it is perhaps surprising that this band is not larger.

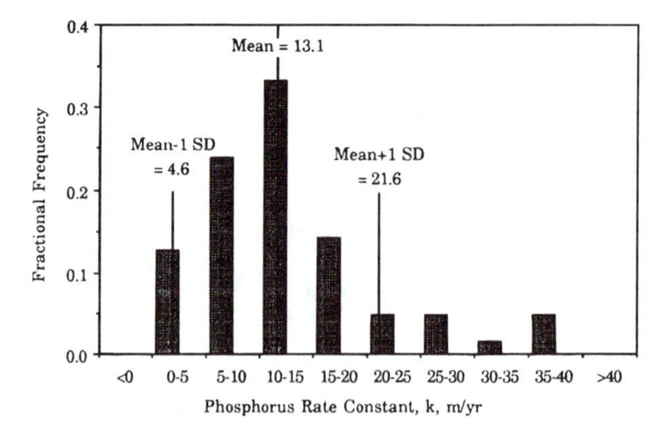

Figure 14-14 Distribution of phosphorus removal rate constants for North American treatment marshes. Sixty-three wetland cells at twenty sites are represented at one point per wetland.

FORESTED WETLANDS

A composite of transect data for the Bellaire, MI Wetland System 1 displayed a decreasing profile of phosphorus with distance (Kadlec, 1983). There are no other data on transects through forested wetlands. It is still possible to calibrate the resulting input/output relation (Equation 14-25) for those systems which report inlet and outlet phosphorus concentrations and hydraulic loading rates.

The rate constant is lower for wetlands with trees. For those systems, k = 3.1 ± 5.2 m/yr. The distribution of values is given in Figure 14-15.

SSF WETLANDS

It is not clear whether a first-order model of phosphorus removal applies to these wetlands. There is very little information available on internal transect profiles of total phosphorus in the flow direction for SSF wetlands, and that data present conflicts. Long-term average performance profiles, after the startup adaptation period, are not available.

Decreasing total phosphorus profiles have been reported by Vymazal (1993) and by TVA (1990a), but in both cases there was no evidence that the startup trends were over. Green and Upton (1993) report decreasing profiles of total phosphorus through the series of SSF cells at Little Stretton, U.K. after the system had been in operation for 3 years. However, transect sampling was conducted over only two 1-week periods.

In contrast, Mann (1990) found no trend of reactive phosphorus with distance through mature planted and unplanted SSF systems. However, as was seen in earlier discussions, phosphorus forms interconvert, and total phosphorus may not follow the same trend.

As a result of this meager database concerning internal phosphorus processes in SSF wetlands, it is not possible to adequately verify a model which predicts an exponential decrease in total phosphorus. The input/output relation (Equation 14-25) is calibrated for those systems which report inlet and outlet phosphorus concentrations and hydraulic loading rates. The rate constant is virtually the same as that for emergent marsh wetlands, k = 11.7 ± 4.2 m/yr. The distribution of values is given in Figure 14-16.

CALIBRATION OF THE FIRST-ORDER MODEL: INTRASITE VARIABILITY

For emergent marshes, there is sufficient information available to begin to understand the origins of the distributions of rate constants. The majority of the rate constant variability is apparently due to stochastic and/or chaotic behavior of the ecosystem.

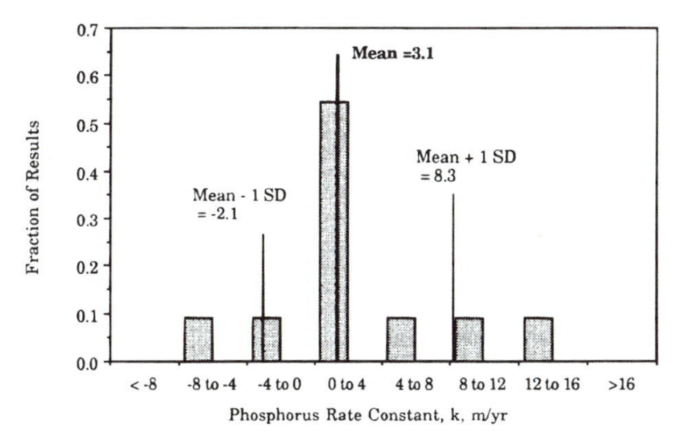

Figure 14-15 Distribution of phosphorus removal rate constants for forested North American treatment wetlands. Eleven wetlands at eight sites are represented at one point per wetland.

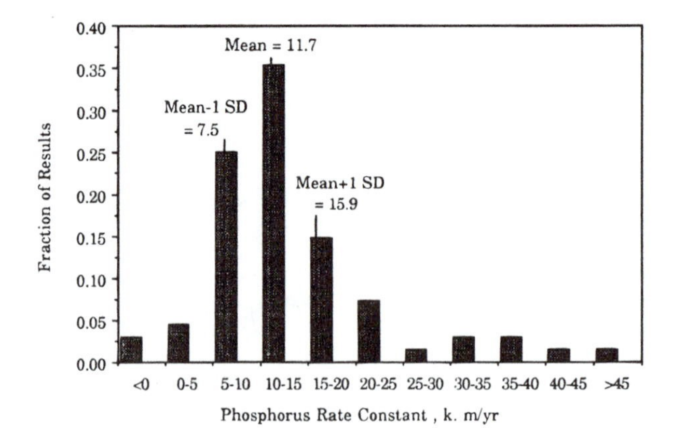

Figure 14-16 Distribution of phosphorus removal rate constants for SSF treatment wetlands. Ninety wetlands are represented at one point per wetland. Most of these wetlands exhibit surface flow at times.

Temporal Variability

Individual cells and systems at the same site display changing values of k with time, due to a number of uncontrollable environmental and ecosystem factors. As an example, consider the Iron Bridge, FL wetland treatment system. This is a multi-cell facility, the first 15 cells of which are emergent marshes. Exponential decreases of phosphorus concentrations have been documented (Kadlec and Newman, 1992). The balance between rain and evapotranspiration is close, and water budgets show little or no infiltration. Added water flows are relatively constant and dominate the water balance. There were no major or sudden changes in inlet phosphorus concentrations. Consequently, the monthly averaging window is acceptable in terms of detention time. The model conditions are reasonably well met, and the variability is due to ecosystem and environmental factors. There is a relatively wide spread in monthly average k values, 9.5 ± 4.1 m/yr (Figure 14-17). Emergent marsh data displays a 65 percent coefficient of variation in k across different wetlands (Figure 14-14), but the temporal variability of k for this single system is also large; CV = 43 percent.

Variability Due to Ecosystem Structure

Individual wetlands at the same site, replicated in aspect ratio and in inlet concentration and similar in species composition, also display variability in k. A good example is the set

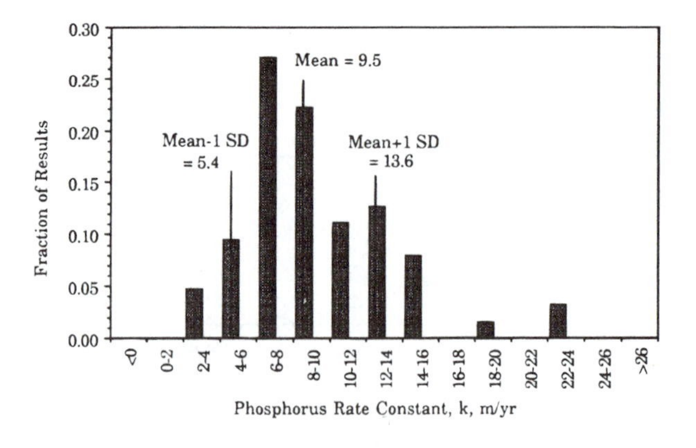

Figure 14-17 Distribution of phosphorus removal rate constants for the Iron Bridge, FL treatment system.

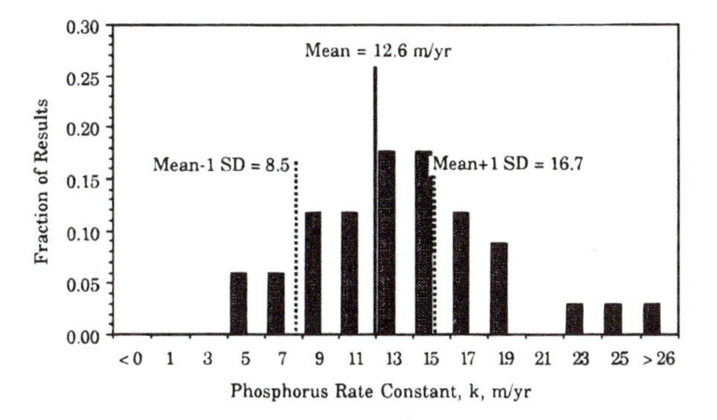

Figure 14-18 Distribution of phosphorus removal rate constants over treatment wetlands at the same site. Inlet concentrations and aspect ratios (length to width ratios) were the same, but there were differences in depth, soils, and vegetation type and density. Seventeen wetlands are represented at one point per wetland per year.

of 17 wetlands in Jackson Bottoms, near Hillsboro, OR (SRI, 1990, 1991). Although these were set side-by-side, there were both planned and unplanned differences in vegetation, soils, and hydrology. There were accompanying differences in k values (Figure 14-18). These differences are not explainable on the basis of depth, soil type, and percent emergent vegetative cover, but some of the variance may be attributed to hydraulic loading rate. A major part of the scatter in the data appears to be attributable to the vagaries of ecosystem function.

Variability Due to Wetland Configuration

Two of the Listowel, Ontario wetlands, numbers 4 and 5, were similar in all respects except aspect ratio. The time history of rate constants follows the same trends, with essentially the same 4-year means (Figure 14-19). The rate constant was not sensitive to aspect ratio.

SOME POTENTIALLY IMPORTANT SITE-SPECIFIC FACTORS

One of the first questions that may be asked is whether the first-order model has adequately accounted for the effects of hydraulic loading rate on the removal of phosphorus. In fact, the rate constants do not depend on HLR on an intersystem basis ($R^2 < 0.06$, one point per wetland; $R^2 < 0.07$, quarterly data each wetland). Trends are also lacking for individual systems after startup adaptation, such as Iron Bridge ($R^2 < 0.07$, monthly data) and Listowel Systems 4 and 5 ($R^2 < 0.08$, monthly data).

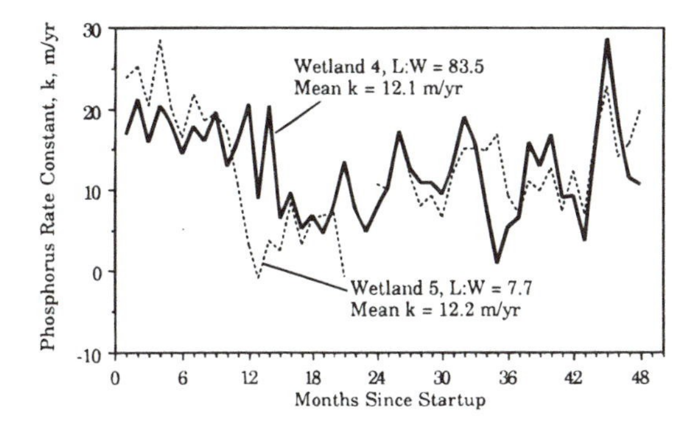

Figure 14-19 Phosphorus removal rate constants for two treatment wetlands at Listowel. These differ in aspect ratio; other system variables were similar.

Part of the difference between wetlands has to do with vegetation density. The potential amount of biological activity clearly depends strongly on the amount of biota available to process phosphorus. Shallow ponds devoid of emergent or submerged vegetation do not show as much phosphorus removal as the vegetated equivalent. This fact has been demonstrated by side-by-side studies at Richmond, NSW, Australia (Bavor et al., 1988). Lakshman (1993) correlated phosphorus removal with vegetation density for the five constructed wetlands in his study and concluded that a direct relation existed. The data from Iron Bridge, FL show a similar stem density effect for a developing constructed wetland.

Filamentous algae often populate a newly constructed wetland with sparse plantings. These organisms can fill a niche and provide phosphorus removal during the early phase of a wetland development. However, a second possibility may also occur: the establishment of a planktonic algal community. This leads to export of planktonic material and essentially zero phosphorus removal.

Insufficient information is presently available to construct regression equations involving plant density. An *ad hoc* approach would be to use a linear interpolation between the uptake rate ($J = kC$) for full plant density and zero uptake at zero plant density.

DEPTH EFFECTS

There is not sufficient data to yield intersystem regression equations for this effect. Most existing SF wetlands are operated within a narrow depth range, and no systematic investigation of depth has been performed for single wetlands.

The 17 Jackson Bottoms wetlands were constructed with varying proportions of deep and shallow areas, ranging from 12 to 96 percent in deep zones. These offer a side-by-side comparison of depth effects. The areal rate constants for the Jackson Bottoms wetlands display no trend with depth (Figure 14-20a).

The Jackson Bottoms data also give a strong denial of the first-order volumetric model for phosphorus removal. The first-order volumetric model was compared with the first-order areal model for BOD in Chapter 12, where it was shown that

$$k_V = k/\varepsilon h \tag{14-26}$$

If areal uptake were the better model, k would be constant and k_v would be inversely proportional to depth. This is apparently the case for Jackson Bottoms (Figure 14-20b).

MIXING EFFECTS

Landscape Scale

This scale encompasses the entire wetland, often measured in hectares.

The landscape scale flow patterns in the wetland, together with depth, determine the fraction of the total wetland surface area that is involved with phosphorus processing. A few wetland treatment systems, such as the Incline Village, NV system, have variable wetted area due to hydrologic conditions, but most have a relatively constant wetted area. In contrast, many wetlands, especially natural ones, have dead zones and channels. The result is that the wetland contains inactive zones, because they are either out of the water or out of the flow pattern. The Lake Wingra, WI wetland, described in Chapter 9, is a good example (Perry et al., 1981). This wetland possessed a high degree of natural channelization and received a point discharge. A large proportion of the delineated wetland area was topographically isolated. Therefore, the rate constant derived using the total delineated area is expected to be low and that is the result: the value of $k = 2.0$ m/yr.

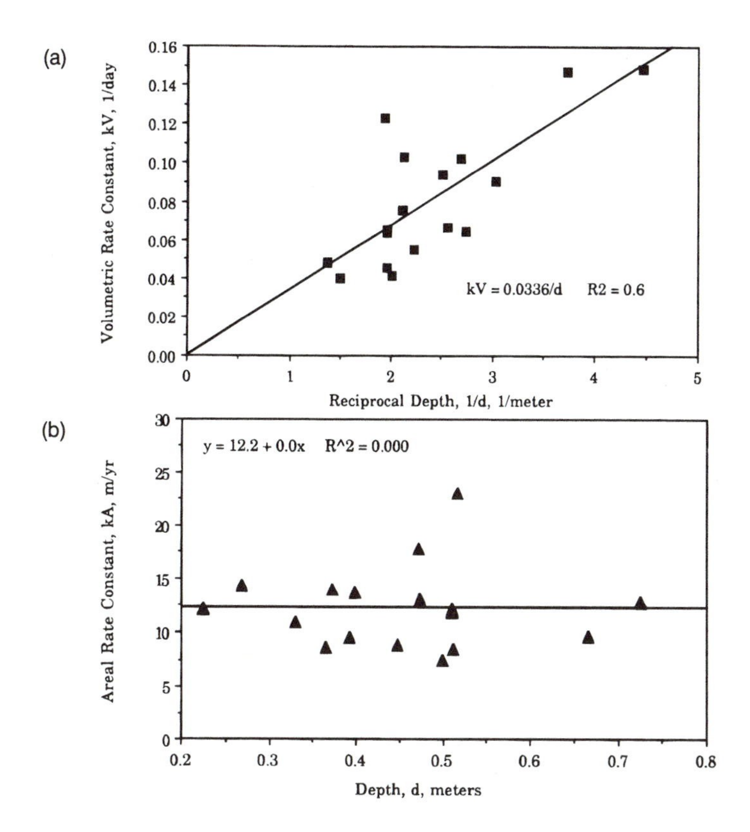

Figure 14-20 (a) Areal phosphorus rate control vs. reciprocal depth for Jackson Bottoms, 1990. (From data in SRI, 1991.) (b) Volumetric phosphorus rate constant vs. reciprocal depth for Jackson Bottoms, 1990. (From data in SRI, 1991.)

Proper analysis of phosphorus removal data from such wetlands should necessarily include determinations of wetted area and flushed volume.

The physical layout and distribution design of a new treatment wetland should clearly seek to minimize macroscopic flow maldistribution. Depending on the methods of flow introduction and redistribution, there may be greater or lesser contacting efficiency.

Plot Scale

The plot scale is defined here to be a size measured in terms of some tens of square meters. Such plots are large enough to be representative of vegetation, as measured by species composition, relative abundance, or biomass.

Within the hydrologically active zones of the total wetland, there exist pockets of stagnant water, clumps of litter (which are mostly water), and zones of water moving faster than average. Boundary layers exist around all submerged objects—plant stems, litter, and soil—and the water in such layers moves slowly if at all. These hydrologic heterogeneities all interact to generate the distribution of transit times discussed in Chapter 9. As noted there, such distributions are not ideal; wetlands are neither in plug flow nor are they well mixed. It may be necessary to acknowledge these effects for phosphorus removal in terms of the rate constant derived from data or the treatment area calculated in design.

The necessary adjustments may be made from a knowledge of the wetland flow patterns. Tracer testing produces the wetland dimensionless variance (σ_θ^2), which may be used with existing chemical reactor theory to determine the rate constant. An illustration follows.

Example 14-1

Wetland EW3 at Des Plaines, IL has been characterized by tracer studies (Kadlec, 1994). The dimensionless variance for EW3 was thus found to be $\sigma_\theta^2 = 0.4$. Phosphorus removal data has been acquired and analyzed with the plug flow assumption to yield $k_{pf} = 31.7$ m/yr at a hydraulic loading rate of 7.85 cm/d (3.65*7.85 = 28.65 m/yr). What is the actual value of k under the known degree of nonideal mixing?

Solution

Use the tanks-in-series (TIS) model described in Chapter 9. The plug flow model for this example gives

$$\frac{(C_o - C^*)}{(C_i - C^*)} = \exp\left(-\frac{k_{pf}}{q}\right) = \exp\left(-\frac{31.7}{28.65}\right) = 0.331 \qquad (9\text{-}103)$$

The number of TIS is

$$\sigma_\theta^2 = \frac{1}{N} = 0.4 \qquad (9\text{-}112)$$

$$N = 2.5$$

The TIS model is

$$0.331 = \frac{(C_o - C^*)}{(C_i - C^*)} = \left[1 + \frac{k}{Nq}\right]^{-N} = \left[1 + \frac{k}{2.5(28.65)}\right]^{-2.5} \qquad (9\text{-}117)$$

Solving yields k = 39.9 m/yr. This is higher than the plug flow value of 31.7 because the partially mixed wetland is less efficient than plug flow.

TEMPERATURE EFFECTS

No temperature trends are to be found for the areal rate constant in the mass balance approach for emergent marshes. It is necessary to utilize data from several replicated annual periods to establish trends that are valid against the background of other stochastic variations. Listowel Systems 4 and 5 are called upon to demonstrate this surprising observation (Figure 14-21). The lack of a temperature effect may be expressed by the lack of correlation

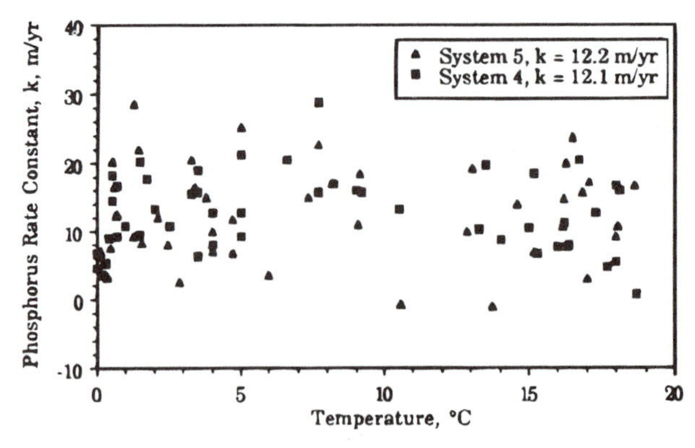

Figure 14-21 There is no temperature effect to be seen in the data from these two Canadian constructed wetlands.

($R^2 < 0.01$) for both these systems over a 4-year period. This behavior is not unique to northern climates; the Iron Bridge, FL wetlands also display no correlation ($R^2 < 0.05$) between k and temperature over the range 10 to 30°C.

The same insensitivity to temperature has also been found for SSF wetland phosphorus performance. For example, Gumbricht (1992) found "negligible temperature dependence" over a 140-week study of an SSF system in Sweden (2 to 21°C).

Forty-seven SSF soil-based wetland systems in Denmark showed no different performance in winter than in summer (Schierup et al., 1990b).

SEASONAL EFFECTS

It was earlier noted that no seasonal trends were to be found in regression equation parameters for the set of emergent marshes. The same is not true for the rate constant in the mass balance approach.

Listowel Systems 4 and 5 are again called upon to demonstrate the phenomena (Figure 14-22). There is a clear seasonal pattern for the phosphorus removal constant: it is higher in the spring and fall and lower in the summer and winter. Again there is a surprise, because the "common wisdom" is that phosphorus removal should be greatest in the summer and lowest in the winter. This view is conditioned by the premise that biological activity is greatest in the summer and lowest in the winter.

A plausible explanation is found when the life style of the vegetation is taken into account. The Listowel wetlands were cattail monocultures (*Typha latifolia*). In a northern environment, *Typha* undergoes maximum growth in the spring. In the fall, it enters a period in which it translocates large amounts of nutrients and biomass to its rhizomes. Thus, the high values of spring and fall are plausible. Summer is indeed a period of high biological activity in the wetland. However, the rate constant is a parameter in a lumped model, which describes a net transfer, which is the difference between uptake processes and release processes to and from the active compartments in the ecosystem. Thus, although plant growth and microbial uptake may be operating in high gear, so are leaching and decomposition processes. High returns of phosphorus from the static compartments counterbalance high rates of uptake. Winter temperatures slow both uptake and return transfers, but the net transfer clearly remains high.

Caution must be exercised; these trends are not necessarily duplicated by other species or in other climates.

No seasonal trends are apparent in the data for forested wetlands. For instance, the Central Slough wetland in South Carolina shows 5-year seasonal averages that vary, but only a little: $k = 2.30 \pm 0.23$ m/yr.

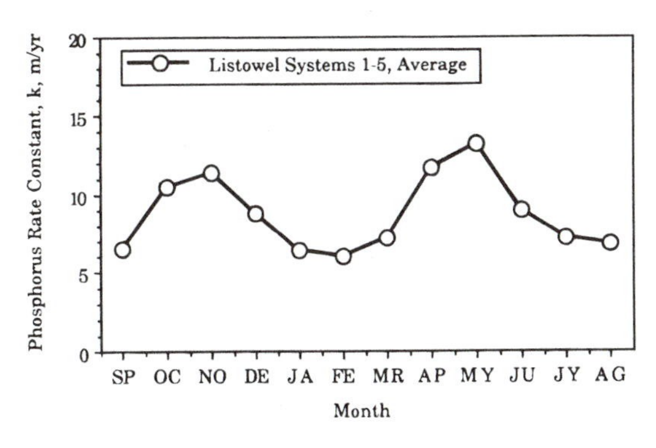

Figure 14-22 There is two-humped seasonal effect to be seen in the data from these five Canadian constructed wetlands. Data span a 4-year period.

Very little seasonal information is available for SSF wetlands. It would be preferrable to have at least 4 years of data past the adaptation period to look for seasonal trends. Tanner (1992) reported performance similar to Listowel for *Schoenoplectus* (*Scirpus*) in gravel, beginning after 1 year of adaptation. He found maximal phosphorus uptake in January (mid-summer in New Zealand) and June (mid-winter).

ADAPTATION TRENDS

The "startup" period for a wetland can extend over varying periods of time, ranging from 1 to 5 years for phosphorus removal. During this startup period, the mass balance model must include the storage of phosphorus on sorption sites and in expanded amounts of biomass. The terms on the left side of chapter 9, Equation 9-141 may not be neglected. Of course, it is still possible to execute the calculation of a rate constant from chapter 9, Equation 9-143, but it will include uptake into, or delivery back from, temporary storages. For phosphorus, the simplified version of the mass balance is

$$\left\{\frac{1}{t_m}\sum_{i=2}^{N-1}\Delta(m_iX_i) + \frac{\Delta(hC)}{t_m}\right\} + J = -q\frac{dC}{dy} = J_U = k_UC \qquad (14\text{-}27)$$

If the term in brackets on the left is negative, then phosphorus is being stripped from static compartments, either from biomass or active soil, leading to decreased uptake. If it is positive, there is extra uptake from water into sorption or expanded biomass. Extra storage in biomass yields a higher value of the uptake rate constant. After the adaptation period passes, the long-term average is attained. An example of this startup period for the Houghton Lake wetland treatment system is shown in Figure 14-23. Initial k values for this natural wetland were very much higher than the long-term average.

Other patterns of startup include the rapid vegetation of a bare soil wetland after an initial planting. If that planting is sparse, the startup period will be characterized by the time for plant fill in plus the time for litter development. That can be relatively brief, especially in warm climates. At Iron Bridge, FL, the startup duration was approximately 24 months,

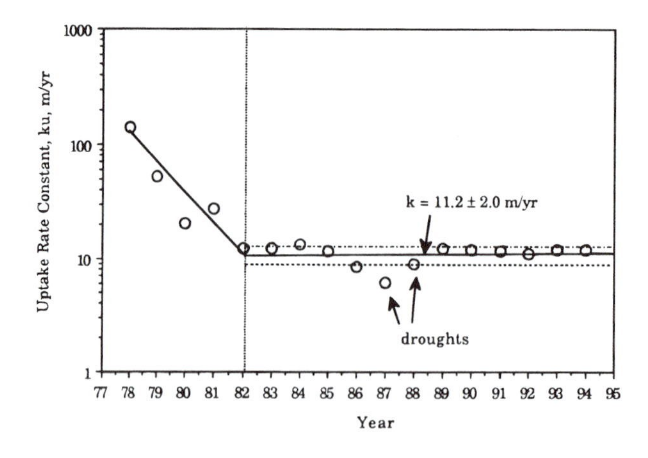

Figure 14-23 Phosphorus removal rate constants over the history of the Houghton Lake Treatment Wetlands. Each point represents the annual average for that year, as determined from transect data and confirmed by input-output data.

during which considerable change in the ecosystem took place. Figure 14-24 shows two of the principal variables: vegetation density and phosphorus rate constant. Note the close parallel between the amount of vegetation and the rate constant during the first 2 years. After that period, the vegetation density levels off, and the rate constant drops to its long-term average value of 13.5 m/yr. The phosphorus requirement for building the new standing crop of biomass leads to a peak uptake rate constant that is roughly double the long-term average value.

Forested wetlands also undergo a period of adaptation. Kadlec (1983) showed that the Bellaire, MI forested wetland provided reasonably good phosphorus removal, but the k value decreased from 5.45 m/yr to zero over a period of 6 years.

SSF wetlands are no exception. Wolstenholme and Bayes (1990) showed a large decrease in the phosphorus k value for the reed beds at Valleyfield, Fife, Scotland. The vegetation reached full density by the end of 2 years. The rate constant calculated from their data started at 60 m/yr and decreased to 13 m/yr over a 3-year period with no evidence of leveling out. Since no North American SSF wetland has reported data for more than 3 years, there is a strong chance that all reported U.S. SSF data represent the startup period. The period of startup adaptation for an SSF system has also been observed to exceed 2 years in Australia (Mann, 1990). However, ten SSF soil-based wetland systems in Denmark showed no adaptation period for phosphorus uptake (Schierup et al., 1990b).

MORE DETAILED MODELS

The global uptake model advocated in the preceeding sections may seem overly simplistic, especially in view of the considerable knowledge about the many component processes and transfers. Several investigators have attempted to develop and calibrate more sophisticated models which involve many ecosystem compartments and transfers (see, for instance, Kadlec and Hammer, 1988, or Mitsch et al., 1993). All such models in the literature to date are *deterministic* models: they contain no stochastic components. These models are difficult to calibrate because of the large number of adjustable parameters—more than 100 for some of the more complicated models (Kadlec, 1990b). Measurements of the biomass and chemical composition of all wetland compartments would ideally be desirable, as indicated in Figures 14-1 and 14-5. This formidable task has been attempted on only a few occasions in a research environment.

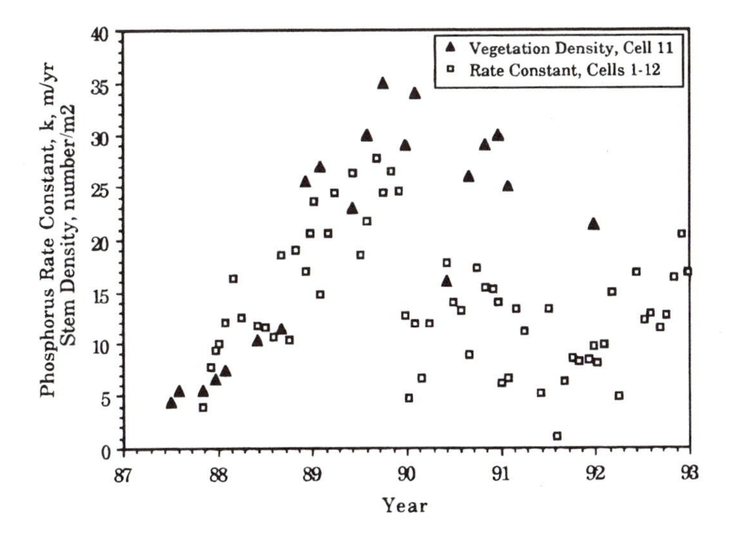

Figure 14-24 Vegetation density and phosphorus rate constant at Iron Bridge FL, during startup.

Detailed models can of course adequately represent the long-term behavior of the wetland if properly calibrated, but they typically are not good over short intervals. As an illustration, the detailed model of Mitsch et al. (1993) is briefly described. Five compartments were considered: water, standing dead macrophytes, detritus, and active and deep sediments (Figure 14-25). Twenty constants were used, eleven calibrated and nine estimated from literature sources. Primary calibration was to correctly represent the total phosphorus removed by the wetland over the course of 1 year (Figure 14-26). For this level of detail, one would expect a reasonable representation of the week-to-week behavior, but that is not the case (Figure 14-27). A possible explanation for the discrepancies between model and data is the inability of a deterministic model to deal with short-term random events in the wetland. A wide variety of stochastic events can temporarily upset the phosphorus mass balance: storms, algal blooms, animal activity, and insect herbivory, for example.

Data from the great majority of treatment wetlands are not sufficiently detailed to support multi-parameter, multi-compartment modeling. The results from the few projects that have

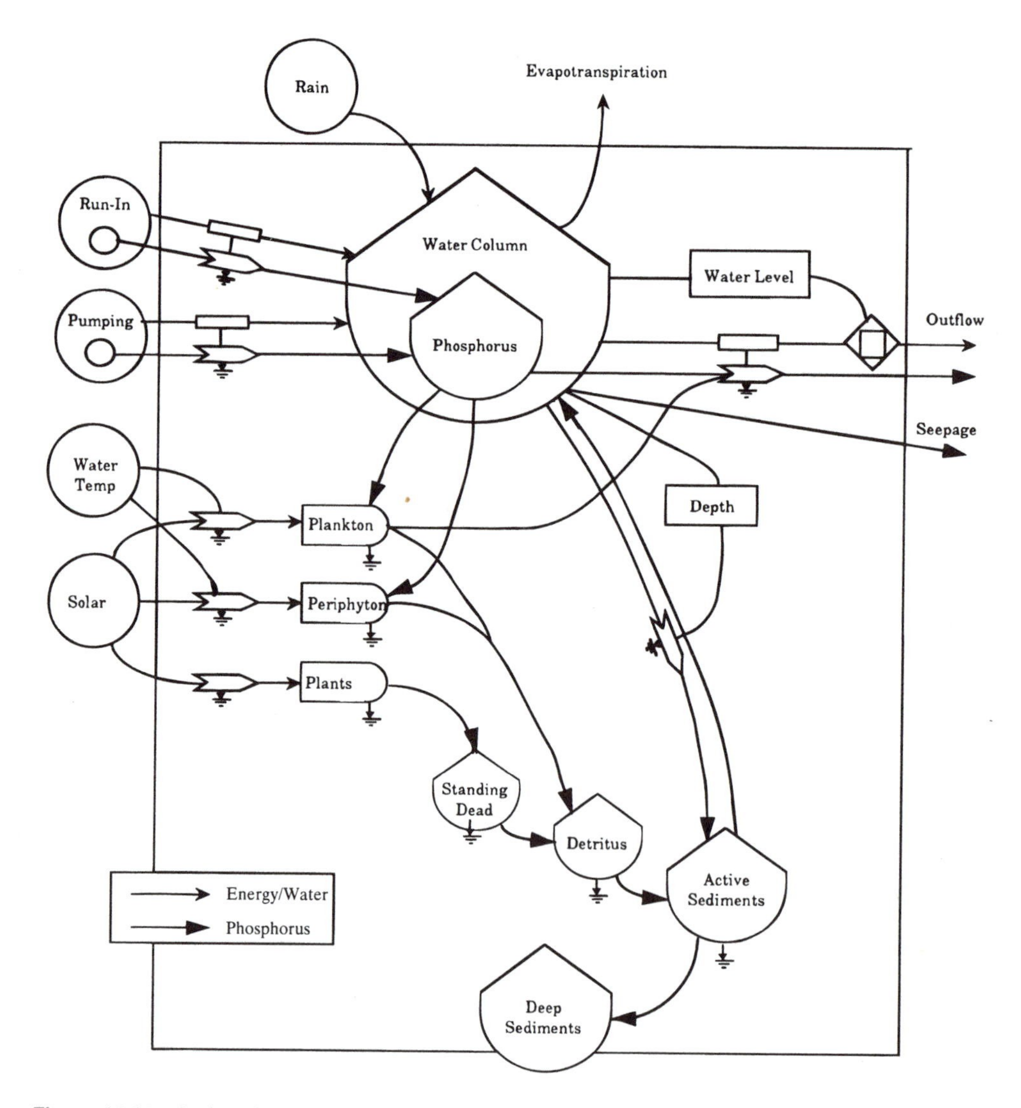

Figure 14-25 A phosphorus model for the Des Plaines River wetlands. (Adapted from Mitsch et al., 1993.) (See Figure 7-15 for an explanation of symbols.)

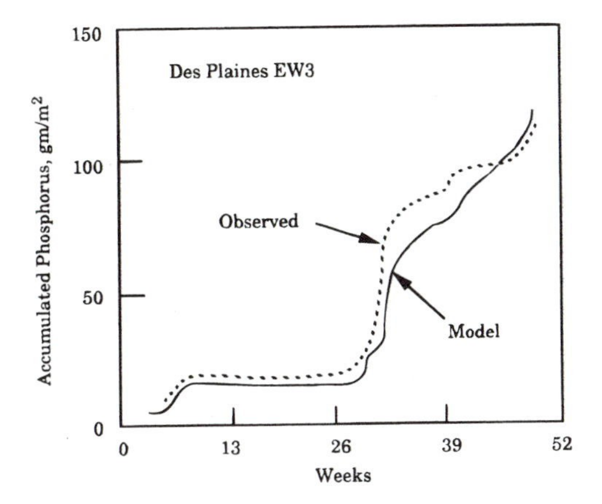

Figure 14-26 Calibration of a detailed phosphorus model to data from Des Plaines wetland EW3. The model is forced to removal of the correct amount of phosphorus over the year. Intermediate storage is reasonably well described. (Adapted from Mitsch et al., 1993.)

developed such models do not provide improved design tools. For these reasons, the global uptake model described in the phosphorus sections represents the appropriate level of detail for the available data.

STOCHASTIC EFFECTS

Field Results

Probabilistic effects are important in the utilization of design models for predicting phosphorus removal performance of treatment wetlands. Regulatory requirements vary from location to location, but often there is a standard other than the long-term average performance described by the model in this chapter. There may be a maximum monthly concentration not to be exceeded or a specified concentration not be exceeded for more than a certain percentage of samples. To meet this type of regulatory requirement, the designer needs information on the variability of wetland outlet concentrations.

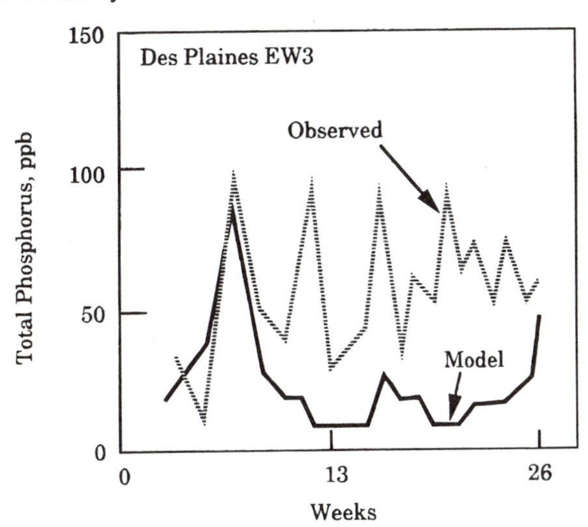

Figure 14-27 Model and data comparison for a detailed phosphorus model of Des Plaines wetland EW3. Weekly effluent concentrations are not well described. (Adapted from Mitsch et al., 1993.)

The data from Listowel, Ontario provides an idea of the variability to be expected with respect to the mean long-term performance for phosphorus removal. These two wetlands received an average inlet concentration of 3.15 mg/L over 4 years, at an average hydraulic loading rate of 2.3 cm/d (nominal detention time = 11.4 days), and produced a mean outlet concentration of 0.80 mg/L. The corresponding plug flow k value was 11.4 m/yr. Figure 14-28 shows a considerable probability band around the mean performance.

The maximum monthly value was 3.9 mg/L at the outlet, five times the average outlet concentration. This maximum did not occur in response to the maximum inlet concentration of 6.8 mg/L, which occurred 11 months earlier. A search on the North American Database reveals that the maximum monthly effluent total phosphorus is on average 1.8 times greater than the annual average effluent total phosphorus for SF wetlands.

Some European countries require performance standards to be met four times out of five for samples at monthly frequency. The 80th percentile for Listowel Systems 4 and 5 was an effluent concentration of 1.14 mg/L, which is only 43 percent greater than the long-term average outlet concentration.

Stochastic-Deterministic Modeling

The mass balance model used for design for long-term average performance may be used to forecast the expected probability band in that performance. Each of the major forcing variables in the model—hydraulic loading rate, inlet concentration, rain, and evapotranspiration—has an associated probability distribution. The rate constant k also may be presumed to vary stochastically in response to effects noted previously. Those random effects may be propagated through the mass balance model to yield the probability distribution of outflow concentrations.

As an illustration, the tanks-in-series (TIS) model and the plug flow model may be applied to historical data available for Water Conservation Area 2A (WCA 2A) in the Florida Everglades. The marsh background concentration C^* is taken to be zero. The models will be applied here on an annual average basis, so that seasonal effects are combined and changes in inventory in the water are minimal. The plug flow model equation is Chapter 9, Equation 9-176. Based on the ideas in Chapter 9, a three-tank model may also be developed. The steady state mass balances for water and phosphorus for Tank 1 are

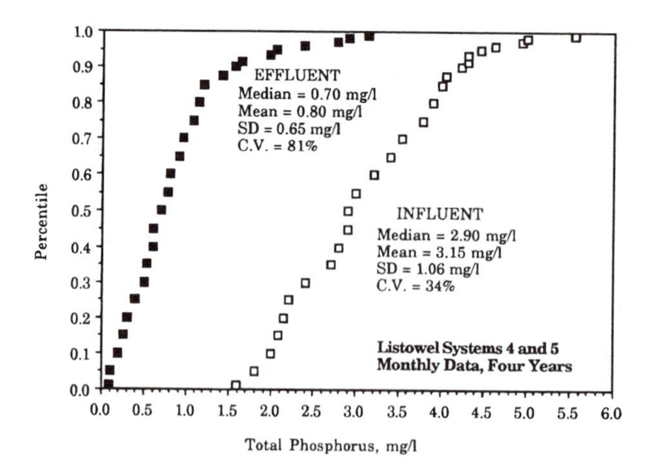

Figure 14-28 Frequency distribution of influent and effluent total phosphorus concentration for two emergent marsh wetlands at Listowel, Ontario.

$$Q_1 = Q_i + PA_1 - ETA_1 \tag{14-28}$$

$$Q_1C_1 = Q_iC_i + PA_1C_p - kA_1C_1 \tag{14-29}$$

These may be solved for the concentration leaving the first tank:

$$C_1 = \frac{Q_iC_i + PA_1C_p}{Q_i + PA_1 - ETA_1 + kA_1} = \frac{q_{1i}C_i + PC_p}{q_{1i} + (P - ET) + k} \tag{14-30}$$

This equation is applied three times in succession to calculate the exit total phosphorus concentration from the last tank.

All the variables on the right sides of Equation 14-30 and Chapter 9, Equation 9-176 have associated probability distributions. The distributions of hydraulic loading rates, rainfall, rainfall phosphorus, and evapotranspiration are all available from data over a 10-year period of record for WCA 2A (Figure 14-29). The values of k were determined to center on a mean value of 10 m/yr, with a standard error of 2 m/yr (Walker, 1992b). In this illustration, k is assumed to be normally distributed with mean = 10 and SD = 2. Rainfall phosphorus was measured and may be represented approximately by a normal distribution with mean = 30 μg/L and SD = 10 μg/L (Figure 14-29).

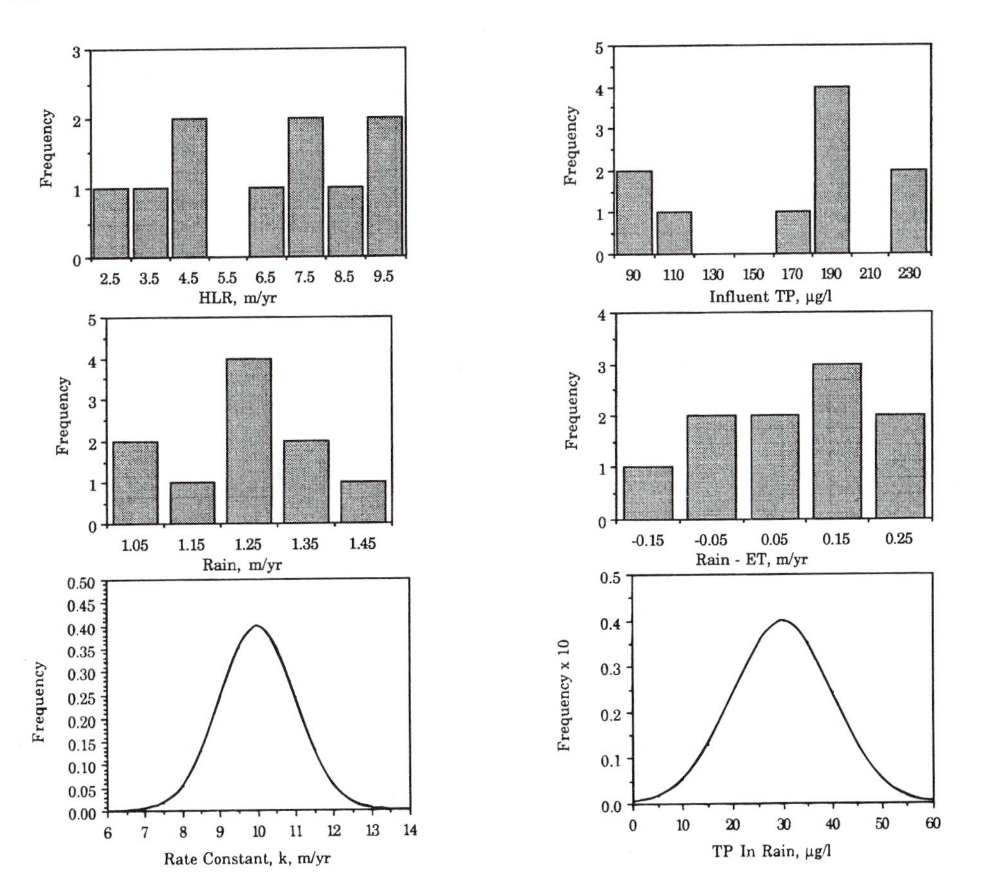

Figure 14-29 Probability distributions used to drive the TIS model for WCA 2A.

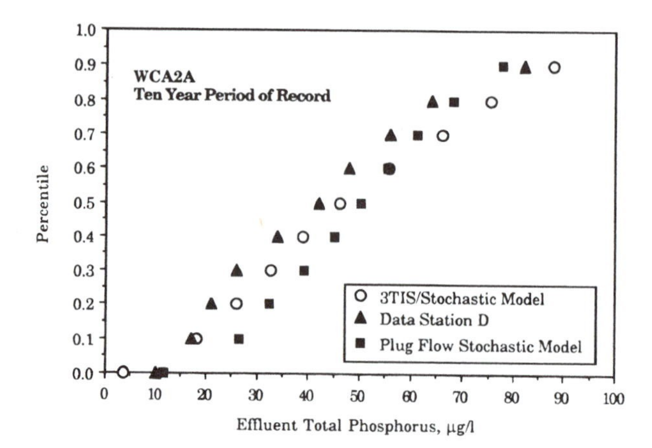

Figure 14-30 Model and data comparison for total phosphorus in WCA 2A. The models are the three TIS flow pattern and the plug flow pattern driven by the probability distributions for wetland variables. The mean total phosphorus over 10 years was measured to be 53 μg/L; the model values were 50 μg/L (three TIS) and 51 μg/L (plug flow).

Calculations using the mass balance models of Equation 14-30 and Chapter 9, Equation 9-176 were repeated 10,000 times for random selections from the probability distributions.* The resulting distribution of exit total phosphorus concentrations is a reasonable estimate of the probability distribution in the measured data (Figure 14-30).

This illustration suggests, but does not confirm, that incorporation of stochastic effects into design models can lead to better definition of the expected wetland performance.

SUMMARY

The first-order areal model has been established as a tool for incorporating water and phosphorus mass balances into design and for acknowledging site-specific factors. The model in its simplest form describes long-term average performance of the wetland for phosphorus removal. Periods of adaptation of a few years preceed this long-term stationary state. This model, under simplifying assumptions, is represented by

$$\ln\left(\frac{C_o}{C_i}\right) = -\frac{k}{q} \tag{9-154}$$

Emergent marshes and SSF wetlands were shown to be equivalent, and these have intersystem mean k values of about 11.5 m/yr. Forested wetlands are not as efficient and are represented by intersystem mean k values of about 3 m/yr.

Corrections can and should be made to properly account for nonideal flow and for wetlands with infiltration, or significant atmospheric inputs and outputs. Temperature is not a consequential factor, nor is aspect ratio. There is no trend in SF k values with depth, but depth considerations strongly suggest that a volume-specific uptake rate is inappropriate. Season does have a minor effect on k values. Uptake is influenced by vegetation density and by the fraction of wetted area. A very significant factor is a wide band of stochastic phosphorus removal behavior exhibited by all wetlands.

* This task is not onerous if a spreadsheet program such as Excel™ is exercised by a Monte Carlo simulator such as Crystal Ball™ (Decisioneering, Inc., Boulder, CO).

CHAPTER 15

Other Substances

INTRODUCTION

In addition to the pollutants discussed in earlier chapters, wastewaters typically contain many other substances. Some of these elements can cause problems when discharged to receiving waters, and their removal must be considered during design.

These additional materials can be categorized as salts, acids, bases, macronutrients, micronutrients, or heavy metals. These categories sometimes overlap because each has a different level of importance for different forms of life. Salts, acids, and bases include compounds that readily dissociate in water to form charged ions that may or may not be used as nutrients for plant and animal growth. Common examples of salts are sodium chloride (NaCl), gypsum ($CaSO_4$), and fluorite (CaF_2). Acids release a hydrogen ion when they dissociate (for example, hydrochloric acid, HCl), and bases release a hydroxl ion (for example, ferric hydroxide, $FeOH_3$). Specific environmental conditions determine whether the cations (positively charged ions) and anions (negatively charged ions), formed when a salt, acid, or base is dissolved in water, are chemically or biologically active.

Nitrogen and phosphorus, discussed in Chapters 13 and 14, are examples of macronutrients. Other plant growth macronutrients are carbon (Chapters 11 and 12); oxygen and hydrogen (Chapter 10); and potassium, silicon, calcium, iron, magnesium, sodium, sulfur, and chlorides (this chapter). Other biologically active elements (such as molybdenum, used in plants for nitrogen fixation and nitrate reduction) are needed in proportions smaller than macronutrients. These trace elements, called micronutrients, are a relatively minor amount of the dry weight (dw) of plants and animals (molybdenum concentration is about 0.1 to 5 mg/kg dw in plants).

Metals are salt cations. Metals are malleable, lustrous elements that tend to lose electrons and are efficient conductors. A certain metal can be either a required micronutrient or toxic, depending on the concentration. For example, copper and zinc are essential elements for plants and animals at low concentrations, but they are toxic to some organisms at elevated concentrations.

Table 15-1 lists the macronutrients and trace elements that are frequently of concern in treatment of wastewater and stormwater. This list does not include all of the elements found in these waters, nor all those that might require treatment. However, some of the elements listed behave analogously to unlisted elements. The use of a wetland treatment system to modify the concentration of elements depends on how the elements interact with the wetland environment and on the wetland designer's knowledge of design factors that can enhance or diminish these processes. Not much is known about how wetland treatment systems affect specific trace elements. Pilot testing and a thorough, updated review of the scientific literature are recommended for project design. This chapter provides background information on substances commonly found in wastewater discharged to wetlands.

Table 15-1 Elements of Interest in Wetland Treatment

Trace Element	Chemical Notation	Atomic Weight[a] (g/mol)	Average Abundance in Biosphere[b] (mg/kg)	Average Abundance in Fresh water[b] (mg/L)	Average Abundance in Earth's Crust[b] (mg/kg)	Biological Significance
Aluminum	Al	26.98	510	0.24	83,600	Nonessential; low solubility; low toxicity
Antimony	Sb	121.75	—	—	—	Nonessential; moderate toxicity
Arsenic	As	74.92	3.1	0.0004	1.8	Nonessential; high toxicity
Barium	Ba	137.33	310	0.054	390	Essential micronutrient; low toxicity
Beryllium	Be	9.01	—	<0.001	2.0	Essential micronutrient; low toxicity
Boron	B	10.81	110	0.013	9.0	Essential micronutrient; low toxicity
Cadmium	Cd	112.41	—	0.00007	0.049	Nonessential; high toxicity
Calcium	Ca	40.08	51,000	15.0	46,600	Essential macronutrient
Carbon	C	12.01	180,000	11.0	180	Essential macronutrient
Chlorine	Cl	35.45	2,100	7.8	126	Essential macronutrient; conservative; low toxicity
Chromium	Cr	52.00	—	0.0002	122	Essential macronutrient; high toxicity
Cobalt	Co	58.93	2.1	0.0009	29	Essential micronutrient; low toxicity
Copper	Cu	63.55	11	0.002	24	Essential micronutrient; variable toxicity
Fluorine	F	19.00	51	0.090	544	Essential micronutrient; low toxicity
Iodine	I	126.90	—	—	0.106	Essential micronutrient; low toxicity
Iron	Fe	55.85	1,100	0.67	62,200	Essential macronutrient; low to moderate toxicity
Lead	Pb	207.2	5.1	0.005	13	Nonessential; high toxicity
Lithium	Li	6.94	1.1	0.0011	18	Nonessential; low toxicity; conservative
Magnesium	Mg	24.30	4,100	4.1	27,640	Essential micronutrient; low toxicity
Manganese	Mn	54.94	110	0.012	1,060	Essential micronutrient; low toxicity
Mercury	Hg	200.59	—	0.00008	0.086	Nonessential; high toxicity
Molybdenum	Mo	95.94	—	—	0.44	Essential micronutrient; low toxicity
Nickel	Ni	58.69	5	0.01	99	Essential micronutrient; high toxicity
Oxygen	O	16.00	780,000	889,000	456,000	Essential macronutrient
Potassium	K	39.10	31,000	2.3	18,400	Essential macronutrient
Selenium	Se	78.96	—	0.0001	0.017	Essential micronutrient; moderate toxicity
Silicon	Si	28.09	21,000	6.5	273,000	Essential micronutrient; low toxicity
Silver	Ag	107.87	1.7	0.0003	0.1	Nonessential; high toxicity
Sodium	Na	22.99	2,100	6.3	22,700	Essential macronutrient; conservative; low toxicity
Sulfur	S	32.06	5,100	3.7	340	Essential micronutrient
Zinc	Zn	65.38	51	0.01	76	Essential micronutrient; moderate toxicity

[a] One mole is equal to 6.02×10^{23} atoms.
[b] Data from: Fortescue (1980), Brownlow (1979), Stephenson (1987), and Fowler (1983).

INTEGRATIVE PARAMETERS

Five chemical parameters are commonly used to indicate the collective concentrations of chemically similar ions: hardness, specific conductance, salinity, chlorinity, and total dissolved solids. These parameters do not specify the distribution of elements being measured, but they are helpful because they take little time and money to analyze. This section describes how these parameters affect the dissolved ion content of wetlands.

HARDNESS (CALCIUM AND MAGNESIUM)

Hardness measures the concentrations of divalent cations in a water sample. The prevalent divalent ions in most surface waters are calcium and magnesium. Rainwater typically has low hardness (soft water), with a calcium concentration between 0.1 and 10 mg/L, a magnesium concentration of about 0.1 mg/L, and a hardness value less than 30 mg/L as $CaCO_3$. Surface water hardness is variable, depending on the soil and rock concentrations of calcium and magnesium and on the degree of contact with rocks, soils, and pollution. Inland surface water hardness varies from 10 to 300 mg/L as $CaCO_3$, with a calcium concentration between 0.3 and 70 mg/L and magnesium concentration between 0.4 and 40 mg/L.

Calcium is biologically active because it is used as a nutrient by invertebrates and vertebrates and because of its role in the carbonate cycle. Calcium is a principal component of calcium carbonate, which is the basic building block for mollusk shells, coral, corallaceous algae, and bone. During photosynthesis, calcium is removed from surface water along with carbon dioxide. During respiration, the carbonate (and the associated calcium) concentration increases as carbon dioxide in the form of carbonic acid is released. The net effect of these processes is generally zero change in water hardness when photosynthesis is in balance with respiration. However, in some systems, the calcium component of hardness may change.

Because there is generally an excess of calcium in surface water and wastewater, calcium concentration does not change significantly in most wetland treatment systems. Richardson (1989) concluded that wetlands are as likely to be calcium sources as they are to be calcium sinks. Outflow calcium mass discharge rates summarized by Richardson from natural wetlands ranged from 0.0126 to 0.16 kg/ha/d. Mass removal efficiencies for calcium varied from −547 to 56 percent. Wetland litter decomposition studies by Davis and van der Valk (1978) and by Chamie and Richardson (1978) concluded that calcium is one of the least mobile elements in wetland plants.

Magnesium is an essential micronutrient because of its role in phosphate energy transfer and because it is a structural component in the chlorophyll molecule. Because magnesium concentration of surface water almost always exceeds the requirements for plant or animal growth, elevated magnesium concentrations are not affected when wastewater travels through wetland treatment systems. In fact, Richardson (1989) found that most natural wetlands act as magnesium sources rather than sinks, with a removal efficiency range of −300 to 36 percent in five systems. Chamie and Richardson (1978) found that magnesium is relatively mobile in wetlands and leaches quickly from dead vegetation.

SPECIFIC CONDUCTANCE

The specific conductance, also called electrical conductivity, of an aqueous solution is the reciprocal of the resistance between two platinum electrodes 1 cm apart and with a surface area of 1 cm^2. The reciprocal of resistance is equal to electrical conductivity and is a function of the total quantity of ionized materials in a surface water sample. Specific conductance usually is reported at a temperature of 25°C and in units of micromhos per centimeter (μmho/cm). Specific conductance is proportional to the total dissolved solids or salinity in many surface waters and is a convenient measure of the salt content of wastewaters. In the SI system of units, the reciprocal ohm (mho) is the siemens (S), and conductivity is reported in mS/m or dS/m. One decisiemens per meter equals one millimho per centimeter.

Total ionic salts in wetlands, as measured by specific conductance, are somewhat altered by biological and physical environmental conditions in wetlands. Therefore, specific conductance is a relatively inaccurate indicator of dilution and concentration effects by rainfall and runoff and evapotranspiration in wetland treatment systems. For instance, Bavor et al. (1988) found 10 to 20 percent loss in conductivity in both surface-flow (SF) and subsurface-flow (SSF) systems in both wet and dry seasons.

The specific conductance of most natural, inland surface waters is between 10 and 300 μmho/cm. Specific conductance in depressional salt lakes and in salt pans can reach levels over 60,000 μmho/cm.

SALINITY

Salinity is a measurement of the mass of total dissolved ions in a water sample. It is used most often in reference to estuarine and marine environments including salt marshes, mangrove wetlands, and seagrass beds. Salinity is equal to the grams of salts dissolved in 1 kg of water and is generally reported in parts per thousand (ppt). In seawater, commonly measured salts usually occur at the following levels: chloride (55.2 percent), sodium (30.4 percent), sulfate (7.7 percent), magnesium (3.7 percent), calcium (1.16 percent), potassium (1.1 percent), and carbonate (0.35 percent). Strontium, bromide, and boric acid also are measured as salts (Reid and Wood, 1976).

In seawater, chlorinity is the named measure of chloride (ionized chlorine) and is related to salinity by (Reid and Wood, 1976)

$$\text{Salinity (ppt)} = 0.030 + 1.805 \text{ Chlorinity (ppt)} \tag{15-1}$$

This equation does not hold for natural or treatment wetlands.

TOTAL DISSOLVED SOLIDS

Total dissolved solids (TDS) are used to quantify the degree of pollution in many industrial wastewater effluents, including textile wastes, food processing wastes, and pulp and paper wastes. When discharged to surface or groundwaters, these dissolved solids may represent a significant pollution source.

The total quantity of dissolved solids in a water sample can be measured by filtration followed by sample evaporation. This quantity contains both inorganic ions and organic compounds. The more volatile organic components can be removed by ignition in a furnace to determine material lost on ignition and the total dissolved inorganic solids. Total dissolved solids are not as conservative in wetlands as specific conductance, salinity, and chlorinity. However, because TDS concentrations are high in many wastewaters and the individual components of these solids greatly exceed the biological requirements for growth, wetlands generally have a negligible effect on this parameter.

CHLORIDE

Chlorine is an essential element for biological systems because of the chloride ion's role in photosynthesis, adenosine triphosphate production, and phosphorylation reactions. Free chlorine, although toxic to most life forms, is one of the most frequently used wastewater disinfectants. Only about half of the free chlorine added during disinfection remains in solution as chlorides or chloramines. The rest is lost to the atmosphere as chlorine gas in a few hours. Because of low biological demand for chlorine, its abundance in surface water, and its high solubility, the total mass of chlorine is relatively constant between the inflows and outflows and storages of a wetland.

Chloride content can be used as a tracer in wetland systems. Table 15-2 provides a sample calculation for estimating flow rate based on the effect of chloride changes in surface water quality in a natural wetland treatment system. Chloride can also be used to determine the extent of wastewater infiltration to the groundwater beneath and adjacent to a wetland treatment system (Knight and Ferda, 1989; CH2M HILL, 1992). Figure 15-1 shows the shallow groundwater chloride concentrations beneath a natural wetland treatment system after 5 years of discharge. The direction of groundwater movement and the effects of dilution and concentration can be seen readily when wastewater chloride levels are elevated compared to background concentrations.

OTHER MACRONUTRIENTS

Sodium, potassium, sulfur, and silicon are essential for life and sometimes are required in large quantities. These four elements occasionally have elevated concentrations in wastewaters, but wetland systems are not commonly used for their control. A brief description of these macronutrients is provided in this section.

Table 15-2 Estimation of Net Dilution by Measurement of Chloride

Problem: Estimate outflow rate when wastewater inflow rate and sample chlorides are known (assumes no water or chloride storages in the wetland system).

Q_2, C_2 Net Rainfall/Runoff
↓
Q_1, C_1 Wastewater Inflow → Wetland → Q_3, C_3 combined Outflow

where Q_1	= wastewater inflow rate	=	3,786 m^3/d
Q_2	= net rainfall/runoff to wetland	=	unknown
Q_3	= surface outflow from wetland	=	unknown
C_1	= wastewater chloride	=	150 mg/L
C_2	= rainfall/runoff chloride	=	15 mg/L
C_3	= outflow chloride	=	60 mg/L

Model: Use mass conservation for chlorides and for water (assumes no groundwater gains/losses).

Chlorides:

$$Q_1C_1 + Q_2C_2 = Q_3C_3$$

Water:

$$Q_1 + Q_2 = Q_3$$

Then:

$$Q_1C_1 + Q_2C_2 = (Q_1 + Q_2)C_3 = Q_1C_3 + Q_2C_3$$

Rearrange:

$$Q_2 = \frac{Q_1(C_1 - C_3)}{(C_3 - C_2)} = 7{,}572 \text{ m}^3/\text{d}$$

and

$$Q_3 = Q_1 + Q_2 = 11{,}358 \text{ m}^3/\text{d}$$

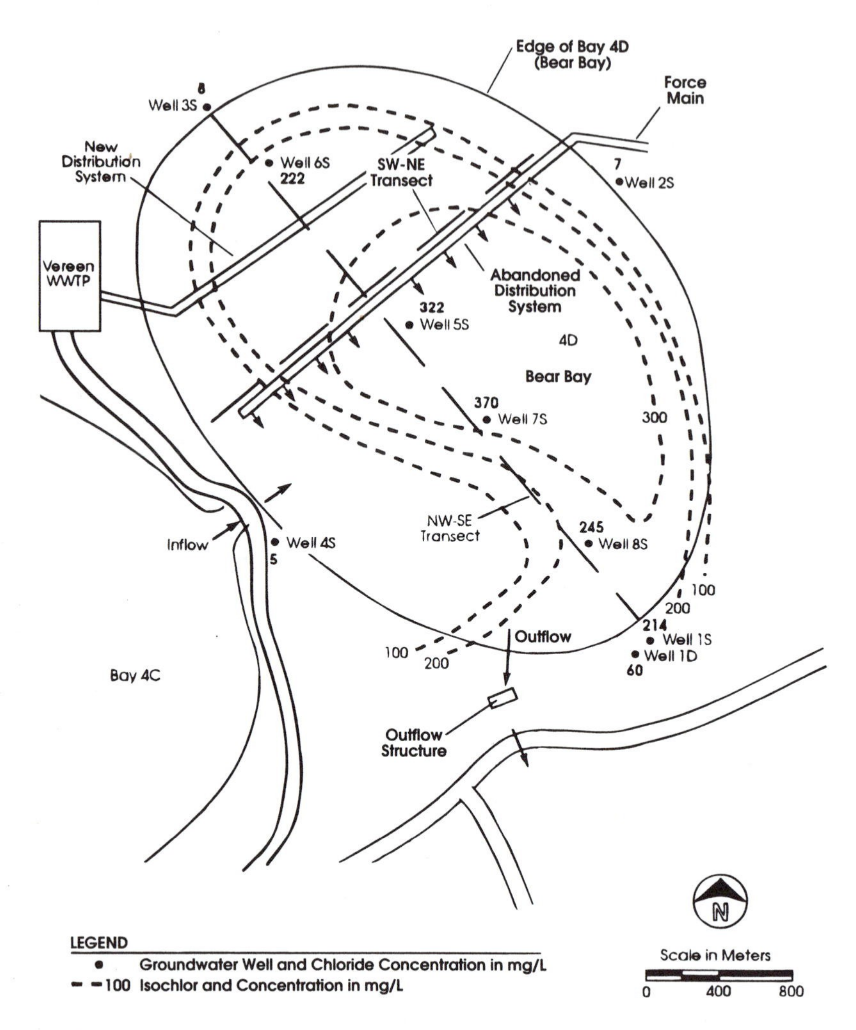

Figure 15-1 Chloride concentrations in the shallow groundwater in the vicinity of Bear Bay, SC in December 1991 after 5 years of treated discharge. The average effluent chloride concentration was 242 mg/L. (Data from CH2M HILL, 1992.)

SODIUM AND POTASSIUM

Sodium and potassium are important in plant and animal physiology and are functionally analogous in some of their properties. Sodium is usually more important for the growth of marine organisms than for freshwater species, and some freshwater organisms can grow in the complete absence of sodium (Brock, 1974). Sodium ions help to regulate osmotic pressure in cells and therefore affect the diffusion of all essential growth nutrients between the external environment and the protoplasm of the living cells. A "sodium pump" fueled by the conversion of energy-bearing adenosine triphosphate maintains internal cell sodium concentrations at optimal levels.

Similar ionic pumps maintain potassium levels in the cells of plant roots at concentrations of 250:1 compared to the concentration of potassium dissolved in the soil water. Potassium regulates the opening and closing of stomates on plant leaves. Stomates are the functional "mouths" of a plant that allow gases inside the plant to be exchanged with the atmosphere. Potassium also is used as an enzyme activator in protein synthesis in most cells. Potassium typically comprises about 2.6 percent of the dry weight of wetland plants.

Because most freshwater wetland species have low sodium requirements, the dissolved sodium content of wastewater passing through wetlands changes little. Thus, sodium concentrations can be used as a conservative tracer for calculating dilution and concentration and for tracking groundwater discharges from wetlands.

Potassium concentrations in surface waters are typically between 0.2 and 33 mg/L (Goldman and Horne, 1983), with an average world river concentration of about 3.4 mg/L (Hutchinson, 1975). Richardson (1989) considers sodium and potassium to be highly mobile in wetlands because they are more exchangeable from wetland soils than nitrogen and phosphorus. Richardson (1989) observed that about 60 percent of the available potassium in swamps, marshes, and fens is associated with peat and detrital litter, 25 to 40 percent is in the plants, and less than 10 percent is in the water. Davis and van der Valk (1978) and Chamie and Richardson (1978) both concluded that potassium is highly leachable from wetland plant litter in herbaceous and forested wetlands.

From his review of the nutrient budgets of a variety of wetland ecosystems, Richardson (1989) concluded that natural wetlands are frequently net exporters of sodium and potassium and that long-term, historic geologic sources (soil weathering) were the suspected sources of these elements. Table 15-3 summarizes some of the mass balance studies for sodium and potassium in wetlands. In four natural wetlands, the sodium mass loading rates ranged from 0.012 to 0.07 kg/ha/d; with removal efficiencies from −78 to 43 percent. In a constructed wetland treatment system, the sodium loading was two orders of magnitude higher at 3.32 kg/ha/d with a 9.1 percent removal efficiency. Mass loadings of potassium to most natural wetlands are lower than sodium (0.003 to 0.03 kg/ha/d), but mass removal efficiencies are also highly variable (−255 to 45 percent). Mass removal efficiencies were positive for two constructed treatment wetlands at much higher loading rates (0.66 to 10.13 kg/ha/d).

Although the mobility of sodium and potassium in wetlands could be temporarily reversed by increasing their mass loading, initial removals caused by the changed equilibria between water and tissue and soil concentrations would decline rapidly after new sediment and tissue concentrations were established.

SULFUR

Sulfur occurs in surface waters in two forms: as sulfate (SO_4) in aerobic waters and as hydrogen sulfide (H_2S) in anaerobic waters. Both forms of sulfur are present in wetlands because of the range of oxidation states found in these systems. Natural surface waters receive sulfur from rainfall—(about 1 to 2 mg/L as sulfate (Hutchinson, 1975) and from weathering of sedimentary rocks such as dolomite and pyrite. Because many sulfur-containing compounds have low solubilities, the sulfate concentration of natural surface waters in open

Table 15-3 Example Sodium and Potassium Nutrient Budgets from a Variety of Natural and Constructed Wetlands

	Sodium			Potassium		
	Mass (kg/ha/d)		Removal Eff (%)	Mass (kg/ha/d)		Removal Eff (%)
Wetland Description	In	Out		In	Out	
Spruce bog, Minnesota	0.012	0.0066	43	0.030	0.017	45
Sphagnum bog, England	0.070	0.125	−78	0.0085	0.030	−255
Maple-gum swamp, North Carolina	0.025	0.023	7.8	0.0030	0.0071	−136
Pocosin, North Carolina	0.051	0.081	−58	0.0052	0.0066	−26
Constructed meadow/marsh/pond, Brookhaven, NY	3.32	3.02	9.1	0.66	0.41	37.5
Subsurface-flow reed bed receiving landfill leachate, New York	—	—	—	10.13	8.33	17.7

Note: Data from Richardson, 1989; Hendrey et al., 1979; and Surface et al., 1993.

basins is generally low. Hutchinson (1975) cites a mean river sulfate concentration of 16 mg/L, and Goldman and Horne (1983) list surface water values between 0.2 and 36 mg/L in lakes and rivers. Natural wetlands typically have sulfate concentrations in this same range. Industrialization has increased the concentration of sulfur dioxide (SO_2) in the atmosphere, which can convert to sulfuric acid (H_2SO_4), increasing rainfall sulfur concentrations and acidifying surface waters.

The sulfur cycle in wetlands, shown in Figure 15-2, is characterized as an interconnected series of oxidation-reduction reactions and biological cycling mechanisms. Sulfate is an essential nutrient because its reduced, sulfhydryl (−SH) form is used in the formation of amino acids. Because there is usually enough sulfate in surface waters to meet the sulfur requirement, sulfate rarely limits overall productivity in wetland systems.

Aerobic organisms excrete sulfur as sulfate. However, upon death and sedimentation, heterotrophic bacteria release the sulfur in detritus in the reduced state, which can result in the accumulation of high levels of hydrogen sulfide in wetland sediments. A second process that transforms sulfate and other oxidized sulfur forms (sulfite, thiosulfate, and elemental sulfur) to hydrogen sulfide in anaerobic sediments is sulfate reduction, mediated by anaerobic, heterotrophic bacteria such as *Desulfovibrio desulphuricans,* which use sulfate as a hydrogen acceptor. Since ferrous sulfide (FeS) is highly insoluble, hydrogen sulfide does not tend to accumulate until the reduced iron is removed from solution. When iron concentrations are low or when sulfate and organic matter concentrations are high, significant hydrogen sulfide concentrations can occur. Several other metal sulfides are also very insoluble, including ZnS, CdS, and others (see Table 15-4). Hydrogen sulfide is a reactive and toxic gas with problematic side effects including a rotten egg odor, corrosion, and acute toxicity.

When it is exposed to air or oxygenated water, hydrogen sulfide may be spontaneously oxidized back to sulfate or may be used sequentially as an energy source by sulfur bacteria such as *Beggiatoa* (oxidation of hydrogen sulfide to elemental sulfur) and *Thiobacillus* (oxidation of elemental sulfur to sulfate). Photosynthetic bacteria, such as purple sulfur bacteria, use hydrogen sulfide as an oxygen acceptor in the reduction of carbon dioxide, resulting in partial or complete oxidation back to sulfate.

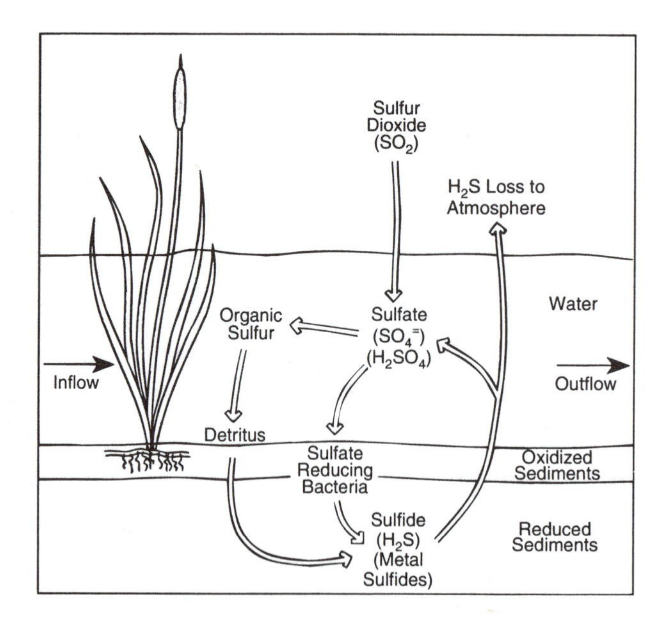

Figure 15-2 Typical wetland sulfur cycle.

Table 15-4 Summary of Certain Aspects of Metal Chemistry Important in Wetland Treatment Systems

Metal	Common Valence States In Natural Water[a]	Common Forms In Surface Water[a]	Insoluble Compounds[a]	Partition Coefficients (L/kg)		
				Soil/Water[b] (L/kg)	Plant/Soil[c]	Fish/Water[d] (L/kg)
Arsenic	As(III), As(V)	Inorganic arsenate and arsenite	Sulfides	1.0–8.3	—	44
Cadmium	Cd(II)	Ionic hydrate, complexes with humics, carbonate, etc.	Carbonate, sulfide, hydroxide	1.26–26.8	1–10	64
Chromium	Cr(III), Cr(VI)		Hydroxide, chloride	1.2–1,800	0.01–01	16
Copper	Cu(II)	Complexes with carbonate, hydroxide, and organics	Carbonate, sulfide, and hydroxide	1.4–333	0.1–1.0	36
Iron	Fe(II), Fe(III)	Hydroxide, organic complexes, sulfides	None			
Lead	Pb(II)	Ionic hydrate, adsorbed on particulates	Sulfides, carbonates	4.5–7,640	0.01–0.1	49
Manganese	Mn(II)	Ionic hydrate adsorbed on particulates	Hydroxide, sulfide	0.2–10,000	—	—
Mercury	Hg(O), Hg(I), Hg(II)	Ionic hydrate, complexed with humics; elemental	Sulfide carbonate	—	0.01–0.1	5,500
Nickel	Ni(II)	Ionic hydrate	Sulfide, hydroxide			47
Selenium	Se(-II), Se(O), Se (IV), Se VI	Ferric selenite; calcium selenate; elemental	—	—	—	6(566–4,000[e,f])
Silver	Ag(I)		Sulfide	10–1,000	—	0.5
Zinc	Zn(II)	Ionic hydrate, carbonate, organic complexes	Sulfide, hydroxide	10–8,000	1–10	47

[a]Rudd (1987).
[b]Dragun (1988).
[c]Bolt et al., 1991.
[d]U.S. EPA (1993a).
[e]Schuler et al., 1990.
[f]Ohlendorf et al., 1986.

Wetlands can function as sulfur sinks through their internal production and release of hydrogen sulfide as a gas, release of elemental sulfur or methyl sulfide gas, precipitation of elemental sulfur, and precipitation and burial of insoluble metallic sulfides. Adams et al. (1981a) measured hydrogen sulfide release from a South Carolina salt marsh as 0.0108 kg/ha/d. Winter and Kickuth (1989a, 1989b) reported that a root-zone, soil-based treatment system receiving textile wastewaters from a facility in Bielefeld, Germany removed from 80 to 85 percent of the sulfur mass at a hydraulic loading rate of 1.14 cm/d for a removal rate of 9.6 kg/ha/d. These authors reported that the majority of this sulfur was largely stored in the wetland soil as elemental sulfur (31 percent) and organic sulfur (25 percent) and that only a small fraction was released by volatilization to the atmosphere or taken up by plants (1 percent).

Sulfate inputs to SF wetlands are frequently lower when input rates are low and are derived primarily from rainfall and runoff (Bayley et al., 1986). Bayley et al. (1986) measured an annual average net retention of 0.017 kg/ha/d for an average removal efficiency of 51 percent in a natural black spruce (*Picea mariana*) and sphagnum fen in Ontario, Canada. The portion of this sulfate stored in the organic form was quickly released on a seasonal basis during dry summer conditions. Since sulfate inputs in SF wetland treatment systems frequently exceed the biological requirements of wetland biota, wetlands generally are not effective for removal of sulfur (Wieder, 1989).

SILICON

Silicon is abundant in the earth's crust and in many soils. The average concentration of silica (SiO_2) in rivers is about 13 mg/L with lake concentrations ranging from 0.5 to 60 mg/L (Goldman and Horne, 1983). Most animals and plants, except for sponges and diatoms, which use silica as a critical structural component, have only a minor need for silicon. Between 25 and 60 percent of the dry weight of diatoms is silica (Goldman and Horne, 1983). Diatoms are extremely productive components of many aquatic and estuarine environments where they may significantly alter silica concentrations. However, diatoms are a relatively insignificant component of the normal algal productivity of wetland ecosystems, and usually silica concentrations are affected minimally by passage of wastewater through wetlands.

TRACE METALS

GENERAL INFORMATION

A number of metals are essential micronutrients at trace concentrations, but some of these metals occur in wastewaters at concentrations that are toxic to sensitive organisms. For a few metals, biochemical transformations and chemical characteristics can lead to "biomagnification," a phenomenon in which increasing concentrations occur in consumers along a food chain. Biomagnification can have devastating effects at top consumer levels, including humans. However, although most metals are more concentrated in biological tissues and soils than they are in surface water, biomagnification does not always occur.

Metals in wastewater must be removed prior to final discharge to protect the environment from toxic effects, but the use of wetlands to accomplish this goal must be examined cautiously. Surface flow wetland treatment systems are usually open to allow free movement of biota between the treatment wetland and adjacent environments. Thus, organisms exposed to potentially dangerous levels of metals in wetland treatment systems may move offsite and contribute to the contamination of natural areas or become a part of the human food chain.

To prevent this problem from occurring, wetland treatment system owners, designers, operators, and regulators should consider reducing influent metal concentrations to noncritical levels by pretreatment before discharging to a wetland treatment system. A second alternative is to minimize the opportunity for ingestion of metals. Subsurface flow wetlands and dense monotypic stands of macrophytes in SF wetlands partially accomplish this purpose.

This section presents general information on the behavior of a variety of metals in wetlands and provides some published information concerning recommended "safe" threshold metal concentrations in various system types. None of these threshold concentrations applies in every instance; the concentrations may be overly conservative in some cases and may not be protective enough in others. In all cases, problems can be avoided by minimizing waste and reducing metals at the original point of use, followed by industrial pretreatment to reduce metal inflow concentrations before discharging to a wetland treatment system. Local, state, and national ordinances concerning allowable concentrations of metals should be consulted whenever they are available.

Metals that have been shown to be essential for the growth of animals are cobalt, chromium, copper, iron, manganese, molybdenum, nickel, selenium, tin, vanadium, and zinc. Metals that have been found to be essential for the growth of plants are boron, copper, iron, manganese, molybdenum, selenium, and zinc. Arsenic and barium may possibly be beneficial for growth. No apparent metabolic functions have been observed yet for cadmium, mercury, and lead (Rudd, 1987). Metals that are considered to be toxic at relatively low concentrations include silver, antimony, arsenic, beryllium, cadmium, cobalt, copper, mercury, nickel, lead, selenium, tin, and zinc (Rudd, 1987). Barium is considered to be toxic, but is relatively inaccessible because of its insolubility.

CHEMISTRY, OCCURRENCE, AND SIGNIFICANCE

Table 15-4 summarizes some of the chemical properties of metals that are important in wetland treatment systems. The table lists the valence states for metals typically encountered in surface waters, the most common forms of metals found in the water column, common insoluble forms usually found in the sediments, and a normal range of partition coefficients and bioconcentration factors.

Partition coefficients are equal to the ratio of the concentrations of the metal in two related media and are reported in units of liter per kilogram for solid to liquid media ratios and unitless for solid to solid media ratios. Partition coefficients provide a method of comparing metals to determine their characteristics for concentration in the environment or in the food chain.

Table 15-5 summarizes published concentrations for a variety of trace metals in soils, surface water, and in biological tissues (plants, invertebrates, fish, and birds). The table also lists some published criteria and advisory levels for these metals in the environment. The table can be used as a reference to determine what metal concentration ranges are typically associated with unaffected and affected aquatic and wetland environments and to help focus attention on those metals that approach or exceed action level concentrations.

WETLAND REMOVAL PERFORMANCE

Information on the effects of wetlands on concentrations of trace elements is widely scattered and has been reported in a variety of formats and level of detail. Consistent measurement of inflow and outflow concentrations as well as mass loads and removal rates will allow successful design criteria for wetlands treatment to be formulated. Unfortunately, the conclusions summarized in this section are somewhat limited because of the current incomplete knowledge in this area.

Table 15-5 Action Levels and Occurrence of Selected Metals in Wetland and Related Surface Waters, Plants, and Soils

Description	Metal													Notes	Ref.
	Al	As	Cd	Cr	Cu	Fe	Pb	Mn	Hg	Ni	Se	Ag	Zn		
Soils and sludge (mg/kg dw)															
Average upland soils	—	5	0.06	100	30	38,000	10	800	0.03	40	8	0.05	80	—	Lindsay, 1979; Lake, 1987
Average U.S. upland soils	72,000	7.2	—	54	25	26,000	19	600	0.089	19	0.39	—	60	—	Shacklette and Boerngen, 1984
Average Arizona upland soils	10,650	9.4	0.4	17.5	16.6	—	7.7	—	0.05	18.2	0.6	0.5	38.9	—	TETC, 1991
Average China upland soils	66,000	11	0.097	61	23	30,000	27	582	0.065	27	0.29	0.13	74	—	Chen et al., 1991
Average sludge	18,300	14	110	2,620	1,210	31,000	1,360	380	9	320	3	225	2,790	—	Lake, 1987
USDA NOAEL	—	100	18	2,000	1,200	—	300	—	15	500	—	—	2,700	—	Chaney, Proposed no observable adverse effect limit
Nonagricultural land maximum	—	36	385	11,000	3,300	—	1,622	—	17	988	—	—	8,600	503 Sludge regulations	U.S. EPA, 1989b
U.K. annual loading limits (kg/ha)	—	0.33	0.17	33.3	9.33	—	33.3	—	0.07	2.33	0.17	—	18.7	—	Lake, 1987
US cumulative loading (kg/ha)	—	14	18	530	46	—	125	—	15	78	—	—	170	503 Sludge regulations	U.S. EPA, 1989b
Unpolluted wetland (U.K.)	—	—	2	—	20	—	40	—	—	—	—	—	35	—	Zhang et al., 1990
Polluted wetland (U.K.)	—	—	12	—	220	—	841	—	—	—	—	—	779	Stormwater municipal effluent	Zhang et al., 1990
Control wetland (China)	—	85.6	1.85	—	20.2	—	55.2	0.17	—	—	—	—	119.8	—	Lan et al., 1990
Lead-zinc mine wetland (China)	—	365	20.9	—	68.9	—	5,980	6.28	—	—	—	—	5,796	Mine wastewater	Lan et al., 1990
Carolina Bay baseline-1986 (South Carolina)	—	—	0.11	3.29	0.93	1,409	—	—	<0.02	1.30	—	0.020	2.87	—	CH2M HILL, 1988
Carolina Bay operational-1991 (South Carolina)	—	—	<0.04	8.6	2.28	2,024	9.6	—	<0.05	3.07	—	<12	2.94	Municipal effluent	CH2M HILL, 1992
Alaska salt marsh	—	—	—	—	—	—	—	95–120	—	—	—	—	45–55	Unpolluted	Nixon and Lee, 1986
Oregon salt marsh	—	—	0.1–0.5	—	9–94	—	8–13	—	—	—	—	—	39–101	Unpolluted	Nixon and Lee, 1986
Everglades peat soils (Florida)	—	—	<2–5	<7–71	<5–36	437–28,400	<16–95	—	—	<5–23	—	—	<1–69	Unpolluted	Delfino et al., 1993
Pond cypress wetland (Florida)	—	—	<1.9	—	1.2	689	5.2	—	<0.07	<5.6	—	<1.9	<1.9	Municipal effluent (8 years)	CH2M HILL, 1994
Riverine swamp forest (South Carolina)	—	<5	<5	9.5	23.2	1,535	24	—	0.30	<16	<25	<5	18	Unpolluted	CH2M HILL, 1988
Okefenokee swamp	—	—	—	—	15	—	20	6.5	0.54	—	—	—	23.5	Unpolluted	Nixon and Lee, 1986

Soils and sludge (mg/kg dw) cont.															
Reed marsh, Nanokita River, (Japan)	—	—	0.1–0.14	—	8–9	—	4–6	—	—	2–2.5	—	—	45–50	Unpolluted	Suzuki et al. 1989
Southeast salt marshes	—	—	0.1–5.0	—	2–30	—	4–49	30–366	0.01–1.7	—	—	—	6–70	37 sites	Nixon and Lee, 1986
Freshwater tidal marsh (Camden, NJ)	—	—	5.5	—	117	—	1,025	—	—	64.4	—	—	452	Delaware River wetlands urban runoff	Simpson et al., 1983
Freshwater marsh, Orlando, FL inlet area	—	—	7	30	92	1,300	1,300	—	—	—	—	—	410	Urban runoff	Schiffer, 1989
Freshwater marsh, Orlando, FL outlet area	—	—	1	9	4	15,000	40	—	—	—	—	—	23	Background	Schiffer, 1989
Swamp forest, Sanford, FL	—	—	2.2	—	8.5	—	48	—	—	—	—	—	40	Urban runoff	Harper and Livingston (unpublished)
Shallow lake with emergent wetland vegetation, Tacoma, WA	—	—	5.9	300	160	—	3,000	1,000	—	100	—	—	900	Urban runoff	Wisseman and Cook, 1977
Freshwater tidal marsh, Hudson River, NY	—	—	2–42,000	—	—	—	—	—	—	—	—	—	—	Nickel-cadmium battery factory	Hazen and Kneip, 1980
Surface waters (μg/L)															
Freshwater (Gold Book) standards	—	190	0.66	Cr(VI):11 Cr(III):117	6.54	1,000	1.32	—	0.012	87.71	5	1.23	58.91	Hardness = 50 mg/L as $CaCO_2$	U.S. EPA, 1986b
California inland surface water (4-day average)	—	190	0.66	11	6.5	—	1.3	—	—	88	5	—	59		
Florida class III criteria	1,500	50	0.8–1.2	50	30	1,000	30	100	0.2	100	25	0.07	30		Florida Administrative Code, 1989
Raw wastewater	—	7	8	167	117	2,250	148	117	1	—	6	22	419	Municipal	Williams, 1982
U.S. limits for agricultural irrigation	5,000	100	10	100	200	5,000	5,000	200	—	200	20	—	2,000		Kirk, 1987
U.K. limits for agricultural irrigation	—	400	—	2,000	500	—	2,000	—	—	150	—	—	1,000		Kirk, 1987
USSR limits for hygenic and domestic purposes	—	50	10	500	100	500	100	—	5	100	1	—	1,000		Kirk, 1987
Suggested discharge limits for waterfowl wetlands	10	100	10	50	1,000	1,000	150	50	7	400	50	3	200		Kaczynski, 1985
World rivers	—	2	0.07	0.5	2	35	0.2	<5	0.01	0.3	0.1	0.3	10		Stephenson, 1987
Swamp forest, Sanford, FL	176	—	3.92	2.78	19.9	105	24.7	3.10	—	2.71	—	—	3.90	Urban runoff	Harper and Livingston, 1985 (unpublished)
Riverine Swamp (South Carolina)	—	<5	<5	<6	6.2	936	<2	—	<0.2	<15	<15	<5	6.8	Predischarge baseline	CH2M HILL, 1995
Rural rainfall	—	—	0.03–0.7	—	0.04–5.4	15–160	2–9	1.3–22	—	—	—	—	4.2–150	Rural locations	Ross, 1987
Biota—plants (mg/kg dw)															
Volta, CA	—	5.4	—	31	14	—	4.5	—		36	0.43	—	17	Unpolluted site	Ohlendorf et al., 1986

Table 15-5 (Continued)

Description	Metal													Notes	Ref.
	Al	As	Cd	Cr	Cu	Fe	Pb	Mn	Hg	Ni	Se	Ag	Zn		
Biota—plants (mg/kg dw) cont.															
Kesterson NWR, California emergent plants	—	1.80	<0.2	3.5	4.92	—	<0.8	—	0.01	3.05	52.1	<0.2	13.7	Agricultural returns	Ohlendorf et al., 1986
Kesterson NWR, California filamentous algae	—	9.6	<0.2–3	3.04	20	—	<2–3	—	0.03	4.0	—	<0.2	31.6	Agricultural returns	Ohlendorf et al., 1986
Inlet Valley, NY *Carex rostrata*	—	—	—	—	5.3–16	—	—	252–616	—	—	—	—	20–28	Unimpacted marsh	Bernard and Bernard, 1989
DUST Marsh, Fremont, CA *Scirpus* and *Typha*	—	—	—	—	—	—	3–16	100–1,200	—	—	—	—	7–41	Urban runoff	Meiorin, 1989
Mannersdorf, Austria *Phragmites*	—	—	—	—	10–38	—	1–48	—	—	—	—	—	18–51	Domestic sewage	Haberl and Perfler, 1989
Lead-zinc mine, China *Typha* and *Phragmites*	—	—	0.3–4.03	—	3–13.9	—	10–444	—	—	—	—	—	30–341	Lead-zinc mine	Lan et al., 1990
U.K. ponds *Typha*	—	—	4–10	—	17–45	—	20–95	—	—	—	—	—	38–170	Urban runoff and treated effluent	Zhang et al., 1990
Tundra biome site—3 spp.	—	—	—	—	6–21	—	—	152–500	—	—	—	—	41–63	Unpolluted marsh	Nixon and Lee, 1986
Landfill leachate treatment marsh, NY *Phragmites australis* shoots	—	—	0.09	—	—	65.8	0.21	68.9	—	—	—	—	17.4	Landfill leachate	Surface et al., 1993
Landfill leachate treatment marsh, NY *Phragmites australis* roots	—	—	0.39	—	—	3,709	8.02	289	—	—	—	—	35.0	Landfill leachate	Surface et al., 1993
Oregon salt marsh—6 spp.	—	—	—	—	8–13	—	2.2–97	30–341	—	—	—	—	23–60	Unpolluted marsh	Nixon and Lee, 1986
Okefenokee Swamp—7 spp. trees	—	—	—	—	<1	—	<6	196	0.26	—	—	—	98	Unpolluted	Nixon and Lee, 1986
Southeastern U.S. saltmarshes	—	—	—	—	0.1–11	—	0.01–55	6–340	0.01–1.13	—	—	—	0.4–686	70 Sites	Nixon and Lee, 1986
Nanakita River *Phragmites*	—	—	0.3	—	8	—	3.2	—	—	2.8	—	—	40	Unpolluted	Suzuki et al., 1989
Northeastern U.S. saltmarshes	—	—	0.00–0.73	—	1–16	—	0.3–26	11–323	0.01–0.15	—	—	—	5–101	Multiple sites	Nixon and Lee, 1986
Sphagnum bogs (remote), Quebec	—	—	—	—	14	1,731	30.5	—	—	—	—	—	33	Unpolluted	Glooschenko et al., 1986
Sphagnum bogs near Cumelter, Quebec	—	—	—	—	83	1,295	217	—	—	—	—	—	115	<20 km from Smelter	Glooschenko et al., 1986
Freshwater tidal marsh, Camden, NJ	—	—	2.13	—	14.5	—	23.6	—	—	8.3	—	—	124	Summer values, urban runoff	Simpson et al., 1983
Cyperus marshes near Detroit, MI	—	—	0.26	—	6	—	7	91	0.034	—	—	—	65	Urban marshes	

Biota—plants (mg/kg dw) cont.															
Wild rice grains, Manitoba	—	—	<0.01–6.2	—	1.6–14.4	—	<0.01–6.7	—	—	—	—	—	—	Lakes	Pip, 1993
Phytotoxic levels	—	20	10	10	20	—	35	—	3	11	30	4	200	Agricultural plants	Lake, 1987
Duckweed, Hamilton, Ontario	—	—	4.0	38	—	—	200	—	—	—	—	—	158	Municipal wastewater	Murdoch and Capobianco, 1979
Glyceria, Hamilton, Ontario	—	—	<1.0	2.2	—	4.2	—	—	—	—	—	—	16	Municipal wastewater	Murdoch and Capobianco, 1979
Sphagnum bog, northern Finland	—	—	—	6.3	231	5.6	—	—	—	—	—	—	34	Unpolluted	Glooschenko et al., 1986
Biota—Invertebrates (mg/kg dw)															
Volta		1.26	0.189	3.03	20.4	—	0.610	—	0.259	2.12	1.29–2.09	0.152	108	Unpolluted site	Ohlendorf et al., 1986
Kesterson NWR, California		2.46	0.29	1.47	15.6	—	0.207	—	0.063	1.25	22.1–175	0.093	81.1	Agricultural returns	Ohlendorf et al., 1986
Red River (New Mexico)	—	0.9	1.9	4.9	43	1,040	0.5	240	—	7.1	0.9	—	320	Upstream	Lynch et al., 1988
Red River (New Mexico)	—	0.5	1.3	2.7	82	1,300	0.9	540	—	13	0.2	—	350	Downstream of mine/mill	Lynch et al., 1988
Constructed marsh		<1.76	<0.07	—	37.7	—	<3.5	112	0.088	—	—	0.69	99.3	Pulp mill effluent	Knight, 1994 (unpublished data)
Biota—Fish (mg/kg dw)															
Concentration protective of human health	—	4–20	43	215	40–400	—	2–40	4,320	4	—	—	215	8,000		U.S. EPA, 1993a
National contaminant biomonitoring program	—	0.72	0.16	—	2.84	—	0.76	—	0.48	—	—	—	89.2		Schmitt and Brumbaugh, 1990[b]
Kesterson NWR, California, mosquitofish	—	0.664–1.54	0.041	0.88	6.76–7.14	—	<0.9–1	—	0.03–0.068	1.05	170	0.122	155–167	Agricultural return flows	Ohlendorf et al., 1986
Volta NWR, California mosquitofish	—	0.426	—	0.389	3.58	—	—	—	0.330	1.09	1.29	0.040	126	Unpolluted site	Ohlendorf et al, 1986
Merced wetlands, California mosquitofish	44	<1.1	<0.027	<0.81	4.9	—	0.22	12	0.32	<2.2	1.0	<0.27	110	Municipal wastewater	Ohlendorf, 1992 (unpublished data)
Constructed wetlands, mosquitofish	—	<10	<0.04	<8	—	—	<2	36	0.08	—	—	0.2	103	Pulp mill effluent	Knight, 1994 (unpublished data)
Constructed wetlands, catfish	57	<8	<0.4	1.4	1.7	95	<4	25	<1.6	<3.2	<16	<0.8	104	Pulp mill effluent	Knight, 1994 (unpublished data)
Biota—birds															
Kesterson NWR, California[b] in livers	—	<0.79	0.362	—	—	—	<3.2	—	1.05	—	28.6–37.2	1.02	105	Agricultural return flows	Ohlendorf et al., 1986
Volta Wildlife Area, California[b] in livers	—	0.251	0.583	—	—	—	0.255	—	1.04	—	4.14–6.1	0.201	120	Unpolluted site	Ohlendorf et al., 1986
Kesterson NWR, California[b] breast	—	—	—	—	—	—	—	—	—	—	9.61–110	—	—	Agricultural return flows	Ohlendorf et al., 1986
Volta wildlife area, California breast	—	—	—	—	—	—	—	—	—	—	2.45–13.2	—	—	Unpolluted site	Ohlendorf et al., 1986

[a]Food and Drug Administration action levels. Calculated from wet weight using 0.25 dry to wet ratio.
[b]Converted from wet weight using 0.25 dry to wet ratio.

As with most constituents discussed in this book, metal removal efficiencies are highly correlated with mass loading rates and influent concentrations. Therefore, average metal removal rates are nearly useless when used outside of the context of surface water concentration and hydraulic loading rates. For this reason, the data provided here should be applied cautiously in design of new wetland treatment systems.

Wetlands interact strongly with trace metals in a number of ways and thus are capable of significant metal removal. Three major mechanisms are operative:

- Binding to soils, sediments, particulates, and soluble organics
- Precipitation as insoluble salts, principally sulfides and oxyhydroxides
- Uptake by plants, including algae, and by bacteria

There are only minor gaseous removal routes (except notably for mercury and selenium). Biological mediation is present in the form of microbial production of the necessary sulfides for precipitation. Uptake by plants and algae may be for purposes of growth enhancement or at higher metal concentrations for protective purposes.

Wetland solids bind metals by cation exchange and chelation. Humic substances, both in solution and as solids, can form bonds with metals. As discussed in Chapter 6, divalent and trivalent metals exchange with hydrogen ions on the cation exchange sites of the sediments and soils. The binding capacity of a particular wetland soil is not often known or predictable; therefore, it must be measured before design (Eger et al., 1993). Typical values of capacities are on the order of 1,000 to 10,000 mg/kg for nickel on peat (Eger et al., 1993) and the same for Cu, Cd, and Ni on peat (Kadlec and Keoleian, 1986).

Sulfide precipitation relies on production of S^{-2} in the sulfide reduction zone of the wetland soil profile. This requires low redox potentials associated with anaerobic conditions, as well as a sufficient source of sulfate to match the metal requirement. For instance, precipitation of 1.0 mg/L cadmium (atomic weight 112.4) requires the reduction of 0.85 mg/L of sulfate (molecular weight 96.1) to obtain the required sulfide.

Metals reach plants via their fine root structure, and most are intercepted there. Some small amounts may find their way to stems, leaves, and rhizomes (Zhang et al., 1990; Sinicrope et al., 1992). Upon root death, some fraction of the metal content may be permanently buried, but there are no data on metal release during root decomposition.

Algae react in different ways to different metals. For example, *Chlorella vulgaris* sequesters cadmium, but not copper or chromium (Shanker and Kadlec, 1993). However, some trace metals are toxic, or inhibitory, to many microorganisms, algae, and invertebrates.

As with other wetland functions, it is not yet possible to piece together the various ecosystem functions to form an integrated metal removal model. Therefore, reliance must be placed on data from microcosm and mesocosm process studies, together with field data on trace metal removal in treatment wetlands.

A number of batch micro- and mesocosm studies have shown an exponential decrease in metal concentrations with time. This occurs for Zn, Cr, and Cd in water hyacinth systems (Delgado et al., 1993) and for Cu and Cr in cattail systems (Srinivasan and Kadlec, 1995). There is no evidence of a return flux of metals, but that may be an artifact of the short duration of the studies. Field studies show removals at very low concentrations, in the range of 10 to 100 μg/L (Noller et al., 1994; Sinicrope et al., 1992). Therefore, possible background concentrations (C^*) are likely to be quite low.

Based on this limited information, the first-order areal model seems appropriate for SF and SSF wetlands, with $C^* \cong 0$. The k values deduced from operating data will likely reflect mass transfer processes, from the water to the solids, and diffusion into the solids. This mechanism has been shown to be appropriate for copper removal to peat (Kadlec and Rathbun, 1984).

OCCURRENCE AND WETLAND TREATMENT

Aluminum

Aluminum occurs naturally in surface waters, to a small extent in the hydrated ionic forms, and to a greater extent complexed with silicates in a colloidal form. Aluminum enters surface waters from a variety of sources because of its widespread anthropogenic use. Aluminum solubility varies with pH. It is least soluble at a pH of 7 and increases in solubility as oxyaluminum ($Al^2O_3H^{2+}$) and aluminohydroxyl ($Al(OH)^{2+}$ ions at lower pH values and as aluminate ion ($HAl_2O_4^-$) at higher pH values (Hutchinson, 1975). Aluminum is not involved in oxidation-reduction reactions, and concentrations in sediments are not directly affected by the presence of aerobic or anaerobic conditions (Reddy, 1991).

Aluminum is associated with phosphorus adsorption capacity in wetland soils (Richardson, 1985). Richardson found that his phosphorus adsorption index, which measures the capacity of wetland soils to adsorb applied phosphorus, correlated highly with the concentration of extractable amorphous (noncrystalline) aluminum (Figure 15-3). It can be hypothesized that this same adsorption mechanism may reduce trace concentrations of aluminum from the water column of wetlands receiving elevated phosphorus concentrations.

Wieder et al. (1988) used a constructed, peat-based wetland for removal of aluminum from runoff and reduced average surface water aluminum concentrations of 6.6 to 35.3 mg/L to an average outflow concentration of 1.5 mg/L. Total aluminum content of the peat increased from 2375 to 13,634 mg/kg dry weight, with the majority bound as organic and oxide compounds. In a subsequent mesocosm study, Wieder et al. (1990) found that although aluminum was initially removed in a wetland, this process was quickly saturated and the wetland began to export aluminum to downstream waters. Also, aluminum concentrations of 10 mg/L were found to be toxic to cattails, leading to their mortality and release of sorbed aluminum. For more information on the fate and effects of aluminum in the environment, refer to Lewis (1989).

Table 15-6 summarizes the published information on aluminum concentrations and mass balances in wetlands. Based on this limited database, it appears that aluminum typically is removed from surface water passing through wetlands.

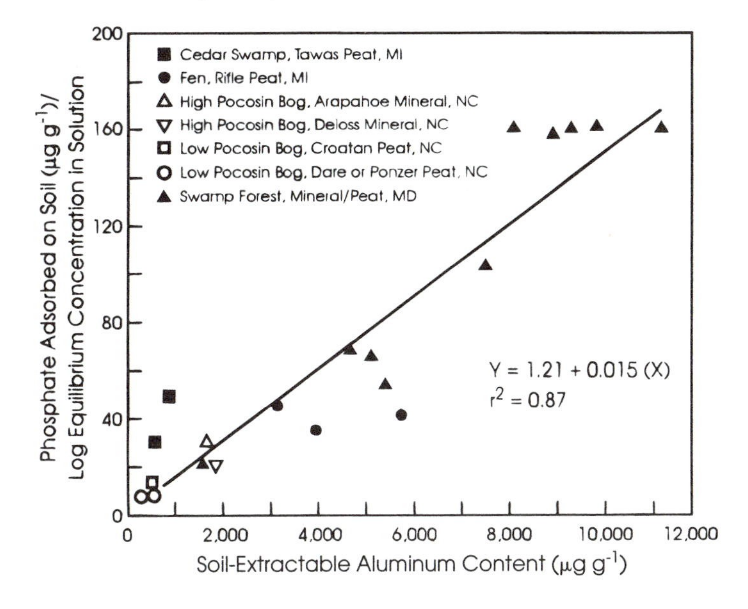

Figure 15-3 The observed relationship between extractable amorphous aluminum and a phosphorus adsorption index in wetland soils. (Data from Richardson, 1985.)

Table 15-6 Aluminum Dynamics in Wetlands

Description	Concentration (μg/L)			Mass Rates (kg/ha/yr)				Ref.
	In	Out	% Removal	In	Out	Removal	% Removal	
Freshwater marsh, Barataria Basin, Louisiana	—	—	—	—	—	413	—	Feijtel et al., 1989
Forested swamp receiving urban run-off, Sanford, FL	176	—	—	21.97	8.11	13.86	63.1	Harper et al., 1986
AMD[a] wetland, Kentucky (Fabius IMP1)	30	40	−33	1.10	1.82	−0.72	−66	Edwards, 1993 (unpublished data, TVA)
AMD wetland, Kentucky (Widows Creek)	300	340	−13	45.26	51.1	−5.84	−13	Edwards, 1993 (unpublished data, TVA)
Natural wetland, Tennessee	110	110	0	36.5	36.5	0	0	Edwards, 1993 (unpublished data, TVA)

[a]AMD-acid-mine drainage.

Arsenic

Arsenic is a naturally occurring element in surface waters, occurring as As(III) and As(V). Arsenic is considered a metalloid because it behaves as a metal and as a nonmetal. Arsenic compounds are mostly insoluble in water, with the arsenate form predominating in aerobic waters and arsenite occurring at low pH and under reducing conditions. Arsenic can be methylated by bacteria, but does not excessively bioconcentrate in biota. Arsenic is of concern in wastewater treatment because it is a known human carcinogen and because it is acutely and chronically toxic to aquatic organisms.

The normal arsenic content of surface waters in Germany was reported as between 2 and 3 μg/L by Hutchinson (1975). The average concentration in the world's rivers is estimated at 2 μg/L (Stephenson, 1987). Background arsenic concentrations in birds and freshwater biota are usually less than 2 mg/kg and are 0.72 mg/kg for fish on a dry-weight basis (Schmitt and Brumbaugh, 1990). Although action levels for arsenic have not been established by the U.S. Food and Drug Administration, legal limits in about 14 other nations range from 1 to 5 mg/kg wet weight (about 4 to 20 mg/kg dw). For more information about the occurrence of arsenic in the environment, see Fowler (1983).

Cadmium

Cadmium is a naturally occurring heavy metal with no known nutritional requirement for biota. Knight (1980) and Giesy et al. (1979) did observe a slight stimulation to aquatic ecosystem productivity at very low cadmium concentrations and increased populations of some periphytic algal species at cadmium concentrations as high as 10 μg/L. In surface waters, cadmium typically occurs as Cd(II) and is most soluble at low pH in waters with low hardness. In solution, cadmium is present as the ion and in a number of complexed, soluble compounds with carbonate, sulfate, chloride, hydroxides, and humates. These cadmium complexes are highly adsorbable on organic particulates. Cadmium may also be removed from solution by formation of cadmium sulfide.

Freshwater biota are sensitive to elevated cadmium concentrations. Birds and mammals are comparatively less sensitive. Giesy et al. (1979) found that aquatic fungi were sensitive to cadmium concentrations as low as 5 μg/L, resulting in decreased nutrient cycling in shallow, wetland-like streams and lower primary productivity. Eisler (1985) estimated that adverse effects of cadmium on fish and wildlife are possible at water concentrations above about 3 μg/L in freshwater or 0.1 mg/kg wet weight in the diet. Phytotoxic cadmium concentration in agricultural plants is about 10 mg/kg dw. Some freshwater algae protect their environment by sequestering cadmium, even though it is toxic. *Chlorella vulgaris,* a single-celled algae, can remove over 10 percent of its dry weight of cadmium (Srinivasan and Kadlec, 1995).

The average cadmium concentration in the world's rivers is about 0.07 μg/L. Natural upland and inland wetland soils typically contain less than 0.1 mg/kg dw. Cadmium concentrations up to 5 mg/kg were reported in salt marsh soils (Nixon and Lee, 1986). Cadmium concentration in wetland sediments in a tidal freshwater marsh on the Hudson River in New York were measured between 2 and 42,000 mg/kg dry weight near the outfall from a nickel-cadmium battery facility (Hazen and Kneip, 1980). The cadmium concentration in surface water at this site averaged about 32 μg/L; plant tissue concentrations ranged from 13 to 513 mg/kg dry weight near the outfall; and invertebrates and fish had muscle tissue concentrations generally between 0.4 and 59 mg/kg dry weight.

Municipal wastewaters have variable cadmium concentrations because of the diversity of industrial influent wastewaters; however, the average cadmium concentration in treated municipal effluents ranges from <5 to 20 μg/L. Wetland plants from unimpacted areas have

Table 15-7 Cadmium Dynamics in Wetlands

Description	Concentration (μg/L)			Mass Rates (kg/ha/yr)				Ref.
	In	Out	% Removal	In	Out	Removal	% Removal	
Creekbank salt marsh, LA	—	—	—	0.08	0.02	0.06	75	DeLaune et al., 1981
Cypress swamp, Waldo, FL	—	—	—	—	—	2.61	82	Best, 1987
Salt marsh, MA	—	—	—	—	—	0.004		Giblin, 1985
Constructed meadow/marsh/ pond, Brookhaven, NY	42.94	0.55	98.7	2.43	0.031	2.40	98.7	Hendrey et al., 1979
Carolina bay receiving munici- pal effluent, Myrtle Beach, SC,	<0.2	<0.2	0	—	—	—	—	CH2M HILL, 1992
Cypress-gum swamp receiving municipal effluent, Conway, SC	0.2	<0.2	—	—	—	—	—	CH2M HILL, 1991
Forested swamp receiving urban runoff, Sanford, FL	3.92	—	—	0.33	0.097	0.23	70.7	Harper et al., 1986
Shallow artificial streams, SC	4.88	4.13	15.3	37.6	31.8	5.79	15.4	Giesy et al., 1979
Shallow artificial streams, SC	9.76	9.18	5.9	75.1	70.7	4.43	5.9	Giesy et al., 1979
Bulrushes in gravel	70	17.5	75	12.6	3.2	9.4	75	Sinicrope et al., 1992
SSF wetlands	70	14.7	79	15.6	3.3	12.3	79	
SF cattail	63	0.19	99.7					Noller et al., 1994

cadmium tissue concentrations generally less than about 0.2 mg/kg dw. Cadmium does not appear to become concentrated in the food chain. Background animal tissue concentrations are generally less than about 0.2 mg/kg dw. For more information on the fate and effects of cadmium in the environment, refer to Stoeppler and Piscator (1988) and Nriagu (1980).

Wetland plants incorporate cadmium into their tissues (Zhang et al., 1990). Uptake is strongly preferential to the roots and rhizomes, where concentrations can reach 100 to 700 mg/kg in *Typha* spp. *Eichhornia crassipes* can strip cadmium from water in a matter of days (Delgado et al., 1993).

Cadmium removal has been investigated in several wetlands receiving municipal wastewaters (Best, 1987; Hendrey et al., 1979; CH2M HILL, 1991, 1992) and in unimpacted wetlands (DeLaune et al., 1981). Cadmium mass reduction efficiencies are usually above 75 percent apparently because of the formation of sulfide and subsequent sedimentation of the metal (Table 15-7). Mass removal rates appear to directly correlate with mass loading rates and were estimated to be between 0.004 and 2.61 kg/ha/yr.

Table 15-7 provides some insights concerning wetland removal of cadmium. Wetlands appear more efficient than streams. The Sinicrope et al. (1992) data for SSF wetlands correspond to removal rate constants in the range $25 < k < 35$ m/yr. The fate of the metal was immobilization in soils (54 percent) and fine roots (44 percent), with the balance in coarse roots, rhizomes, and shoots.

Chromium

Chromium typically occurs in the trivalent [Cr(III)] or hexavalent [Cr(VI)] forms in surface waters. The CR(VI) form is the most toxic and is generally associated with the presence of industrial wastewaters. Cr(VI) is relatively unstable under most environmental conditions and converts to the less toxic trivalent form in surface waters, especially when organic matter is present. Trivalent chromium hydroxides and chlorides are relatively insoluble and their formation may significantly reduce chromium availability to biota.

Although chromium is an essential trace element in animals, it does not biomagnify in the food chain. The sensitivity of biota to chromium varies widely, even among related species (Eisler, 1986). Plants generally have chromium concentrations from 0.01 to 0.1 times the soil concentration. The chromium is transferred from soils and roots to the above-ground plant parts to such a small extent that toxicologically significant concentrations are unlikely (Bolt et al., 1991). A concentration of 10 mg/kg dw of chromium is considered to be a phytotoxic threshold level in agricultural plants. Chromium (VI) is reported to be toxic to algae at concentrations between <20 and 10,000 μg/L (Nriagu and Nieboer, 1988).

The world's rivers have an average chromium concentration of about 0.5 μg/L. Freshwater wetland chromium soil concentrations are generally below 10 mg/kg dw. For more information on the fate and effects of chromium in the environment, refer to Nriagu and Nieboer (1988).

Chromium concentration reduction has been investigated in wetlands receiving municipal wastewater (Hendrey et al., 1979; CH2M HILL, 1991, 1992), urban runoff (Schiffer, 1989), and in unimpacted natural wetlands (Giblin, 1985). Chromium concentration reduction efficiencies are related to chromium inflow concentration with zero reduction at low inflow concentrations to 87.5 percent or greater reduction at an inflow concentration of 160 μg/L (Table 15-8). Chromium removal rates have been measured from 0.026 kg/ha/yr in a wetland that does not receive direct discharges of pollutants to 7.92 kg/ha/yr in a constructed meadow/marsh/pond system receiving municipal wastewater.

Chromium is effectively removed in the SSF environment and in the SF environment. The disappearance data for Cr(III) in Table 15-8 translate to an areal k value of 25 to 35 m/yr in the SSF system (Sinicrope et al., 1992). Surface flow microcosm data indicate values of $20 \leq k \leq 80$ m/yr (Srinivasan and Kadlec, 1995) for Cr(VI).

Table 15-8 Chromium Dynamics in Wetlands

	Concentration (μg/L)			Mass Rates (kg/ha/yr)				
Description	In	Out	% Removal	In	Out	Removal	% Removal	References
Salt marsh, MA	—	—	—	—	—	0.026	—	Giblin, 1985
Constructed meadow/marsh/pond, Brookhaven, NY	160	20	87.5	9.05	1.13	7.92	87.5	Hendrey et al., 1979
Freshwater marsh receiving urban stormwater, Orlando, FL	7.5	4.5	40	—	—	—	—	Schiffer, 1989
Carolina bay receiving municipal effluent, Myrtle Beach, SC	<2	3	—	—	—	—	—	CH2M HILL, 1992
Cypress-gum swamp receiving municipal effluent Conway, SC	15.0	15.0	0	—	—	—	—	CH2M HILL, 1991
Forested swamp receiving urban run-off, Sanford, FL	2.78	—	—	0.23	0.063	0.17	72.5	Harper et al., 1986
Bulrushes in gravel	100	16	84	18.0	2.9	15.1	84	Sinicrope et al., 1992
SSF wetland [Cr(III)]	100	32	68	22.3	7.1	15.2	68	

Copper

Typically, copper is present in surface waters as chelated compounds of Cu(II). The ratio of free ionic to total dissolved copper is about 1 percent and decreases with increasing organic loads and pH above neutrality. Copper forms relatively insoluble complexes with hydroxides, sulfides, and carbonates. When chelated with certain organic compounds, copper may remain relatively soluble.

Copper is an essential micronutrient for plants and animals because it is used for protein synthesis and in blood pigments. Plants and animals require minimal amounts of this element, and deficiencies in nature are rare (Goldman and Horne, 1983). The average copper concentration in the world's rivers is about 2 μg/L.

Copper is a biocide that is commonly used to control algae and other plants. Concentrations of less than 5 to 10 μg/L are toxic to some blue-green algae. Copper has relatively low toxicity to benthic macroinvertebrates and to fish, which may be unaffected by concentrations as high as 500 μg/L. The sensitivity of algae and the tolerance of fish and benthos to copper has resulted in the widespread use of copper sulfate as an algicide in lakes. Under conditions of low pH, copper remains soluble and its algicidal activity continues. Under conditions of high pH and alkalinity, copper is readily chelated as the insoluble, hydrated copper carbonate (malachite) and its algae-control ability is greatly diminished.

Copper does not seem to biomagnify significantly in the freshwater food chain. Background wetland soils contain less than 1 mg/kg dw of copper, and the average copper concentration in upland soils is about 30 mg/kg dw (Table 15-5). Copper concentrations in plants growing in relatively unimpacted wetlands are generally below about 20 mg/kg dw; however, pollution may cause copper concentrations in plant tissues to increase. Zhang et al. (1990) found cattail roots with a copper concentration of 45 mg/kg dw in a wetland receiving urban runoff and treated sewage. The phytotoxic level of copper in agricultural plants is about 20 mg/kg dw (Lake, 1987). Typical wetland invertebrate copper concentrations are less than about 40 mg/kg dw and unimpacted wetland fish have copper concentrations of about 3 mg/kg dw. For more information on the fate and effects of copper in the environment, refer to Nriagu (1979).

Copper concentration reduction has been investigated in wetlands receiving municipal wastewater (Hendrey et al., 1979; Best, 1987; CH2M HILL, 1991, 1992), urban runoff (Schiffer, 1989), and in unimpacted wetlands (DeLaune et al., 1981; Giblin, 1985; Feijtel et al., 1989). Copper concentration reduction efficiency appears to correlate with inflow concentration up to at least 1510 μg/L, where the reduction efficiency was about 96 percent (Table 15-9). Copper mass removal rates are highest in wetlands receiving high copper inputs. Removal rates as high as 82 kg/ha/yr were estimated for the Brookhaven, NY meadow/marsh/pond system. The copper removal rate measured in three natural wetlands that do not directly receive wastewater ranged from 0.025 to 0.6 kg/ha/yr.

Copper is effectively removed in wetlands (Table 15-9). Peats and humic substances have a large affinity for this metal, as indicated by the review of Kadlec and Keoleian (1986). Roots, shoots, and rhizomes of *Typha* spp. can contain up to 200 mg/kg of copper (Zhang et al., 1990).

The Sinicrope et al. (1992) data correspond to uptake k values from 18 to 47 m/yr for an SSF system. Srinivasan and Kadlec (1995) report copper uptake in SF cattail microcosms of $25 < k < 120$ m/yr. In the Sinicrope et al. (1992) study, the removed copper was found mostly in the soil (91 percent), with smaller amounts in fine roots (8 percent), and the balance in coarse roots, rhizomes, and shoots.

Table 15-9 Copper Dynamics in Wetlands

Description	Concentration ($\mu g/L$)			Mass Rates (kg/ha/yr)				Ref.
	In	Out	% Removal	In	Out	Removal	% Removal	
Creekbank salt marsh, LA	—	—	—	0.81	0.21	0.60	74	Delaune et al., 1981
Cypress swamp, Waldo, FL	—	—	—	—	—	13.14	78	Best, 1987
Salt marsh, MA	—	—	—	—	—	0.025	—	Giblin, 1985
Freshwater marsh, Barataria Basin, LA	—	—	—	—	—	0.21	—	Feijtel et al., 1989
Constructed meadow/marsh/pond, Brookhaven, NY	1510	60	96.0	85.42	3.39	82.03	96.0	Hendrey et al., 1979
Freshwater marsh receiving urban stormwater, Orlando, FL	8.0	1.0	87.5	—	—	—	—	Schiffer, 1989
Carolina bay receiving municipal effluent, Myrtle Beach, SC	20.4	6.1	70.1	0.27	0.060	0.21	77.9	CH2M HILL, 1992
Cypress-gum swamp receiving municipal effluent, Conway, SC	12.5	7.8	38	—	—	—	—	CH2M HILL, 1991
Forested swamp receiving urban runoff, Sanford, FL	19.9	—	—	1.38	0.83	0.55	39.9	Harper et al., 1986
Bulrushes in gravel	60	22.2	63	10.8	4.0	6.8	63	Sinicrope et al., 1992
SSF wetland	1200	144	88	268	32	236	88	
Typha SF	1.4	0.9	36					Noller et al., 1994
Typha and Melaleuca SF	13	0.5	96					
Carex SF	280	13					90	Eger et al., 1993
	145	14.5					73	
	160	6					95	

Iron

Iron is a metal that may occur at trace to high concentrations in wetland surface waters and sediments. It is frequently required by plants and animals at significant concentrations. In plants, iron is an essential element in chlorophyll synthesis, cytochromes, and in the enzyme nitrogenase. In animals, iron is important in oxidative metabolism and is a key component in hemoglobin. Iron occurs in aquatic plants at a concentration of about 5000 mg/kg dw, which is about 10 times higher than concentrations in terrestrial plants. Plant roots contain a higher proportion of iron than stems or leaves (Wetzel, 1975).

Oxidation and reduction of iron occurs relatively easily depending on redox potential. Ferric iron, or Fe(III), is the dominant ionic form under oxidized conditions. Ferrous iron, or Fe(II), is the dominant form under reduced conditions in wetlands and other aquatic environments. Ferric iron forms stable complexes with a variety of ligands. It joins with the hydroxide ion in surface waters to form reddish-brown ferric hydroxide ($Fe(OH)_3$) or ocher. Ocher is insoluble and either settles to the bottom or remains in suspension, adsorbed to living and dead organic matter. Other important compounds formed by ferric iron include ferric phosphate ($FePO_4$), iron-humate complexes, and ferric hydroxide-phosphate complexes.

Ferric iron is reduced to the ferrous form under anaerobic conditions. The ferrous iron is more soluble, resulting in the release of dissolved iron and associated anions such as phosphate from anaerobic sediments in wetlands. The formation of this soluble ferrous iron may be controlled somewhat by the concentration of sulfide which forms the relatively insoluble ferrous sulfide (FeS).

As shown in Figure 15-4, a diagram of the wetland iron cycle, wetland biota affect iron concentrations through their nutritional and energy requirements. As discussed earlier, iron is an essential element for all life and may be extracted from water in the form of ferric hydroxide or as soluble ferrous iron. Heterotrophic microbes may create ocher in their use of the organic compounds in iron-organic complexes. Iron bacteria that produce ocher, such as *Leptothrix ochracea* and *Spirophyllum ferrugineum,* derive their energy needs from the oxidation of reduced iron. These bacteria typically multiply in the area where slightly aerated

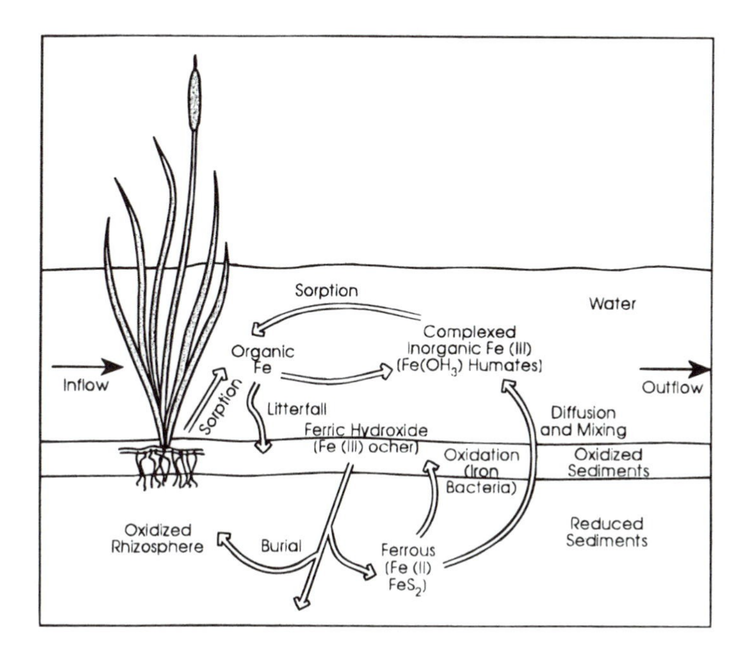

Figure 15-4 Simplified wetland iron cycle.

surface waters and reduced sediments meet. At this location, ferrous iron diffuses upward and is continually available so that these bacteria will grow.

Surface water iron concentrations range from 50 to 200 μg/L in aerated aquatic systems (Hutchinson, 1975). This iron is largely present in insoluble forms that are suspended or adsorbed on small particles. Higher total iron concentrations are found in acidic waters rich in dissolved or colloidal organic matter (Wetzel, 1975).

Low available iron concentrations may limit growth of algae and floating aquatic macrophytes. In one instance in a phytoplankton-dominated area of a treatment wetland with high dissolved oxygen concentrations and high pH, attempts to introduce and grow water hyacinth and pennywort were repeatedly unsuccessful, apparently because of inadequate dissolved iron concentrations (Knight, 1987b). Rooted macrophytes had no problem growing in these same waters. In addition to their effect on the complexation of iron with hydroxide and phosphate at high dissolved oxygen concentrations, some blue-green algae and bacteria species can produce powerful iron chelators called siderochromes that reduce the iron available to other aquatic plants (Goldman and Horne, 1983). Thus, iron limitation may be a nutritional concern in some treatment wetlands.

High concentrations of soluble iron in surface water and wetland systems may result from natural or artificial iron sources, typically as seeps of ferrous iron and iron sulfides (pyrites) from anaerobic groundwaters and as oxidation of iron sulfides exposed during surface mining. A large class of wetland treatment systems is currently used to treat waters that originate as acid-mine drainage (AMD) wastewaters. In many cases, these waters have influent total iron concentrations of 250 mg/L and above that must be reduced to less than an average of 3 mg/L (Wieder, 1989).

Wieder et al. (1990) found that iron retention was similar in peat-based wetland mesocosms with or without *Sphagnum* or cattails. In these studies, iron retention by cation exchange rapidly reached saturation at concentrations of 1000 to 5000 mg/kg dw, organic binding was saturated at about 10,000 to 15,000 mg/kg dw, and oxide formation continued up to a saturated concentration of about 200,000 mg/kg dw.

Information on iron removal in wetlands is available primarily from AMD wetland treatment systems (Wieder, 1989; Kleinmann and Hedin, 1989; Girts et al., 1987), but some iron removal data also have been published for municipal effluents (Hendrey et al., 1979; CH2M HILL, 1991, 1992), landfill leachate (Surface et al., 1993), and unpolluted wetland sites (Giblin, 1985; Feijtel et al., 1989).

Of the 137 AMD wetland treatment sites reviewed by Wieder (1989), 66 percent had influent iron concentrations less than 50 mg/L (Table 15-10). An average total iron concentration of 60.6 mg/L was reduced to an average outflow concentration of 15.4 mg/L, for an average iron removal efficiency of 58.2 percent and a median value of 80.9 percent (Wieder, 1989). This treatment efficiency was found to correlate directly with wetland area and inversely with wetland depth. An average iron removal rate of 100 kg/ha/d at pH 6 in AMD treatment wetlands was given by Kleinmann and Hedin (1989). Because the iron removal rate is correlated with the iron loading rate (Hedin and Nairn, 1990), lower removal rates are expected at low influent concentrations. Recommended hydraulic loading rates for iron removal to meet water quality standards in AMD treatment wetlands range from 9.6 cm/d for systems receiving less than 50 mg/L (Girts et al., 1987) to 29 cm/d (Kleinmann et al., 1986).

In landfill leachate, Surface et al. (1992) measured an iron removal efficiency of 78.6 percent in a SSF wetland planted with common reed. This is equivalent to an iron removal rate of about 963 kg/ha/yr. Hendrey et al. (1979) measured an iron removal rate of about 243 kg/ha/yr and a 66.7 percent removal efficiency in the Brookhaven, NY meadow/marsh/pond pilot wetlands. CH2M HILL (1991, 1992) observed a net release of iron from natural wetlands receiving much lower iron inflow concentrations. The natural streamside wetland

Table 15-10 Iron Dynamics in Wetlands

Description	Concentration (μg/L)			Mass Rates (kg/ha/yr)				Ref.
	In	Out	% Removal	In	Out	Removal	% Removal	
Salt marsh, MA	—	—	—	—	—	11.20	—	Giblin, 1985
Freshwater marsh, Barataria Basin, LA	—	—	—	—	—	161	—	Feijtel et al., 1989
Constructed meadow/marsh/pond, Brookhaven, NY	6,430	2,140	66.7	363.8	121.1	242.7	66.7	Hendrey et al., 1979
Carolina bay, receiving municipal effluent, Myrtle Beach, SC	241.4	765.8	−217	3.18	7.48	−4.29	−135	CH2M HILL, 1992
Cypress-gum swamp receiving municipal effluent, Conway, SC	298	488	−63.8	—	—	—	—	CH2M HILL, 1991
Subsurface-flow reed wetlands receiving landfill leachate, NY	21,700	—	—	1,225	262	963	78.6	Surface et al., 1993
Average for 137-AMD constructed wetlands	60,600	15,400	58.2	—	—	—	—	Wieder, 1989
Average AMD constructed wetlands	—	—	—	—	—	36,500	—	Kleinmann and Hedin, 1989
Forested swamp receiving urban runoff, Sanford, FL	105	—	—	6.08	11.56	−5.48	−90.1	Harper et al., 1986
AMD wetland, KY (Fabius IMP1)	44,000	900	98	1,847	36.5	1,810.5	98	Edwards, 1993 (unpublished TVA data)
AMD wetland, KY (Widows Creek)	205,000	6,300	97	30,828	949	29,879	97	Edwards, 1993 (unpublished TVA data)
Natural wetlands, TN	1,290	1,170	9	449	405	44	10	Edwards, 1993 (unpublished TVA data)

sites studied by Giblin (1985) and Feijtel et al. (1989) removed between 11.2 and 161 kg/ha/yr of iron.

Lead

Lead occurs in surface waters most often in the divalent Pb(II) form, which forms salts with sulfides, carbonates, sulfates, and chlorophosphates. Lead combines with organic ligands to form soluble and colloidal complexes such as sulfhydryl, carboxyl, and amines. Lead solubility is likely to be less than about 1 μg/L at a pH above 8.5, with increasing solubility at lower pH values and lower alkalinities. Lead is largely unavailable in most natural conditions because of its formation of insoluble salts and because the ionic form is adsorbed onto suspended particulates.

The average concentration of lead reported for world rivers is 0.2 μg/L. The average concentrations of lead in agricultural soil is 10 mg/kg dw and less than 40 mg/kg dw in background wetland soil. Sites contaminated by urban runoff or mine drainage may have sediment lead concentrations of more than 7000 mg/kg dw (Lan et al., 1990).

Lead can biomagnify in aquatic biota to concentrations greater than those found in the water. However, concentration in the aquatic food chain does not appear to be a concern (Eisler, 1988). Wetland plants from unimpacted freshwater sites generally have lead concentrations less than 10 mg/kg dw, but saltmarsh plants can have lead concentrations as high as 97 mg/kg dw. According to Lan et al. (1990), the roots of cattails growing in soils contaminated by a lead-zinc mine in China had a lead concentration of 444 mg/kg dw. The average lead concentrations in plants in a cattail and reed marsh in the eutrophic Lake Gandynskie, Poland, were 0.13 to 2.0 mg/kg dw (Kufel, 1991). Wetland invertebrates generally have lead concentrations less than 1 mg/kg dw. The average lead concentration measured in the U.S. in fish was 0.76 mg/kg dw (Table 15-5).

Lead concentration reduction has been investigated in a few wetlands receiving municipal effluent (CH2M HILL, 1991, 1992) and urban runoff (Schiffer, 1989) and in natural systems with no direct pollution sources (DeLaune et al., 1981; Giblin, 1985; Feijtel et al., 1989). As with most other metals, lead concentration removal efficiency increases with increasing inflow concentration in the range of available data (Table 15-11). A removal efficiency of 83.3 percent was estimated for the Island Lake marsh wetland which receives urban runoff near Orlando, FL (Schiffer, 1989), while negative removal efficiencies were measured in natural wetlands receiving low lead concentrations in municipal wastewater in South Carolina (CH2M HILL, 1991, 1992). An average lead removal efficiency of 45 percent in vegetated filter strips and 65 percent in constructed stormwater wetlands was reported by U.S. EPA (1993d). Mass removal rates for lead in unimpacted wetlands were measured in the range of 0.115 to 1.1 kg/ha/yr in stream bank marsh sites in Louisiana and Massachusetts (Table 15-11). Lead removal in wetlands appears to be largely accomplished through formation of insoluble compounds followed by subsequent sedimentation.

The data in Table 15-11 indicate that SF marshes and SSF wetlands effectively remove lead, but forested wetlands may not. The SSF data of Sinicrope et al. (1992) correspond to a k-value of 35 m/yr.

The pattern of allocation to the wetland compartments is similar to that described earlier for other metals. Concentrations in *Typha* are highest in roots and soil (150 to 250 mg/kg) and lower in rhizomes and leaves (30 to 50 mg/kg) (Zhang et al., 1990).

Manganese

Manganese is an essential element that is chemically similar to iron in its behavior in surface waters. Manganese is vital to plant photosynthesis and is used as an enzyme cofactor

Table 15-11 Lead Dynamics in Wetland

Description	Concentration (μg/L)			Mass Rates (kg/ha/yr)				Ref.
	In	Out	% Removal	In	Out	Removal	% Removal	
Creekbank salt marsh, LA	—	—	—	1.08	0.18	0.90	83	DeLaune et al., 1981
Salt marsh, MA	—	—	—	—	—	0.115	—	Giblin, 1985
Fresh water marsh Barataria Basin, LA	—	—	—	—	—	1.1	—	Feijtel et al., 1989
Freshwater marsh receiving urban stormwater, Orlando, FL	18.0	3.0	83.3	—	—	—	—	Schiffer, 1989
Carolina bay receiving municipal effluent, Myrtle Beach, SC	1.96	5.5	−181	0.026	0.054	−0.028	−107	CH2M HILL, 1992
Cypress-gum swamp receiving municipal effluent, Conway, SC	2.8	3.5	−27.3	—	—	—	—	CH2M HILL, 1991
Forested swamp receiving urban runoff, Sanford, FL	24.7	—	—	1.97	0.89	1.08	54.8	Harper et al., 1986
AMD wetland, KY, (Widows Creek)	2.2	1.63	25.9	0.33	0.245	0.085	25.8	Edwards, 1993 (unpublished TVA data)
Bulrushes in gravel	300	42	86	53.9	7.5	46.4	86	Sinicrope et al., 1992
SSF wetland	300	60	80	66.9	13.4	53.5	80	
Typha SF	12	0.2	98					Noller et al., 1994
Typha/Melaleuca SF	9	0.5	94					
Stormwater wetlands (N = 9) median (all SF)							83	Strecker et al., 1992a

for respiration and nitrogen metabolism by plants and animals. Although manganese is toxic to some organisms at elevated concentrations, this situation occurs infrequently and only in response to gross pollution, typically with mining wastes. Manganese concentrations greater than 2 mg/L were found to be toxic to algae in laboratory experiments (Goldman and Horne, 1983).

Manganese is typically present in surface waters as Mn(IV) and the relatively unstable Mn(III), which form insoluble oxides and hydroxides. At low redox potentials and low pH, the predominant form is Mn(II). Bacteria in wetland soils can oxidize this manganous form to manganic hydroxide [$Mn(OH_3)$], which in turn is dismuted to MnO_2 and MnO (Hutchinson, 1975). Manganous manganese can form soluble complexes with bicarbonate, sulfate, and organic compounds. Under reducing conditions, manganese forms insoluble complexes with carbonate, sulfide, and hydroxide.

Total manganese in surface waters is typically less than 100 μg/L. The average for world lakes and rivers is 35 μg/L (Goldman and Horne, 1983). Because soils contain the insoluble forms of manganese described previously, their manganese concentrations are typically higher than those found in surface water. Agricultural soils have average concentrations of 800 mg/kg dry weight, freshwater wetland soils have average concentrations less than 10 mg/kg dw, and saltmarsh soils have concentrations up to about 400 mg/kg dw. Because of their biochemical requirements, plants and animals typically concentrate only a small amount of manganese in their tissues compared to surface water concentrations. Wetland emergent plants from relatively unimpacted sites contain about 200 to 600 mg/kg dw of manganese. Manganese is not observed to bioconcentrate in the wetland food chain.

Wetland reduction of manganese concentrations has been studied at unimpacted natural wetland sites (DeLaune et al., 1981; Giblin, 1985; Feijtel et al., 1989), at sites receiving landfill leachate (Surface et al., 1992), and at sites receiving municipal wastewater (Hendrey et al., 1979; Best, 1987). Manganese removal efficiency increases to at least 68 percent as inflow concentration increases to 100 μg/L, but removal efficiency is 40 percent at inflows between 210 and 5250 μg/L (Table 15-12). Manganese mass removal rates were between 0.105 and 3.75 kg/ha/yr in unimpacted wetlands (Table 15-12), between 5.09 and 26.3 kg/ha/yr in wetlands receiving municipal wastewater, and about 114 kg/ha/yr in a SSF wetland receiving landfill leachate.

Mercury

Mercury is undoubtedly the most unique of the commonly occurring heavy metals in wastewaters. Mercury occurs in the environment in three primary oxidation states: as elemental Hg(0), as relatively insoluble mercurous Hg(I), and as the more soluble mercuric Hg(II). Elemental mercury is formed in reduced sediments and is volatile, resulting in an atmospheric sink for mercury and also subsequent deposition at otherwise remote and unpolluted sites. Because mercury is released from a variety of anthropogenic sources such as solid and medical waste incineration, the combustion of fossil fuels for electric power production, and from paint application, it is sometimes found in elevated concentrations in wetland areas affected by urban air pollution, such as the Florida Everglades (Delfino et al., 1993). Elemental mercury volatilization comprises about 95 percent of the atmospheric transport of this element (Nater and Grigal, 1992). In one study, elemental mercury release was measured as 22 μg/m^2/d in shallow, wetland-like channels receiving a surface water dose of 5 μg/L and 6.7 μg/m^2/d at a mercury dose of 1 μg/L by Kania et al. (1976).

The most common forms of mercury in aquatic systems are mercuric salts such as $HgCl_2$. Under anaerobic sediment conditions, mercuric ions are biomethylated by microorganisms such as *Clostridium cochlearium,* resulting in trace concentrations of mono- and dimethyl mercury, which are more toxic than the more abundant mercuric forms. Methlylated mercury

Table 15-12 Manganese Dynamics in Wetlands

	Concentration (μg/L)			Mass Rates (kg/ha/yr)				
Description	**In**	**Out**	**% Removal**	**In**	**Out**	**Removal**	**% Removal**	**Ref.**
Creekbank salt marsh, LA	—	—	—	7.39	3.64	3.75	51	DeLaune et al., 1981
Cypress swamp, Waldo, FL	—	—	—	—	—	26.3	68	Best, 1987
Salt marsh, MA	—	—	—	—	—	0.105	—	Giblin, 1985
Freshwater marsh, Barataria Basin, LA	—	—	—	—	—	2.7	—	Feijtel et al., 1989
Constructed meadow/marsh/pond, Brookhaven, NY	210	120	42.9	11.88	6.79	5.09	42.9	Hendrey et al., 1979
Subsurface-flow reed wetlands receiving landfill leachate, NY	5250	—	—	291	177	114	39.1	Surface et al., 1993
Forested swamp receiving urban runoff, Sanford, FL	3.10	—	—	0.26	0.24	0.02	7.7	Harper et al., 1986
AMD wetland, KY (Fabius IMP1)	5900	1200	79	244	51.1	193	79	Edwards, 1993 (unpublished TVA data)
AMD wetland, KY (Widows Creek)	7400	3900	47	1113	588	526	47	Edwards, 1993 (unpublished TVA data)
Natural wetland, TN	210	130	40	73	44	29	40	Edwards, 1993 (unpublished TVA data)
Typha SF	600	11	98					Noller et al., 1994
Typha/Melaleuca SF	1300	330	75					

is a serious problem in aquatic ecosystems because it is readily bioconcentrated in the food chain. Mercury is not known to be an essential element for any organism.

The average mercury concentration in the world's rivers is about 0.01 μg/L. Soils have an average mercury content of about 0.03 mg/kg dw. However, Okefenokee Swamp soils had a concentration of 0.54 mg/kg, and saltmarsh concentrations were as high as 1.7 mg/kg dw (Nixon and Lee, 1986). In a detailed study of Everglades mercury concentrations and dynamics, Delfino et al. (1993) found average mercury concentrations of 0.121 mg/kg on a wet-weight basis (approximately 0.6 mg/kg dw). The background total mercury content of plants is typically 0.05 mg/kg dw. However, a concentration of 0.26 mg/kg dw was reported for trees in the Okefenokee Swamp, and some southeastern U.S. saltmarsh plants had concentrations as high as 1.13 mg/kg dw.

Total mercury concentrations are significantly higher in animals than in plants because of the selective storage of methylated mercury compounds in fats and because of low depuration rates. Wetland invertebrates from unpolluted waters typically have mercury concentrations up to 0.1 mg/kg dw and average fish have concentrations of 0.48 mg/kg dw. Older fish and their consumers, which include alligators, piscivorous birds, and mammals, frequently have mercury concentrations over 1 mg/kg dw, even in relatively remote wetland areas such as the Everglades in Florida. Fish from a variety of locations have been found to have mercury bioconcentration factors from 1,000,000 to 3,000,000, compared to aqueous concentrations, due to bioaccumulation of methyl mercury (Zillioux et al., 1993). In polluted waters, invertebrates may have mercury concentrations exceeding 40 mg/kg dw, and fish may have concentrations exceeding 10 mg/kg dw.

Table 15-13 includes data on mercury removal rates from a salt marsh (Giblin, 1985) and for two natural wetlands receiving municipal effluent (CH2M HILL, 1991, 1992). Estimated mass removal rates (0.0001 to 0.0002 kg/ha/yr) and efficiencies (4.8 percent) for mercury are low, probably in response to low influent concentrations. Mercury concentration was observed to increase slightly between the influent and effluent of the two forested wetlands in Table 15-13.

Nickel

The oxidation state for nickel in surface waters is divalent Ni(II). Divalent nickel is hydrated or may be complexed with hydroxide, sulfate, chlorides, or ammonia nitrogen. Under anaerobic conditions, nickel solubility is reduced by sulfide. Much of the nickel in surface waters is present as precipitates associated with suspended particles and organic compounds.

Hutchinson (1975) indicates that the average nickel concentration in surface waters is 5 μg/L. Stephenson (1987) reports the concentration in the world rivers as 0.3 μg/L. Average agricultural soils contain about 40 mg/kg dw of nickel, and the background wetland sites listed in Table 15-5 typically had concentrations less than 25 mg/kg dw. Background wetland peat samples in Minnesota had average nickel concentrations from 154 to 169 mg/kg (Eger, 1993).

Although nickel is an essential micronutrient for the growth of some animals, it also is toxic at elevated concentrations. These concentrations can be elevated in wetland plants at polluted sites. At the unpolluted Volta Wildlife Area in California, emergent wetland plants had about 36 mg/kg dw of nickel in their tissues (Ohlendorf et al., 1986), and at unpolluted sites in Minnesota, the average nickel concentration in cattails ranged from 4.9 mg/kg in the leaves to 18.0 mg/kg in the roots and 7.2 mg/kg in an unidentified grass (Eger, 1993). Following exposure to mine drainage, average nickel concentrations in these plants increased to about 14.8 to 25 mg/kg in the cattail leaves, 195 to 246.6 mg/kg in the cattail roots, and

Table 15-13 Mercury Dynamics in Wetlands

Description	Concentration (μg/L)			Mass Rates (kg/ha/yr)				Ref.
	In	Out	% Removal	In	Out	Removal	% Removal	
Salt marsh, MA	—	—	—	—	—	0.0002	—	Giblin, 1985
Carolina bay receiving municipal effluent, Myrtle Beach, SC	<0.2	0.21	—	0.0021	0.0020	0.0001	4.8	CH2M HILL, 1992
Cypress-gum swamp receiving municipal effluent, Conway, SC	<0.2	0.55	—	—	—	—	—	CH2M HILL, 1991
Shallow artificial streams, SC	1.2	1.0	17.5	10.7	8.9	1.8	17.5	Kania et al., 1976
Shallow artificial streams, SC	5.7	4.7	16.8	51.0	42.0	8.9	16.8	Kania et al., 1976

13 to 26.2 mg/kg in the grass species. Nickel is not biomagnified in the food chain. Invertebrates and fish in wetlands typically have concentrations less than 2 mg/kg dw (Table 15-5).

Nickel concentration and mass reduction have been estimated in a few wetlands receiving municipal effluents (Hendrey et al., 1979; CH2M HILL, 1991, 1992) and a wetland receiving urban runoff (Schiffer, 1989). Within the range of values summarized in Table 15-14, nickel removal efficiency and mass removal rates appear to correlate directly with inflow concentration and mass loading rate. Nickel concentration reduction efficiencies were measured between 25 and 70.7 percent, and mass removal rate was between 0.135 and 1.40 kg/ha/yr. Nickel removal appears effective in several types of wetlands (Table 15-14). Uptake k values are 15 to 18 m/yr for the Sinicrope et al. (1992) data.

Because removal is predominantly to the soils, the largest capacity of a wetland will saturate. Eger et al. (1993) speculate on lifetimes of 3 to 30 years if 20 cm of peat are involved in removing 1.0 mg/L of nickel from about 1.5 cm/d hydraulic loading.

Selenium

Selenium is an essential element for some plants and animals. Selenium is a metalloid because it displays properties of a metal and a nonmetal. The chemistry of selenium resembles that of sulfur, resulting in the presence of four oxidation states: selenide (−II), elemental selenium (O), selenite (+IV), and selenate (+VI). In surface waters, selenium is generally found as ferric selenite ($Fe_2(OH_4SeO_3)$, calcium selenate, or as elemental selenium. Selenium has a variable geographic significance because it occurs in elevated quantities only in regions where seleniferous soils occur. During microbial metabolism, selenium is organically bound as selenide compounds (seleno-amino acids, methyl selenides, and methyl selenones). Upon release from organisms, selenium may be converted to the volatile dimethyl selenide form (Masscheleyn and Patrick, 1993).

The average selenium concentration is 0.1 μg/L in world rivers and 8 mg/kg dw in typical agricultural soil (Table 15-5). At the Kesterson National Wildlife Refuge (NWR) in Merced County, CA, Ohlendorf et al. (1986) found a selenium concentration of 300 μg/L in agricultural irrigation return flows. This elevated selenium level resulted in elevated selenium levels in all trophic levels of the wetland food chain and in severe reproductive impacts to aquatic birds. Emergent plants at Kesterson contained about 52.1 mg/kg dw of selenium compared to 0.43 mg/kg dw at the nearby Volta Wildlife Area, which does not receive agricultural drainage waters. Invertebrates at Kesterson contained from 22.1 to 175 mg/kg dw of selenium compared to 1.29 to 2.09 mg/kg dw at Volta. Mosquitofish contained 170 mg/kg dw at Kesterson compared to 1.29 mg/kg dw at Volta and 1.0 mg/kg dw at a constructed wetland receiving municipal wastewater at Merced, CA. Selenium concentrations in bird eggs at Kesterson average between 9.1 and 81.4 mg/kg dw while bird eggs from Volta had concentrations between 4.14 and 6.1 mg/kg dw. Selenium concentrations in the livers of aquatic birds were even higher (species averages from 28.6 to 130 mg/kg dw) than those in the eggs. Bird breast tissues at Kesterson contained average selenium concentrations between 9.61 and 110 mg/kg dw compared to samples from Volta that had 2.45 to 13.2 mg/kg dw (Ohlendorf et al., 1990).

Selenium in biota at the Kesterson NWR is suspected to be in the organic form (selenomethionine or selenocystine). As with most metals, selenium concentrations in soils and tissues are more than 5000 times those found in surface waters (Schuler et al., 1990). Selenium is different from most metals (except for mercury) because of the importance of organic selenium and its bioconcentration in the aquatic food chain. Biological effects thresholds for selenium are between 30 and 370 mg/kg dw for 300 plankton, benthic invertebrates, and certain forage fish; 4 mg/kg dw for some freshwater and anadromous fish; and 3 to 10 mg/kg dw for some aquatic birds (Lemly, 1993). The "Kesterson Effect" of selenium bioconcentra-

Table 15-14 Nickel Dynamics in Wetlands

Description	Concentration (µg/L)			Mass Rates (kg/ha/yr)				Ref.
	In	Out	% Removal	In	Out	Removal	% Removal	
Constructed meadow/marsh/pond, Brookhaven, NY	35.0	10.27	70.7	1.98	0.58	1.40	70.7	Hendrey et al., 1979
Freshwater marsh receiving urban stormwater, Orlando, FL	4.0	3.0	25.0	—	—	—	—	Schiffer, 1989
Carolina bay receiving municipal effluent, Myrtle Beach, SC	17.0	9.12	46.3	0.224	0.089	0.135	60.2	CH2M HILL, 1992
Cypress-gum receiving municipal effluent, Conway, SC	18.2	13.2	27.5	—	—	—	—	CH2M HILL, 1991
Forested swamp receiving urban runoff, Sanford, FL	2.71	—	—	0.23	0.069	0.16	70.0	Harper et al., 1986
Bulrush in gravel	300	111	63	53.9	19.9	34.0	63	Sinicrope et al., 1992
SSF wetland	300	153	49	66.9	34.1	32.8	49	
Typha/Melaleuca SSF	52	<5	90					Noller et al., 1994
Carex SF	60	7					88	Eger et al., 1993

tion in arid land wetlands receiving agricultural drain waters is considered to be a widespread problem in the American West (Lemly et al., 1993).

Silver

The monovalent form of silver, Ag(I), is dominant in surface water where the silver concentration is largely regulated by the formation of an insoluble sulfide or by adsorption on organic matter. Silver is toxic to all aquatic organisms (Hale, 1977).

The average total silver concentration is 0.3 μg/L in world rivers (Stephenson, 1987) and 0.05 mg/kg dw in average agricultural soils. The background silver concentration in the soils of a South Carolina freshwater wetland was 0.02 mg/kg dw (Table 15-5). Silver concentration in wetland emergent plants at Kesterson NWR in California were <0.2 mg/kg dw, invertebrates had concentrations less than 0.152 mg/kg dw at Volta and Kesterson NWR; and mosquitofish had concentrations less than 0.122 mg/kg dw. Wetland-dependent birds at Volta had silver concentrations in their livers of 0.201 mg/kg, while birds at Kesterson had 1.02 mg/kg dw of silver in their livers.

Silver concentrations have been monitored in two natural wetland treatment systems in South Carolina (Table 15-15) (CH2M HILL, 1991, 1992). As with most other metals, the concentration of silver was not reduced (−11 percent mass change) when near typical ambient levels (0.3 mg/L for world rivers) and was reduced (+76 percent concentration change) when elevated compared to background. Silver is a frequent contaminant in municipal wastewaters and is regulated at very low concentrations in some states. For these reasons, additional research on the fate and effects of silver in treatment wetlands is an important subject for additional research.

Zinc

Zinc is an essential element for both plants and animals. It activates some enzymatic reactions important in respiration and serves as a cofactor in plant photosynthesis and DNA synthesis. Zinc is present in surface waters primarily as the divalent Zn(II) where it forms ionic hydrates, carbonates, and complexes with organics. The sulfide form of zinc is highly insoluble and serves as a sink for zinc in the aquatic environment.

The average concentration of zinc is 10 μg/L in world rivers. Average agricultural soils contain 80 mg/kg dw of zinc, and unpolluted wetland soil concentrations are typically less than 120 mg/kg dw (Table 15-5). Zinc concentrations in contaminated wetland soils have been reported as high as 779 mg/kg dw in a marsh receiving urban runoff and municipal wastewater (Zhang et al., 1990) and 6863 mg/kg dw at a lead-zinc mine impoundment (Lan et al., 1990).

Although zinc concentration in sediments, plants, and animals is typically higher on a dry-weight basis than in water (Table 15-5), zinc does not appear to biomagnify in the food chain. Zinc concentrations in plants from unpolluted wetlands typically range from 10 to 100 mg/kg dw and up to 341 mg/kg in cattail roots at the lead-zinc mine studied by Lan et al. (1990). Above-ground portions of reeds and cattails at this polluted site had 30 to 116 mg/kg dw of zinc, which is comparable to background levels in the Okefenokee Swamp in Georgia. Saltmarsh data summarized by Nixon and Lee (1986) indicated that zinc is sometimes present in these environments to much higher levels (up to 686 mg/kg dw in southeastern U.S. samples). Zinc concentrations in wetland animals from unpolluted sites (Table 15-5) are similar to plant concentrations (less than 120 mg/kg dw).

The effect of wetlands on zinc has been studied at unimpacted sites (DeLaune et al., 1981; Giblin, 1985; Feijtel et al., 1989), sites receiving municipal wastewaters (Best, 1987; Hendrey et al., 1979; CH2M HILL, 1991, 1992), and at sites receiving urban runoff (Schiffer,

Table 15-15 Silver Dynamics in Wetlands

Description	Concentration (μg/L)			Mass Rates (kg/ha/yr)				Reference
	In	Out	% Removal	In	Out	Removal	% Removal	
Carolina bay receiving municipal effluent, Myrtle Beach, SC	0.36	0.53	−48.9	0.0047	0.0052	−0.0005	−10.8	CH2M HILL, 1992
Cypress-gum swamp receiving municipal effluent, Conway, SC	4.0	1.0	75.9	—	—	—	—	CH2M HILL, 1991

Table 15-16 Zinc Dynamics in Wetlands

Description	Concentration (μg/L)			Mass Rates (kg/ha/yr)				Ref.
	In	Out	% Removal	In	Out	Removal	% Removal	
Creekbank salt marsh, LA	—	—	—	2.33	0.21	2.10	91	DeLaune et al., 1981
Cypress swamp, Waldo, FL	—	—	—	—	—	26.3	77	Best, 1987
Salt marsh, MA	—	—	—	—	—	0.055	—	Giblin, 1985
Freshwater marsh, Barataria Basin, LA	—	—	—	—	—	0.47	—	Feijtel et al., 1989
Constructed meadow/marsh/pond, Brookhaven, NY	2200	230	89.5	124.5	13.01	111.5	89.5	Hendrey et al., 1979
Freshwater marsh receiving urban stormwater, Orlando, FL	75.0	25.0	66.7	—	—	—	—	Schiffer, 1989
Carolina bay receiving municipal effluent Myrtle Beach, FL	20.6	5.6	72.7	0.272	0.055	0.217	79.8	CH2M HILL, 1992
Cypress-gum swamp receiving municipal effluent, Conway, SC	20.8	7.0	66.3	—	—	—	—	CH2M HILL, 1991
Forested swamp receiving urban runoff, Sanford, FL	3.90	—	—	0.44	0.26	0.18	40.9	Harper et al., 1986
AMD wetland, KY (Widows Creek)	30	20	33	4.38	2.92	1.46	33	Edwards, 1993 (unpublished TVA data)
Bulrush in gravel	399	63	79	53.9	11.3	42.6	79	Sinicrope et al., 1992
SSF wetland	2500	725	71	558	162	396	71	
Carex SF	1590	41					96	Eger et al., 1993
	1920	44					90	
	800	28					96	
	45	7					33	
Typha/Melaleuca SF	6900	260	96					Noller et al., 1994
	1000	23	98					
Five stormwater SF wetlands			−42					Strecker et al., 1992a

1989). Zinc concentration and mass reduction efficiency appears to correlate with inflow concentration and mass loading rate up to at least 2200 μg/L and 124.5 kg/ha/yr (Table 15-16). Zinc concentration was reduced by 66.3 to 89.5 percent in these wetlands, and mass was reduced by 77 to 91 percent. Outflow concentrations from these wetlands ranged from 5.6 to 230 μg/L, depending on inflow concentration. Zinc mass removal rates were measured between 0.055 and 111.5 kg/ha/yr. An average zinc removal efficiency of 60 percent was reported by U.S. EPA (1993d) for vegetated filter strips and 35 percent for constructed stormwater wetlands.

The removal rate constant for zinc is 28 m/yr for the SSF data of Sinicrope et al. (1992). This is comparable to the results of Delgado et al. (1993) for hyacinths, which give $k \cong 20$ m/yr.

The metal was transferred primarily to soils (80 percent) and fine roots (17 percent) in the Sinicrope et al. (1992) study. This matches the general allocation found by Zhang et al. (1990), who found 300 to 700 mg/kg in *Typha* roots and peat, but only 20 to 100 mg/kg in leaves and rhizomes.

SUMMARY

As industrial wastewater discharges come under increasingly stringent effluent discharge criteria, wetlands are being considered as an option for treating a variety of elements including salts and metals. Design of wetland systems to treat these elements is currently limited by insufficient knowledge of the factors that control removal rates and of the biological effects of concentrating these elements in wetlands.

This chapter summarizes the current knowledge about the fate and effects of these elements in wetlands. From this review, it is clear that wetlands cannot be used for consistent removal of most salts. This limitation exists because the typical surface water concentrations of most salts exceed growth requirements and because salts are soluble and tend to leach from sediments and tissues. In short, there are essentially no removal mechanisms except for nitrate and sulfate reduction.

Wetlands do seem to be effective at retaining significant loads of several trace metals. Sustainable metal uptake occurs primarily in the wetland sediments. However, this storage capacity in wetland sediments and biota will likely eventually be exceeded, and sooner at high influent loadings. In natural wetlands that are subject to state and federal jurisdiction, influent metal levels must be reduced by pretreatment to levels that will not result in chronically toxic conditions to biota. In constructed SF wetland treatment systems, this is not always practical. Extremely high levels of pretreatment, although prudent and necessary, negate the value of the wetlands and can result in potentially worse environmental effects through wastage of chemicals, energy, and creation of harmful sludges. SSF wetlands eliminate most biological contact, but require significant effort to rebuild or clean out when saturated.

In some cases, constructed wetlands can be used like land application sites receiving industrial wastewater and sludge. These land sites are used for a limited time depending on the cumulative loading of one or more metals. Limits on metal loadings to soils are based on accumulated concentrations that are kept below toxicity thresholds of sensitive plants. When the allowable loading limit is reached, these sites are abandoned or cleaned out to an alternate burial site. Constructed wetlands that have sediment or tissue metal levels that are approaching threshold limits, such as the action levels summarized in Table 15-5, should be closed. In many cases, however, calculation of actual metals loading rates indicate that the system's life is on the order of hundreds of years or longer. Actual burial of metals in accreting sediments may further extend wetland treatment system life expectancies.

CHAPTER 16

Organic Compounds

INTRODUCTION

Most of the wetland data for organic compounds has resulted from studies which measure biochemical oxygen demand (BOD) or chemical oxygen demand (COD) in municipal wastewater. However, there is an increasing number of treatment wetlands that target different suites of organic chemicals. These pose a difficult set of problems due to possible toxicity to plants and limitations in aerobic and anaerobic degradation.

Wetlands manufacture and contain a wide spectrum of organic compounds. These range from small molecules such as methane up to humic acids of very high molecular weight. Many wetland soils are organic in nature and possess an affinity for introduced organics, via sorption and other binding mechanisms. Aliphatic hydrocarbons, both straight chain and branched, are present as natural waxes. Aliphatics degrade in the environment, but more slowly when the molecular weight is high (Alexander, 1977). Aromatics follow a similar pattern, with polyaromatic hydrocarbons (PAHs) degrading more slowly than benzene; those with more than three rings cannot support microbial growth (Zander, 1980).

The major routes for removal of hydrocarbons from wetland waters are (1) volatilization, (2) photochemical oxidation, (3) sedimentation, (4) sorption, and (5) biological (microbial) degradation. Three types of microbial processes can contribute: fermentation, aerobic, and anaerobic respiration.

WETLAND PROJECTS

Most of the major petroleum companies in the U.S. have either pilot or full-scale wetland treatment projects, associated with production, refining, and petrochemical wastewaters. Some of that information is proprietary, but a good deal has been published in summary form. Six paper mills are using or piloting treatment wetlands for wastewater polishing. Food processing produces wastewaters with specific types of hydrocarbons. Wetlands treating milk, potato, sugar, egg, spice, and olive oil processing wastewater are all in use. However, the nature of the organic compounds differs from starch to protein to sugar to lignin to oil across these nonmunicipal wastewaters.

A few wetlands are treating wastewaters that contain a small set of solvents, which pose a different problem. The target species is not a group of compounds, such as "BOD," but a single species, such as methanol or acetone.

Perhaps the most challenging, yet the most promising, application of constructed wetlands is for the control of the large number of single organic compounds which are the subject of landfill leachate regulation. The prospect of small pulses of individual organics at erratic frequency and variable concentration requires a broad-spectrum treatment device which is always poised for the next event. Leachate flows are relatively small by most comparative standards, and thus, the land required for natural treatment systems is not a large burden. The passive nature of the constructed wetland fits the task admirably. However, design data are sparse at the present time.

PETROLEUM PROCESSING

REFINERY EFFLUENTS

Refinery wastewaters have received polishing in constructed pond and island wetlands at Amoco's Mandan, ND facility for over 20 years (Litchfield and Schatz, 1989; Litchfield, 1990, 1993). Approximately 2500 m^3/d of water pass through the 16.6-ha system, providing a hydraulic loading rate of 1.5 cm/d (see Chapter 27 for a case history for this wetland). Reasonable removals of many refinery pollutants have been achieved by these wetlands (Table 16-1). The facility also has considerable wildlife value and has received awards for that. High reductions in hexavalent chromium, phenols, oil and grease, and BOD and COD were sustainably maintained.

Chevron utilizes surface-flow (SF) wetlands to polish all the wastewater leaving the Richmond, CA refinery before it enters San Francisco Bay (Duda, 1992). The 36 ha of wetlands treat 9500 m^3/d of water emerging from the oxidation ponds at the refinery. The hydraulic loading is therefore 2.6 cm/d, which is at the light end of the spectrum for treatment wetlands. Consequently, the removals of the ammonium (76%) and nitrate (69%) over a 3-year period were good. BOD dropped 51% and TSS 45%, but both these parameters were fairly low in the influent (*circa* 8 and 20 mg/L, respectively). Return fluxes from the wetland biomass are important in this range. The wetlands also immobilize metals, notably zinc, chromium, and selenium. This is viewed as potentially detrimental to wildlife populations, although there have been no instances of difficulties and toxicity tests on rainbow trout have given 100% clean results.

Wetland wastewater treatment in the petroleum industry is not confined to the U.S. The Jinling Petrochemical Company, Nanjing, Peoples' Republic of China (PRC), reported small reductions in several parameters, including phenol and oil, in water hyacinth wetlands (Tang and Lu, 1993). Dong and Lin (1994a,b) report data and models for both a full-scale facility of 50 ha treating 100,000 m^3/d and a research facility treating lesser amounts of water in ten, 1.5-ha test units. The full-scale facility reduces phenols and oil and grease (Table 16-2). The research facility tested a variety of plants and soil substrates over the seasons. No large effect was found for any variable except the hydraulic loading rate (Table 16-3). A first-order areal uptake model with $C^* = 0$ was proposed for several pollutants, including phenol.

SPILLS AND WASHINGS

Tenneco, Inc., Houston, Texas, utilized a rock-reed wetland to treat wastewaters from a compressor station. The loading was 2.4 cm/d, and treatment efficiency for oil and grease was about 90% (Honig, 1988).

Table 16-1 Performance of the Mandan, ND Wetlands in 1990

Parameter	Wetland Influent	Wetland Effluent
BOD, mg/L	25	3
COD, mg/L	175	37
TSS, mg/L	35	5
Oil and grease, μg/L	2100	130
Phenols, μg/L	80	5
Hex chrome, μg/L	16.0	3.6

Note: Data from Litchfield, 1993.

Table 16-2 Performance of the Yanshan Full-Scale Wetlands at Beijing, PRC, 1992–93

Parameter	Wetland Influent	Wetland Effluent
BOD, mg/L	38	15.3
COD, mg/L	170	47.5
TSS, mg/L	181	41
TP, mg/L	1.51	0.43
TN, mg/L	9.9	5.8
NH_4-N, mg/L	5.8	3.5
DO, mg/L	4.3	7.8
pH	7.8	7.9
Oil and grease, μg/L	840	290
Phenols, μg/L	27	10

Note: Data from Dong and Lin, 1994a.

In South Africa, a number of reed bed wetlands have been established to treat wastewaters generated from truck washing operations at oil industry depots (Wood, 1993). A subsurface-flow (SSF) wetland treats the runoff from a 0.8-ha vehicle yard in Surprise, AZ with 54 to 92 percent removal of oil and grease (Wass and Fox, 1993).

OIL SAND PROCESSING WATER

Suncor Inc., Oil Sands Group in British Columbia, Canada, has built a pilot wetland facility to treat wastewater from the processing plant (Gully and Nix, 1993). The principal hydrocarbon contaminant is naphthoic acid, which was reduced in the wetlands more effectively in the summer than in the winter. Performance was good in terms of reducing toxicity to *Daphnia magna.*

PRODUCED WATER

Studies are under way at Argonne National Laboratory, Argonne, Illinois, to determine the optimal selection of plant types for volume reduction of produced waters and for maximizing metal uptake. The proposed technique capitalizes on the transpiration of water by wetland plants, accompanied by harvest for metal removal (Hinchman et al., 1993). Early projections of water loss seem unrealistically high: 6.1 cm/d for bulrushes, for instance. That is one order of magnitude greater than the solar energy available for evaporation. Water budget data for wetlands indicate water losses approximately equal to lake evaporation (Kadlec et al., 1987; Williams et al., 1987).

A Wyoming pilot project consisted of bacterial ponds followed by a riffle channel followed by SF wetlands (Caswell et al., 1992). It has run in all seasons and reduces Radon 226, benzene, and phenolics.

Table 16-3 Performance of the Fangshan Research Wetlands at Beijing, PRC, 1991–1993

Parameter	Season	% Reduction in Wetland
COD, mg/L	Spring	38.9
	Summer	47.7
	Fall	30.9
	Winter	32.5
TSS, mg/L	Spring	44.6
	Summer	42.1
	Fall	55.6
	Winter	27.8
Phenols, μg/L	Spring	34.2
	Summer	36.7
	Fall	29.2
	Winter	27.8

Note: Data from Dong and Lin, 1994b.

RESEARCH

A variety of highly focused research studies on individual hydrocarbons in wetlands has been reported. For instance, New Mexico State University has conducted microcosm research on the ability of *Scirpus validus* SSF wetlands to degrade a variety of hydrocarbons (Zachritz et al., 1993). Initial results on benzoic acid indicated 99 percent removal at doses up to 40 mg/L.

SURFACTANTS

Linear alkylbenzenesulfonate (LAS), a widely used synthetic surfactant for domestic detergents, was effectively reduced in a *Typha-Phragmites* wetland. The 474-m^2 SF system treated gray water from a 100-inhabitant community. Adsorption and degradation were found to both be operative in the wetland and in companion laboratory research (Inaba, 1992). The k values for the disappearance of these surfactants are in the range of 30 to 80 m/yr, depending upon chain length and isomeric form (Figure 16-1). Seasonal dependence was strong, with a temperature factor of $\theta = 1.087 \pm 0.005$ averaged over all forms of the detergent. The inlet concentration totaled about 5 mg/L and was reduced by over 90 percent in the summer and over 40 percent in the winter at an average hydraulic loading of approximately 5 cm/d. The surfactant partitioned strongly to wetland total suspended solid (TSS), with a partition coefficient of 30 L/kg for chain length 11, isomer 5. Values for other forms of the surfactant were higher, ranging up to 520 L/kg for chain length 13, isomer 2. Sorption was not diminished at cold temperatures. Therefore, the sorption capacity and TSS trapping efficiency of the wetland partially offset the diminished degradation during winter. Sediment-bound surfactant reached 300 mg/kg in mid-winter, but was essentially zero in mid-summer.

FOOD PROCESSING WASTEWATERS

SUGAR REFINING

Wastewaters from sugar processing plants are being treated in both natural (Gambrell et al., 1987) and constructed SF wetlands (American Crystal Sugar). The form of the BOD in

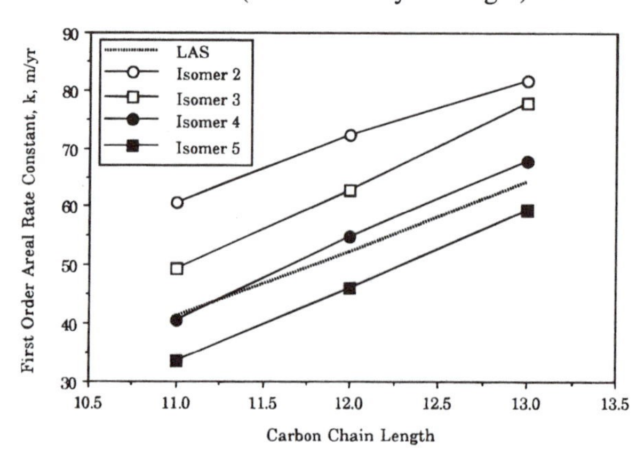

Figure 16-1 First-order disappearance rate constants for surfactants linear alkylbenzenesulfonates in SF wetlands. The data of Inaba (1992) have been analyzed under the plug flow assumption and presuming $C^* = 0$. The isomers result from different positions of the phenylsulfonate substitution; LAS refers to the blend of all chain lengths and isomers. The blend is weighted toward isomer 5 and chain length 11.

these wastewaters is almost exclusively sucrose. It is therefore not surprising that the k-C* model does a good job of fitting the disappearance of soluble total organic carbon (TOC) (Figure 16-2). The rate constant is virtually identical to that obtained from data from numerous domestic wastewater wetlands (36 vs. 34 m/yr in Chapter 12). The rate constant is not sensitive to the multiplier between BOD and TOC, but the value of C* is.

POTATO PROCESSING

A pilot project treating high-starch potato processing water in an SF wetland with 2 to 4 days detention produced 60 to 70 percent reduction in COD (Kadlec et al., 1990b). The corresponding k = 37 m/yr; a value for C* was not determined.

MILK AND EGGS

The addition of milk processing water to domestic treatment lagoons has generated *de facto* data analysis for this particular form of BOD: high in colloids. The Gustine, CA data permit analysis, according to the k-C* model. That analysis is presented in Chapter 21 and leads to k = 23.6 m/yr and C* = 17.5 mg/L. Apparently, milk-derived BOD is harder to degrade than domestic wastewater BOD. Egg breakage can cause extremely high BOD in wastewaters from egg processing facilities. Constructed SF wetlands are proving very effective in treating those waters (Ogden, 1994).

MEAT PROCESSING

Wetlands are in use for treating abattoir wastewaters in Australia (Finlayson et al., 1990) and in New Zealand (van Oostrom and Cooper, 1990). These waters contain significant quantities of blood and other sources of BOD and are well treated in the wetland environment.

DAIRYING

Dairy wastewaters have been successfully treated in SSF bulrush gravel systems (Tanner et al., 1995). The areal k values for CBOD were 20 m/yr with no plants and 25 m/yr with

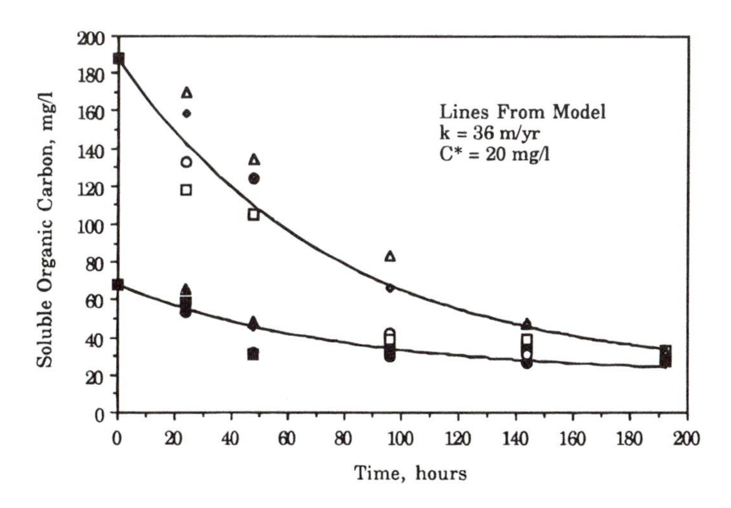

Figure 16-2 Disappearance of soluble carbon in wetland microcosms dosed with sucrose. Data represent two levels of starting concentration, two levels of stirring, and two levels of supplementary nutrient addition. (Data form Gambrell et al., 1987.)

plants. These values are considerably lower than those for domestic wastewater in SSF wetlands (Chapter 12, Table 12-4), but agree with those for a mix containing milk processing water in the SF wetlands at Gustine, CA (Chapter 12, Table 12-3).

LEACHATES

Landfill leachates can contain a large variety of hydrocarbons, including priority pollutants, phenolics, and high BOD of many origins. Reports indicate significant reductions of BOD and TOC, but data are thus far not available on the speciation of the carbon compounds (Surface et al., 1993; Martin and Moshiri, 1994).

PULP AND PAPER WASTEWATERS

SF wetlands are in use in the pulp and paper industry for polishing paper mill effluents (NCASI, 1995; Knight, 1993; Knight, et al., 1994). The standard wastewater parameters of BOD, TSS, and nutrients are used to characterize these waters, but there is a clear difference in the character of the BOD. Lignin-based compounds that survive the pretreatment processes are not found in domestic wastewater and are harder to degrade than starches and sugars. Nevertheless, wetlands have proven to be effective in the reduction of this specialized form of BOD. Data from six operating systems indicate that the k values for degradation may be comparable to those for domestic wastewater.

PESTICIDES

The list of pesticides is very long, and therefore, wetland fate and transport studies lag far behind their introduction and use. A distinction may be drawn between the persistent chemicals used prior to the 1950s and the more degradable substances used since that time.

CHLORINATED ORGANICS

Many of the "old" pesticides, such as DDT, are very persistent in the environment. These substances partition strongly to particulate matter. It is doubtful that wetlands can provide any effective mechanism for degradation, but wetlands can act as a trap for the particulates that carry most of the load. Very little data is available for this class of compounds, but Winter (1991) reports some success treating polychlorinated biphenyls (PCBs) and Lindane™ in SSF systems.

The microcosm research at Krefeld, Germany found "a surprisingly large amount of pentachlorophenol was absorbed" in wetland systems (Seidel, 1976). This result is somewhat unexpected, since this compound is used as a wood preservative.

Studies at the Des Plaines, IL treatment wetlands indicated the presence of some of the old "hard" pesticides. The river and wetland sediment samples all contained quantifiable levels of PCBs, and some of them also had one or more of the chlorinated pesticides or metabolites DDT, DDE, and dieldrin. None of the samples contained gamma-BHC (Lindane), heptachlor, heptachlor epoxide, alpha-endosulfan, or endrin at quantifiable levels, although these chemicals were or are used in the watershed. All identified species were found in the low micrograms per kilogram (parts per billion) range. A brief survey of the residue level literature indicates that these results are typical for midwestern rivers in industrial and agricultural areas. PCBs are present at the 20 to 25 μg/kg level in riverborne sediments throughout most of the year and are in the sediments of wetland EW3 at less than 10 μg/kg. DDT, together with its decomposition product DDE, were present at approximately 2 μg/kg. Wetland cell EW3 sediments contained similar low levels of DDT + DDE. Dieldrin ranged from 0.1 to 3.0 μg/kg in river sediments and 0.5 μ/kg in wetland sediments.

ATRAZINE

Atrazine, a triazine herbicide used on cornfields, exists in many streams in the midwestern part of the U.S. Agricultural practices within the region produce pollution with atrazine, at concentrations which sometimes peak in excess of the federal drinking water standard. The atrazine-wetland interaction is very complex, including removal from the area by convection in the water, loss of chemical identity by hydrolysis to hydroxytriazine and dealkylation, and sorption on wetland sediments and litter.

Atrazine transport, sorption, and identity loss were studied at the Des Plaines site and in accompanying laboratory work. Sorption was effective for soils and sediments, but the more organic materials, such as litter, showed a stronger affinity for atrazine than the mineral base soils of the wetland cells at Des Plaines (Alvord and Kadlec, 1995a,b,c). Atrazine was found to degrade on those sediments according to a first-order rate law, with a half life of 40 to 90 days (Figure 16-3). However, degradation was faster on cattail litter, precluding a rate measurement. The estimated half life on cattail detritus was on the order of 5 days.

Outflows from the Des Plaines wetland cells contained reduced amounts of atrazine compared to the river water inputs. During 1991, atrazine peaked in the river due to two rain events. Only about 25 percent of the incoming atrazine was removed in wetland cell EW3, but 95 percent was removed in wetland cell EW4. The explanation is that the detention time in EW4 is longer than in EW3. Either time-averaged or dynamic models, both embodying a nonideal mixing model, fit the data for wetland EW3 (Figure 16-4). The value of k $\approx$ 14 m/yr calibrated the models to the data.

PHENOL

Phenol is a single compound of considerable industrial importance and a frequent pollutant in industrial effluents. It also has been studied for over 30 years in the wetland treatment context. Early work at the Max Planck Institute in Krefeld, Germany focused on SF microcosms with bulrushes (*Schoenoplectus lacustris*) and resulted in a series of publications on

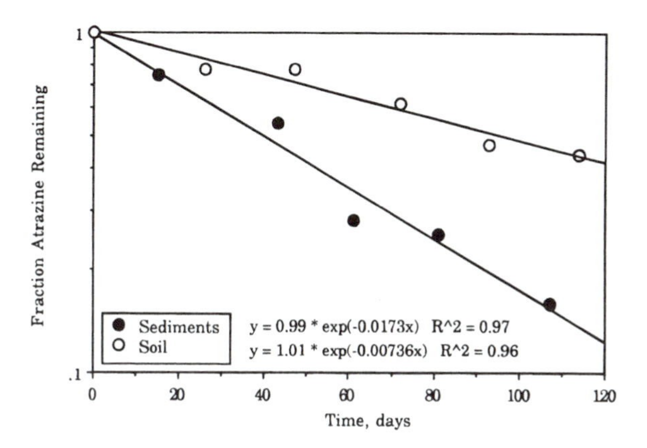

Figure 16-3 Disappearance of atrazine sorbed to wetland soils and sediments from the Des Plaines site.

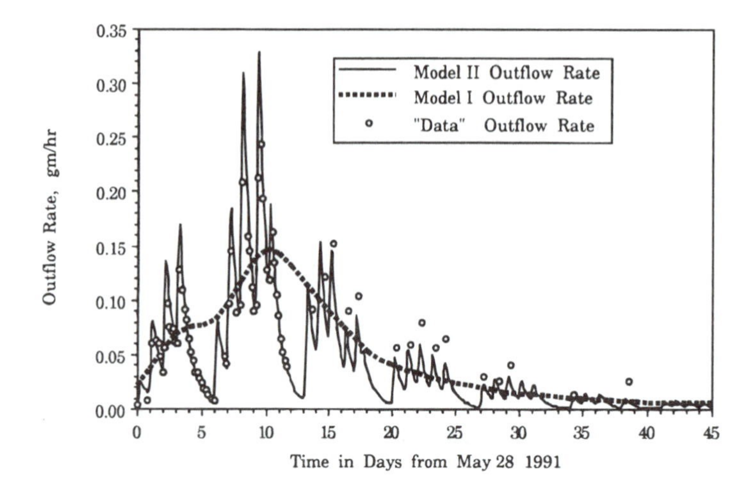

Figure 16-4 EW3 atrazine mass outflow as a function of time. Model I is a steady flow, tracer-based RTD calculation. Model II is a dynamic simulation which fits the hourly changes in flows of water and atrazine. Both models use the same postulated first-order homogeneous reaction rate constant (k = 14.4 m/yr).

removals and effects on plant physiology (for example, Seidel, 1966). That data showed degradation rates of 5 to 20 $g/m^3/d$, and in some instances, an induction period of 1 to 3 days before reduction commenced. The declining data may be modeled by a zero-order reaction (Figure 16-5).

$$\frac{d(VC)}{dt} = -k_0'A = -k_0 \quad (16\text{-}1)$$

where A = wetland area, m^2
C = phenol concentration, mg/m^3
k_0 = zero-order rate constant, mg/d
k_0' = zero-order rate constant, $mg/m^2/d$
V = water volume, m^3
t = time, days

The zero-order rate constant for the Krefeld data was $k_0 = 81 \pm 18$ mg/d in the summer and 34 ± 10 mg/d in the winter. The corresponding temperature coefficient is $\theta = 1.083$.

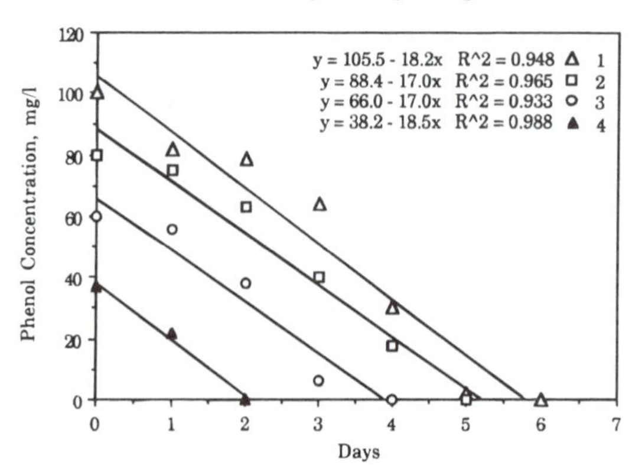

Figure 16-5 The disappearance of phenol in microcosms planted to *Schoenoplectus lacustris.* A zero-order fit to the data is quite appropriate. (Data from Seidel, 1976.)

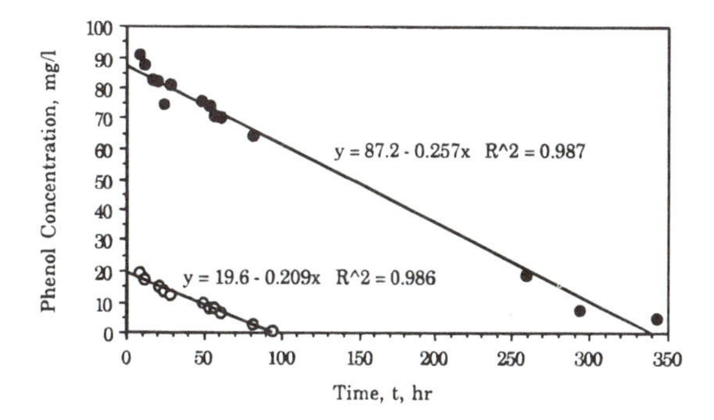

Figure 16-6 Disappearance of phenol in cattail microcosms. A zero-order fit is shown.

Microcosm research on SF wetlands with cattails produced similar results (Srinivasan and Kadlec, 1995). Figure 16-6 shows the decline of phenol in those experimental microcosms. The average value of the zero-order rate constant was 76 $\pm$ 33 mg/d ($k_0' = 0.55$ g/m²/d) for vegetated microcosms and $k_0 = 108 \pm 26$ mg/d ($k_0' = 0.78$ g/m²/d) for unvegetated microcosms. The sizes of these microcosms were comparable to those at Krefeld.

There was little selective loss of phenol from microcosms containing only water and the phenol dose. This study showed no effect of water depth, but a slight effect of soil type, with more organic content promoting phenol disappearance. There were higher evaporative losses in the summer, with phenol being lost with water at the bulk water concentration; there was not significant evaporative concentration nor selective stripping of phenol.

Polprasert and Dan (1994) operated both SSF microcosms and an SSF pilot wetland in Bangkok, Thailand for phenol reduction at high strength. Nearly complete removal of phenol was measured for loading rates up to 300 kg/ha/d. Some phenol was volatilized; some was found in the roots of the cattails. Disappearance followed a first-order model (Figure 16-7).

These microcosm and pilot scale rates are greater than those observed in some full-scale field situations. For instance, the Mandan, ND wetlands do not reduce phenols to zero, as they should, because of a detention time on the order of months. Further, the loss rates for the Jinling project are also much smaller than would be predicted from the microcosm data. Both of these systems receive phenol at concentrations much lower, by a factor of more than 100, than the lab studies. It is clear that pilot scale data is required for determination of design parameters for phenol; microcosms are not adequate. For instance, the areal mixing efficiency of the wetland may be an important determinant of trace amounts of phenol in the effluents from pilot and full-scale wetlands.

The Listowel, Ontario studies included the monitoring of phenols in five wetlands for 4 years (Herskowitz, 1986). These substances were present in the aeration cell and in the

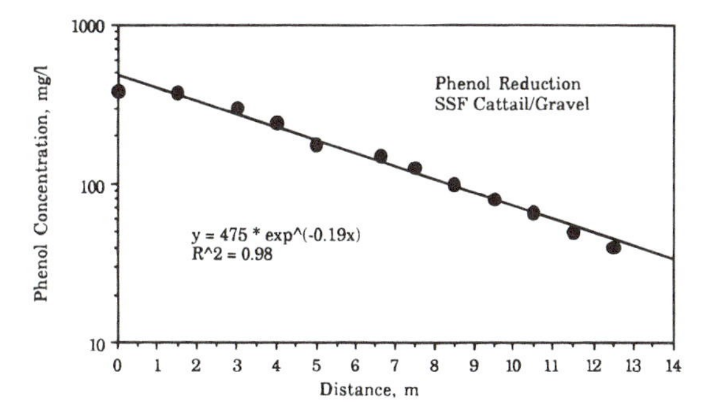

Figure 16-7 Reduction of phenol in a cattail/gravel SSF wetland. (Data from Polprasert and Dan, 1994.) A first order fit is quite good.

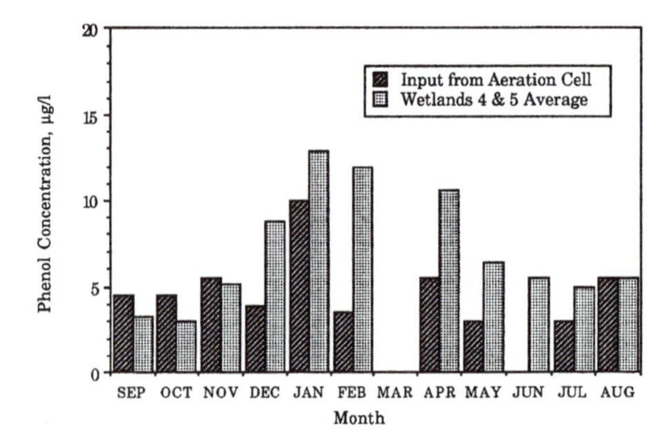

Figure 16-8 Phenol in and out of Listowel wetlands 4 and 5 for 1983-84. There was no chemical addition; the phenol was generated in the treatment system. Data from Herskowitz (1986).

lagoon that provided the wetland influent at concentrations on the order of 10 µg/L. Winter levels were higher than summer, and the wetlands did not provide removal at these low concentration levels (Figure 16-8).

NAPHTHOIC ACID

Cattail microcosms were also used to examine the removal of naphthoic acid in the Srinivasan and Kadlec (1995) project. There is a lag time before removal starts, as observed in the Krefeld experiments with phenol. However, that lag phase is followed by a steady but slow decline of this hydrophilic hydrocarbon (Figure 16-9). The zero-order rate constant is smaller than for phenol, k = 6 mg/d (0.044 $g/m^2/d$).

SUMMARY

Wastewaters containing a wide variety of hydrocarbon constituents are presently being treated in wetland systems. In terms of the lumped measures of BOD, COD, and TOC, performance has been generally good. Surprisingly, the rate constants for domestic wastewater polishing wetlands are not far different from those for other generating sources, such as food and paper processing.

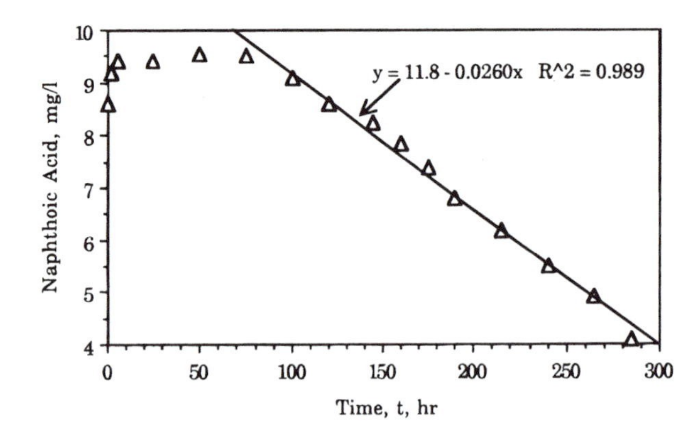

Figure 16-9 Disappearance of napthoic acid in cattail microcosms. A zero-order fit is appropriate after an induction period.

The surfactant work in Japan has shown that individual compounds may behave quite differently, depending upon isomeric form and upon small changes in molecular weight. This is somewhat daunting, because it means preproject investigations for specialty chemicals are a necessity; rate constants may not be inferred from data on related compounds.

The available data on phenol show that microcosms are only qualitative predictors of the performance of full-scale wetland systems. This too is bad news, because it means that pilot scale tests are a necessity for generating design data.

However, there is some removal of nearly every organic compound tested in wetland treatment systems, whether full scale, pilot scale, or microcosm. This includes organics from very important generators: pesticides, petrochemicals, paper and food wastes, and landfill leachates. In each of these categories, there are multiple studies that have initiated the design database. What is known thus far shows great promise for wetland technology for organics control, especially in those situations where these passive systems can be fit into the landscape and accumulated organics will not pose a threat to wetland biota.

CHAPTER 17

Pathogens

INTRODUCTION

Human pathogens are typically present in untreated domestic wastewaters. Although not components of water chemistry, these organisms range from submicroscopic viruses to parasitic worms visible to the unaided eye and represent particulate components of water quality. The efficiencies of conventional treatment technologies to reduce pathogens to noninfective levels have been studied thoroughly, and wastewater treatment plants regularly add processes to accomplish necessary removals (WEF, 1992; Metcalf and Eddy, 1991). The most common add-on treatment or disinfection processes are chlorination, ozonation, and ultraviolet disinfection.

Because of its low cost and proven effectiveness, chlorination has been the disinfection method of choice for many years. However, negative side effects of effluent chlorination have become apparent in the past 20 years. Residual free chlorine harms a variety of aquatic organisms and causes chronic and acute toxicity to microorganisms and fish. Also, when it comes in contact with organic compounds in the wastewater or naturally occurring compounds in the receiving water, free chlorine forms trihalomethanes and other organochlorine compounds known to be carcinogenic. These findings have resulted in the increased use of add-on dechlorination techniques and in the development of the ozonation and ultraviolet disinfection technologies.

Cost, operation and maintenance, and performance problems have limited the use of ozonation and ultraviolet disinfection. Consequently, regulators have developed more information on the effectiveness of low-energy, nonchemical methods of disinfecting wastewater. In some cases where receiving waters are not used for potable water supply or for contact recreation, disinfection requirements have been relaxed.

Most natural treatment technologies have the potential to reduce populations of human pathogens because of natural die-off rates and hostile environmental conditions. Wetlands have been found to reduce pathogen populations with varying but significant degrees of effectiveness. This chapter reviews the pathogens typically found in domestic wastewater and describes the effect wetlands have on pathogen populations passing through them.

REVIEW OF WASTEWATER PATHOGENS

Table 17-1 lists some of the most prevalent human pathogens associated with domestic wastewater. The pathogens are functionally divided into five groups: viruses, bacteria, fungi, protozoans, and helminths. Although the density of these organisms in raw wastewater varies because of the health of the resident population, some typical information on frequency of occurrence, density of colony-forming units, and infective doses is provided when available.

Viruses are submicroscopic, nonliving particles of genetic material that are enclosed in a sheath. Viruses cannot divide and reproduce alone, but they can infect host organisms and

Table 17-1 Some Human Pathogens Typical of Domestic Wastewater

Pathogen	Illness	Human Feces		Infective Dose
		Frequency of Occurrence (%)	Density (cfu/g)	
Viruses				
Adenovirus (31 types)	Respiratory disease	—	—	—
Enteroviruses (67 types)	Diarrhea, respiratory disease, polio	—	—	<10
Hepatitis A	Infectious hepatitis	—	—	—
Norwalk Agent	Gastroenteritis	—	—	—
Rotavirus	Diarrhea	—	10^{10}–10^{11}	—
Reovirus	Gastroenteritis	—	—	—
HIV	AIDS	—	—	—
Bacteria				
Campylobacter jejuni	Diarrhea	100	1.5×10^{11}	10^{6}
Clostridium spp.	Tetanus	76	4×10^{6}	—
Escherichia coli (includes pathogenic strains)	Diarrhea	100	4×10^{8}	—
Klebsiella spp.	Respiratory and urinary tract infections	50	5×10^{4}	—
Leptospira (150 spp.)	Leptospirosis	—	—	Low
Salmonella typhi	Typhoid Fever	—	10^{8}	10^{7}
Salmonella (~1700 spp.)	Salmonellosis	—	10^{6}	10^{5}–10^{8}
Shigella (4 spp.)	Diarrhea, dysentery	—	10^{6}	10^{1}–10^{2}
Vibrio spp.	Cholera, diarrhea	—	10^{6}	10^{8}
Yersinia spp.	Yersiniosis	—	10^{5}	10^{9}
Fungi		53	4×10^{4}	—
Aspergillus fumigatus	Aspergillosis	—	—	—
Candida albicans	Fungal infections	—	—	—
Protozoa				
Balantidium coli	Diarrhea, dysentery	—	—	25–100
Cryptosporidium	Diarrhea	—	—	—
Entamoeba histolytica	Diarrhea, dysentery	—	10^{7}	10–100
Giardia lamblia	Diarrhea	—	10^{5}	25–100
Helminths				
Ascaris lumbricoides	Roundworm	—	10^{4}	Several units
Clonorchis sinensis	Bile duct infection	—	—	—
Diphyllobothrium latum	Fish tapeworm	—	—	—
Enterobius vericularis	Pinworm	—	—	—
Fasciola hepatica	Liver fluke	—	—	Several units
Fasciolopsis buski	Intestinal fluke	—	10^{2}	Several units
Hymenolepis nana	Dwarf tapeworm	—	—	1
Opisthorchis spp.	Bile duct infection	—	—	—
Schistosoma spp.	Schistosomiasis	—	—	—
Taenia spp.	Tapeworm	—	10^{4}	1
Trichuris trichura	Whipworm			

Note: Data from Krishnan and Smith, 1987; Shiaris, 1985; Leclerc et al., 1977; Cabelli, 1977; Metcalf and Eddy, 1991; Prost, 1987.

functionally reproduce to very large populations at the expense of the host organism. Over 100 virus types are known to occur in human feces, with minimum infective doses as low as one organism for some species. Enteroviruses include poliovirus (polio), coxsackievirus (meningitis and colds), and echovirus (meningitis and colds). Viral gastroenteritis is considered to be the most common waterborne illness in the U.S. (Shiaris, 1985). Hepatitis A virus is a major health problem that results in more than 40,000 reported cases of hepatitis in the U.S. annually (Shiaris, 1985). Rotavirus is also a public health concern because of its potentially fatal effects on infants. Human Immunodeficiency Virus (HIV) is the causative agent of Acquired Immune Deficiency Syndrome (AIDS), one of the major health concerns in the world today. HIV may occur in the bodily fluids and excretions of infected individuals, possibly resulting in the presence of HIV in raw domestic wastewater.

Bacteria are universally present in human feces, with normal populations of about 10^{11} organisms per gram (Leclerc et al., 1977). Although most of these organisms live symbiotically with their hosts, a number of species are known human pathogens and occur with great frequency in infected individuals. Table 17-1 lists some of the more important disease-causing bacteria species. In the U.S., *Salmonella* spp. is a major concern to public health because of the more than 30,000 reported cases of salmonella poisoning each year (Shiaris, 1985). But as with a number of infectious bacterial pathogens, *Salmonella* also derives from animals, and infection may be largely from nonwater contact sources. The minimum infective dose for *Salmonella* is considered to be relatively high (10^5 to 10^8 organisms), but *Shigella* spp. causes bacterial dysentery at much lower doses (10 to 100 organisms), resulting in over 15,000 reported cases each year in the U.S. Other waterborne bacterial diseases that now occur less frequently in the U.S. have created significant public health problems in the past and continue to be problematic in other countries.

Human parasites derived from wastewater-related infections include protozoa and helminths. Two common protozoan parasites are *Entamoeba histolytica* and *Giardia lamblia* which both cause diarrhea in infected humans. The phylum Aschelminthes (cavity worms) includes all parasitic worms incapable of adult life without a host organism. A number of helminths, including tapeworms and flukes, are found in human feces from infected individuals and can be spread through wastewater pathways. Wetlands have been shown to reduce these organisms and their cysts and eggs (Rivera et al., 1994).

PATHOGEN REMOVAL PERFORMANCE

Pathogenic organisms live at the expense of their hosts. In many cases, they depend on their host for the environmental conditions necessary for life and reproduction. Wastewater is a hostile environment for pathogenic organisms, and factors such as natural die-off, temperature, ultraviolet light, unfavorable water chemistry, predation, and sedimentation cause pathogen populations to be reduced. Figure 17-1 illustrates generalized percentage removals for viruses, bacteria, and helminths when wastewater is retained in facultative treatment ponds. Many of the processes that reduce pathogen populations in natural pond treatment systems are equally or even more effective in wetland treatment systems.

INDICATOR ORGANISMS

Measurement of human pathogenic organisms in untreated and treated wastewater is expensive and technically challenging. Consequently, environmental engineers have sought indicator organisms that are (1) easy to monitor and (2) correlate with populations of pathogenic organisms. No perfect indicators have been found, but the coliform bacteria group has long been used as the first choice among indicator organisms. In the U.S., coliforms are

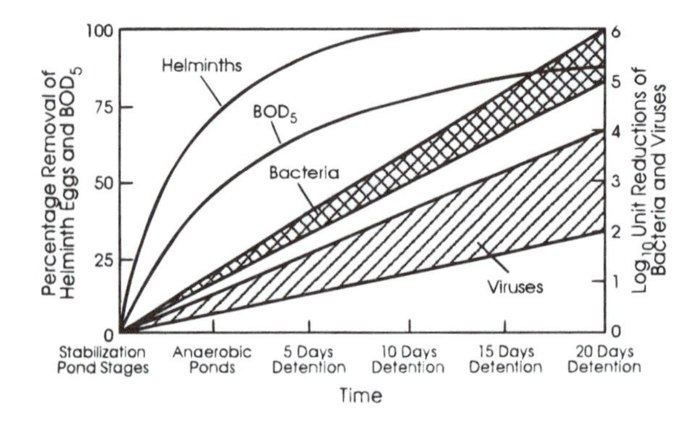

Figure 17-1 Effect of residence time in stabilization ponds on removal efficiency for BOD_5, helminth eggs, bacteria, and viruses. (Data from Krishnan and Smith, 1987.)

monitored most frequently as total or fecal coliforms. Total coliforms include bacterial species that are rod-shaped, stain Gram-negative, do not form spores, are facultatively anaerobic, and ferment lactose with gas production in 48 h at a temperature of 35°C. Fecal coliforms are separated from total coliforms by their ability to ferment lactose with gas production in 24 h at a temperature of 44.5°C. In many other countries, an even narrower group, *Escherichia coli,* is used as the indicator group of choice. The fecal streptococcus group also is used frequently to confirm fecal contamination.

These indicators have been the subject of intense investigation and frequent criticism. Total coliforms are ubiquitous in surface waters, and they include many bacteria that are not derived from human or other animal pollution sources. Thus, the total coliform measurement is the least specific indicator for providing evidence of human fecal contamination. The fecal coliform group is composed largely of fecally derived coliforms, but it also includes free-living bacteria (primarily *Klebsiella* spp.) and bacteria from other warm-blooded animals including birds and mammals. Thus, although the fecal coliform measure is a better indicator of human fecal contamination than total coliform, it is by no means specific. The *E. coli* test can be used to diagnose fecal contamination. However, because these bacteria also originate in other warm-blooded animals, the *E. coli* is not diagnostic of human fecal contamination alone. Typically, *E. coli* constitutes about 20 to 30 percent of the total coliforms found in raw and treated domestic wastewater (Dufour, 1977).

Fecal streptococci are found in the feces of humans and other warm-blooded animals including birds and mammals (Table 17-2). These bacteria are found frequently in waters receiving fecal contamination and are not believed to multiply in natural or polluted waters and soils. Because fecal streptococci bacteria seem to survive longer in receiving waters than fecal coliforms, they are used as a second indicator of fecal contamination and may be a better indicator of the presence of the longer-living viruses originating in wastewater (Clausen et al., 1977; Scheuerman et al., 1989). Some species such as *Streptococcus faecalis* and *S. durans* are considered to be diagnostic of human fecal contamination. Also, the ratio between fecal coliforms and fecal streptococci (FC to FS ratio) is used to distinguish between human and nonhuman coliform contamination. Since animal feces have a higher density of fecal streptococci, the FC to FS ratio for nonhuman waste is typically less than 0.7 and the ratio for human waste is typically greater than 4.0 (Clausen et al., 1977). Because bacterial die-off affects the ratio, it is only applicable to fecal pollution within 24 h of discharge.

It is generally recognized that fecal coliforms are not suitable indicators of the potential for viral contamination in surface water receiving domestic waste because some viruses are more resistant to chlorination and environmental deactivation than bacteria (Kraus, 1977; Gersberg et al., 1987). Bacteriophages (viruses that infect bacteria) have been used as viral

Table 17-2 Bacterial Densities (Number Per Gram) of Fecal Coliforms and Fecal Streptococci in Warm-Blooded Animal Feces

	Densities per gram			
Fecal Source	No. of Samples	Fecal Coliforms	Fecal Streptococci	Ratio FC to FS
Human	43	13,000,000	3,000,000	4.4
Animal pets				
Cat	19	7,900,000	27,000,000	0.3
Dog	24	23,000,000	980,000,000	0.02
Rodents	24	160,000	4,600,000	0.04
Rabbit				0.0004
Chipmunk				0.03
Livestock				
Cow	11	230,000	1,300,000	0.2
Pig	11	3,300,000	84,000,000	0.04
Sheep	10	16,000,000	38,000,000	0.4
Poultry				
Duck	8	33,000,000	54,000,000	0.6
Chicken	10	1,300,000	3,400,000	0.4
Turkey	10	290,000	2,800,000	0.1

Data from Comin, 1994.

indicators, such as coliphage MS-2 in wetland treatment systems (Gersberg et al., 1987; Scheuerman et al., 1989). In a subsurface-flow (SSF) wetland study at Santee, CA, Gersberg et al. (1987) added cultured MS-2 virus to the wetland influent wastewater to determine die-off rates. Enumeration of bacteriophages is technically simpler and more rapid than enumeration of the target pathogenic viruses. Also, MS-2 is nearly the same size as enteroviruses and is more resistant to ultraviolet light, heat, and disinfection than most enteric viruses.

REMOVAL EFFICIENCY

Inflow and outflow densities of fecal and total coliforms and fecal streptococcus bacteria have been monitored at a number of pilot and full-scale wetland treatment systems. Table 17-3 summarizes these data for some of the wetland treatment systems receiving municipal wastewater. This information can be used to make several generalizations concerning removal of indicator bacteria during wetland treatment.

Total and coliform bacteria have been measured in natural wetlands that receive no wastewater. For example, Fox et al. (1984) measured between 109 and 456 col/100 mL of fecal coliforms in cypress wetlands in Florida that were not receiving any outside inputs. Residual populations of fecal coliforms were in the range of 400 to 550 col/100 mL at a surface-flow (SF) constructed wetland in Whangarei, New Zealand. Lake water passing through a wetland in Montreal, Quebec increased to a fecal coliform concentration of 40 to 110 col/100 mL (Vincent, 1992). Bavor et al. (1987) concluded that vegetated wetlands reduce bacterial populations more than oxidation ponds with similar hydraulic retention times and that wetlands with alternating zones of shallow marsh and open water most effectively removed indicator bacteria. In SF wetlands in Listowel, Ontario, Herskowitz (1986) found that indicator bacteria populations were reduced least during the colder winter months.

These natural bacteria populations are generally low, but they may be variable and seasonally high because of wildlife populations. Because natural sources of coliforms and fecal streptococcus bacteria are found in all wetlands open to wildlife, outflow indicator bacteria populations in treatment wetlands can never be consistently reduced to near zero unless disinfection is used. This observation is extremely important during permit writing and negotiation because it is not technically feasible for wetland treatment systems to consis-

Table 17-3 Summary of Indicator Bacteria Data from Wetland Treatment Systems

				Fecal Coliforms (col/100 ml)			
System Name	Location	Type	HLR (cm/d)	In	Out	%	ka (m/d)
Brookhaven meadow/ marsh/ pond	Brookhaven, NY	CON SF	1.55	25,800	473	98.2	0.062
Brookhaven marsh/pond	Brookhaven, NY	CON SF	3.06	25,800	898	96.5	0.103
West Jackson County	Ocean Springs, MS	CON SF	3.18	239	674	−182	—
Lakeland cell no. 1	Lakeland, FL	CON SF	4.37	27,000	115	99.6	0.239
Benton cell no. 1	Benton, KY	CON SF	4.72	4,700	700	85.1	0.007
Benton cell no. 2	Benton, KY	CON SF	4.72	4,700	130	97.2	0.169
Pembroke MPM	Pembroke, KY	CON SF	3.78	166,000	270	99.8	0.243
Listowel no. 3	Listowel, Ontario	CON SF	1.28	2,000	80	96	0.041
Listowel no. 4	Listowel, Ontario	CON SF	1.28	200,000	200	99.9	0.088
Arcata pilot no. 1	Arcata, CA	CON SF	13.3	3,183	440	86.2	0.263
Arcata pilot no. 2	Arcata, CA	CON SF	7.89	12,500	316	97.5	0.290
Cobalt	Cobalt, Ontario	CON SF	1.7	159,301	1,087	99.3	0.085
Iselin	Iselin, PA	CON SF	2.09	1,800,000	150	99.9	0.196
Benton cell no. 3	Benton, KY	CON SSF	5.89	4,700	100	97.9	0.227
Phillips High School	Phillips, AL	CON SSF	3.72	256,000	10	99.9	0.378
Denham Springs	Denham Springs, LA	CON SSF	12.18	52,700	4,150	92.1	0.310
Santee pilot—bulrush	Santee, CA	CON SSF	5	—	—	—	—
Carolina bays	North Myrtle Beach, SC	NAT SF	0.15	66,000	56	99.9	0.011
Central slough	Conway, SC	NAT SF	0.51	857	50	94.2	0.014
Boggy gut	Sea Pines Plantation, SC	NAT SF	3.01	2	236	−117	—
Waldo pilot	Waldo, FL	NAT SF	17.64	7,700,000	270,000	96.5	0.591

Table 17-3 Continued

Total Coliforms (col/100 ml)				Fecal Streptococcus (col/100 ml)					
In	Out	%	ka (m/d)	In	Out	%	ka (m/d)	Notes	Ref.
1,150,000	24,000	97.9	0.060	6,780	739	89.1	0.034	Pilot facility; some fish; 42 months	Hendrey et al., 1979
1,150,000	59,300	94.8	0.091	6,780	2300	66.1	0.033	Pilot facility; some fish; 42 months	Hendrey et al., 1979
—	—	—	—	—	—	—	—	Full scale; abundant nutria and birds; 20 months	Unpublished data
—	—	—	—	—	—	—	—	Full scale; abundant birdlife; 9 months	Unpublished data
—	—	—	—	—	—	—	—	Full scale; cattail cell; little wildlife; 20 months	Watson et al., 1990
—	—	—	—	—	—	—	—	Full scale; wool-grass cell; little wildlife; 20 months	Watson et al., 1990
—	—	—	—	—	—	—	—	Full scale marsh/pond/meadow; little wildlife; 9 months	Watson et al., 1990
—	—	—	—	800	100	87.5	0.027	Pilot facility; lagoon influent; 48 months	Herskowitz, 1986
—	—	—	—	90,000	200	99.8	0.078	Pilot facility; aeration cell influent; 48 months	Herskowitz, 1986
—	—	—	—	—	—	—	—	Pilot facility; 24 months	Gearheart et al., 1989
—	—	—	—	—	—	—	—	Pilot facility; 24 months	Gearheart et al., 1989
—	—	—	—	25,012	3,887	84.5	0.032	Pilot facility; 12 months	Miller, 1989
—	—	—	—	—	—	—	—	Full scale marsh/pond/meadow; 29 months	Watson et al., 1987
—	—	—	—	—	—	—	—	Full scale; bulrush cell; 20 months	Watson et al., 1990
—	—	—	—	—	—	—	—	Full scale; 7 months	NADB, 1993
—	—	—	—	—	—	—	—	Full scale; 18 months	NADB, 1993
67,500,000	608,000	99.1	0.235	—	—	—	—	Pilot facility; 24 months	Gersberg et al., 1989a
89,800	2,640	97.1	0.005	—	—	—	—	Full scale; abundant wildlife; 73 months	CH2M HILL, 1991
—	—	—	—	—	—	—	—	Full scale; abundant wildlife; 36 months	CH2M HILL, 1991
3	2,120	−706	—	2	1972	−985	—	Full scale natural marsh; receives runoff and manure from horse trail and abundant wildlife; 12 months	MacClellan, 1989
15,600,000	100,000	93.6	0.891	—	—	—	—	Cypress wetland; experimental	Scheuerman et al., 1989

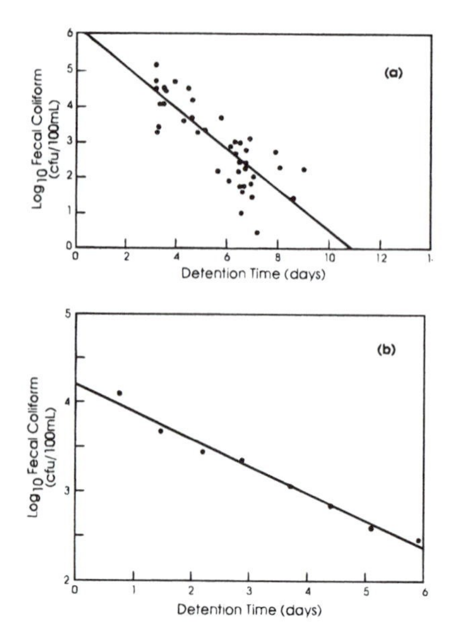

Figure 17-2 Logarithmic reductions of fecal coliforms populations in constructed SF wetlands: (a) combination marsh, open water, gravel system near Sydney, Australia (Bavor et al., 1987) and (b) marsh system of Arcata, CA (Gearheart et al., 1989).

tently attain outflow with total coliform populations less than 10,000 col/100 mL, fecal coliforms less than 500 col/100 mL, and fecal streptococcus less than 500 col/100 mL.

Because of the residual indicator bacteria populations in all wetlands, bacteria removal efficiency is a function of the inflow bacteria population. Removal efficiency typically is high at high inflow populations, but declines to negative efficiencies when inflow populations are lower than the *in situ* bacteria production rates. At the Boggy Gut natural wetland treatment system at Sea Pines Plantation, SC, inflow to the wetland contains essentially no indicator bacteria because of a high level of pretreatment and disinfection; however, the wetland outflow contains elevated coliforms and fecal streptococcus populations due to wetland bird use and stormwater runoff from an adjacent horse trail (Girts and Knight, 1989). A similar negative removal efficiency is observed in the SF constructed wetland at West Jackson County, MS, where inflow fecal coliform populations are reduced due to the high residence time in a facultative pretreatment lagoon (Table 17-3).

When inflow coliform and fecal streptococcus populations are higher and typical of untreated or partially treated municipal wastewater that have not received disinfection, wetland removal efficiencies are nearly always greater than 90 percent for coliforms and greater than 80 percent for fecal streptococcus. Removal efficiency is approximately first order, as long as inflow bacteria populations are high (Figure 17-2).

ZERO-BACKGROUND MODELS

A volume-based, first-order model has been used to estimate die-off of coliform populations in lagoons and wetlands:

$$R = k_{v1}VC = k_{v1}(\varepsilon Ah)C \tag{17-1}$$

where R = death rate, #/day
k_{v1} = volume-based, first-order decay rate, 1/d

V = wetland water volume, m^3
C = bacterial concentration, $\#/m^3$
ε = volume fraction for water
A = wetland area, m^2
h = water depth, m

First-order bacteria decay coefficients have been estimated for total coliforms as 0.86 d^{-1} by Gersberg et al. (1987) in SSF wetlands and 0.74 d^{-1} by Scheuerman et al. (1989) in a Florida cypress wetland. The first-order decay rate was estimated as 0.70 d^{-1} for fecal coliforms and 0.62 d^{-1} for fecal streptococcus in a Florida cypress wetland by Scheuerman et al. (1989), while Gearheart et al. (1989) measured a lower first-order decay rate of 0.29 d^{-1} for fecal coliforms at the Arcata, CA, SF wetland treatment system. These models do not account for the significant background concentrations contributed by local sources.

Direct measurement of some pathogenic bacteria provides results similar to measurement of indicator bacteria species. Gersberg et al. (1989a) measured a 96.1 percent removal of *Salmonella* after 52 h in SSF wetlands at Santee, CA. Scheuerman et al. (1989) measured first-order decay rates in natural cypress wetlands in Florida as 0.72 and 0.91 d^{-1} for *Streptococcus faecalis* and *Salmonella typhimurium,* respectively. At the Listowel, Ontario pilot wetlands, populations of *Yersinia enterocolitica, Pseudomonas aeruginosa, Clostridium perfringens,* and *Salmonella* were significantly reduced at high residence times, even during the colder winter months (Herskowitz, 1986).

Equation 17-1 may also be written on an areal basis:

$$R = k_1 A C \tag{17-2}$$

where k_1 = first-order, zero-background areal rate constant, m/d.

Clearly, comparing Equations 17-1 and 17-2, $k_1 = k_{v1}\epsilon h$. Integration of a pathogen mass balance equation together with Equation 17-2 yields the exponential decline equation (see Chapter 9):

$$\frac{C}{C_0} = \exp\left(-\frac{k_1}{q}\right) \tag{17-3}$$

where C_o = initial bacterial concentration, #/100 mL
q = hydraulic loading rate, m/d

Using Equation 17-3, values for the area-based, first-order rate constant for coliform die-off in wetland treatment systems listed in Table 17-2 were calculated. The area-based, first-order rate constants are summarized in Table 17-2 for fecal coliform bacteria populations being reduced prior to the outflow sampling point. A common value for k_1 not affected by underestimating is 0.3 m/d. This value for k_1 is equivalent to a value of k_{v1} of 1.05 d^{-1} for a free-water depth (ϵh) of 0.3 m. Values of k in Table 17-2 range from 0.005 to 0.89 m/d for total coliforms and 0.03 to 0.08 m/d for fecal streptococcus bacteria. Additional data are needed to better evaluate the performance of wetlands for natural reduction of indicator bacterial populations.

THE k-C* MODEL

An area-based, first-order bacterial die-off model may be more appropriate in wetland treatment systems because surface area for bacterial inactivation does not increase proportionally to volume as water depth exceeds the vegetated zone. The form of this area-based model is

$$R = kA(C - C^*) \tag{17-4}$$

where k = area-based, first-order rate constant, m/d
C^* = background concentration, #/100 mL

This model can also be written as

$$\frac{(C - C^*)}{(C_0 - C^*)} = \exp\left(-\frac{k}{q}\right) = \exp(-k_v\tau) \tag{17-5}$$

where τ = nominal detention time, days.

The area and volume-based rate constants can be converted using the relationship

$$k = k_v\varepsilon h \tag{17-6}$$

Note that the combination $\frac{k}{q}$ = Da, the Damköhler number.

Wetland data are presently insufficient to calibrate both k and C* for individual systems. However, it is clear that C* is often in the range of 10 to 500 #/100 mL for fecal coliforms, based on data from natural and large wetlands. Therefore, it is possible to display the k-C* model results on concentration coordinates, along with wetland treatment data (Figure 17-3). The result indicates that the general trends are correctly predicted.

Analysis of data in Table 17-3 yields preliminary estimates of k = 75 m/yr for SF wetlands, and k = 95 m/yr for SSF wetlands.The designer may utilize Figure 17-3 to project the likely performance range for a new treatment wetland.

REMOVAL EFFICIENCY FOR VIRUSES

The average viral content of domestic sewage in the U.S. is about 7000 particles per liter (Gersberg et al., 1989a). Because the potential infective dose of viruses can be so low (<10 particles), and because viruses are generally more hardy in natural environments than

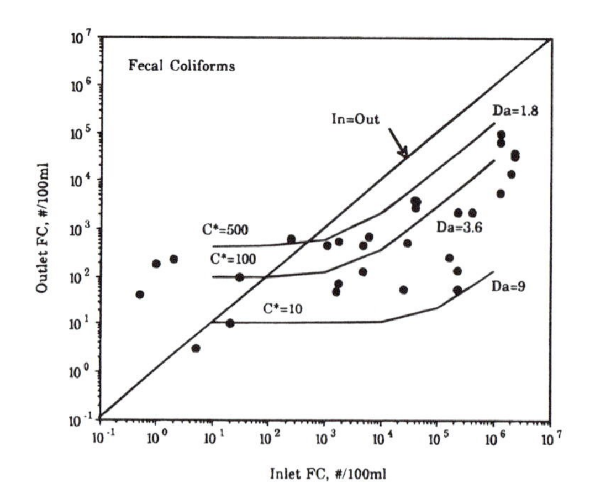

Figure 17-3 I/O data for several marsh systems. Each point is the average for one system. Shown for reference are the predictions of the k-C* model for different values of the Damköhler number and baseline FC.

bacterial pathogens, there has been considerable concern about their fate in wetland and conventional treatment systems. However, because viruses are more costly and difficult to monitor analytically, they have not received as much study as bacterial indicator populations. Existing research indicates that wetlands are generally hostile environments for the survival of viruses. Gersberg et al. (1989a) reported population reductions for MS-2 bacteriophage of 99 percent at the Santee SSF wetlands and 91.5 percent at the Arcata SF wetlands. At Arcata, the wetland channels with low marsh area and high open water pond area were ineffective for virus removal. Virus removal closely correlated with removal of suspended solids. Design criteria that enhance suspended solids removal also are likely to enhance virus removal. These researchers concluded that wetland treatment followed by conventional disinfection was capable of essentially totally removing viruses.

Scheuerman et al. (1989) found that bacteriophages and enteroviruses in a natural cypress wetland declined with distance from an outfall, indicating that wetland virus removals can be effective if residence time is adequate. In this test, bacteriophages were removed at a slower rate than bacterial indicators and enteroviruses. Enteroviruses declined at the fastest rate and were not observed to accumulate in downstream sediments.

Casson et al. (1992) investigated the survival of HIV in unchlorinated domestic wastewater in the laboratory. Inoculated HIV populations in secondary wastewater held at room temperature for 6 h showed no reduction in infectivity. A greater than 1-log reduction in infectivity was observed in 24 h, a 2- to 3-log reduction in infectivity was measured after 48 h, and approximately a 3-log reduction was observed after 72 h. Primary wastewater was a slightly more hostile environment for HIV with a greater than 3-log reduction of infectivity measured after 72 h. HIV survival in wastewater was found to be significantly less than poliovirus survival under similar conditions.

VARIABILITY IN REMOVAL

Because wetland fecal coliform counts are due to internal loading within the wetland as well as declines in populations entering with wastewater, the stochastic variability in effluent concentrations is large. The scatter is often over an order of magnitude (factor of ten). For this reason, excursions around mean performance should be acknowledged in design and regulation.

As an example, consider the performance of the Iron Bridge wetlands (Figure 17-4). Incoming fecal coliforms are very low, but the animal (bird) populations are high. Background geometrical annual averages are 35 and 90/100 mL, but monthly values show frequent exceedances of the permit limit of 100/100 mL.

SUMMARY

Pathogenic organisms are a normal component of domestic wastewater. Their populations are highly variable depending on the health of the human population contributing to the waste. Conventional treatment technologies with short hydraulic residence times depend on disinfection to reduce pathogen populations. Natural wastewater treatment technologies reduce pathogen populations more successfully with longer residence time and land-intensive treatment. Wetland treatment systems that have long residence times (greater than about 10 days) provide some disinfection. Vegetated wetlands appear to be more effective for pathogen removal than facultative ponds and other natural treatment systems that have less physical contact between pathogens and solid surfaces.

Because of the generally low density of actual wastewater pathogens and the difficulty associated with their enumeration, indicator organisms such as fecal coliforms, *E. coli,* and

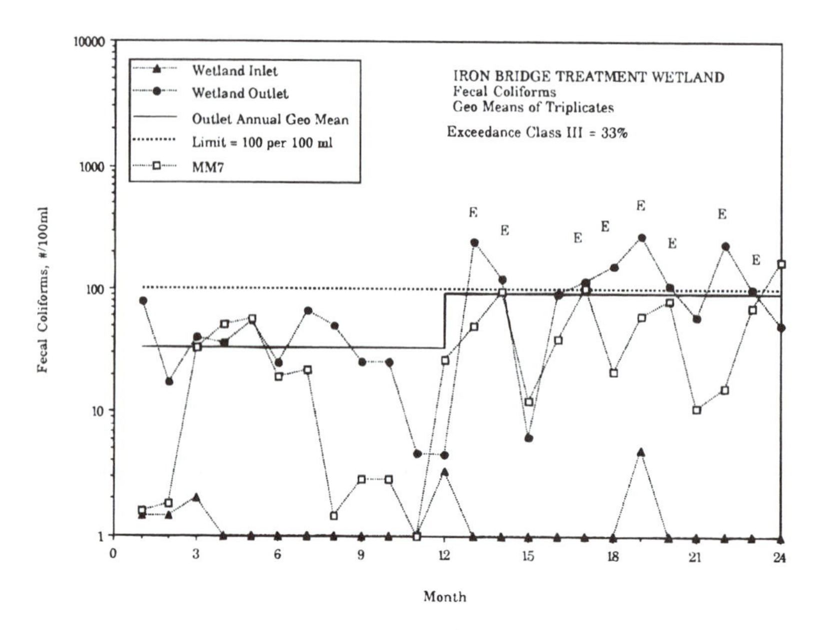

Figure 17-4 The increase of fecal coliforms through the Iron Bridge wetlands. The incoming numbers are very small, are increased at the midpoint (MM7), and are increased still more at the wetland outfall. Exceedances of Florida's Class III standard of 100/100 ml are observed in year two on a monthly basis, but do not occur when annual geometric averages are used.

MS-2 bacteriophage are used to monitor the probable pathogenic activity of wastewater discharged to the environment. Empirical evidence is available that demonstrates that a significant die-off of indicator organisms and pathogenic species occurs in wetland treatment systems. This die-off can be modeled during system design based on the first-order equations given in the text. It is also evident that wetlands contain numerous animals that contribute base populations of these same indicator organisms to surface water quality. These naturally occurring indicator organisms must be recognized during regulatory permitting and review of data from wetland treatment systems.

Because some viral pathogens have low minimum infective doses and because of the difficulty associated with quantification of virus populations, potential problems associated with these pathogens in wetland discharges have led some public health officials to make cautionary statements. For example, Wellings (1986a, b) expressed concern about the potential for greater human incidence of encephalitis because wetlands used for wastewater treatment are likely to stay wet longer and provide better breeding habitat for vectors such as mosquitoes. Also, Bitton et al. (1976) found that at least some virus particles might reach groundwater below natural wetland treatment systems. However, available information indicates that wetlands are at least as effective as other treatment technologies in reducing viral pathogen populations. In their review of the fate of pathogens in wetland treatment systems, Krishnan and Smith (1987) concluded that "while data exist to indicate the potential for public health problems arising from wetland discharges, no incidences of disease resulting directly from such discharges have been identified."

Section 4
Wetland Project Planning and Design

CHAPTER 18

Wastewater Source Characterization

INTRODUCTION

Wastewater quality varies widely among municipal, industrial, agricultural, and stormwater categories. Different wastewater sources have unique mixtures of potential pollutants so that even a single wastewater source category, such as municipal wastewater or urban runoff, may vary depending on local, site-specific circumstances. However, for some chemical constituents, the qualitative and quantitative composition of wastewaters from different sources varies less. For these reasons, any summary of "typical" wastewater concentrations and loads must be considered cautiously.

After careful examination for extenuating circumstances, site-specific wastewater data showing historical flows and mass loads provide the best information for wetland treatment system design. However, because many treatment systems are designed for new facilities or because historical monitoring may be nonexistent or insufficient, it is useful to know the typical concentrations of major constituents in similar wastewaters. This chapter summarizes information from a number of sources on the typical pollutant composition of wastewater applied to engineered wetlands. These "typical" concentrations and loads should only be used when site-specific or better information is not available.

Total wastewater flows for principal industrial and municipal sources in the U.S. in 1990 were estimated at about 69 billion m^3/yr with a total raw BOD_5 load of about 27×10^9 metric tons (Greyson, 1990). Table 18-1 summarizes these flows and loads by industry type. Domestic wastewater contributes about 29 percent of the total flow and 25 percent of the organic load. In terms of pollutant loads, other major industry categories are chemical, pulp and paper, and animal feed products. Some of the key wastewater categories that might benefit from the use of the wetland treatment technology are described in more detail later.

In addition to industrial and municipal wastewaters, nonpoint source pollution contributes over 65 percent of the total pollution load to U.S. inland surface waters (U.S. EPA, 1989a). Sources of nonpoint flows include urban and suburban runoff, diffuse agricultural runoff, forestry activities, runoff from concentrated agricultural activities such as feedlots, mining drainage, and runoff from undisturbed areas. The U.S. EPA (1986a) has estimated that this nonpoint source pollution causes the impairment of two thirds of the impaired water bodies in the U.S., including 332,000 km of rivers, 215,000 ha of lakes, and 1.5×10^6 ha of estuaries (U.S. EPA, 1990a). Agricultural nonpoint source pollution is considered to be the largest single category resulting in these environmental problems. However, in certain areas, urban runoff or other stormwater sources provide the greatest percentage of uncontrolled pollutants.

Some of the pollutants that are common to many of these wastewater sources can be effectively treated by wetland systems. The normal concentration range of these pollutants is an important consideration in evaluating wetland treatment system options. This chapter compares and contrasts these wastewater sources to facilitate this initial alternative evaluation.

Table 18-1 Estimated Wastewater Flows and BOD_5 Loads by Industry Type in the U.S.

Industry	Raw Wastewater Flows (million m^3/yr)	Raw BOD_5 Load (million metric tons/year)
Domestic	20,100	6,600
Chemicals	14,000	8,800
Pulp and paper	7,200	5,400
Animal feed	2,600	3,900
Textiles	530	810
Petroleum and coal	4,900	450
Primary metals	16,300	440
Transportation equipment	910	110
Electrical machinery	340	64
Machinery	190	54
Rubber and plastics	600	36
Other manufacturing	1,700	350

Data from Greyson, 1990.

MUNICIPAL WASTEWATERS

Table 18-2 summarizes the typical quality of medium strength, raw, municipal wastewater in the U.S. and provides a range of values for commonly observed constituents. Municipal wastewater is composed of a variable array of components characterized by the presence of biodegradable organic matter (paper, feces, and food), particulate and dissolved solids, and nutrients. Many municipal wastewaters also receive some component of industrial waste. These flows and residential sources may add trace metals and pesticides to typical municipal wastewater.

Table 18-2 also provides a range of estimated treatment efficiencies for conventional primary and secondary treatment processes. These removal efficiencies vary widely depending on the types of treatment processes. However, it is generally observed that at least 70 percent of the biochemical oxygen demand and total suspended solids are removed from municipal wastewater during primary and secondary treatment. Table 18-2 summarizes the typical quality of secondarily treated municipal wastewaters. This summary can be used as an estimate of the influent water quality to be applied to a wetland system designed for either primary, secondary, or advanced wastewater treatment.

INDUSTRIAL WASTEWATERS

Although industrial wastewater quality varies among industries, it has a fairly consistent intrasystem effluent quality. Table 18-3 summarizes the typical quality of raw wastewater from a number of industries that have used or might consider using a wetland treatment technology. However, raw industrial wastewater usually receives some level of pretreatment before discharge to a wetland treatment system. If total concentrations of biochemical oxygen demand, suspended solids, and ammonia nitrogen in untreated industrial wastewater are in the 100s or 1000s of mg/L concentration range, it is generally not acceptable for wetland discharge without additional pretreatment. Maximum tolerable inlet loads and concentrations for these parameters in wetland treatment systems are discussed in Chapters 12 through 17.

LANDFILL LEACHATES

Treatment and disposal of liquid leachates is one of the most difficult problems associated with the use of sanitary landfills for disposal of solid waste. Leachates are produced when rainfall and percolated groundwater combine with inorganic and organic degraded waste. In

Table 18-2 Typical Composition of Municipal Wastewater and Percent Removals at Various Levels of Treatment

	Raw Wastewater (mg/L)		Percent Removal		Secondary Effluent (mg/L)	
Constituent	Typical	Range	Primary	Secondary	Typical	Range
BOD_5	220	110–400	0–45	65–95	20	10–45
COD	500	250–1000	0–40	60–85	75	35–75
TSS	220	100–350	0–65	60–90	30	15–60
VSS	165	80–275	—	—	—	—
NH_4-N	25	12–50	0–20	8–15	10	<1–20
NO_3 + NO_2-N	0	0	—	—	6	<1–20
Org-N	15	8–35	0–20	15–50	4	2–6
TKN	40	20–85	0–20	20–60	14	10–20
Total N	40	20–85	5–10	10–20	20	10–30
Inorg P	5	4–15	—	—	4	2–8
Org P	3	2–5	—	—	2	0–4
Total P	8	6–20	0–30	10–20	6	4–8
Arsenic	0.007	0.002–0.02	34	28	0.002	—
Cadmium	0.008	<0.005–0.02	38	33–54	0.01	<0.005–6.4
Chromium	0.2	<0.05–3.6	44	58–74	0.09	<0.05–6.8
Copper	0.1	<0.02–0.4	49	28–76	0.05	<0.02–5.9
Iron	0.9	0.10–1.9	43	47–72	0.36	0.10–4.3
Lead	0.1	<0.02–0.2	52	44–69	0.05	<0.02–6.0
Manganese	0.14	0–0.3	20	13–33	0.05	—
Mercury	0.001	<0.0001–0.0045	11	13–83	0.001	<0.0001–0.125
Nickel	0.2	—	—	33	0.02	<0.02–5.4
Silver	0.022	0.004–0.044	55	79	0.002	—
Zinc	1.0	—	36	47–50	0.15	<0.02–20

Note: Partially adapted from WPCF, 1983; Metcalf and Eddy, 1991; Richardson and Nichols, 1985; Krishnan and Smith, 1987; and Williams 1982.

Table 18-3 Typical Pollutant Concentrations in a Variety of Untreated Industrial Wastewaters

Constituent	Units	Pulp and Paper[a]	Landfill Leachate[b]	Non Coal Mine Drainage[c]	Coal Mine Drainage[d]	Petroleum Refinery[e]	Electroplating[f]	Superphosphate Fertilizer[g]	Paint Production[g]	Textile Mills[g]	Starch Production[g]	Breweries[g]
BOD_5	mg/L	100–500	42–10,900	—	—	10–800	—	—	—	75–6,300	1,500–8,000	1,500–3,000
COD	mg/L	600–1,000	40–90,000	—	—	50–600	—	150	19,000	220–31,300	1,500–10,000	800–1,400
TSS	mg/L	500–1,200	100–700	—	—	10–300	4–600	—	—	25–24,500	100–600	100–500
VSS	mg/L	100–250	60–280	—	—	—	30–100	—	16,000	100–400	—	50–500
TDS	mg/L					1,500–3,000	800–5,800	—	—	500–3,000	—	50–350
NH_4+-N	mg/L		0.01–1,000	—	—	0.05–300	—	10	—	—	10–100	—
TN	mg/L		70–1,900	—	—	—	10–120	—	90	10–30	150–600	25–45
TP	mg/L		<0.01–2.7	—	—	1–10	20–50	800	25	—	—	—
pH	Units	6–8	3–7.9	2.1–6.9	3–5.5	8.5–9.5	4–10.5	8–9	6.9	6–12	3.5–8	5–7
Sulfate/sulfides	mg/L		10–260	20–4,000	20–2,000	ND–400	30–120	—	—	—	—	—
Specific conductance	µmho/cm		1,200–16,000	—	—	—	—	—	—	—	—	—
TOC	mg/L		11–8,700	—	—	10–500	—	—	—	—	—	—
Aluminum	mg/L		0.5	18–100	50	—	—	—	—	—	—	—
Arsenic	mg/L		0.011–10,000	<0.001–7	—	—	—	—	—	—	—	—
Barium	mg/L		0.1–2,000	—	—	—	—	—	—	—	—	—
Cadmium	µg/L		5–8,200	<0.01–3	—	—	10,000–50,000	—	—	—	—	—
Chromium	mg/L		0.001–208	—	—	ND–3	10–120	—	—	—	—	—
Iron	mg/L		0.09–678	0.5–700	50–300	—	2–20	—	—	—	—	—
Lead	µg/L		1–19,000	<0.01–0.5	—	—	—	—	—	—	—	—
Manganese	mg/L		0.01–550	1.0–120	20–300	—	—	—	—	—	—	—
Selenium	µg/L		3–590	—	—	—	—	—	—	—	—	—
Silver	µg/L		—	—	—	—	—	—	—	—	—	—
Oil and grease	mg/L		—	—	—	10–700	—	—	—	—	—	—
Phenols	mg/L		<0.003–17	—	—	0.5–100	—	—	—	—	—	—
Cyanide	mg/L						1–50	—	—	—	—	—

Note: ND = Not Detected.

[a]Jorgensen, 1979.
[b]Staubitz et al., 1989; Lema et al., 1988; Bolton and Evans, 1991.
[c]Wildeman and Laudon, 1989.
[d]Girts and Kleinmann, 1986.
[e]Adams et al., 1981b; ANL, 1990.
[f]OECD, 1983.
[g]Cooper, 1978.

unlined landfills, leachates frequently discharge to groundwater or appear as surficial drainage around the base of the landfill. In modern lined landfills, leachates are collected from the lined cells and routed to treatment units. The use of constructed wetlands to treat these landfill leachates is a developing technology, with both subsurface-flow (SSF) wetlands (Cooper and Hobson, 1989; Trautmann et al., 1989; Staubitz et al., 1989; Birkbeck et al., 1990; Surface et al., 1993) and surface-flow (SF) wetlands (Martin and Moshiri, 1994; Martin et al., 1993; Schwartz et al., 1993; Keely et al., 1992).

The highly variable nature of solid waste, differences in age and decomposition, and the diversity of chemical and biological reactions that take place in landfills result in a wide range of chemical quality of leachates. From a review of an "average" landfill, concentrations of chemical oxygen demand, volatile acids, and nitrogenous compounds increase in landfill leachate during the first few years of operation and then decline over 10 or more years. Table 18-3 provides typical ranges encountered in landfill leachates. Flows are generally low, but may vary depending on management and minimization of percolation from rainfall. Clearly, the expected volume and chemical quality of a landfill leachate is highly site specific, may change over time, and must be estimated on a case-by-case basis for wetland treatment system design.

PULP AND PAPER WASTEWATER

The pulp and paper industry converts wood products, including pines, spruce, poplar, beech, birch, and aspen, as well as recycled paper, into liquified cellulose pulp and paper. Raw wood and wood chips are converted to pulp (cellulose fibers) by mechanical grinding (groundwood) or through chemical degradation and leaching (sulfite and Kraft processes). At an increasing number of pulp and paper mills, this pulp is bleached to delignify and decolorize the cellulose fibers before paper manufacture.

About 29 to 34 m^3 of raw wastewater is produced for each metric ton of pulp and paper produced (Britt, 1970). Total wastewater flow for the U.S. pulp and paper industry is about 20×10^6 m^3/d (Greyson, 1990). Table 18-3 summarizes the typical composition of this wastewater, although different manufacturing processes result in different wastewater qualities. This flow is equivalent to a raw BOD_5 load of about 15×10^6 metric tons per day (Greyson, 1990).

Raw wastewater from pulp and paper mills typically receives primary treatment through settling, either in ponds or in primary clarifiers. When required to meet discharge limitations, secondary treatment at most pulp and paper mills includes biological conversion of organic matter (BOD_5) and additional solids settling in aerated lagoons or in conventional activated sludge treatment systems.

To meet reduced effluent limitations, some pulp and paper mills are being required to provide treatment beyond the secondary level. The goals of additional treatment depend on site conditions such as the quality of the effluent after secondary treatment and water quality permit limits in the receiving water. One goal may be to further reduce BOD_5, total suspended solid, nitrogen, phosphorus, color, chlorinated organics (such as adsorbable organic halides or dioxin), and whole effluent toxicity. Constructed and natural wetland treatment systems are being used at an increasing number of pulp and paper mills to provide this advanced secondary or tertiary treatment (NCASI, 1995; Knight et al., 1994; Knight, 1993; Thut, 1989, 1993).

MINE DRAINAGE

Mining provides precious and semiprecious minerals including iron, gold, silver, tin, silver, cadmium, and nickel. During and following mining operations, runoff and leachate from tailings and from abandoned tunnels and shafts dissolve trace metals, contaminating nearby surface waters. Recycling and byproduct recovery can lower the concentrations of polluting metals in discharges offsite, but in most cases some residual trace metals remain (Table 18-3). Also, seeps from small abandoned mines typically receive no treatment. It is estimated that 21,000 km of U.S. rivers have fisheries that are affected by elevated concentrations of copper, zinc, cadmium, lead, and arsenic, and that 2200 km of streams in Colorado alone do not meet water quality standards because of the discharge of metals from past mining activities (Morea et al., 1990).

Constructed wetlands are receiving increasing attention as a technology to reduce metal concentrations in mining wastewater (Eger et al., 1993; Sanders and Brocksen, 1991; Morea et al., 1990). As documented in Chapter 15, metals are sequestered in sediments, plants, and faunal storages in wetlands. Wetland design for metals removal is limited by the need to avoid toxic concentrations in tissues that could subsequently accumulate in the food chain. Information necessary to evaluate the ability of wetlands to provide treatment of these waste products is summarized in Chapter 15 and in the references listed previously. Only rule-of-thumb design approaches are available at this time. Detailed methods to design wetlands to treat mine wastes high in trace metals other than iron and manganese are not included in this book and have not been published elsewhere.

COAL MINE DRAINAGE

During coal mining, iron pyrite and other metal-bearing minerals are exposed to percolating water, which leads to the release of acidic leachates to surface water. These acid mine drainages (AMD) typically have low pH and elevated concentrations of dissolved iron, sulfate, calcium, and magnesium. In addition, the drainages have variable and somewhat elevated concentrations of aluminum, copper, manganese, nickel, and zinc (Table 18-3). It has been estimated that 12,000 km of streams and 12,000 ha of impoundments in the Appalachian coal mining region of the U.S. are affected by acid mine drainage.

Conventional treatment of leachates at these sites includes surface grading and recontouring to reduce or divert flows and chemical buffering and precipitation to improve water quality. Because these processes have relatively high capital and lifetime costs, there has been considerable interest in developing more cost-effective alternatives. Beginning in the early 1980s, research focused on the potential of aerobic wetlands for precipitation of ferric sulfate to neutralize pH and reduce dissolved ferrous iron concentrations. Constructed wetlands are now used at more than 300 sites in the U.S. to increase the pH and reduce concentrations of iron and or manganese at coal mine sites (Kleinmann and Hedin, 1989). The effectiveness of these systems is reviewed briefly in Chapter 15, and the reader is directed to papers by Wieder (1989), Kleinmann and Hedin (1989), and Girts and Kleinmann (1986) for additional information on designing constructed wetlands for treatment of coal mine drainage.

PETROLEUM REFINERY WASTEWATER

Petroleum refineries convert raw oil and other hydrocarbon-bearing petroleum sources (such as natural gas and oil sands) into a variety of end products and intermediate materials for the chemical, fabric, and plastics industries. Because of the diverse processes at refineries and the volume of flammable liquids, land area requirements are large and include many kilometers of piping and hundreds of tanks and storage areas. Wastewater is generated by the topping, cracking, and lube oil manufacturing process; cooling tower blowdown; water and sludge drainage from tanks; and stormwater drainage and runoff (UNEP, 1987).

Typical wastewater pollutants at petroleum refineries include BOD_5, chemical oxygen demand (COD), oil and grease, total suspended solids (TSS), NH_4-N, phenolics, H_2S, trace organics, and heavy metals. Concentrations of many of these pollutants are reduced through source control and preliminary treatments such as sour water stripping, oxidation and neutralization of spent caustics, and cooling tower blowdown treatment. Table 18-3 lists the typical range of pollutant concentrations remaining in raw refinery wastewater. The median concentrations of several principal pollutants at an integrated petroleum refinery follow: BOD_5, 110 mg/L; COD, 260 mg/L; TOC, 52 mg/L; NH_4-N, 14 mg/L; phenols, 2.2 mg/L; sulfides, 1.2 mg/L; oil and grease, 44 mg/L; and total chromium, 0.27 mg/L (ANL, 1990).

Raw wastewater from petroleum refineries typically receives additional treatment including gravity separation of oils and greases; primary clarification; dissolved air flotation; and secondary treatment including oxidation ponds, aerated lagoons, activated sludge, trickling filters, and activated carbon. The API separator process typically removes from 60 to 99 percent of the oil and grease and smaller proportions of other pollutants. Primary treatment removes 20 to 70 percent of the BOD_5 and TSS and 10 to 60 percent of the COD. Secondary treatment will reduce from 40 to 99 percent of the BOD_5, 30 to 95 percent of the COD, 40 to 90 percent of the TOC, 20 to 85 percent of the TSS, 60 to 99 percent of the oil and grease, 60 to 99 percent of the phenol, 9 to 99 percent of the NH_4-N, and 70 to 100 percent of the sulfide (ANL, 1990).

As described in Chapter 16, constructed wetlands are providing advanced secondary and tertiary treatment of process water and stormwater at a growing number of refineries. Amoco's Mandan, ND refinery uses 36 ha of ponds for final effluent polishing and wildlife enhancement (see Chapter 27 for a more detailed description of this facility). Chevron also has a constructed wetland treatment facility at its Richmond, CA refinery. Constructed wetlands typically will reduce remaining concentrations of BOD_5, COD, TSS, NH_4-N, oils and grease, phenols, and metals to advanced treatment levels.

ELECTROPLATING INDUSTRIES

Electroplating is the electrochemical process of applying metal coatings to metallic objects for corrosion protection and for decorative finishing. The primary coating metals are zinc, copper, chromium, and nickel. Secondary metals include iron, cadmium, tin, lead, and various alloys. To a much smaller extent, precious metals including gold, silver, and platinum are used for coatings.

In the metals plating industry, the three main sources of pollutants are waste rinsewater, spent process solutions, and accidental spills and leaks. The pollutants in the wastewater include residual metals, cyanides, and solvents. Typical treatment is chemical precipitation to recover, settle, and recycle byproducts. Pretreated, metal-bearing wastewater typically is discharged to publicly owned treatment works for final treatment and disposal. Wetlands could treat some metal-bearing wastewaters, but this application must be approached cautiously and is not being practiced intentionally on a large scale. Wetlands do have a high potential to remove solids and dissolved metals, but highly contaminated soils and plants would need to be treated as hazardous materials for ultimate disposal. Also, special care would need to be exercised to not attract wildlife or other mobile pollutant vectors.

TEXTILE PRODUCTION

In 1972, 1926 textile mills in the U.S. (Cooper, 1978) used wet processes and discharged 470×10^6 m^3/year of wastewater to municipal or onsite treatment plants. The major pollutants in textile wastewater include solids, BOD_5, COD, nitrogen, phosphorus, phenols, trace metals, oil and grease, sulfides, bacteria, and color. The process wastewater typically is mixed with the sanitary wastewater produced by the high number of mill workers.

Wastewater from textile mills has extremely different qualities depending on the manufacturing process and products (see Table 18-3). BOD_5 in the dyeing process may range from 75 to 340 mg/L, and scouring may produce 4700 to 6300 mg/L. The range of color values recorded in textile mill raw wastewater varies between 325 and 2500 units. The textile industry produces 810 million metric tons of BOD_5 per year in raw wastewater or 0.14 to 0.78 kg of BOD_5 per kilogram of fiber processed (Cooper, 1978). Textile mills use pretreatment and traditional wastewater treatment processes to deal with high waste generation. However, a large proportion of wastewater does not receive adequate treatment. Although no wetland treatment applications for textile mill wastewater currently exist, mill wastewater appears to be compatible with wetland treatment technology.

AGRICULTURAL WASTEWATERS

Table 18-4 summarizes the typical loads per animal and composition of wastes from agricultural waste streams. To estimate total flows and loads to the treatment system, multiply the mass loading in Table 18-4 by the mass of animals or plants at a specific project. Table 18-5 summarizes typical unit loads for agricultural industries and the contribution of total nitrogen and total phosphorus to the dry-weight biomass of several agricultural plant wastes. Table 18-6 provides a summary of the elemental composition of agricultural waste including concentrations of minor and trace nutrients and toxic metals.

Agricultural wastewater contains high BOD_5, COD, TSS, and nutrients and is qualitatively similar to municipal wastewater (Figure 18-1). Mass loadings from animal feed lots and other concentrated agricultural activities require intensive treatment systems to provide environmental protection. Traditional treatment methods such as anaerobic lagoons and spray irrigation are not always adequate to provide high-quality water for offsite discharge. Constructed wetlands are being used in a growing number of cases to receive pretreated dairy and swine wastes (Maddox and Kingsley, 1989; Hammer et al., 1993; DuBowy and Reeves, 1994). These wetland treatment systems must be designed with reasonable organic loadings

Figure 18-1 Each dairy cow produces a raw pollutant load that is roughly equivalent to seven adult humans.

Table 18-4 Typical Waste Characteristics of Selected Agricultural Industries[a]

Parameter	Units[b]	Industry						
		Dairy	Poultry	Cattle	Swine	Fish and Seafood	Cane Sugar	Milling
Flow rate	m^3/1,000 kg/d	0.08–20.3	14–18	0.66	0.076–0.095	1.1–175	12	2.5–41.7
BOD_5	mg/L	—	473	300–12,000	—	84–32,700	180	225–14,633
	kg/1,000 kg/d	1.4	3.0	1.4	8	1.79–210	2.16	0.11–108.4
COD	mg/L	—	722	2500–40,000	—	370–6300	591	473–4901
	kg/1,000 kg/d	13.2	10.9	8.7	20	6.4–410	7.09	0.11–12.5
Total solids	mg/L	—	196	1000–13,400	—	26.2–18,300	375	33–14,824
	kg/1,000 kg/d	10.7	17.8–24.6	7.1	4.4–6	0.70–370	4.50	0.07–109.8
NH_4+-N	mg/L	5.5	—	1–770	—	0.95–50	0–0.46	—
	kg/1,000 kg/d	—	—	—	—	0.0045–0.60	0–0.02	—
TKN	mg/L	—	—	—	—	—	0–1.66	—
	kg/1,000 kg/d	—	—	—	—	—	0–0.08	—
NO_3+-N	mg/L	—	—	0.1–1270	—	—	0–4.3	—
	kg/1,000 kg/d	—	—	—	—	—	0–0.21	—
TP	mg/L	16.2	—	20–480	—	—	0–0.14	5.6–98
	kg/1,000 kg/d	0.11	0.44	0.07	0.13	—	0–2.80	—
TN	mg/L	68	—	—	—	—	—	3.6–13.2
	kg/1,000 kg/d	0.51	1.2	0.39	0.4	—	—	—

[a]Overcash et al., 1983; Middlebrooks, 1979.
[b]All fresh weight basis.

Table 18-5 Typical Unit Loads and Composition of Agricultural Wastes

		Animal Wastes					Plant Wastes			
Constituents	Units	Beef	Dairy	Poultry	Swine	Sheep	Corn	Rice	Alfalfa	Orchard Grass
Wet waste	kg/cap/d	23.6	23.6	0.3	2.7	1.1	—	—	—	—
Dry waste	kg/cap/d	3.9	3.9	0.1	0.9	0.3	—	—	—	—
BOD	% dw	2.3	1.7	2.7	3.3	—				
COD	% dw	5.8	8.8	9.0	10.1	—				
TN	% dw	2.2	2.6	3.9	3.6	4.0	1.2	0.4	4.0	2.9
TP	% dw	0.8	0.6	1.3	1.7	—	0.3	0.2	0.5	0.5

Note: dw: Dry weight. kg/cap/d: Kilograms per animal per day. Data from Reddy, 1981.

to prevent plant mortality, odors, and poor treatment efficiencies. Treatment wetlands are a compatible component of on-farm, total waste management. Their land intensiveness is not a serious limitation in most instances. Farmers typically have the equipment and skills necessary to build their own wetlands and to operate them successfully.

STORMWATER RUNOFF

Concentrations of most parameters in stormwater are time dependent. Stormwater concentrations and loads are cyclic due to periods of dry fall and deposition, followed by the first flush of runoff after rain, followed by exponential decreases in runoff constituent concentrations as storages rinse from the landscape, and finally dry conditions and deposition until the next storm event.

Table 18-7 provides mean concentrations for constituents. The averages are flow weighted to provide realistic estimates of the total constituent load that escapes during multiple storm events. Instantaneous concentrations will be considerably higher than these averages. Pollutant concentrations and loads generally range from low levels from undeveloped and park lands; to low-density residential and commercial; to agricultural (Figure 18-2); to higher-density residential and commercial; and finally to high density commercial, industrial, and agricultural land uses. Mean concentrations per event for BOD_5 vary from 1.45 mg/L for undeveloped lands to 20 mg/L for high-density urban areas. Total suspended solids concentrations vary from 11 mg/L for undeveloped areas up to 150 mg/L for high-density urban areas. Typical reported concentrations for other pollutants occurring in stormwater are summarized in Table 18-7.

The mass loading rates provided in Table 18-7 represent normalized pollutant loads that are somewhat independent of local rainfall amounts. Because pollutant loads per area per time are relatively constant between similar land-use areas, variable local rainfall washes these loads off the land in a few large events or over many smaller events. Urban pollutant loads increase with the imperviousness of the watershed. Although 20 to 40 percent of the material on street surfaces is organic, it does not biodegrade easily because it comes from

Table 18-6 Elemental Composition of Selected Animal Wastes

	Concentration (mg/kg dry weight)			
Parameter	Beef Cattle	Dairy Cattle	Poultry	Swine
Potassium	28,170	39,250	16,980	35,570
Calcium	8030	21,500	51,890	38,810
Magnesium	3240	6360	5190	9700
Sodium	620	7100	2690	9700
Iron	800	1170	1450	1940
Zinc	120	210	310	566
Copper	27	31	54	97–810
Manganese	106	145	250	240
Boron	91	49	87	5
Chloride	28,000	6800	58,000	56,600
Sulfur	3860	3700	11,000	13,740
Cobalt	<15	14	4.3	8
Molybdenum	2.3	5	1.2	0–0.4
Cadmium	—	0.2	0.46	0–1.6
Chromium	—	6.9	—	—
Aluminum	—	1050	2000	570
Nickel	—	18	19	—
Barium	—	16	—	—
Strontium	—	8.5	—	—
Lead	—	—	—	13

Note: Data from Overcash et al., 1983.

Table 18-7 Composition and Mass Loading Rates for Stormwaters

Constituent	Urban Runoff		Industrial Runoff		Residential/Commercial Runoff		Agricultural		Undeveloped	
	Concentration (mg/L)	Load (kg/ha/yr)	Concentration (mg/L)	Load (kg/ha/yr)	Concentration (mg/L)	Load (kg/ha/yr)	Concentration (mg/L)	Load (kg/ha/yr)	Concentration (mg/L)	Load (kg/ha/yr)
BOD_5	20 (7–56)	90	9.6	34–98	3.6–20	31.59–135.2	3.8	11.59	1.45	1.12–2.351
COD	75 (20–275)	—	—	—	—	—	—	—	—	—
TSS	150 (20–2890)	360	93.9	672–954.5	18–140	84.28–797	55.3	24.14	11.1	11.2–18.73
VSS	88 (53–122)	—	—	—	—	—	—	—	—	—
NH_3-N	0.582	—	—	—	—	—	0.33–0.48	—	—	—
TKN	1.4 (0.57–4.2)	—	—	—	—	—	2.16–2.27	—	—	—
TN	2.0 (0.7–20)	11.2	1.79	7.8–18.06	1.1–2.8	9.144–32.18	2.32	10.61	1.25	0.22–2.804
Ortho-P	0.12	—	0.13	1.321	0.05–0.40	0.568–3.302	0.13–0.227	0.942	0.004	0.008
TP	0.36 (0.02–4.3)	3.4	0.31	2.2–3.151	0.14–0.51	1.412–4.85	0.344	1.362	0.053	0.04–0.120
Copper	0.05 (0.01–0.40)	0.049	—	0.077	—	0.045	—	—	—	0.007
Lead	0.18 (0.01–1.20)	0.174	0.202	0.269–2.053	0.065–0.214	0.157–2.431	—	—	—	0.022
Zinc	0.20 (0.01–2.9)	0.630	0.122	0.98–1.240	0.046–0.170	0.218–1.88	—	—	—	0.081
Chromium	—	0.28	—	0.044	—	0.026	—	—	—	0.003
Cadmium	0.0015	0.16	—	0.024	—	0.013	—	—	—	0.002
Iron	8.7	—	—	—	—	—	—	—	—	—
Mercury	0.00005	0.043	—	0.065	—	0.038	—	—	—	0.006
Nickel	0.022	0.032	—	0.030	—	0.029	—	—	—	0.004
Cyanides	0.0025	—	—	—	—	—	—	—	—	—
Total phenols	0.0137	—	—	—	—	—	—	—	—	—
Oil and grease	2.6	—	—	—	—	—	—	—	—	—

Note: Data from Dames and Moore, 1990; U.S. EPA, 1983d; Marsalek and Schroeter, 1989; Bastian, 1986; Lager et al., 1977; Marsalek, 1990; Driscoll, 1986; Shelley and Gaboury, 1986; and Novotny, 1992.

Figure 18-2 Nonpoint source stormwater can be intercepted and treated by wetlands located on the farm.

leaf and wood litter, rubber, and road surface material (Novotny, 1992). The high metal content of highway solids comes from vehicle emissions. Novotny (1992) reported that the average total nitrogen load from urban lands is 5 kg/ha/yr (1 to 38.5 kg/ha/yr), and the total phosphorus load averages 1 kg/ha/yr (0.5 to 6.25 kg/ha/yr). Urban and residential runoff is being treated with wetland detention basins (Kehoe, 1993) and constructed wetlands (Strecker et al., 1992a,b; Baker, 1992; Schueler, 1992).

Table 18-7 summarizes runoff concentrations and loads from general agricultural land uses, and Table 18-8 provides figures for three agricultural activities. Flows and loads are typically highest from areas with altered vegetation, high animal densities, and high fertilization rates. Runoff pollutant concentrations from animal feedlots can be extremely high unless

Table 18-8 Pollutant Concentrations and Mass Loadings for Selected Agricultural Operations

	Animal Feedlots		Fertilized Row Crops		Pasture	
Constituent	Concentration (mg/L)	Load (kg/ha/d)	Concentration (mg/L)	Load (kg/ha/yr)	Concentration (mg/L)	Load (kg/ha/yr)
BOD_5	300–12,000	—	—	20	2.5–107	3.4–220
COD	2,400–40,000	—	—	120	—	48–208
TSS	1,000–13,400	—	—	—	—	—
VSS	100–7,000	—	—	—	—	—
NH_3-N	1–770	—	1.5	—	0.7	—
TKN	—	—	—	—	—	—
TN	128–2,100	—	—	15	0.5–35	1.1–56
Ortho P	5–26	—	0.27	—	0.64–7.1	—
TP	20–480	—	—	4	0.04–8	0.11–2.2
NO_3-N	0.1–1,270	—	2.5	—	1.4	—

Note: mg/L: Milligrams per liter. kg/ha/d: Kilograms per hectare per day.

Data from Overcash et al., 1983; Reddy, 1981; Baker, 1981; and Middlebrooks, 1979.

runoff is collected and treated. Pollutants from feedlot runoff typically include high levels of organic and inorganic solids and associated nutrients. Runoff from row crops and pasture areas may be low or high in mineral solids, depending on farming practices, rainfall intensity, soil types, and topography. Nutrient concentrations and loads from row crops and pastures depend on fertilization practices. Because of the availability of land on many farms, constructed wetlands are being increasingly used to treat runoff from intensive animal operations (Hammer, 1992a) and from row crops (Rodgers and Dunn, 1992; van derValk and Jolly, 1992).

SUMMARY

Design of wetland treatment systems or any other pollution control technology depends on accurate knowledge of influent wastewater characteristics. This chapter summarizes pollutant concentrations and loads from a variety of wastewater categories that might be applied to wetland treatment systems. Raw wastewater generally receives preliminary and sometimes secondary treatment before being discharged to wetlands, so the concentrations and loads summarized in Tables 18-1 to 18-8 typically will be reduced before application. Usually a team of engineers will combine talents to determine the most appropriate balance between pretreatment prior to the wetland and treatment to be accomplished in the wetland. Considerations to be used in selecting the most cost-effective combination of wetland and conventional technologies to meet environmental requirements are discussed in Chapter 19.

CHAPTER 19

Wetland Alternative Analysis

INTRODUCTION

The analysis, comparison, and selection of an appropriate treatment wetland option is a critical aspect of project planning. No amount of careful engineering, construction, and operation can achieve permit compliance and environmental protection unless the selected alternative has the technical, regulatory, and economic feasibility to accomplish the project goals. Alternative selection must be based on the best available information, including clear objectives, wastewater source and flow characterization, technical and permitting feasibility, and site-specific conditions. In some cases, the technical feasibility of a treatment option may not be well known due to unique wastewater characteristics or insufficient technology development. In these cases, a pilot study may be necessary to more fully describe project feasibility. Pilot projects can also help to ascertain regulatory feasibility and public acceptance by serving as a demonstration site for direct scrutiny.

In many cases, the project goals can be met by more than one alternative. In those situations, owner or regulator preference may dictate which of the feasible alternatives will be selected. Economic factors are also important in alternative selection because of the range of capital and operation and management costs associated with different technologies. Lower economic cost benefits not only the owner and consumer, but also the environment because of the reduction of direct and indirect environmental impacts associated with economic expenditures for raw materials and energy. Finally, if all other factors are equal, then ancillary benefits of one alternative may influence selection of the wastewater management technology.

Chapter 3 provides a general description of natural treatment alternatives including wetlands. This chapter describes these wetland options in greater detail to help the reader assess the advantages and disadvantages of each option.

The variables that should guide selection of a wetland treatment option can be grouped into three categories which will be described in this chapter: (1) technical constraints, (2) regulatory constraints, and (3) economic constraints.

BRIEF DESCRIPTION OF WETLAND TREATMENT ALTERNATIVES

Three wetland alternatives are described in this book: constructed wetlands with either surface flow (SF) or subsurface flow (SSF) and natural wetlands with surface flow. A wide variety of design variations exist for each of these alternatives. In addition, these three wetland alternatives can be combined with each other or with other conventional and natural technologies to create hybrid systems that meet specific needs. Each alternative has advantages and disadvantages for different applications. This chapter describes and compares the three major wetland alternatives. Chapters 20 through 22 describe the individual design strategies for each of these three wetland alternatives.

CONSTRUCTED SURFACE-FLOW (SF) WETLANDS

Surface-flow constructed wetlands mimic natural wetlands in that water principally flows above the ground surface, as shallow sheetflow, through a more or less dense growth of emergent wetland plants. Figure 19-1 shows the four features common to all constructed SF wetlands: an inlet device, the wetland basin, the wetland plants, and an outlet device. Various configurations exist for each of these features.

The inlet device initiates the flow of wastewater into the constructed wetland. Constructed wetland inlet devices are designed to initiate and maximize sheetflow of wastewater into the wetland cell.

The size, number, and shape of the constructed wetland basin(s) are important components of design. Basin size should be based on realistic reaction kinetics to provide compliance with regulatory limits. The number of basins is determined by flow rate, land area constraints, and operational redundancy requirements. The shape of the basins will be dictated by site-specific conditions and designer preference.

Wetland plants provide mineral cycling and attachment area for microbial populations that are essential for water quality improvement in SF constructed wetlands. Plants are selected for hardiness under project-specific water quality and hydrologic conditions, by cost and availability, by value as a wildlife cover and food source, and by designer preference.

The SF constructed wetland outlet device recollects surface water from the cell and directs this flow to downstream wetland cells or to the ultimate receiving water system. Outlet devices in SF constructed wetlands frequently are designed to provide water level and flow control as well as the ability to measure outflow rates.

Detailed design methods for SF constructed wetland treatment systems are described in Chapter 20. Typical structural design details and design examples are provided in that chapter.

CONSTRUCTED SUBSURFACE-FLOW (SSF) WETLANDS

Constructed, SSF wetland systems treat wastewater by passing it horizontally or vertically through a permeable media planted with wetland plants. Microbial attachment sites are

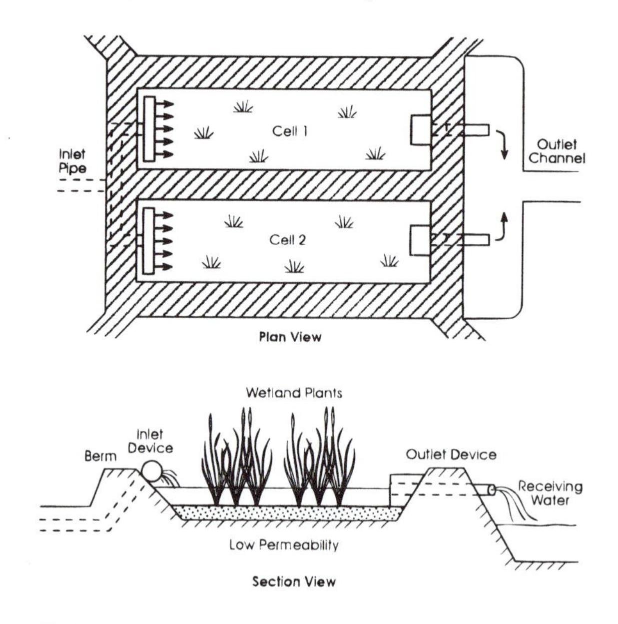

Figure 19-1 Typical configuration of a constructed SF wetland treatment system.

located on the surface of the media and on the roots of the wetland plants. Although SSF wetlands have many features in common with SF wetlands, they also have a number of differences that are important during project planning. As shown in Figure 19-2, the principal components of a constructed SSF wetland are the inlet distribution system, the basin configuration, the bed media, the plants, and the outlet control system.

The inlet wastewater distribution systems and basin configuration in SSF constructed wetlands have the same function as in SF wetlands, but they are designed in a fundamentally different fashion. To operate correctly, a SSF system must initiate and maintain all or most flow, subsurface and horizontally, through a permeable media. The principal design considerations for a SSF system are the media cost and permeability and, for a given permeability, the cross-sectional area necessary to initiate flows into the inlet zone.

Plant selection is frequently quite similar for SF and SSF constructed wetlands. The same small group of emergent wetland plants grows best in both systems.

Outlet devices have the same functions in SSF as in SF constructed wetlands. The principal difference between the two is that SSF outlets must be able to collect water from the base of the media, typically between 0.3 to 0.6 m below the wetland bed (ground) surface.

Chapter 21 provides detailed design guidance for SSF constructed wetlands. As noted previously, many design details apply equally to SF to SSF wetlands, but some design principles are fundamentally different.

NATURAL WETLAND TREATMENT SYSTEMS

From an implementation standpoint, natural wetland treatment systems are simpler than constructed wetlands. The designer must collect and deliver the wastewater, but nature has provided the basin and vegetation. Nevertheless, natural wetland systems include the same components as constructed wetlands, and the basin and vegetation must be evaluated during planning to determine the suitability for project goals. As shown in Figure 19-3, the major components of a natural wetland treatment system include the inlet distribution system, the wetland basin, the natural wetland vegetation, the wetland sediments, and the outlet design features.

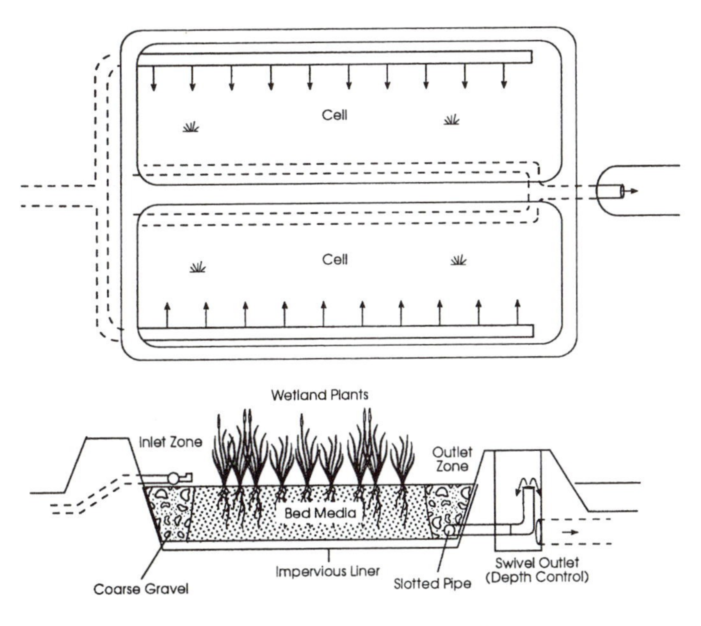

Figure 19-2 Typical configuration of a constructed SSF wetland treatment system.

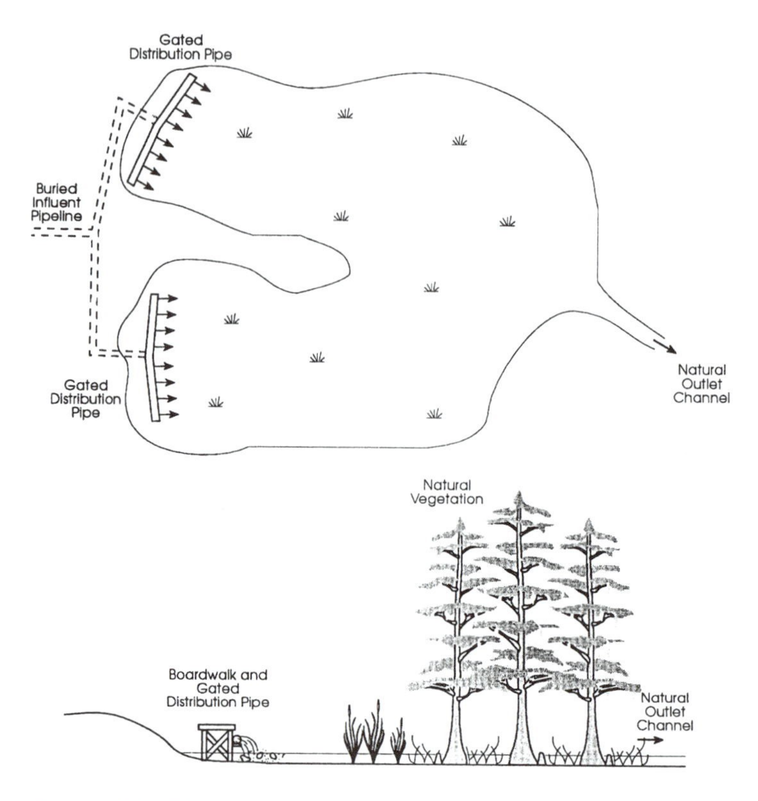

Figure 19-3 Typical components of a natural wetland treatment system.

Design of natural wetland treatment systems is fundamentally the same as for SF constructed wetlands, except that the designer's options are reduced by the existing ecology of natural wetlands that are close enough to the project site to be considered for use. Inlet distribution seeks to create and maintain wastewater sheetflow through the natural wetland. Effective flow distribution in natural wetlands is more of a challenge because of the normal channels and high spots that often are present and because of a general regulatory mandate against significantly altering the wetland's ecology.

Increasingly, regulatory agencies prefer not to alter natural wetland ecology outside allowable changes, which greatly reduces the number of natural wetland sites that can receive treated wastewaters. A limited variety of natural wetland plant communities are compatible with elevated nutrients and altered water regimes.

Outlet structures can be included in some natural wetland treatment systems. However, in most cases, it is preferable to maintain the existing wetland outflow configuration.

Chapter 22 provides a detailed approach to designing natural wetland treatment systems. Typically, design details for structures in natural wetland systems will be similar to those for SF constructed wetlands.

SUMMARY OF WETLAND TREATMENT ALTERNATIVES

The previous descriptions describe the principal components of the three wetland treatment system categories included in this book. When combined with the general performance expectations summarized here, this information can be used to evaluate alternatives and to develop a preliminary conceptual design for a wetland treatment system. Although there are clearly structural differences among these wetland treatment systems, there are also many similarities. Construction cost is directly related to flow control and assimilation rates in most wetlands. The sophistication of engineered system components should be adequate to achieve treatment goals. Reducing construction costs by inadequately distributing flow will result in a "waste" of potentially effective treatment area and may result in a far greater cost due to increased treatment area requirements.

Innovative basin design and interconnection of cells have the potential to reduce wetland system costs. Placing smaller berms between adjacent cells, lowering length-to-width ratios, and minimizing berm height and volume will reduce earthwork, typically the most expensive component of construction costs. When possible, the designer should use gravity flow, minimize pipeline distance and size, use open conveyance channels (that can be vegetated), and use radiating cells to minimize the cost and complexity of the inlet distribution system. In addition to these innovative design methods, the wetland treatment system designer can use experts in other civil engineering subdisciplines (e.g., irrigation, flood drainage, and highway design) to develop new and innovative ways to optimize performance and reduce costs.

Although natural wetlands require less design and construction, they require more initial thought and greater care during site selection. Hydrologic and water quality changes should be minimized in natural wetlands to enhance the survival of existing plant communities. Suitable natural wetlands are available for advanced wastewater treatment in only a few areas. However, when present, they provide an opportunity for environmental enhancement and significant cost savings.

TECHNICAL CONSTRAINTS

The technical feasibility of using wetlands to treat wastewater depends on wastewater characteristics, process performance capabilities, process design, operation and maintenance, discharge standards, and in some cases, site-specific environmental factors. Chapter 18 provided detailed information on the range of wastewater characteristics likely to be discharged to wetland treatment systems. Process performance for wetland alternatives was described for each of the major wastewater constituents in Chapters 10 through 17. Process design for each of the wetland alternatives is described in Chapters 20 through 22, and wetland treatment system operation and maintenance is described in Chapter 25. These chapters should be referred to during project planning to evaluate the technical constraints imposed on the wetland alternatives by each factor. This section compares the advantages and disadvantages of each wetland alternative and describes how general wetland sizing and potential site-specific constraints affect wetland technology selection and implementation.

ASSEMBLY OF DESIGN DATA

Site Conditions

Site conditions dictate the physical, chemical, and biological environment of a wetland treatment system. Conditions that should be evaluated during planning of a wetland treatment system include climate, geography, groundwater, soils and geology, groundwater, rainfall and runoff water chemistry, biology, and socioeconomic factors. The importance of each of these conditions may vary, but all should be investigated to some extent. Detailed studies may be needed to determine the importance of those site conditions that affect technical feasibility.

Climate

Climate cannot typically be controlled when selecting a specific site for a constructed or natural wetland project. However, climate is important during project planning and alternative selection because it affects the type and size of wetland that will be used. Latitude is the most critical determinant of climate because it determines seasonal temperature ranges. Other climatic factors that are important during project planning include rainfall, insolation, wind direction and velocities, and evapotranspiration. In the U.S., climatic information can be obtained from the National Aeronautic and Atmospheric Administration (NOAA) at the following address and telephone number:

National Climatic Data Center
151 Patton Avenue
Room 120
Asheville, North Carolina 28801-5001
(704) 271-4800

Major libraries have the data publications compiled by NOAA. As shown in the design equations provided in earlier chapters, minimum annual average monthly temperature is important when sizing wetland treatment systems for removal of all forms of nitrogen. Figure 19-4 provides a map of these critical temperatures in the U.S. For planning purposes, the long-term average temperature during the coldest month of the year has been found to be a good estimator of the critical low water temperature that will be experienced in a wetland treatment system. For areas where the minimum annual average monthly temperature is less than zero, it can be assumed that the minimum wetland operational temperature will be slightly above zero under an ice cover.

Wetland hydrology is affected not only by the volume of applied wastewater, but also by the net balance between rainfall and evapotranspiration. Wetland designers in areas with high net precipitation must take this incremental water volume into account when deciding on wetland volume, flow control structures, and free board of berms. In areas of high evapotranspiration, wetland design calculations need to consider water loss to the atmosphere. Figure 19-5 provides a map of the annual average net precipitation for the U.S. that can be used with the infiltration estimates discussed later to prepare an approximate water budget. More detailed information concerning the average monthly net precipitation values for the project location should be used to estimate the typical monthly water balance for the wetland under extreme seasonal conditions. Chapter 9 presented examples of water balances for

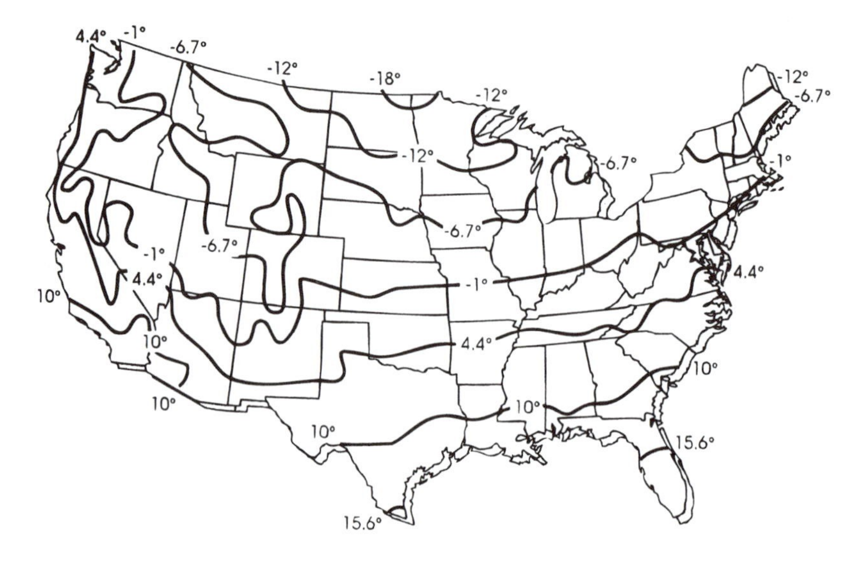

Figure 19-4 Minimum annual average monthly temperatures (°C) in the U.S. (Redrawn from Visher, 1954.)

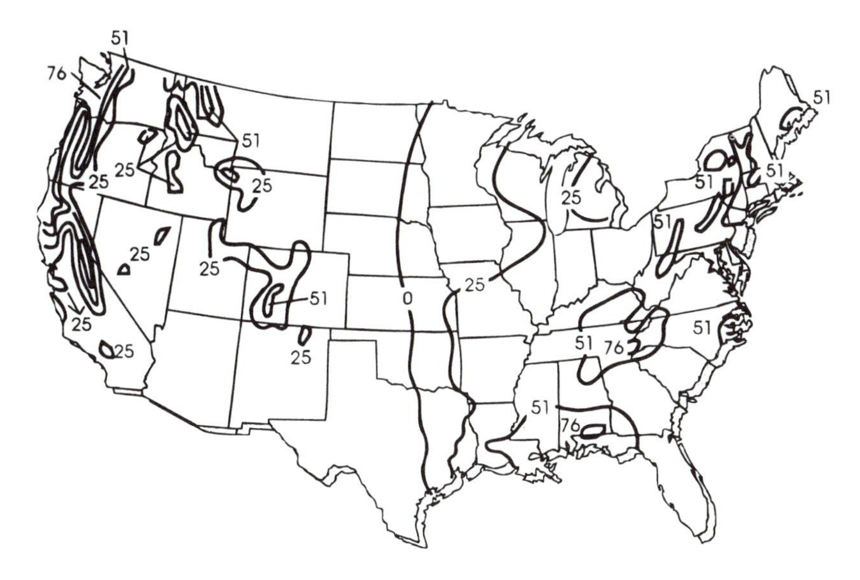

Figure 19-5 Annual average net precipitation (cm) (rainfall minus lake evaporation) for the U.S. (Redrawn from Visher, 1954.)

wetland treatment systems. A water balance is always important in the design of wetland treatment systems.

Geography

The distribution of natural wetlands in the U.S. was described in Chapter 4. If natural wetlands are being considered for a specific project, a preliminary evaluation of their occurrence in the U.S. can be made by reviewing the U.S. Geological Survey's (USGS) National Wetland Inventory maps, USGS topographic maps, USGS National High Altitude Program aerial photographs, and aerial photographs which are often available in the local tax collector's office or from the U.S. Department of Agriculture's (USDA) Natural Resources Conservation Service. Addresses and telephone numbers of some map and aerial photograph sources are as follows:

Earth Science Information Center
United States Geological Survey
507 National Center
Reston, Virginia 22092
800-USA-MAPS
(703) 648-6045

National High Altitude Program
United States Geological Survey
EROS Data Center
Sioux Falls, South Dakota 57198
(605) 594-6151

Figure 19-6 shows some examples of these information sources. The suitability of the vegetation and hydrology of natural wetlands must be confirmed by site data acquisition, including biological assessment.

Preliminary site screening of constructed wetlands should look for relatively level ground (generally less than 6 percent slope) that is not in a conflicting land use. Some sites that

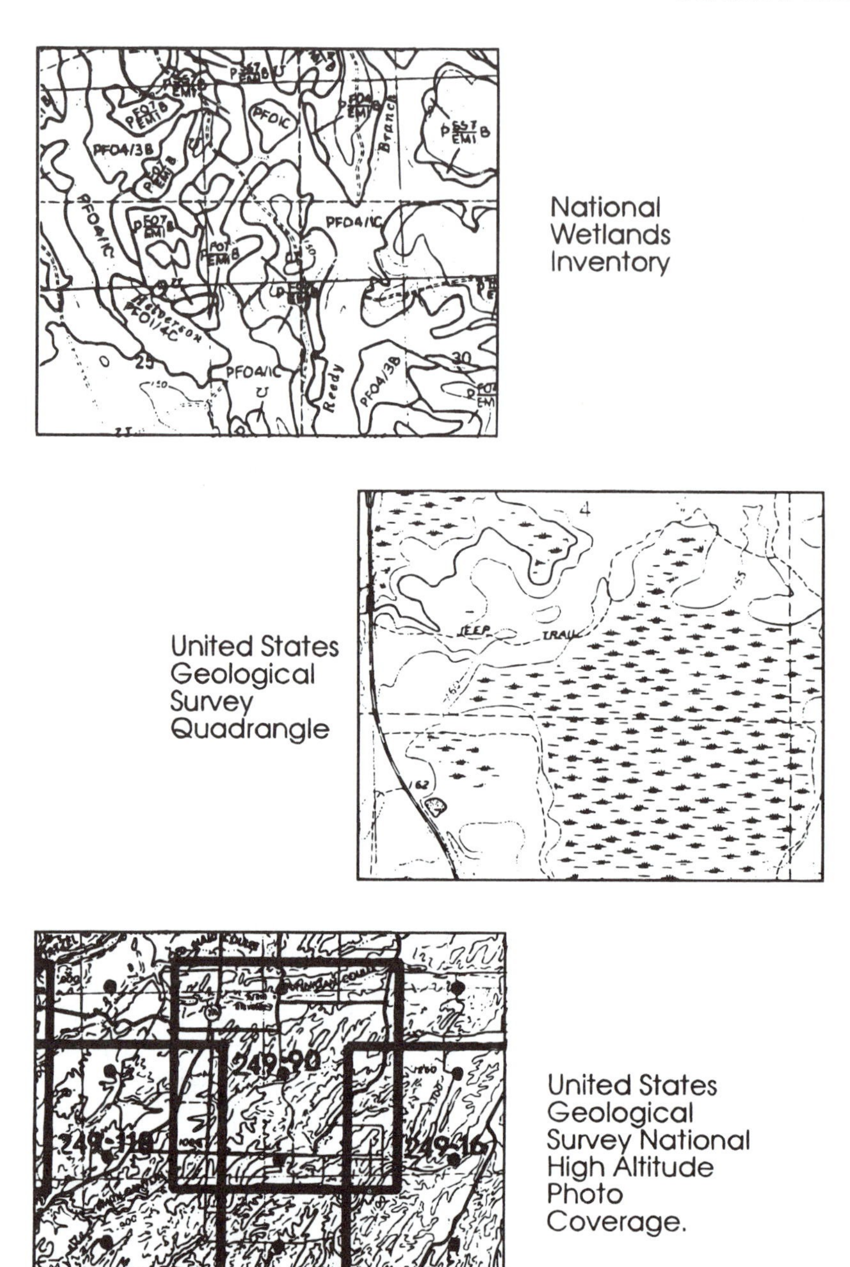

Figure 19-6 Examples of wetland mapping resources.

might be appropriate for wetlands construction are wholly or partially natural, jurisdictional wetland areas that are subject to permitting constraints. Dredging or filling natural wetlands should be avoided to the greatest extent possible.

A candidate site should not include many wet or dry drainage channels because stormwater flows would then need to be redirected. Because piping and pumping can be a significant part of project cost, candidate sites should be close to the wastewater source and alternative pipeline routes should be considered during site selection. Also, because gravity flow can save significant capital and operation and maintenance costs, site suitability is enhanced by the presence of a natural, downhill grade between the source of the pretreated wastewater and the constructed wetland.

Soils and Geology

For planning purposes, site soils can be characterized by using USDA Natural Resources Conservation Service soil surveys which are generally available for many counties of the U.S. Soil surveys include maps of soil types as well as summaries of soil properties, groundwater conditions, climatic information, and plant community information.

Soils are classified by soil scientists based on a complex array of physical and chemical characteristics. Soil information that might be important during project planning includes the presence of hydric soils, which would be a plus for a natural wetland site, but a potential regulatory constraint for a constructed wetland site; soil texture and composition as a suitable medium for berm construction or for impeding leakage to the groundwater; depth to seasonal high groundwater; and depth to confining layers of clays or spodic horizons. In some cases, the sorption potential of the soils will be a design variable, such as for metal removal.

Groundwater

Infiltration of wastewater to the groundwater is important because infiltration affects the wetland water balance and could pose regulatory problems under some conditions. Soil infiltration rates published in soil surveys typically overestimate the actual infiltration rates under sustained, saturated soil conditions and are not reliable for project planning or design. Surface infiltrometer tests or well slug tests provide better estimates of the groundwater leakage that can be expected from a full-scale wetland treatment system. Methods for measuring infiltration rates are described in Hansen et al., 1980 and USBR, 1993. Wetlands can be built on leaky soils as long as regulatory requirements can be met and adequate hydroperiods can be maintained with the wastewater addition and net rainfall. In fact, wetlands have been designed with groundwater recharge as a specific project goal (Ewel and Odum, 1984; Knight and Ferda, 1989). Groundwater infiltration can be eliminated as a project concern for constructed wetlands by using a clay or plastic, impervious liner. Although this approach may not be necessary if the wastewater has received secondary pretreatment, it is recommended when wastewater is less than secondary quality.

Field-scale tests are the most reliable method of estimating groundwater infiltration rates. For constructed wetlands, it may be necessary to construct pilot wetland basins on a proposed site and then instrument inflows and outflows to develop an accurate water balance. For natural and constructed wetlands that do not have significant surface inflows or outflows, infiltration and evapotranspiration can be estimated by continuously measuring changes of water stage (see Figure 19-7). In general, it is wise to understand the directions and flows of regional groundwater under the project site.

Biological Conditions

The addition of any type of water or wastewater will alter biological conditions at a site. Constructed wetlands frequently replace upland habitats with marsh vegetation. The upland habitats that are lost might include plant communities such as grassland, forest, scrub, desert, or agriculture. The environmental values of these upland habitats should be assessed during project planning. Likewise, wastewater discharge to natural wetlands can cause biological changes of varying magnitudes. Existing plant and animal communities in natural wetlands will change, depending on the degree of changes to surface water quality and hydrology.

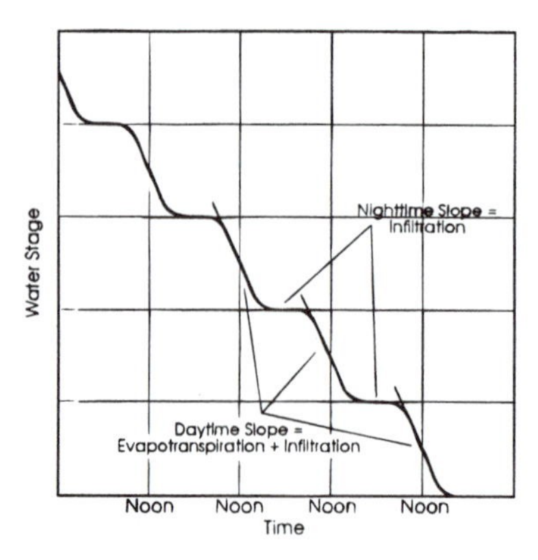

Figure 19-7 Method for estimating infiltration rate and evapotranspiration in natural wetlands based on water stage measurements.

Construction-related impacts will result in replacement of part of the existing vegetation by distribution pipes, boardwalks, and monitoring structures.

For most constructed wetland projects, site-specific biological conditions do not represent a major technical constraint. This is not always the case for projects that use natural wetlands. In planning a natural wetland discharge, the existing wetland biology provides a historical record of hydrology, surface water quality, and climatic factors. This historical water regime can be interpreted by using the information provided in Chapters 4 through 9 and in books such as *Wetlands* by Mitsch and Gosselink (1993). The range of conditions to which the existing natural wetland ecosystem has adapted provides technical bounds for the system's assimilation of wastewater flows. The importance of site-specific biological conditions for project planning is also discussed later in this chapter under Regulatory Constraints.

Characterization of the Water To Be Treated

Flows

The amount and timing of the water to be treated is the first and foremost item of the design basis. This information should include the possible seasonality of flows and the anticipated progression of flows over the life of the design. This is more important for treatment wetland design than for conventional concrete and steel treatment plants because of the implied life cycle of the process and the nature of urban and industrial growth. It is traditional to plan for a 20-year life expectancy for conventional wastewater treatment plants, because mechanical equipment begins to wear out after approximately that time period. But wetlands clearly can continue to function for far longer periods than two decades; for example, there are receiving wetlands that have been in operation for periods of 70 and 90 years (Great Meadows and Brillion). Projecting flow estimates far into the future is risky, so it is necessary to be explicit about flow capacity at the time of design.

Flows, whether municipal, industrial, or stormwater, are often seasonal in character. It is necessary to anticipate those patterns because the wetland must function appropriately under these variable hydraulic conditions. Monthly flow estimates will be required for most point source projects, but stormwater systems may require a definition of the frequencies of events and their magnitude and timing.

The information on water quantities and timing is assembled into the annual and monthly water budgets for the design, including any seasonal or event storage that may be necessary. Such water budgets are easily prepared within the framework of a spreadsheet program on a personal computer (see Table 6-2, Chapter 6). This information is later linked to the computation of the expected reductions in pollutant concentrations.

Quality

The concentrations of the pollutants in the water to be treated are critical to the sizing process and to the prediction of the wetland performance in the face of unknown future variations. A clear definition of the incoming water quality is essential, including the anticipated temporal distribution of concentrations. There are often seasonal fluctuations for point sources and intra-event variability of concentrations for stormwater flows.

Incoming patterns of chemical composition propagate through the wetland and undergo modification, resulting in a spectrum of output compositions. Some of this output variability may be predicted by the design models, namely those variations which represent responses to moderately slow input changes—those which occur on monthly or less time scales. Faster events involve ecosystem processes that are not included in the design models available at the present time and therefore will give the appearance of generating stochastic variations.

Treatment Goals

Receiving Water Standards

Most wetland treatment systems discharge to surface waters and therefore must meet the conditions of a discharge permit. The conditions of the permit dictate the required performance of the wetland and therefore govern its sizing.

Many past permits have been developed, based upon previous technologies, and transferred to wetland treatment systems. There are typically annual averages and monthly and weekly maxima, perhaps adjusted seasonally. The relation between averages and maximum allowable concentrations may have been determined from other technologies and be inappropriate for a wetland system.

At the time of this writing, there is a clear but unmet need for the application of wetland science and technology in the permitting process. If maximum values are set low enough, the existence of unavoidable wetland background values can preclude the use of the wetland technology (see Regulatory Constraints).

Interfacing To Reuse or Further Treatment

Discharge is not always to surface waters, but possibly to groundwater, further treatment elements, or to reuse.

Land application has a long tradition as a means of wastewater disposal, but it is often plagued by a surplus of nitrogen, which escapes crop utilization and nitrifies on its way to the groundwater. Wetlands have the potential to strip excess wastewater nitrogen before land application. In this application, wetland design targets the requirements of the subsequent treatment process (i.e., land application).

Conversely, it is not always optimal to utilize a wetland as the sole element of a natural system to meet final requirements. The integration of the wetland into an engineered natural system often makes better economic sense. The phosphorus binding capacity of soils, and their nitrification potential, can complement the wetland's capabilities for anaerobic processes and denitrification. Ponds can provide sedimentation and water storage in a more land-economical way. The techniques presented in this treatise are applicable to wetlands used as pieces of integrated natural systems.

Groundwater discharges of treated water are feasible in a number of circumstances. The problems of avoidance of eutrophication of surface receiving waters are replaced by problems of ensuring proper quality for the aquifer to be recharged. If the groundwater beneath the wetland is a drinking water source, then attention must be paid to nitrate, pathogens, and metals, as well as to trace organic chemicals. However, many aquifers are not, and will not be, used as a potable water supply. Therefore, wetlands have a role as pretreatment for conventional rapid infiltration basins (RIBs) or as posttreatment for nitrate removal from under-drained RIBs.

Discharge in an increasing number of circumstances is directed to downstream wetlands, which are often a combination of surface and subsurface waters. These wetland water bodies are considered waters of the U.S., and are subject to appropriate regulations. However, the upstream treatment wetland is a treatment system and is regulated according to a different set of rules. The design goals for the treatment wetland therefore become the water quality and quantity desired for the management of the downstream, jurisdictional wetland.

No matter what the receiving ecosystem or postwetland treatment element, proper design requires a clear statement of the required water quality leaving the treatment wetland.

Pretreatment Requirements

Surface-flow wetlands in North America normally receive municipal water of approximately secondary quality or better. This is in contrast to the subsurface technology of northern Europe, which typically treats settled or primary influents. Urban and agricultural runoff and leachates are also relatively dilute influents. Animal wastewaters are normally diluted before wetland treatment. Industrial wastewater treatment wetlands are usually added to the end of a "conventional" treatment process. Some fairly strong food processing wastewaters are being treated with integrated natural systems, but these systems have some form of primary treatment, such as a clarifier, lagoon, or filter.

Table 19-1 compares the pretreatment recommendations for each of the three wetland treatment alternatives. Constructed SF wetlands are not used much at this time for secondary treatment of municipal wastewater. However, there are a few applications for secondary treatment in the U.S. that may serve as a model to judge the success of this application (e.g., Columbia, MO). Constructed SF wetlands can provide advanced secondary and tertiary wastewater treatment of municipal, industrial, and agricultural wastewater. They also can treat urban runoff that has been pretreated with wet or dry stormwater detention basins

Table 19-1 Comparison of Pretreatment Recommendations for Wetland Treatment Alternatives

Wetland Alternative	Pretreatment Goals
Constructed SF	Municipal, agricultural, and industrial wastewater: secondary or greater treatment recommended (can be lagoons, trickling filter, or activated sludge)
	Urban stormwater: wet or dry detention basin with oil and grease skimmer
	Nonpoint source runoff: none recommended
Constructed SSF	Municipal, agricultural, and industrial wastewater: Primary or greater treatment recommended (coarse screening and sedimentation)
	Urban Stormwater and nonpoint source runoff: not recommended for use because of pulse flows
Natural wetlands	Municipal, agricultural, and industrial wastewater: Minimum of advanced secondary recommended (ammonia reduction is desired; also phosphorus and metal removal when necessary)
	Urban stormwater: wet or dry detention basin with oil and grease skimmer
	Nonpoint source runoff: upland vegetated filter strip

used to reduce concentrations of mineral sediments, adsorbed metals, and oils and grease. Constructed SF wetlands can directly capture and treat nonpoint source runoff.

Constructed SSF wetland treatment systems can provide secondary treatment of municipal, industrial, and agricultural wastewater after coarse screening and primary sedimentation. Also, SSF wetlands can provide advanced secondary and tertiary treatment. SSF wetlands for treatment of highly intermittent wastewater, such as urban stormwater or nonpoint source runoff, require off-peak hydrologic management for vegetation maintenance.

Natural wetlands can be used to further polish secondarily treated municipal, industrial, and agricultural wastewater. Adequate pretreatment is necessary before any wastewater is discharged into natural wetlands because of the potential for undesirable environmental changes. Recommended pretreatment prior to natural wetlands discharge includes ammonia removal (nitrification) in addition to standard secondary treatment and industrial pretreatment for metals and organics. If phosphorus in downstream waters is a concern, wastewater that will be discharged to natural wetlands should also be pretreated to an appropriate inflow phosphorus concentration. When natural wetlands are to be used to treat urban stormwater or nonpoint rural stormwaters, a minimum of oil and grease removal and sedimentation of mineral solids is also recommended.

From a regulatory standpoint, there is fairly strong emphasis on creating SF treatment wetlands of a moderately high quality. The unstated principle is one of minimizing the exposure of wildlife to poor-quality waters and habitat. As a result, most of the available SF design data are in the lower ranges of concentration. It implies that incoming water quality is near enough to wetland background that those baseline numbers influence design. In other words, the C^* component of design equations becomes critical.

Sediment accumulation is one of the few processes in wetland treatment that has a foreseeable requirement for maintenance. Sooner or later, buildups will require some sort of removal activity or reconfiguration of the wetland. Based on present experience, such maintenance will be on the schedule of decades unless there is an unusually high content of settleable solids in the incoming water. In that event, a settling basin or pond is a logical pretreatment device. Stormwater wetlands embody such elements.

Phosphorus reduction is one of the least efficient processes in wetland treatment. Low total phosphorus concentrations can be reduced to even lower ones, but large phosphorus load removal requires a large wetland area. Consequently, some pretreatment for the reduction of high total phosphorus concentrations is normally cost effective. Iron and alum addition are the most frequent choices. In the case of SF wetlands, the point of addition needs to be upstream of the wetland because there is not an effective way to provide chemical contacting in the wetland itself. The SSF wetland has an advantage in this regard because the media may be amended with the phosphorus-removing chemicals.

The oxygenation of ammonium nitrogen is more efficient when there is a small diffusional resistance to providing the oxygen to the dissolved or sorbed nitrogen. Neither SSF nor SF wetlands are particularly good in this regard because the reaeration potential of the water sheet is relatively low. But, this tendency toward anaerobiosis is quite beneficial for the reduction of oxidized nitrogen to nitrous oxide and nitrogen gas. Therefore, the greatest efficiency for nitrogen reduction is achieved when the wetland is assisted by some form of nitrification pretreatment. Mechanical nitrification devices, planted or unplanted sand or gravel filters, or vertical alternating flow wetlands are all candidates.

An important prewetland component of northern systems is some form of water storage to avoid winter operation of marsh treatment systems. Although wetlands can be operated under natural or artificial insulation, some processes such as nitrogen removal are not as efficient at cold temperatures. The storage lagoon accomplishes partial treatment while allowing the natural system to be tuned to the requirements of the climate.

Postwetland Requirements

A common regulatory requirement is the reduction of fecal coliform (FC) bacteria to low levels in the wetland effluent. Because SF wetlands are frequented by warm-blooded animals and birds, the background levels of fecal coliforms often do not comply with typical receiving water standards. This particular bacterial indicator does not distinguish the animal of origin, and hence, the conservative regulatory stance is to ascribe the total measure to human origins. Therefore, the attainment of low fecal coliform levels may need to be accomplished by postwetland disinfection, via chlorination, ozonation, or ultraviolet irradiation.

The low amounts of dissolved oxygen found in many SF treatment wetlands also violates receiving water standards in many cases. Some form of postwetland aeration may then be required. If the site topography permits, this can be a simple aeration cascade.

GENERAL WETLAND SIZING

There is a spectrum of procedures that have been used for estimating land area requirements during treatment wetland design. The most simplistic are single-number rules, which have great appeal because they are quickly applied and easily understood. And, there is a popular misconception that wetlands, and other natural systems should be amenable to intuitive quantification. This is certainly not the case, as has been repeatedly proven in wetland science. The structure and function of wetland ecosystems are very complex and require commensurately complex descriptions.

Rule-of-Thumb Approaches

There is some value in a single-number "rule of thumb," provided they are used only to gain a general idea of anticipated performance. Several such rules are popular.

Percentage Removal and Reduction

It is very easy to compare the amounts of a pollutant in the inlet and outlet streams of a wetland and to compute the percentage difference. Unfortunately, this information is of very limited use in design, or in performance predictions, because it reflects none of the features of the ecosystem which are the target of design. By implication, it would be necessary to replicate the wetland that produced the percentage data and to replicate the operating and environmental conditions that prevailed during data acquisition. The second is clearly impossible, and past experience has given strong indications that the first is also difficult.

The literature is replete with review papers that tabulate removals for a selected spectrum of wetlands (for example, Cueto, 1993; Johnston, 1993; Strecker et al., 1992a,b). The implication is that wetlands of a similar type will achieve a similar reduction. While such groups of data begin to elucidate the bounds of performance, the effects of size, loading, flow patterns, depth, and other design variables cannot be deduced from efficiency values alone.

Removal percentage is useful only when correlated with operating conditions and ecosystem characteristics.

Often, the percentage is for a concentration reduction. However, in the case of stormwater wetlands, it may reflect a load reduction instead. These two measures are connected by the alteration in water flow that may occur due to rain, evapotranspiration, and leakage:

$$\frac{Q_o C_o}{Q_i C_i} = \frac{Q_o}{Q_i} \cdot \frac{C_o}{C_i} \tag{19-1}$$

$$\left(1 - \frac{\%L}{100}\right) = \left(1 - \frac{\%W}{100}\right) \cdot \left(1 - \frac{\%C}{100}\right) \qquad (19\text{-}2)$$

where C = pollutant concentration, mg/L
Q = water flow, m^3/d
%C = percent concentration reduction
%L = percent pollutant load reduction
%W = percent flow reduction

The basis for a reduction percentage contains, of necessity, a time period over which input and output data were collected. Therein lie two subtle difficulties. First, the period in question may reflect the filling or emptying of temporary internal wetland storages and, thus, not be representative of long-term sustainable removal capability. Because of this, short-term microcosm and mesocosm studies are susceptible to misinterpretation. Second, wetland outputs are often compared to contemporaneous inputs, thus incurring detention time lag errors. In turn, these fall into two categories: hydrodynamic lag, due solely to the turnover time of the surface water body, and chromatographic lags, due to the filling and emptying of sorption and biomass compartments.

If the input to the wetland fluctuates or is subject to pulses of pollutants, then the interpretation of removal efficiency becomes very difficult. Most treatment wetlands are subject to such events; indeed, stormwater wetlands are solely driven by pulse events. As an illustration, consider a wetland which behaves as a well-mixed unit continuous stirred tank reactor (CSTR) with a detention time of 3 days. It is presumed that there exists a background concentration of $C^* = 5.0$ units. A pulse of pollutant in a constant flow of water is introduced to this hypothetical wetland, with a peak time of 1 day and a peak concentration of 140 units (Figure 19-8). The pulse is 99.7 percent complete in 5 days. The short-term removal behavior of the wetland is presumed to be represented by a first-order areal model with Damköhler number equal to 4.0. Under these conditions, it is easy to calculate the

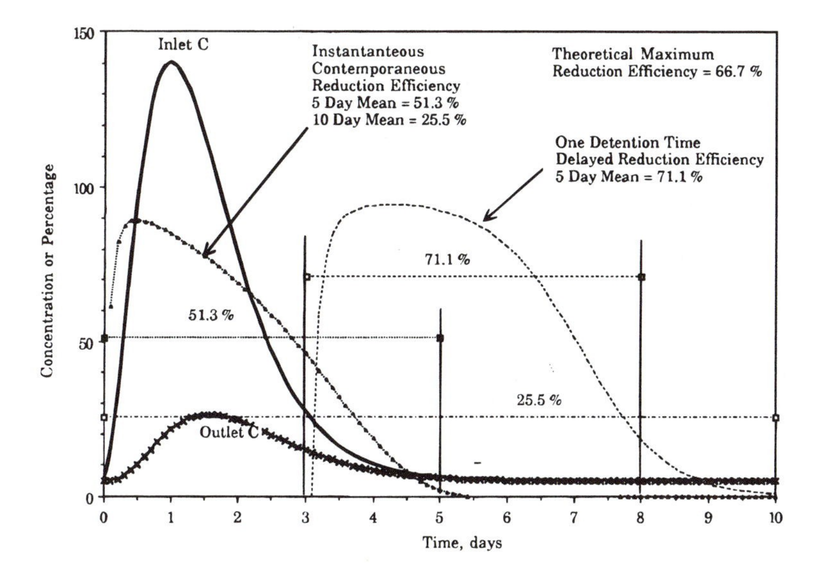

Figure 19-8 Treatment efficiencies associated with a pulse input of pollutant. The wetland is presumed to perform as a CSTR with a detention time of 3 days; the Damköhler number is 4.0. Two thirds of the pollutant pulse is removed under these circumstances. Concentration units are arbitrary.

predicted response of the wetland outflow concentration and to compute the percent reductions in concentration. A 10-day period is considered, starting at the time of pulse initiation.

The reduction of pollutant over the 10-day period is 66.7 percent obtained by adding up all inputs and adding up all outputs and then computing the reduction. It is also possible to compare contemporaneous "samples" of the inlet and outlet and calculate a percent reduction at that particular time. Note that the outlet concentration pulse is delayed only a small amount, occurring 0.6 days after the peak in the inlet pulse. Early in the event, reduction efficiencies are high, approaching 90 percent (Figure 19-8). Later, after the pulse has passed, and background concentrations are restored for both inlet and outlet, there is a zero efficiency. These instantaneous efficiencies time average to 25.5 percent very much lower than the actual reduction.

If the wetland behaved as a plug flow reactor (PFR) with the same 66.7 percent reduction, the required Damköhler number need only be Da = 1.61, due to the higher efficiency of this flow mode. In this case, contemporaneous "samples" of the inlet and outlet cause grave difficulties in interpretation. Whether the instantaneous values are averaged over the 5-day event window or over the 10-day sampling period, a negative reduction occurs. The basic problem is the transit time delay (Figure 19-9). Early in the event, clean water, with background concentrations, is being pushed from the exit regions of the wetland, no matter what the incoming concentration may be. For this example, the incoming pulse has high concentrations compared to background, and the early efficiencies are in the vicinity of 95 percent. Later in the event, the incoming concentrations have dropped to low values, but the unreacted material from the pulse is now exiting the wetland. This results in negative efficiencies, and furthermore, the denominator for those efficiencies is a low value, resulting in high negative values. When these efficiencies are averaged over the event period or the sampling period, the balance is in favor of the negative numbers.

As a result, this PFR wetland would be labeled as a *source* of the contaminant, whereas in fact it is removing two thirds of it.

The fundamental problem is the order of calculations: averaging of efficiencies is to be avoided. Mass average amounts of pollutants should be computed over the period in question, and then a period efficiency should be determined. Stated differently, flow weighting must be used over a period sufficient to accommodate detention time delays.

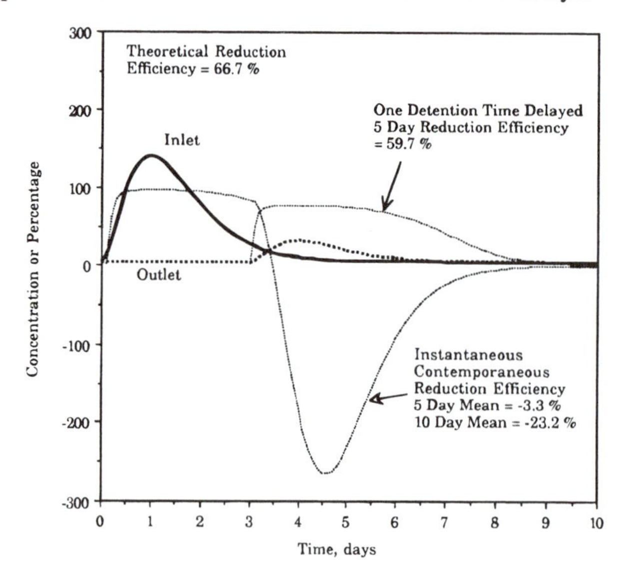

Figure 19-9 Reduction efficiences in a PFR wetland exposed to a pulse of pollutant. The detention time is 3 days; the Damköhler number is 1.61. Two thirds of the pollutant in the pulse is removed under these circumstances. Concentration units are arbitrary.

Low efficiencies always result if the incoming pollutant concentrations are comparable to the wetland background concentration for that constituent. Such data provide no sense of the capacity of the wetland to reduce the concentration of that pollutant under other circumstances.

Detention Time

The idea that more time in the wetland is good for water quality is intuitively very appealing. Early in the history of the technology, there was success for TSS and BOD reduction in wetlands that had 7 to 10 days of nominal detention in SF wetlands and 2 to 4 days in SSF wetlands. The urge to replicate this range is therefore strong, but clearly this basis is inadequate for other constituents and may represent overdesign for TSS and BOD. This attribute of the wetland must be coupled with a knowledge of the irreducible background concentration of the contaminant, as well as other design factors.

Depth is one primary controlling factor for nominal detention time; wetland area is the other:

$$\tau_{nom} = \frac{\varepsilon A h}{Q} \tag{19-3}$$

where A = wetland area, m^2
h = water depth, m
ε = water column void fraction
τ_{nom} = nominal detention time, d

The activity of the wetland in pollutant removal is associated with the emersed sediments and biota. These reactive surfaces dominate the removal processes for all biologically active substances. As a consequence, the rate of removal is strongly dependent on vegetation density: a bare soil, shallow pond has the minimum efficiency; a densely vegetated, fully littered wetland of the same depth has a higher efficiency. For instance, the phosphorus data of Lakshman indicate a linear relation (Figure 19-10).

If the detention time is increased by deeper submergence of these active components, at constant wetland area, further removal activity is typically not proportional to increased depth. In contrast, increasing the area of the wetland at constant depth does in fact increase the biotic material in contact with the water and acts to provide more detention time.

The overall efficacy of a depth increase depends on three factors: the distribution of bottom elevations in the wetland, the distribution of biotic material vertically above the

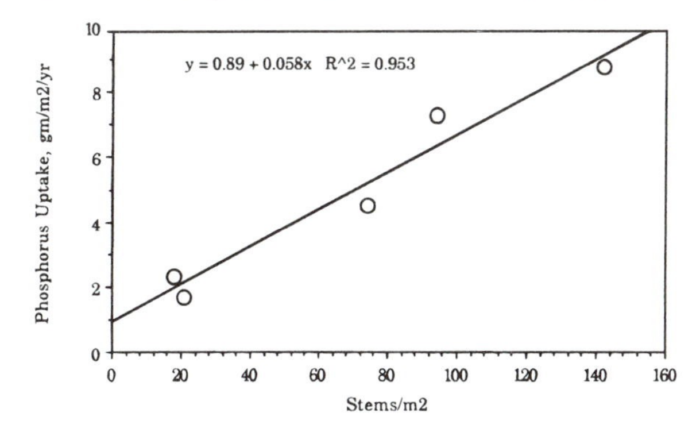

Figure 19-10 The effect of stem density on phosphorus uptake in SF treatment wetlands at Humboldt, Saskatchewan. Cells contained approximately equal amounts of cattail and bulrush. (Adapted from Lakshman, 1981.)

wetland bottom, and the reactive potential of the materials themselves. In general terms, all emersed surfaces are capable of supporting biological activity, although the time scales of uptake may differ, and the degree of support may depend on the degree of decomposition of the biomaterial.

As an illustration, for SF wetlands consider the unaltered Houghton Lake sedge wetland, which possessed a roughly linear distribution of measured bottom elevations (Figure 19-11) and a measured exponentially decreasing amount of biomass surface area with elevation above the bottom (Kadlec, 1990a). The high density of biomaterial near the bottom is the litter layer. If the rate of pollutant removal is first order, and proportional to the immersed area of biomaterial, the rate expression is

$$JA_T = (k_sA_s)A_TC = kA_TC \tag{19-4}$$

where A_T = total wetland area, m^2
A_s = surface area of biomaterial per unit area of total wetland, m^2/m^2
C = concentration of pollutant, g/m^3
J = pollutant removal rate, $g/m^2/yr$
k = first-order areal rate constant, m/yr
k_s = intrinsic first-order rate constant, m/yr

The intrinsic rate constant, k_s, is taken to be a constant, independent of depth, in this discussion.

Alternatively, the first-order rate of pollutant removal may be written as

$$JA_T = (k_sA_v)V_TC = k_vV_TC \tag{19-5}$$

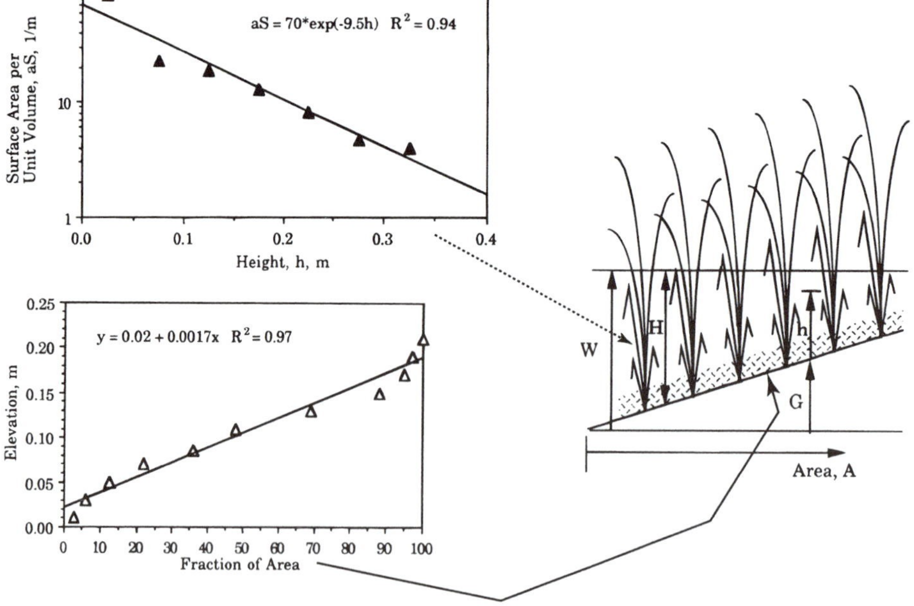

Figure 19-11 Variation in vertical vegetation density and areal ground elevations for a sedge peatland. The mean ground elevation is approximately 10 cm above the arbitrary datum. Shallower water exists in pockets and channels. The entire wetland surface is not submerged until the water elevation is 20 cm above the lowest point. (Adapted from Kadlec, 1990a.)

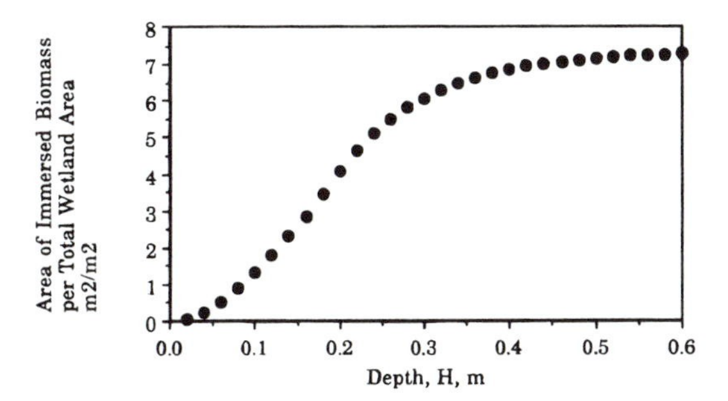

Figure 19-12 The immersed biomass surface area per unit total wetland area for the Houghton Lake sedge area. After the litter layer is covered, the value of A_s is nearly constant.

where A_v = surface area of biomaterial per unit water volume, m^2/m^3
k_v = first-order volumetric rate constant, 1/yr
V_T = volume of water in the wetland, m^3

The information in Figure 19-11 may be used to calculate the values of A_s and A_v. These change with depth, as shown in Figures 19-12 and 19-13. Two distinct types of water regime are possible: a low water level, insufficient to cover all portions of the wetland, and high water levels, which do cover the entire wetland area. In the case of low water, the immersed biomass surface increases with depth, and hence, the uptake rate constant (k) also increases. This effect is amplified by any pollutant releases which may occur under dry conditions.

In deeper water, the immersed area A_s is relatively constant for the case under consideration. Slight increases are due to the submergence of more of the stems, but those surfaces are small compared to that of the litter. The result, via Equation 19-4, is a relatively constant value of k as depth is varied within this regime. This same effect is responsible for a sharp decrease in A_v with increasing depth (Figure 19-13). The result, via Equation 19-5, is a value of k_v that is inversely proportional to depth within the regime of full coverage. The constant of proportionality is the areal uptake rate constant:

$$k_v = \frac{k}{\varepsilon h} \tag{19-6}$$

where ε = water-filled void fraction in wetland area, m^3/m^3.

The previous considerations have important consequences in terms of design based on detention time. There are two ways to increase the detention time of a wetland for a given

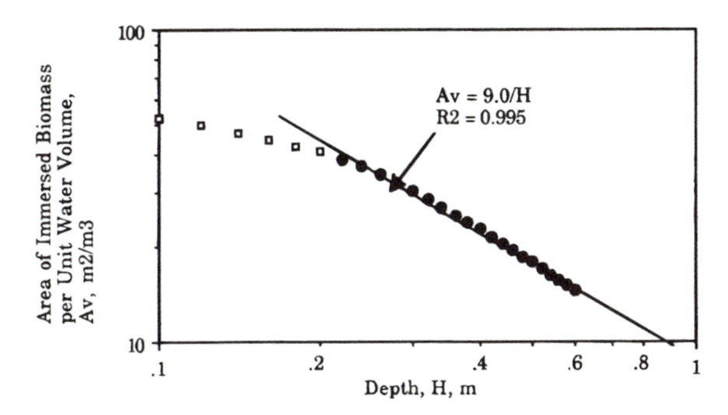

Figure 19-13 The immersed biomass surface area per unit total wetland water volume for the Houghton Lake sedge area. After the litter layer is covered, the value of A_v is inversely proportional to depth.

flow of water: increase the total acreage and/or increase the water depth. In the regime of full inundation, deeper water provides no significant increase in the removal of a pollutant. The amount of contributing biomass is fixed, and therefore, the longer time is utilized in reducing the pollutant in the larger volume of water. Conversely, increasing the wetland acreage at fixed depth does provide for more pollutant removal.

These concepts lead to the conclusion that the more effective method of increasing removal is to decrease the hydraulic loading rate. This has the effect of providing the intuitive increase in detention time without sacrificing efficiency to a depth increase.

The situation for SSF wetlands is less clear. For some pollutants, biofilms on the bed media control removal. For others, cycling of biomass and oxygen transfer are critical. The former are likely volume specific; the latter are area specific. The required research to delineate the best choice has yet to be performed.

Hydraulic Loading Rate

For the reasons stated previously, this single parameter is the most relevant for the sizing of wetland treatment systems. There is another matter of practicality: wetland wetted areas are almost always known more accurately than wetland water volumes. Despite considerable pains in setting bottom profiles during construction, there is always considerable uncertainty in the spatial distribution of bottom elevations and also the mean bottom elevation. Also, effective water depth in a wetland, and consequently water volume, tend to change as a wetland matures and partially fills with roots (SSF) and litter and mineral sediments (SF). It is therefore much easier to achieve a specific design wetted area than a design water volume.

The relation between detention time and hydraulic loading rate is

$$q = \frac{\varepsilon h}{\tau} \tag{19-7}$$

where q = hydraulic loading rate, m/d
ε = volume fraction water in wetland, m^3/m^3
τ = nominal detention time, d

The water volume fraction is normally quite high for SF wetlands, with values typically in the range of 0.90 to 0.95. Consequently, no serious error results from using

$$q \approx \frac{h}{\tau} \tag{19-8}$$

The depth chosen for SF wetlands is usually in the range 0.15 to 0.45 m. The 7 to 10-day detention time range therefore translates to a typical range of hydraulic loading rates from 1.5 to 6.5 cm/d, with a central tendency of about 3.0 cm/d for marshes (Figure 19-14). For SSF wetlands, $\varepsilon \approx 0.4$ and depths range from 0.3 to 0.6 m. The 2- to 4-day detention time translates to loading rates of 8 to 30 cm/d.

Pollutant Loading Rate, Pollutant Uptake Rate

Land application of wastewater has traditionally been designed to meet the water infiltration capacity of the soil together with the nutrient uptake potential of the crop (WPCF, 1990b). The intent is to utilize harvesting for nutrient removal. Further loading criteria are applied to BOD reduction, usually for odor control, and for TSS loading, which reflects

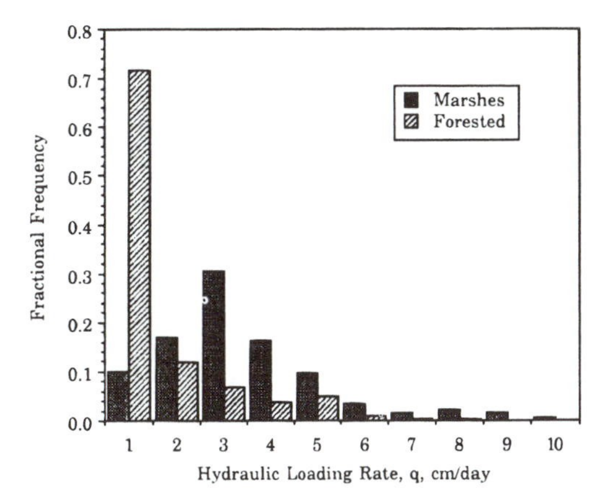

Figure 19-14 Distribution of hydraulic loading rates for SF wetlands. A few marshes (7%) are loaded at rates in excess of 10 cm/d, but no forested wetlands. Data represent mostly point source wetlands, tabulated in the NADB (Knight et al., 1993b).

concerns about soil clogging. These concepts do not transfer to wetland design because there is usually no harvest activity and there are not clogging and conveyance problems.

It is nevertheless instructive to place wetland treatment systems in the overall perspective of land-intensive technologies. Slow-rate land application, designed for pollution control, is usually designed for pollutant loading rates that are comparable to those presently in use in SF wetlands, and marsh removal rates are similar (Table 19-2).

Percentage of the Contributing Watershed

The amount of water and the amounts of pollutants that reach stormwater treatment wetlands from contributing watersheds are not typically known in advance. The number and duration of the events that produce input, together with the interevent spacing, are thought to affect the efficiency of the stormwater wetland. Because of this variability of actual flow rates, detention time and hydraulic loading are difficult to define, and other single-number sizing rules have evolved. One rule-of-thumb states that the wetland size should be a specified fraction of the contributing watershed, usually in the range of 1.0 to 5.0 percent. A little arithmetic shows that this is equivalent to the range of hydraulic loading rates cited earlier for point source treatment wetlands. In a moderate climate,

annual rainfall = 60 cm
average annual rain rate = 60/365 = 0.164 cm/d
watershed runoff coefficient = 0.75
runoff = 0.75 · 0.164 = 0.123 cm/d
watershed/wetland area ratio: WWAR = 1/.04 =25 (for 4% of the watershed)
average annual wetland hydraulic loading rate = 25 · 0.123 = 3.1 cm/d

Table 19-2 Approximate Median Load Removals for Surface-Flow Wetlands and for Land-Based Treatment.

Technology	TSS kg/ha/yr	BOD kg/ha/yr	Total Nitrogen kg/ha/yr	Total Phosphorus kg/ha/yr
Marshes	1340	1050	170	47
Forested wetlands	320	150	320	18
Land application	2000	2000	300	40

Note: These numbers are approximate and should not be used in design.
Data from NADB, 1993; WPCF, 1990b.

Since the average annual hydraulic loading rate (HLR) is close to the mode of the distribution of HLRs for point source-driven marshes (Figure 19-14), it is reasonable to expect that stormwater wetlands designed in this way would perform somewhere near the average for the emergent marsh database set. For instance, the mean reduction for phosphorus in 50 NADB marsh cells was 57 percent at an average HLR = 4.2 cm/d; the mean reduction for several constructed stormwater marshes, with an average WWAR = 4.3 percent, was also 57 percent (Strecker, 1992a).

Design Storm Detention

A stormwater wetland may also be sized to contain a specific volume of water, usually the volume associated with a rain event of a specified return frequency or probability of occurrence. For instance, Schueler (1992a) suggests that the wetland has sufficient volume to fully contain any rain event up to the 90th percentile of the rainstorm quantity distribution. Again, this may be shown to match the loading and detention design ranges used for point-source wetland systems.

In the vicinity of Washington, DC, there are 104 cm of annual rainfall, and the 90th percentile storm is 3.18 cm.

watershed area = 40 ha = 400,000 m^2
watershed runoff coefficient = 0.75
design storm runoff volume = $0.75 \cdot 0.0318 \cdot 400{,}000 = 9540$ m^3
wetland area @ 0.3 m depth = 9540/.3 = 31,800 m^2 = 3.18 ha
WWAR = $100 \cdot (3.18/40) = 8\%$
annual flow = $0.75 \cdot 1.04 \cdot 400{,}000 = 312{,}000$ m^3
average annual detention time = 9540/312,000 = 0.03 yr = 11 d
average annual wetland HLR = 312,000/31,800 = 9.8 m/yr = 2.7 cm/d

This single-number design technique has the advantage of allowing a variable percentage of watershed, depending upon the annual rainfall pattern and annual rainfall total. As in the case of WWAR design, the loading and detention times correspond to the mean values for point source treatment wetlands. It is therefore not surprising that Schueler (1992a) lists pollutant reductions that are in the mid-range for other treatment wetlands. For instance, the total phosphorus removal is projected to be 45 percent, compared to the 57 percent mean for marshes in the NADB.

The utility of these rules-of-thumb lies in the ability to quickly screen out a wetland alternative which has no chance of meeting project requirements. Accordingly, Table 19-3 is provided for that purpose. It should be noted that the central tendencies for stormwater wetlands do not differ much from those for the point source wetlands represented in Table 19-3 (Strecker et al., 1992a; Schueler, 1992a).

Despite the intuitive appeal of these single-number approaches, they are totally inadequate for final design. Most of the important factors, discussed in previous chapters, are ignored by these simplistic approaches. Final design should acknowledge the following additional items:

1. A set of pollution reduction targets
2. Spatial variability of pollutant removal
3. Hydraulic and meteorological constraints
4. Internal depth and vegetation density patterns
5. Internal water flow and mixing
6. Baseline wetland concentration values
7. Seasonality
8. Interaction with other treatment system components

Table 19-3 Baseline and Mean Values of Performance Parameters for Treatment Wetlands

Surface-Flow Marshes[a]	Units	Central Tendency	Wetland Effluent Baseline Lower Limit
Detention time	days	7–14	
Hydraulic loading rate	cm/d	1–5	
TSS	%	66	5–10 mg/L
BOD	%	50	5–10 mg/L
Total nitrogen	%	45	1–2 mg/L
TKN	%	50	1–2 mg/L
Organic nitrogen	%	55	1–2 mg/L
Ammonium nitrogen	%	25	0.05–.50 mg/L
Oxidized nitrogen	%	68	0.0 mg/L
Total phosphorus	%	57	0.02–0.05 mg/L
Bacteria	#/100 ml	Two or three logs	50–500 #/100 ml

		Central Tendencies					
Subsurface-Flow	Units	19 Wetlands[b] Lagoon Influent	42 Wetlands[c] Secondary Influent	5 Wetlands[d] Strong Influent	22 Wetlands[e] Strong Influent	71 Wetlands[f] Strong Influent	Wetland Effluent Baseline Lower Limit
Detention time	days		1–3	5–10	2–14	3–12	
Hydraulic loading rate	cm/d		10–20	3–8	2–15	2–10	
TSS	mg/L	14	4	10	12	27	2–5 mg/L
BOD	mg/L	14	2	5	20	18	2–5 mg/L
Total nitrogen	mg/L		15	19		45	1–2 mg/L
TKN	mg/L		5	18		17	1–2 mg/L
Organic nitrogen	mg/L		1	1		3	1–2 mg/L
Ammonium nitrogen	mg/L	3	4	17		14	1–2 mg/L
Oxidized nitrogen	mg/L		10	1		2	0.0 mg/L
Total phosphorus	mg/L			8	2	6	0.02–0.05 mg/L
Bacteria	#/100 ml	7000		1000			20–200 #/100 ml

Information from: [a]NADB (1993), ca. 50 wetlands; [b]Pennison (1993a); [c]Green (1993), Cooper and Green (1994); [d]Bavor et al. (1988); [e]Saurer (1994); [f]Schierup et al. (1990b).

9. Nature of the regulatory requirements
10. Acceptable level of risk

These items have been discussed in previous chapters, but in the following chapters they are jointly considered in the context of the determination of wetland design.

Mass Balance Design Model

A general wetland design model has been presented previously in Chapter 9, Section 3 and provides a simplified yet quantitative approach to predicting wetland size necessary to reduce an inlet concentration C_1 to an outlet concentration C_2 at a known average flow rate Q.

This wetland design model is presented here again as Equation 19-9 in its generalized form. For an isolated constituent (not involved in a sequential mechanism),

$$A = -\frac{Q}{k}\ln\left(\frac{C_2 - C^*}{C_1 - C^*}\right) \qquad (19\text{-}9)$$

where A = wetland area (m^2)
Q = flow rate (m^3/yr)
k = rate constant (m/yr)
C_1 = inflow concentration (mg/L)
C_2 = outflow concentration (mg/L)
C^* = irreducible background concentration (mg/L)

Values for the area-based rate constant in this model have been derived in previous chapters and are summarized in Table 19-4. The reader should refer to the appropriate preceding chapters for the caveats concerning Table 19-4. These rate constants can be corrected for temperature in some cases and are provided for two principal wetland types discussed in this book: SF and SSF.

On a per area basis, constructed SSF wetlands generally have higher pollutant mass removal rates and reduction efficiencies than constructed SF wetlands; however, there is no apparent difference in performance between these two wetland system types for phosphorus (Table 19-5). Natural wetlands may have lower removal rates and efficiencies than constructed

Table 19-4 Model Parameter Values—Preliminary

Surface Flow

	BOD	TSS	Organic N	Sequential NH_4-N	Sequential NO_x-N	TN	TP	FC
k20, m/yr	34	1000[a]	17	18	35	22	12	75
θ	1.00	1.00	1.05	1.04	1.09	1.05	1	1.00
C*, mg/L	3.5 + 0.053Ci	5.1+0.16Ci	1.50	0.00	0.0	1.50	0.02	300[b]
θ	1.00	1.065	—	—	—	—	1.00	—

Subsurface Flow

	BOD	TSS	Organic N	Sequential NH_4-N	Sequential NO_x-N	TN	TP	FC
k20, m/yr	180	3000[a]	35	34	50	27	12	95
θ	1.00	1.00	1.05	1.04	1.09	1.05	1.00	1.00
C*, mg/L	3.5 + 0.053Ci	7.8 + 0.063Ci	1.50	0.00	0.0	1.50	0.02	10[c]
θ	1.00	1.00	—	—	—	—	1.00	—

[a] Rough unsubstantiated estimate, settling rate determination preferred.
[b] Central tendency of widely variable values.
[c] Rough unsubstantiated estimate.

Table 19-5 Comparison of Estimated Annual Average Treatment Performance Capabilities for Wetland Treatment Alternatives[a]

Wetland Alternative	Typical Performance									
	BOD and TSS		NH_3-N		TN		TP		Metals	
	Removal Rate[b]	Eff[c]	Removal Rate[b]	Eff[c]	Removal Rate[b]	Eff[c]	Removal Rate[b]	Eff[c]	Removal Rate[b]	Eff[c]
Constructed SF	10	67	4.7	62	6.9	69	0.95	48	0.1	50
Constructed SSF	10–12	67–80	6.3	84	7.6	76	0.95	48	0.12	60
Natural Wetlands	4.5–6.7	30–45	4.4	50	5	50	0.4	20	0.06	30

[a] Assumes wetland influent is a typical municipal secondary effluent:

BOD	=	30 mg/L	metals (μg/L)
TSS	=	30 mg/L	Cd - 10
NH_3	=	15 mg/L	Cu - 50
TN	=	20 mg/L	Pb - 50
TP	=	4 mg/L	Zn - 300

Hydraulic loading rate = 5 cm/d.

[b] kg/ha/d.

[c] percent.

wetlands because of less optimal flow control. Specific performance expectations for these alternatives can be obtained by using information from other chapters of this book.

REGULATORY CONSTRAINTS

Wastewater treatment and disposal are regulated by an ever-increasing number of federal, state, and local laws, rules, ordinances, and standards. In some cases, the most challenging part of implementing a wetland treatment project is complying with regulations through the permitting process. In fact, regulations may hinder innovative project design for wetland treatment systems, even when the new design improves upon existing technology. Permitting an innovative project may delay implementation and increase cost with no incremental benefit for environmental protection. As a result of regulatory constraints, designers may opt for conventional technology with well-known deficiencies instead of more risky technology with potentially high environmental benefits.

A detailed knowledge of the pertinent regulations is essential to evaluate the feasibility of a wetland treatment project. This section provides an overview of the regulations that affect the use of wetlands for wastewater treatment in the U.S. as well as a summary of some specific state regulations. An updated and more detailed survey of federal, state, and local ordinances should be conducted to determine those that might be relevant to specific projects.

FEDERAL REGULATIONS

Federal laws related to environmental protection are drafted by congressional committees, congressional aides, and agency advisors; critiqued by professionals in private and public employment; debated, amended, and passed by the U.S. Congress, and enacted by approval of the president. Following enactment, these laws are codified as rules and implemented by government agencies. In addition, federal executive orders, policies, letters of understanding, and memoranda do not have the force of law, but may greatly influence regulatory feasibility of some wetland activities.

The law that most directly affects the permitting and implementation of wetland wastewater treatment systems in the U.S. is the Clean Water Act and its amendments (formerly the Federal Water Pollution Control Act of 1972). The government agency that is primarily responsible for implementing the requirements of the Clean Water Act is the U.S. Environmental Protection Agency (EPA). Other federal regulations that affect wetland treatment system permitting include the National Environmental Policy Act (NEPA) and the Endangered Species Act. NEPA is implemented primarily by EPA, and the Endangered Species Act is implemented primarily by the U.S. Fish and Wildlife Service (USFWS), a division of the Department of Interior. The U.S. Army Corps of Engineers (COE), under the Department of Defense, implements portions of the Clean Water Act and NEPA. In addition to these acts of Congress, presidential executive orders and letters of interpretation from agency representatives shape the overall framework of federal regulations affecting the use of wetlands for wastewater treatment.

Clean Water Act

The Clean Water Act regulates the use of waters of the U.S. for waste treatment and disposal. Waters of the U.S. are defined in Volume 40 of the Code of Federal Regulations (CFR), Part 122.2, to include nearly all naturally occurring surface waters from deep lakes and rivers to intermittently flooded (as little as 2 percent or 7 days/year), shallow, isolated wetlands (see Chapter 4 for the federal wetlands definition). The extent of U.S. waters expanded considerably in the 1980s as intermittent and isolated wetlands increasingly were included in enforcement of the Clean Water Act. Wetland systems constructed for wastewater treatment in upland areas are not considered waters of the U.S. and are not subject to the Clean Water Act (Bastian et al., 1989). However, in some cases, wetlands created to enhance wildlife and abandoned treatment lagoons that have developed dependent wildlife populations have been considered waters of the U.S. The importance of this definition of waters of the U.S. is discussed later. Unfortunately, some uncertainty remains about regulatory interpretation of the definition of waters of the U.S.

National Pollutant Discharge Elimination System (NPDES)

Section 402 of the Clean Water Act created the National Pollutant Discharge Elimination System (NPDES) permitting program. An NPDES permit is required for nearly all point discharges of water or wastewater into waters of the U.S. Municipal, industrial, agricultural, and urban runoff wastewater requires an NPDES permit. NPDES permits specify allowable flows and chemical quality of discharges into waters of the U.S. based on established water quality standards for those receiving waters. NPDES permits also are required for intermittent flows and for some emergency or backup discharges.

The Clean Water Act and EPA (40 CFR 131) guide water quality standards, which are promulgated individually by the states. Water quality standards vary among water bodies within a state and among states, depending on specific water resources. For example, most surface waters are classified for recreation and for maintaining healthy, balanced populations of fish and other aquatic wildlife. However, some surface waters have lower standards for wastewater discharge, and others have higher standards for water supply, shellfish populations, and pristine environmental goals. Consequently, NPDES permit limits are determined from the limitations of the receiving water. These limitations are calculated from effluent characteristics and dilution in the receiving water or from minimum technological limitations associated with treatment of various wastewater categories.

Because wetlands have special ambient water quality and biological conditions, the EPA directed all states to develop wetland water quality standards by the end of fiscal year 1993 (U.S. EPA, 1990b). These standards had to share the components of all water quality standards: designation of uses, antidegradation and backsliding on specific designated uses, and narrative and numeric criteria protecting aesthetic and biological wetland functions. To establish standards to regulate wastewater discharge to natural wetlands, studies must determine the levels of chemical and biological changes which do not affect benefits of the wetlands (U.S. EPA, 1985). To date, only Florida has identified chemical and biological standards for wetlands used for wastewater and stormwater treatment.

Generally, NPDES permits are issued for 5 years and list general and specific conditions. Table 19-6 provides an example of the general and specific conditions in an actual NPDES permit for a discharge of municipal wastewater to a natural wetland treatment system. Preparation, issuance, and enforcement of NPDES permits has been delegated by the EPA to state environmental protection agencies in all states and territories except for Alaska, Arizona, Idaho, Louisiana, Maine, Massachusetts, New Hampshire, New Mexico, Oklahoma, South Dakota, Texas, American Samoa, District of Columbia, Guam, Northern Mariana Islands, Puerto Rico, and the Trust Territory of the Pacific Islands. The EPA and the states have aggressively enforced these permits by levying severe fines for noncompliance. The Clean Water Act also allows criminal penalties for knowledgeable violations of permit conditions including 1 to 15 years of imprisonment for certain types of violations. When NPDES permit conditions are not met, the EPA or the delegated state agency may revoke

Table 19-6 Example NPDES Permit Effluent Limitations, Monitoring Requirements, and Standard Conditions for a Municipal Effluent Discharge to a Natural Wetland

Part I
Specific Standards

Effluent Limitations and Monitoring Requirements

During the period beginning on the effective date and lasting through the expiration date of this permit, the permittee is authorized to discharge from outfall 001, sanitary wastewater. Such discharges shall be limited and monitored by the permittee as specified below:

	Discharge Limitations			Monitoring Requirements		
Parameter	Annual Average	Monthly Average	Daily Maximum	Measurement Frequency	Sample Type	Sampling Point
Flow, millions of gallons per day	Report	Report	Report	Weekdays	Representative	Effluent
Carbonaceous biochemical oxygen demand (5 day), mg/L	20.0	30.0	60.0	Every 2 weeks	Grab	Effluent
Total suspended solids, mg/L	20.0	30.0	60.0	Every 2 weeks	Grab	Effluent
Ammonia as nitrogen, mg/L	—	2.0	—	Every 2 weeks	Grab	Effluent
pH, standard units	6.0–8.5	6.0–8.5	6.0–8.5	Weekdays	Grab	Effluent
Fecal coliform bacteria, N/100 ml	200	Report	800	Every 2 weeks	Grab	Effluent
Total residual chlorine, mg/L	—	—	<0.01	Weekdays	Grab	Effluent
Dissolved oxygen, mg/L		Report daily minimum		Weekdays	Grab	Effluent

Part II
Abbreviated Standard Conditions for NPDES Permits

Section A. General Conditions

1. Permittee must comply with all conditions of the permit. Permit noncompliance constitutes violation of the Clean Water Act.
2. Civil penalties not to exceed $25,000 per day. Criminal penalties of $2,500 to $50,000 per day and/or imprisonment up to 3 years. Administrative penalties up to $10,000 per violation, not to exceed $125,000 total.
3. Permittee will attempt to minimize or prevent discharges.
4. Permit may be modified, terminated, or revoked for various causes.
5. Permit may be automatically modified to comply with more stringent toxic effluent standards.
6–8. Permittee is liable for civil or criminal penalties for noncompliance, legal actions, and state laws.
9. Permit does not convey any property rights.
10. Permit does not authorize any construction in waters of the U.S.
11. Permit provisions are severable.
12. Permittee shall furnish information relevant to the permit.

Section B. Operation and Maintenance of Pollution Controls

1. Permittee shall properly operate and maintain all facilities and systems of treatment and control.
2. The necessity to halt or reduce the discharge to maintain compliance is not a defense during enforcement.
3. Bypass is prohibited unless effluent limitations can be met. Ten days notice shall be given for anticipated bypass and 24-hour notice, after-the-fact, for unanticipated bypass.
4. Upsets are an affirmative defense for noncompliance.
5. Permit does not authorize discharge of treatment byproducts unless specifically listed.

Section C. Monitoring and Records

1. Samples and measurements will be representative.
2. Flow monitoring devices and methods shall be capable of measuring with errors less than ±10 percent of true discharge rates.
3. Monitoring to be conducted according to 40 CFR Part 136.
4. Tampering with a monitoring device is punishable by a fine of up to $10,000 per violation or imprisonment up to 2 years.
5. Permittee will retain all records for at least 3 years.
6. Records will include date, place, time, individual(s), methods, and results.
7. Permit issuing authority may enter, inspect, copy documents, and sample.

Section D. Reporting Requirements

1. Permittee will notify permit issuing authority of any planned physical alterations or additions.
2. Permittee will provide advance notice of changes or activities that may result in noncompliance.
3. Permit can be transferred with prior notification.
4. Routine monitoring is required, including notification of no discharge.
5. Results of any additional monitoring must be reported.
6. Arithmetic means shall be used for averaging unless otherwise specified.
7. Reports of compliance shall be submitted within 14 days.
8. Noncompliance that may endanger health or the environment will be orally reported within 24 hours, and in writing within 5 days.
9. Other noncompliance shall be reported with routine monitoring reports.
10. Certain changes in discharges of toxic substances must be brought to the attention of the permit issuing authority.
11. Reapplication for a permit must be made at least 180 days before expiration.
12. All applications, reports, or information must be signed and certified by a responsible owner or officer.
13. All reports are available for public inspection except for confidential data under 40 CFR Part 2.
14. False statements in a report may be punishable with fines up to $10,000 or imprisonment up to 2 years.

the NPDES permit and issue a Temporary Operating Permit (TOP) or Consent Order that specifies corrective actions and incentives for compliance.

According to the Clean Water Act, nearly all wastewater must receive at least secondary treatment prior to discharge to waters of the U.S. (including natural wetlands). Secondary treatment is defined as "attaining an average effluent quality for both 5-day biochemical oxygen demand and suspended solids of 30 mg/L in a period of 30 consecutive days, an average effluent quality of 45 mg/L for the same pollutants in a period of 7 consecutive days and 85 percent removal of the same pollutants in a period of 30 consecutive days" (Bastian et al., 1989). This definition has been relaxed for discharges from waste stabilization ponds and trickling filters to recognize that these systems discharge some of their solids as new algae and microbes that do not have the same impact as the solids in the raw wastewater. Under these conditions, states may relax secondary standards for TSS to 45 mg/L for a 30-day average, 65 mg/L for a 7-day average, and 65 percent removal efficiency. Because natural wetlands are considered waters of the U.S., these standards also dictate the minimum quality of secondary wastewater that can be discharged to a natural wetland treatment system.

Increasingly, the EPA is requiring treatment beyond secondary standards to protect the designated water quality of surface waters. For example, additional treatment is required to receive an NPDES permit when wastewater discharges use the natural assimilative capacity of a receiving water and when the assimilative capacity of the surface water is limited by natural conditions such as low seasonal flow, little natural aeration, or naturally high oxygen demand. Also, because toxic chemicals that affect wetlands' designated uses have been found in an increasing number of water bodies, most NPDES permits limit effluent toxicity. These limits generally include a provision for no measurable acute toxicity in the effluent at the end of the pipe and no detectable chronic toxicity after complete mixing with the receiving water. Constructed wetlands provide one alternative to increase treatment levels prior to discharge to waters of the U.S.

Construction Activities in Natural Wetlands

Section 404 of the Clean Water Act regulates the discharge of fill materials in waters of the U.S., including wetlands. Although filling of wetlands is not prohibited, a permit is required above certain threshold volumes and for nonexempted activities. A nationwide permit system is used to expedite certain less significant wetland fill activities, such as repairs to existing structures and public utility construction. Otherwise, all permits for filling of wetland areas are evaluated by the U.S. Army Corps of Engineers (COE) on a site-specific basis. Evaluation criteria include issues related to water dependency, environmental impacts, and water quality. If wetland fill activities cannot be avoided and negative impacts can be mitigated by creating new wetland habitat in upland areas, the COE generally will issue a permit for the activity if other federal agencies agree. The following agencies have a specific interest: the USFWS for endangered species, the National Marine Fisheries Service for fisheries in coastal areas, and the state historical preservation office for cultural resources. In addition, the EPA can veto a permit issuance if it determines that the activity will have a detrimental effect on water quality. This veto is called a Section 404c veto.

When choosing a site for wetland wastewater treatment, natural wetlands and other waters of the U.S. should be avoided to the greatest extent possible. Avoiding these waters reduces the need for a Section 404 permit when constructing an outfall to the ultimate receiving water, a requirement needed by any surface-discharging option and one that may be covered by a nationwide permit. In some regions of the U.S., areas with saturated soils cover as much as 40 to 80 percent of the landscape. Since these areas have been included in the definition of waters subject to Section 404 permitting, avoidance has become difficult. If nonwetland sites are not available for wetlands construction, a Section 404 permit can be obtained as long as the environmental values lost by displacing the natural wetland are compensated by mitigation. For example, in a project in Mississippi, destruction of about 7 ha of jurisdictional wet slash pine forest was mitigated by constructing 21 ha of wetlands for wastewater treatment. The wildlife and water quality benefits provided by the constructed wetlands were assessed to be much higher than the values of the wet pine forest. The issue of using constructed SF treatment wetlands to mitigate impacts to jurisdictional wetlands has not been resolved by federal agencies. A favorable resolution of this issue could result in constructed treatment wetlands designed for habitat creation as mitigation banks (Knight, 1991).

Section 404 permitting issues frequently arise during planning and construction of natural wetland treatment systems. In some cases, inflow distribution systems have required Section 404 permits, and in other cases, boardwalk-type pipe support structures were exempted. Any diking that is planned within a natural wetland to isolate or control water flows is subject to Section 404 permitting, although minor diking and placement of water monitoring structures has been included in nationwide permits. As with constructed wetlands, Section 404 permits can be issued for minor fill activities associated with using a natural wetland for wastewater polishing and assimilation.

In a few cases, agency representatives have argued that the pretreated wastewater itself falls within the legal definition of "fill" and should be subject to Section 404 permitting. To the best of our knowledge, this logic has not yet prevailed.

Wetlands Executive Order 11990

Executive Order 11990 directs federal agencies to minimize degradation of wetlands and enhance and protect their natural and beneficial values. This executive order mandates avoidance and mitigation of impacts to wetlands. Before issuing an NPDES permit for a discharge to natural wetlands, this order must be considered, and assurances must be provided that the natural and beneficial values of wetlands will be protected and enhanced by the discharge. Enhancement is easy to show in wetlands that have been degraded by drainage, forestry, or other activities. However, enhancement in unaltered, pristine wetlands can only occur when pretreatment results in high inflow water quality. If a wastewater discharge enhances a natural wetland, there is some concern that ceasing the discharge in the future because of wastewater management changes might not be legally feasible (Rusincovitch, 1985).

Floodplain Management Executive Order 11988

Executive Order 11988 directs federal agencies to avoid direct or indirect support of development within floodplain areas. The purpose of this avoidance is to reduce the risk of economic losses due to flooding, minimize the impact of floods on human safety and health, and preserve the beneficial values of floodplains. This order covers many constructed wetlands and nearly all natural wetland treatment systems because of their dependence on nearby receiving waters. In project planning, this order affects routing and storage of rainfall and runoff to a constructed wetland treatment system. Frequently, storage can be incorporated into wetland system design to deal with extreme rainfall events, and weir design can reduce flooding conditions. If existing watershed flood storage volume is reduced due to construction of new dikes or other structures, adequate replacement storage volume should be provided. Constructed wetlands located in floodplains with extreme flood conditions, such as along major rivers, must have dikes that are designed to allow the passage of floods or that are sized to exclude flood waters. Both approaches have been taken at existing projects.

National Environmental Policy Act (NEPA)

NEPA directs all federal agencies to evaluate the environmental effects of projects that they construct or permit. Federal agency actions covered by Sections 402 and 404 of the Clean Water Act can trigger the NEPA process and consequently trigger the need for an environmental impact assessment. This environmental assessment can take one of two regulatory routes. On smaller projects that have small, localized effects, the environmental assessment is less formal and focuses on the immediate direct and indirect environmental impacts. The project must be modified to avoid or to mitigate affected environmental values until a finding of no significant impact (FONSI) can be issued to allow project implementation.

Larger projects that have the potential to affect regional environmental and socioeconomic resources are subject to the environmental impact statement (EIS) process. A multi-agency and public review determines whether an EIS is required for a proposed project. The EIS process typically takes more than 1 year to complete and is prepared and issued under a lead agency, generally the COE or EPA. To date, this lengthy process has not been applied to a wastewater to wetland project.

Endangered Species Act

Under the Endangered Species Act, the USFWS is mandated to identify and list plant and animal species that are threatened or endangered by extinction or that are considered likely to be threatened in the future. Threatened and endangered species cannot be harmed, killed, or otherwise negatively affected by human activities. The USFWS reviews potential impacts to federally listed species and can veto a Section 404 wetland fill permit or an NPDES permit by issuing a "jeopardy" opinion.

The potential for occurrence of threatened or endangered species inhabiting a project site should be considered during project planning. Typically, a list of threatened and endangered species that could occur at a project location is compiled from federal and state natural resource agency records. Potential project impacts on any of these species are assessed. If negative impacts on a species are anticipated during this review, it may be wise to conduct a field survey to ascertain whether that species actually occurs at the proposed site. This confirmation step is important in eliminating the chance of finding a "fatal flaw" after considerable planning, design, or construction monies have been spent.

Experience indicates that the presence of federally protected species does not always necessitate substantial changes to wetland treatment project siting or design. In a number of cases, the discharge of treated wastewater effluents to constructed wetlands has enhanced the population of listed, wetland-dependent species by creating additional habitat and food resources. For example, populations of bald eagles, wood storks, and snail kites, all federally protected bird species, have been increased in and around the 485-ha Iron Bridge constructed wetland treatment system east of Orlando, FL. Thus, the potential impacts of a project on protected species must be evaluated on a species-by-species basis.

STATE REGULATIONS

All states require permits for municipal and industrial wastewaters and, increasingly, for agricultural wastewaters and stormwaters. The 40 states and territories that have delegated NPDES program authority from the EPA essentially follow Section 402 requirements and EPA guidelines to prepare joint state and federal permits. The 11 states and 6 territories that do not have NPDES delegation require discharge permits in addition to the federally mandated NPDES permit. Each of these states and territories has different permitting requirements that are subject to frequent legislative and rule-making changes. This section reviews some examples of state regulations that specifically concern the use of wetlands for water quality treatment. Slayden and Schwartz (1989) reviewed states' activities and policies concerning the use of wetlands for wastewater treatment. Their findings are updated here and are based on more recent regulatory developments. The reader is cautioned to obtain the most recent rule revisions during project planning.

Alabama

The Alabama Department of Environmental Management (ADEM) prepared a preliminary report on using natural treatment systems to upgrade wastewater treatment (ADEM, 1988). This report provided preliminary design recommendations for constructed wetlands, including multiple, parallel cells followed by cells in series; minimum pretreatment of secondary; inflow distribution; water depth between 15 and 60 cm with 23 to 30 cm optimal; and hydraulic loading rates between 1.2 and 15 cm/d.

Arizona

The State of Arizona has recently published guidance for review of permit applications for constructed wetlands and floating aquatic plant systems (ADEQ, 1995). Most constructed wetlands in Arizona do not have surface discharges and are permitted under the state's Aquifer Protection Permit (APP). In those cases where wetlands have surface discharges, a federal NPDES permit is necessary. NPDES permits in Arizona are generally written to insure the creation of "net ecological benefit" in downstream receiving waters, and the provision of wildlife habitat has been an important consideration for most of the SF constructed treatment wetlands.

Florida

In 1975, Florida passed a regulatory exemption allowing the "Experimental Use of Wetlands For Wastewater Recycling," currently found in Chapter 17-600.120 of the Florida Administrative Code (FAC). This rule allows normal inland surface water quality standards to be exempted to encourage research on applying municipal wastewater to natural wetlands. The rule was a legislative byproduct of the wetlands research being conducted by the University of Florida cypress dome research team (see Chapter 1 for more background on the cypress dome study). Although the rule only provides for an exemption of 5 years at a time, it has been used at seven Florida sites to permit full-scale, natural wetland treatment systems.

In 1984, the Florida legislature passed the Warren S. Henderson Wetlands Protection Act which substantially broadened regulatory requirements for dredging and filling in wetlands. In addition to providing an increased level of protection for natural wetlands, this act also provided economic and environmental balancing by allowing certain compatible uses of natural wetlands. Specifically, the Warren S. Henderson Wetlands Protection Act allowed for the development of rules to permit the use of natural wetlands for receiving and treating municipal wastewaters and stormwaters as long as the "type, nature, and function" of the wetlands was protected. Rules for these uses were developed over 3 years and are now found in Sections 62-611 and 62-25 of the FAC (Schwartz, 1989).

The major features of the Florida wastewater-to-wetlands and stormwater-to-wetlands rules are summarized in Table 19-7. These two rules are quite different in their scope and approaches. The stormwater-to-wetlands rule (Chapter 62-25.042 FAC) applies to the discharge of pretreated urban stormwaters to a small class of natural wetlands (Livingston, 1989). Wetlands that can be used for stormwater treatment include formerly isolated wetlands that have intermittent or artificial connections to waters of the state. Design guidelines are

Table 19-7 Regulatory Criteria Governing Wetland Discharges in Florida

Category	Municipal Wastewater[a]	Stormwater[b]
Wetland types	Natural and constructed wetlands except non-cattail marshes	Natural wetlands connected by an artificial or intermittent channel to state waters
Pretreatment	Minimum of secondary; nitrification and phosphorus removal for natural wetlands Advanced treatment (BOD and TSS <5 mg/L, TN <3 mg/L, TP <1 mg/L) for receiving wetlands	Part of comprehensive storm-water management system; oil and grease removal, grassed swales, dry detention, wet detention
Design criteria	Natural treatment wetlands: TN loading <25 $g/m^2/yr$ TP loading <3.0 $g/m^2/yr$ HLR <2 in/week Hydrologically altered: TN loading <75 $g/m^2/yr$ TP loading <9.0 $g/m^2/yr$ HLR <6 in./week	Treat first 2.5 cm of runoff from rainfall event; maintain normal fluctuation of water levels; promote sheetflow and maximize HRT; bleed down treatment volume in >120 h, with no more than 50% in less than 60 h
Standards within wetland	Exempt from standards for DO Biological standards: Well-balanced fish and wildlife populations Benthic macroinvertebrate diversity reduced less than 50% Fish—biomass less than 10% decrease in sport, commercial, and forage species; less than 25% increase in rough fish Vegetation—less than 25% reduction in importance value for dominant species (species with importance values of at least 15% of total) Toxic substances—must be less than chronic levels	None
Discharge standards	Annual average TN <3 mg/L Un-ionized ammonia <0.02 mg/L Annual average TP <0.2 mg/L or Water quality based effluent limits Cannot cause or contribute to water quality violations in contiguous waters	Cannot cause or contribute to water quality violations in contiguous waters
Monitoring requirements	Baseline water quality, sediment, and biological monitoring Operational water quality, sediment, and biological monitoring	Project specific

[a] Chapter 62-611, Florida Administrative Code.
[b] Chapter 62-25.042, Florida Administrative Code.

based on performance and presume an 80 percent removal of the major contaminants in the stormwater. Pretreatment for oil and grease and sediment removal are an integral part of stormwater treatment. Monitoring is an important aspect of permits for these treatment wetlands.

In Florida, discharges of municipal wastewaters can be permitted to either natural or constructed wetlands. Natural wetlands are designated as either treatment or receiving wetlands depending on the level of pretreatment. They must be forested wetlands or marsh wetlands dominated by cattail species. The minimum pretreatment level for all wastewater discharges to Florida wetlands is secondary. In addition, discharges to natural wetlands must be pretreated by nitrification (treatment wetlands) and denitrification and phosphorus removal (receiving wetlands). Quantitative design criteria specifying maximum allowable total nitrogen and total phosphorus loading rates are provided in the Florida wastewater-to-wetlands rule. Although many of the normal water quality standards are exempted in treatment wetlands, other "biological standards" are specified. These biological standards provide an imaginative approach to protecting the natural wetland's "type, nature, and function" by specifying allowable changes for dominant plant populations, benthic macroinvertebrate diversity, and fish biomass. Normal groundwater standards apply to all wetland treatment systems.

Effluent limits from Florida treatment and receiving wetlands are stringent (TN <3 mg/L and TP <0.2 mg/L) unless other water quality-based standards can be determined from assimilative and dilution aspects of the receiving water. Monitoring requirements are also stringent (Table 19-8) and include a 1-year baseline monitoring period and comprehensive operational monitoring. As shown in Table 19-8, receiving wetlands have reduced operational monitoring requirements compared to treatment wetlands.

Maryland

Maryland developed design guidelines for shallow constructed stormwater wetlands in response to legislation passed in 1982. These guidelines call for establishment of a permanent pool created by an excess of inflow vs. outflow or by a constructed impervious liner; a surface area of 3 percent of the watershed drainage area or a detention time of 24 h for a 1-year storm; about 75 percent of areas with water depths less than 30 cm for emergent vegetation and 25 percent over 1 m deep to encourage submerged aquatic vegetation; energy dissipation devices and sediment collection areas at inflow points; length-to-width ratio of 2:1 and use of baffles and islands to reduce short circuiting; a minimum of 10 cm of soil for vegetation rooting; and five or more planted wetland species. Establishment of cattail species and common reed are specifically prohibited.

Mississippi

The Mississippi Department of Environmental Quality, Office of Pollution Control (OPC) has promulgated limited standards for constructed wetland treatment systems in Chapter 130 of OPC rules. Preliminary design loading rates are 8.6 $m^2/m^3/d$ (12 cm/d) for SF constructed wetlands and 6.4 $m^2/m^3/d$ (16 cm/d) for SSF constructed wetlands. Actual design loading rates are site specific. For SSF wetlands, the inlet cross-sectional area must be designed to prevent hydraulic overloading and plugging by algal solids. The wetland outlet structure must be designed to allow variable water depth in the wetland cell. Vegetation plant spacing must be dense enough to result in permit compliance within the first full growing season. Chapter 100 of the OPC rules provides guidance on levee design and calls for stripping of existing vegetation, compaction to a 90 percent Standard Proctor Density, minimum top width of 2.4 m, minimum slopes of 1:3 (vertical to horizontal), and minimum freeboard not less than 0.6 m for small systems and 1.0 m for larger systems. Constructed wetland cells used for primary treatment must be sealed by using low permeability native soils, bentonite, or synthetic liners so that maximum water loss will be less than 4.7 $m^3/ha/d$ (0.05 cm/d). Constructed wetlands receiving secondary or better effluent do not need to be lined.

South Carolina

South Carolina began permitting experimental discharges to natural wetlands in 1983 at Sea Pines Plantation on Hilton Head Island and in 1985 at the Central Slough wetlands near Conway. Experience at these systems led to permitting the full-scale Carolina Bay natural land treatment system near North Myrtle Beach and an expansion of the Boggy Gut system (White Ibis Marsh). These projects resulted in considerable interest and acceptance of natural

Table 19-8 Summary of Monitoring Requirements for Municipal Wastewater to Wetland Systems in Florida

	Baseline Monitoring[a]			Operational Monitoring		
Parameter	**NTW**	**HATW**	**NRW**	**NTW**	**HATW& CTW**	**NRW**
Water Quality						
Temperature	M(DD)		O(DD)	M(DD)	M(DD)	Q(DD)
Dissolved oxygen	M(DD)		O(DD)	M(DD)	M(DD)	Q(DD)
pH	M	O	O	M	M	Q
Conductivity	M		O	M	M	Q
Color	M			M	M	
Carbonaceous BOD_5	M		O	M	M	Q
Total suspended solids	M		O	M	M	Q
Total phosphorus	M	M	O	M	M	Q
Ortho phosphorus	M			M	M	
Total Kjeldahl nitrogen	M	M	O	M	M	Q
Total ammonia nitrogen	M	M	O	M	M	Q
Nitrate-nitrite nitrogen	M	M	O	M	M	Q
Sulfate sulfur	Q	O	O	Q	Q	Q
Fecal coliforms	M	O	O	M	M	Q
Chlorophyll a	Q		O	Q	Q	Q
Priority pollutants (nonmetallic)	O	O		A	A	
Metals (Hg, Pb, Cd, Cu, Zn, Fe, Ni, Ag)	O	O		SA	SA	
Water stage	C	C	C		C	C
Sediment						
pH	O			A		
Total phosphorus	O	O		A	A	
Total Kjeldahl nitrogen	O			A		
Total ammonia nitrogen	O			A		
Nitrate-nitrite nitrogen	O			A		
Sulfide sulfur	O		O	A	A	A
Metals (Hg, Pb, Cd, Cu, Zn, Fe, Ni, Ag)	O	O		SA	SA	
Biological						
Benthic macroinvertebrates	Q	Q		Q	Q	
Woody vegetation	O	O	O	A	A	A
Herbaceous vegetation	Q	Q	O	Q	Q	Q
Fish	Q	Q	O	Q	Q	O
Mosquitoes	B	B		B	B	
Threatened and endangered species	O	O	O	A	A	A
Woody plant tissues (Hg, Pb, Cd, Cr, Cu, Zn, Fe, Ni, Ag, TKN, TP)	O			F		
Leafy and woody plant tissues (TP, TKN, Fe, Zn)	O			A		

[a] Baseline monitoring for one year in natural treatment wetlands; one event in natural receiving wetlands; none required in constructed wetlands.

Note:

NTW = Natural treatment wetland
HATW = Hydrologically altered treatment wetland
CTW = Constructed treatment wetland
NRW = Natural receiving wetland
M = Monthly
DD = Dawn and dusk for 48 h.
O = Once
Q = Quarterly
C = Continuous
A = Annually
SA = Semiannually
B = Monthly (April to November)
F = One final sample after 5 years

wetland discharges in South Carolina as long as stringent pretreatment standards are met and biological diversity is not significantly altered. Following Florida's lead, South Carolina is currently developing rules for wastewater-to-wetland discharges that rely on conservative hydraulic loading rates, a high level of pretreatment, and biological standards for protection of natural wetland ecosystems (S.C. DHEC 1992).

South Dakota

South Dakota permitted more than 40 constructed wetland treatment systems for municipal discharges between 1987 and 1992 (Dornbush, 1993). Typically, these wetlands are designed to upgrade facultative lagoon (34 kg/ha/d BOD_5 design loading rate to a minimum of two ponds) discharges to meet secondary standards. They usually are seasonal discharge systems with about 180 days of winter storage (pond plus wetland storage). Several of the systems rely on evaporation to eliminate surface discharges during the 180-day winter storage period. Other design recommendations for the constructed wetlands include a maximum hydraulic loading rate of 2.35 cm/d, 34 kg/ha/d of BOD_5, a minimum detention time of 7 to 14 days, a water depth between 15 and 60 cm with 23 cm considered optimum, an inlet distribution header, a length-to-width ratio between 5:1 and 10:1 with irregular shorelines for improved wildlife habitat, maximum bottom slopes of 0.2 percent, and seepage addressed on a site-specific basis. Dikes must be compacted, have a minimum top width of 2.4 m, have a top elevation at least 30 cm higher than the 100-year flood elevation, have side slopes at least 3:1 (horizontal to vertical), and have 60 cm of freeboard. Vegetation should be established from seeds or transplants collected within 160 km of the site, the spacing of transplants should be about 60 cm on centers, and recommended species are cattails, bulrush, or common reeds.

Texas

In 1991, the Texas Water Commission (TWC) published guidelines for the design and construction of constructed wetlands for wastewater treatment (Cueto, 1993). These guidelines set minimal acceptable standards for system design and construction, provide recommendations for project implementation, and demonstrate the acceptance of constructed wetlands as a viable treatment technology.

TREATMENT WETLAND CONCEPTUAL PLAN

Based on the technical and regulatory constraints described previously, one or more wetland treatment alternatives may be able to meet project goals. In this case, it is prudent to develop preliminary conceptual plans for these alternatives to visualize site constraints and estimate project construction costs. A conceptual plan should be based on a specific site to accurately estimate pipe and pumping and cut and fill requirements. The conceptual plan does not need to include all design details, but should include all major dimensions and site components including wetland area, berm cross-sections, number and types of weirs and water control structures, pump stations, and planting density. A realistic conceptual plan is important for assessing the economic constraints on a wetland treatment project. This section provides some ideas for site selection and for development of a conceptual plan.

SITE RANKING

In some instances there is a single preferred site for wetland development because of location (e.g., next to an existing treatment facility) or because of prior ownership by the discharger. In many cases however, there may be multiple possible sites with different political, technical, regulatory, and economic constraints. In these cases, careful site selection is an important component of project planning and can result in significant cost savings during permitting, design, construction, and operation.

Site selection begins with the assembly of existing and new information specific to the potential sites. A minimum of three site locations will generally provide an adequate basis for comparison of implementation constraints. In some cases, all possible sites within a selected distance of the wastewater treatment plant should be screened before a smaller subset of sites is selected for more detailed examination. A preliminary estimate of the required site area to screen potential sites can be determined based on information in Chapter 3 and in this chapter.

Preliminary screening of treatment wetland sites should include the following factors: (1) distance from the wastewater source; (2) land ownership including the number of individual property parcels; (3) presence or absence of sensitive environmental or cultural resources; and (4) site topography, soils, and geology. For natural wetland treatment systems, this list also will include the suitability of the existing physical, chemical, and biological conditions at the various sites. Existing data will provide some information, and preliminary site-specific investigations will evaluate factors such as water quality and biology, geohydrology, and cultural resources.

To rank the alternative sites, evaluation factors should be weighted according to importance. An example of a site-ranking study to select a natural wetland treatment site is presented in Table 19-9. For that project, key site selection factors included distance from the wastewater source, the level of previous disturbances to the sites' biological communities, the absence of threatened or endangered species, the hydroperiod tolerance of the existing plant communities, the absence of open water areas, the presence of peat soils and a clay substratum, and the number of natural outlets that would have to be monitored. Each project will require different factors for site selection, and the importance of these factors will depend on project-specific constraints.

CONCEPTUAL PLAN

Conceptual plans should be developed for each of the treatment and disposal alternatives. In many municipal and industrial wastewater planning studies, treatment options will include both SF and SSF constructed wetland treatment systems and conventional treatment with surface discharge, overland flow, slow-rate land application, and high-rate land application.

Table 19-9 Site Ranking for Natural Wetlands Treatment of Municipal Effluent in Coastal South Carolina

	Criterion	Site No. 1	Site No. 2	Site No. 3	Site No. 4
1.	Distance from wastewater source	9	2	2	8
2.	Prior environmental disturbance	3	9	5	9
3.	Absence of rare plant and animal species	8	1	4	9
4.	Presence of peat soils for nutrient absorption	10	7	8	6
5.	Presence of a continuous circling rim	7	4	7	8
6.	Minimum number of natural surface water outlets	8	6	5	3
7.	Plant community adaptation to extended hydroperiod	10	10	10	8
8.	Presence of low permeability clay subsoils	8	5	0	2
9.	Absence of open water (unvegetated) areas	10	10	10	8
	Total	73[a]	54	51	61[a]

[a] Selected sites.

Note: Each criterion was scored with a number from 1 to 10, with 10 being the most favorable for selection and 1 being least favorable. Data from Knight et al., 1985a.

In a few cases, natural wetlands and other alternatives will be added to the list of options. For stormwater treatment, a more limited number of alternatives are available, including wet and dry detention and wetlands.

Figure 19-15 illustrates a conceptual plan for a constructed wetland treatment system to reduce the biochemical oxygen demand and total nitrogen concentration in municipal wastewater. This plan includes a scaled plan view of the system layout showing the approximate route of the influent force main; the form of pretreatment; a likely effluent distribution system; the size of the wetland cells and berm lengths; preliminary number and size of water control structures; preliminary plant selection and density; liner materials, if any; and outfall structure location and configuration. Table 19-10 provides the conceptual level cost estimate for this 3785 m^3/d treatment wetland. This cost estimate includes quantities and unit costs for land purchase, clearing and grubbing, earthwork, distribution and weir structures, wetland planting, electrical and mechanical devices including pumps and controls, and engineering and permitting tasks.

ALTERNATIVE SELECTION

The goal of project planning is to select the most cost-effective, feasible alternative. Only alternatives that are feasible based on technical, regulatory, legal, and political constraints

Design Criteria	
Design Flow	3,785 m^3/d
Effluent Limits	
Bod_5	30 mg/L
TSS	30 mg/L
Wetland Area	9.7 ha
HLR	3.8 cm/d
Water Depth	0.1 - 0.45 m
HRT	4 - 11 d
Cells	24

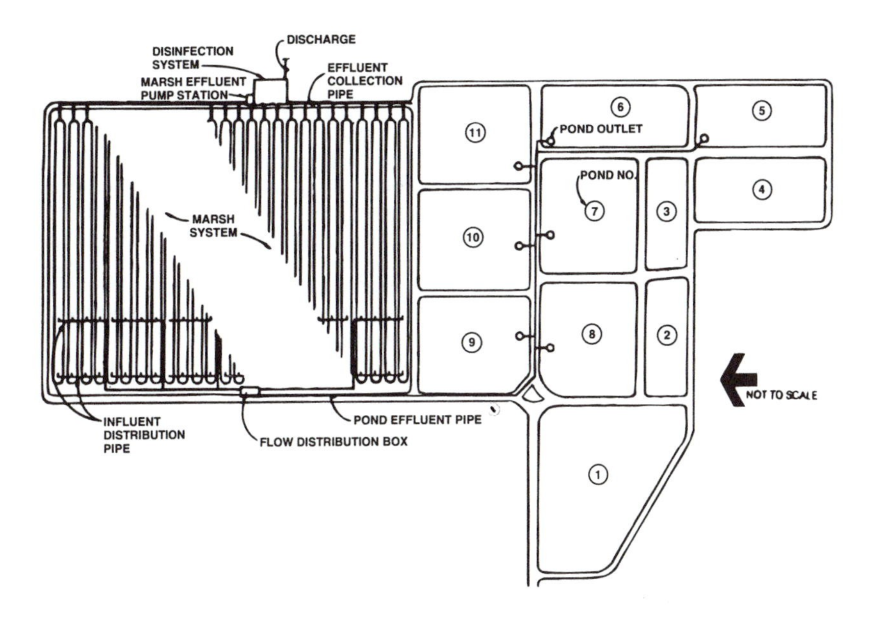

Figure 19-15 Constructed wetland treatment system conceptual plan summary (Gustine, CA). (Modified from U.S. EPA, 1988).

Table 19-10 Example of a Conceptual Level Cost Estimate for a Constructed Wetland Treatment System at Gustine, CA

Item	Cost, $ (August 1985)
Pond effluent piping[a]	192,000
Earthwork[b]	200,000
Flow distribution structure[c]	16,000
Flow distribution piping in marsh[d]	205,000
Marsh cell water level control structures[e]	27,000
Marsh effluent collection piping[f]	83,000
Planting[g]	69,000
Paving[h]	90,000
Total	882,000

[a] Includes 790 m (2,600 ft) of 53-cm (21-in.) PVC gravity piping, five manholes, and seven pond outlet control pipes with wooden access platforms.

[b] Total earthwork volume, approximately 334,000 m^3 (45,000 yd^3). Cost includes clearing and grubbing, extra effort to work in area of very shallow groundwater and to construct a 2-m (6.5-ft) high outer levee to enclose the marsh area and protect it from the 100-year flood.

[c] A concrete structure with V-notch weirs, grating, access stairs, and handrail.

[d] Approximately 850 m (2,600 ft) of 20-cm (8-in.) PVC gravity sewer pipe, 760 m (2,500 ft) of 20-cm (8-in.) gated aluminum pipe, and wooden support structures with concrete base slabs for the gated pipe installed at the one third of length point.

[e] Small concrete structures in each cell with weir board guides and 60-mm (0.24-in.) mesh stainless steel screen.

[f] Approximately 460 m (1,500 ft) of 10 to 38 cm (4 to 5 in.) PVC gravity sewer pipe plus manholes.

[g] Based on mechanical planting of bulrush and cattail rhizomes on 45- and 90-cm (18- and 36-in.) grid, respectively. Total bulrush area of about 2.4 ha (6 ac); 7.2 ha (18 ac) for cattails.

[h] Aggregate base paving of the outer levee and selected inner levees of the marsh area.

Note: Data from the U.S. EPA, 1988.

and that can win public acceptance should be compared during alternative selection. Cost-effectiveness refers to the cost of project implementation and operation as well as the environmental cost, which is harder to quantify. Economic and environmental factors should be weighed carefully against each other. Feasibility also may depend on the owners' philosophical preference. Do the owners wish to have a concrete and steel treatment system that can be kept neat and orderly, or would they prefer a more natural system with ecological aesthetics such as low energy and chemical usage and wildlife habitat enhancement? In a growing number of cases, communities are preferring wetland treatment at an equal or higher cost than equally feasible conventional technologies.

To compare the importance of both cost and environmental considerations in alternative selection, consider a comparison between high-rate land application (infiltration basins) and constructed wetlands for final treatment and disposal of a small, southeastern coastal plain town's municipal wastewater. The present worth (lifetime) cost of the high-rate land application system might be lower than for the wetland with comparable capacity. However, because of the need for highly drained soils, the high-rate system would have to be constructed on a site with a rare, sandhill plant community, while the constructed wetland could be sited in a more common planted pine community with lower permeability soils. Given these environmental concerns, the cost-effectiveness of the two alternatives might be more equal than the economic data would suggest.

When environmental costs and benefits of feasible project alternatives are relatively equal, then lowest lifetime cost or total annual cost becomes the primary consideration for selection. Determining the proposed project's present worth cost requires estimating capital costs (engineering and construction) and operation and maintenance costs using a discount rate over a standardized project life, usually 20 years. Total annual cost adds the amortized capital costs to the annual operation and maintenance costs to estimate the total annual expenditure over the project's life.

Table 19-11 provides an example of a total present worth cost comparison for a hypothetical system required to provide final polishing and disposal of about 20,000 m^3/d of pretreated municipal wastewater. All of the options listed in Table 19-11 require a minimum of secondary treatment prior to discharge. On a cost basis alone, secondary treatment followed by surface discharge would be the preferred alternative where regulations permit. However, in many areas of the country where dilution flow in receiving waters is not adequate, this option is not feasible.

For the hypothetical example in Table 19-11, high-rate land application has the next lowest cost. High-rate systems use higher hydraulic loadings than most constructed wetlands and thus require less land area for implementation. Also, although their per area cost may be higher than constructed wetlands, high-rate land application systems frequently require less effort for permitting and monitoring, and their overall present worth cost may be less. Thus, where the presence of highly permeable soils makes the option technically feasible, high-rate land application systems are frequently the preferred alternative for effluent management.

Natural wetlands are the next lowest cost alternative listed in the hypothetical example in Table 19-11. Although natural wetland treatment systems require greater land areas than constructed wetlands, they have the distinct cost advantage of not requiring extensive design efforts, earthwork, and vegetation planting. Thus, the cost per area for natural wetland treatment systems may be very low. Much of the cost of these systems results from land and pre- or post-treatment costs. Selection of natural wetland treatment systems may be limited by the scarcity of compatible sites, distance from the wastewater source, and by regulatory and political (public acceptability) constraints.

Utilization of a natural wetland unavoidably involves impacts on that ecosystem. Whether those are acceptable impacts or not is a judgement call that will vary widely from state to state. A background study of the natural receiving wetland is a normal precursor to the evaluation of its potential utilization for treatment. This subject will be examined in more detail in Chapter 22. At this juncture, it is simply noted that in many areas of the country, other alternatives would normally need to be precluded before a natural wetland were considered for treatment. Exceptions would be regions of extensive natural wetlands, with some fraction that could be allocated to managing the pollution control needs of local communities, agriculture, and industry.

Constructed wetlands generally have lower present worth costs than equivalent land treatment alternatives, such as slow-rate land application or overland flow. Although these alternatives have similar earthwork costs on a per area basis, slow-rate systems typically have lower hydraulic loading rates and consequently require larger areas. In addition, both slow-rate and overland flow systems have higher piping and pumping costs.

Table 19-11 Example of a Total Present Worth Cost Summary for Effluent Management Alternatives in Florida for a 20,000 m^3/d Municipal Wastewater Facility

Disposal Alternative	Total Present Worth $ (millions)
Slow-rate land application	20.8
High-rate land application	13.4
2° Surface discharge	10.7
Deep well injection	10.4
Natural wetlands	13.0
AWT surface discharge	18.8
Constructed wetlands	15.2

Construction of an SSF treatment wetland would normally be predicated on the need to avoid problems of exposure of humans and wildlife to the waters during treatment. The possible advantages of greater treatment efficiency are not evident in unamended beds and normally cannot overcome the added cost of the media. But, the media can be chemically designed to add to the performance potential. For instance, the addition of iron to the bed can provide for enhanced phosphorus removal. In a wetland system, metals and toxicants are passed primarily from the sediments to the sediment feeding birds and animals. The SSF wetland places these materials out of reach, below ground—when it is not overflooding. When the water is subsurface, mosquito breeding potential is minimized. The design of SSF wetlands is addressed in Chapter 21.

The choice of the constructed SF wetland alternative would normally be predicated on the desire for ancillary benefits, together with the ease of operation and possible residuals management. Part of the general perception of a wetland project has to do with whether or not humans and wildlife think it is a wetland. Both people and ducks expect surface water. Full "green points" are awarded only for projects that look right.

Before selecting an alternative for implementation based on technical and economic feasibility, it is good practice to meet with regulatory agencies to discuss implementation issues and agency preferences. In some cases, regulators have strong preferences, and neither cost nor technical feasibility will dictate which alternative should be selected. Although wetland alternatives are popular in some areas, in other areas, agency personnel may be unfamiliar or have had bad experiences with wetland treatment projects. The rationale for alternative selection should be thoroughly discussed with the agencies to gain their support during project permitting and implementation.

In the event that the project goals admit more than one type of wetland option, the design process can follow parallel paths to the point of documenting the economic, ecological, and technological attributes of the alternatives.

SUMMARY

Careful planning can prevent years of problems for an owner of a wastewater or stormwater treatment system. Because wetlands are not always the best overall alternative for treatment, they should be selected with a full understanding of the cost, permitting, and performance constraints. The treatment wetland alternative should not be selected without the support of owners and regulators, because the support of those individuals will be needed if problems arise during construction or operation. In a fairly small subset of all treatment planning studies, wetlands will be the overall favorite for project implementation. In those cases, it is wise to pursue this alternative with careful evaluation of technology options and conservative design.

This book presents information concerning three different treatment wetland alternatives: natural SF wetlands, constructed SF wetlands, and constructed SSF wetlands. SSF constructed wetlands have advantages in certain applications.

Constructed SF wetlands are useful for treating most types of municipal, industrial, agricultural, and nonpoint source wastewaters. They have almost the same treatment capabilities as SSF wetlands for some parameters and they cost much less to build. Constructed SF wetlands provide the greatest increase in ancillary wetland functional values.

Although natural wetlands may be the lowest cost alternative, they require higher levels of pretreatment, may not be available in many areas, and they are subject to the greatest level of public and agency scrutiny because of the need to protect their natural functions. Natural wetlands should only be used for wastewater treatment after careful evaluation of other alternatives and with adequate pretreatment and conservative design.

CHAPTER 20

Wetland Design: Surface-Flow Wetlands

INTRODUCTION

The design process for surface-flow (SF) wetlands is deceptively simple in outline: determine if there is enough room to do the required treatment, size a shallow basin, and select a means of transporting the water to it. However, the nuances of each situation preclude a "cookbook" approach to design; no two constructed SF wetland systems are alike in all respects. Indeed, the character of the technology precludes packaged designs for SF systems. The principles and scientific data from the preceeding chapters must be invoked as necessary in each new situation.

PRELIMINARY FEASIBILITY

The first cut at SF wetland design addresses the general question of whether a constructed SF wetland fits the geographic and economic constraints of the problem. This involves first estimates of the necessary size, combined with a knowledge of land availability. Then, if the wetland can fit on the site, the question of whether it meets economic constraints must be addressed. In preliminary feasibility, quick and simple calculations are used, although these may be organized on a spreadsheet.

PRELIMINARY SIZING

The goal of these calculations is to obtain a rough idea of the size of wetland required, or whether in fact all the target water quality goals can be met. The tools have been presented in previous chapters in the form of regression equations and areal uptake models.

Wetland size may be limited by geography, a lack of suitable construction sites, or regulatory limitations such as natural upland or wetland areas that cannot be altered. Site topography must be relatively flat to accommodate constructed wetlands without excessive amounts of earthwork to cut and fill the landscape. Wetland size and hydraulic loading rate may be limited in some arid localities because evapotranspiration may exceed total water inflows. Constructed SF treatment wetlands typically have areas from 10 to 100 $m^2/m^3/d$ (4 to 40 ha/3786 m^3/d).

Each regulated parameter gives rise to a wetland area necessary for the reduction of that pollutant to the required level. The required wetland area is the largest of the individual required areas. Calculations based on the k-C* models may be organized by the water quality parameter (Table 20-1). The general form of this model is

$$\ln\left(\frac{C_e - C^*}{C_i - C^*}\right) = -\frac{k}{q} \tag{20-1}$$

where C_e = outlet target concentration, mg/L
C_i = inlet concentration, mg/L
C^* = background concentration, mg/L
k = first-order areal rate constant, m/yr
q = hydraulic loading rate, m/yr

Rearrangement and a unit conversion give the area required for a particular pollutant:

$$A = \left(\frac{0.0365 \cdot Q}{k}\right) \cdot \ln\left(\frac{C_i - C^*}{C_e - C^*}\right) \tag{20-2}$$

where A = required wetland area, ha
Q = water flow rate, m^3/d

Both the hydraulic loading rate and the nominal detention time for a 30-cm depth may then be calculated from their definitions. The concentration of all pollutants is computed from the model using the largest area:

$$C_o = C^* + (C_i - C^*) \exp\left(-\frac{kA}{0.0365\ Q}\right) \tag{20-3}$$

where C_o is the outlet concentration, mg/L.

Table 20-1 illustrates these calculations for a hypothetical set of inlet and target concentrations. In this particular example, a 90 percent reduction in phosphorus is the design "bottleneck," yielding the highest area requirement. The target of 1.0 mg/L for organic nitrogen is below wetland background and cannot be achieved using this technology.

The result of the example feasibility calculations is an estimate of 7.1 ha for the active treatment area. Dikes, buffers, and other peripherals will occupy some extra space, on the order of 25 percent so the wetland project would occupy about 9 ha. If that amount of useable land is available, the project passes the spatial feasibility test.

SITE CHARACTERISTICS

In preliminary feasibility, three site characteristics bear further consideration. The first is site topography, which will determine the amount of earth moving and thus influence the project cost. Topography also determines the need for pumps to move water to and from the site, again an important cost consideration. At the level of preliminary feasibility, these need only be placed in relative categories: no pumping and berm-only earthwork are the cheapest, pumps and land leveling add to the cost.

The second is site ownership. If the land must be purchased, a significant cost is incurred, including the market price of land and ancillary acquisition costs.

The third consideration is the nature of the soils on the site. In some instances, the wetland will be required to be relatively impermeable to protect groundwater. This is true

Table 20-1 Preliminary Feasibility Calculation Sheet

		TSS	BOD	TP	TN	Organic N	FC
Design flow, m^3/d	Q = 1000						
Influent concentration, mg/L	Ci=	60	80	10	30	5	100,000
Target effluent concentration, mg/L	Ce=	15	15	1	5	1	200
Wetland background limit, mg/L (For TSS, $C^* = 5.1 + 0.16\ C_i$) (For BOD, $C^* = 3.5 + 0.053\ C_i$)	C*=	**14.7**	**7.7**	**0.05**	**2**	**1.5**	**100**
Reduction fraction to target	Fe = 1 − Ce/Ci=	0.750	0.813	0.900	0.833	0.800	0.998
Reduction fraction to background	Fb = 1 − C*/Ci=	0.755	0.903	0.995	0.933	0.700	0.999
Areal rate constant, m/yr	k=	**1,000**	**34**	**12**	**22**	**17**	**77**
Required wetland area, ha	A=	0.183	2.467	7.144	3.706	HELP	3.274
$A = \left(\frac{0.0365 \cdot Q}{k}\right) \cdot \ln\left(\frac{C_i - C^*}{C_e - C^*}\right)$			The necessary area = 7.1 ha (HLR = 1.40 cm/day) (HRT @ 30 cm depth = 21.4 days)				
Effluent concentrations, mg/L via k-C* Model	C_o=	14.7	7.8	1.0	2.4	1.6	100
$C_0 = C^* + (C_i - C^*)\exp\left(-\frac{kA}{0.0365Q}\right)$							

Note: TSS, total suspended solids; BOD, biochemical oxygen demand; TP, total phosphorus; TN, total nitrogen; FC, fecal coliform; HLR, hydraulic loading rate; and HRT, hydraulic residence time. Annual averages, constant Q, 20°C, plug flow assumption. Numbers in boxes from designer.

even if the project is designed to recharge groundwater as its method of discharge. If the soils on the site are permeable, then a liner may be required and therefore contribute to the project cost. This is usually a significant cost if native materials are not available onsite. At the feasibility level, the need for a wetland lining is estimated, and the cost factors are increased accordingly.

PRELIMINARY ECONOMICS

The purpose of the preceeding steps is to determine if the wetland concept bears further investigation as an alternative for solving the water quality problem. If the technology can do the job, and the land is available, then economic factors are checked.

Capital

There is a distribution of capital costs for existing constructed SF wetlands, due to facts mentioned previously, in a preliminary feasibility study, the relative position of the proposed project within the distribution is estimated. Figure 20-1 provides capital costs for constructed SF wetlands in 1993 dollars. The reasons for the wide distribution are the presence or absence of pumps, liners, land costs, and other project-specific items.

Investments in land are not subject to normal wear out and replacement charges in a long-term view of the project economics. Therefore, the "salvage" value of the wetland property, which may have appreciated over the project life, is typically much higher than that of a mechanical plant, which may have depreciated. The present worth of the land after the nominal project life should be included as a credit in economic evaluation.

Two items often run up project costs: elaborate water control structures and a perceived need to plant the wetland. The former originate from overengineering and from the need to provide and house expensive monitoring and control equipment. The later originates with the desire to propagate a selected suite of plant species in a short period of time. These items can often be minimized through recognition of the passive, low-tech, self-sustaining character of the most successful projects.

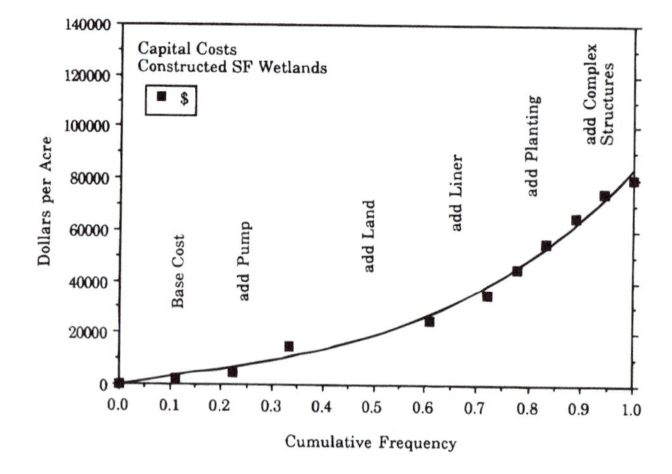

Figure 20-1 Distribution of capital costs for SF constructed wetlands. The base cost, without pumps, land, liners, planting, or complex structures, is on the order of $2000 per acre. The data points are in 1993 dollars. (Data from the NADB.)

Operating and Maintenance (O&M) Costs

Wetland systems have very low intrinsic O&M costs, including pumping energy, compliance monitoring, maintenance of access roads and berms, and mechanical component repair. These basic costs are much lower than those for competing concrete and steel technologies, by a factor of two to ten. Median annual O&M cost for four constructed SF wetlands in the NADB was $400 per acre in 1993 dollars.

However, ancillary research costs — which occur at the directive of the regulatory agencies — have frequently cancelled out this large potential advantage.

Attempts at harvesting or at maintenance of a particular vegetation species composition can prove costly. However, nuisance control, such as mosquito and rodent eradication, have not proven to be unduly expensive.

O&M costs should be capitalized and added to the construction costs to obtain a present worth for the entire project.

If the wetland alternative survives these preliminary operational, site, and economic considerations, then more details must be factored into the design, and the project must be reevaluated.

DETAILED CONCEPTUAL DESIGN

One of the historical difficulties with wetland technology has been the construction of projects based on the feasibility study as outlined in the previous section. Many design considerations still remain at this stage of the process.

NONQUANTIFIED DESIGN PARAMETERS

Relatively few water quality variables have received study in the short history of constructed wetlands technology, most in connection with municipal or domestic wastewater treatment. Those are important and will continue to form the basis for many designs. But for a long list of specific chemicals which might be good targets for wetland treatment, there are insufficient data for design.

If there are positive indications of treatment potential, evidenced from results for analogous compounds, or field or laboratory wetland microcosms, then a pilot treatment project is in order.

In no case should short-term microcosm or mesocosm results be used as the sole basis of a wetland design. These have been notoriously bad predictors of the performance of full-scale ecosystems. The reasons are many: the hydrology is wrong, the proportion of edge effects is too large, environmental factors are constrained or absent, the study period does not carry on to the sustainable limit, short-term capacities are included, and species composition changes are constrained. The value of such tiny scale experiments lies in understanding of process details. Unfortunately, the status of wetland ecological science is not strong enough to synthesize and extrapolate to full-scale ecosystem behavior.

Accordingly, pilot wetlands should be no less than 500 m^2 and should be run long enough to pass the startup transients — normally 1 to 2 years. The subsequent data acquisition period should span all seasons and preferentially more than once. Key design variables include HLR, inlet concentration, depth, and vegetation density. True replication has never been achieved within such pilot projects, but nominal replication serves to more adequately define the system response surface and to define intersystem variability. The cost of complete pilot projects is high, and therefore, designers have opted in some instances for a demonstration project without experimental replication.

The demonstration philosophy suggests that a promising application be tested by building and operating a prototype, usually of medium size. The design is inferred as the best possible from analogous wetlands and applications. If successful, the system is then scaled up, in the expectation of achieving similar performance to the demonstration. Limited options exist for using the demonstration project to optimize future designs. If the potential for application in the particular service is wide enough, the demonstration project may be followed or accompanied by a set of pilot wetlands to gather optimization data.

NONDESIGN PARAMETERS

The models used in feasibility are either inappropriate or uncalibrated for some of the commonly regulated water quality parameters. Chief among these would be dissolved oxygen (DO), pH, temperature, and total dissolved solids (TDS); others may be included in specific circumstances. The expectations for concentrations of the first three of these constituents have been discussed in Chapter 10, and TDS was discussed in Chapter 15. There are no easy methods to alter these in the wetland environment; instead, it is necessary to accept what a wetland will produce in its effluent. In summary, expect

- Low DO averaging about 2.5 mg/L, but lower in more heavily loaded (BOD_5, TKN) wetlands
- Circumneutral pH
- Wetland effluent water temperatures that are at the mean ambient air temperature during the unfrozen seasons
- Pass-through of TDS (including chlorides, sodium, and other salts) without much modification except via rain and evapotranspiration

The prevailing regulatory attitude toward these attributes of wetland water quality treatment is a key ingredient in project evaluation.

MAXIMUM VALUES VS. AVERAGES

All treatment technologies possess a spectrum of effluent concentrations, which is usually predictable only in the probabilistic sense. Therefore, in addition to the mean effluent concentration (which may vary in a deterministic way with temperature and loading), there is an associated band width of concentration. Regulations may constrain both the mean and the maximum of the band, via specification of a limit on the maximum daily, weekly, or monthly value; together with a limit on the average annual value. In this phase of design, care must be taken to select the most restrictive of multiple averaging tests given by the regulation. Table 20-2 can be used as one tool to deal with this selection.

Table 20-2 Ratios of Maximum Monthly Values to Annual Average Values for Wetlands in the NADB

Parameter	Number of Wetlands	Maximum Month/Annual Ratio
Total phosphorus	43	1.8
Dissolved phosphorus	21	1.9
Total nitrogen	30	1.6
TKN	36	1.5
NH_4-N	48	2.5
NO_x-N	46	2.5
Organic nitrogen	22	1.8
BOD	47	1.7
TSS	49	1.9
Fecal coliforms	23	3.0
Dissolved oxygen	32	1.9

The actual relationship between annual average effluent concentration and the maximum average monthly concentration for typically regulated constituents is given in Table 20-2. Exit concentrations fluctuate with an amplitude as high as three times the mean (for fecal coliforms). Some of these excursions from the annual mean are predictable from mass balances and a knowledge of the inlet flows and compositions; some are not. In any case, the effect is an increased wetland size, if there is an unrealistic ratio between maximum month and annual average for regulatory values.

Example 20-1

A wetland is being designed to reduce BOD from 60 mg/L to an annual average of 15 mg/L, with a monthly effluent of 20 mg/L. The flow rate is to be a steady 3786 m^3/d (1.0 mgd). What size SF wetland should be chosen?

Solution

The first-order areal model is chosen (Chapter 13, Equation 13-32), with k = 35 m/yr and C* = 5.8 mg/L.

$$\ln\left(\frac{C_e - C^*}{C_i - C^*}\right) = -\frac{k}{q} = -Da \tag{20-4}$$

$$\ln\left(\frac{C_e - 5.8}{60 - 5.8}\right) = -\frac{35}{q}$$

The annual average, C_e = 15, can be achieved with q = 19.7 m/yr = 5.4 cm/d. The required area is A = Q/q = 70,000 m^2 = 7.0 ha.

The max monthly value, C_e = 20, reflects a departure from a lower annual mean. From Table 20-2,

C max month/C annual average = 1.7

The required annual average is therefore lower, C_e = 20/1.7 = 11.8 mg/L. This requires q = 15.9 m/yr = 4.35 cm/d. The required area is A = Q/q = 87,000 m^2 = 8.7 ha.

Meeting the maximum monthly requirement is more stringent and needs 25 percent more area.

FINAL DESIGN AND LAYOUT

The previous sections have determined the general feasibility of treating the candidate water on the proposed site. Two major tasks remain in the detailed design phase: sizing for the anticipated detailed annual patterns of wastewater parameters, seasonal variables and seasonal regulatory requirements; and determination of the wetland configuration. The first task requires a more detailed exploration of predicted performance; the second involves considerations of water conveyance and control of flow.

LOCALIZING THE DESIGN PARAMETERS

Emphasis has been placed on global values of constants for use in design equations, but it is sometimes necessary to adjust those values for a particular wastewater or a particular regional circumstance. The error bands on correlations presented previously are wide and may require narrowing for final design.

The SF wetland treatment system at Gustine, CA provides an excellent illustration. The wastewater to be treated was lagoon effluent, but those lagoons received raw wastewater from a milk processing plant as well as the municipal sewage from Gustine. At times, some lagoon cells resembled giant milkshakes. Thus, the character of the BOD and total suspended solids (TSS) were not typical of purely municipal effluents. Consequently, a pilot project was conducted to ascertain design parameters for a full-scale wetland polishing system (Nolte and Associates, 1983). The full-scale system also was studied (Walker and Walker, 1990).

Walker and Walker (1990) found that published models, such as the U.S. EPA first-order volumetric model (U.S. EPA, 1988), badly overpredicted BOD removals. Furthermore, this model could not fit the Gustine data: the average R^2 was 0.00.

The k-C* model does fit the Gustine data, but the constants are not the same as the intersystem averages presented earlier (Figure 20-2). The best-fit value of k = 23.6 m/yr was lower than the global average of 35 m/yr and C* = 17.5 mg/L compared to the global average of 5.8 mg/L. Differences are presumably due to the difficulty of biodegrading the milk solids and a greater regeneration of BOD from the wetland engendered by a particular conditioning of the biota.

DETAILED AREA CALCULATIONS

The goal of the calculations is the determination of a wetland area which will meet the design requirements at all times. The primary calculational tools are the k-C* mass balance model and the water budget. It will be assumed that a steady state model is workable for the monthly frequency, which is probably not strictly true. This assumption is more appropriate for constituents that are not stored in soils and vegetation, such as BOD, TSS, and bacteria. It is not appropriate for nutrients such as total phosphorus and some forms of nitrogen, for which there may be large monthly exchanges with those storages. In those cases, seasonal calculations are the highest frequency that should be utilized.

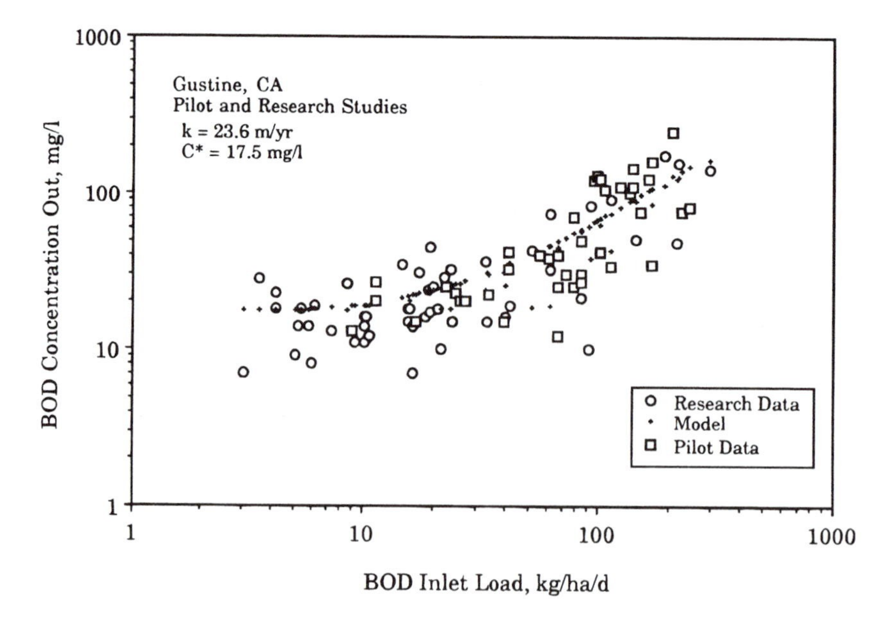

Figure 20-2 The fit of the k-C* model to data acquired at Gustine, CA (Walker and Walker, 1990). The one-cell, 1-ac pilot study ran from January through October 1983; research studies were conducted on six 1-ac cells during March 1989 to March 1990.

Wetland Performance Variability

There are two fundamentally different sources of variability in the performance of the treatment wetland: variations in the inputs and variations within the ecosystem. The first may be quantified from previous information, such as design specifications of the wastewater and climatic data. It is this category that is explored at this stage of design. The second category has to do with stochastic processes within the ecosystem. In sum, these are reflected in variability in the model parameters. It is possible to list many causative factors: algal blooms, insect attacks, vertebrate herbivory, seasonal plant growth rates, and species competition, for instance. But predictive models of these phenomena are beyond the present state of our knowledge, and thus, their cumulative effects on water quality remain as a stochastic variation in the parameters of the simple k-C* model used here.

The sum of both types of deviations fall within the statistical band defined by the NADB and are reflected in the ratio of the maximum monthly value to the long-term or annual average value. The monthly value is typically an average of weekly or more frequent samples, and therefore, some of the high-frequency probabilistic "chatter" has been filtered out of the data. Most of the variability represented in the NADB is not due to changes in the incoming water quantity or quality or to rain or evapotranspiration. Most NADB systems receive relatively constant inflows and are not subject to significant atmospheric augmentation. Therefore, nearly all sources of variability are internal and will not be accounted for in monthly mass balance calculations.

Therefore, each monthly calculation will represent the mean value of a probability distribution, and that distribution is estimated from the NADB. The ratio of the maximum month to the annual average is taken as an estimate of the ratio of the maximum average value observed during any month to the value predicted using the monthly input data. This approach is slightly conservative, since some of the scatter in the NADB is attributable to the seasonal and monthly variations in inlet flows and concentrations.

The calculation procedure is simply a determination of the wetland area so that all regulatory constraints are met. The general ideas will be illustrated via monthly BOD calculations, which represent the highest frequency that can be supported from the information presently available. However, in many instances, it would be more reasonable to consider a seasonal frequency.

A spreadsheet approach is strongly suggested, since it is often desirable to explore the consequences of changing values of flows, concentrations, or environmental variables. It also permits the exploration of stochastic variation in system variables.

The following example illustrates the technique.

Example 20-2

A variation on the Gustine, CA project is considered here. The wastewater flows and quality are approximately those from that project, but the regulatory criterion is different.

An SF wetland is to be built to reduce BOD from an average annual inlet concentration of 150 mg/L to a value not to exceed a maximum monthly value of 40 mg/L in any month of the year. The average annual flow is to be 3786 m^3/d (1.00 mgd). The inlet concentration of BOD is known to vary seasonally, with values given in Table 20-3.

What size wetland is required?

Solution

First, check the predicted annual average performance against the land availability. For this preliminary check, global values of the design parameters are used. The area required to meet an annual average BOD = 40 mg/L is calculated from Equation 20-4;

Global parameters from the pilot project are C* = 17.5 mg/L and k = 23.6 m/yr:

$$\ln\left[\frac{40 - 17.5}{150 - 17.5}\right] = -\frac{23.6}{q}$$

$$q = 13.3 \text{ m/yr} = 3.64 \text{ cm/d}$$

$$A = Q/q = 3786/0.0364 = 103{,}820 \text{ m}^2 = 10.4 \text{ ha}$$

However, it is further necessary to estimate the monthly maximum that accompanies the mean values computed from the k-C* model. For BOD, that multiplier is 1.7 (Table 20-2). Therefore, the calculation must be redone with an average annual BOD = 40/1.7 = 23.5 mg/L:

$$q = 7.63 \text{ m/yr} = 2.09 \text{ cm/d}$$

$$\ln\left[\frac{23.5 - 17.5}{150 - 17.5}\right] = -\frac{23.6}{q}$$

$$q = 7.63 \text{ m/yr} = 2.09 \text{ cm/d}$$

$$A = Q/q = 3786/0.0209 = 181{,}220 \text{ m}^2 = 18.1 \text{ ha}$$

It is now supposed that the site allows for a wetland of that size: perhaps larger, because of the seasonal nature of the wastewater strength. The multiplier of 1.7 is an average value derived for conditions represented by the NADB. It may not apply to the large swings in inlet concentrations in this example. Consequently, the decision is made to compute the monthly performance for the system.

Detailed calculations are shown in Table 20-3. Constant flow is presumed through the entire length of the wetland system during each month for which calculations are made. This is based on the assumption that on average during each month, atmospheric augmentation is small compared to the wastewater loading. This appears reasonable, since rain and evapotranspiration compensate each other, and, on a monthly basis, neither comes close to the mean annual wastewater loading. (If this had not been the case, the techniques of Chapter 9 would be employed, which allow for excessive rain or evapotranspiration components in the water mass balance.)

Each month has its own expected mean effluent BOD concentration. The wetland area is adjusted till the BOD limit is met, which requires 25.6 ha. The required inlet hydraulic loading is 1.48 cm/d. As anticipated, the design bottleneck occurs in January and February when there are excessive inlet concentrations. This procedure yields a larger area than that obtained from a multiplier on the average annual effluent concentration. That is because the variations in the incoming BOD exceed the typical variability in the NADB.

Note that the nominal detention time is checked, not as a design parameter, but as a check on the reasonability of doing monthly calculations. The value of 13.5 days gives about two nominal displacements of the water in the wetland during the month, which is pushing the minimum acceptable time step for the k-C* model for BOD. However, inlet conditions hold constant over 2-month periods, so there are in fact over four detention times represented at each inlet concentration.

Monthly calculations are typically required whenever temporal water quantity management is included as part of a design. Storage pond-wetland combinations used in northern climates are one such circumstance. Other strategies include alternating use of wetland cells and bypassing of extreme flows (off-lining).

A Reality Check

After detailed calculations have been made, it is appropriate to compare the spectrum of anticipated performances against the background of existing data. To that end, a convenient

Table 20-3 Worksheet for Detailed Design, Example 20-2

Parameter = BOD
Annual plug flow Damköhler number = Da = k/q = 4.38
Design area = 25.65 ha
Design depth = 20 cm

Design equation: $\ln\left(\frac{C_o - C^*}{C_i - C^*}\right) = -\frac{k}{q}$

Design parameters:
k = 23.6 m/yr
C* = 17.5 mg/L
Multiplier = 1.7

		Jan	Feb	Mar	Apr	May	Jun	Jul	Aug	Sep	Oct	Nov	Dec	Annual
Flow	m^3/d	3786	3786	3786	3786	3786	3786	3786	3786	3786	3786	3786	3786	3786
Inlet HLR	cm/d	1.48	1.48	1.48	1.48	1.48	1.48	1.48	1.48	1.48	1.48	1.48	1.48	1.48
Detention time	days	13.5	13.5	13.5	13.5	13.5	13.5	13.5	13.5	13.5	13.5	13.5	13.5	13.5
Inlet conc.	mg/L	500	500	250	50	50	50	50	50	50	50	50	150	150
Inlet loading	kg/ha/d	74	74	37	7	7	7	7	7	7	7	7	22	22
Background conc.	mg/L	17.5	17.5	17.5	17.5	17.5	17.5	17.5	17.5	17.5	17.5	17.5	17.5	18
Max. allowable conc.	mg/L	40	40	40	40	40	40	40	40	40	40	40	40	40
Pred. mean outlet conc.	mg/L	23.5	23.5	20.4	17.9	17.9	17.9	17.9	17.9	17.9	17.9	17.9	19.2	19.2
Pred. max. outlet conc.	mg/L	40.0	40.0	34.7	30.4	30.4	30.4	30.4	30.4	30.4	30.4	30.4	32.6	32.6

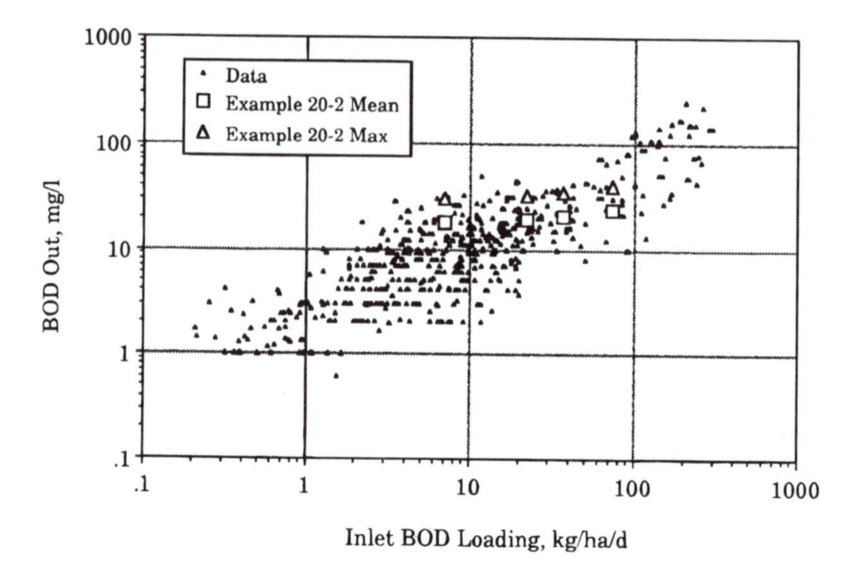

Figure 20-3 Scatterplot of BOD performance of SF wetlands. Data points are quarterly averages from the NADB (NADB, 1993). The mean and maximum performance expectations from Example 20-2 are also shown.

tool is the output concentration-input loading plot. This graph embodies the key variables associated with design: inlet concentration and hydraulic loading (their product is input loading), and outlet concentration. Each operating condition can be located on such a graph. Figures 20-3 through 20-6 show scatterplots of the quarterly data from the NADB and other sources, which form the experience base.

As an example of this procedure, Figure 20-3 contains the calculated results from Example 20-2. It is seen that the mean and maximum predictions are in fact near the mean and maximum of the data cloud. A few points of the database are above the range calculated in Example 20-2, so the design might produce occasional exceedences.

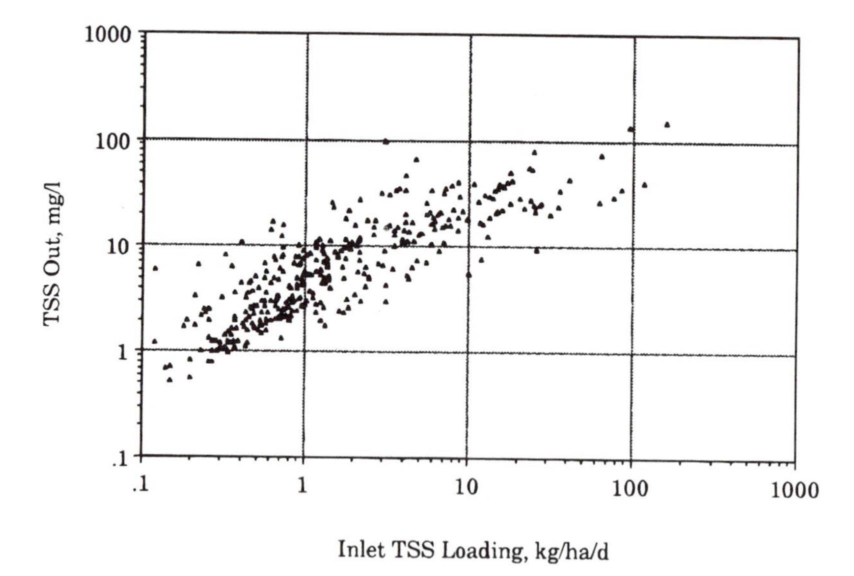

Figure 20-4 Scatterplot of TSS performance of SF wetlands. Data points are quarterly averages from the NADB (NADB, 1993).

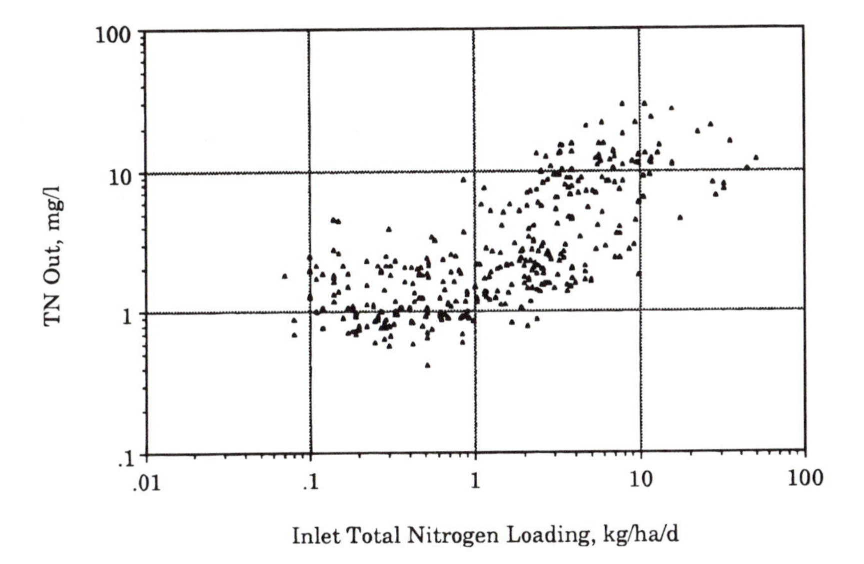

Figure 20-5 Scatterplot of total nitrogen performance of SF wetlands. Data points are quarterly averages from the NADB (NADB, 1993).

Presumed State of Mixing

The premise in global parameter estimation has been that of plug flow, which has been adopted on the premise of conservatism. If that assumption is to be modified, it is in the final sizing calculation. The tools for doing this have been discussed in Chapter 9, but will be reiterated here.

If the designer wishes to take some credit for the acknowledged partial mixing in the database wetlands, he does so by adjusting the rate constant upward, utilizing the presumed imperfect mixing model for the database. One such presumption is that SF wetlands behave

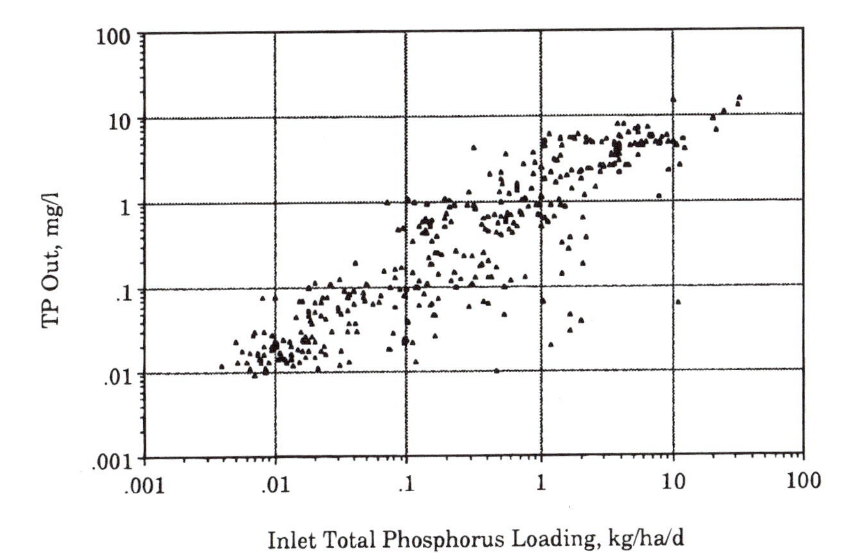

Figure 20-6 Scatterplot of total phosphorus performances of SF wetlands. Data points are quarterly averages from the NADB (NADB, 1993).

more like N well-mixed units in series. Then for the estimation of the k value,

$$\frac{C_e - C^*}{C_i - C^*} = \exp\left(-\frac{k_{pf}}{q}\right) = \left(1 + \frac{k_N}{Nq}\right)^{-N} \qquad (20\text{-}5)$$

where C_e = measured outlet concentration, mg/L
C_i = measured inlet concentration, mg/L
C^* = background concentration, mg/L
k_{pf} = first-order areal rate constant, plug flow, m/yr
k_N = first-order areal rate constant, N CSTRs, m/yr
N = number of well-mixed units assumed
q = measured hydraulic loading rate, m/yr

Therefore,

$$k_N = Nq\left[\exp\left(\frac{k_{pf}}{Nq}\right) - 1\right] \qquad (20\text{-}6)$$

In Example 20-2, suppose the estimate is that the pilot wetland behaved like three continuous stirred tank reactors (CSTRs). (That model produces a better fit to the Gustine BOD data than the plug flow model.) Then at the design hydraulic loading rate of 1.48 cm/d = 5.4 m/yr, the revised k value associated with the pilot project would be

$$k_3 = 3(5.4)\left[\exp\left(\frac{23.6}{3(5.4)}\right) - 1\right] = 53.3 \text{ m/yr} \qquad (20\text{-}7)$$

If the design wetland is presumed to also behave like three CSTRs, then the area calculated is the same as for Example 20-2: this higher rate constant is used in a model with poorer efficiency, the three CSTR model.

But if the designer now decides that hydraulics can be improved so that the design wetland will behave as plug flow, then there is an effect on sizing. The area is reduced from 25.6 to 15.8 ha, as calculated via a spreadsheet like Table 20-3. Note that the k value presumed for the imperfect mixing assumption on the database depends on hydraulic loading rate as well as the plug flow k value. Therefore, the spreadsheet (Table 20-3) must recalculate the correct k_N as the area, and hence the hydraulic loading, is varied.

Other procedures for mixing adjustments are contained in Chapter 9.

CONVEYANCE: ASPECT RATIOS, HEAD LOSS, AND LINEAR VELOCITY

The length-to-width (aspect) ratio is important in basin design because of its effect on flow distribution and hydraulic short circuiting. Theoretically, a constructed wetland with a high aspect ratio is no better for treatment than one with a lower aspect ratio, as long as flow is distributed effectively. However, reality dictates consideration of aspect ratio during design as a trade-off between theory and practice.

Higher aspect ratios increase the area of berms that must be constructed to enclose a given wetland area. Many authors have ventured opinions on the optimal length-to-width ratio. Stowell et al. (1985) recommended a ratio greater than 15:1 for water hyacinth ponds, while Dinges (1978) recommended a ratio greater than 3:1. Gersberg et al. (1984) recommended a ratio of 6:1 for subsurface-flow (SSF) gravel wetlands, but Reed (1990) has recommended ratios less than 1:1. The economic minimum recommended length-to-width ratio for constructed SF wetlands is 2:1 (Knight, 1987a). This ratio is based on the trade-off between enhanced effluent distribution a and the increased earthwork costs. There has also been some conjecture in the literature that long narrow wetlands behave closer to plug flow, but tracer data do not support that speculation (see Chapter 9, Figure 9-42). Other methods for maintaining effective flow redistribution, such as deep zones, are recommended to reduce the need for higher length-to-width ratios.

As the length-to-width ratio is increased, the linear velocity increases, and the head loss increases. At some point, the head loss will produce an inlet water depth that is unacceptably large, as detailed in Chapter 9. The head loss design procedure is detailed in Chapter 9 and illustrated in Examples 9-4 and 9-5. It would usually be used to check an aspect ratio set from site constraints.

Example 20-3

Example 20-2 determined that 25.65 ha would be necessary to treat the 3786 m^3/d to the required level of BOD at a design depth of 20 cm. If the inlet depth is to be no greater than 5 cm deeper than the outlet, what is the maximum length-to-width ratio that may be considered?

Solution

Assume that the outlet depth is fixed at 20 cm, and a horizontal bottom. Then $S_l = 0$. Then the maximum ratio of inlet depth to outlet depth is 25/20 = 1.25. From Chapter 9, Figure 9-14, the parameter grouping M1 must be less than 0.2 (dimensionless). Therefore,

$$M_l = \frac{qL^2}{ah_L^4} \leq 0.2$$

The hydraulic loading is q = 3,786/256,500 = 0.00148 m/d, and $h_L = 0.2$ m. Assume that there will be a dense stand of macrophytes, so that $a = 1.0 \times 10^7\ d^{-1}/m^{-1}$ (Chapter 9). Then,

$$M_l = \frac{0.00148\ L^2}{(1.0 - 10^7)(0.2)^4} \geq 0.2$$

or

$$L \leq 1470\ m$$

which corresponds to W = 175 m. Therefore, the length-to-width ratio should not exceed 1470:175 = 8.4:1.

Linear velocity can become a consideration in design for very large wetlands. For instance, TSS removal depends on sedimentation and trapping within the wetland. Excessive linear velocities lead to large values of shear stress on deposited solids and therefore can lead to the potential for resuspension of those solids. The critical shear stress for resuspension has not been determined for any treatment wetland, as discussed in Chapter 11. As a conservative interim design criteria, it is recommended that linear velocity be kept below a value which would resuspend 15-μm particles that settle at w = 0.1 m/d in a 0.3-m deep flow of Manning's $n = 0.10\ s/m^{1/3}$.

That number is approximately u = 1000 m/d.

However, it is well to bear in mind that existing SF wetlands operate at velocities lower than this, mostly below 100 m/d; hence, there is no field test of the criterion. It may be noted that virtually all SF wetlands do trap sediments effectively, however.

COMPARTMENTALIZATION

At this point in the design procedure, the size and shape limits for the wetland system have been determined. There is next a need to set the compartmentalization of the system. The number of wetland cells in the design of SF wetlands is based on consideration of redundancy, maintenance, and topography. All constructed wetland treatment systems should have at least two cells that can operate in parallel to allow for operational flexibility (cell resting, rotation of flows, or maintenance). Having at least two parallel cells is especially important because of unexpected events such as vegetation die-off, pretreatment failures and subsequent wetland contamination, and berm and other structural failures. Multiple flow paths allow the loading rate to be manipulated to meet varying inflow water quality. Also, parallel flow paths allow cells to be drained for replanting, rodent control, harvesting, burning, leak patching, or other possible operational controls. In the extreme long term, replacement of structures and piping become necessary. Some of the older SF treatment wetlands are now reaching this point in their service life.

The number of cells required must be determined by evaluating the cost of more cells (the ratio of berm area to wetland surface area increases with more cells), site constraints where sloped ground mandates terraced, multi-cell design, and operational flexibility to isolate various fractions of the total wetland treatment area. For example, with two cells, half of the treatment area must be shut off to conduct any maintenance, but with five cells, as little as 20 percent of the treatment area must be turned off. Large systems may profitably incorporate more than two flow paths, for purposes of internal flow control. However, multiplicity of inlet and outlet control structures can add significant cost to the overall project.

Wetlands have a tendency to channelize from points of inlet to points of outlet. If permitted, this operational feature reduces the gross areal efficiency of the wetland (see Chapter 9, Figures 9-34 and 9-35). Control of the bottom elevation and vegetation density can in principle prevent poor flow distribution. But in practice, the bottom of the wetland can neither be constructed nor maintained at tolerances that promote full areal contacting. Nevertheless, care must be taken to degrade any preexisting ditches, roads, or berms on the site because these will exert possibly undesirable flow control in the SF wetland.

The analogous problem exists in concrete and steel processing vessels and is cured by utilization of baffles and redistribution devices. The same is true in treatment wetlands where the baffles are low level berms and the redistribution devices are either structures or transverse deep water ditches.

Deep zones in SF constructed wetlands serve several purposes (Knight and Iverson, 1990). These deeper areas extend below the bottom of the vegetated basin areas by at least 1 m to exclude the development of rooted macrophytes.

Unvegetated cross ditches provide a low resistance path for water to move laterally and provide a nearly constant head across the wetland. They also provide for extra detention time, but in a deep water zone. Such ditches often become covered with duckweed (*Lemna* spp.) and can be used by wetland birds and fish as reliable habitat. These redistribution ditches materially change the overall degree of mixing within the wetland, because high-speed rivulets are intercepted and mixed with slower moving water. However, the redistribution ditch adds a potential for wind mixing that compensates the reduced short circuiting. But water is more effectively distributed over the wetland, improving the gross areal efficiency (Knight et al., 1994).

It has been speculated that long, narrow flow paths are closer to plug flow than short, wide flow paths. This is an intuitive carryover from small-scale devices, in which the size of recirculation patterns is limited by the width of the device. The resulting observations on mixing in pipe flow (without packing) therefore show a dramatic decrease of the dispersion number ($D/\mu l$) with increasing length-to-diameter ratio. But, the interior microchannels of the SF wetland have small dimensions: recirculation eddies are limited by depth (on the order of 0.3 m for a SF wetland) or by the lateral spacing of plant stems or clumps of stems (also a fraction of a meter). Thus, the effective length-to-width ratio is predetermined, and widening the wetland only adds more parallel channels.

As detailed in Chapter 9, observations of wetland mixing do not fit the patterns seen for small-scale devices. In particular, wetland data show that "longer and narrower" does not mean less effect of mixing and hence a closer approximation to plug flow. Rather, D/ul is relatively constant across operating conditions and design parameters, implying that the dispersion constant D is proportioned to length and to velocity. The mechanism of mixing is therefore distributed along the wetland length. The most probable explanation is that of localized storage in quiescent pockets, which exchange material with the adjacent flowing water, as well as the multi-channel effect discussed in Chapter 9.

As a result, there is no mixing pattern advantage for large aspect ratios, as quantified by the wetland dispersion number. However, a large aspect ratio tends to improve gross areal efficiency, quantified by the ratio of actual to theoretical detention time. In other words, it is not possible to get closer to the plug flow model, but it is possible to utilize the entire wetland area.

There are not "hard and fast rules" for design for high areal efficiency. Some general ideas are

1. Avoid blind spots in corners.
2. Provide flow straightener berms interior to an individual wetland cell.
3. Reestablish flow distribution at intermediate points in a flow path.
4. Maintain good bottom uniformity during construction and startup: minimize formation of topographic channels parallel to flow.

These considerations lead to a system configuration that may generally be described as "multiple strings of beads." Figure 20-7 illustrates some of these ideas.

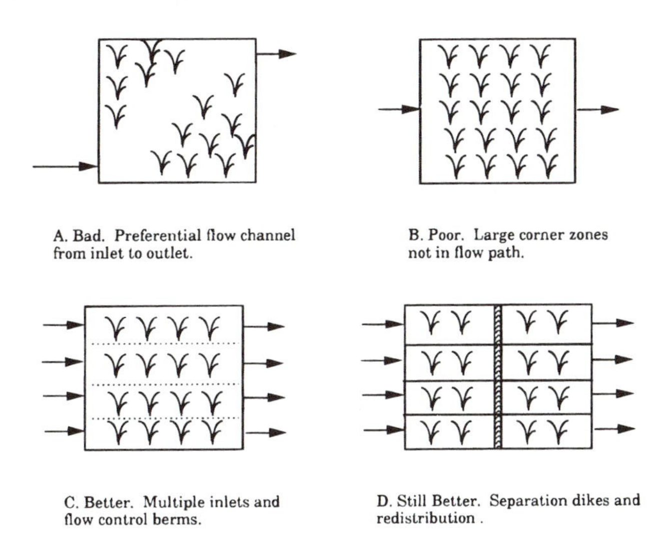

Figure 20-7 Configuration of wetland system elements. Flow path redundancy and within-path redistribution maximize the utilization of the wetland area.

FITTING THE PROJECT TO THE SITE

Given the total required wetland area and the concepts of system configuration, there still remains the placement of the wetland on the site. The principal considerations are adaptation to the boundaries and contours of the site, minimization of intercell conveyance, and minimization of earthmoving.

Site boundaries often determine the external shape of the overall system because there is often no extra land or the ability to choose the shape of the available land. In that event, the various pieces of the overall system must conform to the space available. The topology of the conceptual layout is retained, but shapes and perhaps areas are sacrificed. A good example is the layout of the Columbia, MO treatment wetlands (Metcalf and Eddy, 1990). The conceptual configuration was selected to be comparable to that of Figure 20-7D: three banks of cells in parallel. The available lands were bounded by streams, roads, railroads, and the floodway of the Missouri River. As a consequence, the actual layout was not completely rectangular, nor were the cells all in close proximity (Figures 20-8 and 20-9). Despite the long transfer distances, this complex of cells is gravity driven. A transfer pump is provided at the system outlet to supply the treated water to the Eagle Bluffs wildlife wetlands where it is used to foster habitat.

EARTHMOVING: DIKES, BERMS, AND LEVEES

Berms are designed based on hydraulic and geotechnical considerations. The purpose of berms is to regulate and contain water within specific flow paths. Figure 20-10 shows typical design features of constructed wetland berms. Exterior wetland berms should be kept as small as possible while still providing adequate freeboard to prevent unauthorized flow releases. Interior berms may be used to augment flow distribution, but do not have to be designed to control offsite water releases. Exterior berm freeboard should be adequate to prevent overtopping during sudden storm events (based on a storm event frequency of 10, 25, or more years) and allow overflow of less frequent storm events through controlled and protected emergency overflow points. Berm freeboard should also consider that the wetland will gradually fill with vegetation and with mineral and organic sediments which increase flow resistance and decrease freeboard during system life. Berm height should equal the sum of the maximum desired normal water level (for example, 45 cm), the return storm rainfall amount (for example, 20 cm for a 25-year storm event), and the lifetime loss of freeboard due to sediment and plant accumulation (approximately 1 cm/yr in some wetlands receiving municipal wastewaters). For a 20-year life, this hypothetical wetland should have an emergency spillway height of at least 85 cm with an additional 20 cm or so of berm above that level. Any additional berm height provides additional system life and insures against unauthorized discharges.

Berms should be constructed on the basis of standard geotechnical considerations. The materials that are available dictate how berms will be designed and constructed. Surface liners or internal clay plugs may be required to minimize berm seepage if sandy or other permeable materials are used for berm construction. External seepage collection channels may be necessary if soils are unconsolidated. An exterior slurry wall tied into deeper, low permeability sediments can be used to limit offsite infiltration.

Berm slope is dictated by geotechnical considerations and a slope-stability analysis. Minimum berm slopes typically used in constructed wetlands are 2:1 (horizontal to vertical), and slopes up to 10:1 or 20:1 are used when a shallow littoral shelf is desired to create habitat diversity. Berm width should be adequate for the intended use. For example, if the berm will be used for vehicular traffic, it should be more than 3 m in width.

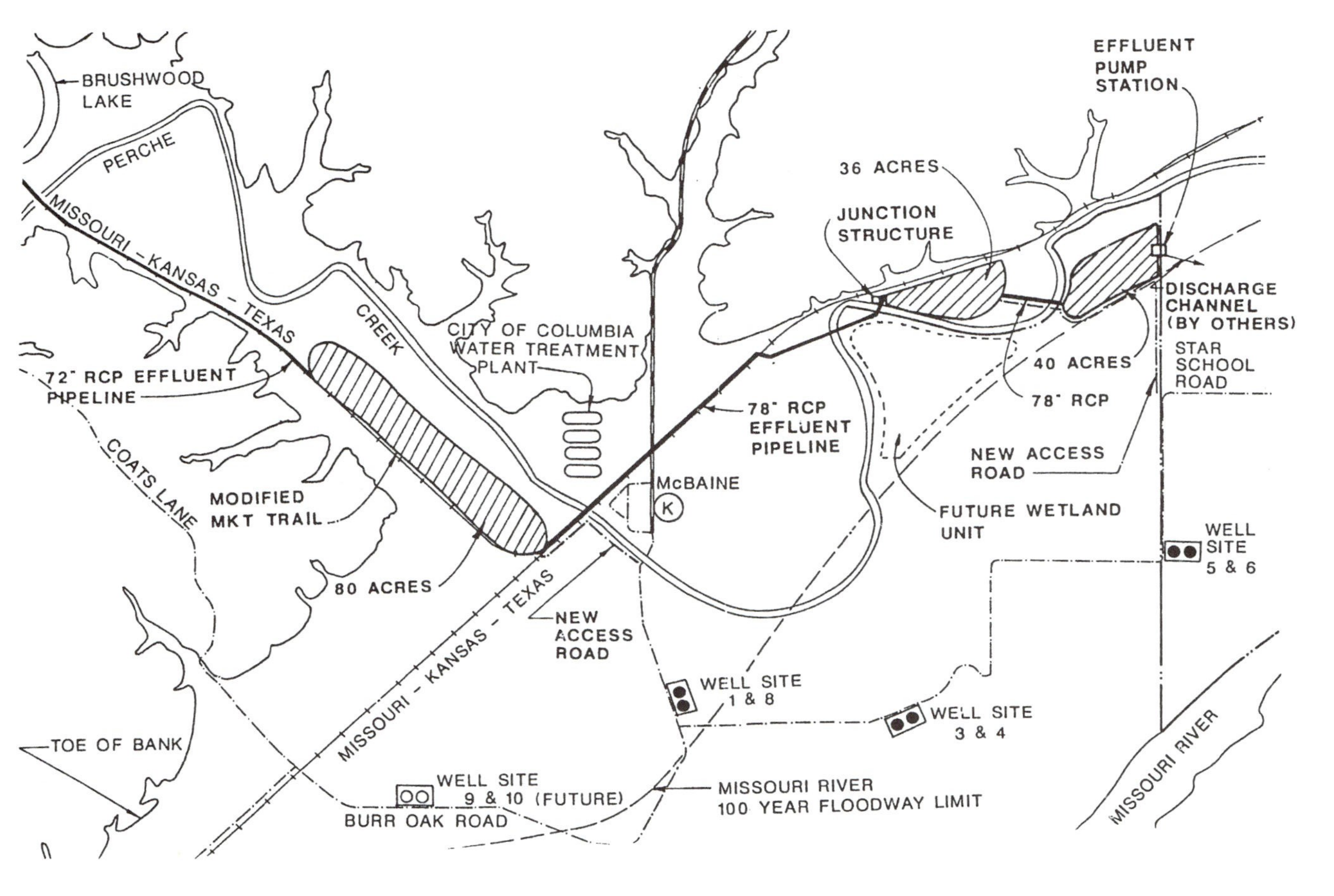

Figure 20-8 Site constraints for the Columbia Treatment Wetlands. The project could not block the Missouri River floodway. The irregular contour is a vertical limestone bluff about 50 ft high. Other constraints include the town McBaine, the water treatment plant, the city drinking water well fields, railroads, roads, and Perche Creek and its floodway.

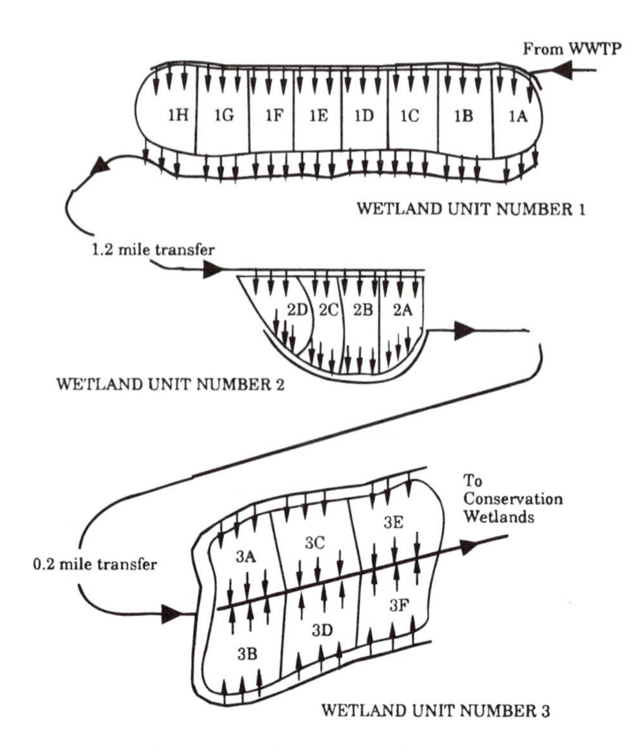

Figure 20-9 Arrangement of cells within the Columbia Wetland Treatment Units. All cells receive water via submerged gated distribution pipes along the inlet edge; each cell has three outlet structures.

The constructed wetland complex of cells and dikes requires earthmoving. The cost for this activity consists of two distinct components: the cost for moving soils within the site boundaries (cut and fill) and the costs for importing or exporting material from the site, from external borrow sources, or to spoil disposal sites. Equal cut and fill is the preferred option, since it avoids import/export costs. Selection of the bottom elevation of the cells, together

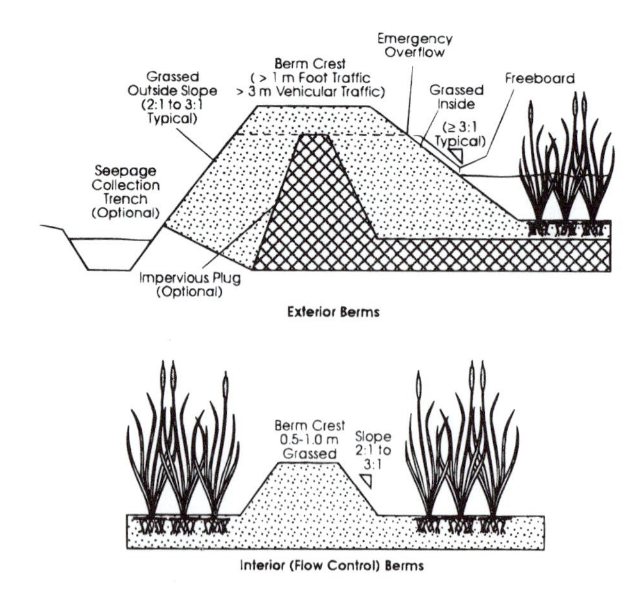

Figure 20-10 Design considerations for constructed wetland berms.

with proper positioning on the site with respect to its topography, generally allow balancing of cut and fill.

The bottom of the wetland, as well as the core of containment dikes, may be formed of compacted clays or bentonite. Locally available clays are preferred from the standpoint of cost reduction. Plastic liners may be feasible for smaller-size wetlands. This clay layer, or other sealant, should not be penetrated by plant roots, so that it retains its integrity. The topsoil from the site should be stockpiled and replaced within the wetland to form a rooting medium. The roots and rhizomes of emergent macrophytes now in use — cattails, bulrushes, and *Phragmites* — usually occupy the top 30 to 40 cm of the soil column. Therefore, a layer of that thickness should be used. The original topsoil from the site may be useable; if so, it should be stockpiled separately from the other soils during construction. This topsoil will contain seeds of the wetland plants of the region, which may assist in vegetating the wetland.

If topsoil is not available at the site, it may need to be imported to optimize plant survival and growth. Final grading to tolerances of about ± 3 cm are necessary to maintain sheetflow conditions in shallow marsh areas. Wetland cells may need to be lined with clay or plastic if regulatory requirements prohibit mixing with groundwater or if natural infiltration rates will make it difficult to maintain surface water wetland conditions.

Dikes are used for access, by walking or driving. A vehicle access dike needs to be more than 3 m wide at the top; interior divider berms designed for pedestrian access may be as narrow as 1 m. Dikes greater than about 5 m in width are less likely to be fully penetrated by muskrats. The side slopes of these are typically at a three to one slope and may be rip-rapped with stone to prevent erosion or rodent burrowing. The interior of a containment dike usually is designed to contain a compacted core and a layer of sealant which extends above water level. Water containment dikes are subject to local dam safety regulations.

FLOOD PROTECTION

A treatment wetland site may be unavoidably in the floodplain of a river. The question of protection of the wetland from flooding then arises. There are two aspects to potential flooding: maintaining the physical integrity of the system and the effect on treatment during flood events.

Physical damage to dikes and structures is not likely to be serious if the system is not placed in the floodway of the river. The main flood currents should not impinge upon the wetlands. Several treatment wetlands have easily survived gentle inundation without any significant damage: Des Plaines, IL; Jackson Bottoms, OR; Tarrant County, TX; and Columbia, MO; for example. Care must be taken to allow uniform flooding to prevent uplift of a sealed bottom by hydrostatic forces. Outlet structures should be designed to allow backflooding (not through-flooding) of wetland cells.

Damage to the wetland vegetation is not likely for inundation up to a week or two. However, prolonged submergence, such as the midwestern U.S. floods of 1993 which lasted for many weeks, will drown the wetland plants.

The treatment efficiency of the wetland will obviously be reduced under flood conditions. Generally, the dilution effect of the flood flow is so great that reduced efficiency can easily be tolerated for the brief duration of the flood. Also, some of the floodwater will receive beneficial detention in the wetland after the flood recedes. However, there is no substantial interruption of function due to brief flooding, as evidenced by little or no change in the removal rate constants (Figure 20-11).

The sides of the dikes must be protected from runoff and wave action and the attendant erosion during a period of wetland vegetation establishment. Emergent vegetation will prevent the wind fetch necessary to cause wave erosion, but those plants are absent during startup and in wetlands with intentional patches of open water that abut the dike. Rainfall runoff

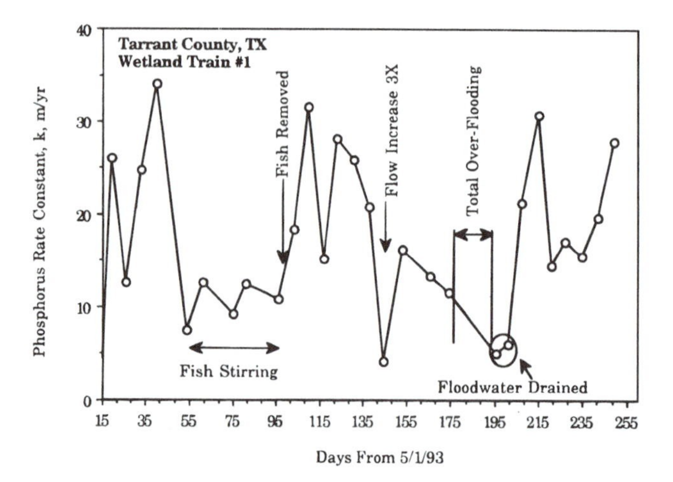

Figure 20-11 The effects of a total immersion flood on the phosphorus removal rate constant for Wetland Train #1, Tarrant County, TX (Alan Plummer and Associates, 1994). No dikes or vegetation were visible above the water during this flood.

can erode newly constructed dikes as well. In such circumstances, it is important to stabilize the bank with an appropriate cover crop (Figure 20-12).

HYDRAULIC PROFILE

Gravity is the preferred means of moving water through the wetlands. Although each cell may have a slight head loss from inlet to outlet, control of the level within each cell must be dominated by the exit structure. In the case of SF wetlands, that structure is normally some form of weir or flume. In any case, there is an associated drop in water level at the structures. A cross-sectional view of the projected water levels through the system of cells is termed the hydraulic profile.

Figure 20-12 Wetland dike erosion occurring due to wave and rain action at Hayward Marsh. At this site, waterfowl and wading birds routinely devoured all plantings, thus frustrating vegetation establishment and dike stabilization. Winds at the site have a clear path all the way across San Francisco Bay to the dike and a long fetch across this large wetland cell.

One of the hydraulic profiles for the Incline Village, NV constructed wetlands is shown in Figure 20-13. The water elevation drops sequentially by 3.05 m over the 2400-m-long section. The flow path is sinuous and totals about 6000 m. The wetland cells are sparsely vegetated, and the entire drop is taken at 18 intercell structures.

PUMPING REQUIREMENTS

Pumps are required for one of four reasons: supplying water to the wetland for treatment, removing water from the wetland to downstream uses, recycling water back to the inlet, and returning groundwater leakage to the wetland. Some systems have none of these; no North American system presently has more than three. Pumping incurs a capital cost, which is repeated on a time scale shorter than the lifespan of the ecosystem. Further, the operations and maintenance costs add to the capitalized total cost of the facility.

In very flat terrain, a low-head inlet pump may be needed to transfer water from the source to the wetlands at an elevation sufficient to move the water cross-country to the outlet. For the same reason, it may be necessary to utilize pumping to lower the water level at the system outlet. Redundancy is usually built into these pumping facilities, so that a portion of the capacity is normally out of service and accessible for maintenance.

If the wetland leaks an appreciable amount of water, it is a common practice to surround the wetland with a perimeter ditch to collect lateral seepage. Seepage water is then pumped back to the wetlands. This is the case in South Florida, where water conservation wetlands and stormwater treatment areas are so configured (Figure 20-14).

Recycle inevitably requires pumping. Treatment advantages that may be gained by this procedure are therefore expensive and often are more economically accomplished by increasing the size of the wetland.

CONTROL STRUCTURES

Inlets

The inlet device initiates the flow of wastewater into the constructed wetland. In cases where water quality treatment is the primary objective, the inlet device initiates a sheetflow of wastewater in the wetland by broadly and evenly distributing the influent. The need for sheetflow can be eliminated by basin configuration or by pretreating the influent wastewater to a level where it will not create an impacted zone of discharge. Multiple inlet devices may be used to distribute flows to large wetlands. The inlet device must be able to shut off during maintenance or resting of the constructed wetland. In some cases, the device is used to monitor flow and water quality.

Inflow can be distributed by gravity flow or by pressurized flow. Gravity flow is desirable because it conserves energy and reduces operation and maintenance costs. Gravity flow distribution systems may require larger pipe sizes to reduce head loss to acceptable levels. Pressurized inflows may result in high orifice velocities that can cause sediment erosion or physical damage to emergent wetland plants. By providing a greater number of individual outlets and greater cross-sectional area, erosion is less likely to result.

Figure 20-15 shows several approaches that have been used to distribute inflows and create sheetflow conditions in constructed wetlands. These include gated distribution pipes, level spreader swales, multiple inlet points, and a single inlet point with appropriate cell design. Selection of one of these alternative inlet configurations is based on wastewater characteristics, cost, and engineer preference. In cold climates, inlet water distribution and its associated plumbing must be kept below the ice layer.

Gated distribution pipes have been used at a number of constructed SF wetland treatment systems. The header pipe material should be selected on the basis of required system life

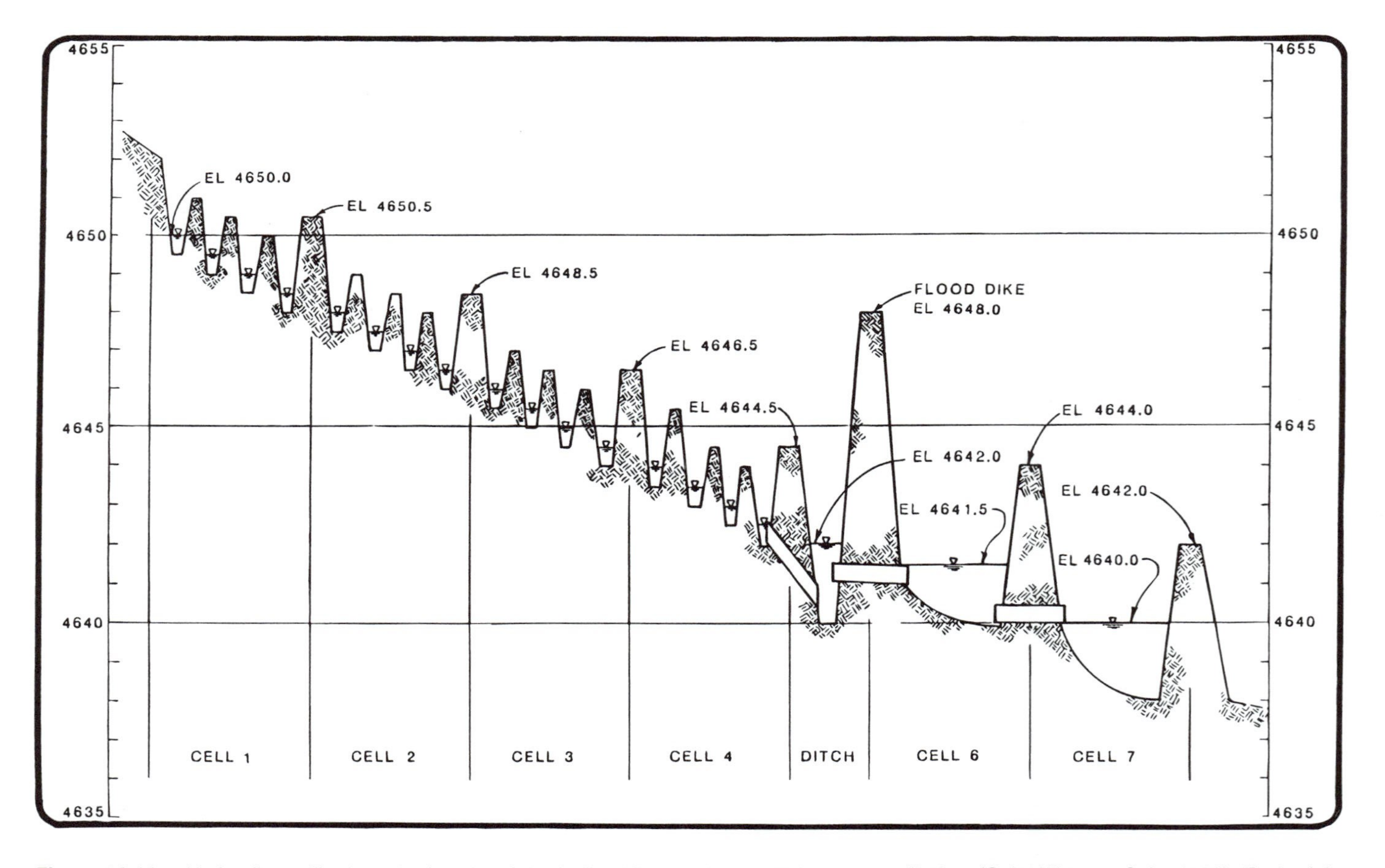

Figure 20-13 Hydraulic profile through six cells of the Incline Village Wetland Enhancement Project (Culp, Wesner, Culp, 1983). Each of the first four cells is divided into four subcells. Elevations are referenced to mean sea level, ft. MSL.

Figure 20-14 The perimeter seepage collection canal (right) and inlet distribution canal (left) for the Everglades Nutrient Removal project in Florida.

and cost. Inlet pipes have been made from a variety of materials including irrigation gauge aluminum pipe, PVC water pipe, and ductile iron. Because the header pipe is exposed to temperature extremes and ultraviolet radiation, PVC may have a limited life expectancy and may break. Aluminum pipe may dent and break. Ductile iron is strong and has a long life, but it corrodes and generally costs more.

The pipe must be sized to control head loss so that flow is distributed evenly over the distribution pipelines. A clean-out should be provided at the end of distribution pipes so that accumulated solids can be removed periodically. Also, pressure relief valves should be installed in pipes where air blocks might occur.

Gated distribution pipes use a variety of outlet configurations depending on specific goals (Figure 20-16). Simple, plastic, sliding outlet gates are used on aluminum irrigation pipe to regulate flow volume (from off to wide open). Screw-type nozzles also can be used on aluminum pipe, but they tend to corrode (oxidize) and quickly lose their effectiveness for flow control. PVC fittings or simple orifices are commonly used on PVC distribution systems. If individual outlet flow control is desired, a valve must be placed at every outlet. This approach has only been used at small facilities. If flow distribution is a concern because of the length of the distribution pipe (potentially a problem when the distribution pipe is more than about 20 m long), swivel outlet pipes can be used to more precisely control the elevation of individual outlets. Swivel outlets also can be used with ductile iron pipe, although they may be more difficult to regulate after they begin to corrode. Simple open-hole orifices made from any of these materials can be used successfully on short, gravity-fed, distribution pipes. Orifices should be at least 5 cm in diameter to minimize plugging by wastewater solids and to reduce the need for frequent maintenance. Orifices are generally placed along the side of the pipe to create a multiple weir-type configuration.

Level spreader swales also provide a gravity-fed distribution system (Figure 20-17). Water can be pumped or flows by gravity to the spreader swale. Outlets from the swale can be hard structures that are precisely leveled during construction or they can be adjustable weir plates. Notched metal weir plates typically used in clarifiers are used in some wetland applications, but these are difficult to install in large projects. The degree of vertical control on positioning is very high.

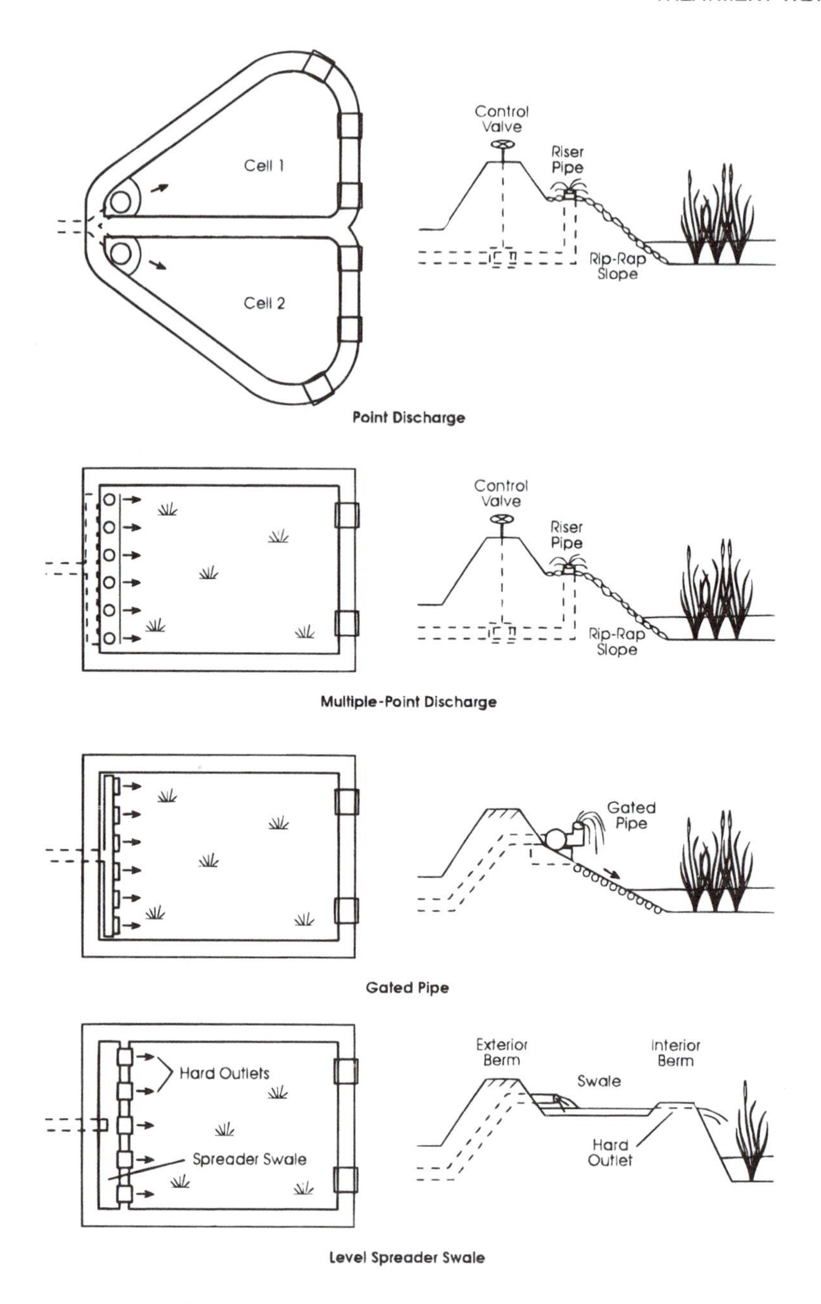

Figure 20-15 Wetland treatment system inlet configuration alternatives.

In some circumstances, single- or multiple-point inlets can distribute wastewater into wetland treatment systems. A single-point discharge distributes water almost as effectively as a linear discharge when the wetland inlet area is relatively narrow. Single-point discharge may work in pie-shaped wetland cells shown in Figure 20-15 or when the wetland is channel like (high length-to-width ratio). Multiple-point discharges distribute the influent over a broader inlet zone. Point discharges may consist of pipes in either a horizontal or vertical configuration. Vertical pipes tend to dissipate energy and reduce water velocities. Because point discharge systems have much higher water flows than diffuse distribution systems such as gated pipes, rip-rap or other hard structures in the inlet area help reduce erosion.

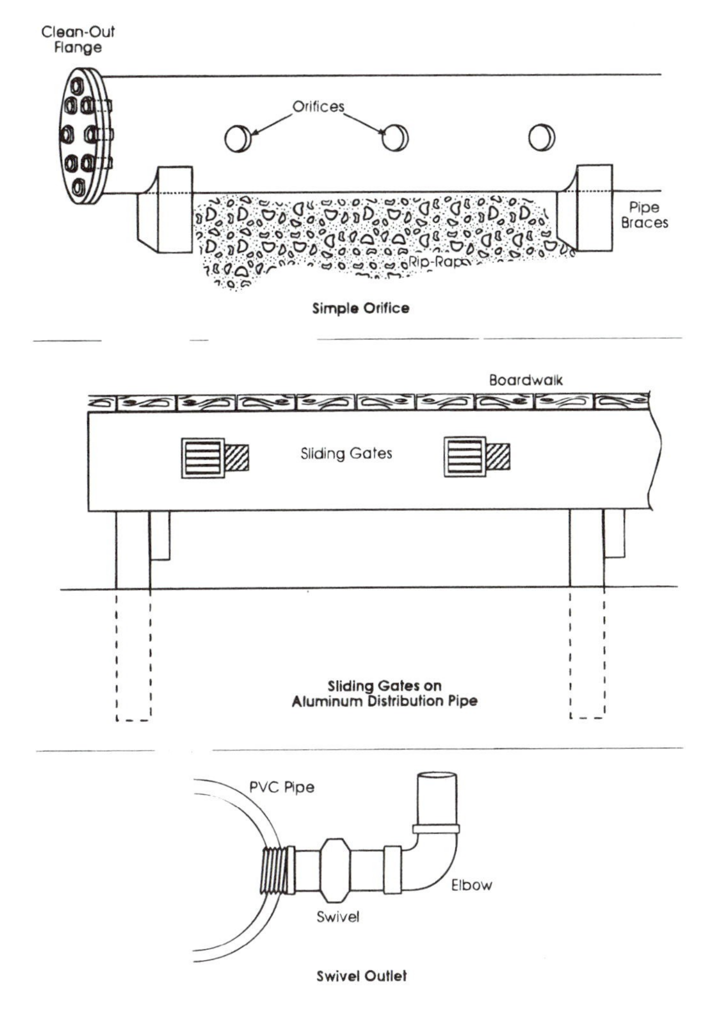

Figure 20-16 Alternative gated distribution pipe configurations.

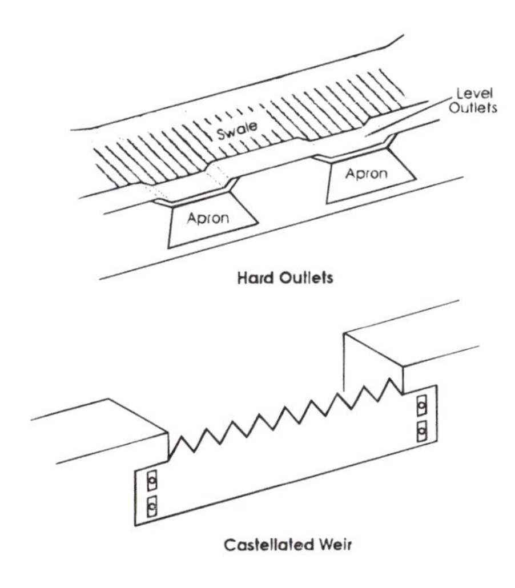

Figure 20-17 Alternative level spreader swale design configurations.

Inlet selection is partly based on the need for fail-safe operation under the prevailing climatic conditions. The need for uniform transverse distribution was discussed earlier. Problems to be avoided include freeze-up in northern climates and clogging by debris or animals.

Freezing can be prevented by use of buried feeder mains followed by point discharge structures or submerged distribution headers. Allowance must be made for ice formation on the surface of the wetland waters.

Incoming waters that originate in lagoons can often contain leaves, sticks, and other debris. These can block small orifices in a distribution system. The collection well in the lagoon system is frequently a favorite habitat for turtles, which can block flow. Screens or other appropriate prevention measures should be considered.

Another consideration in selection is the ability to control the flow distribution across the head of the cell. This can be accomplished by adjustable gates in a distribution pipe or by adjustable height weirs at each inlet point.

If the inlet distribution system should fail, the route of the incoming water should be acceptable for continued treatment and unimpaired system integrity. This is accomplished by incorporation of an emergency overflow at the inlet.

Outlets

Wetland outlet design is important for maintaining sheetflow distribution, for controlling water level, and for monitoring flow and water quality. Many outlet designs have been used in successful constructed wetlands, ranging in complexity from a corrugated metal pipe embedded in a berm to electronically actuated weirs powered by solar cells. To minimize capital, operation, and maintenance costs, outlet designs should be as simple as possible.

As shown in Figure 20-18, flow distribution in a rectangular wetland basin with a single outlet point is compromised, and effective hydraulic retention time is decreased. This situation can be improved by adding outlet weirs along the downstream berm of the wetland or by using a perpendicular deep zone to collect and route flows to a single outlet weir. If a terminal deep zone is used to accomplish this flow control goal, it must be kept as small as possible to discourage a long residence time and subsequent algal growth.

Outlet weirs or pipes do not need to be sized to handle peak wastewater inflow rates because of the amplitude dampening effect of volumetric storage in the wetland cell. However, weirs should be designed to pass maximum average daily or weekly flows plus the volume of rainwater that might fall on the wetland surface from a storm event. Weirs can be designed to pass these storm flows quickly (broad-crested horizontal weir) or slowly (v-notch or vertical notch weirs). Hydraulic flow routing techniques can be used to estimate peak flow events more accurately. Weir design is also important in quantifying wetland outflow rates. Integrating flow meters can be designed to work with almost any weir configuration. If weirs are used solely for water level control, flow monitoring may be accomplished downstream with a flume and flow monitoring device. Figures 20-19 and 20-20 show some typical weir and flume designs used in constructed wetland treatment systems.

One of the most important functions of outlet devices is to provide convenient control of wetland water levels. The required water level is based on requirements for plant growth, hydraulic residence time, and cell maintenance. In most constructed SF wetland treatment systems, water levels should be controllable between 0 (the cell bottom) and 60 cm. Weirs can provide water level control by adjustment and total removal. If adjustments will be infrequent, weir plates can be slotted and bolted to the weir box. If water level adjustments will be frequent or if other factors make the use of anchors and nuts impractical, then variable control weir gates can be designed. For large applications, these weir designs can become quite sophisticated and may rely on electric power and remote control.

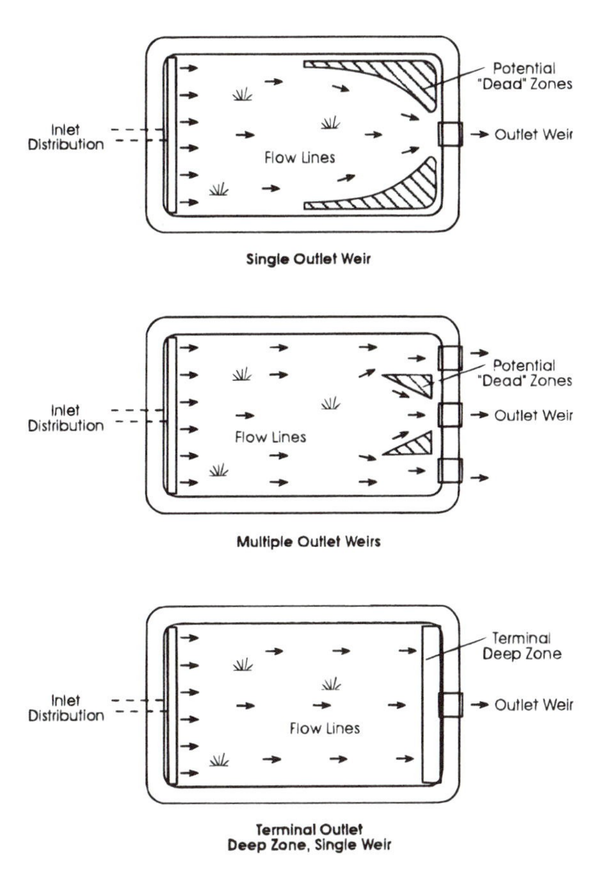

Figure 20-18 Illustration of the effect of outlet design on flow distribution in constructed SF wetlands.

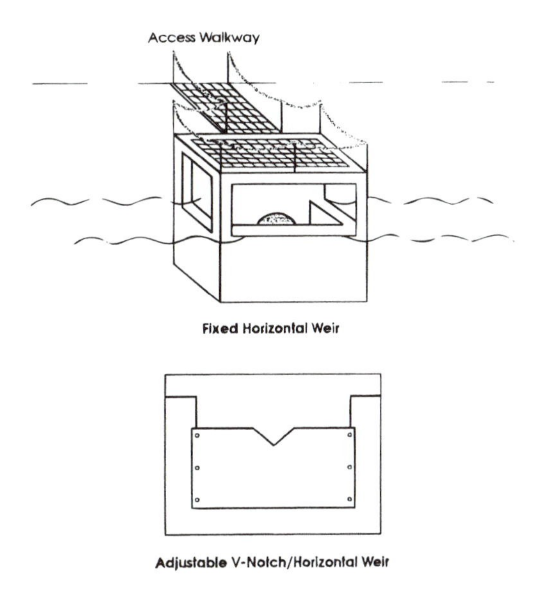

Figure 20-19 Examples of constructed SF wetland outlet weir designs.

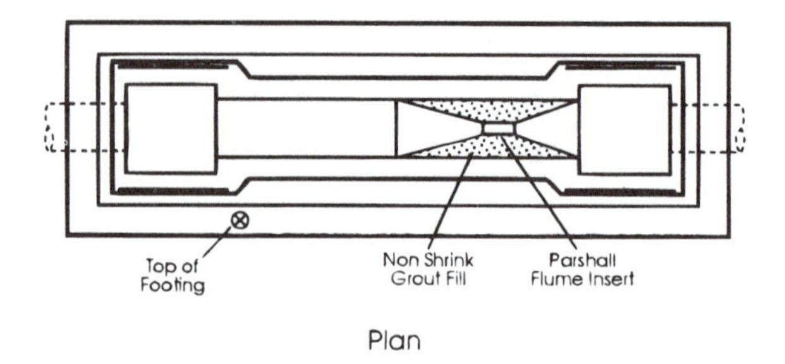

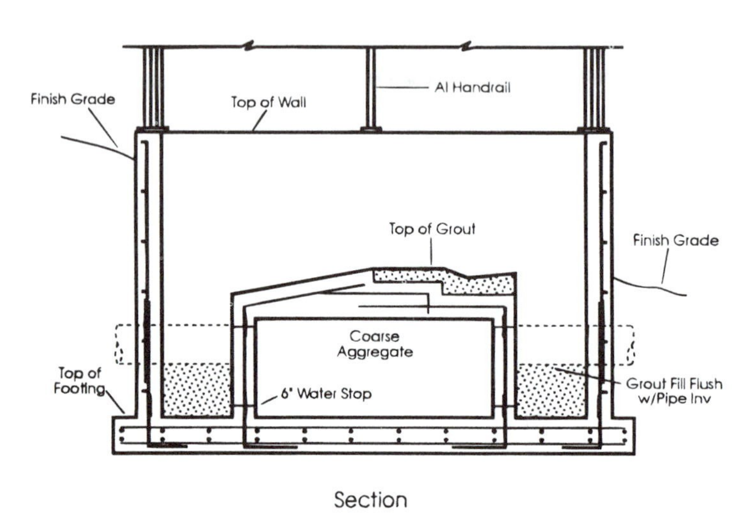

Figure 20-20 Parshall flume structural design details.

Flows at the wetland outlet device can be monitored with a staff gauge or water level recorder calibrated to weir design or by direct or electronic readout from a Parshall-type flume. For a variable outlet configuration or for a hybrid weir design, a stage-discharge curve can be prepared by using periodic volumetric flow measurements at various stage values. After preparing the stage-discharge curve, flow rate can be estimated at any time by reading the water stage with a staff gauge and referring to the stage-discharge graph (Figure 20-21).

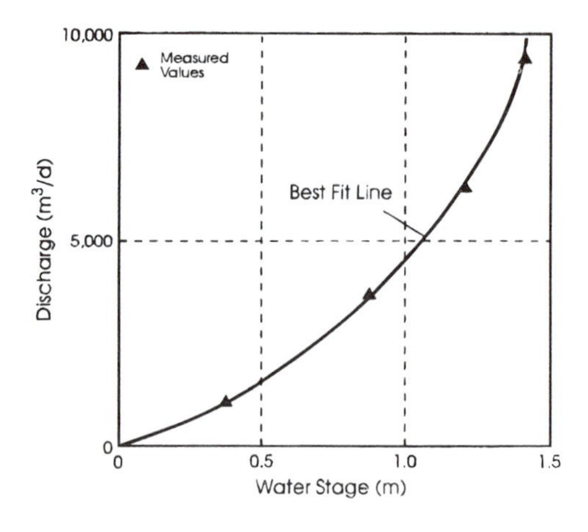

Figure 20-21 Typical empirical stage-discharge relationship for a constructed wetland.

Parshall flumes typically have a graduated scale to enable flow to be estimated instantaneously, or they have a recorder that provides instantaneous and cumulative flows.

Outlet structures are prone to the same difficulties as inlet structures and are more sensitive to accumulation of debris. The combination of water direction and the litter production of the wetland ensures the potential for an accumulation at outlets. A large-mesh debris fence, placed a meter or two from the structure, can alleviate this problem.

Outlet structures are designed to be the primary control on water level within the wetland. Selection of a particular device should include ease of operation as well as ability to set the desired water level. Stop-log weirs are a common choice.

If flood conditions are anticipated, there should be appropriate provisions, possibly including emergency overflow spillways and back-flood gates.

PLANTS AND PLANTING

Constructed wetlands can be planted with a number of adapted, emergent wetland plant species. Wetlands created as part of compensatory mitigation or for wildlife habitat typically include a large number of planted species. However, in constructed SF wetland treatment systems, diversity typically is quite low. Various aspects of wetland plant ecology and potential treatment wetland plant species were summarized in Chapter 7.

Wetland plant species selection should consider the following variables: expected water quality, normal and extreme water depths, climate and latitude, maintenance requirements, and project goals. At this time, there is no evidence that treatment performance is superior or different among the common emergent wetland plant species used in treatment wetlands. The best selection criteria are growth potential, survivability, and cost of planting and maintenance. It is clear that densely vegetated areas are more effective at treating pollutants than sparsely vegetated areas. A corollary to this observation is that plant species that provide structure year-round perform better than species that die below the water line after the onset of cold temperatures. For these reasons, fast-growing emergent species that have high lignin contents and that are adapted to variable water depths are the most ideal for constructed wetland treatment systems. Wetland plant genera that most successfully meet these criteria include *Typha, Scirpus,* and *Phragmites.*

Only a small fraction of the ultimate plant density is planted in the new wetland. Planting densities range from 1000 to 10,000 plants per hectare. Through vegetative reproduction, these plants spread to densities greater than 100,000 plants per hectare. As the first round of plants mature and die, rhizomes send up new shoots, thus maintaining the wetland plant community. Most constructed wetlands also have colonizing plant species around the shallow edges and in unvegetated areas inside the cells. Although these colonizers typically do not provide much cover, they do provide some buffer against plant pathogens, provide habitat diversity important to wildlife, and fill niches that the dominant plant species might otherwise not occupy.

As wetland plants mature and die, they form organic detritus, an essential structural component of a mature constructed wetland treatment system. The standing and fallen dead plants provide a continuing source of organic carbon that is used as substrate by heterotrophic bacteria and fungi. In turn, these microorganisms influence many of the water quality treatment functions important in wetland treatment systems. The organic detritus that is typical of a mature wetland requires from 1 to more than 5 years to develop, depending largely on the nutritive value of the influent wastewater. For this reason, the pollutant transformation functions of wetland treatment systems receiving secondary municipal wastewater typically will mature faster than systems receiving advanced treated, highly renovated and dilute wastewater or runoff from relatively clean watersheds. It is often possible to speed the overall ecosystem development by placement of imported litter such as straw.

If the wetland is to be planted, the cost and availability of plant materials must be addressed early in the design process. The option of establishing an onsite wetland plant nursery must be decided very early, because mature 1- to 2-year-old plants are preferred. These have the energy reserves to survive the transplanting operation. Consequently, the establishment of the nursery must be completed well in advance of other construction.

Another option is to allow natural regrowth of the wetland basins. In southern climates, this process is complete within one growing season, but may require two or more seasons in northern climates. In either climate, the option of transplanting will accelerate the establishment of vegetation. The design decision is based on economics and the regulatory requirements on startup. A 1-year delay in the imposition of permit requirements allows for natural regrowth and can save a considerable amount of money.

The performance of an SF treatment wetland is not sensitive to the particular plant species which populate the wetland. It is difficult to sort this effect from other phenomena in most wetland treatment systems, but there are some side-by-side wetland comparisons that give strong indications of this lack of sensitivity. The Tarrant County, TX facility is one such source of evidence. Three separate trains of three wetlands each were geometric and hydraulic replicates. However, different plant species were established in the three trains: train 1 was bulrush, cattail, arrowhead, and smartweed; train 2 was softrush, pondweed, and water primrose; train 3 was natural regrowth, including Colorado River Hemp, arrowhead, reed canary grass, and smartweed. After a 1-year startup period, which produced full vegetative cover, there were no measured differences in performance, as indicated by the removals in Table 20-4.

It is likely that small performance differences exist due to vegetation type in SF wetlands, but these are often masked by other unavoidable differences in comparison wetlands. At the time of this writing, the case for superiority of a particular plant species had not been proven or disproven. The evidence points toward minimal differences.

From the standpoint of system resiliency, the wetland should probably contain a diverse mix of macrophyte species and thus be in a position to accommodate changes in water quality and timing that may occur. In other words, a polyculture is preferable to a monoculture. Most SF treatment wetlands undergo a process of alteration after an initial planting, with the more robust species gaining dominance, typically cattails, phragmites, and bulrushes. However, in ultrapolishing systems with very high water quality, a very diverse species composition may develop (Schwartz et al., 1992).

DETAILED ECONOMICS

CAPITAL COSTS

Estimating the initial capital cost of the project is a routine exercise in most respects. However, there are two nuances peculiar to SF treatment wetlands that need consideration: the life expectancy of the items purchased and their value (positive or negative) at the end of service life. In many situations, the wetland alternative is to be compared to other types of processes. Traditionally, the life expectancy of a "conventional" treatment

Table 20-4 Performance of the Tarrant County Wetland Systems, November–December 1993

Train	Vegetation	HLR cm/d	TSS Removal %	TP Removal %
1	Planting 1	2.90	95.4 ± 5.2	80.9 ± 18.6
2	Planting 2	4.40	91.8 ± 6.3	73.6 ± 22.2
3	Natural regrowth	2.0	94.6 ± 1.9	81.8 ± 8.5

Note: Data from Alan Plummer and Associates, 1994.

alternative is 20 years, and neither positive nor negative value is assigned to the components after that time.

A treatment wetland has a longer life expectancy than concrete and steel equipment. Although there are no examples of engineered systems with long periods of operation, there are long-lived SF wetland systems that have retained their effectiveness for up to 80 years, based on *ex post facto* monitoring. Both the Brillion Marsh (Spangler et al., 1976) and Great Meadows Marsh (Yonika et al., 1979) operated for over 70 years and, in later years, were shown to have retained treatment efficiency. As fully functional ecosystems, treatment wetlands may be expected to sustain their character for as long as appropriate hydrology is maintained.

It is common practice to claim no salvage value at the end of project life in a feasibility study for mechanical plants; but this does not make sense in the context of a wetland project. Typically, the entire acquisition price is charged to the project up front, and there is no "salvage" value at the end of 20 years. In contrast to the crumbling concrete and rusted steel left after the mechanical process reaches the end of its useful life, the land associated with the wetland project will probably have greater or equal value to that at the time of acquisition. This principal components of the wetland project will have appreciated in value. It may be more accurate to delete land cost from the comparison for that reason.

The breakdown of capital costs includes the major categories discussed in preliminary design; it is generally possible to refine the estimates after final sizing and siting. Still more precise economic estimating is possible after final design drawings have been prepared. A sample of a capital cost estimate based on final conceptual design is shown in Table 20-5. The Incline Village, NV system encompasses 175 ha, and therefore, the estimated cost was $23,700 per ha ($9600 per acre).

O&M

The O&M costs for an SF facility include pumping energy, compliance monitoring, dike maintenance, and equipment replacement and repairs. Dike maintenance consists of mowing and preservation of structural integrity. Equipment replacement and repairs pertain to piping and pipe supports, structures, and pumps.

It is early in the history of constructed SF wetland facilities, so there is not a long track record on frequencies for many of these activities. However, in general terms, pumps and piping may last on the order of 40 years, and repair frequencies are known.

Pumping energy may be accurately quantified, as can the initial level of compliance monitoring, once a permit is issued. Mowing is primarily a matter of aesthetics, with secondary emphasis on visual detection of snakes and alligators. If public use is encouraged, there may be a need to maintain signage, trails, and boardwalks. Nuisance control or removal may be required, most often targeting mosquitoes, burrowing rodents, and bottom-stirring fish.

The sum total of these activities is usually relatively inexpensive. No chemical purchases are involved, and there is not a need for highly trained personnel, nor significant time requirements for the necessary semi-skilled employees. Annual costs range from $5000 to $50,000 per year for small systems. However, ancillary research can greatly increase these expenditures. The estimate for the Incline Village system, made at the time of final conceptual design, was $85,500 per year (Table 20-6).

TOTAL PRESENT WORTH

The total cost of a project at the time of inception is the total of capital costs, engineering services, and the present worth of O&M costs over the project life. This approach to economic

Table 20-5 Estimated Capital Costs for the Incline Village SF Wetland System

Item	Estimated Cost ($)
Site Preparation	
Clearing and grubbing	195,000
Fencing	124,000
Dike Construction	
Stripping	50,000
Flood embankment	450,000[a]
Embankment construction	1,150,000[a]
Erosion control and dike stability requirements	150,000[b]
Gravel roadway	256,000
Water Supply and Distribution	
River crossing	50,000
Outfall pipeline	288,000
Distribution piping	318,500
Overflow structures	105,000
Return-flow system	40,000
Miscellaneous	20,000
Site Improvements	
Operations building	95,000
Chain-link fence	6,000
Access road and parking lot	10,000
Septic tank/leach field	7,500
Potable water well	7,500
Landscaping	15,000
Wetlands vegetation	50,000
Monitoring	
Monitoring wells	32,500[c]
Initial survey	34,000
Subtotal	$3,454,000
Contingencies (20%)	691,000
Total	$4,145,000

[a] Preliminary estimate pending results of more detailed design.
[b] Allowance based on soils investigation.
[c] Abstracted from Facilities Plan, Wetlands Enhancement Addendum pending discussions with state personnel.

Note: This information was developed after the conceptual design was finalized. It does not include engineering costs. (Data from Culp, Wesner, Culp, 1983.)

estimating is required when the alternatives under consideration vary greatly in their life expectancy and in their O&M costs. This is the case for wetlands. The overall project evaluation requires consideration of both capital and O&M costs, and the present worth technique is the appropriate vehicle for combining the two. The present worth of O&M costs, including equipment repairs and replacements, is the money that needs to be set aside now, at the prevailing interest rate, to pay for these future costs.

An alternative comparison is illustrated in Table 20-7, evaluating wetland treatment and chemical treatment to remove phosphorus from agricultural runoff water in southern Florida.

Table 20-6 Estimated O&M Costs for the Incline Village SF Wetland System

Item	Annual Cost[a] ($)
Personnel	50,000
Energy	2,500
Monitoring	21,000
Maintenance materials	12,000
Total	85,500

[a] Preliminary estimate pending discussions with state personnel.

Note: This information was developed after the conceptual design was finalized. It does not include research costs, nor profits derived from hunter use charges. (Data from Culp, Wesner, Culp, 1983.)

Table 20-7 Estimated Cost Comparison for Phosphorus Reduction in Agricultural Runoff

Wetland Alternative		Chemical Treatment Alternative	
Capital Costs		Capital Costs	
Total ex replacement	$129,748	Total ex replacement	$107,770
Land	$34,434	Land	$2,140
Procurement premium	$10,330	Procurement premium	$375
Pump station capital cost	$14,288	Replaceable equipment	
		Pumps piping electrical	$18,670
Pump station replacement (present worth, 8% discount rate) (25% pump station capital replaced @ 25 years)	$522	Mixing through thickening	$51,390
		Equipment replacement (present worth, 8% discount rate)	
		Pumps piping electrical 25% @ 25 years	$682
		Mixing through thickening 100% @ 20 and 40 years	$1,948
Land free capital cost	$95,836	Land free capital cost	$108,260
Operating and Maintenance Costs		**Operating and Maintenance Costs**	
Labor	$592	Labor	$1,060
Materials	$124	Materials	$250
Chemicals	$0	Chemicals	$560
Energy	$228	Energy	$228
Monitoring	$150	Monitoring	$150
Total Annual O&M	$1,094	Total Annual O&M	$2,490
Present worth O&M (50-year life span, 8%)	$33,443	Present worth O&M (50-year life span, 8%)	$76,153
Total present worth capital + O&M	$129,279	Total present worth capital + O&M	$185,637

Note: Base information from Brown and Caldwell, 1993, and Burns and McDonnell, 1992. Dollars in thousands.

The dollar values in this example are large, because the basis is treatment of a very large flow (*circa* 200 mgd). The estimates in this table were developed from information available at the time of final conceptual design and are subject to change during final design and the accompanying modifications. The example is included here to illustrate the unique features of wetland alternatives evaluation.

The chemical treatment alternative is 17 percent cheaper than the treatment wetland on the basis of the capital expenditures needed to build the project. On the surface, this makes chemical treatment the more attractive alternative. This up-front comparison presumes a life-span short enough that equipment does not need to be replaced, nominally 20 years. However, if the lifespan of the project is taken to be 50 years—which is characteristic of constructed wetlands and has been demonstrated in the region—the analysis changes. It becomes necessary to consider the salvage value of worn out components and their replacement costs. In Table 20-7, it is assumed that worn out equipment has zero value: it is unsalable, and there is no charge for disposal. On the other hand, land acquired for the project is assumed to maintain its value; no replacement purchases are necessary. It is therefore logical to exclude land costs from the analysis, because it can be sold at the conclusion of the project with no loss in value.

When these factors are taken into consideration, the treatment wetlands are 13 percent cheaper than chemical treatment. The conclusion of the capital cost analysis is reversed.

Next, the O&M costs are totalled and converted to their present worth. Chemical treatment, as the name implies, requires more energy, materials, labor, and supplies than wetland treatment. Monitoring costs would be the same. In this example, and in virtually all cases like it, O&M costs are higher for the equipment-oriented technology. The annual O&M for chemical treatment is twice as expensive. The present worth of O&M is a significant fraction

Table 20-8 Summary of Design Models

Plug flow areal k-C* model	$\ln\left(\frac{C - C^*}{C_i - C^*}\right) = -\frac{k}{q}y$
Plug flow input/output relation	$\ln\left(\frac{C_O - C^*}{C_i - C^*}\right) = -\frac{k}{q}$
N tanks input/output relation	$\frac{C_O - C^*}{C_i - C^*} = \left(1 + \frac{k}{Nq}\right)^{-N}$

where C = concentration at location y, mg/L
C_i = inlet concentration, mg/L
C_o = outlet concentration, mg/L
C* = background concentration, mg/L
k = first-order areal rate constant, m/yr
N = number of well-mixed units in series
q = hydraulic loading rate, m/yr
(= 3.65 × q in cm/d)
y = fractional distance through wetland

of the capital cost. Consequently, the total present worth of the wetlands project is only 70 percent of the total present worth of the chemical treatment alternative.

The assumed factors in this example will not prevail in all circumstances, but it does serve to indicate that extra care should be taken in economic analysis of a wetland alternative.

SUMMARY

This chapter has integrated the analytical elements of preceeding chapters into a sequential approach to designing and evaluating an SF treatment wetland system. Sizing is the first step in design, for which there are three complementary techniques. The first is the use of rational design equations, for which a summary is presented in Table 20-8. Both tanks-in-series (TIS) and plug flow versions are shown, and both include the possibility of a non-zero background concentration, corresponding to a return flux from the ecosystem.

The recommended values of the constants are derived from information reported to date, analyzed with the plug flow model. The rate constants are therefore the smallest that is possible to infer from the data on the basis of a mixing assumption (Table 20-9). The reader should refer to the appropriate preceding chapter for the caveats concerning Table 20-9. Adjustments may be made for other degrees of mixing via the TIS model. Rate constants and background concentrations vary with temperature in some instances.

Regression equations may also be used to describe the data (Table 20-10); these form a second method of setting wetland size for some variables. These regressions have unsatisfy-

Table 20-9 Model Parameter Values—Preliminary

	BOD	TSS	Organic N	Sequential NH_4-N	Sequential NO_x-N	TN	TP[a]	FC
Surface flow								
k20 m/yr	34	1000[b]	17	18	35	22	12	75
θ	1.00	1.00	1.05	1.04	1.09	1.05	1	1.00
C*, mg/L	3.5 + 0.053Ci	5.1 + 0.16Ci	1.50	0.00	0.0	1.50	0.02	300[c]
θ	1.00	1.065						

[a] Nonforested.
[b] Rough unsubstantiated estimate, settling rate determination preferred.
[c] Central tendency of widely variable values.

Table 20-10 Regression Equations for the Principal Variables

Variable	Equation / Range	Units	Statistic	Source
SF BOD Correlation	$C_o = 0.173\ C_i + 4.70$			(Equation 12-19)
	$R^2 = 0.62$, N = 440			
	Standard error in C_o = 13.6			
	$0.27 < q < 25.4$	cm/d	mean = 3.0	
	$10 < C_i < 680$	mg/L	mean = 57	
	$0.5 < C_o < 227$	mg/L	mean = 15	
SF TSS 20°C Correlation	$C_o = C^* = 0.158\ C_i + 5.1$			(Equation 11-17)
	$R^2 = 0.23$, N = 1,582			
	Standard error in C_o = 15			
	$0.02 < q < 29$	cm/d	mean = 3.0	
	$0.1 < C_i < 807$	mg/L	mean = 26	
	$0.0 < C_o < 290$	mg/L	mean = 9	
SF TSS Temperature Correction	$C^* = 5.1\ \theta^{(T-20)}$			(Equation 11-18)
	$\theta = 1.065$			
SF NH_4-N Correlation	$C_o = 0.336 q^{0.456}\ C_i^{0.728}$			(Table 13-14)
	$R^2 = 0.44$, N = 542			
	Standard error in C_o = 4.4			
	$0.11 < q < 33.3$	cm/d	median = 2.9	
	$0.04 < C_i < 58.5$	mg/L	median = 2.2	
	$0.01 < C_o < 58.4$	mg/L	median = 0.6	
SF NO_3-N Correlation	$C_o = 0.093 C_i^{0.474}\ q^{0.745}$			(Table 13-14)
	$R^2 = 0.35$, N = 553			
	Standard error in C_o = 4.9			
	$0.017 < q < 27.4$	cm/d	median = 2.7	
	$0.01 < C_i < 24.5$	mg/L	median = 1.7	
	$0.01 < C_o < 21.7$	mg/L	median = 0.2	
SF TN Correlation	$C_o = 0.409 C_i + 0.122 q$			(Table 13–14)
	$R^2 = 0.48$, N = 408			
	Standard error in C_o = 3.5			
	$0.2 < q < 28.6$	cm/d	median = 2.5	
	$2.0 < C_i < 39.7$	mg/L	median = 9.1	
	$0.4 < C_o < 29.1$	mg/L	median = 2.2	
Marsh Phosphorus Correlation	$C_o = 0.195\ q_{avg}^{0.53}\ C_i^{0.91}$			(Equation 14-16)
	$R^2 = 0.77$, N = 373			
	Standard error in $\ln C_o$ = 1.00			
	$0.1 < q_{avg} < 33$	cm/d	mean = 4.2	
	$0.02 < C_i < 20$	mg/L	mean = 3.0	
	$0.009 < C_o < 20$	mg/L	mean = 1.6	
Marsh Fecal Coliforms Correlation	$C_o = 6.65\ q_{avg}^{0.513}\ C_i^{0.343}$			
	$R^2 = 0.36$, N = 106			
	Standard error in $\ln C_o$ = 2.15			
	$0.06 < q_{avg} < 16.4$	cm/d	mean = 4.5	
	$1 < C_i < 3{,}180{,}000$	#100/mL	geometric mean = 328	
	$1 < C_o < 263{,}000$	#100/mL	geometric mean = 74	

ingly low correlation coefficients because they span a large number of unquantified system variables.

Scatterplots are a third method of understanding the behavior of existing systems. Figures 20-3 through 20-6 present the means for placing the proposed project on such scatterplots. Designs which fall outside the range of the existing data scatter would require further justification.

The compartmentalization and aspect ratios of the wetland units are then set, with due attention to the hydraulic constraints and site constraints. Structures are selected, a plant establishment strategy is determined, and water conveyance is set. The project is then ready for a more detailed economic estimate, using special techniques that acknowledge the features of wetland systems.

CHAPTER 21

Wetland Design: Subsurface-Flow Wetlands

INTRODUCTION

Design recipes for subsurface-flow (SSF) wetlands abound in the literature (U.S. EPA, 1993b; TVA, 1993; Cooper, 1990; Schierup et al., 1990b). Some of those procedures are rules of thumb, based on intuition and analogs, and others are based upon data analysis and application of physico-chemical principles.

The general concepts for SSF wetland design are not far different from those for surface-flow (SF) wetlands detailed in the preceding chapter, but the details are quite different in some instances. Most importantly, the regressions and rate constants for pollutant removal are different, and the hydraulics are different.

The design procedures offered here are based on data analysis. As was the case for SF wetlands, the rational design model for pollutant removal is the k-C* areal model, which was substantiated in previous chapters for SSF wetlands. In so doing, the SSF wetland is viewed as a vertically integrated unit, consisting of media, water, and plants. Because there are no reports which delineate the effect of saturated media depth, there is presently no way to distinguish between the volume-specific and area-specific versions of the model for SSF systems. However, as detailed in Chapter 12, return fluxes are just as important in SSF wetlands as they are in SF wetlands, and hence, the inclusion of C* is necessary to prevent underdesign.

The first task is assembly of design data, which was discussed in the preceding chapter. It is necessary to carefully define the quantity and quality of the water to be treated and the goals of the treatment. The next task is determination of preliminary feasibility, which follows the same pattern as for SF wetlands.

PRELIMINARY FEASIBILITY

The goal in this step is to obtain a first estimate of the wetland size and to determine if there is sufficient land available for the wetland. Secondarily, the costs of the project are estimated.

PRELIMINARY SIZING

Each regulated water quality parameter will require its own particular wetland area for reduction to the desired level. The maximum of the individual treatment areas is then selected as the necessary size. This is exactly the same procedure as for SF wetlands, but the constants in the model are different for the SSF wetlands (Table 21-1). The reader should refer to the appropriate preceding chapter for the caveats concerning Table 21-1.

An example calculation sheet (Table 21-2) shows that the required areas vary considerably depending on which parameter is being targeted.

PRELIMINARY ECONOMICS

Capital Costs

Gravel beds are more costly on an areal basis than SF wetlands. However, they possess certain advantages, in terms of larger rate constants and of nuisance reduction. Therefore, economics must be evaluated in the context of ancillary benefits and values. The cost of the media is a large fraction of the total cost of gravel bed wetlands, and this added expense must be weighed against the potential advantages of the SSF system.

The distribution of capital costs for SSF wetlands in the NADB is wide (Figure 21-1). However, the median cost of the SSF systems is $145,000 per acre, compared to $20,000 per acre for SF wetlands. The reed beds used in the U.K. average about $400,000 per acre, which includes pumps, liners, land costs, and construction. The land cost is usually not a significant contribution to the total capital cost. Land cost should be excluded from consideration of total capitalized cost, since it will appreciate in value at about the rate of inflation. This is current practice for Severn Trent Water, Ltd. of the U.K. (Green, 1994).

The life expectancy of SSF systems may be limited by the accumulation of mineral solids in the pore space. Blockage by degradable biosolids is also expected, but this is accounted for in hydraulic design. The mineral content of incoming wastewaters is characterized by the non-volatile component of the total suspended solids (NVSS). This material will accumulate in pore spaces, preferentially near the bottom of the gravel bed (Kadlec and Watson, 1993). This process is very slow when the incoming waters have NVSS less than 100 mg/L. At a loading rate of 30 cm/d and NVSS = 100, it would take 37 years to fill half the voids in the bed with mineral residues. Depending on the location of the material in the pore space, the hydraulic conductivity would be reduced by a factor between 2.0 (all on the bottom) and about 16 (uniform pore blockage).

Operating and Maintenance Costs

Experience is very limited, but all indications are that SSF wetlands need little maintenance. Estimates range from $1000 to $2000 per acre per year.

Cost Comparison

The costs of SF and SSF systems depend on the size requirement and the unit areal cost. In turn, the area required is roughly proportional to the inverse of the k value for a given

Table 21-1 SSF Model Parameter Values—Preliminary

	BOD	TSS	Crganic N	Sequential NH_4-N	Sequential NO_x-N	TN	TP	FC
k20, m/yr	180	1000[a]	35	34	50	27	12	95
θ	1.00	1.00	1.05	1.04	1.09	1.05	1.00	1.00
C*, mg/L	3.5 + 0.053 C_i	7.8 + 0.063 C_i	1.50	0.00	0.00	1.50	0.02	10[b]
θ	1.00	1.065						

[a] Rough unsubstantiated estimate, settling rate determination preferred.
[b] Central tendency of widely variable values.

Table 21-2 Preliminary Feasibility Calculation Sheet, SSF Wetlands

		TSS	BOD	TP	TN	FC
Design flow, m^3d	Q = 500	180 L/pe/d (Population Equivalent = 2778)				
Influent concentration, mg/L	C_i=	80	80	10	30	100,000
Target effluent concentration, mg/L	C_e=	15	10	—	—	—
Wetland background limit, mg/L (For TSS, C* = 7.8 + 0.063Ci) (For BOD, C* = 3.5 + 0.053Ci)	C*=	**12.8**	**7.7**	**0.02**	**1.5**	**10**
Reduction fraction to target	$F_e = 1 - C_e/C_1$=	0.813	0.875	0.000	0.000	0.000
Reduction fraction to background	$F_b = 1 - C^*/C_1$=	0.840	0.903	0.998	0.950	1.000
Areal rate constant, m/yr	k=	**3000**	**180**	**12**	**27**	**95**
Required wetland area, ha	A=	0.021	0.351	—	—	—

$$A = \left(\frac{0.0365 \cdot Q}{k}\right) \cdot \ln\left(\frac{C_i - C^*}{C_e - C^*}\right)$$

The area requirement is	1.3	m^2/pe
The necessary area =	0.351	ha
HLR =	14.23	cm/d
HRT @ 60 cm depth and 40% porosity =	1.7	days

		TSS	BOD	TP	TN	FC
Effluent concentrations, mg/L via k-C* Model	C_o=	12.8	10.0	7.9	18.4	16,071

$$C_0 = C^* + (C_i - C^*)\exp\left(-\frac{kA}{0.0365Q}\right)$$

Note: Annual averages, constant Q, 20°C, plug flow assumption. Numbers in boxes from designer.

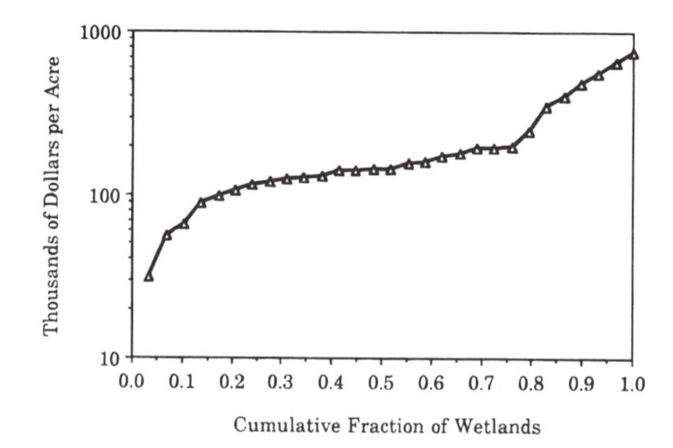

Figure 21-1 Distribution of capital costs for 29 constructed SSF wetlands. The base cost, without pumps, land, liners, planting, or complex structures, is on the order of $30,000. The data points are in 1993 dollars, adjusted from the year of construction at 4%. (Data from the NADB, 1993.)

pollutant. If the median capital costs from the NADB are accepted as norms, the capital cost ratio for a given pollutant may be computed (Table 21-3). On the basis of performance and cost alone, there is no reason to consider an SSF wetland.

DETAILED CONCEPTUAL DESIGN

One of the historical difficulties with SSF wetlands has been the construction of projects based on rules-of-thumb and personal whims rather than on sound and thorough analysis of available data. The feasibility study outlined in the previous section will normally determine the need to pursue the SSF option in more detail. Many design considerations still remain at this stage of the process.

Most of the ideas presented in the preceding chapter on the need for pilot projects also apply to SSF wetlands. An important addition is the need to carefully consider the performance and difficulties associated with operation in cold climates. Operation year-round in cold climates has been successfully conducted (Lemon and Smith, 1993; Jenssen, et al., 1992), but a sufficiently large database is lacking thus far.

A second addition is consideration of the tailoring of the media to provide added sorption capacity for phosphorus and metals. There is an emerging body of knowledge on the use of specialized media to invoke added treatment potential (Jenssen et al., 1992; Brix, 1994a).

Localizing the Design Parameters

Just as in the case of SF wetlands, emphasis has been placed on average values of constants for use in design equations for SSF wetlands, and it may be necessary to adjust those values for a particular wastewater or a particular regional circumstance. The error bands on correlations presented previously are wide and may require narrowing for final design.

Table 21-3 Capital Cost Comparison for SF and SSF Wetlands

	TSS	BOD	TP	NH_4-N	Organic N	NO_x-N	TN	FC
SF k value, m/yr	—	34	12	18	17	35	22	75
SSF k value, m/yr	—	180	12	34	35	50	27	95
SF area/SSF area	1.00	5.29	1.00	1.89	2.06	1.43	1.23	1.27
SF cost/SSF cost	0.14	0.73	0.14	0.26	0.28	0.20	0.17	0.17

Note: TSS is presumed to be reduced to background by both types. Unit costs: $145,000/acre SSF; $20,000/acre SF.

The Severn Trent polishing wetlands provide a good example (Green and Upton, 1992, 1993; Green 1994). In contrast to the fairly strong wastewaters that are treated in other locations, these SSF systems treat nitrified effluents that are already low in biochemical oxygen demand (BOD) and total suspended solids (TSS). Ammonium forms only about half the total Kjeldahl nitrogen (TKN) load, which in turn is small compared to the total oxidized nitrogen (TON) load. Information has been collected at 11 wetlands for 2 years and provides a sufficient basis for reevaluating the operative values for BOD of $k = 160$ m/yr and $C^* = 1.9$ mg/L. These differ from the average values determined from a wider set of data in Chapter 11 ($k = 180$ m/yr and $C^* = 9.8$ mg/L).

The SSF wetlands are subject to the same level of stochastic variability as SF wetlands. Monthly or quarterly calculations, accompanied by appropriate maximum month to annual average ratios (Chapter 20, Table 20-2), are required if there are strong variations within a year.

THE REALITY CHECK

As was the case for SF wetland design, it is appropriate to compare the spectrum of anticipated performances against the background of existing data for the SSF design. The output concentration-input loading plots again serve this purpose. Each operating condition can be located on such a graph. Figures 21-2 through 21-5 show scatterplots of the quarterly data from the NADB and other sources, which form the experience base. Shown for reference are the data for SF wetlands.

The SSF wetland seems to be a better filter for TSS, but this may be an artifact of the higher loading rates. Both types are typically overdesigned for TSS removal.

PRESUMED STATE OF MIXING

As for SF wetlands, the premise in global parameter estimation has been that of plug flow, which has been adopted on the premise of conservatism. If that assumption is to be modified, it is in the final sizing calculation. The tools for doing this have been discussed in Chapters 9, 12, 14, and 20. There is clearly not plug flow in SSF systems, as has been repeatedly demonstrated in tracer tests (Brix and Schierup, 1990; Netter, 1994; Fisher, 1990).

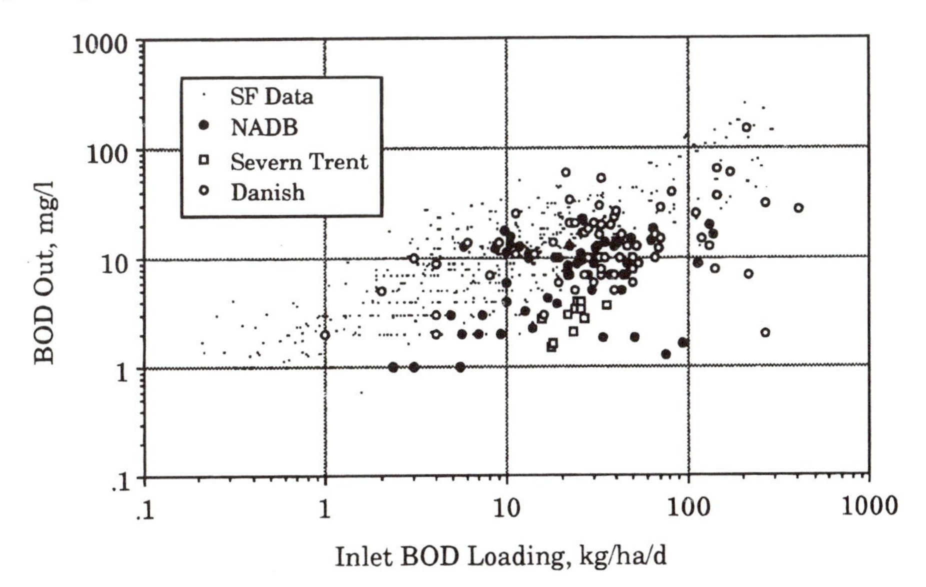

Figure 21-2 Scatterplot of BOD performance of SSF wetlands, with a background of the SF data.

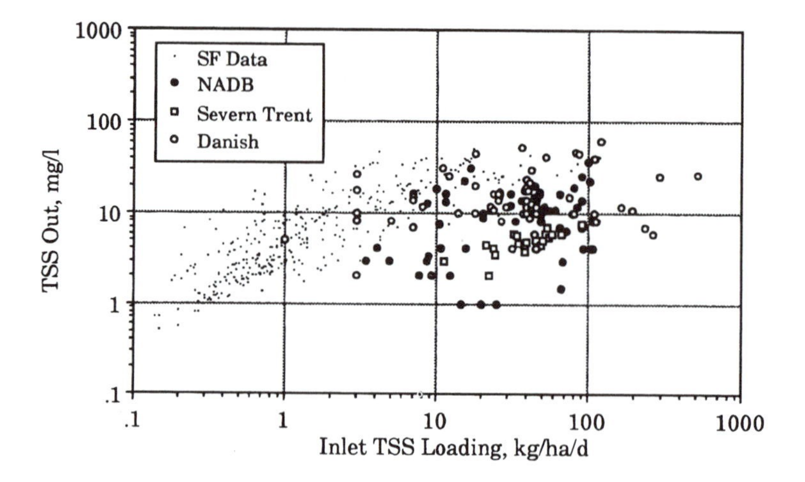

Figure 21-3 Scatterplot of TSS performance of SSF wetlands, with a background of the SF data.

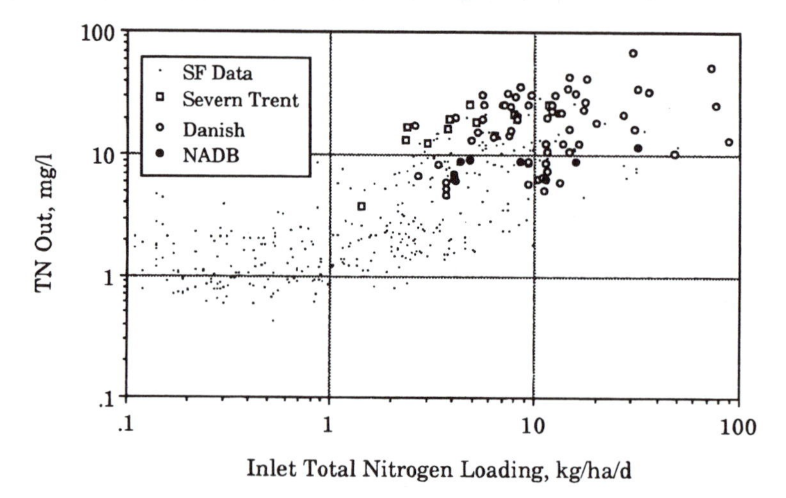

Figure 21-4 Scatterplot of total nitrogen performance of SSF wetlands, with a background of the SF data.

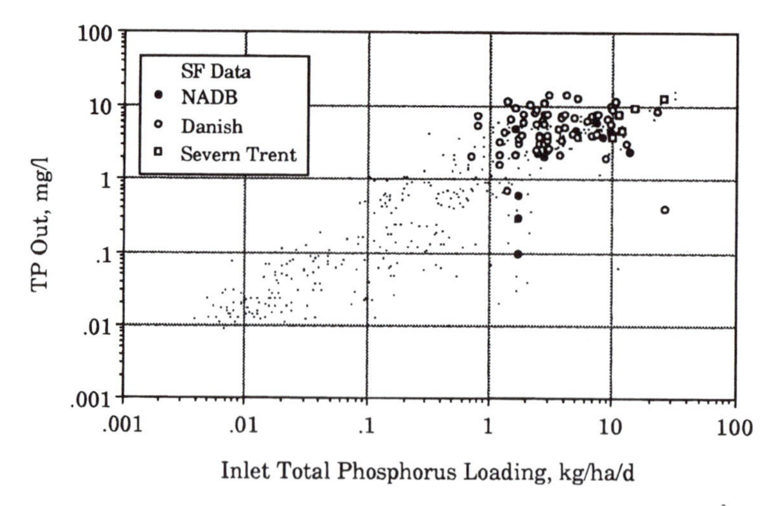

Figure 21-5 Scatterplot of total phosphorus performance of SSF wetlands, with a background of the SF data.

Those tests show that most of the pore volume of a gravel bed is eventually swept by the flowing water. The mean detention time is close to the nominal detention time, and there is a good gross areal efficiency (U.S. EPA, 1993b). However, those same tracer tests show a good deal of internal mixing. In SSF wetlands, mixing is obviously not due to large-scale convection currents. Rather, it is due to complex traffic patterns within the media. Some water follows preferential microchannels and exits sooner; some follows slower paths and exits much later. This leads to the spreading of the tracer pulse during passage. Further, small pockets and crevices within the media are not swept on a flow-through basis; these exchange water and solutes with the flowing water by diffusive and convective processes. When advanced and delayed miniflows are recombined at the wetland outlet, the effect is the same as if large-scale convection currents had been operative.

A popular notion has been that increasing the length-to-width ratio will produce closer-to-plug flow behavior in SSF wetlands. Because "mixing" occurs on a microscale, this notion is not correct. Residence time distributions, with their attendant spreads, do not narrow as length-to-width ratio is increased (Figure 21-6).

As was the case of SF wetlands, it is necessary to avoid gross areal inefficiency by correct selection of inlet distribution and outlet collection. A larger length-to-width ratio promotes greater gross areal efficiency, but will not prevent pocket and profile mixing.

The dispersion numbers for SSF wetlands are approximately the same as those for SF wetlands (Chapter 9, Table 9-5). Accordingly, the effects of nonideal flow may be considerable, especially when the pollutant reduction is targeted close to background. Under those circumstances, the early breakthrough of a small amount of poorly treated water is enough to influence the overall mixed result.

A strategy for accounting for departure from plug flow was presented in the preceding chapter. If designers wish to take some credit for the acknowledged partial mixing in the database wetlands, they do so by adjusting the rate constant upward, utilizing the presumed imperfect mixing model for the database. It is possible to reanalyze the database with a different mixing model, the tanks-in-series (TIS) model, for instance. For example, for BOD reregression of the Severn Trent data for four TIS yields $k = 203$ m/yr and $C^* = 1.7$ mg/L.

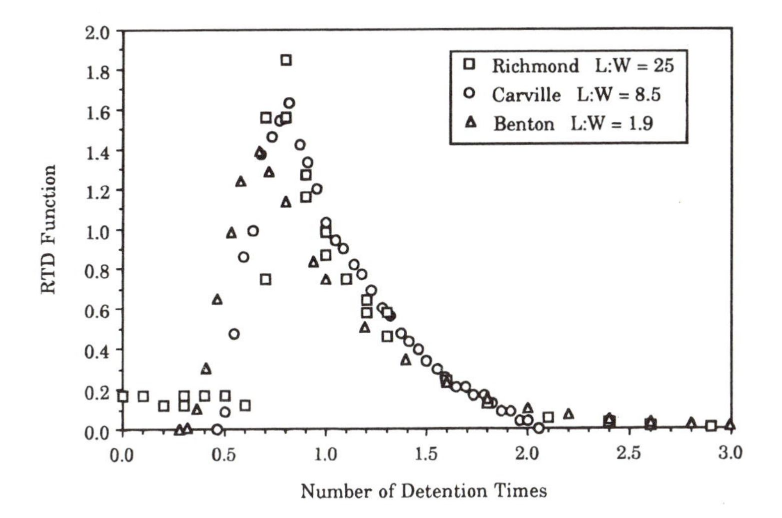

Figure 21-6 The effect of length-to-width ratio on tracer response for SSF wetlands. Note that the breakthrough time is somewhat shorter for the lower aspect ratios, but the spread of the response curve (variance) is similar for all. Data from Bavor et al., 1988 (Richmond, NSW); TVA, 1990b (Benton, KY); and U.S. EPA, 1993b (Carville, LA).

There is an easier way to estimate the TIS rate constant. The reduction toward background given by the two alternative models is

$$\frac{C_e - C^*}{C_i - C^*} = \exp\left(-\frac{k_{pf}}{q}\right) = \left(1 + \frac{k_N}{Nq}\right)^{-N} \tag{21-11}$$

where C_e = measured outlet concentration, mg/L
C_i = measured inlet concentration, mg/L
C^* = background concentration, mg/L
k_{pf} = first-order areal rate constant, plug flow, m/yr
k_N = first-order areal rate constant, N TIS, m/yr
N = number of well-mixed units assumed
q = measured hydraulic loading rate, m/yr

Therefore, provided that C* is the same for both models,

$$k_N = Nq\left[\exp\left(\frac{k_{pf}}{Nq}\right) - 1\right] \tag{21-12}$$

The estimated four TIS value is k = 226 m/yr, which is also the best fit for C* = 1.9 mg/L.
Other procedures for mixing adjustments are contained in Chapters 9, 12, and 14.

CONVEYANCE: ASPECT RATIOS AND HEAD LOSS

The SSF wetland area may be configured in many different length-to-width ratios. As the length-to-width ratio is increased, the linear velocity increases, and the head loss increases. At some point, the head loss will produce overland flow, as detailed in Chapter 9.

The head loss design procedure was detailed in Chapter 9 and illustrated in Examples 9-8 and 9-9. The planar area of the wetland has been selected to provide the correct hydraulic loading, and that in turn sets the detention time via the selection of the bed depth. Since the bed depth is typically 0.5 m, there remain only two questions: what length to width ratio, and what media size?

The question of optimal media size for treatment has been confused with the issue of flooding in most SSF analyses. At present, there is not a rational way to compute the treatment efficiency of a particular type and size of media. Some treatment processes rely simply upon the biofilms on the media or upon the sorptive capacity of the media. For those purposes, it is desirable to use small media which possess more available area per unit volume for those biofilms. Other aspects of treatment rely upon macrophytes and algae, neither of which are sensitive to media size within reasonable limits.

Because the choice of a SSF wetland is normally predicated upon the desire to keep the water below ground (rather than upon economics), the design strategy should be to select both the media size and the length-to-width aspect ratio to control the bed hydraulics.

Therefore, these two variables are selected to meet the design criteria set forth in Chapter 9. Those are repeated here:

1. Anticipated flows must pass through the bed without overland flow or flooding.
2. Anticipated flows must pass through the bed without stranding the plants above water.
3. Operation should remain acceptable in the likely event of changing hydraulic conductivity.
4. The bed should be drainable.
5. The bed should be floodable.

6. Water levels within the system should be fully controllable through the use of inlet and outlet structures.
7. The configuration must fit the site, in terms of project boundaries and of hydraulic profiles.

The case of a rectangular bed may be reduced to a fairly simple set of criteria. The exit water level is presumed to be controlled at 90 percent of the bed depth. The maximum freeboard is 10 percent of the bed depth. When profiles are computed over the full range of values of design variables, the region in which these constraints are met may be identified. The results may be summarized in the constraints:

Design constraint

$$G_1 = \frac{S_b}{(h_o/L)} < 0.1 \qquad (9\text{-}83)$$

Design constraint

$$G_3 = \frac{(q/k)}{(h_o/L)^2} < 0.1 \qquad (9\text{-}88)$$

where h_o = outlet water depth, m
k = hydraulic conductivity of the bed, m/d
L = bed length, m
q = hydraulic loading rate, m/d
S_b = bed slope, dimensionless

These constraints must be met for all anticipated operating conditions, including initial and clogged conductivity, and the range of operating flows expected.

A HYDRAULIC DESIGN EXAMPLE

This hypothetical example is used to illustrate the complete hydraulic design procedure.

Example 21-1

A wetland is to be designed to carry 300 m^3/d of water at full capacity, but only half that amount will be available at startup. The nominal detention time in the media is to be 4 days. A cheap gravel is available with a nominal size of 0.5 cm and a porosity of 40 percent; other sources of media are not as cheap.

Solution

The upper limit on loading sets the wetland volume required, since the largest flow must be treated. The depth of media is selected as 0.6 m to allow for a bottom sediment buildup and to match the rooting requirement of the vegetation. The porosity of the gravel is presumed to be 40 percent. The outlet depth will be set at 0.55 m.

The wetland area will be selected to meet the treatment requirement

$$\tau = \frac{LW\,h_o\varepsilon}{Q}$$

$$4 = \frac{LW(0.55)(0.4)}{300}$$

$$A_T = LW = 5{,}455\ m^2$$

The hydraulic loading is therefore

$$q = Q/LW = 300/5{,}455 = 0.055 \text{ m/d} = 5.5 \text{ cm/d}$$

The length-to-width ratio and media conductivity must meet the hydraulic constraint. A tenfold reduction in the conductivity will be presumed to occur due to clogging; k = 10,500/10 = 1050 m/d.

$$G_3 = \frac{(q/k)}{(h_o/L)^2} < 0.1$$

$$G_3 = \frac{(0.055/k)}{(0.55/L)^2} < 0.1$$

$$\frac{L^2}{k} < 0.55$$

$$L < 24 \text{ m} \qquad \text{units: meters, days}$$

Selecting L = 24 m, the width is then W = 5455/24 = 227 m.

The velocity is u = Q/WH. The depth will be approximately 90 percent of the media depth, or 55 cm. Hence u = 300/(227)(0.55) = 2.4 m/d, which is well within the laminar range. This is established by checking the Reynolds number:

$$Re = \frac{D\rho u}{(1 - \varepsilon)\mu} = \frac{(0.5)(1.0)[(2.4)(100/86{,}400)]}{(1 - 0.4)(0.01)} = 0.23 < 10$$

The drainability condition sets the bottom slope to avoid dryout at reduced loadings:

$$G_1 = \frac{(S_b)}{(h_o/L)} < 0.1$$

$$S_b < 0.1(h_o/L) = 0.1(0.55/24) = 0.0023$$

For the 25-m length, this means a drop in bottom elevation of no more than 5.7 cm. Select a drop of 5 cm (2 in.) to keep the American surveyors happy.

The aspect ratio of 24/227 = 0.105 is conducive to short circuiting and should be corrected by using several cells in parallel. This is consistent with a design strategy which allows for the possibility of cell shutdown. An aspect ratio of 1.0 or greater is reasonable. This leads to nine parallel cells in the current design, each about 25 × 25 m.

A single bed can also be designed if a different media is obtained. Suppose the aspect ratio is to be L/W = 2.0. Then L = 104 m, and W = 52 m. The slope of the bottom can be no higher than 0.00053. The flooding constraint requires

$$k > L^2(0.55) = (104)^2/(0.55) = 19{,}665 \text{ m/d}$$

Allowing for a tenfold reduction due to clogging, the bare media should have k = 197,000 m/d. In the laminar range, conductivity is proportional to the square of the particle size; hence, we need a size of

$$197{,}000 = \left(\frac{D}{0.5}\right)^2 (10{,}500)$$

$$D = 2.2 \text{ cm}$$

A decision must now be made on the media to be used. The extra cost of the larger media must be compared to the extra cost of more separation dikes and more inlet distribution piping. Inlet distribution must be along the entire width of the cell, to avoid corner dead zones. The nine-cell system requires 695 m of diking; the one-cell system requires only 312 m. The nine-cell system requires 227 m of inlet distribution; the one-cell system requires only 52 m. The cost of the media is likely to be more than half the cost of the system, so this decision is economically important. Conductivity measurements on the candidate media are therefore warranted.

Next examine the response of this system to the anticipated variations in operating conditions. Figure 21-7 shows that water level control is virtually completely due to the positioning of the exit water level control point.

Further, the water surface is stable with respect to evaporative losses, as long as the water fed exceeds evapotranspiration. Protection against rainfall events of all magnitudes is not feasible. The system as designed has approximately 5 cm of freeboard at 40 percent porosity, which can contain a 2-cm rain event without surface flow.

INLET DISTRIBUTION SYSTEM

The inlet distribution system of SSF wetlands has the same purpose as it does in SF wetlands: to promote effective flow distribution along the entire edge of the treatment unit. The distribution system designs used are the same as those shown in Chapter 20, Figures 20-15, 20-16, and 20-17. The most common design incorporates a gated or slotted distribution header in or on a coarse gravel inlet area (Cooper et al., 1990). This gravel enhances distribution across the width and depth of the bed media. Although castellated inlet weirs have been used on a number of SSF systems, Cooper et al. (1990) has recommended that they not be used, presumably because of difficulties with leveling and inadequate flow distribution.

BASIN CONFIGURATION

As with SF wetlands, at least two wetland cells are recommended for SSF systems. Systems larger than 3786 m^3/d should have more than two parallel flow paths. These cells must be sized based on peak flows, limiting winter temperatures, and the specific treatment goals. Multiple cells provide system redundancy during periods of maintenance and flexibility with wastewater loading schedules.

SSF wetlands have historically been designed using two different basin configurations. The first of these, which has been used extensively in North America, incorporates length-

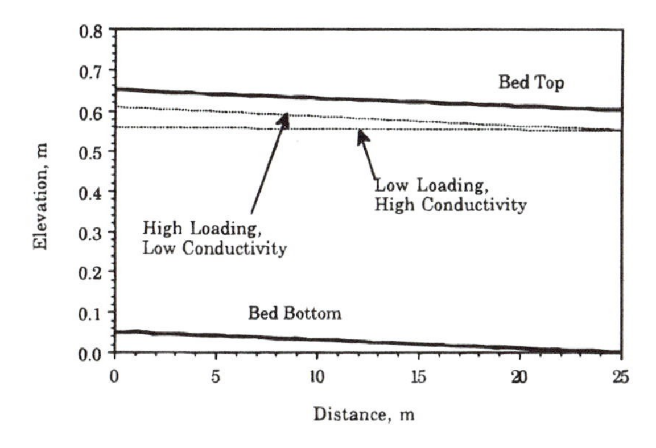

Figure 21-7 Water surface profiles under robust design conditions for Example 21-1. The extremes of operation cause neither flooding nor dryout. Exit level control is at 55 cm.

to-width ratios greater than 2:1. The purpose of this design is supposedly to minimize short circuiting and the subsequent loss of plug flow conditions. Because short circuiting does not typically occur in the subsurface portion of the wetland bed, this design tacitly acknowledges the frequent occurrence of surface flow in these systems and the resulting ease of short circuiting in surface flow when length-to-width ratios are less than one. The second design approach recognizes the importance of inlet surface area on initiating the subsurface flow. This approach generally results in beds that are wider than they are long (length-to-width <1:1). The rational procedure illustrated in the previous section is to be preferred to either of these *ad hoc* procedures.

SSF wetlands can be built with or without slopes on the bed surface and at the base of the bed. A sloped bed surface is more likely to encourage surface flow and is not recommended by Cooper (1990) or in the German design guidelines (Anonymous, 1989). A flat bed surface is useful for flooding as a means of weed control during vegetation establishment in well-drained beds. The bed bottom should not be sloped to enhance the gradient for subsurface flow, but a slight slope provides the capability for full drainage upon shutdown. The hydraulic gradient is also not enhanced by the vertical lowering of the exit collector pipe, as shown in Chapter 9. Normal bed depths have historically been limited to the potential penetration depth of the plant roots, which is generally between 30 and 60 cm.

Berm height in SSF systems is lower than in SF systems because surface water is of limited occurrence. Berm freeboard must provide storage for rainfall events based on a desired return frequency as discussed previously. Emergency overflow areas can also be provided in SSF wetland berms to prevent catastrophic berm failures during storm events.

In SSF wetlands, it may be important to line the system to prevent exchange between the wastewater and groundwater. This consideration is site specific and depends on the quality of the influent wastewater. For constructed, SSF wetland treatment systems that receive raw or primary wastewater, groundwater discharges should be prevented with clay or other impermeable liners. If the influent water has received at least secondary treatment, discharge to the groundwater may not cause or contribute to water quality violations and liners may not be required.

BED MEDIA

The original European design for constructed SSF wetland treatment systems, known as the root zone method or the Kickuth system (Kickuth, 1982, as referenced by Schierup et al., 1990b), used soils as a bed media. These systems are still in use in Germany and in other nations such as South Africa and Denmark. Parallel development of the SSF technology occurred in the U.S. starting in the 1970s (Spangler et al., 1976). These systems have all used various types and grades of sand, gravel, or crushed stone. Plastic trickling filter media have been used for microcosm-scale facilities (Burgoon et al., 1991a, 1991b).

The Kickuth systems relied on an increase in bed media hydraulic conductivity in their soils due to plant root development, but studies in the U.S., U.K., Germany, Austria, and Denmark have found that the hydraulic conductivity may actually decrease over time. For that reason, experts in these countries believe excess hydraulic conductivity should be designed into these systems.

OUTLET CONTROL SYSTEM

The preferred method of outlet control from SSF wetlands is by use of a slotted or perforated pipe buried in a coarse gravel bed along the entire downstream width of the wetland. Water level in the bed can be controlled easily with a swivel-type riser pipe attached to this outlet pipe (Figure 21-8). Water overflowing this riser exits through a culvert at the base of the outlet sump and flows to a system outlet which can be instrumented for flow monitoring and sample collection.

SELECTION OF PLANTS

A gravel bed will require planting, because seed banks are typically lacking, and the media is not optimal for germination. If a portion of the bed remains flooded, a litter layer may develop that is conducive to germination of wetland plant seeds, thus permitting invasion. More frequently, a portion of the bed may remain too dry, permitting invasion by terrestrial species (weeds).

The presence of macrophytes is important for many if not all pollutant removal functions. The three genera of wetland plants that are most frequently used in SF wetland treatment systems are also used in SSF wetlands. The most commonly used plant species worldwide is *Phragmites australis* (common reed). This species has remarkable growth rates, root development, and tolerance to saturated soil conditions. They are known to provide some ancillary benefits in terms of wildlife habitat in the U.K. (Merritt, 1994). The other two species commonly used in U.S. SSF systems are *Typha* spp. and *Scirpus* spp.

Phragmites is planted using rhizomes, seedlings, or field harvested reeds. All of these techniques are effective if the plants are healthy and adequate (but not excessive) soil moisture is maintained during plant establishment. Planting densities between 2 to 6 per square meter (20,000 to 60,000 per hectare) are normally recommended for *Phragmites* (Cooper, 1990; Anonymous, 1989). Planting densities for cattail and bulrush are typically only 1 to 2 per square meter.

The question of which plant may be best is not resolved at the time of this writing. The results of various side-by-side investigations are inconclusive, as will be discussed for ammonium removal. The Santee, CA project ranked *Scirpus* best, *Phragmites* second, and *Typha* a distant third, close to no plants (fourth) (Gersberg et al., 1984). The Lake Buena Vista, FL project ranked *Sagittaria* better than *Scirpus* (DeBusk et al., 1989). The Hamilton, NZ project ranked *Glyceria* better than *Schoenoplectus* (*Scirpus*) better than no plants (van

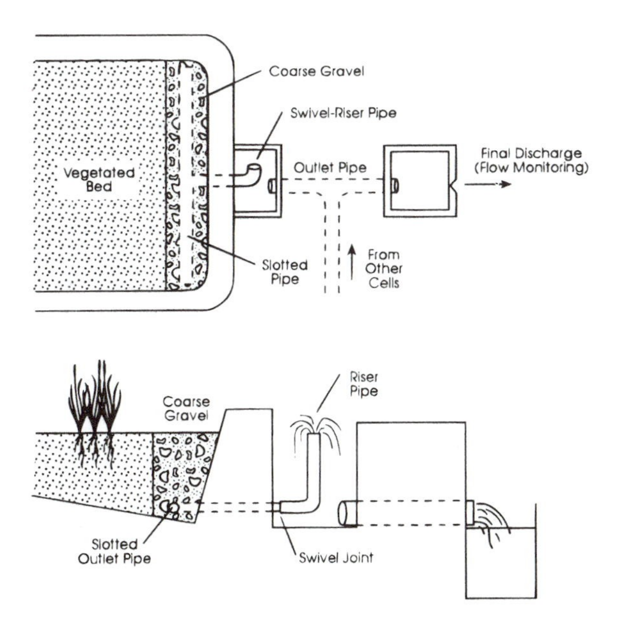

Figure 21-8 Typical outlet control system design for constructed, SSF wetland treatment systems.

Oostrom and Cooper, 1990). The Pretoria, SA project ranked *Phragmites* better than *Scirpus* better than *Typha* for lagoon effluent, but *Scirpus* better than *Typha* better than *Phragmites* for settled sewage (Batchelor et al., 1990). Bavor et al. (1988) found very little difference between *Schoenoplectus* (*Scirpus*) and *Typha* and no plants at Richmond, NSW. At Hardin, KY, *Phragmites* was better than *Scirpus* (NADB).

Table 21-4 Regression Equations for the Principal Variables for SSF Wetlands

Variable	Equation / Range	Units	Statistic	Source
Soil-based SSF BOD Correlation	$C_o = 0.11C_i + 1.87$			(Equation 12-20)
	$R^2 = 0.74$, $N = 73$			
	$1 < C_i < 330$	mg/L	mean = 128	
	$1 < C_o < 50$	mg/L	mean = 18	
	$0.8 < q < 22$	cm/d	mean = 5.2	
US gravel bed SSF BOD Correlation (NADB)	$C_o = 0.33C_i + 1.4$			(Equation 12-21)
	$R^2 = 0.48$, $N = 100$			
	Standard error in $C_o = 5.0$			
	$1 < C_i < 57$	mg/L	mean = 24	
	$1 < C_o < 36$	mg/L	mean = 9	
	$1.9 < q < 11.4$	cm/d	mean = 9.9	
Soil-based SSF TSS Correlation (NADB)	$C_o = 4.7 + 0.09C_i$			(Equation 11-19)
	$R^2 = 0.67$, $N = 77$			
	$0 < C_i < 330$	mg/L	mean = 163	
	$0 < C_o < 60$	mg/L	mean = 27	
	$0.8 < q < 22$	cm/d	mean = 5.2	
SSF TSS Correlation (NADB)	$C_o = 7.8 + 0.063C_i$			(Equation 11-18)
	$R^2 = 0.09$, $N = 107$			
	Standard error in $C_o = 10$			
	$0.1 < C_i < 253$	mg/L	mean = 54	
	$0.1 < C_o < 160$	mg/L	mean = 13	
	$1.9 < q < 44.2$	cm/d	mean = 11.8	
SSF NH4-N Correlation	$C_o = 3.3 + 0.46\ C_i$			(Table 13-14)
	$R^2 = 0.63$, $N = 92$			
	Standard error in $C_o = 4.4$			
	$0.1 < C_i < 43.8$	mg/L	median = 6.7	
	$0.1 < C_o < 26.6$	mg/L	median = 6.1	
	$0.7 < q < 48.5$	cm/d	median = 5.5	
SSF NO_3-N Correlation	$C_o = 0.62\ C_i$			(Table 13-14)
	$R^2 = 0.8$, $N = 95$			
	Standard error in $C_o = 2.4$			
	$0.01 < C_i < 27.0$	mg/L	median = 0.3	
	$0.01 < C_o < 21.0$	mg/L	median = 0.4	
	$0.7 < q < 48.5$	cm/d	median = 5.5	
SSF TN Correlation	$C_o = 2.6 + 0.46\ C_i + 0.124\ q$			(Table 13-14)
	$R^2 = 0.45$, $N = 135$			
	Standard error in $C_o = 6.1$			
	$5.1 < C_i < 58.6$	mg/L	median = 21.0	
	$2.3 < C_o < 37.5$	mg/L	median = 13.6	
	$0.7 < q < 48.5$	cm/d	median = 7.1	
European and NADB SSF Phosphorus Correlation	$C_o = 0.51\ C_i^{1.10}$			(Equation 14-19)
	$R^2 = 0.64$, $N = 90$			
	Standard error in $C_o = 1.8$			
	$0.5 < C_i < 20$	mg/L	mean = 8.1	
	$0.1 < C_o < 15$	mg/L	mean = 5.6	

All of these results read like a set of football game results: the reader may use a sequence of his choice to prove that a particular plant is better than another, just as game scores may be used to establish one team's superiority.

It is therefore not surprising that vegetation has been chosen in the past on the basis of hardiness, cost, and local perceptions rather than treatment optimality.

SUMMARY

This chapter builds upon the preceding SF design techniques by adding those elements which are different for a SSF treatment wetland system. Sizing is the first step in design, for which there are three complementary techniques. The first is the use of rational design equations, for which a summary is presented in Table 21-1. These are applicable to SSF wetlands as well as to SF wetlands.

The recommended values of the constants are derived from information reported to date, analyzed with the plug flow model. The rate constants are therefore the smallest that are possible to infer from the data on the basis of a mixing assumption. Adjustments may be made for other degrees of mixing via the TIS model. For most parameters, the values of the rate constants are not greatly different from those for SF wetlands. The gravel media is typically a large added cost compared to SF pricing, so SSF systems are generally more expensive. The additional cost may be justifiable in terms of ancillary considerations, such as mosquito production, human contact, and wildlife protection.

Regression equations may also be used to describe the data (Table 21-4); these form a second method of setting wetland size for some variables. These regressions have unsatisfyingly low correlation coefficients because they span a large number of unquantified system variables.

Scatterplots are a third method of understanding the behavior of existing systems. Figures 21-2 through 21-5 present the means for placing the proposed project on such scatterplots. Designs which fall outside the range of the existing data scatter would require further justification.

The compartmentalization and aspect ratios of the wetland units are then set with due attention to the hydraulic constraints and site constraints. The hydraulics are of greater concern for the SSF wetland because a premium price is to be paid to keep water below ground. Loading and dimensional criteria must be met if flooding is not to occur and dryout is to be avoided. To that end, the length-to-width ratio and media size are selected.

Vegetation cannot be optimally selected for treatment with the currently existing information, but transplanting is a necessity.

CHAPTER 22

Natural Wetland Systems

INTRODUCTION

Natural wetlands have been used to treat human-generated wastewaters for hundreds if not thousands of years. In China, human waste products have been used for thousands of years to fertilize natural marshes and swamps, which in turn were harvested of fish and birds for human consumption. Wastes from Tenotchitlan, the ancient capital of the Aztecs, were discharged into a nearby lake more than 500 years ago to fertilize their floating gardens. In ancient Egypt, the pharaoh's sewage no doubt filtered through the vast papyrus marshes next to the Nile River. And in every growing metropolitan area in Europe and North America near natural swamps and marshes, it is likely that human and stormwater wastes have been directed to those wetlands at some point during history.

Formal documentation of how these natural wetlands affected wastewater quality began in the 1960s and 1970s. Expanding research efforts described the highly consistent reduction in pollutant concentrations as wastewaters passed through the microbially active and complex flooded portions of these wetlands. By the late 1970s and early 1980s, this research had led to the planning and construction of discharges to natural wetlands at a number of locations in North America, as well as to the implementation of constructed wetland treatment technology around the world.

Although constructed wetlands now dominate the new applications of this ancient technology, the age-old advantages of using natural wetlands for water quality purification still offer potential cost savings. If appropriate natural wetland plant community types are available, if pretreatment is sufficient, and if using natural wetlands for human purposes is publically acceptable, then these natural systems may offer the most cost-effective and environmentally sound approach to provide advanced treatment of municipal wastewater or stormwater.

This chapter describes methods for designing successful natural wetland treatment systems. It focuses on design aspects that are different or unique compared to constructed wetlands described previously in Chapters 20 and 21. This chapter includes guidance for the planner, scientist, or engineer concerning selection of a suitable natural wetland treatment site; determination of appropriate pretreatment requirements and treatment area; system configuration and water control; and design of monitoring structures.

SITE SELECTION

To select an appropriate site for natural wetlands treatment, the designer needs to estimate the wetland area necessary to achieve project goals and then find a suitable wetland plant community type with sufficient area. In addition, a careful analysis must be made of wastewater pretreatment needs, flow control, mass and hydraulic loading rates, and existing environmental resources in the wetland to demonstrate that short- and long-term ecological alterations to the wetland will remain within acceptable limits.

This process of identifying project goals, selecting the required natural wetland area, and selecting the most appropriate natural wetland site may be iterative (Figure 22-1). An initial estimate of the required site area is made based on an estimate of pretreatment, flow rate, and final effluent criteria. If there is not enough natural wetland area available to treat anticipated flow to the necessary final effluent quality, then the designer may wish to repeat the analysis assuming a higher level of pretreatment and a smaller natural wetland treatment area. Success can be measured by the selection of a natural wetland site that achieves the desired advanced treatment with low project cost and high environmental protection. Selection of an unsuitable natural wetland site will eventually lead to problems with permit compliance, increased treatment costs, and public dissatisfaction.

NATURAL WETLAND TREATMENT AREA

The area necessary for natural wetlands treatment can be determined by the level of pretreatment, the flow rate, the final effluent quality goals, and the type of natural wetlands available.

Pollutants that commonly occur in municipal and industrial wastewaters, agricultural wastewaters, and stormwaters should be reduced to below harmful concentrations before being introduced into natural wetlands. Adequate pretreatment must provide consistent influent quality and have built-in safeguards to protect against excursions above acceptable pollutant levels. Table 22-1 summarizes pretreatment levels that have been found to protect most natural wetlands that are otherwise adapted to receiving wastewaters.

To comply with federal standards in the U.S., wastewater must be pretreated to at least secondary standards before discharge to a natural wetland. Additional pretreatment may

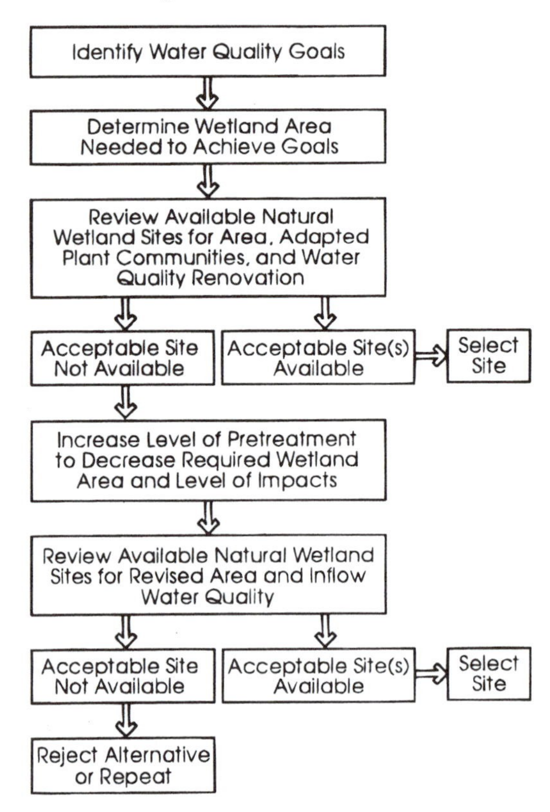

Figure 22-1 Preliminary planning process for analysis of a natural wetland treatment system alternative.

Table 22-1 Summary of Recommended Minimum Wastewater Pretreatment Levels Prior to Natural Wetland Discharge

Constituent	Suggested Pretreatment Level	Potential Harmful Effect
BOD_5	Minimum of secondary (20–30 mg/L)	Oxygen depletion; odor; mosquito production
TSS	Minimum of secondary (30–50 mg/L); mineral solids should be reduced to < 10 mg/L	Oxygen depletion; smothering of plant roots and elimination of woody plant species
NH_4-N	Nitrification is highly desirable; maximum of 5 mg/L	Oxygen depletion; un-ionized ammonia toxicity
Total N	Less than 20 mg/L	Eutrophication; selection of fast-growing species
Total P	Less than 1.0 mg/L	Eutrophication; selection of fast-growing species
TDS	Site specific; dependent on available natural wetland type	Toxic to unadapted plant and animal species
Metals and other toxins	Below chronic toxicity levels	Accumulation to toxic concentrations; food chain biomagnification for a few constituents

include further reductions in 5-day biochemical oxygen demand (BOD_5) and total suspended solid (TSS) concentrations to reduce the impact of organic solids on available dissolved oxygen within the natural wetland. For example, an influent with BOD_5 and TSS concentrations as low as 5 mg/L, loosely defined as advanced treatment standards in Florida and other states, will probably avoid creating a localized zone of impact in the natural wetland. Influent concentration of ammonia nitrogen (and easily decomposable organic nitrogen) also should be low to protect against dissolved oxygen depletion and the potential for un-ionized ammonia toxicity. Florida has a pretreatment standard of 3 mg/L for ammonia nitrogen prior to natural wetlands discharge, and South Carolina has considered a pretreatment standard of 2 mg/L. Natural wetland treatment systems that have received these levels of nitrified municipal wastewaters are showing no significant signs of dissolved oxygen depletion.

Suggested pretreatment to minimize ecological changes in natural treatment wetlands include minimizing the input of mineral solids (sand, silt, and clay) that can smother plant roots and have resulted in tree mortality at a number of locations; reducing the level of total nitrogen to less than 20 mg/L, primarily in the form of nitrate or stable organic nitrogen; reducing total phosphorus to the extent necessary to meet downstream receiving water goals on a sustainable basis; and reducing dissolved solids, trace metals, and organics to levels that will not result in chronic or acute toxicity to the biota in the natural wetland.

The equations and models described in Section 3 and summarized in Chapter 19 (Equation 19-9; Table 19-4) can be used to estimate the minimum natural wetland area required to achieve a given amount of advanced treatment. For example, Equation 19-9 estimates that 21.5 ha of wetlands would be required to reduce an annual average total nitrogen concentration from 20 to 3 mg/L on an annual average basis at an average flow of 3786 m^3/d. This is equivalent to a hydraulic loading rate of 1.76 cm/d. However, for design of a natural wetland treatment system, these estimated areas should generally be increased based on regulatory requirements to protect the biological integrity of the natural wetland ecosystem. To allow for resting periods to reestablish aerated soil conditions and propagate woody species, conservative design of natural wetland treatment systems will incorporate at least twice the area estimated for pollutant removal. This conservative factor will allow at least one half of the natural wetland area to be rested at any given time.

The Florida wastewater-to-wetlands rule establishes a maximum hydraulic application rate of 0.73 cm/d (13.8 ha/1000 m^3/d) for unaltered, woody wetlands and 2.18 cm/d (41.4 ha/1000 m^3/d) for previously impacted, woody wetlands (Schwartz, 1989). A commonly accepted hydraulic loading rate for natural wetlands during the early years of this technology development was 0.36 cm/d (27.6 ha/1000 m^3/d) (Thabaraj, 1982). However, research with AWT (advanced wastewater treatment)-quality wastewaters discharged to natural wetlands (Knight et al., 1987) indicates that much higher hydraulic loading rates (up to at least 7 cm/d) are sometimes appropriate and that excessive concentrations of ammonia nitrogen and BOD_5 are more likely to cause impacts than higher hydraulic loading rates.

The actual natural wetland area necessary to protect against unacceptable biological changes cannot be precisely estimated from existing knowledge. As a result of observations from a small number of natural wetland treatment systems, it is known that the necessary area (and the resulting hydraulic loading rate) is inversely proportional to the level of pretreatment. The definition of an unacceptable change varies from state to state and from one wetland type to another.

Any nontrivial amount of water, with or without nutrients and contaminants, will eventually cause changes to a wetland ecosystem. Changes in hydroperiod can eliminate some species. Elevated nutrient concentrations can allow opportunistic species to crowd out those adapted to nutrient-poor conditions. In many locations, trees have suffered or been eliminated and have been replaced by hardy herbaceous species such as *Typha* spp. The rates of such changes may be very slow. For instance, changes in relative abundance and areal extent are still occuring at the Houghton Lake, MI site after 17 years of wastewater additions.

The extent of change has to do with the starting condition of the wetland as well as the intended water and pollutant loadings. For example, more water and nutrients are unlikely to alter an already-eutrophic cattail wetland.

On a regional basis, some rough guidelines may be set down, based on experiences with treatment wetlands in that region. For instance, some general concepts are emerging from systems in the southeastern U.S.

A conservative recommendation would be that if pretreatment is secondary, without nitrification, the maximum hydraulic loading rate should be about 0.2 cm/d (50 ha/1000 m^3/d). For increased levels of nitrification, this hydraulic loading rate can be increased, resulting in a recommended hydraulic loading rate of 0.5 cm/d (20 ha/1000 m^3/d) for a fully nitrified, secondary effluent. If concentrations of BOD_5, TSS, phosphorus, and other constituents are reduced in pretreatment, the recommended conservative hydraulic loading rate to natural wetland treatment systems is 2.5 cm/d (4 ha/1000 m^3/d). As experience with the design and use of new natural wetland treatment systems grows, these recommended hydraulic loading rates might increase.

Degraded natural wetlands may sometimes be improved by reuse of partially treated wastewater. Hydroperiod restoration is the most frequent benefit, but added nutrients can also be utilized to increase productivity.

SUITABLE NATURAL WETLAND TYPES

As discussed in Chapter 7, many wetland plant species cannot propagate or survive in wetlands receiving increased water flows which change the hydroperiod. In most circumstances, natural wetland hydroperiod will be altered if regular or intermittent wastewater flows are added. Only a relatively small subset of the plant species that occur in natural wetlands are adapted to benefit from or tolerate these hydroperiod alterations. However, discharging wastewater to a natural wetland does not automatically change the water regime. For example, in a natural marsh with fairly constant water levels regulated by an adjacent lake or stream, additional water will not significantly increase the depth or duration of flooding as long as the configuration of the natural outlet remains unchanged. Also, as long as the outlet elevation is not altered during construction, maximum water depths in the wetland generally will not be significantly changed by adding water. This is because maximum depths usually are regulated more by the topographical configuration of the outlet than by the duration of inflows.

Very little vegetation maintenance is practiced in natural wetland treatment systems. The goal is usually to minimize vegetation changes or, where the wetland was previously altered by forestry or drainage, to restore a more natural wetland plant community. Sometimes revegetation can be accelerated by planting adapted herbaceous or tree species. As mentioned earlier, tree clearing near the inlet distribution system may prevent damage during storms. In some natural wetlands containing transitional tree species such as pines and certain hardwoods, the owner may wish to harvest these less-adapted species before the wetland is used for treatment.

Most natural wetland areas (based on the U.S. Army Corps of Engineers' definition of wetlands) are not suitable as water quality treatment systems because of the intolerance of many wetland plant communities to prolonged or deep flooding. Chapter 7 provides detailed information concerning the ability of natural wetland plant communities to withstand changes in water regime (hydroperiod and water depth). Only those plant species in the first five categories in Table 7-6 can withstand continuous inundation. Natural wetlands that are dominated by species in the remaining two categories (less than continuous inundation) will undergo significant alteration in plant dominance if they are inundated with additional water. The extent of the change in plant dominance will be proportional to the frequency, duration, and depth of inundation.

Table 22-2 lists some of the natural wetland plant communities that have been used for wastewater and stormwater recycling. These systems include marshes, scrub-shrub swamps, and forested swamps. The primary adapted species in marshes are in the genera *Typha, Scirpus,* and *Phragmites.* Many other marsh species also occur in monospecific and mixed plant communities more-or-less adapted to continuous flooding (see Chapter 7, Table 7-6). Scrub-shrub swamps dominated by buttonbush (*Cephalanthus occidentalis*), water willow (*Decodon verticillatus*), and water primrose willow (*Ludwidgea* spp.) are generally tolerant of continuous inundation. Wetland tree species generally adapted to continuous inundation are cypresses (*Taxodium* spp.), gums (*Nyssa* spp.), and willows (*Salix* spp.). Other shrub and tree species such as titi (*Cyrilla racemiflora*), red maple (*Acer rubrum*), ashes (*Fraxinus* spp.), palms (*Sabal minor*), and melaleuca (*Melaleuca quinquinerva*) have been included in natural wetland discharge systems with favorable results as long as inflow quality is high and water levels are not significantly increased. Under prolonged loadings of secondary wastewaters, these tree species have shown signs of stress and, in a number of cases, have totally succumbed to be replaced by cattails and duckweed.

A background survey of the site, including soils, hydrology, water chemistry, and ecology, should be conducted. The best technique for predicting impacts is to examine the biological changes that have occurred in similar wetlands receiving wastewaters or other hydrologic

Table 22-2 Natural Wetland Plant Communities that May Be Compatible with Final Polishing of Wastewaters

General Wetland Type	Specific Wetland Communities	Typical Dominants
Marsh	Cattail marsh	*Typha* spp.
	Mixed emergent	*Pontederia* spp.
		Sagittaria spp.
	Bulrush marsh	*Scirpus* spp.
	Wet prairie	Sedges
Scrub shrub	Buttonbush swamp	*Cephalanthus occidentalis*
	Titi swamp	*Cyrilla racemiflora*
Swamps	Cypress/gum	*Taxodium* spp., *Nyssa* spp.
	Palm	*Sabal minor*
	Melaleuca	*Melaleuca quinquinerva*

and water quality changes near the proposed project site. Such a study is not inexpensive and may prejudice the overall project economics.

OTHER POTENTIAL CONSTRAINTS

Other factors that may influence the selection of a natural wetland treatment system site include property ownership, water quality standards, adjacent receiving waters, presence of threatened or endangered plant or animal species, presence of cultural resources, and public or political opposition. An overview of these potential constraints was provided in Chapter 19. In this section, the importance of each of these constraints in the selection of a natural wetland treatment system site is described further.

Publicly owned natural wetlands generally cannot be used as natural wetland treatment systems because of the public's desire to maintain them in an unaltered condition. In some cases, publicly owned wetlands may have been altered previously through drainage or by groundwater withdrawals. In these cases, public entities may be interested in a reliable source of water for wetland restoration or enhancement, and therefore, the publicly owned wetlands might be a candidate site to receive highly pretreated wastewaters.

If natural wetlands are used for treatment, ownership or a lease arrangement may be required to protect the discharger from liability that might occur because of impacts to the wetland. The difficulty of obtaining fee-simple ownership of a natural wetland generally increases as the number of owners increases. For this reason, selection of a natural wetland treatment site might be based on the number of property owners who will be involved and their receptiveness to selling or leasing their property for wastewater discharge.

Natural wetlands are waters of the U.S. and of the state in which they are located. As such, they are classified for certain, sometimes limited, beneficial uses. These classifications can specifically exclude their use for receiving and assimilating wastewaters. A review of a wetland's water quality classification and its allowable uses is an essential part of selecting a natural wetland treatment system site. During this review, the classification of downstream contiguous waters should also be determined because, in most cases, the natural wetland treatment system will discharge to these waters. In some cases, downstream use classifications may further limit the regulatory feasibility of a natural wetland treatment alternative.

State and federal agencies require an assessment of the effects of project development on threatened or endangered species and on cultural resources. Preliminary investigations should determine the potential for the presence of protected or sensitive resources at each of the candidate sites. Detailed site investigations leading to final site selection should thoroughly document the presence of any listed species or cultural resources and assess the likelihood for any detrimental impacts resulting from the proposed project. If the project will harm protected species or degrade cultural resources, the project design and permitting must be modified to mitigate or avoid these impacts.

Public perception and interest concerning a proposed natural wetland treatment system may be important during the permitting process. Some natural wetlands or adjacent streams, rivers, or lakes may have a public constituency that will oppose any plans that appear to alter the resource's perceived value. These constituencies should be considered during site selection. If project impacts will be minimal or may actually enhance the natural wetland resource value, then support of these special interest groups may be solicited to facilitate project permitting. If project impacts might be unacceptable to a special interest group, the discharger may wish to select an alternative natural wetland site or provide mitigation or enhancement of wetland values to offset unavoidable project impacts. An important part of almost every natural wetland treatment system project is a public information program that informs the public of the technology's potential benefits and undesirable impacts and presents a specific site for implementation.

SYSTEM CONFIGURATION

Natural wetlands occur in many shapes and sizes, but unlike constructed wetlands, few have right angles or straight sides. The topography of natural wetlands may vary from highly regular with mostly flat surfaces to highly irregular with undulating or eroded surfaces. One of the challenges in engineering a natural wetland treatment system is designing a water conveyance and distribution system that uses a large portion of the wetland surface area, but is cost-effective and environmentally protective.

Important design aspects include the layout of the distribution system, the provision of an alternate discharge site, and the control of internal flow and water elevation. The influent distribution system must be designed to transport minimum, average, and peak flows to discharge points in the natural wetland in a manner that results in natural sheetflow to most of the wetland area. Alternative discharge sites allow flows to be rotated as may be required for vegetation management. In some cases, internal levees, weirs, and channels may enable the natural wetland to be used for treatment without causing significant environmental impacts. Each of these aspects of a natural wetland treatment system is described here.

INFLUENT DISTRIBUTION

Effective flow distribution in natural wetland treatment systems enhances pollutant assimilation and minimizes the impacts that can result from a localized, intensive discharge. Because natural wetlands typically have more irregular gradients and slopes than constructed wetlands, effective flow distribution is more of a challenge. Point discharges can be used to initiate good distribution only in natural wetlands that have high length-to-width ratios or in systems that are naturally ponded. A linear distribution system is necessary in most natural wetland treatment systems. This linear system frequently uses a gated pipe or a level spreader swale. In some cases, the inlet distribution system can be located outside of the wetland in the adjacent uplands, thus reducing the impacts to the wetland. This approach is only viable if the wetland-upland interface is relatively steep and narrow. Otherwise, the distribution system will be located in a transition zone that may not be adapted to increased water levels.

Inlet distribution system design should try to minimize clearing and filling impacts to natural wetlands. Typically, in warm climates or for seasonal discharges, a boardwalk lifts the distribution pipe off the wetland substrate, thereby reducing the wetland area that is "filled" and providing a level platform for the distribution pipe. Figure 22-2 illustrates some natural wetland distribution systems that use a boardwalk with gated piping.

Erosion is a concern in forested natural wetlands because roots may lose their ability to support the trees. In addition, mineral sediments could cut off oxygen supply to tree roots and cause tree stress or mortality. Inlet systems should promote low velocities and possibly use rip-rap to prevent soil erosion. Trees near the distribution system should be cleared to reduce the damage that might occur by tree fall. In addition, many wetland trees cannot withstand a continuous water spray on their bark even though their roots are in flooded soil. Thus, inlet distribution systems should not spray water onto trees within the natural wetland.

Pretreated wastewater can be conveyed to a natural wetland treatment system with a gravity or pressurized pipeline or with a canal or ditch. In a few hybrid systems, influent to the natural wetland has been conveyed by overland flow from an adjacent upland or constructed wetland. The pipe or channel size necessary depends on minimum, average, and peak flows, pipe material, wastewater characteristics, and site conditions.

Distribution system layout and design is based on a detailed knowledge of site topography. A topographical survey with a 0.3- to 0.6-m (1- to 2-ft) contour interval is normally desired for design. If a detailed survey is not feasible because a closed canopy exists or because of inaccessible conditions, elevational transects should be made across the natural wetland and

Figure 22-2 Two natural wetland treatment system inlet distribution pipes and boardwalks.

information should be collected concerning the observed movement of surface waters in the wetland. The distribution system is laid out along the upgradient edge of the wetland over a distance necessary to initiate sheetflow along the entire width of the wetland. Effective flow distribution may be cost prohibitive in sloped wetlands such as narrow floodplains and bowl-shaped basins.

The precise location of the influent distribution system in a natural wetland treatment system should be determined from a detailed site evaluation of adapted plant communities. If an abrupt transition occurs between adjacent upland plant communities and the adapted wetland plant community, the distribution system can be located on the upland-wetland edge or just inside the wetland margin. If a transitional zone is present because of a more gradual variation in elevation and water depths, the distribution system should be located in the interior of the wetland, some distance from the community that may not be adapted to continuous flooding. This wetland area between the distribution pipe and the unadapted wetland transitional zone serves as a buffer against the mounded water levels that generally extend for many meters upgradient from the discharge point.

Spreader swales are effective for smaller natural wetland treatment systems, but are generally not practical for larger systems. Single-point discharge systems are practical for bowl-shaped and narrow wetlands where the outflow elevation can be set above the normal wetland bottom elevation, resulting in impounded conditions. Gated distribution pipes, either elevated on boardwalks or anchored to the wetland substrate on pipe support structures, are the only effective alternative for distributing inflow over a long, linear discharge zone. These gated pipe systems are flexible and can be designed to use a significant fraction of the wetland area.

BASIN CONFIGURATION AND ALTERNATE DISCHARGE LOCATIONS

Natural wetlands are found in an almost infinite variety of shapes, depths, and slopes. Although most shapes can be used to treat wastewater, some basin configurations are not effective for pollutant assimilation. Figure 22-3 illustrates three marsh types that generally are amenable to wastewater treatment. Narrow natural wetland zones along lakes or streams are relatively ineffective for water quality treatment because it is difficult to eliminate short circuiting through the wetlands. The only way to use a narrow, linear floodplain wetlands is to construct long distribution systems with very low hydraulic loading rates.

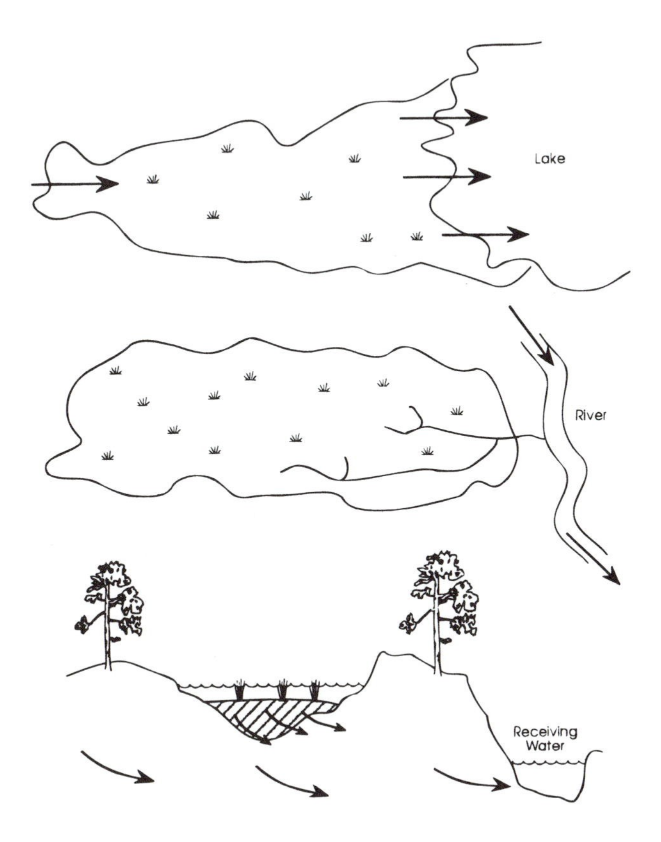

Figure 22-3 Natural marsh wetland basin configurations suitable for engineered water quality treatment.

Bowl-shaped wetlands can be integrated effectively into an effluent assimilation system. As discussed earlier, because natural wetlands have irregular ground surfaces, short circuiting is more likely to occur than in constructed wetlands. Therefore, the designer must use conservative loading rates and higher levels of pretreatment before discharge to natural wetlands. Types of natural wetlands with shapes favorable for water treatment include isolated and flow-through marshes, prairie potholes, cypress domes and strands, Carolina bays, and floodplain oxbows and sloughs (Figure 22-4).

Most rooted, emergent wetland plant species prefer some relatively dry period for growth and propagation. Use of a natural wetland for municipal wastewater treatment does not allow periodic drydowns unless alternative discharge sites are available. Although alternate discharge sites may not be needed more than once a year, they should be included to maintain system stability. However, such dry periods promote sediment oxidation and pollutant release.

Ideally, a natural wetland treatment system will have at least two separate wetland sites that are not hydraulically connected. Each site should be large enough to receive the entire average influent loading for at least 1 year. The two wetland sites might be adjacent areas that drain into a common receiving water or they may be two arms of the same contiguous wetland area that connect at a downgradient location.

While two alternative discharge sites are the minimum number for a conservative design, three or more sites are preferred. With two sites of equal size, one half of the effective treatment area is unavailable whenever one site is rested, resulting in twice the average normal hydraulic loading to the active discharge site. However, if three or more sites exist, the fraction of the entire treatment area that is lost for shutdown of one site decreases from one third to less, depending on how many sites are available. It should be noted that

Cypress dome

Oxbow slough

Carolina bay

Figure 22-4 Natural wetland landforms that are suitable for engineered water quality treatment in the southern U.S. environment.

this redundancy has rarely been implemented because of the increased expense of extra distribution works.

FLOW CONTROL STRUCTURES

In most cases, altering the existing natural wetland outlet should be avoided during project planning and construction. Maintaining the outlet will minimize hydroperiod alterations and reduce direct impacts on the existing natural wetland ecosystem. If the natural outlet is constrained by a culvert or other modification, the engineer must consider increasing the size of this conveyance to handle the increased water flows.

In general, any natural or altered wetland outlet can be used to monitor flow and water quality by using the same methods discussed earlier for constructed wetlands. Precise flow measurement might not be possible. However, accurate inflow measurements augmented by estimated outflows are environmentally more preferable than construction of berms and weirs to obtain precise data. Natural wetlands with multiple outflow points will be more difficult to monitor than those with a single outflow and will therefore be more expensive for treatment. Dams should generally not be constructed in natural wetland outlets unless the projected vegetative impacts in the wetland are within acceptable limits.

Internal berms and fill in natural wetlands are not recommended because of potential harm to existing wetland functions. If natural wetlands were previously degraded by erosion or if drainage and redistribution of flows is desired, internal berms can be used to direct wastewater flows. Long, low berms parallel to the direction of water flow can help maintain sheetflow through natural wetland areas that are otherwise criss-crossed by channels or streams. These low berms do not need to be maintained and can therefore be designed with

Figure 22-5 Example of an outlet water level and flow control weir from natural wetland treatment system.

a low crest elevation (typically only 20 to 30 cm above the average wetland surface elevation). They can be laid out along existing wetland contours to create a terraced effect, or they can be routed perpendicular to the natural contours to minimize short circuiting along the normal flow path.

In some situations, natural wetland areas can be channelized to prevent excessive water levels in areas of unadapted wetland vegetation. For example, if a natural drainage divide separates flows in a wetland during low water levels, a small channel or canal could be constructed to move effluent from one area to the other to prevent buildup of water levels that might harm intermittently flooded plant communities on higher ground.

The ability to control water levels without modifying natural wetland treatment systems is generally minimal. Reliance on natural outlet conditions in a natural wetland treatment system will help to minimize biological changes resulting from changes to the natural water regime. On the other hand, if site topography limits the effective treatment fraction of the natural wetland area or if seasonal water level changes are desired to increase or reduce hydraulic residence time and flow velocities, then outlet water level and flow control might be an advantage. Outlet control can be provided only where a weir or other device can be installed in a natural outlet constriction or where a berm can be constructed across the wetland outlet area. In some cases, beavers have provided natural berms which can be enhanced with a hardened weir structure. The problem with beavers is that they want to shut off the flow through the weir, too. In other cases, a Section 404 permit will be required to construct a berm.

If necessary, an outlet berm should be designed to allow access to and maintenance of water level control and monitoring structures. The berm crest should be at least 0.5 m above the normal high water mark and at least 3 to 4 m wide for light vehicle access, with a minimum of 3:1 slopes to allow for occasional clearing or mowing of vegetation. Shrubs and trees should not be allowed to establish on berms to prevent root penetration from causing piping failure. The outlet berm must include one or more weirs to allow for water level control and flow measurement. Figure 22-5 shows one example of an outlet weir design for a natural wetland treatment system. Weir design generally relies on a broad-crested configuration to allow for highly variable flows without wide variations in water levels. A smaller V-notch can be incorporated in the broad-crested weir to provide more accurate flow monitoring at lower flows.

SUMMARY OF NATURAL WETLAND TREATMENT SYSTEM DESIGN CONSIDERATIONS

Some natural wetland ecosystems have the potential to buffer recreational surface waters from direct discharges of wastewaters and stormwaters. These wetlands can assimilate some residual pollutants in pretreated wastewaters and recycle carbon and nutrients in natural food chains to enhance plant community production and wildlife populations. Successful design of natural wetland treatment systems requires careful site selection, adequate pretreatment, conservative hydraulic loadings, and minimal reliance on structural modifications.

CHAPTER 23

Ancillary Benefits of Wetland Treatment Systems

INTRODUCTION

The physical, chemical, and biological functions that wetlands perform have economic or aesthetic value to society and provide support to plant and animal populations (Gosselink and Turner, 1978; Sather and Smith, 1984; Mitsch and Gosselink, 1993; Hemond and Benoit, 1988; Brinson, 1988; Kusler and Kentula, 1990; and Erwin, 1990a). However, because the term "wetland" includes an area of ecosystems with diverse abiotic and biotic factors, all wetlands are not created equal.

Published summaries of wetland functions show a range of quantitative functional attributes, even for a single wetland plant community type (Mitsch and Gosselink, 1993). For example, Figure 23-1 shows the net primary production (NPP) varies from as low as 50 $g/m^2/yr$ in arctic tundra to 3500 $g/m^2/yr$ in freshwater marshes (Nixon and Lee, 1986). Wetland total nitrogen assimilation rates vary from less than 0.08 to more than 90 $g/m^2/yr$ (Knight, 1986; Nixon and Lee, 1986).

Despite the range of functional attributes, approximate functional levels for individual wetlands can be estimated based on the wetland structure and the forcing functions affecting the wetland (Brinson, 1988; Erwin, 1990a). In the previous example, freshwater marshes display a wide range of NPP rates in North America. Therefore, a knowledge of a specific marsh's latitude, hydrology, water quality, vegetation, and soil type can greatly increase the accuracy of an annual NPP estimate.

The relationships between a wetland's structure and function can be used to predict and enhance the functions of wetlands designed to control pollutants. When wetlands are used to reduce pollutant concentrations and peak stormwater flows, ancillary benefits can be achieved through thoughtful site selection and design. This chapter summarizes the ancillary benefits of wetlands used for pollution control and recommends design features to optimize these benefits. Research and manuals on wetland evaluation techniques (Golet, 1978; Greeson et al., 1978; Richardson, 1981; Kusler and Riexinger, 1986) are used to compare the additional benefits of treatment wetlands with natural wetland functions.

The primary objectives of most wetland pollution control projects include (1) water quality enhancement through assimilation and transformation of sediments, nutrients, and toxic chemicals and (2) water storage and flood attenuation for stormwater projects. Secondary benefits that can be incorporated in wetland treatment designs include (1) photosynthetic production; (2) secondary production of fauna; (3) food chain and habitat diversity; (4) export to adjacent ecosystems; and (5) aesthetic, recreational, commercial, and educational human uses. The potential for including each of these benefits in treatment wetlands are described in this chapter.

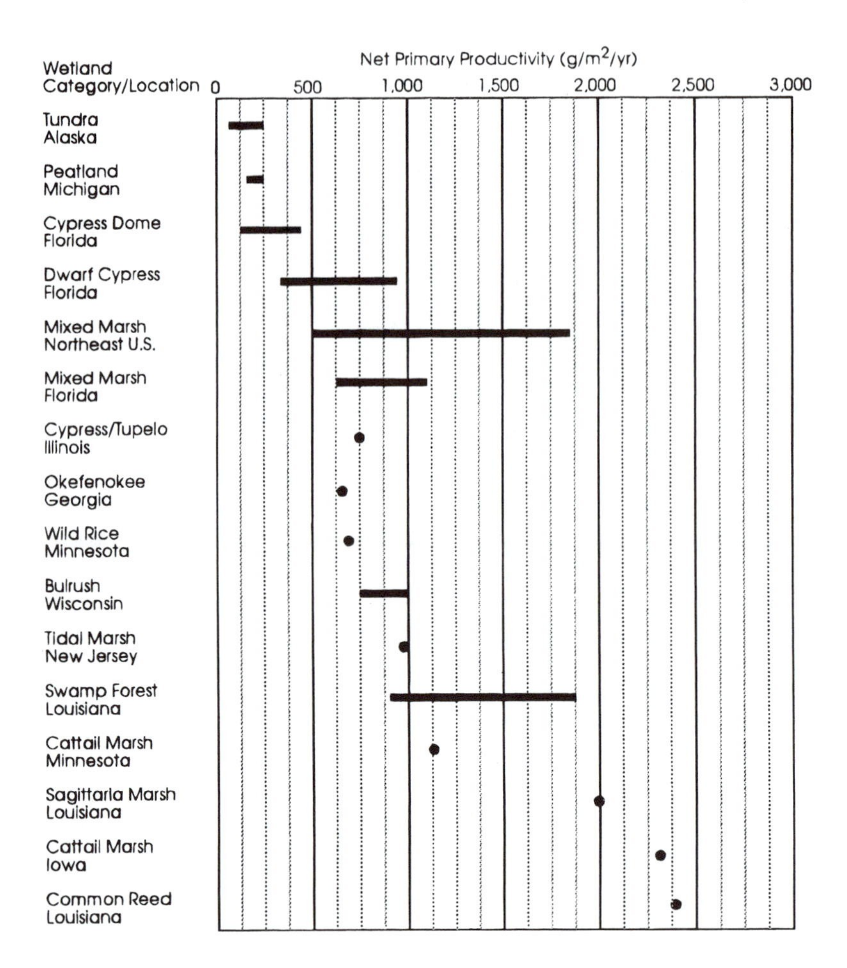

Figure 23-1 Annual net primary productivity of a variety of natural wetlands in North America (Nixon and Lee, 1988).

PRIMARY PRODUCTION AND FOOD CHAIN SUPPORT

All wetlands include photosynthetic production by vascular and nonvascular plants (see Chapter 7). However, because of different environmental forcing functions in individual wetland areas, the magnitude of primary production varies greatly among wetland types and even for a single wetland habitat. Wetlands generally have higher NPP than adjacent uplands (900 to 2700 g/m^2/yr in wetlands compared to about 500 g/m^2/yr in grasslands) because water and nutrients enter the wetlands from the adjacent upland systems (Richardson, 1979).

The project's goals will determine whether or not the primary production of wetlands constructed for pollution control should be enhanced. In constructed wetlands, factors limiting primary production include light, water, macronutrients, and micronutrients. When project goals include generating organic matter as the basis of a food chain leading to domesticated or wild animal populations, then system design and operational control can be used to supply needed limiting factors. For example, in emergent marshes, light is most likely to limit algal production. If reduction of algal suspended solids is a goal, then wetlands can be designed with a densely vegetated emergent zone at the downstream end of the wetland treatment system. If algal productivity is desired to enhance an aquatic food chain (such as fish or shellfish), then deeper, open water areas should be included in design.

In many cases, the addition of nutrients in municipal wastewater to a surface-flow (SF) wetland will greatly stimulate NPP of emergent wetland plants. Since the total insolation available for macrophytic plant growth is roughly the same from year to year, total NPP is generally proportional to nitrogen or phosphorus concentrations. Even in SF wetlands receiving AWT municipal wastewaters, nitrogen and phosphorus are adequate for high NPP and resulting high wildlife production.

In wetlands constructed for stormwater treatment, water may be the factor limiting primary production because of seasonal water supply. To maintain high primary production year-round, an alternate water source should be provided. Highest NPP usually is measured in shallow (<0.3 m), regularly flooded, emergent marshes (Brown et al., 1979). In such shallow or flowing systems, high primary production may result from the availability of water combined with higher sediment dissolved oxygen levels (Gosselink and Turner, 1978). More natural, fluctuating water levels generally result in lower NPP.

Constructed wetlands designed to simulate natural wetland hydroperiods need to include water level control structures to prevent damaging floods and to replicate the slow bleed down of water levels following storms (Livingston, 1989). Prolonged high water in a constructed wetland will result in a rapid change from emergent, shallow water vegetation to an aquatic system dominated by phytoplankton, filamentous algae, or floating aquatics (Guntenspergen et al., 1989). If these prolonged flood events occur several times each year, with dry spells interspersed, the plant community will be highly stressed and is likely to have low annual NPP.

WILDLIFE MANAGEMENT

In some cases, secondary benefits of wetlands constructed for pollution control include management of "domesticated" species. Wengrzynek and Terrell (1990) have developed a generic constructed wetland design for agriculture runoff control that incorporates bait fish or freshwater mussel production as an ancillary benefit. In addition to aquaculture, treatment wetlands dominated by grasses could be productive areas for livestock grazing. Generally, however, greater operational control is required to achieve ancillary benefits such as production of crayfish, fish, or other aquaculture. In some cases, commercial production of wetland aquaculture species will be at odds with goals related to water pollution control.

To many conservationists, the most exciting ancillary benefit of treatment wetlands is wildlife enhancement. In fact, constructed wetlands sometimes provide expanded habitat for threatened or endangered species. Wetland and impoundment design for wildlife enhancement is relatively well known (Weller, 1978, 1990, Smith et al., 1989; Payne, 1992). For example, vegetation and water levels can be managed to provide a range of niches for wildlife. A diversity of plant species and growth forms and varying plant and seed maturity dates create a greater variety of habitats within the wetland. In terms of water, fluctuating water levels create more ecological niches and result in higher wildlife species diversity. Because many wildlife species are attracted to wetlands with perennial water, less frequently flooded areas generally will have lower populations of wetland-dependent wildlife.

Water quality is important to wildlife production because of its influence on primary production. Constructed wetlands receiving nutrient-rich waters generally have high wildlife populations. The abundance of nutrients allows high NPP, and then nutrients move through the food chain from detritus to invertebrates and small fish, reptiles, amphibians, and birds. As long as intermediate consumers are not restricted from colonizing and increasing their populations, they will develop a food base for the more visible avifauna typical of constructed wetlands.

When inflow waters have been extensively pretreated or come from nonpoint sources in less-developed watersheds, the waters carry fewer nutrients. In those cases, primary productivity and therefore food chain support will be lower than in wetlands receiving nutrient-rich discharges. Although wetland secondary production is typically higher at higher nutrient concentrations, faunal diversity may be higher or lower as nutrient concentrations increase. Wetland faunal diversity appears to be related more to niche diversity than to plant diversity, as evidenced by high macroarthropod diversities in near monocultures of *Spartina alterniflora* salt marsh (McMahan et al., 1972) and in *Taxodium ascendus* cypress domes (McMahan and Davis, 1984).

In addition to nutrient levels, physical features of a wetland influence faunal diversity and abundance. For example, waterfowl populations are enhanced if open water areas are interspersed with deep emergent marsh and upland islands. An approximate 1:1 ratio of wetland area devoted to marsh and to open water will provide maximum habitat for a variety of waterfowl species (Weller, 1978). Generally, wetlands adjacent to upland habitats will have greater floral and faunal diversity in proportion to the amount of edge between adjacent ecosystems.

Wading birds, however, prefer a different habitat mix. These species require shallow, sparsely vegetated, littoral areas or perching substrates adjacent to open water areas. To benefit wading bird populations, constructed wetland treatment systems can be designed to provide a broad shelf of emergent marsh with water depth less than 20 to 30 cm. Deep, open water areas next to a shallow marsh provide additional foraging habitat.

Open water areas and the transitional ecotones among marsh, open water and adjacent uplands help promote growth of herptile and fish populations, which in turn provide food for wading and diving birds. Including diverse fish species during system startup or through natural immigration will lengthen the consumer food chain and potentially provide support for raptors such as ospreys, hawks, eagles, and kites.

Depending on regional occurrence and available habitats, other bird varieties will colonize constructed wetlands, too. For example, if living or dead trees are included in the wetland, they will serve as perching and possible nesting sites (Hair et al., 1978). Nesting boxes might attract wood ducks and owls. Upland islands surrounded by open water provide protection for ground-nesting bird species such as waterfowl.

The food and habitat available at a constructed wetland also might support a variety of mammals. Small mammals typically develop large populations on upland areas adjacent to and within constructed wetlands, where they provide forage for raptors and large wading birds. Larger mammals also can be included to add diversity to the system. In some cases, large mammals could be harvested for market (for example, animal skins). Nutria, muskrats, and beaver generally colonize constructed wetlands only when perennial water is available. When present, these large mammals may play important roles in the wetland ecosystem. For example, their ability to rapidly reduce biomass for lodges and food helps maintain patches of high net productivity and early successional marsh areas (Weller, 1978; Hair et al., 1978).

HUMAN USES

Humans appreciate wetlands for their commercial values (plant harvesting, livestock grazing, hunting, and aquaculture) and nonconsumptive values (aesthetics, recreation, and research) (Reimold and Hardisky, 1978; Nash, 1978; Sather and Smith, 1984; and Smardon, 1988).

Consumptive uses of wetlands such as hunting and trapping are more easily quantified than nonconsumptive functions. For example, over 10 million ducks and $35 million worth of furs are harvested from wetlands annually (Chabreck, 1978). Constructed treatment wetlands at Orlando, FL and Incline Village, NV are used for waterfowl hunting on a seasonal basis.

Nonconsumptive uses of wetlands constructed primarily for water quality treatment include recreation, nature study, aesthetics, and education. Increasingly, designs of treatment wetlands have incorporated attractive and informative park-like areas. For instance, stormwater treatment wetlands in urban settings such as Greenwood Park in Orlando, FL and Coyote Hills east of San Francisco Bay, CA are used frequently for field trips and other educational purposes (Figure 23-2). Wetlands constructed for wastewater treatment in Arcata, CA; Hillsboro, OR, and Orlando, FL are vital recreational areas that offer jogging and bird watching. These human uses of wetlands, including the satisfaction of having a wetland and wildlife at the edge of town, are perhaps the most important factors behind public support of protection and enhancement of existing wetlands.

DESIGN FOR ANCILLARY BENEFITS

The previous sections described the secondary features of wetland treatment systems. This section focuses on design considerations to enhance ancillary benefits while optimizing the primary functions of pollutant removal and flood attenuation.

Arcata

Ironbridge

Carolina Bays

Greenwood

Figure 23-2 Wetland treatment systems designed for ancillary benefits.

Wetland design considerations include siting, cell size and configuration, water flow and depth control, planting, and species stocking. Because this book assumes that the treatment wetland's primary function is pollution control for wastewater, the water source itself is not an ancillary design decision. However, the level of pretreatment is important to determine potential for secondary benefits.

WETLAND SITING

Siting wetlands to treat municipal and industrial wastewater is relatively flexible because the effluent can be pumped to the wetland. However, wetlands to control stormwater must be sited close to stormwater sources or further downstream in a watershed, intercepting a tributary. The location of a stormwater treatment wetland will determine the quantity and timing of the influent, which will affect the ancillary functions of the wetland. Generally, wetlands located in headwater areas will receive more irregular and less dependable inflows, resulting in prolonged dry conditions, unless soils are impermeable or groundwater is normally high. This relative lack of flooding will limit ancillary benefits such as primary and secondary production. In those cases, maintaining a healthy stand of wetland-dependent vegetation may be difficult, and upland or transitional species may eventually predominate. This type of system certainly will have some upland values and may support faunal assemblages seasonally. However, overall production of wetland-dependent species likely will be lower than in a perennially flooded wetland.

Siting the constructed stormwater treatment wetland further downstream in the watershed may yield a different constraint: too much water during stormwater runoff periods. To control high stormwater volume, the downstream wetland can be designed to be offline (out of the main flow path) so that it captures only some flood flow, preventing washout of vegetation and berms. A series of offline constructed wetlands, each capturing a portion of the storm flows, can be used to deal with high volumes. Wetlands located downstream in a watershed are more likely than wetlands at headwaters to have perennial water because of more constant base flows and higher groundwater levels. In turn, downstream stormwater treatment wetlands generally will have more wildlife and more food chain support.

Siting of wetland treatment systems to optimize ancillary benefits may also consider the advantages of locations with adjacent donor wetlands (sources of plants and wildlife to colonize the area), with adjacent undeveloped uplands (habitat diversity), or with proximity to humans (aesthetics). These issues depend on project-specific goals.

CELL SIZE AND CONFIGURATION

As discussed previously in Chapters 19 and 20, wetland cell size depends primarily on water quality treatment needs and cost considerations. Because large cells require less berm construction per unit area and fewer inlet and outlet structures, per area project costs are lower. For example, a 100-ha constructed wetland may cost about $10,000 per hectare to construct, while a smaller constructed wetland may cost about $50,000 per hectare. Although cell size may influence the use of wetlands by larger wildlife, it has minimal affect on plant productivity or secondary production of most wetland animals (Sather and Smith, 1984). A higher berm-to-cell area ratio, typical of smaller wetland cells, may result in increased beneficial edge effects. As long as berms are infrequently mowed or visited, the larger edge area provides nesting and feeding habitat for more mammal and bird species.

Islands surrounded by marsh or open water provide excellent habitat for nesting waterfowl. In many wetlands, islands with trees are the preferred nesting habitat for wading bird rookeries. Nesting islands for waterfowl should be about 0.6 m above normal high water, while higher and lower islands may also be valuable for other species for feeding, resting, or nesting.

Open water areas improve the water quality treatment potential of constructed wetlands (Knight and Iverson, 1990) and enhance their ancillary benefits for wildlife. Mallard duck production is maximum in wetlands with about even areas of marsh and open water (Ball and Nudds, 1989). Open water areas can be created by excavating at least 1.5 m below normal water level. Deeper excavations can provide greater hydraulic residence times and fisheries habitat. To prevent hydraulic short circuiting, open water areas should not be connected along the flow path, but rather interspersed with densely vegetated shallow marsh habitat (about 0.3 m average water depth or less). In general terms, areas will be larger when ancillary benefits are desired. Ponds, islands, and edge amenities are not required for treatment.

Cell number and configuration in series or in parallel are major considerations when determining treatment capability and operational flexibility. These design considerations primarily affect ancillary wetland benefits because of their importance in water flow and depth control.

WATER FLOW AND DEPTH CONTROL

Water depth and flow rate are important factors affecting dissolved oxygen in wetlands. Higher flow rates resulting from shallow water tend to provide higher dissolved oxygen concentrations in marshes because of atmospheric reaeration. These higher dissolved oxygen levels generally result in higher secondary production of aquatic invertebrates and vertebrates, thereby increasing ancillary wetland benefits. Although deeper water in a marsh may increase hydraulic residence time, this longer reaction time does not always result in enhanced water quality treatment (oxidation of organic matter and ammonia) because of the resulting reduction of dissolved oxygen concentrations.

Water depth is one of the main factors that affects wetland plant growth. High water levels will stress growth of emergent macrophytes and encourage dominance by floating or submerged plants or algae. The hydrological tolerance range and optimum hydroperiod of any desired vegetation type should be known and closely adhered to when water level control structures are designed. Ideal design allows water levels to be varied from zero (drained) to the maximum depth tolerance of desired wetland plant communities. Stop logs or weir plates should seal against leaks to help maintain water levels during periods of limited inflows. Multiple inlet and outlet weirs between adjacent cells allow greatest hydroperiod control flexibility.

VEGETATION PLANTING

The plant species selected for a constructed wetland will greatly influence ancillary benefits such as primary and secondary productivity. The selection of improper plant species will result in low productivity, and a lengthy adaptive period may be necessary until available plant species, either planted or natural volunteers, rearrange themselves according to hydrologic and water quality factors. High plant diversity can be initially achieved by assisting existing soil seed banks to colonize areas that are graded and shallowly flooded or by spreading muck and propagules from a donor wetland area (Gilbert et al., 1981). Wetland vegetation establishment is most rapid when plants are closely spaced, less than 1 m on centers, and planted during the growing season (Lewis and Bunce, 1980; Broome, 1990). Wildlife use all marsh plant species directly for food or shelter or indirectly through the detritus food chain, so expensive management to exclude "noxious species" or to select favored species may lower overall wildlife use in favor of optimizing specific wildlife species. Burning marshes may be good management for waterfowl species, but may be poor management for fish, small mammals, or other bird species. Burning is ruinous for treatment.

WILDLIFE STOCKING

Stocking constructed wetlands with mosquito fish (*Gambusia affinis*) has been found to control mosquitoes as long as deeper water refuge areas, periodically free of floating vegetation, are available as perennial habitat. Mosquito fish, in turn, are an important forage fish for wildlife. Other forage fish that can be stocked easily, such as shiners, minnows, shad, and sunfish, contribute to a potentially long food chain of sport fish, reptiles, wading birds, waterfowl, and raptors.

Other parts of the food chain may be stocked or allowed to naturally immigrate to the constructed wetland. Significant fur bearer populations (otter, mink, muskrat, nutria) can be supported in highly productive constructed wetlands. As with many wetland-dependent birds, a mixture of open water and marsh habitat is essential for enhancement of these mammal species.

Stocking of constructed wetlands located some distance from existing wetland habitat may be important to quickly establish ancillary benefits for wildlife and aesthetic uses.

INFLOW PRETREATMENT

In addition to wetland hydroperiod, water quality is a key determinant of a wetland's form and function. Primary water quality characteristics that affect wetland plant communities include nutrients (especially nitrogen and phosphorus), suspended solids, salts, pH, and temperature. Except for nutrient concentrations, these same water quality characteristics also influence faunal populations. Inflows of biodegradable solids and the ammonia form of nitrogen can affect wetland flora and fauna indirectly by their impact on dissolved oxygen concentrations.

If high loads of suspended solids are released in the constructed wetland, the solids can smother plant growth in inflow areas (Kuenzler, 1990). Generally, solids loads can be controlled by conventional pretreatment. In stormwater systems, the problem can be minimized with a pretreatment grassed swale, a high maintenance pretreatment wetland cell, or a pond prior to the habitat wetland (Livingston, 1989). If the pretreatment area traps the mineral suspended solids (clays, silt, and sand), the wetland treatment system may need to be maintained less frequently.

Swales and ponds also can reduce nutrient levels, although nutrient reduction may not be desired in constructed wetlands designed to enhance wildlife. As noted earlier, higher nutrient levels generally result in higher primary production of wetland plants, which then support increased wildlife populations.

HUMAN ACCESS

Boardwalks and blinds can greatly enhance the recreation and scientific research benefits of a constructed wetland. Although public access to a created wetland might disturb wildlife populations, disturbances can be minimized with controlled access to certain areas and with design features such as islands for roosting and nesting.

Figure 23-3 shows some of the features that might be incorporated in a constructed wetland to control water pollution and to provide secondary wildlife and recreation benefits. In this conceptual plan, the wetlands are sized and configured to meet specific water quality treatment goals, and operation is based on meeting effluent criteria. However, the wetland also is designed to take advantage of the potential for wildlife habitat creation by incorporating deeper, open water areas in the marsh and providing habitat islands as a refuge from human access and predators. Boardwalks and an interpretive center facilitate nature study and outdoor recreation. Designing constructed wetland treatment systems for a multitude of ancillary

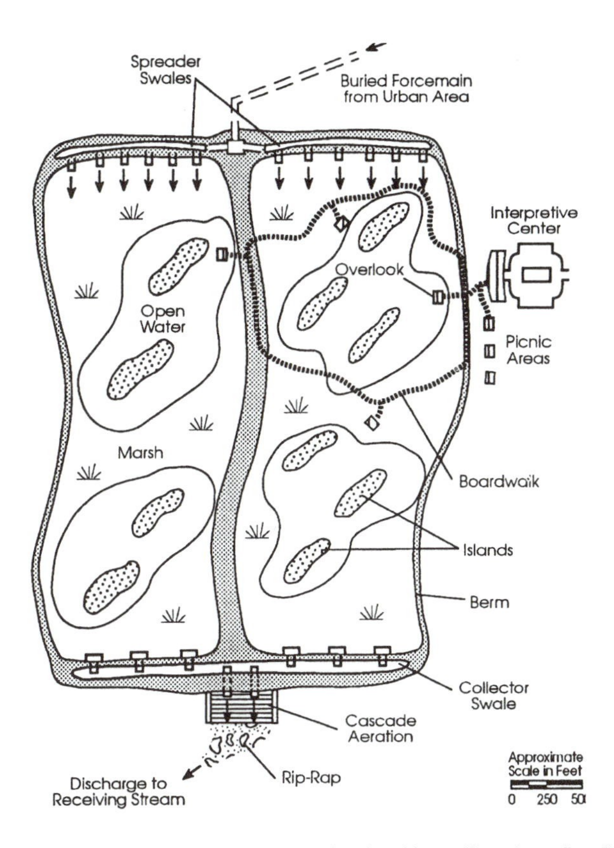

Figure 23-3 Conceptual plan for treatment wetlands with ancillary benefits. (Knight, 1989)

benefits is currently under way at a number of project sites. The use of treatment wetlands to provide mitigation for wetland-dependent construction projects is potentially a win-win situation for society. Municipal wastewaters can provide a dependable water source resulting in high wildlife values in mitigation wetlands, and are likely to be used to establish and maintain large wetland mitigation bank projects in the future.

SUMMARY

Wetlands for water quality improvement are being constructed throughout North America at an accelerating rate. However, many wetland designs do not incorporate ancillary benefits to the extent possible. With thoughtful design, constructed wetlands can provide benefits beyond effective water treatment, such as wildlife enhancement and recreational opportunities. In fact, with the decline of the total area of natural wetlands, constructed treatment wetlands are a viable and cost-efficient way to compensate for the loss of productive wetland habitat.

Pretreated wastewater and stormwater can nourish hydrologically altered natural wetlands or create natural wetland functions in constructed marshes. Wastewater constituents can be controlled to enhance a wetland's ancillary benefits. For example, nutrients promote plant growth at the bottom of the food chain, which then supports other wildlife species. By considering certain design features, such as cell configuration, hydroperiod control, vegetation and wildlife stocking, and human access, wetlands can be designed to meet water treatment goals as well as provide additional benefits.

Section 5
Wetland Treatment System Establishment, Operation, and Maintenance

CHAPTER 24

Wetland Treatment System Establishment

INTRODUCTION

At one extreme, wetland construction can be an enjoyable experience of creation; at the other extreme, it can be a living hell of frustration and financial ruin. Almost any consultant or contractor who has built a wetland will tell you, there is no substitute for actual experience. This statement is true as long as the consultant or contractor meets two conditions: (1) he or she learns from their mistakes, and (2) they get a second chance. The purpose of this chapter is to provide information that will increase chances for successful wetland construction the first time around and to help with understanding and overcoming difficulties that may arise during construction and startup of wetland treatment systems.

Wetland treatment system establishment can be broken into two topics: (1) site civil construction and (2) vegetation establishment. Site civil construction is a broad discipline of engineering and has been refined through centuries of experience. This chapter does not attempt to discuss all aspects of this engineering practice, but rather describes the highlights of site civil construction that are most relevant to wetland treatment system construction. A general discussion of wetland site civil construction is provided by Tomljanovich and Perez (1989).

Establishment of wetland vegetation is currently more of an art than a science and is frequently outside of the experience of many site civil contractors. Therefore, this important aspect of site establishment is generally subcontracted to specialists, frequently nursery owner/operators or wetland scientists. Since vegetation establishment is the most unfamiliar aspect of wetland construction for many individuals, this chapter concentrates on this specialty service.

WETLAND CONSTRUCTION

Site civil construction includes preconstruction activities such as land clearing and preparation, construction of wetland landform and berms, and construction or installation of water control structures. In many cases, wetland treatment system construction also includes installation of instrumentation and control systems.

SITE PREPARATION

Wetland treatment systems may be constructed in nonwetland sites or, rarely, in natural, preexisting wetlands. Construction requirements are somewhat variable at these different types of sites. Wetlands constructed at upland sites are more inclusive of all of the aspects of site preparation and form the basis for much of this discussion. Specific aspects of site preparation that are most important at natural wetland treatment system sites are also mentioned.

The greatest amount of site preparation is typically required when an upland site has existing vegetation or structures. In some cases, these vegetation and/or structures must be removed to allow construction of wetland cells. Clearly, construction costs can be reduced if existing vegetation and structures are avoided during project design. Assuming that design cannot avoid these features, the first step in construction is typically site clearing (Figure 24-1). Site clearing might include demolition and removal of rubble from existing structures or alternatively involve clearing of trees, rocks, or other obstacles present on the site. Tree clearing may involve recovery of valuable timber, followed by uprooting and toppling remaining trees and stumps with a bulldozer.

Some sites, including many natural wetland sites, might require installation of access roads prior to site construction. Access roads should be sited to avoid important environmental resources and should be integrated into the final site access design to avoid unnecessary clearing. Site access to natural wetland sites can be difficult in some cases due to mucky soils and poorly drained conditions. Curduroy or tram roads can utilize tree trunks harvested from the right-of-way to provide better equipment access to some sites.

Following site clearing for a constructed wetland project, it is important to remove any below-ground obstacles that might interfere with site grading or with the integrity of site

Figure 24-1 Preliminary construction of wetland treatment systems typically includes clearing and grubbing to remove larger plant roots.

soils or with plant growth. Grubbing refers to the removal of stumps and roots following removal of trees and shrubby growth on a site. Grubbing should also remove any unwanted buried structures such as utilities and foundations on previously developed sites.

WETLAND LANDFORM AND BERMS

Following site clearing and grubbing, basic landforming can begin (Figure 24-2). Using soil grading and hauling equipment, the contractor will generally remove and stockpile topspoils from the foot print of future wetland cells to be reused for final wetland grading. Following removal of the topsoil, the general contours of the wetland site are graded and all site features are horizontally located by standard surveying methods. Grade stakes are used to identify vertical elevation goals during site grading.

Wetland cell berms are typically constructed in lifts or successive steps of placing and compacting soils. Interior berm cores and any wire meshing needed for muskrat burrow prevention should be placed in the berm during this process. Soil moisture is critical during compaction of berms and internal cell liners, and construction may be delayed by excessively moist weather conditions.

Wetland liners are placed following rough cell grading. Site preparation is very dependent upon the type of liner, but will usually include compaction prior to installation. Following installation of liners, wetland cells should be leak tested prior to further installation of topsoils, media, or wetland vegetation.

Final site grading follows construction of berms, deep zones, and islands. Since a small amount of erosion can negate this work, in rainy climates it is advisable to establish ground cover on berms before final grading. Final grading typically consists of leveling the wetland cell bottom to optimize spreading and sheetflow of wastewaters in the completed wetland. Generally, bottom elevation is constant from side-to-side within the wetland cell and decreases slightly or is flat from the upstream to the downstream end of the wetland. Preliminary grading can be accomplished with a bulldozer and grade stakes, but final grading to elevation tolerances of 3 to 6 cm are typically accomplished by use of a laser leveling system or by hand raking. Hand raking works well in small cells by pulling a level string across the width of the cell and working downgradient. Laser leveling with laser-equipped bulldozers and scrapers works best for larger sites.

Figure 24-2 Landforming is necessary to define the spatial (horizontal and vertical) limits of wetland treatment systems.

As soon as berms have been shaped and dressed with any topsoils that are specified, they should be protected from erosion by use of straw and/or cover crops of grasses. Typically, a mixture of grass species is specified to allow establishment under variable climatic conditions. Hydroseeding with a mixture of grass seeds, shredded newsprint, and fertilizer has been found to be an effective method for rapid cover establishment on wetland berms. Wetland cell floors may also be seeded with a cover crop of grasses if final construction or wetland planting will be delayed due to the need for a seasonal planting window. All cover crops should be irrigated as needed to insure good germination and growth of plants.

Erosion onsite and offsite should be prevented as much as possible during wetland construction. Silt screens can be used to minimize offsite erosion. Silt screens typically consist of polyethylene fabric or hay bales anchored by use of wooden or metal stakes. Geotextiles and fabric cloth can also be used to protect berms and high-flow areas within the constructed wetlands from erosion.

GRAVEL/SAND MEDIA PLACEMENT

In the case of a subsurface-flow (SSF) wetland, the basin is filled with clean (washed) gravel or sand. Care must be taken to avoid mud and debris associated with the transfer vehicles. Leveling is quite important, because it will ultimately set the distance from the surface to the water (bed headspace).

PIPING AND WATER CONTROL STRUCTURES

Wetland water conveyance systems may be fabricated offsite and then brought to the site and installed or in some cases may be totally fabricated onsite. Wastewater conveyance systems typically include piping and pump stations; water control structures such as splitter boxes, weirs, and flumes; and rock collection and dispersal gabions.

Wastewater may be conveyed to a wetland treatment system site by use of a pressurized forcemain, by a gravity pipeline, or by an open channel. Construction of pump stations, pipelines, and open channels are not described in this chapter, and the reader should consult general references on hydraulic design and construction for more information. Once the wastewater is brought to the wetland site, it is typically distributed into the wetland via some distribution system consisting of pipes, channels, or coarse rock beds. Construction related to these wetland distribution systems is briefly discussed.

As described previously, effective effluent distribution is critical to optimizing wetland treatment performance. Wetland distribution system construction is important to increase the ability of the system to achieve this potential. Gravity distribution systems rely on firm foundations and level installation across the width of the wetland cells. Ample footings should be provided for pipe-support structures and for hardened lips on level-spreader swales. If the wetland design allows for outlet height adjustment, then vertical control tolerances are less critical.

Installation of water control structures is often the only significant aspect of construction in natural wetland treatment systems. When the distribution system is located within the natural wetland, extra care should be exercised to minimize environmental effects of clearing and subsequent construction. Because of the relatively small size of these distribution systems, it may be possible to specify that all construction be conducted with light equipment or only with hand tools (Figure 24-3). Almost any traffic will result in significant alteration of sediment elevations in organic wetland soils, resulting in flow paths that may lead to hydraulic short-circuiting. For this reason, pathways and access roads into natural wetlands should not cross areas that will be used for treatment.

Figure 24-3 Hand clearing of vegetation and use of light machinery in natural wetlands minimizes environmental disturbance.

Weirs and flumes should be constructed using standard site, civil construction techniques. Heavy structures require adequate footings on noncompressible soils and adequate reinforcement. If organic or high-clay soils are present in locations where footings are to be constructed, they should be removed and replaced with better geotechnical materials.

VEGETATION ESTABLISHMENT

A healthy stand of emergent macrophytic vegetation is the most important feature affecting the consistent performance of wetland treatment systems. Attaining and maintaining that vegetative cover can be a challenging obstacle for many contractors. The science of effectively establishing wetland vegetation on the first attempt is relatively simple, and yet the knowledge necessary to accomplish this goal has been laboriously relearned on dozens of projects. These trial and error attempts to successfully grow wetlands can delay project implementation and displease clients and regulators.

This state of affairs is understandable considering that, until recently, only a few people had ever tried to create wetlands in upland environments. As the popularity of constructing wetlands for water quality treatment and habitat creation has increased, the number of experienced contractors has increased. However, the number of inexperienced contractors who are awarded wetland construction jobs based on their earthwork capabilities with no consideration of their lack of other relevant qualifications has also increased. Inexperienced contractors and inexperienced engineers writing specifications and providing construction supervision have often killed their wetland plants through insufficient soil moisture, excessive water depths, inadequate soil preparation, damaged plant material, inadequate plant spacing, inappropriate planting methods, and bad timing.

This chapter summarizes the biological requirements and engineering considerations necessary to establish plants successfully on the first try. Success is highly probable as long as a contractor understands the growth requirements of the desired wetland plants and uses information available to most horticulturists (see Chapter 7 for more information on wetland plants). Other aspects of wetland vegetation maintenance are not as simple as initial plant establishment. Perpetuation of dominance by desired species, maintenance of desired plant cover density, and exclusion of undesirable plant species are all complex, problematic goals that cannot always be achieved. This chapter also provides practical considerations for vegetation maintenance through the construction/startup period and suggests preventive methods for dealing with some of these more difficult issues. A number of references are becoming available to provide guidance with wetlands construction, including Allen et al. (1989), Hammer (1992b), and Thunhorst (1993).

PLANT PROPAGULES AND SOURCES

Propagation and sale of wetland plant species has become a big business in several areas of the U.S. and in Great Britain and Europe. Wetland plant nurseries supply thousands of plants which are used to renovate altered landscapes such as phosphate and coal mining areas and to create landscaped wetlands and ponds (aquascaping) for wildlife habitat. Although most of these plants are currently being propagated for use in habitat creation projects, this market has attracted suppliers who can propagate the types and quantities of plants required for constructing large wetland treatment systems.

All of the plant species discussed in earlier chapters as applicable for use in constructed wetland treatment systems are available for purchase. The actual types of plant propagules that can be purchased are more variable. Plant propagules that are frequently used to establish constructed wetlands for wastewater treatment, shown in Figure 24-4, include seeds, bare-root seedlings (sprigs), tissue cultures on agar slants, rhizomes, greenhouse-grown potted seedlings, and field-harvested plants. Each of these plant propagule types has different qualities for wetland planting.

Bare-Root Seedlings

Seedlings are young plants that have been established from fertile seeds that were field collected or collected from nursery brood stock. Both herbaceous and woody wetland plants may be propagated as seedlings. Wetland seed germination is a highly variable process that depends on species-specific dormancy conditions and on some phenotypic differences for the same plant species gathered from different geographical areas.

To produce bare-root seedlings, nurseries plant germinated seeds in flats containing potting soil. The flats may hold dozens of individual plants that are later thinned. Bare root seedlings are harvested from these flats after a few months when they have reached a height of 20 to 50 cm. Their roots are cleaned of potting soil and the plants are wrapped together with moistened paper towels for temporary storage and eventual transportation to the planting site.

Bare-root seedlings are easily planted in the field in shallow individual holes prepared with a shovel, trowel, spike, or dibble (Figure 24-5). The survival rate of bare-root seedlings is significantly higher than for field germination of seeds and can generally be maintained at 80 percent or higher with healthy plant stock and an adequate moisture regime. Use of bare-root seedlings at sufficiently close spacings can insure rapid establishment of a dense, monospecific plant cover in a short time. This success is critical for projects where minimal startup time is of the essence.

Seeds

Wetland plants can be established directly from seeds, given suitable seed stock and suitable soil moisture, light, and temperature conditions. Some species have seeds that can be field harvested in very large numbers, while other species have few seeds. For example, a typical cattail seed head contains thousands of individual seeds, while mature bulrush culms may contain only 20 to 30 seeds each. Most wetland seeds can be broadcast using rotary seeders or by hand and lightly harrowed into the surface soil layer. An estimated

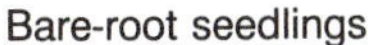
Bare-root seedlings

Container growth

Field harvested

Figure 24-4 Examples of plant propagules used for vegetation establishment in constructed wetlands.

1.2×10^6 seeds per hectare are required for establishment of a *Spartina* salt marsh for an estimated cost of \$2400 per hectare in 1987 dollars (Broome et al., 1988).

A second approach to establishing a wetland plant community from seed is to harvest the seed bank from a neighboring wetland (natural or constructed) that has a plant community similar to the one desired for the new constructed wetland. This seed bank is harvested by scraping the top 10 to 20 cm of topsoil or muck from the donor wetland and then redistributing this muck in strips or over the entire surface of the new wetland. An adequate moisture regime, resulting in saturated, but not flooded soil conditions, is absolutely essential for success of this approach.

Establishing new wetlands from seed has the advantage of lower initial cost. Seeds can generally be harvested and planted for a considerably lower cost than bare-root seedlings. However, there are two potential problems with establishing wetlands from seeds. The first is the time it takes to establish a viable plant community that will provide adequate cover

Trowel

Dibble

Figure 24-5 Some tools and methods used for planting bare-root seedlings in constructed wetlands.

to allow system startup and to initiate water quality treatment. Because bare-root seedlings pass their first several months of growth in greenhouse conditions, they have a comparable time advantage over seed germination in the constructed wetland.

A second potential problem with establishing constructed wetlands from a seed bank or by broadcasting field-harvested seeds is the near impossibility of controlling the plant species that germinate and establish dominance in the treatment wetland. Variations of the initial seed selection, bed preparation, and flood regime can be used with some success to control the resulting plant dominance; however, these controls are only successful after considerable piloting and field studies that may require several growing seasons to perfect. This level of study and planning is most appropriate for large-scale wetland construction projects where the cost savings resulting from use of seeds can offset the cost of preliminary research to perfect propagation methods.

A common result of using seeds to establish wetland plant communities is the development of areas that have poor cover or that are dominated by undesirable species. These undesirable plant species may be upland species that were inadvertently germinated because of slightly higher ground elevation in some parts of the wetland. When the wetland is flooded with wastewater flows, these upland or transitional wetland species will die, leaving unvegetated areas. Several years may pass before these bare or thinly vegetated areas are covered by adapted wetland plants that are spreading slowly by rhizomatous growth.

Field-Harvested Plants

In some cases, herbaceous wetland plant stock can be field harvested and used to plant a new constructed wetland. This is especially true in areas with abundant natural wetlands and high regional water tables, where wetland plants such as cattails and bulrush are common in roadside ditches, in man-made ponds, and along canals. A plant collection permit may be required to legally remove wetland plants from these sources. Therefore, the state department of natural resources or equivalent agency should always be contacted before field harvesting from natural wetlands.

Field harvesting is done by hand digging or by using a back hoe or dragline to scoop wetland plants from the ground and spread them on an open, upland area where they are separated by hand into plantable-sized units. Many wetland herbaceous plant species grow from rhizomes, corms, or woody subterranean tubers that must be cut or pulled apart, each with a viable emergent plant attached. For some species, the rhizome alone is sufficient for propagation. For example, in cattails and the common reed, emergent leaves harvested with the rhizomes and planted in the new wetland will generally die and be replaced by new shoots arising from the rhizomes.

Planting field-harvested plants is more difficult than planting nursery-grown seedlings because of the size of the propagules. Rhizomes and attached shoots and leaves are usually larger than bare-root seedlings and require a larger hole for planting. These holes may be prepared with a shovel, a post-hole digger, or a hand or tractor-mounted auger. In all cases, the hole must be deep enough to allow the rhizome or root material to be completely buried without excessive folding and with the node between rhizome and shoot at the ground surface. Attempts to plant field harvested rhizomes by harrowing or mechanical planting have had varied success at providing plant cover. Field-harvested plants must be protected from excessive heat or cold, kept moist but not wet, and must be used within a few days to minimize mortality.

Field-harvested plants have several distinct advantages over nursery-grown stock. Because herbaceous wetland plants store most of their growth reserves in their roots, rhizomes, corms, and tubers, these plant parts can produce shoots and mature plants faster than newly developing seedlings. In fact, a carefully tended rhizome generally produces numerous daughter plants in a few weeks, each of which in turn produces new rhizomes and a second generation of daughter plants. This growth results in faster development of adequate plant cover so that water quality treatment can begin earlier than otherwise possible. A second advantage of using plants that have been locally harvested is that these plants are more likely to be adapted to local climatic conditions including their phenology and genetic tolerance to temperature and moisture regimes. In a number of cases, nonlocal nursery-grown wetland plants from a distant latitude have not been successful at their new location because of subtle genetic differences.

Another aspect of using field-harvested plants may be an advantage in some cases and a disadvantage in other cases. Invariably, other wetland plant species are introduced into a newly constructed wetland in the rootstock and attached soil of field-harvested plants. These additional plant species come as small seedlings or as seeds in the existing seed bank from the donor wetland. For example, at the West Jackson County, MS constructed wetland near Ocean Springs, use of field-harvested cattails and the existing seed bank in the site soils resulted in the presence of 48 different plant species growing in the constructed wetlands 1 year after construction. This introduction of volunteer plant species is generally an advantage because it provides diversity in a constructed wetland that might otherwise be nearly monospecific and subject to potential pest or climate-induced catastrophes. However, in cases where monospecific plant communities are required for research or because of agency prejudice against wetland "weed" species, this introduction of volunteer species might be unacceptable.

In some cases, the genotype of local field-harvested plants may not be as desirable as nursery-grown plants that have been genetically selected for superior growth characteristics. European research with *Phragmites australis* has developed genotypes that have more robust growth characteristics and greater flooding tolerance than naturally occurring varieties. This may also be the case with *Typha* spp. which are highly variable in growth height, leaf size, and shoot density. Careful examination of local sources of wetland plants for field harvesting should be conducted to verify their acceptability for use in a constructed wetland treatment system.

Potted Seedlings

In some cases, seedlings are planted in containers filled with potting soil to establish older and more robust planting materials. This method is especially useful for establishing woody plants in wetlands because they have a slower growth potential than many herbaceous wetland plant species. If bare-root seedlings of tree species are planted in a new wetland, they may be outcompeted and shaded by herbaceous species. However, when these same tree species are grown in a container for one or more growing seasons and then directly transplanted with their developed root system relatively intact, they can compete with surrounding herbaceous growth and eventually dominate through their ability to grow taller and shade the herbaceous species below their canopy.

Similar to field-harvested wetland plants, potted seedlings have a greater advantage for initial growth than seeds or bare-root seedlings. They also have the disadvantage of higher initial cost, making them economically unattractive for most large-scale wetland plantings. They are not only more expensive to grow, they are also more expensive to transport to the new constructed wetland and within the wetland during planting.

PLANT ESTABLISHMENT

Plant growth is one of the delights of the biological world. With a seed, water, nutrients, and light, something animate appears as if by magic. Frustrated owners waiting for their wetlands to magically appear have often been disappointed and, in more than one case, have concluded that there was too much "hocus pocus" in creating a wetland on dry ground. There is now sufficient knowledge to ensure success in plant establishment.

The key requirements for healthy wetland plant propagules to succeed are water, soil, nutrients, and light. The first two ingredients must be controlled to some extent by the engineer and the contractor, and nature generally provides the other requirements for plant growth. This section describes methods to provide adequate moisture and soil preparation and describes the importance of seasonality for success.

Climatic Factors

Plants have a growing season and are dormant for the rest of the year in temperate and subtropical climates. Annual plants die each fall or winter and must be re-established from seed the following spring or summer. Fortunately, most plants used in treatment wetlands are perennials which lose their above-ground tissues and become dormant during the cold season and regrow from stored reserves in below-ground tissues during the next growing season. In some cases, new growth is timed to take advantage of predictable rainfall and moisture conditions rather than specific day length or temperatures. Knowledge of these climatic influences is important in planting specifications and during wetland planting.

The best time to establish new plants in a constructed wetland is in the beginning or height of the growing season which usually coincides with spring or early summer. During this period, available light is increasing daily, and competition from previously established weeds or damage from pests and pathogens is minimal. Most plants are genetically adapted to grow during this period. Also, planting at the beginning of the growing season provides ample time for full plant development and attainment of suitable plant cover before the onset of cold temperatures and declining plant growth rate. A wetland treatment system can typically begin operation when plant cover is at least 60 to 80 percent. If this percent cover can be obtained during the first growing season after planting, the wetland system can begin operating during the ensuing fall.

A constructed wetland planted with cattails, bulrush, or common reed will vegetate quickly if care is taken when selecting planting materials, planting density (see later), soil preparation and moisture, and seasonal timing. In fact, plant cover in the 60 to 80 percent range is commonly achieved for these species in 3 to 4 months in most temperate climates. To initiate system startup within the first year of planting, it is advisable to plant at least 3 to 4 months before the end of the growing season (before the last frost-free date in most areas). Wetland plants can be planted later in the summer or fall, but their growth will be interrupted by cold weather and decreasing day length, and the below-ground plant organs must in turn be protected from killing frost during the first winter so they can resume growth the next spring.

Soil Preparation

Emergent wetland plants require suitable soil conditions for rapid initial growth and long-term propagation and survival. Loamy soils containing a mixture of sand, silt, and clays are optimal for growth of most plants. These soils have adequate texture and organic matter to retain moisture, allow diffusion of oxygen and carbon dioxide, and retain nutrients for absorption through the plant roots. During design of a constructed wetland, suitable soil conditions should be incorporated to ensure successful plant growth.

While sandy soils provide good gas exchange and allow for relatively unencumbered root development, they do not retain nutrients or moisture well. As long as nutrients and moisture are adequate through continual flooding with enriched wastewaters, sandy soils can be used for constructed wetland development. However, if moisture is unpredictable, as it is in some stormwater treatment wetlands, or nutrients might be in short supply, sandy soils may result in slow or inconsistent wetland plant growth.

The sands and gravels of SSF wetlands present the extreme of this situation. Plants will be growing under near-hydroponic conditions. Therefore, it is important to properly regulate both the water level and its nutrient content. Plant roots must reach the water.

Clays or other fine-textured soils have poor characteristics for gas exchange and root development and may limit water movement to the wetland plant roots. However, clays typically have high cation exchange capacities and provide excellent binding for important plant growth nutrients. In tight clays, the nutrient advantage is overwhelmed by the poor texture for root development, resulting in very slow plant growth.

Highly organic soils may have excellent water holding capacity, but very poor nutrient retention. But when mixed with sands and silts, organic soils are frequently the best medium for growth of wetland plants. Good agricultural topsoils frequently provide an excellent medium for wetland plant growth and should be stockpiled for use during wetland construction.

Constructed surface-flow (SF) wetland design should incorporate a minimum of 20 cm of topsoil as a rooting medium in all areas that will be planted with emergent macrophytes. To increase the rate of plant growth, this topsoil should be amended with agricultural fertilizer at a rate typically used to cultivate a grass crop, which can be determined by soil analyses and recommendations from the local agricultural extension agent. When using nutrient-rich effluent for plant establishment and maintenance, a soil fertilizer amendment is only required initially, but when groundwater or low-nutrient surface water will be used to maintain soil moisture, a subsequent fertilizer application might be considered based on the observed rate of plant cover establishment.

Soil Moisture

As described previously, incorrect control of soil moisture is the most frequent cause of failure to establish wetland plants. Inadequate soil moisture results in desiccation of roots and shoots and causes wetland species to be replaced by weedy upland plant species that may be in the seed bank of the constructed wetland soils. Too much water results in oxygen depletion in the root zone and slow growth or plant death because of insufficient oxygen for root metabolism. The correct amount of moisture can be maintained through adequate planning and attention during the construction period.

To maintain suitable soil moisture during plant establishment, there must be a reliable and adequate supply of water for site irrigation. This requirement can be quite large when constructing wetland systems more than a few hectares in size. When practical, the best water source for site establishment is usually the wastewater that will be treated in the wetland. This water can generally be used if an existing treatment system is being retrofitted or expanded by the addition of a constructed wetland. If the wetland influent water is not available, then another irrigation supply must be provided to optimize chances for planting success. This irrigation source might be a potable water supply, an existing or temporary well, a pond or lake, or a nearby stream or river. Adequate pumps, piping, and sprinklers or hoses must be provided to allow even flow distribution.

The typical sequence for maintaining soil moisture for wetland planting starts with initial saturation of soil by sprinkling or flood irrigation. For optimal plant growth, the soil should be fully or partially saturated with water immediately before planting and should not be allowed to completely dry for any time after planting. High soil moisture must be maintained after planting for the first few weeks without creating flooded conditions for more than a few hours. The best method to maintain soil saturation without excessive flooding is to start planting at the downgradient end of the wetland and continue planting upgradient while gradually raising water levels using the wetland outlet water level control weirs. When planting is complete, water levels can be dropped or raised as needed to maintain saturated soil conditions. Sprinklers can also be used to irrigate evenly over planted areas.

After an entire cell is planted, water levels should be maintained at a level that ensures that all areas of the cell continue to have saturated soil conditions between waterings. This goal can be accomplished by flood irrigating the entire cell with enough water to allow infiltration or evapotranspiration to eliminate the applied surface water within 1 or 2 days, or by distributing water through the inlet distribution structures and allowing this water to resaturate the wetland soils as it sheetflows across the wetland to the outlet. Weirs or outlet water control gates should be removed or left open during plant establishment to prevent flooded conditions from occurring if there is high rainfall or if a sprinkler or irrigator is accidentally left running.

As the wetland plants grow in height, they have an increased ability to transport oxygen to the root zone through their leaves (see Chapter 7). This ability may allow longer flooded conditions after initial plant establishment. However, the best technique for establishing rapid plant cover is to maintain saturated soil conditions without surface flooding. The higher soil oxygen conditions resulting from the absence of flood waters allow maximum root metabolism, effective nutrient use, and rapid development of the original plant propagules and their daughter plants within the wetland. This soil condition should optimally be maintained until the plants achieve complete cover (100 percent) or at least the minimum cover required for system startup (about 60 to 80 percent).

Plant Density

The initial density of plant propagules will greatly influence the rate of establishment of plant cover and the cost of planting. When the goal is establishment of high plant cover (greater than 60 percent) during the first growing season after planting, the minimum density should be about 10,000 plants per hectare (1-m spacing). A higher density of about 18,000 to 28,000 plants per hectare (60- to 75-cm spacing) can be used to ensure even faster plant cover. Much wider spacings have been used on larger-scale wetlands because of the high cost of these numbers of plants. The widest spacing that has much potential to establish a desired wetland plant community is about 2 m or 2500 plants per hectare. This spacing usually provides low cover for two or more growing seasons, resulting in delays in system operation. Also, if wider spacings are used when planting bulrush or broad-leaved emergents (*Pontederia* and *Sagittaria*), there is a higher probability of invasion by unwanted wetland or upland plant species including cattails, duckweed, Johnson grass, and willows.

PLANT INSPECTION AND MAINTENANCE

Design specifications for vegetation establishment in constructed wetlands should clearly delegate responsibility for plant maintenance from the time of planting until system startup. This task may be the responsibility of the contractor or planting subcontractor, the engineer, the owner, or some combination of these parties and should be carefully described in the planting specifications (example planting specifications are given in Table 24-1). Successful plant establishment requires periodic inspections to document soil moisture conditions, plant survival, and plant growth. The frequency of these inspections is project specific, but must be great enough to prevent problems or to detect them relatively quickly after they begin to occur.

Initial plant inspection should examine the viability of the planted propagules, whether seeds, seedlings, or field-harvested mature plants. If plants were planted in rows on specified centers, the surviving plants in a subset of these rows can be counted to determine survival rate. If seeds were used, then random square meter plots can be used to estimate germination success. When counting surviving plants, each of the original seedlings or plant clumps should only be counted once and any daughter plants that may have arisen from rhizomes should not be counted. Planting success can be compared to planting specifications based on these plant counts. In some cases, a contractor may have to fill in some areas with new plants to comply with the requirements of the planting specifications.

Subsequent plant growth is monitored by estimating percent cover and average plant height. These nondestructive techniques are used to ascertain the status of plant development before and during wetland treatment system operation. Plant cover is an estimate of the percentage of the total ground area covered by stems and leaves. This parameter can be estimated by walking through or next to a plant stand and visually determining a cover category for the plants. Typically, seven cover categories are sufficient for this visual estimation method: (1) <1 percent, (2) 1 to 5 percent, (3) 6 to 10 percent, (4) 11 to 25 percent, (5) 26 to 50 percent, (6) 51 to 75 percent, and (7) 76 to 100 percent. Cover estimates can be made on a finer scale by using a square meter frame to delineate specific areas. Cover estimates should be made at enough locations in the constructed wetland to provide reasonable statistical averages for comparison between cells and between dates.

Cover estimates and observations concerning plant health should be a routine part of operational monitoring in a constructed wetland treatment system. Because plants grow slowly and are important for maintaining the performance of wetland treatment systems used for water quality treatment, problems must be anticipated and prevented before they are serious or have progressed too far. Reestablishing a healthy plant community in a natural or constructed wetland is a slow process when the plants have been irrevocably harmed because of operator neglect.

Initial control of water levels in a constructed wetland should optimize plant population growth while achieving water quality goals. Water levels should typically be maintained in the following manner: levels should be kept lowest during late spring and summer and highest during late fall, winter, and early spring. Low water levels during the warmer months optimize new plant growth and maximize cover by reducing oxygen stress in the plant root

Table 24-1 Example of Constructed Wetland Planting Specifications

Part 1 General

1.1 Summary

A. This section describes work necessary to establish a wetland marsh plant community in the constructed wetland cells, including plant species, quality of plants, planting guidelines, and success of plant establishment. No planting shall occur in the deep water areas.

1.2 Submittals

A. Submittal During Construction: The following specific information shall be provided:

1. Samples: Two weeks before the wetland planting, submit to the owner or engineer for approval, samples of the seedlings and/or tublings of the cattails (*Typha latifolia*) and bulrush (either *Scirpus validus* or *S. californicus*).

Part 2 Products

2.1 Plant Materials

A. Plants: Material to be planted shall be tublings (seedlings grown in narrow, tube-shaped containers), bare-root seedlings, or field-harvested cattail (*Typha latifolia*), and bulrush (either *S. validus* or *S. californicus*). Plants shall be live, fresh, healthy, and uninjured at the time of planting. Plants shall have 12 in. of green leaf tissue above the root crown. Plants may be transplants from local wetlands, provided that written documentation of state resource agency approval is presented to the engineer. Plants shall be kept continually moist and shaded until they are planted by covering with wet burlap or by other method approved by the owner or engineer. Any plants which have been permitted to dry out, to become overheated, or for any reason, in the judgment of the Engineer, do not clearly show a viable condition shall be rejected for use.

Part 3 Execution

3.1 Planting Methods

A. Planting Areas: Cattail and bulrush will be established in zones as indicated approximately on the drawings.

B. Planting Schedule: Planting shall be performed as soon as possible after the cells have been constructed. Once started, the planting operation shall continue uninterrupted until all areas have been planted, except for periods when the soil is too wet or too dry to continue the operation. If downtime occurs because of soil condition, planting shall be resumed as soon as the soil is suitable for planting. Planting after August 31 must be approved by the owner or engineer.

C. Planting Method

1. Planting shall be done by hand using a dibble, planting spade, or other method approved by the owner or engineer. The dibble or spade shall be used to create a shallow hole in the moist topsoil for planting. The hole shall be of sufficient depth and width to allow the root to be inserted without breakage or other damage occurring. The plants shall be planted so that the leaf bases are at the soil surface. After placement of the plant in the hole, the hole shall be carefully closed around the plant root by gently applying pressure to the edge of the hole with one's foot or other method approved by the owner or engineer. Plants which have been judged by the owner or engineer to be damaged by mishandling during the planting process shall be replaced at the contractor's expense.

2. The cattail and bulrush shall be planted at a minimum rate of 4900 plants per acre on approximately 3-foot centers in all areas shown on the drawings. The total number of plants shall be equally divided between the two species, cattail and bulrush, unless otherwise approved by the owner or engineer.

3.2 Maintenance of Stand

A. Maintenance: The contractor shall make arrangements and bear all costs of providing adequate water for initial planting and plant maintenance. Treatment effluent from the existing facultative lagoon can be used, if desired by the contractor, as long as offsite runoff is controlled. The contractor shall be responsible for maintaining the plants to ensure optimum growing conditions until accepted by the owner. Initial water depth in the wetlands shall be increased gradually from 0 in. to maximum depth in 2-in. increments over a period of 2 months following completion of planting. All planting shall be done in moist soil without standing water. Maintenance may include watering or dewatering of the wetlands cells to ensure optimum growing conditions.

B. Inspection: Sixty days after the completion of the planting, the owner or engineer will make an inspection to determine if satisfactory stands of cattail and bulrush have been produced. A satisfactory stand is defined as one in which (1) spacing between planted seedlings averages 3 feet or less, (2) the seedling survival averages at least 80 percent (3) no areas of greater than 100 contiguous square feet with a seedling survival rate of less than 50 percent. If satisfactory stands have not been established, another inspection will be made after the contractor has corrected any deficiencies and provided the owner or engineer with written notice that the stands are ready for inspection.

systems. Lower dissolved oxygen solubility in warm water burdens the plant's oxygen supply. This burden can be overcome by keeping the water very shallow to increase water velocity and reduce the barrier to oxygen diffusion to the root zone. The reduced reaction time for microbial transformations is compensated for by higher reaction rates for many microbial reactions in warmer water. During colder months, water levels can be raised to provide longer reaction times without starving the plant roots of oxygen. Also, during the dormant plant growth season, plant roots require a smaller fraction of the oxygen needed for metabolism during the summer.

After the system has operated for several months, the water level control scheme can be varied in individual cells to develop optimum growth conditions for the specific site. Plant cover and water quality changes must be monitored to quantify and document the effect of different water level control regimes. As is the case with all complex water quality treatment technologies, monitoring is one of the key determinants of the success of wetland treatment systems.

TROUBLESHOOTING

Regardless of the cautions and care with which new wetlands are constructed, problems with plant growth will likely develop at some time during the project's life. Most of the existing constructed and natural wetland treatment systems have previously had or currently have plant growth problems. Frequently, these problems can be overcome without jeopardizing treatment performance. Although most of the principles important for optimal plant growth were discussed in Chapter 7 and earlier in this chapter, it may be helpful to some owners and operators of wetland treatment systems to have a troubleshooting guide organized by the typical categories of observed problems.

Table 24-2 lists the most common physical, chemical, or biological factors which may contribute to poor plant growth and survival in constructed wetlands used for wastewater treatment. Each of these factors is discussed briefly and should be examined as a possible contributor to poor plant success. Most of the plant growth observations from a specific wetland treatment system are likely to be a combination of one or more of the factors listed in Table 24-2.

Table 24-2 Summary of Potential Factors Resulting in Wetland Vegetation Maintenance Problems

Problem	Corrective Measures
Water stress (levels too low)	Raise outlet weirs, add more water, or provide supplemental irrigation to maintain adequate soil moisture
Flood stress (levels too high)	Lower outlet weirs or reduce flow to lower water levels
Macronutrient stress (N, P, K)	Fertilize as required to promote healthy plant growth
Micronutrient stress (Fe, Mg, Mo, etc.)	Add micronutrients as required to promote healthy plant growth
Dissolved oxygen stress a. Organic loading b. Ammonia loading c. Smothering (sludge or solids) d. Tight soils	Reduce the input of oxygen demanding substances (BOD_5 and NH_4); lower water levels; reduce the input of solids (mineral solids and sludge); design with loamy topsoil to provide a suitable rooting medium
Pathogens/herbivory a. Insects b. Plant diseases c. Mammals	Tolerate without chemical controls as much as possible. Burn during winter months to reduce insect and pathogen resting stages; trap and remove mammals as necessary
Weather/physical a. Frost b. Heat c. Wind d. Excessive evapotranspiration	Maintain flooded conditions to regulate favorable root temperatures; use suitable topsoil to provide plant stability

Water Stress (Levels Too Low)

If high water levels cause plants to be planted and rooted in the upper few inches of soil or gravel in constructed wetlands, the rapid and prolonged reduction of water levels will result in a hostile root environment and plant death under prolonged dry conditions. If water levels are reduced in a gravel bed to promote root penetration, this reduction must not exceed the growth rate of the roots and should generally be less than about 1 cm/d.

Flood Stress (Levels Too High)

As described in Chapter 7, all plant species including all emergent hydrophytes have some upper tolerance limit for flooding depth. This upper limit is a function of the complex interaction of physical, chemical, and biological factors which affect the available oxygen in the plant roots and the effect of the resulting oxygen concentration (or anoxia) on root metabolism and accumulations of toxic substances.

The upper limit of flood tolerance for a particular plant species is partially a function of dissolved oxygen regime of the water that is flooding the plant. Thus, flood tolerance may be 0.6 m in water with a higher flow rate or lower oxygen demand and may be only 15 to 30 cm in stagnant water with a higher dissolved oxygen demand (see Chapter 10 for a more detailed discussion of the factors influencing dissolved oxygen concentrations in wetland surface waters). Also, individual hydrophytic plant species may have multiple ecotypes with varying flood tolerance (see Grace, 1989 for an example with cattail). A nonadapted ecotype may not survive, but another plant of the same species with an adapted genetic makeup might survive. Overtopping of all emergent plant parts for a prolonged period causes drowning.

Because of this complex balance between physical, chemical, and biological factors, no standard flood tolerance depth can be insured for any hydrophyte species. Emergent hydrophytes generally grow better in saturated soil than in standing water. The depth of standing water on new plants must be increased slowly during early plant growth to allow for the development of morphological adaptations (such as increased numbers of lenticels and shoot length for gaseous diffusion, or the growth of adventitious roots in some species) that also impart flood tolerance in hydrophytes. Plant growth must be observed during this startup phase and monitored in some semiquantitative manner to allow operating water depths to be set at levels that will not exceed the tolerance of the species selected.

Macronutrient Stress (Nitrogen, Phosphorus, and Potassium)

An adequate supply of macronutrients is essential to optimize wetland plant growth, survival, and succession. For emergent macrophytes, these nutrients are most commonly supplied from the wetland soils through the roots rather than through foliar uptake as in some submerged aquatics. Nutrient condition of emergent macrophytes depends not only on the total quantity of nutrients in the soil, but more importantly on the available nutrients in the soil. The overall quantity of one or more of the macronutrients might be in low supply, and the availability of one or more of these nutrients may also be low. Nutrient availability depends on many soil properties including pH and redox potential, two soil properties that may be greatly altered by flooding with wastewaters.

Adequate supplies of macronutrients for wetland plant growth are generally found in domestic wastewaters. Limiting concentrations for nitrogen are typically less than 1 mg/L and for phosphorus are less than 0.1 mg/L. In SSF wetlands, wetland plants obtain macronutrients directly from the hydroponic effluent. Macronutrient limitations in municipal wetland systems are only found under very high levels of pretreatment and then only in the downstream areas of the wetland. Macronutrient concentrations necessary for wetland plant growth may be in short supply or grossly insufficient in wetlands receiving only rainfall, runoff, groundwater, or certain industrial effluents. If nitrogen and/or phosphorus concentrations are below minimal levels for adequate plant growth, these nutrients may need to be added for project success.

Micronutrient Stress

In addition to macronutrients, plants require numerous micronutrients. The availability of micronutrients depends on complex interactions between soil chemistry and the chemistry of the water above the wetland sediments. Vascular plant micronutrient requirements are well documented in plant physiology text books (e.g., Salisbury and Ross, 1978), as are symptoms of specific micronutrient deficiencies. Generally, micronutrients are not likely to limit growth of plants exposed to domestic wastewaters or plants which are rooted in clay or other mineral soils. However, specific micronutrient limitations do occasionally occur and should be considered in optimizing plant condition.

Dissolved Oxygen Stress (Physical Factors)

Multiple physical and chemical factors may result in low to very low dissolved oxygen conditions in flood water (even very shallow flood water) and ultimately in extreme reducing conditions in flooded soils. Physical factors which may limit oxygen diffusion into the root zone include the presence of a sludge "blanket" or coating of aerobically digested solids from an activated sludge treatment facility. Sludge reduces oxygen diffusion and increases reducing conditions because of the presence of increased metabolic activity within these solids. Deposits of mineral sediments entering a natural or constructed wetland through erosion can also smother plant roots, especially wetland tree species. Dense clay soils may also contribute to severely reduced oxygen diffusion rates to the root zone.

Dissolved Oxygen Stress (Chemical Factors)

Chemical constituents dissolved or suspended in the water column may exert a demand on dissolved oxygen in the water. Organic carbon measured as biochemical oxygen demand (BOD) is one of the two principal components of the overall oxygen demand, the second typically being total ammonia which exerts an oxygen demand of about 4.3 mg/L for each milligram per liter of ammonia.

Atmospheric reaeration tends to restore dissolved oxygen to its saturated value. However, in constructed wetlands receiving high loadings of BOD and total ammonia from municipal wastewaters, the atmospheric diffusion rate is often insufficient to fully offset the biological utilization of oxygen in the water column, and resulting dissolved oxygen levels are typically close to zero even near the water surface. At the sediment interface, these processes and the oxygen demand exerted by benthic microbes further reduce oxygen concentration and redox potentials. These conditions result in an anoxic environment which is not conducive to the prolonged presence of any oxygen, including oxygen being transported by hydrophytes to their root zones. Under extreme reducing conditions typical of municipal effluent discharge and high water levels, this balance may be so one sided that plants cannot oxidize any of their root zone and therefore cannot obtain nutrients or avoid toxins in the sediments.

Rooting Problems

In addition to inhibiting oxygen diffusion, stiff clay soils are generally unfavorable for rapid root or rhizome expansion because of their density. Where possible, plants will avoid rooting in this type of media. If unconsolidated, flocculent, higher oxygen sediments exist over a tight clay subsoil, aquatic macrophytes will root in the softer soils. Rooting in softer soils can be detrimental to plants because the soft soil cannot provide support against wind or other stress on the above-ground portion of the plant.

Pathogens/Herbivory

Numerous potential stressors on plant growth exist other than low oxygen or the presence of toxins in the root zone. The most visible of these stressors are herbivores such as geese, muskrats, or nutria which occur in high populations in some constructed and natural wetlands and insects such as army worms (cattail worms) which are commonly found in monospecific cattail marshes (Figure 24-6).

In a review of herbivory in wetlands, Lodge (1991) found that invertebrate and vertebrate grazers can remove between 5 to 83 percent of emergent plant biomass. These organisms generally do not completely eliminate a wetland plant species, but rather contribute an additional stress great enough to eliminate portions of the plant population, resulting in open areas available for weed colonization. This loss of productive biomass adds to the debit side of net production and, in concert with other stressors, can result in a chronic loss of wetland plant cover.

Other stressors which are more difficult to observe are a variety of plant pathogens including rusts, fungi, viruses, and bacteria. These organisms are not considered to be

Figure 24-6 Cattail worms can defoliate a constructed wetland treatment system within a few weeks. Fortunately, cattails will rebound from several infestations per year.

especially virulent in any of the plant species planted in most constructed wetlands, but are common residents in wetland plant populations and add an additional stress to plants already under stress from other factors. Infestations of plant pathogens are often associated with plants that have been previously stressed due to age, nutrient shortages, or excessive water depth.

Weather/Physical

The extremes of hot and cold temperatures, wet and dry conditions, and severe winds or hail may all contribute to poor plant survival in wetlands. One side effect of low water levels in gravel-based, SSF wetlands during summer conditions is the high temperature of the surface gravel exposed to direct sunlight. These temperatures may physically wilt and damage the wetland plants, exacerbating the desiccation resulting from low water levels. Also, evapotranspiration is greatly increased during these conditions, further lowering water levels and stressing the plants.

Cold winter temperatures including freezing weather generally kill all above-ground portions of cattails, grasses, and some of the other commonly used constructed wetland plant species. Bulrush and duckweed may survive winter, but may be stressed severely by cold weather. If above-water portions of wetland plants die and rot to and below the water level, rhizome viability may be compromised. For example, the lack of second year reproduction by some wetland plant species such as woolgrass (*Scirpus cyperinus*) and flag (*Thalia geniculata*) in wetland treatment systems may be caused by prolonged flooding after winter dieback of the aerial plant organs in wetland treatment systems. Subsequent to loss of the aerial plant organs, the rhizomes cannot maintain a positive oxygen balance and succumb to low-oxygen stress before aerial plant tissues can regrow in the spring. Because cattails, bulrush, and reeds maintain above-ground living or dead plant material throughout the winter months, these species can supply enough oxygen to their rhizomes to survive during the winter months even under flooded conditions.

SUMMARY

Wetland construction should be one of the most rewarding and satisfying aspects of implementing a wetland treatment system. As long as a few simple guidelines are followed and common sense is employed by contractors and field engineers, wetland project construction can help insure the success of a good design. Inexperience and lack of common sense are the two most likely causes of project failure during construction. These two problems can, at least in part, be corrected by studying what others have done before moving ahead during wetland construction.

Site civil construction activities, pump stations, forcemains, and open channel conveyance systems are standard experience for most wetland contractors. For wetland systems, it is important to not create conditions that will ultimately limit the wetland treatment system from accomplishing its water quality treatment or habitat creation goals. Contractors should protect biological resources onsite and in the project's vicinity, including seed banks in top soils to be used on the site. Good construction practices will minimize erosion.

The presence of a healthy, dense population of emergent plants is one of the most important requirements for pollutant assimilation in wetland treatment systems. Plants provide the structural surface area occupied by the microbial flora that attacks dissolved pollutants; they provide the physical environment that helps settle particulate pollutants; and they shade the water surface, preventing algal regrowth and dampening temperature fluctuations. In addition, the wetland plants provide a vast amount of fixed organic carbon important for microbial growth and denitrification and are the basis for a complex food web of wetland biota. A wetland without emergent plants is a shallow pond and, like a pond, will have limited potential for dissolved and particulate pollutant reductions.

Wetland plants can be established successfully by a number of different methods that have varying benefits and limitations. Designer preference and planting cost will guide selection of the appropriate plant propagation technique. Care must be taken with the health of the plant propagules, soil and site preparation, planting, and soil moisture maintenance. The time required for vegetation establishment is often critical to meet scheduled system startup.

Emergent plant populations must be kept healthy to achieve consistent treatment performance. This vegetation maintenance requires monitoring of plant health and may require seasonal adjustment of water levels. Troubleshooting of observed plant condition problems will enable corrective action to be made in time to avert reduced treatment effectiveness and permit violations. This troubleshooting may need to be the responsibility of an experienced biologist. A small investment in time and expense to monitor plant condition and to understand the factors that are controlling plant health in a wetland treatment system may save time and expense by averting a disastrous loss of emergent wetland plant cover.

CHAPTER 25

Treatment Wetland Operation and Maintenance

INTRODUCTION

As discussed in Chapters 20 through 22, treatment wetlands should be designed and constructed to require infrequent operational control or maintenance. Through conservative loading rates, high influent water quality, and simple mechanical controls requiring low maintenance, wetland treatment systems will experience minimal ecological changes and will continue to meet final effluent limits indefinitely. Monitoring and adjustment of flows, water levels, water quality, and biological parameters are the only day-to-day activities required to achieve successful performance in treatment wetlands. Other operations and maintenance activities in treatment wetlands such as repair of pumps, berms, and control structures; vegetation management; and removal of accumulated mineral solids must be attended to at much less frequent intervals.

This chapter provides guidance on the appropriate level of operational monitoring necessary for control of wetland treatment systems. Operational efforts may vary widely, depending on whether the wetland is a highly loaded constructed wetland or a natural or seminatural system with low hydraulic and constituent loadings. Suggestions are presented for decision-making related to the few operational controls that can be implemented in these systems, including pollutant loading rates and schedules, vegetation maintenance, water level control, and site rotation.

MONITORING AND CONTROL

Monitoring is the most important factor in successful operation of treatment wetlands. Monitoring information must be collected accurately and consistently and frequently reviewed by a knowledgeable operator to anticipate the need for operational changes. Operating a wetland treatment system is similar to guiding an ocean liner in a confined space with a tugboat. Changing the direction of the ocean liner takes considerable time and energy. Once a new course is established, it takes a large amount of time and energy to return to the original course or to an alternate course. Incorrect operational control decisions and design errors can cause prolonged periods of poor operational performance of constructed wetlands and significant ecological changes in a natural wetland. Early detection of subtle changes in a treatment wetland's water quality and biological resources requires adequate data collection and frequent data analysis.

All wetland treatment systems should be monitored for at least inflow and outflow water quality, water levels, and indicators of biological condition. These parameters are essential for successful system control. Regulatory requirements may also dictate other monitoring requirements. The frequency of operational monitoring for system control is dictated by the size and capacity of the system, the sophistication of the owner's staff and sampling equipment, and site-specific factors related to influent quality variability and climatic factors.

Table 25-1 summarizes a minimum monitoring program for operation of a wetland treatment system. This list includes measurements of the water quality of all major inflows and outflows associated with the treatment wetland. Inflows include the source(s) of pretreated wastewater entering the wetland as well as natural inflow streams that may have a significant effect on the water quality or the hydrologic budget of the natural wetland treatment system. As described in Table 25-1, the parameter list to be tested at all major inflows and outflows at least monthly includes all regulated pollutants and integrative measures such as 5-day biochemical oxygen demand, total suspended solids, pH, dissolved oxygen, water temperature, conductivity, NO_2+NO_3−N, ammonia nitrogen, total Kjeldahl nitrogen, total phosphorus, chloride, and sulfate.

Inflow and outflow stations may also be monitored less frequently for selected heavy metals or organics that might be present in the wastewater and for whole effluent acute and chronic toxicity. If water quality characteristics are highly variable for any of the inflow or outflow locations, or if there are weekly or monthly permit limits, sampling should be more frequent than monthly or quarterly.

These water quality data as well as any other parameters required by permit should be organized and recorded in computerized spreadsheets for visual analysis of variability and trends. Seasonal and successional changes can be detected by examining trend data regularly. Operational controls are called for when trends indicate the potential for future permit violations. Operational modifications should also be made in response to seasonal changes in dissolved oxygen and water temperature.

Flow rate should be measured or estimated at the inflow and outflow locations daily. These data can be collected by installing flow meters at some locations and by collecting stage data and using stage-discharge relationships at noninstrumented locations. Flow estimates are essential for quantifying constituent mass balances in wetland treatment systems.

Rainfall should be monitored at a location next to or near the wetland treatment system. Rainfall measurements are used to estimate the wetland water balance and to anticipate elevated flow conditions at the wetland outflow location(s). Evapotranspiration can be estimated with pan evaporation data (corrected by a factor between 0.7 and 0.8) from a regional weather station. Utilization of rainfall, evapotranspiration, and inflow-outflow measurements

Table 25-1 Typical Minimum Monitoring Requirements for Successful Operation of Wetland Treatment Systems

Recommended Parameters	Recommended Sample Locations	Minimum Sample Frequency
Inflow and outflow water quality		
All systems:		
Temperature, dissolved oxygen, pH, conductivity	Inflow(s) and outflow(s)	Weekly
Municipal systems:		
BOD_5, TSS, Cl^-, SO_4^{2-}	Inflow(s) and outflow(s)	Monthly
Industrial systems:		
COD, TSS	Inflow(s) and outflow(s)	Monthly
Stormwater systems:		
TSS	Inflow(s) and outflow(s)	One storm event per month
Permit parameters as required:		
NO_2+NO_3−N, NH_4−N, TKN, TP	Inflow(s) and outflow(s)	Monthly
Metals, organics, toxicity	Inflow(s) and outflow(s)	Quarterly
Flow	Inflow(s) and outflow(s)	Daily
Rainfall	Adjacent to wetland	Daily
Water stage	Within wetland	Daily
Plant cover for dominant species	Near inflow, near wetland center, near outflow	Annually

Note: BOD_5, 5-day biochemical oxygen demand; TSS, total suspended solids; COD, chemical oxygen demand; TKN, total Kjeldahl nitrogen; TP, total phosphorus.

can be used to maintain a continuing water balance for the wetland to detect groundwater exchanges that might be due to leaks in a liner, if one is present.

Water stage in the wetland should be measured daily near any outflow locations. When combined with a topographic survey of the wetland, stage measurements provide a quantitative tool for assessing the average, maximum, and minimum water depths in the wetland and the frequency with which these depths occur. These data are essential for interpreting tracer measurements of hydraulic residence time and for assessing any detrimental hydroperiod effects on biota.

Biological monitoring within a wetland treatment system provides the operator with information concerning the structural integrity (health) of the vegetation and fauna. Protection of this biological integrity is important from an environmental habitat perspective and because of the biota's control of wetland operational performance.

The status and health of the microbial populations accomplishing most of the pollutant assimilation in wetland treatment systems can be indirectly monitored by measuring changes in water quality. The populations of these microscopic organisms usually vary too much to warrant direct monitoring without incurring significant expense and confusion concerning interpretation of trends. Trends observed for microbial processes such as decomposition of biochemical oxygen demand, nitrification, and denitrification, in addition to changes in dissolved oxygen concentrations, can provide a sensitive indication of microbial activity within the treatment wetland. If declining trends for pollutant assimilation indicate that microbial activity is impaired, operational adjustments to constituent loadings and water depth may be required.

The percent cover of dominant plant species should be recorded in all wetland treatment systems on a quarterly to annual frequency. In addition, quarterly or annual surveys may also be conducted for benthic macroinvertebrate and fish populations at representative stations in wetlands constructed for habitat and in natural treatment wetlands. Quarterly or annual surveys for rare or threatened species may also be conducted when appropriate. This monitoring provides a record of biological changes that occur due to the altered hydrologic regime resulting from the prolonged discharge of pretreated wastewaters. Declining cover for dominant plant species, shifts in macroinvertebrate species composition and density, and changes in fish populations may indicate that hydroperiod changes are excessive and low dissolved oxygen concentrations are impacting plant and animal communities. Trends in use of the treatment wetlands by protected plant and wildlife species may indicate needed changes in system management. Operational control changes may be able to halt and reverse degradation in the wetland biological systems if it is observed early enough.

Routine access to monitoring stations in wetlands may be difficult, and foot traffic over mucky soils may impact wetland biota and create hydraulic short circuits (Figure 25-1). Wading through deep water or walking over roots, fallen trees, and other dense vegetation may be hazardous to operators and may discourage routine sampling. For this reason, sampling stations should be accessible, and boardwalks or catwalks to remote stations should be provided to facilitate sampling. If boardwalks are not a part of system design, foot travel in treatment wetlands should be across and not along the flow path to reduce the chance of channelization.

CONSTITUENT LOADING

Influent water quality to a treatment wetland is generally set by the performance and capacity of the pretreatment process and by the normal fluctuation in inflows and loads from the collection system. This quality is more predictable and controllable for municipal and industrial wastewater flows than for many stormwater flows. For all wetland treatment

Figure 25-1 Operational monitoring in natural wetlands can be facilitated by the use of boardwalks for access. If boardwalks are not present, foot travel should be across and not along the flow path.

systems, maintaining consistent and effective pretreatment is essential for achieving final wetland effluent quality and for protecting the environmental resources in the natural wetland. Types of pretreatment systems to use with various treatment wetland designs have been described in previous chapters. Pretreatment safeguards such as offline storage capacity for nonspec wastes and redundant capacity to handle peak flows are essential to help mitigate upsets that would otherwise be transported to the wetland, potentially resulting in disastrous consequences for the biota in the wetland or in downstream waters.

Under normal conditions of consistent influent water quality, the operator of the treatment wetland can regulate wetland system performance somewhat by regulating inflow rates. Water can be temporarily stored (given this feature in pretreatment system design) before it is discharged to the wetland to dampen flow rate variations. Water can also be discharged to multiple sites in the wetland treatment system to reduce constituent and hydraulic loadings to any one wetland cell or area. Pumping should be avoided during high rainfall periods if possible. Because wetland constituent assimilation performance for almost all parameters correlates with constituent loading rate (see Chapters 9 through 17), the operator should maintain constituent loading rates below design values to achieve permit compliance.

WATER LEVEL AND FLOW CONTROL

Water level and flow control is often the only significant operational variable available to influence the performance of a treatment wetland. Water level variation affects hydraulic residence time, atmospheric oxygen diffusion, plant cover, and all other factors that influence wetland performance. Wastewater flow rate affects hydraulic and constituent loading, hydraulic residence time, water velocity, and longitudinal gradients in water elevation.

Depending on climatic conditions, wetland water levels can be varied on a seasonal basis to prevent freezing in the winter and low sediment oxygen in the summer. In cold climates, wetland treatment system operation during the winter months is possible if water levels are raised in the fall to about 50 cm to allow formation of an insulating cover of ice and snow. Following freeze-up, operational water levels are reduced by lowering the downstream weirs so that water will flow freely under the ice and snow cover. Maintaining a dense stand of emergent vegetation and dead plant litter in cold climate wetlands will help support the ice and insulate the unfrozen water zone (see Chapter 10).

During the summer when water temperatures are elevated and plant productivity is highest, water levels should be lowered to promote better oxygen diffusion to the wetland sediments and plant roots. This guidance is also dictated by the reliance of some microbial treatment processes on adequate dissolved oxygen concentrations and the lower oxygen saturation possible in warmer waters.

Only a few natural wetland treatment systems are able to regulate water levels because altering outlet elevations can hydrologically impact plant communities. In natural wetland treatment systems that have the flexibility for water level control, monitoring information and performance goals should be considered when water level control elevations are selected.

Wastewater flow and water levels are typically controlled by distribution systems with valves, gates, and weirs. These structures must be examined on a regular basis to prevent operational problems. Any flow constriction is likely to become plugged by plant debris, by animals and their activities (e.g., turtles or beavers), or by floating solids in the wastewater itself.

Frequent inspections will also detect vandalism if it occurs. Types of vandalism observed at a single natural wetland system (Houghton Lake, MI) included removal of monitoring wells, axe cuts and bullet holes in distribution header, broken gates and tie-downs removed from the distribution header, stolen boardwalks, stolen engine from monitoring vehicle, water sampling pump stolen, staff gauges pulled up, water level recorder knocked over, aboveground pipe driven over, and fence damaged. An effective public involvement program, adequate security, and sturdy vandal-proof structures will help reduce the incidence of vandalism in treatment wetlands.

DISCHARGE SITE ROTATION

One of the operational controls available with most constructed and some natural wetland treatment systems is moving the point of discharge from one area to another in the wetland or to a separate wetland cell entirely. Since constituent loadings are increased when only a fraction of the total treatment area is used, the ability to rotate discharge sites in a treatment wetland requires conservative system sizing. This redundancy may be invaluable because of the benefits derived from allowing resting periods for the wetland biological communities.

Parallel flow paths are generally provided in constructed treatment wetlands. During normal operation, flows are directed simultaneously to all parallel systems in proportion to the area of each system. Maintenance activities may occasionally require reduction or elimination of flow from a wetland cell for activities such as berm or outlet weir maintenance or to allow lowering of water levels to stimulate increased plant cover. Flow is proportionally increased to all other operational wetland cells during these maintenance activities. Higher flows and greater pollutant loads require the treatment wetland operator to increase vigilance over water control structures, plant community health, and water quality.

Several early natural wetland treatment system operation plans called for discharge site rotation and wetland resting to allow the natural propagation of wetland tree species. This goal is unrealistic in almost every case because of the difficulty in (1) timing a resting period with a natural drought when water levels will naturally be low in the wetland and (2) because it typically requires 2 to 3 years of drought conditions for most wetland tree species to successfully propagate from seed in natural wetlands. These two conditions must occur together, and it is impractical to rely on such an occurrence in most cases. The goal of tree propagation in some natural wetlands used for treatment can be better achieved by planting

2- to 3-year-old, container-grown seedlings that are tall enough to extend well above the water surface.

Periodic resting periods offer natural wetlands several other advantages. The most important advantage is the reduction in the mass of inflow constituents that are creating an oxygen demand within the wetland. This decrease in mass loading may help to restore partially or fully oxygenated conditions in the wetland sediments over a period of 3 to 6 months, allowing a resurgence in plant growth rates that will compensate for a longer period of lowered plant growth rates during system operation.

VEGETATION MANAGEMENT

Since emergent wetland plants provide for microbial growth and pollutant assimilation, their healthy growth and maintenance is a critical component of system operation. The first step in assuring healthy plant cover in a treatment wetland is use of plant species that are tolerant of the normal range of environmental conditions that will occur within the wetland. The environmental conditions that may affect plant survival and growth have been described previously in Chapters 7 and 24. If the plant species specified during design are not tolerant of actual operational conditions, the operator/owner of the system may wish to consider installing more tolerant species to maintain adequate plant cover. In some cases, there will be no species that are more tolerant, and it is necessary to alter system operation to reestablish healthy plant populations in constructed wetlands.

There is generally no need for routine maintenance of the emergent plant species in constructed wetlands as long as water depth and constituent loadings are kept within design criteria. Wetland plant communities are self-maintaining and will grow, die, and regrow year after year in a typical marsh habitat. Plants will naturally spread into unplanted areas that have suitable environmental conditions and will die and be displaced in areas that are environmentally stressful, and a dynamic equilibrium of plant growth and diversity will be established.

Operational and design problems may result in reduction of plant cover and the need to more actively manage plant growth. Management activities will generally be limited to regulating water levels, reducing loadings for short- or long-term periods, eradication of undesirable species through harvesting or herbiciding, and replanting.

The primary goal of vegetation management in most natural wetland treatment systems is to maintain existing plant communities within relatively narrow margins of change. This goal is accomplished by adequate pretreatment and minimal alteration in the wetland hydroperiod. In forested wetland treatment systems, the normal allowance for change usually includes the successional maturation of the existing wetland tree species. Thus, in a forested natural wetland that is not being negatively impacted by a pretreated wastewater discharge, the basal area of trees will increase over time.

Because propagation of most wetland tree species depends on the absence of flooded conditions for several years, a situation that is not likely to occur in a natural wetland receiving additional water flows, tree density (the number of individual trees per area) will decline as trees die. Although tree basal area is a much better indicator of the status of the forested wetland plant community, some circumstances may exist where short-lived tree species lose basal area or where an owner may wish to ensure continued tree propagation in a natural wetland treatment system. In these cases, container-grown wetland trees can be transplanted to maintain tree density and basal area goals.

Other natural wetland treatment systems may use marsh or shrub-scrub wetland plant communities. These systems sometimes exist because of interrupted successional stages resulting from fluctuating water levels and periodic natural fires. With the use of natural wetlands for treatment, normal drydowns and the chance for wildfires may be decreased, potentially leading to increased forested conditions. If this succession is not desirable because of wildlife goals or for preservation of adapted plant and animal species (for example, the Venus fly trap, *Dionaea muscipula,* in pocosin wetlands in the southeastern coastal plain), fire may be an important vegetation management tool.

Fire has become one of the most important tools for managing many of the natural ecosystems under human control, and fire management techniques have been described in the literature (Payne, 1992).

The use of fire management in most wetland treatment systems is not a good idea. Fire mobilizes the stored nutrients and pollutants into water-soluble forms and permits the flushing of the materials the wetland was designed to store. The litter and sediment layers may be destroyed, thus necessitating a new startup period.

Other forms of vegetation management such as herbicide applications, harvesting, or pesticide spraying for insect or pathogen control are not advisable in wetland treatment systems and should be avoided when possible. These controls may alter the ecological functioning of the natural wetland or may negatively impact outflow water quality from the wetlands.

CONTROL OF NUISANCE CONDITIONS IN TREATMENT WETLANDS

Wetlands used for water quality treatment may create nuisance conditions that could negate their positive primary and secondary benefits. These nuisance conditions fall into two general categories: (1) conditions that are a nuisance or hazardous to humans and (2) conditions that are hazardous to plants and wildlife. Nuisance issues that are typically of concern to regulators who permit wetland treatment systems include the potential for odors and increased mosquito problems. Conditions that might represent an environmental problem to plants and animals are increases in parasites, toxicity issues, and impacts to plant communities. Each of these types of potential problems and corrective measures are described here.

POTENTIAL NUISANCES TO SOCIETY

Historically, wetlands were considered to be nuisance areas, harboring disease, poisonous reptiles, and noxious conditions. Wetland drainage saved thousands of lives by preventing malaria and yellow fever. Today, the near eradication of many insect-transmitted diseases and control of biting insect populations through biological and chemical agents has softened society's loathing of wetlands. However, periodic outbreaks of encephalitis still result in warnings from public health officials concerning creation of wetlands for water quality treatment (Wellings, 1986a, 1986b). Mosquito control districts still receive calls after rainy spells because of the increase in biting adult mosquito populations. And golfers and homeowners continue to insist on mowed margins along stormwater ponds in the southern U.S. because of concerns about poisonous snakes.

This section describes problems that can potentially occur in treatment wetlands and their effects on neighboring human populations. Specific problems including mosquitoes and other biting insects, odors, direct contact, and dangerous reptiles are described; the actual occurrence of these problems is summarized; and possible methods to reduce their impact are presented.

Mosquitoes and Biting Insects

Wetlands and other shallowly flooded habitats provide breeding habitat for mosquitoes. These mosquitoes are a nuisance and may transmit debilitating diseases to humans and livestock. Wetlands also provide habitat for bird species that carry encephalitis viruses (eastern, western equine, and St. Louis) which are periodically spread to humans and horses. Yet wetlands are valued for their beneficial functional properties, and arguments for their destruction in the name of mosquito control have nearly been silenced.

Many mosquito species lay their eggs on moist soil and vegetation that is flooded after normal rainfall events. Other species deposit their eggs on the water's surface in shallowly flooded areas including wetlands. Within hours of being flooded or of being laid on the water surface, these eggs hatch and release mosquito larvae into the water column (Figure 25-2). The larvae pass through five growth periods followed by molts (instars) until they are transformed into the pupal stage within about a week of hatching. No additional growth occurs in the pupal stage; however, internal transformations occur that result in the emergence of the adult mosquito in 2 to 3 days. Adult male mosquitoes are nectar feeders and are not a nuisance to society. However, adult female mosquitoes feed on the blood of warm-blooded animals and may be pests because of their biting. In addition, through their multiple blood meals, these mosquitos may transmit infectious diseases such as yellow fever, malaria, and encephalitis. Adult female mosquitoes typically travel less than a mile from their hatching point during their 2- to 3-week lifetime.

Not all mosquito species feed on humans. Many mosquito species do not travel far from their home in search of blood meals. And mosquito-transmitted diseases have become relatively rare in many of the developed parts of the world. For these reasons, the mere breeding of mosquitoes in a treatment wetland does not constitute a problem. However, outbreaks of large populations of some mosquito species continue to cause severe problems

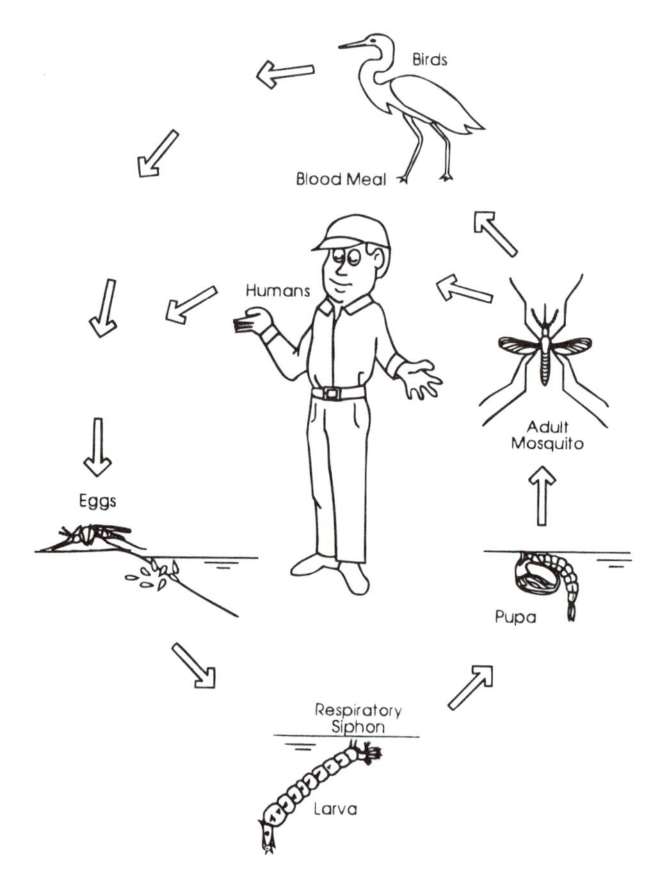

Figure 25-2 Typical mosquito life cycle.

in many parts of the world. Design and operation of natural and constructed wetland treatment systems must take the potential for such outbreaks into account and include measures to minimize their likelihood.

The actual incidence of mosquito-related problems with wetland treatment systems is rare. Studies of natural cypress domes in north central Florida indicated that mosquito populations in domes receiving secondary wastewater were no higher than those in control domes; however, the addition of wastewater may have artificially lengthened the breeding season of these mosquitoes (Davis, 1984). Populations of floodwater-dependent genera such as *Aedes* and *Psorophora* declined when water levels were kept relatively constant. In California, the breeding season of *Culex tarsalis,* an encephalitus carrier, was extended when wastewater was added to wetlands (Dill, 1989); however, mosquito problems in these systems were considered to be minimal and manageable. Studies in flooded mangrove impoundments along the east coast of Florida that are receiving secondary municipal wastewater determined that mosquito production was negligible as long as pretreatment was effective, but that high population densities of *C. nigripalpus* and *C. quinquefasciatus* resulted from the discharge of inadequately treated wastewater after a treatment plant failure (Carlson, 1983; Carlson and Knight, 1987). The Arcata, CA wetland treatment system is believed to have lower mosquito populations than adjacent natural marshes (Crites, 1994). At the Listowel, Ontario constructed wetland pilot facility, the population density of *C. pipiens* was found to be directly related to organic loadings (Wile et al., 1985).

Mosquito problems in natural treatment systems are primarily caused by excessive organic loadings (Stowell et al., 1985; Wilson et al., 1987; Martin and Eldridge, 1989; Wieder et al., 1989). High organic loadings reduce dissolved oxygen levels, limiting the effectiveness of natural aquatic predators such as mosquitofish (*Gambusia affinis*) and aquatic insects (dragonfly and damselfly larvae and beetles). Thick stands of surface vegetation may also limit the access of predaceous fish to mosquito larvae. The abundance of *C. tarsalis* larvae in a wildlife habitat marsh wetland near Los Angeles, CA was found to correlate with the presence of floating mats of dead cattails and was less related to stem density of cattails (*Typha* spp.) or bulrush (*Scirpus californicus*) (Walton et al., 1990). Floating cattail and bulrush root masses protect the developing mosquito larvae from mosquitofish and provide organic nutrition. Mosquitofish effectively controlled larvae in open water areas and at normal cattail and bulrush stem densities of 25 per square meter for cattails and 43.2 stems per square meter for bulrush.

Wilson et al. (1987) found that organic loadings less than 20 kg/ha/d to water hyacinth ponds resulted in no mosquito problems, while loadings more than 80 kg/ha/d limited mosquitofish access and resulted in high mosquito populations. Stowell et al. (1985) recommended that biochemical oxygen demand loadings should be less than 90 kg/ha/d to avoid anaerobic conditions and limit mosquito problems. They also recommended avoidance of dead zones and the use of step feed or sprinkler irrigation to provide uniform influent distribution in hyacinth ponds. It has been suggested that mosquito problems can be avoided by designing for subsurface flow (DeBusk et al., 1989); however, experience at several subsurface-flow systems with surface-flow problems indicates a significant potential for mosquito breeding because mosquitofish usually do not survive in these systems.

Mosquitofish do not survive in the northern climatic zones, and their northern analogs (the top minnows) are not nearly as effective. Control by bacterial and chemical agents has been effective. Disease transmission is virtually nonexistent in northern climates.

Using mosquitofish to control mosquito populations is relatively easy and reliable in constructed, surface-flow wetlands as long as perennial flooded areas exist and strongly anoxic conditions are avoided (Steiner and Freeman, 1989; Martin and Eldridge, 1989; Dill, 1989). Deep water zones that are intermixed with shallow zones provide refuge for fish and other aquatic organisms during fluctuating water level conditions and cold weather (Knight and Iverson, 1990). Wetlands that receive only nonpoint source loadings may periodically go dry, resulting in total loss of mosquitofish populations. If mosquitofish populations are not naturally or intentionally restocked, these constructed wetlands will likely become a public nuisance if they are located near populated areas. Mosquito larvae and mosquitofish populations in wetland treatment systems should be monitored regularly to determine the need for restocking or other operational controls.

Dangerous Reptiles

Wetland-dependent venomous snakes such as the water moccasin (*Ankistrodon pisivorous*) and alligators (*Alligator mississippiensis*) are attracted to created wetlands in the southeastern U.S. because of their high vertebrate and fish productivity (Figure 25-3). Alligator populations as high as one adult per hectare have been observed in Florida constructed wetlands receiving secondary municipal wastewater (Knight, 1987b). While most of the snakes observed in wetland treatment systems are not poisonous, water moccasins are common in some southern natural and constructed wetland treatment systems. No occurrences of the public being injured by these dangerous wetland-dependent reptiles have been reported in wetland treatment systems; however, the danger is real, given increased operator exposure and public use of treatment wetlands for recreation.

Warning signs, boardwalks, and mowed hiking trails can be used to prevent the loss of recreational value because of poisonous snakes and other dangerous reptiles. Operators should be made aware of dangers posed by poisonous snakes and alligators and should use good judgment to avoid injury. Because most poisonous snake bites occur during handling by would-be snake killers, all snakes should be avoided by unknowledgeable individuals. Alligators are most dangerous during nesting. No one should approach an active alligator nest because female alligators may react aggressively. In addition, because of their size, children and dogs should be kept away from the edge of wetlands containing alligator populations.

Pathogen Transmission

Because wetland treatment systems are infrequently used for water contact recreation, direct disease transmission by waterborne pathogens is unlikely (Shiaris, 1985). Wetlands

Alligator

Water mocassin

Figure 25-3 Southeastern wetlands are infamous for their populations of dangerous reptiles.

treating wastewater are expected to pose no greater danger than other points of operator contact. Operators who come into contact with wastewaters should be aware of standard precautions against infectious diseases. Hands should be carefully washed following contact; contact between the hands and mucous membranes of the eyes, nose, and mouth should be avoided; and water quality samples should be stored away from food and potable water.

Wetlands used for water quality treatment are generally not used for human water supply. However, with a high level of pretreatment and adequate monitoring and safeguards, direct potable recycling of wastewaters may occur in the near future. Because wastewaters sometimes contain toxic metals and pesticides, direct potable reuse must be carefully considered.

One potential toxin pathway to humans is consumption of contaminated fish or wildlife from a treatment wetland. As discussed more fully in the next section, toxic metals and organic compounds must not be allowed to accumulate to toxic concentrations in wetland treatment systems to protect wildlife and humans. Forethought in wetland design and water pretreatment and periodic monitoring are necessary to prevent any food chain effects on humans consuming wetland plants or animals.

Odors

Based on the extensive experience of the authors, wetland treatment systems typically operate without problematic odor levels. While all wetlands do have characteristic odors associated with native vegetation and biological processes, these odors are typically no greater in wetlands used for treatment of secondary wastewater effluents. When poor quality wastewater effluents, raw wastewater, or sludges are discharged into wetlands, there is as much potential for odor as there would be by discharge of these inadequately treated waters or solids to a pond system.

Odor-producing compounds are typically associated with anaerobic conditions which occur in both natural and treatment wetlands. The extent of these anaerobic areas is largely dependent upon BOD_5 and ammonia nitrogen loading in treatment wetlands, and hydrogen sulfide is frequently produced in wetlands. The potential for nuisance odor conditions in treatment wetlands can be reduced by reducing loadings of these oxygen-demanding constituents and by interspersing aerobic pools or channels between wetland cells and prior to final discharge. Cascading outfall structures and channels provide an opportunity to dissipate residual odors before they reach nuisance conditions.

POTENTIAL ENVIRONMENTAL PROBLEMS

Environmental problems that may occur when wetlands are used for water quality treatment include effects caused by high loadings of pollutants that are normally subsidies (for example, too much water, organic matter, or nutrients) and effects resulting from metals, pesticides, and other potentially toxic chemicals.

Excess Subsidies

Environmental problems that result from too much water, organic matter, or nutrients generally occur because excess amounts of these subsidies reduce dissolved oxygen concentrations. Drastically lowered dissolved oxygen can result in significant losses of wetland vegetation and fauna. Hydrologic and oxygen-demanding loadings to wetlands can be limited during design by providing appropriate pretreatment levels and constituent loading rates. Although wildlife impacts such as avian cholera and botulism (Friend, 1985) and parasitic nematodes (Spalding, 1990) are possible when oxygen concentrations are reduced, they are not likely to be a widespread threat to the use of wetlands for pollution control.

Environmental Toxins

Other pollutants removed by wetland treatment systems accumulate and are stored in conservative amounts in wetland sediments, plants, and wildlife. Metals and organochlorine compounds are the pollutants that are most likely to accumulate in wetlands receiving various wastewaters. Although toxins can both directly and indirectly affect treatment wetlands, there is little evidence that they are a real limitation on the use of wetlands for water quality management (Chan et al., 1982). However, where inflow concentrations of toxins are a concern, or in environments where indirect toxic or lethal conditions may develop, wetland planning and design must minimize the potential for wildlife impacts.

Heavy metals generally follow one of two behaviors in wetlands (Gardner, 1980; Rudd, 1987). Metals, such as arsenic, cadmium, chromium, nickel, and zinc are quickly concentrated in soils and plants, primarily through direct adsorption and absorption. Rooted plants also acquire some metals via uptake from soils. Bioconcentration factors between water and plant tissues for these metals range from 100 to 1000 times. In spite of elevated concentrations in plants, these metals are not magnified further through the food chain. Dry-weight concentrations decrease at higher levels of the food chain, so that biomagnification factors between water concentrations and fish tissues are less than 100 times (U.S. EPA, 1986c). Essentially, these metals reach saturation levels in tissue based on water concentrations, and additional uptake is matched by tissue metal losses, resulting in a relatively constant body burden. As long as source control or pretreatment prevents high concentrations of these metals in the wetland influent, levels that are toxic to biota will not occur.

Microbially methylated forms of mercury and lead bioaccumulate in plants and also become concentrated through food chain biomagnification. This concentration occurs because of the affinity of these metal-organic complexes with lipids, which accumulate in an organism's tissues during its lifetime. Organochlorines such as DDT and dioxins biomagnify in the wetland food chain because of the same affinity for fats.

As with other metals, excretion and release mechanisms exist for methylated mercury, lead, and organochlorines. Steady state levels that do not result in toxic effects can be reached as long as input concentrations to the wetland are low. Safe input concentrations are unclear, so compliance with published water quality standards is the best recommendation for these compounds. Existing water quality criteria for metals are intended to protect the most sensitive organisms in waters of the U.S. In wetlands, these concentrations protect invertebrates or fish that may reside near the inflow. Because metal concentrations in wastewaters are sometimes above protective criteria, a trade-off is necessary if wetlands are to be used for water quality treatment and if the potential for ancillary benefits are to be realized. Perhaps the most difficult issue is whether creating habitat for ducks and wading birds is ample justification to exceed wetland surface water metal concentrations that could be toxic to invertebrates or larval fish, but that will not be toxic to adult fish or birds. Development of biological criteria for wetlands to replace existing water quality criteria developed for streams and lakes may provide an answer to this regulatory dilemma (U.S. EPA, 1990b).

To date, no conditions in wetlands designed for treatment of municipal wastewater and stormwater have been found to be problematic to propagation of fish or other wildlife populations. The only documented cases of toxicity to wetland wildlife are releases from hazardous waste sites (for examples, see U.S. EPA, 1989d) and discharges of agricultural irrigation return flows in the western U.S. (Willard and Willis, 1988; Deason, 1989). Research with agricultural drainage water at the Kesterson NWR in California has recorded vegetation changes, loss of species, fish die-offs, and acute and chronic effects on birds, primarily from highly concentrated levels of selenium. In the case of the Kesterson Refuge, pretreatment of selenium-rich agricultural return water and rehabilitation of the contaminated wetlands was considered infeasible and the refuge and ponds have been closed (Harris, 1988).

While these examples from Kesterson, Imperial Valley, CA, Stillwater, NV, and other wildlife refuges dependent on agricultural waters were not designed for treatment but rather for habitat, they serve as a poignant example of what must be avoided when new wetland treatment systems are designed.

SUMMARY OF TREATMENT WETLAND OPERATIONAL GUIDELINES

Owners and operators may initially be unfamiliar and uncomfortable with control of their treatment wetlands. Typically, there are no shiny knobs to twirl, buttons to push, or bells to clang. If the new plants begin to die, the wetland operator may have a sinking feeling of helplessness to make changes that will result in rapid system recovery. Once water quality treatment performance declines, months may be required to reestablish control and compliance. Treatment wetlands are a new and innovative technology available for meeting strict regulatory requirements. Treatment wetland owners and operators need to have new values and experience to successfully meet challenges that arise.

This chapter summarizes the anecdotal experience that is available concerning successful wetland operation. Given a conservatively designed wetland treatment system, operation can be a joy. If a treatment wetland is underdesigned or overloaded, operation within tightly defined regulatory limits may not be possible. There are few operational controls in most treatment wetlands, and these controls must be used wisely and with common sense to maintain top performance.

Monitoring is one of the most important aspects of treatment wetlands operation. Monitoring of inflow and outflow water quality provides a system-level barometer of wetland health and performance. Monitoring of the internal wetland structure provides a reference for correlating changes in water quality performance with system structure. Routine monitoring and data analysis are essential for making informed decisions concerning control of operational variables such as water depth and mass loadings. Additional monitoring beyond regulatory requirements may be performed to accomplish specific operational goals. Because of its importance in optimizing system performance and averting unacceptable biological changes, some monitoring should be conducted in all treatment wetlands.

Because they are designated as waters of the U.S. and frequently as waters of the state in which they are found, natural wetlands used for receiving treated wastewaters are protected under a variety of laws and rules and require special attention compared to constructed wetlands. The common basis for these regulatory controls is the protection of the public's interest in clean water, flood control, interstate commerce, migratory wildlife, and threatened or endangered plant and animal species. Treated wastewater discharges cannot be permitted to natural wetlands unless they protect the designated uses of the receiving waters. Additional pretreatment beyond secondary may be required to discharge to natural wetlands to maintain these public values. Use of natural wetlands for providing additional assimilation of wastewater and residual pollutants must embrace these regulations through conservative design and careful monitoring and system operation.

Few operational controls are available in most natural wetland treatment systems. Typically, the only system controls available are variation of inflow loading rates through changes in pretreatment or through storage or diversion of flows, system rotation between alternate discharge locations or wetlands, and some control over water levels within the wetland. Vegetation management might include planting tree species, selective clearing, and the controlled use of fire. Through conservative design and preventive maintenance, natural wetland treatment systems can operate almost care free for many years.

Section 6
Wetland Case Histories

CHAPTER 26

Wetland Treatment System Inventory

INTRODUCTION

Worldwide, designers are recognizing the advantages of using wetlands to treat water. In 1993, North America had more than 200 natural and constructed wetland treatment systems (Knight et al., 1993a, 1993b), and Europe and Great Britain had more than 500 subsurface-flow wetland treatment systems (Brix, 1993a). Dozens of pollution control wetlands exist elsewhere. In addition, hundreds of natural wetlands receive waters after conventional treatment, and thousands of ponds, lagoons, ditches, and shallow lakes receive treated and untreated wastewater and are partially or wholly vegetated by wetland plants.

Although wetland treatment design is becoming more common, good information is limited. In fact, some designs are flawed because of insufficient or dated information or lack of experience. Instead of developing new information to advance treatment technology, pilot studies of design criteria frequently repeat the work of other researchers. Moreover, wetland owners submit data on new systems to regulatory agencies, but the information rarely is available to others. Sometimes, this lack of information wastes time and money, causes system failures, and discourages agencies and owners from using the technology.

To address these problems, this book provides a comprehensive guide to the design and operation wetland treatment systems. Some of the new and revised design criteria in this book are based on projects in the North American Wetland Treatment System Database (Knight, 1994). This database was developed to summarize existing information on engineered wetlands and to provide a quantitative basis for planning and design of new systems. The North American Wetland Treatment System Database is available as an electronic file from

U.S. EPA
Environmental Compliance Laboratory
Cincinnati, OH
Mr. Donald Brown
(513) 569-7630

The structure of the wetland treatment system database has been described elsewhere (Knight et al., 1993a, 1993b); this chapter summarizes its contents and also describes other databases, including the Danish Wetland Treatment System Database initiated by Hans-Henrik Schierup, Hans Brix, and Bent Lorenzen of the Botanical Institute and Aarhus University in Risskov, Denmark.

Although the North American and European databases contain information from many projects, they actually cover only a portion of the information on wetland treatment systems. Also, because the information comes from diverse sources and cannot always be verified by the database authors, the information varies in quality. However, these databases are valuable in that they allow researchers and wetland designers to summarize characteristics and trends in wetland systems and pinpoint possible new advancements. When using these databases, researchers should interpret outlying results with caution. When possible, refer to original project reports and knowledgeable individuals to learn about project-specific details.

This chapter presents information on wetland systems in the aggregate, while Chapter 27 presents detailed case histories. The case studies provide examples of the constraints and responses encountered in treatment wetland projects.

NORTH AMERICAN WETLAND TREATMENT SYSTEMS

Table 26-1 lists some of the engineered wetland treatment systems in the North America Wetland Treatment System Database. This database is restricted to systems with greater than 50,000 gal/d (190 m^3/d). It is incomplete in data content. This list includes 176 wetland treatment sites representing 203 separate wetland treatment systems. Of these systems, 154 were designed to treat municipal wastewater, 9 were designed for industrial wastewater, 6 were designed for agricultural wastewater, and 7 were designed for stormwater. Wieder (1989) reported that in 1988, at least 142 North American wetland treatment systems treated acid-mine drainage wastewater. A recent review of Canadian treatment wetlands documented 67 operational systems (NAWCC, 1995). In addition to constructed wetlands, more than 324 natural wetlands received municipal and industrial wastewater in 1988. All together, by 1994, over 650 natural and constructed wetland systems treated wastewater in North America.

WETLAND TYPES

Of the wetland treatment systems in North America in Table 26-1, 21 percent use natural wetlands and 79 percent are partially or wholly constructed. From 1987 to 1994, the number of new natural wetland treatment systems declined, while the number of new constructed wetlands increased rapidly.

Natural wetlands that are commonly used to treat wastewater are dominated by tree species, including cypress (*Taxodium* spp.), red maple (*Acer rubrum*), willow (*Salix* spp.), black gum (*Nyssa biflora*), and spruce (*Abies* spp.), or by emergent herbaceous species such as cattails (*Typha* spp.) and bulrush (*Scirpus* spp.). The dominant plant species most often established in constructed wetlands are cattails, bulrush, and common reed (*Phragmites australis*).

Of the wetland treatment sites listed in Table 26-1, 120 are surface-flow systems, 48 are subsurface-flow wetlands, and 8 systems include both surface-flow and subsurface-flow components. Surface-flow wetland treatment systems are being used to treat all categories of wastewater and continue to dominate the large new wetland treatment systems in North America. Surface-flow wetland treatment systems are treating municipal, industrial, agricultural, and runoff wastewater.

GEOGRAPHICAL DISTRIBUTION

Wetland treatment systems are located throughout most of the U.S. and in Canada (Figure 26-1). In general, systems are concentrated in areas where research initiatives or local water quality constraints facilitated the use of wetlands to treat water. For example, Florida has many engineered natural wetland treatment systems because of early research at the University of Florida in Gainesville. Mississippi, Louisiana, and Texas, which belong to EPA Regions IV and VI, have most of the existing subsurface-flow constructed wetlands in North America because of research at NASA and by the TVA. A large number of surface-flow constructed wetlands have been implemented in South Dakota because that state developed wetland design guidelines to meet the need for seasonal discharges. In addition, numerous oxidation

Table 26-1 Summary of North American Wetland Treatment Systems

Site Name	City	State	Wastewater Source[a]	Origin[b]	Hydrologic Type[c]	Wetland Area (ha)	Vegetation Type[d]	Number of Cells	Design Flow (m^3/d)	Design HLR (cm/d)
Andrews	Andrews	SC	MUN	NAT	SF	185.00	FOR	1	7,193.00	0.389
Apalachicola	Apalachicola	FL	MUN	NAT	SF	63.70	SHB	1	3,785.00	0.594
Arcata	Arcata	CA	MUN	CON	SF	15.18	MAR	6	8,781.20	5.786
Arlington	Arlington	SD	MUN	CON	SF	3.44	MAR	1	643.45	1.870
Armour	Armour	SD	MUN	CON	SF	3.36	MAR	1		
Armstrong Slough	South Florida	FL	STO	NAT	SF	12.10	MAR	1	41,880.00	34.612
Bellaire	Bellaire	MI	MUN	NAT	SF	66.34	FOR	5	2,445.00	0.369
Belle Fourche	Belle Fourche	SD	MUN	CON	SF	29.34	MAR	13	1,893.00	0.645
Benton	Benton	KY	MUN	CON	SF	3.00	MAR	2	2,800.00	9.333
Bethel	Bethel	MO	MUN	CON	SF	0.34	MAR		56.78	1.690
Biwabik	Biwabik	MN	MUN	NAT	SF	40.50	FOR	1	1,060.00	0.262
Brandt	Brandt	SD	MUN	CON	SF	1.01	MAR	1		
Bridgewater	Bridgewater	SD	MUN	CON	SF	2.02	MAR	2		
Brillion	Brillion	WI	MUN	NAT	SF	156.00	MAR	1	5,400.00	0.346
Bristol	Bristol	SD	MUN	CON	SF	1.01	MAR	1		
Brookhaven	Brookhaven	NY	MUN	CON	SF	0.49	MAR	7	113.56	2.337
Buenaventura Lakes	Buenaventura Lakes	FL	MUN	NAT	SF	68.00	FOR	2	3,029.00	0.445
Canistota	Canistota	SD	MUN	CON	SF	4.57	MAR	1		
Cannon Beach	Cannon Beach	OR	MUN	NAT	SF	7.00	FOR	2	1,174.00	1.677
Cargill/Frank Lake	High River	Alberta, Canada	IND	NAT	SF	1,093.00	MAR	1	5,300.00	0.048
Central	Central	SC	MUN	NAT	SF	31.60	FOR	1	4,543.00	1.438
Chancellor	Chancellor	SD	MUN	CON	SF	0.97	MAR	1		
Clear Lake	Clear Lake	SD	MUN	CON	SF	2.31	MAR	1		
Clermont	Clermont	FL	MUN	NAT	SF	0.60	MAR	3	42.30	0.705
Cobalt	Cobalt	Ontario, Canada	MUN	CON	SF	0.09	MAR	1	17.00	1.828
Cypress Domes	Gainesville	FL	MUN	NAT	SF	1.56	FOR	2	114.00	0.731
Des Plaines	Wadsworth	IL	OTH	CON	SF	10.13	MAR	4	4,635.00	4.576
Doland	Doland	SD	MUN	CON	SF	1.05	MAR	1		
Drummond	Drummond	WI	MUN	NAT	SF	6.00	HYB	1	300.00	0.500
Ecopond	Everglades N.P.	FL	MUN	CON	SF		MAR	1		
Eden	Eden	SD	MUN	CON	SF	0.30	MAR	1		
Ethan	Ethan	SD	MUN	CON	SF	2.83	MAR	2		
Eureka	Eureka	SD	MUN	CON	SF	16.34	HYB	4	1,044.66	0.639

Table 26-1 Continued

Site Name	City	State	Wastewater Source[a]	Origin[b]	Hydrologic Type[c]	Wetland Area (ha)	Vegetation Type[d]	Number of Cells	Design Flow (m^3/d)	Design HLR (cm/d)
Everglades Nutr. Removal	West Palm Beach	FL	OTH	CON	SF	1,406.00	MAR	4	636,208.00	4.525
Fontanges	Fontanges	Quebec, Canada	OTH	NAT	SF	0.50	MAR	2	280.00	5.600
Fort Deposit	Fort Deposit	AL	MUN	CON	SF	6.00	MAR	2	900.00	1.500
Geddes	Geddes	SD	MUN	CON	SF	0.77	MAR	1		
Great Meadows	Concord	MA	MUN	NAT	SF	22.00	MAR	1	2,000.00	0.909
Gustine	Gustine	CA	MUN	CON	SF	9.60	MAR	24	3,785.00	3.943
Gustine	Gustine	CA	MUN	NAT	SF	0.33	MAR	1		
Hamilton Marshes	Hamilton Township	NJ	MUN	NAT	SF	500.00	MAR	3		
Hay River	Hay River	NWT, Canada	MUN	NAT	SF	47.00	MAR	1	1,000.00	0.213
Hayward	Hayward	CA	MUN	CON	SF	58.68	MAR	5	75,720.00	12.904
Hidden Lake	Orlando	FL	STO	NAT	SF	3.00	FOR	1		
Hillsboro	Hillsboro	ND	IND	CON	SF	33.00	MAR	9	5,678.00	1.721
Hillsboro	Hillsboro	OR	IND	CON	SF	35.70	MAR	17		
Hilton Head Plantation	Hilton Head Plantation	SC	MUN	NAT	SF	36.50	FOR	1	1,893.00	0.519
Houghton Lake	Houghton Lake	MI	MUN	NAT	SF	79.00	MAR	2	6,360.00	0.805
Hoven	Hoven	SD	MUN	CON	SF	11.53	HYB	7	359.58	0.312
Huron	Huron	SD	MUN	CON	SF	133.55	MAR	3	9,465.00	0.709
Hurtsboro	Hurtsboro	AL	MUN	NAT	SF	0.16	MAR	2	56.00	3.500
Incline Village	Incline Village	NV	MUN	CON	SF	173.28	MAR	8	5,000.00	0.289
Ironbridge	Orlando	FL	MUN	CON	SF	494.00	HYB	17	75,720.00	1.533
Island Lake	Longwood	FL	STO	NAT	SF	42.00	MAR	1		
Jasper	Jasper	FL	MUN	NAT	SF	24.00	FOR	1		
Johnson City	Johnson City	TX	MUN	CON	SF	0.50	MAR	9	114.00	2.280
Kadoka	Kadoka	SD	MUN	CON	SF	4.98	MAR	2		
Kimball	Kimball	SD	MUN	CON	SF	6.52	MAR	1		
Kinross (Kincheloe)	Kinross	MI	MUN	NAT	SF	110.00	MAR	1	450.00	0.041
Lake Apopka Wetlands Flwy	Apopka	FL	OTH	CON	SF	750.00	MAR	2	733,536.00	9.780
Lake Cochrane San	Lake Cochrane San	SD	MUN	CON	SF	0.61	MAR	1		
Lake Jackson	Tallahassee	FL	STO	CON	SF	2.31	MAR	3		
Lake Preston	Lake Preston	SD	MUN	CON	SF	7.81	MAR	1		
Lakeland	Lakeland	FL	MUN	CON	SF	498.00	MAR	7	52,704.00	1.058
Lakeside	Lakeside	AZ	MUN	CON	SF	38.00	MAR	7	1,540.00	0.405
Leaf River	New Augusta	MS	IND	CON	SF	0.39	MAR	3	698.72	17.916

Listowel Artificial Marsh	Listowel	Ontario, Canada	MUN	CON	SF	0.87	MAR	7	154.00	1.778
Mandan (Amoco)	Mandan	ND	IND	CON	SF	16.60	MAR	11	2,650.00	1.600
Martin	Martin	SD	MUN	CON	SF	2.83	MAR	1		
Mays Chapel	Cockeysville	MD	STO	CON	SF	0.24	MAR	1	160.40	6.683
Mcintosh	Mcintosh	SD	MUN	CON	SF	3.72	HYB	3	223.32	0.600
Mellette	Mellette	SD	MUN	CON	SF	2.45	HYB	3	123.50	0.504
Minot	Minot	ND	MUN	CON	SF	13.58	MAR	4	20,817.50	15.331
Monticello	Monticello	FL	MUN	CON	SF	188.59	HYB	14	3,785.00	0.201
Moodna Basin	Harriman	NY	MUN	CON	SF	0.30	MAR	2	114.00	3.750
Mt Angel	Mt Angel	OR	MUN	CON	SF	4.05	MAR		7,570.00	18.705
Mt. View Sanitary District	Martinez	CA	MUN	CON	SF	37.00	MAR	3	5,300.00	1.432
Murdo	Murdo	SD	MUN	CON	SF	2.43	MAR	2		
Norwalk	Norwalk	IA	MUN	CON	SF	11.70	MAR	2	1,160.00	0.991
Onida	Onida	SD	MUN	CON	SF	2.83	MAR	1		
Orange County	Orlando	FL	MUN	HYB	SF	89.00	HYB	2	13,251.00	1.489
Pembroke	Pembroke	KY	MUN	CON	SF	0.93	HYB	1	340.00	3.656
Plankinton	Plankinton	SD	MUN	CON	SF	1.86	MAR	1		
Poinciana	Poinciana	FL	MUN	NAT	SF	46.60	FOR	1	1,325.00	0.284
Pottsburg	Jacksonville	FL	MUN	NAT	SF	100.00	FOR	1	14,040.00	1.404
Prariewood San	Prariewood San	SD	MUN	CON	SF	0.49	MAR	1		
Presho	Presho	SD	MUN	CON	SF	1.86	MAR	1		
Reedy Creek	Lake Buena Vista	FL	MUN	NAT	SF	82.20	FOR	3	20,066.00	2.441
Reliance	Reliance	SD	MUN	CON	SF	0.28	MAR	1		
Richmond	Richmond	CA	IND	CON	SF	36.00	MAR	2	16,000.00	4.444
Richton	Richton	MS	MUN	CON	SF		MAR	2	1,324.75	
Rosholt	Rosholt	SD	MUN	CON	SF	1.62	MAR	1		
Roslyn	Roslyn	SD	MUN	CON	SF	0.61	MAR	1		
Santa Rosa	Santa Rosa	CA	MUN	CON	SF	4.05	HYB	5	7,570.00	18.691
Sea Pines	Sea Pines	SC	MUN	NAT	SF	20.00	MAR	1	3,786.00	1.893
Seneca Army Depot	Seneca Army Depot	NY	MUN	OTH	SF	2.50	MAR	1	950.00	3.800
Show Low	Show Low	AZ	MUN	CON	SF	54.20	MAR	8	5,299.00	0.978
Silver Springs Shores	Silver Springs Shores	FL	MUN	CON	SF	21.00	MAR	2	3,786.00	1.803
Sisseton	Sisseton	SD	MUN	CON	SF	102.79	MAR	1	2,032.55	0.198
Spencer	Spencer	SD	MUN	CON	SF	1.38	MAR	1	246.03	1.788
St. Joseph	St. Joseph	MN	STO	NAT	SF	18.60	MAR	2	900.00	0.484
Stickney	Stickney	SD	MUN	CON	SF	0.89	MAR	2	257.38	2.892
Tabor	Tabor	SD	MUN	CON	SF	0.49	MAR	2		
Tripp	Tripp	SD	MUN	CON	SF	2.67	MAR	2		
University of Florida	Gainesville	FL	MUN	NAT	SF	33.00	MAR	1	7,500.00	2.273
USDA-NSCS	Orono	ME	OTH	CON	SF		MAR	1		

Table 26-1 Continued

Site Name	City	State	Wastewater Source[a]	Origin[b]	Hydrologic Type[c]	Wetland Area (ha)	Vegetation Type[d]	Number of Cells	Design Flow (m^3/d)	Design HLR (cm/d)
Vereen	Little River	SC	MUN	NAT	SF	229.00	FOR	3	9,466.00	0.413
Vermontville	Vermontville	MI	MUN	CON	SF	4.60	MAR	4	380.00	0.826
Volga	Volga	SD	MUN	CON	SF	6.07	MAR	2	825.13	1.359
Wakonda	Wakonda	SD	MUN	CON	SF	1.62	MAR	1		
Waldo	Waldo	FL	MUN	NAT	SF	2.60	FOR	1	226.00	0.689
Wall Lake San	Wall Lake San	SD	MUN	CON	SF	0.45	MAR	2		
Wessington	Wessington	SD	MUN	CON	SF	0.49	MAR	1		
West Jackson County	Ocean Springs	MS	MUN	CON	SF	22.70	MAR	7	6,057.00	2.668
White Lake	White Lake	SD	MUN	CON	SF	1.50	MAR	2		
Wildwood	Wildwood	FL	MUN	NAT	SF	204.00	FOR	3	3,786.00	0.186
Willow Lake	Willow Lake	SD	MUN	CON	SF	9.71	MAR	6	246.03	0.253
Albany	Albany	LA	MUN	CON	HYB	0.11	MAR	2	132.00	12.000
Cottonwood	Cottonwood	AL	MUN	CON	HYB	0.40	MAR	1	587.00	14.675
Crowley	Crowley	LA	MUN	CON	HYB	17.00	MAR	7	13,248.00	7.793
Degussa Corp.	Theodore	AL	IND	CON	HYB	0.89	MAR	11	2,040.00	22.921
Iselin	Iselin	PA	MUN	CON	HYB	0.22	MAR	3	45.42	2.065
Pelahatchie	Pelahatchie	MS	OTH	CON	HYB	2.63	MAR	5	2,157.00	8.202
Shelbyville	Shelbyville	MO	MUN	CON	HYB	0.16	MAR	4	280.00	17.284
Terry	Terry	MS	MUN	CON	HYB	0.52	MAR	3	378.00	7.269
Benton	Benton	KY	MUN	CON	SSF	1.46	MAR	1	341.00	2.336
Benton	Benton	LA	MUN	CON	SSF	0.48	MAR	1	1,173.00	24.438
Bradford	Bradford	AR	MUN	CON	SSF	1.13	MAR	2	757.00	6.687
Bradley	Bradley	AR	MUN	CON	SSF	0.58	MAR	4	1,134.75	19.464
Carlisle	Carlisle	AR	MUN	CON	SSF	4.35	MAR	4	3,255.11	7.492
Carville	Carville	LA	MUN	CON	SSF	0.26	MAR	1	568.00	21.846
Clarendon	Clarendon	AR	MUN	CON	SSF	0.81	MAR	4	2,649.51	32.549
Denham Springs	Denham Springs	LA	MUN	CON	SSF	6.15	MAR	3	11,355.00	18.463
Dessau Mobile Home Park	Pflugerville	TX	MUN	CON	SSF	0.21	MAR	2	567.75	27.036
Dierks	Dierks	AR	MUN	CON	SSF	0.47	MAR	2	870.55	18.562
Doyline	Doyline	LA	MUN	CON	SSF	0.28	MAR	1	416.00	14.857
Eudora	Eudora	AR	MUN	CON	SSF	1.33	MAR	2	2,271.01	17.037
Foothills Village	Loudon Co.	TN	MUN	CON	SSF	0.10	MAR	2	67.00	6.700
Foreman	Foreman	AR	MUN	CON	SSF	1.03	MAR	4	908.40	8.854
Gillett	Gillett	AR	MUN	CON	SSF	0.95	MAR	4	454.20	4.791
Greenleaves Subdivision	Mandeville	LA	MUN	CON	SSF	0.45	MAR	1	563.97	12.673
Gumdon	Gumdon	AR	MUN	CON	SSF	1.73	MAR	2	3,255.11	18.870

Hammond	Hammond	LA	OTH	CON	SSF	0.13	MAR	1	329.00	26.111
Hardin	Hardin	KY	MUN	CON	SSF	0.64	MAR	2	378.00	5.906
Haughton	Haughton	LA	MUN	CON	SSF	0.62	MAR	1	1,324.00	21.355
Hornbeck	Hornbeck	LA	MUN	CON	SSF	0.09	MAR	1	231.00	25.667
Johnson City	Johnson City	TX	MUN	CON	SSF	0.11	MAR	2	114.00	10.364
Kingston Power Plant	Kingston	TN	MUN	CON	SSF	0.26	MAR	4	76.00	2.923
Lewisville	Lewisville	AR	MUN	CON	SSF	0.70	MAR	2	1,514.00	21.629
Lockesburg	Lockesburg	AR	MUN	CON	SSF	0.32	MAR	2	567.75	17.967
Mandeville	Mandeville	LA	MUN	CON	SSF	2.61	MAR	3	5,678.00	21.755
Marion	Marion	AR	MUN	CON	SSF	2.46	MAR	8	3,785.00	15.386
Mayo Peninsula	Ann Arundel Co.	MD	MUN	CON	SSF	1.53	MAR	4	2,990.00	19.542
McNeil	McNeil	AR	MUN	CON	SSF	0.32	MAR	2	56.78	1.797
Mesquite	Mesquite	NV	MUN	CON	SSF	1.90	MAR	3	1,514.00	7.968
Monterey	Monterey	VA	MUN	CON	SSF	0.02	MAR	1	76.00	33.043
Ola	Ola	AR	MUN	CON	SSF	0.43	MAR	4	757.00	17.812
Paris Landing	Paris Landing State Park	TN	MUN	CON	SSF	0.15	MAR	1	284.00	18.933
Pembroke	Pembroke	KY	MUN	CON	SSF	0.54	MAR	1	340.00	6.296
Phillips High School	Bear Creek	AL	MUN	CON	SSF	0.20	MAR	1	76.00	3.744
Prescott	Prescott	AR	MUN	CON	SSF	0.85	MAR	2	3,217.26	37.939
Provencal	Provencal	LA	MUN	CON	SSF	0.14	MAR	1	344.00	24.571
Rector	Rector	AR	MUN	CON	SSF	1.34	MAR	5	1,324.75	9.916
Roswell	Roswell Correctional Ctr.	NM	MUN	CON	SSF	0.00	MAR	1	15.00	37.500
Shelbyville	Shelbyville	MO	MUN	CON	SSF	0.04	MAR	1	280.00	68.293
Sibley	Sibley	LA	MUN	CON	SSF	0.21	MAR	1	492.00	23.429
Smackover	Smackover	AR	MUN	CON	SSF	2.67	MAR	6	1,892.00	7.081
Swifton	Swifton	AR	MUN	CON	SSF	0.43	MAR	2	416.35	9.705
Thornton	Thornton	AR	MUN	CON	SSF	0.28	MAR	1	378.00	13.357
Tuckerman	Tuckerman	AR	MUN	CON	SSF	2.07	MAR	4	851.63	4.122
Utica, North	Utica	MS	MUN	CON	SSF	0.73	MAR	2	341.00	4.671
Utica, South	Utica	MS	MUN	CON	SSF	0.92	MAR	2	442.00	4.804
Waldo	Waldo	AR	MUN	CON	SSF	0.61	MAR	4	1,324.75	21.825
Averages:										
Natural Wetlands						97.74		2	5,421	2.18
Constructed SF						54.90		4	35,599	3.88
Constructed SSF						1.20		3	1,444	16.08
Median values:										
Natural Wetlands						40.5		1	2737	0.65
Constructed SF						3.4		2	1963	1.78
Constructed SSF						0.5		2	568	15.12

[a] Wastewater source: MUN—municipal, IND—industrial, OTH—other, STO—stormwater; [b] Origin: NAT—natural, CON—constructed, HYB—hybrid; [c] Hydrologic type: SF—surface flow, SSF—subsurface flow, HYB—hybrid; [d] Vegetation type: FOR—forested, MAR—marsh, SHB—shrub, HYB—hybrid.

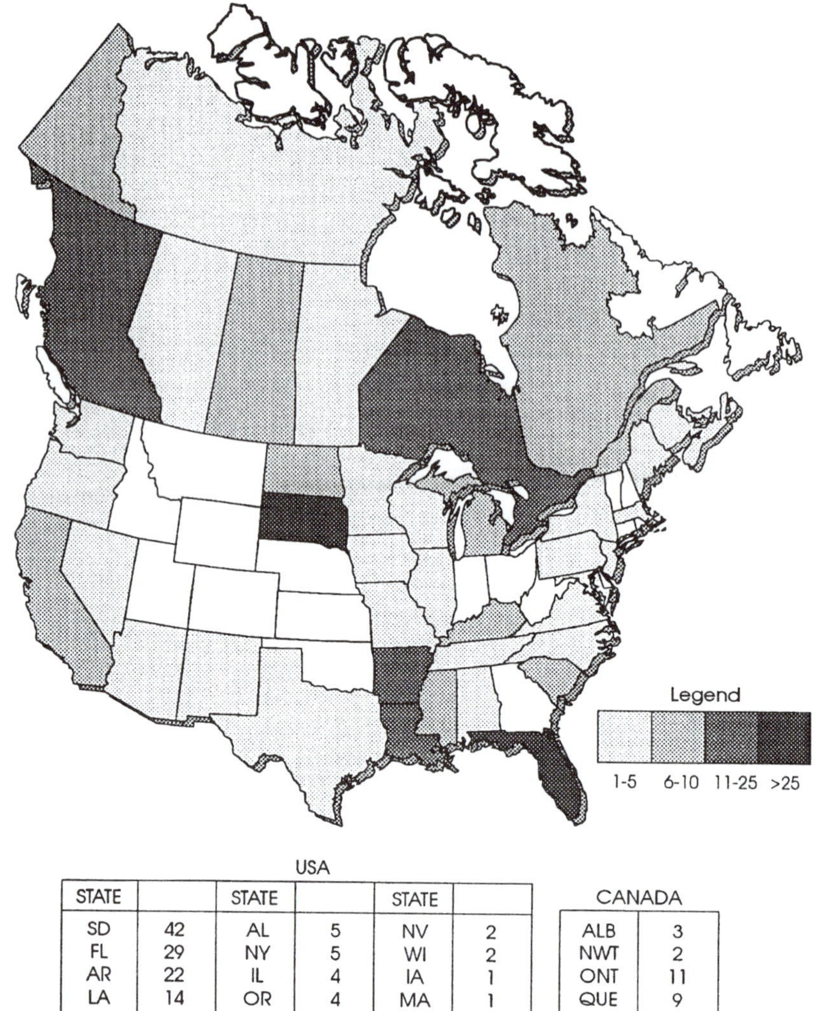

USA

STATE		STATE		STATE	
SD	42	AL	5	NV	2
FL	29	NY	5	WI	2
AR	22	IL	4	IA	1
LA	14	OR	4	MA	1
MS	10	MO	3	ME	1
CA	7	TN	3	NC	1
KY	7	TX	3	NJ	1
MI	7	AZ	2	NM	1
SC	7	MD	2	PA	1
ND	6	MN	2	VA	1
				WA	1

CANADA	
ALB	3
NWT	2
ONT	11
QUE	9
YUK	7
BC	11
SAS	7
MAN	1
NB	1
PEI	1
NS	1

Figure 26-1 Distribution of engineered wetlands in the North American (1993) and Canadian (1995) databases.

pond treatment systems in the southeastern U.S. are being retrofitted with both surface- and subsurface-flow wetlands to provide final treatment of algal solids and nutrients.

WETLAND SYSTEM COSTS

Table 26-2 summarizes capital cost information from the North American Wetland Treatment System Database (Knight et al. 1993) and from Reed (1990). Typically, surface-flow constructed wetlands cost between \$10,000 and \$100,000 per hectare (median \$44,600 per hectare), depending on system size. The major cost in surface-flow wetlands is earthwork. Subsurface-flow constructed wetlands cost about eight times more (median \$358,000 per

Table 26-2 Surface- and Subsurface-Flow Treatment Wetlands Capital Cost Information

System Name	Area (ha)	Flow (m^3/d)	Original Cost	Cost Year	1993 $/Acre (inflated at 4%)	$/ha 1993	$/$m^3$/d
SURFACE FLOW							
Ironbridge, FL	494.00	75,720	$21,020,000	1987	$21,798	$53,840	$351
Mt. View Marsh, CA	37.00	5,300	$90,000	1978	$1,774	$4,381	$31
Show Low, AR	54.20	5,299	$146,750	1980	$1,825	$4,508	$46
Jackson Bottoms, OR	3.31		$185,000	1988	$27,530	$68,000	
Lakeside, AR	38.00	1,540	$286,600	1983	$4,520	$11,164	$275
Everglades ENR, FL	1,406.00	636,208	$14,000,000	1993	$4,031	$9,957	$22
Mandan, ND	16.60		$250,000	1987	$7,715	$19,056	
Eureka, SD	16.34	1,045	$470,000	1990	$13,101	$32,359	$506
Incline Village, NV	173.28	5,000	$5,000,000	1984	$16,628	$41,070	$1,423
Arcata, CA	15.18	8,781	$514,600	1986	$18,065	$44,622	$77
Minot, ND	13.58	20,818	$475,000	1990	$15,930	$39,348	$26
American Crystal Sugar Co., ND	33.00	5,678	$1,600,000	1990	$22,081	$54,539	$317
Fort Deposit, AL	6.00	900	$374,000	1988	$30,704	$75,838	$506
Vermontville, MI	4.60	380	$395,000	1972	$79,221	$195,677	$2,639
Mt. Angel, OR	4.05	7,570	$350,000	1993	$35,014	$86,484	$46
Gustine, CA	9.60	3,785	$882,000	1986	$48,948	$120,901	$307
Mays Chapel, MD	0.24	160	$27,800	1986	$61,712	$152,429	$228
Mcintosh, MD	3.72	223	$530,000	1989	$67,425	$166,539	$2,776
AVERAGE					$26,557	$65,595	$582
MEDIAN					$18,065	$44,622	$275
SUBSURFACE FLOW							
Carlisle, AR	4.35	3,255	$335,430	1993	$31,255	$77,199	$103
Tuckerman, AR	2.07	852	$283,500	1993	$55,555	$137,222	$333
Lewisville, AR	0.70	1,514	$113,000	1993	$65,356	$161,429	$75
Phillips High School, AL	0.20	76	$36,266	1988	$106,274	$262,496	$701
Gurdon, AR	1.73	3,255	$377,411	1993	$88,579	$218,789	$116
Sibley, LA	0.21	492	$48,000	1989	$125,898	$310,969	$133
Gillet, AR	0.95	454	$229,180	1993	$97,875	$241,751	$505
Denham Springs, LA	6.15	11,355	$1,500,000	1988	$145,090	$358,373	$194
Mesquite, NV	1.90	1,514	$515,000	1991	$127,998	$316,156	$397
McNeil, AR	0.32	57	$90,756	1993	$116,276	$287,203	$1,598
Bradford, AR	1.13	757	$335,430	1993	$119,966	$296,316	$443
Smackover, AR	2.67	1,892	$800,000	1992	$130,912	$323,353	$457

Table 26-2 Continued

System Name	Area (ha)	Flow (m^3/d)	Original Cost	Cost Year	1993 $/Acre (inflated at 4%)	$/ha 1993	$/$m^3$/d
Kingston Power Plant, Kingston, TN	0.26	76	$81,000	1989	$171,597	$423,845	$1,450
Foreman, AR	1.03	908	$354,252	1993	$139,787	$345,275	$390
Dierks, AR	0.47	871	$164,758	1993	$142,225	$351,296	$189
Lockesburg, AR	0.32	568	$112,600	1993	$144,263	$356,329	$198
Mandeville, LA	2.61	5,678	$1,000,000	1990	$195,404	$482,648	$222
Carville, LA	0.26	568	$100,000	1987	$247,100	$610,336	$279
Swifton, AR	0.43	416	$165,200	1993	$155,903	$385,082	$397
Clarandon, AR	0.81	2,650	$318,600	1993	$158,462	$391,400	$120
Waldo, AR	0.61	1,325	$248,267	1992	$178,837	$441,727	$202
Eudora, AR	1.33	2,271	$639,619	1993	$194,265	$479,834	$282
Bradley, AR	0.29	568	$145,000	1993	$200,356	$494,881	$255
Benton, LA	0.48	1,173	$262,000	1987	$350,676	$866,169	$354
Hammond, LA	0.13	329	$120,000	1990	$485,719	$1,199,726	$459
Ola, AR	0.43	757	$425,360	1993	$405,201	$1,000,847	$562
Provencal, LA	0.14	344	$152,860	1990	$556,852	$1,375,426	$560
Mandeville, LA	0.45	564	$523,553	1989	$648,035	$1,600,647	$1,263
Hornbeck, LA	0.09	231	$123,870	1989	$758,092	$1,872,486	$730
AVERAGE					$218,752	$540,318	$447
MEDIAN					$145,090	$358,773	$354

Note: Data from Knight et al., 1993 and Reed, 1990.

hectare), because of gravel fill which typicaly represents 50% of the capital cost (Reed and Brown, 1992). Natural wetland treatment system capital costs range from about $5000 to $25,000 per hectare (median $18,000 per hectare). The major costs associated with natural wetlands are land purchase and distribution piping.

LOADING RATES

Reed and Brown (1992) summarized design information from surface- and subsurface-flow constructed wetlands in the U.S. They found that design hydraulic loading rates to surface-flow constructed wetlands ranged from 0.5 to 12 cm/d (0.8 to 20 ha/1000 m^3/d), and loading rates to subsurface-flow constructed wetlands ranged from 3 to 40 cm/d (0.25 to 3.3 ha/1000 m^3/d). The North American Wetland Treatment System Database documented actual hydraulic loading rates ranging from 0.04 to 22.8 cm/d (average 3.16 cm/d) for surface-flow systems and 3.72 to 33.3 cm/d (average 11.1 cm/d) for subsurface-flow wetlands.

Surprisingly, the range of design hydraulic loading rates was equally broad for advanced treatment wetlands (nutrient removal) and advanced secondary treatment wetlands. These ranges are partially a function of pretreatment and, consequently, wetland inflow quality. However, Reed and Brown (1992) also found variable (5-day biochemical oxygen demand BOD_5) mass loading rates to surface-flow systems, ranging from 2 to more than 50 kg/ha/d. Reed and Brown (1992) found no consensus on BOD_5 mass loading rates to subsurface-flow wetlands, which ranged from 10 to nearly 160 kg/ha/d in systems designed according to what the authors termed the "Region VI" concepts developed by Wolverton et al. (1983).

WETLAND PERFORMANCE

To compare individual wetland systems, Table 26-3 summarizes performance data from the North American Wetland Treatment System Database (Knight et al., 1993b). Comparisons of the performance of wetlands receiving different influent concentrations, mass loads, and wastewater characteristics can be misinterpreted because of the importance of inflow quality, loading rate, and climatic factors on removal efficiencies, outflow concentrations, and mass removal rates. For this reason, comparison of individual treatment systems and configurations should be based on multiple indices.

Table 26-4 summarizes the long-term average operational performance of North American wetland treatment systems for several key constituents in wastewater. Many of these data were used in Chapters 10 through 17 to develop empirical design equations for specific wastewater constituents.

OTHER INFORMATION

In addition to the design and operational data described previoulsy, the North American Wetland Treatment System Database includes other types of information, such as individual files for project-specific permit requirements, lists of knowledgeable contacts, and literature citations.

Permit information was available for 80 of the systems in the database. The most commonly permitted parameters in the North American wetland treatment systems are BOD_5 and total suspended solids (TSS). Typical permitted effluent limits are between 10 and 25 mg/L for BOD_5 and between 15 and 30 mg/L for TSS. NH_4-N limits are available for about 45% of the systems with permit information. These NH_4-N effluent limits are generally between 1 and 5 mg/L. Limits for dissolved oxygen, pH, and fecal coliforms are included in many permits, but limits for total nitrogen and phosphorus are relatively uncommon except in Florida.

Table 26-3 Summary of Wetland Operational Data

System Name	Hydrologic Type[a]	Record Type[b]	Area (ha)	Flow (m^3/d)	BOD In (mg/L)	BOD Out (mg/L)	TSS In (mg/L)	TSS Out (mg/L)	NH_4 In (mg/L)	NH_4 Out (mg/L)	TN In (mg/L)	TN Out (mg/L)	TP In (mg/L)	TP Out (mg/L)
Albany, LA	HYB	SA	0.1	21.95		21.4		29.2						
Ann Arundel Co., MD	SSF	SA	0.4	464.50	13.3	9.0	9.6	3.2	9.69	6.48	34.18	16.16	6.01	4.31
Arcata Constructed Wetlands, CA	HYB	SA	12.1	11,350.00	36.2	28.1	28.4	31.8						
Arlington, SD	SF	SA	3.4	1,457.30	14.4	10.8	85.0	81.0	0.13	0.08			0.85	0.63
Armstrong Slough, FL	SF	LT	12.1										0.17	0.13
Bear Bay, SC	SF	LT	28.3	876.63	13.5	1.9	16.4	2.7	2.43	0.27	17.58	2.35	3.88	0.40
Bellaire 1, MI	SF	SA	18.2	1,986.57					8.67	0.70			2.78	0.30
Bellaire 2, MI	SF	SA	10.1						7.30	1.09			1.45	0.06
Benton Cattail, KY	SF	SA	1.5	814.51	25.6	9.7	57.4	10.7	5.04	7.89	13.30	9.81	4.54	4.22
Benton Woolgrass, KY	SF	SA	1.5	818.51	25.6	12.2	57.4	15.6	7.69	6.43	14.05	9.44	4.54	3.98
Benton, LA	SSF	SA	0.5		15.3	6.3	34.6	11.3	5.28	2.74				
Black River Swamp, SC	SF	AN	185.0		65.0									
Boggy Gut, SC	SF	LT	20.2	5,826.78	6.3	3.0	10.7	3.0	4.05	1.96	11.22	3.53	4.26	3.35
Boot, FL	SF	AN	46.6	742.54	2.5	3.5	5.4	12.4	0.99	0.41	6.89	2.92	3.67	1.02
Brillion Marsh, WI	SF	SA	156.0		26.8	6.7	112.0	57.6					3.29	2.56
Cannon Beach, OR	SF	AN	7.0	1,089.54	26.8	5.4	45.2	8.0	6.00	6.00			11.77	5.57
Carville, LA	SSF	SA	0.3	318.23		8.5		3.7		2.51				
Central Slough, SC	SF	AN	32.0	5,198.17	17.1	12.0	21.0	13.1	14.01	3.68	19.51	5.33	4.77	2.43
City of Norwalk, IA	SF	SA	11.7	929.41	31.9	20.9	42.9	69.9	7.19	3.40				
Clermont Plot H, FL	SF	AN	0.2	4.44					1.67	0.26	7.54	2.16		
Clermont Plot L, FL	SF	AN	0.2	11.05					1.67	0.15	7.54	2.09		
Clermont Plot M, FL	SF	AN	0.2	25.32					1.67	0.25	7.54	2.29		
Cobalt I, ONT	SF	SA	0.1	49.42	20.7	4.6	36.2	28.0	2.95	1.04	6.98	3.38	1.68	0.77
Cottonwood, AL	HYB	SA	0.1			13.1		51.8		6.30				
Cypress Domes 1, FL	SF	LT	0.5	24.23							9.39		8.49	
Cypress Domes 2, FL	SF	LT	1.1	39.00							9.29		8.46	
Deer Park, FL	SF	AN	33.7	439.95	8.7	5.0			1.99	0.60	5.63	8.75	2.01	0.33
Degussa Corp., Theodore, AL	HYB	SA	0.9	1,249.00	5.1	4.2	25.5	4.5	5.19	3.02				
Denham Springs, LA	SSF	SA	2.1	2,547.83	25.5	9.8	47.5	14.4		4.33				
Des Plaines 3, IL	SF	SA	2.4	1,870.58			57.5	10.3	0.05	0.06	2.72	1.68	0.10	0.03
Des Plaines 4, IL	SF	SA	2.4	330.16			59.2	6.8	0.05	0.03	2.71	1.26	0.09	0.02
Des Plaines 5, IL	SF	SA	1.9	1,163.48			59.1	5.6	0.05	0.04	2.71	1.53	0.11	0.02
Des Plaines 6, IL	SF	SA	3.4	1,047.46			60.1	5.2	0.05	0.03	2.73	1.34	0.10	0.02
Drummond Bog, WI	SSF	AN	6.0	148.27					0.37	0.32	2.24	1.23	2.88	0.43
Eastern Service Area, FL	SF	LT	121.0	6,689.75	0.7	1.2	0.7	3.0	0.53	0.07	2.30	1.45	0.27	0.09
Fontanges, QUE	SF	SA	0.5		64.8	12.2			12.25	3.35	18.27	5.57	3.57	1.05
Fort Deposit, AL	SF	LT	6.0	673.73	32.8	6.9	91.2	12.6	3.16	0.76				
Great Meadows NWR, MA	SF	SA	22.0	1,796.64	38.2	5.8	27.2	15.6	8.01	1.85	13.07	4.30	2.00	0.51
Gustine, CA	SF	SA	0.4	128.71	173.3	36.7	92.5	28.3	18.72	20.09				

Hammond, LA	SSF	SA	0.1	378.00		1.0		1.0		0.03				
Hardin, KY	SSF	SA	0.3	113.91	108.5	2.5	170.9	4.9	3.00	3.41	24.55	5.82		
Hardin, KY	SSF	SA	0.3	181.49	76.6	2.5	163.1	5.2	3.60	4.97	20.74	7.68	0.00	
Haughton, LA	SSF	SA	0.6		14.3	4.9	34.6	6.1	1.10	3.03				
Hay River, NWT	SF	SA	47.0		117.7	2.7	180.5	5.8	10.30	0.39			11.00	0.26
Hidden Lake S.W., FL	SF	AN	3.0		2.7	3.0	6.4	13.0	0.05	0.05	0.58	0.66	0.07	0.16
Hornbeck, LA	SSF	SA	0.1	137.50		7.9		5.0						
Houghton Lake, MI	SF	SA	4.0	197.00	13.2		17.2		5.00	0.26			3.06	
Houghton Pilot, MI	SF	SA	44.1	4,377.81					1.54	0.30				
Huckleberry Swamp, FL	SF	AN	63.7	3,355.63	2.2	0.3			0.39	0.20			2.48	0.66
Hurtsboro, AL	SF	AN	0.3	37.00	38.1	8.2			0.97	0.92	29.20	14.96		
Incline Village, NV	SF	LT	13.9	1,043.00	18.4	8.5	16.3	16.2	15.96	1.22			23.19	3.62
Iron Bridge, FL	SF	LT	477.5	45,484.99	2.7	2.0	4.3	2.8	1.04	0.18	2.29	0.95	0.26	0.08
Iselin, PA	HYB	LT	0.1	23.27	59.0	14.8	164.7	44.3	16.07	7.17			6.87	3.40
Island Lake, FL	SF	SA	42.0						0.23	0.01			0.23	0.03
Jackson Bottoms, OR	SF	AN	6.3	1,647.38	5.1	3.0	6.1	6.8	9.95	2.94			7.31	4.28
Johnson City, TX	SF	SA	0.1	114.00	20.0	8.0	32.0	7.5	9.00	7.00			7.00	5.50
Johnson City, TX	SSF	SA	0.5	114.00	20.0	8.0	32.0	7.5	9.00	7.00			7.00	5.50
Kelly Farm, CA	SF	AN	0.8	1,363.00			18.9	20.6	1.28	0.56	10.45	8.15	2.88	2.52
Kingston Power Plant, Kingston, TN	SSF	AN	0.1	76.00	56.0	9.0	83.0	3.0	22.00	16.00			3.40	2.10
Kinross, MI	SF	SA	110.0	1,350.00	28.0	10.0	36.0	2.0	10.00	0.24				
Lake Alice, FL	SF	AN	33.0	7,500.00										
Lake Coral, FL	SF	AN	21.0	1,362.68	11.6	2.6	6.0	1.5			20.00	1.60	6.20	5.20
Lakeland, FL	SF	LT	498.0	29,192.00	3.2	2.6	6.7	5.4	0.39	0.29	5.25	2.18	6.70	5.87
Leaf River Pond 1, MS	SF	LT	0.1	84.14	19.2	11.5	80.6	26.8					2.58	1.72
Leaf River Pond 2, MS	SF	LT	0.1	129.35	19.0	12.3	79.8	35.6					2.59	2.07
Leaf River Pond 3, MS	SF	LT	0.1	162.46	19.7	13.8	80.6	40.3					2.58	2.23
Listowel 1, ONT	HYB	LT	0.4	111.50	19.6	8.0	22.8	8.6	7.15	4.85	12.20	7.94	1.05	0.67
Listowel 2, ONT	SF	LT	0.1	26.00	19.6	11.3	22.8	9.0	7.15	5.10	12.20	8.11	1.05	0.76
Listowel 3, ONT	SF	LT	0.1	27.00	19.6	7.6	22.8	9.2	7.15	3.78	12.20	6.28	1.05	0.50
Listowel 4, ONT	SF	LT	0.1	27.00	56.3	9.6	111.1	8.0	8.58	6.13	19.08	8.93	3.18	0.62
Listowel 5, ONT	SF	LT	0.1	24.50	56.3	14.6	111.1	11.6	8.58	7.93	19.08	11.30	3.18	0.99
Mandeville, LA	SSF	SA	2.6	2,839.00		10.0		7.0		0.95				
Mandeville, LA	SSF	SA	0.4	523.94	35.5	12.5	48.5	16.7						
Marion, AR	SSF	SA	2.5	2,369.19	27.0	27.8	28.0	26.0	1.95	4.06	7.11	11.56	3.44	3.94
Mays Chapel, MD	SF	SA	0.2				85.4	33.9	0.23	0.07			0.33	0.19
Meadow Marsh Pond System 1, NY	SF	LT	0.3	48.28	282.3	16.3	588.0	36.4	5.62	2.46	11.86	5.27	11.07	2.33
Mesquite, NV	SSF	SA	1.9			28.9		12.7		10.72		16.37		6.22
Monterey, CA	SSF	SA	0.0	83.00	38.0	15.0	32.0	7.0	9.33	8.67				
Moodna Basin 1, NY	SF	SA	0.2	57.00	88.0	17.2	55.0	12.0	21.15	9.90				
Moodna Basin 2, NY	SF	SA	0.2	57.00	17.5	18.2	12.8	12.2	19.65	12.85				
Mt. View Marsh, CA	SF	SA	4.3	2,820.98	33.4	23.1	21.9	47.9	10.13	6.57			6.72	7.74
Paris Landing State Park, TN	SSF	SA	0.2	169.80		3.7		5.6						
Pembroke FWS 2, KY	HYB	SA	0.9	188.00	67.4	9.4	91.9	8.2	13.80	3.35	32.45	6.03	6.03	3.16

Table 26-3 Continued

System Name	Hydrologic Type[a]	Record Type[b]	Area (ha)	Flow (m^3/d)	BOD In (mg/L)	BOD Out (mg/L)	TSS In (mg/L)	TSS Out (mg/L)	NH_4 In (mg/L)	NH_4 Out (mg/L)	TN In (mg/L)	TN Out (mg/L)	TP In (mg/L)	TP Out (mg/L)
Pflugerville, TX	SSF	SA	0.2	191.14	7.8	2.2	7.7	2.7		2.05				4.00
Phillips High School, AL	SSF	SA	0.2	58.70	15.3	1.0	63.7	2.0	11.00	1.70	51.00	6.67	6.00	0.33
Pottsburg Creek Swamp, FL	SF	LT	100.0	14,040.00					1.03	0.29	5.81	2.43	3.03	1.70
Rector, AR	SSF	SA	0.8	926.88	35.6	19.8	119.3	24.5	0.49	6.00	12.89	9.16	3.98	3.36
Reedy Creek OFWTS, FL	SF	AN	35.2	12,745.55	3.3	1.8	5.1	4.5	3.76	0.08	8.23	1.33	0.69	0.38
Reedy Creek WTS1, FL	SF	LT	5.9	3,533.47	7.0	1.6	9.7	1.7			7.80	1.10	1.70	2.20
Seneca, NY	SF	SA	2.5	622.17	19.0	5.0	11.0	14.3	4.17	1.30				
Shelbyville, MO	HYB	SA	0.2		73.6	31.8			6.84	2.87				
Smackover, AR	SSF	SA	1.6	1,308.72	19.5	17.8	28.0	20.3	3.47	1.85	9.33	8.56	2.78	2.34
Utica, MS	SSF	SA	0.4	140.36	40.4	13.5	51.8	11.3	6.11	2.63	14.07	8.75		
Utica, MS	SSF	SA	0.4	228.68	29.0	9.9	35.3	15.2	5.40	2.83	13.00	6.21		
Vermontville, MI	SF	SA	4.6	247.24		3.3		4.2		0.79				0.25
Waldo, FL	SF	AN	2.6	0.01									1.40	1.30
Waldo, AR	SSF	SA	0.6	875.68	26.6	13.7	61.3	17.2	2.26	3.85	8.26	7.32	2.12	2.00
Wetlands Flowway, FL	SF	AN	750.0	733,536.00										
Weyerhaeuser, MS	SF	SA	101.3	40,878.00	32.0	16.8	81.2	17.9	5.80	6.10	11.51	8.65	3.20	3.30
Whooping Crane, SC	SF	SA	36.5	1,862.00	4.5	3.0	9.0		3.03	0.12				
Wildwood, FL	SF	LT	204.0	946.00									0.40	0.10
WJC System, MS	SF	LT	20.9	5,620.98	25.9	7.4	40.4	14.1	3.85	1.18	11.00	4.39	4.82	3.06
Average			33.3	11,087.41	34.4	10.0	57.5	15.7	5.81	3.17	12.44	5.73	3.76	2.01
Median			1.7	573.05	20.7	8.5	36.0	11.0	5.00	1.96	11.00	5.45	3.03	1.50
Minimum			0.0	0.01	0.7	0.3	0.7	1.0	0.05	0.01	0.58	0.66	0.00	0.02
Maximum			750.0	733,536.00	282.3	36.7	588.0	81.0	22.00	20.09	51.00	16.37	23.19	7.74
N			102.0	88.00	71.0	78.0	71.0	78.0	75.00	83.00	49.00	48.00	65.00	64.00

[a] Hydrologic type: SF—surface flow, SSF—subsurface flow, HYB—hybrid.
[b] Record type: LT—long term, AN—annual average(s), SA—less than annual.

Table 26-4 Summary of North American Wetland Treatment System Operational Performance

Parameter	Type[a]	Concentration (mg/L)			Mass (kg/ha/d)[b]		
		In	Out	Eff (%)	Load	Rem	Eff (%)
BOD_5							
	SF	30.3	8.0	74	7.2	5.1	71
	SSF	27.5	8.6	69	29.2	18.4	63
	ALL	29.8	8.1	73	10.9	7.5	68
TSS							
	SF	45.6	13.5	70	10.4	7.0	68
	SSF	48.2	10.3	79	48.1	35.3	74
	ALL	46.0	13.0	72	16.8	11.9	71
NH_4-N							
	SF	4.88	2.23	54	0.93	0.35	38
	SSF	5.98	4.51	25	7.02	0.62	9
	ALL	4.97	2.41	52	1.46	0.38	26
NO_2 + NO_3-N							
	SF	5.56	2.15	61	0.80	0.40	51
	SSF	4.40	1.35	69	3.10	1.89	61
	ALL	5.49	2.10	62	0.99	0.54	55
ORG-N							
	SF	3.45	1.85	46	0.90	0.51	56
	SSF	10.11	4.03	60	7.28	4.05	56
	ALL	4.01	2.03	49	1.71	0.95	56
TKN							
	SF	7.60	4.31	43	2.20	1.03	47
	SSF	14.21	7.16	50	9.30	3.25	35
	ALL	8.11	4.53	44	2.99	1.29	43
TN							
	SF	9.03	4.27	53	1.94	1.06	55
	SSF	18.92	8.41	56	13.19	5.85	44
	ALL	9.67	4.53	53	2.98	1.52	51
ORTHO-P							
	SF	1.75	1.11	37	0.29	0.12	41
	SSF	ND	ND	ND	ND	ND	ND
	ALL	1.75	1.11	37	0.29	0.12	41
TP							
	SF	3.78	1.62	57	0.50	0.17	34
	SSF	4.41	2.97	32	5.14	1.14	22
	ALL	3.80	1.68	56	0.73	0.22	31

[a] SF—surface flow, SSF—subsurface flow.
[b] kg/ha/d × 0.892 = lb/ac/d.

Note: ND—No data, BOD_5—5-day biochemical oxygen demand; TSS—total suspended solids;—ORG-N—organic nitrogen; TKN—total Kjeldahl nitrogen; ORTHO-P—ortho phosphorus; TP—total phosphorus.

The North American Wetland Treatment System Database includes contact information for 237 individuals with first-hand knowledge of the operations of the wetlands included in the database. When little or no published information is available for a specific wetland treatment system, the reader can contact the operator or manager to see if more information exists.

The database also includes 286 citations to scientific journal articles, system design and data reports, and other documents related to the wetland systems. For example, the database indicates the existence of data related to metals and to biological indices. These data can then be obtained by locating the reports and papers listed in the literature file.

EUROPEAN WETLAND TREATMENT SYSTEMS

Because the constructed wetland technology originated in Europe over 40 years ago, it is not surprising that there are over 500 operational wetland treatment systems with many

Table 26-5 Danish Reed Bed Wetland Treatment System Design and Operational Data

System Name	Community	Type	Year Start	Design Flow (m^3/d)	Area (m^3)	Design HLR (cm/d)	P.E.	Average HLR (cm/d)	BOD_5 in (mg/L)	BOD_5 out (mg/L)	BOD_5 Eff (%)	TSS in (mg/L)
Hjordkaer	Rodekro	SSF	1984	222.00	1341	16.55	600	22.00	121	31	74.4	75
Knudby	Viborg	SSF	1984	24.05	418	5.75	65	6.10	80	8	90.0	63
Borup	Viborg	SSF	1984	74.00	1530	4.84	200	3.60	90	10	88.9	112
Skals 1	Moldrup	SSF	1985		1008			3.90	101	5	95.0	119
Skals 2	Moldrup	SSF	1985					3.90	101	7	93.1	119
Skals 3	Moldrop	SSF	1985					3.90	101	5	95.0	119
Store Binnerup	Norager	SSF	1985	44.40	513	8.65	120	1.90	261	6	97.7	143
Sejerslev	Morso	SSF	1985	555.00	7931	7.00	1500	6.20	54	7	87.0	382
Branderup	Norre Rangstrup	SSF	1985	185.00	3000	6.17	500	5.40	62	8	87.1	72
Frostrup	Hanstholm	SSF	1985	259.00	5229	4.95	700	13.00	92	15	83.7	87
Gimming	Randers	SSF	1985	92.50	1060	8.73	250	3.75	93	40	57.0	49
Dons	Kolding	SSF	1986	37.00	400	9.25	100	6.50	185	12	93.5	112
Omslev	Arhus	SSF	1986	129.50	1845	7.02	350	8.45	43	10	76.7	62
Sabro	Arhus	SSF	1986	740.00	2650	27.92	2000	9.50	44	20	54.5	28
Egense	Sejlfjord	SSF	1986		1000			1.95	51	8	84.3	47
Barmer	Nibe	SSF	1986	92.50	1440	6.42	250	10.20	155	21	86.5	117
Egebaek	Ribe	SSF	1986	555.00	4935	11.25	1500	2.00	70	15	78.6	39
Brondum	Hobro	SSF	1986	53.28	437	12.19	144	1.85	330	16	95.2	392
Valsted		SSF	1986	240.50	3710	6.48	650	1.40	171	16	90.6	219
Rudbol	Hojer	SSF	1986	46.25	2185	2.12	125		191	17	91.1	89
Fjelstervang	Videbaek	SSF	1987	273.80	4275	6.40	740	3.10	38	5	86.8	40
Hjemback	Svinninge	SSF	1987	111.00	2016	5.51	300	2.70	36	11	69.4	35
Mando		SSF	1987	120.25	1094	10.99	325	2.40	121	52	57.0	68
Vrads		SSF	1987	74.00	1504	4.92	200	1.70	38	14	63.2	28
Norby	Samso	SSF	1988	740.00	4430	16.70	2000	1.10	82	11	86.6	43
Osterby	Samso	SSF	1988	37.00	900	4.11	100	3.00	13	2	84.6	13
Ferring	Lemvig	SSF	1984	185.00	2000	9.25	500	1.15	91	7	92.3	178
Fousing Kirkeby	Struer	SSF	1984	74.00	1378	5.37	200	1.85	188	21	88.8	252
Stenhoj	Sindal	SSF	1984	92.50	1427	6.48	250		130	21	83.8	140
Virket	Stubbekobing	SSF	1985	40.70	603	6.75	110	7.20	163	25	84.7	121
Sdr. Thise	Sundsore	SSF	1985	44.40	560	7.93	120	0.80	202	63	68.8	405
Karstoft	Aaskov	SSF	1985	44.40	2400	1.85	120	7.60	135	25	81.5	91
Thise	Sundsore	SSF	1985	51.80	800	6.48	140	7.10	46	10	78.3	52
Gudum	Lemvig	SSF	1985	152.81	2668	5.73	413	8.00	70	10	85.7	60
Jaungyde	Ry	SSF	1985	46.25	800	5.78	125	3.80	53	10	81.1	
Uggerhalne	Alborg	SSF	1985	148.00	2920	5.07	400	6.90	122	13	89.3	114
Iglso	Fjends	SSF	1986	37.00	576	6.42	100	20.55	63	16	74.6	118
Sunby	Thisted	SSF	1986	181.30	2772	6.54	490		197	27	86.3	253
Lidenmark	Skovbo	SSF	1986	55.50	893	6.22	150		83	13	84.3	
Daugbjerg	Fjends	SSF	1986	92.50	1480	6.25	250	3.30	48	6	87.5	422
Svenstrup	Hammel	SSF	1986	92.50	1960	4.72	250	5.45	122	26	78.7	
Fare	Lemvig	SSF	1987	37.00	1500	2.47	100	1.30	30	3	90.0	58
Fjaltring	Lemvig	SSF	1987	111.00	2000	5.55	300	2.10	212	18	91.5	181
Lyngby	Grena	SSF	1987	214.60	4000	5.37	580	1.65	187	23	87.7	108
Hammelev	Grena	SSF	1987	83.25	1500	5.55	225	4.25	135	34	74.8	111
Homa	Grena	SSF	1987	70.30	1700	4.14	190	2.10	75	30	60.0	94
Hobjerg	Grena	SSF	1987		4100				124	17	86.3	199
Saltenskov	Them	SSF	1987	25.90	629	4.12	70	2.70	161	10	93.8	
Borum	Arhus	SSF	1987	185.00	2540	7.28	500		258	29	88.8	282
Lanum	Fjends	SSF	1987	55.50	1058	5.25	150		251	13	94.8	372
Rosmus	Ebeltoft	SSF	1987	37.00	728	5.08	100	3.80	274	8	97.1	
Hals	Alborg	SSF	1988	44.40	720	6.17	120	5.70		7		
DAKA A/S	Hedensted	SSF	1984	673.40	3000	22.45	1820		376	7	98.1	1960
Fovling	Holsted	SSF	1986		2125			5.00	15	2	86.7	852
Vogn	Sindal	SSF	1987	2627.00	8420	31.20	7100	2.00	419	149	64.4	242
Moesgard	Arhus	SSF	1983	66.60	500	13.32	180		92	14	84.8	70
Bredballegard	Arhus	SSF	1984	13.32	150	8.88	36	3.00		35		97
Kalo	Ronde	SSF	1984	81.40	940	8.66	220	0.90	110	20	81.8	
Attruphoj	Alborg	SSF	1985	48.10	900	5.34	130	2.05	230	59	74.3	76
Tjele Camping	Tjele	SSF	1987	74.00	1200	6.17	200	3.05	15	10	33.3	16
Tjele Gods	Tjele	SSF	1987	18.50	400	4.63	50		21	14	33.3	26
Rosmus Skole	Ebeltoft	SSF	1987	37.00	540	6.85	100	16.00	78	9	88.5	
Egeskov	Fredericia	SSF	1984	444.00	3600	12.33	1200		12	6	50.0	11
Lunderskov	Lunderskov	SSF	1984	74.00	1800	4.11	200	3.90	19	7	63.2	11

TSS out (mg/L)	TSS Eff (%)	NH_4-N in (mg/L)	NH_4-N out (mg/L)	NH_4-N Eff (%)	NO_{23}-N in (mg/L)	NO_{23}-N out (mg/L)	NO_{23}-N Eff (%)	TN in (mg/L)	TN out (mg/L)	TN Eff (%)	TP in (mg/L)	TP out (mg/L)	TP Eff (%)
12	84.0	25.5	20.4	20.0	1.5	0.8	46.7	34.8	26.1	25.0	10.4	8.6	17.3
10	84.1	13.5	9.2	31.9	5.1	1.7	66.7	33.1	18.5	44.1	7.8	7.2	7.7
19	83.0	15.4	11.2	27.3	2.1	2.0	4.8	33.9	22.8	32.7	10.6	7.5	29.2
12	89.9	6.4	0.8	87.5	6.3	2.9	54.0	24.0	5.8	75.8	6.5	3.1	52.3
10	91.6	6.4	2.7	57.8	6.3	2.4	61.9	24.0	9.0	62.5	6.3	3.8	39.7
4	96.6	6.4	0.9	85.9	6.3	4.5	28.6	24.0	8.7	63.8	6.3	4.1	34.9
8	94.4	36.0	24.1	33.1				51.0	31.5	38.2	14.5	11.2	22.8
7	98.2	7.3	0.9	87.7	2.8	3.6	−28.6	17.9	5.2	70.9	4.7	2.5	46.8
18	75.0	10.8	9.0	16.7				21.2	12.7	40.1	5.6	2.7	51.8
8	90.8	17.0	12.2	28.2	1.8	1.5	16.7	23.6	16.6	29.7	7.4	5.7	23.0
30	38.8	18.9	23.6	−24.9	2.5	0.3	88.0	29.6	31.2	−5.4	8.7	7.8	10.3
22	80.4	26.2	16.7	36.3	1.0	2.5	−150.0	46.4	23.9	48.5	13.6	3.9	71.3
14	77.4	15.3	10.4	32.0	2.5	3.5	−40.0	22.8	16.3	28.5	5.6	4.9	12.5
11	60.7	30.9	31.2	−1.0				37.8	35.9	5.0	12.4	11.9	4.0
6	87.2	9.3	7.9	15.1	0.6	0.4	33.3	15.7	10.6	32.5	4.9	4.3	12.2
11	90.6	27.9	24.1	13.6	0.3	0.2	33.3	41.6	30.7	26.2	10.6	10.9	−2.8
12	69.2	19.0	21.1	−11.1				26.7	21.5	19.5	7.2	7.8	−8.3
10	97.4							74.5	43.3	41.9	21.0	14.3	31.9
18	91.8	29.0	22.8	21.4	0.0	0.9		38.1	26.3	31.0	12.5	8.4	32.8
26	70.8	43.7	26.5	39.4				53.1	32.5	38.8	10.7	6.7	37.4
14	65.0	5.4	6.6	−22.2	8.9	1.8	79.8	13.2	6.2	53.0	2.5	2.2	12.0
31	11.4	15.5	7.7	50.3	2.3	2.7	−17.4	24.9	15.8	36.5	6.2	6.0	3.2
46	32.4		37.4						41.5			12.0	
7	75.0		14.0					17.0	20.6	−21.2	3.5	7.4	−111.4
14	67.4	33.0	19.6	40.6				42.6	26.1	38.7	11.4	8.0	29.8
5	61.5												
41	77.0	2.4	9.8	−308.3	0.9	1.9	−111.1	20.9	14.2	32.1	5.6	3.2	42.9
12	95.2	32.2	19.2	40.4	2.5	2.6	−4.0	49.9	26.1	47.7	14.1	9.5	32.6
16	88.6												
47	61.2	25.2	22.5	10.7	1.0	0.9	10.0	37.1	26.4	28.8	14.4	10.1	29.9
26	93.6	35.0	25.1	28.3	2.3	3.8	−65.2	50.6	32.8	35.2	13.8	9.2	33.3
16	82.4	25.1	14.2	43.4	1.0	0.5	50.0	33.1	17.6	46.8	10.3	5.5	46.6
10	80.8	7.0	3.3	52.9	6.9	2.4	65.2	15.4	7.5	51.3	3.2	2.5	21.9
13	78.3		10.0					19.4	12.4	36.1	4.0	3.1	22.5
12		7.4	8.5	−14.9	7.9	2.3	70.9	20.8	12.4	40.4	4.8	3.5	27.1
12	89.5							30.6	20.6	32.7	9.9	7.2	27.3
10	91.5		5.4			0.4		16.6	10.7	35.5	4.2	3.4	19.0
	89.3		7.8		1.4	1.1	21.4	43.2	13.5	68.8	6.3	3.1	50.8
11			11.4		1.1	1.4	−27.3	17.4	15.6	10.3	6.5	5.3	18.5
6	98.6		5.9			1.0		16.0	13.2	17.5	4.6	4.1	10.9
16		20.9	17.1	18.2				48.4	32.6	32.6	7.2	5.7	20.8
4	93.1		2.6					19.0	6.3	66.8	2.5	0.7	72.0
16	91.2		26.2					43.4	31.1	28.3	10.6	11.6	−9.4
12	88.9	22.5	17.4	22.7	0.8	0.4	50.0	39.1	21.0	46.3	13.2	7.9	40.2
20	82.0	39.6	31.1	21.5	0.2	0.4	−100.0	52.3	36.8	29.6	18.1	14.6	19.3
24	74.5	19.8	16.3	17.7	5.9	3.5	40.7	32.0	22.6	29.4	6.6	5.8	12.1
9	95.5	23.7	22.0	7.2	0.7	0.7	0.0	36.6	25.3	30.9	11.4	9.1	20.2
10		31.0	9.9	68.1				41.5	20.0	51.8	15.0	6.3	58.0
15	94.7	44.3	36.8	16.9	0.7	0.4	42.9	67.3	42.3	37.1	18.8	13.0	30.9
11	97.0		15.4			0.2		51.2	22.6	55.9	12.8	7.5	41.4
21		26.7	3.0	88.8	0.1	1.6	−1500.0	44.1	6.5	85.3	17.6	2.0	88.6
11						4.4			9.9			1.9	
9	99.5	79.1	46.6	41.1	5.5	8.4	−52.7	127.3	52.3	58.9	2.4	0.7	70.8
750	12.0	2.7	0.8	70.4				33.4	24.3	27.2	0.5	0.2	60.0
63	74.0							61.0	69.4	−13.8	8.2	8.0	2.4
10	85.7	34.8	20.0	42.5	1.2	3.5	−191.7	46.7	26.0	44.3	6.5	4.5	30.8
45	53.6												
11		36.2	28.6	21.0	0.0	2.4		48.5	35.2	27.4	9.8	7.7	21.4
15	80.3							44.0			12.0		
27	−68.8	10.0	10.8	−8.0	9.7	2.0	79.4	27.3	19.9	27.1	3.3	2.1	36.4
12	53.8	5.8	7.6	−31.0	0.8	0.6	25.0	24.7	14.7	40.5	8.8	6.5	26.1
7		23.5	6.3	73.2	0.7	0.1	85.7	33.8	8.8	74.0	5.4	2.2	59.3
10	9.1	1.6	1.5	6.3	6.3	4.4	30.2	8.3	6.0	27.7	4.6	4.3	6.5
8	27.3	7.6	4.2	44.7	3.3	1.9	42.4	13.4	6.8	49.3	4.1	3.6	12.2

Table 26-5 Continued

System Name	Community	Type	Year Start	Design Flow (m^3/d)	Area (m^3)	Design HLR (cm/d)	P.E.	Average HLR (cm/d)	BOD_5 in (mg/L)	BOD_5 out (mg/L)	BOD_5 Eff (%)	TSS in (mg/L)
Skals b1	Moldrup	SSF	1985		1008			3.90	10	3	70.0	7
Skals b2	Moldrup	SSF	1985					3.90	10	9	10.0	7
Skals b3	Moldrop	SSF	1985					8.20	10	2	80.0	7
Stoholm	Fjends	SSF	1985	2220.00	13000	17.08			3	5	−66.7	4
Ingstrup	Tjele	SSF	1984	2.59	100	2.59	7		661	60	90.9	426
Rugballegard	Vejle	SSF	1984	3.70	108	3.43	10		416	36	91.3	108
Host	Grindsted	SSF	1987	17.02	125	13.62	46		194	2	99.0	80
Average				214.67	2037	8.01	580	5	128	18	85.6	163
Median				74.00	1440	6.33	200	4	93	13	85.7	94
Minimum				2.59	100	1.85	7	1	3	2	−66.7	4
Maximum				2627.00	13000	31.20	7100	22	661	149	99.0	1960

Note: Assumes person equivalent = 0.37 m^3/d. Data from Schierup et al., 1990b.

Table 26-6 Summary of Operational Data from Gravel-Based Reed Bed Systems Operating in

System Name	Type	Year Start	Design Flow (m^3/d)	Area (m^3)	Design HLR (cm/d)	P.E.	Sample Year	BOD_5 in (mg/L)	BOD_5 out (mg/L)	BOD_5 Eff (%)	TSS in (mg/L)
Ashby Folville	SSF	1991	156.00	782	19.95	780	1991	15	3	79.3	61
Ashby Folville	SSF	1991	156.00	782	19.95	780	1992	18	3	84.2	26
Himley	SSF	1991	148.00	864	17.13	740	1991	14	2	88.9	39
Himley	SSF	1991	148.00	864	17.13	740	1992	27	4	86.4	39
Leek Wootton	SSF	1990	230.00	900	25.56	1150	1991	13	4	72.7	20
Leek Wootton	SSF	1990	230.00	900	25.56	1150	1992	12	2	82.6	21
Middleton	SSF	1987	80.00	450	17.78	400	1991	12	3	76.7	26
Middleton	SSF	1987	80.00	450	17.78	400	1992	13	2	88.0	26
Thorpe Satchville	SSF	1991	126.00	600	21.00	630	1991	15	4	74.0	35
Thorpe Satchville	SSF	1991	126.00	600	21.00	630	1992	15	3	77.2	34
Average			148.00	719.20	20.28	740.00	1991.50	15.35	2.84	81.01	32.57
Median			148.00	782	19.95	740	1991.5	14	3	81.0	30
Minimum			80.00	450	17.13	400	1991	12	2	72.7	20
Maximum			230.00	900	25.56	1150	1992	27	4	88.9	61

Note: Assumes person equivalent = 0.20 m^3/d.

From Green, M.M. and J. Upton, 1992. *Constructed Reed Beds: A Cost Effective Way to Polish Waste-* Conference & Exposition. New Orleans, LA. September 20–24, 1992. p. 13. With permission.

new systems constructed each year. Interestingly, the subsurface-flow wetland technology has dominated wetland treatment technology in Europe and, more recently, in Great Britian. This slightly different technological evolution results from initial design efforts by Kickuth and others he influenced and from a different overall philosophy concerning municipal wastewater treatment. In Europe, wetlands have typically been added to wastewater facilities that had collection systems with rudimentary levels of primary or secondary treatment. Consequently, the wetlands were designed in most cases to provide secondary and, in a few cases, advanced secondary treatment. Because of the potential for creating nuisance conditions in wetlands receiving poor quality wastewater, designers preferred subsurface flow through soil or sand planted with common reed. However, many of these systems are reported to operate largely by surface flow at normal design loading rates (Reed and Brown, 1992).

Subsurface-flow wetland designs in Great Britain and in the southern U.S. use gravel rather that soil in an attempt to achieve actual subsurface-flow. The mixed results are described elsewhere in this book. The North American surface-flow constructed wetlands and the Danish subsurface-flow wetlands with sandy or loamy soils result in similar kinetic rate constrants for most of the major wastewater constituents, suggesting that the two approaches represent variants of the same technology.

TSS out (mg/L)	TSS Eff (%)	NH_4-N in (mg/L)	NH_4-N out (mg/L)	NH_4-N Eff (%)	NO_{23}-N in (mg/L)	NO_{23} out (mg/L)	NO_{23}-N Eff (%)	TN in (mg/L)	TN out (mg/L)	TN Eff (%)	TP in (mg/L)	TP out (mg/L)	TP Eff (%)
8	−14.3	1.1	0.1	90.9	6.5	3.5	46.2	9.6	4.7	51.0	3.2	1.6	50.0
18	−157.1	1.1	0.1	90.9	6.5	3.7	43.1	9.6	5.3	44.8	3.2	3.3	−3.1
2	71.4	1.1	0.1	90.9	6.5	4.3	33.8	9.6	6.1	36.5	3.2	2.2	31.3
10	−150.0	0.1	0.7	−600.0	3.4	3.0	11.8	4.1	8.4	−104.9	2.2	4.1	−86.4
40	90.6	59.1	15.2	74.3	50.3	0.1	99.8	142.3	28.1	80.3	35.0	6.7	80.9
53	50.9	44.8	26.9	40.0	0.2	0.1	50.0	88.6	43.8	50.6	19.9	4.1	79.4
10	87.5				2.8	3.4	−21.4	72.9	10.5	85.6	19.3	0.4	97.9
27	83.4	21.0	14.1	33.0	4.1	2.0	51.4	36.7	21.0	42.7	9.1	5.8	35.6
12	82.0	19.4	11.3	30.1	2.3	1.9	31.7	33.3	20.0	37.1	7.3	5.5	29.2
2	−157.1	0.1	0.1	−600.0	0.0	0.0	−1500.0	4.1	4.7	−104.9	0.5	0.2	−111.4
750	99.5	79.1	46.6	90.9	50.3	8.4	99.8	142.3	69.4	85.6	35.0	14.6	97.9

Great Britain

TSS out (mg/L)	TSS Eff (%)	TKN in (mg/L)	TKN out (mg/L)	TKN Eff (%)	NH_4N in (mg/L)	NH_4N out (mg/L)	NH_4 Eff (%)	TN in (mg/L)	TN out (mg/L)	TN Eff (%)	TP in (mg/L)	TP out (mg/L)	TP Eff (%)
8	87.6	6.9	5.5	20.3	3.2	4.1	−28.1	17.9	16.7	6.7	6.6	3.9	40.9
4	85.6	11.3	8.6	23.9	8.0	7.0	12.5	15.8	12.2	22.8			
5	86.9	6.6	2.5	62.1	2.5	0.9	64.0	18.5	15.7	15.1	8.6	7.8	9.3
5	87.2	7.1	3.3	53.5	1.4	1.5	−7.1	6.0	0.2	96.7			
4	79.7	8.6	5.3	39.0	5.0	3.1	39.0	33.0	22.8	30.9	13.2	12.8	3.0
5	76.7	9.9	4.0	59.6	5.8	1.9	67.2	32.0	20.8	35.0			
6	77.6	6.1	1.8	70.5	3.4	0.7	79.4	20.8	14.6	29.8	11.2	9.6	14.3
5	82.4	9.3	1.8	80.6	5.1	0.5	90.2	17.2	12.2	29.1			
6	84.4	8.7	6.0	31.0	5.1	4.3	15.7	18.5	13.7	25.9	7.6	4.8	36.8
6	82.8	12.6	8.3	34.1	8.3	6.4	22.9	29.9	25.7	14.0			
5.18	83.10	8.71	4.71	47.47	4.78	3.04	35.57	20.96	15.46	30.61	9.44	7.78	20.87
5	83.6	8.7	4.6	46.2	5.1	2.5	30.95	18.5	15.2	27.5	8.6	7.8	14.3
4	76.7	6.1	1.8	20.3	1.4	0.5	−28.1	6.0	0.2	6.7	6.6	3.9	3.0
8	87.6	12.6	8.6	80.6	8.3	7.0	90.2	33.0	25.7	96.7	13.2	12.8	40.9

water Effluents for Small Communities. Proceedings of the Water Environment Federation 65th Annual

DANISH DATABASE

In 1990, Schierup et al. (1990b) published a comprehensive Danish wetland treatment system database. The database contains information from 109 systems including 67 public systems and 42 private systems. Table 26-5 lists design and operation information for 71 of these systems summarized by Schierup et al. (1990b).

The systems in Table 26-5 have an average area of about 2037 m^2. Assuming a flow of about 0.37 m^3/d per person equivalent, the average design hydraulic loading rate to these systems is about 8.0 cm/d, with a range from 1.8 to 27.9 cm/d. The actual average hydraulic loading for these systems during operation was 5.2 cm/d. These systems are very effective at providing secondary and advanced treatment: they reduce average BOD_5 from 128 to 18 mg/L, TSS from 163 to 27 mg/L, ammonia nitrogen from 21 to 14.1 mg/L, nitrate + nitrite nitrogen from 4.1 to 2.0 mg/L, total nitrogen from 36.7 to 21 mg/L, and total phosphorus from 9.1 to 5.8 mg/L. The authors found minimal seasonal variability for these systems.

The authors summarized operational data from the Danish systems using the same area-based, first-order model used for constituent removals in this book. As would be expected because of the physical, chemical, and microbial processes at work in all wetland treatment systems, the rate constants developed from the Danish wetland treament system database

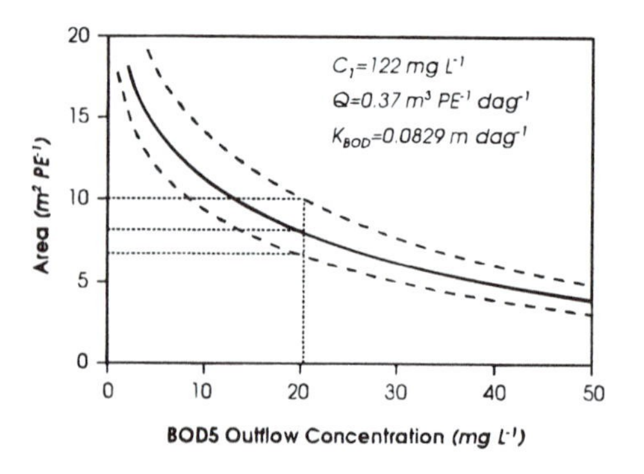

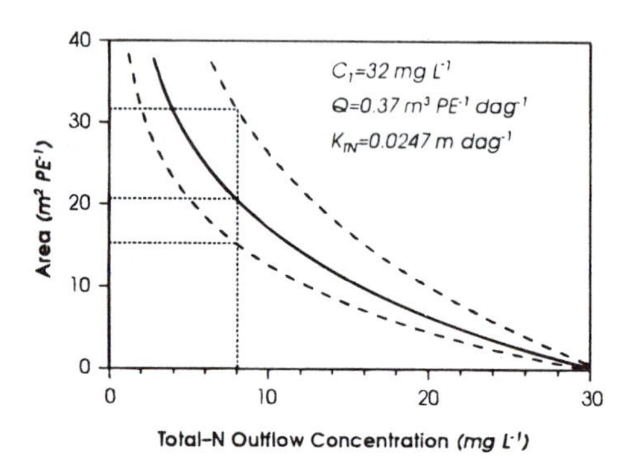

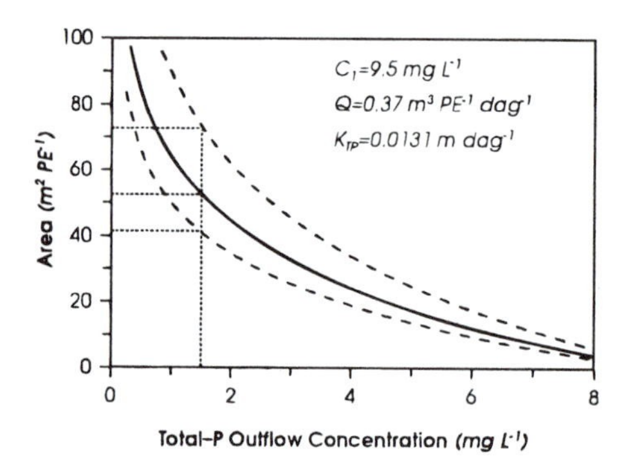

Figure 26-2 Summary of expected performance of soil-based wetland treatment systems planted with *Phragmites communis* (common reed) based on Danish database. (Adapted from Schierup et al., 1990b.)

were similar to those developed for North American wetland teratment systems. Figure 26-2 reproduces the Danish wetland treatment system design developed from analyzing system data.

GREAT BRITAIN

Researchers in Great Britain have been very interested in developing wetland treatment technology since they began to review the performance of German and North American systems in the early and mid-1980s. This interest led to a thorough review of wetland treatment system design approaches, an international conference on the use of wetlands for water quality treatment (Cooper and Findlater, 1990), and publication of design and operations guidelines for reed bed treatment systems for the European community (Cooper, 1990). These guidelines are facilitating the rapid implementation of 200 to 300 reed bed treatment systems for small communities in Great Britain (Green and Upton, 1992; Green, 1993; Cooper and Green, 1994).

Data from a large number of these systems are now available. Table 26-6 presents a summary of design and operational performance of five typical reed bed wetland treatment systems. Data from 42 wetlands is contained in Cooper and Green (1994). Hydraulic loading rates are typically much higher than in Danish systems because of the higher level of pretreatment (usually secondary with trickling filters or rotating biological contactors). These systems have removed BOD_5 and TSS very effectively during initial operation. On the other hand, nitrification and nutrient removal has been relatively minor in these reed beds. Green (1994) reported that 31 of these tertiary treatment reed bed systems served by the Severn Trent Water Company were in place in the Midland counties of England and Wales by mid-1993. Information on 268 European treatment wetlands was summarized by Börner (1992). Saurer (1994) reports on 22 Austrian systems. In this book, data from these numerous worldwide wetlands and reed beds has been used wherever possible.

SUMMARY

All aspects of wetland treatment system planning, design, construction, and operation will benefit from analyzing data from existing wetland systems. Databases can organize many sets of information in a format that allows analysis of ranges, characteristics, operational norms, and trends, and that provides references for further study. The North American Wetland Treatment System Database offers a comprehensive framework and gives an adequate impression of the performance of large surface-flow wetlands for polishing.

Existing inventories of wetland treatment systems indicate that over 1000 engineered wetlands designed for water quality treatment exist worldwide; many hundreds of other wetlands receive wastewater discharges and provide gratuitous water quality treatment benefits. As design guidelines become more standarized and existing wetland systems mature, society's confidence in wetland technology will continue to grow.

CHAPTER 27

Treatment Wetland Case Histories

INTRODUCTION

Treatment wetlands are becoming important tools for assimilating pollutants in municipal and industrial wastewater, agricultural wastewater and urban and nonpoint stormwater. Municipalities and industries are making large investments to implement new wetland treatment systems, which is somewhat surprising, considering the brief history of these treatment works. The rush to implement this technology on a widespread scale is a response to at least two factors: (1) the growing confidence that wetland treatment systems will improve water quality consistently and (2) the shortage of other affordable technologies.

Although the popularity of treatment wetlands is exciting for professionals working in the field, it also is sobering to think how many systems were built before a detailed performance record existed. With this thought, this chapter presents case history information from ten diverse treatment wetlands. These include five constructed surface-flow (SF) wetlands, two subsurface-flow (SSF) systems, and three natural treatment wetlands. The reader may wish to compare and contrast these systems to develop a personal understanding of the types of issues and challenges that arise during wetland project development and implementation.

SURFACE-FLOW CONSTRUCTED TREATMENT WETLANDS

The Incline Village, NV constructed wetland treatment system became operational in 1985 and treats and disposes of pretreated municipal wastewater by evapotranspiration. The West Jackson County, MS constructed wetland treatment system was designed as an intensive, advanced treatment wetland receiving facultative lagoon effluent in a southern, coastal plain climate. The Mandan constructed wetland treatment system in North Dakota treats a combination of process wastewater and stormwater from an oil refinery site, while providing significant wildlife enhancement. The Des Plaines wetlands project near Wadsworth, IL provides offline treatment of a channelized river receiving agricultural runoff and serves as an important recreational site. The Vermontville, MI treatment wetland is an example of an affordable and environmentally protective municipal treatment system for a small community.

Each of these projects has encountered obstacles during permitting, design, and operation. The case histories that follow provide a history gained at significant expense in money and time. Designers of new SF constructed wetland treatment systems should apply these lessons during conceptualization and implementation so that mistakes will not be repeated and so that resources can be used to improve this technology.

INCLINE VILLAGE, NEVADA

Background

Incline Village, NV uses a SF constructed wetland to dispose of secondary effluent. Near an existing, mineralized, warm-water wetland near Minden, NV, the Incline Village General

Improvement District (IVGID) developed a system that uses natural processes to renovate wastewater and benefit wildlife. The goals of this project were to (1) dispose of treated effluent effectively and economically, (2) expand the existing wetland habitat for wildlife in the arid Carson River valley, and (3) provide an educational and recreational experience for visitors.

Until 1975, effluent treated at the IVGID's 11,358 m^3/d, activated sludge plant was exported from the Lake Tahoe Basin and discharged into the Carson River during the winter and used to irrigate hay fields during the summer. A discharge permit issued in 1975 required either more stringent treatment standards or a year-round, land-based disposal system. In 1979, a facility plan recommended meeting a zero surface discharge standard with land application during the growing season and constructed wetland enhancement during the rest of the year (CH2M HILL, 1980). Local agency reviews and public hearings were held, and the wetland concept was finally approved in 1982. The design was completed in 1983, and construction finished in November 1984.

System Description

A 32-km pipeline carries the pretreated effluent from the treatment plant located near the shore of Lake Tahoe to the wetland enhancement facility. Constructed wetland cells, berms, a flood dike, and a distribution ditch are the main components of the system (Figure 27-1). The 312-ha site is made up of several distinct areas:

- Constructed wetlands
- Natural warm-water wetlands
- Seasonal storage and waterfowl areas
- Effluent storage area
- Upland area

Eight constructed wetland cells are the primary disposal area for the treated effluent. The areas of these cells and other design criteria are summarized in Table 27-1. Although treated wastewater flows from cell to cell in series, no surface water is discharged from the overall wetland disposal area because of evaporative water losses and seepage into the ground. The center of each cell has a deeper zone that discourages growth of emergent vegetation and furnishes a landing area for waterfowl. Islands within this deep zone serve as nesting sites.

The adjacent natural warm-water wetland provides a natural habitat for plants and animals and is not part of the disposal process. This perennial habitat is important for resident waterfowl populations, especially during the summer when effluent is used for agricultural irrigation and the constructed wetlands are dry.

The seasonal storage and waterfowl areas store excess water during periods of low evaporation and high rainfall. They are dry during the summer and fall, except for a small ponded area fed by warm-water springs. Three islands in this area provide nesting habitat for waterfowl. Each of the islands was planted to provide food, screened areas, and trees for birds.

The 10,600-m^3 effluent storage area is used only during high flows or heavy rainfall. The 81-ha upland area is used to dispose of effluent by spray irrigation during extended rainy weather.

Operation and Maintenance

The treated effluent passes through the 158-ha system of wetland cells and evaporates, percolates through the soil, and is transpired by plants. The system works in harmony with

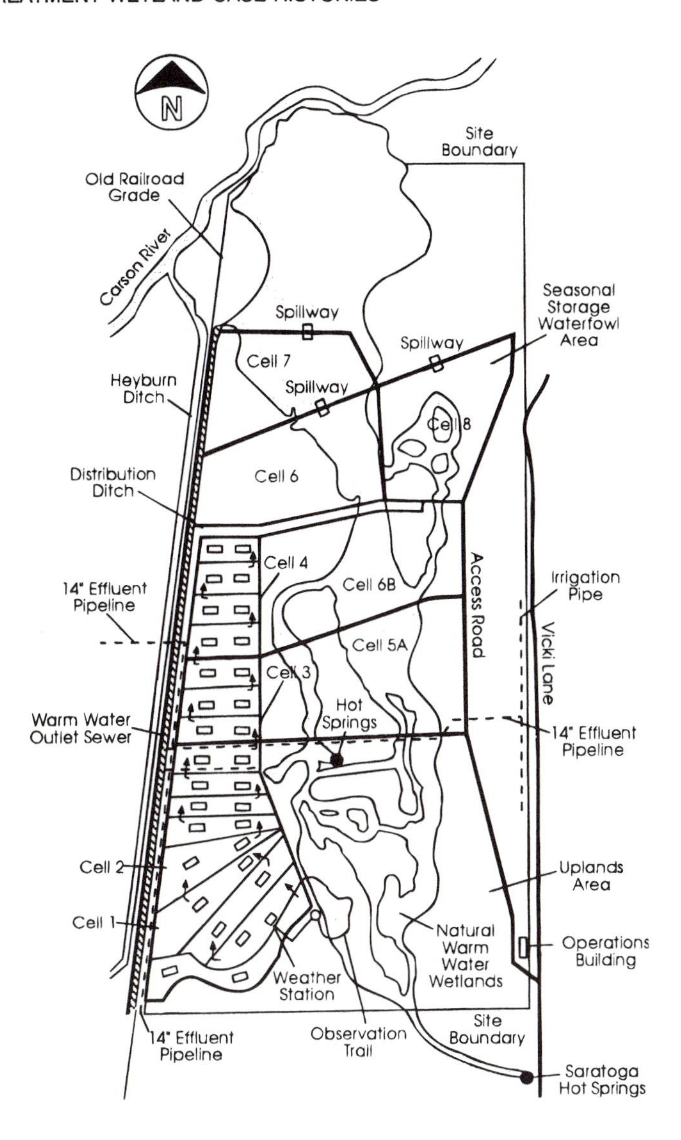

Figure 27-1 Site plan for the Incline Village wetlands enhancement facility near Minden, NV.

the existing warm-water wetlands, adapts well to year-round fluctuations in weather and temperature, meets state and EPA water quality requirements, and avoids surface discharge to the Carson River.

Effluent flows from Cell 1 through Cells 2, 3, and 4 before overflowing to the distribution ditch. Overflows from Cells 3 and 4 are diverted to Cell 5 for storage and evaporation. Water that must be stored is held in Cells 6, 7, and 8.

Plant operators use weather instrumentation and monitoring equipment to determine rainfall, evapotranspiration and percolation rates, and groundwater quality. These data are used to estimate evaporation rates at the site and to determine compliance with groundwater quality standards.

The size of the constructed wetland needed for evapotranspiration and percolation of effluent was determined by calculating the site's water balances. Evaporation rates were estimated with the Penman method using limited data available for the area. Subtracting evapotranspiration and percolation rates from rainfall yielded the net water loss from the site. Dividing the effluent volume by net water loss gave an estimate of the required acreage.

Table 27-1 Design Details for the Incline Village Wetlands Enhancement Facility

Flow, average	6,285 m^3/d
Flow, maximum daily	10,150 m^3/d
Influent quality	
Suspended solids	20 mg/L
BOD_5	20 mg/L
TDS	240 mg/L
Total phosphorus as P	6.5 mg/L
Total nitrogen as N	25 mg/L
Constructed wetland area	
Cell 1	15.3 ha
Cell 2	13.4 ha
Cell 3	11.0 ha
Cell 4	9.5 ha
Cell 5 (overflow area)	47.5 ha
Cells 6 and 7 (floodplain area)	42.7 ha
Cell 8 (seasonal storage)	17.2 ha
Wetland depth	
Emergent marsh	15 cm
Open water	60–90 cm

Note: BOD_5—5-day biochemical oxygen demand and TDS-total dissolved solids.

Percolation is critical to the project's successful operation. At least 2.75 cm of percolation per month is required at the projected flow rate. If percolation occurs at this rate, only 71 ha are needed to treat the effluent. If percolation does not occur, as much as 182 ha would be required.

Performance

Because the Incline Village constructed wetlands treatment system does not discharge to surface water, no surface water quality criteria must be met. However, Kadlec et al. (1990a) summarized many parameters monitored between the wetland cells. Even though all surface water evaporates or is lost to percolation, water quality improvements can be observed as the water passes through the cells in a serial pattern. Table 27-2 summarizes the operational conditions and water quality changes.

Throughout the period of operation of the Incline Village constructed wetland treatment system, nitrogen and phosphorus levels have been reduced in the water, even during the winter. Nutrients in the last cells display only 2 to 6 percent of the concentration values in the incoming wastewater effluent.

The effect of evaporation can be seen in the increases of total dissolved solids (TDS) and chloride ions as water moves through the cells. The evaporites in the original desert soils become rearranged by water movement and increase in concentration in the downstream cells. However, there is no evidence of a buildup of these ions in the downstream cells. Apparently, transport of solutes from upstream to downstream cells has reached a balance with losses to the groundwater.

Ancillary Benefits

Vegetation is essential to the success of the Incline Village constructed wetland. Wetland vegetation includes meadow rush (*Juncus* spp.) and salt grass (*Distichlis spicata*). During 1991, the average percent cover of live and dead vegetation at this site was 86 percent (McAllister, 1992). Upland vegetation consists primarily of sagebrush, rabbitbrush, greasewood, and salt grass, which tolerate the alkaline soils. Floodplain vegetation includes rabbitbrush and salt grass, plants which can exist in saline, silty loam, and clay soils.

Table 27-2 Performance Summary for the Incline Village Constructed Wetland Treatment Facility, 1987–1989 Annual Average (mg/L)

		Wetland Cell									
Parameter	Influent	1	2	3	4	5A	5B	6	7	8	Natural Background[a]
BOD_5	15.1	11.1	9.7	10.8	11.4	7.2	5.6	8.8	6.0	6.4	2.6
Chloride	39	123.3	70.0	95.3	114.4	104.4	121.9	175.0	176.5	173.7	44.2
TDS	269	1408	846	1283	1473	2032	2625	2110	2400	2881	861
Ammonia-N	14.1	2.2	3.8	2.7	1.0	1.0	0.3	0.3	0.2	0.1	0.22
Nitrate-N	41	33.6	30.4	26.3	22.4	3.4	4.5	1.0	0.3	0.6	0.23
Total P	24	10.1	9.7	7.2	6.2	1.2	1.5	1.8	1.2	0.7	0.11

Note: Annual average wastewater inflow was 2247 m^3/d; area of Cells 1 to 8 is about 150 ha.

[a] Onsite warm-water wetland.

Project implementation has allowed existing plant species to flourish. Planting of hundreds of trees and bushes added a new component to the ecosystem, with taller vegetation providing new perching and nesting areas for hawks and eagles.

The Incline Village constructed wetlands provide three types of wildlife habitat: permanent wetlands, seasonal wetlands, and uplands. Many types of aquatic and nonaquatic wildlife coexist at the site. Aquatic invertebrates such as insects, worms, snails, and crayfish eat algae and other plants and serve as food for larger organisms. During a quantitative macroinvertebrate survey in 1991, McAllister (1993a) found 35 species and 5869 individuals in 8.3 person hours of sampling. Fish such as largemouth bass, black bullhead, green sunfish, mosquitofish, and carp were identified before construction and were transferred to several areas within the site.

Birds occupying the Incline Village wetland site include ducks and geese, shorebirds, raptors (hawks and eagles), and passerine birds (such as blackbirds). Many migratory species travel through the Carson Valley and nest on the islands in the seasonal storage and waterfowl area or the grassy areas along the edges of the cells. During a study in 1991, waterfowl surveys over a 75-ha observation area found a maximum density of 32.7 birds per hectare during April with 13 species represented and the lowest density of 2.2 birds per hectare with 4 species represented (McAllister, 1992). At this site, a total of 20 species of birds were observed during counts in the natural wetlands, and 46 species were found in the constructed marshes (Table 27-3). Mammals common to the area include deer, coyote, skunk, mink, muskrat, rabbit, squirrel, and chipmunk.

Although the site is fenced and access is controlled, an observation area at the operations building in the southeast corner of the site encourages the public to enjoy and learn about the constructed and natural environment. Observation trails traverse the warm water and the created wetlands so that visitors can experience the diverse wildlife and vegetation and see how the project operates. Duck blinds are available to the public seasonally, and duck hunting is popular.

Lessons for Wetlands in Arid Regions

A growing number of wetland treatment systems are being constructed in arid regions. The Incline Village system represents an important model for these systems because of its size and the length of its operational history. This case history has described a number of important lessons learned at this constructed wetland treatment system. These lessons include the following: (1) an evaporative system must have some seasonal surface outlet or groundwater outlet to prevent buildup of excessive salt concentrations; (2) although wetland vegetation will die during droughts, many species reestablish quickly from seeds and root structures; (3) wildlife are attracted to arid region wetlands in great numbers and provisions must be made to protect them from excessive predation and exposure to toxic elements; and (4) arid region wetlands are magnets for humans looking for outdoor recreation.

WEST JACKSON COUNTY, MISSISSIPPI

Background and System Description

The West Jackson County, MS constructed wetland treatment system was built to provide additional effluent treatment and disposal capacity for the Mississippi Gulf Coast Regional Wastewater Authority (MGCRWA) land treatment facility. After the land treatment facility's startup in August 1990, soil hydraulic limitations prevented disposing the full design capacity. To provide additional capacity and maintain the innovative wastewater management system, constructed wetlands were added.

Table 27-3 Birds Observed at Incline Village, NV Constructed and Natural Wetlands

Species	Constructed Wetland		Natural Wetland	
	No. of Birds[a]	Frequency (%)[b]	No. of Birds	Frequency (%)
Eared grebe	97	56	0	0
Western grebe	1	6	0	0
Pied-billed grebe	7	17	0	0
American white pelican	30	11	0	0
Cattle egret	1	11	0	0
Snowy egret	1	6	1	6
Tundra swan	14	6	0	0
Canada goose	231	78	44	17
White-fronted goose	18	6	0	0
Mallard	177	94	155	44
Gadwall	254	83	139	28
Northern pintail	266	94	34	22
Green-winged teal	40	33	0	0
Blue-winged teal	10	33	2	6
Cinnamon teal	785	83	147	33
American widgeon	294	89	23	28
Northern shoveler	110	72	10	28
Redhead	74	89	13	17
Ring-necked duck	8	11	0	0
Canvasback	1	11	0	0
Scaup	17	22	0	0
Common goldeneye	11	6	0	0
Bufflehead	22	28	0	0
Ruddy duck	98	83	9	6
Other duck	10	6	0	0
Red-tailed hawk	2	11	1	6
Golden eagle	1	11	0	0
Bald eagle	1	6	0	0
Northern harrier	4	72	4	39
American kestrel	4	11	1	6
Sora	1	11	0	0
Common moorhen	1	6	0	0
American coot	1235	100	81	33
Semipalmated plover	2	6	0	0
Killdeer	31	61	2	11
Common snipe	1	6	0	0
Long-billed curlew	2	22	0	0
Sandpiper spp.	1	6	0	0
Willet	10	67	0	0
Greater yellowlegs	5	6	0	0
Long-billed dowitcher	40	6	0	0
Marbled godwit	0	0	1	6
American avocet	48	72	0	0
Black-necked stilt	72	72	10	22
Phalarope spp.	265	39	0	0
California gull	11	44	0	0
Forster's tern	0	0	4	6
Black tern	0	0	2	6
Common raven	6	6	0	0
Total number of species	46		20	

[a] Maximum number of birds of each species counted on a single survey.
[b] Frequency of observation on 18 counts.
Data from McAllister, 1993a.

Located north of Ocean Springs in coastal Mississippi, the West Jackson County land treatment facility consists of a total area of 145 ha of slow-rate land application fields distributed in three areas and 22.7 ha of constructed wetland treatment cells (Figure 27-2).

The constructed wetland treatment system was designed to treat an average daily flow of 6060 m^3/d and consists of three parallel treatment trains with seven individual wetland

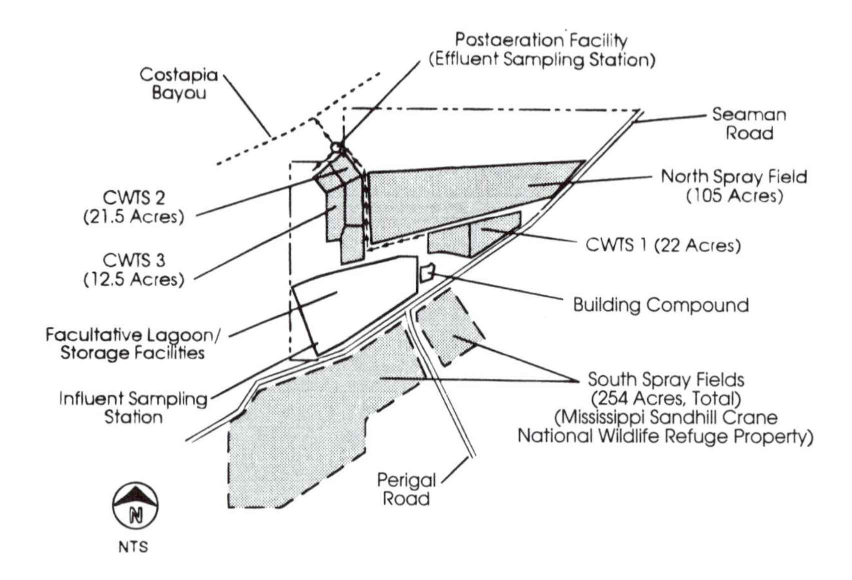

Figure 27-2 West Jackson County, MS regional land treatment facility site plan.

cells (Figure 27-3). The constructed wetland treatment area was constructed in two phases: Phase I consisted of a two-cell constructed wetland treatment system (CWTS1) of 8.9 ha, and Phase II consisted of two parallel systems of three (CWTS2) and two cells (CWTS3) each of 13.8 ha. These three parallel treatment trains are numbered from east to west in Figure 27-3.

Phase I construction of the wetland treatment system began during February 1990, earthwork and planting finished by July 1990, and flows began in August 1990. Construction of Phase II began in June 1990 and ended in February 1991. Influent flows of pretreated wastewater to Phase II began in October 1990. Plant cover was fully established in Phase I by October 1990 and in Phase II by June 1991.

System Design

Table 27-4 summarizes the design considerations incorporated in the West Jackson County constructed wetland treatment system. The area was calculated to achieve an average annual

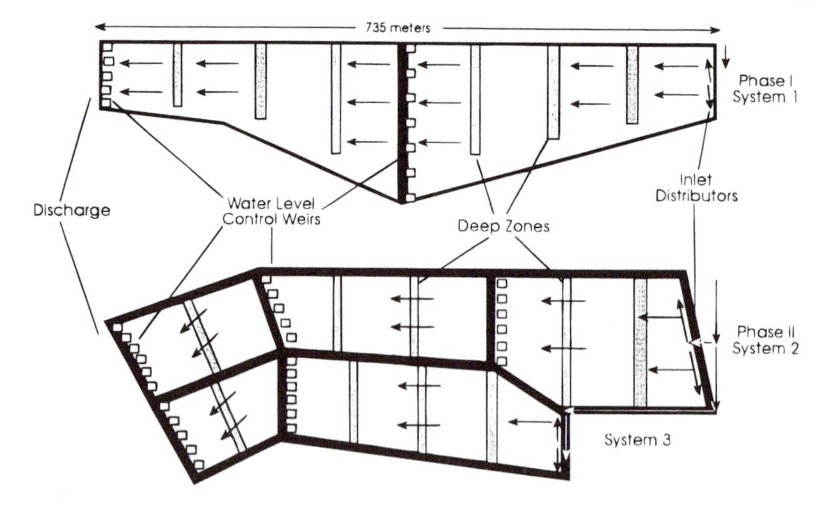

Figure 27-3 West Jackson County constructed wetland treatment system site plan.

Table 27-4 Summary of West Jackson County, MS Constructed Wetland Treatment System Design Parameters

	Phase I			Phase II							
	CWTS1			CWTS2				CWTS3			
Design Parameter	A	B	Comb.	A	B	C	Comb.	A	B	Comb.	Total System
Cell area (ha)	4.9	4.0	8.9	3.9	3.2	1.6	8.7	3.7	1.4	5.1	22.7
Open water (%)	10	10	10	8.4	7.3	6.6	7.7	10	8.7	9.8	9.1
Bottom slope (%)	0.12	0.12	0.19	0.19	0.18	0.37	0.22	0.20	0.48	0.27	0.23
Mean depth (cm)	23	23	23	23	23	23	23	23	23	23	23
Maximum depth (cm)	61	61	61	61	61	61	61	61	61	61	61
Aspect ratio	2.6	2.6	6.0	1.9	2.5	1.5	6.0	3.3	1.1	4.4	—
Volume (1000 m^3)[a]	15.5	12.9	28.4	11.4	8.7	4.5	24.6	10.2	3.8	14.0	67
Hydraulic residence time (d)[a]	6.9	5.6	12.5	4.6	3.6	1.8	10.1	7.7	2.9	10.7	—
Design flow (m^3/d)	2272	2272	2272	2461	2461	2461	2461	1325	1325	1325	6058
Hydraulic loading rate (cm/d)	4.7	5.8	2.5	6.2	7.6	15.2	2.8	3.6	9.8	2.6	2.7
Hydraulic application rate (m/d)[a]	76	88	88	85	107	70	76	61	61	61	73
Plant species[b]	CT/BR	CT/BR	CT/BR	CT	CT	CT	CT	CT	CT	CT	
Deep zones number	3	3	6	2	2	1	5	3	1	4	
Deep zones top width (m)	12	12	—	9	9	9	—	9	9	—	

[a] At normal operating depth.
[b] Cattail-CT; Bulrush-BR.

Note: Design influent quality is BOD −45 mg/L; TN −20 mg/L @ 3786 m^3/d, max loading −76 kg/d; and NH_3 −10 mg/L. Permit criteria are BOD −10 to 13 mg/L, TSS −30 mg/L, NH_3 −2 mg/L, pH −6.0 to 8.5 S.U., DO −6.0 mg/L, and Fecal coliform −2200 col/100 ml.

effluent NH_4-N goal of 2 mg/L, based on an inflow total nitrogen (TN) concentration of 20 mg/L. The equation from Knight and Iverson (1990) was used to estimate a required treatment area of 24 ha for a preliminary design flow if 3786 m^3/d. Because the wetland's size was based on meeting an average annual rather than maximum monthly NH_4-N concentration of 2 mg/L, the actual area needed to achieve maximum monthly goals was probably 1.2 to 1.6 times larger (about 38 ha).

However, because of site constraints, the actual constructed wetland treatment area was reduced to 22.7 ha. In addition, the design flow increased to 6060 m^3/d because of additional limitations discovered for the land treatment facility and because of lower average total Kjeldahl nitrogen (TKN) concentrations in the effluent from the three-cell pretreatment facultative lagoon. These changes opened the possibility of inadequate NH_4-N reduction during high flow periods and high TKN mass loading events.

The West Jackson County constructed wetland cells were designed with sloped bottoms to provide shallow, overland flow conditions in the front of the cells. Although this feature generally has not been used in constructed SF wetlands, it was incorporated based on observations of high nitrification (ammonia transformation) rates in overland flow systems and in natural wetlands with shallow water. Bottom slopes in the West Jackson County constructed wetlands vary from 0.12 to 0.48 percent and typical overland flow slopes are between 2 and 6 percent.

A second innovative feature of this constructed wetland treatment system is the deep water, unvegetated zones perpendicular to the direction of water flow. These "deep zones" enhance atmospheric oxygen diffusion, redistribute flows; provide a deep water refuge for mosquitofish and other wildlife; provide a sump for internal storage of stabilized solids; and provide convenient, intermediate sampling stations for operational research. Deep zones were also part of the Fort Deposit, AL constructed wetland.

In addition to the two design features mentioned previously, several other designs features were modeled after other wetland systems. These features included (1) effective effluent distribution using perforated header pipes, (2) weir control to lower water levels to the ground surface throughout the cells, (3) open water areas in front of outlet weirs to prevent clogging by uprooted plants and algae, (4) densely planted areas adjacent to outlet weirs to provide shading and final filtration of algal growth within the cells, (5) length-to-width ratios (aspect ratio) between 1 and 3 for construction economy and short circuiting control, (6) adequate berm freeboard to provide long system life with minimal maintenance, and (7) emergency overflow points for extreme rainfall.

Lagoon effluent is delivered to the Phase I and Phase II wetlands by different methods. CWTS1 receives all of its daily flow during "power windows" of reduced power cost dictated by the local electric authority. During 1991, influent pumping occurred at a rate of about 145 l/s from 10 p.m. until 5 or 6 a.m. This influent delivery system results in pulsed inflows and outflows at CWTS1. The equivalent loading rate to this wetland is about 12,500 m^3/d (8.2 times the design loading of 1510 m^3/d) during this pumping interval and essentially zero during the remainder of the day. The Phase II wetland cells receive inflow from the lagoon by gravity flow, resulting in continuous loading throughout the day. Flow is passively split between CWTS2 and CWTS3 by horizontal weirs, with 64 percent of the flow going to CWTS2 and 36 percent to CWTS3.

Operation and Performance

This section describes the operation and performance of the West Jackson County constructed wetland treatment system from October 1990 through December 1992. During this period, the combined facultative lagoon/constructed wetland treatment system achieved excellent reductions for 5-day biochemical oxygen demand (BOD_5) and total suspended solids (TSS). TN and total phosphorus (TP) were also reduced significantly. While long-term average outflow NH_4-N concentration was below the 2.0 mg/L goal, monthly averages reached a maximum value exceeding the permit criterion.

As discussed earlier, CWTS1 was vegetated by October 1990, and CWTS2 and CWTS3 were fully vegetated by June 1991. Pretreated wastewater inflows to CWTS1 began on August 13, 1990, and flows to Phase II began on September 28, 1990. Flows to CWTS1 were between 1310 and 3440 m^3/d with a long-term average of 2800 m^3/d. Starting in November 1991, flow rates were varied widely to approximate a constant TKN load to CWTS1 of 2.7 kg/ha/d.

Flows to Phase II were highly variable until the middle of January 1991, when they stabilized at about 2270 m^3/d. The inflow to Phase II was increased in August 1991 to 4540 m^3/d and generally kept above 3786 m^3/d until December 1992. Long-term average flow to Phase II was about 3100 m^3/d.

Routine operational water quality data are collected from the lagoon effluent (CWTS influent) and from the final effluent of the combined CWTS downstream of the three parallel wetland systems. Operational data are summarized in Table 27-5 as monthly averages.

Monthly rainfall varied from 0.2 to 47.6 cm in January 1991 and averaged 15.2 cm per month. Rainfall created a net increase from the average wetland inflow of 5810 m^3/d to an average outflow of 6720 m^3/d. The long-term average combined (average of inflow and outflow volumes) hydraulic loading rate (HLR) for the three parallel wetland treatment systems was about 2.8 cm/d. Annual average hydraulic loading rates were 3.2 cm/d for 1991 and 2.9 cm/d for 1992.

Influent pH typically varied between 7.3 and 9.2 units. Influent pH values were consistently higher than 8.0 from May through October because of algal blooms in the facultative lagoons. CWTS outflow pH generally remained below 8.0 and averaged 7.7 during the period of record. Phase II wetland cells were partially vegetated during the first 7 months of 1991, resulting in high filamentous algal production and higher outflow pH values. Outflow pH values have consistently met permit limits (6.0 to 8.5 units).

Monthly average BOD_5 inflow concentrations ranged from 8.1 to 48 mg/L, with a long-term average of 26 mg/L. The long-term average CWTS outflow BOD_5 concentration was 7.4 mg/L. The average outflow BOD_5 was noticeably lower following plant establishment in Phase II with an average of 5.2 mg/L for the 1992 period. The CWTS outflow has consistently met the summer BOD_5 limit of 10 mg/L since May 1991.

Monthly average TSS inflow concentrations ranged from about 11 to 123 mg/L, with a long-term average of 40 mg/L. The CWTS outflow TSS averaged 14 mg/L for the period reported in Table 27-5 and was increased by high TSS releases from Phase II before full vegetation establishment in June and July 1991. The average outflow TSS for the combined CWTS was 6.5 mg/L for the months after June 1991. The CWTS outflow has consistently met the TSS limit of 30 mg/L after May 1991. Because construction ended in February 1991, these data indicate that about 3 months were necessary for the system to achieve adequate plant cover to meet BOD_5 and TSS goals.

Fecal coliform populations frequently were higher in the CWTS outflow than in the inflow from the facultative lagoons. Averages increased from 243 col/100 ml at the inflow to 337 at the outflow, with peak monthly outflow values of 1200 col/100 ml during the months of July and August. The CWTS outflow consistently met the fecal coliform limit of 2200 col/100 ml. Increases in fecal coliforms in constructed wetlands are not uncommon when wildlife usage is significant. The most likely contributors to this increase are the substantial nutria and bird populations in the wetlands.

The ability of the West Jackson County constructed wetland treatment system to consistently meet the effluent NH_4-N criteria of 2 mg/L was a major concern and closely monitored

Table 27-5 Summary of Operational Performance of the West Jackson County, MS Constructed Wetland Treatment System

Month	Rain (cm)	Flow (m^3/d)		Comb. HLR (cm/d)	BOD_5 (mg/L)		TSS (mg/L)		TKN (mg/L)	NH_4-N (mg/L)	
		In	Out		In	Out	In	Out	In	In	Out
Aug-90	16.0	2074	4129	1.37	24.5	17.7	85.8	32.0	10.00	0.22	0.21
Sep-90	9.0	2278	1159	0.76	36.2	7.0	122.8	6.1	15.50	0.37	0.10
Oct-90	5.0	2765	2765	1.22	31.2	5.5	96.6	11.6	13.50	0.17	0.10
Nov-90	6.4	4317	4317	1.90	18.4	13.9	57.9	34.7	7.00	0.10	0.10
Dec-90	13.0	3936	3936	1.73	13.2	10.2	29.6	39.2	16.50	6.45	0.11
Jan-91	47.6	5353	5859	2.47	8.1	8.8	10.7	23.5	16.90	8.80	0.54
Feb-91	10.5	5766	8559	3.16	16.7	12.6	11.1	33.0	20.20	10.50	0.63
Mar-91	12.5	5818	8355	3.12	47.6	19.6	28.5	32.0	11.50	13.20	0.63
Apr-91	23.8	5598	7982	2.99	18.0	11.6	18.3	30.0	5.40	0.40	0.35
May-91	33.0	8390	9080	3.85	20.1	9.6	28.8	28.3	7.10	2.35	0.45
Jun-91	12.2	7470	7429	3.28	28.2	9.1	40.3	15.5	7.30	0.70	1.20
Jul-91	16.6	6701	8023	3.24	13.2	4.8	41.4	14.6	4.40	0.10	1.33
Aug-91	8.4	7508	7861	3.39	23.4	4.0	48.7	10.4	15.20	1.07	0.96
Sep-91	17.5	7518	8666	3.56	19.4	2.5	34.7	5.1	17.70	2.15	2.32
Oct-91	0.2	7588	6958	3.20	26.9	4.0	35.3	4.5	14.50	6.17	3.54
Nov-91	6.9	4837	5366	2.25	46.3	3.1	35.6	4.0	13.50	5.65	3.92
Dec-91	8.0	7068	7903	3.30	39.3	4.0	29.2	6.6	6.90	0.94	1.26
Jan-92	22.3	5252	7490	2.81	22.6	4.4	17.4	8.3	11.10	6.74	1.42
Feb-92	23.7	6302	8698	3.30	18.8	4.3	11.8	4.1	14.50	11.58	1.58
Mar-92	11.5	5883	7357	2.92	19.2	4.7	16.0	4.8	15.40	10.84	1.69
Apr-92	9.4	5415	5950	2.50	28.5	3.8	18.5	4.1	12.20	6.60	1.19
May-92	5.2	7195	7390	3.21	24.4	4.5	31.0	6.5	6.90	1.78	0.05
Jun-92	17.9	6991	8064	3.32	22.4		37.4		5.20	0.41	0.42
Jul-92	15.1	8650	9040	3.90	23.5	7.0	69.8	4.8	8.60	0.29	1.61
Aug-92	23.2	8012	8909	3.73	26.2	6.5	68.5	3.3	8.90	0.52	2.46
Sep-92	12.0	5848	8341	3.13	29.7	8.8	62.2	5.2	10.70	1.35	3.00
Oct-92	0.3	1121	0	0.25	34.8		32.5		15.50	4.60	
Nov-92	39.6	5120	6590	2.58	43.7	3.4	30.5	4.8	11.30	7.50	1.60
Dec-92	14.9	7734	8815	3.65	27.5	4.3	21	4	5.5	0.2	0.3
Average	15.2	5811	6724	2.76	25.93	7.40	40.41	14.11	11.34	3.85	1.18
Average 1991	16.4	6635	7670	3.15	25.6	7.8	30.2	17.3	11.72	4.34	1.43
Average 1992	16.2	6127	7220	2.94	26.78	5.17	34.72	4.99	10.48	4.37	1.39

during its operational history. To understand the NH_4-N dynamics of the constructed wetlands, all components of the nitrogen cycle were monitored.

TN inflow concentrations were only measured for 10 months, with an average value of 11 mg/L. TN outflow concentrations were measured for 13 months, with an average of 4.4 mg/L which is equivalent to an average TN concentration reduction efficiency of 60 percent.

TKN provides a measure of both dissolved NH_4-N and dissolved and bound organic nitrogen present in partially decomposed waste products. Average annual TKN in the inflow was 11.3 mg/L, with monthly averages as high as 20.2 mg/L and as low as 4.4 mg/L. TKN in the CWTS outflow averaged 3.9 mg/L, with a maximum monthly average of 6.3 mg/L.

NH_4-N concentrations in the wetland inflow were variable reflecting (1) the variability in inflow TKN and (2) the degree to which organic nitrogen had decomposed to NH_4-N in the facultative lagoons. The annual average inflow NH_4-N concentration was 3.85 mg/L, with a range of monthly averages from 0.10 mg/L to a maximum of 13.2 mg/L. Outflow NH_4-N was less than the permitted limit of 2 mg/L for 23 months, but was greater than 2 mg/L for 5 months, typically during late summer and fall. The long-term average NH_4-N outflow concentration from the wetlands was 1.18 mg/L, and annual averages during 1991 and 1992 were 1.43 and 1.39 mg/L.

TP was measured in the wetland inflow throughout the year, but was only measured in the outflow for a few months. The average TP inflow concentration was 4.82 mg/L, and the average outflow concentration was 3.06 mg/L.

Dissolved oxygen profiles were made for one or more of the wetland systems during January, April, July/August, October, and December 1991. Figure 27-4 shows the dissolved oxygen, (DO) profile data from CWTS1. Winter dissolved oxygen concentrations were typically higher than spring and summer values. This relationship was expected because of the inverse relationship between dissolved oxygen saturation potential and water temperature.

Summer dissolved oxygen values were high (up to 13 mg/L) in areas of high filamentous algal productivity and low (less than about 2 mg/L) in the more densely vegetated areas of the wetland cells. Dissolved oxygen levels during late summer were near zero throughout much of the wetland cells because of the duckweed cover and high organic nitrogen loading rates during this period. Winter dissolved oxygen values were generally above 4 mg/L, and areas with high filamentous algae populations generally had values over 10 mg/L.

The dissolved oxygen concentration typically increased 2 to 4 mg/L as the water flowed through weirs when upstream weir plates were installed and about 1 mg/L when the weir plates were off. Deep zones had no consistent effect on increasing or decreasing dissolved oxygen concentrations, even when they were free of floating duckweed populations.

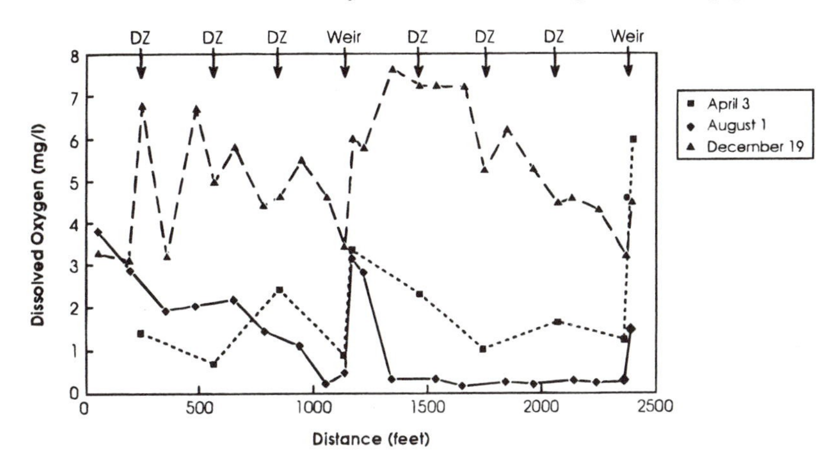

Figure 27-4 Longitudinal dissolved oxygen profile in West Jackson County CWTS1.

Typically, dissolved oxygen concentrations throughout the year decreased from the front of each cell to the downstream end of the cell, with an increase through the weirs and a subsequent decrease through the next cell. While this was observed to be the general trend, it was by no means universal. This pattern lends some support to the hypothesis that dissolved oxygen concentration is inversely proportional to wetland water depth, given that other factors such as water temperature and concentrations of BOD and NH_4-N are similar. However, correlation analysis between depth and dissolved oxygen concentration were negative but inconclusive (low correlation coefficients).

Ammonia Treatment Potential

Final NH_4-N concentration was the limiting constituent at the West Jackson County system as well as at a number of other wetlands throughout the U.S. Achieving a consistently low NH_4-N effluent concentration is important to protect receiving waters from excessive dissolved oxygen depression and to reduce the likelihood of un-ionized ammonia toxicity to invertebrates and fish. A final NH_4-N limit of 2 mg/L was initially set as a default criterion for the West Jackson County system primarily because of the toxicity concern. No relaxed winter NH_4-N limit was provided.

The total TKN loading was the design criterion because most of the nitrogen entering the wetlands as TKN (organic N plus NH_4-N) can be mineralized to NH_4-N in the system and must be transformed or stored to meet the final NH_4-N limit. The design loading for the 22.7-ha West Jackson County system was 76 kg/d (3786 m^3/d at an inflow TKN concentration of 20 mg/L) or 3.35 kg/ha/d.

High monthly average NH_4-N concentrations during the late summer and fall months at the West Jackson County system were correlated with high TKN mass loading rates and low dissolved oxygen concentrations in the wetland cells. It is likely that organic nitrogen stored in the wetland detritus and sediments was released when anaerobic conditions were reached by late summer. Mineralization of the released organic nitrogen then overwhelmed the limited nitrification capacity of the wetlands during this period of low dissolved oxygen. Two operational approaches have been used successfully at other constructed wetlands to deal with this phenomenon. These include lowering water depths throughout the summer months to maintain more aerobic conditions at the sediment surface and reducing overall TKN loadings by increasing use of the land application system. This issue is not a priority at this time because the Mississippi Bureau of Pollution Control has decided that the final NH_4-N limit was not necessary to protect the downstream receiving water.

System Biology

Biological studies at the West Jackson County constructed wetland treatment system have included vegetation monitoring, macroinvertebrate surveys, and bird counts. The U.S. EPA Constructed Wetlands Program conducted a study of the ecological condition of the West Jackson County system during 1991 to compare it to other constructed treatment wetlands around the U.S. (McAllister, 1992).

Although only cattails and bulrush were planted, 48 plant species have been identified in the three constructed wetland treatment systems at West Jackson County. These plants include floating aquatics such as duckweed and pennywort, many emergent herbaceous species, and a small number of shrubs and trees such as willows and red maple. These volunteer species were present in the seed bank of the site soils or were imported with field-harvested cattails. Although the wetland was planted in zones, a mosaic of plant cover has resulted from varying water depths, nutria herbivory, and volunteer plant colonization.

The health of the cattail populations was severely diminished because of a major infestation of cattail caterpillars throughout the summer and early fall of 1991. Caterpillar populations were extremely high and caused significant defoliation of cattails in CWTS2 and CWTS3. Cattails appeared to recover quickly, with new growth even as the caterpillars were feeding.

From observations of reduced plant growth in the bulrush zones during the summer of 1991, it was decided to keep water levels as low as possible by removing all weirs from the wetland cells. Bulrush populations recovered by October. Based on the U.S. EPA's survey during late July 1991, the average emergent plant cover in CWTS1 was 39 percent and floating plant cover was 100 percent.

Semiquantitative macroinvertebrate and fish samples were collected from each of the seven wetland cells at the West Jackson County constructed wetland treatment system in July 1991. A dip net was used to collect samples from all available habitats including floating plants, emergent plant stems, and bottom sediments in each cell, and organisms were picked for a total of 60 min for each cell.

Table 27-6 summarizes the macroinvertebrates and small vertebrates collected in July. Twenty-five invertebrate species and three vertebrate species were collected in these samples. The total number of macroinvertebrate taxa per cell ranged from 13 to 19, and total numbers of organisms collected ranged from 215 to 967. Macroinvertebrate diversity was lowest in Cell 1A and was similar in all other cells. The most common taxa were predaceous diving beetles, backswimmers, midges, and odonates.

Mosquitofish and some unidentified juvenile sunfish species were collected in most of the wetland cells. Tadpoles were collected in all cells in Phase II, but not in the Phase I wetland cells.

Bird counts or observations were made by CH2M HILL in April, July, October, and December 1991. Under contract with the U.S. EPA, Dr. Frank Moore of the University of Southern Mississippi conducted detailed bird counts on five dates between May 29 and June 26, 1991.

Table 27-7 summarizes the bird observations at the West Jackson County system during 1991. A total of 42 bird species were identified in or around the wetlands during 1991. About 26 of these species are considered to be wetland dependent. Bird populations during the winter, spring, and fall seasons were dominated by ducks, sora rails, swamp sparrows, and wading birds which were common prior to full plant establishment in the Phase II wetlands. Summer populations indicated the presence of about 7 nesting bird species and 30 species in and around the wetlands.

From these results, it is clear that the constructed wetlands provide significant cover for feeding and roosting during the nonbreeding seasons for a variety of wetland bird species. The West Jackson County constructed wetlands appear to be less important as a breeding habitat for spring and summer resident bird species.

MANDAN, NORTH DAKOTA

Background

Amoco Oil Company's Mandan Refinery is located on 122 ha of land along the west bank of the Missouri River, north of Mandan, ND (Figure 27-5). During the early 1970s, a stringent National Pollutant Discharge Elimination System (NPDES) permit was issued for this facility which required evaluation of alternatives for wastewater treatment (Litchfield, 1993). A review of the possible technologies for providing compliance with the NPDES permit indicated that a biooxidation pond system together with an existing American Petroleum Institute (API) separator and a 6-ha lagoon was the preferred alternative, based on land availability, past success with the lagoon, and low cost. Approximately 267 ha of land

Table 27-6 Summary of Qualitative Macroinvertebrate and Vertebrate Sampling at West Jackson County CWTS, Summer 1991

	Wetland Cell						
Species	1A	1B	2A	2B	2C	3A	3B
Invertebrates							
Annelida							
Hirudinea	3	6		1			
Ephemeroptera							
Baetis sp.	3	7	22	30		2	11
Trichoptera	4	13	10	1	8	4	5
Coleoptera							
Cybister sp.	5	5	39	12	16	18	13
Hydrophilidae	12	2	16	6	5	6	
Hydrophilus sp.							1
Dytiscus sp.							2
Unknown (beetle larva)		1					
Hydrocanthus sp.	115	43	78	109	35	33	58
Hemiptera							
Belostoma sp.	18	14	24	31	38	15	63
Gerridae				2			3
Ranatra sp.					1		
Notonectidae	4						
Sigara sp.	436	40	12	5	2	48	16
Lethocreus sp.							1
Diptera							
Culicinae		9	10	7	23	5	33
Diptera A					1		
Diptera C							1
Diptera B							2
Chironomidae	353	40	35	8	10	92	54
Psychoda			8	10	9	5	8
Odonata							
Anisoptera spp.	8	6	27	19	13	13	20
Zygoptera spp.	3	12	33	23	31	27	53
Hydracarina							1
Gastropoda							
Neritidae	3	17	38	47	23	42	10
Planorbidae		5	1				
Vertebrates							
Osteicthyes							
Lepomis sp.			2	1		1	1
Gambusia affinis		2	4	8	32	6	18
Amphibia							
Tadpoles			6	7	4	2	12
Number of Organisms	967	217	369	328	251	319	386
Number of Taxa	13	15	17	19	16	16	22
Shannon-Weiner Diversity Index	1.87	3.30	3.61	3.24	3.51	3.18	3.59

surrounds the refinery, and a portion of this area was converted into 11 ponds (total area 35.7 ha), while the rest is managed for wildlife habitat.

System Description

The Mandan Refinery has the capacity to process about 7592 metric tons of crude oil per day and uses about 5700 m^3/d of water from the Missouri River (Litchfield, 1993). Process water is directed to an API separator for primary treatment and then passed through the oxidation lagoon for secondary treatment. Process wastewater and stormwater are then directed through an 0.8-km earthen canal to six of the eleven cascading ponds (16.6 ha) before eventual discharge from Dam 4 to the river. The remaining five ponds (19.1 ha) are reserved for wildlife management or can be used for diversion or additional holding capacity

Table 27-7 West Jackson County CWTS Bird Observations, 1991

	Observation Period				
Common Name	**April**	**May–June**	**July**	**October**	**December**
American bittern	2	—	—	—	1
American coot	9	13	5	—	214
American crow	2	—	—	—	—
American widgeon	—	—	—	—	2
Barn swallow	2	2	—	—	—
Black-crowned night heron	—	1	—	1	—
Black-necked stilt	—	13	2	—	—
Blue-winged teal	133	—	2	50	322
Bobwhite quail	—	—	1	—	1
Brown-headed cowbird	—	1	—	—	—
Chimney swift	2	3	—	—	—
Common gallinule	—	7	2	X[a]	1
Common grackle	—	2	—	—	—
Common merganser	—	—	—	—	2
Common moorhen	—	—	—	—	17
Common nighthawk	—	1	—	—	—
Common snipe	4	—	—	—	126
Common yellowthroat	—	—	—	—	2
Cooper's hawk	—	—	—	—	1
Eastern kingbird	—	2	—	—	—
Eastern meadowlark	—	1	—	—	32
Eastern phoebe	—	—	—	—	1
Great blue heron	—	1	—	—	2
Great egret	—	1	4	5	—
Green-backed heron	—	2	3	—	—
Green-winged teal	—	—	—	—	3
Ground dove	—	—	—	—	2
Hooded merganser	—	—	—	—	2
Killdeer	1	7	—	—	6
Kingbird	2	—	—	—	—
Least tern	—	7	—	—	—
Lesser yellowlegs	35	—	—	X	2
Little blue heron	2	1	—	—	—
Loggerhead shrike	—	—	—	—	1
Long-billed dowitcher	—	—	—	—	1
Long-billed marsh wren	—	—	—	—	10
Mallard	—	3	7	—	3
Marsh hawk	—	—	—	—	1
Mottled duck	2	—	—	X	6
Mourning dove	—	1	1	—	—
Northern mockingbird	—	1	—	—	—
Palm warbler	—	—	—	—	1
Pied-billed grebe	—	3	2	—	2
Purple martin	—	1	200	—	—
Red-breasted merganser	2	—	—	—	—
Red-tailed hawk	—	—	—	—	2
Red-winged blackbird	33	28	104	—	81
Rough-winged swallow	—	1	—	—	1
Sand hill crane	—	—	7	—	—
Savannah sparrow	—	—	—	—	43
Sharp-shinned hawk	—	—	—	—	2
Shoveler	—	—	—	—	28
Snowy egret	—	1	—	—	—
Song sparrow	—	—	—	1	21
Sora rail	69	—	—	—	43
Swamp sparrow	23	—	—	X	78
Tree sparrow	103	—	—	—	108
Turkey vulture	—	1	—	—	—
Virginia rail	—	—	—	—	5
White ibis	—	—	2	—	—
Wood duck	18	2	2	—	—
Yellow-rumped warbler	—	—	—	—	1
Total Species	18	27	15	8	38
Total Individuals	444	107	344	57	1177

[a] X = observed.

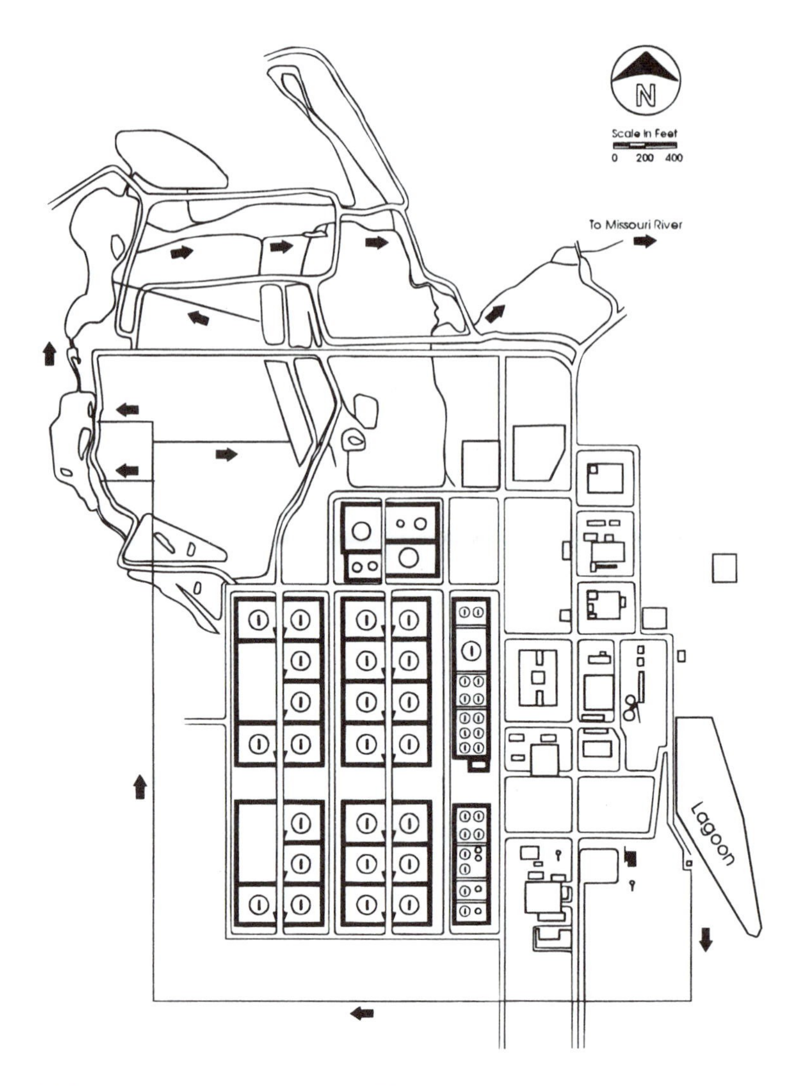

Figure 27-5 Layout of Amoco's Mandan, ND refinery and wetland/cascade ponding system.

during high stormwater runoff conditions or plant upsets. The biooxidation ponds were constructed by damming existing drainages and resulted in average water depths between 1.2 and 1.8 m. Shallower areas along the shorelines, in the earthen canal, and especially in the downstream ponds naturally colonized with emergent wetland plant species including cattails, bulrush, and other species.

Water Quality Performance

Table 27-8 summarizes 3 years of operational data from the Mandan Refinery wastewater treatment system. Performance is given for the oxidation lagoon and for the system of canals-ponds-wetlands between this pond and the final outfall to the Missouri River (Dam 4). Average mass removals occurring in the biooxidation pond system during these 3 years were BOD_5—88 percent; COD—77 percent; NH_4-N—78 percent, TSS—88 percent; sulfides—100 percent; phenols—97 percent; oil and grease—97 percent hexavalent chromium—93 percent; and total chromium—86 percent. This performance has resulted in permit compliance for

Table 27-8 Summary of Operation Performance of Amoco's Mandan Refinery Biooxidation Pond (Constructed Wetland) Treatment

			1987			1989			1990		
Parameter	Units	NPDES Limits	API Sep.	Lagoon	Dam 4	API Sep.	Lagoon	Dam 4	API Sep.	Lagoon	Dam 4
Flow	m^3/d	5700	2411	2650	2542	2433	2690	1976	2000	2500	1900
BOD_5	kg/d	197.7	603.5	79.4	12.4	506.4	72.2	5.5	456.9	61.9	7.2
COD	kg/d	1477.6	1226.5	346.7	101.0	1479.4	369.9	72.5	1393.3	436.5	92.2
NH_4-N	kg/d	131.8	26.3	16.9	2.6	29.5	20.5	6.1	35.5	28.0	5.6
TSS	kg/d	137.8	—	106.1	11.7	—	105.1	11.2	—	88.9	11.4
Sulfides	kg/d	1.3	194.3	0.2	ND[a]	60.3	ND	ND	42.8	0.1	ND
Phenols	kg/d	1.5	6.1	0.2	0.01	5.4	0.3	0.00	6.4	0.2	0.012
Oil and Grease	kg/d	59.9	49.1	21.4	1.0	42.8	1.3	ND	143.0	5.3	0.32
Hex chromium	(kg/d)	0.24	—	0.01	0.00	—	0.2	0.00	—	0.04	0.009
Total chromium	(kg/d)	3.00	—	0.72	0.18	—	0.63	0.03	—	0.10	0.012

[a] ND = not detected.

Note: Data from Litchfield and Schatz (1989) and Litchfield (1990, 1993).

an extended period. The few violations that have occurred were apparently in response to high rainfall or snowmelt events (Litchfield, 1993).

Ancillary Benefits

Following construction of the biooxidation pond-wetland system, wildlife usage of the site increased. A total of 192 bird species have been observed nearby, with about 60 species nesting in the area. Nesting habitat was increased by creating islands within the ponds and planting about 50,000 trees and shrubs. In addition, upland areas are planted in wildlife food crops such as alfalfa, millet, flax, and corn. The ponds were stocked with rainbow trout, largemouth bass, and bluegill. Since 1977, about 1246 giant Canada geese have fledged at the Amoco Mandan site.

Amoco's Mandan wetland project has received significant recognition for its contributions to wildlife conservation, including the 1980 Citizen Participation Award from the U.S. EPA, the 1986 Blue Heron Award from the National Wildlife Federation, and the Environmental Achievement Award from the Renew America Organization.

THE DES PLAINES RIVER WETLAND PROJECT

Introduction and System Description

Four wetlands have been constructed near Wadsworth, IL for purposes of river water quality improvement. The river drains an agricultural and urban watershed and carries a non-point source contaminant load of sediment, nutrients, and agricultural chemicals. It is a "good old muddy midwestern stream." In spring, it regularly floods a large amount of bottomland. In the summer of 1988, a severe drought caused it to dry to a disconnected string of pools. The river flows south, draining 520 km^2 in southern Wisconsin and northeastern Illinois. Eighty percent of the watershed is agricultural, and 20 percent is urban. The river is polluted with nonpoint source contaminants from a variety of land-use activities and point source contaminants from small domestic treatment plants. In support of previous agricultural uses, low-lying portions of the site were drained by means of tiles. Past uses of the site included pasture and a Christmas tree farm, which resulted in the demise of most of the original wetlands and associated fauna and flora.

The Des Plaines River Wetland Demonstration Project was designed to produce the criteria necessary for rebuilding our river systems through the use of wetlands and for developing management programs for the continued operation of the new structures. An ongoing research program is assessing wetland functions through large-scale experimentation, controlled manipulation of flow rates and water depths, testing of soil conditions, and the employment of a wide variety of native plant communities.

The site is located 56 km north of Chicago. It incorporates 4.5 km of the upper Des Plaines River and 182 ha of riparian land. Water is pumped from the river to the wetlands, from a point just south of Wadsworth Road (Figure 27-6). This energy-intensive alternative was necessary because of site constraints and because of the desire to explore a wide range of hydraulic conditions. Gravity diversion would be a preferred alternative in most applications of this technology. Water leaving the wetlands returns to the river via grassy swales. About 15 percent of the variable stream flow is pumped to the wetlands and allowed to return from the wetlands to the river through control structures followed by vegetated channels (Figure 27-7). Native wetland plants have been established, ranging from cattail, bulrushes, water lilies, and arrowhead to duckweed and algae. Pumping began in 1989 and has continued during the ensuing spring, summer, and fall periods. Intensive wetland research began in late summer 1989 and continues to present.

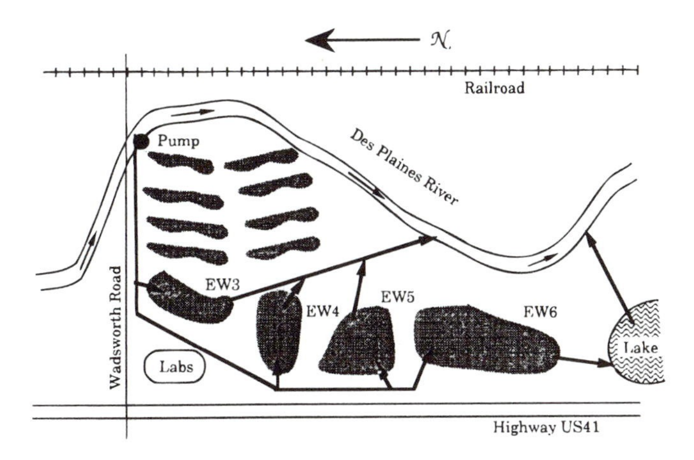

Figure 27-6 Layout of the Des Plaines wetlands. The larger wetlands, EW3-6, were established in 1988–1989. The smaller wetlands were established in 1992 for the purpose of studying sedge meadow construction.

The umbrella organization is Wetlands Research, Inc. (WRI), a not-for-profit corporation. Research has been executed by a number of organizations, including WRI, Iowa State University, Ohio State University, The University of Michigan, Western Illinois University, College of Lake County, D. L. Hey and Associates, M. C. Herp Surveys, North Dakota State University, Northeastern Illinois Planning Commission, Northern Illinois University, Northwestern University, The Illinois State Water Survey, The Illinois Institute of Technology, The Illinois State Geological Survey, The Morton Arboretum, and others. Support for the project has been provided by private individuals (4), charitable trusts (14), private companies (15), and governmental agencies (5). Contributions have been both in-kind and financial.

Hydrology

The hydrology of the wetland complex has been studied extensively. Groundwater investigations showed a relatively complex local flow pattern, with some groundwater interactions

Figure 27-7 Wetlands EW3 and EW4 are encircled by access roads and bordered by US Highway 41 (top) and Wadsworth Road (right). Flow enters EW3 from the right and enters EW4 from the top. Both discharge to a swale (left corner), which is connected to the Des Plaines River (extreme bottom). The sedge meadow wetlands dominate the foreground of this oblique aerial photo.

with the river. Wetland EW5 leaks to groundwater, as does wetland EW5 to a minor extent (Table 27-9). For water year 1990 (October 1989–September 1990), precipitation and evapotranspiration were equal.

Pumping occurred for all weeks in 1990, but was discontinued in the winter in subsequent years. The pump is run on weekdays for a prescheduled period. In water year 1990, it was run 10.5 percent of the time. The experimental design provides for different average hydraulic loading rates, ranging from 0.7 to 8.7 cm/d. Outflow from the wetlands is controlled by weirs. Thus, the hydrologic regime is cyclic, with increasing water levels and flows during the few daily hours of pumping, followed by a lowering of water levels and a slowing of flows during the off hours.

The wetland internal flow patterns are not ideal in any sense of the word. The nominal detention times in the wetlands range from 1 to 3 weeks under moderate to high flow conditions. Some of the pumped water moves quickly toward the outlet and reaches it in about 1 day's time. Other portions of the pumped water are trapped in the litter and floc near the wetland bottom. Still other portions are slowed by plant clumps or blown off course by the wind. The net effect is that some water takes three times as long as the average to find its way out of the wetland.

Tracer studies have been run at Des Plaines using lithium chloride as the tracer material. A sudden dump of dissolved lithium is made into the wetland inflow. The outflow is then analyzed for the lithium, which appears at varying concentrations and at various times after the dump. These tests have shown that the degree of mixing within the wetlands is higher than expected. But surprisingly, there is not a great deal of difference between wetlands, even though they differ in shape.

Water Quality

According to the results of benthic surveys, the stream is classified as semipolluted. A primary water quality problem of the river is associated with turbidity. With a mean concentration of 59 mg/L, over 4500 metric tons of suspended solids enter the site per year via the Des Plaines River and Mill Creek. Seventy-five percent of these solids are inorganic and 95 percent are less than 63 μm in size. Sediment removal efficiencies ranged from 86 to 100 percent for the four cells during the summer and from 38 to 95 percent during the winter, when incoming TSS is very low (Table 27-10).

A fish story developed in 1990. The solids in the wetland effluents were steadily increasing with each passing week. The source of the problem was found: a large number of carp were growing up in the wetlands. These fish foraged in the wetland sediments causing resuspension of solids. They entered as fry in the pumped water and grew to 8 to 10 in. over the first 2 years of the project. The solution was to draw down the wetland water levels, in winter 1990–1991, and freeze out the carp (Figure 27-8). Solids removal returned to the previous high levels of efficiency.

Table 27-9 Annual Average Water Budget Components for the Des Plaines Wetlands in Water Year 1990 (cm/d)

	EW3	EW4	EW5	EW6
Inflows				
Surface inflow	5.36	1.46	5.01	2.78
Precipitation	0.26	0.26	0.26	0.26
Outflows				
Discharge	5.36	1.46	4.80	0.35
Evapotranspiration	0.26	0.26	0.26	0.26
Seepage	0.00	0.00	0.21	2.43

Table 27-10 Suspended Solids In and Out of the Des Plaines Wetlands (mg/L)

	Inlet	EW3	EW4	EW5	EW6
FA 89	8.0	2.0	2.4	2.6	3.0
WI 89	7.1	5.0	3.6	4.2	3.0
SP 90	24.2	5.5	4.5	2.9	3.3
SU 90	47.7	5.7	14.9	4.3	13.9
FA 90	50.1	10.8	7.4	5.4	4.4
SP 91	63.9	5.8	7.4	2.4	6.2
SU 91	123	6.0	6.8	3.2	7.7
FA 91	66.0	10.8	6.7	2	5.8
AVG	48.8	6.5	6.7	4.9	6.1
% Removal		87	86	90	87

The river bears a significant nutrient load, as evidenced by nitrate and phosphorus. These fertilizers peak seasonally, corresponding to runoff timing and land-use practices within the watershed.

Phosphorus removal efficiencies average 65 to 80 percent (Table 27-11). However, efficiency is lower in winter and higher in summer. That is partly because the riverine concentrations of phosphorus are very low in winter. Winter runoff in the watershed is overland, over frozen soils or ice and snow. The result is low phosphorus in the river in winter.

Figure 27-8 Fish were intentionally frozen out of the Des Plaines wetlands to restore high TSS removal efficiency.

Table 27-11 Total Phosphorus Reduction in the Des Plaines Wetlands (mg/L)

	Inlet	EW3	EW4	EW5	EW6
FA 89	0.052	0.018	0.013	0.014	0.018
WI 89	0.073	0.053	0.030	0.058	0.024
SP 90	0.057	0.044	0.015	0.017	0.023
SU 90	0.117	0.038	0.055	0.035	0.062
FA 90	0.131	0.024	0.007	0.017	0.011
SP 91	0.089	0.003	0.002	0.001	0.002
SU 91	0.119	0.010	0.010	0.010	0.009
AVG	0.091	0.027	0.019	0.022	0.021
AVG, %		65	78	73	75

Most phosphorus enters the wetlands associated with mineral suspended solids. These solids settle quickly and may not freely exchange their phosphorus with the wetland waters. In addition, there is a large biotic cycle of growth, death, and decomposition at work, which leaves a residual of organic sedimentary material. The deposition from this cycle exceeds the deposition of incoming river solids by a wide margin. Both processes immobilize phosphorus in these wetlands. During the early years, phosphorus is also tied up in the new biomass associated with these developing ecosystems.

There are a variety of nitrogen forms in the river water. About 0.6 mg/L of organic nitrogen enter the wetlands, and the same amount leaves. Very low ammonium nitrogen concentrations are found in both river and wetland waters: about 0.05 mg/L. Nitrate varies seasonally in the river in response to urban and agricultural practices. High spring and fall concentrations are echoed by similar variations in the nitrate content of the wetland effluent waters (Figure 27-9). However, in the warm seasons, a considerable amount of the incoming nitrate is removed, presumably due to denitrification (Table 27-12). This microbially mediated process appears to be more efficient in the wetlands with lower hydraulic loading rate, which is equivalent to increased detention time, since depths are comparable. Thus, the overall

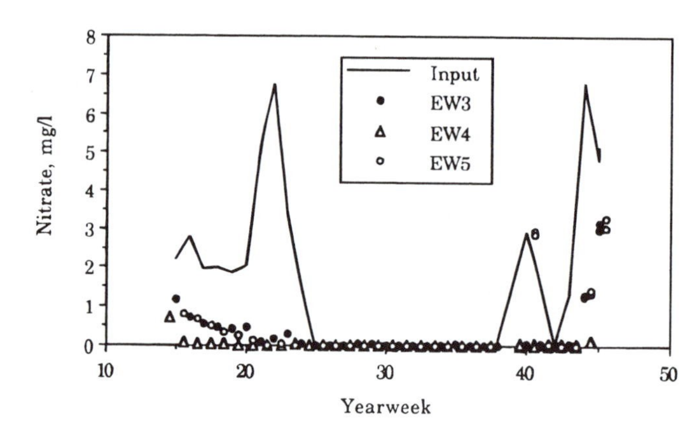

Figure 27-9 Nitrate follows a seasonal pattern in both supply and removal. Wetland EW4 has enough detention time to remove NO_3-N even at cold temperatures.

Table 27-12 Nitrate Nitrogen Reduction in the Des Plaines Wetlands (mg/L)

	Inlet	EW3	EW4	EW5	EW6
FA 89	2.46	1.46	0.04	1.27	0.08
WI 89	2.15	0.67	0.17	1.51	0.25
1990	1.87	0.54	0.24	0.53	0.32
1991	1.22	0.23	0.10	0.18	0.18
AVG	1.80	0.61	0.15	0.70	0.22
AVG, %		66	92	61	88

effect of the wetlands is to control the nitrate in the water when sufficient contact time is available.

Atrazine, a triazine herbicide, exists in many streams in the upper midwestern part of the U.S. including the Des Plaines River. Agricultural practices within the basin produce pollution with atrazine at concentrations which sometimes peak in excess of the federal drinking water standard. The atrazine-wetland interaction is very complex, including removal from the area by convection in the water, loss of chemical identity by hydrolysis to hydroxytriazine and dealkylation, and sorption on wetland sediments and litter. Atrazine transport, sorption, and identity loss were studied at the site and in accompanying laboratory work. Sorption was effective for soils and sediments, but the more organic materials, such as litter, showed a stronger affinity for atrazine than the mineral base soils of the wetland cells at Des Plaines.

Atrazine was found to degrade on those sediments according to a first-order rate law. Therefore, outflows from the Des Plaines wetland cells contained reduced amounts of atrazine compared to the river water inputs. During 1991, atrazine peaked in the river due to two rain events. Only about 25 percent of the incoming atrazine was removed in wetland cell EW3, but 95 percent was removed in wetland cell EW4. The explanation is that the detention time in EW4 is longer than in EW3.

Vegetation

Efforts at vegetation establishment were initially thwarted by the extreme drought conditions of 1988. The planting of white water lily (*Nymphea odorata*) showed small success, and American water lotus (*Nelumbo lutea*) did not survive.

The development of the macrophyte plant communities has been monitored from project startup. Data were acquired on species composition and biomass. Plants were individually measured, and a correlation between dry weight and leaf size was developed. Thus, biomass could be determined nondestructively. There was an overall increase in species as volunteer wetland vegetation replaced the terrestrial vegetation of prepumping. Fourteen species were observed in 1990 that were not present in 1989, and ten species from 1989 did not reappear, the latter being mostly upland species.

The first year of inundation caused the death of many upland species such as cottonwood (*Populus deltoides*). The growing seasons of 1989, 1990, and 1991 all displayed an increase in the amount of cattail (*Typha* spp.). Productivity increased from 200 to 400 dry grams per square meter in 1989 to 600 to 800 in 1990. The growing season of 1990 produced extensive blooms of macrophytic algae, predominantly *Cladophora.*

Animals

Bird populations have grown much larger than in the prewetlands period for the site. For migratory waterfowl, there has been a 500 percent increase in the number of species and a 4500 percent increase in the number of individuals from 1985 to 1990. Forty-seven species of birds nested on the site in 1990, a 27 percent increase over preproject numbers. The fall 1990 bird survey turned up a number of interesting species, including the state endangered pied-billed grebe and black-crowned night heron and also the great egret, American bittern, and the sharp-shinned hawk. The state-endangered yellow-headed blackbird and least bittern nest successfully at the site. The habitat value of the wetland complex is clearly much greater than that of the precursor uplands.

Muskrats have moved in and constructed both dwelling houses and feeding platforms, and beaver are now resident in the wetlands. Beaver chewed off quadrat corner posts—most of the 256 posts initially placed. They attempted to dam the wetland EW3 outflow nearly every night in 1992 and have been active in other wetland cells also (Figure 27-10).

Awards

The project has received the Illinois Society of Professional Engineer's Award for the Outstanding Engineering Achievement of 1991, as well as the Ecological Society of America's Special Recognition Award.

References

Numerous publications have issued from the project. The latest listing may be obtained from

Wetlands Research, Inc.
West Jackson Boulevard
Chicago, IL 60604

VERMONTVILLE, MICHIGAN

Introduction and System Description

Vermontville is a rural community located 40 km southwest of Lansing. The Clean Water Act of the early 1970s dictated that Vermontville upgrade its wastewater treatment capabilities.

Figure 27-10 Beaver are fond of damming any running water, including outlet structures at treatment wetlands.

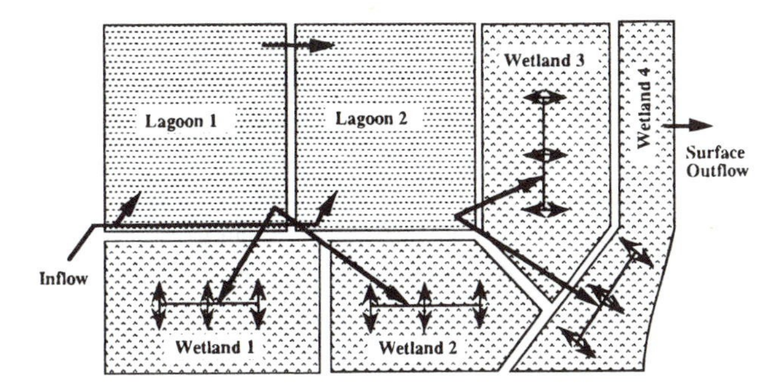

Figure 27-11 Layout of the Vermontville wastewater treatment system. Inflow may be directed to either of the two lagoons. The lagoons are discharged into Wetlands 1, 2, and 3. Wetland 4 no longer receives a direct discharge, but seepage water from the uphill units reemerges into Wetland 4.

In common with many other small communities, Vermontville could not afford to own or operate a "high tech" physical-chemical wastewater treatment plant. But, it was situated to utilize land-intensive natural systems technology and decided to do so. In 1972, the community opted for facultative lagoons followed by seepage beds. Those seepage beds unexpectedly became wetlands, a system which works remarkably well and is liked by the operators.

The municipal wastewater treatment system at Vermontville, MI consists of two facultative stabilization ponds of 4.4 ha, followed by four diked surface (flood) irrigation fields of 4.6 ha constructed on silty-clayey soils (Figure 27-11). The system is located on a hill with the ponds uppermost and the fields at descending elevations (Figure 27-12). After 1991, the 19th year of operation, the fields were totally overgrown with volunteer emergent aquatic vegetation, mainly cattail. The system was designed for 380 m^3/d and a life of 20 years. It is presently operated at about three quarters of design capacity and has shown no signs of "wearing out."

The Vermontville system was intended, in the conceptual stages, to provide phosphorus removal both by harvesting of terrestrial grasses and by soil-water contact as wastewater

Figure 27-12 Wetland 2 is bordered by steep hills next to Wetland 1 and Lagoon 2. Cattails dominate the vegetation, with a few willow shrubs in evidence. Late-summer senescence is in progress, and the cattails are beginning to turn brown.

seeps downward from the irrigation fields. Up to 10 cm of water applied over several hours' time once each week would flood the fields briefly until the water seeped away. The upper pond (Lagoon 1, Figure 27-11) has separate discharge lines into wetlands 1 and 2, and the lower pond (Lagoon 2) has separate discharge lines into wetlands 3 and 4. Wetlands 1 to 4 have all been colonized by volunteer wetland vegetation and are now eutrophic emergent marshes.

Pond-stabilized wastewater is released into each wetland by gravity flow through 0.25 m main and 0.2 m manifold pipe having several ground level outlets in each wetland. The lagoons and wetlands are terraced on a steep hillside, providing ample driving force for gravity flow. Should the water level exceed 15 cm, water would overflow to the next wetland by means of standpipe drain. All applied water was intended to seep into the ground before leaving the treatment area.

The system is operating nearly in this manner today. However, there is a constant surface overflow from the final wetland made up of ground-recycled wastewater which enters the final field at springs. The direct surface overflow from Wetland 3 has been taken out of service. Essentially, the system is a seepage wetland complex and very similar to a conventional flood irrigation facility. The vegetation and relatively small surface overflow from the final wetland provides an established system in which to evaluate the treatment aspects of seepage combined with lateral flow-through wetlands, the potential nutrient removal and wildlife values of these strictly voluntary wastewater wetland, and the economics of the system.

Permits

The facility operates under an NPDES permit issued by Michigan Department of Natural Resources. The outflow from Wetland 4 is to an unnamed tributary of the Thornapple River, which is protected for agricultural uses, navigation, industrial water supply, public water supply at the point of water intake, warm-water fish, and total body contact recreation. There are presently no industrial dischargers. The discharge limitations from the treatment wetlands (Table 27-13) are set for a design flow of 380 m^3/d (0.1 million gal/d). Discharge is limited to the ice-free, high flow periods from May 1 to October 31.

Hydrology

During 1990, approximately 110,000 m^3 of wastewater was introduced into the lagoons. This was a dry year; thus, evaporation exceeded rainfall and snowmelt, leaving only about

Table 27-13 Discharge Limits for the Vermontville Wastewater Treatment Facility

Parameter	Dates	Daily Minimum	Daily Maximum	30-Day Average	7-Day Average
$CBOD_5$	4/15–4/30		25 mg/L	17 mg/L	
				14 lb/d	21 lb/d
	5/1–9/30		10 mg/L	5 mg/L	
				4.2 lb/d	8.3 lb/d
	10/1–10/31		16 mg/L	11 mg/L	
				9.2 lb/d	13.3 lb/d
TSS	4/15–4/30			20 mg/L	30 mg/L
	5/1–10/31			30 mg/L	45 mg/L
NH_4-N	4/15–4/30			7 mg/L	
	5/1–9/30			2.2 mg/L	
	10/1–10/31			5 mg/L	
TP	All year			1.0 mg/L	
				0.83 lb/d	
DO	4/15–4/30	5 mg/L			
	5/1–9/30	6 mg/L			
	10/1–10/31	5 mg/L			
pH	All year	6.5	9.0		

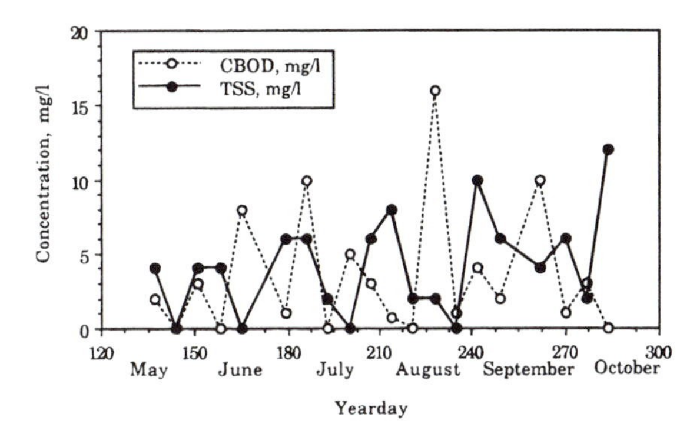

Figure 27-13 Both CBOD and TSS fluctuate in the outflow from the wetlands, but the seasonal averages are quite low: 3.5 mg/L for CBOD and 4.2 mg/L for TSS. (Data are for 1990.)

83,000 m^3 to discharge to Wetlands 1, 2, and 3. There was no lagoon discharge to Wetland 4. About 26,500 m^3 were lost to evaporation in the wetland cells, 49,000 m^3 infiltrated to groundwater, and 7600 m^3 overflowed from Wetland 4 to the receiving stream.

Wetland 4 receives its water from interior springs fed by the groundwater mound under the upgradient wetlands, most importantly Wetland 3. The direct discharge to Wetland 4 was discontinued, since it was in close proximity to the system outflow point and was clearly short circuiting water across Wetland 4. Effluent discharged from the system has therefore passed through the lagoons, through the upper wetlands, through the soils under the site, and finally through the last wetland.

Water Quality

Compliance Monitoring

The overflow from final Wetland 4 contains a fairly constant volume of effluent which has seeped from the higher elevation wetlands, flowed through the ground, and entered wetland springs. This treated effluent is of high quality, as is the groundwater recovered from the project's monitoring wells.

The outflow is monitored weekly. TSS was well within permit limits at all times during 1990 (Figure 27-13), indicating that the wetlands had effectively filtered and settled particulate material.

Carbonaceous biological oxygen demand (CBOD) also remained within 30-day average permit limits in 1990, and only once exceeded the daily permit limit of 10 mg/L. The CBOD load in the surface discharge was less than 10% of that allowed by the permit.

TP in the surface discharge was also well within permit limits, with an average 1990 value of 0.24 mg/L compared to the permit level of 1.0 mg/L (Figure 27-14). The same was true for ammonium nitrogen, which averaged 0.86 mg/L compared to the 2.2 mg/L permit requirement. Both phosphorus and nitrogen display considerable variability, which is characteristic of many wetland systems. The seasonal trends in ammonium nitrogen are likely due to the changing processes of plant uptake and decomposition.

Dissolved oxygen averaged 7.0 mg/L in 1990, with a range from 5.4 to 9.4; in four instances the permit range was exceeded slightly. pH ranged from 6.6 to 7.2, well within the permit range.

Fecal coliform counts (Figure 27-15) are within limits for surface water discharges, but are higher than at other comparable wetland sites.

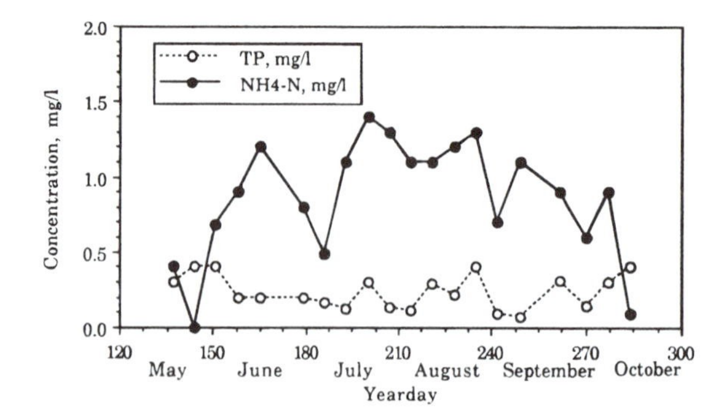

Figure 27-14 The nutrients phosphorus and ammonium nitrogen were well within limits in the wetland outflow in 1990. The seasonal average total phosphorus was 0.24 mg/L; ammonium nitrogen averaged 0.86 mg/L.

Research Results

A thorough study of water quality and other aspects of the system was conducted in 1978 under the sponsorship of The National Science Foundation (Sutherland and Bevis, 1979; Bevis, 1979; Sutherland, 1982).

Some of the more detailed water quality results for 1978 are summarized in Figure 27-16. Greater than twofold dilution across the system was evident in the decreasing chloride concentration from 280 mg/L in the effluent to 124 mg/L in the groundwater. Pond effluent was 25% diluted with respect to influent. Although a few centimeters of precipitation in excess of evaporation from the ponds occurred during the summer, the 25% dilution was more importantly due to excessive snow and ice melt water added to the ponds in spring 1978. The 25% dilution between the pond effluent and the water standing in the wetlands was due principally to a large number of sampling dates coinciding with significant rainfall. Greater than 50 cm (20 in.) of rain fell in the 18 weeks from June to mid-October, which was approximately 50% higher than the normal rate. The decrease in concentration between irrigation fields and groundwater was due to mixing of wastewater with more dilute ambient groundwater.

Phosphorus was removed to the extent of around 97% between the wetland fields and the groundwater, which was sampled from monitoring wells placed at depths ranging from roughly 3.0 to 7.6 m (10 to 25 ft) below the wetland floors. Most removal of phosphorus occurs in the upper 0.9 m (3 ft) of soils judging from a small number of lysimeter samples which averaged 0.11 mg/L total P and 0.06 mg/L ortho-P, with ranges of 0 to 0.3 and 0 to

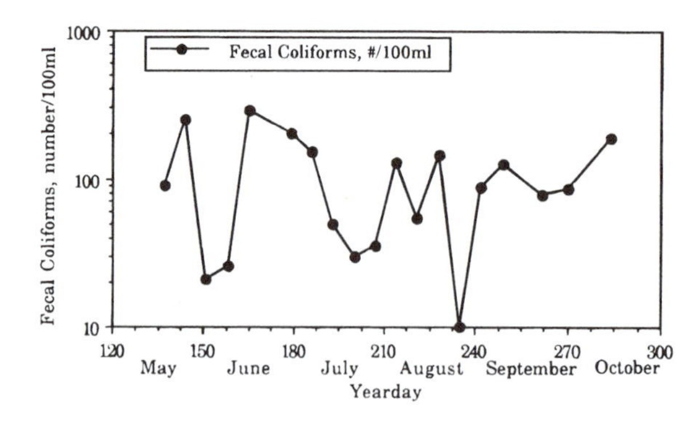

Figure 27-15 Fecal coliform bacteria counts also fluctuate in the outflow from the wetlands, but the seasonal average is quite low; the geometric mean value was 77. (Data are for 1990.)

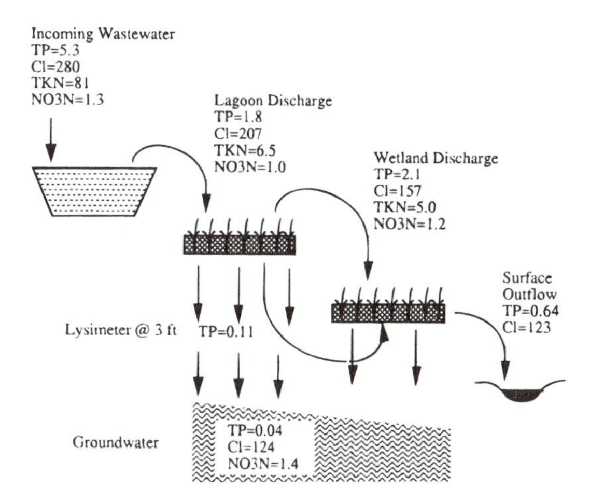

Figure 27-16 Profiles of water quality in 1978. Lagoons and wetlands and soils are functioning to remove nutrients in this system. During the early life of the facility, there were lagoon discharges directly to wetland four, and there was surface overflow directed from wetland three to wetland four. This resulted in some short circuiting to the surface outflow and, consequently, higher phosphorus numbers than in the present mode of operation.

0.2 mg/L, respectively. The average removals of phosphorus affected in the upper 0.9 m (3 ft) of soils were approximately 95%.

Levels of nitrate-nitrogen increased approximately 60% between the pond discharge and the wetland standing water, indicating that aerobic bacteria were at work in the wetland waters. On the other hand, the sediments were anaerobic, as evidenced in the fetid odor which evolved when they were disturbed. Loss of some of the nitrate by denitrification was apparently occurring. Lysimeter samples showed nitrate-nitrogen ranging from 0.0 to 0.9 mg/L, which suggested that denitrification of approximately 60% of the nitrate occurred in the shallow wetland soils. The ambient groundwater contained higher levels of nitrate-nitrogen than did the seeping wastewater, perhaps indicating some further nitrification during passage through the soil.

Levels of TKN and ammonia-nitrogen seemed not to change much between the pond discharge and the wetland waters. But this constancy was likely only apparent, with organic nitrogen and ammonia probably being produced through anaerobic decomposition in the wetland sediments and being consumed in the aerobic wetland waters and plant growth.

Vegetation

The wetlands were observed to contain eight plant communities in 1978. These included areas dominated by grassland, duckweed, cattail, and willow. In 1991, the grassland and duckweed communities were no longer significant. The wetlands are now dominated entirely by cattail and willow shrubs and willow trees.

Standing crops (above-ground plant parts) for the wetlands varied from a minimum of 830 to over 2200 g/m^2 in the wetlands in 1978. Visual estimates in 1991 indicate that the standing crops are presently somewhat higher than that maximum and more uniform. There appears to be approximately 3000 g/m^2 at all locations, not counting trees. Because the wetlands are located on an exposed hillside, winds can and do blow down the cattails. The result is a patchy stand of cattail, about 3 m in height where it is erect and flat on the surface elsewhere.

The phosphorus in the prevailing cattail standing crop is significant compared to the phosphorus released into the wetlands. Cattail harvesting would therefore be a means of reducing effluent phosphorus. But, harvesting is not needed for phosphorus removal in seepage wetland settings where subsurface soil types and volumes are adequate to effect phosphorus removal before effluent groundwater reaches receiving streams. The expense and difficulty of harvesting further preclude its use at Vermontville.

Wildlife

Casual observation reveals the wastewater-grown wetlands have significantly added to the acreage of suitable, adequately isolated habitat for waterfowl and other wildlife in the Vermontville area. Natural, interrupted zones of attached aquatic plant life fringe the nearby Thornapple River, but these are narrow, small, and easily accessible to fisherman and other recreationists. The wastewater wetlands are part of a restricted public access area.

The Vermontville volunteer wetland system created marshland habitat suitable for waterfowl production otherwise not present in the immediate area. Many other types of birds also nest in the marshes, including red-wing blackbirds, American coot, and American goldfinch. Waterfowl (blue-winged teal and mallard), shorebirds (gallinule, killdeer, lesser yellow-legs, and sandpiper), and swallows use the wetland pond system for feeding and/or resting during their migration. Great blue heron, green heron, ring-neck pheasant, and American bittern have also been seen frequenting the wetlands.

These volunteer wetlands are also important habitat for numerous amphibians and reptiles. These include snapping and painted turtles, garter and milk snakes, green and leopard frogs, bullfrogs, and American toads. Muskrats inhabit the wetlands, while raccoon, whitetail deer, and woodchuck are seen feeding in the wetlands.

Operating and Maintenance Activities

Very little wetland maintenance has been required at Vermontville. The berms are mowed three or four times per year for aesthetic reasons only. Water samples are taken on a weekly frequency at the surface outflow. The discharge risers within the wetlands are visited and cleaned periodically during the irrigation season. There is essentially nothing to be vandalized, and there have been no repairs required.

The dikes are monitored for erosion, which has not been a significant problem. Muskrats build lodges and dig holes in the dikes; woodchucks also dig holes in the berms. Therefore, a trapper is allowed on the site to remove these animals periodically. The operator also periodically tears the muskrat lodges apart.

There are no bare soil (tilled) areas to be plugged through siltation caused by rain splash, spray irrigation, or flood suspension of inorganic soils. The Vermontville wetlands showed buildup of about 10 cm of organic residues largely in the form of cattail straw after six irrigation seasons (1972–1978). That litter mat is still of the same thickness today, but is accompanied by a small accretion of new organic sediments and soils. There was one attempt to burn the accumulated detritus, which proved to be difficult and of no value in the system operation or maintenance. The amounts of accreted material have not compromised the freeboard design of the embankments over the system's 20-year operational period. Tree control has not been practiced at Vermontville, and the wetlands now contain willow trees up to several meters in height. No hydraulic problems have been experienced due to these trees or any other cause.

Costs

The Vermontville ponds and wetlands cost $395,000 to build in 1972 (about $99,000 per hectare in 1993 dollars). Much of this expense was incurred for grading because of the uneven topography of the site.

The operating and maintenance costs associated with the wetlands portion of the treatment system are quite low. In 1978, these were approximately $3500 per year, of which $2150 was labor and field costs; the balance was for water quality analytical services. In 1990, these same costs totaled about $4200, including $3400 for labor and field costs.

SUMMARY

SF constructed wetland treatment systems are being successfully used for water quality improvement in many areas. Not all systems have been effective at meeting all permit limitations, but review of design criteria and operating history from these systems can provide useful information for planning new systems. Many wetland treatment systems that have been conservatively designed have had consistently low effluent pollutant concentrations. Constructed wetland treatment systems designed less conservatively have also consistently exceeded effluent quality expectations for conventional wastewater treatment technologies, surprising and pleasing both engineers and regulators. In addition to this overall satisfaction with water quality treatment performance of constructed treatment wetlands, many of these systems have demonstrated a variety of ancillary benefits that increase their overall value for environmental protection and enhancement.

SUBSURFACE-FLOW CONSTRUCTED TREATMENT WETLANDS

INTRODUCTION

Two SSF constructed wetland case histories are presented. The first is for an experimental facility at Richmond, New South Wales, Australia. The second is located at Benton, Kentucky and represents an early application of this technology in the U.S.

RICHMOND, NEW SOUTH WALES, AUSTRALIA

Seven pilot scale treatment wetlands went into operation in January 1984 at Richmond, New South Wales, Australia. The wetlands were located on the campus of the Hawkesbury Agricultural College and received trickling filter effluent (secondary). The objectives of the project were to determine feasibility, to develop design criteria, to develop operational guidelines, and to project possible broader application of the technology.

Three cells were gravel beds: one with cattails (*Typha orientalis*), one with bulrushes (*Schoenoplectus validus*), and one with gravel only. Each unit was 100 m long and 4 m wide. The media was 5 to 10 mm local river gravel, graded and washed and leveled by hand raking. The nominal gravel depth was 50 cm, but operating water depths were lower, approximately 35 to 45 cm.

Two cells contained no media: one unvegetated open water unit and one containing parrot feather (*Myriopyllum aquaticum*), a rooted, partially emergent, mostly submergent macrophyte. These were of the same dimensions and operated in the same depth range.

Two cells were hybrid combinations consisting of cattail-planted gravel sections, followed by open water pools, followed by unplanted gravel sections, in equal proportions. Cell 6 had one such sequence; Cell 7 had two (Figure 27-17).

General Conditions of Operation

The systems were closely monitored from June 1984 through May 1986. Nine student assistants worked with five senior researchers to acquire a large amount of performance data.

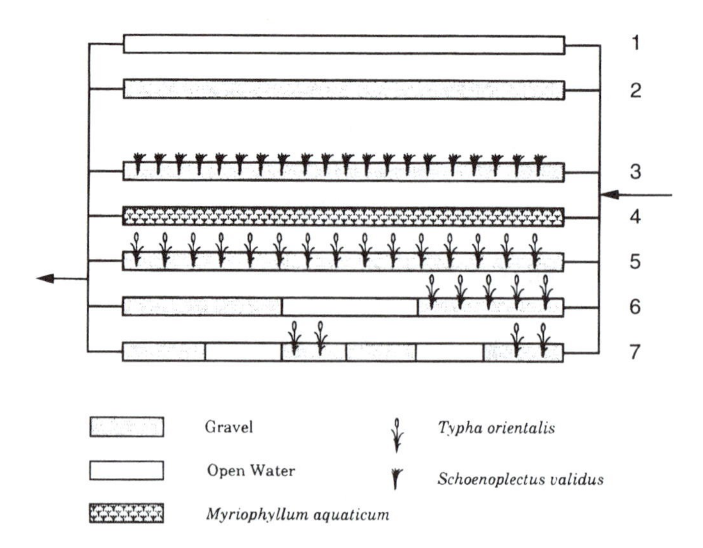

Figure 27-17 Layout of the test wetlands at Richmond, NSW, Australia. Each cell was 100 × 4 × 0.5 m deep. Two cells contained no gravel: one with open water and one with a rooted, mostly submergent macrophyte, parrotfeather (*Myriophyllum*). Three units contained all gravel: one with bulrushes (*Schoenoplectus*), one with cattail (*Typha*), and one unvegetated. Two units were hybrid combinations.

During this period, weekly samples were taken from inlets and outlets and analyzed for a large number of routine water quality indices:

- BOD, TOC, and DOC
- TSS and VSS
- TP and TDP
- TKN, NH_4-N, NO_3-N, NO_2-N
- Fecal and total coliforms
- T, pH, DO

Cross-laboratory checking was conducted to validate analytical procedures. A large number of other parameters were measured as special projects, including detailed microbiology, metals, plant biomass and composition, and sediment accumulation. Diurnal cycles were studied, and transect surveys were done for the major pollutants. Average results for major parameters showed good reductions of pollutants (Table 27-14).

Flow measurements were made daily, and meteorological data were recorded at an onsite weather station. Hydraulic gradients were measured every 6 to 10 weeks at 20 m intervals. Nominal detention times varied from 3 to 19 days, corresponding to hydraulic loading rates in the range of 1.4 to 12.1 cm/d. Dye tracer studies were conducted on several occasions.

During a subsequent research period, the wetlands received primary effluent.

Hydraulics

The front end of all gravel units was subject to a lowered hydraulic conductivity (Fisher, 1990). Flow rates through the gravel beds were reduced to prevent flooding of the inlet zones.

Tracer studies on the open water system indicated a residence time distribution consistent with nearly complete mixing (Figure 27-18a). The *Myriophyllum* trench and the *Schoenoplectus* trench behaved more like two continuous stirred tank reactors (CSTRs) in series with a plug flow element, all three of equal size (Figures 27-18b and c). Tracer profiles taken in

Table 27-14 Mean Performance of the Richmond Wetlands Over a 2-Year Period

	HRT (day)	HLR (cm/d)	BOD (mg/L)	BOD (%)	TSS (mg/L)	TSS (%)	TP (mg/L)	TP (%)	TKN (mg/L)	TKN (%)	NH_4-N (mg/L)	NH_4-N (%)	FC (#/100 ml)	FC (%)
Inlet			51.7		101.4		9.5		43.5		35.2		1,341,000	
Open water	7.45	6.43	22.9	55.8	66.9	34.0	9.4	1.7	24.4	43.9	17.5	50.2	16,640	98.8
Myriophyllum	6.59	7.33	14.3	72.4	23.9	76.4	9.2	3.7	25.6	41.1	24.0	31.7	18,200	98.6
Gravel	4.69	3.85	4.3	91.7	7.0	93.1	7.5	21.1	19.9	54.3	18.8	46.6	2,880	99.8
Typha	5.90	4.61	4.7	91.0	7.0	93.1	7.7	19.2	20.4	53.2	18.8	46.6	11,550	99.1
Bulrush	5.39	5.08	5.8	88.8	8.7	91.4	8.4	12.1	23.8	45.3	19.4	44.8	7,390	99.4
Marsh/pond/gravel	7.79	4.60	4.3	91.8	11.3	78.1	7.8	18.1	16.4	62.4	15.3	56.6	545	100.0
Marsh/pond/gravel/ marsh/pond/gravel	9.21	3.85	4.6	91.1	12.1	76.6	8.3	12.9	11.2	74.2	12.5	64.5	229	100.0

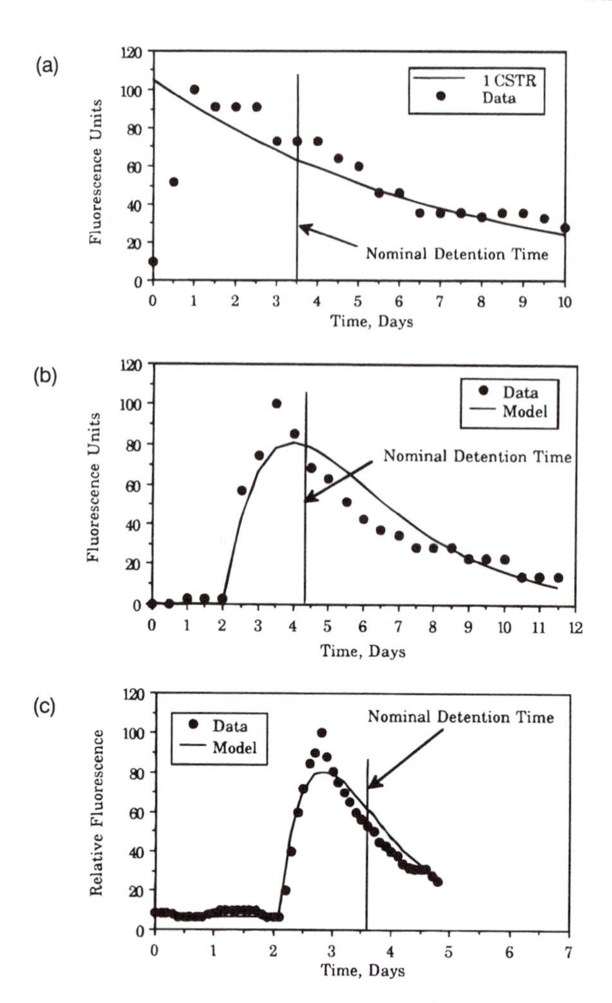

Figure 27-18 (a) Tracer response of the open water cell at Richmond, compared to a single CSTR model. (b) Tracer response of the *Myriophyllum* cell at Richmond, compared to a two CSTR plus plug flow reactor model. (c) Tracer response of the *Schoenoplectus* cell at Richmond, compared to a two CSTR plus plug flow reactor model.

and under the root zone indicated preferential water movement under the roots rather than through that zone (Fisher, 1990).

Evaporative water losses from vegetated trenches was similar to water loss from weather station pans. Open water losses were greater, presumably due to the "clothesline" effect of the long narrow geometry. Bare gravel lost the least amount of water.

The amount of interparticle solids accumulation was low (Bavor and Schulz, 1993). After the 2 years of secondary effluent application, a concentration of about 0.15 g sediments per 100 g of media were found, mostly independent of location within the bed. After switching to primary effluent, bed solids increased to about 0.6 g/100 g of media in vegetated beds and to about 0.3 g/100 g in unvegetated beds. On a mass basis this is not a large amount of solids, and the bulk density of the accumulation was not sufficient to clog an appreciable amount of void space.

The amount of solids found in these gravel beds is much less than the applied or trapped solids loading integrated over the period of operation. A significant fraction (88 percent) of the incoming TSS was volatile, and the presumption is that decomposition processes must have reduced the gross accumulation.

Water Quality

Biochemical Oxygen Demand

Most of the BOD reduction was observed to be in the first few days of retention, as evidenced by the transect measurements of BOD. The transect data revealed that the first-order irreversible model did not fit the data because BOD leveled off to non-zero background in the downstream sections of the trenches. As a result, the k_{V1} values regressed from the data underestimated the effluent concentrations (Bavor et al., 1988). The lack of seasonal variability is shown in Figure 27-19.

Significantly, this research project was one of the first to determine that there was no effect of temperature on the first-order irreversible volumetric rate constants (k_{V1}) for BOD, total organic carbon (TOC), and TSS removal (Bavor et al., 1989). This is mostly due to the fact that these wetlands achieved background BOD concentrations, which are relatively insensitive to temperature.

Total Suspended Solids

TSS was effectively removed in the SSF and hybrid systems, but less effectively in the open water and *Myriophyllum* systems (Table 27-14). Most of the removal occurred in the first 20 m of the wetlands.

Nitrogen Compounds

About 80 percent of the average incoming nitrogen (TN = 44.1 mg/L) was ammonium (NH_4-N = 35.2 mg/L), there was a trace of oxidized nitrogen (NO_x-N = 0.6 mg/L), and the balance was organic nitrogen (ORG-N = 8.3 mg/L). Organic nitrogen was reduced by approximately 90 percent in all wetlands except the open water. Ammonium nitrogen was reduced by a lesser amount, averaging about 50 percent in all systems. Oxidized nitrogen increased slightly in all SSF and hybrid wetlands, to 0.6 to 2.2 mg/L. NO_x-N decreased in the open water and *Myriophyllum* systems.

Profiles indicated that oxidized nitrogen was significant only in the upper layers of the water column in the SSF wetlands, not in the middle zone (rooted) or in the bottom (unrooted) zones. Ammonium declines were measured in the outlet sections of the beds after a peak at

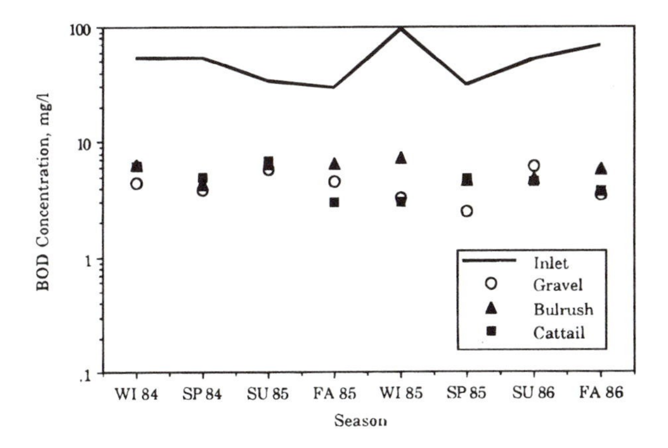

Figure 27-19 Inlet and outlet BOD from the Richmond SSF wetlands. There is no effect of season and, therefore by implication, no effect of temperature over the range 12 to 24°C. This is indicative of little seasonal variation in background BOD.

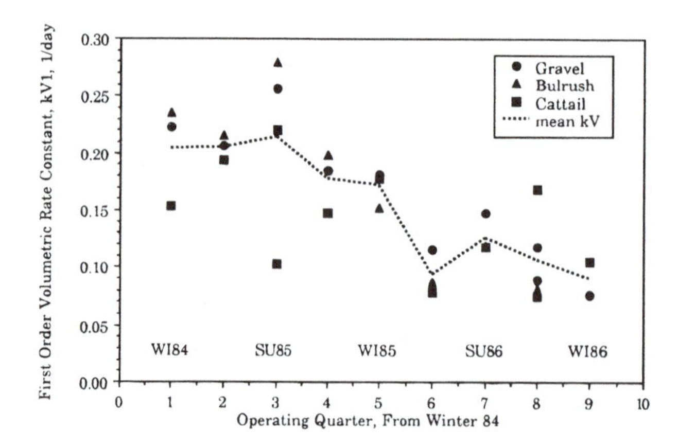

Figure 27-20 Seasonal variation in TKN removal rate constant superimposed on adaption trends.

10 to 20 m from the inlet. Such peaks possibly reflect mineralization of organic nitrogen to ammonium in the inlet zone. Ammonium declines were strongest in the top layer of the media.

Nitrogen performance was found to vary with temperature. Bavor et al. (1989) analyzed both transect and input-output data to obtain a value of k_{V20} and a temperature coefficient for TKN reduction. However, examination of that data shows an adaptation trend (Figure 27-20) to lower values of this rate constant. Accordingly, the last year of data produces a better estimate of the long-term TKN rate constant and $k_{V20} = 0.104\ d^{-1}$ and $\theta = 1.040$. Higher values of the rate constant during early months of operation may have been due to vegetation and sedimentary biomass accumulation.

Phosphorus

The SSF and hybrid systems removed 12 to 21 percent of the incoming total phosphorus, which entered at high concentration (*circa* 9 mg/L). The corresponding first-order areal rate constants were in the range of 3 to 4 m/yr, with performance declining during the adaptation period.

Pathogens

Fecal coliform (FC) bacteria were reduced by two to three logarithms (base 10). Mixed zone systems gave the highest rate constants, while open water and *Myriophyllum* gave the lowest. The plants appear to have little to do with FC reduction (Figure 27-21).

Temperature was found to be a factor in FC reduction. The θ factor for the SSF systems averaged 1.046.

Vegetation

Plant establishment was not a problem, and the systems were considered to have a mature plant cover after the first year. A decreasing gradient of plant biomass density established itself. There were maintenance difficulties. On several occasions, the *Myriophyllum* was attacked by insects and completely defoliated. Coverage was uneven, due to wind effects. This plant produced large amounts of detritus, which contributed to large organic buildups.

Cattails grew vigorously to a height of 3 to 4 m and rooted to a depth of 20 to 30 cm. In contrast, the bulrushes rooted only to a depth of 10 cm. However, the cattails died back in the center of the wetlands, an effect attributed to shading. Harvesting was implemented,

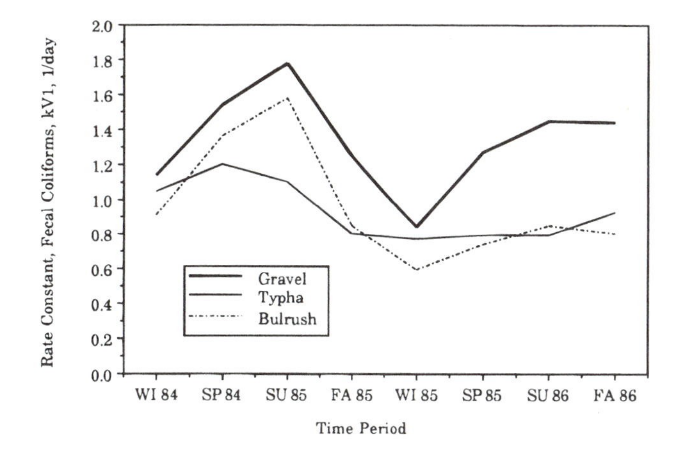

Figure 27-21 First-order irreversible volumetric rate constants for reduction of fecal coliforms. There is evidence of seasonality, especially for the bare gravel bed.

once and twice per year on different sections, in an attempt to enhance regrowth of the beds. These harvests retrieved less than 5 percent of the applied nutrient loads.

Cattails were marginally more effective in water quality improvement than bulrushes, and both were better than the *Myriophyllum.*

BENTON, KENTUCKY

Introduction

Benton, KY (population approximately 5000) is the site of a three-cell constructed wetlands demonstration project designed by the Tennessee Valley Authority (TVA). Wastewater and septage are first treated in a 5-ha (16-ac) primary lagoon and then passed to the wetlands (Figure 27-22). Cells 1 and 2 are SF wetlands; Cell 3 is an SSF wetland, with a

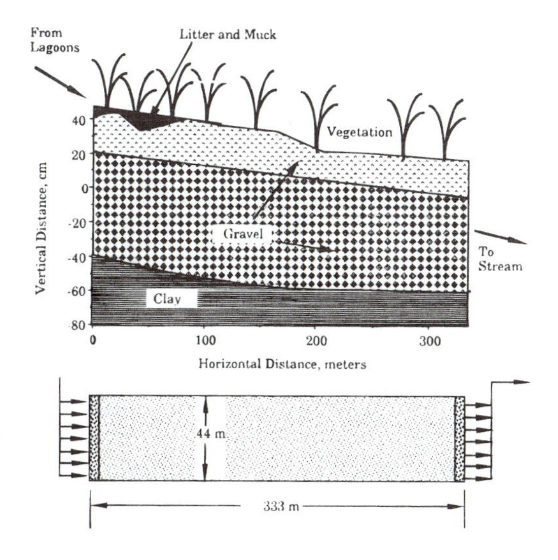

Figure 27-22 Benton cell number three. The bed bottom was sloped at a nominal gradient of 0.1%. The bottom 46 cm was 19- to 25-mm crushed limestone; the top 15 cm was 10- to 19-mm crushed limestone.

two-layer gravel media 0.61 m deep. Each is 44 m wide by 333 m long, and the bottoms are inclined at 0.1% (Steiner et al., 1987). The total flow through the three-cell system was approximately 3785 m^3/d (1.0 million gal/d).

Vegetation establishment was accomplished over the period from fall 1986 through spring 1988. Initial plantings were primarily cattail (*Typha latifolia*) in Cell 1, woolgrass (*Scirpus cyperinus*) in Cell 2, and softstem bulrush (*Scirpus validus*) in Cell 3.

Cell 3 began operation in January 1987; Cells 1 and 2 in January 1988. Cell 3 received all flow during 1987; subsequently, flow was split between all three wetlands.

General Conditions of Operation

The design target of the wetlands was the regulatory requirement (Choate et al., 1990b):

Parameter	Units	Winter November–April	Summer May–October
• BOD_5	mg/L	25	25
• TSS	mg/L	30	30
• Ammonium nitrogen	mg/L	10	4
• Dissolved oxygen	mg/L	7	7
• pH	S.U.	6–9	6–9

The three wetlands were closely monitored from March 1988 through October 1989.

During this period, monthly or bimonthly samples were taken from inlets and outlets, and both inlet and outlet flows were recorded at hourly frequency. On several occasions, a two-dimensional (horizontal) grid of sample stations was sampled for several days, in conjunction with a dye tracer study. Hydraulic gradients were measured on a few occasions.

Chemical analysis was performed for

- BOD and soluble BOD; and TSS
- TP and TDP
- TKN, NH_4-N, organic N, NO_x-N, TN
- Fecal coliforms
- T, pH, DO, redox potential,
- Alkalinity

Fecal coliforms, TN and TP, BOD, and TSS were reduced by all treatment wetlands (Table 27-15). However, dissolved oxygen was also reduced to well below the permit limit.

Table 27-15 Performance of the Benton Wetlands, March 1988 through October 1989

Parameter	Units	Influent	Effluent Cell 1	Effluent Cell 2	Effluent Cell 3
Flow	m^3/d	2801	792	835	765
Hydraulic loading	cm/d	6.37	5.40	5.70	5.22
Temperature	°C	20.3	17.2	18.6	17.6
Dissolved BOD_5	mg/L	8	5	6	5
BOD_5	mg/L	26	10	8	8
TSS	mg/L	57	11	14	5
Organic N	mg/L	9.4	2.9	3.3	1.5
Ammonium N	mg/L	4.8	8	6.7	8.6
TKN	mg/L	14.3	10.9	10	10.1
Oxidized N	mg/L	0.28	0.05	0.16	0.37
TN	mg/L	14.6	10.96	10.2	10.5
Dissolved P	mg/L	3.3	3.7	3.7	3.4
Total P	mg/L	4.5	4.2	4	3.7
Dissolved oxygen	mg/L	8.2	0.8	2	1
pH	S.U.	7.2	6.7	6.9	7
Fecal coliforms	col/100 ml	4,747	698	135	99

Note: Data from Choate et al., 1990b.

On average, all wetlands reduced BOD and TSS to acceptable levels. However, there were a few exceedances of the monthly maximum values dictated by the permit (Table 27-16). Fecal coliforms also showed frequent exceedances, although two of the three wetlands met the permit on average. The winter ammonium nitrogen limit was met on average, but not the summer limit. Monthly exceedances were frequent. pH remained within the permit range.

Hydraulics

With 50% of the flow directed to the SSF cell, there was flooding of the inlet end. Therefore, in June 1989, flow to Cell 3 was reduced to 12% of the total, with the remaining 88% directed to the SF cells. This did not totally cure the flooding problem, which was later found to be due to partial clogging of the inlet zone with biosolids and construction-deposited clays (Kadlec, 1991).

The media size in Cell 3 predicts a hydraulic conductivity of 10,000 to 40,000 m/d (see Chapter 9, Figure 9-18); the measured bare media conductivity was in the range of 20,000 to 27,000 m/d (Choate et al., 1990a and 1990b; Kadlec, 1991). The 0.1% bed slope then projects a maximum hydraulic capacity of 550 m^3/d (0.145 million gal/d) for Cell 3 (see Chapter 9, Equation 9-72). Consequently, flooding during early high flow operation was due to inadequate hydraulic capacity.

Tracer studies revealed that flow patterns were not close to plug flow in either the SF or SSF wetlands. Strong side-to-side variability in tracer response was noted (see Figure 9-37), indicating poor flow distribution.

Water Quality

Biochemical Oxygen Demand

Most of the BOD reduction was observed to be in the first 25% of the wetland length (Figure 27-23). Transect data showed BOD leveled off at non-zero values in the downstream sections of all three wetlands. Approximately half the BOD leaving the wetlands was in dissolved form.

Total Suspended Solids

TSS was removed in both SSF and SF systems, but less effectively in the open water Cell 2. Most of the removal occurred in the first 20 m of the wetlands (Figure 27-23).

Nitrogen Compounds

The wetlands did not live up to design expectations for ammonium nitrogen. Data shows that organic nitrogen was reduced to background levels by all three wetlands (Table 27-15). Transect data shows that this mineralization process occurred in the front end zones of the wetlands. The product of this process is ammonium nitrogen, which increased in all three wetlands from inlet to outlet. Transect data show a maximum in ammonium interior to the wetlands, along the progression from inlet to outlet. Very little (typically less than 0.2 to 0.4 mg/L) oxidized nitrogen appeared in the outflows, and it was below detection (0.01 mg/L) in all but 2 of 30 transect samples.

The load reduction in TKN could be accounted for by the requirement for biomass accumulation in litter, with little or no nitrification and denitrification. This observation led to the exploration of a downflow nitrification bed as a retrofit enhancement to the wetlands.

Table 27-16 Compliance Record for Benton Wetlands

Parameter	Units	Monthly Exceedances SF Cell 1	Compliance on Average SF Cell 1	Monthly Exceedances SF Cell 2	Compliance on Average SF Cell 2	Monthly Exceedances SSF Cell 3	Compliance on Average SSF Cell 3
BOD_5	mg/L	0/20	Yes	0/19	Yes	0/20	Yes
TSS	mg/L	3/20	Yes	1/19	Yes	0/20	Yes
Fecal coliforms	#/100 ml	14/20	No	8/19	Yes	5/20	Yes
Ammonium nitrogen	mg/L						
Summer		11/12	No	9/12	No	11/12	No
Winter		0/8	Yes	1/7	Yes	2/8	Yes
Dissolved oxygen	mg/L	19/20	No	18/19	No	20/20	No
pH	S.U.	0/20	Yes	0/19	Yes	0/20	Yes

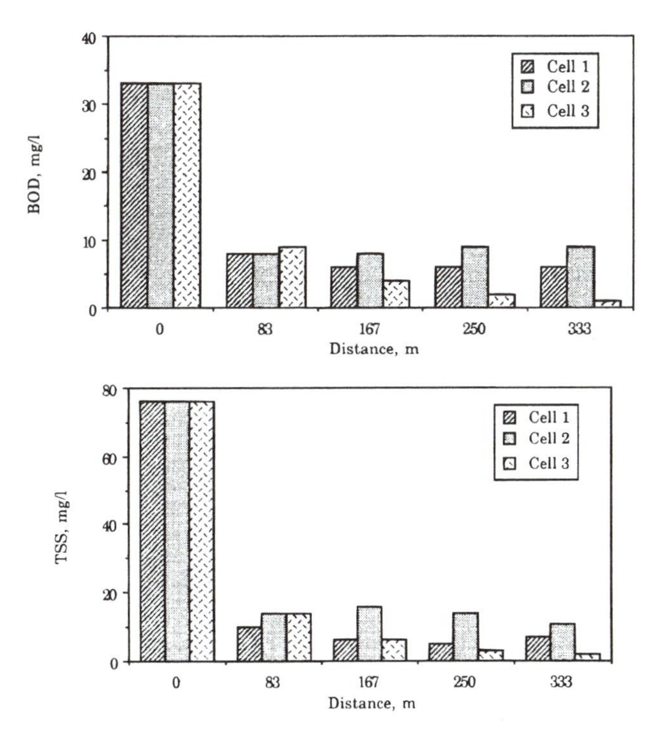

Figure 27-23 Profiles for BOD and TSS for the transect study of June 19–23, 1989. (Data from Choate et al., 1990b.)

This was successful in a pilot test (Watson and Danzig, 1993) and initially successful in full-scale tests (Hines and Reed, 1994).

Phosphorus

The SSF and SF systems removed 7 to 19% of the incoming total phosphorus, which entered at moderate concentration (*circa* 5 mg/L).

Pathogens

Fecal coliform bacteria were reduced by one to two logarithms (base 10). Cells 2 and 3 gave better performance than Cell 1 (Table 27-15).

Mosquitoes

Detailed studies of the species composition and abundance of mosquitoes were conducted (Tennessen, 1993). Chemical control (Abate™) was effective, but damaged the vegetation, which subsequently recovered. Two bacterial agents, *Bacillus thurengiensis israelensis* and *B. sphaericus,* were tested and found to be effective.

Vegetation

Plant establishment presented several problems. The SF wetlands vegetated poorly because of water shortages in the root zone, which were due to poor water absorption by the clay rooting soils. Initial plantings did not survive well and were replaced by volunteer vegetation (Knight, 1991; Choate et al., 1993). Root penetration was slight: 11 cm for cattails in Cell 1, 23 cm for woolgrass in Cell 2, and 7 cm for bulrushes in Cell 3 (Watson et al., 1990).

Bulrushes in Cell 3 disappeared in a large central portion (~75%) of Cell 3 after the flow reduction to alleviate flooding. This central zone became sparsely vegetated with terrestrial weed species (Figure 27-24). This was due to the thick dry top gravel layer caused by the low hydraulic loading. Bulrushes in the inlet zone became rooted in a litter-sediment mat on top of the gravel.

Muskrat herbivory caused cattail loss in Cells 1 and 2.

Deep water zones in Cell 2 lost all emergent macrophytes and became open water areas, sometimes covered with duckweed (*Lemna minor*) (Figure 27-25). A moderately dense population of plankton also developed in these open water zones (Knight, 1991).

SUMMARY

Based on the cost comparison presented earlier in Table 21-3, it would appear that SSF constructed wetlands may have received more consideration than they merit. Nevertheless, many engineers worldwide are faced with the challenge of dealing with primary wastewaters which are more of an environmental nuisance than they are a resource. Soil and gravel-based SSF constructed wetlands are an efficient technology to deal with this quality of wastewater on a local level.

As the Benton, KY case history demonstrates, inappropriate design criteria for these systems can lead to poor regulatory compliance and system failure. On the other hand, the Richmond project as well as numerous gravel-bed SSF systems in England demonstrate that this technology can work hydraulically and meet stringent permit limits.

Figure 27-24 The dry central zone of the gravel bed wetland at Benton, KY.

Figure 27-25 The unvegetated deep water zone near the outlet of Cell 2 at Benton, KY.

NATURAL TREATMENT WETLAND CASE HISTORIES

Natural wetland treatment systems have been engineered for water quality improvement since the early 1970s. For many years before then, natural wetlands also were convenient discharge points for treated and untreated wastewater. Engineered natural wetland treatment systems differ from these opportunistic discharges because they must meet specific water quality goals while minimizing biological impacts that might result from altered hydrology and water quality.

Because of their inherent ecological differences, all of the engineered natural wetland treatment systems in the U.S. have unique permitting and implementation histories. Some systems have consistently met water quality and biological goals, but others have failed. Case histories of several of these systems are presented to illustrate the innovative approaches taken by engineers, scientists, utilities, and regulators to lay the foundation for developing this natural treatment technology.

The Carolina Bay Natural Land Treatment System near Myrtle Beach, SC uses a regionally unique, natural wetland basin for advanced treatment. The southern-most natural wetland treatment system discussed here is the Reedy Creek site at Walt Disney World, FL. Houghton Lake is a natural wetland treatment system in Michigan and is only operated during the summer.

CAROLINA BAYS, SOUTH CAROLINA

Background

Carolina bays are mysterious land features often filled with bay trees (one or more of the species *Gordonia lasianthus, Persea palustris,* or *Magnolia virginianus*) and other wetland vegetation. Because of their oval shape and consistent orientation (Figure 27-26), Carolina bays are considered by some authorities to be the result of a vast meteor shower or comet strike that occurred thousands of years ago (Prouty, 1952). Others think the natural forces of wind and artesian water flow formed lakes which later filled with vegetation (Johnson, 1942; Thom, 1970; Kaczorowski, 1977). The origin of Carolina bays is still a controversial and debated issue among geologists (Savage, 1983).

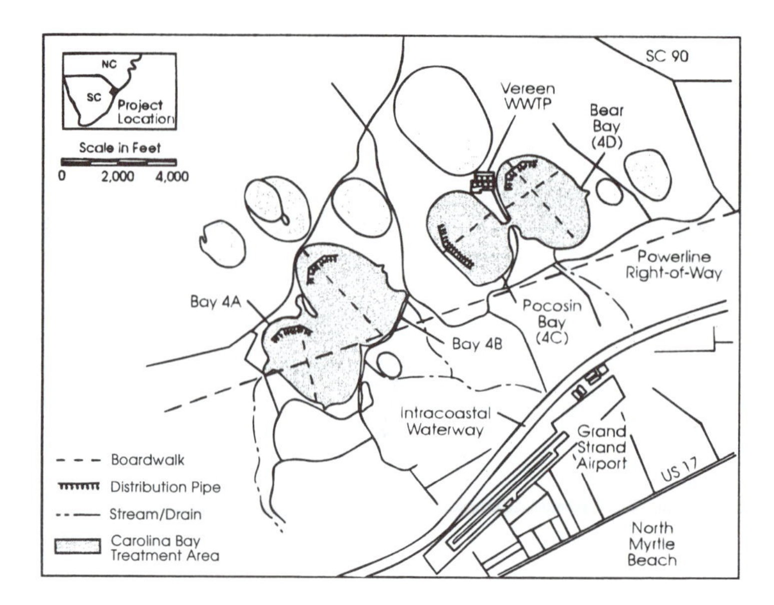

Figure 27-26 Carolina bays selected for advanced wastewater polishing in Horry County, SC, near North Myrtle Beach.

Whatever their origin, more than 500,000 of these shallow basins dot the coastal plain from Georgia to Delaware. Many of them occur in the Carolinas, which accounts for their name. Carolina bays range in size from less than 100 m to more than 10,000 m in length and are mostly swampy or wet areas. Most of the more than 200 Carolina bays in coastal Horry County, SC are vegetated by nearly impenetrable growths of vines, shrubs, and trees, including fetterbush (*Lyonia lucida*), gallberry (*Ilex coriacea*), loblolly bay, pond pine (*Pinus serotina*), and bamboo (*Smilax laurifolia*).

Because of population growth and increased tourism in Horry County in the early 1970s, utility expansion was essential. In the late 1970s, the regional water utility, the Grand Strand Water & Sewer Authority (GSWA), decided to evaluate wastewater treatment and disposal options and to consolidate numerous small wastewater treatment plants into regional plants. Within the Grand Strand, locations to dispose of additional effluent were extremely limited because of sensitive environmental and recreational concerns. It was determined that the slow-moving Waccamaw River and Intracoastal Waterway, into which existing facilities discharged, could only assimilate minor increases in pollutant loadings without adverse effects on water quality and resulting impacts on tourism and recreational activities.

Land-based effluent treatment and disposal were also evaluated and compared to advanced treatment alternatives to alleviate pollutant loading to these adjacent surface waters. Zero discharge treatment systems were found to be impractical because of the limited area of well-drained uplands in the planning area. After the alternatives analysis and subsequent pilot studies, it was recommended that treated effluent from a new 9500 m^3/d wastewater treatment plant be discharged to a group of four Carolina bays.

Permitting

Almost no knowledge existed about the biological effects of releasing treated wastewater into Carolina bays. Projections of vegetation changes that might occur were based on a review of the limited published information on the ecological requirements of the dominant plant species in the selected Carolina bays (Knight et al., 1985a). These proposed changes were closely scrutinized by representatives of the U.S. EPA Region IV, South Carolina

Department of Health and Environmental Control (SCDHEC), the U.S. Fish and Wildlife Service, the U.S. Army Corps of Engineers, the S.C. Wildlife and Marine Resources Commission, the S.C. Coastal Council, the S.C. Water Resources Commission, the S.C. Land Resources Commission, and the Wacammaw Regional Planning Council.

Permitting consensus could not be reached until specific allowable biological changes that might occur within the Carolina bays were proposed and accepted (Schwartz and Knight, 1989). The concept of "biological criteria" had its basis in the widespread use of benthic macroinvertebrate diversity changes for measuring ecological effects in aquatic ecosystems. The Carolina bay permitting was the first attempt to use biological criteria to determine allowable impacts of a wastewater discharge within natural wetlands. Previous natural wetland treatment projects had no criteria for allowable biological changes, so system "failure" was determined from "unacceptable" shifts in plant community composition, an arbitrary and controversial issue. For the Carolina bays, allowable biological changes were negotiated among the agencies and GSWSA and were proportional to previous impact levels documented in the four Carolina bays selected for use.

Biological criteria were negotiated for allowable changes in the following categories: (1) tree basal area (tree cross-sectional area per land area), (2) tree density (number of individual trees per area), (3) percent cover of shrubs and saplings (percentage of ground surface area covered by leaves and woody tissues), and (4) total number of plant species (diversity). Allowable levels of change for each of these criteria were a 50 percent decline in Bay 4D (Bear Bay), part of which had been previously cleared and planted in loblolly pines; a 15 percent decline in Bays 4A and 4B, which had been transected by power transmission lines and peripherally impacted by sand mining; and no change in Bay 4C, which had received no major impacts. These criteria were written to exclude changes that might occur to populations of planted pine trees and any changes resulting from natural (nonwastewater) events such as hurricanes, fires, and pathogens. Use of biological criteria requires that the Carolina bays continue to be monitored biologically in addition to more routine monitoring of inflow and outflow water quality.

Biological criteria were only one part of the permitting process for the Carolina bay system. Water quality criteria were determined for inflow to the bays (waters of the U.S.) and for outflow of the bays. Inflow water quality must be less than 30 mg/L for BOD_5, 50 mg/L for TSS, 20 mg/L for NH_4-N, and 100 fecal coliform colonies per 100 ml. Outflow limits for the Carolina bays were based on an allowable ultimate oxygen demand of 200 kg/d to the Intracoastal Waterway from March to October and 382 kg/d from November to February. This equates to limits of 12 mg/L for BOD_5, 30 mg/L for TSS, and 1.2 mg/L for NH_4-N from March to October, and 5.0 mg/L for NH_4-N from November to February.

System Description and Operation

After 5 years of intensive study to evaluate viable treatment and disposal alternatives, four Carolina bays were selected as treatment sites. Site selection criteria focused on three factors: (1) distance from the wastewater source, (2) available treatment area, and (3) environmental sensitivity. The bays chosen for the GSWSA treatment complex had been previously altered by humans and were the least environmentally sensitive of the bays considered (Knight et al., 1985a). Four bays at Site 4 were selected (Figure 27-27) and were referred to as 4A, 4B, 4C, and 4D. Subsequently, Bays 4C and 4D were renamed as Pocosin Bay and Bear Bay.

Carolina bays 4A and 4B encompass about 158 ha and are dominated by dense, shrubby plant communities with scattered pond pine trees. This plant association is called "pocosin" after an Indian word describing a bog on a hill.

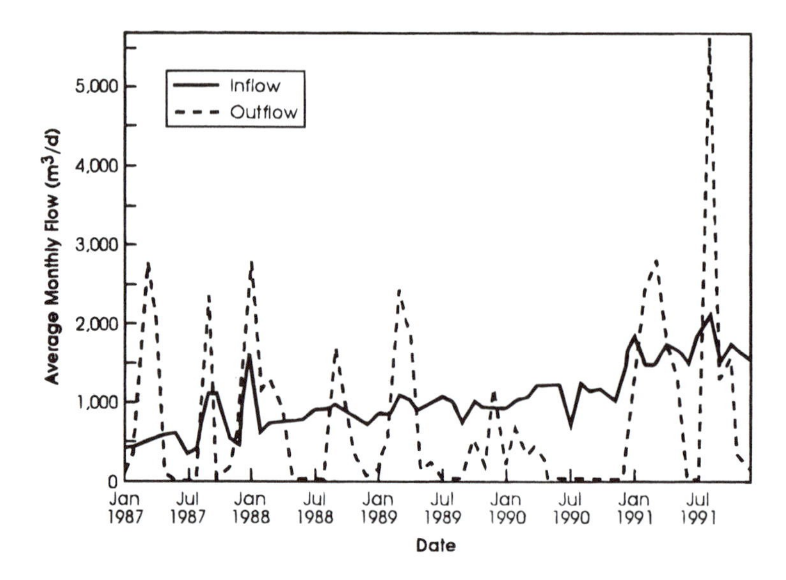

Figure 27-27 Wastewater inflow and surface outflow at Bear Bay, SC.

The 97-ha Pocosin Bay also is dominated by pocosin vegetation and is filled with up to 4.5 m of highly organic peat soils. This bay had received the least amount of prior disturbance and is being used only as a contingency discharge area and as a control site for comparing vegetation changes in an unaffected Carolina bay. Bear Bay covers 69 ha and is somewhat dissimilar from the other bays because portions of it are densely forested by planted pine and wetland hardwood tree species. A large portion of this Carolina bay was cleared for forestry in the mid-1970s, but has since been revegetated with a mixture of upland and wetland plant species.

Because of its diverse plant communities and its history of previous disturbance, Bear Bay was selected as the site of the initial full-scale pilot studies. In January 1987, municipal effluent treated at a neighboring package treatment facility was piped to Bear Bay through a forcemain and discharged into the bay along a 600-m boardwalk and gated aluminum distribution pipe. After one year of monitoring, it was determined that the Carolina bays would meet treatment requirements. In October 1988, construction began on the full-scale, 9500-m^3/d Vereen WWTP and the 364-ha Peter Horry Wildlife Preserve (Carolina Bay Natural Land Treatment System). Operation of the full-scale system began in October 1990, well below design flows, and flow has been increasing ever since. The original 5-year monitoring program continued through 1991 (CH2M HILL, 1992), and intensive monitoring continues as an operational requirement.

Wastewater is pretreated in four aerated lagoons, based on the principles developed by Rich (1988). A total of 143 kW (192 hp) of aeration capacity is available at a design influent BOD_5 concentration of 200 mg/L. The aeration system effluent receives chlorine disinfection before discharge to the Carolina bays.

The treated effluent can be distributed to 364 ha within the four Carolina bays through a series of gated aluminum pipes supported on wooden boardwalks. There are about 2100 m of aluminum distribution pipes installed on boardwalks in the bays. An additional 7000 m of narrower boardwalks provide access to the bays for operational monitoring. Wastewater flow is distributed among the bays, depending on their effluent flow rates and biological conditions. Water levels and outflow rates in Bear Bay can be partially controlled with an adjustable weir gate. Bear Bay is allowed to undergo greater hydrological change than the other bays because of its historical disturbances. Natural surface outlets in the other three bays were not altered by construction of the project to try to reduce the potential magnitude of hydrologic changes.

System Performance

In 1985, after site selection was completed and before wastewater distribution began, baseline studies documented the hydrology, surface water, and groundwater quality and flora and fauna of Bear Bay. Treated effluent was first discharged to the bay in January 1987, and monitoring has continued since then to document variations in the water quality and biological communities. For the period of operation reported here, January 1987 through December 1991, all discharge was to Bear Bay. Baseline and operational data were collected for hydrology, surface water and groundwater quality, soil chemistry, vegetation, and wildlife (CH2M HILL, 1992).

Bear Bay receives water from at least three sources: (1) rainfall and runoff, (2) groundwater discharge where shallow groundwater intersects the bays surface, and (3) pretreated municipal wastewater. Table 27-17 summarizes the annual water balances estimated from 1987 to 1991. The inflow of pretreated wastewater increased from about 570 m^3/d in 1987 to about 1670 m^3/d in 1991, representing from 24 to 74 percent of the annual inflow from rainfall and runoff during those years. On the basis of an estimated area of 69 ha in Bear Bay, these treated wastewater flows are equal to annual average hydraulic loading rates between 0.08 and 0.24 cm/d. However, as noted below, the actual wetted area of Bear Bay is much smaller than 69 ha, and the actual hydraulic loading rates are proportionally higher.

The data in Table 27-17 indicates that surface outflows from Bear Bay varied more than inflows (Figure 27-27). The lowest annual average surface outflow of 120 m^3/d occurred during the year of lowest rainfall. The highest annual average surface outflow rate of 1510 m^3/d coincided with the highest inflow of wastewater and one of the highest rainfall years. During this monitoring period, there were four periods with no surface discharge out of Bear Bay: three 2-month periods and one 7-month period.

Estimated evapotranspiration (0.77 percent of pan evaporation from Charleston, SC) was relatively constant from year to year and may be most subject to inaccuracies. Because most of Bear Bay was not inundated during part or all of this time, evapotranspiration probably was overestimated. The water balance in Table 27-17 indicates a net discharge of groundwater from the bay to evapotranspiration and surface discharge. If evapotranspiration is lower than estimated in Table 27-17, these groundwater exchanges may actually have been positive (net infiltration to the groundwater).

The surface of Bear Bay, and most natural wetlands, is somewhat irregular and consists of numerous small hummocks, pools, and flow paths. Surface water moves in a sheetflow manner; however, under normal water level conditions, this flow path is indirect and somewhat channelized. Figure 27-28 shows the extent of the flow path for a typical conductivity profile from August 1991. The conductivity of the treated wastewater was elevated (about 1300 μmho/cm) because of the nature of the potable water supply (coastal plain groundwater). Rainfall and runoff typically had conductivity values less than 150 μmho/cm. Therefore, this parameter provides a convenient tool to measure dilution and track the wastewater flow path in Bear Bay. As inflow rate increased between 1987 and 1991, and the inflow distribution point moved further upgradient during final construction in 1990, the effective treatment area of Bear Bay increased from about 18 to 28 ha.

A lithium chloride tracer study in Bear Bay in 1991 estimated a hydraulic residence time of 18.8 days at an inflow rate of about 1800 m^3/d. Based on the maximum planned loading to Bear Bay (5000 m^3/d or 0.73 cm/d), the estimated average hydraulic residence time at design flow is expected to be about 9 days.

Table 27-17 Summary of Major Water Inflows and Outflows for the Bear Bay Natural Wetland Treatment System, 1987 to 1991

	Inflows		Outflows			
Year	Pretreated Wastewater	Rainfall and Run In	Surface Outflow	Evapo-transpiration	Storage	Difference (Groundwater Exchange)
1987	570	2390	720	2420	−130	−50
1988	870	2050	760	2490	−40	−290
1989	950	1900	610	2540	+10	−310
1990	1100	1630	120	2880	0	−270
1991	1670	2260	1510	2870	−110	−340
Average	1020	2040	740	2640	−60	−260

Note: All flows are reported in cubic meters per day.

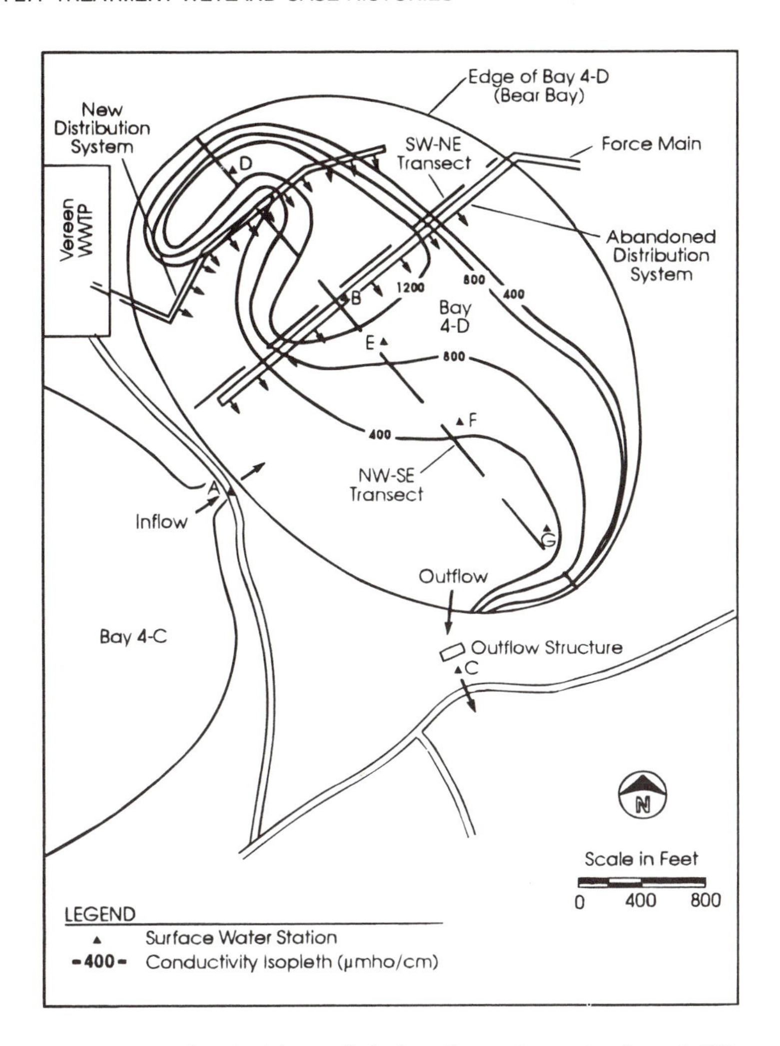

Figure 27-28 Conductivity profile in Bear Bay surface water, August 1991.

Surface water quality has been measured at a number of monitoring points in and around Bear Bay throughout the baseline and operational periods. Table 27-18 summarizes the annual average water quality for Bear Bay's treated wastewater inflow and surface outflow. Concentrations of all major water quality parameters are lowered after passing through the bay. Part of this reduction is due to dilution, as indicated by decreases for conservative parameters such as conductivity, chlorides, and sodium. Concentrations of BOD_5, TSS, TN, TP, and coliforms were also reduced by assimilatory processes described earlier in this book. As the inflow rate has increased over the period of record, outflow concentrations of TKN and TP have increased, apparently because of some saturation of removal mechanisms for these constituents. Concentrations of all other pollutants including BOD_5, TSS, NH_4-N, nitrate + nitrite-N, and coliforms have not increased in spite of the increasing loading rates. These observations confirm results from the pilot studies and from other natural wetland treatment systems showing that constituents are consistently assimilated at reasonable design loadings.

Mass removal efficiencies for Bear Bay from 1987 through 1991 are summarized in Table 27-19. Mass removal efficiencies are annual averages of the mass reduction that occur between surface inflows and outflows and include internal storages, assimilatory processes,

Table 27-18 Summary of Annual Average Surface Water Quality in Bear Bay from 1987 to 1991

Parameter	Units	Inflow (Sta. B)					Outflow (Sta. C)				
		1987	1988	1989	1990	1991	1987	1988	1989	1990	1991
Water depth	ft	0.41	0.37	0.12	0.22	0.55	2.11	2.12	2.12	0.92	1.33
Temperature	°C	18.0	18.7	21.2	20.5	19.5	15.5	11.8	14.4	11.8	16.1
Conductivity	μmho/cm	1,040	1,360	1,469	1,558	1,216	116	278	419	615	674
pH	units	7.2	6.9	7.4	8.2	8.1	5.6	6.4	6.4	7.1	7.3
Dissolved oxygen	mg/L	5.7	7.6	7.4	7.4	7.5	3.5	3.2	3.0	2.5	3.2
BOD_5	mg/L	16.3	10.8	12.2	12.1	16.8	2.2	1.3	1.3	<1.0	2.9
Chloride	mg/L	189	234	284	309	250	27	80	114	176	173
Sodium	mg/L	297	320	354	379	280	19	70	112	179	199
Sulfate	mg/L	21.6	24.3	29.6	31.1	31.1	8.4	7.5	6.1	12.5	3.3
TSS	mg/L	19.2	11.4	14.9	11.7	25.0	4.6	1.1	2.4	<1.0	3.0
Ammonia-N	mg/L	2.91	1.43	0.73	0.23	6.74	0.30	0.37	0.29	0.12	0.23
Nitrate + Nitrite-N	mg/L	11.51	10.39	17.46	14.64	9.75	0.10	0.06	0.07	0.04	0.07
TKN	mg/L	5.84	3.37	2.37	3.42	8.50	1.70	1.79	2.46	2.21	3.59
Total N	mg/L	17.37	14.13	19.83	17.69	18.25	1.70	1.86	2.56	2.26	3.66
Total P	mg/L	5.21	3.06	3.89	3.85	3.38	0.10	0.12	0.26	0.33	1.40
Fecal coliforms	col/100 ml	18,266	227	2,643	553	3	24	15	13	9	6
Total coliforms	col/100 ml	45,262	445	60,000	1,690	56	181	22	61	37	56

Note: Data from CH2M HILL, 1991.

Table 27-19 Summary of Constituent Mass Removal Efficiencies for Bear Bay, 1987 to 1991

	Mass Removal Efficiency (%)					
Year	BOD_5	TSS	NH_3	TN	TP	UOD
1987	87	93	88	85	98	87
1988	92	94	79	89	97	88
1989	90	93	63	93	95	85
1990	99	99	95	98	98	98
1991	82	90	98	82	66	90
Average	88	92	94	89	88	90

and groundwater losses. The 5-year average mass removal efficiencies in Bear Bay were at least 88 percent for BOD_5, TSS, NH_4-N, TN, TP, and UOD (ultimate oxygen demand).

Because of the variable permeability of Bear Bay's peat, sand, and clay soils, groundwater quality changes are an important part of the operational monitoring program. A network of shallow monitoring wells was established in and around the bay. Figure 27-29 shows the observed groundwater distribution of chlorides after 5 years of treated wastewater discharge to this bay. Calculation of possible travel times and observations of the groundwater chloride plume indicate that groundwater constituent concentrations under the bay respond to surface

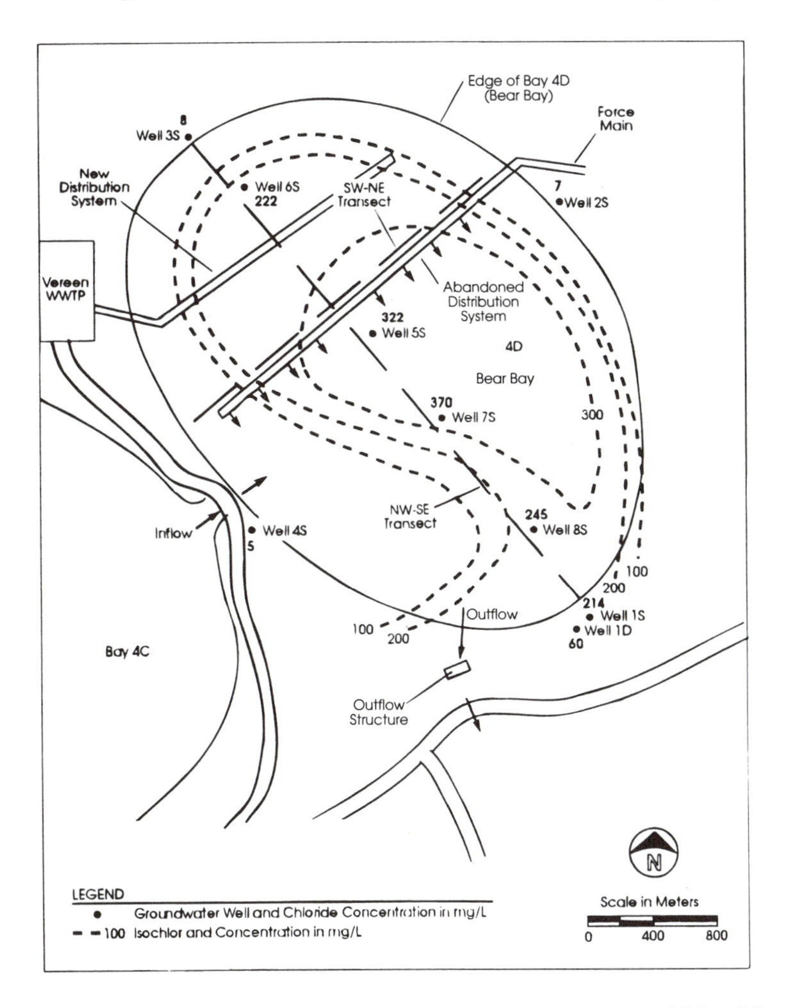

Figure 27-29 Profiles of chloride concentration of shallow groundwater in the vicinity of Bear Bay, SC in 1991.

water constituent concentrations (vertical transport) more rapidly than to longitudinal groundwater transport.

Although wastewater is present in downgradient monitoring wells, as evidenced by elevated chloride concentrations (Table 27-20), no parameters of regulatory concern have reached significantly elevated concentrations in the shallow groundwater directly contiguous with surface water in Bear Bay or in the deeper potable aquifer of concern (compliance monitoring point). Nitrate N concentrations were consistently less than 0.02 mg/L in the compliance well, and fecal coliforms were undetectable. Increased concentrations of TKN and TP were observed in the shallow monitoring wells (2.1 to 5.5 m), but not in the deeper (9.1 m) compliance well.

Biological Changes

Biological changes in the Carolina bays are being monitored to determine how the addition of pretreated wastewater is affecting their flora and fauna. Canopy, subcanopy, and groundcover vegetation species and litterfall and litter decomposition rates are being monitored. Bird counts have been made annually during the breeding season to identify subtle faunal differences that might occur in response to changed hydrology and plant communities.

The vegetation of Bear Bay is dominated by tree species (Table 27-21). The overall average tree basal area for trees greater than 2.5 cm in diameter within 20 10×10 m quadrats in Bear Bay increased from 11.3 to 17.3 m^2/ha between 1986 and 1991. The tree density increased from 695 to 935 individuals per hectare during this operational period. Growth of trees in this successional forest was variable in response to flooding and wastewater effects. Although the effects of these two factors could not be entirely separated, Figure 27-30 shows the apparent different canopy growth rates in response to flooding alone and flooding by wastewater. Canopy growth rates declined in areas where the ground surface was continually flooded with the treated wastewater. This phenomenon was visually evident by the thinning of the tree canopies near the treated wastewater discharge area, due to early and prolonged leaf fall and mortality of susceptible tree species including loblolly pine, sweetgum, American elm, red maple, and water oak.

Decreases in canopy densities in flooded areas of the Carolina bay were offset by increased cover by herbaceous groundcover species such as pennywort (*Hydrocotyle* spp.), duckweed (*Lemna* spp.), and many emergent wetland plant species. Shrub species were somewhat affected by the treated wastewater discharge, with species such as wax myrtle and fetterbush experiencing a decline in coverage.

Table 27-20 Summary of Annual Average Groundwater Quality Around Bear Bay Following 5 Years of Treated Wastewater Discharge

Parameter	Units	Upgradient Well (3S)	Point of Discharge (5S)	Downgradient Wells	
				Shallow (1S)	Deep (1D)
Well depth	m	5.5	2.1	5.5	9.1
Water elevation	ft MSL	34.6	34.0	32.7	27.3
Temperature	°C	19.2	18.2	21.5	17.4
Conductivity	μmho/cm	71	1246	750	200
pH	units	5.9	6.7	5.6	6.1
BOD_5	mg/L	<1	9.0	2.2	1.0
Chloride	mg/L	10.9	360	215	45
Sodium	mg/L	7	331	154	13
Ammonia-N	mg/L	0.09	4.25	0.16	0.24
Nitrate + nitrite-N	mg/L	<0.02	0.06	0.11	0.02
TKN	mg/L	0.12	6.64	6.49	0.80
Total N	mg/L	0.16	6.70	6.60	0.83
Total P	mg/L	0.08	1.12	0.57	0.24
Fecal coliforms	col/100 ml	0	0	0	0

Table 27-21 Summary of Canopy Species Basal Area and Stem Density in Bear Bay During the Period of Wastewater Discharge

	Basal Area[a]						Density[b]					
Species	Oct 1986	Oct 1987	Nov 1988	Nov 1989	Nov 1990	Nov 1991	Oct 1986	Oct 1987	Nov 1988	Nov 1989	Nov 1990	Nov 1991
Acer rubrum	1.03	1.49	1.71	1.86	2.02	2.16	110	145	155	160	165	175
Aralia spinosa	—	—	0.05	0.03	0.03	0.05	—	—	10	5	5	10
Fraxinus caroliniana	0.08	0.11	0.14	0.12	0.13	0.17	15	20	25	20	20	25
Gordonia lasianthus	0.06	0.09	0.10	0.12	0.13	0.14	5	10	10	10	10	10
Liquidambar styraciflua	0.74	0.97	1.26	1.33	1.42	1.34	100	130	170	175	185	175
Liriodendron tulipifera	0.03	0.03	0.03	—	—	—	5	5	5	—	—	—
Magnolia virginiana	—	—	—	—	—	0.33	—	—	—	—	—	5
Myrica cerifera	0.06	0.06	0.09	0.09	0.09	0.12	10	10	15	15	15	20
Nyssa biflora	0.03	0.04	0.04	0.04	0.12	0.13	5	5	5	5	20	20
Persea palustris	0.03	0.18	0.22	0.26	0.32	0.36	5	30	35	40	45	50
Pinus serotina	2.01	2.14	2.24	2.03	2.15	2.35	65	65	65	60	60	65
P. taeda	5.72	6.60	7.13	7.11	7.58	7.71	230	250	255	220	220	205
Quercus laurifolia	0.32	0.50	0.58	0.63	0.66	0.68	25	45	50	50	50	50
Q. nigra	0.06	0.13	0.13	0.14	0.18	0.20	10	20	20	20	20	25
Salix caroliniana	0.79	0.55	0.70	0.72	0.81	0.85	80	40	55	50	45	45
Taxodium ascendens	0.04	0.11	0.20	0.26	0.32	0.38	5	15	25	25	25	25
Ulmus americana	0.32	0.38	0.51	0.57	0.61	0.34	25	25	30	35	40	30
Total	11.33	13.38	15.12	15.31	16.57	17.31	695	815	930	890	925	935

[a] Basal area given as square meters per hectare.
[b] Density given as number of individuals per hectare.

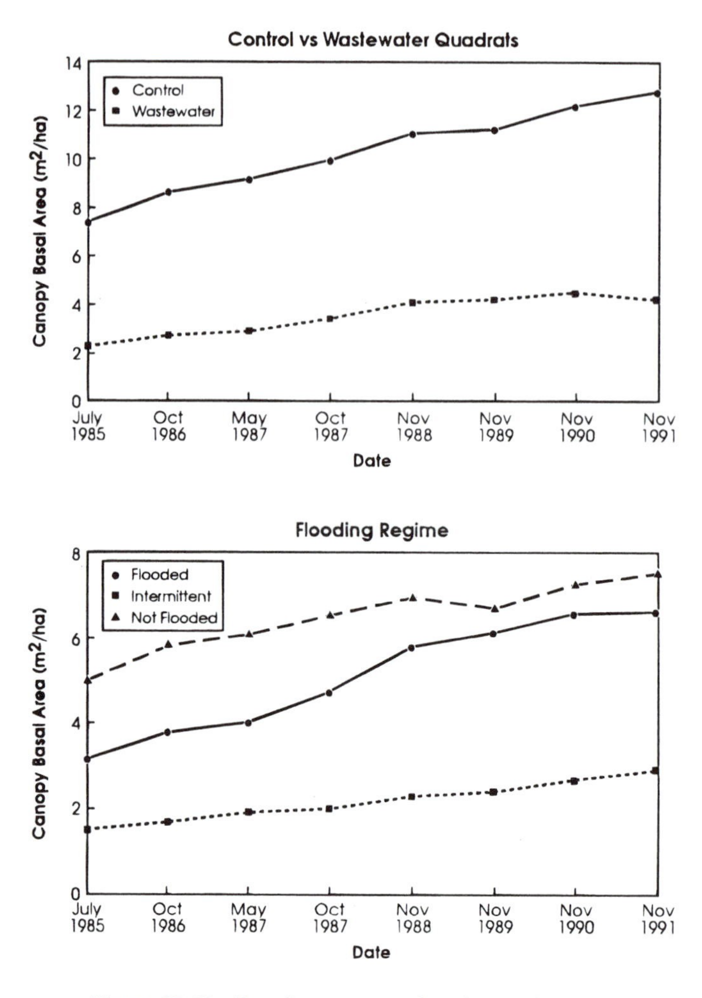

Figure 27-30 Bear Bay overstory basal area changes.

Figure 27-31 illustrates changes in percent cover of woody and herbaceous subcanopy and groundcover species during the operational period in Bear Bay. Increased hydroperiod and wastewater nutrients result in a localized shift from a transitional wetland forest to a scrub-shrub or marsh wetland ecosystem. This change occurs primarily in an area within a few hundred meters of the point of wastewater discharge. Some vegetative stress (canopy thinning) is also evident near the outflow weir from Bear Bay, apparently in response to hydrological changes alone since the water is essentially at background water quality concentrations at this location. No deviation from the previous course of forest succession has yet been observed in the areas of the bay at greater distances from the discharge point. Rotation of treated wastewater discharge to the other Carolina bays may interrupt or partially reverse these vegetational shifts.

Litterfall was measured quarterly in Bear Bay from 1988 through 1991. Average quarterly litterfall values for the whole bay ranged from 0.66 to 2.78 $g/m^2/d$ on a dry-weight basis and largely depended on season with peak litterfall occurring from August to November and minimum litterfall occurring from February to May. Litterfall was highest in the unflooded plots and declined each year in the plots affected by increased wastewater flooding. This trend corroborates the observed decline in canopy and shrub species in the wastewater-affected areas of Bear Bay.

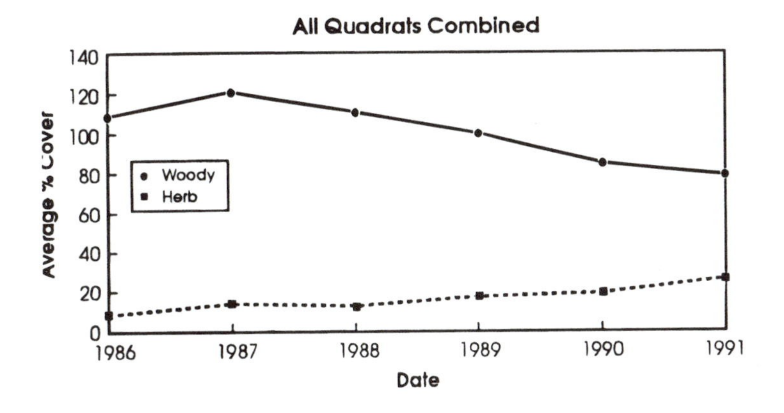

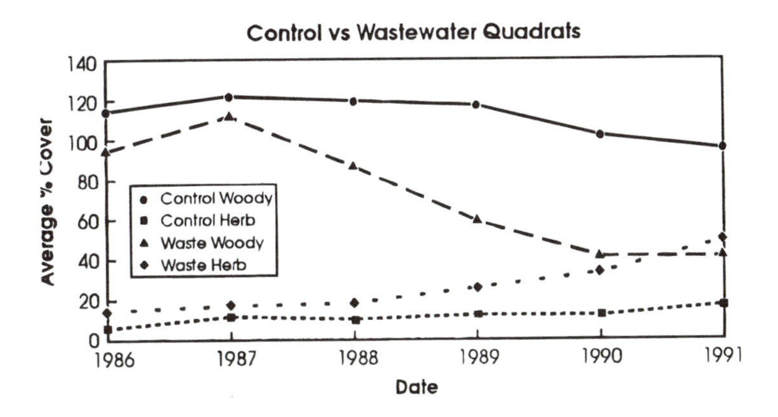

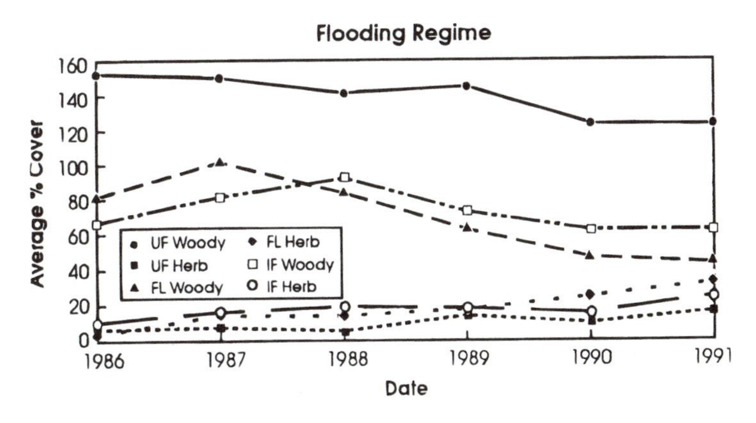

Figure 27-31 Bear Bay woody and herbaceous percent cover changes (UF—unflooded, FL—flooded, IF—intermittently flooded).

Litter decomposition in Bear Bay was measured with a standard indicator species, fetterbush (*Lyonia lucida*). Leaves were incubated *in situ* in fiberglass mesh bags for 3 months. Quarterly average dry-weight losses for the entire bay ranged from 2.14 to 7.33 mg/g/d, with the lowest values during the winter quarter (November to February) and the highest values during the summer (May to August). The presence of surface water derived from the wastewater discharge significantly increased litter decomposition rates, indicating higher nutrient cycling in the effluent-affected area of the Carolina bay.

Bear Bay contains a diverse assemblage of passerine bird species dependent on the mixture of terrestrial and wetland forested conditions. The avifauna was selected as an overall indicator of secondary impacts resulting from indirect changes in hydrology and plant community structure. Bird counts were made during May in each year of operation and compared to a similar count made during the baseline investigations. Table 27-22 summarizes these bird counts. A total of 73 bird species were observed during these breeding season counts, and annual species totals ranged from 36 and 46 species. Some aquatic wading bird species became more common as surface water became more prevalent. Observations of green-backed herons, little blue herons, and wood ducks increased. Bottomland species including the summer tanager, prothonotary warbler, northern parula warbler, and hooded warbler remained abundant. At the same time, species representative of drier pocosin habitats, such as the yellow-breasted chat, worm-eating warbler, and prairie warbler, remained conspicuous throughout the monitoring period.

Biological Criteria

Four biological criteria are being monitored in the Carolina Bay Natural Wetland Treatment System. Bear Bay was operated continuously from system startup in 1987 through the end of 1991 to measure the extent of changes that might occur for each of these biological criteria over a prolonged wastewater discharge period. Figure 27-32 shows the changes observed for the four biological criteria in Bear Bay during this period. These data exclude pine trees from basal area and density calculations, so they are lower than the total values presented previously. Overall canopy basal area and density increased in Bear Bay throughout this operational period because of continuing tree growth and succession. When compared to the canopy declines near the treated wastewater discharge, this figure illustrates the observation that canopy declines in one area were more than offset by increases in the rest of the bay.

Percent cover for the subcanopy and shrub layer increased in Bear Bay during the first 2 years of operation and declined during the remaining period, apparently leveling off about 7 percent below the original level and well above the 30 percent decline criterion after 4 years of continuous discharge. It is not yet known if this downward trend will continue or if it will increase again after the discharge is rotated to the other Carolina bays. The total number of canopy, subcanopy, and shrub species has increased about 28 percent higher than the baseline number of species.

The biological criteria established for the Carolina Bay Natural Wetland Treatment System appear to be an effective way to assess and regulate the biological changes that occur in natural wetlands receiving increased hydrologic and nutrient loadings. The criteria provide a quantitative point of discussion of allowable changes which the regulators and interested parties can agree on to allow cautious, beneficial uses of natural systems.

Ancillary Benefits

The Carolina Bay Natural Land Treatment Program not only serves wastewater management needs, but also plays an important role in protecting the environment. Although the Carolina bays have been recognized as unique, 98 percent of the bays in South Carolina have been disturbed by agricultural activities and ditching. The four bays in the GSWSA treatment program are being conserved in a seminatural ecological condition. These 700 ac of Carolina bays represent one of the largest public holdings of bays in South Carolina.

Carolina bays also provide a critical refuge for rare plants and animals. Amazingly, black bears (*Ursus americanus*) still roam the bays' shrub thickets and forested bottomlands just a few kilometers from the thousands of tourists on South Carolina's beaches. Densely

Table 27-22 Summary of Avifauna Observed in Bear Bay

Common Name	Scientific Name	1986	1987	1988	1989	1990	1991
*Green-backed heron	*Butorides striatus*		+		2	7	7
*Little blue heron	*Florida caerula*		+	1			2
*White ibis	*Eudocimus albus*		+		+		
*Wood duck	*Aix sponsa*			1		10	16
*Black duck	*Anas rubripes*				1		
*Mallard	*A. platyrhynchos*			11			4
Turkey vulture	*Cathartes aura*				1		
Red-shouldered hawk	*Buteo lineatus*		1	2	+	3	1
Red-tailed hawk	*B. jamaicensis*	+		1			
Bobwhite quail	*Colinus virginianus*		1	1		1	
Mourning dove	*Zenaida macroura*	+	5	2	8		+
Yellow-billed cuckoo	*Coccyzus americanus*	5	5	10	7	8	7
Screech owl	*Otus asio*			1			
Barred owl	*Strix varia*	1		+		2	
Chuck will's widow	*Camprimulgus vociferus*	+	+				
Common nighthawk	*Chordeiles minor*	+	1	2	1		
Chimney swift	*Chaetura pelagica*		1	2			10
Ruby-throated hummingbird	*Archilochus colubris*	1	1	2			
Northern flicker	*Colaptes auratus*				1	2	
Pileated woodpecker	*Dryocopus pileatus*	1		2	1		
Red-bellied woodpecker	*Melanerpes carolinus*	4	1	3		2	2
Downy woodpecker	*Picoides pubescens*	+	+	4	1		
Eastern kingbird	*Tyrannus tyrannus*						+
Great crested flycatcher	*Myiarchus crinitus*	8	2	12	7	5	10
*Acadian flycatcher	*Empidonax virescens*	1	3	6	2	5	4
Eastern wood peewee	*Contopus virens*		+	1	+	1	
Barn swallow	*Hirundo rustica*				2		
Tree swallow	*Tachycineta bicolor*						2
Purple martin	*Progne subis*	+	1	3	3		1
Bluejay	*Cyanocitta cristata*	4	1	19	17	12	13
Common crow	*Corvus brachyrhynchos*	1	3		3	5	3
Carolina chickadee	*Parus caroliniensis*	2	2	7	8	5	3
Tufted titmouse	*P. bicolor*	6		2	4	6	9
Brown-headed nuthatch	*Sitta pusilla*			1		1	
Carolina wren	*Thryothorus ludovicianus*	10	9	25	19	8	14
House wren	*Troglodytes aedon*						1
Blue-grey gnatcatcher	*Polioptila caerula*	5	5	9	3	6	8
Ruby-crowned kinglet	*Regulus calendula calendula*	+					
American robin	*Turdus migratorius*	+					
Wood thrush	*Hylocicha mustelina*	1	2	5	2	4	
Eastern bluebird	*Sialia sialis*	1					
Mocking bird	*Mimus polyglottus*				1		
Grey catbird	*Dumatella caroliniensis*		2	1	2	3	4
Brown thrasher	*Toxostoma rufrum*					3	
Red-eyed vireo	*Vireo olivaceus*	8	6	18	16	16	8
White-eyed vireo	*V. griseus*	12	7	17	14	6	6
Yellow-throated vireo	*V. flavifrons*	1	+		2		6
Black-and-White warbler	*Mniotila varia*			1			
*Prothonotary warbler	*Protonotaria citrea*	1	2	14	7	26	26
*Swainson's warbler	*Limnothlypis swainsonii*			4	6	1	2
Worm-eating warbler	*Helmitheros vermivorus*	8	2	2	8	5	4
Northern Parula warbler	*Parula americana*	3	1	7	3	8	3
Myrtle warbler	*Dendroica coronata coronata*	+					
Yellow-throated warbler	*D. dominica*				2		2
Pine warbler	*D. pinus*	5	4	5	7	10	3

Table 27-22 Continued

Common Name	Scientific Name	1986	1987	1988	1989	1990	1991
Prairie warbler	*D. discolor*	2	7	10	3	2	+
*Louisiana water thrush	*Seiurus motacilla*				1		
*Common yellowthroat	*Geothlypis trichas*	5	2	4	9		
Yellow-breasted chat	*Icteria virens*	6	2	6	2	1	1
Hooded warbler	*Wilsonia citrea*	11	16	34	18	18	15
Summer tanager	*Piranga rubra*		+	2	4	3	2
Cardinal	*Richmondena cardinalis*	13	12	31	12	15	12
Blue grosbeak	*Guiraca caerulea*			+			
Indigo bunting	*Passerina cyanea*	1		1			3
*Red-winged blackbird	*Agelaius phoeniceus*	1			1		
Common grackle	*Quisicalus quiscula*	1	1	4	3	15	7
Rufous-sided towhee	*Pipilo erythrophthalmus*	7	7	5	5	9	
Chipping sparrow	*Spizella passerina*					1	6
*Swamp sparrow	*Melospiza georgina*	+					
Total number of species		41	39	46	44	36	39
Total number of detections		146	123	303	222	235	230
Number of wetland-dependent species		5	6	7	9	5	7
Percentage wetland-dependent species		12	15	15	20	14	18

Note: *Wetland-dependent species. +Observed near transect; counted as one detection. (Data from CH2M HILL, 1991.)

vegetated wetlands can provide a variety of ground nests and denning cavities for black bears (Hellgren and Vaughan, 1989), and these dens have been observed in the treatment wetlands. Venus flytraps and pitcher plants, fascinating carnivorous plants that trap trespassing insects, occur naturally in the Carolina bays. In addition, the bays are home to many other interesting plant and animal species.

The Carolina Bay Natural Land Treatment System is currently open to the public through guided tours by GSWSA staff.

Awards

In 1991, the Carolina Bay Natural Land Treatment Program won the Engineering Excellence Award, Best of Show, from the Consulting Engineers of South Carolina.

The American Consulting Engineers Council (ACEC) Grand Conceptor Award, considered the highest national honor in the consulting engineering field, was awarded to CH2M HILL in 1991 for its implementation of the Carolina bays project. ACEC selected the project from a field of 127 national finalist entries, each of which had earlier won in state or regional engineering excellence competitions.

REEDY CREEK, FLORIDA

Background

The Reedy Creek Improvement District (RCID) was established to provide wastewater collection and treatment for Walt Disney World, located west of Orlando, FL. Wastewater generated within RCID received secondary treatment at the Reedy Creek WWTP and was discharged directly to Reedy Creek until 1977. In January 1977, a portion of the effluent from the Reedy Creek WWTP was directed to a percolation pond for final disposal. In the early 1970s, increasing state and federal concern about the potential eutrophication of surface waters led RCID to initiate a long-term program to develop and test wetland, aquatic, and natural land treatment technologies for nutrient removal prior to discharge to Reedy Creek. This testing as well as the success of the Gainesville, FL cypress dome research led RCID

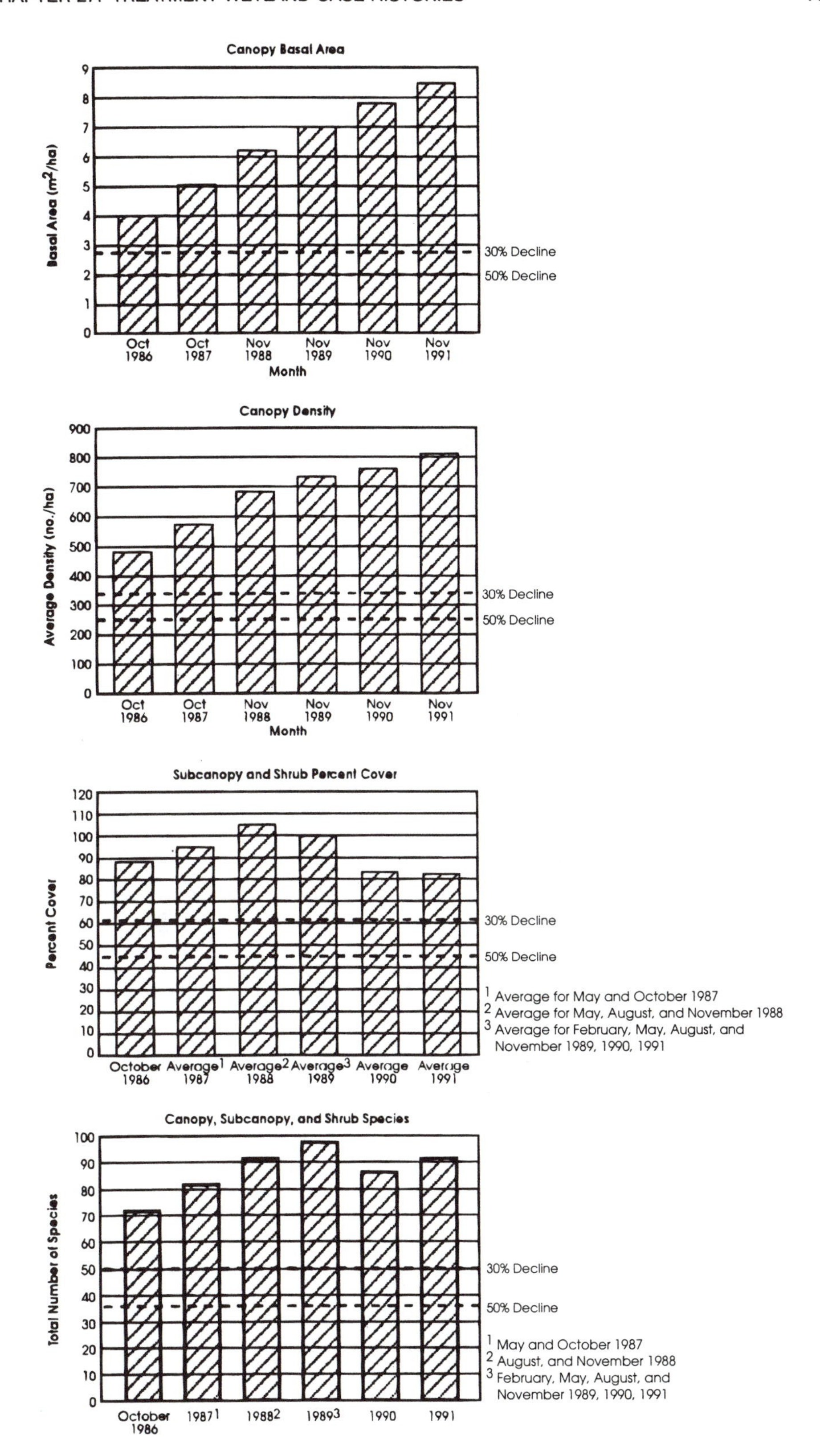

Figure 27-32 Changes in biological criteria indices in Bear Bay from 1986 through 1991.

to incorporate two natural, forested wetland areas in the altered floodplain of Reedy Creek into their overall effluent treatment and management system (McKim, 1982).

The Overland Flow Wetland Treatment System (OFWTS) began receiving effluent either from the percolation ponds or directly from the WWTP in September 1978 (Figure 27-33), and Wetland Treatment System No. 1 (WTS1) began receiving effluent in October 1978. Both systems were shut down in 1992 after nearly continuous use for 13 years. The systems were shut down when a zero discharge option (high-rate land application) was implemented to eliminate all nutrient discharges into the Reedy Creek watershed, which in turn feeds the Kissimmee River-Lake Okeechobee-Everglades watershed.

Beginning in June 1988, a third natural wetland treatment system, Wetland Treatment System No. 2 (WTS2), was operated for 13 months before it was taken out of operation because of water quality criteria compliance problems (Figure 27-33).

The Reedy Creek natural wetland treatment systems received higher loadings and greater monitoring intensity than contemporary engineered natural treatment wetlands. A review and summary of this operational history provides a number of insights concerning the usefulness of natural wetlands for wastewater management.

System Description

Figure 27-33 shows the former locations of the three natural wetland treatment systems at RCID. All three systems were located in the former floodplain of Reedy Creek, a channelized, blackwater low-gradient stream with seasonal flows ranging from 0 to 17 m^3/s. WTS1 included approximately 34 ha of isolated floodplain swamp forest characterized by bald cypress, black gum, red maple, pop ash, bays, and pines (Knight et al., 1987). Pretreated wastewater from the Reedy Creek WWTP was discharged into the northern end of WTS1 and then moved by channelized sheetflow to the south, with eventual discharge over a 6-m horizontal weir into a postaeration basin and from there to the Reedy Creek channel. About 0.9 m of topographic gradient exists along the 1830-m distance from the northern inflow point to the downstream discharge point. Flow was observed in numerous, small braided channels which weave in and out around the buttresses of mature, healthy wetland trees.

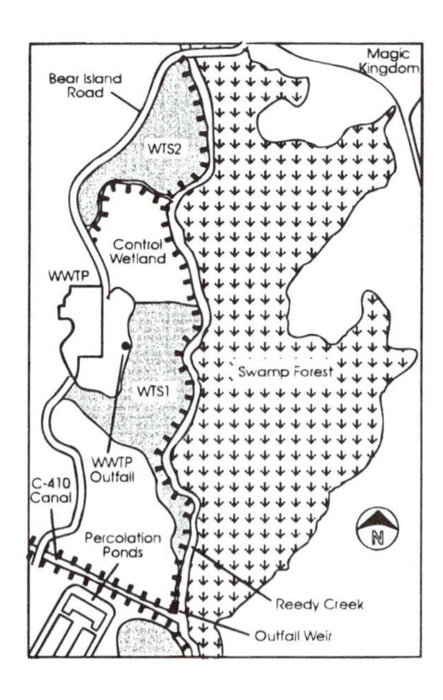

Figure 27-33 Location of wetland treatment systems at RCID.

The average water depth in WTS1 was about 19 cm, and water volume was estimated as 63,000 m^3.

The OFWTS included about 6 ha of floodplain wetlands following an underdrained percolation pond and a small marsh-overland flow area that was created by constructing a level inlet spreader swale and leveling about 0.4 ha of former pine forest. From this point, wastewater entered the natural wetland area as sheetflow through numerous shallow braided channels with an average estimated water depth of about 15 cm and a water volume of 8900 m^3. Water exited the OFWTS about 460 m downstream, where it spilled over a 1.8-m horizontal weir into a small aeration basin before final discharge to the channelized portion of Reedy Creek.

WTS2 included about 36 ha of isolated and drained wetland floodplain swamp forest. This area is surrounded by dikes and roads and received no surface inflows except direct rainfall for about 20 years before its use for discharge of pretreated wastewater in 1988. Significant peat subsidence, leaning cypress trees, and colonization by facultative wetland species was observed in WTS2 before 1988 (Wallace et al., 1990). The consequences of this peat subsidence were not anticipated before the system began operation, but were quickly perceived after system startup as discussed later.

Permitting

Because much of the floodplain wetlands bordering Reedy Creek were hydrologically isolated from the creek when a berm was constructed in 1968, the RCID natural wetland treatment systems were classified as nonjurisdictional wetlands (not waters of the state or of the U.S.) and were permitted under the land disposal regulations of the state of Florida. Surface discharge permits were obtained from both the Florida Department of Environmental Regulation (FDER) and the U.S. EPA to regulate the surface discharge from the wetlands to Reedy Creek. The original Reedy Creek WWTP did not include design for intentional nutrient removal; however, the system's pretreatment capacity was increased so that a significant fraction of TKN was being oxidized and TP concentrations were significantly reduced before discharge to the wetlands.

State and federal permits issued to RCID in 1985 to 1986 allowed the Reedy Creek WWTP to be operated at an annual average flow of about 22,700 m^3/d, with discharge of 3786 m^3/d to the OFWTS and 17,600 m^3/d to WTS1. Discharge limits from these two wetlands were 2 mg/L for TN and 0.5 mg/L for TP, the lowest limits for any wetland treatment system at the time. In 1987, FDER also permitted the use of WTS2 for further polishing of pretreated wastewater effluent. The same discharge standards were used as those for the other wetlands, and an additional flow limit of about 10,800 m^3/d was added for an overall flow capacity of 32,200 m^3/d. In early 1990, treatment difficulties in WTS2 led to a permit modification to cease flows to WTS2 and to allow increased flows to WTS1 and the OFWTS by a total of about 13,250 m^3/d. Increasingly stringent permitting requirements for surface discharge to Reedy Creek eventually led RCID to abandon WTS1 and OFWTS in 1992. All treated effluent from RCID is now discharged to groundwater via rapid infiltration basins or is reused onsite for landscape irrigation.

System Performance

Flows to WTS1 were essentially continuous from October 1978 until May 1992 (Figure 27-34). Average annual daily flows to WTS1 ranged from 6100 to 18,000 m^3/d between 1978 and 1989, resulting in a range of annual average hydraulic loading rates between 1.6 and 4.9 cm/d and estimated hydraulic residence times between 2.9 and 8.8 days. Annual average outflows from WTS1 are very similar to inflows, indicating the influence of the

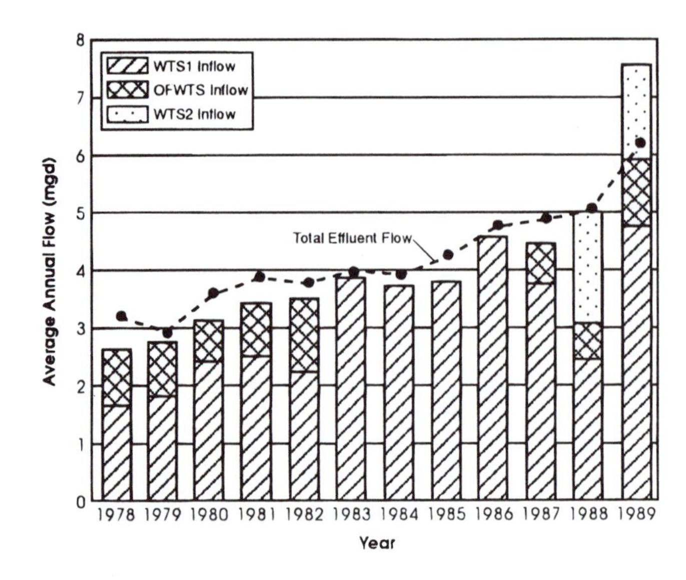

Figure 27-34 Average annual flows from the Reedy Creek WWTP to natural wetland treatment systems.

wastewater loading on system hydrology. Stormwater run-in flows were occasionally high before 1988, when a stormwater inlet was eliminated. Inflows and losses associated with rainfall, run in, and infiltration were a small fraction of the annual water budget for this system.

Table 27-23 summarizes annual average inflow and outflow concentrations, mass removal rates, and mass removal efficiencies for BOD_5, TSS, TN, and TP for WTS1 from 1978 until 1989. A review of the inflow water quality averages in Table 27-23 shows the consistent improvement of the Reedy Creek WWTP effluent quality during this period, except for a 6-month upset at the wastewater treatment plant during 1987. Wetland outflow BOD_5 concentration averaged 2.5 mg/L or less during all years, except during the WWTP upset when the annual average outflow concentration increased to 3.7 mg/L. In spite of the low inflow BOD_5 concentration, the mass removal efficiency averaged 58 percent during the period of record, and mass removal rate averaged about 0.99 kg/ha/d. TSS reductions were similar to those observed for BOD_5, with a range of annual average outflow TSS concentrations between 1 and 4 mg/L over the period of record and a long-term average of 2.4 mg/L. The long-term average mass removal rate for TSS was 1.8 kg/ha/d, for a mass removal efficiency of 61 percent.

The Reedy Creek WTS1 consistently removed nitrogen, including ammonia and nitrate. The long-term average inflow NH_4-N concentration of 2.98 mg/L was reduced to 0.72, with most annual averages less than 0.2 mg/L and a long-term average mass removal efficiency of 83 percent. TN concentration was reduced from about 8.6 to 1.9 mg/L, based on the long-term averages reported in Table 27-23. This concentration is equivalent to an average mass removal rate of 2.09 kg/ha/d and a mass removal efficiency of 78 percent.

TP concentration was never consistently reduced by WTS1. The long-term outflow TP concentration was 1.8 mg/L compared to the inflow concentration of 1.4 mg/L, resulting in a calculated mass removal efficiency of −40 percent. The increase in TP may have resulted from previous stormwater TP inputs to this wetland, resulting in phosphorus-saturated soil conditions and continuing releases as surface water TP concentrations were reduced.

Inflows to the OFWTS began in September 1978 and continued until December 1982 when the flow was ceased. Flow resumed in April 1987 and continued until November 1988 when it was discontinued for 4 months before resuming in March 1989. During its operation, the OFWTS received annual average inflows from the Reedy Creek WWTP directly or via the percolation ponds ranging from 2500 to 5000 m^3/d, resulting in annual average hydraulic

Table 27-23 Summary of Reedy Creek, FL WTS1 Operational Performance

	Flow (m^3/d)		BOD_5				TSS				TN				TP			
Year	In	Out	In	Out	RR	EFF	In	Out	RR	EFF	In	Out	RR	EFF	In	Out	RR	EFF
1978	6,130	4,960	7.9	1.7	1.17	83	14.0	2.0	2.23	88	13.1	1.8	2.09	89	2.5	2.5	0.08	18
1979	6,740	11,850	7.2	1.4	0.93	66	17.0	4.0	1.97	59	11.1	1.4	1.72	78	2.5	3.0	−0.56	−115
1980	9,240	12,720	7.4	1.5	1.45	72	22.0	1.0	5.60	94	10.5	1.0	2.45	86	2.6	3.2	−0.47	−66
1981	9,620	5,600	4.5	2.0	0.95	74	6.0	3.0	1.20	71	9.6	0.8	2.57	95	2.6	3.0	−0.23	32
1982	8,440	8,710	7.3	1.3	1.48	82	10.0	2.0	1.97	79	8.5	1.1	1.84	87	1.8	3.4	−0.43	−95
1983	14,730	18,510	6.3	1.6	1.85	68	8.0	2.0	2.37	69	6.2	0.9	2.16	81	0.7	0.9	−0.23	−80
1984	14,010	15,980	3.2	2.0	0.37	29	5.0	3.0	0.65	32	5.4	1.1	1.71	76	0.6	1.0	−0.20	−78
1985	14,310	14,420	3.9	1.8	0.88	54	4.0	2.0	0.84	50	5.5	1.3	1.76	76	0.6	0.7	−0.07	−25
1986	17,340	14,310	3.5	2.0	0.95	53	3.0	4.0	−0.16	−10	7.3	2.1	2.83	76	0.6	0.7	−0.01	−3
1987	14,540	15,330	5.6	3.7	0.72	30	7.0	2.0	2.09	70	12.8	7.2	2.20	40	1.4	1.5	−0.07	−12
1988	9,280	9,620	3.0	1.4	0.43	52	6.0	1.0	1.35	83	6.2	1.6	1.24	74	0.6	1.0	−0.11	−64
1989	17,980	15,370	3.5	2.5	0.72	39	5.3	3.0	1.44	52	6.5	2.1	2.47	72	0.4	0.5	0.01	6
Average	11,850	12,270	5.3	1.9	0.99	58	8.9	2.4	1.80	61	8.6	1.9	2.09	78	1.4	1.8	−0.15	−40

Note: In = Inflow concentration (mg/L), Out = Outflow concentration (mg/L), RR = Mass removal rate (kg/ha/d), EFF = Mass removal efficiency (%), Area = 35 ha.

Table 27-24 Summary of Reedy Creek, FL OFWTS Operational Performance

	Flow (m^3/d)		BOD_5				TSS				TN				TP			
Year	In	Out	In	Out	RR	EFF	In	Out	RR	EFF	In	Out	RR	EFF	In	Out	RR	EFF
1978	3820	680	7.9	1.4	4.95	97	14.0	2.0	8.89	98	13.1	0.9	8.42	99	2.5	1.1	1.47	92
1979	3600	4690	7.2	1.5	3.25	73	17.0	4.0	7.19	69	11.1	1.6	5.57	82	2.5	1.0	0.70	47
1980	2730	2690	7.4	1.3	2.86	83	22.0	3.0	8.81	87	10.5	0.9	4.41	91	2.6	1.3	0.62	50
1981	3330	3330	4.5	2.1	1.39	53	6.0	2.0	2.24	67	9.6	0.8	5.02	92	2.6	1.0	0.93	63
1982	5000	5340	7.3	1.5	4.87	78	10.0	2.0	6.72	79	8.5	1.7	5.72	78	1.8	1.3	0.39	25
1987	2610	3370	5.6	1.3	1.78	70	7.0	2.0	1.93	63	12.8	1.6	4.79	84	1.4	0.5	0.31	51
1988	2500	3860	3.0	1.3	0.39	33	6.0	1.0	1.93	74	6.2	1.0	2.01	75	0.6	0.4	0.00	8
1989	4240	4390	3.5	1.8	1.16	47	5.3	2.0	2.32	61	6.5	1.4	3.63	77	0.4	0.3	0.08	32
Average	3480	3540	5.8	1.5	2.58	67	10.9	2.2	5.00	75	9.8	1.2	4.94	85	1.8	0.9	0.56	46

Note: In = Inflow concentration (mg/L), Out = Outflow concentration (mg/L), RR = Mass removal rate (kg/ha/d), EFF = Mass removal efficiency (%), Area = 5.8 ha.

loading rates between 3.8 to 8.8 cm/d and estimated annual average hydraulic residence times of 1.7 to 4 days. Since inflow water quality data for the OFWTS do not distinguish which flows went through the percolation ponds and which went directly to the wetland, system performance data are summarized for the combined effect of percolation/natural wetlands.

Table 27-24 summarizes the performance of the combined percolation pond and OFWTS for the years it received pretreated inflow. Treatment in this combination system was generally greater than observed in WTS1 in spite of the typically higher hydraulic loading rates. This result is undoubtedly due to the importance of the percolation pond in attenuating some solids and sorption of associated nutrients. Annual average outflow BOD_5 concentrations were equal to or less than 2.1 mg/L during all years, with a long-term average outflow concentration of 1.5 mg/L and mass removal efficiency of 67 percent. The long-term average TSS outflow concentration from the OFWTS was 2.2 mg/L for a mass removal rate of 5 kg/ha/d and a mass removal efficiency of 75 percent.

The OFWTS was very effective at removing nitrogen and phosphorus. Maximum annual average outflow NH_4-N concentration was 0.3 mg/L, and the long-term average NH_4-N concentration was 0.12 mg/L. This was equivalent to a long-term average NH_4-N mass removal rate of 1.76 kg/ha/d and mass removal efficiency of 95 percent. The maximum annual average TN outflow concentration from the OFWTS was 1.7 mg/L, and the long-term average was 1.2 mg/L. Annual average TN mass removal rates were estimated between 2.01 and 8.42 kg/ha/d, and the long-term TN mass removal efficiency was 85 percent. Unlike WTS1, the OFWTS consistently assimilated TP for a long-term average TP outflow concentration of 0.9 mg/L, a mass removal rate of 0.56 kg/ha/d, and a mass removal efficiency of 46 percent. The two factors that contributed to the treatment efficiency difference between WTS1 and the OFWTS for TP assimilation were the presence of the percolation pond during periods when flow was routed through it and the lack of an historical stormwater input to this natural wetland area.

WTS2 was operated during the period from June 1988 through July 1989 when it was shut down because of its failure to accomplish pollutant concentration and mass reductions. Table 27-25 summarizes operational data from WTS2 during the two annual operational periods. During the entire period of operation, WTS2 received an average inflow of pretreated wastewater from the Reedy Creek WWTP of 11,725 m^3/d, for an average hydraulic loading rate of 3.5 cm/d. Influent quality to WTS2 was very high with the average inflow BOD_5 concentration equal to 2.6 mg/L, TSS equal to 5.3 mg/L, TN 5.97 mg/L, and TP 0.47 mg/L. All major constituent concentrations were increased as the pretreated wastewater passed through WTS2, except for the TSS concentration. The water entering WTS2 was clear with almost no color, while the water exiting the system was highly colored and had elevated concentrations of BOD_5, TOC, NH_4-N, TN, and TP. While the concentration of organic N was increased, the concentration of NO_3-N decreased from about 3.0 to 0.04 mg/L, indicating a high denitrification rate.

The antecedent nutrient loads were flushed from WTS2 over a period of several months (see Figure 14-8). Output phosphorus was less than input TP by the end of 1988. Although long-term stable performance for TP had not been reached, treatment efficiencies had become positive.

RCID and regulatory staff quickly determined that the impaired water quality of WTS2 outflow was caused by the resolubilization of dead plant detritus, powdered organic carbon, and nutrients resulting from subsidence of up to 60 cm of the original peat substrate in this floodplain forest. At the end of 1988, an effort was made to continue the use of WTS2 with reduced inflow to allow complete flushing of these carbon and nutrient deposits before resuming full flow operation. The outflow from WTS2 was routed to the inflow point of WTS1 to allow additional polishing of these accumulated nutrients. It was believed that the

Table 27-25 Summary of Operational Performance for WTS2 at Reedy Creek, FL

			Concentration (mg/L)									
	Flow (m^3/d)		BOD_5		TSS		NH_4-N		TN		TP	
Operational Period	**In**	**Out**	**In**	**Out**	**In**	**Out**	**In**	**Out**	**In**	**Out**	**In**	**Out**
June–Dec 1988	11,400	12,710	2.7	6.8	5.6	2.0	0.97	1.84	5.13	5.71	0.45	0.77
Jan–July 1989	12,050	13,630	2.4	5.5	5.0	4.0	2.62	1.31	6.81	4.85	0.49	0.49
Average	11,725	13,170	2.6	6.2	5.3	3.0	1.80	1.58	5.97	5.28	0.47	0.63

system might achieve steady state performance and treatment levels comparable to WTS1 within a few months.

The actual outcome of lowering flows to WTS2 was opposite of the intended result. Almost immediately the concentration of outflow constituents increased, including BOD_5, TN, and TP. This observation demonstrated that increased constituent concentrations were directly tied to hydraulic residence time (increased due to lower inflow rates) and that the actual time to flush this system and the mass of the nutrients that would be released during that time were unacceptably high from the standpoint of potential downstream water quality impacts. WTS2 was abandoned after July 1989 and has not been used for any further water quality treatment. A fine of several hundred thousand dollars was imposed by U.S. EPA because of permit violations.

Biological Conditions

Biological conditions in the Reedy Creek natural wetland treatment systems were not quantified until 1988, more than 10 years after flows to WTS1 and OFWTS began. In 1988, quarterly sampling of the wetland treatment system biological communities in WTS1 and WTS2 began. This sampling included measurements of tree basal area and density, herbaceous vegetation, fish populations, and mosquitoes. The results of this sampling effort are briefly summarized here.

Table 27-26 summarizes canopy density and basal area information for WTS1 and WTS2 in May 1988, before treated wastewater was discharged to WTS2. These data compare two floodplain forested wetland plant communities that were probably quite similar in 1968 before Reedy Creek was channelized, but which have had very different hydrologic histories since then. WTS1 received only stormwater inflows until 1978, when it also began to receive pretreated wastewaters at relatively high hydraulic loading rates. WTS2 received only direct

Table 27-26 Summary of Canopy Density and Basal Area in Reedy Creek Natural Wetland Treatment Areas, May 1988

	WTS1		WTS2	
Canopy Species	**Density (#/ha)**	**Basal Area (m^2/ha)**	**Density (#/ha)**	**Basal Area (m^2/ha)**
Acer rubrum	320	5.78	455	10.18
Berchemia scandens	4	0.01	—	—
Cephalanthus occidentalis	188	0.26	—	—
Cornus foemina	76	0.10	—	—
Decumaria barbara	6	0.00	3	0.00
Fraxinus caroliniana	1239	3.64	6	0.06
Gordonia lasianthus	—	—	3	0.09
Ilex cassine	74	0.13	33	0.22
Itea virginica	—	—	2	0.00
Magnolia virginiana	721	5.62	188	3.35
Myrica cerifera	60	0.07	—	—
Nyssa biflora	465	7.42	159	3.93
Parthenocissus quinquefolia	9	0.01	3	0.00
Persea palustris	100	1.57	66	0.72
Pinus elliotii	13	0.99	81	5.14
Quercus nigra	—	—	3	0.00
Rhododendron viscosum	2	0.00	—	—
Taxodium ascendens	313	7.29	—	—
Taxodium distichum	87	5.30	405	7.94
Toxicodendron radicans	15	0.01	—	—
Ulmus americana v. floridana	2	0.00	—	—
Vitis rotundifolia	91	0.09	191	0.22
Total	3785	38.29	1598	31.85

Note: Includes all woody stems greater than 2.5 cm diameter at breast height.

rainfall during this 20-year period and was seasonally dry or saturated by groundwater due to its low elevation and position next to the Reedy Creek channel. In spite of about 10 years of wastewater discharges, WTS1 still supported a diverse and robust forested wetland plant community in 1988. Stem density and basal area were both high at 3785 stems per hectare and 38.29 m^2/ha, respectively, and are typical of mature southern coastal plain swamps. Dominant canopy species in WTS1 were black gum, pond cypress, red maple, sweetbay, bald cypress, pop ash, and red bay. A total of 19 canopy species were recorded in six quadrats in WTS1.

By comparison, WTS2 had lower stem density (1598 stems per hectare), basal area (31.85 m^2/ha), and only 14 canopy species. Dominant canopy species in WTS2 were red maple, bald cypress, slash pine, black gum, and sweetbay. The plant community data from WTS1 demonstrate that a healthy natural wetland forest plant community can be maintained while providing advanced wastewater treatment. Successful maintenance of adapted natural wetland plant communities is based on minimizing hydroperiod changes and irrigation with a high-quality, pretreated wastewater.

Drastic vegetation changes occurred in WTS2 when it began to receive treated wastewater inflows in 1988. Much of the understory and herbaceous vegetation died after a normal (nondrained) hydroperiod was reestablished. This response was determined to be the result of the direct and indirect physical effects of inundation and not due to the nutrient or chemical characteristics of the wastewater (Wallace et al., 1990). After peat oxidation and subsidence, numerous plant species had rooted in the deep cavities between hummocks around large trees. When surface water was established to the predrainage level (surface of nonsubsided peat), most of these colonizing species perished. However, less than 1 year after wastewater discharge to WTS2 began, these vegetation changes had stabilized and many of these species were reestablishing populations based on the restored hydrologic regime. After wastewater discharge to WTS2 ceased in 1989, the system was expected to revert to its overdrained condition with declining populations of wetland species and succession to a more upland plant community.

Fish sampling in forested wetlands can be difficult due to the general absence of open water areas suitable for electroshocking or using throw nets. After comparing several sampling methods, a small vertical enclosure method (plastic trash can with an area of 0.13 m^2) was developed and successfully employed in WTS1 to capture the small fish species typical of southern forested wetlands. Minnow traps were also used, but were found to preferentially sample larger fish. Table 27-27 presents the results for fish sampling in WTS1 quarterly between May 1988 and April 1989. Five species of fish were encountered in WTS1 including mosquitofish (*Gambusia affinis*), least killifish (*Heterandria formosa*), Everglades pygmy sunfish (*Elassoma evergladei*), sailfin molly (*Poecilia latipinna*), and spotted sunfish (*Lepomis punctatus*). The average fish density in WTS1 ranged from 6.28 to 47.8 fish per square meter, with an annual average of 23.02 fish per square meter. Fish biomass (wet weight) averaged 4.27 g/m^2 during this 1-year monitoring period. Fish sampling conducted in WTS2 during this period indicated a total absence of fish before wastewater discharge (no surface water) and densities of only about 0.17 fish per square meter after 10 months of discharge.

Mosquito larvae populations were sampled with a dipper in WTS1 and WTS2 quarterly from May 1988 until April 1989. Quarterly average mosquito populations ranged from 0.004 to 0.56 larvae per dip in WTS1 and from 0.68 to 1.56 larvae per dip in WTS2. The annual average populations were 0.12 larvae per dip in WTS1 and 1.24 larvae per dip in WTS2. While neither of these systems had excessive populations of mosquito larvae, the lower populations in WTS1 may have been a direct result of the higher fish populations in that wetland.

Table 27-27 Summary of Fish Density and Biomass for Reedy Creek, FL WTS1

Fish Species	Density (#/m²)					(Biomass (g/m²)[a]				
	May 1988	Sept 1988	Jan 1989	Apr 1989	Average	May 1988	Sept 1988	Jan 1989	Apr 1989	Average
Gambusia affinis	16.56	3.45	5.42	41.60	16.76	4.67	0.62	1.57	2.95	2.45
Heterandria formosa	6.67	1.18	2.28	5.73	3.96	1.25	0.05	0.44	0.60	0.58
Elassoma evergladei	4.16	—	—	0.08	1.06	0.55	—	—	0.01	0.14
Poecilla latipinna	2.35	1.65	0.31	0.39	1.18	2.16	1.18	0.19	0.39	0.98
Lepomis punctatus	0.16	—	0.08	—	0.06	0.41	—	0.04	—	0.11
Totals	29.90	6.28	8.09	47.80	23.02	9.04	1.85	2.24	3.95	4.27

[a] Wet-weight basis.

Summary

The experience at Reedy Creek, FL provides several important lessons for future use of natural wetlands for water quality management. One of the initial goals of this natural wetlands system was to reduce TP as was done at the cypress dome research facility near Gainesville, FL, based on a review of data from other natural wetlands receiving wastewaters in Florida. This goal was never met in WTS1, but was successful in OFWTS, in large measure due to the combination of technologies in that system (percolation pond followed by natural wetlands). Since the time when Reedy Creek was originally designed and permitted, it has become increasingly clear that phosphorus removal in forested wetlands is variable and subject to site-specific hydrology and soil characteristics.

Previous work in Florida and Michigan had led wetland treatment researchers to use and recommend hydraulic loading rates for natural wetland treatment systems less than 0.4 cm/d (Thabaraj, 1982). The successful operational experience from Reedy Creek WTS1 demonstrates that setting an upper level on allowable hydraulic loading rates to natural wetlands is overly simplistic and may be not be warranted to provide maximum water quality benefits while protecting and enhancing environmental benefits. Maximum annual average hydraulic loading rates to WTS1 and OFWTS, both of which were highly successful, following advanced wastewater treatment of BOD_5, TSS, and nitrogen, were 4.9 and 8.8 cm/d, respectively. These hydraulic loading rates did not preclude the long-term maintenance of wetland canopy species diversities, densities, and basal areas as high or higher than most pristine natural wetlands in central Florida.

Three lessons can be derived from these observations: (1) in flow-through wetlands, hydroperiod is less a function of hydraulic loading rate than of existing ground contours and controlling structures such as berms and weirs; (2) long hydroperiods are not detrimental to wetland-adapted tree species as long as water depth is not excessive and water quality is good (adequate dissolved oxygen); and (3) design hydraulic loading rate in natural wetland treatment systems should be based primarily on treatment goals and not on protecting biological systems.

The short-term use of WTS2 also provided new and important information for design of both natural and constructed wetland treatment systems. Antecedent soil conditions may greatly influence the success of wetland treatment systems to meet specific treatment goals. If site soils or existing plant populations contain elevated concentrations of pollutants of concern, they may in turn release these into the water column following reflooding. This condition is not likely in wetlands that have received normal rainfall and surface inflows, but becomes more likely in lands that have been drained or that have been used for disposal of sludge, received agricultural fertilizers, or industrial wastes. Baseline soil chemistry and water quality studies are essential to detect potential fatal flaws with a proposed wetland treatment system site. The history of vegetation impacts in WTS2 is an example of inadequate biological impact assessment prior to using a natural wetland for water quality treatment. Some biological changes always result from changing wetland hydroperiod, and nutrient status and the potential magnitude of these changes must be estimated and agreed upon prior to project initiation to ensure success.

The Reedy Creek natural wetland treatment systems were not a failure; they provided cost-effective, advanced wastewater treatment at Walt Disney World for more than 12 years. Pretreatment at the Reedy Creek WWTP was periodically upgraded to enhance overall system performance throughout the life of the project. Many new and important observations concerning natural wetland treatment systems were documented through an exhaustive monitoring program. Changing political and regulatory goals forced the eventual phasing out of the Reedy Creek natural treatment wetlands and will likely result in their functional decline as they revert to their previously drained condition.

HOUGHTON LAKE, MICHIGAN

Introduction and Project Description

The community of Houghton Lake, located in the central lower peninsula of Michigan, has a seasonally variable population averaging approximately 5000. A 2000 m^3/d (0.5 mgd) sewage treatment plant was built in the early 1970s to help protect the 24-m^2, shallow recreational lake. This treatment facility is operated by the Houghton Lake Sewer Authority (HLSA). Wastewater from this residential community is collected and transported to two 2-ha (5-ac) aerated lagoons, which provide 6 weeks detention. Sludge accumulates on the bottom of these lagoons, below the aeration pipes. Effluent is then stored in a 12-ha (29-ac) pond for summer disposal, resulting in depth variation from 50 cm (fall) to 3 m (spring). Discharge can be to 35 ha of seepage beds, to 35 ha of flood irrigation area, or to a 600 ha peatland. The seepage beds were used until 1978, when the wetland system began operation. The wetland has been used since that time, with only occasional discharges to seepage or flood fields. The annual discharge is currently 775,000 m^3 (200 million gallons, 1993). Secondary wastewater is intermittently discharged to the peatland during May through September at the instantaneous rate of 10,000 m^3/d (2.6 mgd).

Provision for chlorination is available, but has not been used, because of low levels of fecal coliform indicator organisms. Water from the holding pond flows by gravity or is pumped to a 1.2-ha pond which would provide chlorine removal in the event the effluent is ever chlorinated. Wastewater from this pond is pumped through a 30-cm diameter underground force line to the edge of the Porter Ranch peatland (Figure 27-35). There, the transfer line

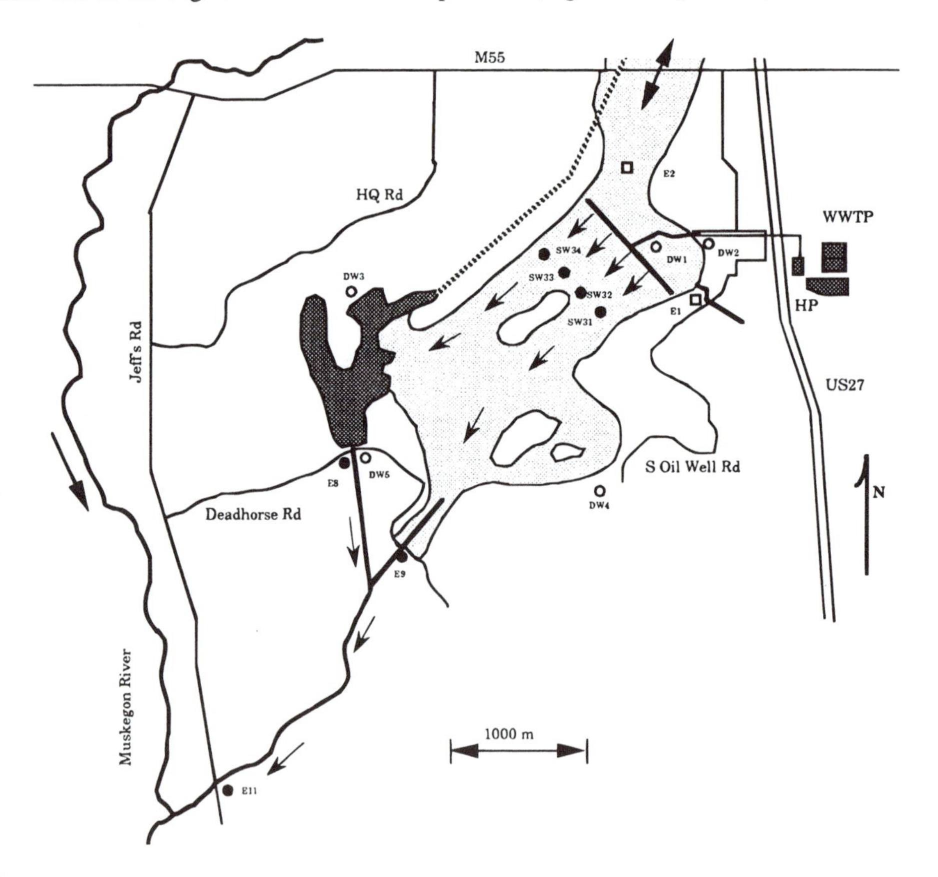

Figure 27-35 Area map of the Porter Ranch wetland vicinity showing sampling stations and wells.

surfaces and runs along a wooden boardwalk for a distance of 760 m to the discharge area in the wetland. The wastewater may be split between two halves of the discharge pipe which runs 500 m in each direction (Figure 27-36). The water is distributed across the width of the peatland through small, gated openings in the discharge pipe. Each of the 100 gates discharge approximately 100 m^3/d under typical conditions, and the water spreads slowly over the peatland. The two branches of the discharge pipe are not used equally in all years.

The wetland provides additional treatment to the wastewater as it progresses eventually to the Muskegon River 10 km away. Small, natural water inflows occur intermittently on the north and east margins of the wetland. These flows are partially controlled by beaver. Interior flow in the wetland occurs by overland flow, proceeding from northeast down a 0.02% gradient to a stream outlet (Deadhorse Dam, Station E8) and beaver dam seepage outflow (Beaver Creek, Station E9), both located about 3 km from the discharge. Wastewater adds to the surface sheetflow. Hydrogeological studies have shown that there is neither recharge or discharge of the shallow groundwater under the wetland because of an impermeable clay layer.

The peatland irrigation site originally supported two distinct vegetation types: a sedge-willow community and a leatherleaf and bog birch community. The edge of the peatland contained alder and willow. Standing water was usually present in the spring and fall, but the wetland had no surface water during dry summers. Soil in the sedge-willow community was 1 to 2 m of highly decomposed sedge peat, while in the leatherleaf-bog community there was 2 to 5 m of medium decomposition sphagnum peat.

Figure 27-36 The distribution pipeline situated on an elevated boardwalk is shown here during the early growing season. It is all but invisible at the peak of the growing season, when the cattail monoculture has grown more than 1 m.

The treated wastewater arriving at the peatland is a good effluent which contains virtually no heavy metals or refractory chemicals, due to the absence of agriculture and industry in the community. Phosphorus and nitrogen are present at 3 to 10 mg/L, mostly as orthophosphate and ammonium. BOD is about 15 mg/L, and solids are about 20 mg/L. Typical levels of chloride are 100 mg/L, pH 8, and conductivity 700 μmho/cm. The character of the water is dramatically altered in its passage through the wetland. After passage through 10 percent of the wetland (60 ha), water quality parameters are at background wetland levels. The system has operated successfully in the treatment of over 8 million m^3 of secondary wastewater over the first 17 years.

This natural wetland treatment system was the cheapest alternative at the time of construction (Table 27-28). It has been very economical to run, although refurbishing of the pipeline and boardwalk will be required as the project nears its projected life of 20 years. Mr. Brett Yardley, operator of the facility, believes "It is a great system. It has low maintenance, and is good for the community." Importantly, he feels that the regulators (Michigan DNR) are "on my side." The public comments he receives are all positive. In 1994, the project was authorized to continue beyond the original 20-year life expectancy.

History

The Porter Ranch peatland has been under study from 1970 to the present. Studies of the background status of the wetland were conducted during the period 1970–1974 under the sponsorship of the Rockefeller Foundation and the National Science Foundation (NSF). The natural peatland, and 6 × 6 m mesocosm plots irrigated with simulated effluent, were studied by an interdisciplinary team from the University of Michigan. This work gave strong indications that water quality improvements would result from wetland processes.

Subsequently, pilot scale (360 m^3/d) wastewater irrigation was conducted for the 3 years, 1975–1977. This system was designed, built, and operated by the Wetland Ecosystem Research Group at the University of Michigan. NSF sponsored this effort, including construction costs and research costs. The pilot study provided the basis for agency approval of the full-scale wetland discharge system.

The full-scale system was designed jointly by Williams and Works, Inc. and the Wetland Ecosystem Research Group at the University of Michigan. Funding for the project included a 75% U.S. EPA construction grant. Construction occurred during the winter and spring of 1978, with the first water discharge in July 1978. Compliance monitoring has been supplemented by full-scale ecosystem studies, spanning 1978 to present, which have focused on

Table 27-28 Economics of the Houghton Lake System

Capital	
Holding pond modification	$38,600
Dechlorination pond	153,200
Pond-wetland water transfer	83,600
Irrigation system	112,800
Monitoring equipment	9,700
1978 dollars	$397,900
1993, @ 4%	**$743,600**
Annual Operating Costs	
Pumping	$2,000
Monitoring	800
Maintenance	500
Research	14,000
1993 dollars	**$17,300**

all aspects of water quality improvement and wetland response. Those studies have been sponsored by NSF and, in a major part, by the Houghton Lake Sewer Authority.

Hydrology

On average, most of the water added to the wetland finds its way to the stream outflows. But in drought years, most of the water evaporates; in wet years, rainfall creates additions to flow (Table 27-29). During most of the drought summers of 1987 and 1988, all the pumped water evaporated in the wetland.

Water flow is strongly depth dependent, because litter and vegetation resistance is the hydrologic control. Doubling the depth causes a tenfold increase in volume flow. Therefore, when the pump is turned on, water depths rise only 1 or 2 in. For similar reasons, a large rainstorm does not flood the peatland to great depths.

There are no manmade outlet control structures, but both man and beaver have relocated the points of outflow, via culvert and dam placements. Inflows at E1 and E2 have ceased (Figure 27-35). The point of principal stream outflow has changed from E8 to E9; and E9 has been relocated three times, twice by beaver and once by man.

The soil elevations in the discharge area were originally extremely flat, with a gentle slope (20 cm/km) toward the outlet. There has developed a significant accumulation of sediment and litter in the irrigation area, which has the effect of an increased soil elevation. This acts as a 10-cm-high dam. As a consequence, the addition of wastewater along the gated irrigation pipe gives rise to a mound of water with the high zone near and upstream of the discharge pipe; in other words, there is a backgradient "pond" (densely vegetated). Depth at the discharge is not greater, but depths are greater at adjacent up- and downstream locations. There is a water flow back into the backgradient pond, which compensates for evaporative losses there, but most water moves downgradient, in a gradually thinning sheet-flow. The hydroperiod of the natural wetland has been altered in the zone of discharge: while wastewater additions only occur between May and October, dryout no longer occurs in the irrigation zone, even under drought conditions.

Permits

The project operates under two permits: an NPDES permit for the surface water discharge and a special use permit for the wetlands.

Table 27-29 Summary of Water Budgets

	P-E	A	Q_i	Q_o	ACC
1978	60	240	0	135	165
1979	−3	384	18	333	66
1980	−103	407	0	304	0
1981	−74	455	30	558	0
1982	−38	404	20	386	0
1983	−110	485	132	487	20
1984	−24	546	73	602	−7
1985	44	379	0	347	76
1986	−11	465	0	412	33
1987	−273	347	0	74	0
1988	−311	425	0	114	0
1989	−153	672	0	522	−3
1990	−43	622	0	628	−49
1991	−100	724	0	660	−36
1992	−263	720	0	498	−41
1993	−79	789	0	702	+8

Note: P-precipitation, E-evapotranspiration, A-added wastewater, Q_o-output, ACC-change in inventory, and Q_i-other inputs. Quantities are in thousands of cubic meters. The interval is the pumping season (May 1 to September 15, 1993), and the area is 1.0 km^2 after 1982, 0.75 km^2 in prior years.

The Michigan Water Resources Commission issues the NPDES permit in compliance with the Federal Clean Water Act. Both the irrigation fields and the wetlands are permitted. The wetland part of the permit establishes three classes of sampling locations: the effluent from the storage or dechlorination ponds, a row of sampling stations approximately 800 m downgradient from the discharge pipeline in the wetland, and steam flows exiting the wetland. At the latter two sets of stations, phosphorus is not to exceed 0.5 mg/L and ammonium nitrogen 3.0 mg/L.

Lagoon discharges are monitored weekly; interior wetland points and stream outflows are measured monthly. The interior wetland stations serve as an early warning line. Background water quality was established in preproject research. Compliance monitoring water chemistry data for the inflows and outflows shows no significant increases in the nitrogen or phosphorus in the wetland waters at these exit locations (Figures 27-37 and 27-38). Target values are set which are the basis for assessing the water quality impacts at the interior stations.

The special use permit was issued by the Wildlife Division of the Michigan Department of Natural Resources. Under this permit, the Roscommon County Department of Public Works is granted permission to maintain a water transporting pipe across state-owned lands, maintain a wooden walkway on the peatlands to support a water distribution pipe, and to

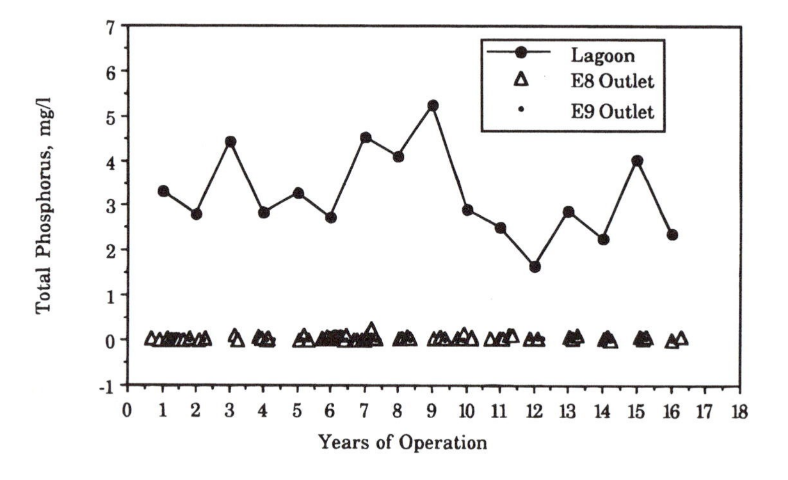

Figure 27-37 There is no evidence of elevated TP at the outlet streams, located about 3 km from the discharge.

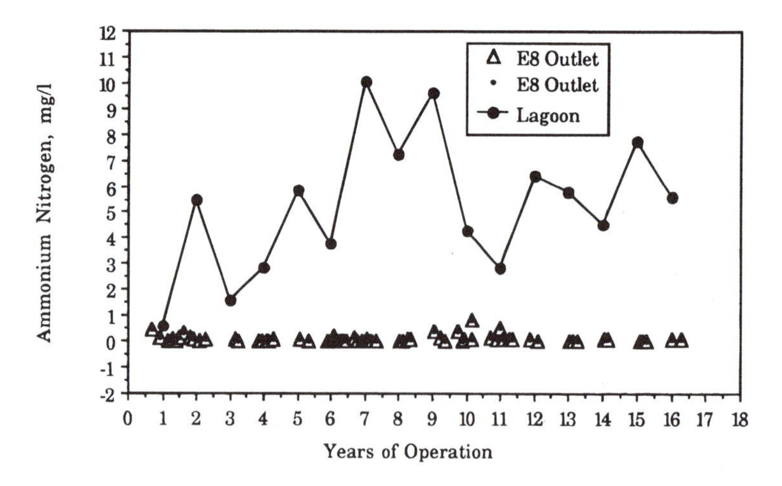

Figure 27-38 Ammonium nitrogen is low at the outlet streams.

distribute secondarily treated effluent onto the peatlands. Under the terms of this permit, if circumstances arise that are detrimental to plant and animal life, the project comes under immediate review. Detrimental circumstances include detection of toxic materials, excessive levels of pathogenic organisms, and excessive water depths. There has not been such an occurrence. This permit also requires monitoring of plant and animal populations, as well as hydrology and water quality.

Water Quality of the Wetland Interior

As the water passes through the ecosystem, both biotic and abiotic interactions occur which reduce the concentration for many parameters, including nitrogen, phosphorus, and sulfur. Surface water samples from the wastewater irrigation area are collected and analyzed throughout the year. The changes in water chemistry as a function of distance from the discharge point are monitored by sampling along lines perpendicular to the discharge pipe, extending to distances up to 1000 m. Such transects are made in the former sedge-willow area along the central axis of the wetland.

The transect concentration profiles are all similar. Water flow carries materials a greater distance in the downgradient (positive) direction than in the upgradient direction. Through the early years of operation, the zone of concentration reduction increased in size; background concentrations are now reached at distances of about 500 m downstream of the discharge, encompassing approximately 60 ha (1994) of the 600-ha (1645-ac) peatland.

Because the irrigation zone is imbedded in a natural wetland of larger extent, care must be taken in the definition of the size of the treatment portion of this larger wetland. A zone extending 300 m upstream and 700 m downstream, spanning the entire 1000-m width of the wetland, encompassed the 1994 treatment zone with room to spare. Nutrient removal is essentially complete within this zone, although background concentrations will always be present in outflows (Table 27-30).

Phosphorus concentrations are represented reasonably well by a first-order areal model (Figure 27-39). The rate constant decreased over the first 4 years of the project, from over 100 m/yr to a stable value of 11.1 ± 2.1 m/yr over the next 12 years. Ammonium also follows a first-order areal model (Figure 27-40). The rate constant stabilized after 4 years to a value of 10.3 ± 4.7 m/yr.

Table 27-30 Affected Area, Mass Average Concentrations, and Nutrient Mass Reductions

		Dissolved Inorganic Nitrogen			Total Phosphorus		
Year	Affected Area (ha)	In (mg/L)	Out (mg/L)	Reduction (%)	In (mg/L)	Out (mg/L)	Reduction (%)
1978	10	0.56	0.100	82	2.85	0.063	97
1979	13	3.68	0.100	97	2.87	0.047	98
1980	17	3.22	0.100	97	4.41	0.068	97
1981	24	2.83	0.094	97	2.83	0.088	96
1982	30	5.85	0.093	98	3.27	0.064	98
1983	55	3.76	0.148	94	2.74	0.066	97
1984	50	10.04	0.078	99	4.52	0.079	97
1985	48	7.64	0.194	98	4.11	0.099	97
1986	46	9.63	0.176	98	5.26	0.063	99
1987	46	4.26	0.244	94	2.90	0.074	97
1988	61	6.26	0.080	99	2.66	0.086	97
1989	54	8.13	0.156	98	1.66	0.047	97
1990	67	8.14	0.119	99	2.93	0.112	96
1991	76	7.80	0.112	99	2.59	0.147	94
1992	76	8.59	0.089	99	4.03	0.140	98
1993	60	7.09	0.457	96	2.63	0.066	98
Average		**6.09**	**0.146**	**97**	**3.27**	**0.082**	**97**

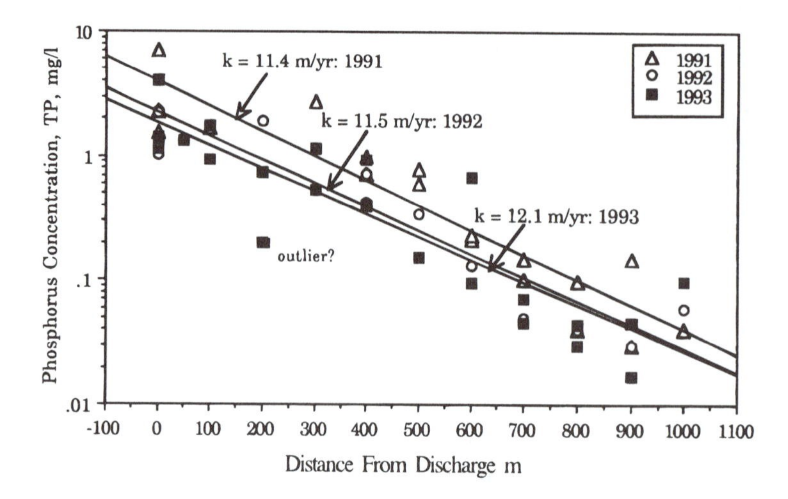

Figure 27-39 Phosphorus disappearance approximately follows a first-order areal rate model. Data are season averages for samples taken on transects parallel to flow. Range of $R^2 = 0.79$ to 0.87.

The reductions in dissolved nutrient concentrations are not due to dilution, as may be seen from the water budgets. There are summers in which rainfall exceeds evapotranspiration, but on average, there are evaporative losses, which would lead to concentration increases in the absence of wetland interactions. There are three major mechanisms by which waterborne substances are removed in this freshwater wetland ecosystem: biomass increases, burial, and gasification. The production of increased biomass due to nutrient stimulation is a non-sustainable storage for nutrients. Cycling occurs between the various biomass storages during the course of the year (Figure 27-41). Accretion of new organic soils represents the permanent, sustainable storage.

Some substances in the wastewater do not interact as strongly with the wetland as others. Chloride, calcium, magnesium, sodium, and potassium all display elevated values in the discharge zone. Chloride, especially, moves freely through the wetland to the outlets. Oxygen levels in the pumped water average approximately 6 mg/L. In the irrigation zone, levels are typically less than 1 mg/L in surface waters. The surrounding, unaffected wetland usually has high dissolved oxygen, representing near-saturation conditions. The zone of depressed oxygen increased in size as the affected area increased. In addition, the diurnal cycle appeared to be suppressed in the irrigation zone. Redox potentials indicate that the sediments are anaerobic in the irrigation area, even at quite shallow depths. Steep gradients occur, leading

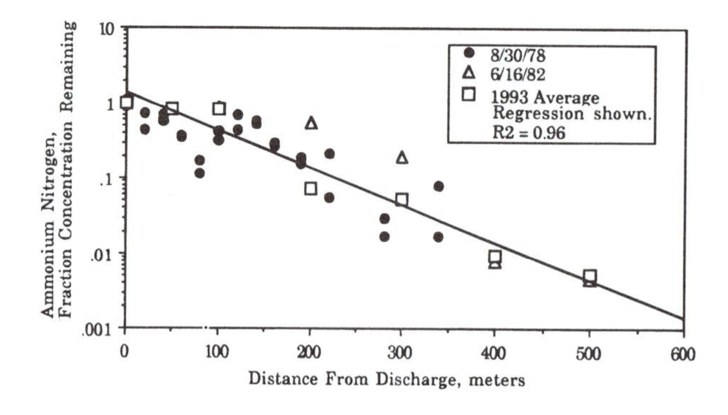

Figure 27-40 Ammonium reduction along transects parallel to flow. A first-order model provides a reasonable fit of the data.

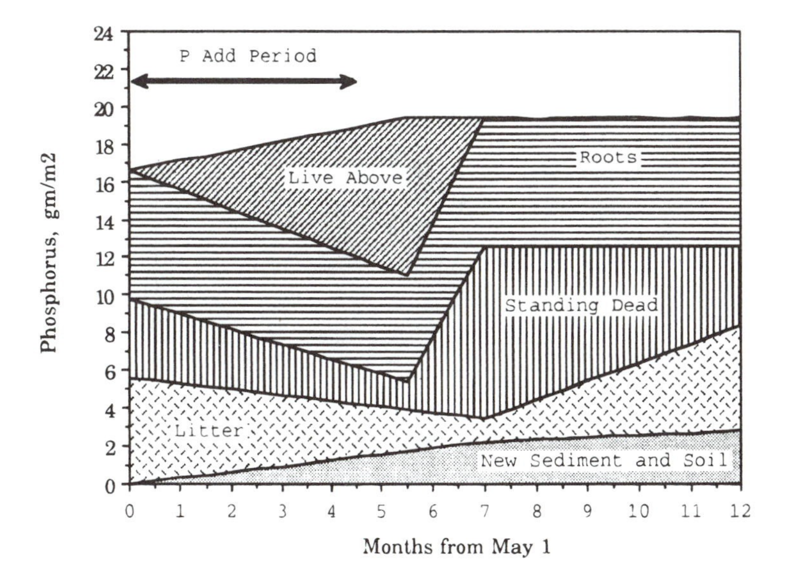

Figure 27-41 Phosphorus pools in the discharge zone. The live, above-ground biomass grows and dies during the summer season, which matches the growth period for the cattails, duckweed, and algae. Standing dead and litter are decomposing during this warm period, leaving a residual of new sediments and soil. In a brief period in fall, live leaves become brown, standing dead. At some time thereafter, depending on snow weight and wind events, the standing dead falls into the water and thus becomes litter. Litter decomposition is very slow in winter. Every 12 months, the cycle repeats, with a net accretion only in new sediments and soil.

to sulfate and nitrate reduction zones and even to a methanogenesis zone, only a few centimeters deep into the sediments and litter.

Soils and Sediments

Wastewater solids are relatively small in amount and deposit near the discharge. Incoming suspended solids average about 25 mg/L, and the wetland functions at levels of about 5 to 10 mg/L. But internal processes in both natural and fertilized wetlands produce large amounts of detrital material, thus complicating the concept of "suspended solids removal."

Some fraction of each year's plant litter does not decompose, but becomes new organic soil. It is joined by detritus from algal and microbial populations. Such organic sediments contain significant amounts of some chemicals in their structural tissues, but in addition can sorb a number of chemicals on their surfaces. The accretion of soils and sediments thus contributes to the effectiveness of the wetland for water purification. The natural Porter Ranch wetland accreted organic soils at the rate of a 2 to 3 mm/yr as determined from carbon-14 and cesium-137 radiotracer techniques. The wastewater addition has stimulated this process to produce a net of 10 mm/yr of new organics in the discharge area. The maximum accumulation rate is located a short distance downflow from the discharge.

Estimated mass balances for particulate, transportable solids indicate a large internal cycle superimposed on net removal for the wetland.

Vegetation

Many changes have occurred in the composition, abundance, and standing crops of the wetland plants in the zone of nutrient removal. There are three observable manifestations of

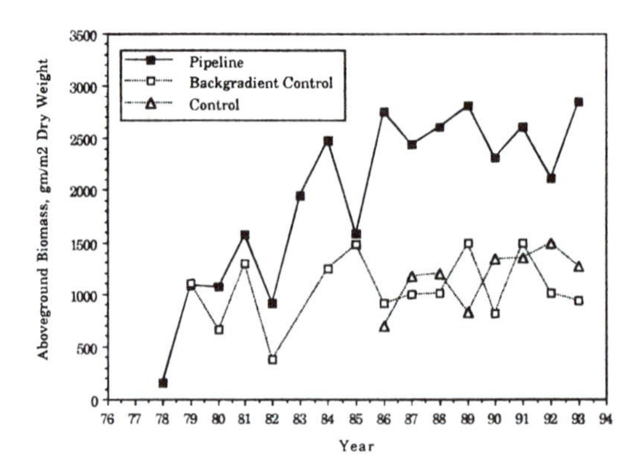

Figure 27-42 Response of the above-ground biomass to wastewater additions.

the wastewater addition: elevated nutrient concentrations in the surface waters; alterations of the size, type and relative abundance of the above-ground vegetation; and physical destabilization of the peatland soil-root mat. Vegetative changes occur in response to changes in hydraulic regime (depth and duration of inundation) and to changes in water nutrient status.

The increased availability of nutrients produces more vegetation during the growing season, which in turn means more litter during the nongrowing season. This litter requires several years to decay, and hence, the total pool of living and dead material grows slowly over a few years to a new and higher value. A significant quantity of nitrogen and phosphorus and other chemical constituents are thus retained, as part of the living and dead tissues, in the wetland. This response at the point of discharge in the Houghton Lake wetland has been slow and large (Figure 27-42). Below-ground biomass responded differently from above-ground biomass, however. Original vegetation required greatly reduced root biomass in the presence of added nutrients, 1500 vs. 4000 g/m^2. However, the sedges initially present were replaced by cattail, which has a root biomass of 4000 g/m^2. The new cattail roots are located in the top 30 cm of the water column, in a floating mat. Because the roots of the sedges retreated prior to their demise, the new root mat is not tied to the original peats below.

Approximately 65 ha of the wetland have been affected in terms of visual vegetative change (Figures 27-43 and 27-44). Some plant species—leatherleaf and sedge—have been nearly all lost in the discharge area, presumably due to shading by other species and the altered water regime. Sedges in the discharge zone went through a large increase followed by a crash to extinction. Species composition within the discharge area is no longer determined

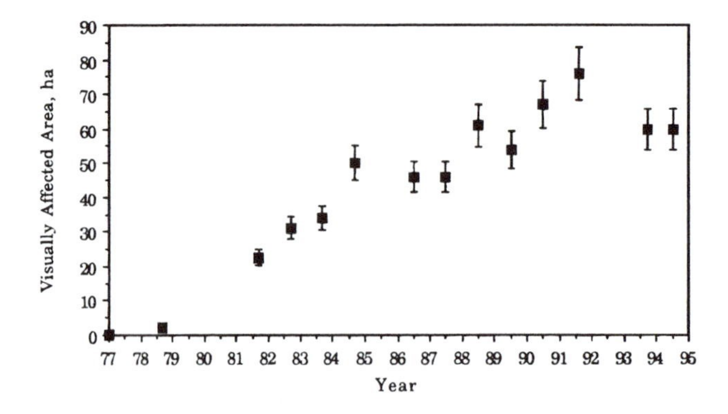

Figure 27-43 The zone of visually impacted vegetation is expanding with time, in spite of the fact that the nutrient removal zone stabilized in about 1982.

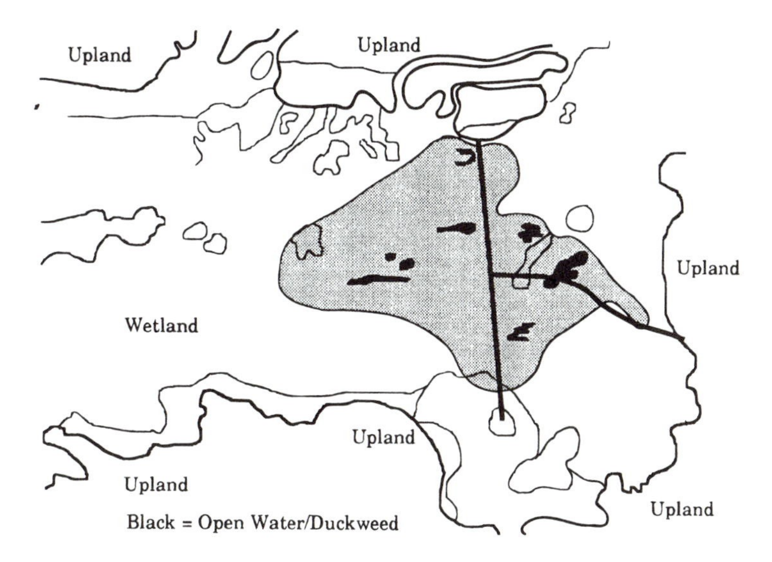

Figure 27-44 The visually impacted area of the Houghton Lake treatment wetland, 1994. The area is one of higher chlorophyll and dominance by cattail.

by earlier vegetative patterns; cattail and duckweed have totally taken over. Cattail has extended its range out to about 600 m along the central water track.

The willows and bog birch decreased in numbers in the irrigation area and are now virtually gone. An aspen community near the pipeline completely succumbed in 1983. A second aspen island, located 500 m downgradient, had also totally died by 1984. The alteration of the water regime has caused tree death along much of the wetland perimeter, in a band up to 50 m wide at a few locations. Long-dead timber at these locations indicates that similar events may have occurred naturally in the past.

Public Use

The project was not designed for purposes of public use, but a set of regular users has evolved. The site serves several organizations as a field classroom. Each year, the sixth grade science classes from the Houghton Lake Schools pay visits—and ask the best questions. Ducks Unlimited and the Michigan United Conservation Clubs also schedule trips to the wetland. The Michigan Department of Natural Resources includes field trips to the system as part of their annual training course. And, Central Michigan University conducts a portion of its wetlands course at the site.

The authorized operating period is set to allow deer hunting; the discharge is stopped in September to permit the wetland to "relax" from the influence of wastewater additions. The bow-and-arrow season in October, and the rifle season in November, both find numerous hunters on and near the wetlands. Those hunters receive a questionnaire, which has demonstrated nearly unanimous acceptance of the project. The only complaint is that the boardwalk allows too easy access to the wetlands.

Duck hunting and muskrat trapping have occurred on an intermittent basis. These activities are new to this wetland, which was formerly too dry to support waterfowl and muskrats.

Animals

In addition to game species, coyotes, bobcats, and raccoons frequent the wetland. Small mammals include a variety of mice, voles, and shrews. The relative numbers have shifted with time in the discharge area; generally, there are now fewer and different small mammals. The number of muskrats has increased greatly in the irrigation zone. Amphibians and reptiles have decreased in abundance.

Bird populations have also changed. The undisturbed wetland (1973) contained 17 species, dominated by swamp sparrows, marsh wrens, and yellowthroats. In 1991, the irrigation zone had 19 species, dominated by tree swallows, red-winged blackbirds, and swamp sparrows.

Insect species and numbers fluctuate from year to year, with no discernible pattern. In some years there are fewer mosquitoes near the discharge; in other years they are more numerous there. There are typically more midges in the discharge zone and fewer mayflies, caddisflies, and dragonflies.

Awards

Clean Waters Award	Michigan Outdoor Writers Association
Award of Merit	Michigan Consulting Engineers Council
Award for Engineering Excellence	American Consulting Engineers Council
Michigan Sesquicentennial Award	Michigan Society of Professional Engineers

Literature

About 100 reports, papers, and dissertations have issued from this project. Each operating year produces three reports: compliance monitoring results; research results for vegetation, hydrology and internal water chemistry; and research results for all types of animals, insects, and invertebrates. Background studies and pilot system performance are contained in several reports and monographs. Over 40 published papers appear in a wide variety of literature sources and involve many authors. Fourteen M.S. and Ph.D. theses have originated from the project.

SUMMARY

Natural wetlands provided the first indications of how wetland ecosystems could effectively modify and improve water quality. Although many natural wetlands receive pretreated wastewaters, few of these systems were actually engineered to provide additional pollutant assimilation. A growing number of natural wetland treatment systems receive pretreated municipal wastewaters and stormwaters and provide cost-effective additional pollutant assimilation in an environmentally acceptable manner. Natural wetland treatment systems are feasible in a small geographical context. In areas where abundant natural wetlands are available, they are worthy of further consideration as polishing systems.

Section 7
Appendices, Glossary, References, and Index

Appendix A Conversion Factors for Metric (SI) and U.S. Units

	U.S. Unit		Equivalent Metric (SI) Unit
Length	1 inch (in)	=	2.54 centimeters (cm)
	1 foot (ft)	=	0.3048 meters (m)
	1 yard (yd)	=	0.9144 meters (m)
	1 mile (mi) (5,280 ft)	=	1.6093 kilometers (km)
Mass	1 ounce (oz)	=	28.35 grams (g)
	1 pound (lb)	=	0.4536 kilograms (kg)
	1 ton (2,000 lb)	=	0.9072 metric tons (MT)
Volume	1 cubic inch (in^3)	=	16.3871 cubic centimeters (cm^3)
	1 cubic foot (ft^3)	=	0.028317 cubic meters (m^3)
	1 cubic yard (yd^3)	=	0.7646 cubic meters (m^3)
	1 acre-foot (ac-ft)	=	1,233.5 cubic meters (m^3)
	1 fluid ounce (oz)	=	29.573 milliliters (mL)
	1 quart (qt)	=	946.336 milliliters (mL)
	1 gallon (gal)	=	3.7854 liters (L)
Area	1 square inch (in^2)	=	6.4516 square centimeters (cm^2)
	1 square foot (ft^2)	=	0.092903 square meters (m^2)
	1 acre (ac) (43,560 ft^2)	=	0.4047 hectares (ha) (1 ha = 10,000 m^2)
	1 square mile (mi^2)	=	2.59 square kilometers (km^2)
Flow	1 cubic foot per second (cfs)	=	0.028317 cubic meters per second (m^3/sec)
	1 gallon per minute (gpm)	=	0.06309 liters per second (L/sec)
	1 million gallons per day (mgd)	=	3785.4 cubic meters per day (m^3/d)
Temperature	degrees Fahrenheit (°F) to degrees Celsius (°C) °C = 0.555 (°F −32)		
	degrees Celsius (°C) to degrees Kelvin (°K) °K = °C + 273		
Some U.S. conversions	1 gallon	=	128 ounces
	1 cubic foot	=	7.48 gallons
	1 pound	=	16 ounces
	1 mile	=	5,280 feet
	1 foot	=	12 inches
	1 yard	=	3 feet
	1 ton	=	2,000 pounds
	1 acre	=	43,560 square feet
	1 square mile	=	640 acres
Some metric conversions	1 centimeter	=	10 millimeters
	1 meter	=	100 centimeters
	1 kilometer	=	1,000 meters
	1 liter	=	1,000 milliliters
	1 cubic meter	=	1,000 liters
	1 hectare	=	10,000 square meters
	1 gram	=	1,000 milligrams
	1 kilogram	=	1,000 grams

Appendix B Atomic Masses of Selected Elements

Element	Symbol	Atomic Number	Atomic Mass[a]	Element	Symbol	Atomic Number	Atomic Mass
Aluminum	Al	13	26.98	Magnesium	Mg	12	24.31
Antimony	Sb	51	121.8	Manganese	Mn	25	54.94
Argon	Ar	18	39.95	Mercury	Hg	80	200.6
Arsenic	As	33	74.92	Molybdenum	Mo	42	95.94
Barium	Ba	56	137.3	Neon	Ne	10	20.18
Beryllium	Be	4	9.012	Nickel	Ni	28	58.70
Boron	B	5	10.81	Nitrogen	N	7	14.01
Bromine	Br	35	79.90	Oxygen	O	8	16.00
Cadmium	Cd	48	112.4	Phosphorus	P	15	30.97
Calcium	Ca	20	40.08	Potassium	K	19	39.10
Carbon	C	6	12.01	Scandium	Sc	21	44.96
Chlorine	Cl	17	35.45	Selenium	Se	34	78.96
Chromium	Cr	24	52.00	Silicon	Si	14	28.09
Cobalt	Co	27	58.93	Silver	Ag	47	107.9
Copper	Cu	29	63.55	Sodium	Na	11	22.99
Fluorine	F	9	19.00	Strontium	Sr	38	87.62
Gold	Au	79	197.0	Sulfur	S	16	32.06
Helium	He	2	4.003	Tin	Sn	50	118.7
Hydrogen	H	1	1.008	Titanium	Ti	22	47.90
Iodine	I	53	126.9	Tungsten	W	74	183.9
Iron	Fe	26	55.85	Vanadium	V	23	50.94
Lead	Pb	82	207.2	Zinc	Zn	30	65.38
Lithium	Li	3	6.941				

[a]Atomic mass in g/mole.

Appendix C Specific Treatment Area as a Function of Hydraulic Loading Rate

Hydraulic Loading Rate		Specific Treatment Area		
(cm/d)	(in/week)	($m^2/m^3/d$)	(ha/1000 m^3/d)	(ac/mgd)
0.20	0.55	500.0	50.0	467.7
0.40	1.10	250.0	25.0	233.9
0.60	1.65	166.7	16.7	155.9
0.80	2.20	125.0	12.5	116.9
1.00	2.76	100.0	10.0	93.5
1.20	3.31	83.3	8.3	78.0
1.40	3.86	71.4	7.1	66.8
1.60	4.41	62.5	6.3	58.5
1.80	4.96	55.6	5.6	52.0
2.00	5.51	50.0	5.0	46.8
2.20	6.06	45.5	4.5	42.5
2.40	6.61	41.7	4.2	39.0
2.60	7.17	38.5	3.8	36.0
2.80	7.72	35.7	3.6	33.4
3.00	8.27	33.3	3.3	31.2
3.50	9.65	28.6	2.9	26.7
4.00	11.02	25.0	2.5	23.4
4.50	12.40	22.2	2.2	20.8
5.00	13.78	20.0	2.0	18.7
5.50	15.16	18.2	1.8	17.0
6.00	16.54	16.7	1.7	15.6
6.50	17.91	15.4	1.5	14.4
7.00	19.29	14.3	1.4	13.4
7.50	20.67	13.3	1.3	12.5
8.00	22.05	12.5	1.3	11.7
8.50	23.43	11.8	1.2	11.0
9.00	24.80	11.1	1.1	10.4
9.50	26.18	10.5	1.1	9.8
10.00	27.56	10.0	1.0	9.4
11.00	30.31	9.1	0.9	8.5
12.00	33.07	8.3	0.8	7.8
13.00	35.83	7.7	0.8	7.2
14.00	38.58	7.1	0.7	6.7
15.00	41.34	6.7	0.7	6.2
16.00	44.09	6.3	0.6	5.8
17.00	46.85	5.9	0.6	5.5
18.00	49.61	5.6	0.6	5.2
19.00	52.36	5.3	0.5	4.9
20.00	55.12	5.0	0.5	4.7

Appendix D Percentage of Total Ammonia in the Un-Ionized Form Based on Water Temperature and pH

Temp. (°C)	pH 6.0	6.5	7.0	7.5	8.0	8.5	9.0	9.5	10.0
0	.01	.03	.08	.26	.82	2.55	7.64	20.7	45.3
1	.01	.03	.09	.28	.89	2.77	8.25	22.1	47.3
2	.01	.03	.10	.31	.97	3.00	8.90	23.6	49.4
3	.01	.03	.11	.34	1.05	3.25	9.60	25.1	51.5
4	.01	.04	.12	.36	1.14	3.52	10.3	26.7	53.5
5	.01	.04	.13	.39	1.23	3.80	11.1	28.3	55.6
6	.01	.04	.14	.43	1.34	4.11	11.9	30.0	57.6
7	.02	.05	.15	.46	1.45	4.44	12.8	31.7	59.5
8	.02	.05	.16	.50	1.57	4.79	13.7	33.5	61.4
9	.02	.05	.17	.54	1.69	5.16	14.7	35.3	63.3
10	.02	.06	.19	.59	1.83	5.56	15.7	37.1	65.1
11	.02	.06	.20	.63	1.97	5.99	16.8	38.9	66.8
12	.02	.07	.22	.68	2.13	6.44	17.9	40.8	68.5
13	.02	.07	.24	.74	2.30	6.92	19.0	42.6	70.2
14	.03	.08	.25	.80	2.48	7.43	20.2	44.5	71.7
15	.03	.09	.27	.86	2.67	7.97	21.5	46.4	73.3
16	.03	.09	.29	.93	2.87	8.54	22.8	48.3	74.7
17	.03	.10	.32	1.00	3.08	9.14	24.1	50.2	76.1
18	.03	.11	.34	1.07	3.31	9.78	25.5	52.0	77.4
19	.04	.12	.37	1.15	3.56	10.5	27.0	53.9	78.7
20	.04	.13	.40	1.24	3.82	11.2	28.4	55.7	79.9
21	.04	.14	.43	1.33	4.10	11.9	29.9	57.5	81.0
22	.05	.15	.46	1.43	4.39	12.7	31.5	59.2	82.1
23	.05	.16	.49	1.54	4.70	13.5	33.0	60.9	83.2
24	.05	.17	.53	1.65	5.03	14.4	34.6	62.6	84.1
25	.06	.18	.57	1.77	5.38	15.3	36.3	64.3	85.1
26	.06	.19	.61	1.89	5.75	16.2	37.9	65.9	85.9
27	.07	.21	.65	2.03	6.15	17.2	39.6	67.4	86.8
28	.07	.22	.70	2.17	6.56	18.2	41.2	68.9	87.5
29	.08	.23	.75	2.32	7.00	19.2	42.9	70.4	88.3
30	.08	.25	.80	2.48	7.46	20.3	44.6	71.8	89.0

Glossary

absorption The movement of a dissolved chemical through a semipermeable membrane into a living organism.

acid A chemical substance that can release excess protons (hydrogen ions).

activated sludge A complex variety of microorganisms growing in sludge in aerated wastewater treatment basins. Following settling, a portion of this microbial and sludge mixture is recycled to the influent of the treatment system, where microbes continue to grow. The remaining activated sludge is removed (wasted) from the treatment system and disposed of by different processes.

adsorption The adherence of a gas, liquid, or dissolved chemical to the surface of a solid.

advanced wastewater treatment (AWT) Treatment of wastewater beyond the secondary treatment level. In some areas, AWT represents treatment to less than 5 mg/L of 5-day biochemical oxygen demand (BOD_5), 5 mg/L of total suspended solids (TSS), 3 mg/L of total nitrogen (TN), and 1 mg/L of total phosphorus (TP).

adventitious roots Roots that grow from the stems of some plants as a response to flooding. Adventitious roots develop on these plants when the plant's normal roots are in oxygen-deficient, flooded soils, and the adventitious roots are in the overlying, oxygen-rich water column.

aeration The addition of air to water, usually for the purpose of providing higher oxygen concentrations for chemical and microbial treatment processes.

aerobic Pertaining to the presence of elemental oxygen.

algae A group of autotrophic plants that are unicellular or multicellular and typically grow in water or humid environments.

alkalinity A measure of the capacity of water to neutralize acids because of the presence of one or more of the following bases in the water: carbonates, bicarbonates, hydroxides, borates, silicates, or phosphates.

allocthonous Pertaining to substances (usually organic carbon) produced outside of and flowing into an aquatic or wetland ecosystem.

ammonification Bacterial decomposition of organic nitrogen to ammonia.

anaerobic Pertaining to the absence of free oxygen.

anion A negatively charged ion.

annual Occurring over a 12-month period.

anoxic Pertaining to the absence of all oxygen (both free oxygen and chemically bound oxygen).

aquaculture The propagation and maintenance of plants or animals by humans in aquatic and wetland environments.

aquatic Pertaining to flooded environments. Over a hydrologic gradient, the aquatic environment is the area waterward from emergent wetlands and is characterized by the growth of floating or submerged plant species.

arenchyma Porous tissues in vascular plants that have large air-filled spaces and thin cell walls. Arenchymous tissues allow gaseous diffusion between above-ground and below-ground plant structures, thus permitting plants to grow in flooded conditions.

aspect ratio Ratio of wetland cell length to width.

autocthonous Pertaining to substances (usually organic carbon) produced internally in an aquatic or wetland ecosystem.

autotrophic The production of organic carbon from inorganic chemicals. Photosynthesis is an example of an autotrophic process.

bacteria Microscopic, unicellular organisms lacking chlorophyll. Most bacteria are heterotrophic (some are chemoautotrophs), and many species perform chemical transformations that are important in nutrient cycling and wastewater treatment.

benthic Pertaining to occurrence on or in the bottom sediments of wetland and aquatic ecosystems.

bioassay The use of plants or animals for testing water quality. Often refers to use of living organisms for testing toxicity of wastewaters.

biomass The total mass of living tissues (plant and animal).

BOD (biochemical oxygen demand) A measure of the oxygen consumed during degradation of organic and inorganic materials in water.

BOD_5 Five-day biochemical oxygen demand.

bog An acidic, freshwater wetland, dominated by mosses, which typically accumulates peat.

bottomland Floodplain wetlands typically dominated by wetland tree species.

brackish water Pertaining to surface or groundwaters containing a salt content greater than 0.5 parts per thousand.

bulk density A measurement of the mass of soil occupying a given volume.

buttress The lower emergent, somewhat conical portion of some trees that grows in response to flooded conditions. The buttress may or may not include distinct ridges that broaden and anchor the base of tree species such as cypress, black gum, and wetland oak species.

carbonate An inorganic chemical compound containing one carbon atom and three oxygen atoms ($-CO_3$).

carnivore A plant or animal that feeds primarily on living animals.

cation A positively charged ion.

$CBOD_5$ Carbonaceous BOD_5.

CEC (cation exchange capacity) A measure of the ability of a soil to bind positively charged ions.

channel A deeper portion of a water flow way that has a faster current and water flow.

channelization The creation of a channel or channels resulting in faster water flow, a reduction in hydraulic residence time, and less contact between waters and solid surfaces within the water body.

chemosynthesis The use of chemically reduced energy for microbial growth.

chlorophyll A green, organic compound produced by plants and used in photosynthesis.

cienega A Spanish term meaning a swamp or marsh typically formed by hillside springs.

clarifier A circular or rectangular sedimentation tank used to remove settled solids in water or wastewater.

COD (chemical oxygen demand) A measure of the oxygen equivalent to the organic matter in water based on reaction with a strong chemical oxidant.

constructed wetland A wetland that is purposely constructed by humans in a nonwetland area.

consumer An animal that derives nutrition from other living organisms. Primary consumers feed on plants, and secondary and higher consumers feed on other animals.

degraded wetland A wetland altered by human action in a way that impairs the wetland's physical or chemical properties, resulting in reduced functions such as habitat value or flood storage.

delineation The process of determining boundaries. Wetlands delineation uses regulatory definitions based on hydrologic, soil, and vegetative indicators to identify these boundaries.

denitrification The anaerobic microbial reduction of oxidized nitrate nitrogen to nitrogen gas.

detritivore An animal that feeds on dead plant material and the associated mass of living bacteria and fungi.

detritus Dead plant material that is in the process of microbial decomposition.

diffusion The transfer of mass through a gas or liquid from a region of high concentration to a region of lower concentration.

disinfection The killing of the majority of microorganisms, including pathogenic bacteria, fungi, and viruses, by using a chemical or physical disinfectant. Disinfection is functionally defined by limits, such as achieving an effluent with no more than 200 colonies of fecal coliform bacteria in 100 milliliter (mL).

dispersion Scattering and mixing within a water or gas volume.

disturbed wetland A wetland directly or indirectly altered by a perturbation, yet retaining some natural wetland characteristics: includes anthropogenic and natural perturbations.

diversity In ecology, diversity refers to the number of species of plants and animals within a defined area. Diversity is measured by a variety of indices that consider the number of species and, in some cases, the distribution of individuals among species.

diurnal Occurring on a daily basis or during the daylight period.

DOC (dissolved organic carbon) Total filterable organic carbon in a water sample.

drained wetland A wetland in which the level or volume of ground or surface water has been reduced or eliminated by artificial means.

ecology The study of the interactions of organisms with their physical environment and with each other and of the results of such interactions.

ecosystem All organisms and the associated nonliving environmental factors with which they interact.

ecotone The boundary between adjacent ecosystem types. An ecotone can include environmental conditions that are common to both neighboring ecosystems and can have higher species diversity.

effluent A liquid or gas that flows out of a process or treatment system. Effluent can be synonymous with wastewater after any level of treatment.

Eh A measure of the reduction-oxidation (redox) potential of a soil based on a hydrogen scale.

emergent plant A rooted, vascular plant that grows in periodically or permanently flooded areas and has portions of the plant (stems and leaves) extending through and above the water plane.

enhanced wetland An existing wetland with certain functional values that have been increased or enhanced by human activity.

estuary An enclosed or open natural, transitional water body between a river and the ocean.

eutrophic Water with an excess of plant growth nutrients that typically result in algal blooms and extreme (high and low) dissolved oxygen concentrations.

evaporation The process by which water in a lake, river, wetland, or other water body becomes a gas.

evapotranspiration The combined processes of evaporation from the water or soil surface and transpiration of water by plants.

exotic species A plant or animal species that has been intentionally or accidentally introduced and that does not naturally occur in a region.

facultative Having the ability to live under different conditions (for example, with or without free oxygen).

fecal Pertaining to feces.

fecal coliform Aerobic and facultative, Gram-negative, nonspore-forming, rod-shaped bacteria capable of growth at 44°C (112°F) and associated with fecal matter of warm-blooded animals.

fen A freshwater wetland occurring on low, poorly drained ground and dominated by herbaceous and shrubby vegetation. Soil is typically organic peat.

flash boards Removable boards used to control water levels.

floating aquatic plant A rooted or nonrooted vascular plant that is adapted to have some plant organs (generally the chlorophyll-bearing leaves) floating on the surface of the water in wetlands, lakes, and rivers.

floodplain Areas that are flooded periodically (usually annually) by the lateral overflow of rivers. In hydrology, the entire area that is flooded at a recurrence interval of 100 years.

food chain or web The interconnected group of plants and animals in an ecosystem. Food chain specifically refers to the progression of trophic levels (for example, primary producer, primary consumer, secondary consumer, tertiary consumer, etc.).

fresh water Water with a total dissolved solids content less than 500 mg/L (0.5 parts per thousand salts).

fungi Microscopic or small nonchlorophyll-bearing, heterotrophic, plant-like organisms that lack roots, stems, or leaves and typically grow in dark and moist environments.

geomorphology The land and submarine relief features of the earth.

grazer An organism that feeds on plants or animals attached to surfaces.

greenway A strip or belt of vegetated land often used for recreation, as a land-use buffer, or to provide a corridor and habitat for wildlife.

groundwater Water that is located below the ground surface.

habitat The environment occupied by individuals of a particular species, population, or community.

heavy metals Metallic elements that are above 21 atomic weight on the periodic table.

herbaceous Plant parts that contain chlorophyll and are nonwoody.

herbivore An animal that feeds primarily on plant tissues.

heterotrophic An organism that derives nutrition from organic carbon compounds.

hydraulic loading rate (HLR) A measure of the application of a volume of water to a land area with units of volume per area per time or simply reduced to applied water depth per time (for example, $m^3/(m^2/d)$ or cm/d).

hydraulic residence time (HRT) A measure of the average time that water occupies a given volume with units of time. The theoretical HRT is calculated as the volume divided by the flow (for example, $m^3/(m^2/d)$). The actual HRT is estimated based on tracer studies using conservative tracers such as lithium or dyes.

hydric soil A soil that is saturated, flooded, or ponded long enough during the growing season to develop anaerobic conditions. Hydric soils that occur in areas having indicators of hydrophytic vegetation and wetland hydrology are wetland soils.

hydrology A science dealing with the properties, distribution, and circulation of water on the land surface and in the soil, underlying rocks, and atmosphere.

hydrograph A record of the rise and fall of water levels during a given time period.

hydroperiod The period of wetland soil saturation or flooding. Hydroperiod is often expressed as a number of days or a percentage of time flooded during an annual period (for example, 25 days or 7 percent).

influent Water, wastewater, or other liquid flowing into a water body or treatment unit.

inorganic All chemicals that do not contain organic carbon.

invertebrate All animals that do not have backbones.

kinetics Pertaining to the rates at which changes occur in chemical, physical, and biological processes.

lacustrine Depressional wetlands with open water.

lagoon Any large holding or detention pond, usually with earthen dikes, used to hold wastewater for sedimentation or biological oxidation.

leachate Liquid that has percolated through permeable solid waste and has extracted soluble dissolved or suspended materials from it.

lentic Pertaining to a lake or other nonflowing water body.

limnetic Relating to or inhabiting the open water portion of a freshwater body with a depth that light penetrates. The area of a wetland without emergent vegetation.

littoral The shoreward zone of a lake or wetland. The area where water is shallow enough to allow the dominance of emergent vegetation.

lotic Pertaining to flowing water bodies such as streams and rivers.

macrophyte Macroscopic (visible to the unassisted eye) vascular plants.

marsh A wetland dominated by herbaceous, emergent plants.

mass loading The total amount, on a mass or mass per area basis, of a constituent entering a system.

mesotrophic Water quality characterized by an intermediate balance of plant growth nutrients.

metabolism The chemical oxidation of organic compounds resulting in the release of energy for maintenance and growth of living organisms.

micronutrient A chemical substance that is required for biological growth in relatively low quantities and in small proportion to the major growth nutrients. Some typical micronutrients include molybdenum, copper, boron, cobalt, iron, and iodine.

microorganism An animal or plant that can only be viewed with the aid of a microscope.

mitigation The replacement of functional values lost when an ecosystem is altered. Mitigation can include replacement, restoration, and enhancement of functional values.

natural wetland A wetland ecosystem that occurs without the aid of humans.

NH_4-N (ammonia nitrogen) A reduced form of nitrogen produced as a byproduct of organic matter decomposition and synthesized from oxidized nitrogen by biological and physical processes.

nitrification Biological transformation (oxidation) of ammonia nitrogen to nitrite and nitrate forms.

nitrogen fixation A microbial process in which atmospheric nitrogen gas is incorporated into the synthesis of organic nitrogen.

NO_3 + NO_2-N (nitrate plus nitrite nitrogen) Oxidized nitrogen.

nutrient A chemical substance that provides a raw material necessary for the growth of a plant or animal.

oligotrophic Water quality characterized by a deficiency of plant growth nutrients.

omnivore An animal that feeds on a mix of plant and animal foods.

organic Pertaining to chemical compounds that contain reduced carbon bonded with hydrogen, oxygen, and a variety of other elements. Organic compounds are typically volatile, combustible, or biodegradable and include proteins, carbohydrates, fats, and oils.

Org-N (organic nitrogen) Nitrogen that is bound in organic compounds.

oxbow A bend in a river channel that over time becomes isolated from the river's main flow and contains water and wetland vegetation.

oxidation A chemical reaction in which the oxidation number (valence) of an element increases because of the loss of one or more electrons. Oxidation of an element is accompanied by the reduction of the other reactant and, in many cases, by the addition of oxygen to the compound.

oxygen sag The decrease in dissolved oxygen measured downstream of a relatively constant addition of an oxygen-consuming wastewater in a flowing water system.

palustrine All nontidal wetlands dominated by trees, shrubs, persistent emergents, emergent mosses, or lichens, and all such tidal wetlands in areas where salinity from ocean-derived salts is below 0.5 parts per thousand.

parasite An organism that lives within or on another organism and derives its sustenance from that organism without providing a useful return to its host.

peat Partially decomposed, but relatively stable organic matter formed from dead plants in flooded environments.

peatland An area where the soil is predominantly peat.

periphyton The community of microscopic plants and animals that grows on the surface of emergent and submergent plants in water bodies.

perennial Persisting for more than 1 year. Perennial plant species persist as woody vegetation from year to year or resprout from their rootstock on an annual basis.

photic zone The area of a water body receiving sunlight.

photosynthesis The biological synthesis of organic matter from inorganic matter in the presence of sunlight and chlorophyll.

phytoplankton Microscopic algae that are suspended in the water column and are not attached to surfaces.

piezometric surface The surface elevation of pressurized groundwater within a well or in a spring.

plant community All of the plant species and individuals occurring in a shared habitat or environment.

plug flow Lock-step flow along the length of a wetland cell.

pocosin A southeastern coastal plain freshwater wetland typically occurring on poorly drained, level lands between stream drainages. Pocosins are dominated by shrubs and trees adapted to periodic fires and have peat soils.

pretreatment (or preliminary treatment) The initial treatment of wastewater to remove substances that might harm downstream treatment processes or to prepare wastewater for subsequent treatment.

primary production The production of organic carbon compounds from inorganic nutrients. The energy source for this production is generally sunlight for chlorophyll-containing plants, but in some cases can be derived from reduced chemicals (chemoautotrophs).

primary treatment The first step in treatment of wastewaters. Primary treatment usually consists of screening and sedimentation of particulate solids.

protozoa Small, one-celled animals including amoebae, ciliates, and flagellates.

receiving water A water body into which wastewater or treated effluent is discharged.

reclaimed wastewater Wastewater that has received treatment sufficient to allow beneficial reuse.

redox potential The potential of a soil to oxidize or reduce chemical substances.

reduction A chemical reaction in which the oxidation state (valence) of a chemical is lowered by the addition of electrons. Reduction of a chemical is simultaneous with the oxidation of another chemical and frequently involves the loss of oxygen.

respiration The intake of oxygen and the release of carbon dioxide as a result of metabolism (biological oxidation of organic carbon).

restoration The return of an ecosystem from a disturbed or altered condition to a previously existing natural condition as a result of human action (for example, by fill removal).

rhizosphere The chemical sphere of influence of plant roots growing in flooded soils. Depending on the overall oxygen balance (availability and consumption), the rhizosphere can be oxidized, resulting in the presence of aerobic soil properties in an otherwise anaerobic soil environment.

riparian Pertaining to a stream or river. Plant communities occurring in association with any spring, lake, river, stream, creek, wash, arroyo, or other body of water or channel having banks and a bed through which waters flow at least periodically.

riverine wetlands Wetlands associated with rivers.

salinity A measure of the total salt content of water. Salinity is usually reported as parts per thousand (ppt). The salinity of normal seawater is about 35 ppt.

saturated soil Soil in which the pore space is filled with water.

secondary production The production of biomass by consumer organisms by feeding on primary producers or lower trophic level consumers.

secondary treatment Generally refers to wastewater treatment beyond initial sedimentation. Secondary treatment typically includes biological reduction in concentrations of particulate and dissolved concentrations of oxygen-demanding pollutants.

sediment Mineral and organic particulate material that has settled from suspension in a liquid.

seed bank The accumulation of viable plant seeds occurring in soils and available for germination under favorable environmental conditions.

SF (surface flow) A treatment wetland category that is designed to have a free-water surface above the ground level.

SSF (subsurface flow) A treatment wetland category that is designed to have the water surface below the level of the ground, with flow through a porous media.

sheet flow Water flow with a relatively thin and uniform depth.

short circuit A faster, channelized water flow route that results in a lower actual hydraulic residence time than the theoretical hydraulic residence time.

shrub swamp Wetlands dominated by woody vegetation less than 6 m (20 ft) tall. Plant species include shrubs, young trees, and trees that are small or stunted because of environmental conditions.

slough A slow-moving creek or stream characterized by herbaceous and woody wetland vegetation.

sludge The accumulated solids separated from liquids, such as water or wastewater, during the treatment process.

soil The upper layer of the earth that can be dug or plowed and in which plants grow.

stabilization pond A type of treatment pond in which biological oxidation of organic matter results by natural or artificially enhanced transfer of oxygen from the atmosphere to the water.

stage-area curve The relationship between the depth of water and the surface area of a wetland or lake.

stage-discharge curve The relationship between water depth and outflow from a body of water.

stemflow Rainfall intercepted by plant leaves and branches and traveling to the ground via stems and the trunk.

submerged plants Aquatic vascular plants or plants that grow below the water surface for all or a majority of their life cycles.

substrate Substances used by organisms for growth in a liquid medium. Surface area of solids or soils used by organisms to attach.

subsurface flow (SSF) Flow of water or wastewater through a porous medium such as soil, sand, or gravel.

succession The temporal changes of plant and animal populations and species in a given area following disturbance.

surface flow (SF) Flow of water or wastewater over the surface of the ground.

swamp A wetland dominated by woody plant species including trees and shrubs.

TDP (total dissolved phosphorus) A measure of the filterable component of total phosphorus (both organic and inorganic) in a water sample.

temperate zone The geographical area in the Northern Hemisphere between the Tropic of Cancer and the Arctic Circle and in the Southern Hemisphere between the Tropic of Capricorn and the Antarctic Circle. Temperate indicates that the climate is moderate and not extremely hot or cold.

terrestrial Living or growing on land that is not normally flooded or saturated.

tertiary treatment Wastewater treatment beyond secondary and often implying the removal of nutrients.

TKN (total Kjeldahl nitrogen) A measure of reduced nitrogen equal to the sum of Org-N and NH_4-N.

TN (total nitrogen) A measure of all organic and inorganic nitrogen forms in a water sample. Functionally, TN is equal to the sum of TKN and $NO_3 + NO_2$-N.

TOC (total organic carbon) A measure of the total reduced carbon in a water sample.

toxicity The adverse effect of a substance on the growth or reproduction of living organisms.

TP (total phosphorus) A measure of the total phosphorus in a water sample including organic and inorganic phosphorus in particulate and soluble forms.

transition zone The area between habitats or ecosystems (see ecotone). Frequently, transition zone is used to refer to the area between uplands and wetlands. In other cases, wetlands are referred to as transitional areas between uplands and aquatic ecosystems.

transpiration The transport of water from the soil to the atmosphere through actively growing plants.

trickling filter A filter with coarse substrate or media to provide secondary treatment of wastewater. Microorganisms attached to the filter media use and reduce concentrations of soluble and particulate organic substances in the wastewater.

trophic level A level of biological organization characterized by a consistent feeding strategy (for example, all primary consumers are in the same trophic level in an ecosystem).

tropical The geographical area between the Tropic of Cancer and the Tropic of Capricorn. An area characterized by little variation in day length and temperature. Most tropical areas have high annual average temperatures. Tropical areas may or may not have seasonably variable rainfall patterns.

TSS (total suspended solids) A measure of the filterable matter in a water sample.

upland Any area that is not an aquatic, wetland, or riparian habitat. An area that does not have the hydrologic regime necessary to support hydrophytic vegetation.

vegetation The accumulation of living plants within an area.

vertebrate An animal characterized by the presence of a spinal cord protected by vertebrae.

volatile Capable of being evaporated at relatively low temperatures.

VSS (volatile suspended solids) A measure of the solids retained on a filter that are lost by ignition in a muffle furnace. An approximate measure of organic and volatile solids in a water sample.

watershed The entire surface drainage area that contributes runoff to a body of water.

water table The upper surface of the groundwater or saturated soil.

weir A device used to control and measure water or wastewater flow.

weir gate Water control device used to adjust water levels and measure flows simultaneously.

wetland An area that is inundated or saturated by surface or groundwater at a frequency, duration, and depth sufficient to support a predominance of emergent plant species adapted to growth in saturated soil conditions.

wetland function A physical, chemical, or biological process occurring in a wetland. Examples of wetland functions include primary production, water quality enhancement, groundwater recharge, organic export, wildlife production, and flood intensity reduction.

wetland mitigation bank A preserved, restored, constructed, or enhanced wetland that has been purposely set aside to provide compensation credits for losses of wetland functions caused by future human development activities as approved by regulatory agencies.

wetland structure The physical, chemical, and biological components of a wetland. Wetland structural components typically include wetland soils, macrophytes, surface water, detritus and microbes, and wetland animal populations.

wetland treatment system A wetland that has been engineered to receive water for the purpose of reducing concentrations of one or more pollutants.

wetland values Structural and functional attributes of wetlands that provide services to humans.

zonation The development of a visible progression of plant or animal communities in response to a gradient of water depth or some other environmental factor.

zooplankton Microscopic and small animals that live suspended in the water column.

References

Adams, D.F., S.O. Farwell, E. Robinson, M.R. Pack, and W. L. Bamesberger. 1981a. Biogenic sulfur source strengths. *Environ. Sci. Technol.,* 15(12):1493–1498.

Adams, C.E., D.L. Ford, and W.W. Eckenfelder, Jr. 1981b. *Development of Design and Operational Criteria for Wastewater Treatment.* Nashville, TN: Enviro Press.

Adamus, P.R. and K. Brandt. 1990. *Impacts on Quality of Inland Wetlands of the United States: A Survey of Indicators, Techniques, and Applications of Community-Level Biomonitoring Data.* U.S. EPA Environmental Research Laboratory. EPA/600/3-90/073.

Adcock, P.W., G.L. Ryan, and P.L. Osborne. 1994. Nutrient partitioning in a clay-based surface flow wetland, in: *Proceedings of the Fourth International Conference on Wetland Systems for Water Pollution Control,* Guangzhou, China, 162–170.

ADEQ. 1995. Arizona Guidance Manual for Constructed Wetlands for Water Quality Improvement. Prepared for the Arizona Department of Environmental Quality by R.L. Knight, R. Randall, M. Girts, J.A. Tress, M. Wilhelm, and R. H. Kadlec.

Adler, P.R., S.T. Summerfelt, D.M. Glenn, and F. Takeda. 1994. Creation of Wetlands Capable of Generating Oligotrophic Water, in: *Proceedings of the 67th Annual Conference.* Alexandria, VA: Water Environment Federation. Volume 8, 37–44.

Agunwamba, J.C., N. Egbunniwe, and J.O. Ademiluyi. 1992. Prediction of the dispersion number in waste stabilization ponds. *Water Res.,* 26(1):85–89.

Ainesworth, G.C., F.K. Sparrow, and A.S. Sussman. 1973. *The Fungi: An Advanced Treatise.* New York: Academic Press.

Alabama Department of Environmental Management (ADEM). 1988. Natural Treatment Systems for Upgrading Secondary Municipal Wastewater Treatment Facilities. Prepared by ADEM and the Department of Civil Engineering, Auburn University, Auburn, AL.

Alan Plummer and Associates. 1994. Pilot-Scale Constructed Wetlands Demonstration Project: Phase I Report. to Tarrant County Water Control and Improvement District, Streetman, TX.

Alexander, M. 1977. *Introduction to Soil Microbiology.* New York: Wiley.

Allen, H.H., G.J. Pierce, and R. Van Wormer. 1989. Considerations and Techniques for Vegetation Establishment in Constructed Wetlands. Chapter 33, pp. 405–415. in D.A. Hammer (Ed.), *Constructed Wetlands for Wastewater Treatment: Municipal, Industrial, and Agricultural.* Chelsea, MI: Lewis Publishers.

Alvord, H.H. and R.H. Kadlec. 1995a. The interaction of atrazine with wetland sorbents. *Ecol. Eng.,* in press.

Alvord, H.H. and R.H. Kadlec. 1995b. Fate and transport of atrazine in the Des Plaines wetlands. *Ecol. Model,* in press.

American Public Health Association (APHA). 1992. *Standard Methods for the Examination of Water and Wastewater.* 18th Edition. Washington, DC: APHA.

ANL. 1990. *Environmental Consequences of, and Control Processes for, Energy Technologies.* Park Ridge, NJ: Argonne National Laboratory. Noyes Data Corporation.

Anonymous. 1992. Unpublished data from the Iron Bridge Wetland Treatment Facility.

Arheimer, B. and H.B. Wittgren. 1994. Modelling the effects of wetlands on regional nitrogen transport. *Ambio,* 23(6):378–386.

Armstrong, W. 1978. Root Aeration in the Wetland Environment. Chapter 9. in D.D. Hook and R.M.M. Crawford (Eds.), *Plant Life in Anaerobic Environments.* Ann Arbor, MI: Ann Arbor Science. pp. 269–297.

Armstrong, W., J. Armstrong, and P.M. Beckett. 1990. Measurement and Modeling of Oxygen Release from Roots of *Phragmites australis,* in P.F. Cooper and B.C. Findlater (Eds.), *Constructed Wetlands in Water Pollution Control,* Oxford, U.K.: Pergamon Press. 41–52.

ASCE (American Society of Civil Engineers). 1975. *Sedimentation Engineering.* V.O. Vanoni, (Ed.). New York: ASCE.

ASCE (American Society of Civil Engineers). 1990. *Evapotranspiration and Irrigation Water Requirements.* M.E. Jensen, R.D. Burman and R.G. Allen (Eds.). ASCE Manuals and Reports on Engineering Practice No. 70, New York: ASCE.

Athanas, C. 1988. Wetlands Creation for Stormwater Treatment. pp. 61–66. in J. Zelazny and J.S. Feierabend (Eds.), *Increasing Our Wetland Resources.* Washington, DC: National Wildilfe Federation-Corporate Conservation Council.

Bagnall, L.O., C.E. Schertz, and D.R. Dubbe. 1987. Harvesting and Handling of Biomass. pp. 599–619. in K.R. Reddy and W.H. Smith (Eds.), *Aquatic Plants for Water Treatment and Resource Recovery.* Orlando: Magnolia Publishing.

Bahlo, K.E. and F.G. Wach. 1990. Purification of Domestic Sewage With and Without Faeces by Vertical Intermittent Filtration in Reed and Rush Beds. pp. 215–221. in P.F. Cooper and B.C. Findlater (Eds.), *Constructed Wetlands in Water Pollution Control.* Oxford, U.K.: Pergamon Press.

Bailey, J.E. and D.F. Ollis. 1986. *Biochemical Engineering Fundamentals.* New York: McGraw-Hill.

Baker, J.L. 1981. Agricultural Areas as Nonpoint Sources of Pollution. pp. 275–310 in M.R. Overcash and J.M. Davidson (Eds.), *Environmental Impact of Nonpoint Source Pollution.* Ann Arbor, MI: Ann Arbor Science.

Baker, L.A. 1992. Introduction to nonpoint source pollution in the United States and prospects for wetland use. *Ecol. Eng.,* 1(1/2):1–26.

Ball, J.P. and T.D. Nudds. 1989. Mallard Habitat Selection: An Experiment and Implications for Management. pp. 659–671. in R.R. Sharitz and J.W. Gibbons (Eds.), *Freshwater Wetlands and Wildlife.* Oak Ridge, TN: U.S. Department of Energy. CONF - 8603101.

Bartlett, M.S., L.C. Brown, N.B. Hanes, and N.H. Nickerson. 1979. Denitrification in freshwater wetland soil. *J. Environ. Qual.,* 8(4):460–464.

Baskett, R.K. 1988. Grand Pass Wildlife Area, Missouri: Modern Wetland Restoration Strategy at Work. pp. 220–224. in J. Zelanzy and J.S. Feierabend (Eds.), *Increasing Our Wetland Resources.* Washington, DC: National Wildilfe Federation-Corporate Conservation Council.

Bastian, R.K. 1981. EPA's Role and Interest in Using Wetlands for Wastewater Treatment. pp. 325–334 in B. Richardson (Ed.), *Selected Proceedings of the Midwest Conference on Wetland Values and Management.* Navarre, MN: Freshwater Society.

Bastian, R.K. 1986. Potential Impacts on Receiving Water. pp. 157–160. in *Urban Runoff Quality—Impact and Quality Enhancement Technology.* New York: American Society for Civil Engineers.

Bastian, R.K. and D.A. Hammer. 1993. The Use of Constructed Wetlands for Wastewater Treatment and Recycling. Chapter 5, pp. 59–68 in G.A. Moshiri (Ed.), *Constructed Wetlands for Water Quality Improvement.* Boca Raton, FL: Lewis Publishers.

Bastian, R.K. and S.C. Reeds (Eds.). 1979. *Proceedings of the Seminar on Aquaculture Systems for Wastewater Treatment.* U.S. EPA Publication No. MCD-67.

Bastian, R.K., P.E. Shanaghan, and B.P. Thompson. 1989. Use of Wetlands for Municipal Wastewater Treatment and Disposal-Regulatory Issues and EPA Policies. Chapter 22, pp. 265–278. in D.A. Hammer (Ed.), *Constructed Wetlands for Wastewater Treatment: Municipal, Industrial, and Agricultural.* Chelsea, MI: Lewis Publishers.

Batchelor, A., W.E. Scott, and A. Wood. 1990. Constructed Wetland Research Programme in South Africa. pp. 373–382. in P.F. Cooper and B.C. Findlater, (Eds.), *Constructed Wetlands in Water Pollution Control.* Oxford, U.K.: Pergamon Press.

Bavor, H.J. and T.J. Schulz. 1993. Sustainable Suspended Solids and Nutrient Removal in Large-Scale, Solid Matrix, Constructed Wetland Systems. pp. 219–225. in G. A. Moshiri (Ed.), *Constructed Wetlands for Water Quality Improvement.* Boca Raton, FL: Lewis Publishers.

Bavor, H.J., D.J. Roser, and S. McKersie. 1987. Nutrient Removal Using Shallow Lagoon-Solid Matrix Macrophyte Systems. pp. 227–235. in K.R. Reddy and W.H. Smith (Eds.), *Aquatic Plants for Water Treatment and Resource Recovery.* Orlando: Magnolia Publishing.

Bavor, H.J., D.J. Roser, P.J. Fisher, and I.C. Smalls. 1989. Performance of Solid-Matrix Wetland Systems, Viewed as Fixed-Film Bioreactors. Chapter 39k, pp. 646–656. in D.A. Hammer (Ed.), *Constructed Wetlands for Wastewater Treatment: Municipal, Industrial, and Agricultural.* Chelsea, MI: Lewis Publishers.

Bavor, H.J., D.J. Roser, S.A. McKersie, and P. Breen. 1988. Treatment of Secondary Effluent. Report to Sydney Water Board, Sydney, NSW, Australia.

Bayley, S.E., R.S. Behr, and C.A. Kelly. 1986. Retention and release of S from a freshwater wetland. *Water Air Soil Pollut.,* 31:101–114.

Behandlung von häuslichen Abwasser in Pflanzenbeeten, ATV. 1989. Gesellschaft zur Föderung der Abwassertechnik e. V., Postfach1160, Markt 71, D-5205 St. Augustin 1, Germany. 1989. English translation of this design guideline document available.

Belanger, T.V. 1981. Benthic oxygen demand in Lake Apopka, FL. *Water Res.,* 15:267–274.

Benforado, J. 1981. Ecological Considerations in Wetland Treatment of Municipal Wastewater. pp. 307–323. in B. Richardson (Ed.), *Selected Proceedings of the Midwest Conference on Wetland Values and Management.* St. Paul, MN: Freshwater Society.

Berg, K.M. and P.C. Kangas. 1989. Effects of Muskrat Mounds on Decomposition in a Wetland Ecosystem. pp. 145–151. in R.R. Sharitz and J.W. Gibbons (Eds.), *Freshwater Wetlands and Wildlife.* Oak Ridge, TN: U.S. Department of Energy. CONF - 8603101.

Bergland, M. 1974. Unpublished data from preproject wetland survey.

Bernard, J.M. and F.A. Bernard. 1989. Seasonal Changes in Copper, Zinc, Manganese, and Iron Levels in *Carex rostrata* (Stokes). pp. 343–350: in R.R. Sharitz and J.W. Gibbons (Eds.), *Freshwater Wetlands and Wildlife.* Oak Ridge, TN: U.S. Dept. of Energy. CONF - 8603101.

Bernatowicz, S., S. Leszczynski, and S. Tyczynska. 1976. The influence of transpiration by emergent plants on the water balance in lakes. *Aquat. Bot.,* 2:275–288.

Best, G.R. 1987. Natural Wetlands—Southern Environment: Wastewater to Wetlands, Where Do We Go from Here? pp. 99–120. in K.R. Reddy and W.H. Smith (Eds.), *Aquatic Plants for Water Treatment and Resource Recovery.* Orlando: Magnolia Publishing.

Bevis, F.B. 1979. Ecological Considerations in the Management of Wastewater-Engendered Volunteer Wetlands. Presented at the Michigan Wetlands Conference, MacMullen Center, Higgins Lake, MI.

Billore, S.K., P. Dass, and H. Hyas. 1994. Ammonia volatilization through plant species in domestic wastewater applied agriculture field and wetland, in: *Proceedings of the Fourth International Conference on Wetland Systems for Water Pollution Control,* Guangzhou, China, 171–179.

Birkbeck, A.E., D. Reil, and R. Hunter. 1990. Application of Natural and Engineered Wetlands for Treatment of Low-Strength Leachate. pp. 411–418. in P.F. Cooper and B.C. Findlater (Eds.), *Proceedings of the International Conference on the Use of Constructed Wetlands in Water Pollution Control.* Oxford, U.K.: Pergamon Press.

Bitton, G., N. Masterson, and G.E. Gifford. 1976. Effect of secondary treated effluent on the movement of viruses through a cypress dome soil. *J. Environ. Qual.,* 5:370–375.

Bolt, G.H., M.F. De Boodt, M.H.B. Hayes, and M.B. McBride (Eds.), 1991. *Interactions at the Soil Colloid-Soil Solution Interface.* Netherlands: Kluwer Academic Publishers.

Bolton, K.A. and L.J. Evans. 1991. Elemental composition and speciation of some landfill leachates with particular reference to cadmium. *Water Air Soil Pollut.,* 60:43–53.

Boltz, J.M. and J.R. Stauffer. 1989. Fish Assemblages of Pennsylvania Wetlands. Chapter 14, pp. 158–170. in S.K. Majumdar, R.P. Brooks, F.J. Brenner, and R.W. Tiner (Eds.), *Wetlands Ecology and Conservation: Emphasis in Pennsylvania.* Easton, PA: The Pennsylvania Academy of Science.

Börner, T. 1992. *Einflussfaktoren für die Leistungsfähigkeit von Pflanzenkläranlagen,* Schriftenreihe Wasserversorgung, Abwasserbeseitigung und Raumplanung, Band 58, *TH Darmstadt.*

Bourbonniere, R.A. 1987. Organic Chemistry of Bog Drainage Waters. pp. 139–146. in C.D.A. Rubec and R.P. Overend, (Eds.), *Wetlands/Peatlands.* Ottawa: Environment Canada.

Bowes, G. and S. Beer. 1987. Physiological Plant processes: Photosynthesis. pp. 311–335. in K. R. Reddy and W.H. Smith (Eds.), *Aquatic Plants for Water Treatment and Resource Recovery.* Orlando: Magnolia Publishing.

Boyd, C.E. 1978. Chemical Composition of Wetland Plants. pp. 155–167. in R. E. Good, D.F. Whigham, and R.L. Simpson (Eds.), *Freshwater Wetlands: Ecological Processes and Management Potential.* New York: Academic Press.

Boyd, C.E. and E. Scarsbrook. 1975. Chemical Composition of Aquatic Weeds. Proceedings of a Symposium on Water Quality Management Through Biological Control. Department of Environmental Engineering Sciences and the U.S. Environmental Protection Agency. January 1975. 144–149.

Bray, J.R. 1962. Estimates of energy budgets for a *Typha* (cattail) marsh. *Science,* 136:1119–1120.

Breen, P.F. 1990. A mass balance method for assessing the potential of artificial wetlands for wastewater treatment. *Water Res.,* 24(6):689–697.

Brennan, K.M. and C.G. Garra. 1981. Wastewater Discharges to Wetlands in Six Midwestern States, pp. 285–293. in B. Richardson (Ed.), *Selected Proceedings of the Midwest Conference on Wetland Values and Management.* Navarre, MN: Freshwater Society.

Brenner, F.J. 1989. Wetland Assessment, Evaluation and Mitigation Procedures. Chapter 21, pp. 249–260. in S.K. Majumdar, R.P. Brooks, F.J. Brenner, and R. W. Tiner (Eds.), *Wetlands Ecology and Conservation: Emphasis in Pennsylvania.* Easton, PA: Pennsylvania Academy of Science.

Brezonik, P.L. 1972. Nitrogen: Sources and Transformations in Natural Waters. Chapter 1, pp. 1–50. in H.E. Allen and J.R. Kramer (Eds.), *Nutrients in Natural Waters.* New York: Wiley-Interscience.

Brightman, R.S. 1984. Benthic Macroinvertebrate Response to Secondarily Treated Wastewater in North-Central Florida Cypress Domes. Chapter 18, pp. 186–196. in K.C. Ewel and H.T. Odum (Eds.), *Cypress Swamps.* Gainesville, University of Florida Press.

Brinson, M.M. 1988. Strategies for assessing the cumulative effects of wetland alteration on water quality. *Environ. Manage.,* 12(5):655–662.

Bristow, J.M. 1974. Nitrogen fixation in the rhizosphere of freshwater angiosperms. *Can. J. Bot.*, Vol. 54, pp. 217–221.

Britt, K.W. 1970. *Handbook of Pulp and Paper Technology.* 2nd Edition. New York: Van Nostrand Reinhold.

Brix, H. 1990. Gas exchange through the soil-atmosphere interphase and through dead culms of *Phragmites australis* in a constructed reed bed receiving domestic sewage. *Water Res.,* 24(2):259–266.

Brix, H. 1994a. Constructed Wetlands for Municipal Wastewater Treatment in Europe. Chapter 20, pp. 325–333. in W.J. Mitsch, (Ed.), *Global Wetlands: Old World and New,* Amsterdam: Elsevier.

Brix, H. 1994b. Constructed Wetlands for Municipal Wastewater Treatment in Europe. pp. 325–333. in W.J. Mitsch, (Ed.), *Global Wetlands: Old World and New.* Amsterdam: Elsevier.

Brix, H. 1994c. Humedales Artificiales. Lectures on wetland treatment, Zaragoza, Spain, 19–30 September, 1994.

Brix, H. 1993a. Wastewater Treatment in Constructed Wetlands: System Design, Removal Processes, and Treatment Performance. Chapter 2, pp. 9–22. in G.A. Moshiri (Ed.), *Constructed Wetlands for Water Quality Improvement.* Boca Raton, FL: Lewis Publishers.

Brix, H. 1993b. Macrophyte-Mediated Oxygen Transfer in Wetlands: Transport Mechanisms and Rates. pp. 391–398. in G.A. Moshiri (Ed.), *Constructed Wetlands for Water Quality Improvement.* Boca Raton, FL: Lewis Publishers.

Brix, H. and H. Schierup. 1989. Sewage treatment in constructed reed beds—Danish experiences. *Water Sci. Technol.,* 21:1665–1668.

Brix, H. and H. Schierup. 1990. Soil Oxygenation in Constructed Reed Beds: The Role of Macrophyte and Soil-Atmosphere Interface Oxygen Transport. pp. 53–66. in P.F. Cooper and B.C. Findlater (Eds.), *Proceedings of the International Conference on the Use of Constructed Wetlands in Water Pollution Control.* Oxford, U.K.: Pergamon Press.

Brock, T.D. 1974. *Biology of Microorganisms.* 2nd Edition. Englewood Clifts, NJ: Prentice-Hall.

Brodie, G.A., D.A. Hammer, and D.A. Tomljanovich. 1988. Constructed Wetlands for Acid Drainage Control in the Tennessee Valley. pp. 173–180. in J. Zelanzy and J. S. Feierabend (Eds.), *Increasing Our Wetland Resources.* Washington, DC: National Wildilfe Federation-Corporate Conservation Council.

Brodrick, S.J., P. Cullen, and W. Maher. 1988. Denitrification in a natural wetland receiving secondary treated effluent. *Water Res.,* 22(4):431–439.

Brooks, R.P. and R.M. Hughes. 1988. Guidelines for Assessing the Biotic Communities of Freshwater Wetlands. pp. 276–282. in J.A. Kusler et al. (Eds.), *Urban Wetlands.* Berne, NY: Association of Wetland Managers.

Broome, S.W. 1990. Creation and Restoration of Tidal Wetlands of the Southeastern United States. pp. 37–72. in Kusler, J.A. and M.E. Kentula (Eds.), *Wetland Creation and Restoration. The Status of the Science.* Washington, DC: Island Press.

Broome, S.W., E.D. Seneca, and W.W. Woodhouse, Jr. 1988. Tidal salt marsh restoration. *Aquat. Bot.* 32:1–22.

Brown, G.G. and Associates. 1956. *Unit Operations.* New York: Wiley.

Brown, S., M.M. Brinson, and A.E. Lugo. 1979. Structure and Function of Riparian Wetlands. pp. 17–31. in R.R. Johnson and J.F. McCormick (Eds.), *Strategies for Protection and Management of Floodplain Wetlands and Other Riparian Ecosystems.* Washington, DC: U.S. Department of Agriculture.

Brown and Caldwell. 1975. *Process Design Manual for Nitrogen Control.* Report to USEPA, No. EPA/625/1–75/007, Brown and Caldwell, Walnut Creek, CA.

Brown and Caldwell. 1993. *Phase II Evaluation of Alternative Treatment Technologies.* Report to the South Florida Water Management District, Feb. 18, 1993.

Brownlow, A.H. 1979. *Geochemistry.* Englewood Cliffs, NJ: Prentice-Hall.

Brunner, C.W. and R.H. Kadlec. 1993. Effluent restores wetland hydrology. *Water Environ. Technol.,* 5(7):60–63.

Buchanan, R.E. and N.E. Gibbons (Eds.). 1974. *Bergey's Manual of Determinative Bacteriology.* 8th Edition. Baltimore, MD: Williams & Wilkins.

Buol, S.W., F.D. Hole, and R.J. McCracken. 1980. *Soil Genesis and Classification.* 2nd Edition. Ames, IA: The Iowa State University Press.

Burgoon, P.S. 1993. Oxidation of Carbon and Nitrogen in the Root Zone of Emergent Macrophytes Grown in Wetland Microcosms. Ph.D. Dissertation, University of Florida, Gainesville.

Burgoon, P.S., T.A. DeBusk, K.R. Reddy, and B. Koopman. 1991a. Vegetated submerged beds with artificial substrates. I: BOD removal. *J. Environ. Eng.,* 117(4):394–407.

Burgoon, P.S., T.A. DeBusk, K.R. Reddy, and B. Koopman. 1991b. Vegetated submerged beds with artificial substrates. II: N and P removal. *J. Environ. Eng.,* 117(4):408–424.

Burka, U. and P.C. Lawrence. 1990. A New Community Approach to Wastewater Treatment with Higher Plants. pp. 359–371. in P.F. Cooper and B.C. Findlater (Eds.), *Constructed Wetlands in Water Pollution Control.* Oxford, U.K.: Pergamon Press.

Burns and McDonnell, 1992. Conceptual Design, Stormwater Treatment Areas. Report to South Florida Water Management District, March 1992.

Bursey, C.R. 1989. Wetland Invertebrates. Chapter 13, pp. 147–157. in S.K. Majumdar, R.P. Brooks, F.J. Brenner and R.W. Tiner (Eds.), *Wetlands Ecology and Conservation: Emphasis in Pennsylvania.* Easton, PA: The Pennsylvania Academy of Science.

Burt, W.H. and R.P. Grossenheider. 1976. *A Field Guide to the Mammals.* 3rd edition. Boston: Houghton Mifflin Co.

Busnardo, M.J., R.M. Gersberg, R. Langis, T.L. Sinicrope, and J.B. Zedler. 1992. Nitrogen and phosphorus removal by wetland mesocosms subjected to different hydroperiods. *Ecological Engineering,* 1(4):287–308.

Butler, J.E., M.G. Ford, E. May, R.F. Ashworth, J.B. Williams, A. Dewedar, M. El-Housseini, and M.M.M. Baghat. 1993. Gravel Bed Hydroponic Sewage Treatment: Performance and Potential. pp. 237–247. in G. A. Moshiri (Ed.), *Constructed Wetlands for Water Quality Improvement.* Boca Raton, FL: Lewis Publishers.

Cabelli, V.J. 1977. Indicators of Recreational Water Quality. pp. 222–238. in A.W. Hoadley and B.J. Dutka (Eds.), *Bacterial Indicators/Health Hazards Associated with Water.* Philadelphia, PA: American Society for Testing and Materials.

Caffrey, J.M. and W.M. Kemp. 1991. Seasonal and spatial patterns of oxygen production, respiration and root-rhizome release in *Potamogeton perfoliatus* L. and *Zostera marina* L. *Aquat. Bot.,* 40:109.

Cairns, J. and W.H. Yongue. 1974. Protozoan colonization rates on artificial substrates suspended at different depths. *Trans. Am. Microsc. Soc.,* 93:206–210.

Call, M.L. 1989. Estimation of Micromixing Parameters from Tracer Concentration Fluctuation Measurements. Ph.D. Thesis. The University of Michigan, Ann Arbor.

Camp, A.R., R.L. Knight, and E.A. McMahan. 1971. Preliminary Comparison of Fiddler Crab Populations in a Salt Marsh Receiving Treated Wastes and in a Control Marsh. pp. 326–332. in H.T. Odum and A.F. Chestnut (Eds.), *Studies of Marine Estuarine Ecosystems Developing With Treated Sewage Wastes.* Annual Report for 1969–1970, Morehead City, NC: Institute of Marine Sciences, University of North Carolina.

Canter, L.W. and R.C. Knox. 1985. *Septic Tank System Effects on Ground Water Quality.* Chelsea, MI: Lewis Publishers.

Choate, K.D., J.T. Watson, and G.R. Steiner. 1993. TVA's Constructed Wetlands Demonstration. pp. 509–516. in G. A. Moshiri (Ed.), *Constructed Wetlands for Water Quality Improvement.* Boca Raton, FL: Lewis Publishers.

Chow, V.T. 1964. *Handbook of Applied Hydrology.* New York: McGraw-Hill.

Christiansen, J.E. 1968. Pan evaporation and evapotranspiration from climatic data. *Trans. Int. Comm. Irrig. Drain.,* III: 23.569–23.596.

Christiansen, J.E. and J.B. Low. 1970. *Water Requirements of Waterfowl Marshlands in Northern Utah.* Utah Division of Fish and Game. Pub. No. 69-12.

Clausen, E.M., B.L. Green, and W. Litsky. 1977. Fecal Streptococci: Indicators of Pollution. pp. 247–264. in A.W. Hoadley and B.J. Dutka (Eds.), *Bacterial Indicators/Health Hazards Associated with Water.* Philadelphia, PA: American Society for Testing and Materials.

Comin, F. 1994. Humedales Artificiales. Lectures on wetland treatment, Zaragoza, Spain, 19–30 September, 1994.

Conant, R. 1975. *A Field Guide to Reptiles and Amphibians of Eastern/Central North America.* The Peterson Field Guide Series. Boston, MA: Houghton Mifflin.

Conley, L.M., R.I. Dick, and L.W. Lion. 1991. An assessment of the root zone method of wastewater treatment. *WPCF Res. J.,* 63(3):239–247.

Conner, W.H. and J.W. Day, Jr. 1976. Productivity and composition of a swamp. *Am. J. Bot.,* 63:1354–1364.

Cooke, J.G. 1994. Nutrient transformations in a natural wetland receiving sewage effluent and the implications for waste treatment. *Wat. Sci. Tech.,* 29(4):209–217.

Coombes, C. 1990. Reed Bed Treatment Systems in Anglian Water. pp. 223–234. in P.F. Cooper and B.C. Findlater (Eds.), *Constructed Wetlands in Water Pollution Control.* Oxford, U.K.: Pergamon Press.

Cooper, A.B. 1992. Coupling Wetland Treatment to Land Treatment: An Innovative Method for Nitrogen Stripping. pp. 37.1–37.9. in *Wetland Systems in Water Pollution Control.* Sydney, Australia: IAWQ.

Cooper, P.F., Ed. 1990. European Design and Operations Guidelines for Reed Bed Treatment Systems. Prepared by EC/EWPCA Emergent Hydropohyte Treatment Systems Expert Contact Group. Swindon, U.K.: Water Research Centre.

Cooper, P.F. and J.A. Hobson. 1989. Sewage Treatment by Reed Beds Systems: The Present Situation in the United Kingdom. Chapter 11, pp. 153–171. in D.A. Hammer (Ed.), *Constructed Wetlands for Wastewater Treatment: Municipal, Industrial, and Agricultural.* Chelsea, MI: Lewis Publishers.

Cooper, P.F. and B.C. Findlater, Eds. 1990. *Constructed Wetlands in Water Pollution Control.* Oxford, U.K.: Pergamon Press.

Cooper, P.F. and M.B. Green. 1994. Reed bed treatment systems for sewage treatment in the United Kingdom—the first ten years experience, in: *Proceedings of the Fourth International Conference on Wetland Systems for Water Pollution Control*, Guangzhou, China, pp. 58–67.

Cooper, P.F., J.A. Hobson, and C. Findlater. 1990. The use of reed bed treatment systems in the UK. *Water Sci. Technol.,* 22(3/4):57–64.

Cooper, S.G. 1978. *The Textile Industry. Environmental Control and Energy Conservation.* Park Ridge, NJ: Noyes Data Corporation.

Correll, D.L. and D.E. Weller. 1989. Factors Limiting Processes in Freshwater Wetlands: An Agricultural Primary Stream Riparian Forest. pp. 9–23. in R.R. Sharitz and J.W. Gibbons (Eds.), *Freshwater Wetlands and Wildlife.* Oak Ridge, TN: U.S. Dept. of Energy. CONF - 8603101.

Coultas, C.L. and M.J. Duever. 1984. Soils of Cypress Swamps. Chapter 5, pp. 51–59. in K.C. Ewel and H.T. Odum (Eds.), *Cypress Swamps.* Gainesville: University of Florida Press.

Cowardin, L.M., V. Carter, F.C. Golet, and E.T. LaRoe. 1979. *Classification of Wetlands and Deep Water Habitats of the United States.* U.S. Fish and Wildlife Service. FWS/OBS-79/31.

Crites, R.W. 1994. Design criteria and practice for constructed wetlands. *Water Science and Technology,* 29(4):1–6.

Crumpton, W.G. and R. Phipps. 1992. Fate of Nitrogen Loads in Experimental Wetlands. Vol. 3, Chapter 5. in *The Des Plaines River Wetlands Demonstration Project.* Chicago, IL: Wetlands Research, Inc.

Crumpton, W.G., T.M. Isenhart, and S.W. Fisher. 1993. The Fate of Non-Point Source Nitrate Loads in Freshwater Wetlands: Results from Experimental Wetland Mesocosms. pp. 283–292. in G.A.

Moshiri (Ed)., *Constructed Wetlands for Water Quality Improvement.* Boca Raton, FL: Lewis Publishers.

Cueto, A.J. 1993. Development of Criteria for the Design and Construction of Engineered Aquatic Treatment Units in Texas. Chapter 9, pp. 99–105. in G.A. Moshiri (Ed.), *Constructed Wetlands for Water Quality Improvement.* Boca Raton, FL: Lewis Publishers.

Culp, Wesner, Culp. 1983. Draft Design Memorandum, Incline Village General Improvement District wetlands Enhancement Project. CWC, 3461 Robin Lane, Cameron Park, CA 95682.

Cussler, E.L. 1984. *Diffusion: Mass Transfer in Fluid Systems.* New York: Cambridge University Press.

Dahl, T.E. 1990. Wetland Losses in the United States 1780's to 1980's. U.S. Department of the Interior, Fish and Wildlife Service, Washington, DC.

Dahl, T.E., C.E. Johnson, and W.E. Frayer. 1991. Wetlands Status and Trends in the Conterminous United States Mid-1970's to mid-1980's. First Update of the National Wetlands Status Report. U.S. Department of the Interior, Fish and Wildlife Service, Washington, DC.

Dames and Moore. 1990. Lakeland Comprehensive Stormwater Management and Lake Pollution Study. Volume I. Report to the City of Lakeland, Florida. May 1990.

Danckwerts, P.V. 1953. Continuous flow systems: distribution of residence times. *Chem. Eng. Sci.,* 2(1):1–13.

Danckwerts, P.V. 1970. *Gas-Liquid Reactions.* New York: McGraw-Hill.

Darcy, H. 1856. *Les Fontaines Publiques de la Ville de Dijon.* Paris: Dalmont.

Das Gupta, A. and G.N. Paudyal. 1985. Characteristics of free surface flow over gravel bed. *J. Irrig. Drain.,* 111(4):299–318.

Davies, T.H. and B.T. Hart. 1990. Use of Aeration to Promote Nitrification in Reed Beds Treating Wastewater. pp. 77–84. in P.C. Cooper and B.C. Findlater (Eds.), *Proceedings of the International Conference on the Use of Constructed Wetlands in Water Pollution Control.* Oxford, U.K.: Pergamon Press.

Davis, C.B. and A.G. van der Valk. 1978. Litter Decomposition in Prairie Glacial Marshes. pp. 99–113. in R.E. Good, R.E. Good, D.F. Whigham, and R.L. Simpson (Eds.), *Freshwater Wetlands: Ecological Processes and Management Potential.* New York: Academic Press.

Davis, D.G. and J.C. Montgomery. 1987. EPA's Regulatory and Policy Considerations on Wetlands and Municipal Wastewater Treatment. pp. 69–79. in K.R. Reddy and W.H. Smith (Eds.), *Aquatic Plants for Water Treatment and Resource Recovery.* Orlando: Magnolia Publishing.

Davis, F.E., Ed. 1989. *Water: Laws and Management.* Bethesda, MD: American Water Research Association.

Davis, H. 1984. Mosquito Populations and Arbovirus Activity in Cypress Domes. Chapter 20, pp. 210–215. in K.C. Ewel and H.T. Odum (Eds.), *Cypress Swamps.* Gainesville: University of Florida Press.

Davis, S.M. 1984. *Cattail Leaf Production, Mortality and Nutrient Flux in Water Conservation Area 2A*, Tech. Publ. 84-8, South Florida Water Management District, West Palm Beach, FL, 40 pp.

Davis, S.M. 1989. Sawgrass and Cattail Production in Relation to Nutrient Supply in the Everglades, in R.R. Sharitz and J.W. Gibbons (Eds.), *Freshwater Wetlands and Wildlife.* Oak Ridge, TN: U.S. Department of Energy. CONF - 8603101.

Davis, S.M. 1990. *Growth, Decomposition and Nutrient Retention of Sawgrass and Cattail in the Everglades.* South Florida Water Management District, West Palm Beach, FL Technical Publication 90-03.

Davis, S.M. 1994. Phosphorus Inputs and Vegetation Sensitivity in the Everglades. Chapter 15. pp. 357–378. in S.M. Davis and J.C. Ogden (Eds.), *Everglades: The Ecosystem and Its Restoration.* Delray Beach, FL: St. Lucie Press.

Davis, S.M. and L.A. Harris. 1978. Marsh Plant Production and Phosphorus Flux in Everglades Conservation Area 2. pp. 105–131. in M.A. Drew (Ed.), A Symposium on Freshwater Wetlands. Sponsored by the Coordinating Council on the Restoration of the Kissimmee River Valley and Taylor Creek-Nubbin Slough Basin. February 28–March 2, 1978. Tallahassee, Florida.

Day, F.P., Jr. 1989. Limits on Decomposition in the Periodically Flooded, Non-Riverine Dismal Swamp. pp. 153–166. in R.R. Sharitz and J.W. Gibbons (Eds.), *Freshwater Wetlands and Wildlife.* Oak Ridge, TN: U.S. Department of Energy. CONF—8603101.

Day, T.J. 1975. Longitudinal dispersion in natural channels. *Water Resour. Res.,* 11(6):909–918.

de la Cruz, A.A. 1978. Primary Production Processes: Summary and Recommendations. pp. 79–86. in R.E. Good, D.F. Whigham, and R.L. Simpson (Eds.), *Freshwater Wetlands: Ecological Processes and Management Potential.* New York: Academic Press.

Deason, J.P. 1989. Impacts of Irrigation Drainwater on Wetlands. pp. 127–138. in D.W. Fisk (Ed.), *Proceedings of the Symposium on Wetlands: Concerns and Successes.* Bethesda, MD: American Water Resources Association.

DeBusk, W.F. and K.R. Reddy. 1987. Removal of floodwater nitrogen in a cypress swamp receiving primary wastewater effluent. *Hydrobiologia,* 153:79–86.

DeBusk, T. and P. Merrick. 1989. Summary of Operational and Research Data for the Reedy Creek Improvement District Wetland-Based Wastewater Treatment System (WTS2). Report of March 1989.

DeBusk, T.A., P.S. Burgoon, and K.R. Reddy. 1989. Secondary Treatment of Domestic Wastewater Using Floating and Emergent Macrophytes. Chapter 38d, pp. 525–529. in D.A. Hammer (Ed.), *Constructed Wetlands for Wastewater Treatment: Municipal, Industrial, and Agricultural.* Chelsea, MI: Lewis Publishers.

DeLaune, R.D., C.M. Reddy, and W.H. Patrick, Jr. 1981. Accumulation of plant nutrients and heavy metals through sedimentation processes and accretion in a Louisiana salt marsh. *Estuaries,* 4:328–334.

DeLaune, R.D., W.H. Patrick, Jr., and R.J. Buresh. 1978. Sedimentation rates determined by Cs-137 dating in a rapidly accreting salt marsh. *Nature,* Vol. 275, pp. 532–533.

Delfino, J.J., T.L. Crisman, and J.F. Gottgens. 1993. Spatial and Temporal Distribution of Mercury in Everglades and Okefenokee Wetland Sediments. Volume 1. Final Project Report. Department of Environmental Engineering Sciences.

Delgado, M., M. Biggeriego, and E. Guardiola. 1993. Uptake of Zn, Cr, and Cd by water hyacinths. *Water Res.,* 27(2):269–272.

De Renzo, D.J., Editor. 1978. *Nitrogen Control and Phosphorus Removal in Sewage Treatment.* Noyes Data Corporation, Park Ridge, NJ, 703 pp.

Dévai, I., L. Felföldy, I. Wittner, and S. Plosz. 1988. Detection of phosphine: New aspects of the phosphorus cycle in the hydrosphere. *Nature,* 333:343–345.

Dierberg, F.E. and P.L. Brezonik. 1984. Nitrogen and Phosphorus Mass Balances in a Cypress Dome Receiving Wastewater. pp. 112–118. in K.C. Ewel and H.T. Odum (eds.). *Cypress Swamps.* Gainesville: University of Florida Press.

Dietrich, W.E. 1982. Settling velocity of natural particles. *Water Resour. Res.,* 18(6):1615–1626.

Dill, C.H. 1989. Wastewater Wetlands: User Friendly Mosquito Habitats. Chapter 39m, pp. 664–667. in D.A. Hammer (Ed.), *Constructed Wetlands for Wastewater Treatment: Municipal, Industrial, and Agricultural.* Chelsea, MI: Lewis Publishers.

Dinges, R. 1978. Upgrading stabilization pond effluent by water hyacinth culture. *J. Water Pollut. Control Fed.,* 50:833–845.

Dixon, K.R. 1974. A Model for Predicting the Effects of Sewage Effluent on a Wetland Ecosystem. Ph.D. Dissertation. University of Michigan, Ann Arbor.

Dolan, T.J., S.E. Bayley, J. Zolteck, and A. Hermann. 1978. The Clermont Project: Renovation of Treated Effluent by a Freshwater Marsh, Biomass Production and Phosphorus Results. pp. 132–152. in M.A. Drew (Ed.), A Symposium on Freshwater Wetlands. Sponsored by the Coordinating Council on the Restoration of the Kissimmee River Valley and Taylor Creek-Nubbin Slough Basin. February 28–March 2, 1978. Tallahassee, Florida.

Dong, K. and C. Lin. 1994a. Treatment of Petrochemical Wastewater by the System of Wetlands and Oxidation Ponds and Engineering Design of the System. Preprint of paper presented at the 4th International Conference on Wetlands for Water Pollution Control, Guangzhou, PRC.

Dong, K. and C. Lin. 1994b. The Purification Mechanism of the System of Wetlands and Oxidation Ponds. Preprint of paper presented at the 4th International Conference on Wetlands for Water Pollution Control, Guangzhou, PRC.

Dornbush, J.N. 1993. Constructed Wastewater Wetlands: The Answer in South Dakota's Challenging Environment. Chapter 63, pp. 569–575. in G.A. Moshiri (Ed.), *Constructed Wetlands for Water Quality Improvement.* Lewis Publishers Boca Raton, FL: Lewis Publishers.

Dorset, J.L., H.H. Prince, and V.J. Brady. 1994. Impact of Wastewater Discharge upon a Northern Michigan Wetland Wildlife Community. Report to Michigan DNR, April, 1994.

Doust, J.L. and L.L. Doust (Eds.). 1988. *Plant Reproductive Ecology. Patterns and Strategies.* New York: Oxford University Press.

Dragun, J. 1988. The fate of hazardous materials in soil (what every geologist and hydrologist should know), *HMC,* Part 2, May/June.

Drew, M.A., Ed. 1978. *Environmental Quality Through Wetlands Utilization. A Symposium on Freshwater Wetlands.* Sponsored by the Coordinating Council on the Restoration of the Kissimmee River Valley and Taylor Creek-Nubbin Slough Basin. February 28–March 2, 1978. Tallahassee, Florida.

Driscoll, E.D. 1986. Lognormality of Point and Non-Point Source Pollutant Concentrations. pp. 438–458. in B. Urbonas and L.A. Roesner (Eds.), *Urban Runoff Quality—Impact and Quality Enhancement Technology.* New York: American Society of Civil Engineers.

DuBowy, P.J. and R.P. Reaves. 1994. *Constructed Wetlands for Animal Waste Management.* Proceedings of Workshop. 4–6 April 1994, LaFayette, Indiana. Department of Forestry and Natural Resources, Purdue University.

Duda, P.J., II. 1992. *Chevron's Richmond Refinery Water Enhancement Wetland.* Report to The Regional Water Quality Control Board, Oakland, CA.

Duellman, W.E. and L. Trueb. 1986. *Biology of Amphibians.* New York: McGraw-Hill.

Duever, M.J., J. McCollom, and L. Neuman. 1988. Plant Community Boundaries and Water Levels at Lake Hatchineha, Florida. pp. 67–72. in J.A. Kusler and G. Brooks (Eds.), *Proceedings of the National Wetland Symposium: Wetland Hydrology.* Berne, NY: Association of State Wetland Managers.

Duffield, J.M. 1986. Waterbird use of a urban stormwater wetland system in central California, USA. *Colonial Waterbirds,* 9(2):227–235.

Dufour, A.P. 1977. *Escherichia coli:* The Fecal Coliform. pp. 48–58. in A.W. Hoadley and B.J. Dutka (Eds.), *Bacterial Indicators/Health Hazards Associated with Water.* Philadelphia, PA: American Society for Testing and Materials.

Durno, S.E. 1961. Evidence regarding the rate of peat growth. *J. Ecol.,* Vol. 49, pp. 347–351.

Eddy, S. and J.C. Underhill. 1978. *How to Know the Freshwater Fishes.* 3rd edition. W.C. Brown, Inc., Dubuque, IA.

Edwards, M.E. 1990. Preliminary Survey of Vegetative Growth and Survival Factors in Constructed Wetlands, Selected TVA Projects. Final Report to TVA. Contract No. TV-81976V.

Edwards, M.E. 1993. Unpublished data from Tennessee Valley Authority.

Eger, P. 1993. Unpublished data from the Minnesota Department of Natural Resources.

Eger, P., G. Melchert, D. Antonson, and J. Wagner. 1993. The Use of Wetland Treatment to Remove Trace Metals from Mine Drainage. pp. 283–292. in G.A. Moshiri (Ed.), *Constructed Wetlands for Water Quality Improvement.* Boca Raton, FL: Lewis Publishers.

Eisenberg, D.M. and J.R. Benneman. 1982. An Overview of Municipal Wastewater Aquaculture. U.S. EPA, Draft Final Report. Contract No. DM41USC252(C).

Eisenlohr, W.S., Jr. 1966. Water loss from a natural pond through transpiration by hydrophytes. *Water Resour. Res.,* 2:443–453.

Eisler, R. 1985. Cadmium hazards to fish, wildlife, and invertebrates: A synoptic review. *U.S. Fish Wild. Serv. Biol. Rep.,* 65(1.2).

Eisler, R. 1986. Chromium hazards to fish, wildlife, and invertebrates: A synoptic review. *U.S. Fish Wild. Serv. Biol. Rep.,* 85(1.6).

Eisler, R. 1988. Lead hazards to fish, wildlife, and invertebrates: A synoptic review. *U.S. Fish Wild. Serv. Biol. Rep.,* 85(1.14).

Elmore, H.L. and T.W. Hayes. 1960. Solubility of atmospheric oxygen in water. Twenty-ninth Progress Report of the Committee on Sanitary Engineering Research. *J. San. Eng. Div.,* 86(SA4):41–53.

Eloubaidy, A.F. and E.J. Plate. 1972. Wind-shear turbulence and reaeration coefficient. *J. Hydraul. Div.,* 98(HY1):153–170.

Ergun, S. 1952. Fluid flow through packed columns. *Chem. Eng. Prog.,* 48:89.

Errington, P.L. 1963. *Muskrat Populations.* Ames: Iowa State University Press.

Erwin, K.L. 1990a. Wetland Evaluation for Restoration and Creation. pp. 429–458. in Kusler, J.A. and M.E. Kentula (Eds.), *Wetland Creation and Restoration. The Status of the Science.* Washington, DC: Island Press.

Erwin, K.L. 1990b. Freshwater Marsh Creation and Restoration in the Southeast. pp. 233–265. in J.A. Kusler and M.E. Kentula (Eds.), *Wetland Creation and Restoration. The Status of the Science.* Washington, DC: Island Press.

Etnier C. and B. Guterstam, Eds. 1991. *Ecological Engineering for Wastewater Treatment.* Gothenburg, Sweden: Bokskogen.

Ewel, K.C. and H.T. Odum, Eds. 1984. *Cypress Swamps.* Gainesville: University of Florida Press.

Faulkner, S.P. and C.J. Richardson. 1989. Physical and Chemical Characteristics of Freshwater Wetland Soils. Chapter 4, pp. 41–72. in D.A. Hammer (Ed.), *Constructed Wetlands for Wastewater Treatment: Municipal, Industrial, and Agricultural.* Chelsea, MI: Lewis Publishers.

Feierabend, J.S. 1989. Wetlands: The Lifeblood of Wildlife. Chapter 7, pp. 107–118. in D.A. Hammer (Ed.), *Constructed Wetlands for Wastewater Treatment: Municipal, Industrial, and Agricultural.* Chelsea, MI: Lewis Publishers.

Feijtel, T.C., R.D. DeLaune, and W.H. Patrick, Jr. 1989. Carbon, Nitrogen, and Micronutrient Dynamics in Gulf Coast Marshes. pp. 47–60. in R.R. Sharitz and J.W. Gibbons (Eds.), *Freshwater Wetlands and Wildlife.* Oak Ridge, TN: U.S. Department of Energy. CONF-8603101.

Fennessy, M.S. 1992. Spatial Patterns of Soil Chemistry Development in Restored Freshwater Wetlands. Volume 4, Chapter 1, in *The Des Plaines River Wetlands Demonstration Project.* Chicago, IL: Wetlands Research, Inc.

Fennessy, M.S., C. Brueske, and W.J. Mitsch. 1992. Sediment Deposition Patterns in Restored Freshwater Wetlands Using Sediment Traps. Volume 4, Chapter 2. in *The Des Plaines River Wetlands Demonstration Project.* Chicago, IL: Wetlands Research, Inc.

Findlater, B.C., J.A. Hobson, and P.F. Cooper. 1990. Reed bed treatment systems: performance evaluation, in: *Constructed Wetlands in Water Pollution Control*, P.F. Cooper and B.C. Findlater, Eds., Pergamon Press, Oxford, UK, 193–204.

Finlayson, C.M., I. Von Oertzen, and A.J., Chick. 1990. Treating Poultry Abattoir and Piggery Effluents in Gravel Trenches. pp. 559–562. in P.F. Cooper and B.C. Findlater (Eds.), *Constructed Wetlands in Water Pollution Control.* Oxford, U.K.: Pergamon Press.

Fischer, H.B., E.J. List, R.C.Y. Koh, J. Imberger, and N.H. Brooks. 1979. *Mixing in Inland and Coastal Waters.* New York: Academic Press.

Fisher, M.M. and K.R. Reddy. 1987. Water Hyacinth for Improving Eutrophic Lake Water: Water Quality and Mass Balance. pp. 969–976. in K.R. Reddy and W.H. Smith (Eds.), in *Aquatic Plants for Water Treatment and Resource Recovery.* Orlando, FL: Magnolia Publishing.

Fisher, P.J. 1990. Hydraulic Characteristics of Constructed Wetlands at Richmond. NSW, Australia. pp 21–32. in P.F. Cooper and B.C. Findlater (Eds.), *Constructed Wetlands in Water Pollution Control.* Oxford, U.K.: Pergamon Press.

Fisk, D.W., (Ed). 1989. *Wetlands: Concerns and Successes.* Bethesda, MD: American Water Resource Association.

Florida Administrative Code. 1989. Florida Department of Environmental Protection. Tallahassee, Florida.

Fogler, H.S. 1992. *Elements of Chemical Reaction Engineering.* 2nd Edition. Englewood Cliffs, NJ: Prentice-Hall.

Font, R. 1991. Analysis of the batch sedimentation test. *Chem. Eng. Sci.*, 46(10):2473–2482.

Fortescue, J.A.C. 1980. *Environmental Geochemistry. A Holistic Approach.* New York: Springer-Verlag.

Fowler, B.A., Ed. 1983. *Biological and Environmental Effects of Arsenic. Topics in Environmental Health.* Volume 6. New York: Elsevier.

Fowler, B.K. and C. Hershner. 1989. Primary Production in Cohoke Swamp, a Tidal Freshwater Wetland in Virginia. pp. 365–374. in R.R. Sharitz and J.W. Gibbons (Eds.), *Freshwater Wetlands and Wildlife.* Oak Ridge, TN: U.S. Dept. of Energy. CONF - 8603101.

Fox, J.L., D.E. Price, and J. Allinson. 1984. Distribution of Fecal Coliform Bacteria in and Around Experimental Cypress Domes. Chapter 22, pp. 225–226. in K.C. Ewel and H.T. Odum (Eds.), *Cypress Swamps.* Gainesville: University of Florida Press.

Freeman, B.J. 1989. Okefenokee Swamp Fishes: Abundance and Production Dynamics in an Aquatic Macrophyte Prairie. pp. 529–540. in R.R. Sharitz and J.W. Gibbons (Eds.), *Freshwater Wetlands and Wildlife.* Oak Ridge, TN: U.S. Dept. of Energy. CONF - 8603101.

Freeze, R.A. and J.A. Cherry. 1979. *Groundwater.* Englewood Ciffs, NJ: Prentice-Hall.

Freney, J.R., R. Leuning, J.R. Simpson, O.T. Denmead, and W.A. Muirhead. 1985. Estimating ammonia volatilization from flooded rice fields by simplified techniques. *Soil Sci. Soc. Am. J.*, Vol. 49, pp. 1049–1054.

French, R.H. 1985. *Open-Channel Hydraulics.* New York: McGraw-Hill.

Friend, M. 1985. Chapter 17. Wildlife Health Implications of Sewage Disposal in Wetlands. pp. 262–269. in P.J. Godfrey, E.R. Kaynor, S. Pelczanski, and J. Benforado (Eds.), *Ecological Considerations in Wetlands Treatment of Municipal Wastewaters.* New York: Van Nostrand Reinhold.

Gale, P.M., K.R. Reddy, and D.A. Graetz. 1993. Nitrogen removal from reclaimed water applied to constructed and natural wetland microcosms. *Water Environment Research*, 65(2):162–168.

Gallagher, J.L. 1978. Decomposition Processes: Summary and Recommendations. pp. 145–151. in R.E. Good, D.F. Whigham, and R.L. Simpson (Eds.), *Freshwater Wetlands: Ecological Processes and Management Potential.* New York: Academic Press.

Gambrell, R.P. and W.H. Patrick, Jr. 1978. Chemical and Microbiological Properties of Anaerobic Soils and Sediments. Chapter 1, pp. 375–423. in D.D. Hook and R.M.M. Crawford (Eds.), *Plant Life in Anaerobic Environments.* Ann Arbor, MI: Ann Arbor Science.

Gambrell, R.P., R.A. Khalid, and W.H. Patrick, Jr. 1987. Capacity of a swamp forest to assimilate the TOC loading from a sugar refinery wastewater stream. *J. WPCF,* 39(10):897–904.

Garbisch, E.W. and L.B. Coleman. 1978. Tidal Freshwater Marsh Establishment in Upper Chesapeake Bay: *Pontederia cordata* and *Peltandra virginica.* pp. 285–298. in R.E. Good, D.F. Whigham, and R.L. Simpson (Eds.), *Freshwater Wetlands: Ecological Processes and Management Potential.* New York: Academic Press.

Gardner, W.S. 1980. Salt Marsh Creation: Impact of Heavy Metals. pp. 126–131. in J.C. Lewis and E.W. Bunce (Eds.), *Rehabilitation and Creation of Selected Coastal Habitats.* U.S. Fish and Wildlife Service. FWS/OBS-80/27.

Gassman, G. and D. Glindemann. 1993. *Angew. Chem.,* 105:749. As reported in *Chem. Eng. News,* May 17, 1993, p 31.

Gaur, S., P.K. Singhal, and S.K. Hasija. 1992. Relative contributions of bacteria and fungi to water hyacinth decomposition. *Aquat. Bot.,* 43:1–15.

Gearheart, R.A. 1990. Nitrogen Removal at the Arcata Constructed Wetlands. Presentation at the Water Pollution Control Federation Meeting, Washington, DC.

Gearheart, R.A. 1992. Use of constructed wetlands to treat domestic wastewater, City of Arcata, California. *Water Sci. Technol.,* 26:1625–1637.

Gearheart, R.A., F. Klopp, and G. Allen. 1989. Constructed Free Surface Wetlands to Treat and Receive Wastewater; Pilot Project to Full Scale. Chapter 8, pp. 121–137. in D.A. Hammer (Ed.), *Constructed Wetlands for Wastewater Treatment: Municipal, Industrial, and Agricultural.* Chelsea, MI: Lewis Publishers.

George, D.B., M.C. Kemp, A.S. Caldwell, S.K. Winfree, and P.J. Tsai. 1994. Design Considerations for Subsurface Flow Wetlands Treating Municipal Wastewater. pp. 13–24. in *Proceedings of the 67th Annual Conference of the Water Environment Federation.* Vol. 8. Alexandria, VA: Water Environment Federation.

Gerritse, R.G. 1993. Prediction of travel times of phosphate in soils at a disposal site for wastewater. *Water Res.,* 27(2):263–267.

Gersberg, R.M., B.V. Elkins, and C.R. Goldman. 1983. Nitrogen removal in artificial wetlands. *Water Research,* 17(9):1009–1014.

Gersberg, R.M., B.V. Elkins, and C.R. Goldman. 1984. Use of artificial wetlands to remove nitrogen from wastewater. *J. Water Pollut. Control Fed.,* 56:152–156.

Gersberg, R.M., B.V. Elkins, S.R. Lyon, and C.R. Goldman. 1986. Role of aquatic plants in wastewater treatment by artificial wetlands, *Water Res.,* 20:363–367.

Gersberg, R.M., R. Brenner, S.R. Lyon, and B.V. Elkins. 1987. Survival of Bacteria and Viruses in Municipal Wastewaters Applied to Artificial Wetlands. pp. 237–245. in K.R. Reddy and

W.H. Smith (Eds.), *Aquatic Plants for Water Treatment and Resource Recovery.* Orlando: Magnolia Publishing.

Gersberg, R.M., R.A. Gearheart, and M. Ives. 1989a. Pathogen Removal in Constructed Wetlands. Chapter 35, pp. 431–445. in D.A. Hammer (Ed.), *Constructed Wetlands for Wastewater Treatment: Municipal, Industrial, and Agricultural.* Chelsea, MI: Lewis Publishers.

Gersberg, R.M., S.R. Lyon, R. Brenner, and B.V. Elkins. 1989b. Integrated Wastewater Treatment Using Artificial Wetlands: A Gravel Marsh Case Study. pp. 145–152. in D.A. Hammer (Ed.), *Constructed Wetlands for Wastewater Treatment.* Chelsea, MI: Lewis Publishers.

Giblin, A.E. 1985. Comparisons of the Processing of Elements by Ecosystems, II: Metals. Chapter 10, pp. 158–179. in P.J. Godfrey et al. (Eds.), *Ecological Considerations in Wetlands Treatment of Municipal Wastewaters.* New York: Van Nostrand Reinhold.

Giesy, J.P., Jr., H.J. Kania, J.W. Bowling, R.L. Knight, S. Mashburn, and S. Clarkin. 1979. *Fate and Biological Effects of Cadmium Introduced into Channel Microcosms.* U.S. EPA. EPA-600/3-79-039.

Gilbert, T., T. King, and B. Barnett. 1981. *An Assessment of Wetland Habitat Establishment at a Central Florida Phosphate Mine Site.* U.S. Fish and Wildlife Service. FWS/OBS-81/38.

Girts, M.A. and R.L. Knight. 1989. Operations Optimization. Chapter 34, pp. 417–429. in D.A. Hammer (Ed.), *Constructed Wetlands for Wastewater Treatment: Municipal, Industrial, and Agricultural.* Chelsea, MI: Lewis Publishers.

Girts, M.A. and R.L.P. Kleinmann. 1986. Constructed Wetlands for Treatment of Acid Mine Drainage: A Preliminary Review. pp. 165–171. in *National Symposium on Mining, Hydrology, Sedimentology, and Reclamation.* Louisville: University of Kentucky Press.

Girts, M.A., R.L. Kleinmann, and P.M. Erickson. 1987. Performance Data on *Typha* and *Sphagnum* Wetlands Constructed to Treat Coal Mine Drainage. in Proceedings of the 8th Annual Surface Mine Drainage Task Force Symposium, Morgantown, WV.

Glaser, P.H. 1987. The Ecology of the Patterned Boreal Peatlands of Northern Minnesota. U.S. FWS Report 85, Washington, DC.

Glass, G.E. and J.E. Podlski. 1975. Interstitial water components and exchange across the water sediment interface of western Lake Superior. *Verh. Int. Verein. Limnol.,* 19:405–420.

Gleason, P.J. 1972. The Origin, Sedimentation, and Stratigraphy of Calcitic Mud Located in the Southern Fresh-Water Everglades. Ph.D. Thesis. The Pennsylvania State University, University Park.

Gleason, P.J. 1974. Chemical Quality of Water in Conservation Area 2A and Associated Canals. Central and Southern Florida Flood Control District Technical Publication 74-1.

Gleason, P.J. and P. Stone, 1994. Age, origin and landscape evolution of the Everglades peatland, in: *Everglades: The Ecosystem and Its Restoration,* S.M. Davis and J.C. Ogden, Editors, St. Lucie Press, Delray Beach, Florida, pp. 149–198.

Glooschenko, W.A., L. Holloway, and N. Arafat. 1986. The use of mires in monitoring the atmospheric deposition of heavy metals. *Aquat. Bot.,* 25:179–190.

Godfrey, R.K. and J.W. Wooten. 1979. *Aquatic and Wetland Plants of Southeastern United States: Monocotyledons.* Athens: University of Georgia Press.

Godfrey, R.K. and J.W. Wooten. 1981. *Aquatic and Wetland Plants of Southeastern United States: Dicotyledons.* Athens: University of Georgia Press.

Godfrey, P.J., E.R. Kaynor, S. Pelczarski, and J. Benforado, Eds. 1985. *Ecological Considerations in Wetlands Treatment of Municipal Wastewaters.* New York, Van Nostrand Reinhold.

Godshalk, G.L. and R.G. Wetzel. 1978. Decomposition in the Littoral Zone of Lakes. pp. 131–143. in R.E. Good, D.F. Whigham, and R.L. Simpson (Eds.), *Freshwater Wetlands: Ecological Processes and Management Potential.* New York: Academic Press.

Goldman, C.R. and A.J. Horne. 1983. *Limnology.* New York: McGraw-Hill.

Golet, F.C. 1978. Rating the Wildlife Value of Northeastern Freshwater Wetlands. pp. 63–73. in Greeson, P.E., J.R. Clark, and J.E. Clark (Eds.), *Wetland Functions and Values: The State of Our Understanding.* Minneapolis, MN: American Water Resources Association.

Good, B.J. and W.H. Patrick, Jr. 1987. Root-Water-Sediment Interface Processes. pp. 359–371. in K.R. Reddy and W.H. Smith (Eds.), *Aquatic Plants for Water Treatment and Resource Recovery.* Orlando, FL: Magnolia Publishing.

Good, R.E., D.F. Whigham, and R.L. Simpson, Eds. 1978. *Freshwater Wetlands: Ecological Processes and Management Potential.* New York: Academic Press.

Gosselink, J.G. and R.E. Turner. 1978. The Role of Hydrology in Freshwater Wetland Ecosystems. pp. 63–78. in R.E. Good, D.F. Whigham, and R.L. Simpson (Eds.), *Freshwater Wetlands: Ecological Processes and Management Potential.* New York: Academic Press.

Grace, J.B. 1989. Effects of water depth on *Typha Latifolia* and *Typha Domingensis. Am. J. Bot.,* 76:762–768.

Graetz, D.A. 1980. *Denitrification in Wetlands as a Means of Water Quality Improvement.* Publication No. 48. Water Resources Research Center. University of Florida Press, Gainesville.

Green, M.B. 1993. Growing Confidence in the Use of Constructed Reed Beds for Polishing Wastewater Effluents. Proceedings of the Water Environment Federation 66th Annual Conference & Exposition. Anaheim, CA. October 3–7, 1993. pp. 85–96.

Green, M.B. 1994. Constructed wetlands are big in small communities. *Water Environ. Technol.,* 6(20)51–55.

Green, M.B. and J. Upton. 1992. Constructed Reed Beds: A Cost Effective Way to Polish Wastewater Effluents for Small Communities. Proceedings of the Water Environment Federation 65th Annual Conference & Exposition. New Orleans, LA. September 20–24, 1992. pp. 13–24.

Green, M.B. and J. Upton, 1993. Reed Bed Treatment for Small Communities: U.K. Experience. pp. 517–524. in G.A. Moshiri (Ed.), *Constructed Wetlands for Water Quality Improvement.* Chelsea, MI: Lewis Publishers.

Greenkorn, R.A. 1983. *Flow Phenomena in Porous Media.* New York: Marcel Dekker.

Greeson, P.E., J.R. Clark, and J.E. Clark, Eds. 1978. *Wetland Functions and Values: The State of our Understanding.* Minneapolis, MN: American Water Resources Association.

Greyson, J. 1990. *Carbon, Nitrogen, and Sulfur Pollutants and Their Determination in Air and Water.* New York: Marcel Dekker.

Gries, C.L., L. Kappen, and R. Lösch. 1990. Mechanism of flood tolerance in reed, *Phragmites australis* (Cav.) Trin. ex Streudel. *New Phytol.,* 114:589.

Gully, J.R. and P.G. Nix. 1993. Wetland Treatment of Oil Sands Operation Waste Waters. Preprint of paper presented at Conference on Biogeochemical Cycling in Wetlands, February, 1993, Baton Rouge, LA.

Gumbricht, T. 1992. Tertiary wastewater treatment using the root zone method in temperate climates. *Ecol. Eng.,* 1:199–212.

Gumbricht, T. 1993. Nutrient removal processes in freshwater submersed macrophyte systems. *Ecol. Eng.,* 2(1):1–30.

Gunderson, L.H. 1989. Historical Hydropatterns in Wetland Communities of Everglades National Park. pp. 1099–1111. in R.R. Sharitz and J.W. Gibbons (Eds.), *Freshwater Wetlands and Wildlife.* Oak Ridge, TN: U.S. Department of Energy. CONF - 8603101.

Guntenspergen, G. and F. Stearns. 1981. Ecological Limitations on Wetland Use for Wastewater Treatment. pp. 273–284. in B. Richardson (Ed.), *Selected Proceedings of the Midwest Conference on Wetland Values and Management.* Navarre, MN: Freshwater Society.

Guntenspergen, G.R., F. Stearns, and J.A. Kadlec. 1989. Wetland Vegetation. Chapter 5, pp. 73–88. in D.A. Hammer (Ed.), *Constructed Wetlands for Wastewater Treatment: Municipal, Industrial, and Agricultural.* Chelsea, MI: Lewis Publishers.

Haack, S.K., G.R. Best, and T.L. Crisman. 1989. Aquatic Macroinvertebrate Communities in a Forested Wetland: Interrelationships with Environmental Gradients. pp. 437–454. in R.R. Sharitz and J.W. Gibbons (Eds.), *Freshwater Wetlands and Wildlife.* Oak Ridge, TN: U.S. Department of Energy. CONF - 8603101.

Haag, K.H. 1987. Potential Impacts of Phytophagous Insects on Aquatic Macrophytes Used in Resource Recovery Systems. pp. 806–807. in K.R. Reddy and W.H. Smith (Eds.), *Aquatic Plants for Water Treatment and Resource Recovery.* Orlando: Magnolia Publishing.

Haag, R.D. 1979. *The Hydrogeology of the Houghton (Lake) Wetland.* Ann Arbor: University of Michigan Press.

Haberl, R. 1994. Unpublished notes for short course, Zaragosa, Spain.

Haberl, R. and R. Perfler. 1989. Root-Zone System: Mannersdorf—New Results. Chapter 39f, pp. 606–621. in D.A. Hammer (Ed.), *Constructed Wetlands for Wastewater Treatment: Municipal, Industrial, and Agricultural.* Chelsea, MI: Lewis Publishers.

Haberl, R. and R. Perfler. 1990. Seven Years of Research Work and Experience with Wastewater Treatment by a Reed Bed System, pp. 205–214. in P.F. Cooper and B.C. Findlater (Eds.), *Constructed Wetlands in Water Pollution Control.* Oxford, U.K.: Pergamon Press.

Haberl, R. and R. Perfler. 1991. Nutrient removal in a reed bed system. *Water Sci. Technol.,* 23:729–737.

Hair, J.D., G.T. Hepp, L.M. Luckett, K.P. Reese, and D.K. Woodward. 1978. Beaver Pond Ecosystems and Their Relationship to Multi-Use Natural Resource Management. pp. 80–92. in R.R. Johnson and J.F. McCormick (Eds.), *Strategies for Protection and Management of Floodplain Wetlands and Other Riparian Ecosystems.* Washington, DC: U.S. Department of Agriculture.

Hale, J.G. 1977. Toxicity of metal mining wastes. *Bull. Environ. Contam. Toxicol.,* 17:66–73.

Hall, B.R. and G.E. Freeman. 1994. Study of Hydraulic Roughness in Wetland Vegetation Takes New Look at Manning's n. *Wetlands Res. Program Bull.,* 4(1):1–4.

Hammer, D.A. 1989a. Constructed Wetlands for Treatment of Agricultural Waste and Urban Stormwater. Chapter 27, pp. 333–348. in S.K. Majumdar, R.P. Brooks, F.J. Brenner, and R.W. Tiner (Eds.), *Wetlands Ecology and Conservation: Emphasis in Pennsylvania.* Easton, PA: The Pennsylvania Academy of Science.

Hammer, D.A., Ed. 1989b. *Constructed Wetlands for Wastewater Treatment: Municipal, Industrial, and Agricultural.* Chelsea, MI: Lewis Publishers.

Hammer, D.A. 1991a. Designing Constructed Wetlands for Localized Rural NPS Sources. in R. Olson (Ed.), Guidelines for the Use of Created and Natural Wetlands in Controlling Nonpoint Source Pollution. Proceedings of a Workshop held in Arlington, VA. June 10–12, 1991.

Hammer, D.A. 1991b. Water Improvement Functions of Natural and Constructed Wetlands. pp. 129–157. in *Proceedings: Protection and Management Issues for South Carolina Wetlands.* Clemson, SC: The Strom Thurmond Institute.

Hammer, D.A. 1992a. Designing constructed wetlands systems to treat agricultural nonpoint source pollution. *Ecol. Eng.,* 1(1/2).

Hammer, D.A. 1992b. *Creating Freshwater Wetlands.* Boca Raton, FL: Lewis Publishers.

Hammer, D.A. and R.L. Knight. 1994. Designing constructed wetlands for nitrogen removal. *Water Science Technology.* 29(4):15–27.

Hammer, D.A., B.P. Pullin, T.A. McCaskey, J. Eason, and V.W.E. Payne. 1993. Treating livestock wastewater with constructed wetlands, in: *Constructed Wetlands for Water Quality Improvement.* G.A. Moshiri (Ed.). Lewis Publishers, Boca Raton, FL. pp. 343–348.

Hammer, D.E. 1984. An Engineering Model of Wetland/Wastewater Interactions. Ph.D. Dissertation. University of Michigan, Ann Arbor.

Hammer, D.E. and R.H. Kadlec. 1980. Ortho-phosphate adsorption on peat. *6th International Peat Congress Proceedings.* Duluth, Minnesota, 563–569.

Hammer, D.E. and R.H. Kadlec. 1983. *Design Principles for Wetland Treatment Systems.* U.S. EPA 600/2-83-026.

Hammer, D.E. and Kadlec, R.H. 1986. A model for wetland surface water dynamics. *Water Resources Research.* 22(13): 1951–1958.

Hancock, S.J. and L. Buddhavarapu. 1993. Control of Algae Using Duckweed (Lemna) Systems. pp. 397–406. in G.A. Moshiri (Ed.), *Constructed Wetlands for Water Quality Improvement.* Boca Raton, FL: Lewis Publishers.

Hansen, B. 1993. Unpublished data.

Hansen, V.E., O.W. Israelsen, and G.E. Stringham. 1980. *Irrigation Principles and Practices.* 4th Edition. New York: John Wiley & Sons.

Hanowski, J.M. and G.J. Niemi. 1993. Effect of sewage effluent on bird abundance and species composition in a northern Minnesota wetland. *J. Minn. Acad. Sci.,* 57(2):5–10.

Harper, H.H. and Livingston. 1985. Unpublished data from Hidden Lake, Orlando, Florida.

Harper, H.H., M.P. Wanielista, B.M. Fries, and D.M. Baker. 1986. Stormwater Treatment by Natural Systems. Florida Department of Environmental Regulation Report 84-026.

Harris, L.D. 1988. The nature of cumulative impacts on biotic diversity of wetland vertebrates. *Environ. Manage.,* 12(5):675–693.

Harris, L.D. and C.R. Vickers. 1984. Some Faunal Community Characteristics of Cypress Ponds and the Changes Induced by Perturbations. Chapter 17, pp. 171–185. in K.C. Ewel and H.T. Odum (Eds.), *Cypress Swamps.* Gainesville: University of Florida Press.

Hayes, T.D., H.R. Isaacson, K.R. Reddy, D.P. Chynoweth, and R. Biljetina. 1987. Water Hyacinth Systems for Water Treatment. pp. 121–140. in K.R. Reddy and W.H. Smith (Eds.), *Aquatic Plants for Water Treatment and Resource Recovery.* Orlando, FL: Magnolia Publishing.

Haynes, R.J. and L. Moore. 1988. Reestablishment of Bottomland Hardwoods Within National Wildlife Refuges in the Southeast. pp. 95–103. in J. Zelanzy and J.S. Feierabend (Eds.), *Increasing Our Wetland Resources.* Washington, DC: National Wildlife Federation-Corporate Conservation Council.

Hazen, R.E. and T.J. Kneip. 1980. Biogeochemical Cycling of Cadmium in a Marsh Ecosystem. Chapter 11, pp. 399–424. in J.O. Nriagu (Ed.), *Cadmium in the Environment. Part I: Ecological Cycling.* New York: Wiley-Interscience.

Hedin, R.S. 1989. Treatment of Coal Mine Drainage with Constructed Wetlands. Chapter 28, pp. 349–362. in S.K. Majumdar, R.P. Brooks, F.J. Brenner, R.W. Tiner (Eds.), *Wetlands Ecology and Conservation: Emphasis in Pennsylvania.* Easton, PA: The Pennsylvania Academy of Science.

Hedin, R.S. and R.W. Nairn. 1990. Sizing and Performance of Constructed Wetlands: Case Studies. Paper Presented at the 1990 Mining and Reclamation Conference and Exhibition, Charleston, WV. April 23–26, 1990.

Hedin, R.S., R. Hammack, and D. Hyman. 1989. Potential Importance of Sulfate Reduction Processes in Wetlands Constructed to Treat Mine Drainage. Chapter 38b, pp. 508–514. in D.A. Hammer (Ed.), *Constructed Wetlands for Wastewater Treatment: Municipal, Industrial, and Agricultural.* Chelsea, MI: Lewis Publishers.

Hellgren, E.C. and M.R. Vaughan. 1989. Denning ecology of black bears in a southeastern wetland. *J. Wildl. Manage.,* 53(2):347–353.

Hemond, H.F. 1983. The Nitrogen Budget of Thoreau's Bog, *Ecology.* 64(1):99–109.

Hemond, H.F. and J. Benoit. 1988. Cumulative impacts on water quality of wetlands. *Environ. Manage.,* 12(5):639–653.

Hendrey, G.R., J. Clinton, K. Blumer, and K. Lewin. 1979. Lowland Recharge Project Operations, Physical, Chemical, and Biological Changes 1975–1978. Final Report to the Town of Brookhaven. Brookhaven National Laboratory, Brookhaven, NY.

Herskowitz, J. 1986. Listowel Artificial Marsh Project Report. Ontario Ministry of the Environment, Water Resources Branch, Toronto.

Herskowitz, J., S. Black, and W. Lewandowski. 1987. Listowel Artificial Marsh Treatment Project. pp. 247–254. in K. Reddy and W.H. Smith (Eds.), *Aquatic Plants for Water Treatment and Resource Recovery.* Orlando, FL: Magnolia Publishing.

Hey, D.L., K. Barrett, and C. Biegen. 1992. The Hydrology of Four Experimental Wetlands. Volume 2, Chapter 1. in *The Des Plaines River Wetlands Demonstration Project.* Chicago, IL: Wetlands Research, Inc.

Hey, D.L., A.L. Kenimer, and K.R. Barrett. 1994a. Water quality improvement by four experimental wetlands. *Ecological Engineering.* 3(4):381–398.

Hey, D.L., K.R. Barrett, and C. Biegen. 1994b. The hydrology of four constructed marshes. *Ecological Engineering.* 3(4):319–344.

Hickman, S. 1994. Improvement of habitat quality for nesting and migrating birds at the Des Plaines River wetlands demonstration site. *Ecol. Eng.,* 3(4):485–494.

Hietz, P. 1992. Decomposition and nutrient dynamics of reed (*Phragmites australis* (Cav.) Trin. ex Steud.) litter in Lake Neusiedl, Austria. *Aquat. Bot.,* 43:211–230.

Hill, B.H. 1987. *Typha* productivity in a Texas pond: Implications for energy and nutrient dynamics in freshwater wetlands. *Aquat. Bot.,* 27:385–394.

Hill, D. 1990. Unpublished operating data from the Lakeland, Florida wetlands.

Hill, D. 1992. Unpublished data from the Lakeland, Florida wetland treatment facility.

Hinchman, R.R., R.G. McNally, and M.C. Negri. 1993. Biotreatment of Produced Waters. Research Project Sponsored by The Gas Research Institute, Chicago, IL; conducted at Argonne National Laboratory, Argonne, IL.

Hines, M. and S.C. Reed. 1994. Constructed Wetlands/Recirculating Gravel Filter System. Abstract Submitted for Fourth International Conference on Wetland Systems for Water Pollution Control, Guangzhou, China, November 1994.

Hokosawa, Y. and T. Horie. 1992. Flow and particulate nutrient removal by wetland with emergent macrophyte. *Sci. Total Environ.*, Supplement; 1271–1282.

Honig, R. 1988. Tenneco's Use of a Rock-Reed Filter at a Natural Gas Pipeline Compressor Station. Poster Paper at International Conference on Constructed Wetlands for Wastewater Treatment, Chattanooga, TN.

Hook, D.D., et al., Eds. 1988. *The Ecology and Management of Wetlands.* Portland, OR: Timber Press.

Hosseini, S.M. and A.G. van der Valk. 1989. The Impact of Prolonged, Above-Normal Flooding on Metaphyton in a Freshwater Marsh. pp. 317–324. in R.R. Sharitz and J.W. Gibbons (Eds.), *Freshwater Wetlands and Wildlife.* Oak Ridge, TN: U.S. Department of Energy. CONF - 8603101.

Hotchkiss, N. 1972. *Common Marsh, Underwater, and Floating-leaved Plants of the United States and Canada.* New York: Dover Publications.

Hu, K.P. 1992. Hydraulic Factors in Constructed Wetlands. pp. 18.1–18.8. in *Wetland Systems in Water Pollution Control.* Sydney, Australia: IAWQ.

Huckabee, J.W., J.W. Elwood, and S.G. Hildebrand. 1979. Accumulation of Mercury in Freshwater Biota. in J.O. Nriagu (Ed.), *The Biogeochemistry of Mercury in the Environment.* New York: Elsevier/North-Holland Biomedical Press.

Hutchinson, G.E. 1975. *A Treatise on Limnology.* Volume 1, Part 1—*Geography and Physics of Lakes.* Part 2—*Chemistry of Lakes* (pp. 1051); Volume 3—*Limnological Botany* (pp. 500). New York: Wiley-Interscience.

Hyde, H.C., et al. 1984. *Technology Assessment of Wetlands for Municipal Wastewater Treatment.* U.S. EPA600/2-84-154, NTIS No. PB 85-106896.

Idelchik, I.E. 1986. *Handbook of Hydraulic Resistance.* New York: Hemisphere Publishing Corp.

IFAS (Institute of Food and Agricultural Sciences). 1991. *Aquatic Plant Identification.* University of Florida, Center for Aquatic Plants, Gainesville, FL, 7 videotapes, 274 total minutes.

Inaba, I. 1992. Quantitative assessment of natural purification in wetland for linear alkylbenzenesulfonates. *Water Res.*, 26(7):893–898.

Ingram, H.A.P. 1983. Hydrology. Vol. 4A, pp. 67–158. in A.J.P. Gore (Ed.), *Mires: Swamp, Bog, Fen and Moor Ecosystems of the World.* Amsterdam: Elsevier Scientific Publications.

Jackson, J.A. 1989. Man-Made Wetlands for Wastewater Treatment: Two Case Studies. pp. 574–580. in D.A. Hammer (Ed.), *Constructed Wetlands for Waswater Treatment. Municipal, Industrial, and Agricultural.* Chelsea, MI: Lewis Publishers.

Jackson, M.B. and M.C. Drew. 1984. Effects of Flooding on Growth and Metabolism of Herbaceous Plants. Chapter 3, pp. 47–128. in T.T. Kozlowski (Ed.), *Flooding and Plant Growth.* Orlando, FL: Academic Press.

Jenssen, P.D., T. Maehlum, and K. Childs. 1994. The Use of Constructed Wetlands in Cold Climates. Paper at Fourth International Conference on Wetland Systems for Water Pollution Control, Guangzhou, P.R. China, November 1994.

Jenssen, P.D., T. Maehlum, and T. Krogstad. 1992. Adapting Constructed Wetlands for Wastewater Treatment to Northern Environments. in W.J. Mitsch, (Ed.), *Global Wetlands: Old World and New.* Amsterdam: Elsevier.

Jiang, Chuncai, Ed. 1994. *Proceedings of the Fourth International Conference on Wetland Systems for Water Pollution Control.* Guangzhou, China: Center for International Development and Research, South China Institute for Environmental Sciences.

Johansen, N.H. 1994. Diseño y Uso de Humedales Artificiales (Design and Use of Constructed Wetlands), Course Notes, IAWQ/IAMZ, Zaragoza, Spain.

Johnson, D.C. 1942. *The Origin of the Carolina Bays.* New York: Columbia University Press.

Johnson, R.R. 1979. The Lower Colorado River: A Western System. pp. 41–55. in R.R. Johnson and J.F. McCoormick (Eds.), *Strategies for Protection and Management of Floodplain Wetlands and Other Riparian Ecosystems.* Washington, DC: U.S. Department of Agriculture General Technical Report WO-12.

Johnson, R.R. and J.F. McCormick, Eds. 1979. *Strategies for Protection and Management of Floodplain Wetlands and Other Riparian Ecosystems.* Washington, DC: U.S. Department of Agriculture General Technical Report WO-12.

Johnston, C.A. 1993. Mechanisms of Wetland-Water Quality Interaction. pp. 293–299. in G.A. Moshiri (Ed.), *Constructed Wetlands for Water Quality Improvement.* Chelsea, MI: Lewis Publishers.

Jordan, T.E. and I. Valiela. 1983. Sedimentation and resuspensions in a New England salt marsh. *Hydrobiologia,* 98:179–184.

Jorgensen, S.E. 1979. Industrial Waste Water Management. *Studies in Environmental Science 5.* New York: Elsevier Scientific Publishing.

Kaczorowski, R.T. 1977. The Carolina Bays: A Comparison with Modern Lakes. Technical Report No. 13-CRD, Coastal Research Division, Department of Geology, University of South Carolina, Columbia.

Kaczynski, V.W. 1985. Considerations for Wetland Treatment of Spent Geothermal Fluids. Chapter 4, pp. 48–65. in P.J. Godfrey et al. (Eds.), *Ecological Considerations in Wetlands Treatment of Municipal Wastewaters.* New York: Van Nostrand Reinhold.

Kadlec, J.A. 1975. Dissolved Nutrients in a Michigan Peatland. pp. 5–67. in R.H. Kadlec, C.J. Richardson and J.A. Kadlec (Eds.), *The Effects of Sewage Effluent on Wetland Ecosystems.* Semi-Annual Report No. 4 to NSF. NTIS PB 2429192.

Kadlec, J.A. 1986. Input-output nutrient budgets for small diked marshes. *Can. J. Fish Aquat. Sci.,* 43(10):2009–2016.

Kadlec, J.A. 1989. Effects of Deep Flooding and Drawdown on Freshwater Marsh Sediments. pp. 127–143. in R.R. Sharitz and J.W. Gibbons (Eds.), *Freshwater Wetlands and Wildlife.* Oak Ridge, TN: U.S. Department of Energy. CONF - 8603101.

Kadlec, R.H. 1983. The Bellaire Wetland: Wastewater alteration and recovery. *Wetlands,* 3:44–63.

Kadlec, R.H. 1987. Wetland Hydrology and Water Pollution Control Functions. pp. 168–173. in J. A. Kusler and G. Brooks (Eds.), *Proceedings of a Symposium on Wetland Hydrology.* Chicago, IL: Association of State Wetland Managers, Berne, NY.

Kadlec, R.H. 1988a. Monitoring Wetland Responses. pp. 114–120. in J. Zelazny and J.S. Feierabend (Eds.), *Increasing Our Wetland Resources.* Washington, DC: National Wildlife Federation.

Kadlec, R.H. 1988b. Unpublished data from the Houghton Lake wetland treatment facility.

Kadlec, R.H. 1988c. Denitrification in wetland treatment systems. Preprint, Session 26, WPCF National Meeting. Dallas, TX.

Kadlec, R.H. 1989a. Wetland Utilization for Management of Community Wastewater. 1988 Operations Summary. Houghton Lake Wetland Treatment System. April 1989. Report to Michigan DNR.

Kadlec, R.H. 1989b. Decomposition in wastewater wetlands, in: *Constructed Wetlands for Wastewater Treatment,* D.A. Hammer, (Ed.). Lewis Publishers, Chelsea, MI, 459–468.

Kadlec, R.H. 1989c. Wetlands for Treatment of Municipal Wastewater. Chapter 25, pp. 300–314. in S.K. Majumdar, R.P. Brooks, F.J. Brenner, and R.W. Tiner (Eds.), *Wetlands Ecology and Conservation: Emphasis in Pennsylvania.* Easton, PA: The Pennsylvania Academy of Science.

Kadlec, R.H. 1990a. Overland flow in wetlands: vegetation resistance. *J. Hydraul. Eng.,* 116(5):691–706.

Kadlec, R.H. 1990b. Modelling nutrient behavior in wetlands used for wastewater treatment. *Utrecht Plant Ecol. News Rep.,* 11:104–129.

Kadlec, R.H. 1991. Analysis of Gravel Cell Number Three Benton, KY Wetlands. Final Report Submitted to the TVA, Water Quality Department.

Kadlec, R.H. 1993a. Natural Wetland Treatment at Houghton Lake, Michigan: The First Fifteen Years. pp. 73–84. in Water Environment Federation Proceedings of the 66th Conference, Anaheim, CA.

Kadlec, R.H. 1993b. Unpublished data from the Houghton Lake Treatment Wetland, 1993 field season.

Kadlec, R.H. 1994. Detention and mixing in free water wetlands. *Ecol. Eng.,* 3(4):1–36.

Kadlec, R.H. and H.H. Alvord, Jr. 1989. Mechanisms of water quality improvement in wetland treatment systems, in: *Wetlands Concerns and Successes,* D.W. Fisk, (Ed.). AWRA, 489–498.

Kadlec, R.H. and W.V. Bastiaens. 1992. The Use of Residence Time Distributions (RTDs) in Wetland Systems. Volume 2, in Chapter 2. *The Des Plaines River Wetlands Demonstration Project.* Chicago, IL: Wetlands Research, Inc.

Kadlec, R.H. and F.B. Bevis. 1990. Wetlands and wastewater: Kinross, Michigan. *Wetlands,* 10(1):77–92.

Kadlec, R.H. and F.B. Bevis. 1992. Baseline Study: Portage Lake Wetlands. Report to Michigan DNR, July 1992.

Kadlec, R.H. and D.E. Hammer, 1981. *Wetland Utilization for Management of Community Wastewater.* 1980 Operations Summary. Report to NSF-ASRA. NTIS PB81-235954.

Kadlec, R.H. and D.E. Hammer. 1985. Simplified Computation of Wetland Vegetation Cycles. Chapter 9, pp. 141–157. in H.H. Prince and F.M. D'Itri (Eds.), *Coastal Wetlands.* Chelsea, MI: Lewis Publishing.

Kadlec, R.H. and D.E. Hammer. 1988. Modelling nutrient behavior in Wetlands. *Ecol. Modell.,* 40:37–66.

Kadlec, R.H. and D.L. Hey. 1994. Constructed wetlands for river water quality improvement. *Water Science and Technology.* 29(4):159–168.

Kadlec, R.H. and G.A. Keoleian. 1986. Metal Ion Exchange on Peat. pp. 61–93. in C.H. Fuchsman (Ed.), *Peat and Water.* New York: Elsevier.

Kadlec, R.H. and S. Newman. 1992. Phosphorus Removal in Wetland Treatment Areas. Report to South Florida Water Management District, No. DRE 321, West Palm Beach, FL.

Kadlec, R.H. and M.A. Rathbun. 1984. Copper Sorption on Peat. pp. 351–364. in C.H. Fuchsman and S.A. Spigarelli (Eds.), *Proceedings of the International Symposium on Peat Utilization.* Minnesota: Bemidji State University.

Kadlec, R.H. and J.A. Robbins. 1984. Sedimentation and sediment accretion in Michigan coastal wetlands (USA). *J. Chem. Geol.,* 44(1/3):119–150.

Kadlec, R.H. and K. Srinivasan. 1994. *Wetlands for Treatment of Oil and Gas Well Wastewaters.* Report to the US D.O.E., Contract No. DE-AC22-92MT92010.

Kadlec, R.H. and J.T. Watson. 1993. Hydraulics and Solids Accumulation in a Gravel Bed Treatment Wetland. pp. 227–236. in G.A. Moshiri (Ed.), *Constructed Wetlands for Water Quality Improvement.* Boca Raton, FL: Lewis Publishers.

Kadlec, R.H., D.E. Hammer, In-Sik Nam, and J.O. Wilkes. 1981. The hydrology of overland flow in wetlands. *Chem. Eng. Commun.,* 9:331–344.

Kadlec, R.H., R.B. Williams, and R.D. Scheffe. 1987. Wetland Evapotranspiration in Temperate and Arid Climates. Chapter 12, pp. 146–160. in D.D. Hook, (Ed.), *The Ecology and Management of Wetlands.* Beckenham: Croom Helm.

Kadlec, R.H., D.E. Hammer, and M.A. Girts. 1990a. A Total Evaporative Constructed Wetland Treatment System. pp. 127–138. in P. F. Cooper and B. C. Findlaters (Eds.), *Constructed Wetlands in Water Pollution Control.* Oxford, U.K.: Pergamon Press.

Kadlec, R.H., K.F. Kirkbride, and R.L. Van Wormer. 1990b. Constructed Wetland Treatment of Potato Processing Wastewater. Paper Presented at PNPCA Conference, October 1990, Bellevue, WA.

Kadlec, R.H., W. Bastiaens, and D.T. Urban. 1993. Hydrological Design of Free Water Surface Treatment Wetlands. pp. 77–86. in G.A. Moshiri, (Ed.), *Constructed Wetlands for Water Quality Improvement.* Boca Raton, FL: Lewis Publishers.

Kania, H.J., R.L. Knight, and R.J. Beyers. 1976. *Fate and Biological Effects of Mercury Introduced into Artificial Streams.* EPA-600/3-76-060.

Kaynor, E.R., et al. 1985. *Ecological Considerations in Wetland Treatment of Municipal Wastewaters.* New York: Van Nostrand Reinhold.

Keely, S.J., L.N. Schwartz, and J.G. Ladner. 1992. Bringing wetlands and landfills into environmental harmony. *World Wastes,* August 1992.

Kehoe, M.J. 1993. *Water Quality Survey of Twenty-Four Stormwater Wet-Detention Ponds.* Final Report. Southwest Florida Water Management District. Brooksville, FL. 84 pp.

Kemp, W.M. and L. Murray. 1986. Oxygen release from roots of the submerged macrophyte *Potamogeton perfoliatus* L.: Regulating factors and ecological implications. *Aquat. Bot.,* 26:349.

Kibby, H.V. 1979. Effects of Wetlands on Water Quality. pp. 289–298. in R.R. Johnson and J.F. McCormick (Eds.), *Strategies for Protection and Management of Floodplain Wetlands and Other Riparian Ecosystems.* Washington, DC: U.S. Department of Agriculture.

Kickuth, R. 1977. Degradation and incorporation of nutrients from rural waste waters by plant rhizosphere under limnic conditions, in: *Utilization of Manure by Land Spreading.* Commission of the European Communities, EUR 5672e. London, UK. pp. 335–343.

Kickuth, R. 1982. Verfahrens- und Dimensionierungsgrundlagen der Wurzelraumentsorgung, Teil I: Die Mineralischen Bodenhorizonte, Unpublished Internal Report, Gesamthochschule Kassel, 67 pp.+ appendices (as referenced by Schierup et al., 1990b).

Kickuth, R. 1984. Das Wurzelraumfahren in der Praxis. Landschaft Stadt. 16:145–153.

King, D. and D.B. Nedwell. 1987. The adaptation of nitrate-reducing bacterial communities in estuarine sediments in response to overlying nitrate load, *Microbiology Ecology*, Vol. 45, pp. 15–20.

Kirk, P.W. 1987. Pollution Control Legislation. Chapter 3, pp. 65–103. in LN. Lester (Ed.), *Heavy Metals in Wastewater and Sludge Treatment Processes. Volume I. Sources, Analysis, and Legislation.* Boca Raton, FL: CRC Press.

Kleinmann, R.L.P. and R. Hedin. 1989. Biological Treatment of Mine Water: An Update. pp. 173–179. in M.E. Chalkley, B.R. Conrad, V.I. Lakshmanan, and K.G. Wheeland (Eds.), *Tailings and Effluent Management.* City, ST: Publisher.

Kleinmann, R.L.P., R. Brooks, B. Huntsman, and B. Pesavento. 1986. Constructing Wetlands for the Treatment of Mine Water. Mini-course Notes. National Symposium on Surface Mining, Hydrology, Sedimentology, and Reclamation, University of Kentucky, Lexington, KY.

Klopatek, J.M. 1978. Nutrient Dynamics of Freshwater Riverine Marshes and the Role of Emergent Macrophytes. pp. 195–216. in R.E. Good, D.F. Whigham, and R.L. Simpson (Eds.), *Freshwater Wetlands: Ecological Processes and Management Potential.* New York: Academic Press.

Knight, R.L. 1980. Energy Basis of Control in Aquatic Ecosystems. Ph.D. Dissertation. Department of Environmental Engineering Science, University of Florida, Gainesville.

Knight, R.L. 1986. Florida Effluent Wetlands, Total Nitrogen. CH2M HILL Wetland Technical Reference Document Series No. 1. Gainesville, Florida.

Knight, R.L. 1987a. Effluent Distribution and Basin Design for Enhanced Pollutant Assimilation by Freshwater Wetlands. pp. 913–921. in K.R. Reddy and W.H. Smith (Eds.), *Aquatic Plants for Water Treatment and Resource Recovery.* Orlando, FL: Magnolia Publishing.

Knight, R.L. 1987b. Unpublished data from the Lake Coral Treatment Wetlands, Silver Springs Shores, FL.

Knight, R.L. 1989. Use of Treated Wastewater for Creation of Mitigation Wetlands. Paper Presented at the 1989 Mid-Year Meeting of the Landscape and Environmental Design Committee of the Transportation Research Board, Rhode Island, July 1989.

Knight, R.L. 1990. Operational Performance of Reedy Creek Wetlands Treatment System and Other Southern Wetlands. pp. 103–108. in U.S. Environmental Protection Agency. Proceedings of the U.S. Environmental Protection Agency Municipal Wastewater Treatment Technology Forum. 1990. March 02–22, 1990. Orlando, FL.

Knight, R.L. 1991. Analysis of Survival and Condition of Planted Vegetation at the Benton, Hardin, and Pembroke, Kentucky Constructed Wetland Treatment Systems. Report to TVA, Chattanooga, TN, November 1991.

Knight, R.L. 1992a. Ancillary benefits and potential problems with the use of wetlands for nonpoint source pollution control. *Ecol. Eng.*, 1:97–113.

Knight, R.L. 1992b. Natural land treatment with Carolina bays. *Water Environment & Technology* 4:13–16.

Knight, R.L. 1993. Operating experience with constructed wetlands for wastewater treatment. *Tappi J.*, 76(1):109–112.

Knight, R.L. 1994a. Unpublished data from a constructed marsh receiving pulp and paper mill effluent.

Knight, R.L. 1994b. Constructed Wetlands for Wastewater Management at Ocean Springs, Mississippi. Vol. 8, pp. 125–132. in *Proceedings of the 67th Annual Conference of the Water Environment Federation.* Alexandria, VA: Water Environment Federation.

Knight, R.L. 1994c. Treatment wetlands database now available. *Water Environ. Technol.*, 6(2):31–33.

Knight, R.L. and K.A. Ferda. 1989. Performance of the Boggy Gut Wetland Treatment System, Hilton Head, South Carolina. pp. 439–450. in D. Fisk (Ed.), *Proceedings of the Symposium on Wetlands: Concerns and Successes.* Bethesda, MD: American Water Resources Association.

Knight, R.L., and M.E. Iverson. 1990. Design of the Fort Deposit, Alabama Constructed Wetlands Treatment System. pp. 521–524. in P.F. Cooper and B.C. Findlater (Eds.), *Constructed Wetlands in Water Pollution Control.* Oxford, U.K.: Pergamon Press.

Knight, R.L., B.H. Winchester, and J.C. Higman. 1985a. Carolina bays—feasibility for effluent advanced treatment and disposal. *Wetlands,* 4:177–203.

Knight, R.L., B.H. Winchester, and J.C. Higman. 1985b. Ecology, hydrology, and advanced wastewater treatment potential of an artificial wetland in north-central Florida. *Wetlands,* 5:167–180.

Knight, R.L., T.W. McKim, and H.R. Kohl. 1987. Performance of a natural wetland treatment system for wastewater management. *J. Water Pollut. Control Fed.*, 59:746–754.

Knight, R.L., J.S. Bays, and F.R. Richardson. 1989. Floral Composition, Soil Relations, and Hydrology of a Carolina Bay in South Carolina. pp. 219–234. in R.R. Sharitz and J.W. Gibbons (Eds.), *Freshwater Wetlands and Wildlife.* Oak Ridge, TN: U.S. Department of Energy. CONF - 8603101.

Knight, R.L., R.H. Kadlec, and S. Reed. 1992. Wetlands Treatment Database. Water Environmental Federation 65th Annual Conference and Exposition. New Orleans, LA. September 20–24, 1992. pp. 25–35.

Knight, R.L., R.W. Ruble, R.H. Kadlec, and S. Reed. 1993a. Wetlands for Wastewater Treatment Performance Database. Chapter 4, pp. 35–58. in G.A. Moshiri (Ed.), *Constructed Wetlands for Water Quality Improvement.* Boca Raton, FL: Lewis Publishers.

Knight, R.L., Ruble, R.W., Kadlec, R.H., and S.C. Reed. 1993b. Database: North American Wetlands for Water Quality Treatment. Phase II Report. Prepared for U.S. EPA. September 1993.

Knight, R.L., J. Hilleke, and S. Grayson. 1994. Design and performance of the Champion pilot-constructed wetland treatment system. *Tappi J.*, 77:240–245.

Koerselman, W. and B. Beltman. 1988. Evapotranspiration from fens in relation to Penman's potential free water evaporation (Eo) and pan evaporation. *Aquat. Bot.*, 31:307–320.

Kozlowski, T.T. 1984. Responses of Woody Plants to Flooding. Chapter 4, pp. 129–163. in T.T. Kozlowski (Ed.), *Flooding and Plant Growth.* Orlando, FL: Academic Press.

Kraus, M.L. 1988. Wetlands: Toxicant Sinks or Reservoirs? pp. 192–196. in J.A. Kusler and G. Brooks (Eds.), *Proceedings of the National Wetland Symposium: Wetland Hydrology.* Berne, NY: Association of State Wetland Managers.

Kraus, M.P. 1977. Bacterial Indicators and Potential Health Hazards of Aquatic Viruses. pp. 196–217. in A.W. Hoadley and B.J. Dutka (Eds.), *Bacterial Indicators/Health Hazards Associated with Water.* Philadelphia, PA: American Society for Testing and Materials.

Krishnan, S.B. and J.E. Smith. 1987. Public Health Issues of Aquatic Systems Used for Wastewater Treatment. pp. 855–878. in K.R. Reddy and W.H. Smith (Eds.), *Aquatic Plants for Water Treatment and Resource Recovery.* Orlando: Magnolia Publishing.

Kroodsma, D.E. 1978. Habitat Values for Nongame Wetland Birds. pp. 320–326. in P.E. Greeson, J.R. Clark, and J.E. Clark (Eds.), *Wetland Functions and Values: The State of Our Understanding.* Minneapolis, MN: American Water Resources Association.

Kuehn, E. and J.A. Moore. 1994. Variability of Treatment Performance in Constructed Wetlands. in *Proceedings of the Fourth International Conference on Wetland Systems for Water Pollution Control.* Guangzhou, P.R. China. Center for International Development and Research, South: China Institute for Environmental Sciences.

Kuenzler, E.J. 1990. Wetlands as Sediment and Nutrient Traps for Lakes. pp. 105–112. in Proceedings of a National Conference on Enhancing the States' Lake and Wetland Management Programs. May 18–19, 1989. Chicago, IL.

Kufel, I. 1991. Lead and molybdenum in reed and cattail—open versus closed type of metal cycling. *Aquat. Bot.* 40:275–288.

Kühl, H. and J-G. Kohl. 1993. Seasonal nitrogen dynamics in reed beds (*Phragmites australis* [Cav.] Trin. ex. Steudel) in relation to productivity, *Hydrobiologia*, Vol. 251, pp. 1–12.

Kushlan, J.A. 1989. Avian Use of Fluctuating Wetlands. pp. 593–604. in R. Sharitz and J.W. Gibbons (Eds.), *Freshwater Wetlands and Wildlife.* Oak Ridge, TN: U.S. Department of Energy. CONF - 8603101.

Kusler, J.A. and G. Brooks. 1988. *Proceedings of the National Wetland Symposium: Wetland Hydrology.* Berne, NY: Association of State Wetland Managers.

Kusler, J.A. and S. Daly, Eds. 1989. *Wetlands and River Corridor Management.* Berne, NY: Association of State Wetland Managers.

Kusler, J.A. and M.E. Kentula. 1990. *Wetland Creation and Restoration. The Status of the Science.* Washington, DC: Island Press.

Kusler, J.A. and P. Riexinger, Eds. 1986. *Proceedings of the National Wetland Assessment Symposium.* Chester, VT: Association of State Wetland Managers.

Kusler, J.A., S. Daly, and G. Brooks, Eds. 1988a. *Urban Wetlands. Proceedings of the National Symposium.* Berne, NY: Association of Wetland Managers.

Kusler, J.A., M.L. Quammen, and G. Brooks, Eds. 1988b. *Proceedings of the National Wetland Symposium: Mitigation of Impacts and Losses.* Berne, NY: Association of State Wetland Managers.

Lack, D.L. 1966. *Population Studies of Birds.* Oxford: Clarendon Press.

Lacki, M.J., W.T. Peneston, K.B. Adams, F.D. Vogt, and J.C. Houppert. 1990. Summer foraging patterns and diet selection of muskrats inhabiting a fen wetland. *Can. J. Zool.,* 68:1163–1167.

Lafleur, P.M. 1990. Evapotranspiration from sedge-dominated wetland surfaces. *Aquat. Bot.,* 37:341–353.

Lager, J.A., W.G. Smith, W.G. Lynard, R.M. Finn, and E.J. Finnemore. 1977. *Urban Stormwater Management and Technology: Update and User's Guide.* U.S. Environmental Protection Agency. Office of Research and Development, Municipal Environmental Research Laboratory, Cincinnati, OH. EPA-600/8-77-014.

Lake, D.L. 1987. Sludge Disposal to Land. Chapter 5, pp. 91–130. in J.N. Lester (Ed.), *Heavy Metals in Wastewater and Sludge Treatment Processes. Volume II. Treatment and Disposal.* Boca Raton, FL: CRC Press.

Lakshman, G. 1993. Design and Operational Limitations of Engineered Wetlands in Cold Climates—Canadian Experience. pp. 399–409. in W. Mitsch (Ed.), *Global Wetlands: Old World and New.* Amsterdam: Elsevier.

Lakshman, G. 1981. *A Demonstration Project at Humboldt to Provide Tertiary Treatment to the Municipal Effluent Using Aquatic Plants.* Saskatchewan Research Council, SRC Publication No. E-820-4-E-81.

Lan, C., G. Chen, L. Li, and M.H. Wong. 1990. Purification of Wastewater from a Pb/Zn Mine Using Hydrophytes. pp. 419–427. in P.F. Cooper and B.C. Findlater (Eds.), *Proceedings of the International Conference on the Use of Constructed Wetlands in Water Pollution Control.* Oxford, U.K.: Pergamon Press.

Lawson, G.J. 1985. Cultivating Reeds (Phragmites australis) for Root Zone Treatment of Sewage. Contract Report to the Water Research Centre, Cumbria, U.K. IRE Project 965.

Leck, M.A., R.L. Simpson, and V.T. Parker. 1989. The Seed Bank of a Freshwater Tidal Wetland and Its Relationship to Vegetation Dynamics. pp. 189–205. in R.R. Sharitz and J.W. Gibbons (Eds.), *Freshwater Wetlands and Wildlife.* Oak Ridge, TN: U.S. Department of Energy. CONF - 8603101.

Leclerc, H., D.A. Mossel, P.A. Trinel, and F. Gavini. 1977. Microbiological Monitoring—A New Test for Fecal Contamination. pp. 23–36. in A.W. Hoadley and B.J. Dutka (Eds.), *Bacterial Indicators/ Health Hazards Associated with Water.* Philadelphia, PA: American Society for Testing and Materials.

Lee, R.E. 1980. *Phycology.* Cambridge, U.K.: Cambridge University Press.

Lema, J.M., R. Mendez, and R. Blazquez. 1988. Characteristics of landfill leachates and alternatives for their treatment: a review. *Water Air Soil Pollut.,* 40:223–250.

Lemly, A.D. 1993. Guidelines for evaluating selenium data from aquatic monitoring and assessment studies. *Environ. Monit. Assess.,* 28:83–100.

Lemly, A.D., S.E. Finger, and M.K. Nelson. 1993. Sources and impacts of irrigation drainwater contaminants in arid wetlands. *Environ. Toxicol. Chem.,* 12:2265–2279.

Lemon, E.R. and I.A. Smith. 1993. Sewage Waste Amendment Marsh Process (SWAMP). Report to Ontario Ministry of Environment and Energy.

Lester, J.N. 1987. *Heavy Metals in Wastewater and Sludge Treatment Processes. Volume I. Sources, Analysis, and Legislation.* Boca Raton, FL: CRC Press.

Leva, M. 1947. Pressure drop through packed tubes. I. A general correlation. *Chem. Eng. Prog.,* 43:549.

Levenspiel, O. 1972. *Chemical Reaction Engineering.* 2nd Edition. New York: John Wiley & Sons.

Levine, D.A. and D.E. Willard. 1990. Regional Analysis of Fringe Wetlands in the Midwest: Creation and Restoration. pp. 299–325. in J.A. Kusler and M.E. Kentula (Eds.), *Wetland Creation and Restoration. The Status of the Science.* Washington, DC: Island Press.

Lewis, T.E. 1989. *Environmental Chemistry and Toxicology of Aluminum.* Chelsea, MI: Lewis Publishers.

Lewis, J.C. and E.W. Bunce, Eds. 1980. *Rehabilitation and Creation of Selected Coastal Habitats: Proceedings of a Workshop.* U.S. Fish and Wildlife Service, loction FWS/OBS-80/27.

Linacre, E.T. 1976. Swamps. pp. 329–347. in J.L. Monteith (Ed.), *Vegetation and Atmosphere, Volume 2: Case Studies.* London: Academic Press.

Lindsay, W.L. 1979. *Chemical Equilibria in Soils.* New York: John Wiley & Sons.

Litchfield, D.K. 1990. Constructed Wetlands for Wastewater Treatment at Amoco Oil Company's Mandan, North Dakota Refinery. pp. 399–402. in P.F. Cooper and B.C. Findlater (Eds.), *Proceedings of the International Conference on the Use of Constructed Wetlands in Water Pollution Control.* Oxford, U.K.: Pergamon Press.

Litchfield, D.K. 1993. Constructed Wetlands for Wastewater Treatment at Amoco Oil Company's Mandan, North Dakota, Refinery. pp. 485–488. in: G.A. Moshiri (Eds.), *Constructed Wetlands for Water Quality Improvement.* Chelsea, MI: Lewis Publishers.

Litchfield, D.K. and D.D. Schatz. 1989. Constructed Wetlands for Wastewater Treatment at Amoco Oil Company's Mandan, North Dakota Refinery. Chapter 18, pp. 233–237. in D.A. Hammer (Ed.), *Constructed Wetlands for Wastewater Treatment: Municipal, Industrial, and Agricultural.* Chelsea, MI: Lewis Publishers.

Livingston, E.H. 1989. Use of Wetlands for Urban Stormwater Management. Chapter 21, pp. 253–262. in D.A. Hammer (Ed.), *Constructed Wetlands for Wastewater Treatment. Municipal, Industrial, and Agricultural.* Chelsea, MI: Lewis Publishers.

Lodge, D.M. 1991. Herbivory on freshwater macrophytes. *Aquat. Bot.,* 41:195–224.

Loxham, M. and W. Burghardt. 1986. Saturated and Unsaturated Permeabilities of North German Peats, in: *Peat and Water,* C.H. Fuchsman (Ed.), Elsevier, New York, 37–59.

Ludwig, J.P. 1991. Impact of Wastewater Discharge upon Northern Michigan Wetland Bird and Wildlife Communities at the Houghton Lake Wetland Treatment Project Discharge Site—1991. Report to Michigan DNR. November 1991.

Lugo, A.E. and S.C. Snedaker. 1974. The ecology of mangroves. *Annu. Rev. Ecol. Syst.,* 5:39–64.

LWAI (Larry Walker Associates, Inc.). 1990. *City of Gustine Marsh Evaluation Study.* Davis, CA: Larry Walker Associates, Inc.

Lynch, T.R., C.J. Popp, and G.Z. Jacobi. 1988. Aquatic insects as environmental monitors of trace metal contamination: Red River, New Mexico. *Water Air Soil Pollut.,* 42:19–31.

MacClellan, D.A. 1989. Boggy Gut Wetland Status Report. Prepared for the Sea Pines Public Service District, Hilton Head Island, SC.

MacVicar, T.K. 1985. *A Wet Season Field Test of Experimental Water Deliveries to Northeast Shark River Slough.* South Florida Water Management District, West Palm Beach, FL. Technical Pub. 85-3.

Maddox, J.J. and J.B. Kingsley. 1989. Waste treatment for confined swine with an integrated artificial wetland and aquaculture system, in: *Constructed Wetlands for Wastewater Treatment.* D.A. Hammer (Ed.), Lewis Publishers, Chelsea, MI. Chapter 14, pp. 191–200.

Madsen, T.V. and K. Sand-Jensen. 1991. Photosynthetic carbon assimilation in aquatic macrophytes. *Aquat. Bot.,* 41:5–40.

Magnuson, J.J., C.A. Paszkowski, F.J. Rahel, and W.M. Tonn. 1989. Fish Ecology in Severe Environments of Small Isolated Lakes in Northern Wisconsin. pp. 487–515. in R.R. Sharitz and J.W. Gibbons (Eds.), *Freshwater Wetlands and Wildlife.* Oak Ridge, TN: U.S. Department of Energy. CONF - 8603101.

Majumdar, S.K., R.P. Brooks, F.J. Brenner, and R.W. Tiner, Eds. 1989. *Wetlands Ecology and Conservation: Emphasis in Pennsylvania.* Easton, PA: The Pennsylvania Academy of Science.

Manci, K.M. and D.H. Rusch. 1989. Waterbird Use of Wetland Habitats Identified by Aerial Photography. pp. 1045–1058. in R.R. Sharitz and J.W. Gibbons (Eds.), *Freshwater Wetlands and Wildlife.* Oak Ridge, TN: U.S. Department of Energy. CONF - 8603101.

Mann, R.A. 1990. Phosphorus Removal by Constructed Wetlands: Substratum Absorption. pp. 97–106. in P.F. Cooper and B.C. Findlater (Eds.), *Constructed Wetlands in Water Pollution Control.* Oxford, U.K.: Pergamon Press.

Marsalek, J. 1990. Evaluation of pollutant loads from urban nonpoint sources. *Water Sci. Technol.,* 22:23–30.

Marsalek, J. and H.O. Schroeter. 1989. Annual loadings of toxic contaminants in urban runoff from the Canadian Great Lakes Basin. *Water Pollut. Res. J. Can.,* 23:360–378.

Marshall, D.E. 1970. Characteristics of *Spartina* Marsh Which Is Receiving Treated Municipal Sewage Wastes. pp. 317–363. in H.T. Odum and A.F. Chestnut (Eds.), Studies of Marine Estuarine

Ecosystems Developing With Treated Sewage Wastes. Annual Report for 1969–1970. Institute of Marine Sciences, University of North Carolina, Morehead City, N.C.

Martin, C.V. and B.F. Eldridge. 1989. California's Experience with Mosquitoes in Aquatic Wastewater Treatment Systems. Chapter 31, pp. 393–398. in D.A. Hammer (Ed.), *Constructed Wetlands for Wastewater Treatment: Municipal, Industrial, and Agricultural.* Chelsea, MI: Lewis Publishers.

Martin, C.D., G.A. Moshiri, and C.C. Miller. 1993. Mitigation of Landfill Leachate Incorporating In-Series Constructed Wetlands of a Closed-Loop Design. Chapter 51, pp. 473–476. in G.A. Moshiri (Ed.), *Constructed Wetlands for Water Quality Improvement.* Boca Raton, FL: Lewis Publishers.

Martin, J.J. 1948. Ph.D. Thesis. Carnegie Institute of Technology, Pittsburgh, PA.

Martin, C.D. and G.A. Moshiri. 1994. The Use of In-Series Surface-Flow Wetlands for Landfill Leachate Treatment. Preprint of Paper Presented at the Fourth International Conference on Wetlands for Water Pollution Control. Guangzhou, PR China.

Masch, F.D. and K.J. Denny. 1966. Grain size distribution and its effect on the permeability of unconsolidated sands. *Water Resour. Res.*, 2:665–677.

Masscheleyn, P.H. and W.H. Patrick, Jr. 1993. Biogeochemical processes affecting selenium cycling in wetlands. *Environ. Toxicol. Chem.*, 12:2235–2243.

Mattingly, G.E. 1977. Experimental study of wind effects on reaeration. *J. Hydraul. Div.*, 103(HY3):311.

May, E., J.E. Butler, M.G. Ford, R. Ashworth, J.B. Williams, and M.M.M. Bahgat. 1990. Chemical and microbiological processes in gravel-bed hydroponic (GBH) systems for sewage treatment, in: *Constructed Wetlands in Water Pollution Control*, P.F. Cooper and B.C. Findlater (Eds.), Pergamon Press, Oxford, UK, pp. 33–40.

McAllister, L.S. 1992. *Habitat Quality Assessment of Two Wetland Treatment Systems in Mississippi—A Pilot Study.* U.S. Environmental Protection Agency. Environmental Research Laboratory, Corvallis, OR. November 1992. EPA/600/R-92/229.

McAllister, L.S. 1993a. *Habitat Quality Assessment of Two Wetland Treatment Systems in the Arid West—Pilot Study.* U.S. Environmental Protection Agency. Environmental Research Laboratory, Corvallis, OR. July 1993. EPA/600/R-93/117.

McAllister, L.S. 1993b. *Habitat Quality Assessment of Two Wetland Treatment Systems in Florida—A Pilot Study.* U.S. Environmental Protection Agency. Environmental Research Laboratory, Corvallis, OR. November 1993. EPA/600/R-93/222.

McKim, T.W. 1982. Advanced Wastewater Treatment at the Walt Disney World Resort Complex. Unpublished Report. Reedy Creek Utilities Co., Inc. Lake Buena Vista, FL.

McMahan, E.A. and L.R. Davis, Jr. 1984. Density and Diversity of Microarthropods in Manipulated and Undisturbed Cypress Domes. Chapter 19, pp. 197–209. in K.C. Ewel and H.T. Odum (Eds.), *Cypress Swamps.* Gainesville: University of Florida Press.

McMahan, E.A., R.L. Knight, and A.R. Camp. 1972. A comparison of microarthropod populations in sewage-exposed and sewage-free *Spartina* salt marshes. *Environ. Entomol.*, 1(2):244–252.

Meiorin, E.C. 1989. Urban Runoff Treatment in a Fresh/Brackish Water Marsh in Fremont, California. Chapter 40b, pp. 677–685. in D.A. Hammer (Ed.), *Constructed Wetlands for Wastewater Treatment: Municipal, Industrial, and Agricultural.* Chelsea, MI: Lewis Publishers.

Melzer, A. and D. Exler. 1982. Nitrate and nitrite reductase activities in aquatic macrophytes, in: *Studies on Aquatic Vascular Plants*, J.J. Symoens, S.S. Hooper, and P. Compère, (Eds.), Royal Botanical Society of Belgium, Brussels, pp. 128–135.

Merritt, A., Ed. 1994. *Wetlands, Industry, and Wildlife.* Slimbridge, Gloucester, U.K.: The Wildfowl and Wetlands Trust.

Merritt, R.W. and D.L. Lawson. 1979. Leaf Litter Processing in Floodplain and Stream Communities. pp. 93–105. in R.R. Johnson and J.F. McCormick (Eds.), *Strategies for Protection and Management of Floodplain Wetlands and Other Riparian Ecosystems.* Washington, DC: U.S. Department of Agriculture.

Messer, J.J. and P.L. Brezonik. 1977. Nitrogen Transformations in Everglades Agricultural Area Soils and Sediments. Report Presented to the Florida Sugar Cane League. Black Crow & Eidsness/ CH2M HILL.

Metcalf & Eddy, Inc. 1990. Preliminary Design Report. Wastewater Treatment Improvements for the City of Columbia, Missouri.

Metcalf & Eddy, Inc. 1991. *Wastewater Engineering, Treatment, Disposal, and Reuse.* Third Edition. Revised by G. Tchobanoglous and F.L. Burton. New York: McGraw-Hill.

Middlebrooks, E.J. 1979. *Industrial Pollution Control. Volume 1: Agro-Industries.* New York: John Wiley & Sons.

Mierau, R. and P. Trimble. 1988. Hydrologic Characteristics of the Kissimmee River Floodplain Boney Marsh Experimental Area. Tech. Memorandum. South Florida Water Management District. September 1988.

Miller, C.C. and G.A. Moshiri. 1993. An Integrated Composite Solid Waste Facility Design Involving Recycling, Volume Reduction, and Wetlands Leachate Treatment. In press.

Miller, G. 1989. Use of Artificial Cattail Marshes to Treat Sewage in Northern Ontario, Canada. Chapter 39, pp. 636–642. in D.A. Hammer (Ed.), *Constructed Wetlands for Wastewater Treatment: Municipal, Industrial, and Agricultural.* Chelsea, MI: Lewis Publishers.

Mir, Z. and G. Laksham. 1987. Voluntary Intake and Digestibility of Complete Diets Containing Varying Levels of Cattail (*Typha latifolia* L.) by Sheep and Cattle. pp. 813. in K.R. Reddy and W.H. Smith (Eds.), *Aquatic Plants for Water Treatment and Resource Recovery.* Orlando FL: Magnolia Publishing.

Mitchell, L.J., R.A. Lancia, R. Lea, and S.A. Gauthreaux. 1989. Effects of Clearcutting and Natural Regeneration on Breeding Bird Communities of a Bald Cypress-Tupelo Wetland in South Carolina. pp. 155–161. in J.A. Kusler and S. Daly (Eds.), *Wetlands and River Corridor Management.* Berne, NY: Association of State Wetland Managers.

Mitsch, W.J. 1979. Interactions Between a Riparian Swamp and a River in Southern Illinois. pp. 63–72. in R.R. Johnson and J.F. McCormick (Eds.), *Strategies for Protection and Management of Floodplain Wetlands and Other Riparian Ecosystems.* Washington, DC: U.S. Department of Agriculture.

Mitsch, W.J. 1991. Landscape Design and the Role of Created, Restored, and Natural Wetlands in Controlling Rural NPS. in R. Olson (Ed.), *Created and Natural Wetlands in Controlling Nonpoint Source Pollution.* Boca Raton, FL: C.K. Smoley, pp. 43–70.

Mitsch, W.J., Ed. 1994. *Global Wetlands: Old World and New.* Amsterdam: Elsevier.

Mitsch, W.J. and J.G. Gosselink. 1993. *Wetlands.* New York: Van Nostrand Reinhold.

Mitsch, W.J., C.L. Dorge, and J.R. Wiemhoff. 1979. Ecosystem dynamics and a phosphorus budget of an alluvial cypress swamp in southern Illinois. *Ecology.* 60:1116–1124.

Mitsch, W.J., X. Wu, and N. Wang, 1993. *Modelling the Des Plaines Experimental Wetlands—An Integrative Approach to Data Management and Ecosystem Prediction.* Report to U.S. EPA Region V, Contract 91044, Project 769641/725753.

Moerassen voor de Zuivering van Water (Wetlands for the Purification of Water). 1990. Post Academic Course. Published as The Utrecht Plant Ecology News Report. University of Utrecht, Lange Nieuwstraat 106, 3512 PN Utrecht, The Netherlands, No. 11. October 1990 (about 50% in English).

Montgomery, J.C. 1987. Institutional Factors Affecting Wastewater Discharge to Wetlands. pp. 889–894. in K.R. Reddy and W.H. Smith (Eds.), *Aquatic Plants for Water Treatment and Resource Recovery.* Orlando: Magnolia Publishing.

Moorhead, K.K. and K.R. Reddy. 1988. Oxygen transport through selected aquatic macrophytes. *J. Environ. Qual.,* 17:138.

Morea, S., R. Olsen, and T. Wilderman. 1990. Passive treatment technology cleans up Colorado mining waste. *Water Environ. Technol.* December: 6–9.

Morris, J.T. and K. Lajtha. 1986. Decomposition and nutrient dynamics of litter from four species of freshwater emergent macrophytes. *Hydrobiologia,* 131:215–223.

Moshiri, G.A., Ed. 1993. *Constructed Wetlands for Water Quality Improvement.* Boca Raton, FL: Lewis Publishers.

Murdroch, A. and J.A. Capobianco. 1979. Effects of treated effluent on a natural marsh. *J. Water Pollut. Control Fed.,* 51:2243–2256.

NADB (North American Treatment Wetland Database). 1993. Electronic database created by R. Knight, R. Ruble, R. Kadlec, and S. Reed for the U.S. Environmental Protection Agency. Copies available from Don Brown, U.S. EPA, (513) 569-7630.

David Nairne & Assoc. and NovaTec Consultants Inc. 1991. *Ross River Wetlands Environmental Impact Assessment.* Report to the Yukon Territorial Government, Whitehorse, Yukon, February, 1991.

Nakata, K. 1989. A simulation of the process of sedimentation of suspended solids in the Yoshii River estuary. *Hydrobiologia,* 176/177:431–438.

Nash, R. 1978. Who Loves a Swamp? pp. 149–156. in R.R. Johnson and J.F. McCormick (Eds.), *Strategies for Protection and Management of Floodplain Wetlands and Other Riparian Ecosystems.* Washington, DC: U.S. Department of Agriculture.

Nater, E.A. and D.F. Grigal. 1992. Regional trends in mercury distribution across the Great Lake states, north central USA. *Nature,* 358:139–141.

NCASI (National Council for Air and Stream Improvement, Inc.). 1978. Interfacial Velocity Effects on the Measurement of Sediment Oxygen Demand. NCASI Technical Bulletin No. 317.

NCASI (National Council for Air and Stream Improvement, Inc.). 1979. Further Studies of Sediment Oxygen Demand Measurement and Its Variability. NCASI Technical Bulletin No. 321.

NCASI (National Council for Air and Stream Improvement, Inc.). 1995. Experience With the Use of Constructed Wetland Effluent Treatment Systems in the Pulp and Paper Industry. National Council of the Paper Industry for Air and Stream Improvement. Gainesville, FL.

Navarra, G.A. 1992. Constructed Wetlands for Extensive Sewage Treatment in the Alps. Paper Presented at Intecol's IV International Wetlands Conference, Columbus, Ohio. September 1992.

Netter, R. 1994. Flow characteristics of planted soil filters. *Water Sci. Technol.,* 29(4):37–44.

Netter, R. and W. Bischofsberger. 1990. Hydraulic Investigations on Planted Soil Filters. pp. 11–20. in P.F. Cooper and B.C. Findlater (Eds.), *Constructed Wetlands in Water Pollution Control.* Oxford, U.K.: Pergamon Press.

Nichols, D.S. 1983. Capacity of natural wetlands to remove nutrients from wastewater. *Water Pollut. Control Fed. J.,* 55:495–505.

Niering, W.A. 1985. *Wetlands.* New York: Knopf.

Nixon, S.W. and V. Lee. 1986. Wetlands and Water Quality. A Regional Review of Recent Research in the United States on the Role of Freshwater and Saltwater Wetlands as Sources, Sinks, and Transformers of Nitrogen, Phosphorus, and Various Heavy Metals. Final Report Prepared for the Department of the Army, U.S. Army Corps of Engineers. Waterways Experiment Station. Technical Report Y-86-2.

NOAA (National Oceanic and Atmospheric Administration). Year. *Climatological Data.* A monthly publication. Asheville, NC: National Climatic Data Center.

Noland, L.E. and M. Gojdics. 1967. Ecology of free-living protozoa. *Res. Protozool.,* 2:215–266.

Noller, B.N., P.H. Woods, and B.J. Ross. 1994. Case studies of wetland filtration of mine waste water in constructed and naturally occurring systems in northern Australia. *Water Sci. Technol.* 29(4):257–265.

Nolte and Associates. 1983. *Marsh System Pilot Study Report, City of Gustine, CA.* EPA Project No. C-06-2824-010.

North American Wetlands Conservation Council (NAWCC). 1995. Wastewater Applications of Wetlands in Canada. Prepared by CH2M HILL Engineering LTD and Canadian Wildlife Service. Sustaining Wetlands Issues Paper No. 1994-1.

Novitzki, R.P. 1989. Wetland Hydrology. Chapter 5, pp. 47–64. in S.K. Majumdar, R.P. Brooks, F.J. Brenner, and R.W. Tiner (Eds.), *Wetlands Ecology and Conservation: Emphasis in Pennsylvania.* Easton, PA: The Pennsylvania Academy of Science.

Novotny, V. 1992. Unit pollutant loads. Their fit in abatement strategies. *Water Environ. Technol.,* January: 40–43.

Nriagu, J.O., Ed. 1979. *Copper in the Environment. Part I. Ecological Cycling.* New York: Wiley-Interscience.

Nriagu, J.O., Ed. 1980. *Cadmium in the Environment. Part I. Ecological Cycling.* New York: Wiley-Interscience.

Nriagu, J.O. and E. Nieboer, Eds. 1988. Chromium in the natural and human environments. *Adv. Environ. Sci. Technol.,* 20:1–571.

O'Connor, D.J. and W.E. Dobbins. 1958. Mechanism of reaeration in natural streams. *ASCE Trans.* Paper No. 2934:641–684.

Odum, E.P. 1971. *Fundamentals of Ecology.* 3rd Edition. Philadelphia, PA: W.B. Sanders.

Odum, E.P. 1989. Wetland Values in Retrospect. pp. 1–8. in R.R. Sharitz and J.W. Gibbons (Eds.), *Freshwater Wetlands and Wildlife.* Oak Ridge, TN: U.S. Department of Energy. CONF - 8603101.

Odum, H.T. 1978. Principles for Interfacing Wetlands with Development. pp. 29–56. in M.A. Drew (Ed.), A Symposium on Freshwater Wetlands. Sponsored by the Coordinating Council on the Restoration of the Kissimmee River Valley and Taylor Creek-Nubbin Slough Basin. February 28–March 2, 1978. Tallahassee, FL.

Odum, H.T. 1983. *Systems Ecology: An Introduction.* New York: John Wiley & Sons.

Odum, H.T. 1985. Self-Organization of Estuarine Ecosystems in Marine Ponds Receiving Treated Sewage. Data From Experimental Pond Studies at Morehead City, North Carolina, 1968–1972. A Data Report. University of North Carolina Sea Grant Publication #UNC-SG-85-04.

Odum W.E. and M.A. Heywood. 1978. Decomposition of Intertidal Freshwater Marsh Plants. pp. 89–97. in R.E. Good, D.F. Whigham, and R.L. Simpson, (Eds.), *Freshwater Wetlands: Ecological Processes and Management Potential.* New York: Academic Press.

Odum, H.T., K.C. Ewel, W.J. Mitsch, and J.W. Ordway. 1977. Recycling Treated Sewage Through Cypress Wetlands in Florida. Chapter 2, pp. 35–67. in F.M. D'Itri (Ed.), *Wastewater Renovation and Reuse.* New York: Marcel Dekker.

OECD. 1983. Emission Control Costs in the Metal Plating Industry. Organization for Economic Cooperation and Development.

Ogden, M.C. 1994. Personal communication.

Ohlendorf, H.M. 1992. Unpublished data from the Merced, CA constructed wetlands.

Ohlendorf, H.M., D.J. Hoffman, M.K. Saiki, and T.W. Aldrich. 1986. Embryonic mortality and abnormalities of aquatic birds: Apparent impacts of selenium from irrigation drainwater. *Sci. Total Environ.,* 52:49–63.

Ohlendorf, H.M., R.L. Hothem, C.M. Bunck, and K.C. Marois. 1990. Bioaccumulation of selenium in birds at Kesterson Reservoir, California. *Arch. Environ. Contam. Toxicol.,* 19:495–507.

Olson, R.K., Ed. 1992. The role of created and natural wetlands in controlling non-point source pollution. *Ecol. Eng.,* 1(1/2): 1–170.

Overcash, M.R., F.J. Humenik, and J.R. Miner. 1983. *Livestock Waste Management.* Volume I. Boca Raton, FL: CRC Press.

Owens, L.P., C.R. Hinkle, and G.R. Best. 1989. Low-Energy Wastewater Recycling Through Wetland Ecosystems: Copper and Zinc in Wetland Microcosms. pp. 1227–1235. in R.R. Sharitz and J.W. Gibbons (Eds.), *Freshwater Wetlands and Wildlife.* Oak Ridge, TN: U.S. Department of Energy. CONF - 8603101.

Palmer, J.F. and R.C. Smardon. 1988. Visual Amenity Value of Wetlands: An Assessment in Juneau, Alaska. pp. 104–107. in J.A. Kusler et al. (Eds.), *Urban Wetlands. Proceedings of the National Symposium.* Berne, NY: Association of Wetland Managers.

Pankow, J.F. 1991. *Aquatic Chemistry Concepts.* Chelsea, MI: Lewis Publishers.

Pano, A. and E.J. Middlebrooks. 1982. Ammonia nitrogen removal in facultative wastewater stabilization ponds. *Journal WPCF,* 54(4):344–351.

Parker, P.E. 1974. A Dynamic Ecosystem Simulator. Ph.D. Thesis. University of Michigan, Ann. Arbor.

Patrick, W.H., Jr. 1994. The effect of projected sea level rise on stability of coastal wetlands. Paper presented at the Third Symposium on the Biogeochemistry of Wetlands. Orlando, Florida, June, 1994.

Patrick, W.H., R.D. Delaune, R.M. Engler, and S. Gotoh. 1976. *Nitrate Removal from Water at the Water-Mud Interface in Wetlands.* EPA-600/3-76-042.

Payne, N.F. 1992. *Techniques for Wildlife Habitat Management of Wetlands.* New York: McGraw-Hill.

PBSJ (Post, Buckley, Schuh and Jernigan). 1989–1993. Orlando Easterly Wetlands: Monitoring Reports. 1988–92. Reports to City of Orlando, FL.

Peat Testing Manual. 1979. Technical Memorandum No. 125. National Research Council of Canada, Ottawa, Ontario.

Pederson, R.L. 1981. Seed Bank Characteristics of the Delta Marsh, Manitoba: Applications for Wetland Management. pp. 61–69. in B. Richardson (Ed.), *Selected Proceedings of the Midwest Conference on Wetland Values and Management.* St. Paul, MN: Freshwater Society.

Penman, H.L. 1948. Natural evapotranspiration from open-water, bare soil and grass. *Proc. Roy. Soc. Acad.,* 193:120–145.

Penman, H.L. 1956. Estimating evaporation. *Trans. Am. Geophys. Union,* 37:43–48.

Penman, H.L. 1963. *Vegetation and Hydrology.*: Commonw. Agric. Bur., Farnham Royal, U.K.

Pennak, R.W. 1978. *Freshwater Invertebrates of the United States.* 2nd Edition. New York: John Wiley & Sons.

Pennison, G.P. 1993a. *Microbial Rock Plant Filters in Louisiana Wastewater Treatment.* M.S. Thesis. Louisiana Tech University, Rushton, LA.

Pennison, G.P. 1993b. Microbial Rock Plant Filters in Louisiana Wastewater Treatment. *Water Environ. Fed.* 9:61–71.

Perry, E.W. and I. Garskof. 1989. Regulatory and Technical Constraints for Wetland Creation and Mitigation. Chapter 23, pp. 276–288. in S.K. Majumdar, R.P. Brooks, F.J. Brenner, and R.W. Tiner (Eds.), *Wetlands Ecology and Conservation: Emphasis in Pennsylvania.* Easton, PA: The Pennsylvania Academy of Science.

Perry, J.J., D.E. Armstrong, and D.D. Huff. 1981. Phosphorus Fluxes in an Urban Marsh During Runoff. pp. 199–211. in B. Richardson (Ed), *Selected Proceedings of the Midwest Conference on Wetland Values and Management.* St. Paul, MN: Minnesota Water Planning Board.

Perry, R.H. and D.W. Green. 1984. *Perry's Chemical Engineers' Handbook.* New York: McGraw-Hill.

Phillips, D.R., M.G. Messina, A. Clark, and D.J. Frederick. 1989. Nutrient concentration prediction equations for wetland trees in the US southern coastal plain. *Biomass.* 19:169–187.

Phipps, R.G. and W.G. Crumpton. 1994. Factors affecting nitrogen loss in experimental wetlands with different hydrologic loads. *Ecological Engineering,* 3(4):399–408.

Pickett, J., H. McKellar, and J. Kelley. 1989. Plant Community Composition, Leaf Mortality, and Aboveground Production in a Tidal Freshwater Marsh. pp. 351–364. in R.R. Sharitz and J.W. Gibbons (Eds.), *Freshwater Wetlands and Wildlife.* Oak Ridge, TN: U.S. Department of Energy. CONF - 8603101.

Pierce, G.J. 1989. Wetland Soils. Chapter 6, pp. 65–74. in S.K. Majumdar, R.P. Brooks, F.J. Brenner, and R.W. Tiner (Eds.), *Wetlands Ecology and Conservation: Emphasis in Pennsylvania.* Easton, PA: The Pennsylvania Academy of Science.

Pip, E. 1993. Cadmium, copper, and lead in wild rice from central Canada. *Arch. Environ. Contam. Toxicol.* 24:179–181.

PLA (P. Lane and Associates). 1992. Global Environmental Facility: Egyptian Engineered Wetlands. Volume I, Project Brief, Summary of Environmental Impact Assessment and Project Implementation. Report Prepared for UNDP.

Platzer, C. and R. Netter. 1992. Factors Affecting Nitrogen Removal in Horizontal Flow Reed Beds. pp. 4.1–4.6. in *Wetland Systems in Water Pollution Control.* Sydney, Australia: IAWQ.

Polprasert, C. and B.K. Agarwalla. 1994. A facultative pond model incorporating biofilm activity. *Water Environ. Res.,* 66(5):725–732.

Polprasert, C. and N.P. Dan. 1994. Phenol Removal in Model Constructed Wetlands Located in the Tropics. Vol. 8, pp. 45–55. in *Proceedings of the 67th Annual Conference of the Water Environment Federation.* Alexandria, VA: Water Environment Federation.

Poole, W.D. and V. Ngo. 1992. Lemna Systems for Wastewater Treatment in Different Climates pp. 30.1–30.10. in *Wetland Systems in Water Pollution Control.* Sydney, Australia: IAWQ.

Portier, R.J. and S.J. Palmer. 1989. Wetlands Microbiology: Form, Function, Processes. Chapter 6, pp. 89–105. in D.A. Hammer (Ed.), *Constructed Wetlands for Wastewater Treatment: Municipal, Industrial, and Agricultural.* Chelsea, MI: Lewis Publishers.

Pratt, D.C. and N.J. Andrews. 1981. Research in Biomass/Special Energy Crop Production in Wetlands. pp. 71–81. in B. Richardson (Ed.), *Selected Proceedings of the Midwest Conference on Wetland Values and Management.* St. Paul, MN: Freshwater Society.

Pratt, J.R. and T. Pluto. 1989. Use of Wetlands for the Treatment of Industrial and Nonpoint Source Pollutants. Chapter 26, pp. 315–332. in S.K. Majumdar, R.P. Brooks, F.J. Brenner, and R.W. Tiner (Eds.), *Wetlands Ecology and Conservation: Emphasis in Pennsylvania.* Easton, PA: The Pennsylvania Academy of Science.

Prentki, R.T., T.D. Gustafson, and M.S. Adams. 1978. Nutrient Movements in Lakeshore Marshes. pp. 169–194. in R.E. Good, D.F. Whigham, and R.L. Simpson (Eds.), *Freshwater Wetlands: Ecological Processes and Management Potential.* New York: Academic Press.

Prescott, G.W. 1951. *Algae of the Western Great Lakes Area.* Dubuque, IA: Wm. C. Brown Co.

Prost, A. 1987. Health risks stemming from wastewater reutilization. *Water Qual. Bull.,* 12(2):73–78.

Prouty, W.F. 1952. Carolina bays and their origin. *Geol. Soc. Am. Bull.,* 63:167–224.

Pullin, B.P. and D.A. Hammer. 1991. Aquatic plants improve wastewater treatment. *Water Environ. Technol.,* March: 36–40.

Qualls, R.G., C.J. Richardson, R. Johnson, and J. Zahina. 1994. Response of Everglades Slough Communities to Increased Concentrations of PO_4. Project Report to Everglades Agricultural Area Environmental Protection District, Duke Wetland Center.

Rabe, M.L. 1979. Impact Assessment of Wastewater Discharge upon a Northern Michigan Wetland Wildlife Community. Report to Michigan DNR.

Rabe, M.L. 1984. Impact of Wastewater Discharge upon a Northern Michigan Wetland Wildlife Community. Report to Michigan DNR.

Rabe, M.L. 1988. Impact of Wastewater Discharge Upon a Northern Michigan Wetland Wildlife Community. Report to Michigan DNR.

Rabe, M.L. 1990. Impact of Wastewater Discharge Upon a Northern Michigan Wetland Wildlife Community. Report to Michigan DNR.

Raven, P.H., R.F. Evert, and H. Curtis. 1981. *Biology of Plants.* 3rd Edition. New York: Worth Publishers.

Reader, R.J. 1978. Primary Production in Northern Bog Marshes. pp. 53–62. in R.E. Good, D.F. Whigham, and R.L. Simpson (Eds.), *Freshwater Wetlands: Ecological Processes and Management Potential.* New York: Academic Press.

Reddy, K.R. 1981. Land Areas Receiving Organic Wastes; Transformations and Transport in Relation to Nonpoint Source Pollution. pp. 243–274. in M.R. Overcash and J.M. Davidson (Eds.), *Environmental Impact of Nonpoint Source Pollution.* Ann Arbor, MI: Ann Arbor Science.

Reddy, K.R. 1982. Mineralization of nitrogen in organic soils. *Soil Sci. Soc. Am. J.,* 46(3):561–566.

Reddy, K.R. 1991. Phosphorus Sorption Capacity of Stream Sediments and Adjacent Wetlands. Preprint Extended Abstract. Presented Before the Division of Environmental Chemistry. American Chemical Society, Atlanta, GA. April 14–19, 1991. pp. 496–499.

Reddy, K.R. and D.A. Graetz. 1988. Carbon and Nitrogen Dynamics in Wetland Soils. pp. 307–318 in D.D. Hook and Others (Eds.), *The Ecology and Management of Wetlands.* London: Croom Helm.

Reddy, K.R. and E.M. D'Angelo. 1994. Soil Processes Regulating Water Quality in Wetlands. pp. 309–324. in W. Mitsch (Ed.), *Global Wetlands: Old World and New.* Amsterdam: Elsevier.

Reddy, K.R. and W.F. DeBusk. 1987. Nutrient Storage Capabilities of Aquatic and Wetland Plants. pp. 337–357. in K.R. Reddy and W.H. Smith (Eds.), *Aquatic Plants for Water Treatment and Resource Recovery.* Orlando, FL: Magnolia Publishing.

Reddy, K.R. and W.H. Patrick. 1984. Nitrogen transformations and loss in flooded soils and sediments. *CRC Crit. Rev. Environ. Control,* 13:273–309.

Reddy, K.R. and W.H. Smith (Eds.). 1987. *Aquatic Plants for Water Treatment and Resource Recovery.* Orlando, FL: Magnolia Publishing.

Reddy, K.R., W.F. DeBusk, Y. Wang, R. DeLaune, and M. Koch. 1991a. Physico-Chemical Properties of Soils in the Water Conservation Area 2 of the Everglades. Report to the South Florida Water Management District, West Palm Beach, FL.

Reddy, K.R., Y. Wang, L. Scinto, M.M. Fisher, and M. Koch. 1991b. Physico-Chemical Properties of Soils in the Holeyland Wildlife Management Area. Final Report to the South Florida Water Management District. Institute of Food and Agricultural Services, University of Florida, Gainesville, FL.

Reddy, K.R., R.D. DeLaune, W.F. DeBusk, and M. Koch. 1993. Long term nutrient accumulation rates in everglades wetlands. *Soil Sci. Soc. Am. J.*, Vol. 57, pp. 1147–1155.

Reed, S.C. 1984. *Nitrogen Removal in Wastewater Ponds.* CRREL Report 84-13, U.S. CRREL, Hanover, NH, June, 1984, 33 pp.

Reed, S.C. 1985. Nitrogen removal in wastewater stabilization ponds. *J. Water Pollut. Control Fed.,* 57(1):39–45.

Reed, S.C. 1990. An Inventory of Constructed Wetlands Used for Wastewater Treatment in the United States. Prepared for the U.S. EPA.

Reed, S.C. 1992. Subsurface Flow Constructed Wetlands for Wastewater Treatment—Status and Prospects. Presented at the International Specialist Conference on Wetland Systems in Water Pollution Control. Sydney, Australia. November 1992.

Reed, S.C. and D.S. Brown. 1992. Constructed wetland design—the first generation. *Water Environ. Res.,* 64(6):776–781.

Reed, S.C., R.W. Crites, and E.J. Middlebrooks. 1995. *Natural Systems for Waste Management and Treatment.* 2nd Ed. New York: McGraw-Hill.

Reeder, B.C. 1992. Personal communication.

Reid, G.K. and R.D. Wood. 1976. *Ecology of Inland Waters and Estuaries.* New York: D. Van Nostrand.

Reimold, R.J. and M.A. Hardisky. 1978. Nonconsumptive Use Values of Wetlands. in P.E. Greeson, J.R. Clark, and J.E. Clark (Eds.), *Wetland Functions and Values: The State of Our Understanding.* Minneapolis, MN: American Water Resources Association.

Rich, L.G. 1988. Dual power-level aerated lagoon systems. *Proc. 1988 Environmental Conference.* TAPPI, Charleston, SC, pp. 197–198.

Richardson, B., Ed. 1981. *Selected Proceedings of the Midwest Conference on Wetland Values and Management.* Navarre, MN: Freshwater Society.

Richardson, C.J. 1979. Primary Productivity Values in Freshwater Wetlands. pp. 131–145. in P.E. Greeson, J.R. Clark, and J.E. Clark (Eds.), *Wetland Functions and Values: The State of Our Understanding.* Minneapolis, MN: American Water Resources Association.

Richardson, C.J. 1985. Mechanisms controlling phosphorus retention capacity in freshwater wetlands. *Science.* 228:1424–1427.

Richardson, C.J. 1989. Freshwater Wetlands: Transformers, Filters, or Sinks? pp. 25–46. in R.R. Sharitz and J.W. Gibbons (Eds.), *Freshwater Wetlands and Wildlife.* Oak Ridge, TN: U.S. Department of Energy. CONF - 8603101.

Richardson, C.J. and C.B. Craft. 1990. Phase One: A Preliminary Assessment of Nitrogen and Phosphorus Accumulation and Surface Water Quality in Water Conservation Areas 2A and 3A of South Florida. Report to the Florida Sugar Cane League. Duke Wetland Center Publication 90-01.

Richardson, C.J. and C.B. Craft. 1993. Effective phosphorus retention in wetlands: fact or fiction? in: *Constructed Wetlands for Water Quality Improvement*, G.A. Moshiri (Ed.), Lewis Publishers, Boca Raton, FL, 271–282.

Richardson, C.J. and P.E. Marshall. 1986. Processes controlling movement, storage, and export of phosphorus in a fen peatland. *Ecol. Monogr.*, 56(4):279–302.

Richardson, C.J. and D.S. Nichols. 1985. Ecological Analysis of Wastewater Management Criteria in Wetland Ecosystems. Chapter 24, pp. 351–391. in P.J. Godfrey et al. (Eds.), *Ecological Considerations in Wetlands Treatment of Municipal Wastewaters.* New York: Van Nostrand Reinhold Company.

Richardson, C.J., D.L. Tilton, J.A. Kadlec, J.P.M. Chamie, and W.A. Wentz. 1978. Nutrient Dynamics of Northern Wetland Ecosystems. in R.E. Good, D.F. Whigham, and R.L. Simpson (Eds.), *Freshwater Wetlands: Ecological Processes and Management Potential.* New York: Academic Press.

Rijs, G.B.J. and S. Veenstra. 1990. Artificial Reed Beds as Post Treatment for Anaerobic Effluents—Urban Sanitation in Developing Countries. pp. 583–586 in P.F. Cooper and B.C. Findlater (Eds.), *Constructed Wetlands in Water Pollution Control.* Oxford, U.K.: Pergamon Press.

Rittman, B.E. and P.L. McCarty. 1980. Evaluation of steady state biofilm kinetics. *Biotechnol. Bioeng.*, 22:2359.

Rivera, F., A. Warren, E. Ramirez, O. Decamp, P. Bonilla, E. Gallegos, A. Calderon, and J.T. Sanchez. 1994. Removal of Pathogens from Wastewaters by the Root Zone Method (RZM). pp. 180–189. in *Proceedings of the Fourth International Conference on Wetland Systems for Water Pollution Control.* Guangzhou, P.R. China: Center for International Development and Research, South China Institute for Environmental Sciences.

RMG (Resource Management Group, Inc.). 1992. *National List of Plant Species That Occur in Wetlands for USF&WS Region 3 (Includes MI, IN, IL, MO, IA, WI, MN).* Grand Haven, MI: Resource Management Group, Incs.

Rodgers, J.H. and A. Dunn. 1992. Developing design guidelines for constructed wetlands to remove pesticides from agricultural runoff, in: R.K. Olson (Ed.). *Created and Natural Wetlands for Controlling Nonpoint Source Pollution.* C.K. Smoley. Boca Raton, FL. 113–125.

Rogers, K.H., P.F. Breen and A.J. Chick. 1991. Nitrogen removal in experimental wetland treatment systems: evidence for the role of aquatic plants. *Research Journal WPCF.* 63(7):934–941.

Roig, L.C. and I.P. King. 1993. Continuum model for flows in emergent marsh vegetation. *J. Hydral. Eng.*, submitted.

Rosendahl, P.C. 1981. The Determination of Manning's Roughness and Longitudinal Dispersion Coefficients in the Everglades Marsh. Paper Presented at the Symposium on Progress in Wetlands Utilization and Management, Orlando, FL. June 1981.

Roser, D.J., S.A. McKersie, P.J. Fisher, P. Breen, and H.J. Bavor. 1987. Sewage treatment using aquatic plants and artificial wetlands. *Water,* 14(3):20.

Rosman, L. 1978. Impact Assessment of Northern Michigan Wetland Invertebrate and Vertebrate Fauna Receiving Secondarily Treated Sewage Effluent. pp. 38–85. in R.H. Kadlec et al. (Eds.), *First Annual Operations Report: Houghton Lake Wetland Treatment Project.*

Ross, H.B. 1987. Trace metals in precipitation in Sweden. *Water Air Soil Pollut.,* 36:349–363.

Roulet, N.T. and M.K. Woo. 1986. Wetland and lake evaporation in the low arctic. *Arct. Alp. Res.,* 18:195–200.

Rudd, T. 1987. Scope of the Problem. Chapter 1, pp. 1–29. in J.N. Lester (Ed.), *Heavy Metals in Wastewater and Sludge Treatment Processes. Volume I Sources, Analysis, and Legislation.* Boca Raton, FL: CRC Press.

Rusincovitch, F. 185. Use of wetlands for wastewater treatment and effluent disposal: institutional constraints, in: P.J. Godfrey, E.R. Kaynor, S. Pelczarski, J. Benforado (Eds.). *Ecological Considerations in Wetlands Treatment of Municipal Wastewaters.* Van Nostrand Reinhold, New York, NY, 427–432.

Ruthven, D.M. 1984. *Principles of Adsorption and Adsorption Processes.* New York: John Wiley and Sons.

S.C. DHEC. 1992. Grant for Development of a Wetlands Classification and Standards System, Effluent Criteria, and Criteria for Other Activities in Wetlands. Final Report. S.C. Department of Health and Environmental Control. Columbia, S.C.

Salisbury, F.B. and C.W. Ross. 1978. *Plant Physiology.* 2nd Edition. Belmont, CA: Wadsworth Publishing Co.

Sand-Jensen, K., C. Prahl, and H. Stokholm. 1982. Oxygen release from roots of submerged aquatic macrophytes. *Oikos,* 38:349.

Sanders, F.S. and R.W. Brocksen. 1991. Constructed Wetlands for Remediation of Metal-Bearing Aqueous Waste Streams. Project Report to Electric Power Research Institute. EPRI Contract No. RP2485-24.

Sanford, W.E., T.S. Steenhuis, J.Y. Parlange, J.M. Surface, and J.H. Peverly. 1995. Hydraulic performance of rock-reed filters. *Ecol. Eng.,* 4(4):321–336.

Sather, J.H. 1989. Chapter 28a. Ancillary Benefits of Wetlands Constructed Primarily for Wastewater Treatment. pp. 353–358. in D.A. Hammer (Ed.), *Constructed Wetlands for Wastewater Treatment, Municipal, Industrial, and Agricultural.* Chelsea, MI: Lewis Publishers.

Sather, J.H. and R.D. Smith. 1984. *An Overview of Major Wetland Functions and Values.* U.S. Fish and Wildlife Service. FWS/OBS-84/18.

Saurer, B., Ed. 1994. *Pflanzenkläranlagen: Abwasserreinigung mit Bepflantzen Bodenkörpern.* Wasserwirtschaft Land Steiermark, Graz, Austria. (in German).

Savage, H. 1983. *The Mysterious Carolina Bays.* Columbia, SC: University of South Carolina Press.

Schalles, J.F. 1989. The Chemical Environment of Wetlands. Chapter 7, pp. 75–92. in S.K. Majumdar, R.P. Brooks, F.J. Brenner, and R.W. Tiner (Eds.), *Wetlands Ecology and Conservation: Emphasis in Pennsylvania.* Easton, PA: The Pennsylvania Academy of Science.

Scheffe, R.D. 1978. Estimation and Prediction of Summer Evapotranspiration from a Northern Wetland. M.S. Thesis. University of Michigan, Ann Arbor, MI.

Scheuerman, P.R., G. Bitton, and S.R. Farrah. 1989. Fate of Microbial Indicators and Viruses in a Forested Wetland. Chapter 391, pp. 657–663. in D.A. Hammer (Ed.), *Constructed Wetlands for Wastewater Treatment: Municipal, Industrial, and Agricultural.* Chelsea, MI: Lewis Publishers.

Schierup, H.H., H. Brix, and B. Lorenzen. 1990a. Wastewater Treatment in Constructed Reed Beds in Denmark—State of the Art. pp. 495–504. in P.F. Cooper and B.C. Findlater (Eds.), *Proceedings of the International Conference on the Use of Constructed Wetlands in Water Pollution Control.* Oxford, U.K.: Pergamon Press.

Schierup H.H., H. Brix, and B. Lorenzen. 1990b. *Spildevandsrensning i rodzoneanlœg.* Botanical Institute, Aarhus University, Denmark.

Schiffer, D.M. 1989. Water-Quality Variability in a Central Florida Wetland Receiving Highway Runoff. in F.E. Davis (Ed.), *Water: Laws and Management.* Bethesda, MD: American Water Research Association.

Schlesinger, W.H. 1978. Community structure, dynamics, and nutrient ecology in the Okefenokee cypress swamp forest. *Ecol. Monogr.,* 48:43–65.

Schmitt, C.J. and W.G. Brumbaugh. 1990. National contaminant biomonitoring program: Concentrations of arsenic, cadmium, copper, lead, mercury, selenium, and zinc in freshwater fish, 1976–1984. *Arch. Environ. Contam. Toxicol.,* 19:731–747.

Schnitzer, M. 1986. Water retention by humic substances, in: *Peat and Water,* C.H. Fuchsman (Ed.), Elsevier, New York, 159–176.

Schueler, T.R. 1992. *Design of Stormwater Wetland Systems: Guidelines for Creating Diverse and Effective Stormwater Wetlands in the Mid-Atlantic Region.* Metropolitan Washington Council of Governments. Washington, DC. 133 pp.

Schuler, C.A., R.G. Anthony, and H.M. Ohlendorf. 1990. Selenium in wetlands and waterfowl foods at Kesterson Reservoir, California, 1984. *Arch. Environ. Contam. Toxicol.,* 19:845–853.

Schwartz, A.L. and R.L. Knight. 1989. Some Ancillary Benefits of a Natural Land Treatment System. Chapter 39j, pp. 643–645. in D.A. Hammer (Ed.), *Constructed Wetlands for Wastewater Treatment. Municipal, Industrial, and Agricultural.* Chelsea, MI: Lewis Publishers.

Schwartz, L.N. 1987. Regulation of Wastewater Discharge to Florida Wetlands. pp. 951–958. in K.R. Reddy and W.H. Smith (Eds.), *Aquatic Plants for Water Treatment and Resource Recovery.* Orlando, FL: Magnolia Publishing.

Schwartz, L.N. 1989. Regulation of Wastewater Discharge to Florida Wetlands. pp. 909–917. in R.R. Sharitz and J.W. Gibbons (Eds.), *Freshwater Wetlands and Wildlife.* Oak Ridge, TN: U.S. Department of Energy. CONF - 8603101.

Schwartz, L.N., J.G. Ladner, and S.J. Keely. 1993. Orange County Florida landfill dilute leachate wetland treatment and restoration system in: H.W. Shen, F.T. Su, and F. Wenu (Eds.), *Proceedings of the Conference on Hydraulic Engineering*, pp. 287–292. American Soc. Civil Eng. NY, NY.

Schwartz, L.N., P.M. Wallace, P.M. Gale, W.F. Smith, J.T. Wittig, and S.L. McCarty. 1994. Orange County Florida Eastern Service Area Reclaimed Water Wetlands Reuse System, *Water Science and Technology*, 29(4):273–282.

Schwegler, B.R. 1978. Effects of Sewage Effluent on Algal Dynamics of a Northern Michigan Wetland. M.S. Thesis. The University of Michigan, Ann Arbor.

Scott, W.B. and E.J. Crossman. 1973. *Freshwater Fishes of Canada.* Bulletin 184. Fisheries Research Board of Canada, Ottawa.

Seidel, K. 1966. Reinigung von Gewässern durch höhere Pflanzen. *Naturwissenschaften,* 53:289–297.

Seidel, K. 1976. Macrophytes and Water Purification. Chapter 14. in J. Tourbier and R.W. Pierson, Jr. (Eds.), *Biological Control of Water Pollution.* Philadelphia, PA: University of Pennsylvania Press.

Seo, I.W. 1990. Laboratory and numerical investigation of longitudinal dispersion in open channels. *Water Resour. Bull.,* 26(5):811–822.

Shacklette, H.T. and J.G. Boerngen. 1984. Element Concentrations in Soils and Other Surficial Materials of the Coterminus United States. USGS Prof. Paper 1270. U.S. Government Printing Office.

Shanker, R. and R.H. Kadlec. 1993. Unpublished data.

Sharitz, R.R. and J.W. Gibbons, Eds. 1989. *Freshwater Wetlands and Wildlife.* Oak Ridge, TN: U.S. Department of Energy. CONF - 8603101.

Shelley, P.E. and D.R. Gaboury. 1986. Estimation of Pollution from Highway Runoff—Initial Results. pp. 459–473. in B. Urbonas and L.A. Roesner (Eds.), *Urban Runoff Quality—Impact and Quality Enhancement Technology.* New York: American Society of Civil Engineers.

Shiaris, M.P. 1985. Public Health Implications of Sewage Applications on Wetlands: Microbiological Aspects. Chapter 16, pp. 243–261. in P.J. Godfrey et al. (Eds.), *Ecological Considerations in Wetlands Treatment of Municipal Wastewaters.* New York: Van Nostrand Reinhold.

Shih, S.F. and G.S. Rahi. 1982. Seasonal variations of Manning's roughness coefficient in a subtropical marsh. *Trans. ASAE,* 25:116–120.

Shih, S.F., A.C. Federico, J.F. Milleson, and M. Rosen. 1979. Sampling programs for evaluating upland marsh to improve water quality. *Trans. ASAE,* 22:828–833.

Shotyk, W. 1987. European Contributions to the Geochemistry of Peatland Waters, 1890–1940. pp. 115–125. in C.D.A. Rubec and R.P. Overend (Eds.), *Wetlands/Peatlands.* Ottawa: Environment Canada.

Siegel, R.A., J.T. Collins, and S.S. Novak, Eds. 1987. *Snakes: Ecology and Evolutionary Biology.* Toronto, Canada: MacMillan Publishing Co.

Sikora, F.J., T. Zhu, L.L. Behrends, S.L. Steinberg, H.S. Coonrod, and L.G. Softley. 1994. Ammonium and phosphorus removal in constructed wetlands with recirculating subsurface flow: removal rates and mechanisms, in: *Proceedings of the Fourth International Conference on Wetland Systems for Water Pollution Control*, Guangzhou, China, 147–161.

Simpson, R.L., Ed. 1978. *Freshwater Wetlands: Ecological Processes and Management Potential.* pp. 217–242. New York: Academic Press.

Simpson, R.L., R.E. Good, R. Walker, and B.R. Frasco. 1983. The role of Delaware River freshwater tidal wetlands in the retention of nutrients and heavy metals. *J. Environ. Qual.,* 12:41–48.

Singer, A. 1966. *Birds of North America.* Western Publishing Co.

Sinicrope, T.L., R. Langis, R.M. Gersberg, M.J. Busanardo, and J.B. Zedler. 1992. Metal removal by wetland mesocosms subjected to different hydroperiods. *Ecol. Eng.,* 1(4):309–322.

Skipper, D. and M. Tittlebaum. 1991. A Computer Model for Predicting Effluent BOD Concentrations from a Rock-Plant Filter. Preprint #AC91-046-002. 64th Annual Conference, Toronto. Water Environment Federation, Alexandria, VA.

Slayden, R.L. and L.N. Schwartz. 1989. States' Activities, Attitudes and Policies Concerning Constructed Wetlands for Wastewater Treatment. Chapter 23, pp. 279–286. in D.A. Hammer (Ed.), *Constructed Wetlands for Wastewater Treatment: Municipal, Industrial, and Agricultural.* Chelsea, MI: Lewis Publishers.

Small, M. and C. Wurm. 1977. Data Report. Meadow/Marsh/Pond System. Brookhaven National Laboratory, BNL 50675.

Small, M.M. 1978. Artificial Wetlands as Non-Point Source Wastewater Treatment Systems. pp. 171–180. in M.A. Drew (Ed.), *A Symposium on Freshwater Wetlands.* Sponsored by the Coordinating Council on the Restoration of the Kissimmee River Valley and Taylor Creek-Nubbin Slough Basin. February 28–March 2, 1978. Tallahassee, FL.

Smardon, R.C. 1988. Aesthetic, Recreational, Landscape Values of Urban Wetlands. pp. 92–95. in J.A. Kusler, S. Daly, and G. Brooks (Eds.), *Proceedings of the National Wetland Symposium on Urban Wetlands.* Berne, NY: Association of Wetland Managers.

Smith, K.L., G.T. Rowe, and S.A. Nichols. 1973. Benthic community respiration near the Woods Hole sewage outfall. *Estuarine Coastal Marine Sci.,* 1:65–70.

Smith, L.M., R.L. Pederson, and R.M. Kaminski. 1989. *Habitat Management for Migrating and Wintering Waterfowl in North America.* Lubbock, TX: Texas Tech University Press.

R.A. Smith and Associates. 1994. Wetlands to Improve Wind Lake Water Quality. Report to Wind Lake Protection Association. December 1994.

South, G.R. and A. Whittick. 1987. *Introduction to Phycology.* Oxford, U.K.: Blackwell Science Publishers.

Spalding, M.G. 1990. Antemortem diagnosis of eustrongylidosis in wading birds (Ciconiiformes). *Colonial Waterbirds,* 13:75–77.

Spangler, F.L., W.E. Sloey, and C.W. Fetter, Jr. 1976. Wastewater Treatment by Natural and Artificial Marshes. Report to the Robert S. Kerr. Environmental Research Laboratory, Ada, OK. EPA-600/2-76-207.

SRI (Scientific Resources, Inc.). 1990. Report on Monitoring of Jackson Bottom Experimental Wetland, 1989. Prepared for Unified Sewerage Agency, Hillsboro, OR.

SRI (Scientific Resources, Inc.). 1991. Operating data from Jackson Bottom Experimental Wetland, 1990.

Srinivasan, K.R. and R.H. Kadlec. 1995. Wetland Treatment of Oil and Gas Well Wastewaters. Report to U.S. Department of Energy. Contract DE-AC22-92MT92010.

Stairs, D.B. 1993. Flow Characteristics of Constructed Wetlands: Tracer Studies of the Hydraulic Regime. M.S. Thesis. Oregon State University, Corvallis, OR.

Staubitz, W.W., J.M. Surface, T.S. Steenhuis, J.H. Peverly, M.J. Lavine, N.C. Weeks, W.E. Sanford, and R.J. Kopka. 1989. Potential Use of Constructed Wetlands to Treat Landfill Leachate. Chapter

41c, pp. 735–742. in D.A. Hammer (Ed.), *Constructed Wetlands for Wastewater Treatment: Municipal, Industrial, and Agricultural.* Chelsea, MI: Lewis Publishers.

Stebbins, R.C. 1966. *A Field Guide to Western Reptiles and Amphibians.* The Peterson Field Guide Series. Boston, MA: Houghton Mifflin.

Steffeck, D.W. and J.A. Blankenship. 1989. Status of Bioassessment Activities in the U.S. Fish and Wildlife Service. pp. 3A-17-3A-29. in F.E. Davis (Ed.), *Water: Laws and Management.* Bethesda, MD: American Water Research Association.

Steiner, G.R. and R.J. Freeman, Jr. 1989. Configuration and Substrate Design Considerations for Constructed Wetlands Wastewater Treatment. Chapter 29, pp. 363–377. in D.A. Hammer (Ed.), *Constructed Wetlands for Wastewater Treatment: Municipal, Industrial, and Agricultural.* Chelsea, MI: Lewis Publishers.

Steiner, G.R., J.T. Watson, D.A. Hammer, and D.F. Harker. 1987. Municipal Wastewater Treatment with Artificial Wetlands. pp. 923–932. in K.R. Reddy and W.H. Smith (Eds.), *Aquatic Plants for Water Treatment and Resource Recovery.* Orlando, FL: Magnolia Publishing.

Stengel, E. 1993. Species-Specific Aeration of Water by Different Vegetation Types in Constructed Wetlands. pp. 427–434. in G.A. Moshiri (Ed.), *Constructed Wetlands for Water Quality Improvement.* Boca Raton, FL: Lewis Publishers.

Stengel, E., W. Carduck, and C. Jebsen. 1987. Evidence for Denitrification in Artificial Wetlands. pp. 543–550. in K.R. Reddy and W.H. Smith (Eds.), *Aquatic Plants for Water Treatment and Resource Recovery.* Orlando, FL: Magnolia Publishing.

Stephens, J.C. 1956. Subsidence of organic soils in the Floridan Everglades. *Soil Sci. Soc. Am. Proc.,* 20:77–80.

Stephenson, M., G. Turner, P. Pope, J. Colt, A. Knight, and G. Tchobanoglous. 1980. The Use and Potential of Aquatic Species for Wastewater Treatment. Appendix A. The Environmental Requirements of Aquatic Plants. Publication No. 65. California State Water Resources Control Board, Sacramento, CA.

Stephenson, T. 1987. Water Treatment and Reuse. Chapter 4, pp. 69–90. in J.N. Lester (Ed.), *Heavy Metals in Wastewater and Sludge Treatment Processes. Volume II Treatment and Disposal.* Boca Raton, FL: CRC Press.

Stiven, A.E. and J.T. Hunter. 1976. Growth and mortality of *Littorina irrorata* (Say) in three North Carolina marshes. *Chesapeake Sci.,* 17:168–176.

Stoeppler, M. and M. Piscator, Eds. 1988. *Cadmium.* New York: Springer-Verlag.

Stowell, R., S. Weber, G. Tchobanoglous, B.A. Wilson, and K.R. Townzen. 1985. Mosquito Considerations in the Design of Wetland Systems for the Treatment of Wastewater. Chapter 3, pp. 38–47. in P.J. Godfrey et al. (Eds.), *Ecological Considerations in Wetlands Treatment of Municipal Wastewaters.* New York: Van Nostrand Reinhold.

Strecker, E., G. Palhegyi, E. Driscoll, J. Duffield, and R. Horner. 1990. The Use of Wetlands for Controlling Stormwater Pollution. Final Report. Prepared by Woodward-Clyde Consultants, Seattle, WA.

Strecker, E.W., J.M. Kersnar, E.D. Driscoll, and R.R. Homer. 1992a. *The Use of Wetlands for Controlling Stormwater Pollution.* Washington, DC: Terrene Institute. EPA/600.

Strecker, E.W., R.R. Horner, and T.E. Davenport. 1992b. *The Use of Wetlands for Controlling Stormwater Pollution.* Washington, DC: The Terrene Institute. 66 pp.

Suberkropp, K., A. Michelis, H.J. Lorch, and J.C.G. Ottow. 1988. Effect of sewage treatment plant effluents on the distribution of aquatic hyphomycetes in the River Erms, Schwabische Alb, F.R.G. *Aquat. Bot.,* 32:141–153.

Surface, J.M., J.H. Peverly, T.S. Steenhuis, and W.E. Sanford. 1993. Effect of Season, Substrate Composition, and Plant Growth on Landfill-Leachate Treatment in a Constructed Wetland. Chapter 50, pp. 461–472. in G.A. Moshiri (Ed.), *Constructed Wetlands for Water Quality Improvement.* Boca Raton, FL: Lewis Publishers.

Sutherland, J.C. 1982. Michigan Wetland Wastewater Tertiary Treatment Systems. Chapter 16. in E.J. Middlebrooks (Ed.), *Water Reuse.* Ann Arbor, MI: Ann Arbor Science Publishers.

Sutherland, J.C. and F.B. Bevis. 1979. Reuse of Municipal Wastewater by Volunteer Fresh-Water Wetlands. Report to the National Science Foundation, Washington, DC.

Sutherland, J.C. and R.H. Kadlec, Eds. 1979. Freshwater Wetlands and Sanitary Wastewater Disposal. Conference Abstracts. Higgins Lake, MI.

Suzuki, T., K. Moriyama, and Y. Kurihara. 1989. Distribution of heavy metals in a reed marsh on a riverbank in Japan. *Aquat. Bot.,* 35:121–127.

Swindell, C.E. and J.A. Jackson. 1990. Constructed Wetlands Design and Operation to Maximize Nutrient Removal Capabilities. pp. 107–114. in P.F. Cooper and B.C. Findlater (Eds.), *Proceedings of the International Conference on the Use of Constructed Wetlands in Water Pollution Control.* Oxford, U.K.: Pergamon Press.

Tang, S-y. and X-w. Lu. 1993. The use of *Eichhornia crassipes* to cleanse oil-refinery wastewater in China. *Ecol. Eng.,* 2:243–251.

Tanner, C.C. 1992. Treatment of dairy farm wastewaters in horizontal and up-flow gravel bed constructed wetlands. *Water Sci. Technol.,* 29(4):85–93.

Tanner, C.C. and J.P. Sukias. 1994. Accumulation of Organic Solids in Gravel-Bed Constructed Wetlands. pp. 617–627. in *Proceedings of the Fourth International Conference on Wetland Systems for Water Pollution Control.* Guangzhou, P.R. China.

Tanner, C.C., J.S. Clayton, and M.P. Updell. 1995. Effect of loading rate and planting on treatment of dairy farm wastewaters in constructed wetlands. I. Removal of oxygen demand, suspended solids and fecal coliforms. *Water Res.,* 291(1):17–26.

Tennessen, K.J. 1993. Production and Suppression of Mosquitoes in Constructed Wetlands. pp. 591–601. in G.A. Moshiri (Ed.), *Constructed Wetlands for Water Quality Improvement.* Boca Raton, FL: Lewis Publishers.

TETC (The Earth Technology Corporation). 1991. Evaluation of Background Metals Concentrations in Arizona Soils. Prepared for Arizona Department of Environmental Quality, Groundwater Hydrology Section, Phoenix, AZ.

Thabaraj, G.J. 1982. Wastewater Discharge to Wetlands: Regulatory Aspects. Paper Presented to Florida Wastewater Management Seminar.

Thackson, E.L., F.D. Shields, Jr., and P.R. Schroeder. 1987. Residence time distributions of shallow basins. *J. Environ. Eng.,* 113(6):1319–1332.

Thom, B.G. 1970. Carolina bays in Horry and Marion Counties, South Carolina. *Geol. Soc. Am. Bull.,* 81:783–814.

Thomann, R.V. 1972. Systems Analysis and Water Quality Measurement. Environmental Research and Applications, Inc., NY.

Thunhorst, G.A. 1993. *Wetland Planting Guide for the Northeastern United States. Plants for Wetland Creation, Restoration, and Enhancement.* St. Michaels, MD: Environmental Concern.

Thut, R.N. 1989. Utilization of Artificial Marshes for Treatment of Pulp Mill Effluents. Chapter 19, pp. 239–244. in D.A. Hammer (Ed.), *Constructed Wetlands for Wastewater Treatment. Municipal, Industrial and Agricultural.* Chelsea, MI: Lewis Publishers.

Thut, R.N. 1993. Feasibility of Treating Pulp Mill Effluent with a Constructed Wetland. pp. 441–447. in G.A. Moshiri (Ed.), *Constructed Wetlands for Water Quality Improvement.* Boca Raton, FL: Lewis Publishers.

Tilton, D.L., R.H. Kadlec, and C.J. Richardson, Eds. 1976. Freshwater Wetlands and Sewage Effluent Disposal. *Proceedings of NSF/RANN Conference.* Ann Arbor, MI: The University of Michigan. NTIS PB259305.

Tiner, R.W. 1984. *Wetlands of the United States: Current Status and Recent Trends.* Washington, DC: U.S. Department of the Interior, Fish and Wildlife Service.

Tomljanovich, D.A. and O. Perez. 1989. Constructing the Wastewater Treatment Wetland—Some Factors to Consider. Chapter 32, pp 399–402. in D.A. Hammer (Ed.), *Constructed Wetlands for Wastewater Treatment. Municipal, Industrial, and Agricultural.* Chelsea, MI: Lewis Publishers.

Toth, L.A. 1990. Effects of Hydrologic Regimes on Lifetime Production and Nutrient Dynamics of Cattail. Technical Publication 88-6. South Florida Water Management District, West Palm Beach, Florida.

Trautmann, N.M., J.H. Martin, Jr., K.S. Porter, and K.C. Hawk, Jr. 1989. Use of Artificial Wetlands for Treatment of Municipal Solid Waste Landfill Leachate. Chapter 20, pp. 245–251. in D.A. Hammer (Ed), *Constructed Wetlands for Wastewater Treatment: Municipal, Industrial, and Agricultural.* Chelsea, MI: Lewis Publishers.

Tupacz, E.G. and F.P. Day. 1990. Decomposition of roots in a seasonally flooded swamp ecosystem. *Aquat. Bot.,* 37:199–214.

Tuschall, J.R., Jr. 1981. Heavy Metal Complexation with Naturally Occurring Organic Ligands in a Wetland Ecosystem. Ph.D. Thesis. University of Florida, Gainesville.

TVA (Tennessee Valley Authority). 1989. Unpublished data.

TVA (Tennessee Valley Authority). 1990a. Unpublished data from Benton, KY and Pembroke, KY.

TVA (Tennessee Valley Authority). 1990b. Unpublished dye study results from September 1990 at Benton KY.

TVA (Tennessee Valley Authority). 1993. General Design, Construction, and Operation Guidelines: Constructed Wetlands Wastewater Treatment System for Small Users Including Individual Residences. Chattanooga, TN: TVA. Technical Report TVA/MW—93/10.

U.S. Bureau of Reclamation (USBR). 1993. *Drainage Manual. A Water Resources Technical Publication.* U.S. Dept. of the Interior. 321 pp.

U.S. Environmental Protection Agency. 1969. Secondary Treatment of Potato Processing Wastes. Report 12060-07/69. Prepared by K. J. Dostal for the Water Quality Office, U.S. EPA, Washington, DC.

U.S. Environmental Protection Agency. 1975. *Process Design Manual for Nitrogen Control.* EPA/625/1-75/007.

U.S. Environmental Protection Agency. 1978. Analysis of Operations & Maintenance Costs for Municipal Wastewater Treatment Systems. Office of Water Programs (WH-547), Washington, DC. EPA 430/9-77-015.

U.S. Environmental Protection Agency. 1980. *Design Manual: Onsite Wastewater Treatment and Disposal Systems.* EPA 625/1-80-012.

U.S. Environmental Protection Agency. 1981. *Process Design Manual for Land Treatment of Municipal Wastewater.* EPA 625/1-81-013.

U.S. Environmental Protection Agency. 1983a. *The Effects of Wastewater Treatment Facilities on Wetlands in the Midwest.* U.S. EPA 905/3-83-002.

U.S. Environmental Protection Agency. 1983b. Construction Costs for Municipal Wastewater Treatment Plants: 1973–1982. Office of Water Program Operations, Washington, DC. EPA/430/9-83-004.

U.S. Environmental Protection Agency. 1983c. Freshwater Wetlands for Wastewater Management. Region IV Environmental Impact Statement. Phase 1 Report. EPA 904/9-83-107.

U.S. Environmental Protection Agency. 1983d. *Results of the Nationwide Urban Runoff Program.* Volume I. Final Report. Water Planning Division, U.S. EPA, NTIS No. P884-185552, Washington, DC.

U.S. Environmental Protection Agency. 1984a. *Process Design Manual for Land Treatment of Municipal Wastewater. Supplement on Rapid Infiltration and Overland Flow.* U.S. Environmental Research Information, Cincinnati, OH. EPA 625/1-81-013a. 121 pp.

U.S. Environmental Protection Agency. 1984b. *The Ecological Impacts of Wastewater on Wetlands: An Annotated Bibliography.* U.S. EPA 905/3-84-002.

U.S. Environmental Protection Agency. 1985a. Rates, Constants, and Kinetics Formulations in Surface Water Quality Modeling, Second Edition. Report EPA/600/3-85/040. Environmental Research Laboratory, Athens, GA.

U.S. Environmental Protection Agency. 1985b. *Freshwater Wetlands for Wastewater Management Handbook.* USEPA 904/9-85-135.

U.S. Environmental Protection Agency. 1986a. National Water Quality Inventory: Report to Congress. U.S. EPA, Washington, DC.

U.S. Environmental Protection Agency. 1986b. Quality Criteria for Water 1986 (Gold Book). U.S. EPA Office of Regulations and Standards. Washington, DC.

U.S. Environmental Protection Agency. 1986c. *Superfund Public Health Evaluation Manual.* EPA 540/1-86/060.

U.S. Environmental Protection Agency. 1987. Report on the Use of Wetlands for Municipal Wastewater Treatment and Disposal. EPA Report 430/09-88-005.

U.S. Environmental Protection Agency. 1988. Design Manual. Constructed Wetlands and Aquatic Plant Systems for Municipal Wastewater Treatment. Office of Research and Development. Center for Environmental Research Information. Cincinnati, OH. EPA/625/1-88/022.

U.S. Environmental Protection Agency. 1989a. Focus on Nonpoint Source Pollution. The Information Broker, Office of Water Regulations and Standards, Nonpoint Sources Control Branch, Washington, DC.

U.S. Environmental Protection Agency. 1989b. Proposed Rule: Standards for Disposal of Sewage Sludge. Part 503 Regulations. *Federal Register.*

U.S. Environmental Protection Agency. 1989c. Nonpoint sources. Agenda for the Future. Office of Water (WH-556), Washington, DC.

U.S. Environmental Protection Agency. 1989d. Water Quality and Toxic Assessment Study, Mangrove Preserve, Munisport Landfill Site, North Miami, Florida. Environmental Services Division, Athens, GA.

U.S. Environmental Protection Agency. 1989e. Short-Term Methods for Estimating the Chronic Toxicity of Effluents and Receiving Waters to Freshwater Organisms. Second Edition. U.S. EPA, Environmental Monitoring and Support Laboratory, Cincinnati, OH. EPA/600/4-89/001.

U.S. Environmental Protection Agency. 1990a. Managing Nonpoint Source Pollution: Final Report to Congress on Section 319 of the Clean Water Act (1989). EPA/506/9-90. U.S. EPA, Washington, DC.

U.S. Environmental Protection Agency. 1990b. Biological Criteria. National Program Guidance for Surface Waters. Criteria and Standards Division. EPA-440/5-90-004.

U.S. Environmental Protection Agency. 1993a. Toxic Substances Spreadsheet. Updated by Region IV Water Quality Standards Section. Section 304(a) of the Clean Water Act.

U.S. Environmental Protection Agency Office of Water. 1993b. Subsurface Flow Constructed Wetlands for Wastewater Treatment: A Technology Assessment. EPA Report 832-R-93-001.

U.S. Environmental Protection Agency. 1993c. 1992 Needs Survey Report to Congress. Office of Water (WH-547), Washington, DC. EPA 832-R-93-002.

U.S. Environmental Protection Agency. 1993d. Guidance Specifying Management Measures for sources of Nonpoint Pollution in Coastal Waters. Issued Under the Authority of Section 6217(g) of the Coastal Zone Act Reauthorization Amendments of 1990. Washington, DC.

U.S. Environmental Protection Agency. 1993e. Nitrogen Control Manual. Office of Research and Development. EPA/625/R-93/010.

U.S. Soil Conservation Service. (SCS). 1975. Soil Taxonomy: A Basic System of Soil Classification for Making and Interpreting Soil Surveys. U.S. Soil Conservation Service Agricultural Handbook 436. Washington, DC.

U.S. Soil Conservation Service. (SCS). 1981. *Soil Survey Manual.* Washington, DC: U.S. Department of Agriculture—Soil Conservation Service.

U.S. Soil Conservation Service. (SCS). 1987. Hydric Soils of the United States. National Technical Committee for Hydric Soils. Washington DC.

UNEP. 1987. Environmental Management Practices in Oil Refineries and Terminals. An Overview. United Nations Environment Program. Industry & Environment Overview Series.

Urban, N.R. and S.J. Eisenreich. 1988. Nitrogen cycling in a forested Minnesota bog. *Can. J. Bot.,* 66:435–449.

van der Valk, A.G. and C.B. Davis. 1978. Primary Production of Prairie Glacial Marshes. pp. 21–37. in R.E. Good, D.F. Whigham, and R.L. Simpson (Eds.), *Freshwater Wetlands: Ecological Processes and Management Potential.* New York: Academic Press.

van der Valk, A.G. and R.W. Jolly. 1992. Recommendations for research to develop guidelines for the use of wetlands to control rural nonpoint source pollution. *Ecol. Eng.,* 1(1/2):115–134.

VanDemark, P.J. and B.L. Batzing. 1987. *The Microbes, an Introduction to Their Nature and Importance.* Menlo Park, CA: Benjamin/Cummings.

van Oostrom, A.J. 1994. Nitrogen Removal in Constructed Wetlands Treating Nitrified Meat Processing Wastewater. pp. 569–579. in *Proceedings of the Fourth International Conference on Wetland Systems for Water Pollution Control.* Guangzhou, P.R. China: Center for International Development and Research, South China Institute for Environmental Sciences.

van Oostrom, A.J. and R.N. Cooper. 1990. "Meat Processing Effluent Treatment in Surface-Flow and Gravel-Bed Constructed Wastewater Wetlands," in: *Constructed Wetlands in Water Pollution Control,* P.F. Cooper and B.C. Findlater, Eds., Pergamon Press, Oxford, UK, pp 321–332.

van Oostrom, A.J. and J.M. Russell. 1994. Denitrification in constructed wastewater wetlands receiving high concentrations of nitrate. *Wat. Sci. Tech.*, 29(4):7–14.

Vartapetian, B.B. 1978. Introduction—Life Without Oxygen. Chapter 1, pp. 1–11. in D.D. Hook and R.M.M. Crawford (Eds.), *Plant Life in Anaerobic Environments.* Ann Arbor, MI: Ann Arbor Science.

Verry, E.S. 1989. Selection and Management of Shallow Water Impoundments for Wildlife. pp. 1177–1194. in R.R. Sharitz and J.W. Gibbons (Eds.), *Freshwater Wetlands and Wildlife.* Oak Ridge, TN: U.S. Department of Energy. CONF - 8603101.

Vile, M.A. and R.K. Weider. 1993. Alkalinity generation by Fe(III) reduction versus sulfate reduction in wetlands constructed for acid mine drainage treatment. *Water, Soil and Air Pollution*, in press.

Virta, J. 1966. Measurement of evapotranspiration and computation of water budget in treeless peatlands in the natural state. *Comment. Phys.-Math. Soc. Sci. Fenn.*, 32(11):1–70.

Visher, S.S. 1954. *Climatic Atlas of the United States.* Cambridge, MA: Harvard University Press.

Vymazal, J. 1993. Constructed Wetlands for Wastewater Treatment in Czechoslovakia: State of the Art. pp. 255–267. in G.A. Moshiri (Ed.), *Constructed Wetlands for Water Quality Improvement.* Chelsea, MI: Lewis Publishers.

Vymazal, J. 1995. *Algae and Element Cycling in Wetlands.* Lewis Publishers, Boca Raton, FL. 689 pp.

Walker, D.R., M.D., Flora, R.G. Rice, and D.J. Scheidt. 1988. Response of the Everglades Marsh to Increased Nitrogen and Phosphorus Loading. Part II: Macrophyte Community Structure and Chemical Composition. Report to the Superintendent, Everglades National Park, Homestead, FL.

Walker, L.P. and M.R. Walker. 1990. *City of Gustine Marsh Evaluation Study.* Davis, CA: Larry Walker Associates, Inc.

Walker, W.W. 1992a. Refinements to Phosphorus Uptake Relationship for Everglades Stormwater Treatment Areas Based upon Data from Water Conservation Area 2A. Paper Prepared for U.S. Department of Justice.

Walker, W.W. 1992b. A Mass-Balance Model for Estimating Phosphorus Settling Rate in Water Conservation Area 2A. Report to U.S. Department of Justice.

Walker, W.W. 1992c. Mass-Balance Models for the S10 Inflow Zone of WCA-2A. Report to U.S. Department of Justice.

Walker, W.W. 1995. Design Basis for Everglades Stormwater Treatment Areas. *Water Resources Bulletin*, in press.

Wallace, P.M., R.A. Garren, D.R. Rich, and A. Hernandez. 1990. An Ecological Study of Wetland Treatment Systems 1 and 2, Reedy Creek Swamp, Walt Disney World, Orange County, Florida. First Annual Report, 1988–1989. Prepared for Reedy Creek Energy Services, Lake Buena Vista, FL.

Walton, W.E., E.T. Schreiber, and M.S. Mulla. 1990. Distribution of *Culex tarsalis* larvae in a freshwater marsh in Orange County, California. *J. Am. Mosq. Control Assoc.*, 6:539–543.

Wass, R.D. and P. Fox. 1993. Constructed Wetlands for Nonpoint Source Control of Wastewater from a Vehicle Maintenance Yard. Proceedings of the Water Environment Federation 66th Annual Conference, Anaheim, CA. Vol. 9, pp. 9–20.

Watson, J.T. 1991. Summary of the Kentucky Constructed Wetlands Demonstration Interagency Task Force Meeting. January 18, 1991.

Watson, J.T. and A.J. Danzig. 1993. Pilot-Scale Nitrification Studies Using Vertical-Flow and Shallow Horizontal-Flow Constructed Wetland Cells pp. 301–313. in G.A. Moshiri (Ed.), *Constructed Wetlands for Water Quality Improvement.* Boca Raton, FL: Lewis Publishers.

Watson, J.T., K.D. Choate, and G.R. Steiner. 1990. Performance of Constructed Wetland Treatment Systems at Benton, Hardin, and Pembroke, Kentucky, During the Early Vegetation Establishment Phase. pp. 171–182. in P.F. Cooper and B.C. Findlater (Eds.), *Proceedings of the International Conference on the Use of Constructed Wetlands in Water Pollution Control.* Oxford, U.K.: Pergamon Press.

Watson, J.T., F.D. Diodata, and M. Lauch. 1987. Design and Performance of the Artificial Wetlands Wastewater Treatment Plant at Iselin, Pennsylvania. pp. 263–270. in K. R. Reddy and W.H. Smith (Eds.), *Aquatic Plants for Water Treatment and Resource Recovery.* Orlando, FL: Magnolia Publishing.

Waughman, G.J. and D.J. Bellamy. 1980. Nitrogen fixation and the nitrogen balance in peatland ecosystems. *Ecology.* 6(5):1185–1198.

Weber, A.S. and G. Tchobanoglous. 1986. Prediction of nitrification in water hyacinth treatment systems. *Journal WPCF.* 58(5):376–380.

WEF (Water Environment Federation) 1992. Design of Municipal Wastewater Treatment Plants. *WEF Manual of Practice No. 76.* Volume I: Chapters 1–12 and Volume II: Chapters 13–20. WEF, Alexandria, VA and American Society of Civil Engineers, New York.

Weller, M.W. 1978. Management of Freshwater Marshes for Wildlife. pp. 267–284. in R.E. Good, D.F. Whigham, and R.L. Simpson (Eds.), *Freshwater Wetlands: Ecological Processes and Management Potential.* New York: Academic Press.

Weller, M.W. 1981. Estimating Wildlife and Wetland Losses Due to Drainage and Other Perturbations. pp. 337–346. in B. Richardson (Ed.), *Selected Proceedings of the Midwest Conference on Wetland Values and Management.* St. Paul, MN: Freshwater Society.

Weller, M.W. 1988. The Influence of Hydrologic Maxima and Minima on Wildlife Habitat and Production Values of Wetlands. pp. 55–60. in J.A. Kusler and G.Brooks (Eds.), *Proceedings of the National Wetland Symposium: Wetland Hydrology.* Berne, NY: Association of State Wetland Managers.

Weller, M.W. 1990. Waterfowl Management Techniques for Wetland Enhancement, Restoration and Creation Useful in Mitigation Procedures. pp. 517–528. in J.A. Kusler and M.E. Kentula (Eds.), *Wetland Creation and Restoration. The Status of the Science.* Washington, DC: Island Press.

Wellings, F.M. 1986a. Letters to the Editor. Wetland, continued: discussion. *Fl. Water Resour. J.* 38(1):38.

Wellings, F.M. 1986b. Letters to the Editor. Wetlands. *Fl. Water Resour. J.* 38(2):22.

Wengrzynek, R.J. and C.R. Terrell. 1990. Using Constructed Wetlands to Control Agricultural Nonpoint Source Pollution. in *Proceedings of the International Conference on the Use of Constructed Wetlands in Water Pollution Control.* Cambridge, U.K.: Churchill College.

Wershaw, R.L., K.A. Thorn, D.J. Pincney, P. MacCarthy, J.A. Rice, and H.F. Hemond. 1986. Application of a membrane model to secondary structure of humic materials in peat, in: *Peat and Water,* C.H. Fuchsman, (Ed.), Elsevier, New York, 133–157.

Wetland Systems in Water Pollution Control. 1994. *Water Sci. Technol.,* 29(4):1–336.

Wetzel, R.G. 1975. *Limnology.* Philadelphia, PA: W.B. Saunders.

Wetzel, R.G. 1978. Foreword and Introduction. pp. xiii–xvii. in R.E. Good, D.F. Whigham, and R.L. Simpson, (Eds.), *Freshwater Wetlands: Ecological Processes and Management Potential.* New York: Academic Press.

Wharton, C.H. and M.M. Brinson. 1979. Characteristics of Southeastern River Systems. pp. 32–40. in R.R. Johnson and J.F. McCormick (Eds.), *Strategies for Protection and Management of Floodplain Wetlands and Other Riparian Ecosystems.* Washington, DC: U.S. Department of Agriculture.

Whigham, D.F., J. McCormick, R.E. Good, and R.L. Simpson. 1978. Biomass and Primary Production in Freshwater Tidal Wetlands of the Middle Atlantic Coast. pp. 3–20. in R.E. Good, D.F. Whigham, and R.L. Simpson, (Eds.), *Freshwater Wetlands: Ecological Processes and Management Potential.* New York: Academic Press.

Whigham, D.F., R.L. Simpson, R.E. Good, and F.A. Sickels. 1989. Decomposition and Nutrient-Metal Dynamics of Litter in Freshwater Tidal Wetlands. pp. 167–188. in R.R. Sharitz and J.W. Gibbons (Eds.), *Freshwater Wetlands and Wildlife.* Oak Ridge, TN: U.S. Department of Energy. CONF-8603101.

Whitlow, T.H. and R.W. Harris. 1979. Flood Tolerance in Plants: A State-Of-The-Art Review. Environmental & Water Quality Operational Studies. Technical Report E-79-2. U.S. Army Engineer Waterways Experiment Station.

Wiberg, P.L. and J.D. Smith. 1989. Model for calculating bed load transport of sediment. *J. Hydraul. Eng.,* 115(1):101–123.

Wieder, R.K. 1989. A survey of constructed wetlands for acid coal mine drainage treatment in the eastern United States. *Wetlands,* 9:299–315.

Wieder, R.K., K.P. Heston, E.M. O'Hara, G.E. Lang, A.E. Whitehouse, and J. Hett. 1988. Aluminum retention in a man-made *Sphagnum* wetland. *Water Air Soil Pollut.,* 37:177–191.

Wieder, R.K., M.N. Linton, and K.P. Heston. 1990. Laboratory mesocosm studies of Fe, Al, Mn, Ca, and Mg dynamics in wetlands exposed to synthetic acid coal mine drainage. *Water Air Soil Pollut.,* 51:181–196.

Wieder, R.K., G. Tchobanoglous, and R.W. Tuttle. 1989. Preliminary Considerations Regarding Constructed Wetlands for Wastewater Treatment. Chapter 25, pp. 297–305. in D.A. Hammer (Ed.), *Constructed Wetlands for Wastewater Treatment: Municipal, Industrial, and Agricultural.* Chelsea, MI: Lewis Publishers.

Wildeman, T.R. and L.S. Laudon. 1989. Use of Wetlands for Treatment of Environmental Problems in Mining: Non-Coal Mining Applications. Chapter 17. in D.A. Hammer (Ed.), *Constructed Wetlands for Wastewater Treatment. Municipal, Industrial and Agricultural.* Chelsea, MI: Lewis Publishers.

Wile, I., G. Miller, and S. Black. 1985. Design and Use of Artificial Wetlands. pp. 26–37. in P.J. Godfrey, E.R. Kaynor, and S. Pelczarski (Eds.), *Ecological Considerations in Wetlands Treatment of Municipal Wastewaters.* New York: Van Nostrand Reinhold.

Wile, I., G. Palmateer, and G. Miller. 1981. Use of Artificial Wetlands for Wastewater Treatment. pp. 255–271. in B. Richardson (Ed.), *Selected Proceedings of the Midwest Conference on Wetland Values and Management.* St. Paul, MN: Freshwater Society.

Wilhelm, M., S.R. Lawry, and D.D. Hardy. 1989. Creation and Management of Wetlands Using Municipal Wastewater in Northern Arizona: A Status Report. Chapter 13a, pp. 179–185. in D.A. Hammer (Ed.), *Constructed Wetlands for Wastewater Treatment. Municipal, Industrial and Agricultural.* Chelsea, MI: Lewis Publishers.

Willard, D.E. and J.A. Willis. 1988. Lessons from Kesterson. pp. 116–121. in J.A. Kusler, S. Daly, and G. Brooks (Eds.), *Proceedings of the National Wetland Symposium on Urban Wetlands.* Berne, NY: Association of Wetland Managers.

Willard, D.E., V.M. Finn, D.A. Levine, and J.E. Klarquist. 1990. Creation and Restoration of Riparian Wetlands in the Agricultural Midwest. pp. 327–350. in J.A. Kusler and M.E. Kentula (Eds.), *Wetland Creation and Restoration. The Status of the Science.* Washington, DC: Island Press.

Williams, J.B., E. May, M.G. Ford, and J.E. Butler. 1994. Nitrogen transformations in gravel bed hydroponic beds used as a tertiary treatment stage for sewage effluents. *Wat. Sci. Tech.*, 29(4):29–36.

Williams, M.C., C.E. Bagley, M.O. Dillard, C.E. Fox, L. Suddereth, D.J. Williams, M.R. McLean, and L.E. Alberts. 1989. Acid Input to an Urban Freshwater Marsh. pp. 1237–1247. in R.R. Sharitz and J.W. Gibbons (Eds.), *Freshwater Wetlands and Wildlife.* Oak Ridge, TN: U.S. Department of Energy. CONF - 8603101.

Williams, R.B. 1982. Wastewater Reuse—An Assessment of the Potential and Technology. Chapter 5, pp. 87–136. in E.J. Middlebrooks (Ed.), *Water Reuse.* Ann Arbor, MI: Ann Arbor Science.

Williams, R.B., Borgerding, J., Richey, D., and R.H. Kadlec. 1987. Start-Up and Operation of an Evaporative Wetlands Facility. pp. 209–216. in K.R. Reddy and W.H. Smith (Eds.), Aquatic Plants for Water Treatment and Resource Recovery. Orlando, FL: Magnolia Publishing.

Wilson, M.F. 1983. *Plant Reproductive Ecology.* New York: John Wiley & Sons.

Wilson, B.A., K.R. Townsend, and T.H. Anderson. 1987. Mosquito and Mosquitofish Responses to Loading of Water Hyacinth Wastewater Treatment Ponds. pp. 807–808. in K.R. Reddy and W.H. Smith (Eds.), *Aquatic Plants for Water Treatment and Resource Recovery.* Orlando, FL: Magnolia Publishing.

Winter, M. 1991. As quoted in: Gilges, K., 1991. For treating wastewater, build your own swamp! *Chem. Eng.* October:56.

Winter, M. and R. Kickuth. 1989a. Elimination of sulphur compounds from wastewater by the root zone process. I. Performance of a large-scale purification plant at a textile finishing industry. *Water Res.,* 23:535–546.

Winter, M. and R. Kickuth. 1989b. Elimination of sulphur compounds from wastewater by the root zone process. II. mode of formation of sulphur deposits. *Water Res.,* 23:547–560.

Wisseman, R. and S. Cook, Jr. 1977. Heavy metal accumulation in the sediments of a Washington Lake. *Bull. Environ. Contam. Toxicol.,* 18.

Wittgren, H.B. and K. Hasselgren. 1993. Naturliga system för avloppsrening och resursutnyttjande i tempererat klimat. VA-VORSK, Stockholm, Sweden, 68 pp. (in Swedish).

Wittgren, H.B. and K. Sundblad. 1990. Removal of Wastewater Nitrogen in an Infiltration Wetland with *Glyceria maxima.* pp. 85–96. in P.F. Cooper and B.C. Findlater (Eds.), *Proceedings of the International Conference on the Use of Constructed Wetlands in Water Pollution Control.* Oxford, U.K.: Pergamon Press.

Wolstenholme, R. and C.D. Bayes. 1990. An Evaluation of Nutrient Removal by the Reed Bed Treatment System at Valleyfield, Fife, Scotland. pp. 139–148. in P.F. Cooper and B.C. Findlater (Eds.), *Constructed Wetlands in Water Pollution Control.* Oxford, U.K.: Pergamon Press.

Wolverton, B.C., R.C. McDonald, and W.R. Duffer. 1983. Microorganisms and higher plants for waste water treatment. *J. Environ. Qual.,* 12(2):236–242.

Wood, A. 1990. Constructed Wetlands for Wastewater Treatment—Engineering and Design Considerations. pp. 481–494. in P.F. Cooper and B.C. Findlater (Eds.), *Proceedings of the International Conference on the Use of Constructed Wetlands in Water Pollution Control.* U.K.: Pergamon Press, Oxford.

Wood, A. 1993. Application of Full Scale Artificial Wetlands for Wastewater Treatment in South Africa. IAWQ Newsletter. Specialist Group on the Use of Macrophytes in Water Pollution Control, No. 9. Department of Plant Ecology, Aarhus University, Aarhus, The Netherlands. pp. 11–16.

WPCF (Water Pollution Control Federation). 1990a. Wastewater Biology: The Microlife. Task Force on Wastewater Biology.

WPCF (Water Pollution Control Federation). 1990b. *Natural Systems for Wastewater Treatment.* Manual of Practice FD-16. Alexandria, VA. Washington, DC: WPCF.

WPCF (Water Pollution Control Federation). 1983. Nutrient Control. Manual of Practice FD-7. Washington, DC: WPCF.

WRI (Wetlands Research, Inc.). 1992. *The Des Plaines River Wetlands Demonstration Project.* Volumes 1–7. Chicago, IL: Wetlands Research Inc.

Wright, J.L. 1982. New evapotranspiration crop coefficients. *J. Irrig. Drain. Div.,* 108(IR2):57–74.

Wright, J.L. 1987. Personal communication cited in ASCE, 1990.

Yin, H. and W. Shen. 1994. Winter Operation of Constructed Infiltration Wetland Treatment System for Wastewater. Paper Presented at Fourth International Conference on Wetland Systems for Water Pollution Control, Guangzhou, P.R. China, November 1994.

Yonika, D., D. Lowry, G. Hollands, W. Mullica, G. Smith, and S. Bigelow. 1979. Feasibility Study of Wetland Disposal of Wastewater Treatment Plant Effluent. Report to Massachusetts Water Resources Commission. Project 78-04.

Zachritz, W.H., II and J.W. Fuller. 1993. Performance of an artificial wetlands filter treating facultative lagoon effluent at Carville, Louisiana. *Water Environ. Res.,* 65(1):46–52.

Zachritz, W.H., II, L. Lundie, Jr., and R. Berghage. 1993. Removal; of Complex Organics Using Artificial Wetland Filters (AWF) Systems. IAWQ Newsletter. Specialist Group on the Use of Macrophytes in Water Pollution Control, No. 8. Department of Plant Ecology, Aarhus University, Aarhus, The Netherlands. p. 24.

Zak, D.R. and D. F. Grigal. 1991. Nitrogen mineralization, nitrification and denitrification in upland and wetland ecosystems. *Oecologia,* 88:189–196.

Zander, M. 1980. Polycyclic Aromatic and Heteroaromatic Hydrocarbons. in O. Hutzinger (Ed.), *Handbook of Environmental Chemistry.* New York: Springer-Verlag.

Zelazny, J. and J.S. Feierabend, Eds. 1988. *Increasing Our Wetland Resources.* Washington, DC: National Wildlife Federation-Corporate Conservation Council.

Zhang, T., J.B. Ellis, D.M. Revitt, and R.B. Shutes. 1990. Metal Uptake and Associated Pollution Control by *Typha latifolia* in Urban Wetlands. pp. 451–459. in P.F. Cooper and B.C. Findlater (Eds.), *Proceedings of the International Conference on the Use of Constructed Wetlands in Water Pollution Control.* Oxford, U.K.: Pergamon Press.

Zhu, T. and F.J. Sikora. 1994. Ammonium and nitrate removal in vegetated and unvegetated gravel bed microcosm wetlands, in: *Proceedings of the Fourth International Conference on Wetland Systems for Water Pollution Control*, Guangzhou, China, 355–366.

Zillioux, E.J., D.B. Porcella, and J.M. Benoit. 1993. Mercury cycling and effects in freshwater wetland ecosystems. *Environ. Toxicol. Chem.,* 12(12):2245–2264.

Zimmerman, J.H. 1988. A Multi-Purpose Wetland Characterization Procedure, Featuring the Hydroperiod. pp. 31–48. in J.A. Kusler and G. Brooks (Eds.), *Proceedings of the National Wetland Symposium: Wetland Hydrology.* Berne, NY: Association of State Wetland Managers.

Zoltek, J., S.E. Bagley, A.J. Hermann, L.R. Tortora, and T.J. Dolan. 1979. Removal of Nutrients from Treated Municipal Wastewater by Freshwater Marshes. Report to City of Clermont, FL. Center for Wetlands, University of Florida, Gainesville.

Index